Beginning Algebra

Beginning Algebra

Fifth Edition

JULIE MILLER

Professor Emerita, Daytona State College

MOLLY O'NEILL

Professor Emerita, Daytona State College

NANCY HYDE

Professor Emerita, Broward College

McGraw Hill Education

BEGINNING ALGEBRA, FIFTH EDITION

Published by McGraw-Hill Education, 2 Penn Plaza, New York, NY 10121. Copyright © 2018 by McGraw-Hill Education. All rights reserved.Printed in the United States of America. Previous editions © 2014, 2011, 2008, 2004. No part of this publication may be reproduced or distributed in any form or by any means, or stored in a database or retrieval system, without the prior written consent of McGraw-Hill Education, including, but not limited to, in any network or other electronic storage or transmission, or broadcast for distance learning.

Some ancillaries, including electronic and print components, may not be available to customers outside the United States.

This book is printed on acid-free paper.

1 2 3 4 5 6 7 8 9 0 LWI 21 20 19 18 17

ISBN 978-1-259-61025-7
MHID 1-259-61025-X

ISBN 978-1-259-93611-1(Annotated Instructor's Edition)
MHID 1-259-93611-2

Chief Product Officer, SVP Products & Markets: *G. Scott Virkler*
Vice President, General Manager, Products & Markets: *Marty Lange*
Vice President, Content Design & Delivery: *Betsy Whalen*
Managing Director: *Ryan Blankenship*
Brand Manager: *Amber Van Namee*
Director, Product Development: *Rose Koos*
Product Developer: *Luke Whalen*
Director of Marketing: *Sally Yagan*
Marketing Coordinator: *Annie Clarke*
Digital Product Analyst: *Ruth Czarnecki-Lichstein*
Digital Product Analyst: *Adam Fischer*
Director, Content Design & Delivery: *Linda Avenarius*
Program Manager: *Lora Neyens*
Content Project Manager: *Peggy J. Selle*
Assessment Project Manager: *Emily Windelborn*
Buyer: *Jennifer Pickel*
Design: *David W. Hash*
Content Licensing Specialists: *Carrie Burger, Melisa Seegmiller*
Cover Image: *© Juanmonino/iStock/Getty Images Plus/Getty Images*
Compositor: *SPi Global*
Typeface: *10/12 pt STIX MathJax Main*
Printer: *LSC Communications*

All credits appearing on page or at the end of the book are considered to be an extension of the copyright page.

Library of Congress Cataloging-in-Publication Data

Names: Miller, Julie, 1962- | O'Neill, Molly, 1953- | Hyde, Nancy.
Title: Beginning algebra/Julie Miller, Daytona State College, Molly
   O'Neill, Daytona State College, Nancy Hyde, Broward College, professor
   emeritus.
Description: Fifth edition. | New York, NY : McGraw-Hill, 2017.
Identifiers: LCCN 2016027890 | ISBN 9781259610257 (alk. paper)
Subjects: LCSH: Algebra—Textbooks.
Classification: LCC QA152.3.M55 2017 | DDC 512.9—dc23
LC record available at https://lccn.loc.gov/2016027890

The Internet addresses listed in the text were accurate at the time of publication. The inclusion of a website does not indicate an endorsement by the authors or McGraw-Hill Education, and McGraw-Hill Education does not guarantee the accuracy of the information presented at these sites.

mheducation.com/highered

The Miller/O'Neill/Hyde Developmental Math Series

Julie Miller, Molly O'Neill, and Nancy Hyde originally wrote their developmental math series because students were entering their College Algebra course underprepared. The students were not mathematically mature enough to understand the concepts of math, nor were they fully engaged with the material. The authors began their developmental mathematics offerings with intermediate algebra to help bridge that gap. This in turn developed into several series of textbooks from Prealgebra through Precalculus to help students at all levels before Calculus.

What sets all of the Miller/O'Neill/Hyde series apart is that they address course issues through an author-created digital package that maintains a consistent voice and notation throughout the program. This consistency—in videos, PowerPoints, Lecture Notes, and Group Activities—coupled with the power of ALEKS and Connect Hosted by ALEKS, ensures that students master the skills necessary to be successful in Developmental Math through Precalculus and prepares them for the calculus sequence.

Developmental Math Series (Hardback)
The hardback series is the more traditional in approach, yet balanced in its treatment of skills and concepts development for success in subsequent courses.

    Beginning Algebra, Fifth Edition
    Beginning & Intermediate Algebra, Fifth Edition
    Intermediate Algebra, Fifth Edition

Developmental Math Series (Softback)
The softback series includes a stronger emphasis on conceptual learning through Skill Practice features and Concept Connections, which are intended to help students with the conceptual meaning of the problems they are solving.

    Basic College Mathematics, Third Edition
    Prealgebra, Second Edition
    Prealgebra & Introductory Algebra, First Edition
    Introductory Algebra, Third Edition
    Intermediate Algebra, Third Edition

College Algebra/Precalculus Series
The Precalculus series serves as the bridge from Developmental Math coursework to setting the stage for future courses, including the skills and concepts needed for Calculus.

    College Algebra, Second Edition
    College Algebra and Trigonometry, First Edition
    Precalculus, First Edition

Acknowledgments

The author team most humbly would like to thank all the people who have contributed to this project.

Special thanks to our team of digital contributors for their thousands of hours of work: to Kelly Jackson, Andrea Hendricks, Jody Harris, Lizette Hernandez Foley, Lisa Rombes, Kelly Kohlmetz, and Leah Rineck for their devoted work on the integrated video and study guides. Thank you as well to Lisa Rombes, J.D. Herdlick, Adam Fischer, and Rob Brieler, the masters of ceremonies for SmartBook with Learning Resources. To Donna Gerken, Nathalie Vega-Rhodes, and Steve Toner: thank you for the countless grueling hours working through spreadsheets to ensure thorough coverage of Connect Math content. To our digital authors, Jody Harris, Linda Schott, Lizette Hernandez Foley, Michael Larkin, and Alina Coronel: thank you for spreading our content to the digital world of Connect Math. We also offer our sincerest appreciation to the outstanding video talent: Jody Harris, Alina Coronel, Didi Quesada, Tony Alfonso, and Brianna Kurtz. So many students have learned from you! To Hal Whipple, Carey Lange, and Julie Kennedy: thank you so much for ensuring accuracy in our manuscripts.

Finally, we greatly appreciate the many people behind the scenes at McGraw-Hill without whom we would still be on page 1. First and foremost, to Luke Whalen, our product developer and newest member of the team. Thanks for being our help desk. You've been a hero filling some big shoes in the day-to-day help on all things math, English, and editorial. To Amber Van Namee, our brand manager and team leader: thank you so much for leading us down this path. Your insight, creativity, and commitment to our project has made our job easier.

To the marketing team, Sally Yagan and Annie Clark: thank you for your creative ideas in making our books come to life in the market. Thank you as well to Mary Ellen Rahn for continuing to drive our long-term content vision through her market development efforts. To the digital content experts, Rob Brieler and Adam Fischer: we are most grateful for your long hours of work and innovation in a world that changes from day to day. And many thanks to the team at ALEKS for creating its spectacular adaptive technology and for overseeing the quality control in Connect Math.

To the production team: Peggy Selle, Carrie Burger, Emily Windelborn, Lora Neyens, and Lorraine Buczek—thank you for making the manuscript beautiful and for keeping the train on the track. You've been amazing. And finally, to Ryan Blankenship, Marty Lange, and Kurt Strand: thank you for supporting our projects for many years and for the confidence you've always shown in us.

Most importantly, we give special thanks to the students and instructors who use our series in their classes.

Julie Miller
Molly O'Neill
Nancy Hyde

Contents

Take a deep breath and know that you aren't alone. Your instructor, fellow students, and we, your authors, are here to help you learn and master the material for this course and prepare you for future courses. You may feel like math just isn't your thing, or maybe it's been a long time since you've had a math class—that's okay!

We wrote the text and all the supporting materials with you in mind. Most of our students aren't really sure how to be successful in math, but we can help with that.

As you begin your class, we'd like to offer some specific suggestions:

1. **Attend class.** Arrive on time and be prepared. If your instructor has asked you to read prior to attending class—do it. How often have you sat in class and thought you understood the material, only to get home and realize you don't know how to get started? By reading and trying a couple of Skill Practice exercises, which follow each example, you will be able to ask questions and gain clarification from your instructor when needed.

2. **Be an *active* learner.** Whether you are at lecture, watching an author lecture or exercise video, or are reading the text, pick up a pencil and work out the examples given. Math is learned only by doing; we like to say, "Math is not a spectator sport." If you like a bit more guidance, we encourage you to use the Integrated Video and Study Guide. It was designed to provide structure and note-taking for lectures and while watching the accompanying videos.

3. **Schedule time to do some math every day.** Exercise, foreign language study, and math are three things that you must do every day to get the results you want. If you are used to cramming and doing all of your work in a few hours on a weekend, you should know that even mathematicians start making silly errors after an hour or so! Check your answers. Skill Practice exercises all have the answer at the bottom of that page. Odd-numbered exercises throughout the text have answers at the back of the text. If you didn't get it right, don't throw in the towel. Try again, revisit an example, or bring your questions to class for extra help.

4. **Prepare for quizzes and exams.** At the end of each chapter is a summary that highlights all the concepts and problem types you need to understand and know how to do. There are additional problem sets at the end of each chapter: a set of review exercises, a chapter test, and a cumulative review. Working through the cumulative review will help keep your skills fresh from previous chapters—one of the key ways to do well on your exams. If you use ALEKS or Connect Hosted by ALEKS, use all of the tools available within the program to test your understanding.

5. **Use your resources.** This text comes with numerous supporting resources designed to help you succeed in this class and your future classes. Additionally, your instructor can direct you to resources within your institution or community. Form a student study group. Teaching others is a great way to strengthen your own understanding and they might be able to return the favor if you get stuck.

We wish you all the best in this class and your educational journey!

Julie Miller
julie.miller.math@gmail.com

Molly O'Neill
molly.s.oneill@gmail.com

Nancy Hyde
nhyde@montanasky.com

Student Guide to the Text

Clear, Precise Writing

Learning from our own students, we have written this text in simple and accessible language. Our goal is to keep you engaged and supported throughout your coursework.

Callouts

Just as your instructor will share tips and math advice in class, we provide callouts throughout the text to offer tips and warn against common mistakes.

- Tip boxes offer additional insight to a concept or procedure.
- Avoiding Mistakes help fend off common student errors.

Examples

- Each example is step-by-step, with thorough annotation to the right explaining each step.
- Following each example is a similar **Skill Practice** exercise to give you a chance to test your understanding. You will find the answer at the bottom of the page—providing a quick check.
- When you see this ⊙ in an example, there is an online dynamic animation within your online materials. Sometimes an animation is worth a thousand words.

Exercise Sets

Each type of exercise is built for your success in learning the materials and showing your mastery on exams.

- **Study Skills Exercises** integrate your studies of math concepts with strategies for helping you grow as a student overall.
- **Vocabulary and Key Concept Exercises** check your understanding of the language and ideas presented within the section.
- **Review Exercises** keep fresh your knowledge of math content already learned by providing practice with concepts explored in previous sections.
- **Concept Exercises** assess your comprehension of the specific math concepts presented within the section.
- **Mixed Exercises** evaluate your ability to successfully complete exercises that combine multiple concepts presented within the section.
- **Expanding Your Skills** challenge you with advanced skills practice exercises around the concepts presented within the section.
- **Problem Recognition Exercises** appear in strategic locations in each chapter of the text. These will require you to distinguish between similar problem types and to determine what type of problem-solving technique to apply.

Calculator Connections

Throughout the text are materials highlighting how you can use a graphing calculator to enhance understanding through a visual approach. Your instructor will let you know if you will be using these in class.

End-of-Chapter Materials

The features at the end of each chapter are perfect for reviewing before test time.

- Section-by-section summaries provide references to key concepts, examples, and vocabulary.
- Chapter review exercises provide additional opportunities to practice material from the entire chapter.
- Chapter tests are an excellent way to test your complete understanding of the chapter concepts.
- Cumulative review exercises are the best preparation to maintain a strong foundation of skills to help you move forward into new material. These exercises cover concepts from all the material covered up to that point in the text and will help you study for your final exam.

How Will Miller/O'Neill/Hyde Help Your Students *Get Better Results?*

Better Clarity, Quality, and Accuracy!

Julie Miller, Molly O'Neill, and Nancy Hyde know what students need to be successful in mathematics. Better results come from clarity in their exposition, quality of step-by-step worked examples, and accuracy of their exercises sets; but it takes more than just great authors to build a textbook series to help students achieve success in mathematics. Our authors worked with a strong mathematical team of instructors from around the country to ensure that the clarity, quality, and accuracy you expect from the Miller/O'Neill/Hyde series was included in this edition.

> "The most complete text at this level in its thoroughness, accuracy, and pedagogical soundness. The best developmental mathematics text I have seen."
>
> —Frederick Bakenhus, *Saint Phillips College*

Better Exercise Sets!

Comprehensive sets of exercises are available for every student level. Julie Miller, Molly O'Neill, and Nancy Hyde worked with a board of advisors from across the country to offer the appropriate depth and breadth of exercises for your students. **Problem Recognition Exercises** were created to improve student performance while testing.

Practice exercise sets help students progress from skill development to conceptual understanding. Student tested and instructor approved, the Miller/O'Neill/Hyde exercise sets will help your students *get better results*.

- ▶ **Problem Recognition Exercises**
- ▶ **Skill Practice Exercises**
- ▶ **Study Skills Exercises**
- ▶ **Mixed Exercises**
- ▶ **Expanding Your Skills Exercises**
- ▶ **Vocabulary and Key Concepts Exercises**

> "This series was thoughtfully constructed with students' needs in mind. The Problem Recognition section was extremely well designed to focus on concepts that students often misinterpret."
>
> —Christine V. Wetzel-Ulrich, *Northampton Community College*

Better Step-By-Step Pedagogy!

Beginning Algebra provides enhanced step-by-step learning tools to help students *get better results*.

- ▶ **Worked Examples** provide an "easy-to-understand" approach, clearly guiding each student through a step-by-step approach to master each practice exercise for better comprehension.
- ▶ **TIPs** offer students extra cautious direction to help improve understanding through hints and further insight.
- ▶ **Avoiding Mistakes** boxes alert students to common errors and provide practical ways to avoid them. All three learning aids will help students get better results by showing how to work through a problem using a clearly defined step-by-step methodology that has been class tested and student approved.

> "The book is designed with both instructors and students in mind. I appreciate that great care was used in the placement of 'Tips' and 'Avoiding Mistakes' as it creates a lot of teachable moments in the classroom."
>
> —Shannon Vinson, *Wake Tech Community College*

Get Better Results

Formula for Student Success

Step-by-Step Worked Examples

▶ Do you get the feeling that there is a disconnection between your students' class work and homework?

▶ Do your students have trouble finding worked examples that match the practice exercises?

▶ Do you prefer that your students see examples in the textbook that match the ones you use in class?

Miller/O'Neill/Hyde's *Worked Examples* offer a clear, concise methodology that replicates the mathematical processes used in the authors' classroom lectures!

Example 5 Solving a Linear Equation

Solve the equation. $2.2y - 8.3 = 6.2y + 12.1$

Solution:

$$2.2y - 8.3 = 6.2y + 12.1$$

Step 1: The right- and left-hand sides are already simplified.

$$-8.3 = 6.2y - 2.2y + 12.1$$
$$-8.3 = 4y + 12.1$$

Step 2: Subtract 2.2y from both sides to collect the variable terms on one side of the equation.

$$-8.3 - 12.1 = 4y + 12.1 - 12.1$$
$$-20.4 = 4y$$

Step 3: Subtract 12.1 from both sides to collect the constant terms on the other side.

$$\frac{-20.4}{4} = \frac{4y}{4}$$

Step 4:

$$-5.1 = y$$
$$y = -5.1$$

Step 5:

$$2.2y$$
$$2.2(-5.1)$$
$$-11.22$$

The solution set is $\{-5.1\}$.

Skill Practice Solve the equation.

5. $1.5t + 2.3 = 3.5t - 1.9$

> "As always, MOH's Worked Examples are so clear and useful for the students. All steps have wonderfully detailed explanations written with wording that the students can understand. MOH is also excellent with arrows and labels making the Worked Examples extremely clear and understandable."
>
> —Kelli Hammer, *Broward College–South*

> "Easy to read step-by-step solutions to sample textbook problems. The 'why' is provided for students, which is invaluable when working exercises without available teacher/tutor assistance."
>
> —Arcola Sullivan, *Copiah-Lincoln Community College*

Classroom Example: p. 282, Exercise 30

Example 3 Solving a System of Linear Equations by Graphing

Solve the system by the graphing method.

$$x - 2y = -2$$
$$-3x + 2y = 6$$

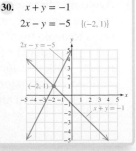

30. $x + y = -1$
$2x - y = -5$ $\{(-2, 1)\}$

Classroom Examples

To ensure that the classroom experience also matches the examples in the text and the practice exercises, we have included references to even-numbered exercises to be used as Classroom Examples. These exercises are highlighted in the Practice Exercises at the end of each section.

Better Learning Tools

TIP and Avoiding Mistakes Boxes

TIP and **Avoiding Mistakes** boxes have been created based on the authors' classroom experiences—they have also been integrated into the **Worked Examples**. These pedagogical tools will help students get better results by learning how to work through a problem using a clearly defined step-by-step methodology.

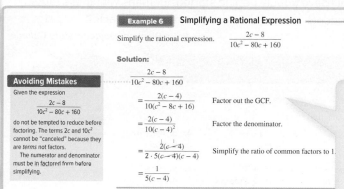

Avoiding Mistakes Boxes

Avoiding Mistakes boxes are integrated throughout the textbook to alert students to common errors and how to avoid them.

"MOH presentation of reinforcement concepts builds students' confidence and provides easy to read guidance in developing basic skills and understanding concepts. I love the visual clue boxes 'Avoiding Mistakes.' Visual clue boxes provide tips and advice to assist students in avoiding common mistakes."

—Arcola Sullivan, *Copiah-Lincoln Community College*

TIP Boxes

Teaching tips are usually revealed only in the classroom. Not anymore! TIP boxes offer students helpful hints and extra direction to help improve understanding and provide further insight.

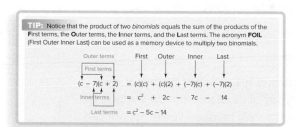

Better Exercise Sets! Better Practice! Better Results!

▶ Do your students have trouble with problem solving?

▶ Do you want to help students overcome math anxiety?

▶ Do you want to help your students improve performance on math assessments?

Get Better Results

Problem Recognition Exercises

Problem Recognition Exercises present a collection of problems that look similar to a student upon first glance, but are actually quite different in the manner of their individual solutions. Students sharpen critical thinking skills and better develop their "solution recall" to help them distinguish the method needed to solve an exercise—an essential skill in developmental mathematics.

Problem Recognition Exercises were tested in the authors' developmental mathematics classes and were created to improve student performance on tests.

> "The PREs are an excellent source of additional mixed problem sets. Frequently students have questions/comments like 'Where do I start?' or 'I know what to do once I get started, but I have trouble getting started.' Perhaps with these PREs, students will be able to overcome this obstacle."
>
> —Erika Blanken, *Daytona State College*

Problem Recognition Exercises

Operations on Polynomials

For Exercises 1–40, perform the indicated operations and simplify.

1. a. $6x^2 + 2x^2$
   b. $(6x^2)(2x^2)$

2. a. $8y^3 + y^3$
   b. $(8y^3)(y^3)$

3. a. $(4x + y)^2$
   b. $(4xy)^2$

4. a. $(2a + b)^2$
   b. $(2ab)^2$

5. a. $(2x + 3) + (4x - 2)$
   b. $(2x + 3)(4x - 2)$

6. a. $(5m^2 + 1) + (m^2 + m)$
   b. $(5m^2 + 1)(m^2 + m)$

7. a. $(3z + 2)^2$
   b. $(3z + 2)(3z - 2)$

8. a. $(6y - 7)^2$
   b. $(6y - 7)(6y + 7)$

9. a. $(2x - 4)(x^2 - 2x + 3)$
   b. $(2x - 4) + (x^2 - 2x + 3)$

10. a. $(3y^2 + 8)(-y^2 - 4)$
    b. $(3y^2 + 8) - (-y^2 - 4)$

11. a. $x + x$
    b. $x \cdot x$

12. a. $2c + 2c$
    b. $2c \cdot 2c$

13. $(7mn)^2$

14. $(8pq)^2$

15. $(-2x^4 - 6x^3 + 8x^2) \div (2x^2)$

16. $(-15m^3 + 12m^2 - 3m) \div (-3m)$

17. $(m^3 - 4m^2 - 6) - (3m^2 + 7m) + (-m^3 - 9m + 6)$

18. $(n^4 + 2n^2 - 3n) + (4n^2 + 2n - 1) - (4n^5 + 6n - 3)$

19. $(8x^3 + 2x + 6) \div (x - 2)$

20. $(-4x^3 + 2x^2 - 5) \div (x - 3)$

21. $(2x - y)(3x^2 + 4xy - y^2)$

22. $(3a + b)(2a^2 - ab + 2b^2)$

24. $(m^2 + 1)(m^4 - m^2 + 1)$

27. $(a^3 + 2b)(a^3 - 2b)$

28. $(y^3 - 6z)(y^3 + 6z)$

31. $\dfrac{12x^3y^7}{3xy^5}$

32. $\dfrac{-18p^2q^4}{2pq^3}$

> "These are so important to test whether a student can recognize different types of problems and the method of solving each. They seem very unique—I have not noticed this feature in many other texts or at least your presentation of the problems is very organized and unique."
>
> —Linda Kuroski, *Erie Community College*

Student-Centered Applications!

The Miller/O'Neill/Hyde Board of Advisors partnered with our authors to bring the *best applications* from every region in the country! These applications include real data and topics that are more relevant and interesting to today's student.

11. A bicyclist rides 24 mi against a wind and returns 24 mi with the same wind. His average speed for the return trip traveling with the wind is 8 mph faster than his speed going out against the wind. If x represents the bicyclist's speed going out against the wind, then the total time, t, required for the round trip is given by

$$t = \frac{24}{x} + \frac{24}{x + 8} \qquad \text{where } t \text{ is measured in hours.}$$

© Royalty Free/Corbis RF

a. Find the time required for the round trip if the cyclist rides 12 mph against the wind.

b. Find the time required for the round trip if the cyclist rides 24 mph against the wind.

Group Activities!

Each chapter concludes with a Group Activity to promote classroom discussion and collaboration—helping students not only to solve problems but to explain their solutions for better mathematical mastery. Group Activities are great for both full-time and adjunct instructors—bringing a more interactive approach to teaching mathematics! All required materials, activity time, and suggested group sizes are provided in the end-of-chapter material.

Group Activity

The Pythagorean Theorem and a Geometric "Proof"

Estimated Time: 25–30 minutes

Group Size: 2

Right triangles occur in many applications of mathematics. By definition, a right triangle is a triangle that contains a 90° angle. The two shorter sides in a right triangle are referred to as the "legs," and the longest side is called the "hypotenuse." In the triangle shown, the legs are labeled as a and b, and the hypotenuse is labeled as c.

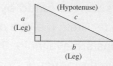

Right triangles have an important property that the sum of the squares of the two legs of a right triangle equals the square of the hypotenuse. This fact is referred to as the Pythagorean theorem. In symbols, the Pythagorean theorem is stated as:

$$a^2 + b^2 = c^2$$

1. The following triangles are right triangles. Verify that $a^2 + b^2 = c^2$. (The units may be left off when performing these calculations.)

$a = 3$
$b = 4$
$c = 5$

$a =$ ____
$b =$ ____
$c =$ ____

$a^2 + b^2 = c^2$

$(3)^2 + (4)^2 \stackrel{?}{=} (5)^2$

$9 + 16 = 25 \ \checkmark$

$a^2 + b^2 = c^2$

$(\_\_)^2 + (\_\_)^2 \stackrel{?}{=} (\_\_)^2$

___ + ___ = ___ $\checkmark$

...an theorem uses addition, subtraction, an... ...e figure. The length of each side of the... ...the large outer square is $(a + b)^2$. ...ound by adding the area of the inner... ...four right triangles (pictured in dark gray...

ght triangles: $4 \cdot \left(\frac{1}{2}\, a\, b\right)$

½ Base · Height

...rea of the large outer square:

Get Better Results

Dynamic Math Animations

The Miller/O'Neill/Hyde author team has developed a series of animations to illustrate difficult concepts where static images and text fall short. The animations leverage the use of on-screen movement and morphing shapes to enhance conceptual learning.

Through their classroom experience, the authors recognize that such media assets are great teaching tools for the classroom and excellent for online learning. The Miller/O'Neill/Hyde animations are interactive and quite diverse in their use. Some provide a virtual laboratory for which an application is simulated and where students can collect data points for analysis and modeling. Others provide interactive question-and-answer sessions to test conceptual learning. For word problem applications, the animations ask students to estimate answers and practice "number sense."

The animations were created by the authors based on over 75 years of combined teaching experience! To facilitate the use of the animations, the authors have placed icons in the text to indicate where animations are available. Students and instructors can access these assets online in either the ALEKS 360 Course product or Connect Math Hosted by ALEKS.

Additional Supplements

SmartBook. . . NOW with Learning Resources!

SmartBook is the first and only adaptive reading experience available for the world of higher education, and facilitates the reading process by identifying what content a student knows and doesn't know. As a student reads, the material continuously adapts to ensure the student is focused on the content he or she needs the most to close specific knowledge gaps. Additionally, new interactive Learning Resources now allow students to explore connections between different representations of problems, and also serve as an added resource right at the moment when a student answers a probe incorrectly and needs help. These Learning Resources—such as videos, interactive activities, and kaleidoscopes—are available at all times to provide support for students, even when they are working late at night or over the weekend and therefore do not have access to an instructor.

NEW Integrated Video and Study Workbooks

The Integrated Video and Study Workbooks were built to be used in conjunction with the Miller/O'Neill/Hyde Developmental Math series online lecture videos. These new video guides allow students to consolidate their notes as they work through the material in the book, and provide students with an opportunity to focus their studies on particular topics that they are struggling with rather than entire chapters at a time. Each video guide contains written examples to reinforce the content students are watching in the corresponding lecture video, along with additional written exercises for extra practice. There is also space provided for students to take their own notes alongside the guided notes already provided. By the end of the academic term, the video guides will not only be a robust study resource for exams, but will serve as a portfolio showcasing the hard work of students throughout the term.

Student Resource Manual

The *Student Resource Manual (SRM),* created by the authors, is a printable, electronic supplement available to students through Connect Math Hosted by ALEKS Corp. Instructors can also choose to customize this manual and package with their course materials. With increasing demands on faculty schedules, this resource offers a convenient means for both full-time and adjunct faculty to promote active learning and success strategies in the classroom.

This manual supports the series in a variety of different ways:

- Discovery-based classroom activities written by the authors for each section
- Worksheets for extra practice written by the authors

- Excel activities that not only provide students with numerical insights into algebraic concepts, but also teach simple computer skills to manipulate data in a spreadsheet
- Additional fun group activities
- Lecture Notes designed to help students organize and take notes on key concepts
- Materials for a student portfolio

Lecture Videos Created by the Authors (Available in ALEKS and Connect Math Hosted by ALEKS Corp.)

Julie Miller began creating these lecture videos for her own students to use when they were absent from class. The student response was overwhelmingly positive, prompting the author team to create the lecture videos for their entire developmental math book series. In these new videos, the authors walk students through the learning objectives using the same language and procedures outlined in the book. Students learn and review right alongside the author! Students can also access the written notes that accompany the videos.

All videos are closed-captioned for the hearing-impaired, and meet the Americans with Disabilities Act Standards for Accessible Design. These videos are available online through Connect Math Hosted by ALEKS Corp. as well as in ALEKS 360.

Exercise Videos (Available in ALEKS and Connect Math Hosted by ALEKS Corp.)

The authors, along with a team of faculty who have used the Miller/O'Neill/Hyde textbooks for many years, have created new exercise videos for designated exercises in the textbook. These videos cover a representative sample of the main objectives in each section of the text. Each presenter works through selected problems, following the solution methodology employed in the text.

Annotated Instructor's Edition

In the *Annotated Instructor's Edition* (*AIE*), answers to all exercises appear adjacent to each exercise in a color used *only* for annotations. The *AIE* also contains Instructor Notes that appear in the margin. These notes offer instructors assistance with lecture preparation. In addition, there are Classroom Examples referenced in the text that are highlighted in the Practice Exercises. Also found in the *AIE* are icons within the Practice Exercises that serve to guide instructors in their preparation of homework assignments and lessons.

Powerpoints (Available in the Resources of Connect Math Hosted by ALEKS Corp.)

The Powerpoints present key concepts and definitions with fully editable slides that follow the textbook. An instructor may project the slides in class or post to a website in an online course.

McGraw-Hill Connect Math Hosted by ALEKS Corp.

Connect Math Hosted by ALEKS Corp. is an exciting new assignment and assessment ehomework platform. Instructors can assign an AI-driven ALEKS Asssessment to identify the strengths and weaknesses of each student at the beginning of the term rather than after the first exam. Assignment creation and navigation is efficient and intuitive. The gradebook, based on instructor feedback, has a straightforward design and allows flexibility to import and export additional grades.

ALEKS Prep for Developmental Mathematics

ALEKS Prep for Beginning Algebra focuses on prerequisite and introductory material for this text. The prep products can be used during the first 3 weeks of a traditional course or in a corequisite course where students need to quickly narrow the gap in their skill and concept base.

ALEKS Prep Course Products Feature:
- Artificial Intelligence Targeting Gaps in Individual Student's Knowledge
- Assessment and Learning Directed Toward Individual Student's Needs
- Open Response Environment with Realistic Input Tools
- Unlimited Online Access—PC and Mac Compatible

Free trial at www.aleks.com/free_trial/instructor

Get Better Results

Instructor's and Student's Solutions Manuals (Available in the Resources of Connect Math Hosted by ALEKS Corp., the Online Learning Center, and at www. mcgrawhillcreate.com, our print-on-demand book-building website)

The *Instructor's Solutions Manual* provides comprehensive, worked-out solutions to all exercises in the Chapter Openers, the Practice Exercises, the Problem Recognition Exercises, the end-of-chapter Review Exercises, the Chapter Tests, and the Cumulative Review Exercises. The *Student's Solutions Manual* provides answers to the odd-numbered exercises in the text.

Instructor's Test Bank (Available in the Resources of Connect Math Hosted by ALEKS Corp.)

Among the supplements is a computerized test bank utilizing algorithm-based testing software to create customized exams quickly. Hundreds of text-specific, open-ended, and multiple-choice questions are included in the question bank. Sample chapter tests are also provided.

Loose-Leaf Text

This three-hole punched version of the traditional printed text allows students to carry it lightly and comfortably in a binder, integrated with notes and workbook pages as desired.

Acknowledgments and Reviewers

Paramount to the development of *Beginning Algebra* was the invaluable feedback provided by the instructors from around the country that reviewed the manuscript or attended a market development event over the course of the several years the text was in development.

Reviewers of Miller/O'Neill/Hyde Developmental Mathematics Series

Maryann Faller, *Adirondack Community College*
Albert Miller, *Ball State University*
Debra Pearson, *Ball State University*
Patricia Parkison, *Ball State University*
Robin Rufatto, *Ball State University*
Melanie Walker, *Bergen Community College*
Robert Fusco, *Bergen Community College*
Latonya Ellis, *Bishop State Community College*
Ana Leon, *Bluegrass Community College & Technical College*
Kaye Black, *Bluegrass Community College & Technical College*
Barbara Elzey, *Bluegrass Community College & Technical College*
Cheryl Grant, *Bowling Green State University*
Beth Rountree, *Brevard College*
Juliet Carl, *Broward College*
Lizette Foley, *Broward College*
Angie Matthews, *Broward College*
Mitchel Levy, *Broward College*
Jody Harris, *Broward College*
Michelle Carmel, *Broward College*
Antonnette Gibbs, *Broward College*
Kelly Jackson, *Camden Community College*
Elizabeth Valentine, *Charleston Southern University*

Adedoyin Adeyiga, *Cheyney University of Pennsylvania*
Dot French, *Community College of Philadelphia*
Brad Berger, *Copper Mountain College*
Donna Troy, *Cuyamaca College*
Brianna Kurtz, *Daytona State College–Daytona Beach*
Jennifer Walsh, *Daytona State College–Daytona Beach*
Marc Campbell, *Daytona State College–Daytona Beach*
Richard Rupp, *Del Mar College*
Joseph Hernandez, *Delta College*
Randall Nichols, *Delta College*
Thomas Wells, *Delta College*
Paul Yun, *El Camino College*
Catherine Bliss, *Empire State College–Saratoga Springs*
Laurie Davis, *Erie Community College*
Linda Kuroski, *Erie Community College*
David Usinski, *Erie Community College*
Ron Bannon, *Essex County College*
David Platt, *Front Range Community College*
Alan Dinwiddie, *Front Range Community College*
Andrea Hendricks, *Georgia Perimeter College*
Shanna Goff, *Grand Rapids Community College*
Betsy McKinney, *Grand Rapids Community College*
Cathy Gardner, *Grand Valley State University*

Reviewers of the Miller/O'Neill/Hyde Developmental Mathematics Series

Jane Mays, *Grand Valley State University*
John Greene, *Henderson State University*
Fred Worth, *Henderson State University*
Ryan Baxter, *Illinois State University*
Angela Mccombs, *Illinois State University*
Elisha Van Meenen, *Illinois State University*
Teresa Hasenauer, *Indian River State College*
Tiffany Lewis, *Indian River State College*
Deanna Voehl, *Indian River State College*
Joe Jordan, *John Tyler Community College*
Sally Copeland, *Johnson County Community College*
Nancy Carpenter, *Johnson County Community College*
Susan Yellott, *Kilgore College*
Kim Miller, *Labette Community College*
Michelle Hempton, *Lansing Community College*
Michelle Whitmer, *Lansing Community College*
Nathalie Vega-Rhodes, *Lone Star College*
Kuen Lee, *Los Angeles Trade Tech*
Nic Lahue, *MCC-Longview Community College*
Jason Pallett, *MCC-Longview Community College*
Janet Wyatt, *MCC-Longview Community College*
Rene Barrientos, *Miami Dade College—Kendall*
Nelson De La Rosa, *Miami Dade College—Kendall*
Jody Balzer, *Milwaukee Area Technical College*
Shahla Razavi, *Mt. San Jacinto College*
Shawna Bynum, *Napa Valley College*
Tammy Ford, *North Carolina A & T University*
Ebrahim Ahmadizadeh, *Northampton Community College*
Christine Wetzel-Ulrich, *Northampton Community College*
Sharon Totten, *Northeast Alabama Community College*
Rodolfo Maglio, *Northeastern Illinios University*
Christine Copple, *Northwest State Community College*
Sumitana Chatterjee, *Nova Community College*
Charbel Fahed, *Nova Community College*
Ken Hirschel, *Orange County Community College*
Linda K. Schott, *Ozarks Technical Community College*
Matthew Harris, *Ozarks Technical Community College*
Daniel Kopsas, *Ozarks Technical Community College*
Andrew Aberle, *Ozarks Technical Community College*
Alan Papen, *Ozarks Technical Community College*
Angela Shreckhise, *Ozarks Technical Community College*
Jacob Lewellen, *Ozarks Technical Community College*
Marylynne Abbott, *Ozarks Technical Community College*
Jeffrey Gervasi, *Porterville College*
Stewart Hathaway, *Porterville College*
Lauran Johnson, *Richard Bland College*
Matthew Nickodemus, *Richard Bland College*
Cameron English, *Rio Hondo College*

Lydia Gonzalez, *Rio Hondo College*
Mark Littrell, *Rio Hondo College*
Matthew Pitassi, *Rio Hondo College*
Wayne Lee, *Saint Philips College*
Paula Looney, *Saint Philips College*
Fred Bakenhus, *Saint Philips College*
Lydia Casas, *Saint Philips College*
Gloria Guerra, *Saint Philips College*
Sounny Slitine, *Saint Philips College*
Jessica Lopez, *Saint Philips College*
Lorraine Lopez, *San Antonio College*
Peter Georgakis, *Santa Barbara City College*
Sandi Nieto-Navarro, *Santa Rosa Junior College*
Steve Drucker, *Santa Rosa Junior College*
Jean-Marie Magnier, *Springfield Tech Community College*
Dave Delrossi, *Tallahassee Community College*
Natalie Johnson, *Tarrant County College South*
Marilyn Peacock, *Tidewater Community College*
Yvonne Aucoin, *Tidewater Community College*
Cynthia Harris, *Triton College*
Jennifer Burkett, *Triton College*
Christyn Senese, *Triton College*
Jennifer Dale, *Triton College*
Patricia Hussey, *Triton College*
Glenn Jablonski, *Triton College*
Myrna La Rosa, *Triton College*
Michael Maltenfort, *Truman College*
Abdallah Shuaibi, *Truman College*
Marta Hidegkuti, *Truman College*
Sandra Wilder, *University of Akron*
Sandra Jovicic, *University of Akron*
Edward Migliore, *University of California–Santa Cruz*
Kelly Kohlmetz, *University of Wisconsin—Milwaukee*
Leah Rineck, *University of Wisconsin—Milwaukee*
Carolann Van Galder, *University of Wisconsin—Rock County*
Claudia Martinez, *Valencia College*
Stephen Toner, *Victor Valley Community College*
David Cooper, *Wake Tech Community College*
Karlata Elliott, *Wake Tech Community College*
Laura Kalbaugh, *Wake Tech Community College*
Kelly Vetter, *Wake Tech Community College*
Jacqui Fields, *Wake Tech Community College*
Jennifer Smeal, *Wake Tech Community College*
Shannon Vinson, *Wake Tech Community College*
Kim Walaski, *Wake Tech Community College*
Lisa Rombes, *Washtenaw Community College*
Maziar Ouliaeinia, *Western Iowa Tech Community College*
Keith McCoy, *Wilbur Wright College*

Also, a special thanks to all instructors who have reviewed previous editions of this series.

Get Better Results

Our Commitment to Market Development and Accuracy

McGraw-Hill's Development Process is an ongoing, never-ending, market-oriented approach to building accurate and innovative print and digital products. We begin developing a series by partnering with authors that desire to make an impact within their discipline to help students succeed. Next, we share these ideas and manuscript with instructors for review for feedback and to ensure that the authors' ideas represent the needs within that discipline. Throughout multiple drafts, we help our authors adapt to incorporate ideas and suggestions from reviewers to ensure that the series carries the same pulse as today's classrooms. With any new series, we commit to accuracy across the series and its supplements. In addition to involving instructors as we develop our content, we also utilize accuracy checks through our various stages of development and production. With our commitment to this process, we are confident that our series has the most developed content the industry has to offer, thus pushing our desire for quality and accurate content that meets the needs of today's students and instructors.

New and Updated Content for Miller/O'Neill/Hyde Beginning Algebra, Fifth Edition:

- New Chapter Openers focused on contextualized learning that introduce the main idea of each chapter in an applied setting
- New SmartBook with Learning Resources digital resource added for students that includes over 1,400 student learning resources, over 500 learning objectives, and over 1,300 student activity probes
- New Integrated Video & Study Guide workbook to accompany the online lecture video series created by the Miller/O'Neill/Hyde author team
- Updated Applications to be timely in all instances where appropriate
- Modularized content for easier course customization and flexibility in a digital or traditional classroom environment
- Over 200 new algorithmic exercises added to Connect Math to better cover developmental math content for students
- New Introduction to Modeling online appendix added to provide students with additional targeted instruction on linear, quadratic, and exponential models

McGraw Hill Education

connect MATH
HOSTED BY ALEKS
+SMARTBOOK

Students lacking confidence in math? Looking for a consistent voice between text and digital?

Problem Solved!

Connect Math Hosted by ALEKS + SmartBook is a complete system that offers everything students and instructors need in one, intuitive platform. ConnectMath is an online homework engine where the problems and solutions are consistent with the textbook authors' approach. SmartBook is an assignable, adaptive eBook and study tool that directs students to the content they don't know and helps them study more efficiently. This combination gives you tools you need to be the teacher you want to be.

"I like that ConnectMath reaches students with different learning styles ... our students are engaged, attend class, and ask excellent questions."
— Kelly Davis, South Texas College

Developed by instructors for instructors to create a seamless transition from text to digital

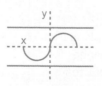

SmartBook integration offers an interactive reading and learning experience

Access to author-developed, text-specific assignments, learning resources, and videos

Because learning changes everything™

mheducation.com

How can ConnectMath + SmartBook help solve your students' math challenges?

I like to learn by _____.

Whether it's reading, watching, discovering, or doing, ConnectMath has something for everyone. Instructors can create assignments that accommodate different learning styles, and students aren't stuck with boring multiple-choice problems. Instead they have a myriad of motivational learning and media resources at their fingertips. SmartBook delivers an interactive reading and learning experience that provides personalized guidance and just-in-time remediation. This helps students to focus on what they need, right when they need it.

I still don't get it. Can you do that problem again?

Because the content in ConnectMath is author-developed and goes through a rigorous quality control process, students hear one voice, one style, and don't get lost moving from text to digital. The high-quality, author-developed videos provide students ample opportunities to master concepts and practice skills that they need extra help with . . . all of which are integrated in the ConnectMath platform and the eBook.

How can ConnectMath + SmartBook help solve your classroom challenges?

I need meaningful data to measure student success!

From helping the student in the back row to tracking learning trends for your entire course, ConnectMath + SmartBook delivers the data you need to make an impactful, meaningful learning experience for students. With easy-to-interpret, downloadable reports, you can analyze learning rates for each assignment, monitor time on task, and learn where students' strengths and weaknesses are in each course area.

We're going with the _____ (flipped classroom, corequisite model, hybrid, etc.) implementation.

ConnectMath + SmartBook is an intuitive digital solution that can be used in any course setup. Each course in ConnectMath comes complete with its own set of text-specific assignments, author-developed videos and learning resources, and an integrated eBook that cater to the needs of your specific course. The easy-to-navigate home page keeps the learning curve at a minimum, but offers an abundance of tutorials and videos to help get you and your colleagues started.

Let's Talk!

Ready to take the next step?

Let's talk. Whether you're looking to redesign your course, change technologies, or just explore, our team of dedicated faculty consultants, reps, and digital specialists are standing by.

http://shop.mheducation.com/
store/paris/user/findltr.html

Looking for stable technology? Connect Math Hosted by ALEKS has an uptime of 99.97%.

Problem Solved!

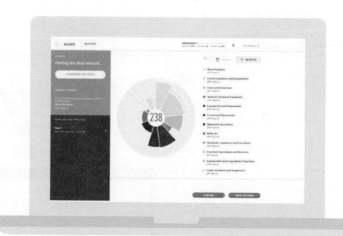

How can ALEKS help solve your students' challenges?

I've never been good at math, so why should I try?

The perceived struggle with math is often too real for many students. ALEKS offers a chance to break from that struggle through its cycle of individualized assessment and learning. Students only work on topics they are ready to learn, which have a proven learning success rate of 93% or higher. Periodic assessments reinforce content mastery and provide targeted remediation. As students watch their progress in the ALEKS® Pie grow, their confidence grows with it.

I did all my homework, so why am I failing my exams?

The purpose of homework is to ensure mastery and prepare students for exams. Yet how well do homework scores correlate to exam scores? ALEKS is the only adaptive learning system that ensures mastery through periodic reassessments and delivers just-in-time remediation to efficiently prepare students. Because of how ALEKS presents lessons and practice, students learn by understanding the core principle of a concept rather than just memorizing a process.

I'm too far behind to catch up. - OR - I've already done this, I'm bored.

No two students are alike. So why start everyone on the same page? ALEKS diagnoses what each student knows and doesn't know, and prescribes an optimized learning path through the curriculum you put forth. Students are only working on what they need, when they need it, rather than focusing on topics they already know or aren't ready for. The frustration of falling behind and the boredom from redundant review is virtually eliminated.

"ALEKS improved my math skills tremendously. What I love the most about it is that it makes sure you understand the material ... Thanks to ALEKS I got the job that I have now."
– *Student, Broward College, FL*

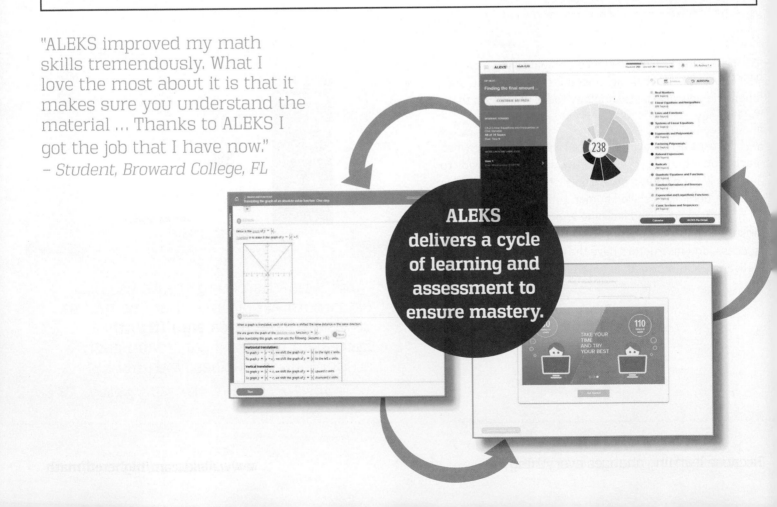

ALEKS delivers a cycle of learning and assessment to ensure mastery.

How can ALEKS help solve your classroom challenges?

I need something that solves the problem of cost, time to completion, and student preparedness.

ALEKS is the perfect solution to the trifecta of these problems. It provides an efficient path to mastery through its individualized cycle of learning and assessment that targets prerequisite gaps and focuses students on what they are ready to learn. Students move through their math requirements more quickly and are better prepared for subsequent courses. This saves both the institution and the student money.

We're going with the _____ (flipped classroom, corequisite, accelerated, etc.) implementation.

No matter your course setup, ALEKS can handle it. ALEKS courses cover a broad curriculum, so that you can easily tailor it to cover just the topics that you need. ALEKS Objectives gives you control over when and in what order students move through the content so that the pacing matches that of the textbook and/or curriculum. The flexibility of ALEKS allows you to use it in conjunction with any textbook or set of resources that you want to use.

My administration and department measure success differently. How can we compare notes?

ALEKS offers the most comprehensive and detailed data analytics on the market. From helping the student in the back row to monitoring pass rates across the department and institution, ALEKS delivers the data needed at all levels. The customizable and intuitive reporting features allow you and your colleagues to easily gather, interpret, and share the data you need, when you need it.

"ALEKS is the only product that diagnoses and prescribes an individualized learning path. The ALEKS management system is intuitive and easy to use. The reports are easy to construct, read and interpret, allowing us to help students focus and succeed."
–Yoshi Yamato and Marie McClendon, Pasadena City College, CA

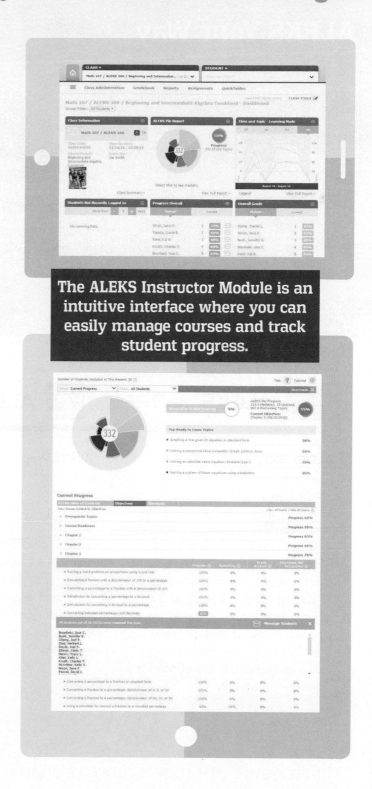

The ALEKS Instructor Module is an intuitive interface where you can easily manage courses and track student progress.

Because learning changes everything™

www.aleks.com/highered/math

ALEKS Your Way

ALEKS offers STEM and non-STEM courses ranging from developmental math through precalculus, including custom state-mandated curriculums, combined courses, and a suite of Prep courses that prepare students up to calculus. Each course covers a broad, flexible curriculum that you can easily customize to match your curriculum and textbook. Select courses include sophisticated graphing tools, and all math courses are bilingual for Spanish-speaking students.

ALEKS 360 combines the power of ALEKS with an integrated, interactive McGraw-Hill eBook. Students have direct access to the text, lecture videos, supplementary resources, and practice examples right when they need it as they work each topic. ALEKS 360 is available with most ALEKS courses, and at an affordable price. For those students who still prefer to hold a page in their hand, they can order a loose-leaf version of the eBook for a significant discount.

ALEKS Prep courses target students' prerequisite knowledge gaps and provide individualized learning and remediation to ensure students are ready to succeed in their math course. It is the perfect solution for the accelerated "bootcamp" program, instructors that need to get students up to speed during the first few weeks of a course, or as the remedial component of a corequisite course.

ALEKS Placement, Preparation, and Learning (ALEKS PPL) is the first and only institution-wide placement solution to both diagnose and remediate, all in one completely personalized program. The adaptive placement assessment places students from Basic Math through Calculus I. Then students get six months of personalized learning and remediation. ALEKS PPL gives institutions the tools to improve college preparedness, course performance, and retention.

"My students are much more prepared because of ALEKS Prep. I saw noticeably fewer errors on exams and fewer Algebra-related questions during class. I can now focus on Calculus material without having to review."

–Jeanette Martin, University of Washington

Because learning changes everything™   **www.aleks.com/highered/math**

Study Tips

In taking a course in algebra, you are making a commitment to yourself, your instructor, and your classmates. Following some or all of the study tips presented here can help you be successful in this endeavor. The features of this text that will assist you are printed in blue.

© Blend Images/Getty Images RF

1. Before the Course

1. Purchase the necessary materials for the course before the course begins or on the first day.
2. Obtain a three-ring binder to keep and organize your notes, homework, tests, and any other materials acquired in the class. We call this type of notebook a portfolio.
3. Arrange your schedule so that you have enough time to attend class and to do homework. A common rule is to set aside at least 2 hours for homework for every hour spent in class. That is, if you are taking a 4-credit-hour course, plan on at least 8 hours a week for homework. A 6-credit-hour course will then take *at least* 12 hours each week—about the same as a part-time job. If you experience difficulty in mathematics, plan for more time.
4. Communicate with your employer and family members the importance of your success in this course so that they can support you.
5. Be sure to find out the type of calculator (if any) that your instructor requires.

2. During the Course

1. Read the section in the text *before* the lecture to familiarize yourself with the material and terminology. It is recommended that you read your math book with paper and pencil in hand. Write a one-sentence preview of what the section is about.
2. Attend every class and be on time. Be sure to bring any materials that are needed for class such as graph paper, a ruler, or a calculator.
3. Take notes in class. Write down all of the examples that the instructor presents. Read the notes after class, and add any comments to make your notes clearer to you. Use an audio recorder to record the lecture if the instructor permits the recording of lectures.
4. Ask questions in class.
5. Read the section in the text *after* the lecture, and pay special attention to the Tip boxes and Avoiding Mistakes boxes.
6. After you read an example, try the accompanying Skill Practice problem. The skill practice problem mirrors the example and tests your understanding of what you have read.
7. Do homework every day. Even if your class does not meet every day, you should still do some work every day to keep the material fresh in your mind.
8. Check your homework with the answers that are supplied in the back of this text. Correct the exercises that do not match, and circle or star those that you cannot correct yourself. This way you can easily find them and ask your instructor, tutor, online tutor, or math lab staff the next day.
9. Be sure to do the Vocabulary and Key Concepts exercises found at the beginning of the Practice Exercises.

10. The Problem Recognition Exercises are located in all chapters. These provide additional practice distinguishing among a variety of problem types. Sometimes the most difficult part of learning mathematics is retaining all that you learn. These exercises are excellent tools for retention of material.

11. Form a study group with fellow students in your class, and exchange phone numbers. You will be surprised by how much you can learn by talking about mathematics with other students.

12. If you use a calculator in your class, read the Calculator Connections boxes to learn how and when to use your calculator.

13. Ask your instructor where you might obtain extra help if necessary.

3. Preparation for Exams

1. Look over your homework. Pay special attention to the exercises you have circled or starred to be sure that you have learned that concept.

2. Begin preparations for exams on the first day of class. As you do each homework assignment, think about how you would recognize similar problems when they appear on a test.

3. Read through the Summary at the end of the chapter. Be sure that you understand each concept and example. If not, go to the section in the text and reread that section.

4. Give yourself enough time to take the Chapter Test uninterrupted. Then check the answers. For each problem you answered incorrectly, go to the Review Exercises and do all of the problems that are similar.

5. To prepare for the final exam, complete the Cumulative Review Exercises at the end of each chapter. If you complete the cumulative reviews after finishing each chapter, then you will be preparing for the final exam throughout the course. The Cumulative Review Exercises are another excellent tool for helping you retain material.

4. Where to Go for Help

1. At the first sign of trouble, see your instructor. Most instructors have specific office hours set aside to help students. Don't wait until after you have failed an exam to seek assistance.

2. Get a tutor. Most colleges and universities have free tutoring available. There may also be an online tutor available.

3. When your instructor and tutor are unavailable, use the Student Solutions Manual for step-by-step solutions to the odd-numbered problems in the exercise sets.

4. Work with another student from your class.

5. Work on the computer. Many mathematics tutorial programs and websites are available on the Internet, including the website that accompanies this text.

© PhotoDisc/Getty Images RF

Group Activity

Becoming a Successful Student

Materials: Computer with Internet access

Estimated Time: 15 minutes

Group Size: 4

Good time management, good study skills, and good organization will help you be successful in this course. Answer the following questions and compare your answers with your group members.

Answers will vary.

1. To motivate yourself to complete a course, it is helpful to have clear reasons for taking the course. List your goals for taking this course and discuss them with your group.

2. For the following week, write down the times each day that you plan to study math.

Monday	Tuesday	Wednesday	Thursday	Friday	Saturday	Sunday

3. Write down the date of your next math test. _____

4. Taking 12 credit-hours is the equivalent of a full-time job. Often students try to work too many hours while taking classes at school.

   a. Write down the number of hours you work per week and the number of credit-hours you are taking this term.

      Number of hours worked per week _____

      Number of credit-hours this term _____

   b. The table gives a recommended limit to the number of hours you should work for the number of credit-hours you are taking at school. (Keep in mind that other responsibilities in your life such as your family might also make it necessary to limit your hours at work even more.) How do your numbers from part (a) compare to those in the table? Are you working too many hours?

Number of Credit-Hours	Maximum Number of Hours of Work per Week
3	40
6	30
9	20
12	10
15	0

5. Discuss with your group members where you can go for extra help in math. Then write down three of the suggestions.

   _____

   _____

   _____

6. Do you keep an organized notebook for this class? Can you think of any suggestions that you can share with your group members to help them keep their materials organized?

7. Look through a chapter and find the page number corresponding to each feature in that chapter. Discuss with your group members how you might use each feature.

   Problem Recognition Exercises: page _____

   Chapter Summary: page _____

   Chapter Review Exercises: page _____

   Chapter Test: page _____

   Cumulative Review Exercises: page _____

8. Look at the Skill Practice exercises. For example, find Skill Practice exercises 1 and 2 in the first section. Where are the answers to these exercises located? Discuss with your group members how you might use the Skill Practice exercises.

9. Do you think that you have math anxiety? Read the following list for some possible solutions. Check the activities that you can realistically try to help you overcome this problem.

   _____   Read a book on math anxiety.

   _____   Search the Web for tips on handling math anxiety.

   _____   See a counselor to discuss your anxiety.

   _____   Talk with your instructor to discuss strategies to manage math anxiety.

   _____   Evaluate your time management to see if you are trying to do too much. Then adjust your schedule accordingly.

10. Some students favor different methods of learning over others. For example, you might prefer:

    - Learning through listening and hearing.

    - Learning through seeing images, watching demonstrations, and visualizing diagrams and charts.

    - Learning by experience through a hands-on approach.

    - Learning through reading and writing.

    Most experts believe that the most effective learning comes when a student engages in *all* of these activities. However, each individual is different and may benefit from one activity more than another. You can visit a number of different websites to determine your "learning style." Try doing a search on the Internet with the key words "*learning styles assessment.*" Once you have found a suitable website, answer the questionnaire and the site will give you feedback on what method of learning works best for you.

11. As you read through Chapter 1, try to become familiar with the features of this textbook. Then match the feature in column B with its description in column A.

    Column A

    1. Allows you to check your work as you do your homework   e
    2. Shows you how to avoid common errors   g
    3. Provides an online tutorial and exercise supplement   b
    4. Outlines key concepts for each section in the chapter   f
    5. Provides exercises that help you distinguish between different types of problems   d
    6. Offers helpful hints and insight   a
    7. Offers practice exercises that go along with each example   c

    Column B

    a. Tips
    b. ConnectMath
    c. Skill Practice exercises
    d. Problem Recognition Exercises
    e. Answers to odd-numbered exercises
    f. Chapter Summary
    g. Avoiding Mistakes

The Set of Real Numbers

1

Numbers in Our World

Imagine a world where the only numbers known are the counting or natural numbers (1, 2, 3, 4, . . .). Now imagine that you want to sell only a fraction of your land to another person, or that you owe twenty dollars and fifteen cents to your bank. How could these values be written without formal numerical symbols? Living in such a world would deter the growth of a complex society like ours.

It is difficult to fathom that the use of zero, fractions, and negative numbers was formally accepted only about a thousand years ago! Before that time, communicating about parts of items, the absence of value, and owing money to a lender was likely done in creative but arduous ways. When we talk about a *third* of a parcel of land, temperatures below *zero* such as *negative* 20 degrees, and the number π, we make use of the many subsets of the **set of real numbers**.

© Zephyr_p/Shutterstock RF

Real numbers enable us to talk about parts of things, to expand our thinking to explain phenomena in precise ways, and to operate with numbers in a consistent and predicable manner. In this chapter we will explore how real numbers are used, and the way they open the door to algebra.

Concepts

1. Basic Definitions

The study of algebra involves many of the operations and procedures used in arithmetic. Therefore, we begin this text by reviewing the basic operations of addition, subtraction, multiplication, and division on fractions and mixed numbers.

We begin with the numbers used for counting:

the **natural numbers:** 1, 2, 3, 4, . . .

and

the **whole numbers:** 0, 1, 2, 3, . . .

Whole numbers are used to count the number of whole units in a quantity. A fraction is used to express part of a whole unit. If a child gains $2\frac{1}{2}$ lb, the child has gained two whole pounds plus a portion of a pound. To express the additional half pound mathematically, we may use the fraction, $\frac{1}{2}$.

> ### A Fraction and Its Parts
>
> **Fractions** are numbers of the form $\frac{a}{b}$, where $\frac{a}{b} = a \div b$ and b does not equal zero.
>
> In the fraction $\frac{a}{b}$, the **numerator** is a, and the **denominator** is b.

The denominator of a fraction indicates how many equal parts divide the whole. The numerator indicates how many parts are being represented. For instance, suppose Jack wants to plant carrots in $\frac{2}{5}$ of a rectangular garden. He can divide the garden into five equal parts and use two of the parts for carrots (Figure 1-1).

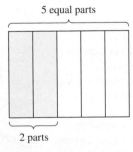

5 equal parts

2 parts

The shaded region represents $\frac{2}{5}$ of the garden.

Figure 1-1

Instructor Note: Tell the students that the word "improper" does not imply that there is anything wrong with a fraction.

> ### Proper Fractions, Improper Fractions, and Mixed Numbers
>
> 1. If the numerator of a fraction is less than the denominator, the fraction is a **proper fraction**. A proper fraction represents a quantity that is less than a whole unit.
> 2. If the numerator of a fraction is greater than or equal to the denominator, then the fraction is an **improper fraction**. An improper fraction represents a quantity greater than or equal to a whole unit.
> 3. A **mixed number** is a whole number added to a proper fraction.

Proper Fractions: $\frac{3}{5}$ $\frac{1}{8}$

Improper Fractions: $\frac{7}{5}$ $\frac{8}{8}$

Mixed Numbers: $1\frac{1}{5}$ $2\frac{3}{8}$

2. Prime Factorization

To perform operations on fractions it is important to understand the concept of a factor. For example, when the numbers 2 and 6 are multiplied, the result (called the **product**) is 12.

$$2 \times 6 = 12$$

factors   product

The numbers 2 and 6 are said to be **factors** of 12. (In this context, we refer only to natural number factors.) The number 12 is said to be factored when it is written as the product of two or more natural numbers. For example, 12 can be factored in several ways:

$$12 = 1 \times 12 \qquad 12 = 2 \times 6 \qquad 12 = 3 \times 4 \qquad 12 = 2 \times 2 \times 3$$

A natural number greater than 1 that has only two factors, 1 and itself, is called a **prime number**. The first several prime numbers are 2, 3, 5, 7, 11, and 13. A natural number greater than 1 that is not prime is called a **composite number**. That is, a composite number has factors other than itself and 1. The first several composite numbers are 4, 6, 8, 9, 10, 12, 14, 15, and 16.

Avoiding Mistakes

The number 1 is neither prime nor composite.

Example 1 **Writing a Natural Number as a Product of Prime Factors**

Write each number as a product of prime factors.

a. 12   **b.** 30

Classroom Example: p. 18, Exercise 36

Solution:

a. $12 = 2 \times 2 \times 3$ Divide 12 by prime numbers until the result is also a prime number.

$$\begin{array}{r} 2\overline{)12} \\ 2\overline{)6} \\ \hline 3 \end{array}$$

Or use a factor tree

b. $30 = 2 \times 3 \times 5$

$$\begin{array}{r} 2\overline{)30} \\ 3\overline{)15} \\ \hline 5 \end{array}$$

Skill Practice Write the number as a product of prime factors.

1. 40   **2.** 72

Answers

1. $2 \times 2 \times 2 \times 5$
2. $2 \times 2 \times 2 \times 3 \times 3$

3. Simplifying Fractions to Lowest Terms

The process of factoring numbers can be used to reduce or simplify fractions to lowest terms. A fractional portion of a whole can be represented by infinitely many fractions. For example, Figure 1-2 shows that $\frac{1}{2}$ is equivalent to $\frac{2}{4}$, $\frac{3}{6}$, $\frac{4}{8}$, and so on.

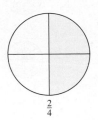 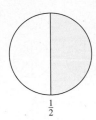 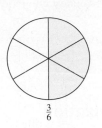 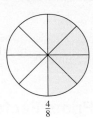

$\frac{1}{2}$ $\frac{2}{4}$ $\frac{3}{6}$ $\frac{4}{8}$

Figure 1-2

The fraction $\frac{1}{2}$ is said to be in **lowest terms** because the numerator and denominator share no common factor other than 1.

To simplify a fraction to lowest terms, we use the following important principle.

Fundamental Principle of Fractions

Suppose that a number, c, is a common factor in the numerator and denominator of a fraction. Then

$$\frac{a \times c}{b \times c} = \frac{a}{b} \times \frac{c}{c} = \frac{a}{b} \times 1 = \frac{a}{b}$$

To simplify a fraction, we begin by factoring the numerator and denominator into prime factors. This will help identify the common factors.

Classroom Example: p. 19,
Exercise 44

Example 2   **Simplifying a Fraction to Lowest Terms**

Simplify $\dfrac{45}{30}$ to lowest terms.

Solution:

$$\frac{45}{30} = \frac{3 \times 3 \times 5}{2 \times 3 \times 5} \qquad \text{Factor the numerator and denominator.}$$

$$= \frac{3}{2} \times \frac{3}{3} \times \frac{5}{5} \qquad \text{Apply the fundamental principle of fractions.}$$

$$= \frac{3}{2} \times 1 \times 1 \qquad \text{Any nonzero number divided by itself is 1.}$$

$$= \frac{3}{2} \qquad \text{Any number multiplied by 1 is itself.}$$

Instructor Note: Students often confuse writing a fraction in lowest terms with writing a mixed number. The fraction $\frac{3}{2}$ is in lowest terms.

Skill Practice Simplify to lowest terms.

3. $\dfrac{20}{50}$

Answer

3. $\dfrac{2}{5}$

In Example 2, we showed numerous steps to reduce fractions to lowest terms. However, the process is often simplified. Notice that the same result can be obtained by dividing out the greatest common factor from the numerator and denominator. (The **greatest common factor** is the largest factor that is common to both numerator and denominator.)

$$\frac{45}{30} = \frac{3 \times 15}{2 \times 15} \qquad \text{The greatest common factor of 45 and 30 is 15.}$$

$$= \frac{3 \times \cancel{15}^{1}}{2 \times \cancel{15}_{1}} \qquad \text{The symbol / is often used to show that a common factor has been divided out.}$$

$$= \frac{3}{2} \qquad \text{Notice that "dividing out" the common factor of 15 has the same effect as dividing the numerator and denominator by 15. This is often done mentally.}$$

$$\frac{\overset{3}{\cancel{45}}}{\underset{2}{\cancel{30}}} = \frac{3}{2} \longleftarrow \text{45 divided by 15 equals 3.} \atop \longleftarrow \text{30 divided by 15 equals 2.}$$

| Example 3 | **Simplifying a Fraction to Lowest Terms** |

Simplify $\frac{14}{42}$ to lowest terms.

Solution:

$$\frac{14}{42} = \frac{1 \times 14}{3 \times 14} \qquad \text{The greatest common factor of 14 and 42 is 14.}$$

$$= \frac{1 \times \cancel{14}^{1}}{3 \times \cancel{14}_{1}}$$

$$= \frac{1}{3} \qquad \frac{\overset{1}{\cancel{14}}}{\underset{3}{\cancel{42}}} = \frac{1}{3} \longleftarrow \text{14 divided by 14 equals 1.} \atop \longleftarrow \text{42 divided by 14 equals 3.}$$

Classroom Example: p. 19, Exercise 48

Avoiding Mistakes

In Example 3, the common factor 14 in the numerator and denominator simplifies to 1. It is important to remember to write the factor of 1 in the numerator. The simplified form of the fraction is $\frac{1}{3}$.

Skill Practice Simplify to lowest terms.

4. $\frac{32}{12}$

4. Multiplying Fractions

Multiplying Fractions

If b is not zero and d is not zero, then

$$\frac{a}{b} \times \frac{c}{d} = \frac{a \times c}{b \times d}$$

To multiply fractions, multiply the numerators and multiply the denominators.

Answer

4. $\frac{8}{3}$

Classroom Example: p. 19, Exercise 58

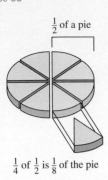

$\frac{1}{2}$ of a pie

$\frac{1}{4}$ of $\frac{1}{2}$ is $\frac{1}{8}$ of the pie

Figure 1-3

| Example 4 | **Multiplying Fractions** |

Multiply the fractions: $\frac{1}{4} \times \frac{1}{2}$

Solution:

$$\frac{1}{4} \times \frac{1}{2} = \frac{1 \times 1}{4 \times 2} = \frac{1}{8}$$ Multiply the numerators. Multiply the denominators.

Notice that the product $\frac{1}{4} \times \frac{1}{2}$ represents a quantity that is $\frac{1}{4}$ of $\frac{1}{2}$. Taking $\frac{1}{4}$ of a quantity is equivalent to dividing the quantity by 4. One-half of a pie divided into four pieces leaves pieces that each represent $\frac{1}{8}$ of the pie (Figure 1-3).

Skill Practice Multiply.

5. $\dfrac{2}{7} \times \dfrac{3}{5}$

Classroom Example: p. 19, Exercise 62

TIP: The same result can be obtained by dividing out common factors *before* multiplying.

$$\frac{\overset{1}{\cancel{7}}}{\underset{2}{\cancel{10}}} \times \frac{\overset{3}{\cancel{15}}}{\underset{2}{\cancel{14}}} = \frac{3}{4}$$

Instructor Note: You can also show multiplication of fractions using shading: The overlapped portion is the product.

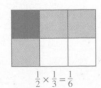

$\frac{1}{2} \times \frac{1}{3} = \frac{1}{6}$

$\frac{1}{2}$

$\frac{1}{3}$

| Example 5 | **Multiplying Fractions** |

Multiply the fractions.

a. $\dfrac{7}{10} \times \dfrac{15}{14}$ b. $\dfrac{2}{13} \times \dfrac{13}{2}$ c. $5 \times \dfrac{1}{5}$

Solution:

a. $\dfrac{7}{10} \times \dfrac{15}{14} = \dfrac{7 \times 15}{10 \times 14}$ Multiply the numerators. Multiply the denominators.

$= \dfrac{\overset{1}{\cancel{7}} \times \overset{3}{\cancel{15}}}{\underset{2}{\cancel{10}} \times \underset{2}{\cancel{14}}}$ Divide out the common factors.

$= \dfrac{3}{4}$ Multiply.

b. $\dfrac{2}{13} \times \dfrac{13}{2} = \dfrac{2 \times 13}{13 \times 2} = \dfrac{\overset{1}{\cancel{2}} \times \overset{1}{\cancel{13}}}{\underset{1}{\cancel{13}} \times \underset{1}{\cancel{2}}} = \dfrac{1}{1} = 1$ Multiply $1 \times 1 = 1$.   Multiply $1 \times 1 = 1$.

c. $5 \times \dfrac{1}{5} = \dfrac{5}{1} \times \dfrac{1}{5}$ The whole number 5 can be written as $\frac{5}{1}$.

$= \dfrac{\overset{1}{\cancel{5}} \times 1}{1 \times \underset{1}{\cancel{5}}} = \dfrac{1}{1} = 1$ Divide out the common factors and multiply.

Skill Practice Multiply.

6. $\dfrac{8}{9} \times \dfrac{3}{4}$ 7. $\dfrac{4}{5} \times \dfrac{5}{4}$ 8. $10 \times \dfrac{1}{10}$

Answers

5. $\dfrac{6}{35}$ 6. $\dfrac{2}{3}$ 7. 1   8. 1

5. Dividing Fractions

Before we divide fractions, we need to know how to find the reciprocal of a fraction. Notice from Example 5 that $\frac{2}{13} \times \frac{13}{2} = 1$ and $5 \times \frac{1}{5} = 1$. The numbers $\frac{2}{13}$ and $\frac{13}{2}$ are said to be reciprocals because their product is 1. Likewise the numbers 5 and $\frac{1}{5}$ are reciprocals.

The Reciprocal of a Number

Two nonzero numbers are **reciprocals** of each other if their product is 1. Therefore, the reciprocal of the fraction

$$\frac{a}{b} \text{ is } \frac{b}{a} \qquad \text{because} \qquad \frac{a}{b} \times \frac{b}{a} = 1$$

Number	Reciprocal	Product
$\dfrac{2}{15}$	$\dfrac{15}{2}$	$\dfrac{2}{15} \times \dfrac{15}{2} = 1$
$\dfrac{1}{8}$	$\dfrac{8}{1}$ (or equivalently 8)	$\dfrac{1}{8} \times 8 = 1$
$6 \left(\text{or equivalently } \dfrac{6}{1}\right)$	$\dfrac{1}{6}$	$6 \times \dfrac{1}{6} = 1$

To understand the concept of dividing fractions, consider a pie that is half-eaten. Suppose the remaining half must be divided among three people, that is, $\frac{1}{2} \div 3$. However, dividing by 3 is equivalent to taking $\frac{1}{3}$ of the remaining $\frac{1}{2}$ of the pie (Figure 1-4).

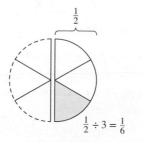

$$\frac{1}{2} \div 3 = \frac{1}{2} \cdot \frac{1}{3} = \frac{1}{6}$$

Figure 1-4

This example illustrates that dividing two numbers is equivalent to multiplying the first number by the reciprocal of the second number.

Dividing Fractions

Let a, b, c, and d be numbers such that b, c, and d are not zero. Then,

$$\frac{a}{b} \div \frac{c}{d} = \frac{a}{b} \times \frac{d}{c}$$

To divide fractions, multiply the first fraction by the reciprocal of the second fraction.

Classroom Examples: p. 19,
Exercises 60 and 64

Example 6   Dividing Fractions

Divide the fractions.

a. $\dfrac{8}{5} \div \dfrac{3}{10}$ **b.** $\dfrac{12}{13} \div 6$

Solution:

a. $\dfrac{8}{5} \div \dfrac{3}{10} = \dfrac{8}{5} \times \dfrac{10}{3}$ Multiply by the reciprocal of $\frac{3}{10}$, which is $\frac{10}{3}$.

$= \dfrac{8 \times \overset{2}{\cancel{10}}}{\underset{1}{\cancel{5}} \times 3} = \dfrac{16}{3}$ Divide out the common factors and multiply.

b. $\dfrac{12}{13} \div 6 = \dfrac{12}{13} \div \dfrac{6}{1}$ Write the whole number 6 as $\frac{6}{1}$.

$= \dfrac{12}{13} \times \dfrac{1}{6}$ Multiply by the reciprocal of $\frac{6}{1}$, which is $\frac{1}{6}$.

$= \dfrac{\overset{2}{\cancel{12}} \times 1}{13 \times \underset{1}{\cancel{6}}} = \dfrac{2}{13}$ Divide out the common factors and multiply.

Avoiding Mistakes

Always check that the final answer is in lowest terms.

Skill Practice Divide.

9. $\dfrac{12}{25} \div \dfrac{8}{15}$ **10.** $\dfrac{1}{4} \div 2$

6. Adding and Subtracting Fractions

Adding and Subtracting Fractions

Two fractions can be added or subtracted if they have a common denominator. Let a, b, and c be numbers such that b does not equal zero. Then,

$$\frac{a}{b} + \frac{c}{b} = \frac{a+c}{b} \qquad \text{and} \qquad \frac{a}{b} - \frac{c}{b} = \frac{a-c}{b}$$

To add or subtract fractions with the same denominator, add or subtract the numerators and write the result over the common denominator.

Classroom Examples: p. 19,
Exercises 78 and 80

Example 7   Adding and Subtracting Fractions with the Same Denominator

Add or subtract as indicated.

a. $\dfrac{1}{12} + \dfrac{7}{12}$ **b.** $\dfrac{13}{5} - \dfrac{3}{5}$

Answers

9. $\dfrac{9}{10}$ **10.** $\dfrac{1}{8}$

Solution:

a. $\dfrac{1}{12} + \dfrac{7}{12} = \dfrac{1+7}{12}$ Add the numerators.

$= \dfrac{8}{12}$

$= \dfrac{2}{3}$ Simplify to lowest terms.

b. $\dfrac{13}{5} - \dfrac{3}{5} = \dfrac{13-3}{5}$ Subtract the numerators.

$= \dfrac{10}{5}$ Simplify.

$= 2$ Simplify to lowest terms.

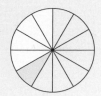

TIP: The sum $\frac{1}{12} + \frac{7}{12}$ can be visualized as the sum of the pink and blue sections of the figure.

Skill Practice Add or subtract as indicated.

11. $\dfrac{2}{3} + \dfrac{5}{3}$ **12.** $\dfrac{5}{8} - \dfrac{1}{8}$

In Example 7, we added and subtracted fractions with the same denominators. To add or subtract fractions with different denominators, we must first become familiar with the idea of the least common multiple between two or more numbers. The **least common multiple (LCM)** of two numbers is the smallest whole number that is a multiple of each number. For example, the LCM of 6 and 9 is 18.

multiples of 6:   6, 12, 18, 24, 30, 36, . . .
multiples of 9:   9, 18, 27, 36, 45, 54, . . .

Listing the multiples of two or more given numbers can be a cumbersome way to find the LCM. Therefore, we offer the following method to find the LCM of two numbers.

Finding the LCM of Two Numbers

Step 1   Write each number as a product of prime factors.
Step 2   The LCM is the product of unique prime factors from *both* numbers. Use repeated factors the maximum number of times they appear in *either* factorization.

Example 8   **Finding the LCM of Two Numbers**

Find the LCM of 9 and 15.

Classroom Example: p. 20, Exercise 82

Solution:

	3's	5's
9 =	3 × 3	
15 =	3 ×	5

LCM $= 3 \times 3 \times 5 = 45$

For the factors of 3 and 5, we circle the greatest number of times each occurs. The LCM is the product.

Skill Practice Find the LCM.

13. 10 and 25

Answers

11. $\dfrac{7}{3}$ or $2\dfrac{1}{3}$ **12.** $\dfrac{1}{2}$ **13.** 50

To add or subtract fractions with *different* denominators, we must first write each fraction as an equivalent fraction with a common denominator. A common denominator may be *any* common multiple of the denominators. However, we will use the least common denominator. The **least common denominator (LCD)** of two or more fractions is the LCM of the denominators of the fractions. The following example uses the fundamental principle of fractions to rewrite fractions with the desired denominator. *Note:* Multiplying the numerator and denominator by the *same* nonzero quantity will not change the value of the fraction.

Classroom Examples: p. 20, Exercises 86 and 90

Example 9   **Writing Equivalent Fractions and Subtracting Fractions**

a. Write each of the fractions $\frac{1}{9}$ and $\frac{1}{15}$ as an equivalent fraction with the LCD as its denominator.

b. Subtract $\frac{1}{9} - \frac{1}{15}$.

Solution:

From Example 8, we know that the LCM for 9 and 15 is 45. Therefore, the LCD of $\frac{1}{9}$ and $\frac{1}{15}$ is 45.

a. $\frac{1}{19} = \frac{}{45}$ $\frac{1 \times 5}{9 \times 5} = \frac{5}{45}$ So, $\frac{1}{9}$ is equivalent to $\frac{5}{45}$.

What number must we multiply 9 by to get 45?   Multiply numerator and denominator by 5.

$\frac{1}{15} = \frac{}{45}$ $\frac{1 \times 3}{15 \times 3} = \frac{3}{45}$ So, $\frac{1}{15}$ is equivalent to $\frac{3}{45}$.

What number must we multiply 15 by to get 45?   Multiply numerator and denominator by 3.

b. $\frac{1}{9} - \frac{1}{15}$

$= \frac{5}{45} - \frac{3}{45}$ Write $\frac{1}{9}$ and $\frac{1}{15}$ as equivalent fractions with the same denominator.

$= \frac{2}{45}$ Subtract.

Skill Practice

14. Write each of the fractions $\frac{5}{8}$ and $\frac{5}{12}$ as an equivalent fraction with the LCD as its denominator.

15. Subtract. $\frac{5}{8} - \frac{5}{12}$

Answers

14. $\frac{5}{8} = \frac{15}{24}$ and $\frac{5}{12} = \frac{10}{24}$ 15. $\frac{5}{24}$

| Example 10 | **Adding and Subtracting Fractions** |

Classroom Example: p. 20, Exercise 98

Simplify. $\dfrac{5}{12}+\dfrac{3}{4}-\dfrac{1}{2}$

Solution:

$\dfrac{5}{12}+\dfrac{3}{4}-\dfrac{1}{2}$

To find the LCD, we have:
LCD $= 2 \times 2 \times 3 = 12$

	2's	3's
12 =	2 × 2	3
4 =	2 × 2	
2 =	2	

$=\dfrac{5}{12}+\dfrac{3\times3}{4\times3}-\dfrac{1\times6}{2\times6}$

Write each fraction as an equivalent fraction with the LCD as its denominator.

$=\dfrac{5}{12}+\dfrac{9}{12}-\dfrac{6}{12}$

$=\dfrac{5+9-6}{12}$

Add and subtract the numerators.

$=\dfrac{8}{12}$

Simplify to lowest terms.

$=\dfrac{2}{3}$

Skill Practice Add.

16. $\dfrac{2}{3}+\dfrac{1}{2}+\dfrac{5}{6}$

7. Operations on Mixed Numbers

Recall that a mixed number is a whole number added to a fraction. The number $3\frac{1}{2}$ represents the sum of three wholes plus a half, that is, $3\frac{1}{2}=3+\frac{1}{2}$. For this reason, any mixed number can be converted to an improper fraction by using addition.

$$3\frac{1}{2}=3+\frac{1}{2}=\frac{6}{2}+\frac{1}{2}=\frac{7}{2}$$

TIP: A shortcut to writing a mixed number as an improper fraction is to multiply the whole number by the denominator of the fraction. Then add this value to the numerator of the fraction, and write the result over the denominator.

$3\frac{1}{2}$ ⟶ Multiply the whole number by the denominator: $3\times2=6$
Add the numerator: $6+1=7$
Write the result over the denominator: $\frac{7}{2}$

To add, subtract, multiply, or divide mixed numbers, we will first write the mixed number as an improper fraction.

Answer

16. 2

Classroom Example: p. 20, Exercise 114

Example 11 **Operations on Mixed Numbers**

Subtract. $5\frac{1}{3} - 2\frac{1}{4}$

Solution:

$5\frac{1}{3} - 2\frac{1}{4}$

$= \frac{16}{3} - \frac{9}{4}$ Write the mixed numbers as improper fractions.

$= \frac{16 \times 4}{3 \times 4} - \frac{9 \times 3}{4 \times 3}$ The LCD is 12. Multiply numerators and denominators by the missing factors from the denominators.

$= \frac{64}{12} - \frac{27}{12}$

$= \frac{37}{12}$ or $3\frac{1}{12}$ Subtract the fractions.

Skill Practice Subtract.

17. $2\frac{3}{4} - 1\frac{1}{3}$

TIP: An improper fraction can also be written as a mixed number. Both answers are acceptable. Note that

$$\frac{37}{12} = \frac{36}{12} + \frac{1}{12} = 3 + \frac{1}{12}, \text{ or } 3\frac{1}{12}$$

This can easily be found by dividing.

$$\frac{37}{12} \longrightarrow 12\overline{)37} \quad \begin{array}{c} \text{quotient} \\ 3 \\ -36 \\ \hline 1 \end{array} \quad 3\frac{1}{12} \quad \begin{array}{c} \text{remainder} \\ \text{divisor} \end{array}$$

Classroom Example: p. 20, Exercise 110

Example 12 **Operations on Mixed Numbers**

Divide. $7\frac{1}{2} \div 3$

Solution:

$7\frac{1}{2} \div 3$

$= \frac{15}{2} \div \frac{3}{1}$ Write the mixed number and whole number as fractions.

$= \frac{\overset{5}{\cancel{15}}}{2} \times \frac{1}{\underset{1}{\cancel{3}}}$ Multiply by the reciprocal of $\frac{3}{1}$, which is $\frac{1}{3}$.

$= \frac{5}{2}$ or $2\frac{1}{2}$ The answer may be written as an improper fraction or as a mixed number.

Avoiding Mistakes

Remember that when dividing (or multiplying) fractions, a common denominator is not necessary.

Answer

17. $\frac{17}{12}$ or $1\frac{5}{12}$

Skill Practice Divide.

18. $5\dfrac{5}{6} \div 3\dfrac{2}{3}$

Answer

18. $\dfrac{35}{22}$ or $1\dfrac{13}{22}$

Section 1.1 Practice Exercises

Study Skills Exercise

For additional exercises, see Classroom Activities 1.1A–1.1E in the *Student's Resource Manual* at www.mhhe.com/moh.

To enhance your learning experience, we provide study skills that focus on eight areas: learning about your course, using your text, taking notes, doing homework, taking an exam (test and math anxiety), managing your time, recognizing your learning style, and studying for the final exam.

Each activity requires only a few minutes and will help you pass this course and become a better math student. Many of these skills can be carried over to other disciplines and help you become a model college student. To begin, write down the following information:

a. Instructor's name

b. Instructor's office number

c. Instructor's telephone number

d. Instructor's e-mail address

e. Instructor's office hours

f. Days of the week that the class meets

g. The room number in which the class meets

h. Is there a lab requirement for this course? How often must you attend lab and where is it located?

Vocabulary and Key Concepts

1. a. A ___product___ is the result of multiplying two or more numbers.

b. The numbers being multiplied in a product are called ___factors___.

c. Given a fraction $\frac{a}{b}$ with $b \neq 0$, the value a is the ___numerator___ and ___b___ is the denominator.

d. A fraction is said to be in ___lowest___ terms if the numerator and denominator share no common factor other than 1.

e. The fraction $\frac{4}{4}$ can also be written as the whole number ___1___, and the fraction $\frac{4}{1}$ can be written as the whole number ___4___.

f. Two nonzero numbers $\frac{a}{b}$ and $\frac{b}{a}$ are ___reciprocals___ because their product is 1.

g. The least common multiple (LCM) of two numbers is the smallest whole number that is a ___multiple___ of both numbers.

h. The ___least___ common denominator of two or more fractions is the LCM of their denominators.

Concept 1: Basic Definitions

For Exercises 2–10, identify the numerator and denominator of each fraction. Then determine if the fraction is a proper fraction or an improper fraction.

2. $\dfrac{7}{8}$ Numerator: 7; denominator: 8; proper

3. $\dfrac{2}{3}$ Numerator: 2; denominator: 3; proper

4. $\dfrac{9}{5}$ Numerator: 9; denominator: 5; improper

5. $\dfrac{5}{2}$ Numerator: 5; denominator: 2; improper

6. $\dfrac{6}{6}$ Numerator: 6; denominator: 6; improper

7. $\dfrac{4}{4}$ Numerator: 4; denominator: 4; improper

8. $\dfrac{12}{1}$ Numerator: 12; denominator: 1; improper

9. $\dfrac{5}{1}$ Numerator: 5; denominator: 1; improper

10. $\dfrac{6}{7}$ Numerator: 6; denominator: 7; proper

 Writing   Translating Expression  Geometry   Scientific Calculator   Video

For Exercises 11–18, write a proper or improper fraction associated with the shaded region of each figure.

11.  $\frac{3}{4}$ **12.**  $\frac{4}{5}$

13.  $\frac{4}{3}$ **14.** 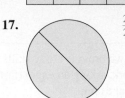 $\frac{5}{4}$

15. $\frac{1}{6}$ **16.** $\frac{1}{8}$ **17.** $\frac{2}{2}$ **18.** $\frac{4}{4}$

   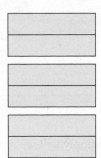   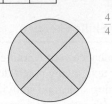   

For Exercises 19–22, write both an improper fraction and a mixed number associated with the shaded region of each figure.

19. $\frac{5}{2}$ or $2\frac{1}{2}$ **20.** $\frac{5}{3}$ or $1\frac{2}{3}$ **21.** $\frac{6}{2}$ or 3   **22.** $\frac{6}{3}$ or 2

23. Explain the difference between the set of whole numbers and the set of natural numbers.
The set of whole numbers includes the number 0 and the set of natural numbers does not.

24. Explain the difference between a proper fraction and an improper fraction. A proper fraction represents a number less than one unit. An improper fraction represents a number greater than or equal to a whole unit.

25. Write a fraction that simplifies to $\frac{1}{2}$. (Answers may vary.)   For example: $\frac{2}{4}$

26. Write a fraction that simplifies to $\frac{1}{3}$. (Answers may vary.)   For example: $\frac{2}{6}$

Concept 2: Prime Factorization

For Exercises 27–34, identify each number as either a prime number or a composite number.

27. 5 Prime   **28.** 9 Composite   **29.** 4 Composite   **30.** 2 Prime

31. 39 Composite   **32.** 23 Prime   **33.** 53 Prime   **34.** 51 Composite

For Exercises 35–42, write each number as a product of prime factors. **(See Example 1.)**

35. 36 $2 \times 2 \times 3 \times 3$ **36.** 70 $2 \times 5 \times 7$ **37.** 42 $2 \times 3 \times 7$ **38.** 35 5×7

39. 110 $2 \times 5 \times 11$ **40.** 136 $2 \times 2 \times 2 \times 17$ **41.** 135 $3 \times 3 \times 3 \times 5$ **42.** 105 $3 \times 5 \times 7$

Concept 3: Simplifying Fractions to Lowest Terms

For Exercises 43–54, simplify each fraction to lowest terms. **(See Examples 2–3.)**

43. $\dfrac{3}{15}$ $\dfrac{1}{5}$

44. $\dfrac{8}{12}$ $\dfrac{2}{3}$

45. $\dfrac{16}{6}$ $\dfrac{8}{3}$ or $2\dfrac{2}{3}$

46. $\dfrac{20}{12}$ $\dfrac{5}{3}$ or $1\dfrac{2}{3}$

47. $\dfrac{42}{48}$ $\dfrac{7}{8}$

48. $\dfrac{35}{80}$ $\dfrac{7}{16}$

49. $\dfrac{48}{64}$ $\dfrac{3}{4}$

50. $\dfrac{32}{48}$ $\dfrac{2}{3}$

51. $\dfrac{110}{176}$ $\dfrac{5}{8}$

52. $\dfrac{70}{120}$ $\dfrac{7}{12}$

53. $\dfrac{200}{150}$ $\dfrac{4}{3}$ or $1\dfrac{1}{3}$

54. $\dfrac{210}{119}$ $\dfrac{30}{17}$ or $1\dfrac{13}{17}$

Concepts 4–5: Multiplying and Dividing Fractions

For Exercises 55–56, determine if the statement is true or false. If it is false, rewrite as a true statement.

55. When multiplying or dividing fractions, it is necessary to have a common denominator.   False: When adding or subtracting fractions, it is necessary to have a common denominator.

56. When dividing two fractions, it is necessary to multiply the first fraction by the reciprocal of the second fraction.   True

For Exercises 57–68, multiply or divide as indicated. **(See Examples 4–6.)**

57. $\dfrac{10}{13} \times \dfrac{26}{15}$ $\dfrac{4}{3}$ or $1\dfrac{1}{3}$

58. $\dfrac{15}{28} \times \dfrac{7}{9}$ $\dfrac{5}{12}$

59. $\dfrac{3}{7} \div \dfrac{9}{14}$ $\dfrac{2}{3}$

60. $\dfrac{7}{25} \div \dfrac{1}{5}$ $\dfrac{7}{5}$ or $1\dfrac{2}{5}$

61. $\dfrac{9}{10} \times 5$ $\dfrac{9}{2}$ or $4\dfrac{1}{2}$

62. $\dfrac{3}{7} \times 14$ 6

63. $\dfrac{12}{5} \div 4$ $\dfrac{3}{5}$

64. $\dfrac{20}{6} \div 5$ $\dfrac{2}{3}$

65. $\dfrac{5}{2} \times \dfrac{10}{21} \times \dfrac{7}{5}$ $\dfrac{5}{3}$ or $1\dfrac{2}{3}$

66. $\dfrac{55}{9} \times \dfrac{18}{32} \times \dfrac{24}{11}$ $\dfrac{15}{2}$ or $7\dfrac{1}{2}$

67. $\dfrac{9}{100} \div \dfrac{13}{1000}$ $\dfrac{90}{13}$ or $6\dfrac{12}{13}$

68. $\dfrac{1000}{17} \div \dfrac{10}{3}$ $\dfrac{300}{17}$ or $17\dfrac{11}{17}$

69. Gus decides to save $\frac{1}{3}$ of his pay each month. If his monthly pay is \$2112, how much will he save each month?   \$704

70. Stephen's take-home pay is \$4200 a month. If he budgeted $\frac{1}{4}$ of his pay for rent, how much is his rent?   \$1050

71. In Professor Foley's Beginning Algebra class, $\frac{5}{6}$ of the students passed the first test. If there are 42 students in the class, how many passed the test?   35 students

72. Shontell had only enough paper to print out $\frac{3}{5}$ of her book report before school. If the report is 10 pages long, how many pages did she print out?   6 pages

73. Marty will reinforce a concrete walkway by cutting a steel rod (called rebar) that is 4 yd long. How many pieces can he cut if each piece must be $\frac{1}{2}$ yd in length?   8 pieces

74. There are 4 cups of oatmeal in a box. If each serving is $\frac{1}{3}$ of a cup, how many servings are contained in the box?   12 servings

75. Anita buys 6 lb of mixed nuts to be divided into decorative jars that will each hold $\frac{3}{4}$ lb of nuts. How many jars will she be able to fill?   8 jars

76. Beth has a $\frac{7}{8}$-in. nail that she must hammer into a board. Each strike of the hammer moves the nail $\frac{1}{16}$ in. into the board. How many strikes of the hammer must she make to drive the nail completely into the board?   Beth must make 14 strikes.

Concept 6: Adding and Subtracting Fractions

For Exercises 77–80, add or subtract as indicated. **(See Example 7.)**

77. $\dfrac{5}{14} + \dfrac{1}{14}$ $\dfrac{3}{7}$

78. $\dfrac{9}{5} + \dfrac{1}{5}$ 2

79. $\dfrac{17}{24} - \dfrac{5}{24}$ $\dfrac{1}{2}$

80. $\dfrac{11}{18} - \dfrac{5}{18}$ $\dfrac{1}{3}$

For Exercises 81–84, find the least common multiple for each list of numbers. **(See Example 8.)**

81. 6, 15 $\quad$ 30

82. 12, 30 $\quad$ 60

83. 20, 8, 4 $\quad$ 40

84. 24, 40, 30 $\quad$ 120

For Exercises 85–100, add or subtract as indicated. **(See Examples 9–10.)**

85. $\dfrac{1}{8} + \dfrac{3}{4}$ $\quad \dfrac{7}{8}$

86. $\dfrac{3}{16} + \dfrac{1}{2}$ $\quad \dfrac{11}{16}$

87. $\dfrac{11}{8} - \dfrac{3}{10}$ $\quad \dfrac{43}{40}$ or $1\dfrac{3}{40}$

88. $\dfrac{12}{35} - \dfrac{1}{10}$ $\quad \dfrac{17}{70}$

89. $\dfrac{7}{26} - \dfrac{2}{13}$ $\quad \dfrac{3}{26}$

90. $\dfrac{25}{24} - \dfrac{5}{16}$ $\quad \dfrac{35}{48}$

91. $\dfrac{7}{18} + \dfrac{5}{12}$ $\quad \dfrac{29}{36}$

92. $\dfrac{3}{16} + \dfrac{9}{20}$ $\quad \dfrac{51}{80}$

93. $\dfrac{5}{4} - \dfrac{1}{20}$ $\quad \dfrac{6}{5}$ or $1\dfrac{1}{5}$

94. $\dfrac{7}{6} - \dfrac{1}{24}$ $\quad \dfrac{9}{8}$ or $1\dfrac{1}{8}$

95. $\dfrac{5}{12} + \dfrac{5}{16}$ $\quad \dfrac{35}{48}$

96. $\dfrac{3}{25} + \dfrac{8}{35}$ $\quad \dfrac{61}{175}$

97. $\dfrac{1}{6} + \dfrac{3}{4} - \dfrac{5}{8}$ $\quad \dfrac{7}{24}$

98. $\dfrac{1}{2} + \dfrac{2}{3} - \dfrac{5}{12}$ $\quad \dfrac{3}{4}$

99. $\dfrac{4}{7} + \dfrac{1}{2} + \dfrac{3}{4}$ $\quad \dfrac{51}{28}$ or $1\dfrac{23}{28}$

100. $\dfrac{9}{10} + \dfrac{4}{5} + \dfrac{3}{4}$ $\quad \dfrac{49}{20}$ or $2\dfrac{9}{20}$

101. For his famous brownie recipe, Chef Alfonso combines $\frac{2}{3}$ cup granulated sugar with $\frac{1}{4}$ cup brown sugar. What is the total amount of sugar in his recipe? $\dfrac{11}{12}$ cup sugar

102. Chef Alfonso eats too many of his brownies and his waistline increased by $\frac{3}{4}$ in. during one month and $\frac{3}{8}$ in. the next month. What was his total increase for the 2-month period? $\dfrac{9}{8}$ in. or $1\dfrac{1}{8}$ in.

103. Currently the most popular smartphone has a thickness of $\frac{9}{25}$ in. The second most popular is $\frac{1}{2}$ in. thick. How much thicker is the second most popular smartphone? $\dfrac{7}{50}$ in.

104. The diameter of a penny is $\frac{3}{4}$ in. while the dime is $\frac{7}{10}$ in. How much larger is the penny than the dime? $\dfrac{1}{20}$ in.

Concept 7: Operations on Mixed Numbers

For Exercises 105–118, perform the indicated operations. **(See Examples 11–12.)**

105. $3\dfrac{1}{5} \times 2\dfrac{7}{8}$ $\quad \dfrac{46}{5}$ or $9\dfrac{1}{5}$

106. $2\dfrac{1}{2} \times 1\dfrac{4}{5}$ $\quad \dfrac{9}{2}$ or $4\dfrac{1}{2}$

107. $1\dfrac{2}{9} \div 7\dfrac{1}{3}$ $\quad \dfrac{1}{6}$

108. $2\dfrac{2}{5} \div 1\dfrac{2}{7}$ $\quad \dfrac{28}{15}$ or $1\dfrac{13}{15}$

109. $1\dfrac{2}{9} \div 6$ $\quad \dfrac{11}{54}$

110. $2\dfrac{2}{5} \div 2$ $\quad \dfrac{6}{5}$ or $1\dfrac{1}{5}$

111. $2\dfrac{1}{8} + 1\dfrac{3}{8}$ $\quad \dfrac{7}{2}$ or $3\dfrac{1}{2}$

112. $1\dfrac{3}{14} + 1\dfrac{1}{14}$ $\quad \dfrac{16}{7}$ or $2\dfrac{2}{7}$

113. $3\dfrac{1}{2} - 1\dfrac{7}{8}$ $\quad \dfrac{13}{8}$ or $1\dfrac{5}{8}$

114. $5\dfrac{1}{3} - 2\dfrac{3}{4}$ $\quad \dfrac{31}{12}$ or $2\dfrac{7}{12}$

115. $1\dfrac{1}{6} + 3\dfrac{3}{4}$ $\quad \dfrac{59}{12}$ or $4\dfrac{11}{12}$

116. $4\dfrac{1}{2} + 2\dfrac{2}{3}$ $\quad \dfrac{43}{6}$ or $7\dfrac{1}{6}$

117. $1 - \dfrac{7}{8}$ $\quad \dfrac{1}{8}$

118. $2 - \dfrac{3}{7}$ $\quad \dfrac{11}{7}$ or $1\dfrac{4}{7}$

119. A board $26\frac{3}{8}$ in. long must be cut into three pieces of equal length. Find the length of each piece.

$8\frac{19}{24}$ in.

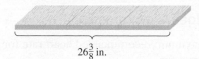

$26\frac{3}{8}$ in.

120. A futon, when set up as a sofa, measures $3\frac{5}{6}$ ft wide. When it is opened to be used as a bed, the width is increased by $1\frac{3}{4}$ ft. What is the total width of this bed? $5\frac{7}{12}$ ft

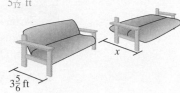

$3\frac{5}{6}$ ft x

121. A plane trip from Orlando to Detroit takes $2\frac{3}{4}$ hr. If the plane traveled for $1\frac{1}{6}$ hr, how much time remains for the flight? $1\frac{7}{12}$ hr

122. Silvia manages a sub shop and needs to prepare smoked turkey sandwiches. She has $3\frac{3}{4}$ lb of turkey in the cooler, and each sandwich requires $\frac{3}{8}$ lb of turkey. How many sandwiches can she make?

10 sandwiches

123. José's catering company plans to prepare two different shrimp dishes for an upcoming event. One dish requires $1\frac{1}{2}$ lb of shrimp and the other requires $\frac{3}{4}$ lb of shrimp. How much shrimp should José order for the two dishes? $2\frac{1}{4}$ lb

124. Ayako took a trip to the store $5\frac{1}{2}$ mi away. If she rode the bus for $4\frac{5}{6}$ mi and walked the rest of the way, how far did she have to walk? $\frac{2}{3}$ mi

125. If Tampa, Florida, averages $6\frac{1}{4}$ in. of rain during each summer month, how much total rain would be expected in June, July, August, and September?

25 in.

126. Pete started working out and found that he lost approximately $\frac{3}{4}$ in. off his waistline every month. How much would he lose around his waist in 6 months? $\frac{9}{2}$ in. or $4\frac{1}{2}$ in.

Introduction to Algebra and the Set of Real Numbers   Section 1.2

1. Variables and Expressions

Concepts

1. **Variables and Expressions**
2. **The Set of Real Numbers**
3. **Inequalities**
4. **Opposite of a Real Number**
5. **Absolute Value of a Real Number**

Doctors promote daily exercise as part of a healthy lifestyle. Aerobic exercise is exercise for the heart. During aerobic exercise, the goal is to maintain a heart rate level between 65% and 85% of an individual's maximum recommended heart rate. The maximum recommended heart rate, in beats per minute, for an adult of age a is given by:

$$\text{Maximum recommended heart rate} = 220 - a$$

In this example, value a is called a **variable**. This is a symbol or letter, such as x, y, z, a, and the like, that is used to represent an unknown number that is subject to change. The number 220 is called a **constant**, because it does not vary. The quantity $220 - a$ is called an algebraic expression. An algebraic **expression** is a collection of variables and constants under algebraic operations. For example, $\frac{3}{x}$, $y + 7$, and $t - 1.4$ are algebraic expressions.

The symbols used in algebraic expressions to show the four basic operations are shown here:

Addition $a + b$

Subtraction $a - b$

Multiplication $a \times b$, $a \cdot b$, $(a)b$, $a(b)$, $(a)(b)$, ab
      (*Note:* We rarely use the notation $a \times b$ because the symbol $\times$ may be confused with the variable x.)

Division $a \div b$, $\dfrac{a}{b}$, a/b, $b\overline{)a}$

 Writing   Translating Expression   Geometry   Scientific Calculator   Video

The value of an algebraic expression depends on the values of the variables within the expression.

Classroom Example: p. 32, Exercise 16

Example 1   **Evaluating an Algebraic Expression**

The expression $220 - a$ represents the maximum recommended heart rate for an adult of age a. Determine the maximum heart rate for:

a. A 20-year-old   **b.** A 45-year-old

Solution:

a. In the expression $220 - a$, the variable, a, represents the age of the individual. To calculate the maximum recommended heart rate for a 20-year-old, we substitute 20 for a in the expression.

$$220 - a$$

$220 - (\ \ )$	When substituting a number for a variable, use parentheses.
$= 220 - (20)$	Substitute $a = 20$.
$= 200$	Subtract.

The maximum recommended heart rate for a 20-year-old is 200 beats per minute.

b. $220 - a$

$220 - (\ \ )$	When substituting a number for a variable, use parentheses.
$= 220 - (45)$	Substitute $a = 45$.
$= 175$	Subtract.

The maximum recommended heart rate for a 45-year-old is 175 beats per minute.

Skill Practice

1. After dining out at a restaurant, the recommended minimum amount for tipping the server is 15% of the cost of the meal. This can be represented by the expression $0.15c$, where c is the cost of the meal. Compute the tip for a meal that costs:
   a. $18   **b.** $46

Classroom Examples: p. 31, Exercises 8 and 10

Example 2   **Evaluating Algebraic Expressions**

Evaluate the algebraic expression when $p = 4$ and $q = \frac{3}{4}$.

a. $100 - p$ **b.** pq

Solution:

a. $100 - p$

$100 - (\ \ )$	When substituting a number for a variable, use parentheses.
$= 100 - (4)$	Substitute $p = 4$ in the parentheses.
$= 96$	Subtract.

Answers

1. **a.** $2.70   **b.** $6.90

b. pq

$= (\quad)(\quad)$ When substituting a number for a variable, use parentheses.

$= (4)\left(\dfrac{3}{4}\right)$ Substitute $p = 4$ and $q = \frac{3}{4}$.

$= \dfrac{\overset{1}{\cancel{4}}}{1} \cdot \dfrac{3}{\underset{1}{\cancel{4}}}$ Write the whole number as a fraction.

$= \dfrac{3}{1}$ Multiply fractions.

$= 3$ Simplify.

Skill Practice Evaluate the algebraic expressions when $x = 5$ and $y = 2$.

   2. $20 - y$ **3.** xy

2. The Set of Real Numbers

Typically, the numbers represented by variables in an algebraic expression are all part of the set of **real numbers**. These are the numbers that we work with on a day-to-day basis. The real numbers encompass zero, all positive, and all negative numbers, including those represented by fractions and decimal numbers. The set of real numbers can be represented graphically on a horizontal number line with a point labeled as 0. Positive real numbers are graphed to the right of 0, and negative real numbers are graphed to the left of 0. Zero is neither positive nor negative. Each point on the number line corresponds to exactly one real number. For this reason, this number line is called the *real number line* (Figure 1-5).

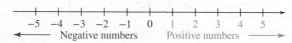

Figure 1-5

Example 3	**Plotting Points on the Real Number Line**

Plot the numbers on the real number line.

   a. -3 **b.** $\dfrac{3}{2}$ **c.** -4.7 **d.** $\dfrac{16}{5}$

Solution:

   a. Because -3 is negative, it lies three units to the left of 0.

   b. The fraction $\frac{3}{2}$ can be expressed as the mixed number $1\frac{1}{2}$ which lies halfway between 1 and 2 on the number line.

   c. The negative number -4.7 lies $\frac{7}{10}$ unit to the left of -4 on the number line.

   d. The fraction $\frac{16}{5}$ can be expressed as the mixed number $3\frac{1}{5}$, which lies $\frac{1}{5}$ unit to the right of 3 on the number line.

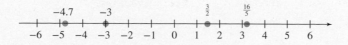

Answers

   2. 18   **3.** 10

Skill Practice Plot the numbers on the real number line.

4. $\{-1, \frac{3}{4}, -2.5, \frac{10}{3}\}$

In mathematics, a well-defined collection of elements is called a **set**. "Well-defined" means the set is described in such a way that it is clear whether an element is in the set. The symbols { } are used to enclose the elements of the set. For example, the set {A, B, C, D, E} represents the set of the first five letters of the alphabet.

Several sets of numbers are used extensively in algebra and are *subsets* (or part) of the set of real numbers.

Natural Numbers, Whole Numbers, and Integers

The set of **natural numbers** is $\{1, 2, 3, \ldots\}$

The set of **whole numbers** is $\{0, 1, 2, 3, \ldots\}$

The set of **integers** is $\{\ldots -3, -2, -1, 0, 1, 2, 3, \ldots\}$

Notice that the set of whole numbers includes the natural numbers. Therefore, every natural number is also a whole number. The set of integers includes the set of whole numbers. Therefore, every whole number is also an integer.

Fractions are also among the numbers we use frequently. A number that can be written as a fraction whose numerator is an integer and whose denominator is a nonzero integer is called a *rational number*.

Rational Numbers

The set of **rational numbers** is the set of numbers that can be expressed in the form $\frac{p}{q}$, where both p and q are integers and q does not equal 0.

We also say that a rational number $\frac{p}{q}$ is a *ratio* of two integers, p and q, where q is not equal to zero.

Example 4   Identifying Rational Numbers

Show that the following numbers are rational numbers by finding an equivalent ratio of two integers.

   a. $\frac{-2}{3}$ **b.** -12 **c.** 0.5 **d.** $0.\overline{6}$

Solution:

   a. The fraction $\frac{-2}{3}$ is a rational number because it can be expressed as the ratio of -2 and 3.

   b. The number -12 is a rational number because it can be expressed as the ratio of -12 and 1, that is, $-12 = \frac{-12}{1}$. In this example, we see that an integer is also a rational number.

Answer

4.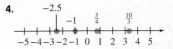

c. The terminating decimal 0.5 is a rational number because it can be expressed as the ratio of 5 and 10, that is, $0.5 = \frac{5}{10}$. In this example, we see that a terminating decimal is also a rational number.

d. The numeral $0.\overline{6}$ represents the nonterminating, repeating decimal $0.6666666\ldots$. The number $0.\overline{6}$ is a rational number because it can be expressed as the ratio of 2 and 3, that is, $0.\overline{6} = \frac{2}{3}$. In this example, we see that a repeating decimal is also a rational number.

Skill Practice Show that each number is rational by finding an equivalent ratio of two integers.

5. $\dfrac{3}{7}$ **6.** -5 **7.** 0.3 **8.** $0.\overline{3}$

> **TIP:** A rational number can be represented by a terminating decimal or by a repeating decimal.

Some real numbers, such as the number π, cannot be represented by the ratio of two integers. These numbers are called irrational numbers and in decimal form are nonterminating, nonrepeating decimals. The value of π, for example, can be approximated as $\pi \approx 3.1415926535897932$. However, the decimal digits continue forever with no repeated pattern. Another example of an irrational number is $\sqrt{3}$ (read as "the positive square root of 3"). The expression $\sqrt{3}$ is a number that when multiplied by itself is 3. There is no rational number that satisfies this condition. Thus, $\sqrt{3}$ is an irrational number.

Irrational Numbers

The set of **irrational numbers** is a subset of the real numbers whose elements cannot be written as a ratio of two integers.

Note: An irrational number cannot be written as a terminating decimal or as a repeating decimal.

The set of real numbers consists of both the rational and the irrational numbers. The relationship among these important sets of numbers is illustrated in Figure 1-6 along with numerical examples.

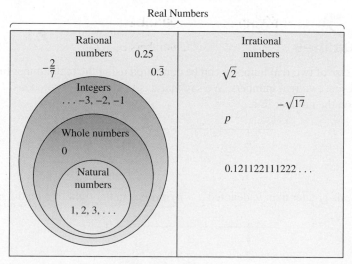

Figure 1-6

Animation

Answers
5. Ratio of 3 and 7
6. Ratio of −5 and 1
7. Ratio of 3 and 10
8. Ratio of 1 and 3

Classroom Examples: p. 33,
Exercises 44 and 46

Example 5   Classifying Numbers by Set

Check the set(s) to which each number belongs. The numbers may belong to more than one set.

	Natural Numbers	Whole Numbers	Integers	Rational Numbers	Irrational Numbers	Real Numbers
5						
$\dfrac{-47}{3}$						
1.48						
$\sqrt{7}$						
0						

Solution:

	Natural Numbers	Whole Numbers	Integers	Rational Numbers	Irrational Numbers	Real Numbers
5	✓	✓	✓	✓ (ratio of 5 and 1)		✓
$\dfrac{-47}{3}$				✓ (ratio of −47 and 3)		✓
1.48				✓ (ratio of 148 and 100)		✓
$\sqrt{7}$					✓	✓
0		✓	✓	✓ (ratio of 0 and 1)		✓

Skill Practice Identify the sets to which each number belongs. Choose from: natural numbers, whole numbers, integers, rational numbers, irrational numbers, real numbers.

9. −4   **10.** $0.\overline{7}$ **11.** $\sqrt{13}$ **12.** 12   **13.** 1

3. Inequalities

The relative size of two real numbers can be compared using the real number line. Suppose a and b represent two real numbers. We say that a is less than b, denoted $a < b$, if a lies to the left of b on the number line.

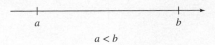

$a < b$

We say that a is greater than b, denoted $a > b$, if a lies to the right of b on the number line.

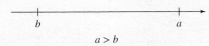

$a > b$

Answers
   9. Integers, rational numbers, real numbers
   10. Rational numbers, real numbers
   11. Irrational numbers, real numbers
   12. Natural numbers, whole numbers, integers, rational numbers, real numbers
   13. Natural numbers, whole numbers, integers, rational numbers, real numbers

Table 1-1 summarizes the relational operators that compare two real numbers a and b.

Table 1-1

Mathematical Expression	Translation	Example
$a < b$	a is less than b.	$2 < 3$
$a > b$	a is greater than b.	$5 > 1$
$a \leq b$	a is less than or equal to b.	$4 \leq 4$
$a \geq b$	a is greater than or equal to b.	$10 \geq 9$
$a = b$	a is equal to b.	$6 = 6$
$a \neq b$	a is not equal to b.	$7 \neq 0$
$a \approx b$	a is approximately equal to b.	$2.3 \approx 2$

The symbols $<$, $>$, $\leq$, $\geq$, and $\neq$ are called *inequality signs*, and the statements $a < b$, $a > b$, $a \leq b$, $a \geq b$, and $a \neq b$ are called **inequalities**.

Example 6 **Ordering Real Numbers**

Classroom Example: p. 33, Exercise 50

The average temperatures (in degrees Celsius) for selected cities in the United States and Canada in January are shown in Table 1-2.

Table 1-2

City	Temp (°C)
Prince George, British Columbia	−12.5
Corpus Christi, Texas	13.4
Parkersburg, West Virginia	−0.9
San Jose, California	9.7
Juneau, Alaska	−5.7
New Bedford, Massachusetts	−0.2
Durham, North Carolina	4.2

Plot a point on the real number line representing the temperature of each city. Compare the temperatures between the following cities, and fill in the blank with the appropriate inequality sign: $<$ or $>$.

Solution:

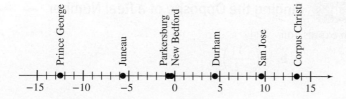

a. Temperature of San Jose $\boxed{<}$ temperature of Corpus Christi

b. Temperature of Juneau $\boxed{>}$ temperature of Prince George

c. Temperature of Parkersburg $\boxed{<}$ temperature of New Bedford

d. Temperature of Parkersburg $\boxed{>}$ temperature of Prince George

Skill Practice Fill in the blanks with the appropriate inequality sign:
< or >.

14. −11 _____ 20   **15.** −3 _____ −6

16. 0 _____ −9   **17.** −6.2 _____ −1.8

4. Opposite of a Real Number

To gain mastery of any algebraic skill, it is necessary to know the meaning of key definitions and key symbols. Two important definitions are the *opposite* of a real number and the *absolute value* of a real number.

> ### The Opposite of a Real Number
>
> Two numbers that are the same distance from 0 but on opposite sides of 0 on the number line are called **opposites** of each other. Symbolically, we denote the opposite of a real number a as $-a$.

Classroom Examples: p. 33,
Exercises 52 and 54

Example 7   **Finding the Opposite of a Real Number**

a. Find the opposite of 5.   **b.** Find the opposite of $-\frac{4}{7}$.

Solution:

a. The opposite of 5 is −5.   **b.** The opposite of $-\frac{4}{7}$ is $\frac{4}{7}$.

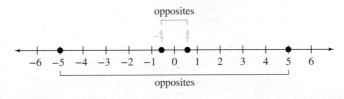

Skill Practice Find the opposite.

18. 224   **19.** −3.4

Classroom Example: p. 34,
Exercise 62

Example 8   **Finding the Opposite of a Real Number**

Evaluate each expression.

a. $-(0.46)$ **b.** $-\left(-\frac{11}{3}\right)$

Solution:

a. $-(0.46) = -0.46$ The expression $-(0.46)$ represents the opposite of 0.46.

b. $-\left(-\frac{11}{3}\right) = \frac{11}{3}$ The expression $-\left(-\frac{11}{3}\right)$ represents the opposite of $-\frac{11}{3}$.

Skill Practice Evaluate.

20. $-(2.8)$ **21.** $-\left(-\frac{1}{5}\right)$

Answers

14. <   **15.** >   **16.** >
17. <   **18.** −224   **19.** 3.4
20. −2.8   **21.** $\frac{1}{5}$

5. Absolute Value of a Real Number

To define the addition of real numbers, we use the concept of absolute value.

> ### Informal Definition of the Absolute Value of a Real Number
>
> The **absolute value** of a real number a, denoted $|a|$, is the distance between a and 0 on the number line.
>
> *Note:* The absolute value of any real number is positive or zero.

Instructor Note: Remind students that because absolute value is a measure of *distance* it will always be nonnegative. We don't measure distance with negative numbers.

For example, $|3| = 3$ and $|-3| = 3$.

| **Example 9** | **Finding the Absolute Value of a Real Number** |

Classroom Examples: p. 34, Exercises 68 and 72

Evaluate the absolute value expressions.

a. $|-4|$　　**b.** $\left|\dfrac{1}{2}\right|$　　**c.** $|-6.2|$　　**d.** $|0|$

Solution:

a. $|-4| = 4$　　　-4 is 4 units from 0 on the number line.

b. $\left|\dfrac{1}{2}\right| = \dfrac{1}{2}$　　　$\dfrac{1}{2}$ is $\dfrac{1}{2}$ unit from 0 on the number line.

c. $|-6.2| = 6.2$　　　-6.2 is 6.2 units from 0 on the number line.

d. $|0| = 0$　　　0 is 0 units from 0 on the number line.

Skill Practice Evaluate.

22. $|-99|$　　**23.** $\left|\dfrac{7}{8}\right|$　　**24.** $|-1.4|$　　**25.** $|1|$

Answers

22. 99　**23.** $\dfrac{7}{8}$

24. 1.4　**25.** 1

The absolute value of a number a is its distance from 0 on the number line. The definition of $|a|$ may also be given symbolically depending on whether a is negative or nonnegative.

Absolute Value of a Real Number

Let a be a real number. Then

1. If a is nonnegative (that is, $a \geq 0$), then $|a| = a$.

2. If a is negative (that is, $a < 0$), then $|a| = -a$.

This definition states that if a is a nonnegative number, then $|a|$ equals a itself. If a is a negative number, then $|a|$ equals the opposite of a. For example:

$|9| = 9$ Because 9 is positive, then $|9|$ equals the number 9 itself.

$|-7| = 7$ Because -7 is negative, then $|-7|$ equals the opposite of -7, which is 7.

Classroom Examples: p. 34,
Exercises 94 and 100

Example 10   **Comparing Absolute Value Expressions**

Determine if the statements are true or false.

  a. $|3| \leq 3$ **b.** $-|5| = |-5|$

Solution:

  a. $|3| \leq 3$ $|3| \overset{?}{\leq} 3$ Simplify the absolute value.

                  $3 \overset{?}{\leq} 3$ True

  b. $-|5| = |-5|$ $-|5| \overset{?}{=} |-5|$ Simplify the absolute values.

                  $-5 \overset{?}{=} 5$ False

Skill Practice   Answer true or false.

  26. $-|4| > |-4|$ **27.** $|-17| = 17$

Answers

26. False   **27.** True

Calculator Connections

Topic: Approximating Irrational Numbers on a Calculator

Scientific and graphing calculators approximate irrational numbers by using rational numbers in the form of terminating decimals. For example, consider approximating π and $\sqrt{3}$.

Scientific Calculator:

Enter: $\boxed{\pi}$ or $\boxed{\text{2nd}}$ $\boxed{\pi}$ **Result:** $\boxed{3.141592654}$

Enter: 3 $\boxed{\sqrt{}}$ **Result:** $\boxed{1.732050808}$

Graphing Calculator:

Enter: $\boxed{\text{2nd}}$ $\boxed{\pi}$ $\boxed{\text{ENTER}}$

Enter: $\boxed{\text{2nd}}$ $\boxed{\sqrt{}}$ 3 $\boxed{\text{ENTER}}$

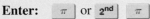

```
π
            3.141592654
√(3)
            1.732050808
```

The symbol $\approx$ is read "is approximately equal to" and is used when writing approximations.

$$\pi \approx 3.141592654 \quad \text{and} \quad \sqrt{3} \approx 1.732050808$$

Calculator Exercises

Use a calculator to approximate the irrational numbers. Remember to use the appropriate symbol, $\approx$, when expressing answers.

1. $\sqrt{12}$ ≈ 3.464101615 **2.** $\sqrt{99}$ ≈ 9.949874371 **3.** 4π ≈ 12.56637061 **4.** $\sqrt{\pi}$ ≈ 1.772453851

Section 1.2   Practice Exercises

Study Skills Exercise

For additional exercises, see Classroom Activities 1.2A–1.2C in the *Student's Resource Manual* at www.mhhe.com/moh.

Look over the notes that you took today. Do you understand what you wrote? If there were any rules, definitions, or formulas, highlight them so that they can be easily found when studying for the test.

Vocabulary and Key Concepts

1. a. A ___variable___ is a symbol or letter used to represent an unknown number.

  b. Values that do not vary are called ___constants___.

  c. In mathematics, a well-defined collection of elements is called a ___set___.

  d. The statements $a < b$, $a > b$, and $a \neq b$ are examples of ___inequalities___.

  e. The statement $a < b$ is read as "___a is less than b___."

  f. The statement $c \geq d$ is read as "___c is greater than or equal to d___."

  g. The statement $5 \neq 6$ is read as "___5 is not equal to 6___."

  h. Two numbers that are the same distance from 0 but on opposite sides of 0 on the number line are called ___opposites___.

  i. The absolute value of a real number, a, is denoted by ___$|a|$___ and is the distance between a and ___0___ on the number line.

Review Exercises

For Exercises 2–4, simplify.

2. $4\frac{1}{2} - 1\frac{5}{6}$ $2\frac{2}{3}$ or $\frac{8}{3}$ **3.** $4\frac{1}{2} \times 1\frac{5}{6}$ $8\frac{1}{4}$ or $\frac{33}{4}$ **4.** $4\frac{1}{2} \div 1\frac{5}{6}$ $2\frac{5}{11}$ or $\frac{27}{11}$

Concept 1: Variables and Expressions

For Exercises 5–14, evaluate the expressions for the given substitution. **(See Examples 1–2.)**

5. $y - 3$ when $y = 18$ 15

6. $3q$ when $q = 5$ 15

7. $\dfrac{15}{t}$ when $t = 5$ 3

8. $8 + w$ when $w = 12$ 20

9. $6d$ when $d = \dfrac{2}{3}$ 4

10. $\dfrac{6}{5}h$ when $h = 10$ 12

 Writing   Translating Expression   Geometry   Scientific Calculator   Video

11. $c - 2 - d$ when $c = 15.4$, $d = 8.1$ 5.3

12. $1.1 + t + s$ when $t = 93.2$, $s = 11.5$ 105.8

13. abc when $a = \dfrac{1}{10}$, $b = \dfrac{1}{4}$, $c = \dfrac{1}{2}$ $\dfrac{1}{80}$

14. $x - y - z$ when $x = \dfrac{7}{8}$, $y = \dfrac{1}{2}$, $z = \dfrac{1}{4}$ $\dfrac{1}{8}$

15. The cost of downloading songs from the Internet can be represented by the expression $1.29s$, where s is the number of songs downloaded. Calculate the cost of downloading:

    a. 3 songs $3.87   **b.** 8 songs $10.32   **c.** 10 songs $12.90

16. The number of calories burned by a 150-lb person by walking 2 mph can be represented by the expression $240h$, where h represents the number of hours spent walking. Calculate the number of calories burned by walking:

    a. 4 hr 960 calories   **b.** $2\frac{1}{2}$ hr 600 calories   **c.** $1\frac{1}{4}$ hr 300 calories

17. Aly is trying to limit her total calorie intake for breakfast and lunch to 850 calories. The number of calories that she can consume for lunch is given by the expression $850 - b$, where b is the number of calories consumed for breakfast. Determine the number of calories allowed for lunch assuming that she had the following number of calories at breakfast:

    a. 475 calories   **b.** 220 calories   **c.** 580 calories

    375 calories   630 calories   270 calories

18. Lorenzo knows that the gas mileage on his car is about 25 miles per gallon. The number of gallons needed to travel a certain distance is given by the expression $\frac{d}{25}$, where d is the distance traveled. Find the number of gallons of fuel needed if Lorenzo drives:

    a. 200 mi   **b.** 450 mi   **c.** 180 mi

    8 gal   18 gal   7.2 gal

Concept 2: The Set of Real Numbers

19. Plot the numbers on the real number line: $\left\{1, -2, -\pi, 0, -\frac{5}{2}, 5.1\right\}$ **(See Example 3.)**

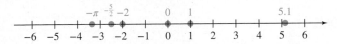

20. Plot the numbers on the real number line: $\left\{3, -4, \frac{1}{8}, -1.7, -\frac{4}{3}, 1.75\right\}$

For Exercises 21–36, describe each number as (a) a terminating decimal, (b) a repeating decimal, or (c) a nonterminating, nonrepeating decimal. Then classify the number as a rational number or as an irrational number. **(See Example 4.)**

21. 0.29 a; rational   **22.** 3.8 a; rational   **23.** $\dfrac{1}{9}$ b; rational   **24.** $\dfrac{1}{3}$ b; rational

25. $\dfrac{1}{8}$ a; rational   **26.** $\dfrac{1}{5}$ a; rational   **27.** 2π c; irrational   **28.** 3π c; irrational

29. -0.125 a; rational   **30.** -3.24 a; rational   **31.** -3 a; rational   **32.** -6 a; rational

33. $0.\overline{2}$ b; rational   **34.** $0.\overline{6}$ b; rational   **35.** $\sqrt{6}$ c; irrational   **36.** $\sqrt{10}$ c; irrational

37. List three numbers that are real numbers but not rational numbers.   For example: $\pi, -\sqrt{2}, \sqrt{3}$

38. List three numbers that are real numbers but not irrational numbers.   For example: $-\frac{1}{2}, 5, 0.\overline{3}$

39. List three numbers that are integers but not natural numbers.   For example: $-4, -1, 0$

40. List three numbers that are integers but not whole numbers.   For example: $-5, -2, -1$

41. List three numbers that are rational numbers but not integers.   For example: $-\frac{3}{4}, \frac{1}{2}, 0.206$

For Exercises 42–48, let $A = \left\{ -\frac{3}{2}, \ \sqrt{11}, \ -4, 0.\overline{6}, 0, \sqrt{7}, 1 \right\}$ **(See Example 5.)**

42. Are all of the numbers in set A real numbers?
Yes

43. List all of the rational numbers in set A. $-\frac{3}{2}, -4, 0.\overline{6}, 0, 1$

44. List all of the whole numbers in set A.
0, 1

45. List all of the natural numbers in set A.   1

46. List all of the irrational numbers in set A.
$\sqrt{11}, \sqrt{7}$

47. List all of the integers in set A. $-4, 0, 1$

48. Plot the real numbers from set A on a number line. (*Hint:* $\sqrt{11} \approx 3.3$ and $\sqrt{7} \approx 2.6$)

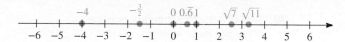

Concept 3: Inequalities

49. The women's golf scores for selected players at an LPGA event are given in the table. Compare the scores and fill in the blank with the appropriate inequality sign: $<$ or $>$. **(See Example 6.)**

   a. Kane's score ___>___ Pak's score.

   b. Sorenstam's score ___>___ Davies's score.

   c. Pak's score ___<___ McCurdy's score.

   d. Kane's score ___>___ Davies's score.

LPGA Golfers	Final Score with Respect to Par
Annika Sorenstam	7
Laura Davies	−4
Lorie Kane	0
Cindy McCurdy	3
Se Ri Pak	−8

50. The elevations of selected cities in the United States are shown in the figure. Compare the elevations and fill in the blank with the appropriate inequality sign: $<$ or $>$. (A negative number indicates that the city is below sea level.)

   a. Elevation of Tucson ___>___ elevation of Cincinnati.

   b. Elevation of New Orleans ___<___ elevation of Chicago.

   c. Elevation of New Orleans ___<___ elevation of Houston.

   d. Elevation of Chicago ___>___ elevation of Cincinnati.

Concept 4: Opposite of a Real Number

For Exercises 51–58, find the opposite of each number. **(See Example 7.)**

51. 18 -18

52. 2 -2

53. -6.1 6.1

54. -2.5 2.5

55. $-\dfrac{5}{8}$ $\dfrac{5}{8}$

56. $-\dfrac{1}{3}$ $\dfrac{1}{3}$

57. $\dfrac{7}{3}$ $-\dfrac{7}{3}$

58. $\dfrac{1}{9}$ $-\dfrac{1}{9}$

The opposite of a is denoted as $-a$. For Exercises 59–66, simplify. **(See Example 8.)**

59. $-(-3)$ 3

60. $-(-5.1)$ 5.1

61. $-\left(\dfrac{7}{3}\right)$ $-\dfrac{7}{3}$

62. $-(-7)$ 7

63. $-(-8)$ 8

64. $-(36)$ -36

65. $-(72.1)$ -72.1

66. $-\left(\dfrac{9}{10}\right)$ $-\dfrac{9}{10}$

Concept 5: Absolute Value of a Real Number

For Exercises 67–78, simplify. **(See Example 9.)**

67. $|-2|$ 2

68. $|-7|$ 7

69. $|-1.5|$ 1.5

70. $|-3.7|$ 3.7

71. $-|-1.5|$ -1.5

72. $-|-3.7|$ -3.7

73. $\left|\dfrac{3}{2}\right|$ $\dfrac{3}{2}$

74. $\left|\dfrac{7}{4}\right|$ $\dfrac{7}{4}$

75. $-|10|$ -10

76. $-|20|$ -20

77. $-\left|-\dfrac{1}{2}\right|$ $-\dfrac{1}{2}$

78. $-\left|-\dfrac{11}{3}\right|$ $-\dfrac{11}{3}$

For Exercises 79–80, answer true or false. If a statement is false, explain why.

79. If n is positive, then $|n|$ is negative.
False, $|n|$ is never negative.

80. If m is negative, then $|m|$ is negative.
False, $|m|$ is never negative.

For Exercises 81–104, determine if the statements are true or false. Use the real number line to justify the answer. **(See Example 10.)**

81. $5 > 2$ True

82. $8 < 10$ True

83. $6 < 6$ False

84. $19 > 19$ False

85. $-7 \geq -7$ True

86. $-1 \leq -1$ True

87. $\dfrac{3}{2} \leq \dfrac{1}{6}$ False

88. $-\dfrac{1}{4} \geq -\dfrac{7}{8}$ True

89. $-5 > -2$ False

90. $6 < -10$ False

91. $8 \neq 8$ False

92. $10 \neq 10$ False

93. $|-2| \geq |-1|$ True

94. $|3| \leq |-1|$ False

95. $\left|-\dfrac{1}{9}\right| = \left|\dfrac{1}{9}\right|$ True

96. $\left|-\dfrac{1}{3}\right| = \left|\dfrac{1}{3}\right|$ True

97. $|7| \neq |-7|$ False

98. $|-13| \neq |13|$ False

99. $-1 < |-1|$ True

100. $-6 < |-6|$ True

101. $|-8| \geq |8|$ True

102. $|-11| \geq |11|$ True

103. $|-2| \leq |2|$ True

104. $|-21| \leq |21|$ True

Expanding Your Skills

105. For what numbers, a, is $-a$ positive?
For all $a < 0$

106. For what numbers, a, is $|a| = a$?
For all $a \geq 0$

 Writing   Translating Expression   Geometry   Scientific Calculator   Video

Exponents, Square Roots, and the Order of Operations

1. Exponential Expressions

In algebra, repeated multiplication can be expressed using exponents. The expression $4 \cdot 4 \cdot 4$ can be written as

$$4^3$$
exponent / base

In the expression 4^3, 4 is the base, and 3 is the exponent, or power. The exponent indicates how many factors of the base to multiply.

Concepts

1. Exponential Expressions
2. Square Roots
3. Order of Operations
3. Translations

Definition of b^n

Let b represent any real number and n represent a positive integer. Then,

$$b^n = \underbrace{b \cdot b \cdot b \cdot b \cdot \ldots b}_{n \text{ factors of } b}$$

b^n is read as "b to the nth power."

b is called the **base**, and n is called the **exponent,** or **power**.

b^2 is read as "b squared," and b^3 is read as "b cubed."

The exponent, n, is the number of times the base, b, is used as a factor.

TIP: A number or variable with no exponent shown implies that there is an exponent of 1. That is, $b = b^1$.

| **Example 1** | **Evaluating Exponential Expressions** |

Translate the expression into words and then evaluate the expression.

a. 2^5 **b.** 5^2 **c.** $\left(\dfrac{3}{4}\right)^3$ **d.** 1^6

Solution:

a. The expression 2^5 is read as "two to the fifth power."
$2^5 = (2)(2)(2)(2)(2) = 32$

b. The expression 5^2 is read as "five to the second power" or "five, squared."
$5^2 = (5)(5) = 25$

c. The expression $\left(\frac{3}{4}\right)^3$ is read as "three-fourths to the third power" or "three-fourths, cubed."

$$\left(\frac{3}{4}\right)^3 = \left(\frac{3}{4}\right)\left(\frac{3}{4}\right)\left(\frac{3}{4}\right) = \frac{27}{64}$$

d. The expression 1^6 is read as "one to the sixth power."
$1^6 = (1)(1)(1)(1)(1)(1) = 1$

Classroom Examples: p. 43, Exercises 24 and 26

Skill Practice Evaluate.

1. 4^3 **2.** 2^4 **3.** $\left(\dfrac{2}{3}\right)^2$ **4.** $(1)^7$

Answers

1. 64 **2.** 16 **3.** $\dfrac{4}{9}$ **4.** 1

2. Square Roots

If we reverse the process of squaring a number, we can find the square roots of the number. For example, finding a square root of 9 is equivalent to asking "what number(s) when squared equals 9?" The symbol, $\sqrt{}$ (called a **radical sign**), is used to find the *principal* square root of a number. By definition, the principal square root of a number is nonnegative. Therefore, $\sqrt{9}$ is the nonnegative number that when squared equals 9. Hence, $\sqrt{9} = 3$ because 3 is nonnegative and $(3)^2 = 9$.

Classroom Examples: p. 43,
Exercises 34 and 42

| Example 2 | **Evaluating Square Roots** |

Evaluate the square roots.

a. $\sqrt{64}$ **b.** $\sqrt{121}$ **c.** $\sqrt{0}$ **d.** $\sqrt{\dfrac{4}{9}}$

Solution:

a. $\sqrt{64} = 8$ Because $(8)^2 = 64$

b. $\sqrt{121} = 11$ Because $(11)^2 = 121$

c. $\sqrt{0} = 0$ Because $(0)^2 = 0$

d. $\sqrt{\dfrac{4}{9}} = \dfrac{2}{3}$ Because $\dfrac{2}{3} \cdot \dfrac{2}{3} = \dfrac{4}{9}$

Skill Practice   Evaluate.

5. $\sqrt{81}$ **6.** $\sqrt{100}$ **7.** $\sqrt{1}$ **8.** $\sqrt{\dfrac{9}{25}}$

A perfect square is a number whose square root is a rational number. If a positive number is not a perfect square, its square root is an irrational number that can be approximated on a calculator.

TIP: To simplify square roots, it is advisable to become familiar with the following perfect squares and square roots.

$$0^2 = 0 \longrightarrow \sqrt{0} = 0 \qquad\qquad 7^2 = 49 \longrightarrow \sqrt{49} = 7$$
$$1^2 = 1 \longrightarrow \sqrt{1} = 1 \qquad\qquad 8^2 = 64 \longrightarrow \sqrt{64} = 8$$
$$2^2 = 4 \longrightarrow \sqrt{4} = 2 \qquad\qquad 9^2 = 81 \longrightarrow \sqrt{81} = 9$$
$$3^2 = 9 \longrightarrow \sqrt{9} = 3 \qquad\qquad 10^2 = 100 \longrightarrow \sqrt{100} = 10$$
$$4^2 = 16 \longrightarrow \sqrt{16} = 4 \qquad\qquad 11^2 = 121 \longrightarrow \sqrt{121} = 11$$
$$5^2 = 25 \longrightarrow \sqrt{25} = 5 \qquad\qquad 12^2 = 144 \longrightarrow \sqrt{144} = 12$$
$$6^2 = 36 \longrightarrow \sqrt{36} = 6 \qquad\qquad 13^2 = 169 \longrightarrow \sqrt{169} = 13$$

3. Order of Operations

When algebraic expressions contain numerous operations, it is important to evaluate the operations in the proper order. Parentheses ( ), brackets [ ], and braces { } are used for grouping numbers and algebraic expressions. It is important to recognize that operations

Answers
5. 9   **6.** 10   **7.** 1   **8.** $\dfrac{3}{5}$

must be done within parentheses and other grouping symbols first. Other grouping symbols include absolute value bars, radical signs, and fraction bars.

Applying the Order of Operations

Step 1 Simplify expressions within parentheses and other grouping symbols first. These include absolute value bars, fraction bars, and radicals. If imbedded parentheses are present, start with the innermost parentheses.

Step 2 Evaluate expressions involving exponents, radicals, and absolute values.

Step 3 Perform multiplication or division in the order that they occur from left to right.

Step 4 Perform addition or subtraction in the order that they occur from left to right.

TIP: Radical signs act as grouping symbols.
$$\sqrt{16 + 9} = \sqrt{(16 + 9)}$$
$$= \sqrt{25}$$
$$= 5$$
Perform operations inside the radical first, then apply the square root.

| **Example 3** | **Applying the Order of Operations** |

Simplify the expressions.

a. $17 - 3 \cdot 2 + 2^2$ **b.** $\dfrac{1}{2}\left(\dfrac{5}{6} - \dfrac{3}{4}\right)$

Solution:

a. $17 - 3 \cdot 2 + 2^2$

$= 17 - 3 \cdot 2 + 4$ Simplify exponents.

$= 17 - 6 + 4$ Multiply before adding or subtracting.

$= 11 + 4$ Add or subtract from left to right.

$= 15$

b. $\dfrac{1}{2}\left(\dfrac{5}{6} - \dfrac{3}{4}\right)$ Subtract fractions within the parentheses.

$= \dfrac{1}{2}\left(\dfrac{10}{12} - \dfrac{9}{12}\right)$ The least common denominator is 12.

$= \dfrac{1}{2}\left(\dfrac{1}{12}\right)$

$= \dfrac{1}{24}$ Multiply fractions.

Classroom Examples: p. 43, Exercises 48 and 54

Skill Practice Simplify the expressions.

9. $14 - 3 \cdot 2 + 3^2$

10. $\dfrac{13}{4} - \dfrac{1}{4}(10 - 2)$

Answers
9. 17   **10.** $\dfrac{5}{4}$

Classroom Examples: pp. 43–44, Exercises 62 and 68

| **Example 4** | **Applying the Order of Operations** |

Simplify the expressions.

a. $25 - 12 \div 3 \cdot 4$

b. $6.2 - |-2.1| + \sqrt{16 + 9}$

c. $28 - 2[(6 - 3)^2 + 4]$

Solution:

Avoiding Mistakes

In Example 4(a), division is performed before multiplication because it occurs first as we read from left to right.

a. $25 - 12 \div 3 \cdot 4$ Multiply or divide in order from left to right.

$= 25 - 4 \cdot 4$ Notice that the operation $12 \div 3$ is performed first (not $4 \cdot 4$).

$= 25 - 16$ Multiply $4 \cdot 4$ before subtracting.

$= 9$ Subtract.

b. $6.2 - |-2.1| + \sqrt{16 + 9}$

$= 6.2 - |-2.1| + \sqrt{25}$ Simplify within the square root.

$= 6.2 - (2.1) + 5$ Simplify the absolute value and square root.

$= 4.1 + 5$ Add or subtract from left to right.

$= 9.1$ Add.

c. $28 - 2[(6 - 3)^2 + 4]$

$= 28 - 2[(3)^2 + 4]$ Simplify within the inner parentheses first.

$= 28 - 2[(9) + 4]$ Simplify exponents.

$= 28 - 2[13]$ Add within the square brackets.

$= 28 - 26$ Multiply before subtracting.

$= 2$ Subtract.

Skill Practice Simplify the expressions.

11. $1 + 2 \cdot 3^2 \div 6$ **12.** $|-20| - \sqrt{20 - 4}$ **13.** $60 - 5[(7 - 4) + 2^2]$

Classroom Example: p. 44, Exercise 72

| **Example 5** | **Applying the Order of Operations** |

Simplify the expression. $\dfrac{32 + 8 \div 2}{2 \cdot 3^2}$

Solution:

$\dfrac{32 + 8 \div 2}{2 \cdot 3^2}$ In this expression, the fraction bar acts as a grouping symbol.

$= \dfrac{32 + 4}{2 \cdot 9}$ First, simplify the expressions above and below the fraction bar using the order of operations.

$= \dfrac{36}{18}$ The last step is to simplify the fraction.

$= 2$

Answers

11. 4   **12.** 16   **13.** 25

Skill Practice Simplify the expression.

14. $\dfrac{60 - 3^2 \cdot 2}{3 + 8 \div 2}$

4. Translations

Algebra is a powerful tool used in science, business, economics, and many day-to-day applications. To apply algebra to a real-world application, we need the important skill of translating an English phrase to a mathematical expression. Table 1-3 summarizes commonly used phrases and expressions.

Table 1-3

Operation	Symbols	Translation
Addition	$a + b$	**sum** of a and b a plus b b added to a b more than a a increased by b the total of a and b
Subtraction	$a - b$	**difference** of a and b a minus b b subtracted from a a decreased by b b less than a a less b
Multiplication	$a \times b,\ a \cdot b,\ a(b),\ (a)b,\ (a)(b),\ ab$ (*Note:* From this point forward we will seldom use the notation $a \times b$ because the symbol, $\times$, might be confused with the variable, x.)	**product** of a and b a times b a multiplied by b
Division	$a \div b,\ \dfrac{a}{b},\ a/b,\ b\overline{)a}$	**quotient** of a and b a divided by b b divided into a ratio of a and b a over b a per b

Example 6   **Writing an English Phrase as an Algebraic Expression**

Classroom Examples: p. 44, Exercises 82 and 92

Translate each English phrase to an algebraic expression.

a. The quotient of x and 5

b. The difference of p and the square root of q

c. Seven less than n

d. Seven less n

e. Eight more than the absolute value of w

f. x subtracted from 18

Answer

14. 6

Solution:

a. $\dfrac{x}{5}$ or $x \div 5$ 　　　The quotient of x and 5

b. $p - \sqrt{q}$ 　　　The difference of p and the square root of q

c. $n - 7$ 　　　Seven less than n

d. $7 - n$ 　　　Seven less n

e. $|w| + 8$ 　　　Eight more than the absolute value of w

f. $18 - x$ 　　　x subtracted from 18

Instructor Note: Show students that 7 less than x would not be written $7 < x$. This expression would be worded as 7 is less than x.

> **Avoiding Mistakes**
>
> Recall that "a less than b" is translated as $b - a$. Therefore, the statement "seven less than n" must be translated as $n - 7$, not $7 - n$.

Skill Practice Translate each English phrase to an algebraic expression.

15. The product of 6 and y
16. The difference of the square root of t and 7
17. Twelve less than x
18. Twelve less x
19. One more than two times x
20. Five subtracted from the absolute value of w

Classroom Examples: p. 45, Exercises 94 and 100

Example 7　**Writing English Phrases as Algebraic Expressions**

Translate each English phrase into an algebraic expression. Then evaluate the expression for $a = 6$, $b = 4$, and $c = 20$.

a. The product of a and the square root of b

b. Twice the sum of b and c

c. The difference of twice a and b

Solution:

a. The product of a and the square root of b
$$a\sqrt{b}$$
$$= ()\sqrt{()} \qquad \text{Use parentheses to substitute a number for a variable.}$$
$$= (6)\sqrt{(4)} \qquad \text{Substitute } a = 6 \text{ and } b = 4.$$
$$= 6 \cdot 2 \qquad \text{Simplify the radical first.}$$
$$= 12 \qquad \text{Multiply.}$$

b. Twice the sum of b and c

$2(b + c)$ 　　　To compute "twice the sum of b and c," it is necessary to take the sum first and then multiply by 2. To ensure the proper order, the sum of b and c must be enclosed in parentheses. The proper translation is $2(b + c)$.

$$= 2(() + ()) \qquad \text{Use parentheses to substitute a number for a variable.}$$
$$= 2((4) + (20)) \qquad \text{Substitute } b = 4 \text{ and } c = 20.$$
$$= 2(24) \qquad \text{Simplify within the parentheses first.}$$
$$= 48 \qquad \text{Multiply.}$$

Answers

15. $6y$ 　　**16.** $\sqrt{t} - 7$
17. $x - 12$ 　　**18.** $12 - x$
19. $2x + 1$ 　　**20.** $|w| - 5$

c. The difference of twice a and b

$2a - b$

$= 2(\ \ )-(\ \ )$ Use parentheses to substitute a number for a variable.

$= 2(6)-(4)$ Substitute $a = 6$ and $b = 4$.

$= 12 - 4$ Multiply first.

$= 8$ Subtract.

Skill Practice Translate each English phrase to an algebraic expression. Then evaluate the expression for $x = 3$, $y = 9$, $z = 10$.

21. The quotient of the square root of y and x

22. One-half the sum of x and y

23. The difference of z and twice x

Answers

21. $\dfrac{\sqrt{y}}{x}$; 1 **22.** $\dfrac{1}{2}(x + y)$; 6

23. $z - 2x$; 4

Calculator Connections

Topic: Evaluating Exponential Expressions on a Calculator

On a calculator, we enter exponents greater than the second power by using the key labeled $\boxed{y^x}$ or $\boxed{\wedge}$. For example, evaluate 2^4 and 10^6:

Scientific Calculator:

Enter: 2 $\boxed{y^x}$ 4 $\boxed{=}$ **Result:** | 16 |

Enter: 10 $\boxed{y^x}$ 6 $\boxed{=}$ **Result:** | 1000000 |

Graphing Calculator:

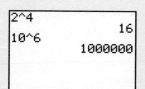

```
2^4
                16
10^6
        1000000
```

Topic: Applying the Order of Operations on a Calculator

Most calculators also have the capability to enter several operations at once. However, it is important to note that fraction bars and radicals require user-defined parentheses to ensure that the proper order of operations is followed. For example, evaluate the following expressions on a calculator:

a. $130 - 2(5 - 1)^3$ **b.** $\dfrac{18 - 2}{11 - 9}$ **c.** $\sqrt{25 - 9}$

Scientific Calculator:

Enter: 130 $\boxed{-}$ 2 $\boxed{\times}$ $\boxed{(}$ 5 $\boxed{-}$ 1 $\boxed{)}$ $\boxed{y^x}$ 3 $\boxed{=}$ **Result:** | 2 |

Enter: $\boxed{(}$ 18 $\boxed{-}$ 2 $\boxed{)}$ $\boxed{\div}$ $\boxed{(}$ 11 $\boxed{-}$ 9 $\boxed{)}$ $\boxed{=}$ **Result:** | 8 |

Enter: $\boxed{(}$ 25 $\boxed{-}$ 9 $\boxed{)}$ $\boxed{\sqrt{\ }}$ **Result:** | 4 |

Graphing Calculator:

```
130-2*(5-1)^3
              2
(18-2)/(11-9)
              8
√(25-9)
              4
```

Calculator Exercises

Simplify each expression without the use of a calculator. Then enter the expression into the calculator to verify your answer.

1. $\dfrac{4+6}{8-3}$ 2

2. $110 - 5(2+1) - 4$ 91

3. $100 - 2(5-3)^3$ 84

4. $3 + (4-1)^2$ 12

5. $(12 - 6 + 1)^2$ 49

6. $3 \cdot 8 - \sqrt{32 + 2^2}$ 18

7. $\sqrt{18-2}$ 4

8. $(4 \cdot 3 - 3 \cdot 3)^3$ 27

9. $\dfrac{20 - 3^2}{26 - 2^2}$ 0.5

Section 1.3   Practice Exercises

For additional exercises, see Classroom Activities 1.3A–1.3C in the *Student's Resource Manual* at www.mhhe.com/moh.

Study Skills Exercise

Sometimes you may run into a problem with homework or you find that you are having trouble keeping up with the pace of the class. A tutor can be a good resource.

a. Does your college offer tutoring?   **b.** Is it free?   **c.** Where would you go to sign up for a tutor?

Vocabulary and Key Concepts

1. **a.** Fill in the blanks with the words *sum, difference, product,* or *quotient.*

   The __quotient__ of 10 and 2 is 5. The __product__ of 10 and 2 is 20.

   The __sum__ of 10 and 2 is 12. The __difference__ of 10 and 2 is 8.

   b. In the expression b^n, the value b is called the __base__ and n is called the __exponent__ or __power__.

   c. The expression __8^2__ is read as "8-squared."

   d. The expression __p^4__ is read as "p to the 4th power."

   e. The symbol $\sqrt{}$ is called a __radical__ sign and is used to find the principal __square__ root of a nonnegative real number.

   f. The set of rules that tell us the order in which to perform operations to simplify an algebraic expression is called the __order of operations__ .

Review Exercises

2. Which of the following are rational numbers? $-4, 5.\overline{6}, \sqrt{29}, 0, \pi, 4.02, \dfrac{7}{9}$ $-4, 5.\overline{6}, 0, 4.02, \dfrac{7}{9}$

3. Evaluate. $|-56|$ 56

4. Evaluate. $-|-14|$ −14

5. Find the opposite of 19.   −19

6. Find the opposite of −34.2.   34.2

Concept 1: Exponential Expressions

For Exercises 7–12, write each product using exponents.

7. $\dfrac{1}{6} \cdot \dfrac{1}{6} \cdot \dfrac{1}{6} \cdot \dfrac{1}{6}$ $\left(\dfrac{1}{6}\right)^4$

8. $10 \cdot 10 \cdot 10 \cdot 10 \cdot 10 \cdot 10$ 10^6

9. $a \cdot a \cdot a \cdot b \cdot b$ $a^3 b^2$

10. $7 \cdot x \cdot x \cdot y \cdot y$ $7x^2 y^2$

11. $5c \cdot 5c \cdot 5c \cdot 5c \cdot 5c$ $(5c)^5$

12. $3 \cdot w \cdot z \cdot z \cdot z \cdot z$ $3wz^4$

13. a. For the expression $5x^3$, what is the base for the exponent 3? x

   b. Does 5 have an exponent? If so, what is it? Yes, 1

14. a. For the expression $2y^4$, what is the base for the exponent 4? y

   b. Does 2 have an exponent? If so, what is it? Yes, 1

For Exercises 15–22, write each expression in expanded form using the definition of an exponent.

15. x^3 $x \cdot x \cdot x$

16. y^4 $y \cdot y \cdot y \cdot y$

17. $(2b)^3$ $2b \cdot 2b \cdot 2b$

18. $(8c)^2$ $8c \cdot 8c$

19. $10y^5$ $10 \cdot y \cdot y \cdot y \cdot y \cdot y$

20. $x^2 y^3$ $x \cdot x \cdot y \cdot y \cdot y$

21. $2wz^2$ $2 \cdot w \cdot z \cdot z$

22. $3a^3 b$ $3 \cdot a \cdot a \cdot a \cdot b$

For Exercises 23–30, simplify each expression. **(See Example 1.)**

23. 6^2 36

24. 5^3 125

25. $\left(\dfrac{1}{7}\right)^2$ $\dfrac{1}{49}$

26. $\left(\dfrac{1}{2}\right)^5$ $\dfrac{1}{32}$

27. $(0.2)^3$ 0.008

28. $(0.8)^2$ 0.64

29. 2^6 64

30. 13^2 169

Concept 2: Square Roots

For Exercises 31–42, simplify the square roots. **(See Example 2.)**

31. $\sqrt{81}$ 9

32. $\sqrt{64}$ 8

33. $\sqrt{4}$ 2

34. $\sqrt{9}$ 3

35. $\sqrt{144}$ 12

36. $\sqrt{49}$ 7

37. $\sqrt{16}$ 4

38. $\sqrt{36}$ 6

39. $\sqrt{\dfrac{1}{9}}$ $\dfrac{1}{3}$

40. $\sqrt{\dfrac{1}{64}}$ $\dfrac{1}{8}$

41. $\sqrt{\dfrac{25}{81}}$ $\dfrac{5}{9}$

42. $\sqrt{\dfrac{49}{100}}$ $\dfrac{7}{10}$

Concept 3: Order of Operations

For Exercises 43–74, use the order of operations to simplify each expression. **(See Examples 3–5.)**

43. $8 + 2 \cdot 6$ 20

44. $7 + 3 \cdot 4$ 19

45. $(8 + 2) \cdot 6$ 60

46. $(7 + 3) \cdot 4$ 40

47. $4 + 2 \div 2 \cdot 3 + 1$ 8

48. $5 + 12 \div 2 \cdot 6 - 1$ 40

49. $81 - 4 \cdot 3 + 3^2$ 78

50. $100 - 25 \cdot 2 - 5^2$ 25

51. $\dfrac{1}{4} \cdot \dfrac{2}{3} - \dfrac{1}{6}$ 0

52. $\dfrac{3}{4} \cdot \dfrac{2}{3} + \dfrac{2}{3}$ $\dfrac{7}{6}$

53. $\left(\dfrac{11}{6} - \dfrac{3}{8}\right) \cdot \dfrac{4}{5}$ $\dfrac{7}{6}$

54. $\left(\dfrac{9}{8} - \dfrac{1}{3}\right) \cdot \dfrac{3}{4}$ $\dfrac{19}{32}$

55. $3[5 + 2(8 - 3)]$ 45

56. $2[4 + 3(6 - 4)]$ 20

57. $10 + |-6|$ 16

58. $18 + |-3|$ 21

59. $21 - |8 - 2|$ 15

60. $12 - |6 - 1|$ 7

61. $2^2 + \sqrt{9} \cdot 5$ 19

62. $3^2 + \sqrt{16} \cdot 2$ 17

63. $3 \cdot 5^2$ 75

64. $10 \cdot 2^3$ 80

65. $\sqrt{9 + 16} - 2$ 3

66. $\sqrt{36 + 13} - 5$ 2

67. $[4^2 \cdot (6 - 4) \div 8] + [7 \cdot (8 - 3)]$ 39

68. $(18 \div \sqrt{4}) \cdot \{[(9^2 - 1) \div 2] - 15\}$ 225

69. $48 - 13 \cdot 3 + [(50 - 7 \cdot 5) + 2]$ 26

70. $80 \div 16 \cdot 2 + (6^2 - |-2|)$ 44

71. $\dfrac{7 + 3(8 - 2)}{(7 + 3)(8 - 2)}$ $\dfrac{5}{12}$

72. $\dfrac{16 - 8 \div 4}{4 + 8 \div 4 - 2}$ $\dfrac{7}{2}$

73. $\dfrac{15 - 5(3 \cdot 2 - 4)}{10 - 2(4 \cdot 5 - 16)}$ $\dfrac{5}{2}$

74. $\dfrac{5(7 - 3) + 8(6 - 4)}{4[7 + 3(2 \cdot 9 - 8)]}$ $\dfrac{9}{37}$

75. A person's debt-to-income ratio is the sum of all monthly installment payments (credit cards, loans, etc.) divided by monthly take-home pay. This number is often considered when one is applying for a loan. Each month, Monica makes credit card payments of $52 and $20, a student loan payment of $65, and a payment for furniture of $43. Her monthly take-home pay is $1500.

   a. Determine Monica's debt-to-income ratio.   0.12

   b. To obtain a car loan, Monica's debt-to-income ratio must be less than 0.20. Does she meet this criteria?
   Yes. 0.12 < 0.20

76. Each month, Jared makes credit card payments of $115, $63, and $95. He also makes a student loan payment of $77 and another loan payment of $100. His monthly take-home pay is $2000.

   a. Use the definition given in Exercise 75 to determine Jared's debt-to-income ratio.   0.225

   b. To obtain a loan for a new home theater system his debt-to-income ratio must be less than 0.15. Does he meet this criteria?   No. 0.225 > 0.15

77. The area of a rectangle is given by $A = lw$, where l is the length of the rectangle and w is the width. Find the area for the rectangle shown.   57,600 ft^2

78. The perimeter of a rectangle is given by $P = 2l + 2w$. Find the perimeter for the rectangle shown.   1040 ft

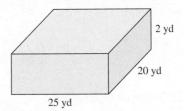

160 ft

360 ft

79. The area of a trapezoid is given by $A = \frac{1}{2}(b_1 + b_2)h$, where b_1 and b_2 are the lengths of the two parallel sides and h is the height. A window is in the shape of a trapezoid. Find the area of the trapezoid with dimensions shown in the figure.   21 ft^2

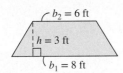

$b_2 = 6$ ft

$h = 3$ ft

$b_1 = 8$ ft

80. The volume of a rectangular solid is given by $V = lwh$, where l is the length of the box, w is the width, and h is the height. Find the volume of the box shown in the figure.   1000 yd^3

2 yd

20 yd

25 yd

Concept 4: Translations

For Exercises 81–92, write each English phrase as an algebraic expression. **(See Example 6.)**

81. The product of 3 and x $3x$

82. The sum of b and 6 $b + 6$

83. The quotient of x and 7 $\dfrac{x}{7}$ or $x \div 7$

84. Four divided by k $\dfrac{4}{k}$ or $4 \div k$

85. The difference of 2 and a $2 - a$

86. Three subtracted from t $t - 3$

87. x more than twice y $2y + x$

88. Nine decreased by the product of 3 and p $9 - 3p$

89. Four times the sum of x and 12 $4(x + 12)$

90. Twice the difference of x and 3 $2(x - 3)$

91. Q less than 3 $3 - Q$

92. Fourteen less than t $t - 14$

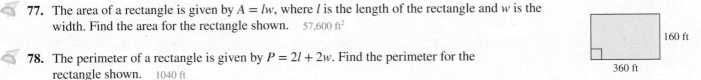

Writing   Translating Expression   Geometry   Scientific Calculator   Video

For Exercises 93–100, write the English phrase as an algebraic expression. Then evaluate each expression for $x = 4$, $y = 2$, and $z = 10$. **(See Example 7.)**

93. Two times y cubed $2y^3$; 16

94. Three times z squared $3z^2$; 300

95. The absolute value of the difference of z and 8
$|z - 8|$; 2

96. The absolute value of the difference of x and 3
$|x - 3|$; 1

97. The product of 5 and the square root of x
$5\sqrt{x}$; 10

98. The square root of the difference of z and 1
$\sqrt{z - 1}$; 3

99. The value x subtracted from the product of y and z
$yz - x$; 16

100. The difference of z and the product of x and y
$z - xy$; 2

Expanding Your Skills

For Exercises 101–104, use the order of operations to simplify each expression.

101. $\dfrac{\sqrt{\frac{1}{9}} + \frac{2}{3}}{\sqrt{\frac{4}{25}} + \frac{3}{5}}$ 1

102. $\dfrac{5 - \sqrt{9}}{\sqrt{\frac{4}{9}} + \frac{1}{3}}$ 2

103. $\dfrac{|-2|}{|-10| - |2|}$ $\frac{1}{4}$

104. $\dfrac{|-4|^2}{2^2 + \sqrt{144}}$ 1

105. Some students use the following common memorization device (mnemonic) to help them remember the order of operations: the acronym PEMDAS or **P**lease **E**xcuse **M**y **D**ear **A**unt **S**ally to remember **P**arentheses, **E**xponents, **M**ultiplication, **D**ivision, **A**ddition, and **S**ubtraction. The problem with this mnemonic is that it suggests that multiplication is done before division and similarly, it suggests that addition is performed before subtraction. Explain why following this acronym may give incorrect answers for the expressions:

a. $36 \div 4 \cdot 3$ $36 \div 4 \cdot 3$ Division must be performed
$= 9 \cdot 3$ before multiplication.
$= 27$

b. $36 - 4 + 3$ $36 - 4 + 3$ Subtraction must be performed
$= 32 + 3$ before addition.
$= 35$

106. If you use the acronym **P**lease **E**xcuse **M**y **D**ear **A**unt **S**ally to remember the order of operations, what must you keep in mind about the last four operations?   Multiplication or division is performed in order from left to right. Addition or subtraction is performed in order from left to right.

107. Explain why the acronym **P**lease **E**xcuse **D**r. **M**ichael **S**mith's **A**unt could also be used as a memory device for the order of operations.   This is acceptable, provided division and multiplication are performed in order from left to right, and subtraction and addition are performed in order from left to right.

Addition of Real Numbers

Section 1.4

1. Addition of Real Numbers and the Number Line

Adding real numbers can be visualized on the number line. To do so, locate the first addend on the number line. Then to add a positive number, move to the right on the number line. To add a negative number, move to the left on the number line. The following example may help to illustrate the process.

On a winter day in Detroit, suppose the temperature starts out at 5 degrees Fahrenheit ($5°F$) at noon, and then drops $12°$ two hours later when a cold front passes through. The resulting temperature can be represented by the expression $5° + (-12°)$. On the number line, start at 5 and count 12 units to the left (Figure 1-7). The resulting temperature at 2:00 P.M. is $-7°F$.

Concepts

1. **Addition of Real Numbers and the Number Line**
2. **Addition of Real Numbers**
3. **Translations**
4. **Applications Involving Addition of Real Numbers**

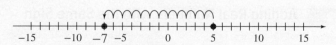

Figure 1-7

Classroom Examples: p. 51,
Exercises 10 and 12

> **Example 1**   **Using the Number Line to Add Real Numbers**

Use the number line to add the numbers.

    a. $-5 + 2$ **b.** $-1 + (-4)$ **c.** $7 + (-4)$

Solution:

a. $-5 + 2 = -3$

Start at -5, and count
2 units to the right.

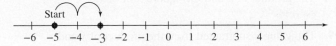

TIP: Note that we move to
the left on the number line
when we add a negative
number. We move to the
right when we add a positive
number.

b. $-1 + (-4) = -5$

Start at -1, and count
4 units to the left.

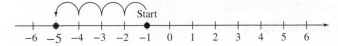

c. $7 + (-4) = 3$

Start at 7, and count
4 units to the left.

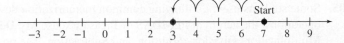

Skill Practice   Use the number line to add the numbers.

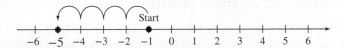

    1. $-2 + 4$ **2.** $-2 + (-3)$ **3.** $5 + (-6)$

2. Addition of Real Numbers

When adding large numbers or numbers that involve fractions or decimals, counting units on the number line can be cumbersome. Study the following example to determine a pattern for adding two numbers with the *same* sign.

     $1 + 4 = 5$

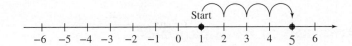

     $-1 + (-4) = -5$

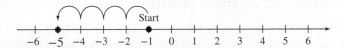

> ### Adding Numbers with the *Same* Sign
>
> To add two numbers with the *same* sign, add their absolute values and apply the common sign.

Classroom Examples: pp. 51–52,
Exercises 22 and 52

> **Example 2**   **Adding Real Numbers with the Same Sign**

Add.

    a. $-12 + (-14)$ **b.** $-8.8 + (-3.7)$ **c.** $-\dfrac{4}{3} + \left(-\dfrac{6}{7}\right)$

Answers

1. 2   **2.** -5 **3.** -1

Solution:

a. $-12 + (-14)$

First find the absolute value of the addends.
$|-12| = 12$ and $|-14| = 14$.

$= -(12 + 14)$

Add their absolute values and apply the common sign (in this case, the common sign is negative).

common sign is negative

$= -26$

The sum is -26.

b. $-8.8 + (-3.7)$

First find the absolute value of the addends.
$|-8.8| = 8.8$ and $|-3.7| = 3.7$.

$= -(8.8 + 3.7)$

Add their absolute values and apply the common sign (in this case, the common sign is negative).

common sign is negative

$= -12.5$

The sum is -12.5.

c. $-\dfrac{4}{3} + \left(-\dfrac{6}{7}\right)$

The least common denominator (LCD) is 21.

$= -\dfrac{4 \cdot 7}{3 \cdot 7} + \left(-\dfrac{6 \cdot 3}{7 \cdot 3}\right)$

Write each fraction with the LCD.

$= -\dfrac{28}{21} + \left(-\dfrac{18}{21}\right)$

Find the absolute value of the addends.

$\left|-\dfrac{28}{21}\right| = \dfrac{28}{21}$ and $\left|-\dfrac{18}{21}\right| = \dfrac{18}{21}$.

$= -\left(\dfrac{28}{21} + \dfrac{18}{21}\right)$

Add their absolute values and apply the common sign (in this case, the common sign is negative).

common sign is negative

$= -\dfrac{46}{21}$

The sum is $-\dfrac{46}{21}$.

Skill Practice Add.

4. $-5 + (-25)$ **5.** $-14.8 + (-9.7)$ **6.** $-\dfrac{1}{2} + \left(-\dfrac{5}{8}\right)$

Study the following example to determine a pattern for adding two numbers with *different* signs.

Instructor Note: Memory Device:
Same Sign ⇒ Sum
Different Sign ⇒ Difference

$1 + (-4) = -3$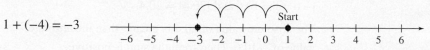

$-1 + 4 = 3$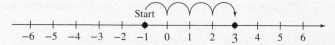

Answers

4. -30 **5.** -24.5 **6.** $-\dfrac{9}{8}$

> ### Adding Numbers with *Different* Signs
> To add two numbers with *different* signs, subtract the smaller absolute value from the larger absolute value. Then apply the sign of the number having the larger absolute value.

Classroom Examples: pp. 51–52, Exercises 20 and 32

Example 3   Adding Real Numbers with Different Signs

Add.

a. $12 + (-17)$ **b.** $-8 + 8$

Solution:

a. $12 + (-17)$ First find the absolute value of the addends.
$|12| = 12$ and $|-17| = 17$.

The absolute value of -17 is greater than the absolute value of 12. Therefore, the sum is negative.

$= -(17 - 12)$ Next, subtract the smaller absolute value from the larger absolute value.

Apply the sign of the number with the larger absolute value.

$= -5$

b. $-8 + 8$ First find the absolute value of the addends.
$|-8| = 8$ and $|8| = 8$.

$= 8 - 8$ The absolute values are equal. Therefore, their difference is 0. The number zero is neither positive

$= 0$ nor negative.

Skill Practice Add.

7. $-15 + 16$ **8.** $6 + (-6)$

Classroom Examples: p. 52, Exercises 46 and 54

Example 4   Adding Real Numbers with Different Signs

Add.

a. $-10.6 + 20.4$ **b.** $\dfrac{2}{15} + \left(-\dfrac{4}{5}\right)$

Solution:

a. $-10.6 + 20.4$ First find the absolute value of the addends.
$|-10.6| = 10.6$ and $|20.4| = 20.4$.

The absolute value of 20.4 is greater than the absolute value of -10.6. Therefore, the sum is positive.

$= +(20.4 - 10.6)$ Next, subtract the smaller absolute value from the larger absolute value.

Apply the sign of the number with the larger absolute value.

$= 9.8$

Answers

7. 1   **8.** 0

b. $\dfrac{2}{15} + \left(-\dfrac{4}{5}\right)$ The least common denominator is 15.

$= \dfrac{2}{15} + \left(-\dfrac{4 \cdot 3}{5 \cdot 3}\right)$ Write each fraction with the LCD.

$= \dfrac{2}{15} + \left(-\dfrac{12}{15}\right)$ Find the absolute value of the addends.

$\left|\dfrac{2}{15}\right| = \dfrac{2}{15}$ and $\left|-\dfrac{12}{15}\right| = \dfrac{12}{15}$.

The absolute value of $-\dfrac{12}{15}$ is greater than the absolute value of $\dfrac{2}{15}$. Therefore, the sum is negative.

$= \underset{\uparrow}{-}\left(\dfrac{12}{15} - \dfrac{2}{15}\right)$ Next, subtract the smaller absolute value from the larger absolute value.

Apply the sign of the number with the larger absolute value.

$= -\dfrac{10}{15}$ Subtract.

$= -\dfrac{2}{3}$ Simplify to lowest terms.

Skill Practice   Add.

9. $27.3 + (-18.1)$ **10.** $-\dfrac{9}{10} + \dfrac{2}{5}$

---

3. Translations

| Example 5 | **Translating Expressions Involving the Addition of Real Numbers** |

Classroom Examples: p. 53, Exercises 86 and 88

Write each English phrase as an algebraic expression. Then simplify the result.

a. The sum of $-12, -8, 9,$ and -1 **b.** Negative three-tenths added to $-\dfrac{7}{8}$

c. The sum of -12 and its opposite

Solution:

a. The sum of $-12, -8, 9,$ and -1

$\underbrace{-12 + (-8)} + 9 + (-1)$

$= \underbrace{-20 + 9} + (-1)$

$= \underbrace{-11 + (-1)}$ Apply the order of operations by adding from left to right.

$= \quad -12$

b. Negative three-tenths added to $-\dfrac{7}{8}$

$-\dfrac{7}{8} + \left(-\dfrac{3}{10}\right)$

$= -\dfrac{35}{40} + \left(-\dfrac{12}{40}\right)$ The common denominator is 40.

$= -\dfrac{47}{40}$ The numbers have the same signs. Add their absolute values and keep the common sign. $-\left(\dfrac{35}{40} + \dfrac{12}{40}\right)$

Answers

9. 9.2 **10.** $-\dfrac{1}{2}$

TIP: The sum of any number and its opposite is 0.

c. The sum of -12 and its opposite

$$-12 + (12)$$

$$= 0 \qquad \text{Add.}$$

Skill Practice Write as an algebraic expression, and simplify the result.

11. The sum of -10, 4, and -6

12. Negative 2 added to $-\frac{1}{2}$

13. -60 added to its opposite

4. Applications Involving Addition of Real Numbers

Classroom Example: p. 53, Exercise 94

© Ryan McVay/Getty Images RF

| Example 6 | **Adding Real Numbers in Applications** |

a. It is common for newborn infants to fluctuate in weight. Elise and Benjamin's baby lost 7 oz the first week after birth and gained 10 oz the second week. Write a mathematical expression to describe this situation and then simplify the result.

b. A student has $120 in her checking account. After depositing her paycheck of $215, she writes a check for $255 to cover her portion of the rent and another check for $294 to cover her car payment. Write a mathematical expression to describe this situation and then simplify the result.

Solution:

a. $-7 + 10$ \qquad A loss of 7 oz can be interpreted as -7 oz.

$\quad = 3$ \qquad The infant had a net gain of 3 oz.

b. $\underbrace{120 + 215} + (-255) + (-294)$ \qquad Writing a check is equivalent to adding a negative amount to the bank account.

$= \underbrace{335 + (-255)} + (-294)$ \qquad Use the order of operations. Add from left to right.

$= \quad 80 + (-294)$

$= \quad -214$ \qquad The student has overdrawn her account by $214.

Skill Practice

14. A stock was priced at $32.00 per share at the beginning of the month. After the first week, the price went up $2.15 per share. At the end of the second week it went down $3.28 per share. Write a mathematical expression to describe the price of the stock and find the price of the stock at the end of the 2-week period.

Answers

11. $-10 + 4 + (-6)$; -12

12. $-\frac{1}{2} + (-2)$; $-\frac{5}{2}$

13. $60 + (-60)$; 0

14. $32.00 + 2.15 + (-3.28)$; $30.87 per share

Section 1.4 Practice Exercises

For additional exercises, see Classroom Activities 1.4A–1.4B in the *Student's Resource Manual* at www.mhhe.com/moh.

Study Skills Exercise

It is very important to attend class every day. Math is cumulative in nature, and you must master the material learned in the previous class to understand today's lesson. Because this is so important, many instructors have an attendance policy that may affect your final grade. Write down the attendance policy for your class.

Vocabulary and Key Concepts

1. a. If a and b are both negative, then $a + b$ will be (choose one: positive or negative).   negative

   b. If a and b have different signs, and if $|b| > |a|$, then the sum will have the same sign as (choose one: a or b). b

Review Exercises

Plot the points in set A on a number line. Then for Exercises 2–7 place the appropriate inequality ($<$, $>$) between the numbers.

$$A = \left\{-2, \frac{3}{4}, -\frac{5}{2}, 3, \frac{9}{2}, 1.6, 0\right\}$$

2. $-2 \,\square\, 0$ $<$

3. $\dfrac{9}{2} \,\square\, \dfrac{3}{4}$ $>$

4. $-2 \,\square\, -\dfrac{5}{2}$ $>$

5. $0 \,\square\, -\dfrac{5}{2}$ $>$

6. $\dfrac{3}{4} \,\square\, 1.6$ $<$

7. $\dfrac{3}{4} \,\square\, -\dfrac{5}{2}$ $>$

8. Evaluate the expressions.

   a. $-(-8)$ 8   **b.** $-|-8|$ -8

Concept 1: Addition of Real Numbers and the Number Line

For Exercises 9–16, add the numbers using the number line. **(See Example 1.)**

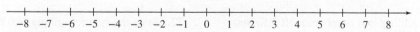

9. $-2 + (-4)$ -6

10. $-3 + (-5)$ -8

11. $-7 + 10$ 3

12. $-2 + 9$ 7

13. $6 + (-3)$ 3

14. $8 + (-2)$ 6

15. $2 + (-5)$ -3

16. $7 + (-3)$ 4

Concept 2: Addition of Real Numbers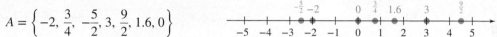

For Exercises 17–70, add. **(See Examples 2–4.)**

17. $-19 + 2$ -17

18. $-25 + 18$ -17

19. $-4 + 11$ 7

20. $-3 + 9$ 6

21. $-16 + (-3)$ -19

22. $-12 + (-23)$ -35

23. $-2 + (-21)$ -23

24. $-13 + (-1)$ -14

25. $0 + (-5)$ -5 **26.** $0 + (-4)$ -4 **27.** $-3 + 0$ -3 **28.** $-8 + 0$ -8

29. $-16 + 16$ 0 **30.** $11 + (-11)$ 0 **31.** $41 + (-41)$ 0 **32.** $-15 + 15$ 0

33. $4 + (-9)$ -5 **34.** $6 + (-9)$ -3 **35.** $7 + (-2) + (-8)$ -3 **36.** $2 + (-3) + (-6)$ -7

37. $-17 + (-3) + 20$ 0 **38.** $-9 + (-6) + 15$ 0

39. $-3 + (-8) + (-12)$ -23 **40.** $-8 + (-2) + (-13)$ -23

41. $-42 + (-3) + 45 + (-6)$ -6 **42.** $36 + (-3) + (-8) + (-25)$ 0

43. $-5 + (-3) + (-7) + 4 + 8$ -3 **44.** $-13 + (-1) + 5 + 2 + (-20)$ -27

45. $23.81 + (-2.51)$ 21.3 **46.** $-9.23 + 10.53$ 1.3 **47.** $-\dfrac{2}{7} + \dfrac{1}{14}$ $-\dfrac{3}{14}$ **48.** $-\dfrac{1}{8} + \dfrac{5}{16}$ $\dfrac{3}{16}$

49. $\dfrac{2}{3} + \left(-\dfrac{5}{6}\right)$ $-\dfrac{1}{6}$ **50.** $\dfrac{1}{2} + \left(-\dfrac{3}{4}\right)$ $-\dfrac{1}{4}$ **51.** $-\dfrac{7}{8} + \left(-\dfrac{1}{16}\right)$ $-\dfrac{15}{16}$ **52.** $-\dfrac{1}{9} + \left(-\dfrac{4}{3}\right)$ $-\dfrac{13}{9}$

53. $-\dfrac{1}{4} + \dfrac{3}{10}$ $\dfrac{1}{20}$ **54.** $-\dfrac{7}{6} + \dfrac{7}{8}$ $-\dfrac{7}{24}$ **55.** $-2.1 + \left(-\dfrac{3}{10}\right)$ **56.** $-8.3 + \left(-\dfrac{9}{10}\right)$

                                                                    -2.4 or $-\dfrac{12}{5}$ -9.2 or $-\dfrac{46}{5}$

57. $\dfrac{3}{4} + (-0.5)$ **58.** $-\dfrac{3}{2} + 0.45$ **59.** $8.23 + (-8.23)$ 0 **60.** $-7.5 + 7.5$ 0

    $\dfrac{1}{4}$ or 0.25 $-\dfrac{21}{20}$ or -1.05

61. $-\dfrac{7}{8} + 0$ $-\dfrac{7}{8}$ **62.** $0 + \left(-\dfrac{21}{22}\right)$ $-\dfrac{21}{22}$ **63.** $-\dfrac{3}{2} + \left(-\dfrac{1}{3}\right) + \dfrac{5}{6}$ -1 **64.** $-\dfrac{7}{8} + \dfrac{7}{6} + \dfrac{7}{12}$ $\dfrac{7}{8}$

65. $-\dfrac{2}{3} + \left(-\dfrac{1}{9}\right) + 2$ $\dfrac{11}{9}$ **66.** $-\dfrac{1}{4} + \left(-\dfrac{3}{2}\right) + 2$ $\dfrac{1}{4}$ **67.** $-47.36 + 24.28$ **68.** $-0.015 + (0.0026)$

                                                                    -23.08 -0.0124

69. $-0.000617 + (-0.0015)$ -0.002117 **70.** $-5315.26 + (-314.89)$ -5630.15

71. State the rule for adding two numbers with different signs.   To add two numbers with different signs, subtract the smaller absolute value from the larger absolute value and apply the sign of the number with the larger absolute value.

72. State the rule for adding two numbers with the same sign.   To add two numbers with the same sign, add their absolute values and apply the common sign.

For Exercises 73–80, evaluate each expression for $x = -3, y = -2,$ and $z = 16.$

73. $x + y + \sqrt{z}$ -1 **74.** $2z + x + y$ 27 **75.** $y + 3\sqrt{z}$ 10 **76.** $-\sqrt{z} + y$ -6

77. $|x| + |y|$ 5 **78.** $z + x + |y|$ 15 **79.** $-x + y$ 1 **80.** $x + (-y) + z$ 15

Concept 3: Translations

For Exercises 81–90, write each English phrase as an algebraic expression. Then simplify the result. **(See Example 5.)**

81. The sum of -6 and -10 $-6 + (-10); -16$ **82.** The sum of -3 and 5 $-3 + 5; 2$

83. Negative three increased by 8 $-3 + 8; 5$ **84.** Twenty-one increased by 4 $21 + 4; 25$

85. Seventeen more than −21   −21 + 17; −4

86. Twenty-four more than −7   −7 + 24; 17

87. Three times the sum of −14 and 20   3(−14 + 20); 18

88. Two times the sum of 6 and −10   2[6 + (−10)]; −8

89. Five more than the sum of −7 and −2
[−7 + (−2)] + 5; −4

90. Negative six more than the sum of 4 and −1
[4 + (−1)] + (−6); −3

Concept 4: Applications Involving Addition of Real Numbers

91. The temperature in Minneapolis, Minnesota, began at −5°F (5° below zero) at 6:00 A.M. By noon, the temperature had risen 13°, and by the end of the day, the temperature had dropped 11° from its noontime high. Write an expression using addition that describes the change in temperatures during the day. Then evaluate the expression to give the temperature at the end of the day.
−5 + 13 + (−11); −3°F

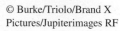

92. The temperature in Toronto, Ontario, Canada, began at 4°F. A cold front went through at noon, and the temperature dropped 9°. By 4:00 P.M., the temperature had risen 2° from its noontime low. Write an expression using addition that describes the changes in temperature during the day. Then evaluate the expression to give the temperature at the end of the day.   4 + (−9) + 2; −3°F

© Burke/Triolo/Brand X Pictures/Jupiterimages RF

93. For 4 months, Amara monitored her weight loss or gain. Her records showed that she lost 8 lb, gained 1 lb, gained 2 lb, and lost 5 lb. Write an expression using addition that describes Amara's total loss or gain and evaluate the expression. Interpret the result. **(See Example 6.)**
−8 + 1 + 2 + (−5) = −10; Amara lost 10 lb.

94. Alan just started an online business. His profit/loss records for the past 5 months show that he had a profit of $200, a profit of $750, a loss of $340, a loss of $290, and a profit of $900. Write an expression using addition that describes Alan's total profit or loss and evaluate the expression. Interpret the result.
200 + 750 + (−340) + (−290) + 900 = 1220; Alan had a profit of $1220.

95. Yoshima has $52.23 in her checking account. She writes a check for groceries for $52.95. **(See Example 6.)**

   a. Write an addition statement that expresses Yoshima's transaction.   52.23 + (−52.95)

   b. Is Yoshima's account overdrawn?   Yes

96. Mohammad has $40.02 in his checking account. He writes a check for a pair of shoes for $40.96.

   a. Write an addition statement that expresses Mohammad's transaction.   40.02 + (−40.96)

   b. Is Mohammad's account overdrawn?   Yes

97. The table gives the golf scores for a top golfer for five rounds of the LPGA Final Qualifying Tournament. Find her total score.   −6; She was 6 below par.

Score per Round	
Round 1	−5
Round 2	0
Round 3	−1
Round 4	−1
Round 5	+1

98. A company that has been in business for 5 years has the following profit and loss record.

   a. Write an expression using addition to describe the company's profit/loss activity.
   −50,000 + (−32,000) + (−5000) + 13,000 + 26,000

   b. Evaluate the expression from part (a) to determine the company's net profit or loss.
   −48,000; The company lost $48,000.

Year	Profit/Loss ($)
1	−50,000
2	−32,000
3	−5,000
4	13,000
5	26,000

Section 1.5 Subtraction of Real Numbers

1. Subtraction of Real Numbers

We have learned the rules for adding real numbers. Subtraction of real numbers is defined in terms of the addition process. For example, consider the following subtraction problem and the corresponding addition problem:

$$6 - 4 = 2 \quad \Leftrightarrow \quad 6 + (-4) = 2$$

In each case, we start at 6 on the number line and move to the left 4 units. That is, adding the opposite of 4 produces the same result as subtracting 4. This is true in general. To subtract two real numbers, add the opposite of the second number to the first number.

Subtracting Real Numbers

If a and b are real numbers, then $a - b = a + (-b)$.

$$\left.\begin{array}{l} 10 - 4 = 10 + (-4) = 6 \\ -10 - 4 = -10 + (-4) = -14 \end{array}\right\} \quad \text{Subtracting 4 is the same as adding } -4.$$

$$\left.\begin{array}{l} 10 - (-4) = 10 + (4) = 14 \\ -10 - (-4) = -10 + (4) = -6 \end{array}\right\} \quad \text{Subtracting } -4 \text{ is the same as adding 4.}$$

Classroom Examples: p. 60, Exercises 16, 18, and 20

Example 1 Subtracting Integers

Subtract the numbers.

  a. $4 - (-9)$ **b.** $-6 - 9$ **c.** $-11 - (-5)$ **d.** $7 - 10$

Solution:

a. $4 - (-9)$

    $= 4 + (9) = 13$

    Take the opposite of −9.

  Change subtraction to addition.

b. $-6 - 9$

    $= -6 + (-9) = -15$

    Take the opposite of 9.

  Change subtraction to addition.

c. $-11 - (-5)$

    $= -11 + (5) = -6$

    Take the opposite of −5.

  Change subtraction to addition.

d. $7 - 10$

    $= 7 + (-10) = -3$

    Take the opposite of 10.

  Change subtraction to addition.

Skill Practice Subtract.

  1. $1 - (-3)$ **2.** $-2 - 2$ **3.** $-6 - (-11)$ **4.** $8 - 15$

Answers

1. 4   **2.** −4   **3.** 5   **4.** −7

Example 2	**Subtracting Real Numbers**

Classroom Examples: p. 60, Exercises 46 and 48.

a. $\dfrac{3}{20} - \left(-\dfrac{4}{15}\right)$　　**b.** $-2.3 - 6.04$

Solution:

a. $\dfrac{3}{20} - \left(-\dfrac{4}{15}\right)$ 　　The least common denominator is 60.

$= \dfrac{9}{60} - \left(-\dfrac{16}{60}\right)$ 　　Write equivalent fractions with the LCD.

$= \dfrac{9}{60} + \left(\dfrac{16}{60}\right)$ 　　Rewrite subtraction in terms of addition.

$= \dfrac{25}{60}$ 　　Add.

$= \dfrac{5}{12}$ 　　Simplify by dividing the numerator and denominator by the GCF of 5.

b. $-2.3 - 6.04$

$-2.3 + (-6.04)$ 　　Rewrite subtraction in terms of addition.

-8.34 　　Add.

Skill Practice Subtract.

5. $\dfrac{1}{6} - \left(-\dfrac{7}{12}\right)$ 　　**6.** $-7.5 - 1.5$

2. Translations

Example 3	**Translating Expressions Involving Subtraction**

Classroom Examples: p. 60, Exercises 64 and 70.

Write an algebraic expression for each English phrase and then simplify the result.

a. The difference of -7 and -5

b. 12.4 subtracted from -4.7

c. -24 decreased by the sum of -10 and 13

d. Seven-fourths less than one-third

Solution:

a. The difference of -7 and -5

$-7 - (-5)$

$= -7 + (5)$ 　　Rewrite subtraction in terms of addition.

$= -2$ 　　Simplify.

Answers

5. $\dfrac{3}{4}$ 　**6.** -9

TIP: Recall that "*b* subtracted from *a*" is translated as *a* − *b*. In Example 3(b), −4.7 is written first and then 12.4.

TIP: In Example 3(c), parentheses must be used around the sum of −10 and 13 so that −24 is decreased by the entire quantity (−10 + 13).

b. 12.4 subtracted from −4.7

$$-4.7 - 12.4$$

$= -4.7 + (-12.4)$	Rewrite subtraction in terms of addition.
$= -17.1$	Simplify.

c. −24 decreased by the sum of −10 and 13

$$-24 - (-10 + 13)$$

$= -24 - (3)$	Simplify inside parentheses.
$= -24 + (-3)$	Rewrite subtraction in terms of addition.
$= -27$	Simplify.

d. Seven-fourths less than one-third

$$\frac{1}{3} - \frac{7}{4}$$

$= \frac{1}{3} + \left(-\frac{7}{4}\right)$	Rewrite subtraction in terms of addition.
$= \frac{4}{12} + \left(-\frac{21}{12}\right)$	The common denominator is 12.
$= -\frac{17}{12}$	

Skill Practice   Write an algebraic expression for each phrase and then simplify.

7. 8 less than −10
8. −7.2 subtracted from −8.2
9. 10 more than the difference of −2 and 3
10. Two-fifths decreased by four-thirds

3. Applications Involving Subtraction

Classroom Example: p. 61, Exercise 72

Example 4   Using Subtraction of Real Numbers in an Application

During one of his turns on *Jeopardy*, Harold selected the category "Show Tunes." He got the $200, $600, and $1000 questions correct, but he got the $400 and $800 questions incorrect. Write an expression that determines Harold's score. Then simplify the expression to find his total winnings for that category.

Solution:

$$200 + 600 + 1000 - 400 - 800$$

$= 200 + 600 + 1000 + (-400) + (-800)$	Add the positive numbers.
$= 1800 + (-1200)$	Add the negative numbers.
$= 600$	Harold won $600.

Skill Practice

11. During Harold's first round on *Jeopardy*, he got the $100, $200, and $400 questions correct but he got the $300 and $500 questions incorrect. Determine Harold's score for this round.

Answers

7. $-10 - 8$; -18
8. $-8.2 - (-7.2)$; -1
9. $(-2 - 3) + 10$; 5
10. $\frac{2}{5} - \frac{4}{3}$; $-\frac{14}{15}$
11. -100, Harold lost $100.

Example 5 — Using Subtraction of Real Numbers in an Application

The highest recorded temperature in North America was 134°F, recorded on July 10, 1913, in Death Valley, California. The lowest temperature of −81°F was recorded on February 3, 1947, in Snag, Yukon, Canada.

Find the difference between the highest and lowest recorded temperatures in North America.

Solution:

$134 - (-81)$

$= 134 + (81)$ Rewrite subtraction in terms of addition.

$= 215$ Add.

The difference between the highest and lowest temperatures is 215°F.

Classroom Example: p. 61, Exercise 74

Skill Practice

12. The record high temperature for the state of Montana occurred in 1937 and was 117°F. The record low occurred in 1954 and was −70°F. Find the difference between the highest and lowest temperatures.

4. Applying the Order of Operations

Example 6 — Applying the Order of Operations

Simplify the expressions.

   a. $-6 + \{10 - [7 - (-4)]\}$ **b.** $5 - \sqrt{35 - (-14)} - 2$

Classroom Examples: p. 61–62, Exercises 86 and 92

Solution:

   a. $-6 + \{10 - [7 - (-4)]\}$ Simplify inside the inner brackets first.

     $= -6 + \{10 - [7 + (4)]\}$ Rewrite subtraction in terms of addition.

     $= -6 + \{10 - (11)\}$ Simplify the expression inside brackets.

     $= -6 + \{10 + (-11)\}$ Rewrite subtraction in terms of addition.

     $= -6 + (-1)$ Add within the braces.

     $= -7$ Add.

   b. $5 - \sqrt{35 - (-14)} - 2$ Simplify inside the radical first.

     $= 5 - \sqrt{35 + (14)} - 2$ Rewrite subtraction in terms of addition.

     $= 5 - \sqrt{49} - 2$ Add within the radical sign.

     $= 5 - 7 - 2$ Simplify the radical.

     $= 5 + (-7) + (-2)$ Rewrite subtraction in terms of addition.

     $= -2 + (-2)$ Add from left to right.

     $= -4$

Skill Practice Simplify the expressions.

13. $-11 - \{8 - [2 - (-3)]\}$ **14.** $(12 - 5)^2 + \sqrt{4 - (-21)}$

Answers

12. 187°F   **13.** −14   **14.** 54

Classroom Examples: p. 62, Exercises 90 and 94

Example 7 **Applying the Order of Operations**

Simplify the expressions.

a. $\left(-\dfrac{5}{8} - \dfrac{2}{3}\right) - \left(\dfrac{1}{8} + 2\right)$
  b. $-6 - |7 - 11| + (-3 + 7)^2$

Solution:

a. $\left(-\dfrac{5}{8} - \dfrac{2}{3}\right) - \left(\dfrac{1}{8} + 2\right)$
  Simplify inside the parentheses first.

$= \left[-\dfrac{5}{8} + \left(-\dfrac{2}{3}\right)\right] - \left(\dfrac{1}{8} + 2\right)$
  Rewrite subtraction in terms of addition.

$= \left[-\dfrac{15}{24} + \left(-\dfrac{16}{24}\right)\right] - \left(\dfrac{1}{8} + \dfrac{16}{8}\right)$
  Get a common denominator in each parentheses.

$= \left(-\dfrac{31}{24}\right) - \left(\dfrac{17}{8}\right)$
  Add fractions in each parentheses.

$= \left(-\dfrac{31}{24}\right) + \left(-\dfrac{17}{8}\right)$
  Rewrite subtraction in terms of addition.

$= -\dfrac{31}{24} + \left(-\dfrac{51}{24}\right)$
  Get a common denominator.

$= -\dfrac{82}{24}$
  Add.

$= -\dfrac{41}{12}$
  Simplify to lowest terms.

b. $-6 - |7 - 11| + (-3 + 7)^2$
  Simplify within absolute value bars and parentheses first.

$= -6 - |7 + (-11)| + (-3 + 7)^2$
  Rewrite subtraction in terms of addition.

$= -6 - |-4| + (4)^2$

$= -6 - (4) + 16$
  Simplify the absolute value and exponent.

$= -6 + (-4) + 16$
  Rewrite subtraction in terms of addition.

$= -10 + 16$
  Add from left to right.

$= 6$

Skill Practice Simplify the expressions.

15. $\left(-1 + \dfrac{1}{4}\right) - \left(\dfrac{3}{4} - \dfrac{1}{2}\right)$

16. $4 - 2|6 + (-8)| + (4)^2$

Answers

15. -1 **16.** 16

Calculator Connections

Topic: Operations with Signed Numbers on a Calculator

Most calculators can add, subtract, multiply, and divide signed numbers. It is important to note, however, that the key used for the negative sign is different from the key used for subtraction. On a scientific calculator, the $\boxed{+/-}$ key or $\boxed{+\circ-}$ key is used to enter a negative number or to change the sign of an existing number. On a graphing calculator, the $\boxed{(-)}$ key is used. These keys should not be confused with the $\boxed{-}$ key which is used for subtraction. For example, try simplifying the following expressions.

a. $-7 + (-4) - 6$ **b.** $-3.1 - (-0.5) + 1.1$

Scientific Calculator:

Enter: $7\ \boxed{+\circ-}\ \boxed{+}\ \boxed{(}\ 4\ \boxed{+\circ-}\ \boxed{)}\ \boxed{-}\ 6\ \boxed{=}$ **Result:** $\boxed{\quad -17 \quad}$

Enter: $3.1\ \boxed{+\circ-}\ \boxed{-}\ \boxed{(}\ 0.5\ \boxed{+\circ-}\ \boxed{)}\ \boxed{+}\ 1.1\ \boxed{=}$ **Result:** $\boxed{\quad -1.5 \quad}$

Graphing Calculator:

```
-7+(-4)-6
              -17
-3.1-(-0.5)+1.1
              -1.5
```

Calculator Exercises

Simplify the expression without the use of a calculator. Then use the calculator to verify your answer.

1. $-8 + (-5)$ -13 **2.** $4 + (-5) + (-1)$ -2 **3.** $627 - (-84)$ 711 **4.** $-0.06 - 0.12$ -0.18

5. $-3.2 + (-14.5)$ -17.7 **6.** $-472 + (-518)$ -990 **7.** $-12 - 9 + 4$ -17 **8.** $209 - 108 + (-63)$ 38

Section 1.5 Practice Exercises

Study Skills Exercise

For additional exercises, see Classroom Activities 1.5A–1.5C in the *Student's Resource Manual* at www.mhhe.com/moh.

Some instructors allow the use of calculators. What is your instructor's policy regarding calculators in class, on the homework, and on tests?

Helpful Hint: If you are not permitted to use a calculator on tests, it is a good idea to do your homework in the same way, without a calculator.

Vocabulary and Key Concepts

1. a. The expression $a - b$ is equal to $a + \underline{\quad -b \quad}$.

   b. If a is positive and b is negative, then the difference $a - b$ will be (choose one: positive or negative).   positive

Review Exercises

For Exercises 2–5, write each English phrase as an algebraic expression.

2. The square root of 6 $\sqrt{6}$ **3.** The square of x x^2

4. Negative seven increased by 10 $-7 + 10$ **5.** Two more than $-b$ $-b + 2$

 Writing   Translating Expression   Geometry   Scientific Calculator   Video

For Exercises 6–8, simplify the expression.

6. $4^2 - 6 \div 2$ 13

7. $1 + 36 \div 9 \cdot 2$ 9

8. $14 - |10 - 6|$ 10

Concept 1: Subtraction of Real Numbers

For Exercises 9–14, fill in the blank to make each statement correct.

9. $5 - 3 = 5 + \underline{}$ −3

10. $8 - 7 = 8 + \underline{}$ −7

11. $-2 - 12 = -2 + \underline{}$ −12

12. $-4 - 9 = -4 + \underline{}$ −9

13. $7 - (-4) = 7 + \underline{}$ 4

14. $13 - (-4) = 13 + \underline{}$ 4

For Exercises 15–60, simplify. **(See Examples 1–2.)**

15. $3 - 5$ −2

16. $9 - 12$ −3

17. $3 - (-5)$ 8

18. $9 - (-12)$ 21

19. $-3 - 5$ −8

20. $-9 - 12$ −21

21. $-3 - (-5)$ 2

22. $-9 - (-5)$ −4

23. $23 - 17$ 6

24. $14 - 2$ 12

25. $23 - (-17)$ 40

26. $14 - (-2)$ 16

27. $-23 - 17$ −40

28. $-14 - 2$ −16

29. $-23 - (-23)$ 0

30. $-14 - (-14)$ 0

31. $-6 - 14$ −20

32. $-9 - 12$ −21

33. $-7 - 17$ −24

34. $-8 - 21$ −29

35. $13 - (-12)$ 25

36. $20 - (-5)$ 25

37. $-14 - (-9)$ −5

38. $-21 - (-17)$ −4

39. $-\dfrac{6}{5} - \dfrac{3}{10}$ $-\dfrac{3}{2}$

40. $-\dfrac{2}{9} - \dfrac{5}{3}$ $-\dfrac{17}{9}$

41. $\dfrac{3}{8} - \left(-\dfrac{4}{3}\right)$ $\dfrac{41}{24}$

42. $\dfrac{7}{10} - \left(-\dfrac{5}{6}\right)$ $\dfrac{23}{15}$

43. $\dfrac{1}{2} - \dfrac{1}{10}$ $\dfrac{2}{5}$

44. $\dfrac{2}{7} - \dfrac{3}{14}$ $\dfrac{1}{14}$

45. $-\dfrac{11}{12} - \left(-\dfrac{1}{4}\right)$ $-\dfrac{2}{3}$

46. $-\dfrac{7}{8} - \left(-\dfrac{1}{6}\right)$ $-\dfrac{17}{24}$

47. $6.8 - (-2.4)$ 9.2

48. $7.2 - (-1.9)$ 9.1

49. $3.1 - 8.82$ −5.72

50. $1.8 - 9.59$ −7.79

51. $-4 - 3 - 2 - 1$ −10

52. $-10 - 9 - 8 - 7$ −34

53. $6 - 8 - 2 - 10$ −14

54. $20 - 50 - 10 - 5$ −45

55. $10 + (-14) + 6 - 22$ −20

56. $-3 - (-8) + (-11) - 6$ −12

57. $-112.846 + (-13.03) - 47.312$ −173.188

58. $-96.473 + (-36.02) - 16.617$ −149.11

59. $0.085 - (-3.14) + 0.018$ 3.243

60. $0.00061 - (-0.00057) + 0.0014$ 0.00258

Concept 2: Translations

↞→ For Exercises 61–70, write each English phrase as an algebraic expression. Then evaluate the expression.
(See Example 3.)

61. Six minus −7 $6 - (-7)$; 13

62. Eighteen minus −1 $18 - (-1)$; 19

63. Eighteen subtracted from 3 $3 - 18$; −15

64. Twenty-one subtracted from 8 $8 - 21$; −13

65. The difference of −5 and −11 $-5 - (-11)$; 6

66. The difference of −2 and −18 $-2 - (-18)$; 16

67. Negative thirteen subtracted from −1 $-1 - (-13)$; 12

68. Negative thirty-one subtracted from −19 $-19 - (-31)$; 12

69. Twenty less than −32 $-32 - 20$; −52

70. Seven less than −3 $-3 - 7$; −10

Concept 3: Applications Involving Subtraction

71. On the game, *Jeopardy,* Jasper selected the category "The Last." He got the first four questions correct (worth $200, $400, $600, and $800) but then missed the last question (worth $1000). Write an expression that determines Jasper's score. Then simplify the expression to find his total winnings for that category.
(See Example 4.) $200 + 400 + 600 + 800 - 1000; \$1000$

72. On Courtney's turn in *Jeopardy,* she chose the category "Birds of a Feather." She already had $1200 when she selected a Double Jeopardy question. She wagered $500 but guessed incorrectly (therefore she lost $500). On her next turn, she got the $800 question correct. Write an expression that determines Courtney's score. Then simplify the expression. $1200 - 500 + 800; \$1500$

73. In Ohio, the highest temperature ever recorded was 113°F and the lowest was −39°F. Find the difference between the highest and lowest temperatures. (*Source: Information Please Almanac*) **(See Example 5.)** 152°F

74. On a recent winter day at the South Pole, the temperature was −52°F. On the same day in Springfield, Missouri, it was a pleasant summer temperature of 75°F. What was the difference in temperatures? The difference in the temperature was 127°F.

75. The highest mountain in the world is Mt. Everest, located in South Asia. Its height is 8848 meters (m). The lowest recorded depth in the ocean is located in the Marianas Trench in the Pacific Ocean. Its "height" relative to sea level is −11,033 m. Determine the difference in elevation, in meters, between the highest mountain in the world and the deepest ocean trench. (*Source: Information Please Almanac*) 19,881 m

© Daniel Prudek/iStockphoto/Getty Images RF

76. The lowest point in North America is located in Death Valley, California, at an elevation of −282 ft. The highest point in North America is Denali, Alaska, at an elevation of 20,320 ft. Find the difference in elevation, in feet, between the highest and lowest points in North America. (*Source: Information Please Almanac*) 20,602 ft

© Comstock Images/Alamy RF

Concept 4: Applying the Order of Operations

For Exercises 77–96, perform the indicated operations. **(See Examples 6–7.)**

77. $6 + 8 - (-2) - 4 + 1$ 13

78. $-3 - (-4) + 1 - 2 - 5$ −5

79. $-1 - 7 + (-3) - 8 + 10$ −9

80. $13 - 7 + 4 - 3 - (-1)$ 8

81. $2 - (-8) + 7 + 3 - 15$ 5

82. $8 - (-13) + 1 - 9$ 13

83. $-6 + (-1) + (-8) + (-10)$ −25

84. $-8 + (-3) + (-5) + (-2)$ −18

85. $-4 - \{11 - [4 - (-9)]\}$ −2

86. $15 - \{25 + 2[3 - (-1)]\}$ −18

87. $-\dfrac{13}{10} + \dfrac{8}{15} - \left(-\dfrac{2}{5}\right)$ $-\dfrac{11}{30}$

88. $\dfrac{11}{14} - \left(-\dfrac{9}{7}\right) - \dfrac{3}{2}$ $\dfrac{4}{7}$

89. $\left(\dfrac{2}{3} - \dfrac{5}{9}\right) - \left(\dfrac{4}{3} - (-2)\right)$ $-\dfrac{29}{9}$ **90.** $\left(-\dfrac{9}{8} - \dfrac{1}{4}\right) - \left(-\dfrac{5}{6} + \dfrac{1}{8}\right)$ $-\dfrac{2}{3}$ **91.** $\sqrt{29 + (-4)} - 7$ -2

92. $8 - \sqrt{98 + (-3)} + 5$ -2 **93.** $|10 + (-3)| - |-12 + (-6)|$ -11 **94.** $|6 - 8| + |12 - 5|$ 9

95. $\dfrac{3 - 4 + 5}{4 + (-2)}$ 2 **96.** $\dfrac{12 - 14 + 6}{6 + (-2)}$ 1

For Exercises 97–104, evaluate each expression for $a = -2$, $b = -6$, and $c = -1$.

97. $(a + b) - c$ -7 **98.** $(a - b) + c$ 3 **99.** $a - (b + c)$ 5 **100.** $a + (b - c)$ -7

101. $(a - b) - c$ 5 **102.** $(a + b) + c$ -9 **103.** $a - (b - c)$ 3 **104.** $a + (b + c)$ -9

Problem Recognition Exercises

Addition and Subtraction of Real Numbers

1. State the rule for adding two negative numbers.
Add their absolute values and apply a negative sign.

2. State the rule for adding a negative number to a positive number. Subtract the smaller absolute value from the larger absolute value. Apply the sign of the number with the larger absolute value.

For Exercises 3–10, perform the indicated operations.

3. a. $14 + (-8)$ 6 **b.** $-14 + 8$ -6 **c.** $-14 + (-8)$ -22 **d.** $14 - (-8)$ 22 **e.** $-14 - 8$ -22

4. a. $-5 - (-3)$ -2 **b.** $-5 + (-3)$ -8 **c.** $-5 - 3$ -8 **d.** $-5 + 3$ -2 **e.** $5 - (-3)$ 8

5. a. $-25 + 25$ 0 **b.** $25 - 25$ 0 **c.** $25 - (-25)$ 50 **d.** $-25 - (-25)$ 0 **e.** $-25 + (-25)$ -50

6. a. $\dfrac{1}{2} + \left(-\dfrac{2}{3}\right)$ $-\dfrac{1}{6}$ **b.** $-\dfrac{1}{2} + \left(\dfrac{2}{3}\right)$ $\dfrac{1}{6}$ **c.** $-\dfrac{1}{2} + \left(-\dfrac{2}{3}\right)$ $-\dfrac{7}{6}$ **d.** $\dfrac{1}{2} - \left(-\dfrac{2}{3}\right)$ $\dfrac{7}{6}$ **e.** $-\dfrac{1}{2} - \dfrac{2}{3}$ $-\dfrac{7}{6}$

7. a. $3.5 - 7.1$ -3.6 **b.** $3.5 - (-7.1)$ 10.6 **c.** $-3.5 + 7.1$ 3.6 **d.** $-3.5 - (-7.1)$ 3.6 **e.** $-3.5 + (-7.1)$ -10.6

8. a. $6 - 1 + 4 - 5$ 4 **b.** $6 - (1 + 4) - 5$ -4 **c.** $6 - (1 + 4 - 5)$ 6 **d.** $(6 - 1) - (4 - 5)$ 6

9. a. $-100 - 90 - 80$ -270 **b.** $-100 - (90 - 80)$ -110 **c.** $-100 + (90 - 80)$ -90 **d.** $-100 - (90 + 80)$ -270

10. a. $-8 - (-10) + 20^2$ 402 **b.** $-8 - (-10 + 20)^2$ -108 **c.** $[-8 - (-10) + 20]^2$ 484 **d.** $[-8 - (-10)]^2 + 20$ 24

| Multiplication and Division of Real Numbers | **Section 1.6** |

1. Multiplication of Real Numbers

Multiplication of real numbers can be interpreted as repeated addition. For example:

$$3(4) = 4 + 4 + 4 = 12 \qquad \text{Add 3 groups of 4.}$$
$$3(-4) = -4 + (-4) + (-4) = -12 \qquad \text{Add 3 groups of } -4.$$

These results suggest that the product of a positive number and a negative number is *negative*. Consider the following pattern of products.

$$
\begin{aligned}
4 \cdot 3 &= 12 \\
4 \cdot 2 &= 8 \\
4 \cdot 1 &= 4 \\
4 \cdot 0 &= 0 \\
4 \cdot (-1) &= -4 \\
4 \cdot (-2) &= -8 \\
4 \cdot (-3) &= -12
\end{aligned}
$$

The pattern decreases by 4 with each row.

Thus, the product of a positive number and a negative number must be *negative* for the pattern to continue.

Now suppose we have a product of two negative numbers. To determine the sign, consider the following pattern of products.

$$
\begin{aligned}
-4 \cdot 3 &= -12 \\
-4 \cdot 2 &= -8 \\
-4 \cdot 1 &= -4 \\
-4 \cdot 0 &= 0 \\
-4 \cdot (-1) &= 4 \\
-4 \cdot (-2) &= 8 \\
-4 \cdot (-3) &= 12
\end{aligned}
$$

The pattern increases by 4 with each row.

Thus, the product of two negative numbers must be *positive* for the pattern to continue.

From the first four rows, we see that the product increases by 4 for each row. For the pattern to continue, it follows that the product of two negative numbers must be *positive*.

We now summarize the rules for multiplying real numbers.

Multiplying Real Numbers

- The product of two real numbers with the *same* sign is positive.

  Examples: $(5)(6) = 30$

  $(-4)(-10) = 40$

- The product of two real numbers with *different* signs is negative.

  Examples: $(-2)(5) = -10$

  $(4)(-9) = -36$

- The product of any real number and zero is zero.

  Examples: $(8)(0) = 0$

  $(0)(-6) = 0$

Classroom Examples: p. 71,
Exercises 8 and 14

Example 1	**Multiplying Real Numbers**

Multiply the real numbers.

  a. $-8(-4)$ **b.** $-2.5(-1.7)$ **c.** $-7(10)$

  d. $\dfrac{1}{2}(-8)$ **e.** $0(-8.3)$ **f.** $-\dfrac{2}{7}\left(-\dfrac{7}{2}\right)$

Solution:

  a. $-8(-4) = 32$ *Same* signs. Product is positive.

  b. $-2.5(-1.7) = 4.25$ *Same* signs. Product is positive.

  c. $-7(10) = -70$ *Different* signs. Product is negative.

  d. $\dfrac{1}{2}(-8) = -4$ *Different* signs. Product is negative.

  e. $0(-8.3) = 0$ The product of any real number and zero is zero.

  f. $-\dfrac{2}{7}\left(-\dfrac{7}{2}\right) = \dfrac{14}{14}$ *Same* signs. Product is positive.

               $= 1$ Simplify.

Skill Practice Multiply.

  1. $-9(-3)$ **2.** $-1.5(-1.5)$ **3.** $-6(4)$

  4. $\dfrac{1}{3}(-15)$ **5.** $0(-4.1)$ **6.** $-\dfrac{5}{9}\left(-\dfrac{9}{5}\right)$

Observe the pattern for repeated multiplications.

$$(-1)(-1) \qquad \underline{(-1)(-1)}(-1) \qquad \underline{(-1)(-1)}(-1)(-1) \qquad \underline{(-1)(-1)}(-1)(-1)(-1)$$

$$\quad = 1 \qquad\qquad = \underline{(1)(-1)} \qquad\qquad = \underline{(1)(-1)}(-1) \qquad\qquad = \underline{(1)(-1)}(-1)(-1)$$

$$\qquad\qquad\qquad\quad = -1 \qquad\qquad\qquad = \underline{(-1)(-1)} \qquad\qquad = \underline{(-1)(-1)}(-1)$$

$$\qquad\qquad\qquad\qquad\qquad\qquad\qquad\qquad = 1 \qquad\qquad\qquad = \underline{(1)(-1)}$$

$$\qquad\qquad\qquad\qquad\qquad\qquad\qquad\qquad\qquad\qquad\qquad\qquad\qquad = -1$$

The pattern demonstrated in these examples indicates that

- The product of an even number of negative factors is positive.
- The product of an odd number of negative factors is negative.

2. Exponential Expressions

Recall that for any real number b and any positive integer, n:

$$b^n = \underbrace{b \cdot b \cdot b \cdot b \cdot \ldots \cdot b}_{n \text{ factors of } b}$$

Be particularly careful when evaluating exponential expressions involving negative numbers. An exponential expression with a negative base is written with parentheses around the base, such as $(-2)^4$.

Answers

1. 27   **2.** 2.25   **3.** -24
4. -5 **5.** 0   **6.** 1

To evaluate $(-2)^4$, the base -2 is used as a factor four times:

$$(-2)^4 = (-2)(-2)(-2)(-2) = 16$$

If parentheses are *not* used, the expression -2^4 has a different meaning:

- The expression -2^4 has a base of 2 (not -2) and can be interpreted as $-1 \cdot 2^4$.

$$-2^4 = -1(2)(2)(2)(2) = -16$$

- The expression -2^4 can also be interpreted as the opposite of 2^4.

$$-2^4 = -(2 \cdot 2 \cdot 2 \cdot 2) = -16$$

TIP: The following expressions are translated as:

$-(-3)$: opposite of negative 3

-3^2: opposite of 3 squared

$(-3)^2$: negative 3, squared

| **Example 2** | **Evaluating Exponential Expressions** |

Simplify.

a. $(-5)^2$ **b.** -5^2 **c.** $(-0.4)^3$ **d.** -0.4^3 **e.** $\left(-\dfrac{1}{2}\right)^3$

Solution:

a. $(-5)^2 = (-5)(-5) = 25$ Multiply two factors of -5.

b. $-5^2 = -1(5)(5) = -25$ Multiply -1 by two factors of 5.

c. $(-0.4)^3 = (-0.4)(-0.4)(-0.4) = -0.064$ Multiply three factors of -0.4.

d. $-0.4^3 = -1(0.4)(0.4)(0.4) = -0.064$ Multiply -1 by three factors of 0.4.

e. $\left(-\dfrac{1}{2}\right)^3 = \left(-\dfrac{1}{2}\right)\left(-\dfrac{1}{2}\right)\left(-\dfrac{1}{2}\right) = -\dfrac{1}{8}$ Multiply three factors of $-\dfrac{1}{2}$.

Classroom Examples: p. 71, Exercises 16, 18, 20 and 22

Avoiding Mistakes

The negative sign is not part of the base unless it is in parentheses with the base. Thus, in the expression -5^2, the exponent applies only to 5 and not to the negative sign.

Skill Practice Simplify.

7. $(-7)^2$ **8.** -7^2 **9.** $\left(-\dfrac{2}{3}\right)^3$ **10.** -0.2^3

3. Division of Real Numbers

Two numbers are *reciprocals* if their product is 1. For example, $-\frac{2}{7}$ and $-\frac{7}{2}$ are reciprocals because $-\frac{2}{7}\left(-\frac{7}{2}\right) = 1$. Symbolically, if a is a nonzero real number, then the reciprocal of a is $\frac{1}{a}$ because $a \cdot \frac{1}{a} = 1$. This definition also implies that a number and its reciprocal have the same sign.

The Reciprocal of a Real Number

Let a be a nonzero real number. Then, the **reciprocal** of a is $\frac{1}{a}$.

Recall that to subtract two real numbers, we add the opposite of the second number to the first number. In a similar way, division of real numbers is defined in terms of multiplication. To divide two real numbers, we multiply the first number by the reciprocal of the second number.

Answers

7. 49   **8.** -49

9. $-\dfrac{8}{27}$ **10.** -0.008

> ### Division of Real Numbers
> Let a and b be real numbers such that $b \neq 0$. Then, $a \div b = a \cdot \dfrac{1}{b}$.

Consider the quotient $10 \div 5$. The reciprocal of 5 is $\frac{1}{5}$, so we have

multiply

$$10 \div 5 = 2 \qquad \text{or equivalently,} \qquad 10 \cdot \frac{1}{5} = 2$$

reciprocal

Because division of real numbers can be expressed in terms of multiplication, then the sign rules that apply to multiplication also apply to division.

> ### Dividing Real Numbers
> - The quotient of two real numbers with the *same* sign is positive.
>
>   Examples: $24 \div 4 = 6$
>
> $-36 \div (-9) = 4$
>
> - The quotient of two real numbers with *different* signs is negative.
>
>   Examples: $100 \div (-5) = -20$
>
> $-12 \div 4 = -3$

Classroom Examples: p. 71, Exercises 24 and 26

Example 3	Dividing Real Numbers

Divide the real numbers.

a. $200 \div (-10)$ **b.** $\dfrac{-48}{16}$ **c.** $\dfrac{-6.25}{-1.25}$ **d.** $\dfrac{-9}{-5}$

Solution:

a. $200 \div (-10) = -20$ *Different* signs. Quotient is negative.

b. $\dfrac{-48}{16} = -3$ *Different* signs. Quotient is negative.

c. $\dfrac{-6.25}{-1.25} = 5$ *Same* signs. Quotient is positive.

d. $\dfrac{-9}{-5} = \dfrac{9}{5}$ *Same* signs. Quotient is positive.

Because 5 does not divide into 9 evenly the answer can be left as a fraction.

> **TIP:** If the numerator and denominator of a fraction are both negative, then the quotient is positive. Therefore, $\frac{-9}{-5}$ can be simplified to $\frac{9}{5}$.

Skill Practice Divide.

11. $-14 \div 7$ **12.** $\dfrac{-18}{3}$ **13.** $\dfrac{-7.6}{-1.9}$ **14.** $\dfrac{-7}{-3}$

Answers

11. -2 **12.** -6 **13.** 4 **14.** $\dfrac{7}{3}$

| Example 4 | **Dividing Real Numbers** |

Classroom Examples: pp. 71–72, Exercises 30 and 78

Divide the real numbers.

a. $15 \div (-25)$ **b.** $-\dfrac{3}{14} \div \dfrac{9}{7}$

Solution:

a. $15 \div (-25)$ *Different* signs. Quotient is negative.

$= \dfrac{15}{-25}$

$= -\dfrac{3}{5}$

b. $-\dfrac{3}{14} \div \dfrac{9}{7}$ *Different* signs. Quotient is negative.

$= -\dfrac{3}{14} \cdot \dfrac{7}{9}$ Multiply by the reciprocal of $\frac{9}{7}$ which is $\frac{7}{9}$.

$= -\dfrac{\overset{1}{\cancel{3}}}{\underset{2}{\cancel{14}}} \cdot \dfrac{\overset{1}{\cancel{7}}}{\underset{3}{\cancel{9}}}$ Divide out common factors.

$= -\dfrac{1}{6}$ Multiply the fractions.

> **TIP:** If the numerator and denominator of a fraction have opposite signs, then the quotient will be negative. Therefore, a fraction has the same value whether the negative sign is written in the numerator, in the denominator, or in front of the fraction.
>
> $$\frac{-3}{5} = \frac{3}{-5} = -\frac{3}{5}$$

Skill Practice Divide.

15. $12 \div (-18)$ **16.** $\dfrac{3}{4} \div \left(-\dfrac{9}{16}\right)$

Multiplication can be used to check any division problem. If $\frac{a}{b} = c$, then $bc = a$ (provided that $b \neq 0$). For example:

$$\frac{8}{-4} = -2 \; \to \; \underline{\text{Check}}: (-4)(-2) = 8 \; \checkmark$$

This relationship between multiplication and division can be used to investigate division problems involving the number zero.

1. The quotient of 0 and any nonzero number is 0. For example:

$$\frac{0}{6} = 0 \qquad \text{because } 6 \cdot 0 = 0 \; \checkmark$$

2. The quotient of any nonzero number and 0 is undefined. For example:

$$\frac{6}{0} = ?$$

Instructor Note: Memory Device: Zero under the line, division undefined.

Finding the quotient $\frac{6}{0}$ is equivalent to asking, "What number times zero will equal 6?" That is, $(0)(?) = 6$. No real number satisfies this condition. Therefore, we say that division by zero is undefined.

3. The quotient of 0 and 0 cannot be determined. Evaluating an expression of the form $\frac{0}{0} = ?$ is equivalent to asking, "What number times zero will equal 0?" That is, $(0)(?) = 0$. Any real number will satisfy this requirement; however, expressions involving $\frac{0}{0}$ are usually discussed in advanced mathematics courses.

Answers

15. $-\dfrac{2}{3}$ **16.** $-\dfrac{4}{3}$

> **Division Involving Zero**
> Let a represent a nonzero real number. Then,
>
> **1.** $\dfrac{0}{a} = 0$ **2.** $\dfrac{a}{0}$ is undefined

4. Order of Operations

Classroom Example: p. 72,
Exercise 96

| Example 5 | **Applying the Order of Operations** |

Simplify. $-8 + 8 \div (-2) \div (-6)$

Solution:

$$-8 + 8 \div (-2) \div (-6)$$

$= -8 + (-4) \div (-6)$ Perform division before addition.

$= -8 + \dfrac{4}{6}$ The quotient of -4 and -6 is positive $\frac{4}{6}$ or $\frac{2}{3}$.

$= -\dfrac{8}{1} + \dfrac{2}{3}$ Write -8 as a fraction.

$= -\dfrac{24}{3} + \dfrac{2}{3}$ Get a common denominator.

$= -\dfrac{22}{3}$ Add.

Skill Practice Simplify.

17. $-36 + 36 \div (-4) \div (-3)$

Classroom Example: p. 72,
Exercise 110

| Example 6 | **Applying the Order of Operations** |

Simplify. $\dfrac{24 - 2[-3 + (5 - 8)]^2}{2|-12 + 3|}$

Solution:

$\dfrac{24 - 2[-3 + (5 - 8)]^2}{2|-12 + 3|}$ Simplify numerator and denominator separately.

$= \dfrac{24 - 2[-3 + (-3)]^2}{2|-9|}$ Simplify within the inner parentheses and absolute value.

$= \dfrac{24 - 2[-6]^2}{2(9)}$ Simplify within brackets, [ ]. Simplify the absolute value.

$= \dfrac{24 - 2(36)}{2(9)}$ Simplify exponents.

$= \dfrac{24 - 72}{18}$ Perform multiplication before subtraction.

$= \dfrac{-48}{18}$ or $-\dfrac{8}{3}$ Simplify to lowest terms.

Answer

17. -33

Skill Practice Simplify.

18. $\dfrac{100 - 3[-1 + (2 - 6)^2]}{|20 - 25|}$

| **Example 7** | **Evaluating Algebraic Expressions** |

Given $y = -6$, evaluate the expressions.

 a. y^2 **b.** $-y^2$

Solution:

 a. y^2

 $= (\ \ )^2$ When substituting a number for a variable, use parentheses.

 $= (-6)^2$ Substitute $y = -6$.

 $= 36$ Square -6, that is, $(-6)(-6) = 36$.

 b. $-y^2$

 $= -(\ \ )^2$ When substituting a number for a variable, use parentheses.

 $= -(-6)^2$ Substitute $y = -6$.

 $= -(36)$ Square -6 within parentheses to get 36.

 $= -36$ Multiply by -1.

Skill Practice Given $a = -7$, evaluate the expressions.

19. a^2 **20.** $-a^2$

Classroom Example: p. 73, Exercise 112

Answers

18. 11 **19.** 49 **20.** -49

Calculator Connections

Topic: Evaluating Exponential Expressions with Positive and Negative Bases

Be particularly careful when raising a negative number to an even power on a calculator. For example, the expressions $(-4)^2$ and -4^2 have different values. That is, $(-4)^2 = 16$ and $-4^2 = -16$. Verify these expressions on a calculator.

Scientific Calculator:

To evaluate $(-4)^2$

Enter: `( 4 +o- ) x²` **Result:** | 16 |

To evaluate -4^2 on a scientific calculator, it is important to square 4 first and then take its opposite.

Enter: `4 x² +o-` **Result:** | -16 |

Graphing Calculator:

```
(-4)²
          16
-4²
         -16
```

The graphing calculator allows for several methods of denoting the multiplication of two real numbers. For example, consider the product of -8 and 4.

```
-8*4
            -32
-8(4)
            -32
(-8)(4)
            -32
```

Calculator Exercises

Simplify the expression without the use of a calculator. Then use the calculator to verify your answer.

1. $-6(5)$ -30 **2.** $\dfrac{-5.2}{2.6}$ -2 **3.** $(-5)(-5)(-5)(-5)$ 625 **4.** $(-5)^4$ 625 **5.** -5^4 -625

6. -2.4^2 -5.76 **7.** $(-2.4)^2$ 5.76 **8.** $(-1)(-1)(-1)$ -1 **9.** $\dfrac{-8.4}{-2.1}$ 4 **10.** $90 \div (-5)(2)$ -36

Section 1.6 Practice Exercises

For additional exercises, see Classroom Activities 1.6A–1.6D in the *Student's Resource Manual* at www.mhhe.com/moh.

Study Skills Exercise

To familiarize yourself with some of the helpful features of the text, look through this section and write down a page number that contains:

a. An Avoiding Mistakes box _____

b. A Tip box _____

c. A key term (shown in bold) _____

Vocabulary and Key Concepts

1. a. If a is a nonzero real number, then the reciprocal of a is ___$\frac{1}{a}$___.

b. If either a or b is zero then the product $ab =$ ___0___.

c. If $a = 0$, and $b \neq 0$, then $\frac{a}{b} =$ ___0___.

d. If $a \neq 0$ and $b = 0$, then $\frac{a}{b}$ is ___undefined___.

e. If a and b have the same sign, then the product ab is (choose one: positive or negative).   positive

f. If a and b have different signs, then the quotient $\frac{a}{b}$ is (choose one: positive or negative).   negative

g. The product of a number and its reciprocal is ___1___. For example $-\frac{2}{3} \cdot (-\frac{3}{2}) = 1$

h. Which of the following expressions represents the product of 2 and x?   All of these

    a. $2x$ **b.** $2 \cdot x$ **c.** $2(x)$ **d.** $(2)x$ **e.** $(2)(x)$

 Writing   Translating Expression   Geometry   Scientific Calculator   Video

Review Exercises

For Exercises 2–6, determine if the expression is true or false.

2. $6 + (-2) > -5 + 6$ True

3. $|-6| + |-14| \leq |-3| + |-17|$ True

4. $\sqrt{36} - |-6| > 0$ False

5. $\sqrt{9} + |-3| \leq 0$ False

6. $14 - |-3| \geq 12 + \sqrt{25}$ False

Concept 1: Multiplication of Real Numbers

For Exercises 7–14, multiply the real numbers. **(See Example 1.)**

7. $8(-7)$ -56

8. $(-3) \cdot 4$ -12

9. $(-11)(-13)$ 143

10. $(-5)(-26)$ 130

11. $(-2.2)(5.8)$ -12.76

12. $(9.1)(-4.5)$ -40.95

13. $\left(-\dfrac{2}{3}\right)\left(-\dfrac{9}{8}\right)$ $\dfrac{3}{4}$

14. $\left(-\dfrac{5}{4}\right)\left(-\dfrac{12}{25}\right)$ $\dfrac{3}{5}$

Concept 2: Exponential Expressions

For Exercises 15–22, simplify the exponential expressions. **(See Example 2.)**

15. $(-6)^2$ 36

16. $(-10)^2$ 100

 17. -6^2 -36

18. -10^2 -100

19. $\left(-\dfrac{3}{5}\right)^3$ $-\dfrac{27}{125}$

20. $\left(-\dfrac{5}{2}\right)^3$ $-\dfrac{125}{8}$

21. $(-0.2)^4$ 0.0016

22. $(-0.1)^4$ 0.0001

Concept 3: Division of Real Numbers

For Exercises 23–30, divide the real numbers. **(See Examples 3–4.)**

23. $\dfrac{54}{-9}$ -6

24. $\dfrac{-27}{3}$ -9

25. $\dfrac{-15}{-17}$ $\dfrac{15}{17}$

26. $\dfrac{-21}{-16}$ $\dfrac{21}{16}$

27. $\dfrac{-14}{-14}$ 1

28. $\dfrac{-21}{-21}$ 1

29. $\dfrac{13}{-65}$ $-\dfrac{1}{5}$

30. $\dfrac{7}{-77}$ $-\dfrac{1}{11}$

For Exercises 31–38, show how multiplication can be used to check the division problems.

31. $\dfrac{14}{-2} = -7$
$(-2)(-7) = 14$

32. $\dfrac{-18}{-6} = 3$
$(-6)(3) = -18$

33. $\dfrac{0}{-5} = 0$
$-5 \cdot 0 = 0$

34. $\dfrac{0}{-4} = 0$
$-4 \cdot 0 = 0$

35. $\dfrac{6}{0}$ is undefined
No number multiplied by zero equals 6.

36. $\dfrac{-4}{0}$ is undefined
No number multiplied by zero equals -4.

37. $-24 \div (-6) = 4$
$(-6)(4) = -24$

38. $-18 \div 2 = -9$
$(2)(-9) = -18$

Mixed Exercises

For Exercises 39–78, multiply or divide as indicated.

39. $2 \cdot 3$ 6

40. $8 \cdot 6$ 48

41. $2(-3)$ -6

42. $8(-6)$ -48

43. $(-24) \div 3$ -8

44. $(-52) \div 2$ -26

45. $(-24) \div (-3)$ 8

46. $(-52) \div (-2)$ 26

47. $-6 \cdot 0$ 0

48. $-8 \cdot 0$ 0

49. $-18 \div 0$ Undefined

50. $-42 \div 0$ Undefined

51. $0\left(-\dfrac{2}{5}\right)$ 0

52. $0\left(-\dfrac{1}{8}\right)$ 0

53. $0 \div \left(-\dfrac{1}{10}\right)$ 0

54. $0 \div \left(\dfrac{4}{9}\right)$ 0

55. $\dfrac{-9}{6}$ $-\dfrac{3}{2}$ **56.** $\dfrac{-15}{10}$ $-\dfrac{3}{2}$ **57.** $\dfrac{-30}{-100}$ $\dfrac{3}{10}$ **58.** $\dfrac{-250}{-1000}$ $\dfrac{1}{4}$

59. $\dfrac{26}{-13}$ -2 **60.** $\dfrac{52}{-4}$ -13 **61.** $1.72(-4.6)$ -7.912 **62.** $361.3(-14.9)$ -5383.37

63. $-0.02(-4.6)$ 0.092 **64.** $-0.06(-2.15)$ 0.129 **65.** $\dfrac{14.4}{-2.4}$ -6 **66.** $\dfrac{50.4}{-6.3}$ -8

67. $\dfrac{-5.25}{-2.5}$ 2.1 **68.** $\dfrac{-8.5}{-27.2}$ 0.3125 **69.** $(-3)^2$ 9 **70.** $(-7)^2$ 49

71. -3^2 -9 **72.** -7^2 -49 **73.** $\left(-\dfrac{4}{3}\right)^3$ $-\dfrac{64}{27}$ **74.** $\left(-\dfrac{1}{5}\right)^3$ $-\dfrac{1}{125}$

75. $(-6.8) \div (-0.02)$ 340 **76.** $(-12.3) \div (-0.03)$ 410 **77.** $\left(-\dfrac{7}{8}\right) \div \left(-\dfrac{9}{16}\right)$ $\dfrac{14}{9}$ **78.** $\left(-\dfrac{22}{23}\right) \div \left(-\dfrac{11}{3}\right)$ $\dfrac{6}{23}$

Concept 4: Order of Operations

For Exercises 79–110, perform the indicated operations. **(See Examples 5–6.)**

79. $(-2)(-5)(-3)$ -30 **80.** $(-6)(-1)(-10)$ -60 **81.** $(-8)(-4)(-1)(-3)$ 96

82. $(-6)(-3)(-1)(-5)$ 90 **83.** $100 \div (-10) \div (-5)$ 2 **84.** $150 \div (-15) \div (-2)$ 5

85. $-12 \div (-6) \div (-2)$ -1 **86.** $-36 \div (-2) \div 6$ 3 **87.** $\dfrac{2}{5} \cdot \dfrac{1}{3} \cdot \left(-\dfrac{10}{11}\right)$ $-\dfrac{4}{33}$

88. $\left(-\dfrac{9}{8}\right) \cdot \left(-\dfrac{2}{3}\right) \cdot \left(1\dfrac{5}{12}\right)$ $\dfrac{17}{16}$ **89.** $\left(1\dfrac{1}{3}\right) \div 3 \div \left(-\dfrac{7}{9}\right)$ $-\dfrac{4}{7}$ **90.** $-\dfrac{7}{8} \div \left(3\dfrac{1}{4}\right) \div (-2)$ $\dfrac{7}{52}$

91. $12 \div (-2)(4)$ -24 **92.** $(-6) \cdot 7 \div (-2)$ 21 **93.** $\left(-\dfrac{12}{5}\right) \div (-6) \cdot \left(-\dfrac{1}{8}\right)$ $-\dfrac{1}{20}$

94. $10 \cdot \dfrac{1}{3} \div \dfrac{25}{6}$ $\dfrac{4}{5}$ **95.** $8 - 2^3 \cdot 5 + 3 - (-6)$ -23 **96.** $-14 \div (-7) - 8 \cdot 2 + 3^3$ 13

97. $-(2 - 8)^2 \div (-6) \cdot 2$ 12 **98.** $-(3 - 5)^2 \cdot 6 \div (-4)$ 6 **99.** $\dfrac{6(-4) - 2(5 - 8)}{-6 - 3 - 5}$ $\dfrac{9}{7}$

100. $\dfrac{3(-4) - 5(9 - 11)}{-9 - 2 - 3}$ $\dfrac{1}{7}$ **101.** $\dfrac{-4 + 5}{(-2) \cdot 5 + 10}$ Undefined   **102.** $\dfrac{-3 + 10}{2(-4) + 8}$ Undefined

103. $-4 - 3[2 - (-5 + 3)] - 8 \cdot 2^2$ **104.** $-6 - 5[-4 - (6 - 12)] + (-5)^2$ **105.** $-|-1| - |5|$ -6
-48 9

106. $-|-10| - |6|$ -16 **107.** $\dfrac{|2 - 9| - |5 - 7|}{10 - 15}$ -1 **108.** $\dfrac{|-2 + 6| - |3 - 5|}{13 - 11}$ 1

109. $\dfrac{6 - 3[2 - (6 - 8)]^2}{-2|2 - 5|}$ 7 **110.** $\dfrac{12 - 4[-6 - (5 - 8)]^2}{4|6 - 10|}$ $-\dfrac{3}{2}$

For Exercises 111–116, evaluate the expression for $x = -2$, $y = -4$, and $z = 6$. **(See Example 7.)**

111. $-x^2$ -4

112. x^2 4

113. $4(2x - z)$ -40

114. $6(3x + y)$ -60

115. $\dfrac{3x + 2y}{y}$ $\dfrac{7}{2}$

 **116.** $\dfrac{2z - y}{x}$ -8

117. Is the expression $\dfrac{10}{5x}$ equal to 10/5x? Explain.
No. The first expression is equivalent to $10 \div (5x)$.
The second is $10 \div 5 \cdot x$.

118. Is the expression 10/(5x) equal to $\dfrac{10}{5x}$? Explain.
Yes, the parentheses indicate that the divisor is the quantity $5x$.

For Exercises 119–126, write each English phrase as an algebraic expression. Then evaluate the expression.

119. The product of -3.75 and 0.3
$-3.75(0.3)$; -1.125

120. The product of -0.4 and -1.258
$(-0.4)(-1.258)$; 0.5032

121. The quotient of $\frac{16}{5}$ and $\left(-\frac{8}{9}\right)$ $\dfrac{16}{5} \div \left(-\dfrac{8}{9}\right)$; $-\dfrac{18}{5}$

122. The quotient of $\left(-\frac{3}{14}\right)$ and $\frac{1}{7}$ $-\dfrac{3}{14} \div \dfrac{1}{7}$; $-\dfrac{3}{2}$

123. The number -0.4 plus the quantity 6 times -0.42
$-0.4 + 6(-0.42)$; -2.92

124. The number 0.5 plus the quantity -2 times 0.125
$0.5 + (-2)(0.125)$; 0.25

125. The number $-\frac{1}{4}$ minus the quantity 6 times $-\frac{1}{3}$
$-\dfrac{1}{4} - 6\left(-\dfrac{1}{3}\right)$; $\dfrac{7}{4}$

126. Negative five minus the quantity $\left(-\frac{5}{6}\right)$ times $\frac{3}{8}$
$-5 - \left(-\dfrac{5}{6}\right)\dfrac{3}{8}$; $-\dfrac{75}{16}$

127. For 3 weeks, Jim pays $2 a week for lottery tickets. Jim has one winning ticket for $3. Write an expression that describes his net gain or loss. How much money has Jim won or lost? $3(-2) + 3 = -3$; loss of $3

128. Stephanie pays $2 a week for 6 weeks for lottery tickets. Stephanie has one winning ticket for $5. Write an expression that describes her net gain or loss. How much money has Stephanie won or lost? $-2(6) + 5 = -7$; loss of $7

© Buena Vista Images/Getty Images RF

129. Lorne sells ads for the local newspaper and has a quota for the number of ads he must sell each week. For the first two days of the week, Lorne was 5 above his daily quota. For the last three days of the week, Lorne was 3 below his daily quota. Write an expression that describes the net amount he was above or below his quota. Interpret the result. $2(5) + 3(-3) = 1$; Lorne was 1 sale above quota for the week.

130. Valerie trades stocks each day and analyzes her results at the end of the week. The first day of the week she had a profit of $650. The next two days she had a loss of $400 each day and the last two days she had a loss of $150 each day. Write an expression that describes Valerie's profit or loss for the week. Interpret the result.
$650 + 2(-400) + 2(-150) = -450$; Valerie lost $450 for the week.

131. A pediatrician analyzed the effect of a special baby food on five toddlers. He recorded the amount of weight lost or gained (in ounces) by each child for one month as shown in the table.

Baby	A	B	C	D	E
Gain/Loss (oz)	12	−15	4	−9	3

$\dfrac{12 + (-15) + 4 + (-9) + 3}{5} = -1$;
The average loss was 1 oz.

Write an expression to calculate the average gain or loss in weight of the children and then simplify. Interpret your answer. (*Hint*: To find the *average* (mean) of a set of values, add the values and divide by the number of values.)

132. At the end of a 2-month class, a personal trainer analyzed the effect of her new exercise routine by recording the gain or loss (in inches) in waist measurements of her clients. The results are shown in the table.

Client	A	B	C	D	E	F
Gain/Loss (in.)	−1.5	−2	1.5	2.5	−3	−0.5

Write an expression to calculate the average gain or loss in waist measurement and then simplify. Interpret your answer.
$\dfrac{-1.5 + (-2) + 1.5 + 2.5 + (-3) + (-0.5)}{6} = -0.5$; The average loss was 0.5 in.

133. Evaluate the expressions in parts (a) and (b).

    a. $-4 - 3 - 2 - 1$ -10

    b. $-4(-3)(-2)(-1)$ 24

    c. Explain the difference between the operations in parts (a) and (b). In part (a), we subtract; in part (b), we multiply.

134. Evaluate the expressions in parts (a) and (b).

    a. $-10 - 9 - 8 - 7$ -34

    b. $-10(-9)(-8)(-7)$ 5040

    c. Explain the difference between the operations in parts (a) and (b). In part (a), we subtract; in part (b), we multiply.

Problem Recognition Exercises

Adding, Subtracting, Multiplying, and Dividing Real Numbers

Perform the indicated operations.

1. a. $-8 - (-4)$ -4 **b.** $-8(-4)$ 32 **c.** $-8 + (-4)$ -12 **d.** $-8 \div (-4)$ 2

2. a. $12 + (-2)$ 10 **b.** $12 - (-2)$ 14 **c.** $12(-2)$ -24 **d.** $12 \div (-2)$ -6

3. a. $-36 + 9$ -27 **b.** $-36(9)$ -324 **c.** $-36 \div 9$ -4 **d.** $-36 - 9$ -45

4. a. $27 - (-3)$ 30 **b.** $27 + (-3)$ 24 **c.** $27(-3)$ -81 **d.** $27 \div (-3)$ -9

5. a. $-5(-10)$ 50 **b.** $-5 + (-10)$ -15 **c.** $-5 \div (-10)$ $\frac{1}{2}$ **d.** $-5 - (-10)$ 5

6. a. $-20 \div 4$ -5 **b.** $-20 - 4$ -24 **c.** $-20 + 4$ -16 **d.** $-20(4)$ -80

7. a. $-4(-16)$ 64 **b.** $-4 - (-16)$ 12 **c.** $-4 \div (-16)$ $\frac{1}{4}$ **d.** $-4 + (-16)$ -20

8. a. $-21 \div 3$ -7 **b.** $-21 - 3$ -24 **c.** $-21(3)$ -63 **d.** $-21 + 3$ -18

9. a. $80(-5)$ -400 **b.** $80 - (-5)$ 85 **c.** $80 \div (-5)$ -16 **d.** $80 + (-5)$ 75

10. a. $-14 - (-21)$ 7 **b.** $-14(-21)$ 294 **c.** $-14 \div (-21)$ $\frac{2}{3}$ **d.** $-14 + (-21)$ -35

11. a. $|-6| + |2|$ 8 **b.** $|-6 + 2|$ 4 **c.** $|-6| - |2|$ 4 **d.** $|-6 - 2|$ 8

12. a. $-|9| - |-7|$ -16 **b.** $|-9| - |-7|$ 2 **c.** $-|9 - 7|$ -2 **d.** $|-9 - 7|$ 16

Properties of Real Numbers and Simplifying Expressions

1. Commutative Properties of Real Numbers

When getting dressed, it makes no difference whether you put on your left shoe first and then your right shoe, or vice versa. This example illustrates a process in which the order does not affect the outcome. Such a process or operation is said to be *commutative*.

In algebra, the operations of addition and multiplication are commutative because the order in which we add or multiply two real numbers does not affect the result. For example:

$$10 + 5 = 5 + 10 \quad \text{and} \quad 10 \cdot 5 = 5 \cdot 10$$

© C. Borland/PhotoLink/
Getty Images RF

Concepts

1. **Commutative Properties of Real Numbers**
2. **Associative Properties of Real Numbers**
3. **Identity and Inverse Properties of Real Numbers**
4. **Distributive Property of Multiplication over Addition**
5. **Algebraic Expressions**

Commutative Properties of Real Numbers

If a and b are real numbers, then

1. $a + b = b + a$ **commutative property of addition**
2. $ab = ba$ **commutative property of multiplication**

It is important to note that although the operations of addition and multiplication are commutative, subtraction and division are *not* commutative. For example:

$$\underline{10 - 5} \neq \underline{5 - 10} \quad \text{and} \quad \underline{10 \div 5} \neq \underline{5 \div 10}$$
$$5 \quad \neq \quad -5 \qquad\qquad 2 \quad \neq \quad \frac{1}{2}$$

Example 1 **Applying the Commutative Property of Addition**

Use the commutative property of addition to rewrite each expression.

a. $-3 + (-7)$ **b.** $3x^3 + 5x^4$

Solution:

a. $-3 + (-7) = -7 + (-3)$

b. $3x^3 + 5x^4 = 5x^4 + 3x^3$

Skill Practice Use the commutative property of addition to rewrite each expression.

1. $-5 + 9$ **2.** $7y + x$

Classroom Example: p. 85, Exercise 16

Recall that subtraction is not a commutative operation. However, if we rewrite $a - b$, as $a + (-b)$, we can apply the commutative property of addition. This is demonstrated in Example 2.

Answers
1. $9 + (-5)$ **2.** $x + 7y$

Classroom Example: p. 85, Exercise 24

Example 2 **Applying the Commutative Property of Addition**

Rewrite the expression in terms of addition. Then apply the commutative property of addition.

a. $5a - 3b$ **b.** $z^2 - \dfrac{1}{4}$

Solution:

a. $5a - 3b$

$\quad = 5a + (-3b)$ Rewrite subtraction as addition of $-3b$.

$\quad = -3b + 5a$ Apply the commutative property of addition.

b. $z^2 - \dfrac{1}{4}$

$\quad = z^2 + \left(-\dfrac{1}{4}\right)$ Rewrite subtraction as addition of $-\frac{1}{4}$.

$\quad = -\dfrac{1}{4} + z^2$ Apply the commutative property of addition.

Skill Practice Rewrite each expression in terms of addition. Then apply the commutative property of addition.

3. $8m - 2n$ **4.** $\dfrac{1}{3}x - \dfrac{3}{4}$

Classroom Example: p. 85, Exercise 22

Example 3 **Applying the Commutative Property of Multiplication**

Use the commutative property of multiplication to rewrite each expression.

a. $12(-6)$ **b.** $x \cdot 4$

Solution:

a. $12(-6) = -6(12)$

b. $x \cdot 4 = 4 \cdot x$ (or simply $4x$)

Skill Practice Use the commutative property of multiplication to rewrite each expression.

5. $-2(5)$ **6.** $y \cdot 6$

2. Associative Properties of Real Numbers

The associative property of real numbers states that the manner in which three or more real numbers are grouped under addition or multiplication will not affect the outcome. For example:

$$(5 + 10) + 2 = 5 + (10 + 2) \qquad \text{and} \qquad (5 \cdot 10)2 = 5(10 \cdot 2)$$
$$15 + 2 = 5 + 12 \qquad\qquad\qquad (50)2 = 5(20)$$
$$17 = 17 \qquad\qquad\qquad\qquad 100 = 100$$

Answers

3. $8m + (-2n); \; -2n + 8m$

4. $\dfrac{1}{3}x + \left(-\dfrac{3}{4}\right); \; -\dfrac{3}{4} + \dfrac{1}{3}x$

5. $5(-2)$ **6.** $6y$

> **Associative Properties of Real Numbers**
>
> If a, b, and c represent real numbers, then
>
> **1.** $(a + b) + c = a + (b + c)$ **associative property of addition**
>
> **2.** $(ab)c = a(bc)$ **associative property of multiplication**

Example 4 **Applying the Associative Property**

Classroom Examples: p. 85, Exercises 28 and 30

Use the associative property of addition or multiplication to rewrite each expression. Then simplify the expression if possible.

a. $(y + 5) + 6$ **b.** $4(5z)$ **c.** $-\dfrac{3}{2}\left(-\dfrac{2}{3}w\right)$

Solution:

a. $(y + 5) + 6$

$= y + (5 + 6)$ Apply the associative property of addition.

$= y + 11$ Simplify.

b. $4(5z)$

$= (4 \cdot 5)z$ Apply the associative property of multiplication.

$= 20z$ Simplify.

c. $-\dfrac{3}{2}\left(-\dfrac{2}{3}w\right)$

$= \left[-\dfrac{3}{2}\left(-\dfrac{2}{3}\right)\right]w$ Apply the associative property of multiplication.

$= 1w$ Simplify.

$= w$

Note: In most cases, a detailed application of the associative property will not be shown. Instead, the process will be written in one step, such as

$$(y + 5) + 6 = y + 11, \quad 4(5z) = 20z, \quad \text{and} \quad -\frac{3}{2}\left(-\frac{2}{3}w\right) = w$$

Skill Practice Use the associative property of addition or multiplication to rewrite each expression. Simplify if possible.

7. $(x + 4) + 3$ **8.** $-2(4x)$ **9.** $\dfrac{5}{4}\left(\dfrac{4}{5}t\right)$

3. Identity and Inverse Properties of Real Numbers

The number 0 has a special role under the operation of addition. Zero added to any real number does not change the number. Therefore, the number 0 is said to be the *additive identity* (also called the *identity element of addition*). For example:

$$-4 + 0 = -4 \quad 0 + 5.7 = 5.7 \quad 0 + \frac{3}{4} = \frac{3}{4}$$

Answers

7. $x + (4 + 3)$; $x + 7$

8. $(-2 \cdot 4)x$; $-8x$

9. $\left(\dfrac{5}{4} \cdot \dfrac{4}{5}\right)t$; t

The number 1 has a special role under the operation of multiplication. Any real number multiplied by 1 does not change the number. Therefore, the number 1 is said to be the *multiplicative identity* (also called the *identity element of multiplication*). For example:

$$(-8)1 = -8 \quad 1(-2.85) = -2.85 \quad 1\left(\frac{1}{5}\right) = \frac{1}{5}$$

Identity Properties of Real Numbers

If a is a real number, then

1. $a + 0 = 0 + a = a$ **identity property of addition**
2. $a \cdot 1 = 1 \cdot a = a$ **identity property of multiplication**

The sum of a number and its opposite equals 0. For example, $-12 + 12 = 0$. For any real number, a, the opposite of a (also called the *additive inverse* of a) is $-a$ and $a + (-a) = -a + a = 0$. The inverse property of addition states that the sum of any number and its additive inverse is the identity element of addition, 0. For example:

Number	Additive Inverse (Opposite)	Sum
9	-9	$9 + (-9) = 0$
-21.6	21.6	$-21.6 + 21.6 = 0$
$\frac{2}{7}$	$-\frac{2}{7}$	$\frac{2}{7} + \left(-\frac{2}{7}\right) = 0$

If b is a nonzero real number, then the reciprocal of b (also called the *multiplicative inverse* of b) is $\frac{1}{b}$. The inverse property of multiplication states that the product of b and its multiplicative inverse is the identity element of multiplication, 1. Symbolically, we have $b \cdot \frac{1}{b} = \frac{1}{b} \cdot b = 1$. For example:

Number	Multiplicative Inverse (Reciprocal)	Product
7	$\frac{1}{7}$	$7 \cdot \frac{1}{7} = 1$
3.14	$\frac{1}{3.14}$	$3.14\left(\frac{1}{3.14}\right) = 1$
$-\frac{3}{5}$	$-\frac{5}{3}$	$-\frac{3}{5}\left(-\frac{5}{3}\right) = 1$

Inverse Properties of Real Numbers

If a is a real number and b is a nonzero real number, then

1. $a + (-a) = -a + a = 0$ **inverse property of addition**
2. $b \cdot \dfrac{1}{b} = \dfrac{1}{b} \cdot b = 1$ **inverse property of multiplication**

4. Distributive Property of Multiplication over Addition

The operations of addition and multiplication are related by an important property called the **distributive property of multiplication over addition**. Consider the expression $6(2 + 3)$. The order of operations indicates that the sum $2 + 3$ is evaluated first, and then the result is multiplied by 6:

$$6(2 + 3)$$
$$= 6(5)$$
$$= 30$$

Notice that the same result is obtained if the factor of 6 is multiplied by each of the numbers 2 and 3, and then their products are added:

$6(2 + 3)$ ⠀⠀⠀⠀⠀⠀ The factor of 6 is *distributed* to the numbers 2 and 3.

$$= 6(2) + 6(3)$$
$$= \quad 12 + 18$$
$$= \quad 30$$

The distributive property of multiplication over addition states that this is true in general.

> **TIP:** The mathematical definition of the distributive property is consistent with the everyday meaning of the word *distribute*. To distribute means to "spread out from one to many." In the mathematical context, the factor a is distributed to both b and c in the parentheses.

Distributive Property of Multiplication over Addition

If a, b, and c are real numbers, then

$$a(b + c) = ab + ac \quad \text{and} \quad (b + c)a = ab + ac$$

Example 5 **Applying the Distributive Property**

Apply the distributive property. $2(a + 6b + 7)$

Solution:

$2(a + 6b + 7)$

$= 2(a + 6b + 7)$

$= 2(a) + 2(6b) + 2(7)$ ⠀⠀⠀ Apply the distributive property.

$= 2a + 12b + 14$ ⠀⠀⠀ Simplify.

Skill Practice Apply the distributive property.

10. $7(x + 4y + z)$

Classroom Example: p. 85, Exercise 48

Answer

10. $7x + 28y + 7z$

Because the difference of two expressions $a - b$ can be written in terms of addition as $a + (-b)$, the distributive property can be applied when the operation of subtraction is present within the parentheses. For example:

$$5(y - 7)$$

$$= 5[y + (-7)] \qquad \text{Rewrite subtraction as addition of } -7.$$

$$= 5[y + (-7)] \qquad \text{Apply the distributive property.}$$

$$= 5(y) + 5(-7)$$

$$= 5y + (-35), \text{ or } 5y - 35 \qquad \text{Simplify.}$$

Classroom Example: p. 85, Exercise 60

Example 6	**Applying the Distributive Property**

Use the distributive property to rewrite each expression.

   a. $-(-3a + 2b + 5c)$ **b.** $-6(2 - 4x)$

Solution:

a. $-(-3a + 2b + 5c)$

> **TIP:** Notice that a negative factor preceding the parentheses changes the signs of all the terms to which it is multiplied.
>
> $$-1(-3a + 2b + 5c)$$
> $$\downarrow \quad \downarrow \quad \downarrow$$
> $$= +3a - 2b - 5c$$

$$= -1(-3a + 2b + 5c) \qquad \text{The negative sign preceding the parentheses can be interpreted as taking the opposite of the quantity that follows or as } -1(-3a + 2b + 5c)$$

$$= -1(-3a + 2b + 5c)$$

$$= -1(-3a) + (-1)(2b) + (-1)(5c) \qquad \text{Apply the distributive property.}$$

$$= 3a + (-2b) + (-5c) \qquad \text{Simplify.}$$

$$= 3a - 2b - 5c$$

b. $-6(2 - 4x)$

$$= -6[2 + (-4x)] \qquad \text{Change subtraction to addition of } -4x.$$

$$= -6[2 + (-4x)] \qquad \text{Apply the distributive property. Notice that multiplying by } -6 \text{ changes the signs of all terms to which it is applied.}$$

$$= -6(2) + (-6)(-4x)$$

$$= -12 + 24x \qquad \text{Simplify.}$$

Skill Practice Use the distributive property to rewrite each expression.

11. $-(12x + 8y - 3z)$ **12.** $-6(-3a + 7b)$

Note: In most cases, the distributive property will be applied without as much detail as shown in Examples 5 and 6. Instead, the distributive property will be applied in one step.

$$2(a + 6b + 7) \qquad\qquad -(3a + 2b + 5c) \qquad\qquad -6(2 - 4x)$$
$$\text{1 step} = 2a + 12b + 14 \qquad \text{1 step} = -3a - 2b - 5c \qquad \text{1 step} = -12 + 24x$$

Answers

11. $-12x - 8y + 3z$

12. $18a - 42b$

5. Algebraic Expressions

A **term** is a constant or the product or quotient of constants and variables. An algebraic expression is the sum of one or more terms. For example, the expression

$$-7x^2 + xy - 100 \quad \text{or} \quad -7x^2 + xy + (-100)$$

consists of the terms $-7x^2$, xy, and -100.

The terms $-7x^2$ and xy are **variable terms** and the term -100 is called a **constant term**. It is important to distinguish between a term and the factors within a term. For example, the quantity xy is one term, and the values x and y are factors within the term. The constant factor in a term is called the *numerical coefficient* (or simply **coefficient**) of the term. In the terms $-7x^2$, xy, and -100, the coefficients are -7, 1, and -100, respectively.

Terms are *like* **terms** if they each have the same variables, and the corresponding variables are raised to the same powers. For example:

Like **Terms**		*Unlike* **Terms**		
$-3b$ and $5b$		$-5c$ and $7d$		(different variables)
$9p^2q^3$ and p^2q^3		$4p^2q^3$ and $8p^3q^2$		(different powers)
$5w$ and $2w$		$5w$ and 2		(different variables)

Example 7 **Identifying Terms, Factors, Coefficients, and *Like* Terms**

Classroom Examples: p. 86, Exercises 82 and 84

a. List the terms of the expression $5x^2 - 3x + 2$.

b. Identify the coefficient of the term $6yz^3$.

c. Which of the pairs are *like* terms: $8b, 3b^2$ or $4c^2d, -6c^2d$?

Solution:

a. The terms of the expression $5x^2 - 3x + 2$ are $5x^2$, $-3x$, and 2.

b. The coefficient of $6yz^3$ is 6.

c. $4c^2d$ and $-6c^2d$ are *like* terms.

Skill Practice

13. List the terms in the expression. $4xy - 9x^2 + 15$

14. Identify the coefficients of each term in the expression. $2a - b + c - 80$

15. Which of the pairs are *like* terms? $5x^3, 5x$ or $-7x^2, 11x^2$

Two terms can be added or subtracted only if they are *like* terms. To add or subtract *like* terms, we use the distributive property as shown in Example 8.

Answers
13. $4xy, -9x^2, 15$
14. $2, -1, 1, -80$
15. $-7x^2$ and $11x^2$ are *like* terms.

Classroom Example: p. 87,
Exercise 90

Example 8   Using the Distributive Property to Add and Subtract *Like* Terms

Add or subtract as indicated.

a. $7x + 2x$ **b.** $-2p + 3p - p$

Solution:

a. $7x + 2x$

$= (7 + 2)x$ Apply the distributive property.

$= 9x$ Simplify.

b. $-2p + 3p - p$

$= -2p + 3p - 1p$ Note that $-p$ equals $-1p$.

$= (-2 + 3 - 1)p$ Apply the distributive property.

$= (0)p$ Simplify.

$= 0$

Skill Practice   Simplify by adding *like* terms.

16. $8x + 3x$ **17.** $-6a + 4a + a$

Although the distributive property is used to add and subtract *like* terms, it is tedious to write each step. Observe that adding or subtracting *like* terms is a matter of adding or subtracting the coefficients and leaving the variable factors unchanged. This can be shown in one step, a shortcut that we will use throughout the text. For example:

$$7x + 2x = 9x \qquad -2p + 3p - 1p = 0p = 0 \qquad -3a - 6a = -9a$$

Classroom Examples: p. 87,
Exercises 92 and 94

Instructor Note: Ask students why $2x^2 + 3x^2$ is not $5x^4$.

Example 9   Combining *Like* Terms

Simplify by combining *like* terms.

a. $3yz + 5 - 2yz + 9$ **b.** $1.2w^3 + 5.7w^3$

Solution:

a. $3yz + 5 - 2yz + 9$

$= 3yz - 2yz + 5 + 9$ Arrange *like* terms together. Notice that constants such as 5 and 9 are *like* terms.

$= 1yz + 14$ Combine *like* terms.

$= yz + 14$

b. $1.2w^3 + 5.7w^3$

$= 6.9w^3$ Combine *like* terms.

Skill Practice   Simplify by combining *like* terms.

18. $4pq - 7 + 5pq - 8$ **19.** $8.3x^2 + 5.1x^2$

Answers

16. $11x$ **17.** $-a$

18. $9pq - 15$ **19.** $13.4x^2$

When we apply the distributive property, the parentheses are removed. Sometimes this is referred to as *clearing parentheses*. In Examples 10 and 11, we clear parentheses and combine *like* terms.

Example 10 Clearing Parentheses and Combining *Like* Terms

Simplify by *clearing parentheses* and combining *like* terms. $5 - 2(3x + 7)$

Classroom Example: p. 87, Exercises 104

Solution:

$5 - 2(3x + 7)$ The order of operations indicates that we must perform multiplication before subtraction.

It is important to understand that a factor of -2 (not 2) will be multiplied to all terms within the parentheses. To see why this is so, we rewrite the subtraction in terms of addition.

$= 5 + (-2)(3x + 7)$ Change subtraction to addition.

$= 5 + (-2)(3x + 7)$ A factor of -2 is to be distributed to terms in the parentheses.

$= 5 + (-2)(3x) + (-2)(7)$ Apply the distributive property.

$= 5 + (-6x) + (-14)$ Simplify.

$= 5 + (-14) + (-6x)$ Arrange *like* terms together.

$= -9 + (-6x)$ Combine *like* terms.

$= -9 - 6x$ Simplify by changing addition of the opposite to subtraction.

Skill Practice Clear the parentheses and combine *like* terms.

20. $9 - 5(2x - 7)$

Example 11 Clearing Parentheses and Combining *Like* Terms

Simplify by clearing parentheses and combining *like* terms.

Classroom Examples: p. 87, Exercises 108 and 116

a. $\dfrac{1}{4}(4k + 2) - \dfrac{1}{2}(6k + 1)$ **b.** $-(4s - 6t) - (3t + 5s) - 2s$

Solution:

a. $\dfrac{1}{4}(4k + 2) - \dfrac{1}{2}(6k + 1)$

$= \dfrac{4}{4}k + \dfrac{2}{4} - \dfrac{6}{2}k - \dfrac{1}{2}$ Apply the distributive property. Notice that a factor of $-\frac{1}{2}$ is distributed through the second parentheses and changes the signs.

$= k + \dfrac{1}{2} - 3k - \dfrac{1}{2}$ Simplify fractions.

$= k - 3k + \dfrac{1}{2} - \dfrac{1}{2}$ Arrange *like* terms together.

$= -2k + 0$ Combine *like* terms.

$= -2k$

Answer

20. $-10x + 44$

b. $-(4s - 6t) - (3t + 5s) - 2s$

$= -1(4s - 6t) - 1(3t + 5s) - 2s$ Notice that a factor of -1 is distributed through each parentheses.

$= -4s + 6t - 3t - 5s - 2s$ Apply the distributive property.

$= -4s - 5s - 2s + 6t - 3t$ Arrange *like* terms together.

$= -11s + 3t$ Combine *like* terms.

Skill Practice Clear the parentheses and combine *like* terms.

21. $\dfrac{1}{2}(8x + 4) + \dfrac{1}{3}(3x - 9)$ **22.** $-4(x + 2y) - (2x - y) - 5x$

Classroom Example: p. 87,
Exercise 124

| **Example 12** | **Clearing Parentheses and Combining *Like* Terms** |

Simplify by clearing parentheses and combining *like* terms.

$$-7a - 4[3a - 2(a + 6)] - 4$$

Solution:

$-7a - 4[3a - 2(a + 6)] - 4$

$= -7a - 4[3a - 2a - 12] - 4$ Apply the distributive property to clear the innermost parentheses.

$= -7a - 4[a - 12] - 4$ Simplify within brackets by combining *like* terms.

$= -7a - 4a + 48 - 4$ Apply the distributive property to clear the brackets.

$= -11a + 44$ Combine *like* terms.

Avoiding Mistakes

First clear the innermost parentheses and combine *like* terms within the brackets. Then use the distributive property to clear the brackets.

Skill Practice Clear the parentheses and combine *like* terms.

23. $6 - 5[-2y - 4(2y - 5)]$

Answers

21. $5x - 1$ **22.** $-11x - 7y$
23. $50y - 94$

Section 1.7   Practice Exercises

For additional exercises, see Classroom Activities 1.7A–1.7D in the
Student's Resource Manual at www.mhhe.com/moh.

Study Skills Exercise

Write down the page number(s) for the Chapter Summary for this chapter. Describe one way in which you can use the Summary found at the end of each chapter.

Vocabulary and Key Concepts

1. a. Given the expression $12y + ab - 2x + 18$, the terms $12y$, ab, and $-2x$ are variable terms, whereas 18 is a ___constant___ term.

b. The constant factor in a term is called the ___coefficient___ of the term.

c. Given the expression x, the value of the coefficient is ___1___, and the exponent is ___1___.

d. Terms that have the same variables, with corresponding variables raised to the same powers, are called ___like___ terms.

Writing   Translating Expression   Geometry   Scientific Calculator   Video

Review Exercises

For Exercises 2–14, perform the indicated operations.

2. $(-6) + 14$ 8

3. $(-2) + 9$ 7

4. $-13 - (-5)$ -8

5. $-1 - (-19)$ 18

6. $18 \div (-4)$ $-\frac{9}{2}$ or -4.5

7. $-27 \div 5$ $-\frac{27}{5}$ or -5.4

8. $-3 \cdot 0$ 0

9. $0(-15)$ 0

10. $\frac{1}{2} + \frac{3}{8}$ $\frac{7}{8}$

11. $\frac{25}{21} - \frac{6}{7}$ $\frac{1}{3}$

12. $\left(-\frac{3}{5}\right)\left(\frac{4}{27}\right)$ $-\frac{4}{45}$

13. $\left(-\frac{11}{12}\right) \div \left(-\frac{5}{4}\right)$ $\frac{11}{15}$

14. $25 \cdot \left(-\frac{4}{5}\right)$ -20

Concept 1: Commutative Properties of Real Numbers

For Exercises 15–22, rewrite each expression using the commutative property of addition or the commutative property of multiplication. **(See Examples 1 and 3.)**

15. $5 + (-8)$ $-8 + 5$

16. $7 + (-2)$ $-2 + 7$

17. $8 + x$ $x + 8$

18. $p + 11$ $11 + p$

19. $5(4)$ $4(5)$

20. $10(8)$ $8(10)$

 21. $x(-12)$ $-12x$

22. $y(-23)$ $-23y$

For Exercises 23–26, rewrite each expression using addition. Then apply the commutative property of addition.
(See Example 2.)

23. $x - 3$
$x + (-3); -3 + x$

24. $y - 7$
$y + (-7); -7 + y$

25. $4p - 9$
$4p + (-9); -9 + 4p$

26. $3m - 12$
$3m + (-12); -12 + 3m$

Concept 2: Associative Properties of Real Numbers

For Exercises 27–38, use the associative property of addition or multiplication to rewrite each expression. Then simplify the expression if possible. **(See Example 4.)**

27. $(x + 4) + 9$
$x + (4 + 9); x + 13$

28. $-3 + (5 + z)$
$(-3 + 5) + z; 2 + z$

29. $-5(3x)$
$(-5 \cdot 3)x; -15x$

30. $-12(4z)$
$(-12 \cdot 4)z; -48z$

31. $\frac{6}{11}\left(\frac{11}{6}x\right)$ $\left(\frac{6}{11} \cdot \frac{11}{6}\right)x; x$

32. $\frac{3}{5}\left(\frac{5}{3}x\right)$ $\left(\frac{3}{5} \cdot \frac{5}{3}\right)x; x$

 33. $-4\left(-\frac{1}{4}t\right)$ $\left(-4 \cdot -\frac{1}{4}\right)t; t$

34. $-5\left(-\frac{1}{5}w\right)$ $\left(-5 \cdot -\frac{1}{5}\right)w; w$

35. $-8 + (2 + y)$
$(-8 + 2) + y; -6 + y$

36. $[x + (-5)] + 7$
$x + (-5 + 7); x + 2$

37. $-5(2x)$
$(-5 \cdot 2)x; -10x$

38. $-10(6t)$
$(-10 \cdot 6)t; -60t$

Concept 3: Identity and Inverse Properties of Real Numbers

39. What is another name for multiplicative inverse?
Reciprocal

40. What is another name for additive inverse?
Opposite

41. What is the additive identity? 0

42. What is the multiplicative identity? 1

Concept 4: Distributive Property of Multiplication over Addition

For Exercises 43–62, use the distributive property to clear parentheses. **(See Examples 5–6.)**

43. $6(5x + 1)$
$30x + 6$

44. $2(x + 7)$
$2x + 14$

45. $-2(a + 8)$
$-2a - 16$

46. $-3(2z + 9)$
$-6z - 27$

47. $3(5c - d)$
$15c - 3d$

48. $4(w - 13z)$
$4w - 52z$

49. $-7(y - 2)$
$-7y + 14$

50. $-2(4x - 1)$
$-8x + 2$

51. $-\frac{2}{3}(x - 6)$ $-\frac{2}{3}x + 4$

52. $-\frac{1}{4}(2b - 8)$ $-\frac{1}{2}b + 2$

53. $\frac{1}{3}(m - 3)$ $\frac{1}{3}m - 1$

54. $\frac{2}{5}(n - 5)$ $\frac{2}{5}n - 2$

55. $-(2p + 10)$
$-2p - 10$

56. $-(7q + 1)$
$-7q - 1$

57. $-2(-3w - 5z + 8)$
$6w + 10z - 16$

58. $-4(-7a - b - 3)$
$28a + 4b + 12$

59. $4(x + 2y - z)$
$4x + 8y - 4z$

60. $-6(2a - b + c)$
$-12a + 6b - 6c$

 61. $-(-6w + x - 3y)$
$6w - x + 3y$

62. $-(-p - 5q - 10r)$
$p + 5q + 10r$

Writing   Translating Expression   Geometry   Scientific Calculator   Video

Mixed Exercises

For Exercises 63–70, use the associative property or distributive property to clear parentheses.

63. $2(3 + x)$ $6 + 2x$ **64.** $5(4 + y)$ $20 + 5y$ **65.** $4(6z)$ $24z$ **66.** $8(2p)$ $16p$

67. $-2(7x)$ $-14x$ **68.** $3(-11t)$ $-33t$ **69.** $-4(1 + x)$ $-4 - 4x$ **70.** $-9(2 + y)$ $-18 - 9y$

For Exercises 71–79, match each statement with the property that describes it.

71. $6 \cdot \dfrac{1}{6} = 1$ b   **a.** Commutative property of addition

72. $7(4 \cdot 9) = (7 \cdot 4) 9$ f   **b.** Inverse property of multiplication

73. $2(3 + k) = 6 + 2k$ i   **c.** Commutative property of multiplication

74. $3 \cdot 7 = 7 \cdot 3$ c   **d.** Associative property of addition

75. $5 + (-5) = 0$ g   **e.** Identity property of multiplication

76. $18 \cdot 1 = 18$ e   **f.** Associative property of multiplication

77. $(3 + 7) + 19 = 3 + (7 + 19)$ d   **g.** Inverse property of addition

78. $23 + 6 = 6 + 23$ a   **h.** Identity property of addition

79. $3 + 0 = 3$ h   **i.** Distributive property of multiplication over addition

Concept 5: Algebraic Expressions

For Exercises 80–83, for each expression list the terms and their coefficients. **(See Example 7.)**

80. $3xy - 6x^2 + y - 17$

Term	Coefficient
$3xy$	3
$-6x^2$	-6
y	1
-17	-17

81. $2x - y + 18xy + 5$

Term	Coefficient
$2x$	2
$-y$	-1
$18xy$	18
5	5

82. $x^4 - 10xy + 12 - y$

Term	Coefficient
x^4	1
$-10xy$	-10
12	12
$-y$	-1

83. $-x + 8y - 9x^2y - 3$

Term	Coefficient
$-x$	-1
$8y$	8
$-9x^2y$	-9
-3	-3

84. Explain why $12x$ and $12x^2$ are not *like* terms.
The exponents on x are different.

85. Explain why $3x$ and $3xy$ are not *like* terms.
The variable factors are different.

86. Explain why $7z$ and $\sqrt{13}z$ are *like* terms.
The variables are the same and raised to the same power.

87. Explain why πx and $8x$ are *like* terms.
The variables are the same and raised to the same power.

88. Write three different *like* terms.
For example: $5y, -2y, y$

89. Write three terms that are not *like*.
For example: $5y, -2x, 6$

  Writing   ⟷ Translating Expression   Geometry   Scientific Calculator   Video

For Exercises 90–98, simplify by combining *like* terms. **(See Examples 8–9.)**

90. $5k - 10k$ $-5k$

91. $-4p - 2p$ $-6p$

92. $-7x^2 + 14x^2$ $7x^2$

93. $2y^2 - 5y^2 - 3y^2$ $-6y^2$

94. $2ab + 5 + 3ab - 2$ $5ab + 3$

95. $8x^3y + 3 - 7 - x^3y$ $7x^3y - 4$

96. $\frac{1}{4}a + b - \frac{3}{4}a - 5b$ $-\frac{1}{2}a - 4b$

97. $\frac{2}{5} + 2t - \frac{3}{5} + t - \frac{6}{5}$ $3t - \frac{7}{5}$

98. $2.8z - 8.1z + 6 - 15.2$ $-5.3z - 9.2$

For Exercises 99–126, simplify by clearing parentheses and combining *like* terms. **(See Examples 10–12.)**

99. $-3(2x - 4) + 10$ $-6x + 22$

100. $-2(4a + 3) - 14$ $-8a - 20$

101. $4(w + 3) - 12$ $4w$

102. $5(2r + 6) - 30$ $10r$

103. $5 - 3(x - 4)$
$-3x + 17$

104. $4 - 2(3x + 8)$
$-6x - 12$

105. $-3(2t + 4w) + 8(2t - 4w)$
$10t - 44w$

106. $-5(5y + 9z) + 3(3y + 6z)$
$-16y - 27z$

107. $2(q - 5u) - (2q + 8u)$
$-18u$

108. $6(x + 3y) - (6x - 5y)$ $23y$

109. $-\frac{1}{3}(6t + 9) + 10$
$-2t + 7$

110. $-\frac{3}{4}(8 + 4q) + 7$
$-3q + 1$

111. $10(5.1a - 3.1) + 4$
$51a - 27$

112. $100(-3.14p - 1.05) + 212$
$-314p + 107$

113. $-4m + 2(m - 3) + 2m$
-6

114. $-3b + 4(b + 2) - 8b$
$-7b + 8$

115. $\frac{1}{2}(10q - 2) + \frac{1}{3}(2 - 3q)$
$4q - \frac{1}{3}$

116. $\frac{1}{5}(15 - 4p) - \frac{1}{10}(10p + 5)$
$-\frac{9}{5}p + \frac{5}{2}$

117. $7n - 2(n - 3) - 6 + n$ $6n$

 118. $8k - 4(k - 1) + 7 - k$
$3k + 11$

119. $6(x + 3) - 12 - 4(x - 3)$
$2x + 18$

120. $5(y - 4) + 3 - 6(y - 7)$
$-y + 25$

121. $0.2(6c - 1.6) + c$
$2.2c - 0.32$

122. $-1.1(5 + 8x) - 3.1$
$-8.8x - 8.6$

123. $6 + 2[-8 - 3(2x + 4)] + 10x$
$-2x - 34$

 124. $-3 + 5[-3 - 4(y + 2)] - 8y$
$-28y - 58$

125. $1 - 3[2(z + 1) - 5(z - 2)]$
$9z - 35$

126. $1 - 6[3(2t + 2) - 8(t + 2)]$
$12t + 61$

Expanding Your Skills

For Exercises 127–134, determine if the expressions are equivalent. If two expressions are not equivalent, state why.

127. $3a + b, b + 3a$
Equivalent

128. $4y + 1, 1 + 4y$
Equivalent

129. $2c + 7, 9c$
Not equivalent. The terms are not *like* terms and cannot be combined.

130. $5z + 4, 9z$
Not equivalent. The terms are not *like* terms and cannot be combined.

131. $5x - 3, 3 - 5x$
Not equivalent; subtraction is not commutative.

132. $6d - 7, 7 - 6d$
Not equivalent; subtraction is not commutative.

133. $5x - 3, -3 + 5x$
Equivalent

134. $8 - 2x, -2x + 8$
Equivalent

135. As a small child in school, the great mathematician Karl Friedrich Gauss (1777–1855) was said to have found the sum of the integers from 1 to 100 mentally:

$$1 + 2 + 3 + 4 + \cdots + 99 + 100$$

Rather than adding the numbers sequentially, he added the numbers in pairs:

$$(1 + 99) + (2 + 98) + (3 + 97) + \cdots + 100$$

a. Use this technique to add the integers from 1 to 10.   55

$$1 + 2 + 3 + 4 + 5 + 6 + 7 + 8 + 9 + 10$$

b. Use this technique to add the integers from 1 to 20.   210

Group Activity

Evaluating Formulas Using a Calculator

Materials: A calculator

Estimated Time: 20 minutes

Group Size: 2

In this chapter, we learned one of the most important concepts in mathematics—the order of operations. The proper order of operations is required whenever we evaluate any mathematical expression. The following formulas are taken from applications from science, mathematics, statistics, and business. These are just some samples of what you may encounter as you work your way through college.

For Exercises 1–8, substitute the given values into the formula. Then use a calculator and the proper order of operations to simplify the result. Round to three decimal places if necessary.

1. $F = \dfrac{9}{5}C + 32$ (biology) $C = 35$ $F = 95$

2. $V = \dfrac{nRT}{P}$ (chemistry) $n = 1.00,\ R = 0.0821,\ T = 273.15,\ P = 1.0$ $V = 22.426$

3. $R = k\left(\dfrac{L}{r^2}\right)$ (electronics) $k = 0.05,\ L = 200,\ r = 0.5$ $R = 40$

4. $m = \dfrac{y_2 - y_1}{x_2 - x_1}$ (mathematics) $x_1 = -8.3,\ x_2 = 3.3,\ y_1 = 4.6,\ y_2 = -9.2$ $m = -1.190$

5. $z = \dfrac{\bar{x} - \mu}{\frac{\sigma}{\sqrt{n}}}$ (statistics) $\bar{x} = 69,\ \mu = 55,\ \sigma = 20,\ n = 25$ $z = 3.5$

6. $S = R\left[\dfrac{(1 + i)^n - 1}{i}\right]$ (finance) $R = 200,\ i = 0.08,\ n = 30$ $S = 22{,}656.642$

7. $x = \dfrac{-b + \sqrt{b^2 - 4ac}}{2a}$ (mathematics) $a = 2,\ b = -7,\ c = -15$ $x = 5$

8. $h = \dfrac{1}{2}gt^2 + v_0 t + h_0$ (physics) $g = -32,\ t = 2.4,\ v_0 = 192,\ h_0 = 288$ $h = 656.64$

/ Writing   ←→ Translating Expression   Geometry   Scientific Calculator   Video

Chapter 1   Summary

Section 1.1   Fractions

Key Concepts	Examples

Key Concepts

Simplifying Fractions

Divide the numerator and denominator by their greatest common factor.

Multiplication of Fractions

$$\frac{a}{b} \times \frac{c}{d} = \frac{a \times c}{b \times d}$$

Division of Fractions

$$\frac{a}{b} \div \frac{c}{d} = \frac{a}{b} \times \frac{d}{c}$$

Addition and Subtraction of Fractions

$$\frac{a}{b} + \frac{c}{b} = \frac{a+c}{b} \quad \text{and} \quad \frac{a}{b} - \frac{c}{b} = \frac{a-c}{b}$$

To perform operations on mixed numbers, convert to improper fractions.

Examples

Example 1

$$\frac{60}{84} = \frac{5 \times \cancel{12}^{1}}{7 \times \cancel{12}_{1}} = \frac{5}{7}$$

Example 2

$$\frac{25}{108} \times \frac{27}{40} = \frac{\cancel{25}^{5}}{\cancel{108}_{4}} \times \frac{\cancel{27}^{1}}{\cancel{40}_{8}}$$

$$= \frac{5 \times 1}{4 \times 8} = \frac{5}{32}$$

Example 3

$$\frac{95}{49} \div \frac{65}{42} = \frac{\cancel{95}^{19}}{\cancel{49}_{7}} \times \frac{\cancel{42}^{6}}{\cancel{65}_{13}}$$

$$= \frac{19 \times 6}{7 \times 13} = \frac{114}{91}$$

Example 4

$$\frac{8}{9} + \frac{2}{15} = \frac{8 \times 5}{9 \times 5} + \frac{2 \times 3}{15 \times 3}$$

$$= \frac{40}{45} + \frac{6}{45} = \frac{46}{45}$$

The least common denominator (LCD) of 9 and 15 is 45.

Example 5

$$2\frac{5}{6} - 1\frac{1}{3} = \frac{17}{6} - \frac{4}{3}$$

The LCD is 6.

$$= \frac{17}{6} - \frac{4 \times 2}{3 \times 2} = \frac{17}{6} - \frac{8}{6}$$

$$= \frac{9}{6}$$

$$= \frac{3}{2} \text{ or } 1\frac{1}{2}$$

Section 1.2 Introduction to Algebra and the Set of Real Numbers

Key Concepts

A **variable** is a symbol or letter used to represent an unknown number.

A **constant** is a value that is not variable.

An algebraic **expression** is a collection of variables and constants under algebraic operations.

Natural numbers: $\{1, 2, 3, \ldots\}$

Whole numbers: $\{0, 1, 2, 3, \ldots\}$

Integers: $\{\ldots -3, -2, -1, 0, 1, 2, 3, \ldots\}$

Rational numbers: The set of numbers that can be expressed in the form $\frac{p}{q}$, where p and q are integers and q does not equal 0. In decimal form, rational numbers are terminating or repeating decimals.

Irrational numbers: A subset of the real numbers whose elements cannot be written as a ratio of two integers. In decimal form, irrational numbers are nonterminating, nonrepeating decimals.

Real numbers: The set of both the rational numbers and the irrational numbers.

$a < b$ "a is less than b."

$a > b$ "a is greater than b."

$a \leq b$ "a is less than or equal to b."

$a \geq b$ "a is greater than or equal to b."

Two numbers that are the same distance from zero but on opposite sides of zero on the number line are called **opposites**. The opposite of a is denoted $-a$.

The **absolute value** of a real number, a, denoted $|a|$, is the distance between a and 0 on the number line.

If $a \geq 0$, $|a| = a$

If $a < 0$, $|a| = -a$

Examples

Example 1

Variables: x, y, z, a, b

Constants: $2, -3, \pi$

Expressions: $2x + 5, 3a + b^2$

Example 2

-5, 0, and 4 are integers.

$-\frac{5}{2}$, -0.5, and $0.\overline{3}$ are rational numbers.

$\sqrt{7}$, $-\sqrt{2}$, and π are irrational numbers.

Example 3

All real numbers can be located on the real number line.

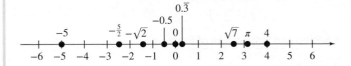

Example 4

$5 < 7$ "5 is less than 7."

$-2 > -10$ "-2 is greater than -10."

$y \leq 3.4$ "y is less than or equal to 3.4."

$x \geq \frac{1}{2}$ "x is greater than or equal to $\frac{1}{2}$."

Example 5

5 and -5 are opposites.

Example 6

$|7| = 7$

$|-7| = 7$

Section 1.3   Exponents, Square Roots, and the Order of Operations

Key Concepts

$b^n = \underbrace{b \cdot b \cdot b \cdot b \cdot \ldots b}_{n \text{ factors of } b}$ b is the **base**,
                        n is the **exponent**

$\sqrt{x}$ is the positive **square root** of x.

The Order of Operations

1. Simplify expressions within parentheses and other grouping symbols first.
2. Evaluate expressions involving exponents, radicals, and absolute values.
3. Perform multiplication or division in the order that they occur from left to right.
4. Perform addition or subtraction in the order that they occur from left to right.

Examples

Example 1

$5^3 = 5 \cdot 5 \cdot 5 = 125$

Example 2

$\sqrt{49} = 7$

Example 3

$$10 + 5(3 - 1)^2 - \sqrt{5 - 1}$$
$$= 10 + 5(2)^2 - \sqrt{4} \quad \text{Work within grouping symbols.}$$
$$= 10 + 5(4) - 2 \quad \text{Simplify exponents and radicals.}$$
$$= 10 + 20 - 2 \quad \text{Perform multiplication.}$$
$$= 30 - 2 \quad \text{Add and subtract, left to right.}$$
$$= 28$$

Section 1.4   Addition of Real Numbers

Key Concepts

Addition of Two Real Numbers

Same Signs. Add the absolute values of the numbers and apply the common sign to the sum.

Different Signs. Subtract the smaller absolute value from the larger absolute value. Then apply the sign of the number having the larger absolute value.

Examples

Example 1

$-3 + (-4) = -7$

$-1.3 + (-9.1) = -10.4$

Example 2

$-5 + 7 = 2$

$\dfrac{2}{3} + \left(-\dfrac{7}{3}\right) = -\dfrac{5}{3}$

| **Section 1.5** | **Subtraction of Real Numbers** |

Key Concepts

Subtraction of Two Real Numbers

Add the opposite of the second number to the first number. That is,

$$a - b = a + (-b)$$

Examples

Example 1

$$7 - (-5) = 7 + (5) = 12$$

$$-3 - 5 = -3 + (-5) = -8$$

$$-11 - (-2) = -11 + (2) = -9$$

| **Section 1.6** | **Multiplication and Division of Real Numbers** |

Key Concepts

Multiplication and Division of Two Real Numbers

Same Signs

Product is positive.

Quotient is positive.

Different Signs

Product is negative.

Quotient is negative.

The **reciprocal** of a nonzero number a is $\dfrac{1}{a}$.

Multiplication and Division Involving Zero

The product of any real number and 0 is 0.

The quotient of 0 and any nonzero real number is 0.

The quotient of any nonzero real number and 0 is undefined.

Examples

Example 1

$$(-5)(-2) = 10 \qquad \frac{-20}{-4} = 5$$

Example 2

$$(-3)(7) = -21 \qquad \frac{-4}{8} = -\frac{1}{2}$$

Example 3

$$-3^4 = -1(3 \cdot 3 \cdot 3 \cdot 3) = -81$$

$$(-3)^4 = (-3)(-3)(-3)(-3) = 81$$

Example 4

The reciprocal of -6 is $-\frac{1}{6}$.

Example 5

$$4 \cdot 0 = 0$$

$$0 \div 4 = 0$$

$$4 \div 0 \text{ is undefined.}$$

Section 1.7	**Properties of Real Numbers and Simplifying Expressions**

Key Concepts

Properties of Real Numbers

Commutative Properties

$a + b = b + a$

$ab = ba$

Associative Properties

$(a + b) + c = a + (b + c)$

$(ab)c = a(bc)$

Identity Properties

$0 + a = a$

$1 \cdot a = a$

Inverse Properties

$a + (-a) = 0$

$b \cdot \dfrac{1}{b} = 1$ for $b \neq 0$

Distributive Property of Multiplication over Addition

$a(b + c) = ab + ac$

A **term** is a constant or the product or quotient of constants and variables. The **coefficient** of a term is the numerical factor of the term.

Like **terms** have the same variables, and the corresponding variables have the same powers.

Terms can be added or subtracted if they are *like* terms. Sometimes it is necessary to clear parentheses before adding or subtracting *like* terms.

Examples

Example 1

$(-5) + (-7) = (-7) + (-5)$

$3 \cdot 8 = 8 \cdot 3$

Example 2

$(2 + 3) + 10 = 2 + (3 + 10)$

$(2 \cdot 4) \cdot 5 = 2 \cdot (4 \cdot 5)$

Example 3

$0 + (-5) = -5$

$1(-8) = -8$

Example 4

$1.5 + (-1.5) = 0$

$6 \cdot \dfrac{1}{6} = 1$

Example 5

$-2(x - 3y) = (-2)x + (-2)(-3y)$
$$= -2x + 6y$$

Example 6

$-2x$ is a term with coefficient -2.
yz^2 is a term with coefficient 1.

$3x$ and $-5x$ are *like* terms.
$4a^2b$ and $4ab$ are not *like* terms.

Example 7

$-2w - 4(w - 2) + 3$
$= -2w - 4w + 8 + 3$ Clear parentheses.
$= -6w + 11$ Combine *like* terms.

Chapter 1   Review Exercises

Section 1.1

For Exercises 1–4, identify as a proper or improper fraction.

1. $\dfrac{14}{5}$ Improper

2. $\dfrac{1}{6}$ Proper

3. $\dfrac{3}{3}$ Improper

4. $\dfrac{7}{1}$ Improper

5. Write 112 as a product of primes. $2 \times 2 \times 2 \times 2 \times 7$

6. Simplify. $\dfrac{84}{70}$ $\dfrac{6}{5}$

For Exercises 7–12, perform the indicated operations.

7. $\dfrac{2}{9} + \dfrac{3}{4}$ $\dfrac{35}{36}$

8. $\dfrac{7}{8} - \dfrac{1}{16}$ $\dfrac{13}{16}$

9. $\dfrac{21}{24} \times \dfrac{16}{49}$ $\dfrac{2}{7}$

10. $\dfrac{68}{34} \div \dfrac{20}{12}$ $\dfrac{6}{5}$ or $1\dfrac{1}{5}$

11. $5\dfrac{1}{3} \div 1\dfrac{7}{9}$ 3

12. $3\dfrac{4}{5} - 2\dfrac{1}{10}$ $\dfrac{17}{10}$ or $1\dfrac{7}{10}$

13. The surface area of the Earth is approximately 510 million km^2. If water covers about $\frac{7}{10}$ of the surface, how many square kilometers of the Earth is covered by water?   357 million km^2

Section 1.2

14. Given the set $\left\{7, \frac{1}{3}, -4, 0, -\sqrt{3}, -0.\overline{2}, \pi, 1\right\}$,

   a. List the natural numbers.   7, 1

   b. List the integers.   7, −4, 0, 1

   c. List the whole numbers.   7, 0, 1

   d. List the rational numbers. $7, \frac{1}{3}, -4, 0, -0.\overline{2}, 1$

   e. List the irrational numbers. $-\sqrt{3}, \pi$

   f. List the real numbers. $7, \frac{1}{3}, -4, 0, -\sqrt{3}, -0.\overline{2}, \pi, 1$

For Exercises 15–18, determine the absolute value.

15. $\left|\dfrac{1}{2}\right|$ $\dfrac{1}{2}$

16. $|-6|$ 6

17. $|-\sqrt{7}|$ $\sqrt{7}$

18. $|0|$ 0

For Exercises 19–27, identify whether the inequality is true or false.

19. $-6 > -1$ False

20. $0 < -5$ False

21. $-10 \le 0$ True

22. $5 \ne -5$ True

23. $7 \ge 7$ True

24. $7 \ge -7$ True

25. $0 \le -3$ False

26. $-\dfrac{2}{3} \le -\dfrac{2}{3}$ True

27. $|-3| > -|3|$ True

For Exercises 28–31, evaluate each expression for $x = 8, y = 4,$ and $z = 1.$

28. $x - 2y$ 0

29. $x^2 - y$ 60

30. $\sqrt{x + z}$ 3

31. $\sqrt{x + 2y}$ 4

Section 1.3

For Exercises 32–37, write each English phrase as an algebraic expression.

32. The product of x and $\dfrac{2}{3}$ $x \cdot \dfrac{2}{3}$ or $\dfrac{2}{3}x$

33. The quotient of 7 and y $\dfrac{7}{y}$ or $7 \div y$

34. The sum of 2 and $3b$ $2 + 3b$

35. The difference of a and 5 $a - 5$

36. Two more than $5k$ $5k + 2$

37. Seven less than $13z$ $13z - 7$

For Exercises 38–43, simplify the expressions.

38. 6^3 216

39. 15^2 225

40. $\sqrt{36}$ 6

41. $\dfrac{1}{\sqrt{100}}$ $\dfrac{1}{10}$

42. $\left(\dfrac{1}{4}\right)^2$ $\dfrac{1}{16}$

43. $\left(\dfrac{3}{2}\right)^3$ $\dfrac{27}{8}$

For Exercises 44–47, perform the indicated operations.

44. $15 - 7 \cdot 2 + 12$ 13

45. $|-11| + |5| - (7 - 2)$ 11

46. $4^2 - (5 - 2)^2$ 7

47. $22 - 3(8 \div 4)^2$ 10

Section 1.4

For Exercises 48–60, add.

48. $-6 + 8$ 2

49. $14 + (-10)$ 4

50. $21 + (-6)$ 15

51. $-12 + (-5)$ −17

52. $\dfrac{2}{7} + \left(-\dfrac{1}{9}\right)$ $\dfrac{11}{63}$

53. $\left(-\dfrac{8}{11}\right) + \left(\dfrac{1}{2}\right)$ $-\dfrac{5}{22}$

54. $\left(-\dfrac{1}{10}\right) + \left(-\dfrac{5}{6}\right)$ $-\dfrac{14}{15}$

55. $\left(-\dfrac{5}{2}\right) + \left(-\dfrac{1}{5}\right)$ $-\dfrac{27}{10}$

56. $-8.17 + 6.02$ −2.15

57. $2.9 + (-7.18)$ −4.28

 Writing   Translating Expression   Geometry   Scientific Calculator   Video

58. $13 + (-2) + (-8)$ 3

59. $-5 + (-7) + 20$ 8

60. $2 + 5 + (-8) + (-7) + 0 + 13 + (-1)$ 4

61. Under what conditions will the expression $a + b$ be negative?   When a and b are both negative or when a and b have different signs and the number with the larger absolute value is negative.

62. Richard's checkbook was overdrawn by $45 (that is, his balance was -45). He deposited $117 but then wrote a check for $80. Was the deposit enough to cover the check? Explain.
No. He is still overdrawn by $8.

Section 1.5

For Exercises 63–75, subtract.

63. $13 - 25$ –12

64. $31 - (-2)$ 33

65. $-8 - (-7)$ –1

66. $-2 - 15$ –17

67. $\left(-\dfrac{7}{9}\right) - \dfrac{5}{6}$ $-\dfrac{29}{18}$

68. $\dfrac{1}{3} - \dfrac{9}{8}$ $-\dfrac{19}{24}$

69. $7 - 8.2$ –1.2

70. $-1.05 - 3.2$ –4.25

71. $-16.1 - (-5.9)$ –10.2

72. $7.09 - (-5)$ 12.09

73. $\dfrac{11}{2} - \left(-\dfrac{1}{6}\right) - \dfrac{7}{3}$ $\dfrac{10}{3}$

74. $\dfrac{4}{5} - \dfrac{7}{10} - \left(-\dfrac{13}{20}\right)$ $\dfrac{17}{20}$

75. $6 - 14 - (-1) - 10 - (-21) - 5$ –1

76. Under what conditions will the expression $a - b$ be negative?   If $a < b$

For Exercises 77–81, write an algebraic expression and simplify.

77. -18 subtracted from -7
$-7 - (-18)$; 11

78. The difference of -6 and 41
$-6 - 41$; –47

79. Seven decreased by 13
$7 - 13$; –6

80. Five subtracted from the difference of 20 and -7
$[20 - (-7)] - 5$; 22

81. The sum of 6 and -12, decreased by 21
$[6 + (-12)] - 21$; –27

82. In Nevada, the highest temperature ever recorded was $125°$F and the lowest was $-50°$F. Find the difference between the highest and lowest temperatures. (*Source: Information Please Almanac*)
175°F

Section 1.6

For Exercises 83–100, multiply or divide as indicated.

83. $10(-17)$ –170

84. $(-7)13$ –91

85. $(-52) \div 26$ –2

86. $(-48) \div (-16)$ 3

87. $\dfrac{7}{4} \div \left(-\dfrac{21}{2}\right)$ $-\dfrac{1}{6}$

88. $\dfrac{2}{3}\left(-\dfrac{12}{11}\right)$ $-\dfrac{8}{11}$

89. $-\dfrac{21}{5} \cdot 0$ 0

90. $\dfrac{3}{4} \div 0$ Undefined

91. $0 \div (-14)$ 0

92. $(-0.45)(-5)$ 2.25

93. $\dfrac{-21}{14}$ $-\dfrac{3}{2}$

94. $\dfrac{-13}{-52}$ $\dfrac{1}{4}$

95. $(5)(-2)(3)$ –30

96. $(-6)(-5)(15)$ 450

97. $\left(-\dfrac{1}{2}\right)\left(\dfrac{7}{8}\right)\left(-\dfrac{4}{7}\right)$ $\dfrac{1}{4}$

98. $\left(\dfrac{12}{13}\right)\left(-\dfrac{1}{6}\right)\left(\dfrac{13}{14}\right)$ $-\dfrac{1}{7}$

99. $40 \div 4 \div (-5)$ –2

100. $\dfrac{10}{11} \div \dfrac{7}{11} \div \dfrac{5}{9}$ $\dfrac{18}{7}$

For Exercises 101–106, perform the indicated operations.

101. $9 - 4[-2(4 - 8) - 5(3 - 1)]$ 17

102. $\dfrac{8(-3) - 6}{-7 - (-2)}$ 6

103. $\dfrac{2}{3} - \left(\dfrac{3}{8} + \dfrac{5}{6}\right) \div \dfrac{5}{3}$ $-\dfrac{7}{120}$

104. $5.4 - (0.3)^2 \div 0.09$ 4.4

105. $\dfrac{5 - [3 - (-4)^2]}{36 \div (-2)(3)}$ $-\dfrac{1}{3}$

106. $|-8 + 5| - \sqrt{5^2 - 3^2}$ –1

For Exercises 107–110, evaluate each expression given the values $x = 4$ and $y = -9$.

107. $3(x + 2) \div y$ –2

108. $\sqrt{x} - y$ 11

109. $-xy$ 36

110. $3x + 2y$ –6

111. In statistics the formula $x = \mu + z\sigma$ is used to find cutoff values for data that follow a bellshaped curve Find x if $\mu = 100$, $z = -1.96$, and $\sigma = 15$.   70.6

Writing   ← → Translating Expression   Geometry   Scientific Calculator   Video

For Exercises 112–118, answer true or false. If a statement is false, explain why.

112. If n is positive, then $-n$ is negative   True

113. If m is negative, then m^4 is negative.   False, any nonzero real number raised to an even power is positive.

114. If m is negative, then m^3 is negative.   True

115. If $m > 0$ and $n > 0$, then $mn > 0$.   True

116. If $p < 0$ and $q < 0$, then $pq < 0$.   False, the product of two negative numbers is positive.

117. A number and its reciprocal have the same signs.   True

118. A nonzero number and its opposite have different signs.   True

Section 1.7

For Exercises 119–126, answers may vary.

119. Give an example of the commutative property of addition.
For example: $2 + 3 = 3 + 2$

120. Give an example of the associative property of addition.
For example: $(2 + 3) + 4 = 2 + (3 + 4)$

121. Give an example of the inverse property of addition.
For example: $5 + (-5) = 0$

122. Give an example of the identity property of addition.
For example: $7 + 0 = 7$

123. Give an example of the commutative property of multiplication.
For example: $5 \cdot 2 = 2 \cdot 5$

124. Give an example of the associative property of multiplication.
For example: $(8 \cdot 2)10 = 8(2 \cdot 10)$

125. Give an example of the inverse property of multiplication.
For example: $3 \cdot \frac{1}{3} = 1$

126. Give an example of the identity property of multiplication.   For example: $8 \cdot 1 = 8$

127. Explain why $5x - 2y$ is the same as $-2y + 5x$.
$5x - 2y = 5x + (-2y)$, then use the commutative property of addition.

128. Explain why $3a - 9y$ is the same as $-9y + 3a$.
$3a - 9y = 3a + (-9y)$, then use the commutative property of addition.

129. List the terms of the expression:
$3y + 10x - 12 + xy$
$3y, 10x, -12, xy$

130. Identify the coefficients for the terms listed in Exercise 129.
$3, 10, -12, 1$

For Exercises 131–132, simplify by combining *like* terms.

131. $3a + 3b - 4b + 5a - 10$
$8a - b - 10$

132. $-6p + 2q + 9 - 13q - p + 7$
$-7p - 11q + 16$

For Exercises 133–134, use the distributive property to clear the parentheses.

133. $-2(4z + 9)$
$-8z - 18$

134. $5(4w - 8y + 1)$
$20w - 40y + 5$

For Exercises 135–140, simplify each expression.

135. $2p - (p + 5w) + 3w$ $p - 2w$

136. $6(h + 3m) - 7h - 4m$ $-h + 14m$

137. $\frac{1}{2}(-6q) + q - 4\left(3q + \frac{1}{4}\right)$ $-14q - 1$

138. $0.3b + 12(0.2 - 0.5b)$ $-5.7b + 2.4$

139. $-4[2(x + 1) - (3x + 8)]$ $4x + 24$

140. $5[(7y - 3) + 3(y + 8)]$ $50y + 105$

Chapter 1   Test

1. Simplify. $\frac{135}{36}$ $\frac{15}{4}$

2. Add and subtract. $\frac{5}{4} - \frac{5}{12} + \frac{2}{3}$ $\frac{3}{2}$

3. Divide. $4\frac{1}{12} \div 1\frac{1}{3}$ $3\frac{1}{16}$

4. Subtract. $4\frac{1}{4} - 1\frac{7}{8}$ $2\frac{3}{8}$

5. Is $0.\overline{315}$ a rational number or an irrational number? Explain your reasoning.
Rational, all repeating decimals are rational numbers.

Writing   Translating Expression   Geometry   Scientific Calculator   Video

6. Use the definition of exponents to expand the expressions:

a. $(4x)^3$ $(4x)(4x)(4x)$ **b.** $4x^3$ $4 \cdot x \cdot x \cdot x$

7. Plot the points on a number line: $|3|$, 0, -2, 0.5, $\left|-\frac{3}{2}\right|$, $\sqrt{16}$.

8. Use the number line in Exercise 7 to identify whether the statements are true or false.

a. $|3| < -2$ False   **b.** $0 \le \left|-\frac{3}{2}\right|$ True

c. $-2 < 0.5$ True   **d.** $|3| \ge \left|-\frac{3}{2}\right|$ True

9. Identify the property that justifies each statement.

a. $6(-8) = (-8)6$
Commutative property
of multiplication

b. $5 + 0 = 5$
Identity property of
addition

c. $(2 + 3) + 4 = 2 + (3 + 4)$
Associative property of addition

d. $\frac{1}{7} \cdot 7 = 1$
Inverse property
of multiplication

e. $8[7(-3)] = (8 \cdot 7)(-3)$
Associative property
of multiplication

10. Write each expression as an English phrase.

a. $2(a - b)$.   Twice the difference of a and b

b. $2a - b$. (Answers may vary.)
b subtracted from twice a

11. Write the phrase as an algebraic expression: "The quotient of the square root of c and the square of d."
$\frac{\sqrt{c}}{d^2}$ or $\sqrt{c} \div d^2$

For Exercises 12–14, write each English statement as an algebraic expression. Then simplify the expression.

12. Subtract -4 from 12 $12 - (-4)$; 16

13. Find the difference of 6 and 8 $6 - 8$; -2

14. The quotient of 10 and -12 $\frac{10}{-12}$; $-\frac{5}{6}$

For Exercises 15–29, perform the indicated operations.

15. $-\frac{1}{8} + \left(-\frac{3}{4}\right)$ $-\frac{7}{8}$ **16.** $-84 \div 7$ -12

17. $21 - (-7)$ 28 **18.** $-15 - (-3)$ -12

19. $-14 + (-2) - 16$ -32

20. $(-16)(-2)(-1)(-3)$ 96

21. $-22 \cdot 0$ 0

22. $38 \div 0$ Undefined

23. $18 + (-12)$ 6

24. $-10.06 - (-14.72)$ 4.66

25. $7(-4)$ -28

26. $\frac{2}{5} \div \left(-\frac{7}{10}\right) \cdot \left(-\frac{7}{6}\right)$ $\frac{2}{3}$

27. $\frac{\sqrt{5^2 - 4^2}}{|-12 + 3|}$ $\frac{1}{3}$

28. $8 - [(2 - 4) - (8 - 9)]$ 9

29. $(8 - 10) \cdot \frac{3}{2} + (-5)$ -8

30. The average high temperature in January for Nova Scotia, Canada, is $-1.2°C$. The average low is $-10.7°C$. Find the difference between the average high and the average low.
The difference is $9.5°C$.

31. In the third quarter of a football game, a quarterback made a 5-yd gain, a 2-yd gain, a 10-yd loss, and then a 4-yd gain.

a. Write an expression using addition to describe the quarterback's movement.
$5 + 2 + (-10) + 4$

b. Evaluate the expression from part (a) to determine the quarterback's net gain or loss in yards.
He gained 1 yd.

For Exercises 32–36, simplify each expression.

32. $3k - 20 + (-9k) + 12$ $-6k - 8$

33. $-5x - 4y + 3 - 7x + 6y - 7$ $-12x + 2y - 4$

34. $4(p - 5) - (8p + 3)$ $-4p - 23$

35. $-3(4m + 8p - 7)$ $-12m - 24p + 21$

36. $\frac{1}{2}(12p - 4) + \frac{1}{3}(2 - 6p)$ $4p - \frac{4}{3}$

For Exercises 37–40, evaluate each expression given the values $x = 4$ and $y = -3$ and $z = -7$.

37. $y^2 - x$ 5 **38.** $3x - 2y$ 18

39. $y(x - 2)$ -6 **40.** $-y^2 - 4x + z$ -32

Linear Equations and Inequalities

2

Mathematics as a Language

Languages make use of symbols to represent sounds and other conventions used in speech and writing. The letters of the alphabet and punctuation marks are all examples of these symbols. Mathematicians cleverly adopted the use of symbols as a way to give a temporary nickname to *unknowns* in order to simplify the problem-solving process.

Suppose that the maximum recommended heart rate for an adult is given by 220 minus the age (in years) of the adult. If we let a be the age of an adult in years, then the expression $220 - a$ represents the adult's maximum recommended heart rate.

© Blend Images/Ariel Skelley/Getty Images RF

If Alan is 60 years old, his maximum heart rate is found by substituting 60 for a. Thus, Alan's maximum recommended heart rate is $220 - 60$, which is 160 beats per minute.

If we know that Ben's recommended heart rate is 178, then we can solve an equation to determine his age, a.

$$220 - a = 178$$
$$-a = 178 - 220$$
$$-a = -42$$
$$a = 42 \quad \text{Ben is 42 years old.}$$

Section 2.1 Addition, Subtraction, Multiplication, and Division Properties of Equality

1. Definition of a Linear Equation in One Variable

An **equation** is a statement that indicates that two expressions are equal. The following are equations.

$$x = 5 \qquad y + 2 = 12 \qquad -4z = 28$$

All equations have an equal sign. Furthermore, notice that the equal sign separates the equation into two parts, the left-hand side and the right-hand side. A **solution to an equation** is a value of the variable that makes the equation a true statement. Substituting a solution into an equation for the variable makes the right-hand side equal to the left-hand side.

Equation	Solution	Check	
$x = 5$	5	$x = 5$ $\downarrow$ $5 = 5 \checkmark$	Substitute 5 for x. Right-hand side equals left-hand side.
$y + 2 = 12$	10	$y + 2 = 12$ $\downarrow$ $10 + 2 = 12 \checkmark$	Substitute 10 for y. Right-hand side equals left-hand side.
$-4z = 28$	-7	$-4z = 28$ $\downarrow$ $-4(-7) = 28 \checkmark$	Substitute -7 for z. Right-hand side equals left-hand side.

Avoiding Mistakes

Be sure to notice the difference between solving an equation versus simplifying an expression. For example, $2x + 1 = 7$ is an equation, whose solution is 3, while $2x + 1 + 7$ is an expression that simplifies to $2x + 8$.

Classroom Examples: p. 110, Exercises 10 and 14

Example 1 Determining Whether a Number Is a Solution to an Equation

Determine whether the given number is a solution to the equation.

a. $4x + 7 = 5$; $-\frac{1}{2}$ **b.** $-4 = 6w - 14$; 3

Solution:

a.
$$4x + 7 = 5$$
$$4\left(-\tfrac{1}{2}\right) + 7 \stackrel{?}{=} 5 \qquad \text{Substitute } -\tfrac{1}{2} \text{ for } x.$$
$$-2 + 7 \stackrel{?}{=} 5 \qquad \text{Simplify.}$$
$$5 \stackrel{?}{=} 5 \checkmark \qquad \text{Right-hand side equals the left-hand side.}$$

Thus, $-\frac{1}{2}$ *is a solution* to the equation $4x + 7 = 5$.

b. $-4 = 6w - 14$
$$-4 \stackrel{?}{=} 6(3) - 14 \qquad \text{Substitute 3 for } w.$$
$$-4 \stackrel{?}{=} 18 - 14 \qquad \text{Simplify.}$$
$$-4 \neq 4 \qquad \text{Right-hand side does not equal left-hand side.}$$

Thus, 3 *is not a solution* to the equation $-4 = 6w - 14$.

Skill Practice Determine whether the given number is a solution to the equation.

1. $4x - 1 = 7$; 3   **2.** $9 = -2y + 5$; -2

Answers

1. No   **2.** Yes

The set of all solutions to an equation is called the **solution set** and is written with set braces. For example, the solution set for Example 1(a) is $\{-\frac{1}{2}\}$.

In the study of algebra, you will encounter a variety of equations. In this chapter, we will focus on a specific type of equation called a linear equation in one variable.

Definition of a Linear Equation in One Variable

Let a, b, and c be real numbers such that $a \neq 0$. A **linear equation in one variable** is an equation that can be written in the form

$$ax + b = c$$

Note: A linear equation in one variable is often called a first-degree equation because the variable x has an implied exponent of 1.

Examples	Notes
$2x + 4 = 20$	$a = 2, b = 4, c = 20$
$-3x - 5 = 16$ can be written as $-3x + (-5) = 16$	$a = -3, b = -5, c = 16$
$5x + 9 - 4x = 1$ can be written as $x + 9 = 1$	$a = 1, b = 9, c = 1$

---

2. Addition and Subtraction Properties of Equality

If two equations have the same solution set, then the equations are equivalent. For example, the following equations are equivalent because the solution set for each equation is $\{6\}$.

Equivalent Equations	Check the Solution 6
$2x - 5 = 7$ $\longrightarrow$	$2(6) - 5 \overset{?}{=} 7$ $\Rightarrow 12 - 5 \overset{?}{=} 7 \checkmark$
$2x = 12$ $\longrightarrow$	$2(6) \overset{?}{=} 12$ $\Rightarrow$ $12 \overset{?}{=} 12 \checkmark$
$x = 6$ $\longrightarrow$	$6 \overset{?}{=} 6$ $\Rightarrow$ $6 \overset{?}{=} 6 \checkmark$

To solve a linear equation, $ax + b = c$, the goal is to find *all* values of x that make the equation true. One general strategy for solving an equation is to rewrite it as an equivalent but simpler equation. This process is repeated until the equation can be written in the form $x = $ number. We call this "isolating the variable." The addition and subtraction properties of equality help us isolate the variable.

Addition and Subtraction Properties of Equality

Let a, b, and c represent algebraic expressions.

1. Addition property of equality: If $a = b$,

then $a + c = b + c$

2. *Subtraction property of equality: If $a = b$,

then $a - c = b - c$

*The subtraction property of equality follows directly from the addition property, because subtraction is defined in terms of addition.

If $a + (-c) = b + (-c)$

then, $a - c = b - c$

The addition and subtraction properties of equality indicate that adding or subtracting the same quantity on each side of an equation results in an equivalent equation. This means that if two equal quantities are increased or decreased by the same amount, then the resulting quantities will also be equal (Figure 2-1).

$$50 = 50$$
$$50 + 20 = 50 + 20$$
$$70 = 70$$

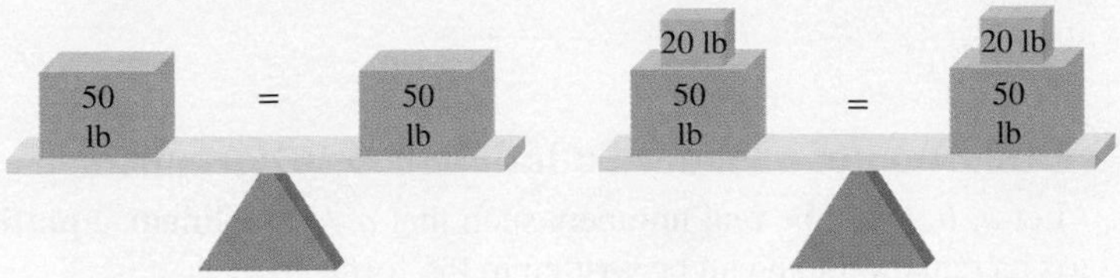

Figure 2-1

Classroom Examples: p. 110, Exercises 16 and 18

Example 2   **Applying the Addition and Subtraction Properties of Equality**

Solve the equations.

a. $p - 4 = 11$ **b.** $w + 5 = -2$

Solution:

In each equation, the goal is to isolate the variable on one side of the equation. To accomplish this, we use the fact that the sum of a number and its opposite is zero and the difference of a number and itself is zero.

a.

$$p - 4 = 11$$
$$p - 4 + 4 = 11 + 4$$ To isolate p, add 4 to both sides ($-4 + 4 = 0$).
$$p + 0 = 15$$ Simplify.
$$p = 15$$ Check by substituting $p = 15$ into the original equation.

Check: $p - 4 = 11$
$$15 - 4 \overset{?}{=} 11$$

The solution set is $\{15\}$. $11 \overset{?}{=} 11 ✓$ True

b.

$$w + 5 = -2$$
$$w + 5 - 5 = -2 - 5$$ To isolate w, subtract 5 from both sides. ($5 - 5 = 0$).
$$w + 0 = -7$$ Simplify.
$$w = -7$$ Check by substituting $w = -7$ into the original equation.

Check: $w + 5 = -2$
$$-7 + 5 \overset{?}{=} -2$$

The solution set is $\{-7\}$. $-2 \overset{?}{=} -2 ✓$ True

Skill Practice   Solve the equations.

3. $v - 7 = 2$ **4.** $x + 4 = 4$

Answers

3. $\{9\}$ **4.** $\{0\}$

Classroom Examples: p. 110,
Exercises 30 and 32

| Example 3 | **Applying the Addition and Subtraction Properties of Equality** |

Solve the equations.

a. $\dfrac{9}{4} = q - \dfrac{3}{4}$ **b.** $-1.2 + z = 4.6$

Solution:

a. $\dfrac{9}{4} = q - \dfrac{3}{4}$

$\dfrac{9}{4} + \dfrac{3}{4} = q - \dfrac{3}{4} + \dfrac{3}{4}$ To isolate q, add $\frac{3}{4}$ to both sides $\left(-\frac{3}{4} + \frac{3}{4} = 0\right)$.

$\dfrac{12}{4} = q + 0$ Simplify.

$3 = q$ or equivalently, $q = 3$

> **TIP:** The variable may be isolated on either side of the equation.

Check: $\dfrac{9}{4} = q - \dfrac{3}{4}$

$\dfrac{9}{4} \overset{?}{=} 3 - \dfrac{3}{4}$ Substitute $q = 3$.

$\dfrac{9}{4} \overset{?}{=} \dfrac{12}{4} - \dfrac{3}{4}$ Common denominator

The solution set is $\{3\}$. $\dfrac{9}{4} \overset{?}{=} \dfrac{9}{4}$ ✓   True

b. $-1.2 + z = 4.6$

$-1.2 + 1.2 + z = 4.6 + 1.2$ To isolate z, add 1.2 to both sides.

$0 + z = 5.8$

$z = 5.8$ Check: $-1.2 + z = 4.6$

$-1.2 + 5.8 \overset{?}{=} 4.6$ Substitute $z = 5.8$.

The solution set is $\{5.8\}$. $4.6 \overset{?}{=} 4.6$ ✓   True

Skill Practice   Solve the equations.

5. $\dfrac{1}{4} = a - \dfrac{2}{3}$ **6.** $-8.1 + w = 11.5$

3. Multiplication and Division Properties of Equality

Adding or subtracting the same quantity to both sides of an equation results in an equivalent equation. In a similar way, multiplying or dividing both sides of an equation by the same nonzero quantity also results in an equivalent equation. This is stated formally as the multiplication and division properties of equality.

Answers

5. $\left\{\dfrac{11}{12}\right\}$ **6.** $\{19.6\}$

Multiplication and Division Properties of Equality

Let a, b, and c represent algebraic expressions, $c \neq 0$.

1. **Multiplication property of equality:**   If $a = b$,

       then $ac = bc$

2. ***Division property of equality:**   If $a = b$

       then $\dfrac{a}{c} = \dfrac{b}{c}$

*The division property of equality follows directly from the multiplication property because division is defined as multiplication by the reciprocal.

$$\text{If} \quad a \cdot \frac{1}{c} = b \cdot \frac{1}{c}$$

$$\text{then,} \quad \frac{a}{c} = \frac{b}{c}$$

To understand the multiplication property of equality, suppose we start with a true equation such as $10 = 10$. If both sides of the equation are multiplied by a constant such as 3, the result is also a true statement (Figure 2-2).

$$10 = 10$$
$$3 \cdot 10 = 3 \cdot 10$$
$$30 = 30$$

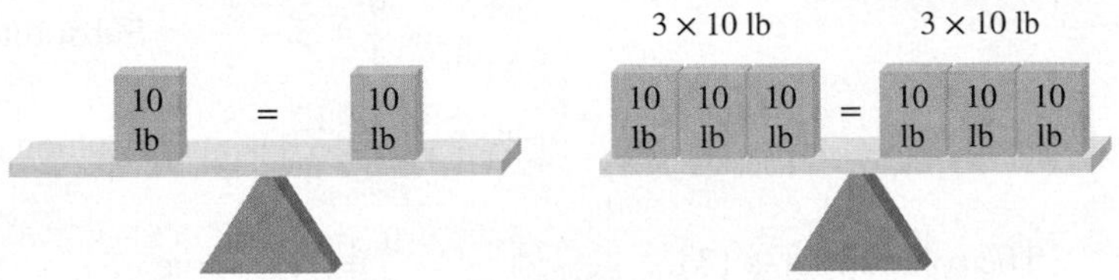

Figure 2-2

Similarly, if both sides of the equation are divided by a nonzero real number such as 2, the result is also a true statement (Figure 2-3).

$$10 = 10$$
$$\frac{10}{2} = \frac{10}{2}$$
$$5 = 5$$

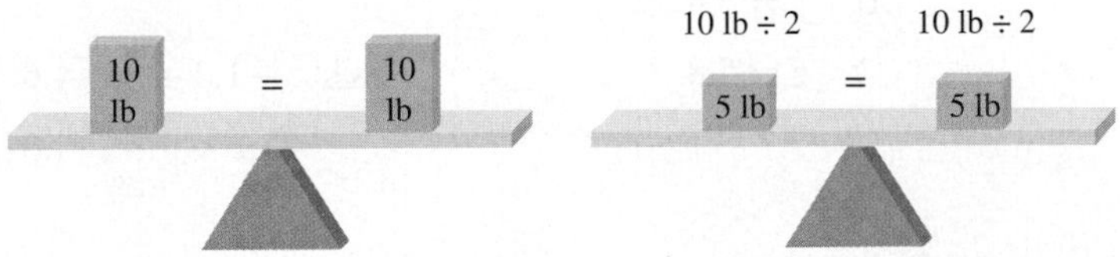

Figure 2-3

TIP: The product of a number and its reciprocal is always 1. For example:

$$\frac{1}{5}(5) = 1$$

$$-\frac{7}{2}\left(-\frac{2}{7}\right) = 1$$

To solve an equation in the variable x, the goal is to write the equation in the form $x = $ number. In particular, notice that we desire the coefficient of x to be 1. That is, we want to write the equation as $1x = $ number. Therefore, to solve an equation such as $5x = 15$, we can multiply both sides of the equation by the reciprocal of the x-term coefficient. In this case, multiply both sides by the reciprocal of 5, which is $\frac{1}{5}$.

$$5x = 15$$
$$\frac{1}{5}(5x) = \frac{1}{5}(15) \qquad \text{Multiply by } \frac{1}{5}.$$
$$1x = 3 \qquad \text{The coefficient of the } x\text{-term is now 1.}$$
$$x = 3$$

The division property of equality can also be used to solve the equation $5x = 15$ by dividing both sides by the coefficient of the x-term. In this case, divide both sides by 5 to make the coefficient of x equal to 1.

$$5x = 15$$

$$\frac{5x}{5} = \frac{15}{5} \qquad \text{Divide by 5.}$$

$$1x = 3 \qquad \text{The coefficient of the } x\text{-term is now 1.}$$

$$x = 3$$

TIP: The quotient of a nonzero real number and itself is always 1. For example:

$$\frac{5}{5} = 1$$

$$\frac{-3.5}{-3.5} = 1$$

Example 4 **Applying the Division Property of Equality**

Solve the equations using the division property of equality.

a. $12x = 60$ **b.** $48 = -8w$ **c.** $-x = 8$

Solution:

a. $12x = 60$

$$\frac{12x}{12} = \frac{60}{12} \qquad \text{To obtain a coefficient of 1 for the } x\text{-term,}$$
$$\text{divide both sides by 12.}$$

$$1x = 5 \qquad \text{Simplify.}$$

$$x = 5 \qquad \underline{\text{Check:}} \quad 12x = 60$$

$$12(5) \overset{?}{=} 60$$

The solution set is $\{5\}$. $60 \overset{?}{=} 60 \checkmark$ True

b. $48 = -8w$

$$\frac{48}{-8} = \frac{-8w}{-8} \qquad \text{To obtain a coefficient of 1 for the } w\text{-term, divide}$$
$$\text{both sides by } -8.$$

$$-6 = 1w \qquad \text{Simplify.}$$

$$-6 = w \qquad \underline{\text{Check:}} \; 48 = -8w$$

$$48 \overset{?}{=} -8(-6)$$

The solution set is $\{-6\}$. $48 \overset{?}{=} 48 \checkmark$ True

c. $-x = 8$ Note that $-x$ is equivalent to $-1 \cdot x$.

$$-1x = 8$$

$$\frac{-1x}{-1} = \frac{8}{-1} \qquad \text{To obtain a coefficient of 1 for the } x\text{-term, divide}$$
$$\text{by } -1.$$

$$x = -8 \qquad \underline{\text{Check:}} \; -x = 8$$

$$-(-8) \overset{?}{=} 8$$

The solution set is $\{-8\}$. $8 \overset{?}{=} 8 \checkmark$ True

Classroom Examples: p. 110, Exercises 38 and 50

TIP: In Example 4(c), we could also have *multiplied* both sides by -1 to create a coefficient of 1 on the x-term.

$$-x = 8$$
$$(-1)(-x) = (-1)8$$
$$x = -8$$

Skill Practice   Solve the equations.

7. $4x = -20$ **8.** $100 = -4p$ **9.** $-y = -11$

Answers

7. $\{-5\}$ **8.** $\{-25\}$ **9.** $\{11\}$

Classroom Example: p. 110,
Exercise 46

Example 5   **Applying the Multiplication Property of Equality**

Solve the equation by using the multiplication property of equality.

$$-\frac{2}{9}q = \frac{1}{3}$$

Solution:

$$-\frac{2}{9}q = \frac{1}{3}$$

$$\left(-\frac{9}{2}\right)\left(-\frac{2}{9}q\right) = \frac{1}{3}\left(-\frac{9}{2}\right)$$ To obtain a coefficient of 1 for the q-term, multiply by the reciprocal of $-\frac{2}{9}$, which is $-\frac{9}{2}$.

$$1q = -\frac{3}{2}$$ Simplify. The product of a number and its reciprocal is 1.

$$q = -\frac{3}{2}$$ Check: $-\frac{2}{9}q = \frac{1}{3}$

$$-\frac{2}{9}\left(-\frac{3}{2}\right) \overset{?}{=} \frac{1}{3}$$

The solution set is $\left\{-\frac{3}{2}\right\}$. $\frac{1}{3} \overset{?}{=} \frac{1}{3}$ ✓   True

Skill Practice   Solve the equation.

10. $-\frac{2}{3}a = \frac{1}{4}$

TIP: When applying the multiplication or division property of equality to obtain a coefficient of 1 for the variable term, we will generally use the following convention:

- If the coefficient of the variable term is expressed as a fraction, we will usually multiply both sides by its reciprocal, as in Example 5.
- If the coefficient of the variable term is an integer or decimal, we will divide both sides by the coefficient itself, as in Example 6.

Classroom Example: p. 110,
Exercise 52

Example 6   **Applying the Division Property of Equality**

Solve the equation by using the division property of equality.

$$-3.43 = -0.7z$$

Solution:

$$-3.43 = -0.7z$$

$$\frac{-3.43}{-0.7} = \frac{-0.7z}{-0.7}$$ To obtain a coefficient of 1 for the z-term, divide by -0.7.

$$4.9 = 1z$$ Simplify.

$$4.9 = z$$

$$z = 4.9$$ Check: $-3.43 = -0.7z$

$$-3.43 \overset{?}{=} -0.7(4.9)$$

The solution set is $\{4.9\}$. $-3.43 \overset{?}{=} -3.43$ ✓   True

Answer

10. $\left\{-\frac{3}{8}\right\}$

Skill Practice   Solve the equation.

11. $6.82 = 2.2w$

| **Example 7** | **Applying the Multiplication Property of Equality** |

Classroom Example: p. 110, Exercise 42

Solve the equation by using the multiplication property of equality.

$$\frac{d}{6} = -4$$

Solution:

$$\frac{d}{6} = -4$$

$$\frac{1}{6}d = -4 \qquad\qquad \frac{d}{6} \text{ is equivalent to } \frac{1}{6}d.$$

$$\frac{6}{1} \cdot \frac{1}{6}d = -4 \cdot \frac{6}{1} \qquad \text{To obtain a coefficient of 1 for the } d\text{-term,}$$
$$\text{multiply by the reciprocal of } \tfrac{1}{6}, \text{ which is } \tfrac{6}{1}.$$

$$1d = -24 \qquad\qquad \text{Simplify.}$$

$$d = -24 \qquad\qquad \underline{\text{Check:}}\ \frac{d}{6} = -4$$

$$\frac{-24}{6} \overset{?}{=} -4$$

The solution set is $\{-24\}$. $\qquad\qquad -4 \overset{?}{=} -4\ \checkmark\quad$ True

Instructor Note: Remind students that dividing by a number is the same as multiplying by its reciprocal. Thus, $\dfrac{d}{6} = \dfrac{1}{6}d$

Skill Practice   Solve the equation.

12. $\dfrac{x}{5} = -8$

It is important to distinguish between cases where the addition or subtraction properties of equality should be used to isolate a variable versus those in which the multiplication or division property of equality should be used. Remember the goal is to isolate the variable term and obtain a coefficient of 1. Compare the equations:

$$5 + x = 20 \qquad \text{and} \qquad 5x = 20$$

In the first equation, the relationship between 5 and x is addition. Therefore, we want to reverse the process by subtracting 5 from both sides. In the second equation, the relationship between 5 and x is multiplication. To isolate x, we reverse the process by dividing by 5 or equivalently, multiplying by the reciprocal, $\frac{1}{5}$.

$$5 + x = 20 \qquad\qquad \text{and} \qquad 5x = 20$$

$$5 - 5 + x = 20 - 5 \qquad\qquad \frac{5x}{5} = \frac{20}{5}$$

$$x = 15 \qquad\qquad\qquad x = 4$$

4. Translations

We have already practiced writing an English sentence as a mathematical equation. Recall that several key words translate to the algebraic operations of addition, subtraction, multiplication, and division.

Classroom Examples: p. 110, Exercises 60 and 62

Example 8	**Translating to a Linear Equation**

Write an algebraic equation to represent each English sentence. Then solve the equation.

   a. The quotient of a number and 4 is 6.

   b. The product of a number and 4 is 6.

   c. Negative twelve is equal to the sum of −5 and a number.

   d. The value 1.4 subtracted from a number is 5.7.

Solution:

For each case we will let x represent the unknown number.

   a. The quotient of a number and 4 is 6.

$$\frac{x}{4} = 6$$

$$4 \cdot \frac{x}{4} = 4 \cdot 6 \qquad \text{Multiply both sides by 4.}$$

$$\frac{4}{1} \cdot \frac{x}{4} = 4 \cdot 6$$

$$x = 24 \qquad \text{Check: } \frac{24}{4} \overset{?}{=} 6 \checkmark \quad \text{True}$$

The number is 24.

   b. The product of a number and 4 is 6.

$$4x = 6$$

$$\frac{4x}{4} = \frac{6}{4} \qquad \text{Divide both sides by 4.}$$

$$x = \frac{3}{2} \qquad \text{Check: } 4\left(\frac{3}{2}\right) \overset{?}{=} 6 \checkmark \quad \text{True}$$

The number is $\frac{3}{2}$.

   c. Negative twelve is equal to the sum of −5 and a number.

$$-12 = -5 + x$$

$$-12 + 5 = -5 + 5 + x \qquad \text{Add 5 to both sides.}$$

$$-7 = x \qquad \text{Check: } -12 \overset{?}{=} -5 + (-7) \checkmark \quad \text{True}$$

The number is −7.

d. The value 1.4 subtracted from a number is 5.7.

$$x - 1.4 = 5.7$$
$$x - 1.4 + 1.4 = 5.7 + 1.4 \qquad \text{Add 1.4 to both sides.}$$
$$x = 7.1 \qquad \text{Check: } \ 7.1 - 1.4 \stackrel{?}{=} 5.7 \checkmark \quad \text{True}$$

The number is 7.1.

Skill Practice   Write an algebraic equation to represent each English sentence. Then solve the equation.

13. The quotient of a number and -2 is 8.

14. The product of a number and -3 is -24.

15. The sum of a number and 6 is -20.

16. 13 is equal to 5 subtracted from a number.

Answers

13. $\dfrac{x}{-2} = 8$; The number is -16.

14. $-3x = -24$; The number is 8.

15. $y + 6 = -20$; The number is -26.

16. $13 = x - 5$; The number is 18.

Section 2.1   Practice Exercises

Study Skills Exercise

For additional exercises, see Classroom Activities 2.1A–2.1D in the *Student's Resource Manual* at www.mhhe.com/moh.

After getting a test back, it is a good idea to correct the test so that you do not make the same errors again. One recommended approach is to use a clean sheet of paper, and divide the paper down the middle vertically as shown. For each problem that you missed on the test, rework the problem correctly on the left-hand side of the paper. Then give a written explanation on the right-hand side of the paper. To reinforce the correct procedure, do four more problems of that type.

Take the time this week to make corrections from your last test.

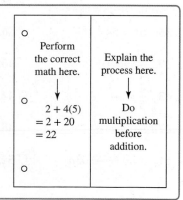

Perform the correct math here.	Explain the process here.
$2 + 4(5)$ $= 2 + 20$ $= 22$	Do multiplication before addition.

Vocabulary and Key Concepts

1. a. An ___equation___ is a statement that indicates that two expressions are equal.

  b. A ___solution___ to an equation is a value of the variable that makes the equation a true statement.

  c. An equation that can be written in the form $ax + b = c \ (a \neq 0)$ is called a ___linear___ equation in one variable.

  d. The set of all solutions to an equation is called the ___solution set___.

Concept 1: Definition of a Linear Equation in One Variable

For Exercises 2–6, identify the following as either an expression or an equation.

2. $2 - 8x + 10$
  Expression

3. $x - 4 + 5x$
  Expression

4. $8x + 2 = 7$
  Equation

5. $9 = 2x - 4$
  Equation

6. $3x^2 + x = -3$
  Equation

7. Explain how to determine if a number is a solution to an equation.
  Substitute the value into the equation and determine if the right-hand side is equal to the left-hand side.

8. Explain why the equations $6x = 12$ and $x = 2$ are *equivalent equations*.
  They are equivalent equations because their solution sets are the same. The solution set is {2} in both cases.

Writing   Translating Expression   Geometry   Scientific Calculator   Video

For Exercises 9–14, determine whether the given number is a solution to the equation. **(See Example 1.)**

9. $x - 1 = 5$; 4   No

10. $x - 2 = 1$; −1   No

11. $5x = -10$;   −2   Yes

12. $3x = 21$;   7   Yes

13. $3x + 9 = 3$;   −2   Yes

14. $2x - 1 = -3$;   −1   Yes

Concept 2: Addition and Subtraction Properties of Equality

For Exercises 15–34, solve each equation using the addition or subtraction property of equality. Be sure to check your answers. **(See Examples 2–3.)**

15. $x + 6 = 5$ {−1}

16. $x - 2 = 10$ {12}

 17. $q - 14 = 6$ {20}

18. $w + 3 = -5$ {−8}

19. $2 + m = -15$ {−17}

20. $-6 + n = 10$ {16}

21. $-23 = y - 7$ {−16}

22. $-9 = -21 + b$ {12}

23. $4 + c = 4$ {0}

24. $-13 + b = -13$ {0}

25. $4.1 = 2.8 + a$ {1.3}

26. $5.1 = -2.5 + y$ {7.6}

27. $5 = z - \dfrac{1}{2}$

$\left\{\dfrac{11}{2}\right\}$ or $\left\{5\dfrac{1}{2}\right\}$

28. $-7 = p + \dfrac{2}{3}$

$\left\{-\dfrac{23}{3}\right\}$ or $\left\{-7\dfrac{2}{3}\right\}$

29. $x + \dfrac{5}{2} = \dfrac{1}{2}$ {−2}

30. $\dfrac{7}{3} = x - \dfrac{2}{3}$ {3}

31. $-6.02 + c = -8.15$ {−2.13}

32. $p + 0.035 = -1.12$ {−1.155}

33. $3.245 + t = -0.0225$ {−3.2675}

34. $-1.004 + k = 3.0589$ {4.0629}

Concept 3: Multiplication and Division Properties of Equality

For Exercises 35–54, solve each equation using the multiplication or division property of equality. Be sure to check your answers. **(See Examples 4–7.)**

35. $6x = 54$ {9}

36. $2w = 8$ {4}

37. $12 = -3p$ {−4}

38. $6 = -2q$ {−3}

39. $-5y = 0$ {0}

40. $-3k = 0$ {0}

41. $-\dfrac{y}{5} = 3$ {−15}

42. $-\dfrac{z}{7} = 1$ {−7}

43. $\dfrac{4}{5} = -t$ $\left\{-\dfrac{4}{5}\right\}$

 44. $-\dfrac{3}{7} = -h$ $\left\{\dfrac{3}{7}\right\}$

45. $\dfrac{2}{5}a = -4$ {−10}

46. $\dfrac{3}{8}b = -9$ {−24}

47. $-\dfrac{1}{5}b = -\dfrac{4}{5}$ {4}

48. $-\dfrac{3}{10}w = \dfrac{2}{5}$ $\left\{-\dfrac{4}{3}\right\}$

49. $-41 = -x$ {41}

50. $32 = -y$ {−32}

51. $3.81 = -0.03p$ {−127}

 52. $2.75 = -0.5q$ {−5.5}

53. $5.82y = -15.132$ {−2.6}

54. $-32.3x = -0.4522$ {0.014}

Concept 4: Translations

⟵ ⟶ For Exercises 55–66, write an algebraic equation to represent each English sentence. (Let x represent the unknown number.) Then solve the equation. **(See Example 8.)**

 55. The sum of negative eight and a number is forty-two. $-8 + x = 42$; The number is 50.

56. The sum of thirty-one and a number is thirteen. $31 + x = 13$; The number is −18.

57. The difference of a number and negative six is eighteen. $x - (-6) = 18$; The number is 12.

58. The sum of negative twelve and a number is negative fifteen. $-12 + x = -15$; The number is −3.

59. The product of a number and seven is the same as negative sixty-three. $x \cdot 7 = -63$ or $7x = -63$; The number is −9.

60. The product of negative three and a number is the same as twenty-four. $-3x = 24$; The number is −8.

61. The value 3.2 subtracted from a number is 2.1. $x - 3.2 = 2.1$; The number is 5.3.

62. The value −3 subtracted from a number is 4. $x - (-3) = 4$; The number is 1.

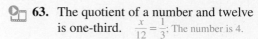 **63.** The quotient of a number and twelve is one-third. $\frac{x}{12}=\frac{1}{3}$; The number is 4.

64. Eighteen is equal to the quotient of a number and two. $18=\frac{x}{2}$; The number is 36.

65. The sum of a number and $\frac{5}{8}$ is $\frac{13}{8}$. $x+\frac{5}{8}=\frac{13}{8}$; The number is 1.

66. The difference of a number and $\frac{2}{3}$ is $\frac{1}{3}$. $x-\frac{2}{3}=\frac{1}{3}$; The number is 1.

Mixed Exercises

For Exercises 67–90, solve each equation using the appropriate property of equality.

67. a. $x-9=1$ {10}

b. $-9x=1$ $\left\{-\frac{1}{9}\right\}$

68. a. $k-2=-4$ {−2}

b. $-2k=-4$ {2}

69. a. $-\frac{2}{3}h=8$ {−12}

b. $\frac{2}{3}+h=8$ $\left\{\frac{22}{3}\right\}$

70. a. $\frac{3}{4}p=15$ {20}

b. $\frac{3}{4}+p=15$ $\left\{\frac{57}{4}\right\}$

71. $\frac{r}{3}=-12$ {−36}

72. $\frac{d}{-4}=5$ {−20}

73. $k+16=32$ {16}

74. $-18=-9+t$ {−9}

75. $16k=32$ {2}

76. $-18=-9t$ {2}

77. $7=-4q$ $\left\{-\frac{7}{4}\right\}$

78. $-3s=10$ $\left\{-\frac{10}{3}\right\}$

79. $-4+q=7$ {11}

80. $s-3=10$ {13}

81. $-\frac{1}{3}d=12$ {−36}

82. $-\frac{2}{5}m=10$ {−25}

83. $4=\frac{1}{2}+z$ $\left\{\frac{7}{2}\right\}$

84. $3=\frac{1}{4}+p$ $\left\{\frac{11}{4}\right\}$

85. $1.2y=4.8$ {4}

86. $4.3w=8.6$ {2}

87. $4.8=1.2+y$ {3.6}

88. $8.6=w-4.3$ {12.9}

89. $0.0034=y-0.405$ {0.4084}

90. $-0.98=m+1.0034$ {−1.9834}

For Exercises 91–98, determine if the equation is a linear equation in one variable. Answer yes or no.

91. $4p+5=0$ Yes

92. $3x-5y=0$ No

93. $4+2a^2=5$ No

94. $-8t=7$ Yes

95. $x-4=9$ Yes

96. $2x^3+y=0$ No

97. $19b=-3$ Yes

98. $13+x=19$ Yes

Expanding Your Skills

For Exercises 99–104, construct an equation with the given solution set. Answers will vary.

99. {6} For example: $y+9=15$

100. {2} For example: $1+x=3$

101. {−4} For example: $2p=-8$

102. {−10} For example: $-4t=40$

103. {0} For example: $5a+5=5$

104. {1} For example: $6k+1=7$

For Exercises 105–108, simplify by collecting the *like* terms. Then solve the equation.

105. $5x-4x+7=8-2$ {−1}

106. $2+3=2y+1-y$ {4}

107. $6p-3p=15+6$ {7}

108. $12-20=2t+2t$ {−2}

Section 2.2 Solving Linear Equations

Concepts

1. Linear Equations Involving Multiple Steps
2. Procedure for Solving a Linear Equation in One Variable
3. Conditional Equations, Identities, and Contradictions

1. Linear Equations Involving Multiple Steps

Previously we studied a one-step process to solve linear equations by using the addition, subtraction, multiplication, and division properties of equality. In Example 1, we solve the equation $-2w - 7 = 11$. Solving this equation will require multiple steps. To understand the proper steps, always remember that the ultimate goal is to isolate the variable. Therefore, we will first isolate the *term* containing the variable before dividing both sides by -2.

Classroom Example: p. 119, Exercise 14

Example 1 Solving a Linear Equation

Solve the equation. $-2w - 7 = 11$

Solution:

$$-2w - 7 = 11$$

$$-2w - 7 + 7 = 11 + 7$$ Add 7 to both sides of the equation. This isolates the w-term.

$$-2w = 18$$

$$\frac{-2w}{-2} = \frac{18}{-2}$$ Next, apply the division property of equality to obtain a coefficient of 1 for w. Divide by -2 on both sides.

$$w = -9$$

Check:

$$-2w - 7 = 11$$

$$-2(-9) - 7 \stackrel{?}{=} 11$$ Substitute $w = -9$ in the original equation.

$$18 - 7 \stackrel{?}{=} 11$$

$$11 \stackrel{?}{=} 11 \checkmark$$ True

The solution set is $\{-9\}$.

Skill Practice Solve the equation.

1. $-5y - 5 = 10$

Classroom Example: p. 119, Exercise 22

Example 2 Solving a Linear Equation

Solve the equation. $2 = \frac{1}{5}x + 3$

Solution:

$$2 = \frac{1}{5}x + 3$$

$$2 - 3 = \frac{1}{5}x + 3 - 3$$ Subtract 3 from both sides. This isolates the x-term.

$$-1 = \frac{1}{5}x$$ Simplify.

Answer

1. $\{-3\}$

$$5(-1) = 5 \cdot \left(\frac{1}{5}x\right)$$

Next, apply the multiplication property of equality to obtain a coefficient of 1 for x.

$$-5 = 1x$$

$$-5 = x$$ Simplify. The answer checks in the original equation.

The solution set is $\{-5\}$.

Skill Practice Solve the equation.

2. $2 = \frac{1}{2}a - 7$

In Example 3, the variable x appears on both sides of the equation. In this case, apply the addition or subtraction property of equality to collect the variable terms on one side of the equation and the constant terms on the other side. Then use the multiplication or division property of equality to get a coefficient equal to 1 on the variable term.

| **Example 3** | **Solving a Linear Equation** |

Classroom Example: p. 119, Exercise 30

Solve the equation. $6x - 4 = 2x - 8$

Solution:

$$6x - 4 = 2x - 8$$

$$6x - 2x - 4 = 2x - 2x - 8$$ Subtract $2x$ from both sides leaving $0x$ on the right-hand side.

$$4x - 4 = 0x - 8$$ Simplify.

$$4x - 4 = -8$$ The x-terms have now been combined on one side of the equation.

$$4x - 4 + 4 = -8 + 4$$ Add 4 to both sides of the equation. This combines the constant terms on the *other* side of the equation.

$$4x = -4$$

$$\frac{4x}{4} = \frac{-4}{4}$$ To obtain a coefficient of 1 for x, divide both sides of the equation by 4.

$$x = -1$$ The answer checks in the original equation.

The solution set is $\{-1\}$.

Skill Practice Solve the equation.

3. $10x - 3 = 4x - 2$

Answers

2. $\{18\}$ **3.** $\left\{\frac{1}{6}\right\}$

TIP: It is important to note that the variable may be isolated on either side of the equation. We will solve the equation from Example 3 again, this time isolating the variable on the right-hand side.

$$6x - 4 = 2x - 8$$

$$6x - 6x - 4 = 2x - 6x - 8 \qquad \text{Subtract } 6x \text{ on both sides.}$$

$$0x - 4 = -4x - 8$$

$$-4 = -4x - 8$$

$$-4 + 8 = -4x - 8 + 8 \qquad \text{Add 8 to both sides.}$$

$$\boxed{4 = -4x}$$

$$\frac{4}{-4} = \frac{-4x}{-4} \qquad \text{Divide both sides by } -4.$$

$$-1 = x \quad \text{or equivalently } x = -1$$

2. Procedure for Solving a Linear Equation in One Variable

In some cases, it is necessary to simplify both sides of a linear equation before applying the properties of equality. Therefore, we offer the following steps to solve a linear equation in one variable.

Solving a Linear Equation in One Variable

Step 1   Simplify both sides of the equation.
- Clear parentheses
- Combine *like* terms

Step 2   Use the addition or subtraction property of equality to collect the variable terms on one side of the equation.

Step 3   Use the addition or subtraction property of equality to collect the constant terms on the other side of the equation.

Step 4   Use the multiplication or division property of equality to make the coefficient of the variable term equal to 1.

Step 5   Check your answer.

Classroom Example: p. 119, Exercise 44

Example 4   Solving a Linear Equation

Solve the equation. $7 + 3 = 2(p - 3)$

Solution:

$$7 + 3 = 2(p - 3)$$

$$10 = 2p - 6 \qquad \textbf{Step 1:} \quad \text{Simplify both sides of the equation by clearing parentheses and combining } like \text{ terms.}$$

$$\textbf{Step 2:} \quad \text{The variable terms are already on one side.}$$

$$10 + 6 = 2p - 6 + 6 \qquad \textbf{Step 3:} \quad \text{Add 6 to both sides to collect the constant terms on the other side.}$$

$$16 = 2p$$

$$\frac{16}{2} = \frac{2p}{2}$$

Step 4: Divide both sides by 2 to obtain a coefficient of 1 for p.

$$8 = p$$

Step 5: Check:

$$7 + 3 = 2(p - 3)$$

$$10 \overset{?}{=} 2(8 - 3)$$

$$10 \overset{?}{=} 2(5)$$

The solution set is $\{8\}$.

$$10 \overset{?}{=} 10 \checkmark \quad \text{True}$$

Skill Practice Solve the equation.

4. $12 + 2 = 7(3 - y)$

Example 5 Solving a Linear Equation

Solve the equation. $2.2y - 8.3 = 6.2y + 12.1$

Classroom Example: p. 119, Exercise 46

Solution:

$$2.2y - 8.3 = 6.2y + 12.1$$

Step 1: The right- and left-hand sides are already simplified.

$$2.2y - 2.2y - 8.3 = 6.2y - 2.2y + 12.1$$
$$-8.3 = 4y + 12.1$$

Step 2: Subtract $2.2y$ from both sides to collect the variable terms on one side of the equation.

$$-8.3 - 12.1 = 4y + 12.1 - 12.1$$
$$-20.4 = 4y$$

Step 3: Subtract 12.1 from both sides to collect the constant terms on the other side.

$$\frac{-20.4}{4} = \frac{4y}{4}$$

Step 4: To obtain a coefficient of 1 for the y-term, divide both sides of the equation by 4.

$$-5.1 = y$$

$$y = -5.1$$

Step 5: Check:

$$2.2y - 8.3 = 6.2y + 12.1$$

$$2.2(-5.1) - 8.3 \overset{?}{=} 6.2(-5.1) + 12.1$$

$$-11.22 - 8.3 \overset{?}{=} -31.62 + 12.1$$

The solution set is $\{-5.1\}$.

$$-19.52 \overset{?}{=} -19.52 \checkmark \text{ True}$$

TIP: In Examples 5 and 6 we collected the variable terms on the right side to avoid negative coefficients on the variable term.

Skill Practice Solve the equation.

5. $1.5t + 2.3 = 3.5t - 1.9$

Answers
4. $\{1\}$ **5.** $\{2.1\}$

Classroom Example: p. 119, Exercise 50

| **Example 6** | **Solving a Linear Equation** |

Solve the equation. $2 + 7x - 5 = 6(x + 3) + 2x$

Solution:

$$2 + 7x - 5 = 6(x + 3) + 2x$$
$$-3 + 7x = 6x + 18 + 2x$$

Step 1: Add *like* terms on the left. Clear parentheses on the right.

$$-3 + 7x = 8x + 18$$

Combine *like* terms.

$$-3 + 7x - 7x = 8x - 7x + 18$$

Step 2: Subtract $7x$ from both sides.

$$-3 = x + 18$$

Simplify.

$$-3 - 18 = x + 18 - 8$$

Step 3: Subtract 18 from both sides.

$$-21 = x$$

Step 4: Because the coefficient of the x term is already 1, there is no need to apply the multiplication or division property of equality.

$$x = -21$$

The solution set is $\{-21\}$.

Step 5: The check is left to the reader.

Skill Practice Solve the equation.

6. $4(2y - 1) + y = 6y + 3 - y$

Classroom Example: p. 119, Exercise 54

| **Example 7** | **Solving a Linear Equation** |

Solve the equation. $9 - (z - 3) + 4z = 4z - 5(z + 2) - 6$

Solution:

$$9 - (z - 3) + 4z = 4z - 5(z + 2) - 6$$
$$9 - z + 3 + 4z = 4z - 5z - 10 - 6$$

Step 1: Clear parentheses.

$$12 + 3z = -z - 16$$

Combine *like* terms.

$$12 + 3z + z = -z + z - 16$$

Step 2: Add z to both sides.

$$12 + 4z = -16$$

$$12 - 12 + 4z = -16 - 12$$

Step 3: Subtract 12 from both sides.

$$4z = -28$$

$$\frac{4z}{4} = \frac{-28}{4}$$

Step 4: Divide both sides by 4.

$$z = -7$$

Step 5: The check is left for the reader.

The solution set is $\{-7\}$.

Skill Practice Solve the equation.

7. $10 - (x + 5) + 3x = 6x - 5(x - 1) - 3$

> **Avoiding Mistakes**
>
> When distributing a negative number through a set of parentheses, be sure to change the signs of every term within the parentheses.

Answers

6. $\left\{\dfrac{7}{4}\right\}$ **7.** $\{-3\}$

3. Conditional Equations, Identities, and Contradictions

The solutions to an equation are the values of x that make the equation a true statement. A linear equation in one variable has one unique solution. Some types of equations, however, have no solution while others have infinitely many solutions.

I. Conditional Equations

An equation that is true for some values of the variable but false for other values is called a **conditional equation**. The equation $x + 4 = 6$, for example, is true on the condition that $x = 2$. For other values of x, the statement $x + 4 = 6$ is false.

$$x + 4 = 6$$
$$x + 4 - 4 = 6 - 4$$
$$x = 2 \quad \text{(Conditional equation)} \quad \text{Solution set: } \{2\}$$

II. Contradictions

Some equations have no solution, such as $x + 1 = x + 2$. There is no value of x, that when increased by 1 will equal the same value increased by 2. If we try to solve the equation by subtracting x from both sides, we get the contradiction $1 = 2$. This indicates that the equation has no solution. An equation that has no solution is called a **contradiction**. The solution set is the empty set. The **empty set** is the set with no elements and is denoted by $\{\ \}$.

> **TIP:** The empty set is also called the null set and can be expressed by the symbol $\varnothing$.

$$x + 1 = x + 2$$
$$x - x + 1 = x - x + 2$$
$$1 = 2 \quad \text{(Contradiction)} \quad \text{Solution set: } \{\ \}$$

III. Identities

An equation that has all real numbers as its solution set is called an **identity**. For example, consider the equation, $x + 4 = x + 4$. Because the left- and right-hand sides are *identical*, any real number substituted for x will result in equal quantities on both sides. If we subtract x from both sides of the equation, we get the identity $4 = 4$. In such a case, the solution is the set of all real numbers.

> **Avoiding Mistakes**
>
> There are two ways to express the empty set: $\{\ \}$ or $\varnothing$. Be sure that you do not use them together. It would be incorrect to write $\{\varnothing\}$.

$$x + 4 = x + 4$$
$$x - x + 4 = x - x + 4$$
$$4 = 4 \quad \text{(Identity)} \qquad \text{Solution set: } \text{The set of real numbers.}$$

Example 8   **Identifying Conditional Equations, Contradictions, and Identities**

Classroom Examples: p. 120, Exercises 60, 62, and 64

Solve the equation. Identify each equation as a conditional equation, a contradiction, or an identity.

a. $4k - 5 = 2(2k - 3) + 1$ **b.** $2(b - 4) = 2b - 7$ **c.** $3x + 7 = 2x - 5$

Solution:

a.
$$4k - 5 = 2(2k - 3) + 1$$
$$4k - 5 = 4k - 6 + 1 \qquad \text{Clear parentheses.}$$
$$4k - 5 = 4k - 5 \qquad \text{Combine } like \text{ terms.}$$
$$4k - 4k - 5 = 4k - 4k - 5 \qquad \text{Subtract } 4k \text{ from both sides.}$$
$$-5 = -5 \quad \text{(Identity)}$$

This is an identity. Solution set: The set of real numbers.

b. $2(b - 4) = 2b - 7$

$\quad\quad 2b - 8 = 2b - 7$ Clear parentheses.

$2b - 2b - 8 = 2b - 2b - 7$ Subtract $2b$ from both sides.

$\quad\quad\quad\quad -8 = -7$ (Contradiction)

This is a contradiction. Solution set: $\{\ \}$

c. $3x + 7 = 2x - 5$

$3x - 2x + 7 = 2x - 2x - 5$ Subtract $2x$ from both sides.

$\quad\quad x + 7 = -5$ Simplify.

$x + 7 - 7 = -5 - 7$ Subtract 7 from both sides.

$\quad\quad\quad x = -12$ (Conditional equation)

This is a conditional equation. The solution set is $\{-12\}$. (The equation is true only on the condition that $x = -12$.)

Skill Practice Solve the equation. Identify the equation as a conditional equation, a contradiction, or an identity.

8. $4(2t + 1) - 1 = 8t + 3$ **9.** $3x - 5 = 4x + 1 - x$ **10.** $6(v - 2) = 2v - 4$

Answers

8. The set of real numbers; identity
9. $\{\ \}$; contradiction
10. $\{2\}$; conditional equation

Section 2.2   Practice Exercises

For additional exercises, see Classroom Activities 2.2A–2.2C in the *Student's Resource Manual* at www.mhhe.com/moh.

Study Skills Exercise

Several strategies are given here about taking notes. Check the activities that you routinely do and discuss how the other suggestions may improve your learning.

_____ Read your notes after class and fill in details.

_____ Highlight important terms and definitions.

_____ Review your notes from the previous class.

_____ Bring pencils (more than one) and paper to class.

_____ Sit in class where you can clearly read the board and hear your instructor.

_____ Turn off your cell phone and keep it off your desk to avoid distraction.

Vocabulary and Key Concepts

1. a. A ___conditional___ equation is true for some values of the variable, but false for other values.

    b. An equation that has no solution is called a ___contradiction___.

    c. The set containing no elements is called the ___empty or null___ set.

    d. An equation that has all real numbers as its solution set is called an ___identity___.

Review Exercises

For Exercises 2–5, simplify each expression by clearing parentheses and combining *like* terms.

2. $5z + 2 - 7z - 3z$ $-5z + 2$ **3.** $10 - 4w + 7w - 2 + w$ $4w + 8$

4. $-(-7p + 9) + (3p - 1)$ $10p - 10$ **5.** $8y - (2y + 3) - 19$ $6y - 22$

Writing   Translating Expression   Geometry   Scientific Calculator   Video

6. Explain the difference between simplifying an expression and solving an equation.
To simplify an expression, clear parentheses and combine *like* terms. To solve an equation, use the addition, subtraction, multiplication, and division properties of equality to isolate the variable.

For Exercises 7–12, solve each equation using the addition, subtraction, multiplication, or division property of equality.

7. $7 = p - 12$ {19}

8. $5w = -30$ {−6}

9. $-7y = 21$ {−3}

10. $x + 8 = -15$ {−23}

11. $z - 23 = -28$ {−5}

12. $-\dfrac{9}{8} = -\dfrac{3}{4}k$ $\left\{\dfrac{3}{2}\right\}$

Concept 1: Linear Equations Involving Multiple Steps

For Exercises 13–36, solve each equation using the steps outlined in the text. **(See Examples 1–3.)**

13. $6z + 1 = 13$ {2}

14. $5x + 2 = -13$ {−3}

15. $3y - 4 = 14$ {6}

16. $-7w - 5 = -19$ {2}

17. $-2p + 8 = 3$ $\left\{\dfrac{5}{2}\right\}$

18. $2b - \dfrac{1}{4} = 5$ $\left\{\dfrac{21}{8}\right\}$

19. $0.2x + 3.1 = -5.3$ {−42}

20. $-1.8 + 2.4a = -6.6$ {−2}

21. $\dfrac{5}{8} = \dfrac{1}{4} - \dfrac{1}{2}p$ $\left\{-\dfrac{3}{4}\right\}$

22. $\dfrac{6}{7} = \dfrac{1}{7} + \dfrac{5}{3}r$ $\left\{\dfrac{3}{7}\right\}$

23. $7w - 6w + 1 = 10 - 4$ {5}

24. $5v - 3 - 4v = 13$ {16}

25. $11h - 8 - 9h = -16$ {−4}

26. $6u - 5 - 8u = -7$ {1}

27. $3a + 7 = 2a - 19$ {−26}

28. $6b - 20 = 14 + 5b$ {34}

29. $-4r - 28 = -58 - r$ {10}

30. $-6x - 7 = -3 - 8x$ {2}

31. $-2z - 8 = -z$ {−8}

32. $-7t + 4 = -6t$ {4}

33. $\dfrac{5}{6}x + \dfrac{2}{3} = -\dfrac{1}{6}x - \dfrac{5}{3}$ $\left\{-\dfrac{7}{3}\right\}$

34. $\dfrac{3}{7}x - \dfrac{1}{4} = -\dfrac{4}{7}x - \dfrac{5}{4}$ {−1}

35. $3y - 2 = 5y - 2$ {0}

36. $4 + 10t = -8t + 4$ {0}

Concept 2: Procedure for Solving a Linear Equation in One Variable

For Exercises 37–58, solve each equation using the steps outlined in the text. **(See Examples 4–7.)**

37. $4q + 14 = 2$ {−3}

38. $6 = 7m - 1$ {1}

39. $-9 = 4n - 1$ {−2}

40. $-\dfrac{1}{2} - 4x = 8$ $\left\{-\dfrac{17}{8}\right\}$

41. $3(2p - 4) = 15$ $\left\{\dfrac{9}{2}\right\}$

42. $4(t + 15) = 20$ {−10}

43. $6(3x + 2) - 10 = -4$ $\left\{-\dfrac{1}{3}\right\}$

44. $4(2k + 1) - 1 = 5$ $\left\{\dfrac{1}{4}\right\}$

45. $3.4x - 2.5 = 2.8x + 3.5$ {10}

46. $5.8w + 1.1 = 6.3w + 5.6$ {−9}

47. $17(s + 3) = 4(s - 10) + 13$ {−6}

48. $5(4 + p) = 3(3p - 1) - 9$ {8}

49. $6(3t - 4) + 10 = 5(t - 2) - (3t + 4)$ {0}

50. $-5y + 2(2y + 1) = 2(5y - 1) - 7$ {1}

51. $5 - 3(x + 2) = 5$ {−2}

52. $1 - 6(2 - h) = 7$ {3}

53. $3(2z - 6) - 4(3z + 1) = 5 - 2(z + 1)$ $\left\{-\dfrac{25}{4}\right\}$

54. $-2(4a + 3) - 5(2 - a) = 3(2a + 3) - 7$ {−2}

55. $-2[(4p + 1) - (3p - 1)] = 5(3 - p) - 9$ $\left\{\dfrac{10}{3}\right\}$ **56.** $5 - (6k + 1) = 2[(5k - 3) - (k - 2)]$ $\left\{\dfrac{3}{7}\right\}$

57. $3(-0.9n + 0.5) = -3.5n + 1.3$ $\{-0.25\}$ **58.** $7(0.4m - 0.1) = 5.2m + 0.86$ $\{-0.65\}$

Concept 3: Conditional Equations, Identities, and Contradictions

For Exercises 59–64, solve each equation. Identify as a conditional equation, an identity, or a contradiction. **(See Example 8.)**

59. $2(k - 7) = 2k - 13$
{ }; contradiction

60. $5h + 4 = 5(h + 1) - 1$
The set of real numbers; identity

61. $7x + 3 = 6(x - 2)$
$\{-15\}$; conditional equation

62. $3y - 1 = 1 + 3y$
{ }; contradiction

63. $3 - 5.2p = -5.2p + 3$
The set of real numbers; identity

64. $2(q + 3) = 4q + q - 9$
$\{5\}$; conditional equation

65. A conditional linear equation has (choose one): one solution, no solution, or infinitely many solutions. One solution

66. An equation that is a contradiction has (choose one): one solution, no solution, or infinitely many solutions. No solution

67. An equation that is an identity has (choose one): one solution, no solution, or infinitely many solutions. Infinitely many solutions

68. If the only solution to a linear equation is 5, then is the equation a conditional equation, an identity, or a contradiction? Conditional equation

Mixed Exercises

For Exercises 69–92, solve each equation.

69. $4p - 6 = 8 + 2p$ $\{7\}$

70. $\dfrac{1}{2}t - 2 = 3$ $\{10\}$

71. $2k - 9 = -8$ $\left\{\dfrac{1}{2}\right\}$

72. $3(y - 2) + 5 = 5$ $\{2\}$

73. $7(w - 2) = -14 - 3w$ $\{0\}$

74. $0.24 = 0.4m$ $\{0.6\}$

75. $2(x + 2) - 3 = 2x + 1$
The set of real numbers

76. $n + \dfrac{1}{4} = -\dfrac{1}{2}$ $\left\{-\dfrac{3}{4}\right\}$

77. $0.5b = -23$ $\{-46\}$

78. $3(2r + 1) = 6(r + 2) - 6$
{ }

79. $8 - 2q = 4$ $\{2\}$

80. $\dfrac{x}{7} - 3 = 1$ $\{28\}$

81. $2 - 4(y - 5) = -4$ $\left\{\dfrac{13}{2}\right\}$

82. $4 - 3(4p - 1) = -8$ $\left\{\dfrac{5}{4}\right\}$

83. $0.4(a + 20) = 6$ $\{-5\}$

84. $2.2r - 12 = 3.4$ $\{7\}$

85. $10(2n + 1) - 6 = 20(n - 1) + 12$
{ }

86. $\dfrac{2}{5}y + 5 = -3$ $\{-20\}$

87. $c + 0.123 = 2.328$ $\{2.205\}$

88. $4(2z + 3) = 8(z - 3) + 36$
The set of real numbers

89. $\dfrac{4}{5}t - 1 = \dfrac{1}{5}t + 5$ $\{10\}$

90. $6g - 8 = 4 - 3g$ $\left\{\dfrac{4}{3}\right\}$

91. $8 - (3q + 4) = 6 - q$ $\{-1\}$

92. $6w - (8 + 2w) = 2(w - 4)$ $\{0\}$

Expanding Your Skills

93. Suppose the solution set for x in the equation $x + a = 10$ is $\{-5\}$. Find the value of a.
$a = 15$

94. Suppose the solution set for x in the equation $x + a = -12$ is $\{6\}$. Find the value of a.
$a = -18$

95. Suppose the solution set for x in the equation $ax = 12$ is $\{3\}$. Find the value of a.
$a = 4$

96. Suppose the solution set for x in the equation $ax = 49.5$ is $\{11\}$. Find the value of a.
$a = 4.5$

97. Write an equation that is an identity.
Answers may vary.
For example: $5x + 2 = 2 + 5x$

98. Write an equation that is a contradiction.
Answers may vary.
For example: $4x - 3 = 4x + 1$

Section 2.3

1. Linear Equations Containing Fractions

Linear equations that contain fractions can be solved in different ways. The first procedure, illustrated here, uses the method previously outlined.

$$\frac{5}{6}x - \frac{3}{4} = \frac{1}{3}$$

$$\frac{5}{6}x - \frac{3}{4} + \frac{3}{4} = \frac{1}{3} + \frac{3}{4} \qquad \text{To isolate the variable term, add } \frac{3}{4} \text{ to both sides.}$$

$$\frac{5}{6}x = \frac{4}{12} + \frac{9}{12} \qquad \text{Find the common denominator on the right-hand side.}$$

$$\frac{5}{6}x = \frac{13}{12} \qquad \text{Simplify.}$$

$$\frac{6}{5}\left(\frac{5}{6}x\right) = \frac{\overset{1}{6}}{5}\left(\frac{13}{\underset{2}{12}}\right) \qquad \text{Multiply by the reciprocal of } \frac{5}{6}, \text{ which is } \frac{6}{5}.$$

$$x = \frac{13}{10} \qquad \text{The solution set is } \left\{\frac{13}{10}\right\}.$$

Sometimes it is simpler to solve an equation with fractions by eliminating the fractions first by using a process called **clearing fractions**. To clear fractions in the equation $\frac{5}{6}x - \frac{3}{4} = \frac{1}{3}$, we can apply the multiplication property of equality to multiply both sides of the equation by the least common denominator (LCD). In this case, the LCD of $\frac{5}{6}x$, $-\frac{3}{4}$, and $\frac{1}{3}$ is 12. Because each denominator in the equation is a factor of 12, we can simplify common factors to leave integer coefficients for each term.

| **Example 1** | **Solving a Linear Equation by Clearing Fractions** |

Solve the equation by clearing fractions first. $\dfrac{5}{6}x - \dfrac{3}{4} = \dfrac{1}{3}$

Solution:

$$\frac{5}{6}x - \frac{3}{4} = \frac{1}{3} \qquad \text{The LCD of } \frac{5}{6}x, -\frac{3}{4}, \text{ and } \frac{1}{3} \text{ is 12.}$$

$$12\left(\frac{5}{6}x - \frac{3}{4}\right) = 12\left(\frac{1}{3}\right) \qquad \text{Multiply both sides of the equation by the LCD, 12.}$$

$$\frac{\overset{2}{12}}{1}\left(\frac{5}{6}x\right) - \frac{\overset{3}{12}}{1}\left(\frac{3}{4}\right) = \frac{\overset{4}{12}}{1}\left(\frac{1}{3}\right) \qquad \text{Apply the distributive property (recall that } 12 = \frac{12}{1}\text{).}$$

$$2(5x) - 3(3) = 4(1) \qquad \text{Simplify common factors to clear the fractions.}$$

$$10x - 9 = 4$$

$$10x - 9 + 9 = 4 + 9 \qquad \text{Add 9 to both sides.}$$

$$10x = 13$$

$$\frac{10x}{10} = \frac{13}{10} \qquad \text{Divide both sides by 10.}$$

$$x = \frac{13}{10} \qquad \text{The solution set is } \left\{\frac{13}{10}\right\}.$$

Concepts

1. **Linear Equations Containing Fractions**
2. **Linear Equations Containing Decimals**

Classroom Example: p. 126, Exercise 18

TIP: Recall that the multiplication property of equality indicates that multiplying both sides of an equation by a nonzero constant results in an equivalent equation.

TIP: The fractions in this equation can be eliminated by multiplying both sides of the equation by *any* common multiple of the denominators. These include 12, 24, 36, 48, and so on. We chose 12 because it is the *least* common multiple.

Skill Practice   Solve the equation by clearing fractions.

1. $\dfrac{2}{5}y + \dfrac{1}{2} = -\dfrac{7}{10}$

In this section, we combine the process for clearing fractions and decimals with the general strategies for solving linear equations. To solve a linear equation, it is important to follow these steps.

Solving a Linear Equation in One Variable

Step 1   Simplify both sides of the equation.
- Clear parentheses
- Consider clearing fractions and decimals (if any are present) by multiplying both sides of the equation by a common denominator of all terms
- Combine *like* terms

Step 2   Use the addition or subtraction property of equality to collect the variable terms on one side of the equation.

Step 3   Use the addition or subtraction property of equality to collect the constant terms on the other side of the equation.

Step 4   Use the multiplication or division property of equality to make the coefficient of the variable term equal to 1.

Step 5   Check your answer.

Classroom Example: p. 126, Exercise 22

Example 2   **Solving a Linear Equation Containing Fractions**

Solve the equation. $\dfrac{1}{6}x - \dfrac{2}{3} = \dfrac{1}{5}x - 1$

Solution:

$$\frac{1}{6}x - \frac{2}{3} = \frac{1}{5}x - 1$$ The LCD of $\frac{1}{6}x$, $-\frac{2}{3}$, $\frac{1}{5}x$, and $\frac{-1}{1}$ is 30.

$$30\left(\frac{1}{6}x - \frac{2}{3}\right) = 30\left(\frac{1}{5}x - 1\right)$$ Multiply by the LCD, 30.

$$\frac{\overset{5}{30}}{1}\cdot\frac{1}{\cancel{6}}x - \frac{\overset{10}{30}}{1}\cdot\frac{2}{\cancel{3}} = \frac{\overset{6}{30}}{1}\cdot\frac{1}{\cancel{5}}x - 30(1)$$ Apply the distributive property (recall $30 = \frac{30}{1}$).

$$5x - 20 = 6x - 30$$ Clear fractions.

$$5x - 6x - 20 = 6x - 6x - 30$$ Subtract $6x$ from both sides.

$$-x - 20 = -30$$

Answers

1. $\{-3\}$

$$-x - 20 + 20 = -30 + 20 \qquad \text{Add 20 to both sides.}$$

$$-x = -10$$

$$\frac{-x}{-1} = \frac{-10}{-1} \qquad \text{Divide both sides by } -1.$$

$$x = 10 \qquad \text{The check is left to the reader.}$$

The solution set is $\{10\}$.

Skill Practice   Solve the equation.

2. $\dfrac{2}{5}x - \dfrac{1}{2} = \dfrac{7}{4} + \dfrac{3}{10}x$

| **Example 3** | **Solving a Linear Equation Containing Fractions** |

Classroom Example: p. 126, Exercise 26

Solve the equation. $\dfrac{1}{3}(x + 7) - \dfrac{1}{2}(x + 1) = 4$

Solution:

$$\frac{1}{3}(x + 7) - \frac{1}{2}(x + 1) = 4$$

$$\frac{1}{3}x + \frac{7}{3} - \frac{1}{2}x - \frac{1}{2} = 4 \qquad \text{Clear parentheses.}$$

$$6\left(\frac{1}{3}x + \frac{7}{3} - \frac{1}{2}x - \frac{1}{2}\right) = 6(4) \qquad \begin{array}{l}\text{The LCD of} \\ \frac{1}{3}x, \frac{7}{3}, -\frac{1}{2}x, -\frac{1}{2}, \text{ and } \frac{4}{1} \text{ is 6.}\end{array}$$

$$\frac{\overset{2}{6}}{1}\cdot\frac{1}{\underset{1}{3}}x + \frac{\overset{2}{6}}{1}\cdot\frac{7}{\underset{1}{3}} + \frac{\overset{3}{6}}{1}\left(-\frac{1}{\underset{1}{2}}x\right) + \frac{\overset{3}{6}}{1}\left(-\frac{1}{\underset{1}{2}}\right) = 6(4) \qquad \begin{array}{l}\text{Apply the distributive} \\ \text{property.}\end{array}$$

$$2x + 14 - 3x - 3 = 24$$

$$-x + 11 = 24 \qquad \text{Combine } like \text{ terms.}$$

$$-x + 11 - 11 = 24 - 11 \qquad \text{Subtract 11.}$$

$$-x = 13$$

$$\frac{-x}{-1} = \frac{13}{-1} \qquad \text{Divide by } -1.$$

$$x = -13 \qquad \begin{array}{l}\text{The check is left to the} \\ \text{reader.}\end{array}$$

The solution set is $\{-13\}$.

TIP: In Example 3 both parentheses and fractions are present within the equation. In such a case, we recommend that you clear parentheses first. Then clear the fractions.

Avoiding Mistakes

When multiplying an equation by the LCD, be sure to multiply all terms on both sides of the equation, including terms that are not fractions.

Skill Practice   Solve the equation.

3. $\dfrac{1}{5}(z + 1) + \dfrac{1}{4}(z + 3) = 2$

Answers

2. $\left\{\dfrac{45}{2}\right\}$ **3.** $\left\{\dfrac{7}{3}\right\}$

Classroom Example: p. 127, Exercise 32

| **Example 4** | Solving a Linear Equation Containing Fractions |

Solve the equation. $\dfrac{x-2}{5} - \dfrac{x-4}{2} = 2$

Solution:

$$\frac{x-2}{5} - \frac{x-4}{2} = \frac{2}{1}$$ The LCD of $\frac{x-2}{5}$, $\frac{x-4}{2}$, and $\frac{2}{1}$ is 10.

$$10\left(\frac{x-2}{5} - \frac{x-4}{2}\right) = 10\left(\frac{2}{1}\right)$$ Multiply both sides by 10.

$$\frac{\overset{2}{\cancel{10}}}{1} \cdot \left(\frac{x-2}{\cancel{5}}\right) - \frac{\overset{5}{\cancel{10}}}{1} \cdot \left(\frac{x-4}{\cancel{2}}\right) = \frac{10}{1} \cdot \left(\frac{2}{1}\right)$$ Apply the distributive property.

$$2(x-2) - 5(x-4) = 20$$ Clear fractions.

$$2x - 4 - 5x + 20 = 20$$ Apply the distributive property.

$$-3x + 16 = 20$$ Simplify both sides of the equation.

$$-3x + 16 - 16 = 20 - 16$$ Subtract 16 from both sides.

$$-3x = 4$$

$$\frac{-3x}{-3} = \frac{4}{-3}$$ Divide both sides by -3.

$$x = -\frac{4}{3}$$ The check is left to the reader.

The solution set is $\left\{-\dfrac{4}{3}\right\}$.

> **Avoiding Mistakes**
>
> In Example 4, several of the fractions in the equation have two terms in the numerator. It is important to enclose these fractions in parentheses when clearing fractions. In this way, we will remember to use the distributive property to multiply the factors shown in blue with both terms from the numerator of the fractions.

Skill Practice Solve the equation.

4. $\dfrac{x+1}{4} + \dfrac{x+2}{6} = 1$

2. Linear Equations Containing Decimals

The same procedure used to clear fractions in an equation can be used to **clear decimals**. For example, consider the equation

$$2.5x + 3 = 1.7x - 6.6$$

Recall that any terminating decimal can be written as a fraction. Therefore, the equation can be interpreted as

$$\frac{25}{10}x + 3 = \frac{17}{10}x - \frac{66}{10}$$

A convenient common denominator of all terms is 10. Therefore, we can multiply the original equation by 10 to clear decimals.

$$10(2.5x + 3) = 10(1.7x - 6.6)$$
$$25x + 30 = 17x - 66$$

Multiplying by the appropriate power of 10 moves the decimal points so that all coefficients become integers.

Answer

4. $\{1\}$

Example 5 Solving a Linear Equation Containing Decimals

Solve the equation by clearing decimals. $2.5x + 3 = 1.7x - 6.6$

Classroom Example: p. 127, Exercise 40

Solution:

$$2.5x + 3 = 1.7x - 6.6$$

$$10(2.5x + 3) = 10(1.7x - 6.6)$$ Multiply both sides of the equation by 10.

$$25x + 30 = 17x - 66$$ Apply the distributive property.

$$25x - 17x + 30 = 17x - 17x - 66$$ Subtract $17x$ from both sides.

$$8x + 30 = -66$$

$$8x + 30 - 30 = -66 - 30$$ Subtract 30 from both sides.

$$8x = -96$$

$$\frac{8x}{8} = \frac{-96}{8}$$ Divide both sides by 8.

$$x = -12$$ The check is left to the reader.

The solution set is $\{-12\}$.

TIP: Notice that multiplying a decimal number by 10 has the effect of moving the decimal point one place to the right. Similarly, multiplying by 100 moves the decimal point two places to the right, and so on.

Skill Practice Solve the equation.

5. $1.2w + 3.5 = 2.1 + w$

Example 6 Solving a Linear Equation Containing Decimals

Solve the equation by clearing decimals. $0.2(x + 4) - 0.45(x + 9) = 12$

Classroom Example: p. 127, Exercise 46

Solution:

$$0.2(x + 4) - 0.45(x + 9) = 12$$

$$0.2x + 0.8 - 0.45x - 4.05 = 12$$ Clear parentheses first.

$$100(0.2x + 0.8 - 0.45x - 4.05) = 100(12)$$ Multiply both sides by 100.

$$20x + 80 - 45x - 405 = 1200$$ Apply the distributive property.

$$-25x - 325 = 1200$$ Simplify both sides.

$$-25x - 325 + 325 = 1200 + 325$$ Add 325 to both sides.

$$-25x = 1525$$

$$\frac{-25x}{-25} = \frac{1525}{-25}$$ Divide both sides by -25.

$$x = -61$$ The check is left to the reader.

The solution set is $\{-61\}$.

TIP: The terms with the most digits following the decimal point are $-0.45x$ and -4.05. Each of these is written to the hundredths place. Therefore, we multiply both sides by 100.

Skill Practice Solve the equation.

6. $0.25(x + 2) - 0.15(x + 3) = 4$

Answers

5. $\{-7\}$ **6.** $\{39.5\}$

Section 2.3   Practice Exercises

For additional exercises, see Classroom Activities 2.3A–2.3B in the *Student's Resource Manual* at www.mhhe.com/moh.

Study Skills Exercise

Instructors vary in what they emphasize on tests. For example, test material may come from the textbook, notes, handouts, or homework. What does your instructor emphasize?

Vocabulary and Key Concepts

1. **a.** The process of eliminating fractions in an equation by multiplying both sides of the equation by the LCD is called __clearing__ __fractions__.

   b. The process of eliminating decimals in an equation by multiplying both sides of the equation by a power of 10 is called __clearing__ __decimals__.

Review Exercises

For Exercises 2–8, solve each equation.

2. $-5t - 17 = -2t + 49$ {−22}

3. $5(x + 2) - 3 = 4x + 5$ {−2}

4. $-2(2x - 4x) = 6 + 18$ {6}

5. $3(2y + 3) - 4(-y + 1) = 7y - 10$ {−5}

6. $-(3w + 4) + 5(w - 2) - 3(6w - 8) = 10$ {0}

7. $7x + 2 = 7(x - 12)$ { }

8. $2(3x - 6) = 3(2x - 4)$ The set of real numbers

Concept 1: Linear Equations Containing Fractions

For Exercises 9–14, determine which of the values could be used to clear fractions or decimals in the given equation.

9. $\dfrac{2}{3}x - \dfrac{1}{6} = \dfrac{x}{9}$

   Values: 6, 9, 12, 18, 24, 36   18, 36

10. $\dfrac{1}{4}x - \dfrac{2}{7} = \dfrac{1}{2}x + 2$

    Values: 4, 7, 14, 21, 28, 42   28

11. $0.02x + 0.5 = 0.35x + 1.2$

    Values: 10; 100; 1000; 10,000

    100; 1000; 10,000

12. $0.003 - 0.002x = 0.1x$

    Values: 10; 100; 1000; 10,000

    1000; 10,000

13. $\dfrac{1}{6}x + \dfrac{7}{10} = x$

    Values: 3, 6, 10, 30, 60

    30, 60

14. $2x - \dfrac{5}{2} = \dfrac{x}{3} - \dfrac{1}{4}$

    Values: 2, 3, 4, 6, 12, 24

    12, 24

For Exercises 15–36, solve each equation. **(See Examples 1–4.)**

15. $\dfrac{1}{2}x + 3 = 5$ {4}

16. $\dfrac{1}{3}y - 4 = 9$ {39}

17. $\dfrac{2}{15}z + 3 = \dfrac{7}{5}$ {−12}

18. $\dfrac{1}{6}y + 2 = \dfrac{5}{12}$ $\left\{-\dfrac{19}{2}\right\}$

19. $\dfrac{1}{3}q + \dfrac{3}{5} = \dfrac{1}{15}q - \dfrac{2}{5}$ $\left\{-\dfrac{15}{4}\right\}$

20. $\dfrac{3}{7}x - 5 = \dfrac{24}{7}x + 7$ {−4}

21. $\dfrac{12}{5}w + 7 = 31 - \dfrac{3}{5}w$ {8}

22. $-\dfrac{1}{9}p - \dfrac{5}{18} = -\dfrac{1}{6}p + \dfrac{1}{3}$ {11}

23. $\dfrac{1}{4}(3m - 4) - \dfrac{1}{5} = \dfrac{1}{4}m + \dfrac{3}{10}$ {3}

24. $\dfrac{1}{25}(20 - t) = \dfrac{4}{25}t - \dfrac{3}{5}$ {7}

25. $\dfrac{1}{6}(5s + 3) = \dfrac{1}{2}(s + 11)$ {15}

26. $\dfrac{1}{12}(4n - 3) = \dfrac{1}{4}(2n + 1)$ {−3}

27. $\dfrac{2}{3}x + 4 = \dfrac{2}{3}x - 6$ { }

28. $-\dfrac{1}{9}a + \dfrac{2}{9} = \dfrac{1}{3} - \dfrac{1}{9}a$
{ }

29. $\dfrac{1}{6}(2c - 1) = \dfrac{1}{3}c - \dfrac{1}{6}$
The set of real numbers

30. $\dfrac{3}{2}b - 1 = \dfrac{1}{8}(12b - 8)$
The set of real numbers

31. $\dfrac{2x + 1}{3} + \dfrac{x - 1}{3} = 5$ {5}

32. $\dfrac{4y - 2}{5} - \dfrac{y + 4}{5} = -3$ {−3}

33. $\dfrac{3w - 2}{6} = 1 - \dfrac{w - 1}{3}$ {2}

34. $\dfrac{z - 7}{4} = \dfrac{6z - 1}{8} - 2$ $\left\{\dfrac{3}{4}\right\}$

35. $\dfrac{x + 3}{3} - \dfrac{x - 1}{2} = 4$ {−15}

36. $\dfrac{5y - 1}{2} - \dfrac{y + 4}{5} = 1$ {1}

Concept 2: Linear Equations Containing Decimals

For Exercises 37–54, solve each equation. **(See Examples 5–6.)**

37. $9.2y - 4.3 = 50.9$
{6}

38. $-6.3x + 1.5 = -4.8$
{1}

39. $0.05z + 0.2 = 0.15z - 10.5$
{107}

40. $21.1w + 4.6 = 10.9w + 35.2$
{3}

41. $0.2p - 1.4 = 0.2(p - 7)$
The set of real numbers

42. $0.5(3q + 87) = 1.5q + 43.5$
The set of real numbers

43. $0.20x + 53.60 = x$
{67}

44. $z + 0.06z = 3816$
{3600}

45. $0.15(90) + 0.05p = 0.1(90 + p)$
{90}

46. $0.25(60) + 0.10x = 0.15(60 + x)$
{120}

47. $0.40(y + 10) - 0.60(y + 2) = 2$
{4}

48. $0.75(x - 2) + 0.25(x + 4) = 0.5$
{1}

49. $0.12x + 3 - 0.8x = 0.22x - 0.6$
{4}

50. $0.4x + 0.2 = -3.6 - 0.6x$
{−3.8}

51. $0.06(x - 0.5) = 0.06x + 0.01$
{ }

52. $0.125x = 0.025(5x + 1)$
{ }

53. $-3.5x + 1.3 = -0.3(9x - 5)$
{−0.25}

54. $x + 4 = 2(0.4x + 1.3)$
{−7}

Mixed Exercises

For Exercises 55–64, solve each equation.

55. $0.2x - 1.8 = -3$ {−6}

56. $9.8h + 2 = 3.8h + 20$ {3}

57. $\dfrac{1}{4}(x + 4) = \dfrac{1}{5}(2x + 3)$ $\left\{\dfrac{8}{3}\right\}$ or $\left\{2\dfrac{2}{3}\right\}$

58. $\dfrac{2}{3}(y - 1) = \dfrac{3}{4}(3y - 2)$ $\left\{\dfrac{10}{19}\right\}$

59. $0.05(2t - 1) - 0.03(4t - 1) = 0.2$ {−11}

60. $0.3(x + 6) - 0.7(x + 2) = 4$ {−9}

61. $\dfrac{2k + 5}{4} = 2 - \dfrac{k + 2}{3}$ $\left\{\dfrac{1}{10}\right\}$

62. $\dfrac{3d - 4}{6} + 1 = \dfrac{d + 1}{8}$ $\left\{-\dfrac{5}{9}\right\}$

63. $\dfrac{1}{8}v + \dfrac{2}{3} = \dfrac{1}{6}v + \dfrac{3}{4}$ {−2}

64. $\dfrac{2}{5}z - \dfrac{1}{4} = \dfrac{3}{10}z + \dfrac{1}{2}$ $\left\{\dfrac{15}{2}\right\}$

Expanding Your Skills

For Exercises 65–68, solve each equation.

65. $\dfrac{1}{2}a + 0.4 = -0.7 - \dfrac{3}{5}a$ {−1}

66. $\dfrac{3}{4}c - 0.11 = 0.23(c - 5)$ {−2}

67. $0.8 + \dfrac{7}{10}b = \dfrac{3}{2}b - 0.8$ {2}

68. $0.78 - \dfrac{1}{25}h = \dfrac{3}{5}h - 0.5$ {2}

Writing   Translating Expression   Geometry   Scientific Calculator   Video

Problem Recognition Exercises

Equations vs. Expressions

For Exercises 1–32, identify each exercise as an expression or an equation. Then simplify the expression or solve the equation.

1. $2b + 23 - 6b - 5$
Expression; $-4b + 18$

2. $10p - 9 + 2p - 3 + 8p - 18$
Expression; $20p - 30$

3. $\dfrac{y}{4} = -2$ Equation; $\{-8\}$

4. $-\dfrac{x}{2} = 7$ Equation; $\{-14\}$

5. $3(4h - 2) - (5h - 8) = 8 - (2h + 3)$ Equation; $\left\{\dfrac{1}{3}\right\}$

6. $7y - 3(2y + 5) = 7 - (10 - 10y)$
Equation; $\left\{-\dfrac{4}{3}\right\}$

7. $3(8z - 1) + 10 - 6(5 + 3z)$
Expression; $6z - 23$

8. $-5(1 - x) - 3(2x + 3) + 5$
Expression; $-x - 9$

9. $6c + 3(c + 1) = 10$
Equation; $\left\{\dfrac{7}{9}\right\}$

10. $-9 + 5(2y + 3) = -7$
Equation; $\left\{-\dfrac{13}{10}\right\}$

11. $0.5(2a - 3) - 0.1 = 0.4(6 + 2a)$
Equation; $\{20\}$

12. $0.07(2v - 4) = 0.1(v - 4)$
Equation; $\{-3\}$

13. $-\dfrac{5}{9}w + \dfrac{11}{12} = \dfrac{23}{36}$ Equation; $\left\{\dfrac{1}{2}\right\}$

14. $\dfrac{3}{8}t - \dfrac{5}{8} = \dfrac{1}{2}t + \dfrac{1}{8}$
Equation; $\{-6\}$

15. $\dfrac{3}{4}x + \dfrac{1}{2} - \dfrac{1}{8}x + \dfrac{5}{4}$
Expression; $\dfrac{5}{8}x + \dfrac{7}{4}$

16. $\dfrac{7}{3}(6 - 12t) + \dfrac{1}{2}(4t + 8)$
Expression; $-26t + 18$

17. $2z - 7 = 2(z - 13)$
Equation; $\{\ \}$

18. $-6x + 2(x + 1) = -2(2x + 3)$
Equation; $\{\ \}$

19. $\dfrac{2x - 1}{4} + \dfrac{3x + 2}{6} = 2$
Equation; $\left\{\dfrac{23}{12}\right\}$

20. $\dfrac{w - 4}{6} - \dfrac{3w - 1}{2} = -1$
Equation; $\left\{\dfrac{5}{8}\right\}$

21. $4b - 8 - b = -3b + 2(3b - 4)$
Equation; The set of real numbers

22. $-k - 41 - 2 - k = -2(20 + k) - 3$
Equation; The set of real numbers

23. $\dfrac{4}{3}(6y - 3) = 0$
Equation; $\left\{\dfrac{1}{2}\right\}$

24. $\dfrac{1}{2}(2c - 4) + 3 = \dfrac{1}{3}(6c + 3)$
Equation; $\{0\}$

25. $3(x + 6) - 7(x + 2) - 4(1 - x)$
Expression; 0

26. $-10(2k + 1) - 4(4 - 5k) + 25$
Expression; -1

27. $3 - 2[4a - 5(a + 1)]$
Expression; $2a + 13$

28. $-9 - 4[3 - 2(q + 3)]$
Expression; $8q + 3$

29. $4 + 2[8 - (6 + x)] = -2(x - 1) - 4 + x$
Equation; $\{10\}$

30. $-1 - 5[2 + 3(w - 2)] = 5(w + 4)$ Equation; $\left\{-\dfrac{1}{20}\right\}$

31. $\dfrac{1}{6}y + y - \dfrac{1}{3}(4y - 1)$
Expression; $-\dfrac{1}{6}y + \dfrac{1}{3}$

32. $\dfrac{1}{2} - \dfrac{1}{5}\left(x + \dfrac{1}{2}\right) + \dfrac{9}{10}x$ Expression; $\dfrac{7}{10}x + \dfrac{2}{5}$

Section 2.4   Applications of Linear Equations: Introduction to Problem Solving

Concepts

1. **Problem-Solving Strategies**
2. **Translations Involving Linear Equations**
3. **Consecutive Integer Problems**
4. **Applications of Linear Equations**

1. Problem-Solving Strategies

Linear equations can be used to solve many real-world applications. However, with "word problems," students often do not know where to start. To help organize the problem-solving process, we offer the following guidelines:

/ Writing   ⇐⇒ Translating Expression   Geometry   Scientific Calculator   Video

Problem-Solving Flowchart for Word Problems

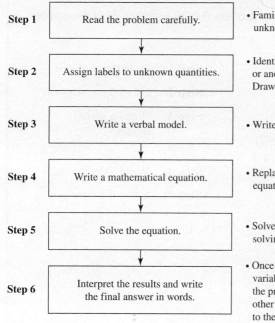

Step 1 | Read the problem carefully. | • Familiarize yourself with the problem. Identify the unknown, and if possible, estimate the answer.

Step 2 | Assign labels to unknown quantities. | • Identify the unknown quantity or quantities. Let x or another variable represent one of the unknowns. Draw a picture and write down relevant formulas.

Step 3 | Write a verbal model. | • Write an equation in *words*.

Step 4 | Write a mathematical equation. | • Replace the verbal model with a mathematical equation using x or another variable.

Step 5 | Solve the equation. | • Solve for the variable using the steps for solving linear equations.

Step 6 | Interpret the results and write the final answer in words. | • Once you have obtained a numerical value for the variable, recall what it represents in the context of the problem. Can this value be used to determine other unknowns in the problem? Write an answer to the word problem in *words*.

Instructor Note: Remind students that x is a convenient variable, but any letter can be used. For example, for an unknown width use w.

Avoiding Mistakes

Once you have reached a solution to a word problem, verify that it is reasonable in the context of the problem.

2. Translations Involving Linear Equations

We have already practiced translating an English sentence to a mathematical equation. Recall that several key words translate to the algebraic operations of addition, subtraction, multiplication, and division.

Example 1   Translating to a Linear Equation

The sum of a number and negative eleven is negative fifteen. Find the number.

Solution:

Let x represent the unknown number.

$$\underset{\text{the sum of}}{\text{(a number)}} + \underset{\text{is}}{(-11)} = (-15)$$

$$x + (-11) = -15$$

$$x + (-11) + 11 = -15 + 11$$

$$x = -4$$

The number is -4.

Step 1: Read the problem.

Step 2: Label the unknown.

Step 3: Write a verbal model.

Step 4: Write an equation.

Step 5: Solve the equation.

Step 6: Write the final answer in words.

Classroom Example: p. 136, Exercise 10

Skill Practice

1. The sum of a number and negative seven is 12. Find the number.

Answer

1. The number is 19.

Classroom Example: p. 136, Exercise 14

Example 2   Translating to a Linear Equation

Forty less than five times a number is fifty-two less than the number. Find the number.

Solution:

	Step 1: Read the problem.
Let x represent the unknown number.	**Step 2:** Label the unknown.

$$\begin{pmatrix} 5 \text{ times} \\ \text{a number} \end{pmatrix} \overset{\text{less}}{-} (40) \overset{\text{is}}{=} \begin{pmatrix} \text{the} \\ \text{number} \end{pmatrix} \overset{\text{less}}{-} (52)$$

Step 3: Write a verbal model.

$$5x \quad - 40 \;=\; x \quad - 52$$

Step 4: Write an equation.

$$5x - 40 = x - 52$$

Step 5: Solve the equation.

$$5x - x - 40 = x - x - 52$$
$$4x - 40 = -52$$
$$4x - 40 + 40 = -52 + 40$$
$$4x = -12$$
$$\frac{4x}{4} = \frac{-12}{4}$$
$$x = -3$$

The number is -3.

Step 6: Write the final answer in words.

> **Avoiding Mistakes**
>
> It is important to remember that subtraction is not a commutative operation. Therefore, the order in which two real numbers are subtracted affects the outcome. The expression "forty less than five times a number" must be translated as: $5x - 40$ (not $40 - 5x$). Similarly, "fifty-two less than the number" must be translated as: $x - 52$ (not $52 - x$).

Skill Practice

2. Thirteen more than twice a number is 5 more than the number. Find the number.

Classroom Example: p. 136, Exercise 18

Example 3   Translating to a Linear Equation

Twice the sum of a number and six is two more than three times the number. Find the number.

Solution:

	Step 1: Read the problem.
Let x represent the unknown number.	**Step 2:** Label the unknown.

$$\overset{\text{twice}}{2} \cdot \overset{\text{the sum}}{(x+6)} \overset{\text{is}}{=} \overset{\text{2 more than}}{3x + 2}$$

$$\underset{\substack{\text{three times} \\ \text{a number}}}{}$$

Step 3: Write a verbal model.

Step 4: Write an equation.

Answer

2. The number is -8.

$$2(x + 6) = 3x + 2$$

Step 5: Solve the equation.

$$2x + 12 = 3x + 2$$

$$2x - 2x + 12 = 3x - 2x + 2$$

$$12 = x + 2$$

$$12 - 2 = x + 2 - 2$$

$$10 = x$$

The number is 10.

Step 6: Write the final answer in words.

> ### Avoiding Mistakes
>
> It is important to enclose "the sum of a number and six" within parentheses so that the entire quantity is multiplied by 2. Forgetting the parentheses would imply that only the x-term is multiplied by 2.
>
> Correct: $2(x + 6)$
>
> Incorrect: $2x + 6$

Skill Practice

3. Three times the sum of a number and eight is 4 more than the number. Find the number.

3. Consecutive Integer Problems

The word *consecutive* means "following one after the other in order without gaps." The numbers 6, 7, and 8 are examples of three **consecutive integers**. The numbers $-4, -2, 0$, and 2 are examples of **consecutive even integers**. The numbers 23, 25, and 27 are examples of **consecutive odd integers**.

Notice that any two consecutive integers differ by 1. Therefore, if x represents an integer, then $(x + 1)$ represents the next larger consecutive integer (Figure 2-4).

> **Instructor Note:** Explain to students that we could let x be the larger of two consecutive integers. Then $x - 1$ is the next smaller integer. However, using x and $x + 1$ generally causes less confusion for students.

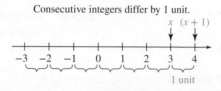

Figure 2-4

Any two consecutive even integers differ by 2. Therefore, if x represents an even integer, then $(x + 2)$ represents the next consecutive larger even integer (Figure 2-5).

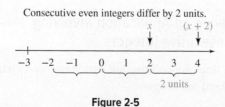

Figure 2-5

Likewise, any two consecutive odd integers differ by 2. If x represents an odd integer, then $(x + 2)$ is the next larger odd integer (Figure 2-6).

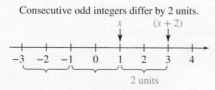

Figure 2-6

Answer

3. The number is -10.

Classroom Example: p. 136,
Exercise 22

Example 4 **Solving an Application Involving Consecutive Integers**

The sum of two consecutive odd integers is −188. Find the integers.

Solution:

In this example we have two unknown integers. We can let x represent either of the unknowns.

	Step 1: Read the problem.
Suppose x represents the first odd integer.	**Step 2:** Label the unknowns.

Then $(x + 2)$ represents the second odd integer.

$$\left(\begin{array}{c}\text{First}\\\text{integer}\end{array}\right) + \left(\begin{array}{c}\text{second}\\\text{integer}\end{array}\right) = (\text{total}) \qquad \textbf{Step 3:} \text{ Write a verbal model.}$$

$$x \quad + \quad (x+2) \quad = -188 \qquad \textbf{Step 4:} \text{ Write a mathematical equation.}$$

$$x + (x + 2) = -188$$

$$2x + 2 = -188 \qquad \textbf{Step 5:} \text{ Solve for } x.$$

$$2x + 2 - 2 = -188 - 2$$

$$2x = -190$$

$$\frac{2x}{2} = \frac{-190}{2}$$

$$x = -95$$

The first integer is $x = -95$. **Step 6:** Interpret the results and write the answer in words.

The second integer is $x + 2 = -95 + 2 = -93$.

The two integers are −95 and −93.

TIP: With word problems, it is advisable to check that the answer is reasonable.

The numbers −95 and −93 are consecutive odd integers. Furthermore, their sum is −188 as desired.

Skill Practice

4. The sum of two consecutive even integers is 66. Find the integers.

Classroom Example: p. 136,
Exercise 30

Example 5 **Solving an Application Involving Consecutive Integers**

Ten times the smallest of three consecutive integers is twenty-two more than three times the sum of the integers. Find the integers.

Solution:

	Step 1: Read the problem.
Let x represent the first integer.	**Step 2:** Label the unknowns.

$x + 1$ represents the second consecutive integer.

$x + 2$ represents the third consecutive integer.

Answer

4. The integers are 32 and 34.

$$\begin{pmatrix} 10 \text{ times} \\ \text{the first} \\ \text{integer} \end{pmatrix} = \begin{pmatrix} 3 \text{ times} \\ \text{the sum of} \\ \text{the integers} \end{pmatrix} + 22$$

Step 3: Write a verbal model.

10 times
the first   is   3 times
integer

22 more
than

$$10x = 3\underbrace{[(x) + (x + 1) + (x + 2)]}_{\text{the sum of the integers}} + 22$$

Step 4: Write a mathematical equation.

$$10x = 3(x + x + 1 + x + 2) + 22$$

Step 5: Solve the equation.

$$10x = 3(3x + 3) + 22$$

Clear parentheses.

$$10x = 9x + 9 + 22$$

Combine *like* terms.

$$10x = 9x + 31$$

$$10x - 9x = 9x - 9x + 31$$

Isolate the x-terms on one side.

$$x = 31$$

The first integer is $x = 31$.

Step 6: Interpret the results and write the answer in words.

The second integer is $x + 1 = 31 + 1 = 32$.

The third integer is $x + 2 = 31 + 2 = 33$.

The three integers are 31, 32, and 33.

Skill Practice

5. Five times the smallest of three consecutive integers is 17 less than twice the sum of the integers. Find the integers.

---

4. Applications of Linear Equations

Example 6 **Using a Linear Equation in an Application**

A carpenter cuts a 6-ft board in two pieces. One piece must be three times as long as the other. Find the length of each piece.

Classroom Example: p. 137, Exercise 32

Solution:

In this problem, one piece must be three times as long as the other. Thus, if x represents the length of one piece, then $3x$ can represent the length of the other.

Step 1: Read the problem completely.

x represents the length of the smaller piece.
$3x$ represents the length of the longer piece.

Step 2: Label the unknowns. Draw a figure.

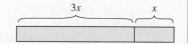

Answer

5. The integers are 11, 12, and 13.

$$\begin{pmatrix} \text{Length of} \\ \text{one piece} \end{pmatrix} + \begin{pmatrix} \text{length of} \\ \text{other piece} \end{pmatrix} = \begin{pmatrix} \text{total length} \\ \text{of the board} \end{pmatrix}$$

Step 3:   Write a verbal model.

$$x \quad + \quad 3x \quad = \quad 6$$

Step 4:   Write an equation.

$$4x = 6$$

Step 5:   Solve the equation.

$$\frac{4x}{4} = \frac{6}{4}$$

$$x = 1.5$$

TIP: The variable can represent either unknown. In Example 6, if we let x represent the length of the longer piece of board, then $\frac{1}{3}x$ would represent the length of the smaller piece. The equation would become $x + \frac{1}{3}x = 6$. Try solving this equation and interpreting the result.

The smaller piece is $x = 1.5$ ft.

Step 6:   Interpret the results.

The longer piece is $3x$ or $3(1.5 \text{ ft}) = 4.5$ ft.

Skill Practice

6. A plumber cuts a 96-in. piece of pipe into two pieces. One piece is five times longer than the other piece. How long is each piece?

Classroom Example: p. 137, Exercise 38

Example 7   **Using a Linear Equation in an Application**

In a recent Olympics, the United States won the greatest number of overall medals, followed by China. The United States won 16 more medals than China, and together they brought home a total of 192 medals. How many medals did each country win?

Solution:

Source: U.S. Department of Defense

In this example, we have two unknowns. The variable x can represent either quantity. However, the number of medals won by the United States is given in terms of the number won by China.

Step 1:   Read the problem.

Let x represent the number of medals won by China.

Step 2:   Label the variables.

Then let $x + 16$ represent the number of medals won by the United States.

$$\begin{pmatrix} \text{Number of} \\ \text{medals won} \\ \text{by China} \end{pmatrix} + \begin{pmatrix} \text{Number of medals} \\ \text{won by the} \\ \text{United States} \end{pmatrix} = \begin{pmatrix} \text{Total} \\ \text{number} \\ \text{of medals} \end{pmatrix}$$

Step 3:   Write a verbal model.

$$x \quad + \quad (x + 16) \quad = \quad 192$$

Step 4:   Write an equation.

$$2x + 16 = 192$$

Step 5:   Solve the equation.

$$2x = 176$$

$$x = 88$$

- Medals won by China, $x = 88$
- Medals won by the United States, $x + 16 = (88) + 16 = 104$

China won 88 medals and the United States won 104 medals.

Answer

6. One piece is 80 in. and the other is 16 in.

Skill Practice

7. There are 40 students in an algebra class. There are 4 more women than men. How many women and how many men are in the class?

Answer

7. There are 22 women and 18 men.

Section 2.4 Practice Exercises

For additional exercises, see Classroom Activities 2.4A–2.4C in the *Student's Resource Manual* at www.mhhe.com/moh.

Study Skills Exercise

After doing a section of homework, check the answers to the odd-numbered exercises in the back of the text. Choose a method to identify the exercises that gave you trouble (i.e., circle the number or put a star by the number). List some reasons why it is important to label these problems.

Vocabulary and Key Concepts

1. a. Integers that follow one after the other without "gaps" are called ____consecutive____ integers.

   b. The integers −2, 0, 2, and 4 are examples of consecutive ____even____ integers.

   c. The integers −3, −1, 1, and 3 are examples of consecutive ____odd____ integers.

   d. Two consecutive integers differ by ____1____.

   e. Two consecutive odd integers differ by ____2____.

   f. Two consecutive even integers differ by ____2____.

Concept 2: Translations Involving Linear Equations

For Exercises 2–8, write an expression representing the unknown quantity.

2. In a math class, the number of students who received an "A" in the class was 5 more than the number of students who received a "B." If x represents the number of "B" students, write an expression for the number of "A" students. $x + 5$

3. There are 5,682,080 fewer men than women on a particular social media site. If x represents the number of women using that site, write an expression for the number of men using that site. $x - 5,682,080$

4. At a recent motorcycle rally, the number of men exceeded the number of women by 216. If x represents the number of women, write an expression for the number of men. $x + 216$

5. There are 10 times as many users of a social media site than there are of a social news site. If x represents the number of users of the news site, write an expression for the number of users of the social media site. $10x$

6. Rebecca downloaded twice as many songs as Nigel. If x represents the number of songs downloaded by Nigel, write an expression for the number downloaded by Rebecca. $2x$

7. Sidney made $20 less than three times Casey's weekly salary. If x represents Casey's weekly salary, write an expression for Sidney's weekly salary. $3x - 20$

8. David scored 26 points less than twice the number of points Rich scored in a video game. If x represents the number of points scored by Rich, write an expression representing the number of points scored by David. $2x - 26$

© McGraw-Hill Education/ Mark Dierker

Writing   Translating Expression   Geometry   Scientific Calculator   Video

For Exercises 9–18, use the Problem-Solving Flowchart for Word Problems. **(See Examples 1–3.)**

9. Six less than a number is −10. Find the number.
The number is −4.

10. Fifteen less than a number is 41. Find the number.   The number is 56.

11. Twice the sum of a number and seven is eight. Find the number.   The number is −3.

12. Twice the sum of a number and negative two is sixteen. Find the number.   The number is 10.

13. A number added to five is the same as twice the number. Find the number.
The number is 5.

14. Three times a number is the same as the difference of twice the number and seven. Find the number.   The number is −7.

15. The sum of six times a number and ten is equal to the difference of the number and fifteen. Find the number.   The number is −5.

16. The difference of fourteen and three times a number is the same as the sum of the number and negative ten. Find the number.   The number is 6.

17. If the difference of a number and four is tripled, the result is six less than the number. Find the number.   The number is 3.

18. Twice the sum of a number and eleven is twenty-two less than three times the number. Find the number.   The number is 44.

Concept 3: Consecutive Integer Problems

19. a. If x represents the smallest of three consecutive integers, write an expression to represent each of the next two consecutive integers.
$x + 1, x + 2$

20. a. If x represents the smallest of three consecutive odd integers, write an expression to represent each of the next two consecutive odd integers.
$x + 2, x + 4$

b. If x represents the largest of three consecutive integers, write an expression to represent each of the previous two consecutive integers.
$x − 1, x − 2$

b. If x represents the largest of three consecutive odd integers, write an expression to represent each of the previous two consecutive odd integers. $x − 2, x − 4$

For Exercises 21–30, use the Problem-Solving Flowchart for Word Problems. **(See Examples 4–5.)**

21. The sum of two consecutive integers is −67. Find the integers.   The integers are −34 and −33.

22. The sum of two consecutive odd integers is 52. Find the integers.   The integers are 25 and 27.

23. The sum of two consecutive odd integers is 28. Find the integers.   The integers are 13 and 15.

24. The sum of three consecutive even integers is 66. Find the integers.   The integers are 20, 22, and 24.

25. The perimeter of a pentagon (a five-sided polygon) is 80 in. The five sides are represented by consecutive integers. Find the lengths of the sides.
The sides are 14 in., 15 in., 16 in., 17 in., and 18 in.

26. The perimeter of a triangle is 96 in. The lengths of the sides are represented by consecutive integers. Find the lengths of the sides.   The sides are 31 in., 32 in., and 33 in.

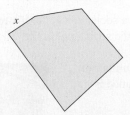

27. The sum of three consecutive even integers is 48 less than twice the smallest of the three integers. Find the integers.   The integers are −54, −52, and −50.

28. The sum of three consecutive odd integers is 89 less than twice the largest integer. Find the integers.   The integers are −87, −85, and −83.

29. Eight times the sum of three consecutive odd integers is 210 more than ten times the middle integer. Find the integers.

30. Five times the sum of three consecutive even integers is 140 more than ten times the smallest of the integers. Find the integers.   The integers are 22, 24, and 26.

Concept 4: Applications of Linear Equations

For Exercises 31–42, use the Problem-Solving Flowchart for Word Problems to solve the problems.

31. A board is 86 cm in length and must be cut so that one piece is 20 cm longer than the other piece. Find the length of each piece. **(See Example 6.)** The lengths of the pieces are 33 cm and 53 cm.

x $x + 20$

32. A rope is 54 in. in length and must be cut into two pieces. If one piece must be twice as long as the other, find the length of each piece. The lengths of the pieces are 18 in. and 36 in.

x $2x$

33. Karen's music library contains 12 fewer playlists than Clarann's music library. The total number of playlists for both music libraries is 58. Find the number of playlists in each person's music library. Karen's music library has 23 playlists and Claran's library has 35 playlists.

34. Maria has 15 fewer apps on her phone than Orlando. If the total number of apps on both phones is 29, how many apps are on each phone? Maria has 7 apps and Orlando has 22 apps.

35. For a recent year, 31 more Democrats than Republicans were in the U.S. House of Representatives. If the total number of representatives in the House from these two parties was 433, find the number of representatives from each party. There were 201 Republicans and 232 Democrats.

36. For a recent year, the number of men in the U.S. Senate totaled 4 more than five times the number of women. Find the number of men and the number of women in the Senate given that the Senate has 100 members. There were 84 men and 16 women.

37. A car dealership sells SUVs and passenger cars. For a recent year, 40 more SUVs were sold than passenger cars. If a total of 420 vehicles were sold, determine the number of each type of vehicle sold. **(See Example 7.)** There were 190 passenger cars and 230 SUVs sold.

38. Two of the largest Internet retailers are eBay and Amazon. Recently, the estimated U.S. sales of eBay were $0.1 billion less than twice the sales of Amazon. Given the total sales of $5.6 billion, determine the sales of eBay and Amazon. eBay's sales were $3.7 billion and Amazon's sales were $1.9 billion.

39. The longest river in Africa is the Nile. It is 2455 km longer than the Congo River, also in Africa. The sum of the lengths of these rivers is 11,195 km. What is the length of each river? The Congo River is 4370 km long, and the Nile River is 6825 km.

Nile

Congo

40. The average depth of the Gulf of Mexico is three times the depth of the Red Sea. The difference between the average depths is 1078 m. What is the average depth of the Gulf of Mexico and the average depth of the Red Sea? The average depth of the Red Sea is 539 m and that of the Gulf of Mexico is 1617 m.

41. Asia and Africa are the two largest continents in the world. The land area of Asia is approximately 14,514,000 km² larger than the land area of Africa. Together their total area is 74,644,000 km². Find the land area of Asia and the land area of Africa. The area of Africa is 30,065,000 km². The area of Asia is 44,579,000 km².

42. Mt. Everest, the highest mountain in the world, is 2654 m higher than Mt. McKinley, the highest mountain in the United States. If the sum of their heights is 15,042 m, find the height of each mountain. Mt. McKinley is 6194 m high. Mt. Everest is 8848 m high.

© J. Luke/PhotoLink/Getty Images RF

Mixed Exercises

43. A group of hikers walked from Hawk Mt. Shelter to Blood Mt. Shelter along the Appalachian Trail, a total distance of 20.5 mi. It took 2 days for the walk. The second day the hikers walked 4.1 mi less than they did on the first day. How far did they walk each day? They walked 12.3 mi on the first day and 8.2 mi on the second.

© Corbis/age fotostock RF

Writing   Translating Expression   Geometry   Scientific Calculator   Video

44. $120 is to be divided among three restaurant servers. Angie made $10 more than Marie. Gwen, who went home sick, made $25 less than Marie. How much money should each server get?   Marie made $45, Angie made $55, and Gwen made $20.

© Comstock Images/
Masterfile RF

45. A 4-ft piece of PVC pipe is cut into three pieces. The longest piece is 5 in. shorter than three times the shortest piece. The middle piece is 8 in. longer than the shortest piece. How long is each piece?   The pieces are 9 in., 17 in., and 22 in.

46. A 6-ft piece of copper wire must be cut into three pieces. The shortest piece is 16 in. less than the middle piece. The longest piece is twice as long as the middle piece. How long is each piece?   The pieces are 6 in., 22 in., and 44 in.

47. Three consecutive integers are such that three times the largest exceeds the sum of the two smaller integers by 47. Find the integers.   The integers are 42, 43, and 44.

48. Four times the smallest of three consecutive odd integers is 236 more than the sum of the other two integers. Find the integers.   The integers are 121, 123, and 125.

49. The winner and runner-up of a TV music contest had lucrative earnings immediately after the show's finale. The runner-up earned $2 million less than half of the winner's earnings. If their combined earnings totaled $19 million, how much did each person make?   The winner earned $14 million and the runner-up earned $5 million.

50. One TV series ran 97 fewer episodes than twice the number of a second TV series. If the total number of episodes is 998, determine the number of each show produced.   There were 633 episodes of the first series and 365 episodes of the second series.

51. Five times the difference of a number and three is four less than four times the number. Find the number.   The number is 11.

52. Three times the difference of a number and seven is one less than twice the number. Find the number.   The number is 20.

53. The sum of the page numbers on two facing pages in a book is 941. What are the page numbers?   The page numbers are 470 and 471.

54. Three raffle tickets are represented by three consecutive integers. If the sum of the three integers is 2,666,031, find the numbers.   The ticket numbers are 888,676; 888,677; and 888,678.

55. If three is added to five times a number, the result is forty-three more than the number. Find the number.   The number is 10.

56. If seven is added to three times a number, the result is thirty-one more than the number. Find the number.   The number is 12.

57. The deepest point in the Pacific Ocean is 676 m more than twice the deepest point in the Arctic Ocean. If the deepest point in the Pacific is 10,920 m, how many meters is the deepest point in the Arctic Ocean?   The deepest point in the Arctic Ocean is 5122 m.

58. The area of Greenland is 201,900 km^2 less than three times the area of New Guinea. What is the area of New Guinea if the area of Greenland is 2,175,600 km^2?   The area of New Guinea is 792,500 km^2.

59. The sum of twice a number and $\frac{3}{4}$ is the same as the difference of four times the number and $\frac{1}{8}$. Find the number.   The number is $\frac{7}{16}$.

60. The difference of a number and $-\frac{11}{12}$ is the same as the difference of three times the number and $\frac{1}{6}$. Find the number.   The number is $\frac{13}{24}$.

61. The product of a number and 3.86 is equal to 7.15 more than the number. Find the number.   The number is 2.5.

62. The product of a number and 4.6 is 33.12 less than the number. Find the number.   The number is −9.2.

 Writing   Translating Expression   Geometry   Scientific Calculator   Video

Applications Involving Percents

1. Basic Percent Equations

Recall that the word *percent* as meaning "per hundred."

Concepts

1. **Basic Percent Equations**
2. **Applications Involving Simple Interest**
3. **Applications Involving Discount and Markup**

Percent	Interpretation
63% of homes have a computer	63 out of 100 homes have a computer.
5% sales tax	5¢ in tax is charged for every 100¢ in merchandise.
15% commission	$15 is earned in commission for every $100 sold.

Percents come up in a variety of applications in day-to-day life. Many such applications follow the basic percent equation:

$$\text{Amount} = (\text{percent})(\text{base}) \qquad \text{Basic percent equation}$$

In Example 1, we apply the basic percent equation to compute sales tax.

Example 1 **Computing Sales Tax**

A new digital camera costs $429.95.

a. Compute the sales tax if the tax rate is 4%.

b. Determine the total cost, including tax.

Solution:

Classroom Example: p. 144, Exercise 18

© BrandX/Punchstock RF

a. Let x represent the amount of tax.	**Step 1:** Read the problem.
	Step 2: Label the variable.
Amount = (percent) · (base)	**Step 3:** Write a verbal model. Apply the percent equation to compute sales tax.
Sales tax = (tax rate)(price of merchandise)	
$x = (0.04)(\$429.95)$	**Step 4:** Write a mathematical equation.
$x = \$17.198$	**Step 5:** Solve the equation.
$x = \$17.20$	Round to the nearest cent.
The tax on the merchandise is $17.20.	**Step 6:** Interpret the results.

Avoiding Mistakes

Be sure to use the decimal form of a percent within an equation.

$$4\% = 0.04$$

b. The total cost is found by:

total cost = cost of merchandise + amount of tax

Therefore, the total cost is $429.95 + $17.20 = $447.15.

In Example 2, we solve a problem in which the percent is unknown.

Example 2   **Finding an Unknown Percent**

A group of 240 college men were asked what intramural sport they most enjoyed playing. The results are in the graph. What percent of the men surveyed preferred tennis?

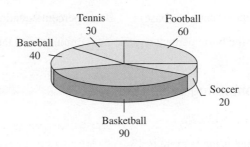

Solution:

	Step 1: Read the problem.
Let x represent the unknown percent (in decimal form).	**Step 2:** Label the variable.

The problem can be rephrased as:

$$30 \text{ is what percent of } 240?$$

$$30 = x \cdot 240$$ **Step 3:** Write a verbal model.

Step 4: Write a mathematical equation.

$$30 = 240x$$ **Step 5:** Solve the equation.

$$\frac{30}{240} = \frac{240x}{240}$$ Divide both sides by 240.

$$0.125 = x$$

$$0.125 \times 100\% = 12.5\%$$

Step 6: Interpret the results. Change the value of x to a percent form by multiplying by 100%.

In this survey, 12.5% of men prefer tennis.

Skill Practice   Refer to the graph in Example 2.

3. What percent of the men surveyed prefer basketball as their favorite intramural sport?

Answers

1. The amount of tax is $5.34.
2. The total cost is $94.34.
3. 37.5% of the men surveyed prefer basketball.

Example 3	Solving a Percent Equation with an Unknown Base

Andrea spends 20% of her monthly paycheck on rent each month. If her rent payment is $950, what is her monthly paycheck?

Solution:

Step 1: Read the problem.

Let x represent the amount of Andrea's monthly paycheck.

Step 2: Label the variables.

The problem can be rephrased as:

$950 is 20% of what number?
↓   ↓   ↓   ↓   ↓

Step 3: Write a verbal model.

$950 = 0.20 \cdot x$

Step 4: Write a mathematical equation.

$950 = 0.20x$

Step 5: Solve the equation.

$$\frac{950}{0.20} = \frac{0.20x}{0.20}$$

Divide both sides by 0.20.

$4750 = x$

Andrea's monthly paycheck is $4750.

Step 6: Interpret the results.

© Stockbyte/Getty Images RF

Skill Practice

4. In order to pass an exam, a student must answer 70% of the questions correctly. If answering 42 questions correctly results in a 70% score, how many questions are on the test?

2. Applications Involving Simple Interest

One important application of percents is in computing simple interest on a loan or on an investment.

Simple interest is interest that is earned or owed on principal (the original amount of money invested or borrowed). The following formula is used to compute simple interest.

$$\left(\begin{array}{c}\text{Simple}\\\text{interest}\end{array}\right) = (\text{principal})\left(\begin{array}{c}\text{annual}\\\text{interest rate}\end{array}\right)\left(\begin{array}{c}\text{time}\\\text{in years}\end{array}\right)$$

This formula is often written symbolically as $I = Prt$. In this formula, I represents the simple interest, P represents the principal, r represents the annual interest rate, and t is the time of the investment in years.

For example, to find the simple interest earned on $2000 invested at 7.5% interest for 3 years, we have $P = \$2000$, $r = 0.075$, and $t = 3$. Thus,

$$I = Prt$$
$$\text{Interest} = (\$2000)(0.075)(3)$$
$$= \$450$$

Answer

4. There are 60 questions on the test.

Classroom Example: p. 144, Exercise 28

© Ingram Publishing RF

Avoiding Mistakes

The interest is computed on the original principal, P, not on the total amount $20,250. That is, the interest is $P(0.025)(5)$, not $(\$20,250)(0.025)(5)$.

Example 4   Applying Simple Interest

Jorge wants to save money to buy a car in 5 years. If Jorge needs to have $20,250 at the end of 5 years, how much money would he need to invest in a certificate of deposit (CD) at a 2.5% interest rate?

Solution:

	Step 1: Read the problem.
Let P represent the original amount invested.	**Step 2:** Label the variables.

$$\left(\begin{array}{c}\text{Original}\\\text{principal}\end{array}\right) + (\text{interest}) = (\text{total})$$ **Step 3:** Write a verbal model.

$$\begin{array}{ccccl} P & + & Prt & = & \text{total} \end{array}$$ **Step 4:** Write a mathematical equation.

$$P + P(0.025)(5) = 20{,}250$$

$$P + 0.125P = 20{,}250$$ **Step 5:** Solve the equation.

$$1.125P = 20{,}250$$

$$\frac{1.125P}{1.125} = \frac{20{,}250}{1.125}$$

$$P = 18{,}000$$

The original investment should be $18,000.   **Step 6:** Interpret the results and write the answer in words.

Skill Practice

5. Cassandra invested some money in her bank account, and after 10 years at 4% simple interest, it has grown to $7700. What was the initial amount invested?

---

3. Applications Involving Discount and Markup

Applications involving percent increase and percent decrease are abundant in many real-world settings. Sales tax, for example, is essentially a markup by a state or local government. It is important to understand that percent increase or decrease is always computed on the original amount given.

In Example 5, we illustrate an example of percent decrease in an application where merchandise is discounted.

Classroom Example: p. 145, Exercise 36

Example 5   Applying Percents to a Discount Problem

After a 38% discount, a used treadmill costs $868 on eBay. What was the original cost of the treadmill?

Solution:

	Step 1: Read the problem.
Let x be the original cost of the treadmill.	**Step 2:** Label the variables.

Answer

5. The initial investment was $5500.

$$\begin{pmatrix}\text{Original}\\\text{cost}\end{pmatrix} - (\text{discount}) = \begin{pmatrix}\text{sale}\\\text{price}\end{pmatrix}$$

$$x \quad - \quad 0.38(x) \quad = \quad 868$$

Step 3: Write a verbal model.

Step 4: Write a mathematical equation. The discount is a percent of the *original* amount.

$$x - 0.38x = 868$$
$$0.62x = 868$$
$$\frac{0.62x}{0.62} = \frac{868}{0.62}$$
$$x = 1400$$

Step 5: Solve the equation.

Combine *like* terms.

Divide by 0.62.

The original cost of the treadmill was $1400.

Step 6: Interpret the result.

© Comstock Images/Alamy RF

Skill Practice

6. A camera is on sale for $151.20. This is after a 20% discount. What was the original cost?

Answer

6. The camera originally cost $189.

Study Skills Exercise

For additional exercises, see Classroom Activities 2.5A–2.5D in the *Student's Resource Manual* at www.mhhe.com/moh.

> It is always helpful to read the material in a section and make notes before it is presented in class. Writing notes ahead of time will free you to listen more in class and to pay special attention to the concepts that need clarification. Refer to your class syllabus and identify the next two sections that will be covered in class. Then determine a time when you can read these sections before class.

Vocabulary and Key Concepts

1. a. Interest that is earned on principal is called ___simple___ interest.

   b. 82% means 82 out of ___100___.

Review Exercises

For Exercises 2–4, use the steps for problem solving to find the unknown quantities.

2. The difference of four times a number and 17 is 5 less than the number. Find the number.
   The number is 4.

3. Find two consecutive integers such that three times the larger is the same as 45 more than the smaller.
   The numbers are 21 and 22.

4. The height of the Great Pyramid of Giza is 17 m more than twice the height of the pyramid found in Saqqara. If the difference in their heights is 77 m, find the height of each pyramid.
   The pyramid at Saqqara is 60 m high, and the Great Pyramid is 137 m high.

 Writing   Translating Expression   Geometry   Scientific Calculator    Video

Concept 1: Basic Percent Equations

For Exercises 5–16, find the missing values.

5. 45 is what percent of 360?   12.5%

6. 338 is what percent of 520?   65%

 7. 544 is what percent of 640?   85%

8. 576 is what percent of 800?   72%

9. What is 0.5% of 150?   0.75

 10. What is 9.5% of 616?   58.52

11. What is 142% of 740?   1050.8

12. What is 156% of 280?   436.8

 13. 177 is 20% of what number?   885

14. 126 is 15% of what number?   840

15. 275 is 12.5% of what number?   2200

16. 594 is 45% of what number?   1320

17. A drill is on sale for $99.99. If the sales tax rate is 7%, how much will Molly have to pay for the drill? **(See Example 1.)**   Molly will have to pay $106.99.

18. Patrick purchased four new tires that were regularly priced at $94.99 each, but are on sale for $20 off per tire. If the sales tax rate is 6%, how much will be charged to Patrick's VISA card?   Patrick will have to pay $317.96.

For Exercises 19–22, use the graph showing the distribution for leading forms of cancer in men. (*Source:* Centers for Disease Control)

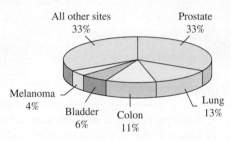

Percent of Cancer Cases by Type (Men)

19. If there are 700,000 cases of cancer in men in the United States, approximately how many are prostate cancer?
Approximately 231,000 cases

20. Approximately how many cases of lung cancer would be expected in 700,000 cancer cases among men in the United States?
Approximately 91,000 cases

21. There were 14,000 cases of cancer of the pancreas diagnosed out of 700,000 cancer cases. What percent is this? **(See Example 2.)**   2%

22. There were 21,000 cases of leukemia diagnosed out of 700,000 cancer cases. What percent is this?   3%

23. Javon is in a 28% tax bracket for his federal income tax. If the amount of money that he paid for federal income tax was $23,520, what was his taxable income? **(See Example 3.)**
Javon's taxable income was $84,000.

24. In a recent survey of college-educated adults, 155 indicated that they regularly work more than 50 hr a week. If this represents 31% of those surveyed, how many people were in the survey?
There were 500 people in the survey.

Concept 2: Applications Involving Simple Interest

25. Aidan is trying to save money and has $1800 to set aside in some type of savings account. He checked his bank one day, and found that the rate for a 12-month CD had an annual percentage yield (APY) of 1.25%. The interest rate on his savings account was 0.75% APY. How much more simple interest would Aidan earn if he invested in a CD for 12 months rather than leaving the $1800 in a regular savings account?
Aidan would earn $9 more in the CD.

26. How much interest will Roxanne have to pay if she borrows $2000 for 2 years at a simple interest rate of 4%?
Roxanne will have to pay $160.

 27. Bob borrowed money for 1 year at 5% simple interest. If he had to pay back a total of $1260, how much did he originally borrow? **(See Example 4.)**   Bob borrowed $1200.

28. Andrea borrowed some money for 2 years at 6% simple interest. If she had to pay back a total of $3640, how much did she originally borrow?   Andrea borrowed $3250.

29. If $1500 grows to $1950 after 5 years, find the simple interest rate.   The rate is 6%.

30. If $9000 grows to $10,440 in 2 years, find the simple interest rate.   The rate is 8%.

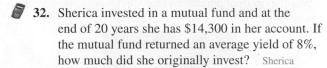 **31.** Perry is planning a vacation to Europe in 2 years. How much should he invest in an account that pays 3% simple interest to get the $3500 that he needs for the trip? Round to the nearest dollar.   Perry needs to invest $3302.

32. Sherica invested in a mutual fund and at the end of 20 years she has $14,300 in her account. If the mutual fund returned an average yield of 8%, how much did she originally invest?   Sherica invested $5500.

Concept 3: Applications Involving Discount and Markup

33. A hands-free kit for a car costs $62. An electronics store has it on sale for 12% off with free installation.

   a. What is the discount on the hands-free kit? $7.44
   b. What is the sale price? $54.56

34. A tablet originally selling for $550 is on sale for 10% off.

   a. What is the discount on the tablet? $55
   b. What is the sale price? $495

35. A digital camera is on sale for $400. This price is 15% off the original price. What was the original price? Round to the nearest cent. **(See Example 5.)**   The original price was $470.59.

36. A Blu-ray disc is on sale for $18. If this represents an 18% discount rate, what was the original price?   The original price was $21.95.

37. The original price of an Audio Jukebox was $250. It is on sale for $220. What percent discount does this represent?   The discount rate is 12%.

38. During its first year, a gaming console sold for $425 in stores. This product was in such demand that it sold for $800 online. What percent markup does this represent? (Round to the nearest whole percent.)   The markup rate is 88%.

39. A doctor ordered a dosage of medicine for a patient. After 2 days, she increased the dosage by 20% and the new dosage came to 18 cc. What was the original dosage?   The original dosage was 15 cc.

40. In one area, the cable company marked up the monthly cost by 6%. The new cost is $63.60 per month. What was the cost before the increase?   The original cost was $60.

Mixed Exercises

41. Sun Lei bought a laptop computer for $1800. The total cost, including tax, came to $1890. What is the tax rate?   The tax rate is 5%.

© Creatas/PictureQuest RF

42. Jamie purchased a beach umbrella and paid $32.04, including tax. If the price before tax is $29.80, what is the sales tax rate (round to the nearest tenth of a percent)?   The tax rate is 7.5%.

© Tero Hakala/123RF

43. To discourage tobacco use and to increase state revenue, several states tax tobacco products. One year, the state of New York increased taxes on tobacco, resulting in a 11% increase in the retail price of a pack of cigarettes. If the new price of a pack of cigarettes is $12.85, what was the cost before the increase in tax?   The original cost was $11.58 per pack.

44. A hotel room rented for 5 nights costs $706.25 including 13% in taxes. Find the original price of the room (before tax) for the 5 nights. Then find the price per night.   The 5 nights (without tax) costs $625. The price per night is $125.

Writing   Translating Expression   Geometry   Scientific Calculator   Video

45. Deon purchased a house and sold it for a 24% profit. If he sold the house for $260,400, what was the original purchase price? The original price was $210,000.

© Comstock/PunchStock RF

46. To meet the rising cost of energy, the yearly membership at a YMCA had to be increased by 12.5% from the past year. The yearly membership fee is currently $450. What was the cost of membership last year? The cost was $400.

© Image100/Corbis RF

47. Alina earns $1600 per month plus a 12% commission on pharmaceutical sales. If she sold $25,000 in pharmaceuticals one month, what was her salary that month? Alina made $4600 that month.

48. Dan sold a beachfront home for $650,000. If his commission rate is 4%, what did he earn in commission? Dan earned $26,000.

49. Diane sells women's sportswear at a department store. She earns a regular salary and, as a bonus, she receives a commission of 4% on all sales over $200. If Diane earned an extra $25.80 last week in commission, how much merchandise did she sell over $200? Diane sold $645 over $200 worth of merchandise.

50. For selling software, Tom received a bonus commission based on sales over $500. If he received $180 in commission for selling a total of $2300 worth of software, what is his commission rate? Tom's commission rate is 10%.

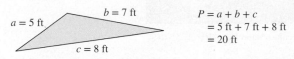

Section 2.6   Formulas and Applications of Geometry

Concepts

1. Literal Equations and Formulas
2. Geometry Applications

1. Literal Equations and Formulas

A *literal equation* is an equation that has more than one variable. A formula is a literal equation with a specific application. For example, the perimeter of a triangle (distance around the triangle) can be found by the formula $P = a + b + c$, where a, b, and c are the lengths of the sides (Figure 2-7).

$b = 7$ ft
$a = 5$ ft
$c = 8$ ft

$P = a + b + c$
$= 5$ ft $+ 7$ ft $+ 8$ ft
$= 20$ ft

Figure 2-7

In this section, we will learn how to rewrite formulas to solve for a different variable within the formula. Suppose, for example, that the perimeter of a triangle is known and two of the sides are known (say, sides a and b). Then the third side, c, can be found by subtracting the lengths of the known sides from the perimeter (Figure 2-8).

 Writing   Translating Expression   Geometry   Scientific Calculator   Video

If the perimeter is 20 ft, then
$c = P - a - b$
$= 20 \text{ ft} - 5 \text{ ft} - 7 \text{ ft}$
$= 8 \text{ ft}$

Figure 2-8

To solve a formula for a different variable, we use the same properties of equality outlined in the earlier sections of this chapter. For example, consider the two equations $2x + 3 = 11$ and $wx + y = z$. Suppose we want to solve for x in each case:

$2x + 3 = 11$		$wx + y = z$	
$2x + 3 - 3 = 11 - 3$ Subtract 3.		$wx + y - y = z - y$ Subtract y.
$2x = 8$		$wx = z - y$
$\dfrac{2x}{2} = \dfrac{8}{2}$ Divide by 2.		$\dfrac{wx}{w} = \dfrac{z - y}{w}$ Divide by w.
$x = 4$		$x = \dfrac{z - y}{w}$

The equation on the left has only one variable and we are able to simplify the equation to find a numerical value for x. The equation on the right has multiple variables. Because we do not know the values of w, y, and z, we are not able to simplify further. The value of x is left as a formula in terms of w, y, and z.

Example 1   Solving for an Indicated Variable

Classroom Examples: p. 153, Exercises 16 and 24

Solve for the indicated variable.

a. $d = rt$ for t **b.** $5x + 2y = 12$ for y

Solution:

a. $d = rt$ for t The goal is to isolate the variable t.

$\dfrac{d}{r} = \dfrac{rt}{r}$ Because the relationship between r and t is multiplication, we reverse the process by dividing both sides by r.

$\dfrac{d}{r} = t$, or equivalently $t = \dfrac{d}{r}$

b. $5x + 2y = 12$ for y The goal is to solve for y.

$5x - 5x + 2y = 12 - 5x$ Subtract $5x$ from both sides to isolate the y term.

$2y = -5x + 12$ $-5x + 12$ is the same as $12 - 5x$.

$\dfrac{2y}{2} = \dfrac{-5x + 12}{2}$ Divide both sides by 2 to isolate y.

$y = \dfrac{-5x + 12}{2}$

Avoiding Mistakes

In the expression $\dfrac{-5x + 12}{2}$ do not try to divide the 2 into the 12. The divisor of 2 is dividing the entire quantity, $-5x + 12$ (not just the 12).

We may, however, apply the divisor to each term individually in the numerator. That is, $y = \dfrac{-5x + 12}{2}$ can be written in several different forms. Each is correct.

$$y = \frac{-5x + 12}{2} \quad \text{or} \quad y = \frac{-5x}{2} + \frac{12}{2} \quad \Rightarrow \quad y = -\frac{5}{2}x + 6$$

Skill Practice   Solve for the indicated variable.

1. $A = lw$ for l **2.** $-2a + 4b = 7$ for a

Classroom Example: p. 153, Exercise 36

Example 2   **Solving for an Indicated Variable**

The formula $C = \frac{5}{9}(F - 32)$ is used to find the temperature, C, in degrees Celsius for a given temperature expressed in degrees Fahrenheit, F. Solve the formula $C = \frac{5}{9}(F - 32)$ for F.

Solution:

$$C = \frac{5}{9}(F - 32)$$

$$C = \frac{5}{9}F - \frac{5}{9} \cdot 32 \qquad \text{Clear parentheses.}$$

$$C = \frac{5}{9}F - \frac{160}{9} \qquad \text{Multiply: } \frac{5}{9} \cdot \frac{32}{1} = \frac{160}{9}$$

$$9(C) = 9\left(\frac{5}{9}F - \frac{160}{9}\right) \qquad \text{Multiply by the LCD to clear fractions.}$$

$$9C = \frac{9}{1} \cdot \frac{5}{9}F - \frac{9}{1} \cdot \frac{160}{9} \qquad \text{Apply the distributive property.}$$

$$9C = 5F - 160 \qquad \text{Simplify.}$$

$$9C + 160 = 5F - 160 + 160 \qquad \text{Add 160 to both sides.}$$

$$9C + 160 = 5F$$

$$\frac{9C + 160}{5} = \frac{5F}{5} \qquad \text{Divide both sides by 5.}$$

$$\frac{9C + 160}{5} = F$$

The answer may be written in several forms:

$$F = \frac{9C + 160}{5} \quad \text{or} \quad F = \frac{9C}{5} + \frac{160}{5} \quad \Rightarrow \quad F = \frac{9}{5}C + 32$$

Answers

1. $l = \dfrac{A}{w}$

2. $a = \dfrac{7 - 4b}{-2}$ or $a = \dfrac{4b - 7}{2}$ or

   $a = 2b - \dfrac{7}{2}$

3. $x = 3y + 7$

Skill Practice   Solve for the indicated variable.

3. $y = \dfrac{1}{3}(x - 7)$ for x.

2. Geometry Applications

In Examples 3 through 6 and the related exercises, we use facts and formulas from geometry.

Classroom Example: p. 153, Exercise 46

Example 3   ## Solving a Geometry Application Involving Perimeter

The length of a rectangular lot is 1 m less than twice the width. If the perimeter is 190 m, find the length and width.

Solution:

Let x represent the width of the rectangle.

Then $2x - 1$ represents the length.

	Step 1: Read the problem.
	Step 2: Label the variables.

$$P = 2l + 2w$$

$$190 = 2(2x - 1) + 2(x)$$

$$190 = 4x - 2 + 2x$$

$$190 = 6x - 2$$

$$192 = 6x$$

$$\frac{192}{6} = \frac{6x}{6}$$

$$32 = x$$

Step 3: Write the formula for perimeter.

Step 4: Write an equation in terms of x.

Step 5: Solve for x.

The width is $x = 32$.

The length is $2x - 1 = 2(32) - 1 = 63$.

Step 6: Interpret the results and write the answer in words.

The width of the rectangular lot is 32 m and the length is 63 m.

Skill Practice

4. The length of a rectangle is 10 ft less than twice the width. If the perimeter is 178 ft, find the length and width.

Recall some facts about angles.

- Two angles are complementary if the sum of their measures is 90°.
- Two angles are supplementary if the sum of their measures is 180°.
- The sum of the measures of the angles within a triangle is 180°.
- The measures of vertical angles are equal.

Answer

4. The length is 56 ft, and the width is 33 ft.

Example 4   Solving a Geometry Application Involving Complementary Angles

Two complementary angles are drawn such that one angle is 4° more than seven times the other angle. Find the measure of each angle.

Solution:

Let x represent the measure of one angle.

Then $7x + 4$ represents the measure of the other angle.

The angles are complementary, so their sum must be 90°.

$$\left(\begin{array}{c}\text{Measure of}\\\text{first angle}\end{array}\right) + \left(\begin{array}{c}\text{measure of}\\\text{second angle}\end{array}\right) = 90°$$

$$x \quad + \quad 7x + 4 \quad = 90$$

$$8x + 4 = 90$$

$$8x = 86$$

$$\frac{8x}{8} = \frac{86}{8}$$

$$x = 10.75$$

One angle is $x = 10.75$.

The other angle is $7x + 4 = 7(10.75) + 4 = 79.25$.

The angles are 10.75° and 79.25°.

Step 1: Read the problem.

Step 2: Label the variables.

Step 3: Write a verbal model.

Step 4: Write a mathematical equation.

Step 5: Solve for x.

Step 6: Interpret the results and write the answer in words.

Skill Practice

5. Two complementary angles are constructed so that one measures 1° less than six times the other. Find the measures of the angles.

Example 5   Solving a Geometry Application Involving Angles in a Triangle

One angle in a triangle is twice as large as the smallest angle. The third angle is 10° more than seven times the smallest angle. Find the measure of each angle.

Solution:

Let x represent the measure of the smallest angle.

Then $2x$ and $7x + 10$ represent the measures of the other two angles.

The sum of the angles must be 180°.

Step 1: Read the problem.

Step 2: Label the variables.

Answer

5. The angles are 13° and 77°.

$$\begin{pmatrix} \text{Measure of} \\ \text{first angle} \end{pmatrix} + \begin{pmatrix} \text{measure of} \\ \text{second angle} \end{pmatrix} + \begin{pmatrix} \text{measure of} \\ \text{third angle} \end{pmatrix} = 180°$$

Step 3: Write a verbal model.

$$x \quad + \quad 2x \quad + \quad (7x + 10) \quad = 180$$

Step 4: Write a mathematical equation.

$$x + 2x + 7x + 10 = 180$$

Step 5: Solve for x.

$$10x + 10 = 180$$
$$10x = 170$$
$$x = 17$$

Step 6: Interpret the results and write the answer in words.

The smallest angle is $x = 17$.

The other angles are $2x = 2(17) = 34$

$$7x + 10 = 7(17) + 10 = 129$$

The angles are $17°$, $34°$, and $129°$.

Skill Practice

6. In a triangle, the measure of the first angle is $80°$ greater than the measure of the second angle. The measure of the third angle is twice that of the second. Find the measures of the angles.

Example 6 **Solving a Geometry Application Involving Circumference**

Classroom Example: p. 155, Exercise 66

The distance around a circular garden is 188.4 ft. Find the radius to the nearest tenth of a foot. Use 3.14 for π.

Solution:

$$C = 2\pi r \qquad \text{Use the formula for the circumference of a circle.}$$
$$188.4 = 2\pi r \qquad \text{Substitute 188.4 for } C.$$
$$\frac{188.4}{2\pi} = \frac{2\pi r}{2\pi} \qquad \text{Divide both sides by } 2\pi.$$
$$\frac{188.4}{2\pi} = r$$
$$r \approx \frac{188.4}{2(3.14)}$$
$$= 30.0$$

$C = 188.4$ ft

The radius is approximately 30.0 ft.

Skill Practice

7. The circumference of a drain pipe is 12.5 cm. Find the radius. Round to the nearest tenth of a centimeter.

Answers

6. The angles are $25°$, $50°$, and $105°$.
7. The radius is 2.0 cm.

Calculator Connections

Topic: Using the π Key on a Calculator

In Example 6 we could have obtained a more accurate result if we had used the π key on a calculator.

   Note that parentheses are required to divide 188.4 by the quantity 2π. This guarantees that the calculator follows the implied order of operations. Without parentheses, the calculator would divide 188.4 by 2 and then multiply the result by π.

Scientific Calculator

Enter: 188.4 ÷ (2 × π) =   **Result:** 29.98479128   correct

Enter: 188.4 ÷ 2 × π =   **Result:** 295.938028   incorrect

Graphing Calculator

```
188.4/(2π)
          29.98479128    ◄—— Correct
188.4/2π
          295.938028     ◄—— Incorrect
```

Calculator Exercises

Approximate the expressions with a calculator. Round to three decimal places if necessary.

1. $\dfrac{880}{2\pi}$ 140.056

2. $\dfrac{1600}{\pi(4)^2}$ 31.831

3. $\dfrac{20}{5\pi}$ 1.273

4. $\dfrac{10}{7\pi}$ 0.455

Section 2.6   Practice Exercises

For additional exercises, see Classroom Activities 2.6A–2.6C in the *Student's Resource Manual* at www.mhhe.com/moh.

Study Skills Exercise

A good technique for studying for a test is to choose four problems from each section of the chapter and write the problems along with the directions on 3×5 cards. On the back of each card, put the page number where you found that problem. Then shuffle the cards and test yourself on the procedure to solve each problem. If you find one that you do not know how to solve, look at the page number and do several of that type. Write four problems you would choose for this section.

Review Exercises

For Exercises 1–8, solve the equation.

1. $3(2y + 3) - 4(-y + 1) = 7y - 10$ {−5}

2. $-(3w + 4) + 5(w - 2) - 3(6w - 8) = 10$ {0}

3. $\dfrac{1}{2}(x - 3) + \dfrac{3}{4} = 3x - \dfrac{3}{4}$ {0}

4. $\dfrac{5}{6}x + \dfrac{1}{2} = \dfrac{1}{4}(x - 4)$ $\left\{-\dfrac{18}{7}\right\}$

5. $0.5(y + 2) - 0.3 = 0.4y + 0.5$ {−2}

6. $0.25(500 - x) + 0.15x = 75$ {500}

7. $8b + 6(7 - 2b) = -4(b + 1)$ { }

8. $2 - 5(t - 3) + t = 7t - (6t + 8)$ {5}

 Writing   Translating Expression   Geometry   Scientific Calculator   Video

Concept 1: Literal Equations and Formulas

For Exercises 9–40, solve for the indicated variable. **(See Examples 1–2.)**

9. $P = a + b + c$ for a
$a = P - b - c$

10. $P = a + b + c$ for b
$b = P - a - c$

11. $x = y - z$ for y
$y = x + z$

12. $c + d = e$ for d
$d = e - c$

13. $p = 250 + q$ for q
$q = p - 250$

14. $y = 35 + x$ for x
$x = y - 35$

15. $A = bh$ for b $b = \dfrac{A}{h}$

16. $d = rt$ for r $r = \dfrac{d}{t}$

17. $PV = nrt$ for t $t = \dfrac{PV}{nr}$

18. $P_1 V_1 = P_2 V_2$ for V_1 $V_1 = \dfrac{P_2 V_2}{P_1}$

19. $x - y = 5$ for x $x = 5 + y$

20. $x + y = -2$ for y $y = -2 - x$

21. $3x + y = -19$ for y
$y = -3x - 19$

22. $x - 6y = -10$ for x
$x = 6y - 10$

23. $2x + 3y = 6$ for y
$y = \dfrac{-2x + 6}{3}$ or $y = -\dfrac{2}{3}x + 2$

24. $7x + 3y = 1$ for y
$y = \dfrac{-7x + 1}{3}$ or $y = -\dfrac{7}{3}x + \dfrac{1}{3}$

25. $-2x - y = 9$ for x
$x = \dfrac{y + 9}{-2}$ or $x = -\dfrac{1}{2}y - \dfrac{9}{2}$

26. $3x - y = -13$ for x
$x = \dfrac{y - 13}{3}$ or $x = \dfrac{1}{3}y - \dfrac{13}{3}$

27. $4x - 3y = 12$ for y
$y = \dfrac{-4x + 12}{-3}$ or $y = \dfrac{4}{3}x - 4$

28. $6x - 3y = 4$ for y
$y = \dfrac{-6x + 4}{-3}$ or $y = 2x - \dfrac{4}{3}$

29. $ax + by = c$ for y
$y = \dfrac{-ax + c}{b}$ or $y = -\dfrac{a}{b}x + \dfrac{c}{b}$

30. $ax + by = c$ for x
$x = \dfrac{-by + c}{a}$ or $x = -\dfrac{b}{a}y + \dfrac{c}{a}$

31. $A = P(1 + rt)$ for t
$t = \dfrac{A - P}{Pr}$ or $t = \dfrac{A}{Pr} - \dfrac{1}{r}$

32. $P = 2(L + w)$ for L
$L = \dfrac{P - 2w}{2}$ or $L = \dfrac{P}{2} - w$

33. $a = 2(b + c)$ for c
$c = \dfrac{a - 2b}{2}$ or $c = \dfrac{a}{2} - b$

34. $3(x + y) = z$ for x
$x = \dfrac{z - 3y}{3}$ or $x = \dfrac{z}{3} - y$

35. $Q = \dfrac{x + y}{2}$ for y
$y = 2Q - x$

36. $Q = \dfrac{a - b}{2}$ for a
$a = 2Q + b$

37. $M = \dfrac{a}{S}$ for a
$a = MS$

38. $A = \dfrac{1}{3}(a + b + c)$ for c
$c = 3A - a - b$

39. $P = I^2 R$ for R $R = \dfrac{P}{I^2}$

40. $F = \dfrac{GMm}{d^2}$ for m $m = \dfrac{Fd^2}{GM}$

Concept 2: Geometry Applications

For Exercises 41–62, use the Problem-Solving Flowchart for Word Problems.

41. The perimeter of a rectangular garden is 24 ft. The length is 2 ft more than the width. Find the length and the width of the garden. **(See Example 3.)**
The length is 7 ft and the width is 5 ft.

42. In a small rectangular wallet photo, the width is 7 cm less than the length. If the perimeter of the photo is 34 cm, find the length and width.
The length is 12 cm and the width is 5 cm.

43. The length of a rectangular parking area is four times the width. The perimeter is 300 yd. Find the length and width of the parking area. The length is 120 yd and the width is 30 yd.

44. The width of Jason's workbench is $\frac{1}{2}$ the length. The perimeter is 240 in. Find the length and the width of the workbench.   The length is 80 in. and the width is 40 in.

45. A builder buys a rectangular lot of land such that the length is 5 m less than two times the width. If the perimeter is 590 m, find the length and the width.   The length is 195 m and the width is 100 m.

$2w - 5$

46. The perimeter of a rectangular pool is 140 yd. If the length is 20 yd less than twice the width, find the length and the width.
The length is 40 yd and the width is 30 yd.

w

$2w - 20$

47. A triangular parking lot has two sides that are the same length, and the third side is 5 m longer. If the perimeter is 71 m, find the lengths of the sides.   The sides are 22 m, 22 m, and 27 m.

48. The perimeter of a triangle is 16 ft. One side is 3 ft longer than the shortest side. The third side is 1 ft longer than the shortest side. Find the lengths of the sides.   The sides are 4 ft, 5 ft, and 7 ft.

49. Sometimes memory devices are helpful for remembering mathematical facts. Recall that the sum of two complementary angles is 90°. That is, two complementary angles when added together form a right angle or "corner." The words *Complementary* and *Corner* both start with the letter "*C*." Derive your own memory device for remembering that the sum of two supplementary angles is 180°. "Adjacent supplementary angles form a straight angle." The words *Supplementary* and *Straight* both begin with the same letter.

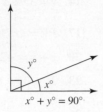

$$x° + y° = 90°$$

Complementary angles form a "Corner"

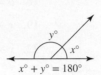

$$x° + y° = 180°$$

Supplementary angles . . .

50. Two angles are complementary. One angle is 20° less than the other angle. Find the measures of the angles.
The angles are 55° and 35°.

51. Two angles are complementary. One angle is 4° less than three times the other angle. Find the measures of the angles. **(See Example 4.)**
The angles are 23.5° and 66.5°.

52. Two angles are supplementary. One angle is three times as large as the other angle. Find the measures of the angles. The angles are 45° and 135°.

53. Two angles are supplementary. One angle is 6° more than four times the other. Find the measures of the angles. The angles are 34.8° and 145.2°.

54. Refer to the figure. The angles, ∠a and ∠b, are vertical angles.

   a. If the measure of ∠a is 32°, what is the measure of ∠b?

   b. What is the measure of the supplement of ∠a?
   a. The measure of ∠b is 32°.
   b. The supplement is 148°.

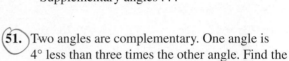

Figure for Exercise 54

55. Find the measures of the vertical angles labeled in the figure by first solving for *x*.
x = 20; the vertical angles measure 37°.

$$(2x - 3)°$$
$$(x + 17)°$$

56. Find the measures of the vertical angles labeled in the figure by first solving for *y*.
y = 40; the vertical angles measure 146°.

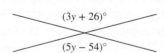

$$(3y + 26)°$$
$$(5y - 54)°$$

57. The largest angle in a triangle is three times the smallest angle. The middle angle is two times the smallest angle. Given that the sum of the angles in a triangle is 180°, find the measure of each angle. **(See Example 5.)**
The measures of the angles are 30°, 60°, and 90°.

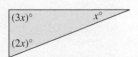

$$(3x)°$$ $$x°$$
$$(2x)°$$

Animation

58. The smallest angle in a triangle measures 90° less than the largest angle. The middle angle measures 60° less than the largest angle. Find the measure of each angle.
The measures of the angles are 20°, 50°, and 110°.

$$x°$$

59. The smallest angle in a triangle is half the largest angle. The middle angle measures 30° less than the largest angle. Find the measure of each angle.
The measures of the angles are 42°, 54°, and 84°.

60. The largest angle of a triangle is three times the middle angle. The smallest angle measures 10° less than the middle angle. Find the measure of each angle.
The measures of the angles are 38°, 28°, and 114°.

61. Find the value of x and the measure of each angle labeled in the figure.

$x = 17$; the measures of the angles are 34° and 56°.

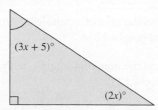

$(3x + 5)°$

$(2x)°$

62. Find the value of y and the measure of each angle labeled in the figure.

$y = 26.5$; the measures of the angles are 28.5° and 61.5°.

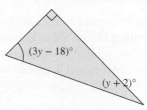

$(3y − 18)°$

$(y + 2)°$

63. a. A rectangle has length l and width w. Write a formula for the area.

$A = lw$

b. Solve the formula for the width, w. $w = \dfrac{A}{l}$

c. The area of a rectangular volleyball court is 1740.5 ft² and the length is 59 ft. Find the width.

The width is 29.5 ft.

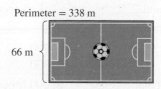

Area = 1740.5 ft²

59 ft

64. a. A parallelogram has height h and base b. Write a formula for the area.

$A = bh$

b. Solve the formula for the base, b. $b = \dfrac{A}{h}$

c. Find the base of the parallelogram pictured if the area is 40 m².

The base is 8 m.

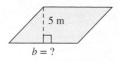

5 m

$b = ?$

65. a. A rectangle has length l and width w. Write a formula for the perimeter. $P = 2l + 2w$

b. Solve the formula for the length, l. $l = \dfrac{P - 2w}{2}$

c. The perimeter of the soccer field at Giants Stadium is 338 m. If the width is 66 m, find the length. The length is 103 m.

Perimeter = 338 m

66 m

66. a. A triangle has height h and base b. Write a formula for the area. $A = \dfrac{1}{2}bh$

b. Solve the formula for the height, h. $h = \dfrac{2A}{b}$

c. Find the height of the triangle pictured if the area is 12 km².

The height is 4 km.

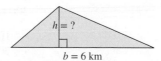

$h = ?$

$b = 6$ km

67. a. A circle has a radius of r. Write a formula for the circumference. **(See Example 6.)** $C = 2\pi r$

b. Solve the formula for the radius, r. $r = \dfrac{C}{2\pi}$

c. The circumference of the circular Buckingham Fountain in Chicago is approximately 880 ft. Find the radius. Round to the nearest foot.

The radius is approximately 140 ft.

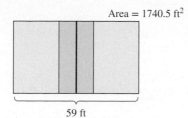

© Brand X Pictures/Getty Images RF

68. a. The length of each side of a square is s. Write a formula for the perimeter of the square. $P = 4s$

b. Solve the formula for the length of a side, s. $s = \dfrac{P}{4}$

c. The Pyramid of Khufu (known as the Great Pyramid) at Giza has a square base. If the distance around the bottom is 921.6 m, find the length of the sides at the bottom of the pyramid.

The length of each side at the bottom of the pyramid is 230.4 m.

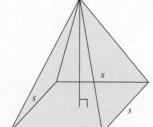

s

s

s

s

Expanding Your Skills

For Exercises 69–70, find the indicated area or volume. Be sure to include the proper units and round each answer to two decimal places if necessary.

69. a. Find the area of a circle with radius 11.5 m. Use the π key on the calculator.   415.48 m²
b. Find the volume of a right circular cylinder with radius 11.5 m and height 25 m.   10,386.89 m³

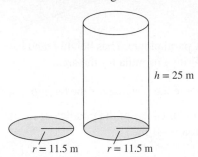

$h = 25$ m

$r = 11.5$ m $r = 11.5$ m

70. a. Find the area of a parallelogram with base 30 in. and height 12 in.
360 in.²
b. Find the area of a triangle with base 30 in. and height 12 in.
180 in.²
c. Compare the areas found in parts (a) and (b).
The area of the triangle is one-half the area of the parallelogram.

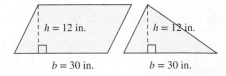

$h = 12$ in. $h = 12$ in.

$b = 30$ in. $b = 30$ in.

Section 2.7   Mixture Applications and Uniform Motion

Concepts

1. Applications Involving Cost
2. Applications Involving Mixtures
3. Applications Involving Uniform Motion

1. Applications Involving Cost

In Examples 1 and 2, we will look at different kinds of mixture problems. The first example "mixes" two types of theater tickets, adult tickets that sell for \$12 and children's tickets that sell for \$9. Furthermore, there were 300 tickets sold for a total revenue of \$3060. Before attempting the problem, we should try to gain some familiarity. Let's try a few combinations to see how many of each type of ticket might have been sold.

Suppose 100 adult tickets were sold and 200 children's tickets were sold (a total of 300 tickets).

- 100 adult tickets at \$12 each gives   100(\$12) = \$1200
- 200 children's tickets at \$9 each gives   200(\$9) = \$1800

Total revenue: \$3000 (not enough)

Suppose 150 adult tickets were sold and 150 children's tickets were sold (a total of 300 tickets).

- 150 adult tickets at \$12 each gives   150(\$12) = \$1800
- 150 children's tickets at \$9 each gives   150(\$9) = \$1350

Total revenue: \$3150 (too much)

As you can see, the trial-and-error process can be tedious and time consuming. Therefore, we will use algebra to determine the correct combination of each type of ticket.

Suppose we let x represent the number of adult tickets. Then the number of children's tickets is the *total minus x*. That is,

$$\left(\begin{array}{c}\text{Number of}\\\text{children's tickets}\end{array}\right) = \left(\begin{array}{c}\text{total number}\\\text{of tickets}\end{array}\right) - \left(\begin{array}{c}\text{number of}\\\text{adult tickets, } x\end{array}\right)$$

Number of children's tickets = $300 - x$.

Notice that the number of tickets sold times the price per ticket gives the revenue:

- x adult tickets at \$12 each gives a revenue of: $x(\$12)$ or simply $12x$.
- $300 - x$ children's tickets at \$9 each gives: $(300 - x)(\$9)$ or $9(300 - x)$.

This will help us set up an equation in Example 1.

Example 1   **Solving a Mixture Problem Involving Ticket Sales**

At a community theater, 300 tickets were sold. Adult tickets cost \$12 and tickets for children cost \$9. If the total revenue from ticket sales was \$3060, determine the number of each type of ticket sold.

© Uppercut/Getty Images RF

Classroom Example: p. 162, Exercise 14

Solution:

                                                    Step 1: Read the problem.

Let x represent the number of adult tickets sold.   **Step 2:** Label the
$300 - x$ is the number of children's tickets.   variables.

	\$12 Tickets	\$9 Tickets	Total
Number of tickets	x	$300 - x$	300
Revenue	$12x$	$9(300 - x)$	3060

$$\left(\begin{array}{c}\text{Revenue from} \\ \text{adult tickets}\end{array}\right) + \left(\begin{array}{c}\text{revenue from} \\ \text{children's tickets}\end{array}\right) = \left(\begin{array}{c}\text{total} \\ \text{revenue}\end{array}\right)$$

Step 3: Write a verbal model.

$$12x \quad + \quad 9(300 - x) \quad = \quad 3060$$

Step 4: Write a mathematical equation.

$$12x + 9(300 - x) = 3060$$ **Step 5:** Solve the equation.

$$12x + 2700 - 9x = 3060$$

$$3x + 2700 = 3060$$

$$3x = 360$$

$$x = 120$$ **Step 6:** Interpret the results.

There were 120 adult tickets sold.
The number of children's tickets is $300 - x$ which is 180.

Avoiding Mistakes

Check that the answer is reasonable. 120 adult tickets and 180 children's tickets makes 300 total tickets.

Furthermore, 120 adult tickets at \$12 each amounts to \$1440, and 180 children's tickets at \$9 each amounts to \$1620. The total revenue is \$3060 as expected.

Skill Practice

1. At a Performing Arts Center, seats in the orchestra section cost \$18 and seats in the balcony cost \$12. If there were 120 seats sold for one performance, for a total revenue of \$1920, how many of each type of seat were sold?

Answer

1. There were 80 seats in the orchestra section, and there were 40 in the balcony.

2. Applications Involving Mixtures

Classroom Example: p. 163, Exercise 24

| Example 2 | Solving a Mixture Application |

How many liters (L) of a 60% antifreeze solution must be added to 8 L of a 10% antifreeze solution to produce a 20% antifreeze solution?

Solution:

The information can be organized in a table. Notice that an algebraic equation is derived from the second row of the table. This relates the number of liters of pure antifreeze in each container.

Step 1: Read the problem.

	60% Antifreeze	10% Antifreeze	Final Mixture: 20% Antifreeze
Number of liters of solution	x	8	$(8 + x)$
Number of liters of pure antifreeze	$0.60x$	$0.10(8)$	$0.20(8 + x)$

Step 2: Label the variables.

> **TIP:** To determine the amount of pure antifreeze in a given solution, multiply the concentration rate by the amount of solution. For example, 8 L of a 10% solutions means that
>
> $(0.10)(8\ \text{L}) = 0.8\ \text{L}$
>
> of the solution is pure antifreeze. The rest is something else such as water.

The amount of pure antifreeze in the final solution equals the sum of the amounts of antifreeze in the first two solutions.

$$\left(\begin{array}{c}\text{Pure antifreeze} \\ \text{from solution 1}\end{array}\right) + \left(\begin{array}{c}\text{pure antifreeze} \\ \text{from solution 2}\end{array}\right) = \left(\begin{array}{c}\text{pure antifreeze} \\ \text{in the final solution}\end{array}\right)$$

Step 3: Write a verbal model.

$$0.60x \quad + \quad 0.10(8) \quad = \quad 0.20(8 + x)$$

$$0.60x + 0.10(8) = 0.20(8 + x)$$

Step 4: Write a mathematical equation.

$$0.6x + 0.8 = 1.6 + 0.2x$$

Step 5: Solve the equation.

$$0.6x - 0.2x + 0.8 = 1.6 + 0.2x - 0.2x$$ Subtract $0.2x$.

$$0.4x + 0.8 = 1.6$$

$$0.4x + 0.8 - 0.8 = 1.6 - 0.8$$ Subtract 0.8.

$$0.4x = 0.8$$

$$\frac{0.4x}{0.4} = \frac{0.8}{0.4}$$ Divide by 0.4.

$$x = 2$$ **Step 6:** Interpret the result.

Therefore, 2 L of 60% antifreeze solution is necessary to make a final solution that is 20% antifreeze.

Skill Practice

2. How many gallons of a 5% bleach solution must be added to 10 gallons (gal) of a 20% bleach solution to produce a solution that is 15% bleach?

Answer

2. 5 gal is needed.

3. Applications Involving Uniform Motion

The formula (distance) = (rate)(time) or simply, $d = rt$, relates the distance traveled to the rate of travel and the time of travel.

For example, if a car travels at 60 mph for 3 hours, then

$$d = (60 \text{ mph})(3 \text{ hours})$$
$$= 180 \text{ miles}$$

If a car travels at 60 mph for x hours, then

$$d = (60 \text{ mph})(x \text{ hours})$$
$$= 60x \text{ miles}$$

© Adam Gault/Getty Images RF

| **Example 3** | **Solving an Application Involving Distance, Rate, and Time** |

Classroom Example: p. 163, Exercise 32

One bicyclist rides 4 mph faster than another bicyclist. The faster rider takes 3 hr to complete a race, while the slower rider takes 4 hr. Find the speed for each rider.

Solution:

Step 1: Read the problem.

The problem is asking us to find the speed of each rider.

Let x represent the speed of the slower rider. Then $(x + 4)$ is the speed of the faster rider.

Step 2: Label the variables and organize the information given in the problem. A distance-rate-time chart may be helpful.

	Distance	Rate	Time
Faster rider	$3(x + 4)$	$x + 4$	3
Slower rider	$4(x)$	x	4

To complete the first column, we can use the relationship, $d = rt$.

Because the riders are riding in the same race, their distances are equal.

$$\begin{pmatrix} \text{Distance} \\ \text{by faster rider} \end{pmatrix} = \begin{pmatrix} \text{distance} \\ \text{by slower rider} \end{pmatrix}$$

Step 3: Write a verbal model.

$$3(x + 4) = 4(x)$$

Step 4: Write a mathematical equation.

$$3x + 12 = 4x$$

Step 5: Solve the equation.

$$12 = x$$

Subtract $3x$ from both sides.

The variable x represents the slower rider's rate. The quantity $x + 4$ is the faster rider's rate. Thus, if $x = 12$, then $x + 4 = 16$.

The slower rider travels 12 mph and the faster rider travels 16 mph.

Avoiding Mistakes

Check that the answer is reasonable. If the slower rider rides at 12 mph for 4 hr, he travels 48 mi. If the faster rider rides at 16 mph for 3 hr, he also travels 48 mi as expected.

Skill Practice

3. An express train travels 25 mph faster than a cargo train. It takes the express train 6 hr to travel a route, and it takes 9 hr for the cargo train to travel the same route. Find the speed of each train.

Answer

3. The express train travels 75 mph, and the cargo train travels 50 mph.

Classroom Example: p. 164,
Exercise 38

Example 4 **Solving an Application Involving Distance, Rate, and Time**

Two families that live 270 mi apart plan to meet for an afternoon picnic at a park that is located between their two homes. Both families leave at 9.00 A.M., but one family averages 12 mph faster than the other family. If the families meet at the designated spot $2\frac{1}{2}$ hr later, determine

© BananaStock/PictureQuest RF

   a. The average rate of speed for each family.

   b. The distance each family traveled to the picnic.

Solution:

For simplicity, we will call the two families, Family A and Family B. Let Family A be the family that travels at the slower rate (Figure 2-9).

Step 1: Read the problem and draw a sketch.

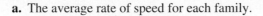

270 miles

Family A   Family B

Figure 2-9

Let x represent the rate of Family A.

Then $(x + 12)$ is the rate of Family B.

Step 2: Label the variables.

	Distance	Rate	Time
Family A	$2.5x$	x	2.5
Family B	$2.5(x + 12)$	$x + 12$	2.5

To complete the first column, we can multiply rate and time: $d = rt$.

   To set up an equation, recall that the total distance between the two families is given as 270 mi.

$$\begin{pmatrix} \text{Distance} \\ \text{traveled by} \\ \text{Family A} \end{pmatrix} + \begin{pmatrix} \text{distance} \\ \text{traveled by} \\ \text{Family B} \end{pmatrix} = \begin{pmatrix} \text{total} \\ \text{distance} \end{pmatrix}$$

Step 3: Write a verbal model.

$$2.5x \quad + \quad 2.5(x + 12) \quad = \quad 270$$

Step 4: Write a mathematical equation.

$$2.5x + 2.5(x + 12) = 270$$

Step 5: Solve for x.

$$2.5x + 2.5x + 30 = 270$$

$$5.0x + 30 = 270$$

$$5x = 240$$

$$x = 48$$

   a. Family A traveled 48 mph.

Step 6: Interpret the results and write the answer in words.

   Family B traveled $x + 12 = 48 + 12 = 60$ mph.

b. To compute the distance each family traveled, use $d = rt$.

Family A traveled $(48 \text{ mph})(2.5 \text{ hr}) = 120 \text{ mi}$.

Family B traveled $(60 \text{ mph})(2.5 \text{ hr}) = 150 \text{ mi}$.

Skill Practice

4. A Piper airplane has an average air speed that is 10 mph faster than a Cessna 150 airplane. If the combined distance traveled by these two small planes is 690 mi after 3 hr, what is the average speed of each plane?

Answer

4. The Cessna's speed is 110 mph, and the Piper's speed is 120 mph.

Section 2.7 Practice Exercises

For additional exercises, see Classroom Activities 2.7A–2.7C in the *Student's Resource Manual* at www.mhhe.com/moh.

Study Skills Exercise

The following is a list of steps to help you solve word problems. Check those that you follow on a regular basis when solving a word problem. Place an asterisk next to the steps that you need to improve.

_____ Read through the entire problem before writing anything down.

_____ Write down exactly what you are being asked to find.

_____ Write down what is known and assign variables to what is unknown.

_____ Draw a figure or diagram if it will help you understand the problem.

_____ Highlight key words like total, sum, difference, etc.

_____ Translate the word problem to a mathematical problem.

_____ After solving, check that your answer makes sense.

Review Exercises

For Exercises 1–3, solve for the indicated variable.

1. $ax - by = c$ for x $x = \dfrac{c + by}{a}$

2. $cd = r$ for c $c = \dfrac{r}{d}$

3. $7x + xy = 18$ for y $y = \dfrac{18 - 7x}{x}$

For Exercises 4–6, solve each equation.

4. $-2d + 11 = 4 - d$ {7}

5. $3(2y + 5) - 8(y - 1) = 3y + 3$ {4}

6. $0.02x + 0.04(10 - x) = 1.26$ {−43}

Concept 1: Applications Involving Cost

For Exercises 7–12, write an algebraic expression as indicated.

7. Two numbers total 200. Let t represent one of the numbers. Write an algebraic expression for the other number. $200 - t$

8. The total of two numbers is 43. Let s represent one of the numbers. Write an algebraic expression for the other number. $43 - s$

Writing   Translating Expression   Geometry   Scientific Calculator   Video

9. Olivia needs to bring 100 cookies to her friend's party. She has already baked x cookies. Write an algebraic expression for the number of cookies Olivia still needs to bake. $100 - x$

© Comstock Images/Jupiter Images RF

10. Rachel needs a mixture of 55 pounds (lb) of nuts consisting of peanuts and cashews. Let p represent the number of pounds of peanuts in the mixture. Write an algebraic expression for the number of pounds of cashews that she needs to add. $55 - p$

© McGraw-Hill Education/Jill Braaten, photographer

11. Max has a total of $3000 in two bank accounts. Let y represent the amount in one account. Write an algebraic expression for the amount in the other account. $3000 - y$

12. Roberto has a total of $7500 in two savings accounts. Let z represent the amount in one account. Write an algebraic expression for the amount in the other account. $7500 - z$

13. A church had an ice cream social and sold tickets for $3 and $2. When the social was over, 81 tickets had been sold totaling $215. How many of each type of ticket did the church sell? **(See Example 1.)**
53 tickets were sold at $3 and 28 tickets were sold at $2.

	$3 Tickets	$2 Tickets	Total
Number of tickets			
Cost of tickets			

14. Anna is a teacher at an elementary school. She purchased 72 tickets to take the first-grade children and some parents on a field trip to the zoo. She purchased children's tickets for $10 each and adult tickets for $18 each. She spent a total of $856. How many of each type of ticket did she buy?
Anna purchased 17 adult tickets and 55 children's tickets.

	Adults	Children	Total
Number of tickets			
Cost of tickets			

15. Josh downloaded 25 songs from an online site. Some songs cost $1.29 and some cost $1.49. He spent a total of $33.85. How many songs at each price were purchased? There were 17 songs purchased at $1.29 and 8 songs purchased at $1.49.

16. During the past year, Kris purchased 30 books at a wholesale club store. She purchased softcover books for $4.50 each and hardcover books for $13.50 each. The total cost of the books was $216. How many of each type of book did she purchase? Kris purchased 21 softcover books and 9 hardcover books.

© Royalty Free/Corbis RF

17. During the past year, Amber purchased 32 books for her e-reader. She purchased some books for $6.99 and others for $9.99. If she spent a total of $256.68, how many books from each price category did she buy?
Amber bought 21 books at $6.99 and 11 books at $9.99.

18. Steven wants to buy some candy with his birthday money. He can choose from jelly beans that sell for $6.99 per pound and a variety mix that sells for $3.99. He likes to have twice the amount of jelly beans as the variety mix. If he spent a total of $53.91, how many pounds of each type of candy did he buy?
Steven bought 6 lb of jelly beans and 3 lb of the variety mix.

Concept 2: Applications Involving Mixtures

For Exercises 19–22, write an algebraic expression as indicated.

19. A container holds 7 ounces (oz) of liquid. Let x represent the number of ounces of liquid in another container. Write an expression for the total amount of liquid. $x + 7$

20. A bucket contains 2.5 L of a bleach solution. Let n represent the number of liters of bleach solution in a second bucket. Write an expression for the total amount of bleach solution. $n + 2.5$

 Writing   Translating Expression   Geometry   Scientific Calculator   Video

21. If Charlene invests $2000 in a certificate of deposit and d dollars in a stock, write an expression for the total amount she invested. $d + 2000$

22. James has $5000 in one savings account. Let y represent the amount he has in another savings account. Write an expression for the total amount of money in both accounts. $y + 5000$

23. How much of a 5% ethanol fuel mixture should be mixed with 2000 gal of 10% ethanol fuel mixture to get a mixture that is 9% ethanol. **(See Example 2.)** Mix 500 gal of 5% ethanol fuel mixture.

	5% Ethanol	10% Ethanol	Final Mixture: 9% Ethanol
Number of gallons of fuel mixture			
Number of gallons of pure ethanol			

24. How many ounces of a 50% antifreeze solution must be mixed with 10 oz of an 80% antifreeze solution to produce a 60% antifreeze solution? Mix 20 oz of 50% antifreeze solution.

	50% Antifreeze	80% Antifreeze	Final Mixture: 60% Antifreeze
Number of ounces of solution			
Number of ounces of pure antifreeze			

25. A pharmacist needs to mix a 1% saline (salt) solution with 24 milliliters (mL) of a 16% saline solution to obtain a 9% saline solution. How many milliliters of the 1% solution must she use? The pharmacist needs to use 21 mL of the 1% saline solution.

26. A landscaper needs to mix a 75% pesticide solution with 30 gal of a 25% pesticide solution to obtain a 60% pesticide solution. How many gallons of the 75% solution must he use? The landscaper needs to use 70 gal of the 75% pesticide solution.

27. To clean a concrete driveway, a contractor needs a solution that is 30% acid. How many ounces of a 50% acid solution must be mixed with 15 oz of a 21% solution to obtain a 30% acid solution? The contractor needs to mix 6.75 oz of 50% acid solution.

28. A veterinarian needs a mixture that contains 12% of a certain medication to treat an injured bird. How many milliliters of a 16% solution should be mixed with 6 mL of a 7% solution to obtain a solution that is 12% medication? The veterinarian needs to use 7.5 mL of the 16% solution.

Concept 3: Applications Involving Uniform Motion

29. **a.** If a car travels 60 mph for 5 hr, find the distance traveled. 300 mi

b. If a car travels at x miles per hour for 5 hr, write an expression that represents the distance traveled. $5x$

c. If a car travels at $x + 12$ mph for 5 hr, write an expression that represents the distance traveled. $5(x + 12)$ or $5x + 60$

30. **a.** If a plane travels 550 mph for 2.5 hr, find the distance traveled. 1375 mi

b. If a plane travels at x miles per hour for 2.5 hr, write an expression that represents the distance traveled. $2.5x$

c. If a plane travels at $x - 100$ mph for 2.5 hr, write an expression that represents the distance traveled. $2.5(x - 100)$ or $2.5x - 250$

31. A woman can walk 2 mph faster down a trail to Cochita Lake than she can on the return trip uphill. It takes her 2 hr to get to the lake and 4 hr to return. What is her speed walking down to the lake? **(See Example 3.)** She walks 4 mph to the lake.

	Distance	Rate	Time
Downhill to the lake			
Uphill from the lake			

32. A car travels 20 mph slower in a bad rain storm than in sunny weather. The car travels the same distance in 2 hr in sunny weather as it does in 3 hr in rainy weather. Find the speed of the car in sunny weather. The car travels 60 mph in sunny weather.

	Distance	Rate	Time
Rain storm			
Sunny weather			

33. Bryan hiked up to the top of City Creek in 3 hr and then returned down the canyon to the trailhead in another 2 hr. His speed downhill was 1 mph faster than his speed uphill. How far up the canyon did he hike? Bryan hiked 6 mi up the canyon.

34. Laura hiked up Lamb's Canyon in 2 hr and then ran back down in 1 hr. Her speed running downhill was 2.5 mph faster than her speed hiking uphill. How far up the canyon did she hike? Laura hiked 5 mi up the canyon.

35. Hazel and Emilie fly from Atlanta to San Diego. The flight from Atlanta to San Diego is against the wind and takes 4 hr. The return flight with the wind takes 3.5 hr. If the wind speed is 40 mph, find the speed of the plane in still air. The plane travels 600 mph in still air.

36. A boat on the Potomac River travels the same distance downstream with the current in $\frac{2}{3}$ hr as it does going upstream against the current in 1 hr. If the speed of the current is 3 mph, find the speed of the boat in still water. The speed of the boat is 15 mph in still water.

37. Two cars are 200 mi apart and travel toward each other on the same road. They meet in 2 hr. One car travels 4 mph faster than the other. What is the speed of each car? **(See Example 4.)** The slower car travels 48 mph and the faster car travels 52 mph.

38. Two cars are 238 mi apart and travel toward each other along the same road. They meet in 2 hr. One car travels 5 mph slower than the other. What is the speed of each car? The cars are traveling 62 mph and 57 mph.

39. After Hurricane Katrina, a rescue vehicle leaves a station at noon and heads for New Orleans. An hour later a second vehicle traveling 10 mph faster leaves the same station. By 4:00 P.M., the first vehicle reaches its destination, and the second is still 10 mi away. How fast is each vehicle? The speeds of the vehicles are 40 mph and 50 mph.

40. A truck leaves a truck stop at 9:00 A.M. and travels toward Sturgis, Wyoming. At 10:00 A.M., a motorcycle leaves the same truck stop and travels the same route. The motorcycle travels 15 mph faster than the truck. By noon, the truck has traveled 20 mi farther than the motorcycle. How fast is each vehicle? The speed of the motorcycle is 65 mph and the speed of the truck is 50 mph.

41. In the Disney Marathon, Jeanette's speed running is twice Sarah's speed walking. After 2 hr, Jeanette is 7 mi ahead of Sarah. Find Jeanette's speed and Sarah's speed. Sarah walks 3.5 mph and Jeanette runs 7 mph.

42. Two canoes travel down a river, starting at 9:00 A.M. One canoe travels twice as fast as the other. After 3.5 hr, the canoes are 5.25 mi apart. Find the speed of each canoe. The canoes travel 1.5 mph and 3 mph.

Mixed Exercises

43. A certain granola mixture is 10% peanuts.

  a. If a container has 20 lb of granola, how many pounds of peanuts are there? 2 lb

  b. If a container has x pounds of granola, write an expression that represents the number of pounds of peanuts in the granola. $0.10x$

  c. If a container has $x + 3$ lb of granola, write an expression that represents the number of pounds of peanuts. $0.10(x + 3) = 0.10x + 0.30$

44. A certain blend of coffee sells for $9.00 per pound.

  a. If a container has 20 lb of coffee, how much will it cost? $180

  b. If a container has x pounds of coffee, write an expression that represents the cost. $9x$

  c. If a container has $40 - x$ pounds of this coffee, write an expression that represents the cost. $9(40 - x) = 360 - 9x$

45. The Coffee Company mixes coffee worth $12 per pound with coffee worth $8 per pound to produce 50 lb of coffee worth $8.80 per pound. How many pounds of the $12 coffee and how many pounds of the $8 coffee must be used? Mix 10 lb of coffee sold at $12 per pound and 40 lb of coffee sold at $8 per pound.

	$12 Coffee	$8 Coffee	Total
Number of pounds			
Value of coffee			

46. The Nut House sells pecans worth $4 per pound and cashews worth $6 per pound. How many pounds of pecans and how many pounds of cashews must be mixed to form 16 lb of a nut mixture worth $4.50 per pound? Mix 12 lb of pecans and 4 lb of cashews.

	$4 Pecans	$6 Cashews	Total
Number of pounds			
Value of nuts			

 Writing  Translating Expression   Geometry  Scientific Calculator   Video

47. A boat in distress, 21 nautical miles from a marina, travels toward the marina at 3 knots (nautical miles per hour). A coast guard cruiser leaves the marina and travels toward the boat at 25 knots. How long will it take for the boats to reach each other?   The boats will meet in $\frac{3}{4}$ hr (45 min).

© Dennis MacDonald/Alamy RF

48. An air traffic controller observes a plane heading from New York to San Francisco traveling at 450 mph. At the same time, another plane leaves San Francisco and travels 500 mph to New York. If the distance between the airports is 2850 mi, how long will it take for the planes to pass each other?   The planes will pass in 3 hr.

49. Surfer Sam purchased a total of 21 items at the surf shop. He bought wax for $3.00 per package and sunscreen for $8.00 per bottle. He spent a total of $88.00. How many of each item did he purchase?   Sam purchased 16 packages of wax and 5 bottles of sunscreen.

50. Tonya Toast loves jam. She purchased 30 jars of gourmet jam for $178.50. She bought raspberry jam for $6.25 per jar and strawberry jam for $5.50 per jar. How many jars of each did she purchase?   Tonya purchased 18 jars of raspberry and 12 jars of strawberry.

51. How many quarts of 85% chlorine solution must be mixed with 5 quarts of 25% chlorine solution to obtain a 45% chlorine solution?   2.5 quarts of 85% chlorine solution

52. How many liters of a 58% sugar solution must be added to 14 L of a 40% sugar solution to obtain a 50% sugar solution?   17.5 L of 58% sugar solution.

Expanding Your Skills

53. How much pure water must be mixed with 12 L of a 40% alcohol solution to obtain a 15% alcohol solution? (*Hint:* Pure water is 0% alcohol.)   20 L of water must be added.

54. How much pure water must be mixed with 10 oz of a 60% alcohol solution to obtain a 25% alcohol solution?   14 oz of water must be added.

55. Amtrak Acela Express is a high-speed train that runs in the United States between Washington, D.C. and Boston. In Japan, a bullet train along the Sanyo line operates at an average speed of 60 km/hr faster than the Amtrak Acela Express. It takes the Japanese bullet train 2.7 hr to travel the same distance as the Acela Express can travel in 3.375 hr. Find the speed of each train.   The Japanese bullet train travels 300 km/hr and the Acela Express travels 240 km/hr.

56. Amtrak Acela Express is a high-speed train along the northeast corridor between Washington, D.C. and Boston. Since its debut, it cuts the travel time from 4 hr 10 min to 3 hr 20 min. On average, if the Acela Express is 30 mph faster than the old train, find the speed of the Acela Express. (*Hint:* 4 hr 10 min = $4\frac{1}{6}$ hr.)   The speed of the Acela Express in 150 mph.

Linear Inequalities

Section 2.8

1. Graphing Linear Inequalities

Consider the following two statements.

$$2x + 7 = 11 \quad \text{and} \quad 2x + 7 < 11$$

The first statement is an equation (it has an = sign). The second statement is an inequality (it has an inequality symbol, <). In this section, we will learn how to solve linear *inequalities*, such as $2x + 7 < 11$.

> ### A Linear Inequality in One Variable
>
> A **linear inequality in one variable**, x, is any inequality that can be written in the form:
>
> $$ax + b < c, \ ax + b \leq c, \ ax + b > c, \ \text{or} \ ax + b \geq c, \text{where } a \neq 0.$$

Concepts

1. **Graphing Linear Inequalities**
2. **Set-Builder Notation and Interval Notation**
3. **Addition and Subtraction Properties of Inequality**
4. **Multiplication and Division Properties of Inequality**
5. **Inequalities of the Form** $a < x < b$
6. **Applications of Linear Inequalities**

Writing   Translating Expression   Geometry   Scientific Calculator   Video

The following inequalities are linear equalities in one variable.

$$2x - 3 < 11 \qquad -4z - 3 > 0 \qquad a \leq 4 \qquad 5.2y \geq 10.4$$

The number line is a useful tool to visualize the solution set of an equation or inequality. For example, the solution set to the equation $x = 2$ is $\{2\}$ and may be graphed as a single point on the number line.

$$x = 2 \qquad \underset{-6\ -5\ -4\ -3\ -2\ -1\ \ 0\ \ 1\ \ 2\ \ 3\ \ 4\ \ 5\ \ 6}{\longleftarrow\!\!\!\mid\!\!\mid\!\!\mid\!\!\mid\!\!\mid\!\!\mid\!\!\mid\!\!\blacklozenge\!\!\mid\!\!\mid\!\!\mid\!\!\mid\!\!\longrightarrow}$$

The solution set to an inequality is the set of real numbers that make the inequality a true statement. For example, the solution set to the inequality $x \geq 2$ is all real numbers 2 or greater. Because the solution set has an infinite number of values, we cannot list all of the individual solutions. However, we can graph the solution set on the number line.

$$x \geq 2 \qquad \underset{-6\ -5\ -4\ -3\ -2\ -1\ \ 0\ \ 1\ \ 2\ \ 3\ \ 4\ \ 5\ \ 6}{\longleftarrow\!\!\!\mid\!\!\mid\!\!\mid\!\!\mid\!\!\mid\!\!\mid\!\!\mid\!\!\mid\!\![\!\!\mid\!\!\mid\!\!\mid\!\!\mid\!\!\longrightarrow}$$

The square bracket symbol, [, is used on the graph to indicate that the point $x = 2$ is included in the solution set. By convention, square brackets, either [or], are used to *include* a point on a number line. Parentheses, (or), are used to *exclude* a point on a number line.

The solution set of the inequality $x > 2$ includes the real numbers greater than 2 but not equal to 2. Therefore, a "(" symbol is used on the graph to indicate that $x = 2$ is not included.

$$x > 2 \qquad \underset{-6\ -5\ -4\ -3\ -2\ -1\ \ 0\ \ 1\ \ 2\ \ 3\ \ 4\ \ 5\ \ 6}{\longleftarrow\!\!\!\mid\!\!\mid\!\!\mid\!\!\mid\!\!\mid\!\!\mid\!\!\mid\!\!\mid\!\!(\!\!\mid\!\!\mid\!\!\mid\!\!\mid\!\!\longrightarrow}$$

In Example 1, we demonstrate how to graph linear inequalities. To graph an inequality means that we graph its solution set. That is, we graph all of the values on the number line that make the inequality true.

Classroom Examples: p. 176,
Exercises 6 and 10

| **Example 1** | **Graphing Linear Inequalities** |

Graph the solution sets.

a. $x > -1$ **b.** $c \leq \dfrac{7}{3}$ **c.** $3 > y$

Solution:

a. $x > -1$ 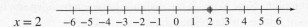

The solution set is the set of all real numbers strictly greater than -1. Therefore, we graph the region on the number line to the right of -1. Because $x = -1$ is not included in the solution set, we use the "(" symbol at $x = -1$.

Instructor Note: Students sometimes confuse the meaning of $x > 1$ to think it means the set $\{2, 3, 4 \ldots\}$. They don't think about the real number line. They don't see that 1.00000001 is a solution.

b. $c \leq \frac{7}{3}$ is equivalent to $c \leq 2\frac{1}{3}$.   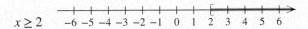

The solution set is the set of all real numbers less than or equal to $2\frac{1}{3}$. Therefore, graph the region on the number line to the left of and including $2\frac{1}{3}$. Use the symbol] to indicate that $c = 2\frac{1}{3}$ is included in the solution set.

c. $3 > y$ This inequality reads "3 is greater than y." This is equivalent to saying, "y is less than 3." The inequality $3 > y$ can also be written as $y < 3$.

$y < 3$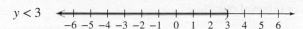

The solution set is the set of real numbers less than 3. Therefore, graph the region on the number line to the left of 3. Use the symbol ")" to denote that the endpoint, 3, is not included in the solution.

Skill Practice Graph the solution sets.

1. $y < 0$ **2.** $x \geq -\dfrac{5}{4}$ **3.** $5 \geq a$

TIP: Some textbooks use a closed circle or an open circle (● or ○) rather than a bracket or parenthesis to denote inclusion or exclusion of a value on the real number line. For example, the solution sets for the inequalities $x > -1$ and $c \leq \frac{7}{3}$ are graphed here.

$x > -1$

$c \leq \frac{7}{3}$

A statement that involves more than one inequality is called a **compound inequality**. One type of compound inequality is used to indicate that one number is between two others. For example, the inequality $-2 < x < 5$ means that $-2 < x$ and $x < 5$. In words, this is easiest to understand if we read the variable first: x is greater than -2 and x is less than 5. The numbers satisfied by these two conditions are those between -2 and 5.

| **Example 2** | **Graphing a Compound Inequality** |

Graph the solution set of the inequality: $-4.1 < y \leq -1.7$

Solution:

$-4.1 < y \leq -1.7$ means that

$-4.1 < y$ and $y \leq -1.7$

Shade the region of the number line greater than -4.1 and less than or equal to -1.7.

Classroom Example: p. 176, Exercise 12

Skill Practice Graph the solution set.

4. $0 \leq y \leq 8.5$

Answers

1.

2.

3.

4.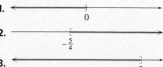

2. Set-Builder Notation and Interval Notation

Graphing the solution set to an inequality is one way to define the set. Two other methods are to use **set-builder notation** or **interval notation**.

Set-Builder Notation

The solution to the inequality $x \geq 2$ can be expressed in set-builder notation as follows:

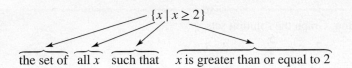

$$\{x \mid x \geq 2\}$$

the set of   all x such that x is greater than or equal to 2

Interval Notation

To understand interval notation, first think of a number line extending infinitely far to the right and infinitely far to the left. Sometimes we use the infinity symbol, ∞, or negative infinity symbol, $-\infty$, to label the far right and far left ends of the number line (Figure 2-10).

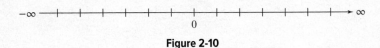

$$-\infty \quad \xrightarrow{\hspace{3cm}} \quad \infty$$
$$0$$

Figure 2-10

To express the solution set of an inequality in interval notation, sketch the graph first. Then use the endpoints to define the interval.

Inequality	Graph	Interval Notation
$x \geq 2$		$[2, \infty)$

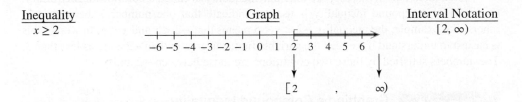

The graph of the solution set $x \geq 2$ begins at 2 and extends infinitely far to the right. The corresponding interval notation begins at 2 and extends to ∞. Notice that a square bracket [is used at 2 for both the graph and the interval notation. A parenthesis is always used at ∞ and for $-\infty$, because there is no endpoint.

Using Interval Notation

- The endpoints used in interval notation are always written from left to right. That is, the smaller number is written first, followed by a comma, followed by the larger number.
- A parenthesis, (or), indicates that an endpoint is *excluded* from the set.
- A square bracket, [or], indicates that an endpoint is *included* in the set.
- A parenthesis, (or), is always used with $-\infty$ or ∞, respectively.

In Table 2-1, we present examples of eight different scenarios for interval notation and the corresponding graph.

Table 2-1

Interval Notation	Graph	Interval Notation	Graph
(a, ∞)	a	$[a, \infty)$	a
$(-\infty, a)$	a	$(-\infty, a]$	a
(a, b)	a b	$[a, b]$	a b
$(a, b]$	a b	$[a, b)$	a b

Example 3 ## Using Set-Builder Notation and Interval Notation

Complete the chart.

Classroom Examples: pp. 176–177, Exercises 20, 24, and 30

Set-Builder Notation	Graph	Interval Notation
	$-6\ -5\ -4\ -3\ -2\ -1\ \ 0\ \ 1\ \ 2\ \ 3\ \ 4\ \ 5\ \ 6$	
		$[-\frac{1}{2}, \infty)$
$\{y \mid -2 \le y < 4\}$		

Solution:

Set-Builder Notation	Graph	Interval Notation
$\{x \mid x < -3\}$	$-6\ -5\ -4\ -3\ -2\ -1\ \ 0\ \ 1\ \ 2\ \ 3\ \ 4\ \ 5\ \ 6$	$(-\infty, -3)$
$\{x \mid x \ge -\frac{1}{2}\}$	$-6\ -5\ -4\ -3\ -2\ -1\ \ 0\ \ 1\ \ 2\ \ 3\ \ 4\ \ 5\ \ 6$ $-\frac{1}{2}$	$[-\frac{1}{2}, \infty)$
$\{y \mid -2 \le y < 4\}$	$-6\ -5\ -4\ -3\ -2\ -1\ \ 0\ \ 1\ \ 2\ \ 3\ \ 4\ \ 5\ \ 6$	$[-2, 4)$

Skill Practice Express each of the following in set-builder notation and interval notation.

5. -2

6. $x < \dfrac{3}{2}$

7. -3 1

3. Addition and Subtraction Properties of Inequality

The process to solve a linear inequality is very similar to the method used to solve linear equations. Recall that adding or subtracting the same quantity to both sides of an equation results in an equivalent equation. The addition and subtraction properties of inequality state that the same is true for an inequality.

Answers

5. $\{x \mid x \ge -2\}$; $[-2, \infty)$
6. $\left\{x \mid x < \dfrac{3}{2}\right\}$; $\left(-\infty, \dfrac{3}{2}\right)$
7. $\{x \mid -3 < x \le 1\}$; $(-3, 1]$

> ### Addition and Subtraction Properties of Inequality
>
> Let a, b, and c represent real numbers.
>
> **1.** *Addition Property of Inequality: If $a < b$,
>   then $a + c < b + c$
>
> **2.** *Subtraction Property of Inequality: If $a < b$,
>   then $a - c < b - c$
>
> *These properties may also be stated for $a \le b$, $a > b$, and $a \ge b$.

To illustrate the addition and subtraction properties of inequality, consider the inequality $5 > 3$. If we add or subtract a real number such as 4 to both sides, the left-hand side will still be greater than the right-hand side. (See Figure 2-11.)

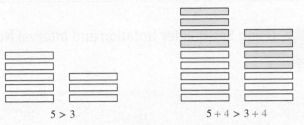

$$5 > 3 \qquad\qquad\qquad 5 + 4 > 3 + 4$$

Figure 2-11

Classroom Example: p. 178,
Exercise 52

> **Example 4**   **Solving a Linear Inequality**
>
> Solve the inequality and graph the solution set. Express the solution set in set-builder notation and in interval notation.
>
> $$-2p + 5 < -3p + 6$$
>
> **Solution:**
>
> $$-2p + 5 < -3p + 6$$
>
> | | |
> |---|---|
> | $-2p + 3p + 5 < -3p + 3p + 6$ | Addition property of inequality (add $3p$ to both sides). |
> | $p + 5 < 6$ | Simplify. |
> | $p + 5 - 5 < 6 - 5$ | Subtraction property of inequality. |
> | $p < 1$ | |
>
> **Instructor Note:** Ask students whether 1 is a solution to the inequality.
>
> Graph: ←————————————)————————→
> -6 -5 -4 -3 -2 -1 $\ 0$ $\ 1$ $\ 2$ $\ 3$ $\ 4$ $\ 5$ $\ 6$
>
> Set-builder notation: $\{p \mid p < 1\}$
>
> Interval notation: $(-\infty, 1)$

Skill Practice Solve the inequality and graph the solution set. Express the solution set in set-builder notation and interval notation.

   8. $2y - 5 < y - 11$

Answer

8.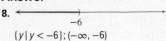
$\{y \mid y < -6\}; (-\infty, -6)$

TIP: The solution to an inequality gives a set of values that make the original inequality true. Therefore, you can test your final answer by using *test points*. That is, pick a value in the proposed solution set and verify that it makes the original inequality true. Furthermore, any test point picked outside the solution set should make the original inequality false. For example,

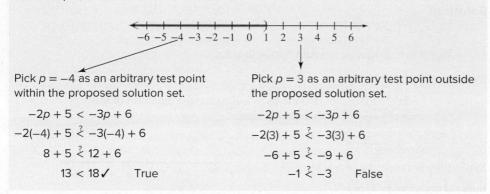

Pick $p = -4$ as an arbitrary test point within the proposed solution set.

$$-2p + 5 < -3p + 6$$
$$-2(-4) + 5 \overset{?}{<} -3(-4) + 6$$
$$8 + 5 \overset{?}{<} 12 + 6$$
$$13 < 18 \checkmark \quad \text{True}$$

Pick $p = 3$ as an arbitrary test point outside the proposed solution set.

$$-2p + 5 < -3p + 6$$
$$-2(3) + 5 \overset{?}{<} -3(3) + 6$$
$$-6 + 5 \overset{?}{<} -9 + 6$$
$$-1 \overset{?}{<} -3 \quad \text{False}$$

4. Multiplication and Division Properties of Inequality

Multiplying both sides of an equation by the same quantity results in an equivalent equation. However, the same is not always true for an inequality. If you multiply or divide an inequality by a negative quantity, the direction of the inequality symbol must be reversed.

For example, consider multiplying or dividing the inequality, $4 < 5$ by -1.

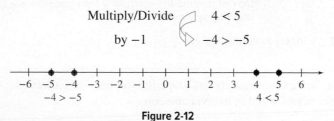

Figure 2-12

Instructor Note: Show that this works for other situations. Start with two negative numbers or one positive and one negative. Show that multiplying or dividing by -1 changes the direction of the inequality sign.

The number 4 lies to the left of 5 on the number line. However, -4 lies to the right of -5 (Figure 2-12). Changing the sign of two numbers changes their relative position on the number line. This is stated formally in the multiplication and division properties of inequality.

Multiplication and Division Properties of Inequality

Let a, b, and c represent real numbers, $c \neq 0$.

*If c is positive and $a < b$, then $ac < bc$ and $\dfrac{a}{c} < \dfrac{b}{c}$

*If c is negative and $a < b$, then $ac > bc$ and $\dfrac{a}{c} > \dfrac{b}{c}$

The second statement indicates that if both sides of an inequality are multiplied or divided by a negative quantity, the inequality sign must be reversed.

*These properties may also be stated for $a \leq b$, $a > b$, and $a \geq b$.

Classroom Example: p. 178, Exercise 74

Example 5   Solving a Linear Inequality

Solve the inequality and graph the solution set. Express the solution set in set-builder notation and in interval notation.

$$-5x - 3 \leq 12$$

Solution:

$$-5x - 3 \leq 12$$

$$-5x - 3 + 3 \leq 12 + 3 \qquad \text{Add 3 to both sides.}$$

$$-5x \leq 15$$

$$\frac{-5x}{-5} \geq \frac{15}{-5} \qquad \text{Divide by } -5. \text{ Reverse the direction of the}$$
$$\text{inequality sign.}$$

$$x \geq -3$$

Set-builder notation: $\{x \mid x \geq -3\}$

Interval notation: $[-3, \infty)$

TIP: The inequality $-5x - 3 \leq 12$, could have been solved by isolating x on the right-hand side of the inequality. This would create a positive coefficient on the variable term and eliminate the need to divide by a negative number.

$$-5x - 3 \leq 12$$

$$-3 \leq 5x + 12$$

$$-15 \leq 5x \qquad \text{Notice that the coefficient of } x \text{ is positive.}$$

$$\frac{-15}{5} \leq \frac{5x}{5} \qquad \text{Do not reverse the inequality sign because we are}$$
$$\text{dividing by a positive number.}$$

$$-3 \leq x, \text{ or equivalently, } x \geq -3$$

Skill Practice Solve the inequality and graph the solution set. Express the solution set in set-builder notation and in interval notation.

9. $-5p + 2 > 22$

Classroom Example: p. 178, Exercise 86

Example 6   Solving a Linear Inequality

Solve the inequality and graph the solution set. Express the solution set in set-builder notation and in interval notation.

$$12 - 2(y + 3) < -3(2y - 1) + 2y$$

Solution:

$$12 - 2(y + 3) < -3(2y - 1) + 2y$$

$$12 - 2y - 6 < -6y + 3 + 2y \qquad \text{Clear parentheses.}$$

$$-2y + 6 < -4y + 3 \qquad \text{Combine } like \text{ terms.}$$

$$-2y + 4y + 6 < -4y + 4y + 3 \qquad \text{Add } 4y \text{ to both sides.}$$

$$2y + 6 < 3 \qquad \text{Simplify.}$$

$$2y + 6 + (-6) < 3 + (-6) \qquad \text{Add } -6 \text{ to both sides.}$$

$$2y < -3 \qquad \text{Simplify.}$$

$$\frac{2y}{2} < \frac{-3}{2} \qquad \begin{array}{l} \text{Divide by 2. The direction of the} \\ \text{inequality sign is } not \text{ reversed because} \\ \text{we divided by a positive number.} \end{array}$$

$$y < -\frac{3}{2}$$

Answer

9.

$\{p \mid p < -4\}; (-\infty, -4)$

Set-builder notation: $\left\{ y \mid y < -\dfrac{3}{2} \right\}$

Interval notation: $\left(-\infty, -\dfrac{3}{2} \right)$

Skill Practice Solve the inequality and graph the solution set. Express the solution set in set-builder notation and in interval notation.

10. $-8 - 8(2x - 4) > -5(4x - 5) - 21$

Example 7 **Solving a Linear Inequality**

Classroom Example: p. 178, Exercise 90

Solve the inequality and graph the solution set. Express the solution set in set-builder notation and in interval notation.

$$-\frac{1}{4}k + \frac{1}{6} \le 2 + \frac{2}{3}k$$

Solution:

$$-\frac{1}{4}k + \frac{1}{6} \le 2 + \frac{2}{3}k$$

$$12\left(-\frac{1}{4}k + \frac{1}{6} \right) \le 12\left(2 + \frac{2}{3}k \right)$$
Multiply both sides by 12 to clear fractions. (Because we multiplied by a positive number, the inequality sign is not reversed.)

$$\frac{12}{1}\left(-\frac{1}{4}k \right) + \frac{12}{1}\left(\frac{1}{6} \right) \le 12(2) + \frac{12}{1}\left(\frac{2}{3}k \right)$$
Apply the distributive property.

$$-3k + 2 \le 24 + 8k$$
Simplify.

$$-3k - 8k + 2 \le 24 + 8k - 8k$$
Subtract $8k$ from both sides.

$$-11k + 2 \le 24$$

$$-11k + 2 - 2 \le 24 - 2$$
Subtract 2 from both sides.

$$-11k \le 22$$

$$\frac{-11k}{-11} \ge \frac{22}{-11}$$
Divide both sides by -11. Reverse the inequality sign.

$$k \ge -2$$

Set-builder notation: $\{ k \mid k \ge -2 \}$

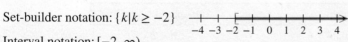

Interval notation: $[-2, \infty)$

Skill Practice Solve the inequality and graph the solution set. Express the solution set in set-builder notation and in interval notation.

11. $\dfrac{1}{5}t + 7 \le \dfrac{1}{2}t - 2$

Answers

10.

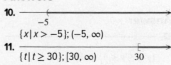

$\{ x \mid x > -5 \}$; $(-5, \infty)$

11.

$\{ t \mid t \ge 30 \}$; $[30, \infty)$

5. Inequalities of the Form $a < x < b$

To solve a compound inequality of the form $a < x < b$ we can work with the inequality as a three-part inequality and isolate the variable, x, as demonstrated in Example 8.

Classroom Example: p. 178, Exercise 70

Example 8   **Solving a Compound Inequality of the Form $a < x < b$**

Solve the inequality and graph the solution set. Express the solution set in set-builder notation and in interval notation.

$$-3 \leq 2x + 1 < 7$$

Solution:

To solve the compound inequality $-3 \leq 2x + 1 < 7$ isolate the variable x in the middle. The operations performed on the middle portion of the inequality must also be performed on the left-hand side and right-hand side.

$$-3 \leq 2x + 1 < 7$$

$$-3 - 1 \leq 2x + 1 - 1 < 7 - 1 \qquad \text{Subtract 1 from all three parts of the inequality.}$$

$$-4 \leq 2x < 6 \qquad \text{Simplify.}$$

$$\frac{-4}{2} \leq \frac{2x}{2} < \frac{6}{2} \qquad \text{Divide by 2 in all three parts of the inequality.}$$

$$-2 \leq x < 3$$

Set-builder notation: $\{x \mid -2 \leq x < 3\}$

Interval notation: $[-2, 3)$

Skill Practice Solve the inequality and graph the solution set. Express the solution set in set-builder notation and in interval notation.

12. $-3 \leq -5 + 2y < 11$

6. Applications of Linear Inequalities

Table 2-2 provides several commonly used translations to express inequalities.

Table 2-2

English Phrase	Mathematical Inequality
a is less than b	$a < b$
a is greater than b a exceeds b	$a > b$
a is less than or equal to b a is at most b a is no more than b	$a \leq b$
a is greater than or equal to b a is at least b a is no less than b	$a \geq b$

Answer

12. $\{y \mid 1 \leq y < 8\}; [1, 8)$

| Example 9 | **Translating Expressions Involving Inequalities** |

Classroom Examples: p. 179, Exercises 106 and 108

Write the English phrases as mathematical inequalities.

  a. Claude's annual salary, s, is no more than $40,000.

  b. A citizen must be at least 18 years old to vote. (Let a represent a citizen's age.)

  c. An amusement park ride has a height requirement between 48 in. and 70 in. (Let h represent height in inches.)

Solution:

  a. $s \le 40{,}000$ Claude's annual salary, s, is no more than $40,000.

  b. $a \ge 18$ A citizen must be at least 18 years old to vote.

  c. $48 < h < 70$ An amusement park ride has a height requirement between 48 in. and 70 in.

Skill Practice   Write the English phrase as a mathematical inequality.

13. Bill needs a score of at least 92 on the final exam. Let x represent Bill's score.

14. Fewer than 19 cars are in the parking lot. Let c represent the number of cars.

15. The heights, h, of women who wear petite size clothing are typically between 58 in. and 63 in., inclusive.

Linear inequalities are found in a variety of applications. Example 10 can help you determine the minimum grade you need on an exam to get an A in your math course.

| Example 10 | **Solving an Application with Linear Inequalities** |

Classroom Example: p. 179, Exercise 112

To earn an A in a math class, Alsha must average at least 90 on all of her tests. Suppose Alsha has scored 79, 86, 93, 90, and 95 on her first five math tests. Determine the minimum score she needs on her sixth test to get an A in the class.

Solution:

Let x represent the score on the sixth exam.   Label the variable.

$$\left(\begin{array}{c}\text{Average of}\\ \text{all tests}\end{array}\right) \ge 90$$ Create a verbal model.

$$\frac{79 + 86 + 93 + 90 + 95 + x}{6} \ge 90$$ The average score is found by taking the sum of the test scores and dividing by the number of scores.

$$\frac{443 + x}{6} \ge 90$$ Simplify.

$$6\left(\frac{443 + x}{6}\right) \ge (90)6$$ Multiply both sides by 6 to clear fractions.

$$443 + x \ge 540$$ Solve the inequality.

$$x \ge 540 - 443$$ Subtract 443 from both sides.

$$x \ge 97$$ Interpret the results.

Alsha must score at least 97 on her sixth exam to receive an A in the course.

Skill Practice

16. To get at least a B in math, Simon must average at least 80 on all tests. Suppose Simon has scored 60, 72, 98, and 85 on the first four tests. Determine the minimum score he needs on the fifth test to receive a B.

Answers
13. $x \ge 92$
14. $c < 19$
15. $58 \le h \le 63$
16. Simon needs at least 85.

Section 2.8 Practice Exercises

For additional exercises, see Classroom Activities 2.8A–2.8D in the *Student's Resource Manual* at www.mhhe.com/moh.

Study Skills Exercise

Find the page numbers for the Chapter Review Exercises, the Chapter Test, and the Cumulative Review Exercises for this chapter.

Chapter Review Exercises _____   Chapter Test _____

Cumulative Review Exercises _____

Compare these features and state the advantages of each.

Vocabulary and Key Concepts

1. **a.** A relationship of the form $ax + b > c$ or $ax + b < c$ $(a \neq 0)$ is called a ___linear inequality___ in one variable.

   b. A statement that involves more than one ___inequality___ is called a compound inequality.

   c. The notation $\{x \mid x > -9\}$ is an example of ___set-builder___ notation, whereas $(-9, \infty)$ is an example of ___interval___ notation.

Review Problems

For Exercises 2–4, solve the equation.

2. $10y - 7(y + 8) + 13 = 13 - 6(2y + 1)$ $\left\{\frac{10}{3}\right\}$

3. $3(x + 2) - (2x - 7) = -(5x - 1) - 2(x + 6)$ $\{-3\}$

4. $6 - 8(x + 3) + 5x = 5x - (2x - 5) + 13$ $\{-6\}$

Concept 1: Graphing Linear Inequalities

For Exercises 5–16, graph the solution set of each inequality. **(See Examples 1–2.)**

5. $x > 5$

6. $x \geq -7.2$

7. $x \leq \frac{5}{2}$

8. $x < -1$

9. $13 > p$

10. $-12 \geq t$

11. $2 \leq y \leq 6.5$

12. $-3 \leq m \leq \frac{8}{9}$

13. $0 < x < 4$

14. $-4 < y < 1$

15. $1 < p \leq 8$

16. $-3 \leq t < 3$

Concept 2: Set-Builder Notation and Interval Notation

For Exercises 17–22, graph each inequality and write the solution set in interval notation. **(See Example 3.)**

Set-Builder Notation	Graph	Interval Notation
17. $\{x \mid x \geq 6\}$		$[6, \infty)$
18. $\left\{x \mid \frac{1}{2} < x \leq 4\right\}$		$\left(\frac{1}{2}, 4\right]$
19. $\{x \mid x \leq 2.1\}$		$(-\infty, 2.1]$
20. $\left\{x \mid x > \frac{7}{3}\right\}$		$\left(\frac{7}{3}, \infty\right)$
21. $\{x \mid -2 < x \leq 7\}$		$(-2, 7]$
22. $\{x \mid x < -5\}$		$(-\infty, -5)$

For Exercises 23–28, write each set in set-builder notation and in interval notation. **(See Example 3.)**

Set-Builder Notation	Graph	Interval Notation
23. $\left\{x \mid x > \frac{3}{4}\right\}$	$\frac{3}{4}$	$\left(\frac{3}{4}, \infty\right)$
24. $\{x \mid x \le -0.3\}$	-0.3	$(-\infty, -0.3]$
25. $\{x \mid -1 < x < 8\}$	$-1 \quad 8$	$(-1, 8)$
26. $\{x \mid x \ge 0\}$	0	$[0, \infty)$
27. $\{x \mid x \le -14\}$	-14	$(-\infty, -14]$
28. $\{x \mid 0 < x \le 9\}$	$0 \quad 9$	$(0, 9]$

For Exercises 29–34, graph each set and write the set in set-builder notation. **(See Example 3.)**

Set-Builder Notation	Graph	Interval Notation
29. $\{x \mid x \ge 18\}$	18	$[18, \infty)$
30. $\{x \mid -10 \le x \le -2\}$	$-10 \quad -2$	$[-10, -2]$
31. $\{x \mid x < -0.6\}$	-0.6	$(-\infty, -0.6)$
32. $\left\{x \mid x < \frac{5}{3}\right\}$	$\frac{5}{3}$	$\left(-\infty, \frac{5}{3}\right)$
33. $\{x \mid -3.5 \le x < 7.1\}$	$-3.5 \quad 7.1$	$[-3.5, 7.1)$
34. $\{x \mid x \ge -10\}$	-10	$[-10, \infty)$

Concepts 3–4: Properties of Inequality

For Exercises 35–42, solve the equation in part (a). For part (b), solve the inequality and graph the solution set. Write the answer in set-builder notation and interval notation. **(See Examples 4–7.)**

35. a. $x + 3 = 6$ {3}

   b. $x + 3 > 6$

     3

     $\{x \mid x > 3\}; (3, \infty)$

36. a. $y - 6 = 12$ {18}

   b. $y - 6 \ge 12$

     18

     $\{y \mid y \ge 18\}; [18, \infty)$

37. a. $p - 4 = 9$ {13}

   b. $p - 4 \le 9$

     $\{p \mid p \le 13\}; (-\infty, 13]$ 13

38. a. $k + 8 = 10$ {2}

   b. $k + 8 < 10$

     2

     $\{k \mid k < 2\}; (-\infty, 2)$

39. a. $4c = -12$ {−3}

   b. $4c < -12$

     -3

     $\{c \mid c < -3\}; (-\infty, -3)$

40. a. $5d = -35$ {−7}

   b. $5d > -35$

     -7

     $\{d \mid d > -7\}; (-7, \infty)$

41. a. $-10z = 15$ $\left\{-\dfrac{3}{2}\right\}$

   b. $-10z \le 15$

     $-\dfrac{3}{2}$

     $\left\{z \mid z \ge -\dfrac{3}{2}\right\}; \left[-\dfrac{3}{2}, \infty\right)$

42. a. $-2w = 14$ {−7}

   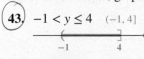 **b.** $-2w < 14$

     -7

     $\{w \mid w > -7\}; (-7, \infty)$

Concept 5: Inequalities of the Form $a < x < b$

For Exercises 43–48, graph the solution and write the set in interval notation. **(See Example 8.)**

43. $-1 < y \le 4$ $(-1, 4]$

   $-1 \quad 4$

44. $2.5 \le t < 5.7$ $[2.5, 5.7)$

   $2.5 \quad 5.7$

45. $0 < x + 3 < 8$ $(-3, 5)$

   $-3 \quad 5$

46. $-2 \le x - 4 \le 3$ $[2, 7]$

   $2 \quad 7$

47. $8 \le 4x \le 24$ $[2, 6]$

   $2 \quad 6$

48. $-9 < 3x < 12$ $(-3, 4)$

   $-3 \quad 4$

Writing   Translating Expression   Geometry   Scientific Calculator   Video

Mixed Exercises

For Exercises 49–96, solve the inequality and graph the solution set. Write the solution set in (a) set-builder notation and (b) interval notation. **(See Exercises 4–8.)**

49. $x + 5 \leq 6$

　　a. $\{x|x \leq 1\}$　**1**　**b.** $(-\infty, 1]$

50. $y - 7 < 6$

　　a. $\{y|y < 13\}$　**b.** $(-\infty, 13)$　**13**

51. $3q - 7 > 2q + 3$

　　10　**a.** $\{q|q > 10\}$　**b.** $(10, \infty)$

52. $5r + 4 \geq 4r - 1$

　　−5　**a.** $\{r|r \geq -5\}$　**b.** $[-5, \infty)$

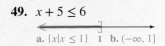　**53.** $4 < 1 + x$

　　3　**a.** $\{x|x > 3\}$　**b.** $(3, \infty)$

54. $3 > z - 6$

　　a. $\{z|z < 9\}$　**9**　**b.** $(-\infty, 9)$

55. $3c > 6$

　　2　**a.** $\{c|c > 2\}$　**b.** $(2, \infty)$

56. $4d \leq 12$

　　3　**a.** $\{d|d \leq 3\}$　**b.** $(-\infty, 3]$

57. $-3c > 6$

　　−2　**a.** $\{c|c < -2\}$　**b.** $(-\infty, -2)$

58. $-4d \leq 12$

　　−3　**a.** $\{d|d \geq -3\}$　**b.** $[-3, \infty)$

59. $-h \leq -14$

　　14　**a.** $\{h|h \geq 14\}$　**b.** $[14, \infty)$

60. $-q > -7$

　　a. $\{q|q < 7\}$　**7**　**b.** $(-\infty, 7)$

61. $12 \geq -\dfrac{x}{2}$

　　−24
　　a. $\{x|x \geq -24\}$　**b.** $[-24, \infty)$

62. $6 < -\dfrac{m}{3}$

　　−18
　　a. $\{m|m < -18\}$　**b.** $(-\infty, -18)$

63. $-2 \leq p + 1 < 4$

　　−3　**3**
　　a. $\{p|-3 \leq p < 3\}$　**b.** $[-3, 3)$

64. $0 < k + 7 < 6$

　　a. $\{k|-7 < k < -1\}$　**b.** $(-7, -1)$

　　−7　**−1**

65. $-3 < 6h - 3 < 12$

　　a. $\left\{h\middle|0 < h < \dfrac{5}{2}\right\}$　**b.** $\left(0, \dfrac{5}{2}\right)$

　　0　$\frac{5}{2}$

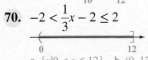　**66.** $-6 \leq 4a - 2 \leq 12$

　　a. $\left\{a\middle|-1 \leq a \leq \dfrac{7}{2}\right\}$　**b.** $\left[-1, \dfrac{7}{2}\right]$

　　−1　$\frac{7}{2}$

67. $-24 < -2x < -20$

　　a. $\{x|10 < x < 12\}$　**b.** $(10, 12)$

　　10　**12**

68. $-12 \leq -3x \leq 6$

　　a. $\{x|-2 \leq x \leq 4\}$　**b.** $[-2, 4]$

　　−2　**4**

69. $-3 \leq \dfrac{1}{4}x - 1 < 5$

　　−8　**24**
　　a. $\{x|-8 \leq x < 24\}$　**b.** $[-8, 24)$

70. $-2 < \dfrac{1}{3}x - 2 \leq 2$

　　0　**12**
　　a. $\{x|0 < x \leq 12\}$　**b.** $(0, 12]$

71. $-\dfrac{2}{3}y < 6$

　　−9
　　a. $\{y|y > -9\}$　**b.** $(-9, \infty)$

72. $\dfrac{3}{4}x \leq -12$

　　−16
　　a. $\{x|x \leq -16\}$　**b.** $(-\infty, -16]$

73. $-2x - 4 \leq 11$

　　$-\frac{15}{2}$　**a.** $\left\{x\middle|x \geq -\dfrac{15}{2}\right\}$　**b.** $\left[-\dfrac{15}{2}, \infty\right)$

74. $-3x + 1 > 0$

　　a. $\left\{x\middle|x < \dfrac{1}{3}\right\}$

　　$\frac{1}{3}$　**b.** $\left(-\infty, \dfrac{1}{3}\right)$

75. $-12 > 7x + 9$

　　a. $\{x|x < -3\}$

　　−3　**b.** $(-\infty, -3)$

76. $8 < 2x - 10$

　　a. $\{x|x > 9\}$　**9**　**b.** $(9, \infty)$

77. $-7b - 3 \leq 2b$

　　a. $\left\{b\middle|b \geq -\dfrac{1}{3}\right\}$

　　$-\frac{1}{3}$　**b.** $\left[-\dfrac{1}{3}, \infty\right)$

78. $3t \geq 7t - 35$

　　a. $\left\{t\middle|t \leq \dfrac{35}{4}\right\}$

　　$\frac{35}{4}$　**b.** $\left(-\infty, \dfrac{35}{4}\right]$

79. $4n + 2 < 6n + 8$

　　−3
　　a. $\{n|n > -3\}$　**b.** $(-3, \infty)$

80. $2w - 1 \leq 5w + 8$

　　−3
　　a. $\{w|w \geq -3\}$　**b.** $[-3, \infty)$

81. $8 - 6(x - 3) > -4x + 12$

　　7
　　a. $\{x|x < 7\}$　**b.** $(-\infty, 7)$

82. $3 - 4(h - 2) > -5h + 6$

　　a. $\{h|h > -5\}$　**b.** $(-5, \infty)$

　　−5

83. $3(x + 1) - 2 \leq \dfrac{1}{2}(4x - 8)$

　　a. $\{x|x \leq -5\}$　**b.** $(-\infty, -5]$

　　−5

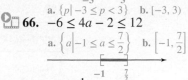　**84.** $8 - (2x - 5) \geq \dfrac{1}{3}(9x - 6)$

　　a. $\{x|x \leq 3\}$　**b.** $(-\infty, 3]$

　　3

85. $4(z - 1) - 6 \geq 6(2z + 3) - 12$

　　a. $\{z|z \leq -2\}$　**b.** $(-\infty, -2]$

　　−2

86. $3(2x + 5) + 2 < 5(2x + 2) + 3$

　　a. $\{x|x > 1\}$　**b.** $(1, \infty)$

　　1

87. $2a + 3(a + 5) > -4a - (3a - 1) + 6$

　　a. $\left\{a\middle|a > -\dfrac{2}{3}\right\}$　**b.** $\left(-\dfrac{2}{3}, \infty\right)$

　　$-\frac{2}{3}$

88. $13 + 7(2y - 3) \leq 12 + 3(3y - 1)$

　　$\frac{17}{5}$
　　a. $\left\{y\middle|y \leq \dfrac{17}{5}\right\}$　**b.** $\left(-\infty, \dfrac{17}{5}\right]$

89. $\dfrac{7}{6}p + \dfrac{4}{3} \geq \dfrac{11}{6}p - \dfrac{7}{6}$

　　$\frac{15}{4}$
　　a. $\left\{p\middle|p \leq \dfrac{15}{4}\right\}$　**b.** $\left(-\infty, \dfrac{15}{4}\right]$

90. $\dfrac{1}{3}w - \dfrac{1}{2} \leq \dfrac{5}{6}w + \dfrac{1}{2}$

　　−2
　　a. $\{w|w \geq -2\}$　**b.** $[-2, \infty)$

91. $\dfrac{y-6}{3} > y+4$

　　⟵———————→
　　　　　−9　**a.** $\{y|y<-9\}$　**b.** $(-\infty,-9)$

92. $\dfrac{5t+7}{2} < t-4$

　　⟵———————→
　　　　−5　**a.** $\{t|t<-5\}$　**b.** $(-\infty,-5)$

93. $-1.2a-0.4 < -0.4a+2$

　　⟵———————→
　　　　　−3　**a.** $\{a|a>-3\}$　**b.** $(-3,\infty)$

94. $-0.4c+1.2 > -2c-0.4$

　　⟵———————→
　　　　　−1
　　a. $\{c|c>-1\}$　**b.** $(-1,\infty)$

95. $-2x+5 \geq -x+5$

　　⟵———————→
　　　　　0
　　a. $\{x|x\leq 0\}$　**b.** $(-\infty,0]$

96. $4x-6 < 5x-6$

　　⟵———————→
　　　　　0
　　a. $\{x|x>0\}$　**b.** $(0,\infty)$

For Exercises 97–100, determine whether the given number is a solution to the inequality.

97. $-2x+5 < 4$;　$x=-2$　No

98. $-3y-7 > 5$;　$y=6$　No

99. $4(p+7)-1 > 2+p$;　$p=1$　Yes

100. $3-k < 2(-1+k)$;　$k=4$　Yes

Concept 6: Applications of Linear Inequalities

For Exercises 101–110, write each English phrase as a mathematical inequality. **(See Example 9.)**

101. The length of a fish, L, was at least 10 in.
$L \geq 10$

102. Tasha's average test score, t, exceeded 90.
$t > 90$

103. The wind speed, w, exceeded 75 mph.　$w > 75$

104. The height of a cave, h, was no more than 2 ft.
$h \leq 2$

105. The temperature of the water in Blue Spring, t, is no more than 72°F.　$t \leq 72$

106. The temperature on the tennis court, t, was no less than 100°F.　$t \geq 100$

107. The length of the hike, L, was no less than 8 km.　$L \geq 8$

108. The depth, d, of a certain pool was at most 10 ft.　$d \leq 10$

109. The snowfall, h, in Monroe County is between 2 in. and 5 in.　$2 < h < 5$

110. The cost, c, of carpeting a room is between $300 and $400.　$300 < c < 400$

© Royalty Free/Corbis RF

111. The average summer rainfall for Miami, Florida, for June, July, and August is 7.4 in. per month. If Miami receives 5.9 in. of rain in June and 6.1 in. in July, how much rain is required in August to exceed the 3-month summer average? **(See Example 10.)**
More than 10.2 in. of rain is needed.

112. The average winter snowfall for Burlington, Vermont, for December, January, and February is 18.7 in. per month. If Burlington receives 22 in. of snow in December and 24 in. in January, how much snow is required in February to exceed the 3-month winter average?　More than 10.1 in. of snow is needed.

113. To earn a B in chemistry, Trevor's average on his five tests must be at least 80. Suppose that Trevor has scored 85, 75, 72, and 82 on his first four chemistry tests. Determine the minimum score needed on his fifth test to get a B in the class.
Trevor needs at least 86 to get a B in the course.

114. In speech class, Carolyn needs at least a B+ to keep her financial aid. To earn a B+, the average of her four speeches must be at least an 85. On the first three speeches she scored 87, 75, and 82. Determine the minimum score on her fourth speech to get a B+.　Carolyn needs at least 96 on her last speech.

115. An artist paints wooden birdhouses. She buys the birdhouses for $9 each. However, for large orders, the price per birdhouse is discounted by a percentage off the original price. Let x represent the number of birdhouses ordered. The corresponding discount is given in the table.

Size of Order	Discount
$x \leq 49$	0%
$50 \leq x \leq 99$	5%
$100 \leq x \leq 199$	10%
$x \geq 200$	20%

a. If the artist places an order for 190 birdhouses, compute the total cost.　$1539

b. Which costs more: 190 birdhouses or 200 birdhouses? Explain your answer.
200 birdhouses cost $1440. It is cheaper to purchase 200 birdhouses because the discount is greater.

116. A wholesaler sells T-shirts to a surf shop at $8 per shirt. However, for large orders, the price per shirt is discounted by a percentage off the original price. Let x represent the number of shirts ordered. The corresponding discount is given in the table.

Number of Shirts Ordered	Discount
$x \leq 24$	0%
$25 \leq x \leq 49$	2%
$50 \leq x \leq 99$	4%
$100 \leq x \leq 149$	6%
$x \geq 150$	8%

   a. If the surf shop orders 50 shirts, compute the total cost. $384

   b. Which costs more: 148 shirts or 150 shirts? Explain your answer.
     It costs $1112.96 for 148 shirts and $1104.00 for 150 shirts. 150 shirts cost less than 148 shirts because the discount is greater.

117. To print a flyer for a new business, Company A charges $39.99 for the design plus $0.50 per flyer. Company B charges $0.60 per flyer but has no design fee. For how many flyers would Company A be a better deal?
Company A is better if more than 400 flyers are printed.

118. Melissa runs a landscaping business. She has equipment and fuel expenses of $313 per month. If she charges $45 for each lawn, how many lawns must she service to make a profit of at least $600 a month?
Melissa must service at least 21 lawns.

119. Madison is planning a 5-night trip to Cancun, Mexico, with her friends. The airfare is $475, her share of the hotel room is $54 per night, and her budget for food and entertainment is $350. She has $700 in savings and has a job earning $10 per hour babysitting. What is the minimum number of hours of babysitting that Madison needs so that she will have enough money to take the trip?   Madison needs to babysit a minimum of 39.5 hr.

120. Luke and Landon are both tutors. Luke charges $50 for an initial assessment and $25 per hour for each hour he tutors. Landon charges $100 for an initial assessment and $20 per hour for tutoring. After how many hours of tutoring will Luke surpass Landon in earnings?   After 10 hr of tutoring, Luke's earnings will surpass Landon's earnings.

Group Activity

Computing Body Mass Index (BMI)

Materials: Calculator

Estimated Time: 10 minutes

Group Size: 2

Body mass index is a statistical measure of an individual's weight in relation to the person's height. It is computed by

$$\text{BMI} = \frac{703W}{h^2}$$
     where W is a person's weight in *pounds*.
         h is the person's height in *inches*.

The National Institutes of Health (NIH) categorizes body mass indices as shown in the table.

Body Mass Index (BMI)	Weight Status
$18.5 \leq \text{BMI} \leq 24.9$	considered ideal
$25.0 \leq \text{BMI} \leq 29.9$	considered overweight
$\text{BMI} \geq 30.0$	considered obese

1. Compute the body mass index for a person 5'4" tall weighing 160 lb. Is this person's weight considered ideal?
27.5; No, the person is considered overweight.

2. At the time that basketball legend Michael Jordan played for the Chicago Bulls, he was 210 lb and stood 6'6" tall. What was Michael Jordan's body mass index?   24.3

3. For a fixed height, body mass index is a function of a person's weight only. For example, for a person 72 in. tall (6 ft), solve the following inequality to determine the person's ideal weight range.   136.4 lb $\leq W \leq$ 183.6 lb

$$18.5 \leq \frac{703W}{(72)^2} \leq 24.9$$

Writing   Translating Expression   Geometry   Scientific Calculator   Video

4. At the time that professional bodybuilder, Jay Cutler, won the Mr. Olympia contest he was 260 lb and stood 5′10″ tall.

  a. What was Jay Cutler's body mass index?   37.3

  b. As a bodybuilder, Jay Cutler has an extraordinarily small percentage of body fat. Yet, according to the chart, would he be considered overweight or obese? Why do you think that the formula is not an accurate measurement of Mr. Cutler's weight status?   Jay Cutler's BMI was 37.3 which placed him in the "obese" category. However, BMI is meant to be used as a simple means of classifying physically inactive individuals with an average body composition. The BMI formula does not account for other factors affecting a person's weight such as fitness level, muscle mass, bone structure, and gender.

Chapter 2   Summary

Section 2.1   Addition, Subtraction, Multiplication, and Division Properties of Equality

Key Concepts

An equation is an algebraic statement that indicates two expressions are equal. A **solution to an equation** is a value of the variable that makes the equation a true statement. The set of all solutions to an equation is the solution set of the equation.

A **linear equation in one variable** can be written in the form $ax + b = c$, where $a \neq 0$.

Addition Property of Equality:

If $a = b$, then $a + c = b + c$

Subtraction Property of Equality:

If $a = b$, then $a - c = b - c$

Multiplication Property of Equality:

If $a = b$, then $ac = bc$ $(c \neq 0)$

Division Property of Equality:

If $a = b$, then $\dfrac{a}{c} = \dfrac{b}{c}$ $(c \neq 0)$

Examples

Example 1

$2x + 1 = 9$ is an equation with solution set $\{4\}$.

Check: $2(4) + 1 \stackrel{?}{=} 9$

$8 + 1 \stackrel{?}{=} 9$

$9 \stackrel{?}{=} 9$ ✓   True

Example 2

$x - 5 = 12$
$x - 5 + 5 = 12 + 5$
$x = 17$ The solution set is $\{17\}$.

Example 3

$z + 1.44 = 2.33$
$z + 1.44 - 1.44 = 2.33 - 1.44$
$z = 0.89$ The solution set is $\{0.89\}$.

Example 4

$\dfrac{3}{4}x = 12$

$\dfrac{4}{3} \cdot \dfrac{3}{4}x = 12 \cdot \dfrac{4}{3}$

$x = 16$ The solution set is $\{16\}$.

Example 5

$16 = 8y$

$\dfrac{16}{8} = \dfrac{8y}{8}$

$2 = y$ The solution set is $\{2\}$.

Section 2.2	Solving Linear Equations

Key Concepts

Steps for Solving a Linear Equation in One Variable:

1. Simplify both sides of the equation.
   - Clear parentheses
   - Combine *like* terms
2. Use the addition or subtraction property of equality to collect the variable terms on one side of the equation.
3. Use the addition or subtraction property of equality to collect the constant terms on the other side of the equation.
4. Use the multiplication or division property of equality to make the coefficient of the variable term equal to 1.
5. Check your answer.

A **conditional equation** is true for some values of the variable but is false for other values.

An equation that has all real numbers as its solution set is an **identity**.

An equation that has no solution is a **contradiction**.

Examples

Example 1

$$5y + 7 = 3(y - 1) + 2$$

$5y + 7 = 3y - 3 + 2$	Clear parentheses.
$5y + 7 = 3y - 1$	Combine *like* terms.
$2y + 7 = -1$	Collect the variable terms.
$2y = -8$	Collect the constant terms.
$y = -4$	Divide both sides by 2.

Check:

$$5(-4) + 7 \stackrel{?}{=} 3[(-4) - 1] + 2$$
$$-20 + 7 \stackrel{?}{=} 3(-5) + 2$$
$$-13 \stackrel{?}{=} -15 + 2$$

The solution set is $\{-4\}$. $-13 \stackrel{?}{=} -13$ ✓   True

Example 2

$x + 5 = 7$ is a conditional equation because it is true only on the condition that $x = 2$.

Solution set: $\{2\}$

Example 3

$$x + 4 = 2(x + 2) - x$$
$$x + 4 = 2x + 4 - x$$
$$x + 4 = x + 4$$
$$4 = 4 \quad \text{is an identity.}$$

Solution set: The set of real numbers.

Example 4

$$y - 5 = 2(y + 3) - y$$
$$y - 5 = 2y + 6 - y$$
$$y - 5 = y + 6$$
$$-5 = 6 \quad \text{is a contradiction.}$$

Solution set: $\{\ \}$

Section 2.3   Linear Equations: Clearing Fractions and Decimals

Key Concepts

Steps for Solving a Linear Equation in One Variable:

1. Simplify both sides of the equation.
   - Clear parentheses
   - Consider clearing fractions or decimals (if any are present) by multiplying both sides of the equation by a common denominator of all terms
   - Combine *like* terms
2. Use the addition or subtraction property of equality to collect the variable terms on one side of the equation.
3. Use the addition or subtraction property of equality to collect the constant terms on the other side of the equation.
4. Use the multiplication or division property of equality to make the coefficient of the variable term equal to 1.
5. Check your answer.

Examples

Example 1

$$\frac{1}{2}x - 2 - \frac{3}{4}x = \frac{7}{4}$$

$$\frac{4}{1}\left(\frac{1}{2}x - 2 - \frac{3}{4}x\right) = \frac{4}{1}\left(\frac{7}{4}\right) \quad \text{Multiply by the LCD.}$$

$$\frac{4}{1}\left(\frac{1}{2}x\right) - \frac{4}{1}\left(\frac{2}{1}\right) - \frac{4}{1}\left(\frac{3}{4}x\right) = \frac{4}{1}\left(\frac{7}{4}\right)$$

$$2x - 8 - 3x = 7 \qquad \text{Apply distributive property.}$$

$$-x - 8 = 7 \qquad \text{Combine } like \text{ terms.}$$

$$-x = 15 \qquad \text{Add 8 to both sides.}$$

$$x = -15 \qquad \text{Divide by } -1.$$

The solution set is $\{-15\}$.

Example 2

$$-1.2x - 5.1 = 16.5$$

$$10(-1.2x - 5.1) = 10(16.5) \qquad \text{Multiply both sides by 10.}$$

$$-12x - 51 = 165$$

$$-12x = 216$$

$$\frac{-12x}{-12} = \frac{216}{-12}$$

$$x = -18$$

The solution set is $\{-18\}$.

Section 2.4	**Applications of Linear Equations:**
	Introduction to Problem Solving

Key Concepts

Problem-Solving Steps for Word Problems:

1. Read the problem carefully.
2. Assign labels to unknown quantities.
3. Write a verbal model.
4. Write a mathematical equation.
5. Solve the equation.
6. Interpret the results and write the answer in words.

Examples

Example 1

The perimeter of a triangle is 54 m. The lengths of the sides are represented by three consecutive even integers. Find the lengths of the three sides.

1. Read the problem.

2. Let x represent one side, $x + 2$ represent the second side, and $x + 4$ represent the third side.

3. $\begin{pmatrix} \text{Length of} \\ \text{first side} \end{pmatrix} + \begin{pmatrix} \text{length of} \\ \text{second side} \end{pmatrix} + \begin{pmatrix} \text{length of} \\ \text{third side} \end{pmatrix}$

   $= \text{perimeter}$

4. $x + (x + 2) + (x + 4) = 54$

5. $3x + 6 = 54$
   $ 3x = 48$
   $ x = 16$

6. $x = 16$ represents the length of the shortest side. The lengths of the other sides are given by $x + 2 = 18$ and $x + 4 = 20$.

   The lengths of the three sides are 16 m, 18 m, and 20 m.

Section 2.5   Applications Involving Percents

Key Concepts

The following formula will help you solve basic percent problems.

$$\text{Amount} = (\text{percent})(\text{base})$$

One common use of percents is in computing **sales tax**.

Another use of percents is in computing **simple interest** using the formula:

$$\left(\begin{array}{c}\text{Simple}\\\text{interest}\end{array}\right) = (\text{principal})\left(\begin{array}{c}\text{annual}\\\text{interest}\\\text{rate}\end{array}\right)\left(\begin{array}{c}\text{time in}\\\text{years}\end{array}\right)$$

or $I = Prt$.

Examples

Example 1

A flat screen television costs $1260.00 after a 5% sales tax is included. What was the price before tax?

$$\left(\begin{array}{c}\text{Price}\\\text{before tax}\end{array}\right) + (\text{tax}) = \left(\begin{array}{c}\text{total}\\\text{price}\end{array}\right)$$

$$x \quad\quad + 0.05x = 1260$$
$$1.05x = 1260$$
$$x = 1200$$

The television costs $1200 before tax.

Example 2

John Li invests $5400 at 2.5% simple interest. How much interest does he earn after 5 years?

$$I = Prt$$
$$I = (\$5400)(0.025)(5)$$
$$I = \$675$$

Section 2.6   Formulas and Applications of Geometry

Key Concepts

A **literal equation** is an equation that has more than one variable. Often such an equation can be manipulated to solve for different variables.

Examples

Example 1

$$P = 2a + b, \text{ solve for } a.$$

$$P - b = 2a + b - b$$
$$P - b = 2a$$
$$\frac{P-b}{2} = \frac{2a}{2}$$
$$\frac{P-b}{2} = a \quad \text{or} \quad a = \frac{P-b}{2}$$

Example 2

Find the length of a side of a square whose perimeter is 28 ft.

Use the formula $P = 4s$. Substitute 28 for P and solve:

$$P = 4s$$
$$28 = 4s$$
$$7 = s$$

The length of a side of the square is 7 ft.

Examples

Example 1 illustrates a mixture problem.

Example 1

How much 80% disinfectant solution should be mixed with 8 L of a 30% disinfectant solution to make a 40% solution?

	80% Solution	30% Solution	40% Solution
Amount of Solution	x	8	$x + 8$
Amount of Pure Disinfectant	$0.80x$	$0.30(8)$	$0.40(x + 8)$

$0.80x + 0.30(8) = 0.40(x + 8)$

$\quad 0.80x + 2.4 = 0.40x + 3.2$

$\quad\quad 0.40x + 2.4 = 3.2$ Subtract $0.40x$.

$\quad\quad\quad 0.40x = 0.80$ Subtract 2.4.

$\quad\quad\quad\quad x = 2$ Divide by 0.40.

2 L of 80% solution is needed.

Examples

Example 2 illustrates a uniform motion problem.

Example 2

Jack and Diane participate in a bicycle race. Jack rides the first half of the race in 1.5 hr. Diane rides the second half at a rate 5 mph slower than Jack and completes her portion in 2 hr. How fast does each person ride?

	Distance	Rate	Time
Jack	$1.5x$	x	1.5
Diane	$2(x - 5)$	$x - 5$	2

$$\begin{pmatrix} \text{Distance} \\ \text{Jack rides} \end{pmatrix} = \begin{pmatrix} \text{distance} \\ \text{Diane rides} \end{pmatrix}$$

$\quad 1.5x \quad\quad = 2(x - 5)$

$\quad\quad 1.5x = 2x - 10$

$\quad\quad -0.5x = -10$ Subtract $2x$.

$\quad\quad\quad x = 20$ Divide by -0.5.

Jack's speed is x. Jack rides 20 mph. Diane's speed is $x - 5$, which is 15 mph.

Key Concepts

A **linear inequality in one variable**, x, is any relationship in the form: $ax + b < c$, $ax + b > c$, $ax + b \leq c$, or $ax + b \geq c$, where $a \neq 0$.

    The solution set to an inequality can be expressed as a graph or in **set-builder notation** or in **interval notation**.

    When graphing an inequality or when writing interval notation, a parenthesis, (or), is used to denote that an endpoint is *not included* in a solution set. A square bracket, [or], is used to show that an endpoint *is included* in a solution set. A parenthesis (or) is always used with $-\infty$ and ∞, respectively.

    The inequality $a < x < b$ is used to show that x is greater than a and less than b. That is, x is *between* a and b.

    Multiplying or dividing an inequality by a negative quantity requires the direction of the inequality sign to be reversed.

Example

Example 1

$\quad -2x + 6 \geq 14$

$-2x + 6 - 6 \geq 14 - 6$ Subtract 6.

$\quad\quad -2x \geq 8$ Simplify.

$\quad\quad \dfrac{-2x}{-2} \leq \dfrac{8}{-2}$ Divide by -2. Reverse the inequality sign.

$\quad\quad\quad x \leq -4$

Graph:

                -4

Set-builder notation: $\{x \mid x \leq -4\}$

Interval notation: $(-\infty, -4]$

Chapter 2 Review Exercises

Section 2.1

1. Label the following as either an expression or an equation:

   a. $3x + y = 10$
   Equation
   b. $9x + 10x - 2xy$
   Expression
   c. $4(x + 3) = 12$
   Equation
   d. $-5x = 7$
   Equation

2. Explain how to determine whether an equation is linear in one variable.   A linear equation can be written in the form $ax + b = c,\ a \neq 0$.

3. Determine if the given equation is a linear equation in one variable. Answer yes or no.

   a. $4x^2 + 8 = -10$
   No
   b. $x + 18 = 72$
   Yes
   c. $-3 + 2y^2 = 0$
   No
   d. $-4p - 5 = 6p$
   Yes

4. For the equation, $4y + 9 = -3$, determine if the given numbers are solutions.

   a. $y = 3$ No
   b. $y = -3$ Yes

For Exercises 5–12, solve each equation using the addition property, subtraction property, multiplication property, or division property of equality.

5. $a + 6 = -2$ $\{-8\}$
6. $6 = z - 9$ $\{15\}$

7. $-\dfrac{3}{4} + k = \dfrac{9}{2}$ $\left\{\dfrac{21}{4}\right\}$
8. $0.1r = 7$ $\{70\}$

9. $-5x = 21$ $\left\{-\dfrac{21}{5}\right\}$
10. $\dfrac{t}{3} = -20$ $\{-60\}$

11. $-\dfrac{2}{5}k = \dfrac{4}{7}$ $\left\{-\dfrac{10}{7}\right\}$
12. $-m = -27$ $\{27\}$

13. The quotient of a number and negative six is equal to negative ten. Find the number.
   The number is 60.

14. The difference of a number and $-\frac{1}{8}$ is $\frac{5}{12}$. Find the number.   The number is $\dfrac{7}{24}$.

15. Four subtracted from a number is negative twelve. Find the number.
   The number is -8.

16. The product of a number and $\frac{1}{4}$ is $-\frac{1}{2}$. Find the number.   The number is -2.

Section 2.2

For Exercises 17–28, solve each equation.

17. $4d + 2 = 6$ $\{1\}$
18. $5c - 6 = -9$ $\left\{-\dfrac{3}{5}\right\}$

19. $-7c = -3c - 8$ $\{2\}$
20. $-28 = 5w + 2$ $\{-6\}$

21. $\dfrac{b}{3} + 1 = 0$ $\{-3\}$
22. $\dfrac{2}{3}h - 5 = 7$ $\{18\}$

23. $-3p + 7 = 5p + 1$ $\left\{\dfrac{3}{4}\right\}$
24. $4t - 6 = 12t + 18$ $\{-3\}$

25. $4a - 9 = 3(a - 3)$ $\{0\}$
26. $3(2c + 5) = -2(c - 8)$ $\left\{\dfrac{1}{8}\right\}$

27. $7b + 3(b - 1) + 3 = 2(b + 8)$ $\{2\}$

28. $2 + (18 - x) + 2(x - 1) = 4(x + 2) - 8$ $\{6\}$

29. Explain the difference between an equation that is a contradiction and an equation that is an identity.   A contradiction has no solution and an identity is true for all real numbers.

For Exercises 30–35, label each equation as a conditional equation, a contradiction, or an identity.

30. $x + 3 = 3 + x$
   Identity
31. $3x - 19 = 2x + 1$
   Conditional equation

32. $5x + 6 = 5x - 28$
   Contradiction
33. $2x - 8 = 2(x - 4)$
   Identity

34. $-8x - 9 = -8(x - 9)$
   Contradiction
35. $4x - 4 = 3x - 2$
   Conditional equation

Section 2.3

For Exercises 36–53, solve each equation.

36. $\dfrac{x}{8} - \dfrac{1}{4} = \dfrac{1}{2}$ $\{6\}$
37. $\dfrac{y}{15} - \dfrac{2}{3} = \dfrac{4}{5}$ $\{22\}$

38. $\dfrac{x + 5}{2} - \dfrac{2x + 10}{9} = 5$ $\{13\}$

39. $\dfrac{x - 6}{3} - \dfrac{2x + 8}{2} = 12$ $\{-27\}$

40. $\dfrac{1}{10}p - 3 = \dfrac{2}{5}p$
   $\{-10\}$
41. $\dfrac{1}{4}y - \dfrac{3}{4} = \dfrac{1}{2}y + 1$
   $\{-7\}$

42. $-\dfrac{1}{4}(2 - 3t) = \dfrac{3}{4}$ $\left\{\dfrac{5}{3}\right\}$
43. $\dfrac{2}{7}(w + 4) = \dfrac{1}{2}$ $\left\{-\dfrac{9}{4}\right\}$

44. $17.3 - 2.7q = 10.55$ $\{2.5\}$

45. $4.9z + 4.6 = 3.2z - 2.2$ $\{-4\}$

46. $5.74a + 9.28 = 2.24a - 5.42$ $\{-4.2\}$

47. $62.84t - 123.66 = 4(2.36 + 2.4t)$ $\{2.5\}$

48. $0.05x + 0.10(24 - x) = 0.75(24)$ $\{-312\}$

49. $0.20(x + 4) + 0.65x = 0.20(854)$ $\{200\}$

50. $100 - (t - 6) = -(t - 1)$ $\{\ \}$

51. $3 - (x + 4) + 5 = 3x + 10 - 4x$ $\{\ \}$

52. $5t - (2t + 14) = 3t - 14$ The set of real numbers

53. $9 - 6(2x + 1) = -3(4z - 1)$
The set of real numbers

Section 2.4

54. Twelve added to the sum of a number and two is forty-four. Find the number.   The number is 30.

55. Twenty added to the sum of a number and six is thirty-seven. Find the number.   The number is 11.

56. Three times a number is the same as the difference of twice the number and seven. Find the number.   The number is -7.

57. Eight less than five times a number is forty-eight less than the number. Find the number.
The number is -10.

58. Three times the largest of three consecutive even integers is 76 more than the sum of the other two integers. Find the integers.
The integers are 66, 68, and 70.

59. Ten times the smallest of three consecutive integers is 213 more than the sum of the other two integers. Find the integers.
The integers are 27, 28, and 29.

60. The perimeter of a triangle is 78 in. The lengths of the sides are represented by three consecutive integers. Find the lengths of the sides of the triangle.   The sides are 25 in., 26 in., and 27 in.

61. The perimeter of a pentagon (a five-sided polygon) is 190 cm. The lengths of the sides are represented by consecutive integers. Find the lengths of the sides.   The sides are 36 cm, 37 cm, 38 cm, 39 cm, and 40 cm.

62. Minimum salaries of major league baseball players soared after a new ruling in 1975. In 2010, the minimum salary for a major league player was $400,000. This is 25 times the minimum salary in 1975. Find the minimum salary in 1975.
In 1975 the minimum salary was $16,000.

63. The state of Indiana has approximately 2.1 million more people than Kentucky. Together their populations total 10.3 million. Approximately how many people are in each state?   Indiana has 6.2 million people and Kentucky has 4.1 million.

Section 2.5

 For Exercises 64–69, solve each problem involving percents.

64. What is 35% of 68?
23.8

65. What is 4% of 720?
28.8

66. 53.5 is what percent of 428?   12.5%

67. 68.4 is what percent of 72?   95%

68. 24 is 15% of what number?   160

69. 8.75 is 0.5% of what number?   1750

70. A couple spent a total of $50.40 for dinner. This included a 20% tip and 6% sales tax on the price of the meal. What was the price of the dinner before tax and tip?   The dinner was $40 before tax and tip.

71. Anna Tsao invested $3000 in an account paying 8% simple interest.

    a. How much interest will she earn in $3\frac{1}{2}$ years?
       $840

    b. What will her balance be at that time?
       $3840

72. Eduardo invested money in an account earning 4% simple interest. At the end of 5 years, he had a total of $14,400. How much money did he originally invest?   He invested $12,000.

73. A novel is discounted 30%. The sale price is $20.65. What was the original price?   The novel originally cost $29.50.

Section 2.6

For Exercises 74–81, solve for the indicated variable.

74. $C = K - 273$ for K $K = C + 273$

75. $K = C + 273$ for C $C = K - 273$

76. $P = 4s$ for s $s = \dfrac{P}{4}$

77. $P = 3s$ for s $s = \dfrac{P}{3}$

78. $y = mx + b$ for x $x = \dfrac{y - b}{m}$

79. $a + bx = c$ for x $x = \dfrac{c - a}{b}$

80. $2x + 5y = -2$ for y $y = \dfrac{-2x - 2}{5}$

81. $4(a + b) = Q$ for b $b = \dfrac{Q - 4a}{4}$ or $b = \dfrac{Q}{4} - a$

✏️ Writing   ↔ Translating Expression   ◿ Geometry   🖩 Scientific Calculator   🎬 Video

For Exercises 82–88, use an appropriate geometry formula to solve the problem.

82. Find the height of a parallelogram whose area is 42 m² and whose base is 6 m.   The height is 7 m.

83. The volume of a cone is given by the formula $V = \frac{1}{3}\pi r^2 h$.

a. Solve the formula for h. $h = \frac{3V}{\pi r^2}$

b. Find the height of a right circular cone whose volume is 47.8 in.³ and whose radius is 3 in. Round to the nearest tenth of an inch.
The height is 5.1 in.

84. The smallest angle of a triangle is 2° more than $\frac{1}{4}$ of the largest angle. The middle angle is 2° less than the largest angle. Find the measure of each angle.   The angles are 22°, 78°, and 80°.

85. A carpenter uses a special saw to cut an angle on a piece of framing. If the angles are complementary and one angle is 10° more than the other, find the measure of each angle.   The angles are 50° and 40°.

86. A rectangular window has width 1 ft less than its length. The perimeter is 18 ft. Find the length and the width of the window.
The length is 5 ft and the width is 4 ft.

87. Find the measure of the vertical angles by first solving for x. $x = 20$. The angle measure is 65°.

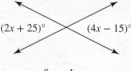

$(2x + 25)°$ $(4x - 15)°$

88. Find the measure of angle y.
The measure of angle y is 53°.

37°

y

Section 2.7

89. In stormy conditions, a delivery truck can travel a route in 14 hr. In good weather, the same trip can be made in 10.5 hr because the truck travels 15 km/hr faster. Find the speed of the truck in stormy weather and the speed in good weather.
The truck travels 45 km/hr in bad weather and 60 km/hr in good weather.

90. Winston and Gus ride their bicycles in a relay. Each person rides the same distance. Winston rides 3 mph faster than Gus and finishes the course in 2.5 hr. Gus finishes in 3 hr. How fast does each person ride?   Gus rides 15 mph and Winston rides 18 mph.

91. Two cars leave a rest stop on Interstate I-10 at the same time. One heads east and the other heads west. One car travels 55 mph and the other 62 mph. How long will it take for them to be 327.6 mi apart?   The cars will be 327.6 mi apart after 2.8 hr (2 hr and 48 min).

92. Two hikers begin at the same time at opposite ends of a 9-mi trail and walk toward each other. One hiker walks 2.5 mph and the other walks 1.5 mph. How long will it be before they meet?
They meet in 2.25 hr (2 hr and 15 min).

93. How much ground beef with 24% fat should be mixed with 8 lb of ground sirloin that is 6% fat to make a mixture that is 9.6% fat?   2 lb of 24% fat content beef is needed.

94. A soldering compound with 40% lead (the rest is tin) must be combined with 80 lb of solder that is 75% lead to make a compound that is 68% lead. How much solder with 40% lead should be used?   20 lb of the 40% solder should be used.

Section 2.8

For Exercises 95–97, graph each inequality and write the set in interval notation.

95. $\{x \mid x > -2\}$ $(-2, \infty)$

-2

96. $\left\{x \mid x \leq \frac{1}{2}\right\}$ $\left(-\infty, \frac{1}{2}\right]$

$\frac{1}{2}$

97. $\{x \mid -1 < x \leq 4\}$

-1 4

$(-1, 4]$

98. A landscaper buys potted geraniums from a nursery at a price of $5 per plant. However, for large orders, the price per plant is discounted by a percentage off the original price. Let x represent the number of potted plants ordered. The corresponding discount is given in the table.

Number of Plants	Discount
$x \leq 99$	0%
$100 \leq x \leq 199$	2%
$200 \leq x \leq 299$	4%
$x \geq 300$	6%

a. Find the cost to purchase 130 plants. $637

b. Which costs more, 300 plants or 295 plants? Explain your answer.   300 plants cost $1410, and 295 plants cost $1416. 295 plants costs more.

For Exercises 99–109, solve the inequality. Graph the solution set and write the answer in set-builder notation and interval notation.

99. $c + 6 < 23$

$\{c | c < 17\}; (-\infty, 17)$

100. $3w - 4 > -5$

$\{w | w > -\frac{1}{3}\}; (-\frac{1}{3}, \infty)$

101. $-2x - 7 \geq 5$

$\{x | x \leq -6\}; (-\infty, -6]$

102. $5(y + 2) \leq -4$

$\{y | y \leq -\frac{14}{5}\}; (-\infty, -\frac{14}{5}]$

103. $-\frac{3}{7}a \leq -21$

$\{a | a \geq 49\}; [49, \infty)$

104. $1.3 > 0.4t - 12.5$

$\{t | t < 34.5\}; (-\infty, 34.5)$

105. $4k + 23 < 7k - 31$

$\{k | k > 18\}; (18, \infty)$

106. $\frac{6}{5}h - \frac{1}{5} \leq \frac{3}{10} + h$

$\{h | h \leq \frac{5}{2}\}; (-\infty, \frac{5}{2}]$

107. $-5x - 2(4x - 3) + 6 > 17 - 4(x - 1)$

$\{x | x < -1\}; (-\infty, -1)$

108. $-6 < 2b \leq 14$

$\{b | -3 < b \leq 7\}; (-3, 7]$

109. $-2 \leq z + 4 \leq 9$

$\{z | -6 \leq z \leq 5\}; [-6, 5]$

110. The summer average rainfall for Bermuda for June, July, and August is 5.3 in. per month. If Bermuda receives 6.3 in. of rain in June and 7.1 in. in July, how much rain is required in August to exceed the 3-month summer average?   More than 2.5 in. is required.

111. Collette has $15.00 to spend on dinner. Of this, 25% will cover the tax and tip, resulting in $11.25 for her to spend on food. If Collette wants veggies and blue cheese, fries, and a drink, what is the maximum number of chicken wings she can get?   Collette can have at most 18 wings.

Wing Special

25¢ each

5:00–7:00 P.M.

*Add veggies and blue cheese for $2.50

*Add fries for $2.50

*Add a drink for $1.75

Chapter 2   Test

1. Which of the equations have $x = -3$ as a solution?

   a. $4x + 1 = 10$ **b.** $6(x - 1) = x - 21$

   c. $5x - 2 = 2x + 1$ **d.** $\frac{1}{3}x + 1 = 0$ b, d

2. a. Simplify: $3x - 1 + 2x + 8$ $5x + 7$

   b. Solve: $3x - 1 = 2x + 8$ $\{9\}$

For Exercises 3–13, solve each equation.

3. $-3x + 5 = -2$ $\{\frac{7}{3}\}$ **4.** $3h + 1 = 3(h + 1)$ $\{\ \}$

5. $t + 3 = -13$ $\{-16\}$

6. $2 + d = 2 - 3(d - 5) - 2$ $\{\frac{13}{4}\}$

7. $8 = p - 4$ $\{12\}$

8. $\frac{3}{7} + \frac{2}{5}x = -\frac{1}{5}x + 1$ $\{\frac{20}{21}\}$ **9.** $\frac{t}{8} = -\frac{2}{9}$ $\{-\frac{16}{9}\}$

10. $2(p - 4) = p + 7$ $\{15\}$

11. $-5(x + 2) + 8x = -2 + 3x - 8$

The set of real numbers

12. $\frac{3x + 1}{2} - \frac{4x - 3}{3} = 1$ $\{-3\}$

13. $0.5c - 1.9 = 2.8 + 0.6c$ $\{-47\}$

14. Solve the equation for y: $3x + y = -4$

$y = -3x - 4$

15. Solve $C = 2\pi r$ for r. $r = \frac{C}{2\pi}$

16. 13% of what is 11.7?   90

17. Graph the inequalities and write the sets in interval notation.

   a. $\{x \mid x < 0\}$ $(-\infty, 0)$

   b. $\{x \mid -2 \leq x < 5\}$ $[-2, 5)$

For Exercises 18–21, solve the inequality. Graph the solution and write the solution set in set-builder notation and interval notation.

18. $5x + 14 > -2x$
   $\{x \mid x > -2\}; (-2, \infty)$

19. $2(3 - x) \geq 14$
   $\{x \mid x \leq -4\}; (-\infty, -4]$

20. $3(2y - 4) + 1 > 2(2y - 3) - 8$
   $\left\{ y \mid y > -\dfrac{3}{2} \right\}; \left(-\dfrac{3}{2}, \infty \right)$

21. $-13 \leq 3p + 2 \leq 5$
   $\{p \mid -5 \leq p \leq 1\}; [-5, 1]$

 22. The total bill for a pair of basketball shoes (including sales tax) is $87.74. If the tax rate is 7%, find the cost of the shoes before tax.
   The cost was $82.00.

23. Clarita borrowed money at a 6% simple interest rate. If she paid back a total of $8000 at the end of 10 yr, how much did she originally borrow?
   Clarita originally borrowed $5000.

24. One number is four plus one-half of another. The sum of the numbers is 31. Find the numbers.
   The numbers are 18 and 13.

25. Two families leave their homes at the same time to meet for lunch. The families live 210 mi apart, and one family drives 5 mph slower than the other. If it takes them 2 hr to meet at a point between their homes, how fast does each family travel?
   One family travels 55 mph and the other travels 50 mph.

26. The average winter snowfall for Syracuse, New York, for December, January, and February is 27.5 in. per month. If Syracuse receives 24 in. of snow in December and 32 in. in January, how much snow is required in February to exceed the 3-month average?   More than 26.5 in. is required.

27. The perimeter of a pentagon (a five-sided polygon) is 315 in. The lengths of the sides are represented by consecutive integers. Find the measures of the sides.
   The sides are 61 in., 62 in., 63 in., 64 in., and 65 in.

28. Matthew mixes macadamia nuts that cost $9.00 per pound with 50 lb of peanuts that cost $5.00 per pound. How many pounds of macadamia nuts should he mix to make a nut mixture that costs $6.50 per pound?
   Matthew needs 30 lb of macadamia nuts.

29. A couple purchased two hockey tickets and two basketball tickets for $153.92. A hockey ticket cost $4.32 more than a basketball ticket. What were the prices of the individual tickets?   Each basketball ticket was $36.32, and each hockey ticket was $40.64.

30. Two angles are complementary. One angle is 26° more than the other angle. What are the measures of the angles?   The measures of the angles are 32° and 58°.

31. The length of a soccer field for international matches is 40 m less than twice its width. If the perimeter is 370 m, what are the dimensions of the field?   The field is 110 m long and 75 m wide.

32. Given the triangle, find the measure of each angle by first solving for y. $y = 30$; The measures of the angles are 30°, 39°, and 111°.

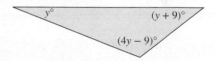

Chapters 1-2   Cumulative Review Exercises

For Exercises 1–5, perform the indicated operations.

1. $\left| -\dfrac{1}{5} + \dfrac{7}{10} \right|$ $\dfrac{1}{2}$

2. $5 - 2[3 - (4 - 7)]$ -7

3. $-\dfrac{2}{3} + \left(\dfrac{1}{2} \right)^2$ $-\dfrac{5}{12}$

4. $-3^2 + (-5)^2$ 16

5. $\sqrt{5 - (-20)} - 3^2$ 4

For Exercises 6–7, translate the mathematical expressions and simplify the results.

6. The square root of the difference of five squared and nine $\sqrt{5^2 - 9}$; 4

7. The sum of -14 and 12 $-14 + 12$; -2

8. List the terms of the expression:
   $-7x^2y + 4xy - 6$ $-7x^2y, 4xy, -6$

9. Simplify: $-4[2x - 3(x + 4)] + 5(x - 7)$
   $9x + 13$

For Exercises 10–15, solve each equation.

10. $8t - 8 = 24$ $\{4\}$

11. $-2.5x - 5.2 = 12.8$ $\{-7.2\}$

12. $-5(p - 3) + 2p = 3(5 - p)$ The set of real numbers

13. $\dfrac{x + 3}{5} - \dfrac{x + 2}{2} = 2$ $\{-8\}$

14. $\dfrac{2}{9}x - \dfrac{1}{3} = x + \dfrac{1}{9}$ $\left\{-\dfrac{4}{7}\right\}$

15. $-0.6w = 48$ $\{-80\}$

16. The sum of two consecutive odd integers is 156. Find the integers.   The numbers are 77 and 79.

17. The cost of a smartphone speaker dock (including sales tax) is $374.50. If the tax rate is 7%, find the cost of the speaker dock before tax.
The cost before tax was $350.00.

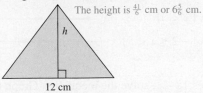

18. The area of a triangle is 41 cm². Find the height of the triangle if the base is 12 cm.

The height is $\frac{41}{6}$ cm or $6\frac{5}{6}$ cm.

h

12 cm

For Exercises 19–20, solve the inequality. Graph the solution set on a number line and express the solution in set-builder notation and interval notation.

19. $-3x - 3(x + 1) < 9$
$\{x \mid x > -2\}; (-2, \infty)$

-2

20. $-6 \leq 2x - 4 \leq 14$
$\{x \mid -1 \leq x \leq 9\}; [-1, 9]$

$-1 \quad 9$

Graphing Linear Equations in Two Variables

3

Mathematics in Gaming

Imagine that you are playing a game of *Battleship*. You say "E-5" to your opponent and his face saddens as he realizes you have sunk his ship. In this game, the "E" and the "5" represent a row or column in a grid, thus defining a *location* on a map. In mathematics we use **ordered pairs**, two numbers of the form (x, y), to determine the location of a point. Furthermore, an equation involving x and y defines a line or curve in a plane that might be the path of a battleship or other object in a computer game.

© ERproductions Ltd/Getty Images RF

For example, suppose that a battleship follows the path defined by $y = 2x + 1$. The value of y depends on the value of x, where y is 1 more than double the value of x. We can determine the value of y for any given value of x by using substitution.

If $x = 1$, then $y = 2(1) + 1 = 3$. Thus, the ordered pair $(1, 3)$ represents a point on the path of the ship.

If $x = 2$, then $y = 2(2) + 1 = 5$. Thus, the ordered pair $(2, 5)$ represents a point on the path of the ship.

In this chapter, we will study equations in x and y and their related graphs. This is important content for a variety of applications, including computer gaming.

1. Interpreting Graphs

Mathematics is a powerful tool used by scientists and has directly contributed to the highly technical world in which we live. Applications of mathematics have led to advances in the sciences, business, computer technology, and medicine.

One fundamental application of mathematics is the graphical representation of numerical information (or data). For example, Table 3-1 represents the number of clients admitted to a drug and alcohol rehabilitation program over a 12-month period.

Table 3-1

	Month	Number of Clients
Jan.	1	55
Feb.	2	62
March	3	64
April	4	60
May	5	70
June	6	73
July	7	77
Aug.	8	80
Sept.	9	80
Oct.	10	74
Nov.	11	85
Dec.	12	90

In table form, the information is difficult to picture and interpret. It appears that on a monthly basis, the number of clients fluctuates. However, when the data are represented in a graph, an upward trend is clear (Figure 3-1).

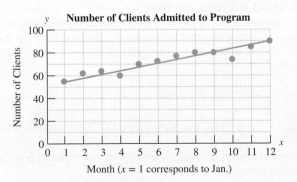

Figure 3-1

From the increase in clients shown in this graph, management for the rehabilitation center might make plans for the future. If the trend continues, management might consider expanding its facilities and increasing its staff to accommodate the expected increase in clients.

Classroom Example: p. 199, Exercise 4

Example 1 | **Interpreting a Graph**

Refer to Figure 3-1 and Table 3-1.

a. For which month was the number of clients the greatest?

b. How many clients were served in the first month (January)?

c. Which month corresponds to 60 clients served?

d. Between which two consecutive months did the number of clients decrease?

e. Between which two consecutive months did the number of clients remain the same?

Solution:

a. Month 12 (December) corresponds to the highest point on the graph. This represents the greatest number of clients, 90.

b. In month 1 (January), there were 55 clients served.

c. Month 4 (April).

d. The number of clients decreased between months 3 and 4 and between months 9 and 10.

e. The number of clients remained the same between months 8 and 9.

Skill Practice Refer to Figure 3-1 and Table 3-1.

1. How many clients were served in October?

2. Which month corresponds to 70 clients?

3. What is the difference between the number of clients in month 12 and month 1?

4. For which month was the number of clients the least?

2. Plotting Points in a Rectangular Coordinate System

The data in Table 3-1 represent a relationship between two variables—the month number and the number of clients. The graph in Figure 3-1 enables us to visualize this relationship. In picturing the relationship between two quantities, we often use a graph with two number lines drawn at right angles to each other (Figure 3-2). This forms a **rectangular coordinate system**. The horizontal line is called the **x-axis**, and the vertical line is called the **y-axis**. The point where the lines intersect is called the **origin**. On the x-axis, the numbers to the right of the origin are positive and the numbers to the left are negative. On the y-axis, the numbers above the origin are positive and the numbers below are negative. The x- and y-axes divide the graphing area into four regions called **quadrants**.

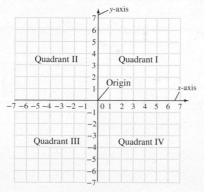

Figure 3-2

Answers

1. 74 clients
2. Month 5 (May)
3. 35 clients
4. Month 1 (January)

Points graphed in a rectangular coordinate system are defined by two numbers as an **ordered pair**, (x, y). The first number (called the **x-coordinate**, or the abscissa) is the horizontal position from the origin. The second number (called the **y-coordinate**, or the ordinate) is the vertical position from the origin. Example 2 shows how points are plotted in a rectangular coordinate system.

Classroom Example: p. 200, Exercise 10

Example 2   **Plotting Points in a Rectangular Coordinate System**

Plot the points.

a. $(4, 5)$ b. $(-4, -5)$ c. $(-1, 3)$ d. $(3, -1)$

e. $\left(\dfrac{1}{2}, -\dfrac{7}{3}\right)$ f. $(-2, 0)$ g. $(0, 0)$ h. $(\pi, 1.1)$

Solution:

See Figure 3-3.

a. The ordered pair $(4, 5)$ indicates that $x = 4$ and $y = 5$. Beginning at the origin, move 4 units in the positive x-direction (4 units to the right), and from there move 5 units in the positive y-direction (5 units up). Then plot the point. The point is in Quadrant I.

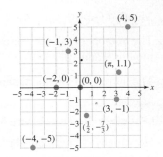

Figure 3-3

b. The ordered pair $(-4, -5)$ indicates that $x = -4$ and $y = -5$. Move 4 units in the negative x-direction (4 units to the left), and from there move 5 units in the negative y-direction (5 units down). Then plot the point. The point is in Quadrant III.

TIP: Notice that changing the order of the x- and y-coordinates changes the location of the point. For example, the point $(-1, 3)$ is in Quadrant II, whereas $(3, -1)$ is in Quadrant IV (Figure 3-3). This is why points are represented by *ordered* pairs. The order of the coordinates is important.

c. The ordered pair $(-1, 3)$ indicates that $x = -1$ and $y = 3$. Move 1 unit to the left and 3 units up. The point is in Quadrant II.

d. The ordered pair $(3, -1)$ indicates that $x = 3$ and $y = -1$. Move 3 units to the right and 1 unit down. The point is in Quadrant IV.

e. The improper fraction $-\dfrac{7}{3}$ can be written as the mixed number $-2\frac{1}{3}$. Therefore, to plot the point $\left(\frac{1}{2}, -\frac{7}{3}\right)$ move to the right $\frac{1}{2}$ unit, and down $2\frac{1}{3}$ units. The point is in Quadrant IV.

f. The point $(-2, 0)$ indicates $y = 0$. Therefore, the point is on the x-axis.

Avoiding Mistakes

Points that lie on either of the axes do not lie in any quadrant.

g. The point $(0, 0)$ is at the origin.

h. The irrational number, π, can be approximated as 3.14. Thus, the point $(\pi, 1.1)$ is located approximately 3.14 units to the right and 1.1 units up. The point is in Quadrant I.

Answer

5.

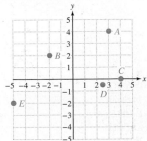

Skill Practice

5. Plot the points.

$A(3, 4)$ $B(-2, 2)$ $C(4, 0)$ $D\left(\dfrac{5}{2}, -\dfrac{1}{2}\right)$ $E(-5, -2)$

3. Applications of Plotting and Identifying Points

The effective use of graphs for mathematical models requires skill in identifying points and interpreting graphs.

Example 3 Determining Points from a Graph

A map of a national park is drawn so that the origin is placed at the ranger station (Figure 3-4). Four fire observation towers are located at points A, B, C, and D. Estimate the coordinates of the fire towers relative to the ranger station (all distances are in miles).

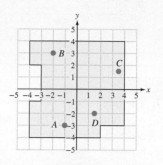

Solution:

Point A: $(-1, -3)$

Point B: $(-2, 3)$

Point C: $(3\frac{1}{2}, 1\frac{1}{2})$ or $(\frac{7}{2}, \frac{3}{2})$ or $(3.5, 1.5)$

Point D: $(1\frac{1}{2}, -2)$ or $(\frac{3}{2}, -2)$ or $(1.5, -2)$

Figure 3-4

Classroom Example: p. 201, Exercise 26

Animation

Skill Practice

6. Towers are located at points A, B, C, and D. Estimate the coordinates of the towers.

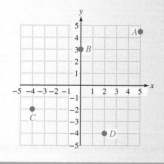

Example 4 Plotting Points in an Application

The daily low temperatures (in degrees Fahrenheit) for one week in January for Sudbury, Ontario, Canada, are given in Table 3-2.

a. Write an ordered pair for each row in the table using the day number as the x-coordinate and the temperature as the y-coordinate.

b. Plot the ordered pairs from part (a) on a rectangular coordinate system.

Table 3-2

Day Number, x	Temperature (°F), y
1	-3
2	-5
3	1
4	6
5	5
6	0
7	-4

Classroom Example: p. 202, Exercise 28

Solution:

a. Each ordered pair represents the day number and the corresponding low temperature for that day.

$(1, -3)$ $(2, -5)$ $(3, 1)$ $(4, 6)$ $(5, 5)$ $(6, 0)$ $(7, -4)$

Answer

6. $A(5, 4\frac{1}{2})$
   $B(0, 3)$
   $C(-4, -2)$
   $D(2, -4)$

b.

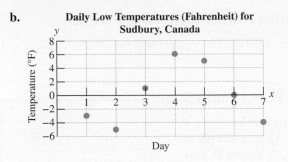

Daily Low Temperatures (Fahrenheit) for Sudbury, Canada

TIP: The graph in Example 4(b) shows only Quadrants I and IV because all *x*-coordinates are positive.

Answer

7.

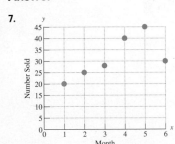

Skill Practice

7. The table shows the number of homes sold in a certain town for a 6-month period. Plot the ordered pairs.

Month, *x*	Number Sold, *y*
1	20
2	25
3	28
4	40
5	45
6	30

Section 3.1   Practice Exercisess

For additional exercises, see Classroom Activities 3.1A–3.1B in the *Student's Resource Manual* at www.mhhe.com/moh.

Study Skills Exercise

> Before you begin the exercises in this section, make your test corrections for the previous test.

Vocabulary and Key Concepts

1. **a.** In a rectangular coordinate system, two number lines are drawn at right angles to each other. The horizontal line is called the ___*x*___-axis, and the vertical line is called the ___*y*-axis___.

b. A point in a rectangular coordinate system is defined by an ___ordered___ pair, (*x*, *y*).

c. In a rectangular coordinate system, the point where the *x*- and *y*-axes intersect is called the ___origin___ and is represented by the ordered pair ___(0, 0)___.

d. The *x*- and *y*-axes divide the coordinate plane into four regions called ___quadrants___.

e. A point with a positive *x*-coordinate and a ___negative___ *y*-coordinate is in Quadrant IV.

f. In Quadrant ___III___, both the *x*- and *y*-coordinates are negative.

Concept 1: Interpreting Graphs

For Exercises 2–6, refer to the graphs to answer the questions.
(See Example 1.)

2. The number of Botox® injection procedures (in millions) in the United States over a 6-yr period is shown in the graph.

a. For which year was the number of procedures the greatest?
Year 3

b. Approximately how many procedures were performed in year 5?   Approximately 2.9 million

c. Which year corresponds to 2.3 million procedures?   Year 1

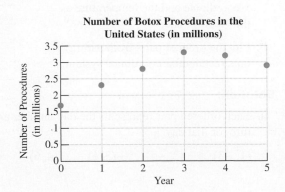

Number of Botox Procedures in the United States (in millions)

3. The number of patients served by a certain hospice care center for the first 12 months after it opened is shown in the graph.

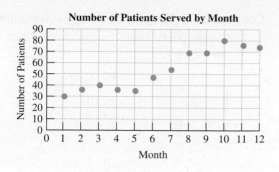

Number of Patients Served by Month

a. For which month was the number of patients greatest?
Month 10

b. How many patients did the center serve in the first month?
30

c. Between which months did the number of patients decrease?
Between months 3 and 5 and between months 10 and 12

d. Between which two months did the number of patients remain the same?   Months 8 and 9

e. Which month corresponds to 40 patients served?   Month 3

f. Approximately how many patients were served during the 10th month?   80

4. Recently the number of housing permits (in thousands) issued by a county in Texas between year 0 and year 6 is shown in the graph.

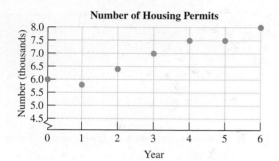

Number of Housing Permits

a. For which year was the number of permits greatest?   Year 6

b. How many permits did the county issue in year 0?
6 thousand or 6000

c. Between which years did the number of permits decrease?
Between year 0 and year 1

d. Between which two years did the number of permits remain the same?   Between year 4 and year 5

e. Which year corresponds to 7000 permits issued?   Year 3

5. The price per share of a stock (in dollars) over a period of 5 days is shown in the graph.

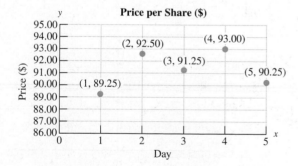

Price per Share ($)

a. Interpret the meaning of the ordered pair (1, 89.25).
On day 1 the price per share was $89.25.

b. What was the change in price between day 3 and day 4?
$1.75

c. What was the change in price between day 4 and day 5?
−$2.75

6. The price per share of a stock (in dollars) over a period of 5 days is shown in the graph.

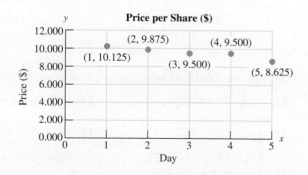

Price per Share ($)

a. Interpret the meaning of the ordered pair (1, 10.125).
On day 1 the price per share was $10.125.

b. What was the change between day 4 and day 5?
−$0.875

c. What is the change between day 1 and day 5?   −$1.50

 Writing   Translating Expression   Geometry   Scientific Calculator   Video

Concept 2: Plotting Points in a Rectangular Coordinate System

7. Plot the points on a rectangular coordinate system.
(See Example 2.)

   a. $(2, 6)$ **b.** $(6, 2)$ **c.** $(-7, 3)$

   d. $(-7, -3)$ **e.** $(0, -3)$ **f.** $(-3, 0)$

   g. $(6, -4)$ **h.** $(0, 5)$

8. Plot the points on a rectangular coordinate system.

   a. $(4, 5)$ **b.** $(-4, 5)$ **c.** $(-6, 0)$

   d. $(6, 0)$ **e.** $(4, -5)$ **f.** $(-4, -5)$

   g. $(0, -2)$ **h.** $(0, 0)$

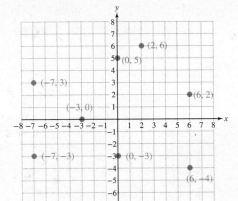

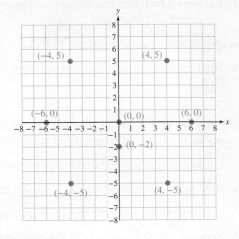

9. Plot the points on a rectangular coordinate system.

   a. $(-1, 5)$ **b.** $(0, 4)$ **c.** $\left(-2, -\dfrac{3}{2}\right)$

   d. $(2, -1.75)$ **e.** $(4, 2)$ **f.** $(-6, 0)$

10. Plot the points on a rectangular coordinate system.

   a. $(7, 0)$ **b.** $(-3, -2)$ **c.** $\left(6\tfrac{3}{5}, 1\right)$

   d. $(0, 1.5)$ **e.** $\left(\dfrac{7}{2}, -4\right)$ **f.** $\left(-\dfrac{7}{2}, 4\right)$

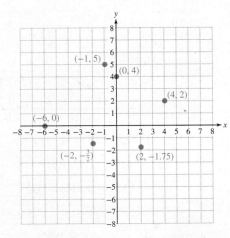

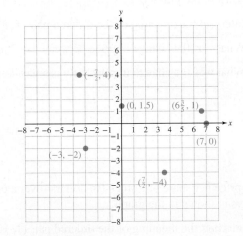

For Exercises 11–18, identify the quadrant in which the given point is located.

11. $(13, -2)$ IV   **12.** $(25, 16)$ I   **13.** $(-8, 14)$ II   **14.** $(-82, -71)$ III

15. $(-5, -19)$ III   **16.** $(-31, 6)$ II   **17.** $\left(\dfrac{5}{2}, \dfrac{7}{4}\right)$ I   **18.** $(9, -40)$ IV

19. Explain why the point $(0, -5)$ is *not* located in Quadrant IV.
$(0, -5)$ lies on the y-axis.

20. Explain why the point $(-1, 0)$ is *not* located in Quadrant II.
$(-1, 0)$ lies on the x-axis.

21. Where is the point $\left(\tfrac{7}{8}, 0\right)$ located?
$\left(\tfrac{7}{8}, 0\right)$ is located on the x-axis.

22. Where is the point $\left(0, \tfrac{6}{5}\right)$ located?
$\left(0, \tfrac{6}{5}\right)$ is located on the y-axis.

Concept 3: Applications of Plotting and Identifying Points

For Exercises 23–24, refer to the graph. **(See Example 3.)**

23. Estimate the coordinates of the points $A, B, C, D, E,$ and F.
$A(-4, 2), B(\frac{1}{2}, 4), C(3, -4), D(-3, -4), E(0, -3), F(5, 0)$

24. Estimate the coordinates of the points $G, H, I, J, K,$ and L.
$G(-5, -2), H(1, -5), I(3\frac{1}{2}, 2), J(-1\frac{1}{2}, 3), K(-2, 0), L(0, 2)$

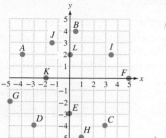

25. A map of a park is laid out with the visitor center located at the origin. Five visitors are in the park located at points $A, B, C, D,$ and E. All distances are in meters.

a. Estimate the coordinates of each visitor. **(See Example 3.)**
$A(400, 200), B(200, -150), C(-300, -200), D(-300, 250), E(0, 450)$

b. How far apart are visitors C and D?   450 m

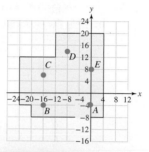

26. A townhouse has a sprinkler system in the backyard. With the water source at the origin, the sprinkler heads are located at points $A, B, C, D,$ and E. All distances are in feet.

a. Estimate the coordinates of each sprinkler head.
$A(0, -4), B(-16, -4), C(-16, 6), D(-8, 14), E(0, 8)$

b. How far is the distance from sprinkler head B to C?   10 ft

27. A movie theater has kept records of popcorn sales versus movie attendance.

a. Use the table to write the corresponding ordered pairs using the movie attendance as the x-variable and sales of popcorn as the y-variable. Interpret the meaning of the first ordered pair. **(See Example 4.)**
$(250, 225), (175, 193), (315, 330), (220, 209), (450, 570), (400, 480), (190, 185)$; the ordered pair $(250, 225)$ means that 250 people produce $225 in popcorn sales.

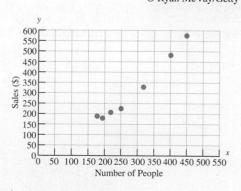

© Ryan McVay/Getty Images RF

b. Plot the data points on a rectangular coordinate system.

Movie Attendance (Number of People)	Sales of Popcorn ($)
250	225
175	193
315	330
220	209
450	570
400	480
190	185

28. The age and systolic blood pressure (in millimeters of mercury, mm Hg) for eight different women are given in the table.

   a. Write the corresponding ordered pairs using the woman's age as the *x*-variable and the systolic blood pressure as the *y*-variable. Interpret the meaning of the first ordered pair.   (57, 149), (41, 120), (71, 158), (36, 115), (64, 151), (25, 110), (40, 118), (77, 165); the ordered pair (57, 149) means that a 57-year-old woman has a systolic blood pressure of 149 mm Hg.

   b. Plot the data points on a rectangular coordinate system.

Age (Years)	Systolic Blood Pressure (mm Hg)
57	149
41	120
71	158
36	115
64	151
25	110
40	118
77	165

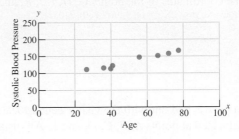

29. The following table shows the average temperature in degrees Celsius for Montreal, Quebec, Canada, by month.

   a. Write the corresponding ordered pairs, letting $x = 1$ correspond to the month of January.   (1, −10.2), (2, −9.0), (3, −2.5), (4, 5.7), (5, 13.0), (6, 18.3), (7, 20.9), (8, 19.6), (9, 14.8), (10, 8.7), (11, 2.0), (12, −6.9)

   b. Plot the ordered pairs on a rectangular coordinate system.

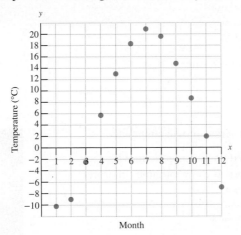

Month, *x*		Temperature (°C), *y*
Jan.	1	−10.2
Feb.	2	−9.0
March	3	−2.5
April	4	5.7
May	5	13.0
June	6	18.3
July	7	20.9
Aug.	8	19.6
Sept.	9	14.8
Oct.	10	8.7
Nov.	11	2.0
Dec.	12	−6.9

30. The table shows the average temperature in degrees Fahrenheit for Fairbanks, Alaska, by month.

   a. Write the corresponding ordered pairs, letting $x = 1$ correspond to the month of January.   (1, −12.8), (2, −4.0), (3, 8.4), (4, 30.2), (5, 48.2), (6, 59.4), (7, 61.5), (8, 56.7), (9, 45.0), (10, 25.0), (11, 6.1), (12, −10.1).

   b. Plot the ordered pairs on a rectangular coordinate system.

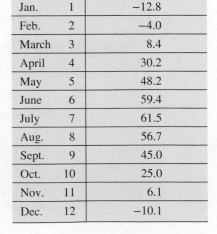

Month, *x*		Temperature (°F), *y*
Jan.	1	−12.8
Feb.	2	−4.0
March	3	8.4
April	4	30.2
May	5	48.2
June	6	59.4
July	7	61.5
Aug.	8	56.7
Sept.	9	45.0
Oct.	10	25.0
Nov.	11	6.1
Dec.	12	−10.1

Expanding Your Skills

31. The data in the table give the percent of males and females who have completed 4 or more years of college education for selected years. Let x represent the number of years since 1960. Let y represent the percent of men and the percent of women that completed 4 or more years of college.

Year	x	Percent, y Men	Percent, y Women
1960	0	9.7	5.8
1970	10	13.5	8.1
1980	20	20.1	12.8
1990	30	24.4	18.4
2000	40	27.8	23.6
2005	45	28.9	26.5
2010	50	29.9	28.8

a. Plot the data points for men and for women on the same graph.

b. Is the percentage of men with 4 or more years of college increasing or decreasing?   Increasing

c. Is the percentage of women with 4 or more years of college increasing or decreasing?   Increasing

32. Use the data and graph from Exercise 31 to answer the questions.

a. In which year was the difference in percentages between men and women with 4 or more years of college the greatest?   1980

b. In which year was the difference in percentages between men and women the least?   2010

c. If the trend continues beyond the data in the graph, does it seem possible that in the future, the percentage of women with 4 or more years of college will be greater than or equal to the percentage of men?   Yes

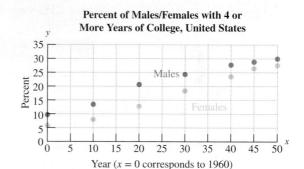

Percent of Males/Females with 4 or More Years of College, United States

---

Linear Equations in Two Variables

1. Definition of a Linear Equation in Two Variables

Recall that an equation in the form $ax + b = c$, where $a \neq 0$, is called a linear equation in one variable. A solution to such an equation is a value of x that makes the equation a true statement. For example, $3x + 5 = -1$ has a solution of -2.

   In this section, we will look at linear equations in *two* variables.

Concepts

1. **Definition of a Linear Equation in Two Variables**
2. **Graphing Linear Equations in Two Variables by Plotting Points**
3. **x- and y-Intercepts**
4. **Horizontal and Vertical Lines**

Linear Equation in Two Variables

Let A, B, and C be real numbers such that A and B are not both zero. Then, an equation that can be written in the form:

$$Ax + By = C$$

is called a **linear equation in two variables**.

 Writing   Translating Expression    Geometry   Scientific Calculator   Video

The equation $x + y = 4$ is a linear equation in two variables. A solution to such an equation is an ordered pair (x, y) that makes the equation a true statement. Several solutions to the equation $x + y = 4$ are listed here:

Solution:	Check:
(x, y)	$x + y = 4$
$(2, 2)$	$(2) + (2) = 4$ ✓
$(1, 3)$	$(1) + (3) = 4$ ✓
$(4, 0)$	$(4) + (0) = 4$ ✓
$(-1, 5)$	$(-1) + (5) = 4$ ✓

By graphing these ordered pairs, we see that the solution points line up (Figure 3-5).

Notice that there are infinitely many solutions to the equation $x + y = 4$ so they cannot all be listed. Therefore, to visualize all solutions to the equation $x + y = 4$, we draw the line through the points in the graph. Every point on the line represents an ordered pair solution to the equation $x + y = 4$, and the line represents the set of *all* solutions to the equation.

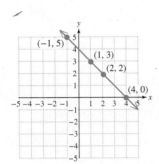

Figure 3-5

Classroom Examples: p. 213, Exercises 10 and 12

Example 1	**Determining Solutions to a Linear Equation**

For the linear equation, $6x - 5y = 12$, determine whether the given ordered pair is a solution.

a. $(2, 0)$ **b.** $(3, 1)$ **c.** $\left(1, -\dfrac{6}{5}\right)$

Solution:

a. $6x - 5y = 12$

$6(2) - 5(0) \overset{?}{=} 12$ Substitute $x = 2$ and $y = 0$.

$\quad\quad 12 - 0 \overset{?}{=} 12$ ✓ True   The ordered pair $(2, 0)$ is a solution.

b. $6x - 5y = 12$

$6(3) - 5(1) \overset{?}{=} 12$ Substitute $x = 3$ and $y = 1$.

$\quad\quad 18 - 5 \neq 12$ The ordered pair $(3, 1)$ is *not* a solution.

c. $6x - 5y = 12$

$6(1) - 5\left(-\dfrac{6}{5}\right) \overset{?}{=} 12$ Substitute $x = 1$ and $y = -\dfrac{6}{5}$.

$\quad\quad 6 + 6 \overset{?}{=} 12$ ✓ True   The ordered pair $\left(1, -\dfrac{6}{5}\right)$ is a solution.

Skill Practice Given the equation $3x - 2y = -12$, determine whether the given ordered pair is a solution.

1. $(4, 0)$ **2.** $(-2, 3)$ **3.** $\left(1, \dfrac{15}{2}\right)$

2. Graphing Linear Equations in Two Variables by Plotting Points

In this section, we will graph linear equations in two variables.

Answers

1. No   **2.** Yes   **3.** Yes

The Graph of an Equation in Two Variables

The graph of an equation in two variables is the graph of all ordered pair solutions to the equation.

The word *linear* means "relating to or resembling a line." It is not surprising then that the solution set for any linear equation in two variables forms a line in a rectangular coordinate system. Because two points determine a line, to graph a linear equation it is sufficient to find two solution points and draw the line between them. We will find three solution points and use the third point as a check point. This process is demonstrated in Example 2.

Example 2   Graphing a Linear Equation

Graph the equation $x - 2y = 8$.

Classroom Example: p. 213, Exercise 18

Solution:

We will find three ordered pairs that are solutions to $x - 2y = 8$. To find the ordered pairs, choose an arbitrary value of x or y. Three choices are recorded in the table. To complete the table, individually substitute each choice into the equation and solve for the missing variable. The substituted value and the solution to the equation form an ordered pair.

x	y	
2		→ (2,)
	−1	→ (, −1)
0		→ (0,)

TIP: Usually we try to choose arbitrary values that will be convenient to graph.

From the first row, substitute $x = 2$:	From the second row, substitute $y = -1$:	From the third row, substitute $x = 0$:
$x - 2y = 8$	$x - 2y = 8$	$x - 2y = 8$
$(2) - 2y = 8$	$x - 2(-1) = 8$	$(0) - 2y = 8$
$-2y = 6$	$x + 2 = 8$	$-2y = 8$
$y = -3$	$x = 6$	$y = -4$

The completed table is shown with the corresponding ordered pairs.

x	y	
2	−3	→ (2, −3)
6	−1	→ (6, −1)
0	−4	→ (0, −4)

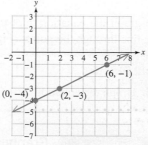

Figure 3-6

To graph the equation, plot the three solutions and draw the line through the points (Figure 3-6).

Avoiding Mistakes

Only two points are needed to graph a line. However, in Example 2, we found a third ordered pair, (0, −4). Notice that this point "lines up" with the other two points. If the three points do not line up, then we know that a mistake was made in solving for at least one of the ordered pairs.

Answer

4.

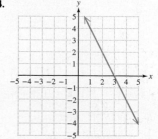

Skill Practice

4. Graph the equation $2x + y = 6$.

In Example 2, the original values for x and y given in the table were chosen arbitrarily by the authors. It is important to note, however, that once you choose an arbitrary value for x, the corresponding y-value is determined by the equation. Similarly, once you choose an arbitrary value for y, the x-value is determined by the equation.

Classroom Example: p. 213, Exercise 22

Example 3   **Graphing a Linear Equation**

Graph the equation $4x + 3y = 15$.

Solution:

We will find three ordered pairs that are solutions to the equation $4x + 3y = 15$. In the table, we have selected arbitrary values for x and y and must complete the ordered pairs. Notice that in this case, we are choosing zero for x and zero for y to illustrate that the resulting equation is often easy to solve.

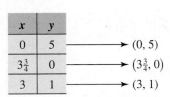

x	y	
0		→ (0,)
	0	→ (, 0)
3		→ (3,)

From the first row, substitute $x = 0$:	From the second row, substitute $y = 0$:	From the third row, substitute $x = 3$:
$4x + 3y = 15$	$4x + 3y = 15$	$4x + 3y = 15$
$4(0) + 3y = 15$	$4x + 3(0) = 15$	$4(3) + 3y = 15$
$3y = 15$	$4x = 15$	$12 + 3y = 15$
$y = 5$	$x = \dfrac{15}{4}$ or $3\dfrac{3}{4}$	$3y = 3$
		$y = 1$

The completed table is shown with the corresponding ordered pairs.

x	y	
0	5	→ (0, 5)
$3\frac{3}{4}$	0	→ $(3\frac{3}{4}, 0)$
3	1	→ (3, 1)

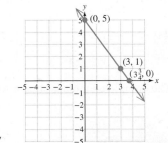

To graph the equation, plot the three solutions and draw the line through the points (Figure 3-7).

Figure 3-7

Skill Practice

5. Graph the equation $2x + 3y = 12$.

Answer

5.

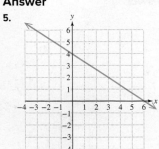

Example 4   **Graphing a Linear Equation**

Graph the equation $y = -\dfrac{1}{3}x + 1$.

Solution:

Because the y-variable is isolated in the equation, it is easy to substitute a value for x and simplify the right-hand side to find y. Since any number for x can be

Classroom Example: p. 215, Exercise 38

chosen, select numbers that are multiples of 3. These will simplify easily when multiplied by $-\frac{1}{3}$.

x	y
3	
0	
−3	

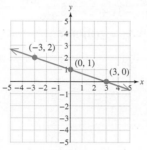

$$y = -\frac{1}{3}x + 1$$

Let $x = 3$:

$$y = -\frac{1}{3}(3) + 1$$
$$y = -1 + 1$$
$$y = 0$$

Let $x = 0$:

$$y = -\frac{1}{3}(0) + 1$$
$$y = 0 + 1$$
$$y = 1$$

Let $x = -3$:

$$y = -\frac{1}{3}(-3) + 1$$
$$y = 1 + 1$$
$$y = 2$$

x	y	
3	0	→ (3, 0)
0	1	→ (0, 1)
−3	2	→ (−3, 2)

The line through the three ordered pairs (3, 0), (0, 1), and (−3, 2) is shown in Figure 3-8. The line represents the set of all solutions to the equation $y = -\frac{1}{3}x + 1$.

Figure 3-8

Skill Practice

6. Graph the equation $y = \frac{1}{2}x + 3$.

3. x- and y-Intercepts

The x- and y-intercepts are the points where the graph intersects the x- and y-axes, respectively. From Example 4, we see that the x-intercept is at the point (3, 0) and the y-intercept is at the point (0, 1). See Figure 3-8. Notice that a y-intercept is a point on the y-axis and must have an x-coordinate of 0. Likewise, an x-intercept is a point on the x-axis and must have a y-coordinate of 0.

> ### Definitions of x- and y-Intercepts
> An **x-intercept** of a graph is a point $(a, 0)$ where the graph intersects the x-axis.
>
> A **y-intercept** of a graph is a point $(0, b)$ where the graph intersects the y-axis.

In some applications, an x-intercept is defined as the x-coordinate of a point of intersection that a graph makes with the x-axis. For example, if an x-intercept is at the point (3, 0), it is sometimes stated simply as 3 (the y-coordinate is assumed to be 0). Similarly, a y-intercept is sometimes defined as the y-coordinate of a point of intersection that a graph makes with the y-axis. For example, if a y-intercept is at the point (0, 7), it may be stated simply as 7 (the x-coordinate is assumed to be 0).

Although any two points may be used to graph a line, in some cases it is convenient to use the x- and y-intercepts of the line. To find the x- and y-intercepts of any two-variable equation in x and y, follow these steps:

> ### Finding x- and y-Intercepts
> - Find the x-intercept(s) by substituting $y = 0$ into the equation and solving for x.
> - Find the y-intercept(s) by substituting $x = 0$ into the equation and solving for y.

Answer

6.

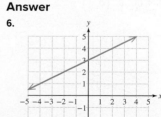

Classroom Example: p. 216, Exercise 50

Example 5   Finding the *x*- and *y*-Intercepts of a Line

Given the equation $-3x + 2y = 8$,

  a. Find the *x*-intercept.

  b. Find the *y*-intercept.

  c. Graph the equation.

Solution:

a. To find the *x*-intercept, substitute $y = 0$.

$$-3x + 2y = 8$$
$$-3x + 2(0) = 8$$
$$-3x = 8$$
$$\frac{-3x}{-3} = \frac{8}{-3}$$
$$x = -\frac{8}{3}$$

The *x*-intercept is $\left(-\frac{8}{3}, 0\right)$.

b. To find the *y*-intercept, substitute $x = 0$.

$$-3x + 2y = 8$$
$$-3(0) + 2y = 8$$
$$2y = 8$$
$$y = 4$$

The *y*-intercept is $(0, 4)$.

Avoiding Mistakes

Be sure to write the *x*- and *y*-intercepts as two separate ordered pairs: $\left(-\frac{8}{3}, 0\right)$ and $(0, 4)$.

c. The line through the ordered pairs $\left(-\frac{8}{3}, 0\right)$ and $(0, 4)$ is shown in Figure 3-9. Note that the point $\left(-\frac{8}{3}, 0\right)$ can be written as $\left(-2\frac{2}{3}, 0\right)$.

The line represents the set of all solutions to the equation $-3x + 2y = 8$.

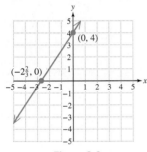

Figure 3-9

Skill Practice   Given the equation $x - 3y = -4$,

7. Find the *x*-intercept.   **8.** Find the *y*-intercept.   **9.** Graph the equation.

Classroom Example: p. 216, Exercise 56

Example 6   Finding the *x*- and *y*-Intercepts of a Line

Given the equation $4x + 5y = 0$,

  a. Find the *x*-intercept.

  b. Find the *y*-intercept.

  c. Graph the equation.

Solution:

Answers

7. $(-4, 0)$ **8.** $\left(0, \frac{4}{3}\right)$

9.

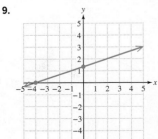

a. To find the *x*-intercept, substitute $y = 0$.

$$4x + 5y = 0$$
$$4x + 5(0) = 0$$
$$4x = 0$$
$$x = 0$$

The *x*-intercept is $(0, 0)$.

b. To find the *y*-intercept, substitute $x = 0$.

$$4x + 5y = 0$$
$$4(0) + 5y = 0$$
$$5y = 0$$
$$y = 0$$

The *y*-intercept is $(0, 0)$.

c. Because the x-intercept and the y-intercept are the same point (the origin), one or more additional points are needed to graph the line. In the table, we have arbitrarily selected additional values for x and y to find two more points on the line.

x	y
-5	
	2

Let $x = -5$:

$$4x + 5y = 0$$
$$4(-5) + 5y = 0$$
$$-20 + 5y = 0$$
$$5y = 20$$
$$y = 4$$

$(-5, 4)$ is a solution.

Let $y = 2$:

$$4x + 5y = 0$$
$$4x + 5(2) = 0$$
$$4x + 10 = 0$$
$$4x = -10$$
$$x = -\frac{10}{4}$$
$$x = -\frac{5}{2}$$

$\left(-\frac{5}{2}, 2\right)$ is a solution.

The line through the ordered pairs $(0, 0)$, $(-5, 4)$, and $\left(-\frac{5}{2}, 2\right)$ is shown in Figure 3-10. Note that the point $\left(-\frac{5}{2}, 2\right)$ can be written as $\left(-2\frac{1}{2}, 2\right)$.

The line represents the set of all solutions to the equation $4x + 5y = 0$.

x	y
-5	4
$-\frac{5}{2}$	2
0	0

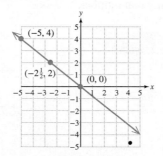

Figure 3-10

Skill Practice Given the equation $2x - 3y = 0$,

10. Find the x-intercept.   **11.** Find the y-intercept.

12. Graph the equation. (*Hint:* You may need to find an additional point.)

4. Horizontal and Vertical Lines

Recall that a linear equation can be written in the form $Ax + By = C$, where A and B are not both zero. However, if A or B is 0, then the line is either horizontal or vertical. A horizontal line either lies on the x-axis or is parallel to the x-axis. A vertical line either lies on the y-axis or is parallel to the y-axis.

Answers

10. $(0, 0)$ **11.** $(0, 0)$

12.

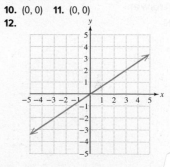

Equations of Vertical and Horizontal Lines

1. A **vertical line** can be represented by an equation of the form $x = k$, where k is a constant.

2. A **horizontal line** can be represented by an equation of the form $y = k$, where k is a constant.

Classroom Example: p. 217, Exercise 68

Example 7 **Graphing a Horizontal Line**

Graph the equation $y = 3$.

Solution:

Because this equation is in the form $y = k$, the line is horizontal and must cross the y-axis at $y = 3$ (Figure 3-11).

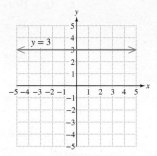

Figure 3-11

> **TIP:** Notice that a horizontal line has a y-intercept, but does not have an x-intercept (unless the horizontal line is the x-axis itself).

Alternative Solution:

Create a table of values for the equation $y = 3$. The choice for the y-coordinate must be 3, but x can be any real number.

x	y
0	3
1	3
2	3

x can be any number.　　y must be 3.

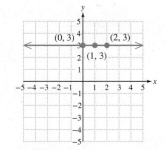

Skill Practice

13. Graph the equation. $y = -2$

Classroom Example: p. 217, Exercise 70

Example 8 **Graphing a Vertical Line**

Graph the equation $7x = -14$.

Solution:

Because the equation does not have a y-variable, we can solve the equation for x.

$$7x = -14 \quad \text{is equivalent to} \quad x = -2$$

This equation is in the form $x = k$, indicating that the line is vertical and must cross the x-axis at $x = -2$ (Figure 3-12).

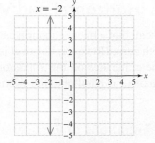

Figure 3-12

Answer

13.

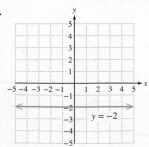

Alternative Solution:

Create a table of values for the equation $x = -2$. The choice for the x-coordinate must be -2, but y can be any real number.

x	y
-2	0
-2	3
-2	-4

x must be -2. y can be any number.

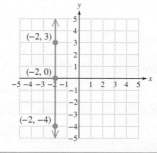

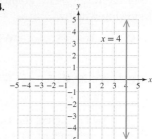

> **TIP:** Notice that a vertical line has an x-intercept but does not have a y-intercept (unless the vertical line is the y-axis itself).

Answer

14.

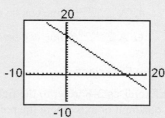

Skill Practice

14. Graph the equation. $3x = 12$

Topic: Graphing Linear Equations on an Appropriate Viewing Window

A viewing window of a graphing calculator shows a portion of a rectangular coordinate system. The standard viewing window for many calculators shows the x-axis between -10 and 10 and the y-axis between -10 and 10 (Figure 3-13). Furthermore, the scale defined by the "tick" marks on both the x- and y-axes is usually set to 1.

The "Standard Viewing Window"

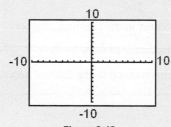

Figure 3-13

To graph an equation in x and y on a graphing calculator, the equation must be written with the y-variable isolated. For example, to graph the equation $x + 3y = 3$, we solve for y by applying the steps for solving a literal equation. The result, $y = -\frac{1}{3}x + 1$, can now be entered into a graphing calculator. To enter the equation $y = -\frac{1}{3}x + 1$, use parentheses around the fraction $\frac{1}{3}$. The *Graph* option displays the graph of the line.

Sometimes the standard viewing window does not provide an adequate display for the graph of an equation. For example, the graph of $y = -x + 15$ is visible only in a small portion of the upper right corner of the standard viewing window.

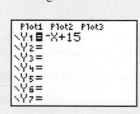

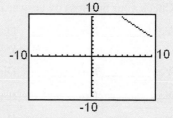

To see where this line crosses the x- and y-axes, we can change the viewing window to accommodate larger values of x and y. Most calculators have a *Range* feature or *Window* feature that allows the user to change the minimum and maximum x- and y-values.

To get a better picture of the equation $y = -x + 15$, change the minimum x-value to -10 and the maximum x-value to 20. Similarly, use a minimum y-value of -10 and a maximum y-value of 20.

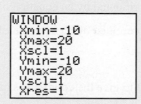

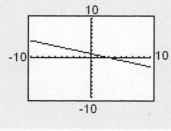

Calculator Exercises

For Exercises 1–8, graph the equations on the standard viewing window.

1. $y = -2x + 5$ **2.** $y = 3x - 1$

3. $y = \dfrac{1}{2}x - \dfrac{7}{2}$

4. $y = -\dfrac{3}{4}x + \dfrac{5}{3}$

5. $4x - 7y = 21$

6. $2x + 3y = 12$

7. $-3x - 4y = 6$

8. $-5x + 4y = 10$

For Exercises 9–12, graph the equations on the given viewing window.

9. $y = 3x + 15$ Window: $-10 \le x \le 10$
 $-5 \le y \le 20$

10. $y = -2x - 25$ Window: $-30 \le x \le 30$
 $-30 \le y \le 30$

 Xscl = 3 (sets the x-axis tick marks to increments of 3)

 Yscl = 3 (sets the y-axis tick marks to increments of 3)

11. $y = -0.2x + 0.04$
 Window: $-0.1 \le x \le 0.3$
 $-0.1 \le y \le 0.1$

 Xscl = 0.01 (sets the x-axis tick marks to increments of 0.01)

 Yscl = 0.01 (sets the y-axis tick marks to increments of 0.01)

12. $y = 0.3x - 0.5$
 Window: $-1 \le x \le 3$
 $-1 \le y \le 1$

 Xscl = 0.1 (sets the x-axis tick marks to increments of 0.1)

 Yscl = 0.1 (sets the y-axis tick marks to increments of 0.1)

Section 3.2 Practice Exercises

For additional exercises, see Classroom Activities 3.2A–3.2C in the *Student's Resource Manual* at www.mhhe.com/moh.

Study Skills Exercise

Check your progress by answering these questions.

Yes _____ No _____ Did you have sufficient time to study for the test in the previous chapter? If not, what could you have done to create more time for studying?

Yes _____ No _____ Did you work all of the assigned homework problems in the previous chapter?

Yes _____ No _____ If you encountered difficulty, did you see your instructor or tutor for help?

Yes _____ No _____ Have you taken advantage of the textbook supplements such as the *Student Solutions Manual?*

Vocabulary and Key Concepts

1. **a.** A linear equation in two variables is an equation that can be written in the form $\underline{Ax + By = C}$ where A and B are not both zero.

 b. A point where a graph intersects the x-axis is called a(n) $\underline{x\text{-intercept}}$.

 c. A point where a graph intersects the y-axis is called a(n) $\underline{y\text{-intercept}}$.

 d. A $\underline{\text{vertical}}$ line can be represented by an equation of the form $x = k$, where k is a constant.

 e. A $\underline{\text{horizontal}}$ line can be represented by an equation of the form $y = k$, where k is a constant.

Review Exercises

For Exercises 2–8, refer to the figure to give the coordinates of the labeled points, and state the quadrant or axis where the point is located.

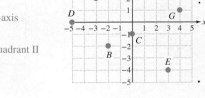

2. A (2, 4); quadrant I

3. B (−2, −2); quadrant III

4. C (0, −1); y-axis

5. D (−5, 0); x-axis

6. E (3, −4); quadrant IV

7. F (−3, 2); quadrant II

8. G (4, 1); quadrant I

 Writing Translating Expression  Geometry Scientific Calculator Video

Concept 1: Definition of a Linear Equation in Two Variables

For Exercises 9–17, determine whether the given ordered pair is a solution to the equation. **(See Example 1.)**

9. $x - y = 6$; $(8, 2)$ Yes

10. $y = 3x - 2$; $(1, 1)$ Yes

 11. $y = -\frac{1}{3}x + 3$; $(-3, 4)$ Yes

12. $y = -\frac{5}{2}x + 5$; $\left(\frac{4}{5}, -3\right)$ No

13. $4x + 5y = 1$; $\left(\frac{1}{4}, -\frac{2}{5}\right)$ No

14. $y = 7$; $(0, 7)$ Yes

15. $y = -2$; $(-2, 6)$ No

16. $x = 1$; $(0, 1)$ No

17. $x = -5$; $(-5, 6)$ Yes

Concept 2: Graphing Linear Equations in Two Variables by Plotting Points

For Exercises 18–31, complete each table, and graph the corresponding ordered pairs. Draw the line defined by the points to represent all solutions to the equation. **(See Examples 2–4.)**

18. $x + y = 3$

x	y
2	1
0	3
-1	4
3	0

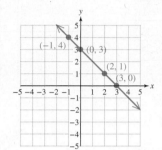

19. $x + y = -2$

x	y
1	-3
-2	0
-3	1
-4	2

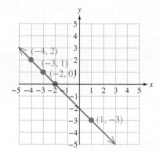

20. $y = 5x + 1$

x	y
1	6
0	1
-1	-4

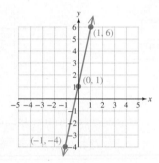

21. $y = -3x - 3$

x	y
-2	3
-1	0
-4	9

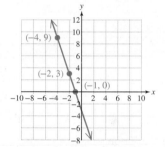

22. $2x - 3y = 6$

x	y
0	-2
3	0
2	-2/3

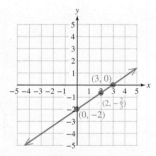

23. $4x + 2y = 8$

x	y
0	4
2	0
3	-2

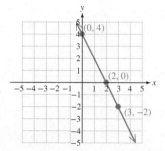

24. $y = \frac{2}{7}x - 5$

x	y
7	-3
-7	-7
0	-5

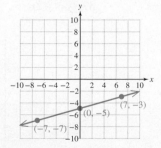

25. $y = -\frac{3}{5}x - 2$

x	y
0	-2
5	-5
10	-8

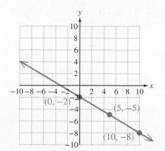

26. $y = 3$

x	y
2	3
0	3
−1	3

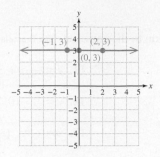

27. $y = -2$

x	y
0	−2
−3	−2
5	−2

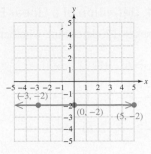

28. $x = -4$

x	y
−4	1
−4	−2
−4	4

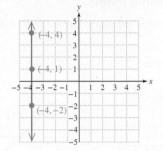

29. $x = \frac{3}{2}$

x	y
3/2	−1
3/2	2
3/2	−3

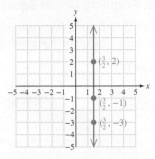

30. $y = -3.4x + 5.8$

x	y
0	5.8
1	2.4
2	−1

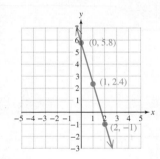

31. $y = -1.2x + 4.6$

x	y
0	4.6
1	3.4
2	2.2

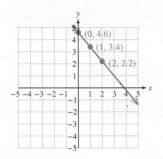

For Exercises 32–43, graph each line by making a table of at least three ordered pairs and plotting the points. **(See Example 4.)**

32. $x - y = 2$

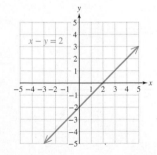

33. $x - y = 4$

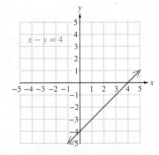

34. $-3x + y = -6$

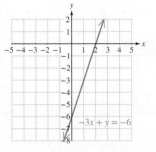

35. $2x - 5y = 10$

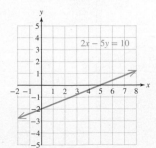

36. $y = 4x$

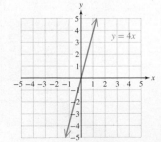

37. $y = -2x$

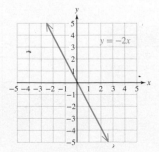

38. $y = -\dfrac{1}{2}x + 3$

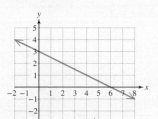

39. $y = \dfrac{1}{4}x - 2$

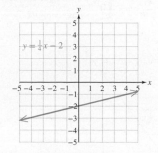

40. $x + y = 0$

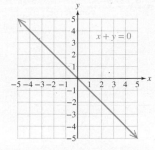

41. $-x + y = 0$

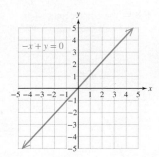

42. $50x - 40y = 200$

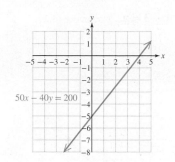

43. $-30x - 20y = 60$

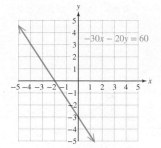

Concept 3: *x*- and *y*-Intercepts

44. The *x*-intercept is on which axis?   *x*-axis

45. The *y*-intercept is on which axis?   *y*-axis

For Exercises 46–49, estimate the coordinates of the *x*- and *y*-intercepts.

46.

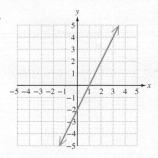

x-intercept: (1, 0);
y-intercept: (0, −2)

47.

x-intercept: (−1, 0);
y-intercept: (0, −3)

48.

x-intercept: (3, 0);
y-intercept: (0, 1)

49.

x-intercept: (−4, 0);
y-intercept: (0, 1)

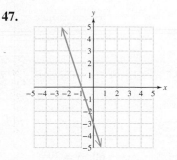

/ Writing   ← → Translating Expression   Geometry   Scientific Calculator   Video

For Exercises 50–61, find the *x*- and *y*-intercepts (if they exist), and graph the line. **(See Examples 5–6.)**

50. $5x + 2y = 5$ *x*-intercept: $(1, 0)$;
y-intercept: $\left(0, \dfrac{5}{2}\right)$

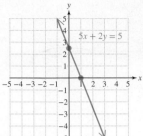

51. $4x - 3y = -9$ *x*-intercept: $\left(-\dfrac{9}{4}, 0\right)$;
y-intercept: $(0, 3)$

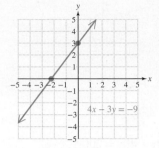

52. $y = \dfrac{2}{3}x - 1$ *x*-intercept: $\left(\dfrac{3}{2}, 0\right)$;
y-intercept: $(0, -1)$

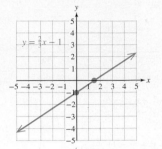

53. $y = -\dfrac{3}{4}x + 2$ *x*-intercept: $\left(\dfrac{8}{3}, 0\right)$;
y-intercept: $(0, 2)$

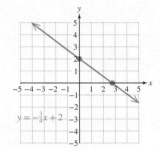

54. $x - 3 = y$ *x*-intercept: $(3, 0)$;
y-intercept: $(0, -3)$

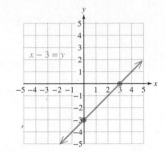

55. $2x + 8 = y$ *x*-intercept: $(-4, 0)$;
y-intercept: $(0, 8)$

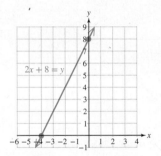

56. $-3x + y = 0$ *x*-intercept: $(0, 0)$;
y-intercept: $(0, 0)$

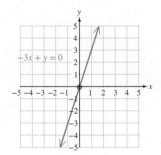

57. $2x - 2y = 0$ *x*-intercept: $(0, 0)$;
y-intercept: $(0, 0)$

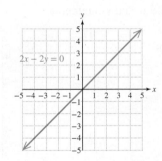

58. $25y = 10x + 100$ *x*-intercept: $(-10, 0)$;
y-intercept: $(0, 4)$

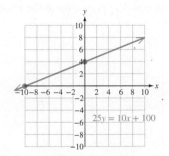

59. $20x = -40y + 200$
x-intercept: $(10, 0)$; *y*-intercept: $(0, 5)$

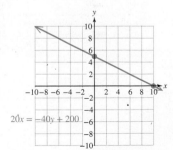

60. $x = 2y$
x-intercept: $(0, 0)$; *y*-intercept: $(0, 0)$

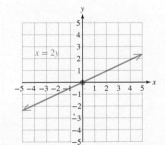

61. $x = -5y$
x-intercept: $(0, 0)$; *y*-intercept: $(0, 0)$

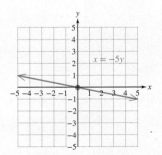

Writing   Translating Expression   Geometry   Scientific Calculator   Video

Concept 4: Horizontal and Vertical Lines

For Exercises 62–65, answer true or false. If the statement is false, rewrite it to be true.

62. The line defined by $x = 3$ is horizontal.
    False, $x = 3$ is vertical.

63. The line defined by $y = -4$ is horizontal.   True

64. A line parallel to the y-axis is vertical.   True

65. A line parallel to the x-axis is horizontal.   True

For Exercises 66–74,

a. Identify the equation as representing a horizontal or vertical line.

b. Graph the line.

c. Identify the x- and y-intercepts if they exist. **(See Examples 7–8.)**

66. $x = 3$ a. Vertical line c. x-intercept: $(3, 0)$; no y-intercept

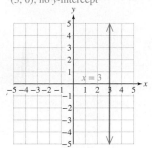

67. $y = -1$ a. Horizontal line c. no x-intercept; y-intercept: $(0, -1)$

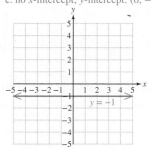

68. $-2y = 8$ a. Horizontal line c. no x-intercept; y-intercept: $(0, -4)$

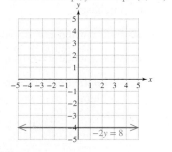

69. $5x = 20$ a. Vertical line c. x-intercept: $(4, 0)$; no y-intercept

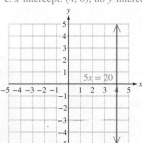

70. $x - 3 = -7$ a. Vertical line c. x-intercept: $(-4, 0)$; no y-intercept

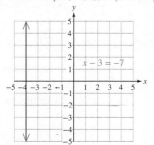

71. $y + 8 = 11$ a. Horizontal line c. no x-intercept; y-intercept: $(0, 3)$

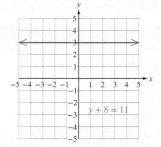

72. $3y = 0$ a. Horizontal line c. All points on the x-axis are x-intercepts; y-intercept: $(0, 0)$

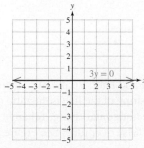

73. $5x = 0$ a. Vertical line c. All points on the y-axis are y-intercepts; x-intercept: $(0, 0)$

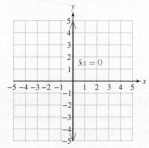

74. $2x + 7 = 10$ a. Vertical line c. x-intercept: $(\frac{3}{2}, 0)$; no y-intercept

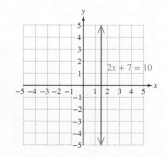

75. Explain why not every line has both an x- and a y-intercept.   A horizontal line may not have an x-intercept. A vertical line may not have a y-intercept.

76. Which of the lines has an x-intercept?

    a. $2x - 3y = 6$ **b.** $x = 5$ **c.** $2y = 8$ **d.** $-x + y = 0$ a, b, d

77. Which of the lines has a y-intercept?

    a. $y = 2$ **b.** $x + y = 0$ **c.** $2x - 10 = 2$ **d.** $x + 4y = 8$ a, b, d

Writing   Translating Expression   Geometry   Scientific Calculator   Video

Expanding Your Skills

78. The store "CDs R US" sells all compact discs for $13.99. The following equation represents the revenue, y, (in dollars) generated by selling x CDs.

$$y = 13.99x \quad (x \geq 0)$$

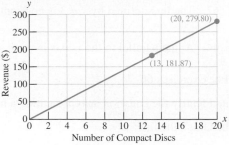

a. Find y when $x = 13$. $y = 181.87$

b. Find x when $y = 279.80$. $x = 20$

c. Write the ordered pairs from parts (a) and (b), and interpret their meaning in the context of the problem.   (13, 181.87) Selling 13 compact discs yields $181.87 in revenue. (20, 279.80) Selling 20 compact discs yields $279.80 in revenue.

d. Graph the ordered pairs and the line defined by the points.

79. The value of a car depreciates once it is driven off of the dealer's lot. For a certain sub-compact car, the value of the car is given by the equation $y = -1025x + 12,215$ $(x \geq 0)$ where y is the value of the car in dollars x years after its purchase.

a. Find y when $x = 1$. $y = 11,190$

b. Find x when $y = 9140$. $x = 3$

c. Write the ordered pairs from parts (a) and (b), and interpret their meaning in the context of the problem.   (1, 11,190) One year after purchase the value of the car is $11,190. (3, 9140) Three years after purchase the value of the car is $9140.

Section 3.3   **Slope of a Line and Rate of Change**

Concepts

1. **Introduction to Slope**
2. **Slope Formula**
3. **Parallel and Perpendicular Lines**
4. **Applications of Slope: Rate of Change**

1. Introduction to Slope

The x- and y-intercepts represent the points where a line crosses the x- and y-axes. Another important feature of a line is its slope. Geometrically, the slope of a line measures the "steepness" of the line. For example, two hiking trails are depicted by the lines in Figure 3-14.

Park Trail   Mt. Dora Trail

Figure 3-14

By visual inspection, Mt. Dora Trail is "steeper" than Park Trail. To measure the slope of a line quantitatively, consider two points on the line. The **slope** of the line is the ratio of the vertical change (change in y) between the two points and the horizontal change (change in x). As a memory device, we might think of the slope of a line as "rise over run." See Figure 3-15.

$$\text{Slope} = \frac{\text{change in } y}{\text{change in } x} = \frac{\text{rise}}{\text{run}}$$

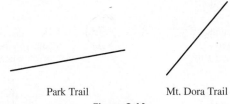

Figure 3-15

To move from point A to point B on Park Trail, rise 2 ft and move to the right 6 ft (Figure 3-16).

To move from point A to point B on Mt. Dora Trail, rise 5 ft and move to the right 4 ft (Figure 3-17).

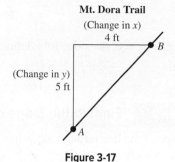

Mt. Dora Trail
(Change in x)
4 ft

(Change in y)
5 ft

B

A

Figure 3-17

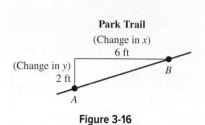

Park Trail
(Change in x)
6 ft

(Change in y)
2 ft

B

A

Figure 3-16

> **TIP:** To find the slope, you can use any two points on the line. The ratio of rise to run will be the same.

$$\text{Slope} = \frac{\text{change in } y}{\text{change in } x} = \frac{2 \text{ ft}}{6 \text{ ft}} = \frac{1}{3}$$

$$\text{Slope} = \frac{\text{change in } y}{\text{change in } x} = \frac{5 \text{ ft}}{4 \text{ ft}} = \frac{5}{4}$$

The slope of Mt. Dora Trail is greater than the slope of Park Trail, confirming the observation that Mt. Dora Trail is steeper. On Mt. Dora Trail there is a 5-ft change in elevation for every 4 ft of horizontal distance (a 5:4 ratio). On Park Trail there is only a 2-ft change in elevation for every 6 ft of horizontal distance (a 1:3 ratio).

Example 1 **Finding Slope in an Application**

Classroom Example: p. 226, Exercise 10

Determine the slope of the ramp up the stairs.

Solution:

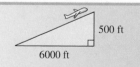

$$\text{Slope} = \frac{\text{change in } y}{\text{change in } x} = \frac{8 \text{ ft}}{16 \text{ ft}}$$

$$\frac{8}{16} = \frac{1}{2} \qquad \text{Write the ratio for the slope and simplify.}$$

The slope is $\frac{1}{2}$.

8 ft

16 ft

Skill Practice

1. Determine the slope of the aircraft's takeoff path. (Figure is not drawn to scale.)

500 ft

6000 ft

---

2. Slope Formula

The slope of a line may be found using any two points on the line—call these points (x_1, y_1) and (x_2, y_2). The numbers to the right and below the variables are called *subscripts*. In this instance, the subscript 1 indicates the coordinates of the first point, and the subscript 2 indicates the coordinates of the second point. The change in y between the points can be found by taking the difference of the y values: $y_2 - y_1$. The change in x can be found by taking the difference of the x values in the same order: $x_2 - x_1$ (Figure 3-18).

The slope of a line is often symbolized by the letter m and is given by the following formula.

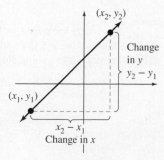

(x_2, y_2)

Change in y
$y_2 - y_1$

(x_1, y_1)

$x_2 - x_1$
Change in x

Figure 3-18

Answer

1. $\dfrac{500}{6000} = \dfrac{1}{12}$

Slope Formula

The slope of a line passing through the distinct points (x_1, y_1) and (x_2, y_2) is

$$m = \frac{y_2 - y_1}{x_2 - x_1} \quad \text{provided } x_2 - x_1 \neq 0.$$

Note: If $x_2 - x_1 = 0$, the slope is undefined.

Classroom Example: p. 228, Exercise 36

Example 2   **Finding the Slope of a Line Given Two Points**

Find the slope of the line through the points $(-1, 3)$ and $(-4, -2)$.

Solution:

To use the slope formula, first label the coordinates of each point and then substitute the coordinates into the slope formula.

$$\underset{(x_1, y_1)}{(-1, 3)} \quad \text{and} \quad \underset{(x_2, y_2)}{(-4, -2)} \qquad \text{Label the points.}$$

$$m = \frac{y_2 - y_1}{x_2 - x_1} = \frac{(-2) - (3)}{(-4) - (-1)} \qquad \text{Apply the slope formula.}$$

$$= \frac{-5}{-3}$$

$$= \frac{5}{3} \qquad \text{Simplify to lowest terms.}$$

Avoiding Mistakes

When calculating slope, always write the change in y in the numerator.

The slope of the line can be verified from the graph (Figure 3-19).

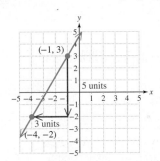

Figure 3-19

Skill Practice Find the slope of the line through the given points.

2. $(-5, 2)$ and $(1, 3)$

TIP: The slope formula is not dependent on which point is labeled (x_1, y_1) and which point is labeled (x_2, y_2). In Example 2, reversing the order in which the points are labeled results in the same slope.

$$\underset{(x_2, y_2)}{(-1, 3)} \quad \text{and} \quad \underset{(x_1, y_1)}{(-4, -2)} \qquad \text{Label the points.}$$

$$m = \frac{(3) - (-2)}{(-1) - (-4)} = \frac{5}{3} \qquad \text{Apply the slope formula.}$$

Answer

2. $\dfrac{1}{6}$

When you apply the slope formula, you will see that the slope of a line may be positive, negative, zero, or undefined.

- Lines that increase, or rise, from left to right have a positive slope.
- Lines that decrease, or fall, from left to right have a negative slope.
- Horizontal lines have a slope of zero.
- Vertical lines have an undefined slope.

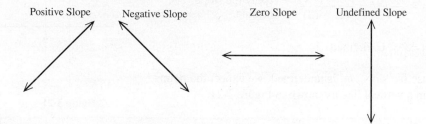

Positive Slope   Negative Slope   Zero Slope   Undefined Slope

---

Example 3	**Finding the Slope of a Line Given Two Points**

Find the slope of the line passing through the points $\left(-5, \frac{1}{2}\right)$ and $\left(2, -\frac{3}{2}\right)$.

Solution:

$$\underset{(x_1, y_1)}{\left(-5, \frac{1}{2}\right)} \quad \text{and} \quad \underset{(x_2, y_2)}{\left(2, -\frac{3}{2}\right)} \qquad \text{Label the points.}$$

$$m = \frac{y_2 - y_1}{x_2 - x_1} = \frac{\left(-\frac{3}{2}\right) - \left(\frac{1}{2}\right)}{(2) - (-5)} \qquad \text{Apply the slope formula.}$$

$$= \frac{-\frac{4}{2}}{2 + 5} \qquad \text{Simplify.}$$

$$= \frac{-2}{7} \quad \text{or} \quad -\frac{2}{7}$$

By graphing the points $\left(-5, \frac{1}{2}\right)$ and $\left(2, -\frac{3}{2}\right)$, we can verify that the slope is $-\frac{2}{7}$ (Figure 3-20). Notice that the line slopes downward from left to right.

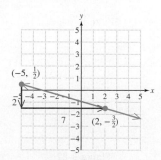

Figure 3-20

> **Classroom Example:** p. 228, Exercise 44

> **Avoiding Mistakes**
>
> When applying the slope formula, y_2 and x_2 are taken from the same ordered pair. Likewise y_1 and x_1 are taken from the same ordered pair.

Skill Practice Find the slope of the line through the given points.

3. $\left(\frac{2}{3}, 0\right)$ and $\left(-\frac{1}{6}, 5\right)$

Answer

3. -6

Classroom Example: p. 228, Exercise 42

Example 4   Determining the Slope of a Vertical Line

Find the slope of the line passing through the points $(2, -1)$ and $(2, 4)$.

Solution:

$$\underset{(x_1, y_1)}{(2, -1)} \quad \text{and} \quad \underset{(x_2, y_2)}{(2, 4)} \qquad \text{Label the points.}$$

$$m = \frac{y_2 - y_1}{x_2 - x_1} = \frac{(4) - (-1)}{(2) - (2)} \qquad \begin{array}{l}\text{Apply the slope} \\ \text{formula.}\end{array}$$

$$m = \frac{5}{0} \quad \text{Undefined}$$

Because the slope, m, is undefined, we expect the points to form a vertical line as shown in Figure 3-21.

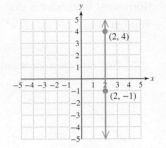

Figure 3-21

Skill Practice Find the slope of the line through the given points.

4. $(5, 6)$ and $(5, -2)$

Classroom Example: p. 228, Exercise 40

Example 5   Determining the Slope of a Horizontal Line

Find the slope of the line passing through the points $(3.4, -2)$ and $(-3.5, -2)$.

Solution:

$$\underset{(x_1, y_1)}{(3.4, -2)} \quad \text{and} \quad \underset{(x_2, y_2)}{(-3.5, -2)} \qquad \text{Label the points.}$$

$$m = \frac{y_2 - y_1}{x_2 - x_1} = \frac{(-2) - (-2)}{(-3.5) - (3.4)} \qquad \text{Apply the slope formula.}$$

$$= \frac{-2 + 2}{-3.5 - 3.4} = \frac{0}{-6.9} = 0 \qquad \text{Simplify.}$$

Because the slope is 0, we expect the points to form a horizontal line, as shown in Figure 3-22.

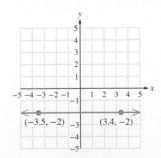

Figure 3-22

Skill Practice Find the slope of the line through the given points.

5. $(3, 8)$ and $(-5, 8)$

Answers
4. Undefined   **5.** 0

3. Parallel and Perpendicular Lines

Lines in the same plane that do not intersect are called **parallel lines**. Parallel lines have the same slope and different y-intercepts (Figure 3-23).

Lines that intersect at a right angle are **perpendicular lines**. If two lines are perpendicular then the slope of one line is the opposite of the reciprocal of the slope of the other line (provided neither line is vertical) (Figure 3-24).

Instructor Note: If you use calculators for graphing lines, try graphing two lines with slopes that are similar ($y = x$ and $y = \frac{19}{20}x + 1$). They will appear parallel, but clearly we can show they are not by inspecting the equations. Sometimes students don't value what we do by hand until they see it does not always appear correctly on the calculator.

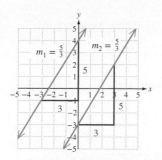

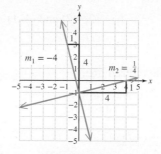

Animation

Figure 3-23   **Figure 3-24**

Slopes of Parallel Lines

If m_1 and m_2 represent the slopes of two parallel (nonvertical) lines, then

$$m_1 = m_2.$$

See Figure 3-23.

Slopes of Perpendicular Lines

If $m_1 \neq 0$ and $m_2 \neq 0$ represent the slopes of two perpendicular lines, then

$$m_1 = -\frac{1}{m_2} \text{ or equivalently, } m_1 m_2 = -1. \text{ See Figure 3-24.}$$

Instructor Note: Ask students how we know that all vertical lines are parallel and all horizontal lines are parallel.

Example 6 **Determining the Slope of Parallel and Perpendicular Lines**

Classroom Example: p. 228, Exercise 52

Suppose a given line has a slope of -6.

a. Find the slope of a line parallel to the line with the given slope.

b. Find the slope of a line perpendicular to the line with the given slope.

Solution:

a. Parallel lines must have the same slope. The slope of a line parallel to the given line is $m = -6$.

b. For perpendicular lines, the slope of one line must be the opposite of the reciprocal of the other. The slope of a line perpendicular to the given line is $m = -\left(\frac{1}{-6}\right) = \frac{1}{6}$.

Skill Practice A given line has a slope of $\frac{5}{3}$.

6. Find the slope of a line parallel to the given line.

7. Find the slope of a line perpendicular to the given line.

Answers

6. $\frac{5}{3}$ **7.** $-\frac{3}{5}$

If the slopes of two lines are known, then we can compare the slopes to determine if the lines are parallel, perpendicular, or neither.

Classroom Example: p. 228, Exercise 64

Example 7 Determining If Lines Are Parallel, Perpendicular, or Neither

Lines l_1 and l_2 pass through the given points. Determine if l_1 and l_2 are parallel, perpendicular, or neither.

$$l_1: \quad (2, -7) \text{ and } (4, 1) \qquad l_2: \quad (-3, 1) \text{ and } (1, 0)$$

Solution:

Find the slope of each line.

$$l_1: \quad \underset{(x_1, y_1)}{(2, -7)} \quad \text{and} \quad \underset{(x_2, y_2)}{(4, 1)} \qquad\qquad l_2: \quad \underset{(x_1, y_1)}{(-3, 1)} \quad \text{and} \quad \underset{(x_2, y_2)}{(1, 0)}$$

TIP: You can check that two lines are perpendicular by checking that the product of their slopes is −1.

$$4\left(-\frac{1}{4}\right) = -1$$

$$m_1 = \frac{1 - (-7)}{4 - 2} \qquad\qquad m_2 = \frac{0 - 1}{1 - (-3)}$$

$$m_1 = \frac{8}{2} \qquad\qquad m_2 = \frac{-1}{4}$$

$$m_1 = 4 \qquad\qquad m_2 = -\frac{1}{4}$$

One slope is the opposite of the reciprocal of the other slope. Therefore, the lines are perpendicular.

Skill Practice Determine if lines l_1 and l_2 are parallel, perpendicular, or neither.

8. $l_1: \quad (-2, -3) \text{ and } (4, -1)$
 $l_2: \quad (0, 2) \text{ and } (-3, 1)$

4. Applications of Slope: Rate of Change

In many applications, the interpretation of slope refers to the *rate of change* of the *y*-variable to the *x*-variable.

Classroom Example: p. 229, Exercise 72

Example 8 Interpreting Slope in an Application

The annual median income for males in the United States for selected years is shown in Figure 3-25. The trend is approximately linear. Find the slope of the line and interpret the meaning of the slope in the context of the problem.

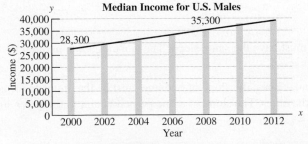

Figure 3-25

Source: U.S. Department of the Census

Answer

8. Parallel

Solution:

To determine the slope we need to know two points on the line. From the graph, the median income for males in the year 2000 was approximately \$28,300. This gives us the ordered pair (2000, 28,300). In the year 2008, the income was \$35,300. This gives the ordered pair (2008, 35,300).

$$\underset{(x_1, y_1)}{(2000, 28{,}300)} \quad \text{and} \quad \underset{(x_2, y_2)}{(2008, 35{,}300)} \qquad \text{Label the points.}$$

$$m = \frac{y_2 - y_1}{x_2 - x_1} = \frac{35{,}300 - 28{,}300}{2008 - 2000} \qquad \text{Apply the slope formula.}$$

$$= \frac{7000}{8}$$

$$= 875 \qquad \text{Simplify.}$$

The slope is 875. This tells us the rate of change of the y-variable (income) to the x-variable (years). This means that men's median income in the United States increased at a rate of \$875 per year during this time period.

Skill Practice

9. In the year 2000, the population of Alaska was approximately 630,000. By 2010, it had grown to 700,000. Use the ordered pairs (2000, 630,000) and (2010, 700,000) to determine the slope of the line through the points. Then interpret the meaning in the context of this problem.

Answer

9. $m = 7000$; The population of Alaska increased at a rate of 7000 people per year.

Section 3.3 Practice Exercises

For additional exercises, see Classroom Activities 3.3A–3.3C in the *Student's Resource Manual* at www.mhhe.com/moh.

Study Skills Exercise

Each day after finishing your homework, choose two or three odd-numbered problems or examples from that section. Write the problem with the directions on one side of a 3 × 5 card. On the back write the section, page, and problem number along with the answer. Each week, shuffle your cards and pull out a few at random, to give yourself a review of $\frac{1}{2}$-hr or more.

Vocabulary and Key Concepts

1. **a.** The ratio of the vertical change and the horizontal change between two distinct points (x_1, y_1) and (x_2, y_2) on a line is called the __slope__ of the line. The slope can be computed from the formula $m = $ _____. $\frac{y_2 - y_1}{x_2 - x_1}$

   b. Lines in the same plane that do not intersect are called __parallel__ lines.

   c. Two lines are perpendicular if they intersect at a __right__ angle.

   d. If m_1 and m_2 represent the slopes of two nonvertical perpendicular lines then $m_1 \cdot m_2 = $ __-1__.

   e. The slope of a vertical line is __undefined__. The slope of a __horizontal__ line is 0.

Review Exercises

For Exercises 2–6, find the *x*- and *y*-intercepts (if they exist). Then graph the line.

2. $x - 5 = 2$ *x*-intercept: (7, 0);
          y-intercept: none

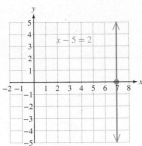

3. $x - 3y = 6$ *x*-intercept: (6, 0);
          y-intercept: (0, −2)

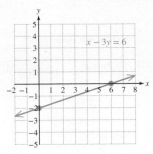

4. $y = \dfrac{2}{3}x$ *x*-intercept: (0, 0);
          y-intercept: (0, 0)

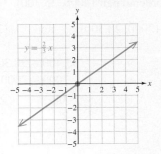

5. $2y - 3 = 0$ *x*-intercept: none;
          y-intercept: $\left(0, \frac{3}{2}\right)$

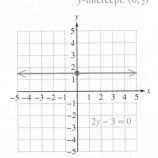

6. $2x = 4y$ *x*-intercept: (0, 0);
          y-intercept: (0, 0)

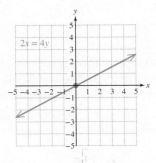

Concept 1: Introduction to Slope

7. Determine the slope of the roof.
   (See Example 1.)

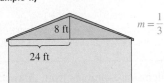

$m = \dfrac{1}{3}$

8. Determine the slope of the stairs.

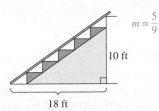

$m = \dfrac{5}{9}$

9. Calculate the slope of the handrail.

5.5 ft
© Ryan McVay/Getty Images RF

$m = \dfrac{6}{11}$

10. Determine the slope of the treadmill

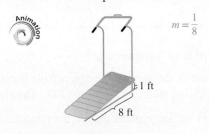

$m = \dfrac{1}{8}$

Concept 2: Slope Formula

For Exercises 11–14, fill in the blank with the appropriate term: *zero*, *negative*, *positive*, or *undefined*.

11. The slope of a line parallel to the *y*-axis is _undefined_.

12. The slope of a horizontal line is _zero_.

13. The slope of a line that rises from left to right is _positive_.

14. The slope of a line that falls from left to right is _negative_.

For Exercises 15–23, determine if the slope is positive, negative, zero, or undefined.

15. Negative

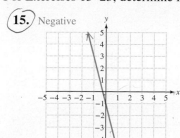

16. Undefined

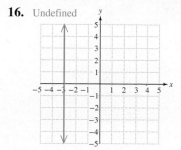

17. Zero

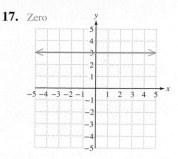

18. Negative

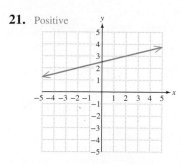

19. Undefined

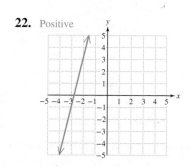

20. Zero

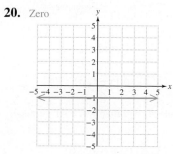

21. Positive

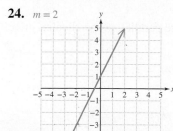

22. Positive

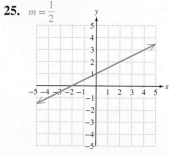

23. Negative

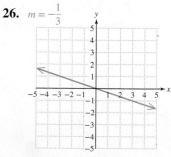

For Exercises 24–32, determine the slope by using the slope formula and any two points on the line. Check your answer by drawing a right triangle, where appropriate, and labeling the "rise" and "run."

24. $m = 2$

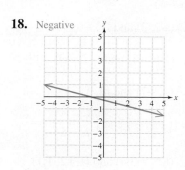

25. $m = \dfrac{1}{2}$

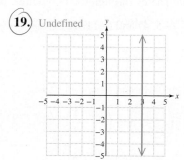

26. $m = -\dfrac{1}{3}$

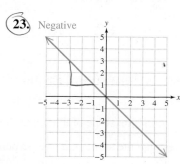

27. $m = -3$

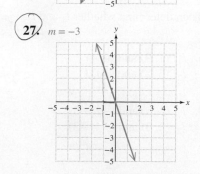

28. $m = 0$

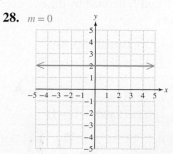

29. $m = 0$

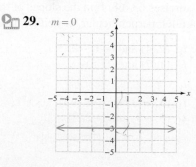

 Writing Translating Expression   Geometry   Scientific Calculator   Video

30. 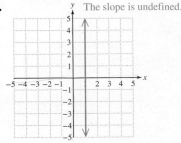 The slope is undefined.

31. 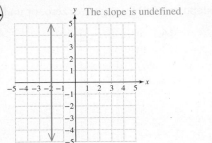 The slope is undefined.

32. $m = -1$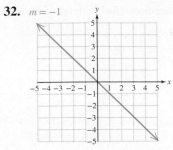

For Exercises 33–50, find the slope of the line that passes through the two points. **(See Example 2–5.)**

33. $(2, 4)$ and $(-4, 2)$ $\dfrac{1}{3}$

34. $(-5, 4)$ and $(-11, 12)$ $-\dfrac{4}{3}$

35. $(-2, 3)$ and $(1, -6)$ -3

36. $(-3, -4)$ and $(1, -5)$ $-\dfrac{1}{4}$

37. $(1, 5)$ and $(-4, 2)$ $\dfrac{3}{5}$

38. $(-6, -1)$ and $(-2, -3)$ $-\dfrac{1}{2}$

39. $(5, 3)$ and $(-2, 3)$ Zero

40. $(0, -1)$ and $(-4, -1)$ Zero

41. $(2, -7)$ and $(2, 5)$ Undefined

42. $(-4, 3)$ and $(-4, -4)$ Undefined

43. $\left(\dfrac{1}{2}, \dfrac{3}{5}\right)$ and $\left(\dfrac{1}{4}, -\dfrac{4}{5}\right)$ $\dfrac{28}{5}$

44. $\left(-\dfrac{2}{7}, \dfrac{1}{3}\right)$ and $\left(\dfrac{8}{7}, -\dfrac{5}{6}\right)$ $-\dfrac{49}{60}$

45. $(3, -1)$ and $(-5, 6)$ $-\dfrac{7}{8}$

46. $(-6, 5)$ and $(-10, 4)$ $\dfrac{1}{4}$

47. $(6.8, -3.4)$ and $(-3.2, 1.1)$
  -0.45 or $-\dfrac{9}{20}$

48. $(-3.15, 8.25)$ and $(6.85, -4.25)$
  -1.25 or $-\dfrac{5}{4}$

49. $(1994, 3.5)$ and $(2000, 2.6)$
  -0.15 or $-\dfrac{3}{20}$

50. $(1988, 4.65)$ and $(1998, 9.25)$
  0.46 or $\dfrac{23}{50}$

Concept 3: Parallel and Perpendicular Lines

For Exercises 51–56, the slope of a line is given. **(See Example 6.)**

a. Determine the slope of a line parallel to the line with the given slope.

b. Determine the slope of a line perpendicular to the line with the given slope.

51. $m = -2$ a. -2 b. $\dfrac{1}{2}$

52. $m = \dfrac{2}{3}$ a. $\dfrac{2}{3}$ b. $-\dfrac{3}{2}$

53. $m = 0$ a. 0 b. undefined

54. The slope is undefined.
  a. undefined   b. 0

55. $m = \dfrac{4}{5}$ a. $\dfrac{4}{5}$ b. $-\dfrac{5}{4}$

56. $m = -4$ a. -4 b. $\dfrac{1}{4}$

For Exercises 57–62, let m_1 and m_2 represent the slopes of two lines. Determine if the lines are parallel, perpendicular, or neither. **(See Example 6.)**

57. $m_1 = -2,\ m_2 = \dfrac{1}{2}$
Perpendicular

58. $m_1 = \dfrac{2}{3},\ m_2 = \dfrac{3}{2}$
Neither

59. $m_1 = 1,\ m_2 = \dfrac{4}{4}$
Parallel

60. $m_1 = \dfrac{3}{4},\ m_2 = -\dfrac{8}{6}$
Perpendicular

61. $m_1 = \dfrac{2}{7},\ m_2 = -\dfrac{2}{7}$
Neither

62. $m_1 = 5,\ m_2 = 5$
Parallel

For Exercises 63–68, find the slopes of the lines l_1 and l_2 defined by the two given points. Then determine whether l_1 and l_2 are parallel, perpendicular, or neither. **(See Example 7.)**

63. l_1: $(2, 4)$ and $(-1, -2)$
l_2: $(1, 7)$ and $(0, 5)$
l_1: $m = 2$, l_2: $m = 2$; parallel

64. l_1: $(0, 0)$ and $(-2, 4)$
l_2: $(1, -5)$ and $(-1, -1)$
l_1: $m = -2$, l_2: $m = -2$; parallel

65. l_1: $(1, 9)$ and $(0, 4)$
l_2: $(5, 2)$ and $(10, 1)$
l_1: $m = 5$, l_2: $m = -\dfrac{1}{5}$; perpendicular

66. l_1: $(3, -4)$ and $(-1, -8)$
l_2: $(5, -5)$ and $(-2, 2)$
l_1: $m = 1$, l_2: $m = -1$; perpendicular

67. l_1: $(4, 4)$ and $(0, 3)$
l_2: $(1, 7)$ and $(-1, -1)$
l_1: $m = \dfrac{1}{4}$, l_2: $m = 4$; neither

68. l_1: $(3, 5)$ and $(-2, -5)$
l_2: $(2, 0)$ and $(-4, -3)$
l_1: $m = 2$, l_2: $m = \dfrac{1}{2}$; neither

Concept 4: Applications of Slope: Rate of Change

69. For a recent year, the average earnings for male workers between the ages of 25 and 34 with a high school diploma was $32,000. Comparing this value in constant dollars to the average earnings 15 yr later showed that the average earnings have decreased to $29,600. Find the average rate of change in dollars per year for this time period. [*Hint:* Use the ordered pairs (0, 32,000) and (15, 29,600).]   The average rate of change is −$160 per year.

70. In 1985, the U.S. Postal Service charged $0.22 for first class letters and cards up to 1 oz. By 2015, the price had increased to $0.49. Let x represent the year, and y represent the cost for 1 oz of first class postage. Find the average rate of change of the cost per year.
The cost increased at a rate of $0.009 per year.

71. In 1985, there were 539 thousand male inmates in federal and state prisons. By 2010, the number increased to 1714 thousand. Let x represent the year, and let y represent the number of prisoners (in thousands). **(See Example 8.)**

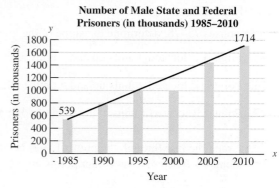

Number of Male State and Federal Prisoners (in thousands) 1985–2010

(*Source:* U.S. Bureau of Justice Statistics)

a. Using the ordered pairs (1985, 539) and (2010, 1714), find the slope of the line. $m = 47$

b. Interpret the slope in the context of this problem.
The number of male prisoners increased at a rate of 47 thousand per year during this time period.

72. In the year 1985, there were 30 thousand female inmates in federal and state prisons. By 2010, the number increased to 120 thousand. Let x represent the year, and let y represent the number of prisoners (in thousands).

a. Using the ordered pairs (1985, 30) and (2010, 120), find the slope of the line. $m = 3.6$

b. Interpret the slope in the context of this problem.
The number of female prisoners increased at a rate of 3.6 thousand per year during this time period.

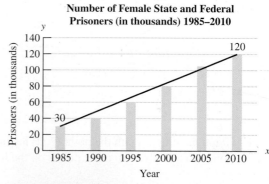

Number of Female State and Federal Prisoners (in thousands) 1985–2010

(*Source:* U.S. Bureau of Justice Statistics)

73. The distance, d (in miles), between a lightning strike and an observer is given by the equation $d = 0.2t$, where t is the time (in seconds) between seeing lightning and hearing thunder.

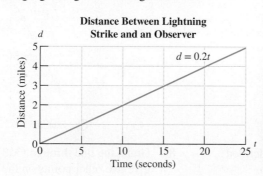

Distance Between Lightning Strike and an Observer

© Royalty Free/Corbis RF

a. If an observer counts 5 sec between seeing lightning and hearing thunder, how far away was the lightning strike?   1 mi

b. If an observer counts 10 sec between seeing lightning and hearing thunder, how far away was the lightning strike?   2 mi

c. If an observer counts 15 sec between seeing lightning and hearing thunder, how far away was the lightning strike?   3 mi

d. What is the slope of the line? Interpret the meaning of the slope in the context of this problem.
$m = 0.2$; The distance between a lightning strike and an observer increases by 0.2 mi for every additional second between seeing lightning and hearing thunder.

74. Michael wants to buy an efficient Smart car that according to the latest EPA standards gets 33 mpg in the city and 40 mpg on the highway. The car that Michael picked out costs $12,600. His dad agreed to purchase the car if Michael would pay it off in equal monthly payments for the next 60 months. The equation $y = -210x + 12,600$ represents the amount, y (in dollars), that Michael owes his father after x months.

© Erica Simone Leeds

   a. How much does Michael owe his dad after 5 months?
     Michael owes $11,550 after 5 months.

   b. Determine the slope of the line and interpret its meaning in the context of this problem.
     $m = -210$; The amount Michael owes his father decreases by $210 per month.

Mixed Exercises

For Exercises 75–78, determine the slope of the line passing through points A and B.

75. Point A is located 3 units up and 4 units to the right of point B. $m = \dfrac{3}{4}$

76. Point A is located 2 units up and 5 units to the left of point B. $m = -\dfrac{2}{5}$

77. Point A is located 5 units to the right of point B. $m = 0$

78. Point A is located 3 units down from point B.
The slope is undefined.

79. Graph the line through the point $(1, -2)$ having slope $-\frac{2}{3}$. Then give two other points on the line.

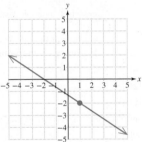

For example: $(4, -4)$ and $(-2, 0)$

80. Graph the line through the point $(1, 2)$ having slope $-\frac{3}{4}$. Then give two other points on the line.

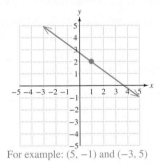

For example: $(5, -1)$ and $(-3, 5)$

81. Graph the line through the point $(2, 2)$ having slope 3. Then give two other points on the line.

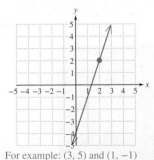

For example: $(3, 5)$ and $(1, -1)$

82. Graph the line through the point $(-1, 3)$ having slope 2. Then give two other points on the line.

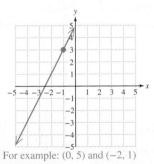

For example: $(0, 5)$ and $(-2, 1)$

83. Graph the line through $(-3, -2)$ with an undefined slope. Then give two other points on the line.

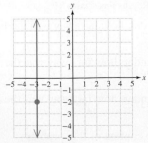

For example: $(-3, 1)$ and $(-3, 4)$

84. Graph the line through $(3, 3)$ with a slope of 0. Then give two other points on the line.

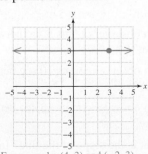

For example: $(4, 3)$ and $(-2, 3)$

For Exercises 85–90, draw a line as indicated. Answers may vary.

85. Draw a line with a positive slope and a positive
y-intercept.

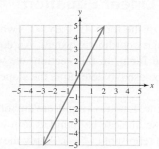

86. Draw a line with a positive slope and a negative
y-intercept.

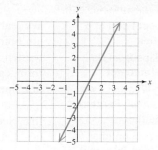

87. Draw a line with a negative slope and a negative
y-intercept.

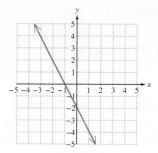

88. Draw a line with a negative slope and positive
y-intercept.

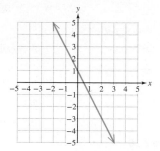

89. Draw a line with a zero slope and a positive
y-intercept.

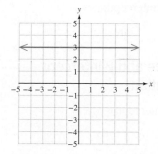

90. Draw a line with undefined slope and a negative
x-intercept.

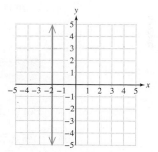

Expanding Your Skills

91. Determine the slope between the points $(a + b, 4m - n)$ and $(a - b, m + 2n)$. $\dfrac{3m - 3n}{2b}$ or $\dfrac{-3m + 3n}{-2b}$

92. Determine the slope between the points $(3c - d, s + t)$ and $(c - 2d, s - t)$. $\dfrac{2t}{2c + d}$ or $\dfrac{-2t}{-2c - d}$

93. Determine the *x*-intercept of the line $ax + by = c$. $\left(\dfrac{c}{a}, 0\right)$

94. Determine the *y*-intercept of the line $ax + by = c$. $\left(0, \dfrac{c}{b}\right)$

95. Find another point on the line that contains the point $(2, -1)$ and has a slope of $\frac{2}{5}$.   For example: $(7, 1)$

96. Find another point on the line that contains the point $(-3, 4)$ and has a slope of $\frac{1}{4}$.   For example: $(1, 5)$

Writing   Translating Expression   Geometry   Scientific Calculator   Video

Section 3.4 Slope-Intercept Form of a Linear Equation

1. Slope-Intercept Form of a Linear Equation

We learned that the solutions to an equation of the form $Ax + By = C$ (where A and B are not both zero) represent a line in a rectangular coordinate system. An equation of a line written in this way is said to be in **standard form**. In this section, we will learn a new form, called **slope-intercept form**, that can be used to determine the slope and y-intercept of a line.

Let $(0, b)$ represent the y-intercept of a line. Let (x, y) represent any other point on the line. See Figure 3-26. Then the slope, m, of the line can be found as follows:

Let $(0, b)$ represent (x_1, y_1), and let (x, y) represent (x_2, y_2). Apply the slope formula.

$$m = \frac{y_2 - y_1}{x_2 - x_1} \rightarrow m = \frac{y - b}{x - 0} \qquad \text{Apply the slope formula.}$$

$$m = \frac{y - b}{x} \qquad \text{Simplify.}$$

$$mx = \left(\frac{y - b}{x}\right)x \qquad \text{Multiply by } x \text{ to clear fractions.}$$

$$mx = y - b$$

$$mx + b = y - b + b \qquad \text{To isolate } y, \text{ add } b \text{ to both sides.}$$

$$mx + b = y \qquad \text{or} \qquad y = mx + b \qquad \text{The equation is in slope-intercept form.}$$

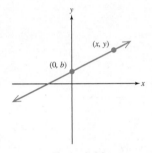

Figure 3-26

> **Slope-Intercept Form of a Linear Equation**
>
> $y = mx + b$ is the slope-intercept form of a linear equation.
> m is the slope and the point $(0, b)$ is the y-intercept.

Classroom Examples: p. 238, Exercises 12 and 16

Example 1 **Identifying the Slope and y-Intercept From a Linear Equation**

For each equation, identify the slope and y-intercept.

   a. $y = 3x - 1$ **b.** $y = -2.7x + 5$ **c.** $y = 4x$

Solution:

Each equation is written in slope-intercept form, $y = mx + b$. The slope is the coefficient of x, and the y-intercept is determined by the constant term.

   a. $y = 3x - 1$ The slope is 3.   The y-intercept is $(0, -1)$.

   b. $y = -2.7x + 5$ The slope is -2.7.   The y-intercept is $(0, 5)$.

   c. $y = 4x$ can be written as $y = 4x + 0$.   The slope is 4.
                                           The y-intercept is $(0, 0)$.

Skill Practice Identify the slope and the y-intercept.

   1. $y = 4x + 6$ **2.** $y = 3.5x - 4.2$ **3.** $y = -7$

Answers

1. slope: 4; y-intercept: $(0, 6)$
2. slope: 3.5; y-intercept: $(0, -4.2)$
3. slope: 0; y-intercept: $(0, -7)$

Given an equation of a line, we can write the equation in slope-intercept form by solving the equation for the y-variable. This is demonstrated in Example 2.

Example 2 **Identifying the Slope and y-Intercept From a Linear Equation**

Classroom Example: p. 238, Exercise 22

Given the equation $-5x - 2y = 6$,

a. Write the equation in slope-intercept form.

b. Identify the slope and y-intercept.

Solution:

a. Write the equation in slope-intercept form, $y = mx + b$, by solving for y.

$-5x - 2y = 6$

$\quad -2y = 5x + 6$ Add $5x$ to both sides.

$\quad \dfrac{-2y}{-2} = \dfrac{5x + 6}{-2}$ Divide both sides by -2.

$\quad\quad y = \dfrac{5x}{-2} + \dfrac{6}{-2}$ Divide each term by -2 and simplify.

$\quad\quad y = -\dfrac{5}{2}x - 3$ Slope-intercept form

b. The slope is $-\frac{5}{2}$, and the y-intercept is $(0, -3)$.

Skill Practice Given the equation $2x - 6y = -3$.

4. Write the equation in slope-intercept form.

5. Identify the slope and the y-intercept.

2. Graphing a Line from Its Slope and y-Intercept

Slope-intercept form is a useful tool to graph a line. The y-intercept is a known point on the line. The slope indicates the direction of the line and can be used to find a second point. Using slope-intercept form to graph a line is demonstrated in Examples 3 and 4.

Animation

Example 3 **Graphing a Line Using the Slope and y-Intercept**

Classroom Example: p. 240, Exercise 44

Graph the equation $y = -\frac{5}{2}x - 3$ by using the slope and y-intercept.

Solution:

First plot the y-intercept, $(0, -3)$.

The slope $m = -\frac{5}{2}$ can be written as

$m = \dfrac{-5}{2}$ ⟵ The change in y is -5.
   ⟵ The change in x is 2.

To find a second point on the line, start at the y-intercept and move down 5 units and to the right 2 units. Then draw the line through the two points (Figure 3-27).

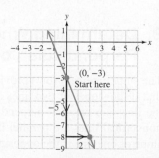

Figure 3-27

Answers

4. $y = \dfrac{1}{3}x + \dfrac{1}{2}$

5. slope is $\dfrac{1}{3}$; y-intercept is $\left(0, \dfrac{1}{2}\right)$

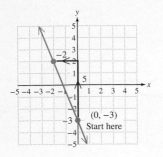

Figure 3-28

Similarly, the slope can be written as

$$m = \frac{5}{-2}$$

The change in y is 5.

The change in x is -2.

To find a second point, start at the y-intercept and move up 5 units and to the left 2 units. Then draw the line through the two points (Figure 3-28).

Skill Practice

6. Graph the equation by using the slope and y-intercept. $y = 2x - 3$

Classroom Example: p. 240, Exercise 50

Example 4   **Graphing a Line Using the Slope and y-Intercept**

Graph the equation $y = 4x$ by using the slope and y-intercept.

Solution:

The equation can be written as $y = 4x + 0$. Therefore, we can plot the y-intercept at (0, 0). The slope $m = 4$ can be written as

$$m = \frac{4}{1}$$

The change in y is 4.

The change in x is 1.

To find a second point on the line, start at the y-intercept and move up 4 units and to the right 1 unit. Then draw the line through the two points (Figure 3-29).

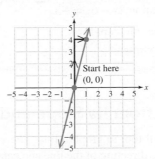

Figure 3-29

Skill Practice

7. Graph the equation by using the slope and y-intercept. $y = -\frac{1}{4}x$

Answers
6–7.

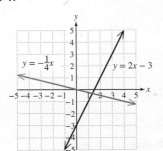

3. Determining Whether Two Lines Are Parallel, Perpendicular, or Neither

The slope-intercept form provides a means to find the slope of a line by inspection. Recall that if the slopes of two lines are known, then we can compare the slopes to determine if the lines are parallel, perpendicular, or neither parallel nor perpendicular. (Two distinct nonvertical lines are parallel if their slopes are equal. Two lines are perpendicular if the slope of one line is the opposite of the reciprocal of the slope of the other line.)

Example 5	**Determining If Two Lines Are Parallel, Perpendicular, or Neither**

Classroom Examples: p. 240, Exercises 54 and 56

For each pair of lines, determine if they are parallel, perpendicular, or neither.

a. l_1: $y = 3x - 5$ **b.** l_1: $y = \frac{3}{2}x + 2$

  l_2: $y = 3x + 1$ l_2: $y = \frac{2}{3}x + 1$

Solution:

a. l_1: $y = 3x - 5$ The slope of l_1 is 3.

  l_2: $y = 3x + 1$ The slope of l_2 is 3.

  Because the slopes are the same, the lines are parallel.

b. l_1: $y = \frac{3}{2}x + 2$ The slope of l_1 is $\frac{3}{2}$.

  l_2: $y = \frac{2}{3}x + 1$ The slope of l_2 is $\frac{2}{3}$.

  The slopes are not the same. Therefore, the lines are not parallel. The values of the slopes are reciprocals, but they are not opposite in sign. Therefore, the lines are not perpendicular. The lines are neither parallel nor perpendicular.

Skill Practice For each pair of lines determine if they are parallel, perpendicular, or neither.

8. $y = 3x - 5$ **9.** $y = \dfrac{5}{6}x - \dfrac{1}{2}$

  $y = -3x - 15$

               $y = \dfrac{5}{6}x + \dfrac{1}{2}$

Example 6	**Determining If Two Lines Are Parallel, Perpendicular, or Neither**

Classroom Examples: p. 240, Exercises 62 and 64

For each pair of lines, determine if they are parallel, perpendicular, or neither.

a. l_1: $x - 3y = -9$ **b.** l_1: $x = 2$

  l_2: $3x = -y + 4$ l_2: $2y = 8$

Solution:

a. First write the equation of each line in slope-intercept form.

  l_1: $x - 3y = -9$ l_2: $3x = -y + 4$

      $-3y = -x - 9$ $3x + y = 4$

      $\dfrac{-3y}{-3} = \dfrac{-x}{-3} - \dfrac{9}{-3}$ $y = -3x + 4$

        $y = \dfrac{1}{3}x + 3$

  l_1: $y = \frac{1}{3}x + 3$ The slope of l_1 is $\frac{1}{3}$.

  l_2: $y = -3x + 4$ The slope of l_2 is -3.

  The slope of $\frac{1}{3}$ is the opposite of the reciprocal of -3. Therefore, the lines are perpendicular.

Answers

8. Neither   **9.** Parallel

b. The equation $x = 2$ represents a vertical line because the equation is in the form $x = k$.

The equation $2y = 8$ can be simplified to $y = 4$, which represents a horizontal line.

In this example, we do not need to analyze the slopes because vertical lines and horizontal lines are perpendicular.

Skill Practice For each pair of lines, determine if they are parallel, perpendicular, or neither.

10. $x - 5y = 10$　　　**11.** $y = -5$

　　$5x - 1 = -y$　　　　　$x = 6$

4. Writing an Equation of a Line Using Slope-Intercept Form

The slope-intercept form of a linear equation can be used to write an equation of a line when the slope is known and the y-intercept is known.

Classroom Example: p. 241, Exercise 70

> **Example 7**　Writing an Equation of a Line Using Slope-Intercept Form
>
> Write an equation of the line whose slope is $\frac{2}{3}$ and whose y-intercept is $(0, 8)$.
>
> **Solution:**
>
> The slope is given as $m = \frac{2}{3}$, and the y-intercept $(0, b)$ is given as $(0, 8)$. Substitute the values $m = \frac{2}{3}$ and $b = 8$ into the slope-intercept form of a line.
>
> $$y = mx + b$$
> $$y = \frac{2}{3}x + 8$$

Skill Practice

12. Write an equation of the line whose slope is -4 and y-intercept is $(0, -10)$.

Classroom Example: p. 241, Exercise 74

> **Example 8**　Writing an Equation of a Line Using Slope-Intercept Form
>
> Write an equation of the line having a slope of 2 and passing through the point $(-3, 1)$.
>
> **Solution:**
>
> To find an equation of a line using slope-intercept form, it is necessary to find the value of m and b. The slope is given in the problem as $m = 2$. Therefore, the slope-intercept form becomes
>
> $$y = mx + b$$
> $$y = 2x + b$$

Answers

10. Perpendicular　　**11.** Perpendicular

12. $y = -4x - 10$

Because the point $(-3, 1)$ is on the line, it is a solution to the equation. Therefore, to find b, substitute the values of x and y from the ordered pair $(-3, 1)$ and solve the resulting equation for b.

$$y = 2x + b$$
$$1 = 2(-3) + b \quad \text{Substitute } y = 1 \text{ and } x = -3.$$
$$1 = -6 + b \quad \text{Simplify and solve for } b.$$
$$7 = b$$

Now with m and b known, the slope-intercept form is $y = 2x + 7$.

TIP: The equation from Example 8 can be checked by graphing the line $y = 2x + 7$. The slope $m = 2$ can be written as $m = \frac{2}{1}$. Therefore, to graph the line, start at the y-intercept $(0, 7)$ and move up 2 units and to the right 1 unit.

The graph verifies that the line passes through the point $(-3, 1)$ as it should.

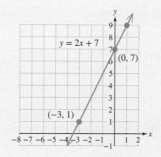

Skill Practice

13. Write an equation of the line having a slope of -3 and passing through the point $(-2, -5)$.

Answer

13. $y = -3x - 11$

Calculator Connections

Topic: Using the *ZSquare* Option in Zoom

In Example 6(a) we found that the equations $y = \frac{1}{3}x + 3$ and $y = -3x + 4$ represent perpendicular lines. We can verify our results by graphing the lines on a graphing calculator.

Notice that the lines do not appear perpendicular in the calculator display on the standard viewing window. That is, they do not appear to form a right angle at the point of intersection. Because many calculators have a rectangular screen, the standard viewing window is elongated in the horizontal direction. To eliminate this distortion, try using a *ZSquare* option, which is located under the Zoom menu. This feature will set the viewing window so that equal distances on the display denote an equal number of units on the graph.

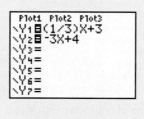

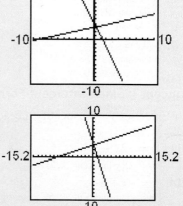

Calculator Exercises

For each pair of lines, determine if the lines are parallel, perpendicular, or neither. Then use a square viewing window to graph the lines on a graphing calculator to verify your results.

1. $x + y = 1$

  $x - y = -3$ Perpendicular

2. $3x + y = -2$

  $6x + 2y = 6$ Parallel

3. $2x - y = 4$

  $3x + 2y = 4$ Neither

4. Graph the lines defined by $y = x + 1$ and $y = 0.99x + 3$. Are these lines parallel? Explain.
The lines may appear parallel; however, they are not parallel because the slopes are different.

5. Graph the lines defined by $y = -2x - 1$ and $y = -2x - 0.99$. Are these lines the same? Explain.
The lines may appear to coincide on a graph; however, they are not the same line because the y-intercepts are different.

6. Graph the line defined by $y = 0.001x + 3$. Is this line horizontal? Explain.
The line may appear to be horizontal, but it is not. The slope is 0.001 rather than 0.

Section 3.4 Practice Exercises

For additional exercises, see Classroom Activities 3.4A–3.4C in the *Student's Resource Manual* at www.mhhe.com/moh.

Study Skills Exercise

When taking a test, go through the test and do all the problems that you know first. Then go back and work on the problems that were more difficult. Give yourself a time limit for how much time you spend on each problem (maybe 3 to 5 min the first time through). Circle the importance of each statement.

	not important	somewhat important	very important
a. Read through the entire test first.	1	2	3
b. If time allows, go back and check each problem.	1	2	3
c. Write out all steps instead of doing the work in your head.	1	2	3

Vocabulary and Key Terms

1. **a.** Consider a line with slope m and y-intercept $(0, b)$. The slope-intercept form of an equation of the line is __$y = mx + b$__.

    b. An equation of a line written in the form $Ax + By = C$ where A and B are not both zero is said to be in __standard__ form.

Review Exercises

2. For each equation given, determine if the line is horizontal, vertical, or slanted.

    a. $3x = 6$ **b.** $y + 3 = 6$ **c.** $x + y = 6$
        vertical   horizontal   slanted

For Exercises 3–10, determine the x- and y-intercepts, if they exist.

3. $x - 5y = 10$
    x-intercept: $(10, 0)$;
    y-intercept: $(0, -2)$

4. $3x + y = -12$
    x-intercept: $(-4, 0)$;
    y-intercept: $(0, -12)$

5. $3y = -9$
    x-intercept: none;
    y-intercept: $(0, -3)$

6. $2 + y = 5$
    x-intercept: none;
    y-intercept: $(0, 3)$

7. $-4x = 6y$
    x-intercept: $(0, 0)$;
    y-intercept: $(0, 0)$

8. $-x + 3 = 8$
    x-intercept: $(-5, 0)$;
    y-intercept: none

9. $5x = 20$
    x-intercept: $(4, 0)$;
    y-intercept: none

10. $y = \dfrac{1}{2}x$
    x-intercept: $(0, 0)$;
    y-intercept: $(0, 0)$

Concept 1: Slope-Intercept Form of a Linear Equation

For Exercises 11–30, identify the slope and y-intercept, if they exist. **(See Examples 1–2.)**

11. $y = -2x + 3$
    $m = -2$; y-intercept: $(0, 3)$

12. $y = \dfrac{2}{3}x + 5$
    $m = \dfrac{2}{3}$; y-intercept: $(0, 5)$

13. $y = x - 2$
    $m = 1$; y-intercept: $(0, -2)$

14. $y = -x + 6$
    $m = -1$; y-intercept: $(0, 6)$

15. $y = -x$
    $m = -1$; y-intercept: $(0; 0)$

16. $y = -5x$
    $m = -5$; y-intercept: $(0, 0)$

17. $y = \dfrac{3}{4}x - 1$
    $m = \dfrac{3}{4}$; y-interncept: $(0, -1)$

18. $y = x - \dfrac{5}{3}$
    $m = 1$; y-intercept: $\left(0, -\dfrac{5}{3}\right)$

19. $2x - 5y = 4$
    $m = \dfrac{2}{5}$; y-intercept: $\left(0, -\dfrac{4}{5}\right)$

20. $3x + 2y = 9$
    $m = -\dfrac{3}{2}$; y-intercept: $\left(0, \dfrac{9}{2}\right)$

21. $3x - y = 5$
    $m = 3$; y-intercept: $(0, -5)$

22. $7x - 3y = -6$
    $m = \dfrac{7}{3}$; y-intercept: $(0, 2)$

23. $x + y = 6$
    $m = -1$; y-intercept: $(0, 6)$

24. $x - y = 1$
    $m = 1$; y-intercept: $(0, -1)$

25. $x + 6 = 8$
    Undefined slope; no y-intercept

26. $-4 + x = 1$
    Undefined slope; no y-intercept

27. $-8y = 2$
    $m = 0$; y-intercept: $\left(0, -\dfrac{1}{4}\right)$

28. $1 - y = 9$
    $m = 0$; y-intercept: $(0, -8)$

29. $3y - 2x = 0$
    $m = \dfrac{2}{3}$; y-intercept: $(0, 0)$

30. $5x = 6y$
    $m = \dfrac{5}{6}$; y-intercept: $(0, 0)$

Concept 2: Graphing a Line from Its Slope and *y*-Intercept

For Exercises 31–34, graph the line using the slope and *y*-intercept. **(See Examples 3–4.)**

31. Graph the line through the point $(0, 2)$, having a slope of -4.

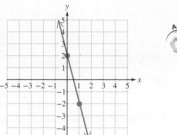

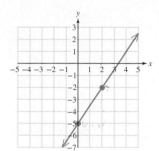

32. Graph the line through the point $(0, -1)$, having a slope of -3.

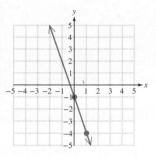

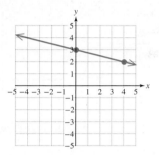

33. Graph the line through the point $(0, -5)$, having a slope of $\frac{3}{2}$.

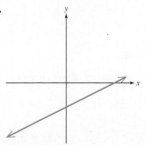

34. Graph the line through the point $(0, 3)$, having a slope of $-\frac{1}{4}$.

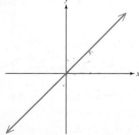

For Exercises 35–40, match the equation with the graph (a–f) by identifying if the slope is positive or negative and if the *y*-intercept is positive, negative, or zero.

35. $y = 2x + 3$ b

36. $y = -3x - 2$ d

37. $y = -\frac{1}{3}x + 3$ e

38. $y = \frac{1}{2}x - 2$ a

39. $y = x$ c

40. $y = -2x$ f

a.

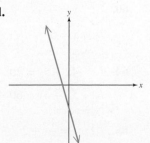

b.

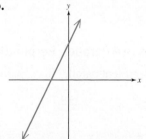

c.

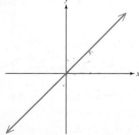

d.

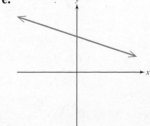

e.

f.

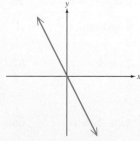

For Exercises 41–52, write each equation in slope-intercept form (if possible) and graph the line. **(See Examples 3–4.)**

41. $2x + y = 9$
$y = -2x + 9$

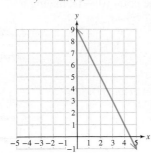

42. $-6x + y = 8$
$y = 6x + 8$

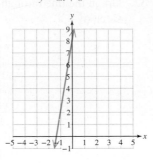

43. $x - 2y = 6$
$y = \frac{1}{2}x - 3$

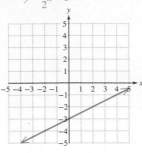

44. $5x - 2y = 2$
$y = \frac{5}{2}x - 1$

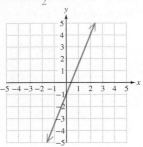

45. $2x = -4y + 6$
$y = -\frac{1}{2}x + \frac{3}{2}$

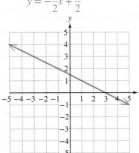

46. $6x = 2y - 14$
$y = 3x + 7$

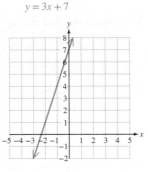

47. $x + y = 0$
$y = -x$

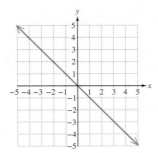

48. $x - y = 0$
$y = x$

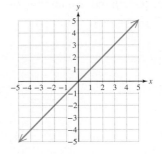

49. $5y = 4x$
$y = \frac{4}{5}x$

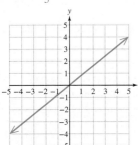

50. $-2x = 5y$
$y = -\frac{2}{5}x$

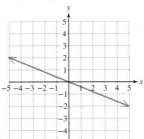

51. $3y + 2 = 0$
$y = -\frac{2}{3}$

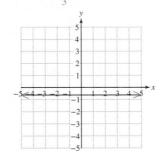

52. $1 + 5y = 6$
$y = 1$

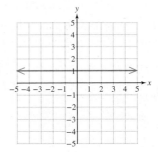

Concept 3: Determining Whether Two Lines Are Parallel, Perpendicular, or Neither

For Exercises 53–68, determine if the equations represent parallel lines, perpendicular lines, or neither. **(See Examples 5–6.)**

53. l_1: $y = -2x - 3$

l_2: $y = \frac{1}{2}x + 4$
Perpendicular

54. l_1: $y = \frac{4}{3}x - 2$

l_2: $y = -\frac{3}{4}x + 6$
Perpendicular

55. l_1: $y = \frac{4}{5}x - \frac{1}{2}$

l_2: $y = \frac{5}{4}x - \frac{2}{3}$
Neither

56. l_1: $y = \frac{1}{5}x + 1$

l_2: $y = 5x - 3$
Neither

57. l_1: $y = -9x + 6$

l_2: $y = -9x - 1$
Parallel

58. l_1: $y = 4x - 1$

l_2: $y = 4x + \frac{1}{2}$
Parallel

59. l_1: $x = 3$

l_2: $y = \frac{7}{4}$
Perpendicular

60. l_1: $y = \frac{2}{3}$

l_2: $x = 6$
Perpendicular

61. l_1: $2x = 4$

l_2: $6 = x$
Parallel

62. l_1: $2y = 7$

l_2: $y = 4$
Parallel

63. l_1: $2x + 3y = 6$

l_2: $3x - 2y = 12$
Perpendicular

64. l_1: $4x + 5y = 20$

l_2: $5x - 4y = 60$
Perpendicular

65. l_1: $4x + 2y = 6$

l_2: $4x + 8y = 16$
Neither

66. l_1: $3x + y = 5$

l_2: $x + 3y = 18$
Neither

67. l_1: $y = \frac{1}{5}x - 3$

l_2: $2x - 10y = 20$
Parallel

68. l_1: $y = \frac{1}{3}x + 2$

l_2: $-x + 3y = 12$
Parallel

Concept 4: Writing an Equation of a Line Using Slope-Intercept Form

For Exercises 69–80, write an equation of the line given the following information. Write the answer in slope-intercept form if possible. **(See Examples 7–8.)**

69. The slope is $-\frac{1}{3}$, and the y-intercept is (0, 2).

$$y = -\frac{1}{3}x + 2$$

70. The slope is $\frac{2}{3}$, and the y-intercept is (0, −1).

$$y = \frac{2}{3}x - 1$$

71. The slope is 10, and the y-intercept is (0, −19).

$y = 10x - 19$

72. The slope is −14, and the y-intercept is (0, 2).

$y = -14x + 2$

73. The slope is 6, and the line passes through the point (1, −2). $y = 6x - 8$

74. The slope is −4, and the line passes through the point (4, −3). $y = -4x + 13$

75. The slope is $\frac{1}{2}$, and the line passes through the point (−4, −5). $y = \frac{1}{2}x - 3$

76. The slope is $-\frac{2}{3}$, and the line passes through the point (3, −1). $y = -\frac{2}{3}x + 1$

77. The slope is 0, and the y-intercept is −11. $y = -11$

78. The slope is 0, and the y-intercept is $\frac{6}{7}$. $y = \frac{6}{7}$

79. The slope is 5, and the line passes through the origin. $y = 5x$

80. The slope is −3, and the line passes through the origin. $y = -3x$

Expanding Your Skills

For Exercises 81–86, write an equation of the line that passes through two points by following these steps:

Step 1: Find the slope of the line using the slope formula, $m = \dfrac{y_2 - y_1}{x_2 - x_1}$.

Step 2: Using the slope from Step 1 and either given point, follow the procedure given in Example 8 to find an equation of the line in slope-intercept form.

81. (2, −1) and (0, 3)

$y = -2x + 3$

82. (4, −8) and (0, −4)

$y = -x - 4$

83. (3, 1) and (−3, 3) $y = -\frac{1}{3}x + 2$

84. (2, −3) and (4, −2) $y = \frac{1}{2}x - 4$

85. (1, 3) and (−2, −9)

$y = 4x - 1$

86. (1, 7) and (−2, 4)

$y = x + 6$

87. The number of reported cases of Lyme disease in the United States can be modeled by the equation $y = 1203x + 10{,}006$. In this equation, x represents the number of years since 1993, and y represents the number of cases of Lyme disease.

a. What is the slope of this line and what does it mean in the context of this problem? $m = 1203$; The slope represents the rate of increase in the number of cases of Lyme disease per year.

b. What is the y-intercept, and what does it mean in the context of this problem? (0, 10006); In the year 1993 there were 10,006 cases reported.

c. Use the model to estimate the number of cases of Lyme disease in the year 2010.   30,457 cases

d. During what year would the predicted number of cases be 42,487? $x = 27$; the year 2020

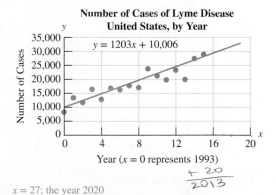

Number of Cases of Lyme Disease United States, by Year

$y = 1203x + 10{,}006$

Number of Cases — Year ($x = 0$ represents 1993)

$\dfrac{+\ 20}{2013}$

88. A phone bill is determined each month by a $16.95 flat fee plus $0.10/min of long distance. The equation, $C = 0.10x + 16.95$ represents the total monthly cost, C, for x minutes of long distance.

a. Identify the slope. Interpret the meaning of the slope in the context of this problem. $m = 0.10$. The cost increases by $0.10/min of long distance.

b. Identify the C-intercept. Interpret the meaning of the C-intercept in the context of this problem.   (0, 16.95). The monthly bill for 0 min is $16.95.

c. Use the equation to determine the total cost of 234 min of long distance.   The cost is $40.35.

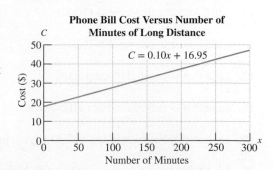

Phone Bill Cost Versus Number of Minutes of Long Distance

$C = 0.10x + 16.95$

Cost ($) — Number of Minutes

Writing ⟷ Translating Expression   Geometry   Scientific Calculator   Video

89. A linear equation is written in standard form if it can be written as $Ax + By = C$, where A and B are not both zero. Write the equation $Ax + By = C$ in slope-intercept form to show that the slope is given by the ratio, $-\frac{A}{B}$. ($B \neq 0$.) $y = -\frac{A}{B}x + \frac{C}{B}$; the slope is $-\frac{A}{B}$.

For Exercises 90–93, use the result of Exercise 89 to find the slope of the line.

90. $2x + 5y = 8$ $m = -\dfrac{2}{5}$ **91.** $6x + 7y = -9$ $m = -\dfrac{6}{7}$ **92.** $4x - 3y = -5$ $m = \dfrac{4}{3}$ **93.** $11x - 8y = 4$ $m = \dfrac{11}{8}$

Problem Recognition Exercises

Linear Equations in Two Variables

For Exercises 1–20, choose the equation(s) from the column on the right whose graph satisfies the condition described. Give all possible answers.

1. Line whose slope is positive.   a, c, d

2. Line whose slope is negative.   b, f, h

3. Line that passes through the origin.   a

4. Line that contains the point $(3, -2)$.   f

5. Line whose y-intercept is $(0, 4)$.   b, f

6. Line whose y-intercept is $(0, -5)$.   c

7. Line whose slope is $\dfrac{1}{2}$.   c, d

8. Line whose slope is -2.   f

9. Line whose slope is 0.   e

10. Line whose slope is undefined.   g

11. Line that is parallel to the line with equation $y = -\dfrac{2}{3}x + 4$.   b

12. Line perpendicular to the line with equation $y = 2x + 9$.   h

13. Line that is vertical.   g

14. Line that is horizontal.   e

15. Line whose x-intercept is $(10, 0)$.   c

16. Line whose x-intercept is $(6, 0)$.   b, h

17. Line that is parallel to the x-axis.   e

18. Line that is perpendicular to the y-axis.   e

19. Line with a negative slope and positive y-intercept.   b, f, h

20. Line with a positive slope and negative y-intercept.   c, d

a. $y = 5x$

b. $2x + 3y = 12$

c. $y = \dfrac{1}{2}x - 5$

d. $3x - 6y = 10$

e. $2y = -8$

f. $y = -2x + 4$

g. $3x = 1$

h. $x + 2y = 6$

 Writing   Translating Expression   Geometry   Scientific Calculator   Video

Point-Slope Formula

1. Writing an Equation of a Line Using the Point-Slope Formula

The slope-intercept form of a line can be used as a tool to construct an equation of a line. Another useful tool to determine an equation of a line is the point-slope formula. The point-slope formula can be derived from the slope formula as follows.

Suppose a line passes through a given point (x_1, y_1) and has slope m. If (x, y) is any other point on the line, then the slope is given by

$$m = \frac{y - y_1}{x - x_1} \qquad \text{Slope formula}$$

$$m(x - x_1) = \frac{y - y_1}{x - x_1}(x - x_1) \qquad \text{Clear fractions.}$$

$$m(x - x_1) = y - y_1$$

$$y - y_1 = m(x - x_1) \qquad \text{Point-slope formula}$$

Point-Slope Formula

The **point-slope formula** is given by

$$y - y_1 = m(x - x_1)$$

where m is the slope of the line and (x_1, y_1) is any known point on the line.

Example 1 demonstrates how to use the point-slope formula to find an equation of a line when a point on the line and slope are given.

Example 1 Writing an Equation of a Line Using the Point-Slope Formula

Use the point-slope formula to write an equation of the line having a slope of 3 and passing through the point $(-2, -4)$. Write the answer in slope-intercept form.

Solution:

The slope of the line is given: $\quad m = 3$

A point on the line is given: $\quad (x_1, y_1) = (-2, -4)$

The point-slope formula:

$$y - y_1 = m(x - x_1)$$
$$y - (-4) = 3[x - (-2)] \qquad \text{Substitute } m = 3, x_1 = -2, \text{ and } y_1 = -4.$$
$$y + 4 = 3(x + 2) \qquad \text{Simplify. Because the final answer is required in slope-intercept form, simplify the equation and solve for } y.$$
$$y + 4 = 3x + 6 \qquad \text{Apply the distributive property.}$$
$$y = 3x + 2 \qquad \text{Slope-intercept form}$$

Concepts

1. Writing an Equation of a Line Using the Point-Slope Formula
2. Writing an Equation of a Line Given Two Points
3. Writing an Equation of a Line Parallel or Perpendicular to Another Line
4. Different Forms of Linear Equations: A Summary

Classroom Example: p. 248, Exercise 12

Instructor Note: Show students an alternative method using slope-intercept form.

$$y = mx + b \quad \text{replace } m, x, y$$
$$-4 = 3(-2) + b$$
$$-4 = -6 + b$$
$$2 = b$$

$$y = mx + b \quad \text{replace } m, b$$
$$y = 3x + 2$$

 Writing   Translating Expression   Geometry   Scientific Calculator   Video

Skill Practice

1. Use the point-slope formula to write an equation of the line having a slope of -4 and passing through $(-1, 5)$. Write the answer in slope-intercept form.

The equation $y = 3x + 2$ from Example 1 is graphed in Figure 3-30. Notice that the line does indeed pass through the point $(-2, -4)$.

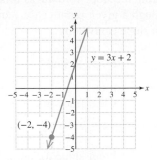

Figure 3-30

2. Writing an Equation of a Line Given Two Points

Example 2 is similar to Example 1; however, the slope must first be found from two given points.

Classroom Example: p. 249, Exercise 20

Example 2   **Writing an Equation of a Line Given Two Points**

Use the point-slope formula to find an equation of the line passing through the points $(-2, 5)$ and $(4, -1)$. Write the final answer in slope-intercept form.

Solution:

Given two points on a line, the slope can be found with the slope formula.

$$(-2, 5) \quad \text{and} \quad (4, -1)$$
$$(x_1, y_1) \qquad\qquad (x_2, y_2) \qquad \text{Label the points.}$$

$$m = \frac{y_2 - y_1}{x_2 - x_1} = \frac{(-1) - (5)}{(4) - (-2)} = \frac{-6}{6} = -1$$

To apply the point-slope formula, use the slope, $m = -1$ and either given point. We will choose the point $(-2, 5)$ as (x_1, y_1).

$$y - y_1 = m(x - x_1)$$
$$y - 5 = -1[x - (-2)] \qquad \text{Substitute } m = -1, x_1 = -2, \text{ and } y_1 = 5.$$
$$y - 5 = -1(x + 2) \qquad \text{Simplify.}$$
$$y - 5 = -x - 2$$
$$y = -x + 3$$

TIP: The point-slope formula can be applied using either given point for (x_1, y_1). In Example 2, using the point $(4, -1)$ for (x_1, y_1) produces the same result.

$$y - y_1 = m(x - x_1)$$
$$y - (-1) = -1(x - 4)$$
$$y + 1 = -x + 4$$
$$y = -x + 3$$

Skill Practice

2. Use the point-slope formula to write an equation of the line passing through the points $(1, -1)$ and $(-1, -5)$.

Answers

1. $y = -4x + 1$ **2.** $y = 2x - 3$

The solution to Example 2 can be checked by graphing the line $y = -x + 3$ using the slope and y-intercept. Notice that the line passes through the points $(-2, 5)$ and $(4, -1)$ as expected. See Figure 3-31.

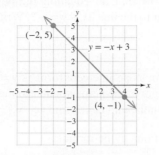

Figure 3-31

3. Writing an Equation of a Line Parallel or Perpendicular to Another Line

To write an equation of a line using the point-slope formula, the slope must be known. If the slope is not explicitly given, then other information must be used to determine the slope. In Example 2, the slope was found using the slope formula. Examples 3 and 4 show other situations in which we might find the slope.

| Example 3 | **Writing an Equation of a Line Parallel to Another Line** |

Classroom Example: p. 249, Exercise 30

Use the point-slope formula to find an equation of the line passing through the point $(-1, 0)$ and parallel to the line $y = -4x + 3$. Write the final answer in slope-intercept form.

Solution:

Figure 3-32 shows the line $y = -4x + 3$ (pictured in black) and a line parallel to it (pictured in blue) that passes through the point $(-1, 0)$. The equation of the given line, $y = -4x + 3$, is written in slope-intercept form, and its slope is easily identified as -4. The line parallel to the given line must also have a slope of -4.

Apply the point-slope formula using $m = -4$ and the point $(x_1, y_1) = (-1, 0)$.

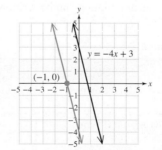

Figure 3-32

$$y - y_1 = m(x - x_1)$$
$$y - 0 = -4[x - (-1)]$$
$$y = -4(x + 1)$$
$$y = -4x - 4$$

TIP: When writing an equation of a line, slope-intercept form or standard form is usually preferred. For instance, the solution to Example 3 can be written as follows.

Slope-intercept form:
$y = -4x - 4$

Standard form:
$4x + y = -4$

Skill Practice

3. Use the point-slope formula to write an equation of the line passing through $(8, 2)$ and parallel to the line $y = \frac{3}{4}x - \frac{1}{2}$.

Answer

3. $y = \frac{3}{4}x - 4$

Classroom Example: p. 249, Exercise 34

| Example 4 | **Writing an Equation of a Line Perpendicular to Another Line** |

Use the point-slope formula to find an equation of the line passing through the point $(-3, 1)$ and perpendicular to the line $3x + y = -2$. Write the final answer in slope-intercept form.

Solution:

The given line can be written in slope-intercept form as $y = -3x - 2$. The slope of this line is -3. Therefore, the slope of a line perpendicular to the given line is $\frac{1}{3}$.

Apply the point-slope formula with $m = \frac{1}{3}$, and $(x_1, y_1) = (-3, 1)$.

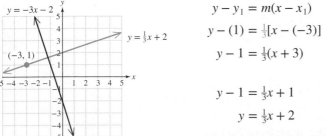

$$y - y_1 = m(x - x_1) \qquad \text{Point-slope formula}$$

$$y - (1) = \tfrac{1}{3}[x - (-3)] \qquad \text{Substitute } m = \tfrac{1}{3}, x_1 = -3, \text{ and } y_1 = 1.$$

$$y - 1 = \tfrac{1}{3}(x + 3) \qquad \text{To write the final answer in slope-intercept form, simplify the equation and solve for } y.$$

$$y - 1 = \tfrac{1}{3}x + 1 \qquad \text{Apply the distributive property.}$$

$$y = \tfrac{1}{3}x + 2 \qquad \text{Add 1 to both sides.}$$

Figure 3-33

A sketch of the perpendicular lines $y = \frac{1}{3}x + 2$ and $y = -3x - 2$ is shown in Figure 3-33. Notice that the line $y = \frac{1}{3}x + 2$ passes through the point $(-3, 1)$ as expected.

Skill Practice

4. Write an equation of the line passing through the point $(10, 4)$ and perpendicular to the line $x + 2y = 1$.

4. Different Forms of Linear Equations: A Summary

A linear equation can be written in several different forms, as summarized in Table 3-3.

Table 3-3

Form	Example	Comments
Standard Form $Ax + By = C$	$4x + 2y = 8$	A and B must not both be zero.
Horizontal Line $y = k$ (k is constant)	$y = 4$	The slope is zero, and the y-intercept is $(0, k)$.
Vertical Line $x = k$ (k is constant)	$x = -1$	The slope is undefined, and the x-intercept is $(k, 0)$.
Slope-Intercept Form $y = mx + b$ the slope is m y-intercept is $(0, b)$	$y = -3x + 7$ Slope $= -3$ y-intercept is $(0, 7)$	Solving a linear equation for y results in slope-intercept form. The coefficient of the x-term is the slope, and the constant defines the location of the y-intercept.
Point-Slope Formula $y - y_1 = m(x - x_1)$	$m = -3$ $(x_1, y_1) = (4, 2)$ $y - 2 = -3(x - 4)$	This formula is typically used to build an equation of a line when a point on the line is known and the slope of the line is known.

Answer

4. $y = 2x - 16$

Although standard form and slope-intercept form can be used to express an equation of a line, often the slope-intercept form is used to give a *unique* representation of the line. For example, the following linear equations are all written in standard form, yet they each define the same line.

$$2x + 5y = 10$$

$$-4x - 10y = -20$$

$$6x + 15y = 30$$

$$\frac{2}{5}x + y = 2$$

The line can be written uniquely in slope-intercept form as: $y = -\frac{2}{5}x + 2$.

Although it is important to understand and apply slope-intercept form and the point-slope formula, they are not necessarily applicable to all problems, particularly when dealing with a horizontal or vertical line.

Example 5 **Writing an Equation of a Line**

Find an equation of the line passing through the point $(2, -4)$ and parallel to the x-axis.

Classroom Example: p. 250, Exercise 48

Solution:

Because the line is parallel to the x-axis, the line must be horizontal. Recall that all horizontal lines can be written in the form $y = k$, where k is a constant. A quick sketch can help find the value of the constant. See Figure 3-34.

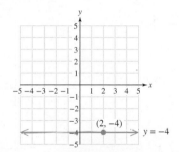

Figure 3-34

Because the line must pass through a point whose y-coordinate is -4, then the equation of the line must be $y = -4$.

Skill Practice

5. Write an equation for the vertical line that passes through the point $(-7, 2)$.

Answer

5. $x = -7$

Section 3.5 Practice Exercises

For additional exercises, see Classroom Activity 3.5A in the *Student's Resource Manual* at www.mhhe.com/moh.

Study Skills Exercise

Prepare a one-page summary sheet with the most important information that you need for the test. On the day of the test, look at this sheet several times to refresh your memory instead of trying to memorize new information.

 Writing   Translating Expression   Geometry    Scientific Calculator   Video

Vocabulary and Key Concepts

1. **a.** The standard form of an equation of a line is ___$Ax + By = C$___, where A and B are not both zero and C is a constant.

    b. A line defined by an equation $y = k$, where k is a constant is a (horizontal/vertical) line.   horizontal

    c. A line defined by an equation $x = k$, where k is a constant is a (horizontal/vertical) line.   vertical

    d. Given the slope-intercept form of an equation of a line, $y = mx + b$, the value of m is the ___slope___ and b is the ___y-intercept___.

    e. Given a point (x_1, y_1) on a line with slope m, the point-slope formula is given by $\underline{y - y_1 = m(x - x_1)}$.

Review Exercises

For Exercises 2–6, graph each equation.

2. $-5x - 15 = 0$

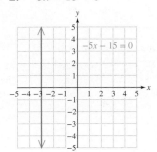

3. $2x - 3y = -3$

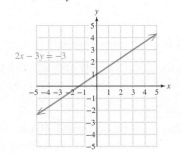

4. $y = -2x$

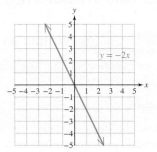

5. $3 - y = 9$

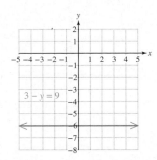

6. $y = \dfrac{4}{5}x$

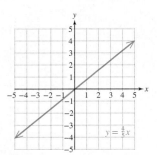

For Exercises 7–10, find the slope of the line that passes through the given points.

7. $(1, -3)$ and $(2, 6)$ 9

8. $(2, -4)$ and $(-2, 4)$ -2

9. $(-2, 5)$ and $(5, 5)$ 0

10. $(6.1, 2.5)$ and $(6.1, -1.5)$ Undefined

Concept 1: Writing an Equation of a Line Using the Point-Slope Formula

For Exercises 11–16, use the point-slope formula (if possible) to write an equation of the line given the following information. **(See Example 1.)**

11. The slope is 3, and the line passes through the point $(-2, 1)$.
    $y = 3x + 7$ or $3x - y = -7$

12. The slope is -2, and the line passes through the point $(1, -5)$.
    $y = -2x - 3$ or $2x + y = -3$

13. The slope is -4, and the line passes through the point $(-3, -2)$.
    $y = -4x - 14$ or $4x + y = -14$

14. The slope is 5, and the line passes through the point $(-1, -3)$.
    $y = 5x + 2$ or $5x - y = -2$

15. The slope is $-\dfrac{1}{2}$, and the line passes through $(-1, 0)$.
    $y = -\dfrac{1}{2}x - \dfrac{1}{2}$ or $x + 2y = -1$

16. The slope is $-\dfrac{3}{4}$, and the line passes through $(2, 0)$.
    $y = -\dfrac{3}{4}x + \dfrac{3}{2}$ or $3x + 4y = 6$

 Writing    Translating Expression    Geometry   Scientific Calculator   Video

Concept 2: Writing an Equation of a Line Given Two Points

For Exercises 17–22, use the point-slope formula to write an equation of the line given the following information. **(See Example 2.)**

17. The line passes through the points $(-2, -6)$ and $(1, 0)$. $y = 2x - 2$ or $2x - y = 2$

18. The line passes through the points $(-2, 5)$ and $(0, 1)$. $y = -2x + 1$ or $2x + y = 1$

19. The line passes through the points $(0, -4)$ and $(-1, -3)$. $y = -x - 4$ or $x + y = -4$

20. The line passes through the points $(1, -3)$ and $(-7, 2)$. $y = -\frac{5}{8}x - \frac{19}{8}$ or $5x + 8y = -19$

21. The line passes through the points $(2.2, -3.3)$ and $(12.2, -5.3)$.
$y = -0.2x - 2.86$ or $20x + 100y = -286$

22. The line passes through the points $(4.7, -2.2)$ and $(-0.3, 6.8)$.
$y = -1.8x + 6.26$ or $180x + 100y = 626$

For Exercises 23–28, find an equation of the line through the given points. Write the final answer in slope-intercept form.

23.
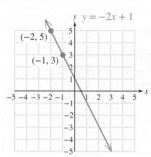
$y \quad y = -2x + 1$
$(-2, 5)$
$(-1, 3)$

24.
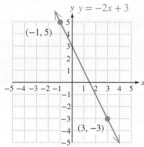
$y \quad y = -2x + 3$
$(-1, 5)$
$(3, -3)$

25.
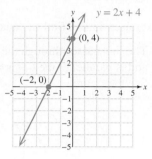
$y \quad y = 2x + 4$
$(0, 4)$
$(-2, 0)$

26.
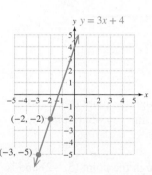
$y \quad y = 3x + 4$
$(-2, -2)$
$(-3, -5)$

27.
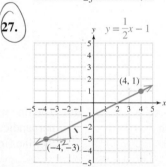
$y \quad y = \frac{1}{2}x - 1$
$(4, 1)$
$(-4, -3)$

28.
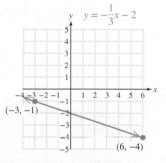
$y \quad y = -\frac{1}{3}x - 2$
$(-3, -1)$
$(6, -4)$

Concept 3: Writing an Equation of a Line Parallel or Perpendicular to Another Line

For Exercises 29–36, use the point-slope formula to write an equation of the line given the following information. **(See Examples 3–4.)**

29. The line passes through the point $(-3, 1)$ and is parallel to the line $y = 4x + 3$.
$y = 4x + 13$ or $4x - y = -13$

30. The line passes through the point $(4, -1)$ and is parallel to the line $y = 3x + 1$.
$y = 3x - 13$ or $3x - y = 13$

31. The line passes through the point $(4, 0)$ and is parallel to the line $3x + 2y = 8$. $y = -\frac{3}{2}x + 6$ or $3x + 2y = 12$

32. The line passes through the point $(2, 0)$ and is parallel to the line $5x + 3y = 6$. $y = -\frac{5}{3}x + \frac{10}{3}$ or $5x + 3y = 10$

33. The line passes through the point $(-5, 2)$ and is perpendicular to the line $y = \frac{1}{2}x + 3$.
$y = -2x - 8$ or $2x + y = -8$

34. The line passes through the point $(-2, -2)$ and is perpendicular to the line $y = \frac{1}{3}x - 5$.
$y = -3x - 8$ or $3x + y = -8$

35. The line passes through the point $(0, -6)$ and is perpendicular to the line $-5x + y = 4$.
$y = -\frac{1}{5}x - 6$ or $x + 5y = -30$

36. The line passes through the point $(0, -8)$ and is perpendicular to the line $2x - y = 5$.
$y = -\frac{1}{2}x - 8$ or $x + 2y = -16$

 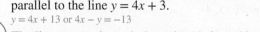

Concept 4: Different Forms of Linear Equations: A Summary

For Exercises 37–42, match the form or formula on the left with its name on the right.

37. $x = k$ iv

i.   Standard form

38. $y = mx + b$ v

ii.   Point-slope formula

39. $m = \dfrac{y_2 - y_1}{x_2 - x_1}$ vi

iii.   Horizontal line

40. $y - y_1 = m(x - x_1)$ ii

iv.   Vertical line

41. $y = k$ iii

v.   Slope-intercept form

42. $Ax + By = C$ i

vi.   Slope formula

For Exercises 43–48, find an equation for the line given the following information. **(See Example 5.)**

43. The line passes through the point $(3, 1)$ and is parallel to the line $y = -4$. See the figure.

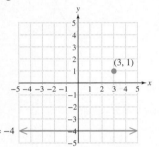

44. The line passes through the point $(-1, 1)$ and is parallel to the line $y = 2$. See the figure.

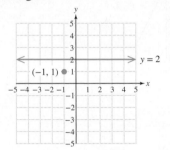

45. The line passes through the point $(2, 6)$ and is perpendicular to the line $y = 1$. (*Hint:* Sketch the line first.) $x = 2$

46. The line passes through the point $(0, 3)$ and is perpendicular to the line $y = -5$. (*Hint:* Sketch the line first.) $x = 0$

47. The line passes through the point $(2, 2)$ and is perpendicular to the line $x = 0$. $y = 2$

48. The line passes through the point $(5, -2)$ and is perpendicular to the line $x = 0$. $y = -2$

Mixed Exercises

For Exercises 49–60, write an equation of the line given the following information.

49. The slope is $\frac{1}{4}$, and the line passes through the point $(-8, 6)$. $y = \frac{1}{4}x + 8$ or $x - 4y = -32$

50. The slope is $\frac{2}{5}$, and the line passes through the point $(-5, 4)$. $y = \frac{2}{5}x + 6$ or $2x - 5y = -30$

51. The line passes through the point $(4, 4)$ and is parallel to the line $3x - y = 6$.
$y = 3x - 8$ or $3x - y = 8$

52. The line passes through the point $(-1, -7)$ and is parallel to the line $5x + y = -5$.
$y = -5x - 12$ or $5x + y = -12$

53. The slope is 4.5, and the line passes through the point $(5.2, -2.2)$. $y = 4.5x - 25.6$ or $45x - 10y = 256$

54. The slope is -3.6, and the line passes through the point $(10.0, 8.2)$. $y = -3.6x + 44.2$ or $36x + 10y = 442$

55. The slope is undefined, and the line passes through the point $(-6, -3)$. $x = -6$

56. The slope is undefined, and the line passes through the point $(2, -1)$. $x = 2$

57. The slope is 0, and the line passes through the point $(3, -2)$. $y = -2$

58. The slope is 0, and the line passes through the point $(0, 5)$. $y = 5$

59. The line passes through the points $(-4, 0)$ and $(-4, 3)$. $x = -4$

60. The line passes through the points $(1, 3)$ and $(1, -4)$. $x = 1$

Expanding Your Skills

For Exercises 61–64, write an equation in slope-intercept form for the line shown.

61.

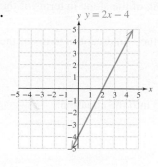

$y = 2x - 4$

62.

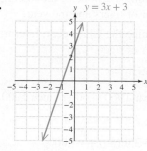

$y = 3x + 3$

63.

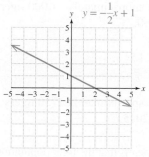

$y = -\dfrac{1}{2}x + 1$

64.

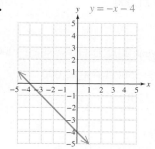

$y = -x - 4$

Applications of Linear Equations and Modeling

Section 3.6

1. Interpreting a Linear Equation in Two Variables

Linear equations can often be used to describe (or model) the relationship between two variables in a real-world event.

Concepts

1. Interpreting a Linear Equation in Two Variables
2. Writing a Linear Model Using Observed Data Points
3. Writing a Linear Model Given a Fixed Value and a Rate of Change

Example 1 Interpreting a Linear Equation

Since the year 1900, the tiger population in India has decreased linearly. A recent study showed this decrease can be approximated by the equation $y = -350x + 42,000$. The variable y represents the number of tigers left in India, and x represents the number of years since 1900.

Classroom Example: p. 256, Exercise 10

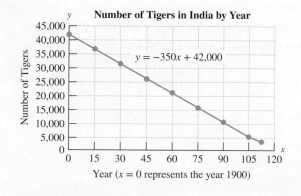

Number of Tigers in India by Year

$y = -350x + 42,000$

Year ($x = 0$ represents the year 1900)

a. Use the equation to estimate the number of tigers in 1960.

b. Use the equation to estimate the number of tigers in 2015.

c. Determine the slope of the line. Interpret the meaning of the slope in terms of the number of tigers and the year.

d. Determine the x-intercept. Interpret the meaning of the x-intercept in terms of the number of tigers.

Solution:

a. The year 1960 is 60 yr since 1900. Substitute $x = 60$ into the equation.

$$y = -350x + 42{,}000$$

$$y = -350(60) + 42{,}000$$

$$= 21{,}000$$ There were approximately 21,000 tigers in India in 1960.

b. The year 2015 is 115 yr since 1900. Substitute $x = 115$.

$$y = -350(115) + 42{,}000$$

$$= 1750$$ There were approximately 1750 tigers in India in 2015.

c. The slope is -350.   The slope means that the tiger population is decreasing by 350 tigers per year.

d. To find the x-intercept, substitute $y = 0$.

$$y = -350x + 42{,}000$$

$$0 = -350x + 42{,}000$$ Substitute 0 for y.

$$-42{,}000 = -350x$$

$$120 = x$$

The x-intercept is (120, 0). This means that 120 yr after the year 1900, the tiger population would be expected to reach zero. That is, in the year 2020, there will be no tigers left in India if this linear trend continues.

Skill Practice

1. The cost y (in dollars) for a local move by a small moving company is given by $y = 60x + 100$, where x is the number of hours required for the move.

   a. How much would be charged for a move that requires 3 hr?

   b. How much would be charged for a move that requires 8 hr?

   c. What is the slope of the line and what does it mean in the context of this problem?

   d. Determine the y-intercept and interpret its meaning in the context of this problem.

Answers

1. a. $280   **b.** $580
   c. 60; This means that for each additional hour of service, the cost of the move goes up by $60.
   d. (0, 100); The $100 charge is a fixed fee in addition to the hourly rate.

2. Writing a Linear Model Using Observed Data Points

Example 2   **Writing a Linear Model from Observed Data Points**

Classroom Example: p. 258, Exercise 18

The monthly sales of hybrid cars sold in the United States are given for a recent year. The sales for the first 8 months of the year are shown in Figure 3-35. The value $x = 0$ represents January, $x = 1$ represents February, and so on.

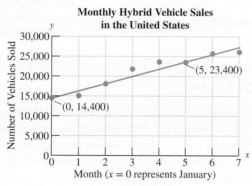

Monthly Hybrid Vehicle Sales in the United States

Month ($x = 0$ represents January)

Figure 3-35

a. Use the data points from Figure 3-35 to find a linear equation that represents the monthly sales of hybrid cars in the United States. Let x represent the month number and let y represent the number of vehicles sold.

b. Use the linear equation in part (a) to estimate the number of hybrid vehicles sold in month 7 (August).

Solution:

a. The ordered pairs (0, 14,400) and (5, 23,400) are given in the graph. Use these points to find the slope.

$$\underset{(x_1, y_1)}{(0, 14{,}400)} \quad \text{and} \quad \underset{(x_2, y_2)}{(5, 23{,}400)} \qquad \text{Label the points.}$$

$$m = \frac{y_2 - y_1}{x_2 - x_1} = \frac{23{,}400 - 14{,}400}{5 - 0}$$

$$= \frac{9000}{5}$$

$$= 1800 \qquad \qquad \text{The slope is 1800. This indicates that sales increased by approximately 1800 per month during this time period.}$$

With $m = 1800$, and the y-intercept given as (0, 14,400), we have the following linear equation in slope-intercept form.

$$y = 1800x + 14{,}400$$

b. To approximate the sales in month number 7, substitute $x = 7$ into the equation from part (a).

$$y = 1800(7) + 14{,}400 \qquad \qquad \text{Substitute } x = 7.$$

$$= 27{,}000$$

The monthly sales for August (month 7) would be 27,000 vehicles.

Skill Practice

2. Soft drink sales at a concession stand at a softball stadium have increased linearly over the course of the summer softball season.

   a. Use the given data points to find a linear equation that relates the sales, y, to week number, x.

   b. Use the equation to predict the number of soft drinks sold in week 10.

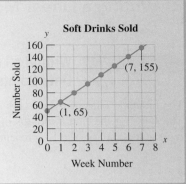

3. Writing a Linear Model Given a Fixed Value and a Rate of Change

Another way to look at the equation $y = mx + b$ is to identify the term mx as the variable term and the term b as the constant term. The value of the term mx will change with the value of x (this is why the slope, m, is called a *rate of change*). However, the term b will remain constant regardless of the value of x. With these ideas in mind, we can write a linear equation if the rate of change and the constant are known.

Classroom Example: p. 258, Exercise 20

Example 3   Writing a Linear Model

A stack of posters to advertise a production by the theater department costs $19.95 plus $1.50 per poster at the printer.

   a. Write a linear equation to compute the cost, c, of buying x posters.

   b. Use the equation to compute the cost of 125 posters.

Solution:

   a. The constant cost is $19.95. The variable cost is $1.50 per poster. If m is replaced with 1.50 and b is replaced with 19.95, the equation is

   $$c = 1.50x + 19.95 \qquad \text{where } c \text{ is the cost (in dollars) of buying } x \text{ posters.}$$

   b. Because x represents the number of posters, substitute $x = 125$.

   $$c = 1.50(125) + 19.95$$
   $$= 187.5 + 19.95$$
   $$= 207.45$$

The total cost of buying 125 posters is $207.45.

Skill Practice

3. The monthly cost for a "minimum use" cellular phone is $19.95 plus $0.10 per minute for all calls.

   a. Write a linear equation to compute the cost, c, of using t minutes.

   b. Use the equation to determine the cost of using 150 minutes.

Answers

2. a. $y = 15x + 50$ b. 200 soft drinks
3. a. $c = 0.10t + 19.95$ b. $34.95

Calculator Connections

Topic: Using the Evaluate Feature on a Graphing Calculator

In Example 3, the equation $c = 1.50x + 19.95$ was used to represent the cost, c, to buy x posters. To graph this equation on a graphing calculator, first replace the variable c by y.

$$y = 1.50x + 19.95$$

We enter the equation into the calculator and set the viewing window.

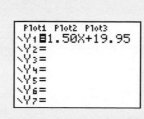 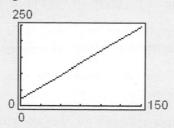

To evaluate the equation for a user-defined value of x, use the *Value* feature in the CALC menu. In this case, we entered $x = 125$, and the calculator returned $y = 207.45$.

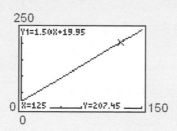

Calculator Exercises

Use a graphing calculator to graph the lines on an appropriate viewing window. Evaluate the equation at the given values of x.

1. $y = -4.6x + 27.1$ at $x = 3$ 13.3

2. $y = -3.6x - 42.3$ at $x = 0$ −42.3

3. $y = 40x + 105$ at $x = 6$ 345

4. $y = 20x - 65$ at $x = 8$ 95

Section 3.6 Practice Exercises

For additional exercises, see Classroom Activities 3.6A–3.6C in the *Student's Resource Manual* at www.mhhe.com/moh.

Study Skills Exercise

On test day, take a look at any formulas or important points that you had to memorize before you enter the classroom. Then when you sit down to take your test, write these formulas on the test or on scrap paper. This is called a memory dump. For practice write down the formulas involving slopes and equations of lines.

Review Exercises

1. Determine the slope of the line defined by $5x + 2y = -6$. $m = -\dfrac{5}{2}$

2. Determine the slope of the line defined by $2x - 8y = 15$. $m = \dfrac{1}{4}$

For Exercises 3–8, find the x- and y-intercepts of the lines, if possible.

3. $5x + 6y = 30$ x-intercept: $(6, 0)$; y-intercept: $(0, 5)$

4. $3x + 4y = 1$ x-intercept: $(\frac{1}{3}, 0)$; y-intercept: $(0, \frac{1}{4})$

5. $y = -2x - 4$ x-intercept: $(-2, 0)$; y-intercept: $(0, -4)$

6. $y = 5x$ x-intercept: $(0, 0)$; y-intercept: $(0, 0)$

7. $y = -9$ x-intercept: none; y-intercept: $(0, -9)$

8. $x = 2$ x-intercept: $(2, 0)$; y-intercept: none

 Writing Translating Expression Geometry Scientific Calculator Video

Concept 1: Interpreting a Linear Equation in Two Variables

9. The minimum hourly wage, y (in dollars per hour), in the United States can be approximated by the equation $y = 0.14x + 1.60$. In this equation, x represents the number of years since 1970 ($x = 0$ represents 1970, $x = 5$ represents 1975, and so on). **(See Example 1.)**

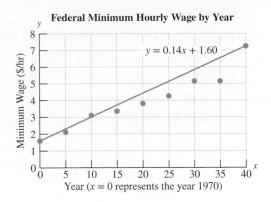

Federal Minimum Hourly Wage by Year

a. Use the equation to approximate the minimum wage in the year 1980. $3.00

b. Use the equation to estimate the minimum wage in 2015. $7.90

c. Determine the y-intercept. Interpret the meaning of the y-intercept in the context of this problem.   The y-intercept is (0, 1.6). This indicates that the minimum wage was $1.60 per hour in the year 1970.

d. Determine the slope. Interpret the meaning of the slope in the context of this problem.   The slope is 0.14. This indicates that the minimum wage has risen approximately $0.14 per year during this period.

10. The average daily temperature in January for cities along the eastern seaboard of the United States and Canada generally decreases for cities farther north. A city's latitude in the northern hemisphere is a measure of how far north it is on the globe.

The average temperature, y (measured in degrees Fahrenheit), can be described by the equation

$y = -2.333x + 124.0$ where x is the latitude of the city.

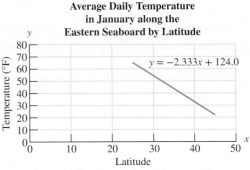

Average Daily Temperature in January along the Eastern Seaboard by Latitude

(*Source:* U.S. National Oceanic and Atmospheric Administration)

a. Use the equation to predict the average daily temperature in January for Philadelphia, Pennsylvania, whose latitude is 40.0°N. Round to one decimal place.   30.7°

b. Use the equation to predict the average daily temperature in January for Edmundston, New Brunswick, Canada, whose latitude is 47.4°N. Round to one decimal place.   13.4°

c. What is the slope of the line? Interpret the meaning of the slope in terms of latitude and temperature.
$m = -2.333$. The average temperature in January decreases at a rate of 2.333° per 1° of latitude.

d. From the equation, determine the value of the x-intercept. Round to one decimal place. Interpret the meaning of the x-intercept in terms of latitude and temperature.
(53.2, 0). At 53.2° latitude, the average temperature in January is 0°.

11. Veterinarians keep records of the weights of animals that are brought in for examination. Grindel, the cat, weighed 42 oz when she was 70 days old. She weighed 46 oz when she was 84 days old. Her sister, Frisco, weighed 40 oz when she was 70 days old and 48 oz at 84 days old.

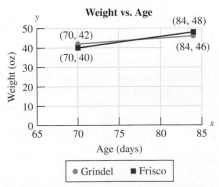

Weight vs. Age

Courtesy Rick Iossi

a. Compute the slope of the line representing Grindel's weight. $m = \frac{2}{7}$

b. Compute the slope of the line representing Frisco's weight. $m = \frac{4}{7}$

c. Interpret the meaning of each slope in the context of this problem. $m = \frac{2}{7}$ means that Grindel's weight increased at a rate of 2 oz in 7 days. $m = \frac{4}{7}$ means that Frisco's weight increased at a rate of 4 oz in 7 days.

d. Which cat gained weight more rapidly during this time period?   Frisco gained weight more rapidly.

Writing   Translating Expression   Geometry   Scientific Calculator   Video

12. The graph depicts the rise in the number of jail inmates in the United States since 2000. Two linear equations are given: one to describe the number of female inmates and one to describe the number of male inmates by year.

    Let y represent the number of inmates (in thousands). Let x represent the number of years since 2000.

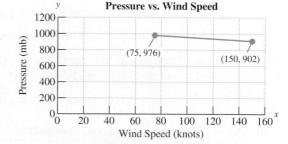

Number of Jail Inmates by Year

$y = 19.2x + 544$

$y = 4x + 71$

Year
($x = 0$ corresponds to 2000)

● Male Inmates
■ Female Inmates

(*Source:* U.S. Bureau of Justice Statistics)

a. What is the slope of the line representing the number of female inmates? Interpret the meaning of the slope in the context of this problem. $m = 4$. The number of female inmates has increased by 4 thousand per year since 2000.

b. What is the slope of the line representing the number of male inmates? Interpret the meaning of the slope in the context of this problem. $m = 19.2$. The number of male inmates has increased by 19.2 thousand per year since 2000.

c. Which group, males or females, has the larger slope? What does this imply about the rise in the number of male and female prisoners?   Males. The number of male inmates is increasing at a faster rate than the number of female inmates.

d. Assuming this trend continues, use the equation to predict the number of female inmates in 2020.
151 thousand or 151,000

13. The electric bill charge for a certain utility company is $0.095 per kilowatt-hour plus a fixed monthly tax of $11.95. The total cost, y, depends on the number of kilowatt-hours, x, according to the equation $y = 0.095x + 11.95$, $x \geq 0$.

a. Determine the cost of using 1000 kilowatt-hours. $106.95

b. Determine the cost of using 2000 kilowatt-hours. $201.95

c. Determine the y-intercept. Interpret the meaning of the y-intercept in the context of this problem.
(0, 11.95). For 0 kilowatt-hours used, the cost consists of only the fixed monthly tax of $11.95.

d. Determine the slope. Interpret the meaning of the slope in the context of this problem.
$m = 0.095$. The cost increases at a rate of $0.095 for each kilowatt-hour used.

14. For a recent year, children's admission to a State Fair was $8. Ride tickets were $0.75 each. The equation $y = 0.75x + 8$ represented the cost, y, in dollars to be admitted to the fair and to purchase x ride tickets.

a. Determine the slope of the line represented by $y = 0.75x + 8$. Interpret the meaning of the slope in the context of this problem.
$m = 0.75$; The slope means that the cost increases at a rate of 75¢ per ride.

b. Determine the y-intercept. Interpret its meaning in the context of this problem.
(0, 8); The cost was $8 if 0 rides were purchased.

c. Use the equation to determine how much money a child need for admission and to ride 10 rides.
$15.50

Concept 2: Writing a Linear Model Using Observed Data Points

15. Meteorologists often measure the intensity of a tropical storm or hurricane by the maximum sustained wind speed and the minimum pressure. The relationship between these two quantities is approximately linear. Hurricane Katrina had a maximum sustained wind speed of 150 knots and a minimum pressure of 902 mb (millibars). Hurricane Ophelia had maximum sustained winds of 75 knots and a pressure of 976 mb. **(See Example 2.)**

Pressure vs. Wind Speed

(75, 976)

(150, 902)

Wind Speed (knots)

a. Find the slope of the line between these two points. Round to one decimal place. $m = -1.0$

b. Using the slope found in part (a) and the point (75, 976), find a linear model that represents the minimum pressure of a hurricane, y, versus its maximum sustained wind speed, x. $y = -1.0x + 1051$

c. Hurricane Dennis had a maximum wind speed of 130 knots. Using the equation found in part (b), predict the minimum pressure.   The minimum pressure was approximately 921 mb.

16. The figure depicts a relationship between a person's height, y (in inches), and the length of the person's arm, x (measured in inches from shoulder to wrist).

   a. Use the points (17, 57.75) and (24, 82.25) to find a linear equation relating height to arm length.
$y = 3.5x - 1.75$

   b. What is the slope of the line? Interpret the slope in the context of this problem. $m = 3.5$. For each additional inch in length of a person's arm, the person's height increases by 3.5 in.

   c. Use the equation from part (a) to estimate the height of a person whose arm length is 21.5 in.
73.5 in. or 6 ft $1\frac{1}{2}$ in.

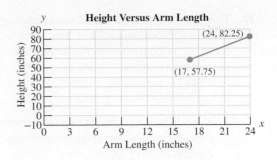

Height Versus Arm Length

(24, 82.25)

(17, 57.75)

Height (inches)

Arm Length (inches)

17. Wind energy is one type of renewable energy that does not produce dangerous greenhouse gases as a by-product. The graph shows the consumption of wind energy in the United States for selected years. The variable y represents the amount of wind energy in trillions of Btu, and the variable x represents the number of years since 2000.

   a. Use the points (0, 57) and (4, 143) to determine the slope of the line. $m = 21.5$
The slope means that the consumption of wind energy in the

   b. Interpret the slope in the context of this problem?
United States increased by 21.5 trillion Btu per year.

   c. Use the points (0, 57) and (4, 143) to find a linear equation relating the consumption of wind energy, y, to the number of years, x, since 2000. $y = 21.5x + 57$

   d. If this linear trend continues beyond the observed data values, use the equation in part (c) to estimate the consumption of wind energy in the year 2010.   272 trillion Btu

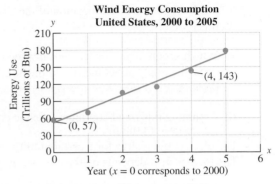

Wind Energy Consumption
United States, 2000 to 2005

(4, 143)

(0, 57)

Energy Use
(Trillions of Btu)

Year ($x = 0$ corresponds to 2000)

(*Source*: United States Department of Energy)

Animation

18. The graph shows the average height for boys based on age. Let x represent a boy's age, and let y represent his height (in inches).

   a. Find a linear equation that represents the height of a boy versus his age. $y = 2.5x + 31$

   b. Use the linear equation found in part (a) to predict the average height of a 5-year-old boy.   43.5 in.

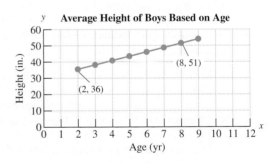

Average Height of Boys Based on Age

(8, 51)

(2, 36)

Height (in.)

Age (yr)

(*Source*: National Parenting Council)

Concept 3: Writing a Linear Model Given a Fixed Value and a Rate of Change

19. The owner of a restaurant franchise pays the parent company a monthly fee of $5000 plus 10% of sales. **(See Example 3.)**

   a. Write a linear model to compute the monthly payment, y, that the restaurant owner must pay the parent company for x dollars in sales. $y = 0.10x + 5000$

   b. Use the equation to compute the amount that the owner has to pay if monthly sales are $11,300. $6130

20. Anabel lives in New York and likes to keep in touch with her family in Texas. She uses 10-10-987 to call them. The cost of a long distance call is $0.83 plus $0.06 per minute.

   a. Write an equation that represents the cost, C, of a long distance call that is x minutes long. $C = 0.06x + 0.83$

   b. Use the equation from part (a) to compute the cost of a long distance phone call that lasted 32 minutes. $2.75

21. The cost to rent a 10 ft by 10 ft storage space is $90 per month plus a nonrefundable deposit of $105.

    a. Write a linear equation to compute the cost, y, of renting a 10 ft by 10 ft space for x months. $y = 90x + 105$

    b. What is the cost of renting such a storage space for 1 year (12 months)? $1185.00

22. An air-conditioning and heating company has a fixed monthly cost of $5000. Furthermore, each service call costs the company $25.

    a. Write a linear equation to compute the total cost, y, for 1 month if x service calls are made. $y = 25x + 5000$

    b. Use the equation to compute the cost for 1 month if 150 service calls are made. $8750.00

23. A bakery that specializes in bread rents a booth at a flea market. The daily cost to rent the booth is $100. Each loaf of bread costs the bakery $0.80 to produce.

    a. Write a linear equation to compute the total cost, y, for 1 day if x loaves of bread are produced. $y = 0.8x + 100$

    b. Use the equation to compute the cost for 1 day if 200 loaves of bread are produced. $260.00

© Nick Gunderson/Getty Images RF

24. A beverage company rents a booth at an art show to sell lemonade. The daily cost to rent a booth is $35. Each lemonade costs $0.50 to produce.

    a. Write a linear equation to compute the total cost, y, for 1 day if x lemonades are produced. $y = 0.5x + 35$

    b. Use the equation to compute the cost for 1 day if 350 lemonades are produced. $210.00

Group Activity

Modeling a Linear Equation

Materials: Yardstick or other device for making linear measurements

Estimated Time: 15–20 minutes

Group Size: 3

  1. The members of each group should measure the length of their arms (in inches) from elbow to wrist. Record this measurement as x and the person's height (in inches) as y. Write these values as ordered pairs for each member of the group. Then write the ordered pairs on the board.

  2. Next, copy the ordered pairs collected from all groups in the class and plot the ordered pairs. (This is called a "scatter diagram.")   Answers will vary throughout this exercise.

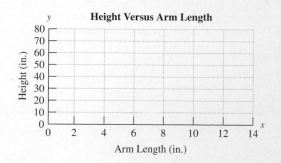

3. Select two ordered pairs that seem to follow the upward trend of the data. Using these data points, determine the slope of the line.

Slope: _____

4. Using the data points and slope from question 3, find an equation of the line through the two points. Write the equation in slope-intercept form, $y = mx + b$.

Equation: _____

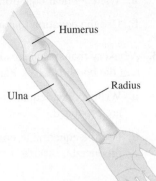

Humerus

Radius

Ulna

5. Using the equation from question 4, estimate the height of a person whose arm length from elbow to wrist is 8.5 in.

6. Suppose a crime scene investigator uncovers a partial skeleton and identifies a bone as a human ulna (the ulna is one of two bones in the forearm and extends from elbow to wrist). If the length of the bone is 12 in., estimate the height of the person before death. Would you expect this person to be male or female?

Chapter 3   Summary

Section 3.1   Rectangular Coordinate System

Key Concepts

Graphical representation of numerical data is often helpful to study problems in real-world applications.

A **rectangular coordinate system** is made up of a horizontal line called the **x-axis** and a vertical line called the **y-axis**. The point where the lines meet is the **origin**. The four regions of the plane are called **quadrants**.

The point (x, y) is an **ordered pair**. The first element in the ordered pair is the point's horizontal position from the origin. The second element in the ordered pair is the point's vertical position from the origin.

Example

Example 1

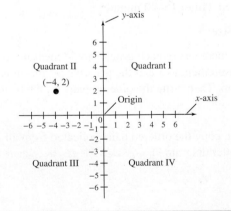

Section 3.2   Linear Equations in Two Variables

Key Concepts

An equation written in the form $Ax + By = C$ (where A and B are not both zero) is a **linear equation in two variables**.

A solution to a linear equation in x and y is an ordered pair (x, y) that makes the equation a true statement. The graph of the set of all solutions of a linear equation in two variables is a line in a rectangular coordinate system.

A linear equation can be graphed by finding at least two solutions and graphing the line through the points.

Examples

Example 1

Graph the equation $2x + y = 2$.

Select arbitrary values of x or y such as those shown in the table. Then complete the table to find the corresponding ordered pairs.

x	y	
0	2	→ $(0, 2)$
−1	4	→ $(-1, 4)$
1	0	→ $(1, 0)$

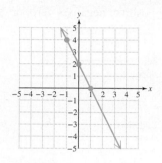

An **x-intercept** of a graph is a point $(a, 0)$ where the graph intersects the x-axis.

To find the x-intercept, let $y = 0$ and solve for x.

A **y-intercept** of a graph is a point $(0, b)$ where the graph intersects the y-axis.

To find the y-intercept, let $x = 0$ and solve for y.

Example 2

For the line $2x + y = 2$, find the x- and y-intercepts.

$\underline{x\text{-intercept}}$ $\underline{y\text{-intercept}}$

$2x + (0) = 2$ $2(0) + y = 2$

$\quad 2x = 2$ $\quad 0 + y = 2$

$\quad\; x = 1$ $\quad\;\; y = 2$

$\quad (1, 0)$ $\quad (0, 2)$

A **vertical line** can be represented by an equation of the form $x = k$.

A **horizontal line** can be represented by an equation of the form $y = k$.

Example 3

$x = 3$ represents a vertical line

$y = 3$ represents a horizontal line

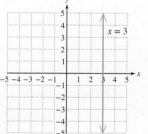
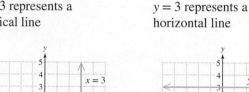

Section 3.3   Slope of a Line and Rate of Change

Key Concepts	**Examples**

Key Concepts

The **slope**, m, of a line between two points (x_1, y_1) and (x_2, y_2) is given by

$$m = \frac{y_2 - y_1}{x_2 - x_1} \quad \text{or} \quad \frac{\text{change in } y}{\text{change in } x}$$

The slope of a line may be positive, negative, zero, or undefined.

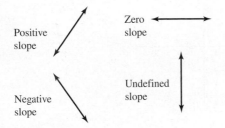

If m_1 and m_2 represent the slopes of two **parallel lines** (nonvertical), then $m_1 = m_2$.

If $m_1 \neq 0$ and $m_2 \neq 0$ represent the slopes of two nonvertical **perpendicular lines**, then

$$m_1 = -\frac{1}{m_2} \text{ or equivalently, } m_1 m_2 = -1.$$

Examples

Example 1

Find the slope of the line between $(1, -5)$ and $(-3, 7)$.

$$m = \frac{7 - (-5)}{-3 - 1} = \frac{12}{-4} = -3$$

Example 2

The slope of the line $y = -2$ is 0 because the line is horizontal.

Example 3

The slope of the line $x = 4$ is undefined because the line is vertical.

Example 4

The slopes of two distinct lines are given. Determine whether the lines are parallel, perpendicular, or neither.

a. $m_1 = -7$ and $m_2 = -7$ Parallel

b. $m_1 = -\dfrac{1}{5}$ and $m_2 = 5$ Perpendicular

c. $m_1 = -\dfrac{3}{2}$ and $m_2 = -\dfrac{2}{3}$ Neither

Section 3.4   Slope-Intercept Form of a Linear Equation

Key Concepts

The **slope-intercept form** of a linear equation is

$$y = mx + b$$

where m is the slope of the line and $(0, b)$ is the y-intercept.

Slope-intercept form is used to identify the slope and y-intercept of a line when the equation is given.

Slope-intercept form can also be used to graph a line.

Examples

Example 1

Find the slope and y-intercept.

$$7x - 2y = 4$$

$$-2y = -7x + 4 \qquad \text{Solve for } y.$$

$$\frac{-2y}{-2} = \frac{-7x}{-2} + \frac{4}{-2}$$

$$y = \frac{7}{2}x - 2$$

The slope is $\dfrac{7}{2}$. The y-intercept is $(0, -2)$.

Example 2

Graph the line.

$$y = \frac{7}{2}x - 2$$

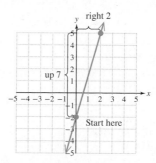

Section 3.5   Point-Slope Formula

Key Concepts

The **point-slope formula** is used primarily to construct an equation of a line given a point and the slope.

Equations of Lines—A Summary:

Standard form: $Ax + By = C$
Horizontal line: $y = k$
Vertical line: $x = k$
Slope-intercept form: $y = mx + b$
Point-slope formula: $y - y_1 = m(x - x_1)$

Example

Example 1

Find an equation of the line passing through the point $(6, -4)$ and having a slope of $-\frac{1}{2}$.

Label the given information:
$m = -\frac{1}{2}$ and $(x_1, y_1) = (6, -4)$

$$y - y_1 = m(x - x_1)$$

$$y - (-4) = -\tfrac{1}{2}(x - 6)$$

$$y + 4 = -\tfrac{1}{2}x + 3$$

$$y = -\tfrac{1}{2}x - 1$$

| Section 3.6 | Applications of Linear Equations and Modeling |

Key Concepts

Linear equations can often be used to describe or model the relationship between variables in a real-world event. In such applications, the slope may be interpreted as a rate of change.

Example

Example 1

The number of drug-related arrests for a small city has been growing approximately linearly since 1985.

Let y represent the number of drug arrests, and let x represent the number of years after 1985.

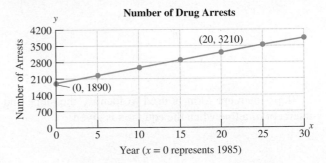

Number of Drug Arrests

Year ($x = 0$ represents 1985)

a. Use the ordered pairs (0, 1890) and (20, 3210) to find an equation of the line shown in the graph.

$$m = \frac{y_2 - y_1}{x_2 - x_1} = \frac{3210 - 1890}{20 - 0}$$

$$= \frac{1320}{20} = 66$$

The slope is 66, indicating that the number of drug arrests is increasing at a rate of 66 per year.
$m = 66$, and the y-intercept is (0, 1890). Thus,

$$y = mx + b \quad \Rightarrow \quad y = 66x + 1890$$

b. Use the equation in part (a) to estimate the number of drug-related arrests in the year 2015. (The year 2015 is 30 years after 1985. Hence, $x = 30$.)

$$y = 66(30) + 1890$$

$$y = 3870$$

For the year 2015, the number of drug arrests was approximately 3870.

Chapter 3 Review Exercises

Section 3.1

1. Graph the points on a rectangular coordinate system.

   a. $\left(\frac{1}{2}, 5\right)$ b. $(-1, 4)$ c. $(2, -1)$

   d. $(0, 3)$ e. $(0, 0)$ f. $\left(-\frac{8}{5}, 0\right)$

   g. $(-2, -5)$ h. $(3, 1)$

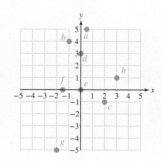

2. Estimate the coordinates of the points A, B, C, D, E, and F.

   $A(4, -3)$; $B(-3, -2)$; $C(\frac{5}{2}, 5)$; $D(-4, 1)$; $E(-\frac{1}{2}, 0)$; $F(0, -5)$

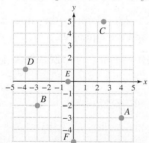

For Exercises 3–8, determine the quadrant in which the given point is located.

3. $(-2, -10)$ III   4. $(-4, 6)$ II

5. $(3, -5)$ IV   6. $\left(\frac{1}{2}, \frac{7}{5}\right)$ I

7. $(\pi, -2.7)$ IV   8. $(-1.2, -6.8)$ III

9. On which axis is the point $(2, 0)$ located?   x-axis

10. On which axis is the point $(0, -3)$ located?   y-axis

11. The price per share of a stock (in dollars) over a period of 5 days is shown in the graph.

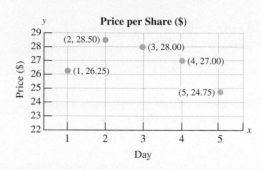

   a. Interpret the meaning of the ordered pair $(1, 26.25)$.   On day 1, the price was $26.25.

   b. On which day was the price the highest?
      Day 2

   c. What was the increase in price between day 1 and day 2? $2.25

12. The number of space shuttle launches for selected years is given by the ordered pairs. Let x represent the number of years since 1995. Let y represent the number of launches.

   $(1, 7)$ $(2, 8)$ $(3, 5)$ $(4, 3)$

   $(5, 5)$ $(6, 6)$ $(7, 5)$ $(8, 1)$

   a. Interpret the meaning of the ordered pair $(8, 1)$.
      In 2003 (8 years after 1995), there was only one space shuttle launch. (This was the year that the Columbia and its crew were lost.)

   b. Plot the points on a rectangular coordinate system.

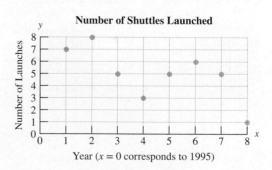

Section 3.2

For Exercises 13–16, determine if the given ordered pair is a solution to the equation.

13. $5x - 3y = 12;$ $(0, 4)$ No

14. $2x - 4y = -6;$ $(3, 0)$ No

15. $y = \frac{1}{3}x - 2;$ $(9, 1)$ Yes

16. $y = -\frac{2}{5}x + 1;$ $(-10, 5)$ Yes

For Exercises 17–20, complete the table and graph the corresponding ordered pairs. Graph the line through the points to represent all solutions to the equation.

17. $3x - y = 5$

x	y
2	1
3	4
1	-2

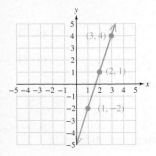

18. $\frac{1}{2}x + 3y = 6$

x	y
0	2
-2	7/3
-6	3

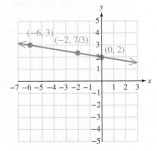

19. $y = \frac{2}{3}x - 1$

x	y
0	-1
3	1
-6	-5

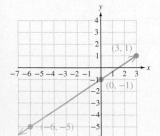

20. $y = -2x - 3$

x	y
0	-3
-3	3
1	-5

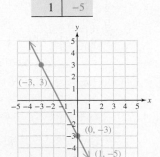

For Exercises 21–24, graph the equation.

21. $x + 2y = 4$

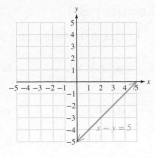

22. $x - y = 5$

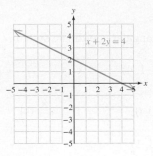

23. $y = 3x$

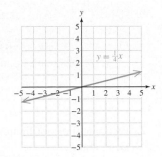

24. $y = \frac{1}{4}x$

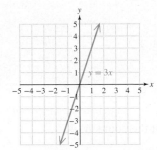

For Exercises 25–28, identify the line as horizontal or vertical. Then graph the equation.

25. $3x - 2 = 10$ Vertical

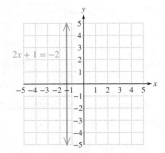

26. $2x + 1 = -2$ Vertical

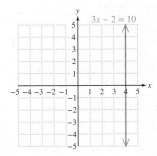

27. $6y + 1 = 13$ Horizontal

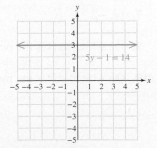

28. $5y - 1 = 14$ Horizontal

For Exercises 29–36, find the x- and y-intercepts if they exist.

29. $-4x + 8y = 12$ x-intercept: $(-3, 0)$; y-intercept: $\left(0, \frac{3}{2}\right)$

30. $2x + y = 6$ x-intercept: $(3, 0)$; y-intercept: $(0, 6)$

31. $y = 8x$ x-intercept: $(0, 0)$; y-intercept: $(0, 0)$

32. $5x - y = 0$ x-intercept: $(0, 0)$; y-intercept: $(0, 0)$

33. $6y = -24$ x-intercept: none; y-intercept: $(0, -4)$

34. $2y - 3 = 1$ x-intercept: none; y-intercept: $(0, 2)$

35. $2x + 5 = 0$ x-intercept: $\left(-\frac{5}{2}, 0\right)$; y-intercept: none

36. $-3x + 1 = 0$ x-intercept: $\left(\frac{1}{3}, 0\right)$; y-intercept: none

Section 3.3

37. What is the slope of the ladder leaning up against the wall? $m = \frac{12}{5}$

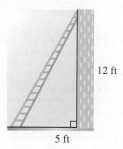

12 ft

5 ft

38. Point A is located 4 units down and 2 units to the right of point B. What is the slope of the line through points A and B? -2

39. Determine the slope of the line that passes through the points $(7, -9)$ and $(-5, -1)$. $-\frac{2}{3}$

40. Determine the slope of the line that has x- and y-intercepts of $(-1, 0)$ and $(0, 8)$.   8

41. Determine the slope of the line that passes through the points $(3, 0)$ and $(3, -7)$.   Undefined

42. Determine the slope of the line given by $y = -1$.   0

43. A given line has a slope of -5.

  a. What is the slope of a line parallel to the given line? -5

  b. What is the slope of a line perpendicular to the given line? $\frac{1}{5}$

44. A given line has a slope of 0.

  a. What is the slope of a line parallel to the given line?   0

  b. What is the slope of a line perpendicular to the given line?   Undefined

For Exercises 45–48, find the slopes of the lines l_1 and l_2 from the two given points. Then determine whether l_1 and l_2 are parallel, perpendicular, or neither.

45. l_1: $(3, 7)$ and $(0, 5)$ $m_1 = \frac{2}{3}$

    l_2: $(6, 3)$ and $(-3, -3)$ $m_2 = \frac{2}{3}$; parallel

46. l_1: $(-2, 1)$ and $(-1, 9)$ $m_1 = 8$

    l_2: $(0, -6)$ and $(2, 10)$ $m_2 = 8$; parallel

47. l_1: $\left(0, \frac{5}{6}\right)$ and $(2, 0)$ $m_1 = -\frac{5}{12}$

    l_2: $\left(0, \frac{6}{5}\right)$ and $\left(-\frac{1}{2}, 0\right)$ $m_2 = \frac{12}{5}$; perpendicular

48. l_1: $(1, 1)$ and $(1, -8)$ m_1 is undefined

    l_2: $(4, -5)$ and $(7, -5)$ $m_2 = 0$; perpendicular

49. Carol's electric bill had an initial reading of 35,955 kilowatt-hours at the beginning of the month. At the end of the month the reading was 37,005 kilowatt-hours. Let x represent the day of the month and y represent the reading on the meter in kilowatt-hours.

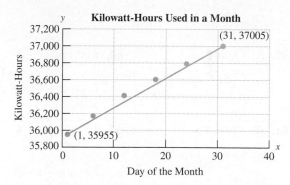

  a. Using the ordered pairs $(1, 35955)$ and $(31, 37005)$, find the slope of the line. $m = 35$

  b. Interpret the slope in the context of this problem.   The number of kilowatt-hours increased at a rate of 35 kilowatt-hours per day.

50. New car sales were recorded over a 5-yr period in Maryland. Let x represent the year and y represent the number of new cars sold.

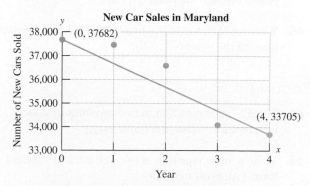

  a. Using the ordered pairs $(0, 37682)$ and $(4, 33705)$, find the slope of the line. Round to the nearest whole unit. $m = -994$

  b. Interpret the slope in the context of this problem.   The number of new cars sold in Maryland decreased at a rate of 994 cars per year.

Section 3.4

For Exercises 51–56, write each equation in slope-intercept form. Identify the slope and the y-intercept.

51. $5x - 2y = 10$ $y = \frac{5}{2}x - 5;$
$m = \frac{5}{2};$ y-intercept: $(0, -5)$

52. $3x + 4y = 12$ $y = -\frac{3}{4}x + 3;$
$m = -\frac{3}{4};$ y-intercept: $(0, 3)$

53. $x - 3y = 0$ $y = \frac{1}{3}x;$
$m = \frac{1}{3};$ y-intercept: $(0, 0)$

54. $5y - 8 = 4$ $y = \frac{12}{5};$
$m = 0;$ y-intercept: $\left(0, \frac{12}{5}\right)$

55. $2y = -5$
$y = -\frac{5}{2}; m = 0;$ y-intercept: $\left(0, -\frac{5}{2}\right)$

56. $y - x = 0$
$y = x;\ m = 1;$ y-intercept: $(0, 0)$

For Exercises 57–62, determine whether the equations represent parallel lines, perpendicular lines, or neither.

57. l_1: $y = \frac{3}{5}x + 3$

    l_2: $y = \frac{5}{3}x + 1$
    Neither

58. l_1: $2x - 5y = 10$

    l_2: $5x + 2y = 20$
    Perpendicular

59. l_1: $3x + 2y = 6$

    l_2: $-6x - 4y = 4$
    Parallel

60. l_1: $y = \frac{1}{4}x - 3$

    l_2: $-x + 4y = 8$
    Parallel

61. l_1: $2x = 4$

    l_2: $y = 6$ Perpendicular

62. l_1: $y = \frac{2}{9}x + 4$

    l_2: $y = \frac{9}{2}x - 3$
    Neither

63. Write an equation of the line whose slope is $-\frac{4}{3}$ and whose y-intercept is $(0, -1)$.
$y = -\frac{4}{3}x - 1$ or $4x + 3y = -3$

64. Write an equation of the line that passes through the origin and has a slope of 5.
$y = 5x$ or $5x - y = 0$

65. Write an equation of the line with slope $-\frac{4}{3}$ that passes through the point $(-6, 2)$.
$y = -\frac{4}{3}x - 6$ or $4x + 3y = -18$

66. Write an equation of the line with slope 5 that passes through the point $(-1, -8)$.
$y = 5x - 3$ or $5x - y = 3$

Section 3.5

67. Write a linear equation in two variables in slope-intercept form. (Answers may vary.)
For example: $y = 3x + 2$

68. Write a linear equation in two variables in standard form. (Answers may vary.)
For example: $5x + 2y = -4$

69. Write the slope formula to find the slope of the line between the points (x_1, y_1) and (x_2, y_2).
$m = \frac{y_2 - y_1}{x_2 - x_1}$

70. Write the point-slope formula. $y - y_1 = m(x - x_1)$

71. Write an equation of a vertical line (answers may vary).   For example: $x = 6$

72. Write an equation of a horizontal line (answers may vary).   For example: $y = -5$

For Exercises 73–78, write an equation of a line given the following information.

73. The slope is -6, and the line passes through the point $(-1, 8)$. $y = -6x + 2$ or $6x + y = 2$

74. The slope is $\frac{2}{3}$, and the line passes through the point $(5, 5)$. $y = \frac{2}{3}x + \frac{5}{3}$ or $2x - 3y = -5$

75. The line passes through the points $(0, -4)$ and $(8, -2)$. $y = \frac{1}{4}x - 4$ or $x - 4y = 16$

76. The line passes through the points $(2, -5)$ and $(8, -5)$. $y = -5$

77. The line passes through the point $(5, 12)$ and is perpendicular to the line $y = -\frac{5}{6}x - 3$.
$y = \frac{6}{5}x + 6$ or $6x - 5y = -30$

78. The line passes through the point $(-6, 7)$ and is parallel to the line $4x - y = 0$. $y = 4x + 31$ or $4x - y = -31$

Section 3.6

79. The graph shows the average height for girls based on age (*Source:* National Parenting Council). Let x represent a girl's age, and let y represent her height (in inches).

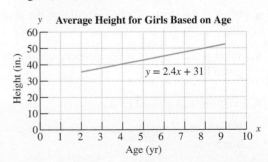

Average Height for Girls Based on Age

$y = 2.4x + 31$

a. Use the equation to estimate the average height of a 7-year-old girl.   47.8 in.

b. What is the slope of the line? Interpret the meaning of the slope in the context of the problem.
The slope is 2.4 and indicates that the average height for girls increases at a rate of 2.4 in. per year.

80. The number of drug prescriptions increased between 2000 and 2010 (see graph). Let x represent the number of years since 2000. Let y represent the number of prescriptions (in millions).

Number of Drug Prescriptions

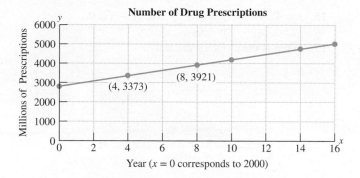

Year ($x = 0$ corresponds to 2000)

a. Using the ordered pairs (4, 3373) and (8, 3921) find the slope of the line. $m = 137$

b. Interpret the meaning of the slope in the context of this problem. The number of prescriptions increased by 137 million per year during this time period.

c. Find a linear equation that represents the number of prescriptions, y, versus the number of years, x, since 2000. $y = 137x + 2825$

d. Use the equation from part (c) to estimate the number of prescriptions for the year 2015. 4880 million

81. A water purification company charges $20 per month and a $55 installation fee.

a. Write a linear equation to compute the total cost, y, of renting this system for x months. $y = 20x + 55$

b. Use the equation from part (a) to determine the total cost to rent the system for 9 months. $235

82. A small cleaning company has a fixed monthly cost of $700 and a variable cost of $8 per service call.

a. Write a linear equation to compute the total cost, y, of making x service calls in one month. $y = 8x + 700$

b. Use the equation from part (a) to determine the total cost of making 80 service calls. $1340

Chapter 3 Test

1. In which quadrant is the given point located?

a. $\left(-\dfrac{7}{2}, 4\right)$ **b.** (4.6, −2) **c.** (−37, −45)

II   IV   III

2. What is the y-coordinate for a point on the x-axis?
0

3. What is the x-coordinate for a point on the y-axis?
0

4. Bamboo is the fastest growing woody plant on earth. At a bamboo farm, the height of a black bamboo plant (phyllostachys nigra) is measured for selected days. Let x represent the day number and y represent the height of the plant.

Day, x	Height (inches), y
4	14
8	28
12	42
16	56
20	70

a. Write the data as ordered pairs and interpret the meaning of the first ordered pair.
(4, 14) After 4 days the bamboo plant is 14 in. tall. (8, 28), (12, 42), (16, 56), (20, 70)

b. Graph the ordered pairs on a rectangular coordinate system.

Height of Bamboo Plant per Day

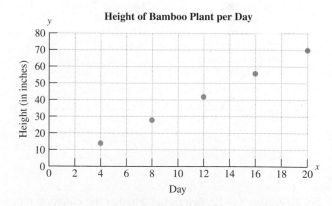

Day

c. From the graph, estimate the height of the bamboo plant after 10 days. Approximately 35 in.

5. Determine whether the ordered pair is a solution to the equation $2x - y = 6$.

a. (0, 6) No   **b.** (4, 2) Yes

c. (3, 0) Yes   **d.** $\left(\dfrac{9}{2}, 3\right)$ Yes

6. Given the equation $y = \frac{1}{4}x - 2$, complete the table. Plot the ordered pairs and graph the line through the points to represent the set of all solutions to the equation.

x	y
0	-2
4	-1
6	$-\frac{1}{2}$

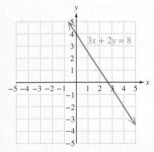

For Exercises 7–10, graph the equations.

7. $y = 3x + 2$

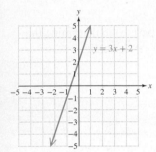

8. $2x + 5y = 0$

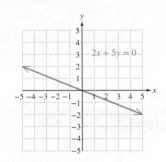

9. $3x + 2y = 8$

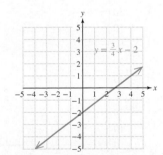

10. $y = \frac{3}{4}x - 2$

For Exercises 11–12, determine whether the equation represents a horizontal or vertical line. Then graph the line.

11. $-6y = 18$ Horizontal

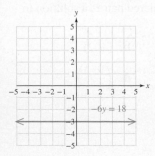

12. $5x + 1 = 8$ Vertical

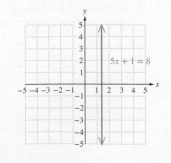

For Exercises 13–16, determine the x- and y-intercepts if they exist.

13. $-4x + 3y = 6$
x-intercept: $\left(-\frac{3}{2}, 0\right)$; y-intercept: $(0, 2)$

14. $2y = 6x$
x-intercept: $(0, 0)$;
y-intercept: $(0, 0)$

15. $x = 4$ x-intercept: $(4, 0)$;
y-intercept: none

16. $y - 3 = 0$
x-intercept: none;
y-intercept: $(0, 3)$

17. What is the slope of the hill?
$\frac{2}{5}$

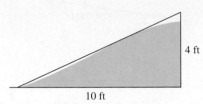

4 ft

10 ft

18. a. Find the slope of the line that passes through the points $(-2, 0)$ and $(-5, -1)$. $\frac{1}{3}$

  b. Find the slope of the line $4x - 3y = 9$. $\frac{4}{3}$

19. a. What is the slope of a line parallel to the line $x + 4y = -16$? $-\frac{1}{4}$

  b. What is the slope of a line perpendicular to the line $x + 4y = -16$? 4

20. a. What is the slope of the line $x = 5$?   Undefined

  b. What is the slope of the line $y = -3$? 0

21. Carlos called a local truck rental company and got quotes for renting a truck. He was told that it would cost $41.95 to rent a truck for one day to travel 20 miles. It costs $89.95 to rent the truck for one day to travel 100 miles. Let x represent the number of miles driven and y represent the cost of the rental.

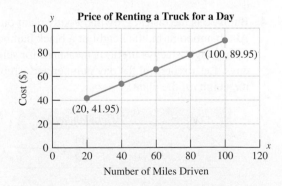

Price of Renting a Truck for a Day

$(100, 89.95)$

$(20, 41.95)$

Cost ($)

Number of Miles Driven

  a. Using the ordered pairs $(20, 41.95)$ and $(100, 89.95)$, find the slope of the line. $m = 0.6$

  b. Interpret the slope in the context of this problem.   The cost of renting a truck increases at a rate of $0.60 per mile.

22. Determine whether the lines through the given points are parallel, perpendicular, or neither.

l_1: (1, 4), (−1, −2) l_2: (0, −5), (−2, −11)
Parallel

23. Determine whether the equations represent parallel lines, perpendicular lines, or neither.

l_1: $2y = 3x − 3$ l_2: $4x = −6y + 1$
Perpendicular

24. Write an equation of the line that passes through the point (3, 0) and is parallel to the line $2x + 6y = −5$.
$y = −\frac{1}{3}x + 1$ or $x + 3y = 3$

25. Write an equation of the line that passes through the points (2, 8) and (4, 1). $y = −\frac{7}{2}x + 15$ or $7x + 2y = 30$

26. Write an equation of the line that has y-intercept $(0, \frac{1}{2})$ and slope $\frac{1}{4}$. $y = \frac{1}{4}x + \frac{1}{2}$ or $x − 4y = −2$

27. Write an equation of the line that passes through the point (−3, −1) and is perpendicular to the line $x + 3y = 9$. $y = 3x + 8$ or $3x − y = −8$

28. Write an equation of the line that passes through the point (2, −6) and is parallel to the x-axis. $y = −6$

29. Write an equation of the line that has slope −1 and passes through the point (−5, 2).
$y = −x − 3$ or $x + y = −3$

30. To attend a state fair, the cost is $10 per person to cover exhibits and musical entertainment. There is an additional cost of $1.50 per ride.

 a. Write an equation that gives the total cost, y, of visiting the state fair and going on x rides.
$y = 1.5x + 10$

 b. Use the equation from part (a) to determine the cost of going to the state fair and going on 10 rides. $25

31. The number of medical doctors for selected years is shown in the graph. Let x represent the number of years since 1980, and let y represent the number of medical doctors (in thousands) in the United States.

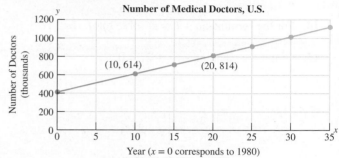

Number of Medical Doctors, U.S.

 a. Find the slope of the line shown in the graph. Interpret the meaning of the slope in the context of this problem. $m = 20$; The slope indicates that there is an increase of 20 thousand medical doctors per year.

 b. Find an equation of the line.
$y = 20x + 414$

 c. Use the equation from part (b) to estimate the number of medical doctors in the United States for the year 2015.
1114 thousand or, equivalently, 1,114,000

Chapters 1–3   Cumulative Review Exercises

1. Identify the number as rational or irrational.

 a. −3   **b.** $\frac{5}{4}$ **c.** $\sqrt{10}$ **d.** 0
 Rational   Rational   Irrational   Rational

2. Write the opposite and the absolute value for each number.

 a. $−\frac{2}{3}$ $\frac{2}{3}; \frac{2}{3}$ **b.** 5.3   −5.3; 5.3

3. Simplify the expression using the order of operations. $32 \div 2 \cdot 4 + 5$ 69

4. Add. $3 + (−8) + 2 + (−10)$ −13

5. Subtract. $16 − 5 − (−7)$ 18

For Exercises 6–7, translate the English phrase into an algebraic expression. Then evaluate the expression.

6. The quotient of $\frac{3}{4}$ and $−\frac{7}{8}$ $\frac{3}{4} \div \left(−\frac{7}{8}\right); −\frac{6}{7}$

7. The product of −2.1 and −6 $(−2.1)(−6); 12.6$

8. Name the property that is illustrated by the following statement. $6 + (8 + 2) = (6 + 8) + 2$ The associative property of addition

Writing   Translating Expression   Geometry   Scientific Calculator   Video

For Exercises 9–12, solve each equation.

9. $6x - 10 = 14$
{4}

10. $3(m + 2) - 3 = 2m + 8$
{5}

11. $\dfrac{2}{3}y - \dfrac{1}{6} = y + \dfrac{4}{3}$
$\left\{-\dfrac{9}{2}\right\}$

12. $1.7z + 2 = -2(0.3z + 1.3)$
{−2}

13. The area of Texas is 267,277 mi². If this is 712 mi² less than 29 times the area of Maine, find the area of Maine.   9241 mi²

14. For the formula $3a + b = c$, solve for a. $a = \dfrac{c - b}{3}$

15. Graph the equation $-6x + 2y = 0$.

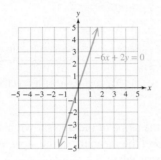

16. Find the x- and y-intercepts of $-2x + 4y = 4$.
x-intercept: $(-2, 0)$; y-intercept: $(0, 1)$

17. Write the equation in slope-intercept form. Then identify the slope and y-intercept. $3x + 2y = -12$
$y = -\dfrac{3}{2}x - 6$; slope: $-\dfrac{3}{2}$; y-intercept: $(0, -6)$

18. Explain why the line $2x + 3 = 5$ has only one intercept. $2x + 3 = 5$ can be written as $x = 1$, which represents a vertical line. A vertical line of the form $x = k$ $(k \neq 0)$ has an x-intercept of $(k, 0)$ and no y-intercept.

19. Find an equation of a line passing through $(2, -5)$ with slope -3. $y = -3x + 1$ or $3x + y = 1$

20. Find an equation of the line passing through $(0, 6)$ and $(-3, 4)$. $y = \frac{2}{3}x + 6$ or $2x - 3y = -18$

Systems of Linear Equations in Two Variables

CHAPTER OUTLINE

Mathematics in Business

Suppose that you have been invited to an end-of-semester party. You ask your friend for directions and she says, "It's on Earl Street and 10th Avenue." If we think of Earl Street and 10th Avenue as lines, then we know that the house is located where these lines intersect. In mathematics, we call these intersections **solutions** to **systems of linear equations**. Furthermore, the applications of systems of linear equations are numerous.

Imagine that the total price of buying a shirt and a tie is normally $42. If the items are on sale and priced at 60% and 90% of the original values, respectively, then the total cost is $30. To determine the original cost s of a single shirt and the original cost t of a single tie, we can set up a system of two equations.

© Lissa Harrison

Original Price: $s + t = 42$
Discounted Price: $0.60s + 0.90t = 30$

With techniques you will learn in this chapter, you can determine that the solution to this system is (26, 16). This means that the ordered pair (26, 16) satisfies both equations and that the point (26, 16) is a point of intersection of the lines defined by the equations in the system. The solution also tells us that the original cost of a shirt is $26 and the original cost of a tie is $16.

In this chapter, you will learn that some systems of linear equations have no solution, indicating that the related lines never intersect. This is similar to parallel streets that never meet. Other systems of linear equations may have infinitely many solutions. This occurs if the equations represent the same line.

Concepts

1. **Solutions to a System of Linear Equations**
2. **Solving Systems of Linear Equations by Graphing**

1. Solutions to a System of Linear Equations

Recall that a linear equation in two variables has an infinite number of solutions. The set of all solutions to a linear equation forms a line in a rectangular coordinate system. Two or more linear equations form a **system of linear equations**. For example, here are three systems of equations:

$$x - 3y = -5 \qquad\qquad y = \tfrac{1}{4}x - \tfrac{3}{4} \qquad\qquad 5a + \ b = 4$$
$$2x + 4y = 10 \qquad\qquad -2x + 8y = -6 \qquad\qquad -10a - 2b = 8$$

A **solution to a system of linear equations** is an ordered pair that is a solution to *both* individual linear equations.

Classroom Examples: p. 280, Exercises 4 and 6

Example 1   Determining Solutions to a System of Linear Equations

Determine whether the ordered pairs are solutions to the system.

$$x + y = 4$$
$$-2x + y = -5$$

a. $(3, 1)$ **b.** $(0, 4)$

Solution:

a. Substitute the ordered pair $(3, 1)$ into both equations:

$$x + y = 4 \longrightarrow (3) + (1) \overset{?}{=} 4 \checkmark \qquad \text{True}$$
$$-2x + y = -5 \longrightarrow -2(3) + (1) \overset{?}{=} -5 \checkmark \qquad \text{True}$$

Avoiding Mistakes

It is important to test an ordered pair in *both* equations to determine if the ordered pair is a solution.

Because the ordered pair $(3, 1)$ is a solution to both equations, it is a solution to the *system* of equations.

b. Substitute the ordered pair $(0, 4)$ into both equations.

$$x + y = 4 \longrightarrow (0) + (4) \overset{?}{=} 4 \checkmark \qquad \text{True}$$
$$-2x + y = -5 \longrightarrow -2(0) + (4) \overset{?}{=} -5 \qquad \text{False}$$

Because the ordered pair $(0, 4)$ is not a solution to the second equation, it is *not* a solution to the system of equations.

Skill Practice   Determine whether the ordered pair is a solution to the system.

$$5x - 2y = 24$$
$$2x + \ y = 6$$

1. $(6, 3)$ **2.** $(4, -2)$

A solution to a system of two linear equations can be interpreted graphically as a point of intersection of the two lines. Using slope-intercept form to graph the lines from Example 1, we have

Answers

1. No
2. Yes

$$l_1: \qquad x + y = 4 \longrightarrow y = -x + 4$$
$$l_2: \quad -2x + y = -5 \longrightarrow y = 2x - 5$$

All points on l_1 are solutions to the equation $y = -x + 4$.

All points on l_2 are solutions to the equation $y = 2x - 5$.

The point of intersection (3, 1) is the only point that is a solution to both equations. (See Figure 4-1).

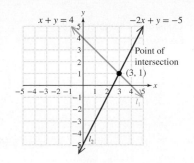

Figure 4-1

When two lines are drawn in a rectangular coordinate system, three geometric relationships are possible:

1. Two lines may intersect at *exactly one point*.

2. Two lines may intersect at *no point*. This occurs if the lines are parallel.

3. Two lines may intersect at *infinitely many points* along the line. This occurs if the equations represent the same line (the lines coincide).

If a system of linear equations has one or more solutions, the system is said to be **consistent**. If a system of linear equations has no solution, it is said to be **inconsistent**.

If two equations represent the same line, the equations are said to be **dependent equations**. In this case, all points on the line are solutions to the system. If two equations represent two different lines, the equations are said to be **independent equations**. In this case, the lines either intersect at one point or are parallel.

Solutions to Systems of Linear Equations in Two Variables

One Unique Solution	No Solution	Infinitely Many Solutions
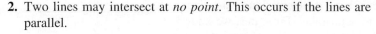		
One point of intersection	Parallel lines	Coinciding lines
• System is consistent.	• System is inconsistent.	• System is consistent.
• Equations are independent.	• Equations are independent.	• Equations are dependent.

2. Solving Systems of Linear Equations by Graphing

One way to find a solution to a system of equations is to graph the equations and find the point (or points) of intersection. This is called the *graphing method* to solve a system of equations.

Example 2 | Solving a System of Linear Equations by Graphing

Solve the system by the graphing method. $\quad y = 2x$
$$y = 2$$

Solution:

The equation $y = 2x$ is written in slope-intercept form as $y = 2x + 0$. The line passes through the origin, with a slope of 2.

The line $y = 2$ is a horizontal line and has a slope of 0.

Because the lines have different slopes, the lines must be different and nonparallel. From this, we know that the lines must intersect at exactly one point. Graph the lines to find the point of intersection (Figure 4-2).

The point $(1, 2)$ appears to be the point of intersection. This can be confirmed by substituting $x = 1$ and $y = 2$ into both original equations.

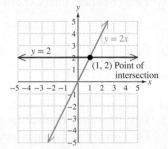

Figure 4-2

$$y = 2x \quad (2) \overset{?}{=} 2(1) \checkmark \quad \text{True}$$

$$y = 2 \quad (2) \overset{?}{=} 2 \checkmark \quad \quad \text{True}$$

The solution set is $\{(1, 2)\}$.

Skill Practice Solve the system by the graphing method.

3. $y = -3x$

$\quad x = -1$

Example 3 | Solving a System of Linear Equations by Graphing

Solve the system by the graphing method.

$$x - 2y = -2$$
$$-3x + 2y = 6$$

Solution:

One method to graph the lines is to write each equation in slope-intercept form, $y = mx + b$.

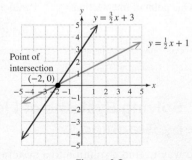

Figure 4-3

Answer

3. $\{(-1, 3)\}$

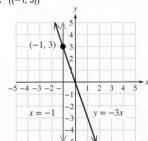

Equation 1	**Equation 2**
$x - 2y = -2$	$-3x + 2y = 6$
$-2y = -x - 2$	$2y = 3x + 6$
$\dfrac{-2y}{-2} = \dfrac{-x}{-2} - \dfrac{2}{-2}$	$\dfrac{2y}{2} = \dfrac{3x}{2} + \dfrac{6}{2}$
$y = \dfrac{1}{2}x + 1$	$y = \dfrac{3}{2}x + 3$

From their slope-intercept forms, we see that the lines have different slopes, indicating that the lines are different and nonparallel. Therefore, the lines must intersect at exactly one point. Graph the lines to find that point (Figure 4-3).

The point $(-2, 0)$ appears to be the point of intersection. This can be confirmed by substituting $x = -2$ and $y = 0$ into both equations.

$$x - 2y = -2 \longrightarrow (-2) - 2(0) \overset{?}{=} -2 \checkmark \quad \text{True}$$

$$-3x + 2y = 6 \longrightarrow -3(-2) + 2(0) \overset{?}{=} 6 \checkmark \quad \text{True}$$

The solution set is $\{(-2, 0)\}$.

Skill Practice   Solve the system by the graphing method.

4. $y = 2x - 3$
$6x + 2y = 4$

TIP: In Examples 2 and 3, the lines could also have been graphed by using the x- and y-intercepts or by using a table of points. However, the advantage of writing the equations in slope-intercept form is that we can compare the slopes and y-intercepts of the two lines.

  1. If the slopes differ, the lines are different and nonparallel and must intersect at exactly one point.

  2. If the slopes are the same and the y-intercepts are different, the lines are parallel and will not intersect.

  3. If the slopes are the same and the y-intercepts are the same, the two equations represent the same line.

Example 4   **Graphing an Inconsistent System**

Classroom Example: p. 282, Exercise 34

Solve the system by graphing.

$$-x + 3y = -6$$
$$6y = 2x + 6$$

Solution:

To graph the lines, write each equation in slope-intercept form.

Answers

4. $\{(1, -1)\}$

Equation 1	**Equation 2**
$-x + 3y = -6$	$6y = 2x + 6$
$3y = x - 6$	
$\dfrac{3y}{3} = \dfrac{x}{3} - \dfrac{6}{3}$	$\dfrac{6y}{6} = \dfrac{2x}{6} + \dfrac{6}{6}$
$y = \dfrac{1}{3}x - 2$	$y = \dfrac{1}{3}x + 1$

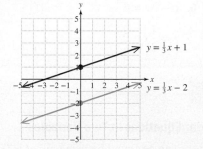

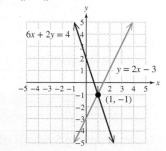

Figure 4-4

Because the lines have the same slope but different y-intercepts, they are parallel (Figure 4-4). Two parallel lines do not intersect, which implies that the system has no solution. Therefore, the solution set is the empty set, { }. The system is inconsistent.

Skill Practice Solve the system by graphing.

  5. $4x + y = 8$
  $y = -4x + 3$

5.

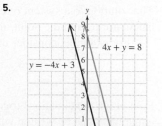

{ } The lines are parallel. The system is inconsistent.

Example 5   Graphing a System of Dependent Equations

Solve the system by graphing. $x + 4y = 8$

$$y = -\frac{1}{4}x + 2$$

Solution:

Write the first equation in slope-intercept form. The second equation is already in slope-intercept form.

TIP: The solution set to a system of dependent equations uses set-builder notation to describe the common line of intersection. Any form of the equation can be used. For example, in Example 5, we show the equation written in slope-intercept form and in standard form.

$$\left\{(x, y) \,\middle|\, y = -\frac{1}{4}x + 2\right\}$$
or
$$\{(x, y) \,|\, x + 4y = 8\}$$

Equation 1

$x + 4y = 8$

$4y = -x + 8$

$\dfrac{4y}{4} = \dfrac{-x}{4} + \dfrac{8}{4}$

$y = -\dfrac{1}{4}x + 2$

Equation 2

$y = -\dfrac{1}{4}x + 2$

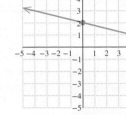

Figure 4-5

Notice that the slope-intercept forms of the two lines are identical. Therefore, the equations represent the same line (Figure 4-5). The equations are dependent, and the solution to the system of equations is the set of all points on the line.

Because there are infinitely many points on the line, the ordered pairs in the solution set cannot all be listed. Therefore, we can write the solution in set-builder notation: $\{(x, y) \,|\, y = -\frac{1}{4}x + 2\}$. This can be read as "the set of all ordered pairs (x, y) such that the ordered pairs satisfy the equation $y = -\frac{1}{4}x + 2$."

In summary:

- There are infinitely many solutions to the system of equations.
- The solution set is $\{(x, y) \,|\, y = -\frac{1}{4}x + 2\}$, or equivalently $\{(x, y) \,|\, x + 4y = 8\}$.
- The equations are dependent.

Answer

6.

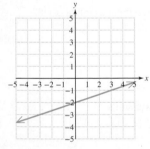

$$\left\{(x, y) \,\middle|\, y = \frac{1}{3}x - 2\right\}$$

The equations are dependent.

Skill Practice Solve the system by graphing.

6. $x - 3y = 6$

$$y = \frac{1}{3}x - 2$$

Calculator Connections

Topic: Graphing Systems of Linear Equations in Two Variables

The solution to a system of equations can be found by using either a *Trace* feature or an *Intersect* feature on a graphing calculator to find the point of intersection of two graphs.

For example, consider the system:

$$-2x + y = 6$$
$$5x + y = -1$$

First graph the equations together on the same viewing window. Recall that to enter the equations into the calculator, the equations must be written with the *y* variable isolated.

$$-2x + y = 6 \xrightarrow{\text{Isolate } y.} y = 2x + 6$$
$$5x + y = -1 \xrightarrow{\phantom{\text{Isolate } y.}} y = -5x - 1$$

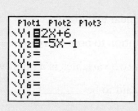

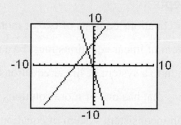

By inspection of the graph, it appears that the solution is $(-1, 4)$. The *Trace* option on the calculator may come close to $(-1, 4)$ but may not show the exact solution (Figure 4-6). However, an *Intersect* feature on a graphing calculator may provide the exact solution (Figure 4-7).

Using *Trace*

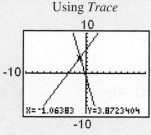

Figure 4-6

Using *Intersect*

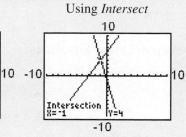

Figure 4-7

Calculator Exercises

Use a graphing calculator to graph each pair of linear equations on the same viewing window. Use a *Trace* or *Intersect* feature to find the point(s) of intersection. Then write the solution set.

1. $y = 2x - 3$ $\{(2, 1)\}$
$y = -4x + 9$

2. $y = -\dfrac{1}{2}x + 2$ $\{(6, -1)\}$
$y = \dfrac{1}{3}x - 3$

3. $x + y = 4$ (Example 1)
$-2x + y = -5$ $\{(3, 1)\}$

4. $x - 2y = -2$ (Example 3)
$-3x + 2y = 6$
$\{(-2, 0)\}$

5. $-x + 3y = -6$ (Example 4)
$6y = 2x + 6$
$\{\ \}$

6. $x + 4y = 8$ (Example 5)
$y = -\dfrac{1}{4}x + 2$ $\left\{ (x, y) \,\middle|\, y = -\dfrac{1}{4}x + 2 \right\}$

Section 4.1   Practice Exercises

Study Skills Exercise

For additional exercises, see Classroom Activity 4.1A, in the *Student's Resource Manual* at www.mhhe.com/moh.

It is important to keep track of your grade throughout the semester. Take a minute to compute your grade at this point. Are you earning the grade that you want? If not, maybe organizing a study group would help.

In a study group, check the activities that you might try to help you learn and understand the material.

_____ Quiz each other by asking each other questions.

_____ Practice teaching each other.

_____ Share and compare class notes.

_____ Support and encourage each other.

_____ Work together on exercises and sample problems.

Writing   Translating Expression   Geometry   Scientific Calculator   Video

Vocabulary and Key Concepts

1. **a.** A _system_ of linear equations consists of two or more linear equations.

    b. A _solution_ to a system of linear equations must be a solution to both individual equations in the system.

    c. Graphically, a solution to a system of linear equations in two variables is a point where the lines _intersect_.

    d. A system of equations that has one or more solutions is said to be _consistent_.

    e. The solution set to an inconsistent system of equations is _{ }_.

    f. Two equations in a system of linear equations in two variables are said to be _dependent_ if they represent the same line.

    g. Two equations in a system of linear equations in two variables are said to be _independent_ if they represent different lines.

Concept 1: Solutions to a System of Linear Equations

For Exercises 2–10, determine if the given point is a solution to the system. **(See Example 1.)**

2. $6x - y = -9$ $(-1, 3)$
    $x + 2y = 5$ Yes

3. $3x - y = 7$ $(2, -1)$
    $x - 2y = 4$ Yes

4. $x - y = 3$ $(4, 1)$
    $x + y = 5$ Yes

5. $4y = -3x + 12$ $(0, 4)$
    $y = \dfrac{2}{3}x - 4$ No

6. $y = -\dfrac{1}{3}x + 2$ $(9, -1)$
    $x = 2y + 6$ No

7. $3x - 6y = 9$ $\left(4, \dfrac{1}{2}\right)$
    $x - 2y = 3$ Yes

8. $x - y = 4$ $(6, 2)$
    $3x - 3y = 12$ Yes

9. $\dfrac{1}{3}x = \dfrac{2}{5}y - \dfrac{4}{5}$ $(0, 2)$
    $\dfrac{3}{4}x + \dfrac{1}{2}y = 2$ No

10. $\dfrac{1}{4}x + \dfrac{1}{2}y = \dfrac{3}{2}$ $(4, 1)$
    $y = \dfrac{3}{2}x - 6$ No

For Exercises 11–14, match the graph of the system of equations with the appropriate description of the solution.

11. b
    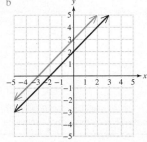

    a. The solution set is {(1, 3)}.

    b. { }

    c. There are infinitely many solutions.

    d. The solution set is {(0, 0)}.

12. c
    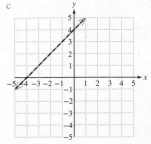

13. d
    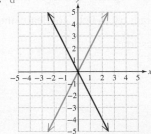

14. a
    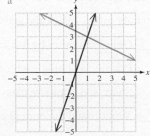

15. Graph each system of equations.

    a. $y = 2x - 3$

        $y = 2x + 5$

    b. $y = 2x + 1$

        $y = 4x - 1$

    c. $y = 3x - 5$

        $y = 3x - 5$

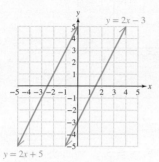 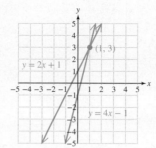 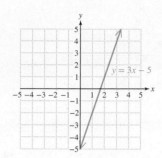

For Exercises 16–26, determine which system of equations (a, b, or c) makes the statement true. (*Hint:* Refer to the graphs from Exercise 15.)

a. $y = 2x - 3$

   $y = 2x + 5$

b. $y = 2x + 1$

   $y = 4x - 1$

c. $y = 3x - 5$

   $y = 3x - 5$

16. The lines are parallel. a

17. The lines coincide. c

18. The lines intersect at exactly one point. b

19. The system is inconsistent. a

20. The equations are dependent. c

21. The lines have the same slope but different y-intercepts. a

22. The lines have the same slope and same y-intercept. c

23. The lines have different slopes. b

24. The system has exactly one solution. b

25. The system has infinitely many solutions. c

26. The system has no solution. a

Concept 2: Solving Systems of Linear Equations by Graphing

For Exercises 27–50, solve the system by graphing. For systems that do not have one unique solution, also state the number of solutions and whether the system is inconsistent or the equations are dependent. **(See Examples 2–5.)**

27. $y = -x + 4$

    $y = x - 2$ $\{(3, 1)\}$

28. $y = 3x + 2$

    $y = 2x$ $\{(-2, -4)\}$

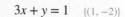**29.** $2x + y = 0$

    $3x + y = 1$ $\{(1, -2)\}$

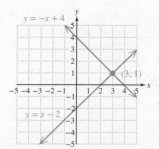 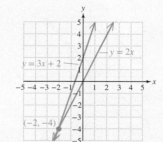 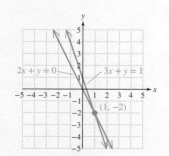

30. $x + y = -1$
$2x - y = -5$ $\{(-2, 1)\}$

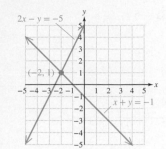

31. $2x + y = 6$
$x = 1$ $\{(1, 4)\}$

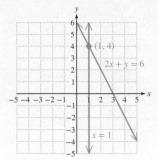

32. $4x + 3y = 9$
$x = 3$ $\{(3, -1)\}$

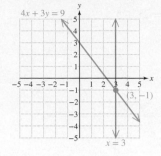

33. $-6x - 3y = 0$ No solution;
$4x + 2y = 4$ { };
inconsistent system

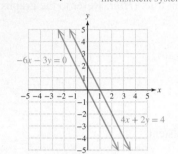

34. $2x - 6y = 12$ No solution;
$-3x + 9y = 12$ { };
inconsistent system

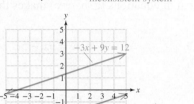

35. $-2x + y = 3$ Infinitely many solutions;
$6x - 3y = -9$ $\{(x, y) \mid -2x + y = 3\}$;
dependent equations

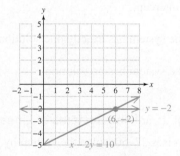

36. $x + 3y = 0$ Infinitely many solutions;
$-2x - 6y = 0$ $\{(x, y) \mid x + 3y = 0\}$;
dependent equations

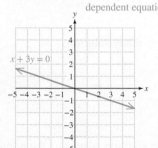

37. $y = 6$
$2x + 3y = 12$ $\{(-3, 6)\}$

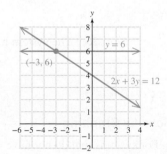

38. $y = -2$
$x - 2y = 10$ $\{(6, -2)\}$

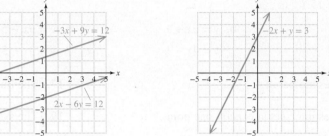

39. $x = 4 + y$
$3y = -3x$ $\{(2, -2)\}$

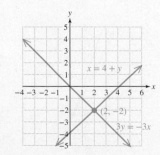

40. $3y = 4x$
$x - y = -1$ $\{(3, 4)\}$

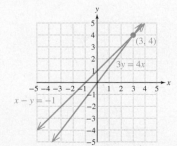

41. $-x + y = 3$ No solution;
$4y = 4x + 6$ { };
inconsistent system

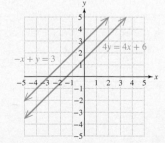

42. $x - y = 4$ No solution;
$3y = 3x + 6$ { };
inconsistent system

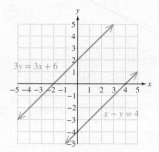

43. $x = 4$
$2y = 4$ {(4, 2)}

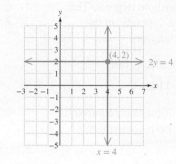

44. $-3x = 6$
$y = 2$ {(−2, 2)}

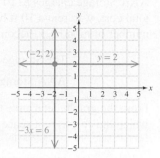

45. $2x + y = 4$
$4x - 2y = 0$ {(1, 2)}

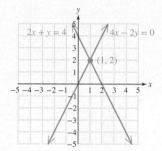

46. $3x + 3y = 3$
$2x - y = 5$ {(2, −1)}

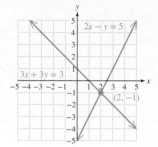

47. $y = 0.5x + 2$ Infinitely many
solutions;
$-x + 2y = 4$ {(x, y) | y = 0.5x + 2};
dependent equations

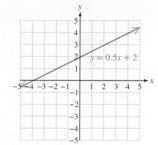

48. $3x - 4y = 6$ Infinitely many
solutions;
$-6x + 8y = -12$ {(x, y) | 3x − 4y = 6};
dependent equations

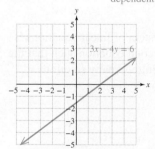

49. $x - 3y = 0$
$y = -x - 4$ {(−3, −1)}

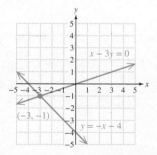

50. $-6x + 3y = -6$
$4x + y = -2$ {(0, −2)}

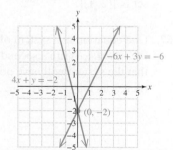

51. A wholesale club offers two types of memberships. The Executive Membership is $100 per year, including an annual 2% reward. The Business Membership is $50 per year without a reward. The total cost for membership, y, depends on the amount of money spent on merchandise, x, and can be represented by the following equations:

Executive Membership: $y = 100 - 0.02x$

Business Membership: $y = 50$

According to the graph, how much money spent on merchandise would result in the same cost for each membership?
The same cost occurs when $2500 of merchandise is purchased.

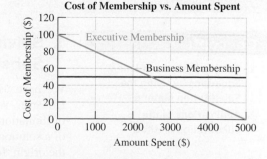

Cost of Membership vs. Amount Spent

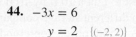 Writing   Translating Expression   Geometry   Scientific Calculator   Video

52. The cost to rent a 10 ft by 10 ft storage space is different for two different storage companies. The Storage Bin charges $90 per month plus a nonrefundable deposit of $120. AAA Storage charges $110 per month with no deposit. The total cost, y, to rent a 10 ft by 10 ft space depends on the number of months, x, according to the equations

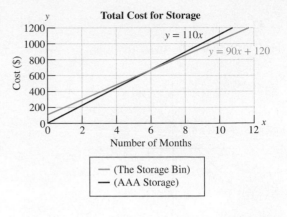

The Storage Bin: $y = 90x + 120$

AAA Storage: $y = 110x$

From the graph, determine the number of months required for which the cost to rent space is equal for both companies.
For 6 months, the cost is $660 for each company.

For the systems graphed in Exercises 53–54, explain why the ordered pair cannot be a solution to the system of equations.

53. $(-3, 1)$ The point of intersection is below the x-axis and cannot have a positive y-coordinate.

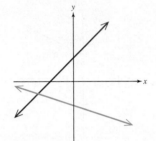

54. $(-1, -4)$ The point of intersection is above the x-axis and cannot have a negative y-coordinate.

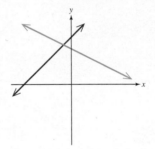

Expanding Your Skills

55. Write a system of linear equations whose solution set is $\{(2, 1)\}$.   For example: $4x + y = 9$
$-2x - y = -5$

56. Write a system of linear equations whose solution set is $\{(1, 4)\}$.   For example: $x - 3y = -11$
$3x + y = 7$

57. One equation in a system of linear equations is $x + y = 4$. Write a second equation such that the system will have no solution. (Answers may vary.)   For example: $2x + 2y = 1$

58. One equation in a system of linear equations is $x - y = 3$. Write a second equation such that the system will have infinitely many solutions. (Answers may vary.)   For example: $3x - 3y = 9$

Section 4.2 Solving Systems of Equations by the Substitution Method

Concepts

1. Solving Systems of Linear Equations by Using the Substitution Method
2. Applications of the Substitution Method

1. Solving Systems of Linear Equations by Using the Substitution Method

We have used the graphing method to find the solution set to a system of equations. However, sometimes it is difficult to determine the solution using this method because of limitations in the accuracy of the graph. This is particularly true when the coordinates of a solution are not integer values or when the solution is a point not sufficiently close to the origin. Identifying the coordinates of the point $\left(\frac{3}{17}, -\frac{23}{9}\right)$ or $(-251, 8349)$, for example, might be difficult from a graph.

In this section, we will cover the substitution method to solve systems of equations. This is an algebraic method that does not require graphing the individual equations. We demonstrate the substitution method in Examples 1 through 5.

| **Example 1** | **Solving a System of Linear Equations by Using the Substitution Method** |

Classroom Example: p. 292, Exercise 14

Solve the system by using the substitution method.

$$x = 2y - 3$$
$$-4x + 3y = 2$$

Solution:

The variable x has been isolated in the first equation. The quantity $2y - 3$ is equal to x and therefore can be substituted for x in the second equation. This leaves the second equation in terms of y only.

First equation: $x = \underline{2y - 3}$

Second equation: $-4x + 3y = 2$

$-4(2y - 3) + 3y = 2$ This equation now contains only one variable.

$-8y + 12 + 3y = 2$ Solve the resulting equation.

$-5y + 12 = 2$

$-5y = -10$

$y = 2$

To find x, substitute $y = 2$ back into the first equation.

$$x = 2y - 3$$
$$x = 2(2) - 3$$
$$x = 1$$

Check the ordered pair $(1, 2)$ in both original equations.

$x = 2y - 3 \longrightarrow 1 \overset{?}{=} 2(2) - 3$ ✓   True

$-4x + 3y = 2 \longrightarrow -4(1) + 3(2) \overset{?}{=} 2$ ✓   True

The solution set is $\{(1, 2)\}$.

Avoiding Mistakes

Remember to solve for *both* variables in the system.

Skill Practice Solve the system by using the substitution method.

1. $2x + 3y = -2$
    $y = x + 1$

Instructor Note: Remind students that they can also graph a system of equations to determine if an answer is reasonable.

In Example 1, we eliminated the x variable from the second equation by substituting an equivalent expression for x. The resulting equation was relatively simple to solve because it had only one variable. This is the premise of the substitution method.

Answer

1. $\{(-1, 0)\}$

The substitution method can be summarized as follows.

> ### Solving a System of Equations by Using the Substitution Method
> **Step 1** Isolate one of the variables from one equation.
> **Step 2** Substitute the expression found in step 1 into the other equation.
> **Step 3** Solve the resulting equation.
> **Step 4** Substitute the value found in step 3 back into the equation in step 1 to find the value of the remaining variable.
> **Step 5** Check the ordered pair in both original equations.

Classroom Example: p. 292, Exercise 22

Example 2 **Solving a System of Linear Equations by Using the Substitution Method**

Solve the system by using the substitution method.

$$x + y = 4$$
$$-5x + 3y = -12$$

Solution:

The x or y variable in the first equation is easy to isolate because the coefficients are both 1. While either variable can be isolated, we arbitrarily choose to solve for the x variable.

$x + y = 4 \longrightarrow x = \underbrace{4 - y}$ **Step 1:** Isolate x in the first equation.

$-5(4 - y) + 3y = -12$ **Step 2:** Substitute $4 - y$ for x in the other equation.

$-20 + 5y + 3y = -12$ **Step 3:** Solve for y.
$-20 + 8y = -12$
$8y = 8$
$y = 1$

$x = 4 - y$ **Step 4:** Substitute $y = 1$ into the equation $x = 4 - y$.
$x = 4 - 1$
$x = 3$

Step 5: Check the ordered pair (3, 1) in both original equations.

$$x + y = 4 \qquad (3) + (1) \overset{?}{=} 4 \checkmark \qquad \text{True}$$
$$-5x + 3y = -12 \qquad -5(3) + 3(1) \overset{?}{=} -12 \checkmark \qquad \text{True}$$

The solution set is $\{(3, 1)\}$.

Skill Practice Solve the system by using the substitution method.

2. $x + y = 3$
    $-2x + 3y = 9$

Avoiding Mistakes

Although we solved for y first, be sure to write the x-coordinate first in the ordered pair. Remember that (1, 3) is not the same as (3, 1).

TIP: The solution to a system of linear equations can be confirmed by graphing. The system from Example 2 is graphed here.

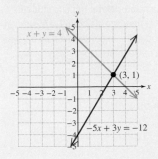

Answer
2. $\{(0, 3)\}$

| Example 3 | Solving a System of Linear Equations by Using the Substitution Method |

Classroom Example: p. 292, Exercise 24

Solve the system by using the substitution method.

$$3x + 5y = 17$$
$$2x - y = -6$$

Solution:

The y variable in the second equation is the easiest variable to isolate because its coefficient is -1.

$3x + 5y = 17$

$2x - y = -6 \longrightarrow -y = -2x - 6$

$\qquad\qquad\qquad y = \underline{2x + 6}$ **Step 1:** Isolate y in the second equation.

$3x + 5(2x + 6) = 17$ **Step 2:** Substitute the quantity $2x + 6$ for y in the other equation.

$3x + 10x + 30 = 17$ **Step 3:** Solve for x.

$\qquad 13x + 30 = 17$

$\qquad\qquad 13x = 17 - 30$

$\qquad\qquad 13x = -13$

$\qquad\qquad\quad x = -1$

$y = 2x + 6$ **Step 4:** Substitute $x = -1$ into the equation $y = 2x + 6$.

$y = 2(-1) + 6$

$y = -2 + 6$

$y = 4$

Step 5: The ordered pair $(-1, 4)$ can be checked in the original equations to verify the answer.

$3x + 5y = 17 \longrightarrow 3(-1) + 5(4) \overset{?}{=} 17 \longrightarrow -3 + 20 \overset{?}{=} 17 \checkmark$ True

$2x - y = -6 \longrightarrow 2(-1) - (4) \overset{?}{=} -6 \longrightarrow -2 - 4 \overset{?}{=} -6 \checkmark$ True

The solution set is $\{(-1, 4)\}$.

Avoiding Mistakes

Do not substitute $y = 2x + 6$ into the same equation from which it came. This mistake will result in an identity:

$$2x - y = -6$$
$$2x - (2x + 6) = -6$$
$$2x - 2x - 6 = -6$$
$$-6 = -6$$

Skill Practice Solve the system by using the substitution method.

3. $x + 4y = 11$
   $2x - 5y = -4$

Answer

3. $\{(3, 2)\}$

Recall that a system of linear equations may represent two parallel lines. In such a case, there is no solution to the system.

Classroom Example: p. 292, Exercise 26

> ### Example 4   Solving an Inconsistent System Using Substitution
>
> Solve the system by using the substitution method.
>
> $$2x + 3y = 6$$
> $$y = -\tfrac{2}{3}x + 4$$
>
> **Solution:**
>
> $$2x + 3y = 6$$
> $$y = -\tfrac{2}{3}x + 4$$
>
> **Step 1:** The variable y is already isolated in the second equation.
>
> $$2x + 3\left(-\tfrac{2}{3}x + 4\right) = 6$$
>
> **Step 2:** Substitute $y = -\tfrac{2}{3}x + 4$ from the second equation into the first equation.
>
> $$2x - 2x + 12 = 6$$
>
> **Step 3:** Solve the resulting equation.
>
> $$12 = 6 \quad \text{(contradiction)}$$
>
> The equation results in a contradiction. There are no values of x and y that will make 12 equal to 6. Therefore, the solution set is { }, and the system is inconsistent.
>
> **Skill Practice** Solve the system by using the substitution method.
>
> **4.** $y = -\dfrac{1}{2}x + 3$
>
> $\ 2x + 4y = 5$

Instructor Note: Remind students that when we solved equations in one variable, a contradiction had no solution. For example, the equation $x + 3 = x + 2$ has no solution.

TIP: The answer to Example 4 can be verified by writing each equation in slope-intercept form and graphing the lines.

Equation 1

$$2x + 3y = 6$$
$$3y = -2x + 6$$
$$\frac{3y}{3} = \frac{-2x}{3} + \frac{6}{3}$$
$$y = -\frac{2}{3}x + 2$$

Equation 2

$$y = -\frac{2}{3}x + 4$$

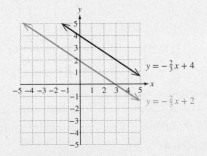

The equations indicate that the lines have the same slope but different y-intercepts. Therefore, the lines must be parallel. There is no point of intersection, indicating that the system has no solution, { }.

Answer

4. { }

Recall that a system of two linear equations may represent the same line. In such a case, the solution is the set of all points on the line.

| Example 5 | **Solving a System of Dependent Equations Using Substitution** |

Classroom Example: p. 293, Exercise 40

Solve the system by using the substitution method.

$$\frac{1}{2}x - \frac{1}{4}y = 1$$

$$6x - 3y = 12$$

Solution:

$\frac{1}{2}x - \frac{1}{4}y = 1$ To make the first equation easier to work with, we have the option of clearing fractions.

$6x - 3y = 12$

$\frac{1}{2}x - \frac{1}{4}y = 1 \xrightarrow{\text{Multiply by 4.}} 4\left(\frac{1}{2}x\right) - 4\left(\frac{1}{4}y\right) = 4(1) \longrightarrow 2x - y = 4$

Now the system becomes:

$2x - y = 4$ The y variable in the first equation is the easiest to isolate because its coefficient is -1.

$6x - 3y = 12$

$2x - y = 4 \xrightarrow{\text{Solve for } y.} -y = -2x + 4 \rightarrow y = \underline{2x - 4}$ **Step 1:** Isolate one of the variables.

$6x - 3y = 12$

$6x - 3(2x - 4) = 12$ **Step 2:** Substitute $y = 2x - 4$ from the first equation into the second equation.

$6x - 6x + 12 = 12$ **Step 3:** Solve the resulting equation.

$12 = 12$ (identity)

Because the equation produces an identity, all values of x make this equation true. Thus, x can be any real number. Substituting any real number, x, into the equation $y = 2x - 4$ produces an ordered pair on the line $y = 2x - 4$. Hence, the solution set to the system of equations is the set of all ordered pairs on the line $y = 2x - 4$. This can be written as $\{(x, y) \mid y = 2x - 4\}$. The equations are dependent.

Skill Practice Solve the system by using the substitution method.

5. $2x + \frac{1}{3}y = -\frac{1}{3}$

$12x + 2y = -2$

Answer

5. Infinitely many solutions;
$\{(x, y) \mid 12x + 2y = -2\}$; dependent equations

TIP: The solution to Example 5 can be verified by writing each equation in slope-intercept form and graphing the lines.

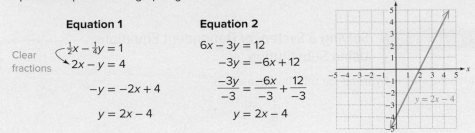

Equation 1	**Equation 2**
Clear fractions $\begin{cases} \frac{1}{2}x - \frac{1}{4}y = 1 \\ 2x - y = 4 \end{cases}$	$6x - 3y = 12$
	$-3y = -6x + 12$
$-y = -2x + 4$	$\dfrac{-3y}{-3} = \dfrac{-6x}{-3} + \dfrac{12}{-3}$
$y = 2x - 4$	$y = 2x - 4$

Notice that the slope-intercept forms for both equations are identical. The equations represent the same line, indicating that they are dependent. Each point on the line is a solution to the system of equations.

The following summary reviews the three different geometric relationships between two lines and the solutions to the corresponding systems of equations.

Interpreting Solutions to a System of Two Linear Equations

- The lines may intersect at one point (yielding one unique solution).
- The lines may be parallel and have no point of intersection (yielding no solution). This is detected algebraically when a contradiction (false statement) is obtained (for example, $0 = -3$ or $12 = 6$).
- The lines may be the same and intersect at all points on the line (yielding an infinite number of solutions). This is detected algebraically when an identity is obtained (for example, $0 = 0$ or $12 = 12$).

---

2. Applications of the Substitution Method

We have already encountered word problems using one linear equation and one variable. In this chapter, we investigate application problems with two unknowns. In such a case, we can use two variables to represent the unknown quantities. However, if two variables are used, we must write a system of *two* distinct equations.

Classroom Example: p. 293, Exercise 52

Example 6   Applying the Substitution Method

One number is 3 more than 4 times another. Their sum is 133. Find the numbers.

Solution:

We can use two variables to represent the two unknown numbers.

Let x represent one number.
Let y represent the other number.   Label the variables.

We must now write two equations. Each of the first two sentences gives a relationship between x and y:

One number is 3 more than 4 times another. ⟶ $x = 4y + 3$ (first equation)

Their sum is 133. ⟶ $x + y = 133$ (second equation)

$$(4y + 3) + y = 133$$ Substitute $x = 4y + 3$ into the second equation, $x + y = 133$.

$$5y + 3 = 133$$ Solve the resulting equation.

$$5y = 130$$

$$y = 26$$

$$x = 4y + 3$$

$$x = 4(26) + 3$$ To solve for x, substitute $y = 26$ into the equation $x = 4y + 3$.

$$x = 104 + 3$$

$$x = 107$$

One number is 26, and the other is 107.

Avoiding Mistakes

Notice that the answer for an application is not represented by an ordered pair.

TIP: Check that the numbers 26 and 107 meet the conditions of Example 6.

- 4 times 26 is 104. Three more than 104 is 107. ✓

- The sum of the numbers should be 133: $26 + 107 = 133$ ✓

Skill Practice

6. One number is 16 more than another. Their sum is 92. Use a system of equations to find the numbers.

Example 7   **Using the Substitution Method in a Geometry Application**

Classroom Example: p. 293, Exercise 56

Two angles are supplementary. The measure of one angle is 15° more than twice the measure of the other angle. Find the measures of the two angles.

Solution:

Let x represent the measure of one angle.
Let y represent the measure of the other angle.

The sum of the measures of supplementary angles is 180°. ⟶ $x + y = 180$

The measure of one angle is 15° more than twice the other angle. ⟶ $x = 2y + 15$

$$x + y = 180$$

$$x = 2y + 15$$ The x variable in the second equation is already isolated.

$$(2y + 15) + y = 180$$ Substitute $2y + 15$ into the first equation for x.

$$2y + 15 + y = 180$$ Solve the resulting equation.

$$3y + 15 = 180$$

$$3y = 165$$

$$y = 55$$

$$x = 2y + 15$$ Substitute $y = 55$ into the equation $x = 2y + 15$.

$$x = 2(55) + 15$$

$$x = 110 + 15$$

$$x = 125$$

One angle is 55°, and the other is 125°.

TIP: Check that the angles 55° and 125° meet the conditions of Example 7.

- Because $55° + 125° = 180°$, the angles are supplementary. ✓

- The angle 125° is 15° more than twice 55°: $125° = 2(55°) + 15°$ ✓

Answer

6. One number is 38, and the other number is 54.

Skill Practice

7. The measure of one angle is 2° less than 3 times the measure of another angle. The angles are complementary. Use a system of equations to find the measures of the two angles.

Answer

7. The measures of the angles are 23° and 67°.

Section 4.2 Practice Exercises

For additional exercises, see Classroom Activities 4.2A–4.2C in the *Student's Resource Manual* at www.mhhe.com/moh.

Review Exercises

For Exercises 1–6, write each pair of lines in slope-intercept form. Then identify whether the lines intersect in exactly one point or if the lines are parallel or coinciding.

1. $2x - y = 4$ $y = 2x - 4$
   $-2y = -4x + 8$
   $y = 2x - 4$; coinciding lines

2. $x - 2y = 5$ $y = \frac{1}{2}x - \frac{5}{2}$
   $3x = 6y + 15$
   $y = \frac{1}{2}x - \frac{5}{2}$; coinciding lines

3. $2x + 3y = 6$ $y = -\frac{2}{3}x + 2$
   $x - y = 5$
   $y = x - 5$; intersecting lines

4. $x - y = -1$ $y = x + 1$
   $x + 2y = 4$ $y = -\frac{1}{2}x + 2$;
   intersecting lines

5. $2x = \frac{1}{2}y + 2$ $y = 4x - 4$
   $4x - y = 13$ $y = 4x - 13$; parallel lines

6. $4y = 3x$ $y = \frac{3}{4}x$
   $3x - 4y = 15$ $y = \frac{3}{4}x - \frac{15}{4}$;
   parallel lines

Concept 1: Solving Systems of Linear Equations by Using the Substitution Method

For Exercises 7–10, solve each system by using the substitution method. **(See Example 1.)**

7. $3x + 2y = -3$
   $y = 2x - 12$
   $\{(3, -6)\}$

8. $4x - 3y = -19$
   $y = -2x + 13$
   $\{(2, 9)\}$

9. $x = -4y + 16$
   $3x + 5y = 20$
   $\{(0, 4)\}$

10. $x = -y + 3$
    $-2x + y = 6$
    $\{(-1, 4)\}$

11. Given the system: $4x - 2y = -6$
    $\qquad\qquad\qquad 3x + y = 8$

    a. Which variable from which equation is easiest to isolate and why? y in the second equation is easiest to isolate because its coefficient is 1.

    b. Solve the system by using the substitution method.
    $\{(1, 5)\}$

12. Given the system: $x - 5y = 2$
    $\qquad\qquad\qquad 11x + 13y = 22$

    a. Which variable from which equation is easiest to isolate and why? x in the first equation is easiest to isolate because its coefficient is 1.

    b. Solve the system by using the substitution method.
    $\{(2, 0)\}$

For Exercises 13–48, solve the system by using the substitution method. For systems that do not have one unique solution, also state the number of solutions and whether the system is inconsistent or the equations are dependent. **(See Examples 1–5.)**

13. $x = 3y - 1$ $\{(5, 2)\}$
    $2x - 4y = 2$

14. $2y = x + 9$ $\{(-1, 4)\}$
    $y = -3x + 1$

15. $-2x + 5y = 5$ $\{(10, 5)\}$
    $x = 4y - 10$

16. $y = -2x + 27$ $\{(11, 5)\}$
    $3x - 7y = -2$

17. $4x - y = -1$ $\left\{\left(\frac{1}{2}, 3\right)\right\}$
    $2x + 4y = 13$

18. $5x - 3y = -2$ $\left\{\left(\frac{1}{5}, 1\right)\right\}$
    $10x - y = 1$

19. $4x - 3y = 11$ $\{(5, 3)\}$
    $x = 5$

20. $y = -3x - 9$ $\{(-7, 12)\}$
    $y = 12$

21. $4x = 8y + 4$ $\{(1, 0)\}$
    $5x - 3y = 5$

22. $3y = 6x - 6$ $\{(2, 2)\}$
    $-3x + y = -4$

23. $x - 3y = -11$ $\{(1, 4)\}$
    $6x - y = 2$

24. $-2x - y = 9$ $\{(-6, 3)\}$
    $x + 7y = 15$

25. $3x + 2y = -1$
    $\frac{3}{2}x + y = 4$
    No solution; $\{\ \}$; inconsistent system

26. $5x - 2y = 6$
    $-\frac{5}{2}x + y = 5$
    No solution; $\{\ \}$; inconsistent system

27. $10x - 30y = -10$
    $2x - 6y = -2$
    Infinitely many solutions; $\{(x, y)\,|\,2x - 6y = -2\}$; dependent equations

28. $3x + 6y = 6$
    $-6x - 12y = -12$
    Infinitely many solutions; $\{(x, y)\,|\,3x + 6y = 6\}$; dependent equations

29. $2x + y = 3$
$y = -7$
$\{(5, -7)\}$

30. $-3x = 2y + 23$
$x = -1$
$\{(-1, -10)\}$

31. $x + 2y = -2$ $\left\{\left(-5, \frac{3}{2}\right)\right\}$
$4x = -2y - 17$

32. $x + y = 1$ $\left\{\left(-\frac{1}{3}, \frac{4}{3}\right)\right\}$
$2x - y = -2$

33. $y = -\frac{1}{2}x - 4$ $\{(2, -5)\}$
$y = 4x - 13$

34. $y = \frac{2}{3}x - 3$ $\{(3, -1)\}$
$y = 6x - 19$

35. $y = 6$ $\{(-4, 6)\}$
$y - 4 = -2x - 6$

36. $x = 9$ $\{(9, -1)\}$
$x - 3 = 6y + 12$

37. $3x + 2y = 4$
$2x - 3y = -6$
$\{(0, 2)\}$

38. $4x + 3y = 4$
$-2x + 5y = -2$
$\{(1, 0)\}$

39. $y = 0.25x + 1$
Infinitely many solutions;
$-x + 4y = 4$
$\{(x, y) \mid y = 0.25x + 1\}$;
dependent equations

40. $y = 0.75x - 3$
Infinitely many solutions;
$-3x + 4y = -12$
$\{(x, y) \mid -3x + 4y = -12\}$;
dependent equations

41. $11x + 6y = 17$
$5x - 4y = 1$
$\{(1, 1)\}$

42. $3x - 8y = 7$
$10x - 5y = 45$
$\{(5, 1)\}$

43. $x + 2y = 4$
No solution;
$4y = -2x - 8$ $\{\ \}$;
inconsistent system

44. $-y = x - 6$
No solution;
$2x + 2y = 4$ $\{\ \}$;
inconsistent system

45. $2x = 3 - y$
$x + y = 4$ $\{(-1, 5)\}$

46. $2x = 4 + 2y$
$3x + y = 10$ $\{(3, 1)\}$

47. $\frac{x}{3} + \frac{y}{2} = -4$
$x - 3y = 6$ $\{(-6, -4)\}$

48. $x - 2y = -5$
$\frac{2x}{3} + \frac{y}{3} = 0$ $\{(-1, 2)\}$

Concept 2: Applications of the Substitution Method

For Exercises 49–58, set up a system of linear equations and solve for the indicated quantities. **(See Examples 6–7.)**

49. Two numbers have a sum of 106. One number is 10 less than the other. Find the numbers.
The numbers are 48 and 58.

50. Two positive numbers have a difference of 8. The larger number is 2 less than 3 times the smaller number. Find the numbers. The numbers are 13 and 5.

51. The difference between two positive numbers is 26. The larger number is 3 times the smaller. Find the numbers. The numbers are 13 and 39.

52. The sum of two numbers is 956. One number is 94 less than 6 times the other. Find the numbers.
The numbers are 150 and 806.

53. Two angles are supplementary. One angle is 15° more than 10 times the other angle. Find the measure of each angle.
The angles are 165° and 15°.

54. Two angles are complementary. One angle is 1° less than 6 times the other angle. Find the measure of each angle. The angles are 13° and 77°.

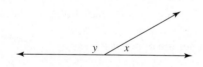

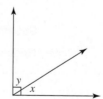

55. Two angles are complementary. One angle is 10° more than 3 times the other angle. Find the measure of each angle. The angles are 70° and 20°.

56. Two angles are supplementary. One angle is 5° less than twice the other angle. Find the measure of each angle. The angles are $118\frac{1}{3}°$ and $61\frac{2}{3}°$.

57. In a right triangle, one of the acute angles is 6° less than the other acute angle. Find the measure of each acute angle. The angles are 42° and 48°.

58. In a right triangle, one of the acute angles is 9° less than twice the other acute angle. Find the measure of each acute angle. The angles are 57° and 33°.

Writing · Translating Expression · Geometry · Scientific Calculator · Video

Expanding Your Skills

59. The following system consists of dependent equations and therefore has infinitely many solutions. Find three ordered pairs that are solutions to the system of equations.

$$y = 2x + 3$$
$$-4x + 2y = 6$$

For example: $(0, 3), (1, 5), (-1, 1)$

60. The following system consists of dependent equations and therefore has infinitely many solutions. Find three ordered pairs that are solutions to the system of equations.

$$y = -x + 1$$
$$2x + 2y = 2$$

For example: $(0, 1), (1, 0), (-1, 2)$

Section 4.3   Solving Systems of Equations by the Addition Method

Concepts

1. Solving Systems of Linear Equations by Using the Addition Method
2. Summary of Methods for Solving Systems of Linear Equations in Two Variables

Classroom Example: p. 301, Exercise 10

1. Solving Systems of Linear Equations by Using the Addition Method

In this section, we present another algebraic method to solve a system of linear equations. This method is called the *addition method* and its underlying principle is to add multiples of the given equations to eliminate a variable from the system. For this reason, the addition method is sometimes called the *elimination method*.

Example 1   Solving a System of Linear Equations by Using the Addition Method

Solve the system by using the addition method.

$$x + y = -2$$
$$x - y = -6$$

Solution:

Notice that the coefficients of the y variables are opposites:

Coefficient is 1.

$$x + 1y = -2$$
$$x - 1y = -6$$

Coefficient is -1.

Because the coefficients of the y variables are opposites, we can add the two equations to eliminate the y variable.

$$x + y = -2$$
$$\underline{x - y = -6}$$
$$2x = -8 \quad \longleftarrow \text{ After adding the equations, we have one equation and one variable.}$$

$$2x = -8 \qquad \text{Solve the resulting equation.}$$
$$x = -4$$

To find the value of y, substitute $x = -4$ into *either* of the original equations.

$$x + y = -2 \qquad \text{First equation}$$

$$(-4) + y = -2$$

$$y = -2 + 4$$

$$y = 2 \qquad \text{The ordered pair is } (-4, 2).$$

Check:

$$x + y = -2 \longrightarrow (-4) + (2) \overset{?}{=} -2 \longrightarrow -2 \overset{?}{=} -2 \checkmark \quad \text{True}$$

$$x - y = -6 \longrightarrow (-4) - (2) \overset{?}{=} -6 \longrightarrow -6 \overset{?}{=} -6 \checkmark \quad \text{True}$$

The solution set is $\{(-4, 2)\}$.

Skill Practice Solve the system by using the addition method.

1. $x + y = 13$
   $2x - y = 2$

TIP: In Example 1, notice that the value $x = -4$ could have been substituted into the second equation, to obtain the same value for y.

$$x - y = -6$$

$$(-4) - y = -6$$

$$-y = -6 + 4$$

$$-y = -2$$

$$y = 2$$

It is important to note that the addition method works on the premise that the two equations have *opposite* values for the coefficients of one of the variables. Sometimes it is necessary to manipulate the original equations to create two coefficients that are opposites. This is accomplished by multiplying one or both equations by an appropriate constant. The process is outlined as follows.

Solving a System of Equations by Using the Addition Method

Step 1 Write both equations in standard form: $Ax + By = C$.

Step 2 Clear fractions or decimals (optional).

Step 3 Multiply one or both equations by nonzero constants to create opposite coefficients for one of the variables.

Step 4 Add the equations from step 3 to eliminate one variable.

Step 5 Solve for the remaining variable.

Step 6 Substitute the known value from step 5 into one of the original equations to solve for the other variable.

Step 7 Check the ordered pair in both equations.

Instructor Note: Point out that in step 3 it does not matter whether we create opposites for the x variables or the y variables.

Answer

1. $\{(5, 8)\}$

Classroom Example: p. 301, Exercise 14

Example 2 **Solving a System of Linear Equations by Using the Addition Method**

Solve the system by using the addition method.

$$3x + 5y = 17$$
$$2x - y = -6$$

Solution:

$3x + 5y = 17$	**Step 1:** Both equations are already written in standard form.
$2x - y = -6$	**Step 2:** There are no fractions or decimals.

Notice that neither the coefficients of x nor the coefficients of y are opposites. However, multiplying the second equation by 5 creates the term $-5y$ in the second equation. This is the opposite of the term $+5y$ in the first equation.

Avoiding Mistakes

Remember to multiply the chosen constant on *both* sides of the equation.

$$3x + 5y = 17 \qquad\qquad\qquad 3x + 5y = 17$$
$$2x - y = -6 \xrightarrow{\text{Multiply by 5.}} \underline{10x - 5y = -30}$$
$$13x = -13$$

Step 3: Multiply the second equation by 5.

Step 4: Add the equations.

$$13x = -13$$

Step 5: Solve the equation.

$$x = -1$$

TIP: In Example 2, we could have eliminated the x variable by multiplying the first equation by 2 and the second equation by -3.

$$3x + 5y = 17 \qquad \text{First equation}$$
$$3(-1) + 5y = 17$$
$$-3 + 5y = 17$$
$$5y = 20$$
$$y = 4$$

Step 6: Substitute $x = -1$ into one of the original equations.

Step 7: Check $(-1, 4)$ in both original equations.

Check:

$$3x + 5y = 17 \longrightarrow 3(-1) + 5(4) \stackrel{?}{=} 17 \longrightarrow -3 + 20 \stackrel{?}{=} 17 \checkmark \quad \text{True}$$
$$2x - y = -6 \longrightarrow 2(-1) - (4) \stackrel{?}{=} -6 \longrightarrow -2 - 4 \stackrel{?}{=} -6 \checkmark \quad \text{True}$$

The solution set is $\{(-1, 4)\}$.

Skill Practice Solve the system by using the addition method.

2. $4x + 3y = 3$
$x - 2y = 9$

In Example 3, the system of equations uses the variables a and b instead of x and y. In such a case, we will write the solution as an ordered pair with the variables written in alphabetical order, such as (a, b).

Answer

2. $\{(3, -3)\}$

| Example 3 | **Solving a System of Linear Equations by Using the Addition Method** |

Classroom Example: p. 302, Exercise 18

Solve the system by using the addition method.

$$5b = 7a + 8$$
$$-4a - 2b = -10$$

Solution:

Step 1: Write the equations in standard form.

The first equation becomes: $\qquad\qquad 5b = 7a + 8 \longrightarrow -7a + 5b = 8$

The system becomes: $\qquad\qquad -7a + 5b = 8$
$$-4a - 2b = -10$$

Step 2: There are no fractions or decimals.

Step 3: We need to obtain opposite coefficients on either the a or b term.

Notice that neither the coefficients of a nor the coefficients of b are opposites. However, it is possible to change the coefficients of b to 10 and -10 (this is because the LCM of 5 and 2 is 10). This is accomplished by multiplying the first equation by 2 and the second equation by 5.

$-7a + 5b = 8 \xrightarrow{\text{Multiply by 2.}} -14a + 10b = 16$

$-4a - 2b = -10 \xrightarrow{\text{Multiply by 5.}} \dfrac{-20a - 10b = -50}{-34a = -34}$ **Step 4:** Add the equations.

$$-34a = -34$$ **Step 5:** Solve the resulting equation.
$$\frac{-34a}{-34} = \frac{-34}{-34}$$
$$a = 1$$

$5b = 7a + 8$ $\qquad$ First equation $\qquad$ **Step 6:** Substitute $a = 1$ into one of the original equations.

$$5b = 7(1) + 8$$
$$5b = 15$$
$$b = 3$$ $\qquad\qquad\qquad\qquad$ **Step 7:** Check $(1, 3)$ in the original equations.

Check:

$5b = 7a + 8 \longrightarrow 5(3) \overset{?}{=} 7(1) + 8 \longrightarrow 15 \overset{?}{=} 7 + 8 \checkmark \qquad$ True

$-4a - 2b = -10 \longrightarrow -4(1) - 2(3) \overset{?}{=} -10 \longrightarrow -4 - 6 \overset{?}{=} -10 \checkmark \qquad$ True

The solution set is $\{(1, 3)\}$.

Skill Practice Solve the system by using the addition method.

3. $8n = 4 - 5m$
   $7m + 6n = -10$

Classroom Example: p. 302, Exercise 22

| Example 4 | **Solving a System of Linear Equations by Using the Addition Method** |

Solve the system by using the addition method.

$$34x - 22y = 4$$

$$17x - 88y = -19$$

Solution:

The equations are already in standard form. There are no fractions or decimals to clear.

$$34x - 22y = 4 \longrightarrow 34x - 22y = 4$$

$$17x - 88y = -19 \xrightarrow{\text{Multiply by } -2.} \underline{-34x + 176y = 38}$$

$$154y = 42$$

Solve for y. $154y = 42$

$$\frac{154y}{154} = \frac{42}{154}$$

Simplify. $y = \dfrac{3}{11}$

To find the value of x, we normally substitute y into one of the original equations and solve for x. In this example, we will show an alternative method for finding x. By repeating the addition method, this time eliminating y, we can solve for x. This approach enables us to avoid substitution of the fractional value for y.

$$34x - 22y = 4 \xrightarrow{\text{Multiply by } -4.} -136x + 88y = -16$$

$$17x - 88y = -19 \longrightarrow \underline{17x - 88y = -19}$$

$$-119x = -35$$

Solve for x. $-119x = -35$

$$\frac{-119x}{-119} = \frac{-35}{-119}$$

Simplify. $x = \dfrac{5}{17}$

The ordered pair $\left(\frac{5}{17}, \frac{3}{11}\right)$ can be checked in the original equations.

$$34x - 22y = 4 \qquad\qquad\qquad 17x - 88y = -19$$

$$34\left(\frac{5}{17}\right) - 22\left(\frac{3}{11}\right) \overset{?}{=} 4 \qquad\qquad 17\left(\frac{5}{17}\right) - 88\left(\frac{3}{11}\right) \overset{?}{=} -19$$

$$10 - 6 \overset{?}{=} 4 \checkmark \text{ True} \qquad\qquad 5 - 24 \overset{?}{=} -19 \checkmark \text{ True}$$

The solution set is $\left\{\left(\frac{5}{17}, \frac{3}{11}\right)\right\}$.

Skill Practice Solve the system by using the addition method.

4. $15x - 16y = 1$

    $45x + 4y = 16$

Answer

4. $\left\{\left(\frac{1}{3}, \frac{1}{4}\right)\right\}$

| Example 5 | Solving an Inconsistent System by the Addition Method |

Classroom Example: p. 302, Exercise 32

Solve the system by using the addition method.

$$2x - 5y = 10$$

$$\frac{1}{2}x - \frac{5}{4}y = 1$$

Solution:

$2x - 5y = 10$

$\frac{1}{2}x - \frac{5}{4}y = 1$ **Step 1:** The equations are in standard form.

                    Step 2: Multiply both sides of the second equation by 4 to clear fractions.

$$\frac{1}{2}x - \frac{5}{4}y = 1 \longrightarrow 4\left(\frac{1}{2}x - \frac{5}{4}y\right) = 4(1) \longrightarrow 2x - 5y = 4$$

Now the system becomes $2x - 5y = 10$

                               $2x - 5y = 4$

To make either the x coefficients or y coefficients opposites, multiply either equation by -1.

$2x - 5y = 10 \xrightarrow{\text{Multiply by } -1.} -2x + 5y = -10$ **Step 3:** Create opposite coefficients.

$2x - 5y = 4 \longrightarrow \underline{2x - 5y = 4}$

                                    $0 = -6$ **Step 4:** Add the equations.

Because the result is a contradiction, the solution set is { }, and the system of equations is inconsistent. Writing each equation in slope-intercept form verifies that the lines are parallel (Figure 4-8).

$$2x - 5y = 10 \xrightarrow{\text{slope-intercept form}} y = \frac{2}{5}x - 2$$

$$\frac{1}{2}x - \frac{5}{4}y = 1 \xrightarrow{\text{slope-intercept form}} y = \frac{2}{5}x - \frac{4}{5}$$

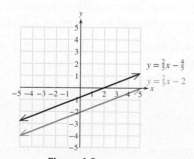

Figure 4-8

Skill Practice Solve the system by using the addition method.

5. $\frac{2}{3}x - \frac{3}{4}y = 2$

     $8x - 9y = 6$

Answer

5. { }

Classroom Example: p. 302, Exercise 34

Example 6 Solving a System of Dependent Equations by the Addition Method

Solve the system by using the addition method.

$$3x - y = 4$$
$$2y = 6x - 8$$

Solution:

$3x - y = 4 \longrightarrow 3x - y = 4$ **Step 1:** Write the equations in standard form.

$2y = 6x - 8 \longrightarrow -6x + 2y = -8$ **Step 2:** There are no fractions or decimals.

Notice that the equations differ exactly by a factor of -2, which indicates that these two equations represent the same line. Multiply the first equation by 2 to create opposite coefficients for the variables.

$$\begin{array}{rl} 3x - y = 4 & \xrightarrow{\text{Multiply by 2.}} \quad 6x - 2y = 8 \\ -6x + 2y = -8 & \phantom{\xrightarrow{\text{Multiply by 2.}}} \quad \underline{-6x + 2y = -8} \\ & \phantom{\xrightarrow{\text{Multiply by 2.}} \quad 6x - 2y = } 0 = 0 \end{array}$$

Step 3: Create opposite coefficients.

Step 4: Add the equations.

Because the resulting equation is an identity, the original equations represent the same line. This can be confirmed by writing each equation in slope-intercept form.

$$3x - y = 4 \quad \longrightarrow \quad -y = -3x + 4 \quad \longrightarrow \quad y = 3x - 4$$
$$-6x + 2y = -8 \quad \longrightarrow \quad 2y = 6x - 8 \quad \longrightarrow \quad y = 3x - 4$$

The solution is the set of all points on the line, or equivalently, $\{(x, y) \mid y = 3x - 4\}$.

Skill Practice Solve the system by using the addition method.

6. $3x = 3y + 15$
   $2x - 2y = 10$

2. Summary of Methods for Solving Systems of Linear Equations in Two Variables

If no method of solving a system of linear equations is specified, you may use the method of your choice. However, we recommend the following guidelines:

1. If one of the equations is written with a variable isolated, the substitution method is a good choice. For example:

$$2x + 5y = 2 \qquad \text{or} \qquad y = \frac{1}{3}x - 2$$
$$x = y - 6 \qquad \qquad\qquad x - 6y = 9$$

2. If both equations are written in standard form, $Ax + By = C$, where none of the variables has coefficients of 1 or -1, then the addition method is a good choice.

$$4x + 5y = 12$$
$$5x + 3y = 15$$

3. If both equations are written in standard form, $Ax + By = C$, and at least one variable has a coefficient of 1 or -1, then either the substitution method or the addition method is a good choice.

Answer

6. $\{(x, y) \mid 2x - 2y = 10\}$

Study Skills Exercise

For additional exercises, see Classroom Activities 4.3A–4.3B in the *Student's Resource Manual* at www.mhhe.com/moh.

Now that you have learned three methods of solving a system of linear equations with two variables, choose a system and solve it all three ways. There are two advantages to this. One is to check your answer (you should get the same answer using all three methods). The second advantage is to show you which method is the easiest for you to use.

Solve the system by using the graphing method, the substitution method, and the addition method.

$$2x + y = -7$$
$$x - 10 = 4y$$

Review Exercises

For Exercises 1–5, check whether the given ordered pair is a solution to the system.

1. $-\dfrac{3}{4}x + 2y = -10$ $(8, -2)$ No

$x - \dfrac{1}{2}y = 7$

2. $x + y = 8$ $(5, 3)$ Yes

$y = x - 2$

3. $x = y + 1$ $(3, 2)$ No

$-x + 2y = 0$

4. $3x + 2y = 14$ $(5, -2)$ No

$5x - 2y = 29$

5. $x = 2y - 11$ $(-3, 4)$ Yes

$-x + 5y = 23$

Concept 1: Solving Systems of Linear Equations by Using the Addition Method

For Exercises 6–7, answer as true or false.

6. Given the system $5x - 4y = 1$
$$7x - 2y = 5$$

a. To eliminate the y variable using the addition method, multiply the second equation by 2.
False, multiply by −2.

b. To eliminate the x variable using the addition method, multiply the first equation by 7 and the second equation by −5.   True

7. Given the system $3x + 5y = -1$
$$9x - 8y = -26$$

a. To eliminate the x variable using the addition method, multiply the first equation by −3.
True

b. To eliminate the y variable using the addition method, multiply the first equation by 8 and the second equation by −5.   False, multiply the second equation by 5.

8. Given the system $3x - 4y = 2$
$$17x + y = 35$$

a. Which variable, x or y, is easier to eliminate using the addition method? y would be easier.

b. Solve the system using the addition method.
$\{(2, 1)\}$

9. Given the system $-2x + 5y = -15$
$$6x - 7y = 21$$

a. Which variable, x or y, is easier to eliminate using the addition method? x would be easier.

b. Solve the system using the addition method.
$\{(0, -3)\}$

For Exercises 10–24, solve each system using the addition method. **(See Examples 1–4.)**

10. $x + 2y = 8$ $\{(2, 3)\}$
$5x - 2y = 4$

11. $2x - 3y = 11$ $\{(4, -1)\}$
$-4x + 3y = -19$

12. $a + b = 3$ $\{(5, -2)\}$
$3a + b = 13$

13. $-2u + 6v = 10$ $\{(4, 3)\}$
$-2u + v = -5$

14. $-3x + y = 1$ $\{(0, 1)\}$
$-6x - 2y = -2$

15. $5m - 2n = 4$ $\{(2, 3)\}$
$3m + n = 9$

 Writing   Translating Expression   Geometry   Scientific Calculator   Video

16. $3x - 5y = 13$ $\{(1, -2)\}$

$x - 2y = 5$

17. $7a + 2b = -1$ $\{(1, -4)\}$

$3a - 4b = 19$

18. $6c - 2d = -2$ $\{(1, 4)\}$

$5c = -3d + 17$

19. $2s + 3t = -1$ $\{(1, -1)\}$

$5s = 2t + 7$

20. $6y - 4z = -2$ $\{(3, 5)\}$

$4y + 6z = 42$

21. $4k - 2r = -4$ $\{(-4, -6)\}$

$3k - 5r = 18$

22. $2x + 3y = 6$ $\left\{\left(\frac{21}{5}, -\frac{4}{5}\right)\right\}$

$x - y = 5$

23. $6x + 6y = 8$ $\left\{\left(\frac{7}{9}, \frac{5}{9}\right)\right\}$

$9x - 18y = -3$

24. $2x - 5y = 4$ $\left\{\left(\frac{8}{9}, -\frac{4}{9}\right)\right\}$

$3x - 3y = 4$

25. In solving a system of equations, suppose you get the statement $0 = 5$. How many solutions will the system have? What can you say about the graphs of these equations?
The system will have no solution. The lines are parallel.

26. In solving a system of equations, suppose you get the statement $0 = 0$. How many solutions will the system have? What can you say about the graphs of these equations?
There are infinitely many solutions. The lines coincide.

27. In solving a system of equations, suppose you get the statement $3 = 3$. How many solutions will the system have? What can you say about the graphs of these equations?
There are infinitely many solutions. The lines coincide.

28. In solving a system of equations, suppose you get the statement $2 = -5$. How many solutions will the system have? What can you say about the graphs of these equations?
The system will have no solution. The lines are parallel.

29. Suppose in solving a system of linear equations, you get the statement $x = 0$. How many solutions will the system have? What can you say about the graphs of these equations?
The system will have one solution. The lines intersect at a point whose x-coordinate is 0.

30. Suppose in solving a system of linear equations, you get the statement $y = 0$. How many solutions will the system have? What can you say about the graphs of these equations?
The system will have one solution. The lines intersect at a point whose y-coordinate is 0.

For Exercises 31–42, solve the system by using the addition method. For systems that do not have one unique solution, also state the number of solutions and whether the system is inconsistent or the equations are dependent. **(See Examples 5–6.)**

31. $-2x + y = -5$

$8x - 4y = 12$

No solution; { }; inconsistent system

32. $x - 3y = 2$

$-5x + 15y = 10$

No solution; { }; inconsistent system

33. $x + 2y = 2$

$-3x - 6y = -6$

Infinitely many solutions; $\{(x, y) \mid x + 2y = 2\}$; dependent equations

34. $4x - 3y = 6$

$-12x + 9y = -18$

Infinitely many solutions; $\{(x, y) \mid 4x - 3y = 6\}$; dependent equations

35. $3a + 2b = 11$ $\{(1, 4)\}$

$7a - 3b = -5$

36. $4y + 5z = -2$

$5y - 3z = 16$

$\{(2, -2)\}$

37. $3x - 5y = 7$

$5x - 2y = -1$

$\{(-1, -2)\}$

38. $4s + 3t = 9$ $\{(0, 3)\}$

$3s + 4t = 12$

39. $2x + 2 = -3y + 9$

$3x - 10 = -4y$

$\{(2, 1)\}$

40. $-3x + 6 + 7y = 5$ $\{(5, 2)\}$

$5y = 2x$

41. $4x - 5y = 0$

$8(x - 1) = 10y$

No solution; { }; inconsistent system

42. $y = 2x + 1$

$-3(2x - y) = 0$

No solution; { }; inconsistent system

Concept 2: Summary of Methods for Solving Systems of Linear Equations in Two Variables

For Exercises 43–63, solve each system of equations by either the addition method or the substitution method.

43. $5x - 2y = 4$ $\{(2, 3)\}$

$y = -3x + 9$

44. $-x = 8y + 5$ $\{(-5, 0)\}$

$4x - 3y = -20$

45. $0.1x + 0.1y = 0.6$ $\{(3.5, 2.5)\}$

$0.1x - 0.1y = 0.1$

46. $0.1x + 0.1y = 0.2$ $\{(2.5, -0.5)\}$

$0.1x - 0.1y = 0.3$

47. $3x = 5y - 9$ $\left\{\left(\frac{1}{3}, 2\right)\right\}$

$2y = 3x + 3$

48. $10x - 5 = 3y$ $\left\{\left(\frac{1}{2}, 0\right)\right\}$

$4x + 5y = 2$

Writing   Translating Expression   Geometry   Scientific Calculator   Video

49. $\frac{1}{10}y = -\frac{1}{2}x - \frac{1}{2}$ $\{(-2, 5)\}$

$\frac{3}{2}x - \frac{3}{4} = -\frac{3}{4}y$

50. $x + \frac{5}{4}y = -\frac{1}{2}$ $\{(-3, 2)\}$

$\frac{3}{4}x = -\frac{1}{2}y - \frac{5}{4}$

51. $x = -\frac{1}{2}$ $\left\{\left(-\frac{1}{2}, 1\right)\right\}$

$6x - 5y = -8$

52. $4x - 2y = 1$ $\left\{\left(\frac{7}{4}, 3\right)\right\}$

$y = 3$

53. $0.02x + 0.04y = 0.12$ $\{(0, 3)\}$

$0.03x - 0.05y = -0.15$

54. $-0.04x + 0.03y = 0.03$ $\{(0, 1)\}$

$-0.06x - 0.02y = -0.02$

55. $8x - 16y = 24$

$2x - 4y = 0$

$\{\ \}$

56. $y = -\frac{1}{2}x - 5$

$2x + 4y = -8$

$\{\ \}$

57. $\frac{m}{2} + \frac{n}{5} = \frac{13}{10}$

$3m - 3n = m - 10$

$\{(1, 4)\}$

58. $\frac{a}{4} - \frac{3b}{2} = \frac{15}{2}$ $\{(0, -5)\}$

$a + 2b = -10$

59. $2m - 6n = m + 4$ $\{(4, 0)\}$

$3m + 8 = 5m - n$

60. $m - 3n = 10$ $\{(4, -2)\}$

$3m + 12n = -12$

61. $9a - 2b = 8$

$18a + 6 = 4b + 22$

Infinitely many solutions;

$\{(a, b) \mid 9a - 2b = 8\}$

62. $a = 5 + 2b$ Infinitely many solutions;

$3a - 6b = 15$ $\{(a, b) \mid a = 5 + 2b\}$

63. $6x - 5y = 7$ $\left\{\left(\frac{7}{16}, -\frac{7}{8}\right)\right\}$

$4x - 6y = 7$

For Exercises 64–69, use a system of linear equations, and solve for the indicated quantities.

64. The sum of two positive numbers is 26. Their difference is 14. Find the numbers.
The numbers are 20 and 6.

65. The difference of two positive numbers is 2. The sum of the numbers is 36. Find the numbers.
The numbers are 17 and 19.

66. Eight times the smaller of two numbers plus 2 times the larger number is 44. Three times the smaller number minus 2 times the larger number is zero. Find the numbers.
The numbers are 4 and 6.

67. Six times the smaller of two numbers minus the larger number is −9. Ten times the smaller number plus five times the larger number is 5. Find the numbers.
The numbers are −1 and 3.

68. Twice the difference of two angles is 64°. If the angles are complementary, find the measures of the angles. The angles are 61° and 29°.

69. The difference of an angle and twice another angle is 42°. If the angles are supplementary, find the measures of the angles.
The angles are 46° and 134°.

For Exercises 70–72, solve the system by using each of the three methods: (a) the graphing method, (b) the substitution method, and (c) the addition method.

70. $2x + y = 1$

$-4x - 2y = -2$

Infinitely many solutions;

$\{(x, y) \mid 2x + y = 1\}$

71. $3x + y = 6$

$-2x + 2y = 4$

$\{(1, 3)\}$

72. $2x - 2y = 6$

$5y = 5x + 5$

$\{\ \}$

Expanding Your Skills

73. Explain why a system of linear equations cannot have exactly two solutions. One line within the system of equations would have to "bend" for the system to have exactly two points of intersection. This is not possible.

74. The solution to the system of linear equations is $\{(1, 2)\}$. Find A and B.

$Ax + 3y = 8$ $A = 2, B = -4$

$x + By = -7$

75. The solution to the system of linear equations is $\{(-3, 4)\}$. Find A and B.

$4x + Ay = -32$ $A = -5, B = 2$

$Bx + 6y = 18$

Problem Recognition Exercises

Systems of Equations

For Exercises 1–6 determine the number of solutions to the system without solving the system. Explain your answers.

1. $y = -4x + 2$

$y = -4x + 2$
Infinitely many solutions. The equations represent the same line.

2. $y = -4x + 6$

$y = -4x + 1$
No solution. The equations represent parallel lines.

3. $y = 4x - 3$

$y = -4x + 5$
One solution. The equations represent intersecting lines.

4. $y = 7$

$2x + 3y = 1$
One solution. The equations represent intersecting lines.

5. $2x + 3y = 1$

$2x + 3y = 8$
No solution. The equations represent parallel lines.

6. $8x - 2y = 6$

$12x - 3y = 9$
Infinitely many solutions. The equations represent the same line.

For Exercises 7–10, a method of solving has been suggested for each system of equations. Explain why that method was suggested for the system and then solve the system using the method given.

7. $2x - 5y = -11$ **Addition Method**

$7x + 5y = -16$ The y variable will easily be eliminated by adding the equations. The solution set is $\{(-3, 1)\}$.

8. $4x + 11y = 56$ **Addition Method**

$-2x - 5y = -26$ The x variable will easily be eliminated by multiplying the second equation by 2 and then adding the equations. The solution set is $\{(3, 4)\}$.

9. $x = -3y + 4$ **Substitution Method**

$5x + 4y = -2$ Because the variable x is already isolated in the first equation, the expression $-3y + 4$ can be easily substituted for x in the second equation. The solution set is $\{(-2, 2)\}$.

10. $2x + 3y = 16$ **Substitution Method**

$y = x - 8$ Because the variable y is already isolated in the second equation, the expression $x - 8$ can be easily substituted for y in the first equation. The solution set is $\{(8, 0)\}$.

For Exercises 11–30, solve each system using the method of your choice. For systems that do not have one unique solution, also state the number of solutions and whether the system is inconsistent or the equations are dependent.

11. $x = -2y + 5$ $\{(5, 0)\}$

$2x - 4y = 10$

12. $y = -3x - 4$ $\{(1, -7)\}$

$2x - y = 9$

13. $3x - 2y = 22$ $\{(4, -5)\}$

$5x + 2y = 10$

14. $-4x + 2y = -2$ $\{(2, 3)\}$

$4x - 5y = -7$

15. $\frac{1}{3}x + \frac{1}{2}y = \frac{2}{3}$ $\{(2, 0)\}$

$-\frac{2}{3}x + y = -\frac{4}{3}$

16. $\frac{1}{4}x + \frac{2}{5}y = 6$ $\{(8, 10)\}$

$\frac{1}{2}x - \frac{1}{10}y = 3$

17. $2c + 7d = -1$ $\left\{\left(2, -\frac{5}{7}\right)\right\}$

$c = 2$

18. $-3w + 5z = -6$ $\left\{\left(-\frac{14}{3}, -4\right)\right\}$

$z = -4$

19. $y = 0.4x - 0.3$

$-4x + 10y = 20$
No solution; { }; inconsistent system

20. $x = -0.5y + 0.1$

$-10x - 5y = 2$
No solution; { }; inconsistent system

21. $3a + 7b = -3$ $\{(-1, 0)\}$

$-11a + 3b = 11$

22. $2v - 5w = 10$ $\{(5, 0)\}$

$9v + 7w = 45$

23. $y = 2x - 14$

$4x - 2y = 28$ Infinitely many solutions; $\{(x, y) \mid y = 2x - 14\}$; dependent equations

24. $x = 5y - 9$

$-2x + 10y = 18$ Infinitely many solutions; $\{(x, y) \mid x = 5y - 9\}$; dependent equations

25. $x + y = 3200$ $\{(2200, 1000)\}$

$0.06x + 0.04y = 172$

26. $x + y = 4500$

$0.07x + 0.05y = 291$
$\{(3300, 1200)\}$

27. $3x + y - 7 = x - 4$ $\{(5, -7)\}$

$3x - 4y + 4 = -6y + 5$

28. $7y - 8y - 3 = -3x + 4$

$10x - 5y - 12 = 13$ $\{(2, -1)\}$

29. $3x - 6y = -1$ $\left\{\left(\frac{2}{3}, \frac{1}{2}\right)\right\}$

$9x + 4y = 8$

30. $8x - 2y = 5$ $\left\{\left(\frac{1}{4}, -\frac{3}{2}\right)\right\}$

$12x + 4y = -3$

Writing   Translating Expression   Geometry   Scientific Calculator   Video

Applications of Linear Equations in Two Variables

1. Applications Involving Cost

We have solved several applied problems by setting up a linear equation in one variable. When solving an application that involves two unknowns, sometimes it is convenient to use a system of linear equations in two variables.

Concepts

1. **Applications Involving Cost**
2. **Applications Involving Principal and Interest**
3. **Applications Involving Mixtures**
4. **Applications Involving Distance, Rate, and Time**

Classroom Example: p. 311, Exercise 10

| Example 1 | **Using a System of Linear Equations Involving Cost** |

At a movie theater a couple buys one large popcorn and two small drinks for $12.50. A group of teenagers buys two large popcorns and five small drinks for $28.50. Find the cost of one large popcorn and the cost of one small drink.

© Burke/Triolo/Brand X Pictures RF

Solution:

In this application we have two unknowns, which we can represent by x and y.

Let x represent the cost of one large popcorn.
Let y represent the cost of one small drink.

We must now write two equations. Each of the first two sentences in the problem gives a relationship between x and y:

$$\begin{pmatrix} \text{Cost of 1} \\ \text{popcorn} \end{pmatrix} + \begin{pmatrix} \text{cost of 2} \\ \text{drinks} \end{pmatrix} = \begin{pmatrix} \text{total} \\ \text{cost} \end{pmatrix} \longrightarrow x + 2y = 12.50$$

$$\begin{pmatrix} \text{Cost of 2} \\ \text{popcorns} \end{pmatrix} + \begin{pmatrix} \text{cost of 5} \\ \text{drinks} \end{pmatrix} = \begin{pmatrix} \text{total} \\ \text{cost} \end{pmatrix} \longrightarrow 2x + 5y = 28.50$$

To solve this system, we may either use the substitution method or the addition method. We will use the substitution method by solving for x in the first equation.

$x + 2y = 12.50 \longrightarrow x = -2y + 12.50$ Isolate x in the first equation.

$2x + 5y = 28.50$

$2(-2y + 12.50) + 5y = 28.50$ Substitute $x = -2y + 12.50$ into the other equation.

$-4y + 25.00 + 5y = 28.50$ Solve for y.

$y + 25.00 = 28.50$

$y = 3.50$

$x = -2y + 12.50$

$x = -2(3.50) + 12.50$ Substitute $y = 3.50$ into the equation $x = -2y + 12.50$.

$x = -7.00 + 12.50$

$x = 5.50$

The cost of one large popcorn is $5.50 and the cost of one small drink is $3.50.

Check by verifying that the solution meets the specified conditions.

1 popcorn + 2 drinks = 1($5.50) + 2($3.50) = $12.50 ✓   True

2 popcorns + 5 drinks = 2($5.50) + 5($3.50) = $28.50 ✓   True

Classroom Example: p. 311, Exercise 18

Skill Practice

1. Lynn went to a fast-food restaurant and spent $20.00. She purchased 4 hamburgers and 5 orders of fries. The next day, Ricardo went to the same restaurant and purchased 10 hamburgers and 7 orders of fries. He spent $41.20. Use a system of equations to determine the cost of a burger and the cost of an order of fries.

2. Applications Involving Principal and Interest

Simple interest is interest computed on the principal amount of money invested (or borrowed). Simple interest, I, is found by using the formula

$$I = Prt$$ where P is the principal,
r is the annual interest rate, and
t is the time in years.

In Example 2, we apply the concept of simple interest to two accounts to produce a desired amount of interest after 1 year.

Example 2 **Using a System of Linear Equations Involving Investments**

Joanne has a total of $6000 to deposit in two accounts. One account earns 3.5% simple interest and the other earns 2.5% simple interest. If the total amount of interest at the end of 1 year is $195, find the amount she deposited in each account.

Solution:

Let x represent the principal deposited in the 2.5% account.
Let y represent the principal deposited in the 3.5% account.

	2.5% Account	3.5% Account	Total
Principal	x	y	6000
Interest ($I = Prt$)	$0.025x(1)$	$0.035y(1)$	195

Each row of the table yields an equation in x and y:

$$\begin{pmatrix} \text{Principal} \\ \text{invested} \\ \text{at } 2.5\% \end{pmatrix} + \begin{pmatrix} \text{principal} \\ \text{invested} \\ \text{at } 3.5\% \end{pmatrix} = \begin{pmatrix} \text{total} \\ \text{principal} \end{pmatrix} \longrightarrow x + y = 6000$$

$$\begin{pmatrix} \text{Interest} \\ \text{earned} \\ \text{at } 2.5\% \end{pmatrix} + \begin{pmatrix} \text{interest} \\ \text{earned} \\ \text{at } 3.5\% \end{pmatrix} = \begin{pmatrix} \text{total} \\ \text{interest} \end{pmatrix} \longrightarrow 0.025x + 0.035y = 195$$

We will choose the addition method to solve the system of equations. First multiply the second equation by 1000 to clear decimals.

Answer

1. The cost of a burger is $3.00, and the cost of an order of fries is $1.60.

Multiply by −25.

$$x + y = 6000 \longrightarrow \qquad x + y = 6000 \qquad \longrightarrow \quad -25x - 25y = -150,000$$

$$0.025x + 0.035y = 195 \longrightarrow \quad 25x + 35y = 195,000 \longrightarrow \quad \underline{25x + 35y = \quad 195,000}$$

Multiply by 1000.

$$10y = \quad 45,000$$

$10y = 45,000$ After eliminating the x variable, solve for y.

$$\frac{10y}{10} = \frac{45,000}{10}$$

$y = 4500$ The amount invested in the 3.5% account is $4500.

$x + y = 6000$ Substitute $y = 4500$ into the equation $x + y = 6000$.

$x + 4500 = 6000$

$x = 1500$ The amount invested in the 2.5% account is $1500.

Joanne deposited $1500 in the 2.5% account and $4500 in the 3.5% account.

To check, verify that the conditions of the problem have been met.

1. The sum of $1500 and $4500 is $6000 as desired. ✓   True

2. The interest earned on $1500 at 2.5% is: $0.025(\$1500) = \$37.50$
   The interest earned on $4500 at 3.5% is: $\underline{0.035(\$4500) = \$157.50}$
                                       Total interest: $195.00 ✓   True

Skill Practice

2. Addie has a total of $8000 in two accounts. One pays 5% interest, and the other pays 6.5% interest. At the end of one year, she earned $475 interest. Use a system of equations to determine the amount invested in each account.

3. Applications Involving Mixtures

Example 3   **Using a System of Linear Equations in a Mixture Application**

Classroom Example: p. 312, Exercise 22

According to new hospital standards, a certain disinfectant solution needs to be 20% alcohol instead of 10% alcohol. There is a 40% alcohol disinfectant available to adjust the mixture. Determine the amount of 10% solution and the amount of 40% solution to produce 30 L of a 20% solution.

Solution:

Each solution contains a percentage of alcohol plus some other mixing agent such as water. Before we set up a system of equations to model this situation, it is helpful to have background understanding of the problem. In Figure 4-9, the liquid depicted in blue is pure alcohol and the liquid shown in gray is the mixing agent (such as water). Together these liquids form a solution. (Realistically the mixture may not separate as shown, but this image may be helpful for your understanding.)

Let x represent the number of liters of 10% solution.
Let y represent the number of liters of 40% solution.

Answer

2. $3000 is invested at 5%, and $5000 is invested at 6.5%.

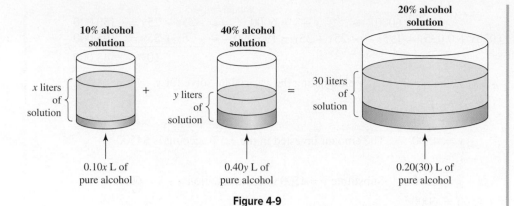

Figure 4-9

The information given in the statement of the problem can be organized in a chart.

	10% Alcohol	40% Alcohol	20% Alcohol
Number of liters of solution	x	y	30
Number of liters of pure alcohol	$0.10x$	$0.40y$	$0.20(30) = 6$

From the first row, we have

$$\left(\begin{array}{c}\text{Amount of}\\\text{10\% solution}\end{array}\right) + \left(\begin{array}{c}\text{amount of}\\\text{40\% solution}\end{array}\right) = \left(\begin{array}{c}\text{total amount}\\\text{of 20\% solution}\end{array}\right) \rightarrow x + y = 30$$

From the second row, we have

$$\left(\begin{array}{c}\text{Amount of}\\\text{alcohol in}\\\text{10\% solution}\end{array}\right) + \left(\begin{array}{c}\text{amount of}\\\text{alcohol in}\\\text{40\% solution}\end{array}\right) = \left(\begin{array}{c}\text{total amount of}\\\text{alcohol in}\\\text{20\% solution}\end{array}\right) \rightarrow 0.10x + 0.40y = 6$$

We will solve the system with the addition method by first clearing decimals.

$$
\begin{array}{lcll}
x + y = 30 & \xrightarrow{\hspace{1cm}} & x + y = 30 & \xrightarrow{\text{Multiply by } -1.} \quad -x - y = -30 \\
0.10x + 0.40y = 6 & \xrightarrow[\text{Multiply by 10.}]{} & x + 4y = 60 & \xrightarrow{\hspace{1cm}} \quad \underline{x + 4y = 60} \\
& & & 3y = 30
\end{array}
$$

$3y = 30$ After eliminating the x variable, solve for y.

$y = 10$ 10 L of 40% solution is needed.

$x + y = 30$ Substitute $y = 10$ into either of the original equations.

$x + (10) = 30$

$x = 20$ 20 L of 10% solution is needed.

10 L of 40% solution must be mixed with 20 L of 10% solution.

Skill Practice

3. How many ounces of 20% and 35% acid solution should be mixed together to obtain 15 oz of 30% acid solution?

Answer

3. 10 oz of the 35% solution, and 5 oz of the 20% solution.

4. Applications Involving Distance, Rate, and Time

The following formula relates the distance traveled to the rate and time of travel.

$$d = rt \qquad \text{distance} = \text{rate} \cdot \text{time}$$

For example, if a car travels 60 mph for 3 hr, then

$$d = (60 \text{ mph})(3 \text{ hr})$$
$$= 180 \text{ mi}$$

If a car travels 60 mph for x hr, then

$$d = (60 \text{ mph})(x \text{ hr})$$
$$= 60x \text{ mi}$$

The relationship $d = rt$ is used in Example 4.

| **Example 4** | **Using a System of Linear Equations in a Distance, Rate, and Time Application** |

Classroom Example: p. 312, Exercise 30

A plane travels with the wind from Kansas City, Missouri, to Denver, Colorado, a distance of 600 mi in 2 hr. The return trip against the same wind takes 3 hr. Find the speed of the plane in still air, and find the speed of the wind.

Solution:

Let p represent the speed of the plane in still air.
Let w represent the speed of the wind.

Notice that when the plane travels with the wind, the net speed is $p + w$. When the plane travels against the wind, the net speed is $p - w$.

The information given in the problem can be organized in a chart.

© Stockbyte/Punchstock
Images RF

	Distance	**Rate**	**Time**
With the wind	600	$p + w$	2
Against the wind	600	$p - w$	3

To set up two equations in p and w, recall that $d = rt$.

From the first row, we have

$$\begin{pmatrix} \text{Distance} \\ \text{with the wind} \end{pmatrix} = \begin{pmatrix} \text{rate with} \\ \text{the wind} \end{pmatrix} \begin{pmatrix} \text{time traveled} \\ \text{with the wind} \end{pmatrix} \longrightarrow 600 = (p + w) \cdot 2$$

From the second row, we have

$$\begin{pmatrix} \text{Distance} \\ \text{against the wind} \end{pmatrix} = \begin{pmatrix} \text{rate against} \\ \text{the wind} \end{pmatrix} \begin{pmatrix} \text{time traveled} \\ \text{against the wind} \end{pmatrix} \longrightarrow 600 = (p - w) \cdot 3$$

Using the distributive property to clear parentheses produces the following system:

$$2p + 2w = 600$$
$$3p - 3w = 600$$

The coefficients of the w variable can be changed to 6 and -6 by multiplying the first equation by 3 and the second equation by 2.

$$2p + 2w = 600 \xrightarrow{\text{Multiply by 3.}} 6p + 6w = 1800$$
$$3p - 3w = 600 \xrightarrow[\text{Multiply by 2.}]{} \underline{6p - 6w = 1200}$$
$$12p \qquad = 3000$$

$$12p = 3000$$

$$\frac{12p}{12} = \frac{3000}{12}$$

$$p = 250 \qquad \text{The speed of the plane in still air is 250 mph.}$$

> **TIP:** To create opposite coefficients on the w variables, we could have divided the first equation by 2 and divided the second equation by 3:
>
> $$2p + 2w = 600 \xrightarrow{\text{Divide by 2.}} p + w = 300$$
> $$3p - 3w = 600 \xrightarrow[\text{Divide by 3.}]{} \underline{p - w = 200}$$
> $$2p \qquad = 500$$
> $$p = 250$$

$$2p + 2w = 600 \qquad \text{Substitute } p = 250 \text{ into the first equation.}$$

$$2(250) + 2w = 600$$

$$500 + 2w = 600$$

$$2w = 100$$

$$w = 50 \qquad \text{The speed of the wind is 50 mph.}$$

The speed of the plane in still air is 250 mph. The speed of the wind is 50 mph.

Skill Practice

4. Dan and Cheryl paddled their canoe 40 mi in 5 hr with the current and 16 mi in 8 hr against the current. Find the speed of the current and the speed of the canoe in still water.

Answer

4. The speed of the canoe in still water is 5 mph. The speed of the current is 3 mph.

Section 4.4   Practice Exercises

Review Exercises

For additional exercises, see Classroom Activities 4.4A–4.4D in the *Student's Resource Manual* at www.mhhe.com/moh.

For Exercises 1–4, solve each system of equations by three different methods:

   a. Graphing method   **b.** Substitution method   **c.** Addition method

1. $-2x + y = 6$
     $2x + y = 2$ $\{(-1, 4)\}$

2. $x - y = 2$
     $x + y = 6$ $\{(4, 2)\}$

3. $y = -2x + 6$
     $4x - 2y = 8$ $\left\{\left(\frac{5}{2}, 1\right)\right\}$

4. $2x = y + 4$
     $4x = 2y + 8$ Infinitely many solutions;
               $\{(x, y) \mid 2x = y + 4\}$

 Writing   Translating Expression   Geometry   Scientific Calculator   Video

For Exercises 5–8, set up a system of linear equations in two variables and solve for the unknown quantities.

5. One number is eight more than twice another. Their sum is 20. Find the numbers.
   The numbers are 4 and 16.

6. The difference of two positive numbers is 264. The larger number is three times the smaller number. Find the numbers.
   The numbers are 132 and 396.

 7. Two angles are complementary. The measure of one angle is 10° less than nine times the measure of the other. Find the measure of each angle.
   The angles are 80° and 10°.

 8. Two angles are supplementary. The measure of one angle is 9° more than twice the measure of the other angle. Find the measure of each angle.
   The angles are 123° and 57°.

Concept 1: Applications Involving Cost

9. An online store sells old video games and DVDs as a bundle. A bundle of two video games and three DVDs can be purchased for $88. A bundle of one video game and two DVDs can be purchased for $51.50. Find the cost of one video game and the cost of one DVD in the bundle. **(See Example 1.)**
   A video game costs $21.50 and a DVD costs $15.

10. Tanya bought three adult tickets and one children's ticket to a movie for $32.00. Li bought two adult tickets and five children's tickets for $49.50. Find the cost of one adult ticket and the cost of one children's ticket.   Adult tickets cost $8.50 each, and children's tickets cost $6.50 each.

11. Nora bought 100 shares of a technology stock and 200 shares of a mutual fund for $3800. Her sister, Erin, bought 300 shares of the technology stock and 50 shares of the same mutual fund for $5350. Find the cost per share of the technology stock, and the cost per share of the mutual fund.
    Technology stock costs $16 per share, and the mutual fund costs $11 per share.

12. Eight students in Ms. Reese's class decided to purchase their textbooks from two different sources. Some students purchased the textbook from the college bookstore for $95.50. The other students purchased the textbook from an online discount store for $65 per book. If the total amount spent by the eight students is $611.50, how many students purchased the book online?   Five students purchased the book online.

13. Mylee is a stamp collector and buys commemorative stamps. Suppose she buys a combination of 47-cent stamps and 34-cent stamps at the post office. If she spends exactly $21.55 on 50 stamps, how many of each type did she buy?
    Mylee bought thirty-five 47-cent stamps and fifteen 34-cent stamps.

14. Zoey purchased some beef and some chicken for a family barbeque. The beef cost $6.00 per pound and the chicken cost $4.50 per pound. She bought a total of 18 lb of meat and spent $96. How much of each type of meat did she purchase?   She bought 10 lb of beef and 8 lb of chicken.

Concept 2: Applications Involving Principal and Interest

15. Shanelle invested $10,000, and at the end of 1 year, she received $805 in interest. She invested part of the money in an account earning 10% simple interest and the remaining money in an account earning 7% simple interest. How much did she invest in each account? **(See Example 2.)**
    Shanelle invested $3500 in the 10% account and $6500 in the 7% account.

	10% Account	7% Account	Total
Principal invested			
Interest earned			

16. $12,000 was borrowed from two sources, one that charges 12% simple interest and the other that charges 8% simple interest. If the total interest at the end of 1 year was $1240, how much money was borrowed from each source?
    $7000 was borrowed from the 12% account and $5000 was borrowed from the 8% account.

	12% Account	8% Account	Total
Principal borrowed			
Interest earned			

17. Troy borrowed a total of $12,000 in two different loans to help pay for his new truck. One loan charges 9% simple interest, and the other charges 6% simple interest. If he is charged $810 in interest after 1 year, find the amount borrowed at each rate.
    $9000 was borrowed at 6%, and $3000 was borrowed at 9%.

 18. Blake has a total of $4000 to invest in two accounts. One account earns 2% simple interest, and the other earns 5% simple interest. How much should be invested in each account to earn exactly $155 at the end of 1 year?
    $1500 must be invested in the 2% account and $2500 invested in the 5% account.

19. Suppose a rich uncle dies and leaves you an inheritance of $30,000. You decide to invest part of the money in a relatively safe bond fund that returns 8%. You invest the rest of the money in a riskier stock fund that you hope will return 12% at the end of 1 year. If you need $3120 at the end of 1 year to make a down payment on a car, how much should you invest at each rate? Invest $12,000 in the bond fund and $18,000 in the stock fund.

20. As part of his retirement strategy, John plans to invest $200,000 in two different funds. He projects that the moderately high risk investments should return, over time, about 9% per year, while the low risk investments should return about 4% per year. If he wants a supplemental income of $12,000 a year, how should he divide his investments? He should invest $80,000 at 9% and $120,000 at 4%.

Concept 3: Applications Involving Mixtures

21. How much 50% disinfectant solution must be mixed with a 40% disinfectant solution to produce 25 gal of a 46% disinfectant solution? **(See Example 3.)** 15 gal of the 50% mixture should be mixed with 10 gal of the 40% mixture.

	50% Mixture	40% Mixture	46% Mixture
Amount of solution			
Amount of disinfectant			

22. How many gallons of 20% antifreeze solution and a 10% antifreeze solution must be mixed to obtain 40 gal of a 16% antifreeze solution? 24 gal of the 20% mixture should be mixed with 16 gal of the 10% mixture.

	20% Mixture	10% Mixture	16% Mixture
Amount of solution			
Amount of antifreeze			

23. How much 45% disinfectant solution must be mixed with a 30% disinfectant solution to produce 20 gal of a 39% disinfectant solution? 12 gal of the 45% disinfectant solution should be mixed with 8 gal of the 30% disinfectant solution.

24. How many gallons of a 25% antifreeze solution and a 15% antifreeze solution must be mixed to obtain 15 gal of a 23% antifreeze solution? 12 gal of the 25% antifreeze solution should be mixed with 3 gal of the 15% antifreeze solution.

25. A chemist needs 50 mL of a 16% salt solution for an experiment. She can only find a 13% salt solution and an 18% salt solution in the supply room. How many milliliters of the 13% solution should be mixed with the 18% solution to produce the desired amount of the 16% solution? She should mix 20 mL of the 13% solution with 30 mL of the 18% solution.

26. Meadowsilver Dairy keeps two kinds of milk on hand, skim milk that has 0.3% butterfat and whole milk that contains 3.3% butterfat. How many gallons of each type of milk does the company need to produce 300 gal of 1% milk for the P&A grocery store? It needs 230 gal of skim milk and 70 gal of whole milk.

27. The cooling system in most cars requires a mixture that is 50% antifreeze. How many liters of pure antifreeze and how many liters of 40% antifreeze solution should Chad mix to obtain 6 L of 50% antifreeze solution? (*Hint:* Pure antifreeze is 100% antifreeze.) Chad needs 1 L of pure antifreeze and 5 L of the 40% solution.

28. Silvia wants to mix a 40% apple juice drink with pure apple juice to make 2 L of a juice drink that is 80% apple juice. How much pure apple juice should she use? Silvia should use $1\frac{1}{3}$ L of pure apple juice.

Concept 4: Applications Involving Distance, Rate, and Time

29. It takes a boat 2 hr to go 16 mi downstream with the current and 4 hr to return against the current. Find the speed of the boat in still water and the speed of the current. **(See Example 4.)** The speed of the boat in still water is 6 mph, and the speed of the current is 2 mph.

	Distance	Rate	Time
Downstream			
Upstream			

30. A boat takes 1.5 hr to go 12 mi upstream against the current. It can go 24 mi downstream with the current in the same amount of time. Find the speed of the current and the speed of the boat in still water. The speed of the boat is 12 mph, and the speed of the current is 4 mph.

	Distance	Rate	Time
Upstream			
Downstream			

31. A plane can fly 960 mi with the wind in 3 hr. It takes the same amount of time to fly 840 mi against the wind. What is the speed of the plane in still air and the speed of the wind? The speed of the plane in still air is 300 mph, and the wind is 20 mph.

32. A plane flies 720 mi with the wind in 3 hr. The return trip against the wind takes 4 hr. What is the speed of the wind and the speed of the plane in still air? The speed of the plane in still air is 210 mph, and the wind is 30 mph.

33. Tony Markins flew from JFK Airport to London. It took him 6 hr to fly with the wind, and 8 hr on the return flight against the wind. If the distance is approximately 3600 mi, determine the speed of the plane in still air and the speed of the wind. The speed of the plane in still air is 525 mph and the speed of the wind is 75 mph.

34. A riverboat cruise upstream on the Mississippi River from New Orleans, Louisiana, to Natchez, Mississippi, takes 10 hr and covers 140 mi. The return trip downstream with the current takes only 7 hr. Find the speed of the riverboat in still water and the speed of the current. The speed of the boat in still water is 17 mph and the speed of the current is 3 mph.

© Royalty Free/Corbis RF

© S. Solum/PhotoLink/Getty Images RF

Mixed Exercises

 35. Debi has $2.80 in a collection of dimes and nickels. The number of nickels is five more than the number of dimes. Find the number of each type of coin. There are 17 dimes and 22 nickels.

36. A child collects state quarters and new $1 coins. If she has a total of 25 coins, and the number of quarters is nine more than the number of dollar coins, how many of each type of coin does she have? She has eight $1 coins and 17 quarters.

37. How many quarts of water should be mixed with a 30% vinegar solution to obtain 12 qt of a 25% vinegar solution? (*Hint:* Water is 0% vinegar.) 2 qt of water should be mixed.

38. How much water should be mixed with an 8% plant fertilizer solution to make a half gallon of a 6% plant fertilizer solution? $\frac{1}{8}$ (or 0.125) gal of water should be mixed.

39. In the 1961–1962 NBA basketball season, Wilt Chamberlain of the Philadelphia Warriors made 2432 baskets. Some of the baskets were free throws (worth 1 point each) and some were field goals (worth 2 points each). The number of field goals was 762 more than the number of free throws.

   a. How many field goals did he make and how many free throws did he make? 835 free throws and 1597 field goals

   b. What was the total number of points scored? 4029 points

   c. If Wilt Chamberlain played 80 games during this season, what was the average number of points per game? Approximately 50 points per game

40. In the 1971–1972 NBA basketball season, Kareem Abdul-Jabbar of the Milwaukee Bucks made 1663 baskets. Some of the baskets were free throws (worth 1 point each) and some were field goals (worth 2 points each). The number of field goals he scored was 151 more than twice the number of free throws.

   a. How many field goals did he make and how many free throws did he make? 504 free throws and 1159 field goals

   b. What was the total number of points scored? 2822 points

   c. If Kareem Abdul-Jabbar played 81 games during this season, what was the average number of points per game? Approximately 35 points per game

41. A small plane can fly 350 mi with the wind in $1\frac{3}{4}$ hr. In the same amount of time, the same plane can travel only 210 mi against the wind. What is the speed of the plane in still air and the speed of the wind? The speed of the plane in still air is 160 mph, and the wind is 40 mph.

42. A plane takes 2 hr to travel 1000 mi with the wind. It can travel only 880 mi against the wind in the same amount of time. Find the speed of the wind and the speed of the plane in still air. The speed of the plane in still air is 470 mph, and the wind is 30 mph.

 Writing   Translating Expression   Geometry   Scientific Calculator   Video

43. A total of $60,000 is invested in two accounts, one that earns 5.5% simple interest, and one that earns 6.5% simple interest. If the total interest at the end of 1 year is $3750, find the amount invested in each account. $15,000 was invested in the 5.5% account, and $45,000 was invested in the 6.5% account.

44. Jacques borrows a total of $15,000. Part of the money is borrowed from a bank that charges 12% simple interest per year. Jacques borrows the remaining part of the money from his sister and promises to pay her 7% simple interest per year. If Jacques' total interest for the year is $1475, find the amount he borrowed from each source. Jacques borrowed $8500 at 12%, and $6500 at 7%.

45. At the holidays, Erica likes to sell a candy/nut mixture to her neighbors. She wants to combine candy that costs $1.80 per pound with nuts that cost $1.20 per pound. If Erica needs 20 lb of mixture that will sell for $1.56 per pound, how many pounds of candy and how many pounds of nuts should she use? 12 lb of candy should be mixed with 8 lb of nuts.

© McGraw-Hill Higher Education/
Jill Braaten, photographer

46. Mary Lee's natural food store sells a combination of teas. The most popular is a mixture of a tea that sells for $3.00 per pound with one that sells for $4.00 per pound. If she needs 40 lb of tea that will sell for $3.65 per pound, how many pounds of each tea should she use? 14 lb of $3 per pound tea should be mixed with 26 lb of $4 per pound tea.

© Steve Mason/Getty Images RF

47. In the 1994 Super Bowl, the Dallas Cowboys scored four more points than twice the number of points scored by the Buffalo Bills. If the total number of points scored by both teams was 43, find the number of points scored by each team. Dallas scored 30 points, and Buffalo scored 13 points.

48. In the 1973 Super Bowl, the Miami Dolphins scored twice as many points as the Washington Redskins. If the total number of points scored by both teams was 21, find the number of points scored by each team. Miami scored 14 points, and Washington scored 7 points.

© herreid/Getty Images RF

Expanding Your Skills

49. In a survey conducted among 500 college students, 340 said that the campus lacked adequate lighting. If $\frac{4}{5}$ of the women and $\frac{1}{2}$ of the men said that they thought the campus lacked adequate lighting, how many men and how many women were in the survey? There were 300 women and 200 men in the survey.

50. During a 1-hr television program, there were 22 commercials. Some commercials were 15 sec and some were 30 sec long. Find the number of 15-sec commercials and the number of 30-sec commercials if the total playing time for commercials was 9.5 min. There are six 15-sec commercials and sixteen 30-sec commercials.

Section 4.5 Linear Inequalities and Systems of Inequalities in Two Variables

Concepts

1. Graphing Linear Inequalities in Two Variables
2. Graphing Systems of Linear Inequalities in Two Variables

1. Graphing Linear Inequalities in Two Variables

A **linear inequality in two variables** x and y is an inequality that can be written in one of the following forms: $Ax + By < C$, $Ax + By > C$, $Ax + By \leq C$, or $Ax + By \geq C$ where A and B are not both zero.

A solution to a linear inequality in two variables is an ordered pair that makes the inequality true. For example, solutions to the inequality $x + y < 3$ are ordered pairs (x, y)

such that the sum of the x- and y-coordinates is less than 3. Several such examples are $(0, 0)$, $(-2, -2)$, $(3, -2)$, and $(-4, 1)$. There are actually infinitely many solutions to this inequality, and therefore it is convenient to express the solution set as a graph. The shaded area in Figure 4-10 represents all solutions (x, y), whose coordinates total less than 3.

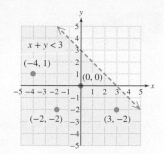

To graph a linear inequality in two variables we will use a process called the test point method. To use the test point method, first graph the related equation. In this case, the related equation represents a line in the xy-plane. Then choose a test point *not* on the line to determine which side of the line to shade. This process is demonstrated in Example 1.

Figure 4-10

| **Example 1** | **Graphing a Linear Inequality in Two Variables** |

Graph the solution set. $2x + y \leq 3$

Classroom Example: p. 323, Exercise 18

Solution:

$2x + y \leq 3 \longrightarrow 2x + y = 3$ **Step 1:** Set up the related equation.

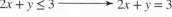

Figure 4-11

Step 2: Graph the related equation.

Graph the line by either setting up a table of points, or by using the slope-intercept form (Figure 4-11).

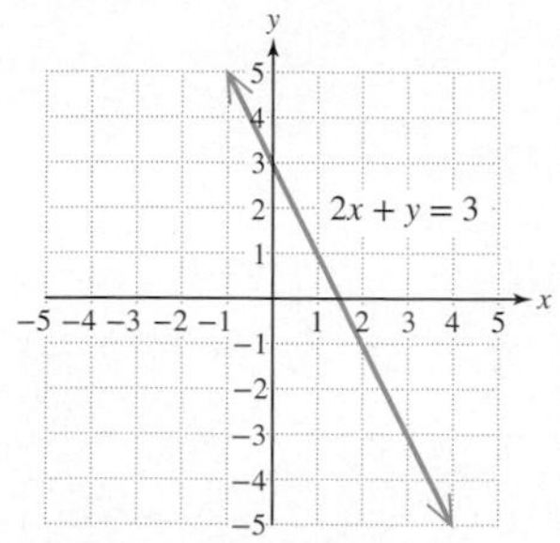

Table:

x	y
1	1
0	3
$\frac{3}{2}$	0

Slope-intercept form:

$2x + y = 3$

$y = -2x + 3$

Step 3: The solution to $2x + y \leq 3$ includes points for which $2x + y$ is less than *or equal to* 3. Because equality is included, points on the line $2x + y = 3$ are included. A solid line shows that the points on the line are included.

Now we must determine which side of the line to shade. To do so, we choose an arbitrary test point *not* on the line. The point $(0, 0)$ is a convenient choice.

Test point: $(0, 0)$

$$2x + y \leq 3$$

$$2(0) + (0) \overset{?}{\leq} 3$$

$$0 \overset{?}{\leq} 3 \checkmark \quad \text{True}$$

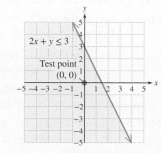

Figure 4-12

The test point $(0, 0)$ is true in the original inequality. This means that the region from which the test point was taken is part of the solution set. Therefore, shade below the line (Figure 4-12).

TIP: If a point above the line is selected as a test point, notice that it will *not* make the original inequality true. For example, test the point (2, 2).

$$2x + y \leq 3$$

$$2(2) + (2) \overset{?}{\leq} 3$$

$$6 \overset{?}{\leq} 3 \quad \text{False}$$

A false result tells us to shade the *other* side of the line.

Skill Practice Graph the solution set.

1. $3x + 2y \geq -6$

Now suppose the inequality from Example 1 had the strict inequality symbol, $<$. That is, consider the inequality $2x + y < 3$. The boundary line $2x + y = 3$ is *not* included in the solution set, because the expression $2x + y$ must be *strictly less than* 3 (not equal to 3). To show that the boundary line is *not* included in the solution set, we draw a dashed line (Figure 4-13).

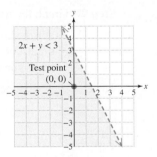

Figure 4-13

The test point method to graph linear inequalities in two variables is summarized as follows:

Avoiding Mistakes

Although one test point is sufficient to select a region to shade, you can choose two test points: one above the line and one below the line. The second point can serve as a check.

Answer

1.

Test Point Method

Step 1 Set up the related equation.

Step 2 Graph the related equation from step 1. The equation will be a boundary line in the *xy*-plane.

- If the original inequality is a strict inequality, $<$ or $>$, then the line is *not* part of the solution set. Graph the line as a *dashed line*.
- If the original inequality is not strict, $\leq$ or $\geq$, then the line *is* part of the solution set. Graph the line as a *solid line*.

Step 3 Choose a point not on the line and substitute its coordinates into the original inequality.

- If the test point makes the inequality true, then the region it represents is part of the solution set. Shade that region.
- If the test point makes the inequality false, then the other region is part of the solution set and should be shaded.

Example 2	**Graphing a Linear Inequality in Two Variables**

Graph the solution set. $\quad 4x - 2y > 6$

Classroom Example: p. 323, Exercise 24

Solution:

$4x - 2y > 6 \longrightarrow 4x - 2y = 6$ **Step 1:** Set up the related equation.

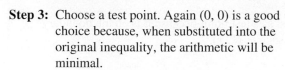

Figure 4-14

Step 2: Graph the equation. Draw a dashed line because the inequality is strict, $>$ (Figure 4-14).

Table:

x	y
0	-3
$\frac{3}{2}$	0
2	1

Slope-intercept form:

$4x - 2y = 6$

$-2y = -4x + 6$

$y = 2x - 3$

Step 3: Choose a test point. Again $(0, 0)$ is a good choice because, when substituted into the original inequality, the arithmetic will be minimal.

$$4x - 2y > 6$$

$$4(0) - 2(0) \overset{?}{>} 6$$

$$0 \overset{?}{>} 6 \quad \text{False}$$

The test point from above the line does not check in the original inequality. Therefore, shade below the line (Figure 4-15).

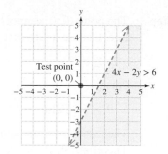

Figure 4-15

Skill Practice Graph the solution set.

2. $6x - 2y < -6$

TIP: An inequality can also be graphed by first solving the inequality for y. Then,

- Shade *below* the line if the inequality is of the form $y < mx + b$ or $y \le mx + b$.
- Shade *above* the line if the inequality is of the form $y > mx + b$ or $y \ge mx + b$.

From Example 2, we have

$4x - 2y > 6$

$\quad -2y > -4x + 6$

$\dfrac{-2y}{-2} < \dfrac{-4x}{-2} + \dfrac{6}{-2}$ $\quad$ Reverse the inequality sign.

$\quad y < 2x - 3$ $\quad$ Shade below the line.

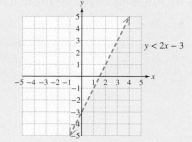

Answer

2.

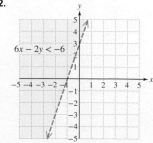

Classroom Example: p. 324,
Exercise 32

Example 3 Graphing a Linear Inequality in Two Variables

Graph the solution set. $\quad 2y \geq 5x$

Solution:

$2y \geq 5x \xrightarrow{\hspace{1cm}} 2y = 5x$ **Step 1:** Set up the related equation.

Step 2: Graph the equation. Draw a solid line because the symbol $\geq$ is used (Figure 4-16).

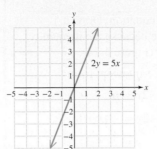

Figure 4-16

Table:

x	y
0	0
2	5
−2	−5

Slope-intercept form:

$2y = 5x$

$y = \dfrac{5}{2}x$

Step 3: The point $(0, 0)$ cannot be used as a test point because it is on the boundary line. Choose a different point such as $(1, 1)$.

$$2y \geq 5x$$

$$2(1) \overset{?}{\geq} 5(1)$$

$$2 \overset{?}{\geq} 5 \quad \text{False}$$

The test point from below the line does not check in the original inequality. Therefore, shade above the line (Figure 4-17).

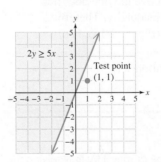

Figure 4-17

Skill Practice Graph the solution set.

  3. $-2y < x$

Answer

3.

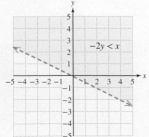

| Example 4 | Graphing a Linear Inequality in Two Variables |

Classroom Example: p. 324, Exercise 26

Graph the solution set. $2x > -4$

Solution:

$2x > -4 \longrightarrow 2x = -4$

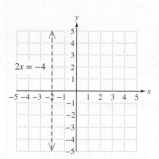

Figure 4-18

Step 1: Set up the related equation.

Step 2: Graph the equation. The equation represents a vertical line.

$$2x = -4$$

$$x = -2$$

Draw a dashed vertical line (Figure 4-18).

Step 3: Choose a test point such as $(0, 0)$.

$$2x > -4$$

$$2(0) \overset{?}{>} -4$$

$$0 \overset{?}{>} -4 \checkmark \quad \text{True}$$

The test point from the right of the line checks in the original inequality. Therefore, shade to the right of the line (Figure 4-19).

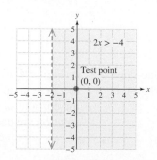

Figure 4-19

Skill Practice Graph the solution set.

4. $4x \geq 12$

2. Graphing Systems of Linear Inequalities in Two Variables

Thus far in this chapter, we have studied systems of linear equations in two variables. Graphically, a solution to such a system is a point of intersection of two lines. In this section, we will study systems of linear *inequalities* in two variables. Graphically, the solution set to such a system is the intersection (or "overlap") of the shaded regions of the two inequalities.

Answer

4.

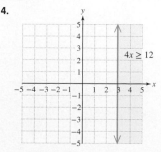

Example 5   Graphing a System of Linear Inequalities

Graph the solution set. $y > \dfrac{1}{2}x - 2$

$$x + y \leq 1$$

Solution:

Sketch each inequality.

$y > \dfrac{1}{2}x - 2 \xrightarrow{\text{Related equation}} y = \dfrac{1}{2}x - 2$ $x + y \leq 1 \xrightarrow{\text{Related equation}} x + y = 1$

The line $y = \dfrac{1}{2}x - 2$ is drawn in red in Figure 4-20. Substituting the test point $(0, 0)$ into the inequality results in a true statement. Therefore, we shade above the line.

The line $x + y = 1$ is drawn in blue in Figure 4-21. Substituting the test point $(0, 0)$ into the inequality results in a true statement. Therefore, we shade below the line.

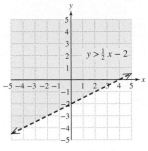

Figure 4-20

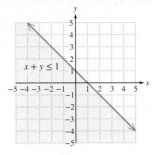

Figure 4-21

Next, we draw these regions on the same graph. The intersection ("overlap") is shown in purple (Figure 4-22).

In Figure 4-23, we show the solution to the system of inequalities. Notice that the portions of the lines not bounding the solution are dashed.

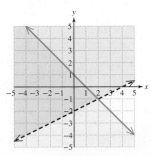

Figure 4-22

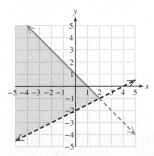

Figure 4-23

Answer

5.

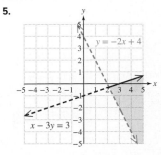

Skill Practice   Graph the solution set.

5. $x - 3y \geq 3$

$\; y > -2x + 4$

Example 6 **Graphing a System of Linear Inequalities**

Graph the solution set.
$$y \leq 3$$
$$2x - y < 2$$

Classroom Example: p. 326, Exercise 54

Solution:

Sketch each inequality.

$$y \leq 3 \xrightarrow{\text{Related equation}} y = 3$$

$$2x - y < 2 \xrightarrow{\text{Related equation}} 2x - y = 2$$

The line $y = 3$ is drawn in red in Figure 4-24. Substituting (0, 0) into the inequality results in a true statement. Therefore, shade below the red line.

The line $2x - y = 2$ is drawn in blue in Figure 4-24. Substituting (0, 0) into the inequality results in a true statement. Therefore, shade above the blue line.

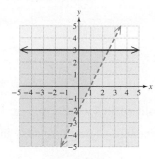

Figure 4-24

In Figure 4-25, we show the solution to the system of inequalities. Notice that the portions of the lines not bounding the solution set are dashed. This is because points not adjacent to the shaded region are not part of the solution set.

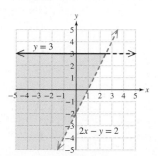

Figure 4-25

Answer

6.

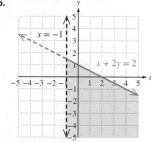

Skill Practice Graph the solution set.

6. $x > -1$
$x + 2y \leq 2$

Section 4.5 Practice Exercises

For additional exercises, see Classroom Activities 4.5A–4.5B in the *Student's Resource Manual* at www.mhhe.com/moh.

Vocabulary and Key Concepts

1. a. An inequality that can be written in the form $Ax + By > C$ is called a ___linear___ inequality in two variables.

b. The solution to a system of linear inequalities is the region where the graphs ___intersect or overlap___ .

Review Exercises

For Exercises 2–4, graph each equation.

2. $x = -3$

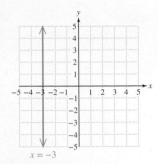

$x = -3$

3. $y = \frac{3}{5}x + 2$

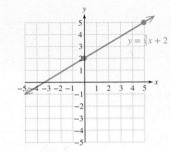

$y = \frac{3}{5}x + 2$

4. $y = -\frac{4}{3}x$

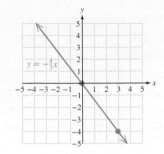

$y = -\frac{4}{3}x$

Concept 1: Graphing Linear Inequalities in Two Variables

5. When is a solid line used in the graph of a linear inequality in two variables?
When the inequality symbol is ≤ or ≥

6. When is a dashed line used in the graph of a linear inequality in two variables?
When the inequality symbol is < or >

7. What does the shaded region represent in the graph of a linear inequality in two variables?
All of the points in the shaded region are solutions to the inequality.

8. When graphing a linear inequality in two variables, how do you determine which side of the boundary line to shade? Choose a test point not on the line. Substitute the coordinates of the test point into the inequality. If this results in a true statement, shade the region represented by the test point. If the test point makes a false statement, shade the other region.

9. Which is the graph of $-2x - y \leq 2$? a

a.

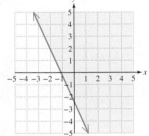

b.

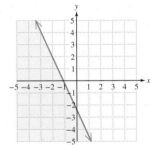

c.

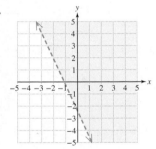

10. Which is the graph of $-3x + y > -1$? c

a.

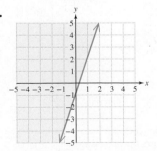

b.

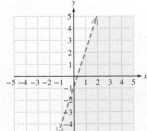

c.

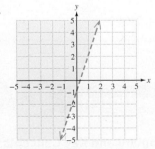

For Exercises 11–16, answer true or false.

11. The point $(3, -1)$ is a solution
   to $3x + 2y > 1$.   True

12. The point $(-2, -2)$ is a solution
   to $-2x + y > 9$.   False

13. The point $(2, 0)$ is a solution
   $y < -2x + 4$.   False

14. The point $(0, 4)$ is a solution
   to $3x + y \leq 4$.   True

15. The point $(-3, 0)$ is a solution
   to $x + 10y < 1$.   True

16. The point $(1, 1)$ is a solution
   to $y \geq x - 4$.   True

For Exercises 17–22, graph each solution set. Then write three ordered pairs that are solutions to the inequality.
(See Examples 1–4.)

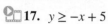 **17.** $y \geq -x + 5$

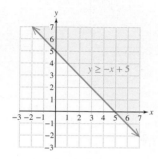

For example: $(0, 5), (2, 7), (-1, 8)$

18. $y \leq 2x - 1$

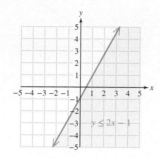

For example: $(0, -1), (2, -2), (-1, -5)$

 **19.** $y < 4x$

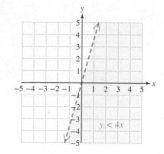

For example: $(1, -1), (3, 0), (-2, -9)$

20. $y > -5x$

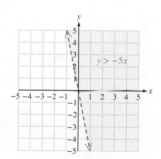

For example: $(2, 1), (0, 4), (-1, 8)$

21. $3x + 7y \leq 14$

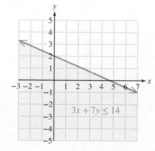

For example: $(0, 0), (0, 2), (-1, -3)$

22. $5x - 6y \geq 18$

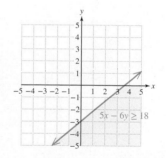

For example: $(0, -3), (4, -4), (-1, -6)$

For Exercises 23–40, graph each solution set. **(See Examples 1–4.)**

23. $x - y > 6$

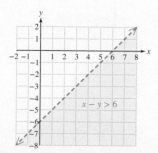

24. $x + y < 5$

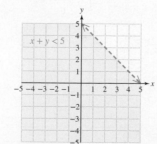

 25. $x \geq -1$

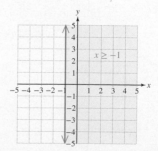

26. $x \le 6$

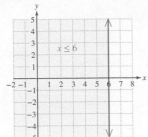

27. $y < 3$

 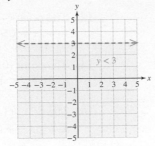

28. $y > -3$

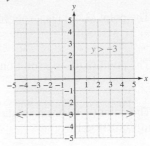

29. $y \le -\dfrac{3}{4}x + 2$

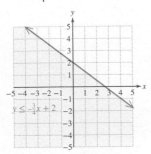

30. $y \ge \dfrac{2}{3}x + 1$

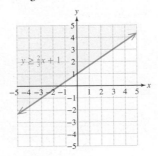

31. $y - 2x > 0$

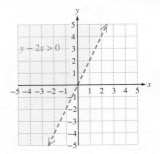

32. $y + 3x < 0$

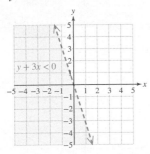

33. $x \le 0$

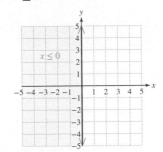

34. $y \le 0$

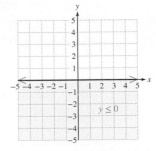

35. $y \ge 0$

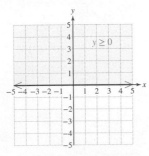

36. $x \ge 0$

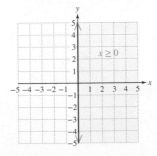

37. $-x \le \dfrac{1}{2}y - 2$

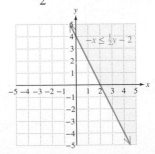

38. $-3 + 2x \le -y$

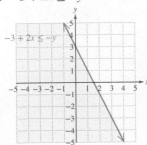

39. $2x > 3y$

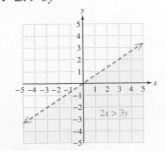

40. $-4x > 5y$

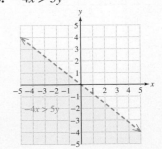

  Writing   ←→ Translating Expression   Geometry   Scientific Calculator   Video

41. a. Describe the graph of the inequality $x + y > 4$. Find three solutions to the inequality (answers will vary).

The set of ordered pairs above the line $x + y = 4$, for example: $(6, 3)$, $(-2, 8)$, $(0, 5)$

b. Describe the graph of the equation $x + y = 4$. Find three solutions to the equation (answers will vary).

The set of ordered pairs on the line $x + y = 4$, for example: $(0, 4)$, $(4, 0)$, $(2, 2)$

c. Describe the graph of the inequality $x + y < 4$. Find three solutions to the inequality (answers will vary).

The set of ordered pairs below the line $x + y = 4$, for example: $(0, 0)$, $(-2, 1)$, $(3, 0)$

42. a. Describe the graph of the inequality $x + y < 3$. Find three solutions to the inequality (answers will vary).

The set of ordered pairs below the line $x + y = 3$, for example: $(0, 2)$, $(-1, -1)$, $(3, -2)$

b. Describe the graph of the equation $x + y = 3$. Find three solutions to the equation (answers will vary).

The set of ordered pairs on the line $x + y = 3$, for example: $(0, 3)$, $(3, 0)$, $(1, 2)$

c. Describe the graph of the inequality $x + y > 3$. Find three solutions to the inequality (answers will vary).

The set of ordered pairs above the line $x + y = 3$, for example: $(4, 0)$, $(-1, 6)$, $(2, 2)$

Concept 2: Graphing Systems of Linear Inequalities in Two Variables

For Exercises 43–60, graph each solution set. **(See Examples 5–6.)**

43. $2x + y < 3$

$y \geq x + 3$

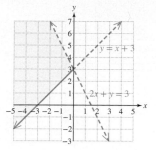

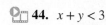

44. $x + y < 3$

$y - x \geq 0$

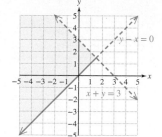

45. $x + y \geq -3$

$x - 2y \geq 6$

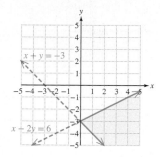

46. $y \geq -3x + 4$

$x + y \leq 4$

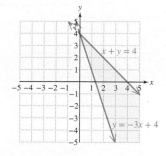

47. $2x + 3y < 6$

$3x + y > -5$

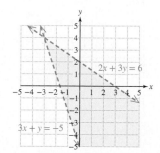

48. $-2x - y < 5$

$x + 2y \geq 2$

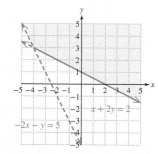

49. $y > 2x$

$y > -4x$

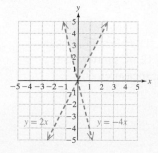

50. $2y \geq 6x$

$y \leq x$

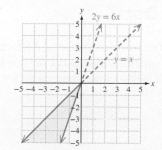

51. $y < \dfrac{1}{2}x - 1$

$x + y \leq -4$

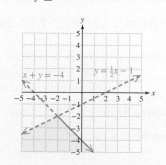

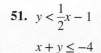

Writing   Translating Expression   Geometry   Scientific Calculator   Video

52. $y \geq \dfrac{1}{3}x + 2$

$4x + y < -2$

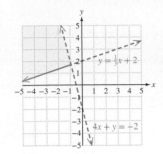

53. $y < 4$

$4x + 3y \geq 12$

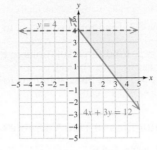

54. $x \geq -3$

$2x + 4y < 4$

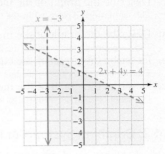

55. $x > -4$

$y \leq 3$

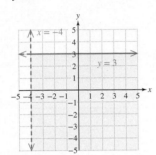

56. $x \leq 3$

$y > 1$

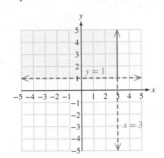

57. $2x \geq 5$

$6 > 3y$

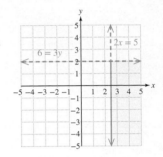

58. $4y \geq 6$

$8 > 2x$

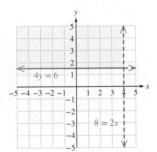

59. $x \geq -4$

$x \leq 1$

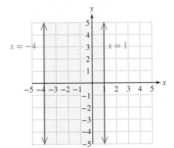

60. $y \geq -2$

$y \leq 3$

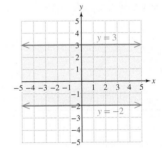

Group Activity

Creating Linear Models from Data

Materials: Two pieces of rope for each group. The ropes should be of different thicknesses. The piece of thicker rope should be between 4 and 5 ft long. The thinner piece of rope should be 8 to 12 in. shorter than the thicker rope. You will also need a yardstick or other device for making linear measurements.

Estimated Time: 30–35 minutes

Group Size: 4 (2 pairs)

 Writing   Translating Expression   Geometry   Scientific Calculator    Video

1. Each group of 4 should divide into two pairs, and each pair will be given a piece of rope. Each pair will measure the initial length of rope. Then students will tie a series of knots in the rope and measure the new length after each knot is tied. (*Hint:* Try to tie the knots with an equal amount of force each time. Also, as the ropes are straightened for measurement, try to use the same amount of tension in the rope.) The results should be recorded in the table.

Thick Rope			Thin Rope	
Number of Knots, x	Length (in.), y		Number of Knots, x	Length (in.), y
0			0	
1			1	
2			2	
3			3	
4			4	

Answers will vary throughout this exercise.

2. Graph each set of data points. Use a different color pen or pencil for each set of points. Does it appear that each set of data follows a linear trend? For each data set, draw a representative line.

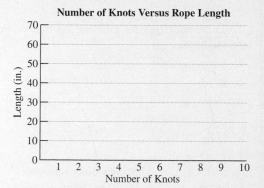

Number of Knots Versus Rope Length

3. Each time a knot is tied, the rope decreases in length. Using the results from question 1, compute the average amount of length lost per knot tied.

   For the thick rope, the length decreases by _____ inches per knot tied.

   For the thin rope, the length decreases by _____ inches per knot tied.

4. For each set of data points, find an equation of the line through the points. Write the equation in slope-intercept form, $y = mx + b$.

   Equation for the thick rope: _____

   Equation for the thin rope: _____

   What does the slope of each line represent? _____

   What does the y-intercept for each line represent? _____

5. Next, you will try to predict the number of knots that you need to tie in each rope so that the ropes will be equal in length. To do this, solve the system of equations in question 4.

   Solution to the system of equations: (_____, _____)
                                         number of knots, x length, y

   Interpret the meaning of the ordered pair in terms of the number of knots tied and the lengths of the ropes.

6. Check your answer from question 5 by actually tying the required number of knots in each rope. After doing this, are the ropes the same length? What is the length of each rope? Does this match the length predicted from question 5?

Chapter 4   Summary

Key Concepts

A **system of two linear equations** can be solved by graphing.

A **solution to a system of linear equations** is an ordered pair that satisfies both equations in the system. Graphically, this represents a point of intersection of the lines.

There may be one solution, infinitely many solutions, or no solution.

One solution
Consistent
Independent

Infinitely many solutions
Consistent
Dependent

No solution
Inconsistent
Independent

A system of equations is **consistent** if there is at least one solution. A system is **inconsistent** if there is no solution.

If two equations represent the same line, the equations are said to be **dependent equations**. In this case, all points on the line are solutions to the system. If two equations represent two different lines, the equations are said to be **independent equations**. In this case, the lines either intersect at one point or are parallel.

Examples

Example 1

Solve by using the graphing method.

$x + y = 3$

$2x - y = 0$

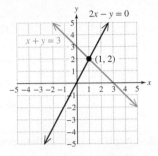

The solution set is $\{(1, 2)\}$.

Example 2

Solve by using the graphing method.

$3x - 2y = 2$

$-6x + 4y = 4$

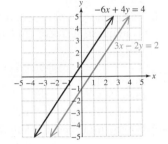

There is no solution, $\{\ \}$. The system is inconsistent.

Example 3

Solve by using the graphing method.

$x + 2y = 2$

$-3x - 6y = -6$

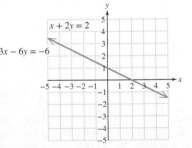

The equations are dependent, and the solution set consists of all points on the line, given by
$\{(x, y)\,|\,x + 2y = 2\}$.

Section 4.2   Solving Systems of Equations by the Substitution Method

Key Concepts

Solving a System of Equations by Using the Substitution Method:

1. Isolate one of the variables from one equation.
2. Substitute the expression found in step 1 into the other equation.
3. Solve the resulting equation.
4. Substitute the value found in step 3 back into the equation in step 1 to find the remaining variable.
5. Check the ordered pair in both original equations.

An inconsistent system has no solution and is detected algebraically by a contradiction (such as $0 = 3$).

If two linear equations represent the same line, the equations are dependent. This is detected algebraically by an identity (such as $0 = 0$).

Examples

Example 1

Solve by using the substitution method.

$$x + 4y = -11$$
$$3x - 2y = -5$$

Isolate x in the first equation: $x = -4y - 11$
Substitute into the second equation.

$$3(-4y - 11) - 2y = -5 \qquad \text{Solve the}$$
$$-12y - 33 - 2y = -5 \qquad \text{equation.}$$
$$-14y = 28$$
$$y = -2$$

$$x = -4y - 11 \qquad \substack{\text{Substitute} \\ y = -2.}$$
$$x = -4(-2) - 11 \qquad \text{Solve for } x.$$
$$x = -3$$

The ordered pair $(-3, -2)$ checks in the original equations. The solution set is $\{(-3, -2)\}$.

Example 2

Solve by using the substitution method.

$$3x + y = 4$$
$$-6x - 2y = 2$$

Isolate y in the first equation: $y = -3x + 4$.
Substitute into the second equation.

$$-6x - 2(-3x + 4) = 2$$
$$-6x + 6x - 8 = 2$$
$$-8 = 2 \qquad \text{Contradiction}$$

The system is inconsistent and has no solution, { }.

Example 3

Solve by using the substitution method.

$$y = x + 2 \qquad y \text{ is already isolated.}$$
$$x - y = -2$$

$$x - (x + 2) = -2 \qquad \substack{\text{Substitute } y = x + 2 \text{ into the} \\ \text{second equation.}}$$
$$x - x - 2 = -2$$
$$-2 = -2 \qquad \text{Identity}$$

The equations are dependent. The solution set is all points on the line $y = x + 2$ or $\{(x, y) \mid y = x + 2\}$.

Section 4.3	**Solving Systems of Equations by the Addition Method**

Key Concepts

Solving a System of Linear Equations by Using the Addition Method:

1. Write both equations in standard form: $Ax + By = C$.
2. Clear fractions or decimals (optional).
3. Multiply one or both equations by a nonzero constant to create opposite coefficients for one of the variables.
4. Add the equations to eliminate one variable.
5. Solve for the remaining variable.
6. Substitute the known value into one of the original equations to solve for the other variable.
7. Check the ordered pair in both equations.

Examples

Example 1

Solve by using the addition method.

$$5x = -4y - 7 \qquad \text{Write the first equation in standard form.}$$
$$6x - 3y = 15$$

$$5x + 4y = -7 \xrightarrow{\text{Multiply by 3.}} 15x + 12y = -21$$
$$6x - 3y = 15 \xrightarrow[\text{Multiply by 4.}]{} \underline{24x - 12y = 60}$$
$$39x = 39$$
$$x = 1$$

$$5x = -4y - 7$$
$$5(1) = -4y - 7$$
$$5 = -4y - 7$$
$$12 = -4y$$
$$-3 = y \qquad \text{The ordered pair } (1, -3) \text{ checks in both original equations. The solution set is } \{(1, -3)\}.$$

Section 4.4 Applications of Linear Equations in Two Variables

Examples

Example 1

A riverboat travels 36 mi with the current in 2 hr. The return trip takes 3 hr against the current. Find the speed of the current and the speed of the boat in still water.

Let x represent the speed of the boat in still water.
Let y represent the speed of the current.

	Distance	Rate	Time
Against current	36	$x - y$	3
With current	36	$x + y$	2

Distance = (rate)(time)

$36 = (x - y) \cdot 3 \longrightarrow 36 = 3x - 3y$

$36 = (x + y) \cdot 2 \longrightarrow 36 = 2x + 2y$

$36 = 3x - 3y \xrightarrow{\text{Multiply by 2.}} 72 = 6x - 6y$

$36 = 2x + 2y \xrightarrow{\text{Multiply by 3.}} \underline{108 = 6x + 6y}$

$\phantom{36 = 2x + 2y \xrightarrow{\text{Multiply by 3.}}} 180 = 12x$

$\phantom{36 = 2x + 2y \xrightarrow{\text{Multiply by 3.}}} 15 = x$

$36 = 2(15) + 2y$

$36 = 30 + 2y$

$6 = 2y$

$3 = y$

The speed of the boat in still water is 15 mph, and the speed of the current is 3 mph.

Example 2

Diane borrows a total of $15,000. Part of the money is borrowed from a lender that charges 8% simple interest. She borrows the rest of the money from her mother and will pay back the money at 5% simple interest. If the total interest after 1 year is $900, how much did she borrow from each source?

	8%	5%	Total
Principal	x	y	15,000
Interest	$0.08x$	$0.05y$	900

$x + y = 15,000$

$0.08x + 0.05y = 900$

Substitute $x = 15,000 - y$ into the second equation.

$0.08(15,000 - y) + 0.05y = 900$

$1200 - 0.08y + 0.05y = 900$

$1200 - 0.03y = 900$

$-0.03y = -300$

$y = 10,000$

$x = 15,000 - 10,000$

$ = 5,000$

The amount borrowed at 8% is $5,000.
The amount borrowed from her mother is $10,000.

Linear Inequalities and Systems of Inequalities in Two Variables

Key Concepts

A **linear inequality in two variables** can be written in one of the forms: $Ax + By < C$, $Ax + By > C$, $Ax + By \leq C$, or $Ax + By \geq C$.

Steps for Using the Test Point Method to Solve a Linear Inequality in Two Variables:

1. Set up the related *equation*.
2. Graph the related equation. This will be a line in the xy-plane.
   - If the original inequality is a strict inequality, $<$ or $>$, then the line is *not* part of the solution set. Therefore, graph the boundary as a dashed line.

   - If the original inequality is not strict, $\leq$ or $\geq$, then the line *is* part of the solution set. Therefore, graph the boundary as a solid line.

3. Choose a point not on the line and substitute its coordinates into the original inequality.
   - If the test point makes the inequality true, then the region it represents is part of the solution set. Shade that region.
   - If the test point makes the inequality false, then the other region is part of the solution set and should be shaded.

Example

Example 1

Graph the solution set. $2x - y < 4$

1. The related equation is $2x - y = 4$.
2. Graph the equation $2x - y = 4$ (dashed line).

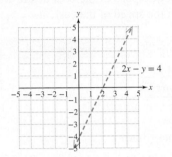

3. Choose an arbitrary test point not on the line such as $(0, 0)$.

$$2x - y < 4$$
$$2(0) - (0) \overset{?}{<} 4$$
$$0 \overset{?}{<} 4 \checkmark \quad \text{True}$$

Shade the region represented by the test point (in this case, above the line).

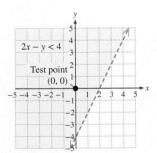

Chapter 4   Review Exercises

Section 4.1

For Exercises 1–4, determine if the ordered pair is a solution to the system.

1. $x - 4y = -4$ $(4, 2)$
   $x + 2y = 8$ Yes

2. $x - 6y = 6$ $(12, 1)$
   $-x + y = 4$ No

3. $3x + y = 9$ $(1, 3)$
   $y = 3$ No

4. $2x - y = 8$ $(2, -4)$
   $x = 2$ Yes

For Exercises 5–10, identify whether the system represents intersecting lines, parallel lines, or coinciding lines by comparing slopes and y-intercepts.

5. $y = -\dfrac{1}{2}x + 4$
   $y = x - 1$
   Intersecting lines (The lines have different slopes.)

6. $y = -3x + 4$
   $y = 3x + 4$
   Intersecting lines (The lines have different slopes.)

7. $y = -\dfrac{4}{7}x + 3$
   $y = -\dfrac{4}{7}x - 5$
   Parallel lines (The lines have the same slope but different y-intercepts.)

8. $y = 5x - 3$
   $y = \dfrac{1}{5}x - 3$
   Intersecting lines (The lines have different slopes.)

9. $y = 9x - 2$
   $9x - y = 2$
   Coinciding lines (The lines have the same slope and same y-intercept.)

10. $x = -5$
    $y = 2$
    Intersecting lines (The lines have different slopes.)

For Exercises 11–18, solve the system by graphing. For systems that do not have one unique solution, also state the number of solutions and whether the system is inconsistent or the equations are dependent.

11. $y = -\dfrac{2}{3}x - 2$
    $-x + 3y = -6$
    $\{(0, -2)\}$

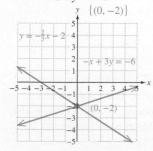

12. $y = -2x - 1$ $\{(-2, 3)\}$
    $x + 2y = 4$

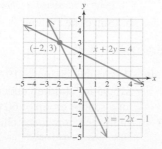

13. $4x = -2y + 10$
    $2x + y = 5$

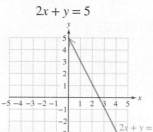

Infinitely many solutions; $\{(x, y) \mid 2x + y = 5\}$; dependent equations

14. $10y = 2x - 10$
    $-x + 5y = -5$

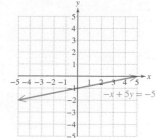

Infinitely many solutions; $\{(x, y) \mid -x + 5y = -5\}$; dependent equations

15. $6x - 3y = 9$
    $y = -1$ $\{(1, -1)\}$

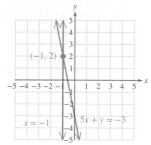

16. $5x + y = -3$
    $x = -1$ $\{(-1, 2)\}$

17. $x - 7y = 14$
    $-2x + 14y = 14$

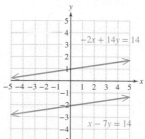

No solution; $\{\ \}$; inconsistent system

18. $y = -5x + 4$
    $10x + 2y = -4$

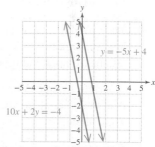

No solution; $\{\ \}$; inconsistent system

Section 4.2

19. One phone company charges $0.15 a minute for calls but adds a $3.90 charge each month. Another company does not have a monthly fee but charges $0.25 per minute. The cost per month, y_1 (in $) for the first company is given by the equation:

    $y_1 = 0.15x + 3.90$ where x represents the number of minutes used.

    The cost per month, y_2, (in $) for the second company is given by the equation:

    $y_2 = 0.25x$ where x represents the number of minutes used.   Using 39 minutes per month, the cost per month for each company is $9.75.

    Find the number of minutes at which the cost per month for each company is the same.

Writing   ←→ Translating Expression   Geometry   Scientific Calculator   Video

For Exercises 20–23, solve each system using the substitution method.

20. $6x + y = 2$ $\left\{\left(\frac{2}{3}, -2\right)\right\}$ **21.** $2x + 3y = -5$ $\{(-4, 1)\}$
$y = 3x - 4$ $\qquad$ $x = y - 5$

22. $2x + 6y = 10$ $\{\ \}$ **23.** $4x + 2y = 4$
$x = -3y + 6$ $\qquad$ Infinitely many solutions;
$\qquad\qquad\qquad\qquad$ $y = -2x + 2$
$\qquad\qquad\qquad\qquad$ $\{(x, y) \mid y = -2x + 2\}$

24. Given the system: $\qquad x + 2y = 11$
$\qquad\qquad\qquad\qquad 5x + 4y = 40$

   a. Which variable from which equation is easiest to isolate and why? $\quad$ *x* in the first equation is easiest to isolate because its coefficient is 1.

   b. Solve the system using the substitution method.
   $\left\{\left(6, \frac{5}{2}\right)\right\}$

25. Given the system: $\qquad 4x - 3y = 9$
$\qquad\qquad\qquad\qquad 2x + y = 12$

   a. Which variable from which equation is easiest to isolate and why? $\quad$ *y* in the second equation is easiest to isolate because its coefficient is 1.

   b. Solve the system using the substitution method.
   $\left\{\left(\frac{9}{2}, 3\right)\right\}$

For Exercises 26–29, solve each system using the substitution method.

26. $3x - 2y = 23$ $\{(5, -4)\}$ **27.** $x + 5y = 20$ $\{(0, 4)\}$
$x + 5y = -15$ $\qquad$ $3x + 2y = 8$

28. $x - 3y = 9$ **29.** $-3x + y = 15$ $\{\ \}$
Infinitely many solutions;
$5x - 15y = 45$ $\qquad\qquad$ $6x - 2y = 12$
$\{(x, y) \mid x - 3y = 9\}$

30. The difference of two positive numbers is 42. The larger number is 2 more than 6 times the smaller number. Find the numbers. $\quad$ The numbers are 50 and 8.

31. In a right triangle, one of the acute angles is 8° less than the other acute angle. Find the measure of each acute angle. $\quad$ The angles are 41° and 49°.

32. Two angles are supplementary. One angle measures 14° less than two times the other angle. Find the measure of each angle. $\quad$ The angles are $115\frac{1}{3}°$ and $64\frac{2}{3}°$.

Section 4.3

33. Consider the system. $\qquad -2x + 7y = 30$
$\qquad\qquad\qquad\qquad\quad 4x + 5y = 16$

   a. Which variable, *x* or *y*, is easier to eliminate using the addition method? (Answers may vary.)

   b. Solve the system using the addition method. $\quad \{(-1, 4)\}$

34. Given the system: $\qquad 3x - 5y = 1$
$\qquad\qquad\qquad\qquad\quad 2x - y = -4$

a. Which variable, *x* or *y*, is easier to eliminate using the addition method? (Answers may vary.)

b. Solve the system using the addition method.
$\{(-3, -2)\}$

35. Given the system: $\qquad 9x - 2y = 14$
$\qquad\qquad\qquad\qquad\quad 4x + 3y = 14$

   a. Which variable, *x* or *y*, is easier to eliminate using the addition method? (Answers may vary.)

   b. Solve the system using the addition method. $\quad \{(2, 2)\}$

For Exercises 36–43, solve each system using the addition method.

36. $2x + 3y = 1$ $\{(2, -1)\}$ **37.** $x + 3y = 0$ $\{(-6, 2)\}$
$x - 2y = 4$ $\qquad\qquad$ $-3x - 10y = -2$

38. $8x + 8 = -6y + 6$ **39.** $12x = 5y + 5$
$10x = 9y - 8$ $\left\{\left(-\frac{1}{2}, \frac{1}{3}\right)\right\}$ $\quad 5y = -1 - 4x$ $\left\{\left(\frac{1}{4}, -\frac{2}{5}\right)\right\}$

40. $-4x - 6y = -2$ **41.** $-8x - 4y = 16$ $\{\ \}$
Infinitely many solutions;
$6x + 9y = 3$ $\qquad\qquad$ $10x + 5y = 5$
$\{(x, y) \mid -4x - 6y = -2\}$

42. $\frac{1}{2}x - \frac{3}{4}y = -\frac{1}{2}$ **43.** $0.5x - 0.2y = 0.5$
$\qquad\qquad\qquad\qquad\qquad\quad 0.4x + 0.7y = 0.4$
$\frac{1}{3}x + y = -\frac{10}{3}$ $\qquad\qquad$ $\{(1, 0)\}$
$\{(-4, -2)\}$

44. Given the system: $\qquad 4x + 9y = -7$
$\qquad\qquad\qquad\qquad\quad y = 2x - 13$

   a. Which method would you choose to solve the system, the substitution method or the addition method? Explain your choice. $\quad$ Use the substitution method because *y* is already isolated in the second equation.

   b. Solve the system $\quad \{(5, -3)\}$

45. Given the system: $\qquad 5x - 8y = -2$
$\qquad\qquad\qquad\qquad\quad 3x - 7y = 1$

   a. Which method would you choose to solve the system, the substitution method or the addition method? Explain your choice. $\quad$ Use the addition method. The substitution method would be cumbersome because isolating either variable from either equation would result in fractional coefficients.

   b. Solve the system $\quad \{(-2, -1)\}$

Section 4.4

46. Zoo Miami charges $19.95 for adult admission and $15.95 for children under 13. The total bill before tax for a school group of 60 people is $989. How many adults and how many children were admitted? $\quad$ There were 8 adult tickets and 52 children's tickets sold.

47. As part of his retirement strategy Winston plans to invest $600,000 in two different funds. He projects that the high-risk investments should return, over time, about 12% per year, while the low-risk investments should return about 4% per year. If he wants a supplemental income of $30,000 a year, how should he divide his investments? He should invest $75,000 at 12% and $525,000 at 4%.

48. Suppose that whole milk with 4% fat is mixed with 1% low fat milk to make a 2% reduced fat milk. How much of the whole milk should be mixed with the low fat milk to make 60 gal of 2% reduced fat milk? 20 gal of whole milk should be mixed with 40 gal of low fat milk.

49. A boat travels 80 mi downstream with the current in 4 hr and 80 mi upstream against the current in 5 hr. Find the speed of the current and the speed of the boat in still water. The speed of the boat is 18 mph, and that of the current is 2 mph.

50. A plane travels 870 mi against the wind in 3 hr. Traveling with the wind, the plane travels 700 mi in 2 hr. Find the speed of the plane in still air and the speed of the wind. The plane's speed in still air is 320 mph. The wind speed is 30 mph.

51. At a sports arena, the total cost of a soft drink and a hot dog is $8.00. The price of the hot dog is $1.00 more than the cost of the soft drink. Find the cost of a soft drink and the cost of a hot dog. A hot dog costs $4.50 and a drink costs $3.50.

52. Ray played two rounds of golf at Pebble Beach for a total score of 154. If his score in the second round is 10 more than his score in the first round, find the scores for each round.

© Royalty Free/Corbis RF

The score was 72 on the first round and 82 on the second round.

Section 4.5

For Exercises 53–56, graph each solution set. Then write three ordered pairs that are in the solution set (answers may vary).

53. $y < 3x - 1$

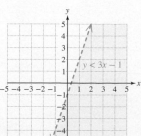

54. $y > -2x + 6$

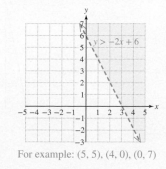

For example: (1, −1), (0, −4), (2, 0)   For example: (5, 5), (4, 0), (0, 7)

55. $-2x - 3y \geq 8$

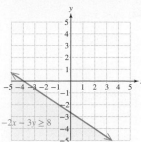

For example:
(−4, 0), (−2, −2), (1, −4)

56. $4x - 2y \leq 10$

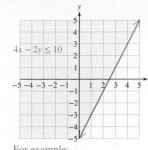

For example:
(0, 0), (0, −5), (−1, 1)

For Exercises 57–62, graph each solution set.

57. $x - 5y \geq 0$

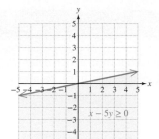

58. $7x - y \leq 0$

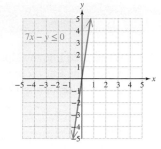

59. $x > 5$

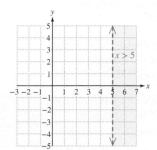

60. $y < -4$

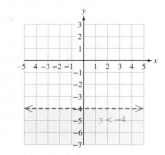

61. $y \geq 0$

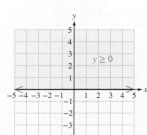

62. $x \geq 0$

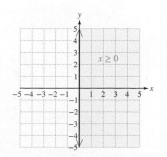

For Exercises 63–66, graph each solution set.

63. $2x - y \geq 8$
$x + y \leq 3$

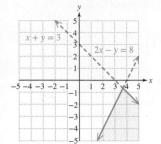

64. $y \leq x - 1$
$x + 2y \geq 4$

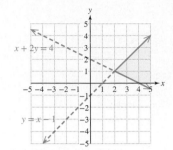

65. $y \leq 2x$
$-2x - y > -3$

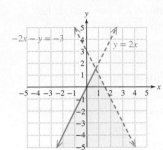

66. $y \leq 4$
$2x - y < 1$

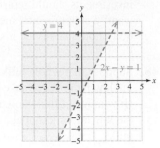

Chapter 4   Test

1. Write each line in slope-intercept form. Then determine if the lines represent intersecting lines, parallel lines, or coinciding lines.

$5x + 2y = -6$ $y = -\frac{5}{2}x - 3$

$-\frac{5}{2}x - y = -3$ $y = -\frac{5}{2}x + 3$

Parallel lines

2. a. How many solutions does a system of two linear equations have if the equations represent parallel lines?   No solution

b. How many solutions does a system of two linear equations have if the equations represent coinciding lines?   Infinitely many solutions

c. How many solutions does a system of two linear equations have if the equations represent intersecting lines?   One solution

For Exercises 3–4, solve each system by graphing.

3. $y = 2x - 4$
$-2x + 3y = 0$
$\{(3, 2)\}$

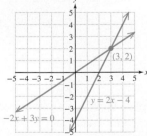

4. $2x + 4y = 12$
$2y - 6 = -x$

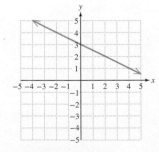

Infinitely many solutions;
$\{(x, y) \mid 2x + 4y = 12\}$

5. Solve the system using the substitution method.

$x = 5y - 2$
$2x + y = -4$ $\{(-2, 0)\}$

6. Solve the system using the addition method.

$3x - 6y = 8$ $\left\{\left(2, -\frac{1}{3}\right)\right\}$
$2x + 3y = 3$

For Exercises 7–12, solve each system using any method.

7. $\frac{1}{3}x + y = \frac{7}{3}$

$x = \frac{3}{2}y - 11$
$\{(-5, 4)\}$

8. $2x - 12 = y$

$2x - \frac{1}{2}y = x + 5$
$\{\ \}$

9. $3x - 4y = 29$ $\{(3, -5)\}$
$2x + 5y = -19$

10. $2x = 6y - 14$ $\{(-1, 2)\}$
$2y = 3 - x$

11. $-0.25x - 0.05y = 0.2$
Infinitely many solutions;
$10x + 2y = -8$
$\{(x, y) \mid 10x + 2y = -8\}$

12. $3x + 3y = -2y - 7$
$-3y = 10 - 4x$
$\{(1, -2)\}$

13. Graph the solution set. $5x - y \geq -6$

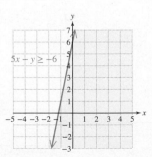

14. Graph the solution set.

$$2x + y > 1$$

$$x + y < 2$$

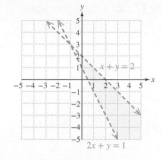

15. In an early WNBA (basketball) season, the league's leading scorer was Sheryl Swoopes from the Houston Comets. Swoopes scored 17 points more than the second leading scorer, Lauren Jackson from the Seattle Storm. Together they scored a total of 1211 points. How many points did each player score?   Swoopes scored 614 points and Jackson scored 597.

16. Latrell buys four CDs and two DVDs for $54 from the sale rack. Kendra buys two CDs and three DVDs from the same rack for $49. What is the price per CD and the price per DVD?   CDs cost $8 each and DVDs cost $11 each.

17. How many milliliters of a 50% acid solution and how many milliliters of a 20% acid solution must be mixed to produce 36 mL of a 30% acid solution?   12 mL of the 50% acid solution should be mixed with 24 mL of the 20% solution.

18. The cost to ride a certain trolley one way is $2.25. Kelly and Hazel had to buy eight tickets for their group.

   a. What was the total amount of money required? $18 was required.

   b. Kelly and Hazel had only quarters and $1 bills. They also determined that they used twice as many quarters as $1 bills. How many quarters and how many $1 bills did they use?   They used 24 quarters and 12 $1 bills.

19. Suppose a total of $5000 is borrowed from two different loans. One loan charges 10% simple interest, and the other charges 8% simple interest.

How much was borrowed at each rate if $424 in interest is charged at the end of 1 year? $1200 was borrowed at 10%, and $3800 was borrowed at 8%.

20. Mark needs to move to a new apartment and is trying to find the most affordable moving truck. He will only need the truck for one day. After checking the AAA Movers website, he finds that he can rent a 10-ft truck for $20.95 a day plus $1.89 per mile. He then checks the website of a local moving company and finds the charge to be $37.95 a day plus $1.19 per mile for the same size truck. Determine the number of miles for which the cost to rent from either company would be the same. Round the answer to the nearest mile.   They would be the same cost for a distance of approximately 24 mi.

21. A plane travels 910 mi in 2 hr against the wind and 1090 mi in 2 hr with the same wind. Find the speed of the plane in still air and the speed of the wind.   The plane travels 500 mph in still air, and the wind speed is 45 mph.

22. The number of calories in a piece of cake is 20 less than 3 times the number of calories in a scoop of ice cream. Together, the cake and ice cream have 460 calories. How many calories are in each?   The cake has 340 calories, and the ice cream has 120 calories.

© McGraw-Hill Education/Jill Braaten, photographer

23. How much 10% acid solution should be mixed with a 25% acid solution to create 100 mL of a 16% acid solution?   60 mL of 10% solution and 40 mL of 25% solution

Chapters 1–4   Cumulative Review Exercises

1. Simplify. $\dfrac{|2 - 5| + 10 \div 2 + 3}{\sqrt{10^2 - 8^2}}$ $\dfrac{11}{6}$

2. Solve for x. $\dfrac{1}{3}x - \dfrac{3}{4} = \dfrac{1}{2}(x + 2)$ $\left\{-\dfrac{21}{2}\right\}$

3. Solve for a. $-4(a + 3) + 2 = -5(a + 1) + a$ { }

4. Solve for y. $3x - 2y = 6$ $y = \dfrac{3}{2}x - 3$

5. Solve for x. $z = \dfrac{x - m}{5}$ $x = 5z + m$

6. Solve for z. Graph the solution set on a number line and write the solution in interval notation.

$$-2(3z + 1) \le 5(z - 3) + 10$$

$\left[\dfrac{3}{11}, \infty\right)$

 Writing   Translating Expression    Geometry   Scientific Calculator   Video

7. The largest angle in a triangle is 110°. Of the remaining two angles, one is 4° less than the other angle. Find the measure of the three angles.

The angles are 37°, 33°, and 110°.

8. Two hikers start at opposite ends of an 18-mi trail and walk toward each other. One hiker walks predominately down hill and averages 2 mph faster than the other hiker. Find the average rate of each hiker if they meet in 3 hr.

The rates of the hikers are 2 mph and 4 mph.

9. Jesse Ventura became the 38th governor of Minnesota by receiving 37% of the votes. If approximately 2,060,000 votes were cast, how many did Mr. Ventura get?

Jesse Ventura received approximately 762,200 votes.

10. The YMCA wants to raise $2500 for its summer program for disadvantaged children. If the YMCA has already raised $900, what percent of its goal has been achieved?

36% of the goal has been achieved.

11. Two angles are complementary. One angle measures 17° more than the other angle. Find the measure of each angle.   The angles are 36.5° and 53.5°.

12. Find the slope and y-intercept of the line $5x + 3y = -6$.   Slope: $-\dfrac{5}{3}$; y-intercept: $(0, -2)$

13. The slope of a given line is $-\frac{2}{3}$.

a. What is the slope of a line parallel to the given line? $-\dfrac{2}{3}$

b. What is the slope of a line perpendicular to the given line? $\dfrac{3}{2}$

14. Find an equation of the line passing through the point $(2, -3)$ and having a slope of -3. Write the final answer in slope-intercept form. $y = -3x + 3$

15. Sketch the following equations on the same graph.

a. $2x + 5y = 10$

b. $2y = 4$

c. Find the point of intersection and check the solution in each equation.

$\{(0, 2)\}$

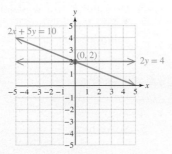

16. Solve the system of equations by using the substitution method.

$$2x + 5y = 10$$
$$2y = 4$$

$\{(0, 2)\}$

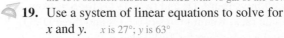

17. a. Graph the equation $2x + y = 3$.   **b.** Graph the solution set $2x + y < 3$.

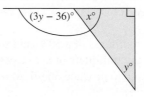

c. Explain the difference between the graphs in parts (a) and (b).   Part (a) represents the solutions to an equation. Part (b) represents the solutions to a strict inequality.

18. How many gallons of a 15% antifreeze solution should be mixed with a 60% antifreeze solution to produce 60 gal of a 45% antifreeze solution?   20 gal of the 15% solution should be mixed with 40 gal of the 60% solution.

19. Use a system of linear equations to solve for x and y. x is 27°; y is 63°

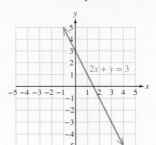

20. In 1920, the average speed for the winner of the Indianapolis 500 car race was 88.6 mph. In 1990, a track record was reached with the speed of 186.0 mph.

a. Find the slope of the line shown in the figure. Round to one decimal place.   1.4

b. Interpret the meaning of the slope in the context of this problem.   Between 1920 and 1990, the winning speed in the Indianapolis 500 increased on average by 1.4 mph per year.

Polynomials and Properties of Exponents

5

CHAPTER OUTLINE

Mathematics to Compute Cost

Trevor is a dance instructor and wants to host a one-day dance event. He has a fixed cost of $200 to rent the dance studio. In addition, to host the event, he has the following variable costs, which depend on the number of participants.

- $1.00 per person for coffee and breakfast snacks
- $7.00 per person for lunch
- $1.20 per person for a step-sheet booklet

If n represents the number of participants, then Trevor's cost for the event is given by

$$\text{Cost} = 1.00n + 7.00n + 1.20n + 200$$

This expression is called a **polynomial**. Terms of a polynomial are separated by addition, and sometimes a polynomial can be simplified by adding *like* **terms**. In this case, the terms containing the variable n are *like* terms. Thus, the polynomial representing cost can be simplified as

$$\text{Cost} = (1.00 + 7.00 + 1.20)n + 200$$
$$= 9.20n + 200$$

The terms $1.00n$, $7.00n$, and $1.20n$ are combined to form $9.20n$ because we're effectively consolidating the costs for breakfast, lunch, and the booklet into one overall cost per person.

As you study polynomials in this chapter, you will encounter other applications including those involving profit and cost.

© McGraw-Hill Education/John Flournoy

Multiplying and Dividing Expressions with Common Bases

Concepts

Classroom Example: p. 347, Exercise 2

1. Review of Exponential Notation

Recall that an **exponent** is used to show repeated multiplication of the **base**.

Definition of b^n

Let b represent any real number and n represent a positive integer. Then,

$$b^n = \underbrace{b \cdot b \cdot b \cdot b \cdot \ldots \cdot b}_{n \text{ factors of } b}$$

Example 1 Evaluating Expressions with Exponents

For each expression, identify the exponent and base. Then evaluate the expression.

a. 6^2 **b.** $\left(-\dfrac{1}{2}\right)^3$ **c.** 0.8^4

Solution:

Expression	Base	Exponent	Result
a. 6^2	6	2	$(6)(6) = 36$
b. $\left(-\dfrac{1}{2}\right)^3$	$-\dfrac{1}{2}$	3	$\left(-\dfrac{1}{2}\right)\left(-\dfrac{1}{2}\right)\left(-\dfrac{1}{2}\right) = -\dfrac{1}{8}$
c. 0.8^4	0.8	4	$(0.8)(0.8)(0.8)(0.8) = 0.4096$

Skill Practice For each expression, identify the base and exponent.

1. 8^3 **2.** $\left(-\dfrac{1}{4}\right)^2$ **3.** 0.2^4

Note that if no exponent is explicitly written for an expression, then the expression has an implied exponent of 1. For example, $x = x^1$.

Consider an expression such as $3y^6$. The factor 3 has an exponent of 1, and the factor y has an exponent of 6. That is, the expression $3y^6$ is interpreted as $3^1 y^6$.

2. Evaluating Expressions with Exponents

Particular care must be taken when evaluating exponential expressions involving negative numbers. An exponential expression with a negative base is written with parentheses around the base, such as $(-3)^2$.

To evaluate $(-3)^2$, we have: $(-3)^2 = (-3)(-3) = 9$

If no parentheses are present, the expression -3^2 is the *opposite* of 3^2, or equivalently, $-1 \cdot 3^2$.

$$-3^2 = -1(3^2) = -1(3)(3) = -9$$

Answers

1. Base 8; exponent 3
2. Base $-\dfrac{1}{4}$; exponent 2
3. Base 0.2; exponent 4

| Example 2 | **Evaluating Expressions with Exponents** |

Classroom Examples: p. 347, Exercises 28 and 30

Evaluate each expression.

a. -5^4 **b.** $(-5)^4$ **c.** $(-0.2)^3$ **d.** -0.2^3

Solution:

a. -5^4

$= -1 \cdot 5^4$ 5 is the base with exponent 4.

$= -1 \cdot 5 \cdot 5 \cdot 5 \cdot 5$ Multiply -1 by four factors of 5.

$= -625$

b. $(-5)^4$

$= (-5)(-5)(-5)(-5)$ Parentheses indicate that -5 is the base with exponent 4.

$= 625$ Multiply four factors of -5.

c. $(-0.2)^3$ Parentheses indicate that -0.2 is the base with exponent 3.

$= (-0.2)(-0.2)(-0.2)$ Multiply three factors of -0.2.

$= -0.008$

d. -0.2^3

$= -1 \cdot 0.2^3$ 0.2 is the base with exponent 3.

$= -1(0.2)(0.2)(0.2)$ Multiply -1 by three factors of 0.2.

$= -0.008$

Skill Practice Evaluate each expression.

4. -2^4 **5.** $(-2)^4$ **6.** $(-0.1)^3$ **7.** -0.1^3

| Example 3 | **Evaluating Expressions with Exponents** |

Classroom Examples: p. 348, Exercises 52 and 56

Evaluate each expression for $a = 2$ and $b = -3$.

a. $5a^2$ **b.** $(5a)^2$ **c.** $5ab^2$ **d.** $(b + a)^2$

Solution:

a. $5a^2$

$= 5(\ )^2$ Use parentheses to substitute a number for a variable.

$= 5(2)^2$ Substitute $a = 2$.

$= 5(4)$ Simplify exponents before multiplying.

$= 20$

b. $(5a)^2$

$= [5(\ )]^2$ Use parentheses to substitute a number for a variable. The original parentheses are replaced with brackets.

$= [5(2)]^2$ Substitute $a = 2$.

$= (10)^2$ Simplify inside the parentheses first.

$= 100$

Answers
4. -16 **5.** 16
6. -0.001 **7.** -0.001

c. $5ab^2$

$\qquad = 5(2)(-3)^2$ $\qquad$ Substitute $a = 2$, $b = -3$.

$\qquad = 5(2)(9)$ $\qquad\qquad$ Simplify exponents before multiplying.

$\qquad = 90$ $\qquad\qquad\quad$ Multiply.

d. $(b + a)^2$

$\qquad = [(-3) + (2)]^2$ $\qquad$ Substitute $b = -3$ and $a = 2$.

$\qquad = (-1)^2$ $\qquad\qquad\quad$ Simplify within the parentheses first.

$\qquad = 1$

Skill Practice Evaluate each expression for $x = 2$ and $y = -5$.

8. $6x^2$ $\qquad$ **9.** $(6x)^2$ $\qquad$ **10.** $2xy^2$ $\qquad$ **11.** $(y - x)^2$

3. Multiplying and Dividing Expressions with Common Bases

In this section, we investigate the effect of multiplying or dividing two quantities with the same base. For example, consider the expressions: $x^5 x^2$ and $\frac{x^5}{x^2}$. Simplifying each expression, we have:

$$x^5 x^2 = (x \cdot x \cdot x \cdot x \cdot x)(x \cdot x) = \overbrace{x \cdot x \cdot x \cdot x \cdot x \cdot x \cdot x}^{7 \text{ factors of } x} = x^7$$

$$\frac{x^5}{x^2} = \frac{x \cdot x \cdot x \cdot \overset{1}{\cancel{x}} \cdot \overset{1}{\cancel{x}}}{x \cdot x} = \frac{x \cdot x \cdot x}{1} = x^3$$

These examples suggest that to multiply two quantities with the same base, we add the exponents. To divide two quantities with the same base, we subtract the exponent in the denominator from the exponent in the numerator. These rules are stated formally in the following two properties.

Multiplication of Expressions with Like Bases

Assume that b is a real number and that m and n represent positive integers. Then,

$$b^m b^n = b^{m+n}$$

Division of Expressions with Like Bases

Assume that $b \neq 0$ is a real number and that m and n represent positive integers. Then,

$$\frac{b^m}{b^n} = b^{m-n}$$

Answers

8. 24 $\quad$ **9.** 144 $\quad$ **10.** 100

11. 49

| Example 4 | **Simplifying Expressions with Exponents** |

Simplify the expressions.　　**a.** $w^3 w^4$　　**b.** $2^3 \cdot 2^4$

Classroom Examples: p. 348, Exercises 68 and 70

Solution:

a. $w^3 w^4$　　　$(w \cdot w \cdot w)(w \cdot w \cdot w \cdot w)$

　　$= w^{3+4}$　　To multiply expressions with like bases, add the exponents.

　　$= w^7$

b. $2^3 \cdot 2^4$　　　$(2 \cdot 2 \cdot 2)(2 \cdot 2 \cdot 2 \cdot 2)$

　　$= 2^{3+4}$　　To multiply expressions with like bases, add the exponents (the base is unchanged).

　　$= 2^7$ or 128

Avoiding Mistakes

When we multiply expressions with like bases, we add the exponents. The base does not change. In Example 4(b), notice that the base 2 does not change. $2^3 \cdot 2^4 = 2^7$.

Skill Practice Simplify the expressions.

12. $q^4 q^8$　　　**13.** $8^4 \cdot 8^8$

| Example 5 | **Simplifying Expressions with Exponents** |

Simplify the expressions.　　**a.** $\dfrac{t^6}{t^4}$　　**b.** $\dfrac{5^6}{5^4}$

Classroom Examples: p. 348, Exercises 80 and 86

Solution:

a. $\dfrac{t^6}{t^4}$　　　$\dfrac{t \cdot t \cdot t \cdot t \cdot t \cdot t}{t \cdot t \cdot t \cdot t}$

　　$= t^{6-4}$　　To divide expressions with like bases, subtract the exponents.

　　$= t^2$

b. $\dfrac{5^6}{5^4}$　　　$\dfrac{5 \cdot 5 \cdot 5 \cdot 5 \cdot 5 \cdot 5}{5 \cdot 5 \cdot 5 \cdot 5}$

　　$= 5^{6-4}$　　To divide expressions with like bases, subtract the exponents (the base is unchanged).

　　$= 5^2$ or 25

Skill Practice Simplify the expressions.

14. $\dfrac{y^{15}}{y^8}$　　　**15.** $\dfrac{3^{15}}{3^8}$

Answers

12. q^{12}　　**13.** 8^{12}
14. y^7　　**15.** 3^7

Classroom Example: p. 348, Exercise 90

> **Example 6**   Simplifying Expressions with Exponents

Simplify the expressions.   **a.** $\dfrac{z^4 z^5}{z^3}$ **b.** $\dfrac{10^7}{10^2 \cdot 10}$

Solution:

a. $\dfrac{z^4 z^5}{z^3}$

$= \dfrac{z^{4+5}}{z^3}$ Add the exponents in the numerator (the base is unchanged).

$= \dfrac{z^9}{z^3}$

$= z^{9-3}$ Subtract the exponents.

$= z^6$

b. $\dfrac{10^7}{10^2 \cdot 10}$

$= \dfrac{10^7}{10^2 \cdot 10^1}$ Note that 10 is equivalent to 10^1.

$= \dfrac{10^7}{10^{2+1}}$ Add the exponents in the denominator (the base is unchanged).

$= \dfrac{10^7}{10^3}$

$= 10^{7-3}$ Subtract the exponents.

$= 10^4$ or 10,000   Simplify.

Skill Practice Simplify the expressions.

16. $\dfrac{a^3 a^8}{a^7}$ **17.** $\dfrac{5^9}{5^2 \cdot 5^5}$

4. Simplifying Expressions with Exponents

Classroom Examples: pp. 348–349, Exercises 98 and 106

> **Example 7**   Simplifying Expressions with Exponents

Use the commutative and associative properties of real numbers and the properties of exponents to simplify the expressions.

   a. $(-3p^2 q^4)(2pq^5)$ **b.** $\dfrac{16 w^9 z^3}{4 w^8 z}$

Avoiding Mistakes

To simplify the expression in Example 7(a) we multiply the coefficients. However, to multiply expressions with like bases, we add the exponents.

Solution:

a. $(-3p^2 q^4)(2pq^5)$

$= (-3 \cdot 2)(p^2 p)(q^4 q^5)$ Apply the associative and commutative properties of multiplication to group coefficients and like bases.

$= (-3 \cdot 2) p^{2+1} q^{4+5}$ Add the exponents when multiplying expressions with like bases.

$= -6 p^3 q^9$ Simplify.

Answers

16. a^4 **17.** 5^2 or 25

b. $\dfrac{16w^9z^3}{4w^8z}$

$= \left(\dfrac{16}{4}\right)\left(\dfrac{w^9}{w^8}\right)\left(\dfrac{z^3}{z}\right)$ Group coefficients and like bases.

$= 4w^{9-8}z^{3-1}$ Subtract the exponents when dividing expressions with like bases.

$= 4wz^2$ Simplify.

> **Avoiding Mistakes**
>
> In Example 7(b) we divide the coefficients. However, to divide expressions with like bases, we subtract the exponents.

Skill Practice Simplify the expressions.

18. $(-4x^2y^3)(3x^5y^7)$ **19.** $\dfrac{81x^4y^7}{9xy^3}$

5. Applications of Exponents

Simple interest on an investment or loan is computed by the formula $I = Prt$, where P is the amount of principal, r is the annual interest rate, and t is the time in years. Simple interest is based only on the original principal. However, in most day-to-day applications, the interest computed on money invested or borrowed is compound interest. **Compound interest** is computed on the original principal and on the interest already accrued.

Suppose $1000 is invested at 8% interest for 3 years. Compare the total amount in the account if the money earns simple interest versus if the interest is compounded annually.

Simple Interest

The simple interest earned is given by $I = Prt$

$$= (\$1000)(0.08)(3)$$

$$= \$240$$

The total amount, A, at the end of 3 years is $A = P + I$

$$= \$1000 + \$240$$

$$= \$1240$$

Compound Annual Interest

The total amount, A, in an account earning compound annual interest may be computed using the following formula:

$A = P(1 + r)^t$ where P is the amount of principal, r is the annual interest rate (expressed in decimal form), and t is the number of years.

For example, for $1000 invested at 8% interest compounded annually for 3 years, we have $P = 1000$, $r = 0.08$, and $t = 3$.

$$A = P(1 + r)^t$$

$$A = 1000(1 + 0.08)^3$$

$$= 1000(1.08)^3$$

$$= 1000(1.259712)$$

$$= 1259.712$$

Rounding to the nearest cent, we have $A = \$1259.71$.

Answers

18. $-12x^7y^{10}$ **19.** $9x^3y^4$

Classroom Example: p. 349,
Exercise 116

Example 8 **Using Exponents in an Application**

Find the amount in an account after 8 years if the initial investment is $7000, invested at 2.25% interest compounded annually.

Solution:

Identify the values for each variable.

$P = 7000$

$r = 0.0225$ Note that the decimal form of a percent is used for calculations.

$t = 8$

$A = P(1 + r)^t$

$= 7000(1 + 0.0225)^8$ Substitute.

$= 7000(1.0225)^8$ Simplify inside the parentheses.

$\approx 7000(1.194831142)$ Approximate $(1.0225)^8$.

≈ 8363.82 Multiply (round to the nearest cent).

The amount in the account after 8 years is $8363.82.

Skill Practice

20. Find the amount in an account after 3 years if the initial investment is $4000 invested at 5% interest compounded annually.

Answer

20. $4630.50

Calculator Connections

Topic: Review of Evaluating Exponential Expressions on a Calculator

In Example 8, it was necessary to evaluate the expression $(1.0225)^8$. Recall that the $\boxed{\wedge}$ or $\boxed{y^x}$ key may be used to enter expressions with exponents.

Scientific Calculator

Enter: 1.0225 $\boxed{y^x}$ 8 $\boxed{=}$ **Result:** $\boxed{\quad 1.194831142 \quad}$

Graphing Calculator

Calculator Exercises

Use a calculator to evaluate the expressions.

1. $(1.06)^5$ 1.338225578

2. $(1.02)^{40}$ 2.208039664

3. $5000(1.06)^5$ 6691.127888

4. $2000(1.02)^{40}$ 4416.079327

5. $3000(1 + 0.06)^2$ 3370.8

6. $1000(1 + 0.05)^3$ 1157.625

Section 5.1   Practice Exercises

For this exercise set, assume all variables represent nonzero real numbers.   For additional exercises, see Classroom Activities 5.1A–5.1D in the *Student's Resource Manual* at www.mhhe.com/moh.

Vocabulary and Key Concepts

1. **a.** A(n) ___exponent___ is used to show repeated multiplication of the base.

    b. Given the expression b^n, the value b is the ___base___ and n is the ___exponent___.

    c. Given the expression x, the value of the exponent on x is understood to be ___1___.

    d. The formula to compute simple interest is ___$I = Prt$___.

    e. Interest that is computed on the original principal and on the accrued interest is called ___compound interest___.

Concept 1: Review of Exponential Notation

For Exercises 2–13, identify the base and the exponent. **(See Example 1.)**

2. c^3 Base: c; exponent: 3

3. x^4 Base: x; exponent: 4

4. 5^2 Base: 5; exponent: 2

5. 3^5 Base: 3; exponent: 5

6. $(-4)^8$ Base: -4; exponent: 8

7. $(-1)^4$ Base: -1; exponent: 4

8. x Base: x; exponent: 1

9. 13   Base: 13; exponent: 1

10. -4^2 Base: 4; exponent: 2

11. -10^3 Base: 10; exponent: 3

12. $-y^5$ Base: y; exponent: 5

13. $-t^6$ Base: t; exponent: 6

14. What base corresponds to the exponent 5 in the expression $x^3y^5z^2$? y

15. What base corresponds to the exponent 2 in the expression w^3v^2? v

16. What is the exponent for the factor of 2 in the expression $2x^3$?   1

17. What is the exponent for the factor of p in the expression pq^7?   1

For Exercises 18–26, write the expression using exponents.

18. $(4n)(4n)(4n)$ $(4n)^3$

19. $(-6b)(-6b)$ $(-6b)^2$

 20. $4 \cdot n \cdot n \cdot n$ $4n^3$

21. $-6 \cdot b \cdot b$ $-6b^2$

22. $(x-5)(x-5)(x-5)$ $(x-5)^3$

23. $(y+2)(y+2)(y+2)(y+2)$ $(y+2)^4$

24. $\dfrac{4}{x \cdot x \cdot x \cdot x \cdot x}$ $\dfrac{4}{x^5}$

25. $\dfrac{-2}{t \cdot t \cdot t}$ $\dfrac{-2}{t^3}$

26. $\dfrac{5 \cdot x \cdot x \cdot x}{(y-7)(y-7)}$ $\dfrac{5x^3}{(y-7)^2}$

Concept 2: Evaluating Expressions with Exponents

For Exercises 27–34, evaluate the two expressions and compare the answers. Do the expressions have the same value?
(See Example 2.)

27. -5^2 and $(-5)^2$
    No; $-5^2 = -25$ and $(-5)^2 = 25$

28. -3^4 and $(-3)^4$
    No; $-3^4 = -81$ and $(-3)^4 = 81$

29. -2^5 and $(-2)^5$
    Yes; -32

30. -5^3 and $(-5)^3$
    Yes; -125

31. $\left(\dfrac{1}{2}\right)^3$ and $\dfrac{1}{2^3}$ Yes; $\dfrac{1}{8}$

32. $\left(\dfrac{1}{5}\right)^2$ and $\dfrac{1}{5^2}$ Yes; $\dfrac{1}{25}$

33. $-(-2)^4$ and $-(2)^4$
    Yes; -16

34. $-(-3)^3$ and $-(3)^3$
    No; $-(-3)^3 = 27$ and $-(3)^3 = -27$

For Exercises 35–42, evaluate each expression. **(See Example 2.)**

35. 16^1 16

36. 20^1 20

37. $(-1)^{21}$ -1

38. $(-1)^{30}$ 1

39. $\left(-\dfrac{1}{3}\right)^2$ $\dfrac{1}{9}$

40. $\left(-\dfrac{1}{4}\right)^3$ $-\dfrac{1}{64}$

41. $-\left(\dfrac{2}{5}\right)^2$ $-\dfrac{4}{25}$

42. $-\left(\dfrac{3}{5}\right)^2$ $-\dfrac{9}{25}$

For Exercises 43–50, simplify using the order of operations.

43. $3 \cdot 2^4$ 48

44. $2 \cdot 0^5$ 0

45. $-4(-1)^7$ 4

46. $-3(-1)^4$ -3

47. $6^2 - 3^3$ 9

48. $4^3 + 2^3$ 72

49. $2 \cdot 3^2 + 4 \cdot 2^3$ 50

50. $6^2 - 3 \cdot 1^3$ 33

For Exercises 51–62, evaluate each expression for $a = -4$ and $b = 5$. **(See Example 3.)**

51. $-4b^2$ -100

52. $3a^2$ 48

53. $(-4b)^2$ 400

54. $(3a)^2$ 144

55. $(a+b)^2$ 1

56. $(a-b)^2$ 81

57. $a^2 + 2ab + b^2$ 1

58. $a^2 - 2ab + b^2$ 81

59. $-10ab^2$ 1000

60. $-6a^3b$ 1920

61. $-10a^2b$ -800

62. $-a^2b$ -80

Concept 3: Multiplying and Dividing Expressions with Common Bases

63. Expand the following expressions first. Then simplify using exponents.

  a. $x^4 \cdot x^3$
    $(x \cdot x \cdot x \cdot x)(x \cdot x \cdot x) = x^7$

  b. $5^4 \cdot 5^3$
    $(5 \cdot 5 \cdot 5 \cdot 5)(5 \cdot 5 \cdot 5) = 5^7$

64. Expand the following expressions first. Then simplify using exponents.

  a. $y^2 \cdot y^4$
    $(y \cdot y)(y \cdot y \cdot y \cdot y) = y^6$

  b. $3^2 \cdot 3^4$
    $(3 \cdot 3)(3 \cdot 3 \cdot 3 \cdot 3) = 3^6$

For Exercises 65–76, simplify each expression. Write the answers in exponent form. **(See Example 4.)**

65. $z^5 z^3$ z^8

66. $w^4 w^7$ w^{11}

67. $a \cdot a^8$ a^9

68. $p^4 p$ p^5

69. $4^5 \cdot 4^9$ 4^{14}

70. $6^7 \cdot 6^5$ 6^{12}

71. $\left(\dfrac{2}{3}\right)^3 \left(\dfrac{2}{3}\right)$ $\left(\dfrac{2}{3}\right)^4$

72. $\left(\dfrac{1}{x}\right)\left(\dfrac{1}{x}\right)^2$ $\left(\dfrac{1}{x}\right)^3$

73. $c^5 c^2 c^7$ c^{14}

74. $b^7 b^2 b^8$ b^{17}

75. $x \cdot x^4 \cdot x^{10} \cdot x^3$ x^{18}

76. $z^7 \cdot z^{11} \cdot z^{60} \cdot z$ z^{79}

77. Expand the expressions. Then simplify.

  a. $\dfrac{p^8}{p^3}$

  b. $\dfrac{8^8}{8^3}$

$\dfrac{p \cdot p \cdot p \cdot p \cdot p \cdot p \cdot p \cdot p}{p \cdot p \cdot p} = p^5$ $\dfrac{8 \cdot 8 \cdot 8 \cdot 8 \cdot 8 \cdot 8 \cdot 8 \cdot 8}{8 \cdot 8 \cdot 8} = 8^5$

78. Expand the expressions. Then simplify.

  a. $\dfrac{w^5}{w^2}$

  b. $\dfrac{4^5}{4^2}$

$\dfrac{w \cdot w \cdot w \cdot w \cdot w}{w \cdot w} = w^3$ $\dfrac{4 \cdot 4 \cdot 4 \cdot 4 \cdot 4}{4 \cdot 4} = 4^3$

For Exercises 79–94, simplify each expression. Write the answers in exponent form. **(See Examples 5–6.)**

79. $\dfrac{x^8}{x^6}$ x^2

80. $\dfrac{z^5}{z^4}$ z

81. $\dfrac{a^{10}}{a}$ a^9

82. $\dfrac{b^{12}}{b}$ b^{11}

83. $\dfrac{7^{13}}{7^6}$ 7^7

84. $\dfrac{2^6}{2^4}$ 2^2

85. $\dfrac{5^8}{5}$ 5^7

86. $\dfrac{3^5}{3}$ 3^4

87. $\dfrac{y^{13}}{y^{12}}$ y

88. $\dfrac{w^7}{w^6}$ w

89. $\dfrac{h^3 h^8}{h^7}$ h^4

90. $\dfrac{n^5 n^4}{n^2}$ n^7

91. $\dfrac{7^2 \cdot 7^6}{7}$ 7^7

92. $\dfrac{5^3 \cdot 5^8}{5}$ 5^{10}

93. $\dfrac{10^{20}}{10^3 \cdot 10^8}$ 10^9

94. $\dfrac{3^{15}}{3^2 \cdot 3^{10}}$ 3^3

Concept 4: Simplifying Expressions with Exponents (Mixed Exercises)

For Exercises 95–114, use the commutative and associative properties of real numbers and the properties of exponents to simplify. **(See Example 7.)**

95. $(2x^3)(3x^4)$ $6x^7$

96. $(10y)(2y^3)$ $20y^4$

97. $(5a^2b)(8a^3b^4)$ $40a^5b^5$

98. $(10xy^3)(3x^4y)$ $30x^5y^4$

99. $s^3 \cdot t^5 \cdot t \cdot t^{10} \cdot s^6$
  $s^9 t^{16}$

100. $c \cdot c^4 \cdot d^2 \cdot c^3 \cdot d^3$
  $c^8 d^5$

101. $(-2v^2)(3v)(5v^5)$
  $-30v^8$

102. $(10q^5)(-3q^8)(q)$
  $-30q^{14}$

Writing   ⬌ Translating Expression   Geometry   Scientific Calculator   Video

103. $\left(\frac{2}{3}m^{13}n^8\right)(24m^7n^2)$ $16m^{20}n^{10}$

104. $\left(\frac{1}{4}c^6d^6\right)(28c^2d^7)$ $7c^8d^{13}$

105. $\frac{14c^4d^5}{7c^3d}$ $2cd^4$

106. $\frac{36h^5k^2}{9h^3k}$ $4h^2k$

107. $\frac{z^3z^{11}}{z^4z^6}$ z^4

108. $\frac{w^{12}w^2}{w^4w^5}$ w^5

109. $\frac{25h^3jk^5}{12h^2k}$ $\frac{25hjk^4}{12}$

110. $\frac{15m^5np^{12}}{4mp^9}$ $\frac{15m^4np^3}{4}$

111. $(-4p^6q^8r^4)(2pqr^2)$ $-8p^7q^9r^6$

112. $(-5a^4bc)(-10a^2b)$ $50a^6b^2c$

113. $\frac{-12s^2tu^3}{4su^2}$ $-3stu$

114. $\frac{15w^5x^{10}y^3}{-15w^4x}$ $-wx^9y^3$

Concept 5: Applications of Exponents

 Use the formula $A = P(1 + r)^t$ for Exercises 115–118. **(See Example 8.)**

115. Find the amount in an account after 2 years if the initial investment is \$5000, invested at 7% interest compounded annually.   \$5724.50

116. Find the amount in an account after 5 years if the initial investment is \$2000, invested at 4% interest compounded annually.   \$2433.31

117. Find the amount in an account after 3 years if the initial investment is \$4000, invested at 6% interest compounded annually.   \$4764.06

118. Find the amount in an account after 4 years if the initial investment is \$10,000, invested at 5% interest compounded annually.   \$12,155.06

 For Exercises 119–122, use appropriate geometry formulas.

119. Find the area of the pizza shown in the figure. Round to the nearest square inch.   201 in.^{2}

16 in.

© Brand X Pictures/Alamy RF

120. Find the volume of the sphere shown in the figure. Round to the nearest cubic centimeter.

113 cm^3

$r = 3$ cm

121. Find the volume of a spherical balloon that is 8 in. in diameter. Round to the nearest cubic inch.

268 in.^{3}

122. Find the area of a circular pool 50 ft in diameter. Round to the nearest square foot.

1963 ft^2

Expanding Your Skills

For Exercises 123–130, simplify each expression using the addition or subtraction rules of exponents. Assume that a, b, m, and n represent positive integers.

123. x^nx^{n+1} x^{2n+1}

124. y^ay^{2a} y^{3a}

125. $p^{3m+5}p^{-m-2}$ p^{2m+3}

126. $q^{4b-3}q^{-4b+4}$ q

127. $\frac{z^{b+1}}{z^b}$ z

128. $\frac{w^{5n+3}}{w^{2n}}$ w^{3n+3}

129. $\frac{r^{3a+3}}{r^{3a}}$ r^3

130. $\frac{t^{3+2m}}{t^{2m}}$ t^3

Concepts

1. **Power Rule for Exponents**
2. **The Properties** $(ab)^m = a^m b^m$ **and** $\left(\dfrac{a}{b}\right)^m = \dfrac{a^m}{b^m}$

1. Power Rule for Exponents

The expression $(x^2)^3$ indicates that the quantity x^2 is cubed.

$$(x^2)^3 = (x^2)(x^2)(x^2) = (x \cdot x)(x \cdot x)(x \cdot x) = x^6$$

From this example, it appears that to raise a base to successive powers, we multiply the exponents and leave the base unchanged. This is stated formally as the power rule for exponents.

> ### Power Rule for Exponents
> Assume that b is a real number and that m and n represent positive integers. Then,
>
> $$(b^m)^n = b^{m \cdot n}$$

Classroom Examples: p. 353, Exercises 12 and 22

Example 1   Simplifying Expressions with Exponents

Simplify the expressions.

   a. $(s^4)^2$ **b.** $(3^4)^2$ **c.** $(x^2 x^5)^4$

Solution:

   a. $(s^4)^2$

        $= s^{4 \cdot 2}$ Multiply exponents (the base is unchanged).

        $= s^8$

   b. $(3^4)^2$

        $= 3^{4 \cdot 2}$ Multiply exponents (the base is unchanged).

        $= 3^8$ or 6561

   c. $(x^2 x^5)^4$

        $= (x^7)^4$ Simplify inside the parentheses by adding exponents.

        $= x^{7 \cdot 4}$ Multiply exponents (the base is unchanged).

        $= x^{28}$

Skill Practice   Simplify the expressions.

   1. $(y^3)^5$ **2.** $(2^8)^{10}$ **3.** $(q^5 q^4)^3$

2. The Properties $(ab)^m = a^m b^m$ and $\left(\dfrac{a}{b}\right)^m = \dfrac{a^m}{b^m}$

Consider the following expressions and their simplified forms:

$$(xy)^3 = (xy)(xy)(xy) = (x \cdot x \cdot x)(y \cdot y \cdot y) = x^3 y^3$$

$$\left(\frac{x}{y}\right)^3 = \left(\frac{x}{y}\right)\left(\frac{x}{y}\right)\left(\frac{x}{y}\right) = \left(\frac{x \cdot x \cdot x}{y \cdot y \cdot y}\right) = \frac{x^3}{y^3}$$

The expressions are simplified using the commutative and associative properties of multiplication. The simplified forms for each expression could have been reached in one step by applying the exponent to each factor inside the parentheses.

Answers

1. y^{15} **2.** 2^{80} **3.** q^{27}

Power of a Product and Power of a Quotient

Assume that a and b are real numbers. Let m represent a positive integer. Then,

$$(ab)^m = a^m b^m$$

$$\left(\frac{a}{b}\right)^m = \frac{a^m}{b^m}, \quad b \neq 0$$

Applying these properties of exponents, we have

$$(xy)^3 = x^3 y^3 \quad \text{and} \quad \left(\frac{x}{y}\right)^3 = \frac{x^3}{y^3}$$

Example 2   **Simplifying Expressions with Exponents**

Simplify the expressions.

a. $(-2xyz)^4$ **b.** $(5x^2y^7)^3$ **c.** $\left(\frac{2}{5}\right)^3$ **d.** $\left(\frac{1}{3xy^4}\right)^2$

Classroom Examples: p. 354, Exercises 28 and 32

Solution:

a. $(-2xyz)^4$

$= (-2)^4 x^4 y^4 z^4$ Raise each factor within parentheses to the fourth power.

$= 16x^4 y^4 z^4$

b. $(5x^2y^7)^3$

$= 5^3 (x^2)^3 (y^7)^3$ Raise each factor within parentheses to the third power.

$= 125x^6 y^{21}$ Multiply exponents and simplify.

Instructor Note: Remind students that they can write exponents of 1 where appropriate to avoid confusion. For example, $(5x^2y^7)^3 = (5^1x^2y^7)^3$.

c. $\left(\frac{2}{5}\right)^3$

$= \frac{(2)^3}{(5)^3}$ Raise each factor within parentheses to the third power.

$= \frac{8}{125}$ Simplify.

d. $\left(\frac{1}{3xy^4}\right)^2$

$= \frac{1^2}{3^2 x^2 (y^4)^2}$ Square each factor within parentheses.

$= \frac{1}{9x^2 y^8}$ Multiply exponents and simplify.

Skill Practice Simplify the expressions.

4. $(3abc)^5$ **5.** $(-2t^2w^4)^3$ **6.** $\left(\frac{3}{4}\right)^3$ **7.** $\left(\frac{2x^3}{y^5}\right)^2$

Answers

4. $3^5 a^5 b^5 c^5$ or $243a^5 b^5 c^5$

5. $-8t^6 w^{12}$ **6.** $\frac{27}{64}$ **7.** $\frac{4x^6}{y^{10}}$

The properties of exponents can be used along with the properties of real numbers to simplify complicated expressions.

Classroom Example: p. 354,
Exercise 56

| Example 3 | **Simplifying Expressions with Exponents** |

Simplify the expression. $\dfrac{(x^2)^6(x^3)}{(x^7)^2}$

Solution:

$\dfrac{(x^2)^6(x^3)}{(x^7)^2}$ Clear parentheses by applying the power rule.

$= \dfrac{x^{2 \cdot 6}x^3}{x^{7 \cdot 2}}$ Multiply exponents.

$= \dfrac{x^{12}x^3}{x^{14}}$

$= \dfrac{x^{12+3}}{x^{14}}$ Add exponents in the numerator.

$= \dfrac{x^{15}}{x^{14}}$

$= x^{15-14}$ Subtract exponents.

$= x$ Simplify.

Skill Practice Simplify the expression.

8. $\dfrac{(k^5)^2 k^8}{(k^2)^4}$

Classroom Example: p. 354,
Exercise 60

| Example 4 | **Simplifying Expressions with Exponents** |

Simplify the expression. $(3cd^2)(2cd^3)^3$

Solution:

$(3cd^2)(2cd^3)^3$ Clear parentheses by applying the power rule.

$= 3cd^2 \cdot 2^3 c^3 d^9$ Raise each factor in the second parentheses to the third power.

$= 3 \cdot 2^3 cc^3 d^2 d^9$ Group like factors.

$= 3 \cdot 8c^{1+3}d^{2+9}$ Add exponents on like bases.

$= 24c^4 d^{11}$ Simplify.

Skill Practice Simplify the expression.

9. $(4x^4y)(2x^3y^4)^4$

Answers

8. k^{10} 9. $64x^{16}y^{17}$

Example 5 **Simplifying Expressions with Exponents**

Simplify the expression. $\left(\dfrac{x^7 y z^4}{8xz^3}\right)^2$

Classroom Example: p. 354, Exercise 68

Solution:

$\left(\dfrac{x^7 y z^4}{8xz^3}\right)^2$

$= \left(\dfrac{x^{7-1} y z^{4-3}}{8}\right)^2$ First simplify inside the parentheses by subtracting exponents on like bases.

$= \left(\dfrac{x^6 y z}{8}\right)^2$

$= \dfrac{(x^6)^2 y^2 z^2}{8^2}$ Apply the power rule of exponents.

$= \dfrac{x^{12} y^2 z^2}{64}$

Skill Practice Simplify the expression.

10. $\left(\dfrac{w^2 x y^4}{6xy^3}\right)^2$

Answer
10. $\dfrac{w^4 y^2}{36}$

Section 5.2   Practice Exercises

For this exercise set assume all variables represent nonzero real numbers.

For additional exercises, see Classroom Activity 5.2A in the *Student's Resource Manual* at www.mhhe.com/moh.

Review Exercises

For Exercises 1–8, simplify.

1. $4^2 \cdot 4^7$ 4^9

2. $5^8 \cdot 5^3 \cdot 5$ 5^{12}

3. $a^{13} \cdot a \cdot a^6$ a^{20}

4. $y^{14} y^3$ y^{17}

5. $\dfrac{d^{13} d}{d^5}$ d^9

6. $\dfrac{3^8 \cdot 3}{3^2}$ 3^7

7. $\dfrac{7^{11}}{7^5}$ 7^6

8. $\dfrac{z^4}{z^3}$ z

9. Explain when to add exponents versus when to multiply exponents.   When multiplying expressions with the same base, add the exponents. When raising an expression with an exponent to a power, multiply the exponents.

10. Explain when to add exponents versus when to subtract exponents.   When multiplying expressions with the same base, add the exponents. When dividing expressions with the same base, subtract the exponents.

Concept 1: Power Rule for Exponents

For Exercises 11–22, simplify and write answers in exponent form. **(See Example 1.)**

11. $(5^3)^4$ 5^{12}

12. $(2^8)^7$ 2^{56}

13. $(12^3)^2$ 12^6

14. $(6^4)^4$ 6^{16}

15. $(y^7)^2$ y^{14}

16. $(z^6)^4$ z^{24}

17. $(w^5)^5$ w^{25}

18. $(t^3)^6$ t^{18}

19. $(a^2 a^4)^6$ a^{36}

20. $(z \cdot z^3)^2$ z^8

21. $(y^3 y^4)^2$ y^{14}

22. $(w^5 w)^4$ w^{24}

 Writing   Translating Expression   Geometry   Scientific Calculator   Video

23. Evaluate the two expressions and compare the answers: $(2^2)^3$ and $(2^3)^2$ They are both equal to 2^6.

24. Evaluate the two expressions and compare the answers: $(4^4)^2$ and $(4^2)^4$ They are both equal to 4^8.

25. Evaluate the two expressions and compare the answers. Which expression is greater? Why?

$$4^{3^2} \quad \text{and} \quad (4^3)^2$$

$4^{3^2} = 4^9$; $(4^3)^2 = 4^6$; the expression 4^{3^2} is greater than $(4^3)^2$.

26. Evaluate the two expressions and compare the answers. Which expression is greater? Why?

$$3^{5^2} \quad \text{and} \quad (3^5)^2$$

$3^{5^2} = 3^{25}$; $(3^5)^2 = 3^{10}$; the expression 3^{5^2} is greater than $(3^5)^2$.

Concept 2: The Properties $(ab)^m = a^m b^m$ and $\left(\frac{a}{b}\right)^m = \frac{a^m}{b^m}$

For Exercises 27–42, use the appropriate property to clear the parentheses. **(See Example 2.)**

27. $(5w)^2$ $25w^2$

28. $(4y)^3$ $64y^3$

29. $(srt)^4$ $s^4 r^4 t^4$

30. $(wxy)^6$ $w^6 x^6 y^6$

31. $\left(\frac{2}{r}\right)^4$ $\frac{16}{r^4}$

32. $\left(\frac{1}{t}\right)^8$ $\frac{1}{t^8}$

33. $\left(\frac{x}{y}\right)^5$ $\frac{x^5}{y^5}$

34. $\left(\frac{w}{z}\right)^7$ $\frac{w^7}{z^7}$

35. $(-3a)^4$ $81a^4$

36. $(2x)^5$ $32x^5$

37. $(-3abc)^3$ $-27a^3 b^3 c^3$

38. $(-5xyz)^2$ $25x^2 y^2 z^2$

39. $\left(-\frac{4}{x}\right)^3$ $-\frac{64}{x^3}$

40. $\left(-\frac{1}{w}\right)^4$ $\frac{1}{w^4}$

41. $\left(-\frac{a}{b}\right)^2$ $\frac{a^2}{b^2}$

42. $\left(-\frac{r}{s}\right)^3$ $-\frac{r^3}{s^3}$

Mixed Exercises

For Exercises 43–74, simplify. **(See Examples 3–5.)**

43. $(6u^2 v^4)^3$
$6^3 u^6 v^{12}$ or $216 u^6 v^{12}$

44. $(3a^5 b^2)^6$
$3^6 a^{30} b^{12}$ or $729 a^{30} b^{12}$

45. $5(x^2 y)^4$ $5x^8 y^4$

46. $18(u^3 v^4)^2$ $18u^6 v^8$

47. $(-h^4)^7$ $-h^{28}$

48. $(-k^6)^3$ $-k^{18}$

49. $(-m^2)^6$ m^{12}

50. $(-n^3)^8$ n^{24}

51. $\left(\frac{4}{rs^4}\right)^5$ $\frac{4^5}{r^5 s^{20}}$ or $\frac{1024}{r^5 s^{20}}$

52. $\left(\frac{2}{h^7 k}\right)^3$ $\frac{8}{h^{21} k^3}$

53. $\left(\frac{3p}{q^3}\right)^5$ $\frac{3^5 p^5}{q^{15}}$ or $\frac{243 p^5}{q^{15}}$

54. $\left(\frac{5x^2}{y^3}\right)^4$ $\frac{5^4 x^8}{y^{12}}$ or $\frac{625 x^8}{y^{12}}$

55. $\frac{y^8 (y^3)^4}{(y^2)^3}$ y^{14}

56. $\frac{(w^3)^2 (w^4)^5}{(w^4)^2}$ w^{18}

57. $(x^2)^5 (x^3)^7$ x^{31}

58. $(y^3)^4 (y^2)^5$ y^{22}

59. $(2a^2 b)^3 (5a^4 b^3)^2$ $200a^{14} b^9$

60. $(4c^3 d^5)^2 (3cd^3)^2$ $144c^8 d^{16}$

61. $(-2p^2 q^4)^4$ $16p^8 q^{16}$

62. $(-7x^4 y^5)^2$ $49x^8 y^{10}$

63. $(-m^7 n^3)^5$ $-m^{35} n^{15}$

64. $(-a^3 b^6)^7$ $-a^{21} b^{42}$

65. $\frac{(5a^3 b)^4 (a^2 b)^4}{(5ab)^2}$ $25a^{18} b^6$

66. $\frac{(6s^3)^2 (s^4 t^5)^2}{(3s^4 t^2)^2}$ $4s^6 t^6$

67. $\left(\frac{2c^3 d^4}{3c^2 d}\right)^2$ $\frac{4c^2 d^6}{9}$

68. $\left(\frac{x^3 y^5 z}{5xy^2}\right)^2$ $\frac{x^4 y^6 z^2}{25}$

69. $(2c^3 d^2)^5 \left(\frac{c^6 d^8}{4c^2 d}\right)^3$ $\frac{c^{27} d^{31}}{2}$

70. $\left(\frac{s^5 t^6}{2s^2 t}\right)^2 (10s^3 t^3)^2$ $25s^{12} t^{16}$

71. $\left(\frac{-3a^3 b}{c^2}\right)^3$ $-\frac{27a^9 b^3}{c^6}$

72. $\left(\frac{-4x^2}{y^4 z}\right)^3$ $-\frac{64x^6}{y^{12} z^3}$

73. $\frac{(-8b^6)^2 (b^3)^5}{4b}$ $16b^{26}$

74. $\frac{(-6a^2)^2 (a^3)^4}{9a}$ $4a^{15}$

Expanding Your Skills

For Exercises 75–82, simplify each expression. Assume that a, b, m, n, and x represent positive integers.

75. $(x^m)^2$ x^{2m}

76. $(y^3)^n$ y^{3n}

77. $(5a^{2n})^3$ $125a^{6n}$

78. $(3b^4)^m$ $3^m b^{4m}$

79. $\left(\frac{m^2}{n^3}\right)^b$ $\frac{m^{2b}}{n^{3b}}$

80. $\left(\frac{x^5}{y^3}\right)^m$ $\frac{x^{5m}}{y^{3m}}$

81. $\left(\frac{3a^3}{5b^4}\right)^n$ $\frac{3^n a^{3n}}{5^n b^{4n}}$

82. $\left(\frac{4m^6}{3n^2}\right)^x$ $\frac{4^x m^{6x}}{3^x n^{2x}}$

Definitions of b^0 and b^{-n}

Section 5.3

We have learned several rules that enable us to manipulate expressions containing *positive* integer exponents. In this section, we present definitions that can be used to simplify expressions with negative exponents or with an exponent of zero.

1. Definition of b^0

To begin, consider the following pattern.

$3^3 = 27$ $\searrow$ Divide by 3.
$3^2 = 9$ $\searrow$ Divide by 3.   As the exponents decrease by
$3^1 = 3$ $\searrow$ Divide by 3.   1, the resulting expressions are
$3^0 = 1$ $\longleftarrow$ divided by 3.

For the pattern to continue, we define $3^0 = 1$.

This pattern suggests that we should define an expression with a zero exponent as follows.

Concepts

1. Definition of b^0
2. Definition of b^{-n}
3. Properties of Integer Exponents: A Summary

Definition of b^0

Let b be a nonzero real number. Then, $b^0 = 1$.

Note: The value of 0^0 is not defined by this definition because the base b must not equal zero.

Example 1 Simplifying Expressions with a Zero Exponent

Simplify. Assume that $z \neq 0$.

a. 4^0 **b.** $(-4)^0$ **c.** -4^0

d. z^0 **e.** $-4z^0$ **f.** $(4z)^0$

Solution:

a. $4^0 = 1$ By definition

b. $(-4)^0 = 1$ By definition

c. $-4^0 = -1 \cdot 4^0 = -1 \cdot 1 = -1$ The exponent 0 applies only to 4.

d. $z^0 = 1$ By definition

e. $-4z^0 = -4 \cdot z^0 = -4 \cdot 1 = -4$ The exponent 0 applies only to z.

f. $(4z)^0 = 1$ The parentheses indicate that the exponent, 0, applies to both factors 4 and z.

Classroom Examples: p. 362, Exercises 14, 18, and 24

Skill Practice Evaluate the expressions. Assume that $x \neq 0$ and $y \neq 0$.

1. 7^0 **2.** $(-7)^0$ **3.** -5^0
4. y^0 **5.** $-2x^0$ **6.** $(2x)^0$

Answers
1. 1 **2.** 1 **3.** −1
4. 1 **5.** −2 **6.** 1

The definition of b^0 is consistent with the other properties of exponents learned thus far. For example, we know that $1 = \frac{5^3}{5^3}$. If we subtract exponents, the result is 5^0.

Subtract exponents.

$$1 = \frac{5^3}{5^3} = 5^{3-3} = 5^0 \qquad \text{Therefore, } 5^0 \text{ must be defined as 1.}$$

2. Definition of b^{-n}

To understand the concept of a *negative* exponent, consider the following pattern.

$3^3 = 27$

$\qquad$ Divide by 3.

$3^2 = 9$

$\qquad$ Divide by 3. $\quad$ As the exponents decrease by

$3^1 = 3$ $\qquad\qquad\qquad$ 1, the resulting expressions are

$\qquad$ Divide by 3. $\quad$ divided by 3.

$3^0 = 1$

$\qquad$ Divide by 3.

$3^{-1} = \frac{1}{3}$ $\qquad\qquad$ For the pattern to continue, we define $3^{-1} = \frac{1}{3^1} = \frac{1}{3}$.

$3^{-2} = \frac{1}{9}$ $\qquad\qquad$ For the pattern to continue, we define $3^{-2} = \frac{1}{3^2} = \frac{1}{9}$.

$3^{-3} = \frac{1}{27}$ $\qquad\qquad$ For the pattern to continue, we define $3^{-3} = \frac{1}{3^3} = \frac{1}{27}$.

This pattern suggests that $3^{-n} = \frac{1}{3^n}$ for all integers, n. In general, we have the following definition involving negative exponents.

Definition of b^{-n}

Let n be an integer and b be a nonzero real number. Then,

$$b^{-n} = \left(\frac{1}{b}\right)^n \quad \text{or} \quad \frac{1}{b^n}$$

The definition of b^{-n} implies that to evaluate b^{-n}, take the reciprocal of the base and change the sign of the exponent.

Change the sign of $\qquad\qquad\qquad\qquad$ Change the sign of

the exponent. $\qquad\qquad\qquad\qquad\qquad$ the exponent.

$$4^{-2} = \left(\frac{1}{4}\right)^2 \quad \text{or} \quad \frac{1}{4^2} \qquad\qquad \left(\frac{a}{b}\right)^{-n} = \left(\frac{b}{a}\right)^n$$

Reciprocal of the base $\qquad\qquad\qquad\qquad$ Reciprocal of the base

Example 2   **Simplifying Expressions with Negative Exponents**

Classroom Examples: pp. 362–363, Exercises 34 and 44

Simplify. Assume that $c \neq 0$.

a. c^{-3} **b.** 5^{-1} **c.** $(-3)^{-4}$

Solution:

a. $c^{-3} = \dfrac{1}{c^3}$ By definition

b. $5^{-1} = \dfrac{1}{5^1}$ By definition

       $= \dfrac{1}{5}$ Simplify.

Avoiding Mistakes

A negative exponent does *not* affect the sign of the base.

c. $(-3)^{-4} = \dfrac{1}{(-3)^4}$ The base is -3 and must be enclosed in parentheses.

         $= \dfrac{1}{81}$ Simplify. Note that $(-3)^4 = (-3)(-3)(-3)(-3) = 81$.

Skill Practice Simplify. Assume that $p \neq 0$.

7. p^{-4} **8.** 3^{-3} **9.** $(-5)^{-2}$

Example 3   **Simplifying Expressions with Negative Exponents**

Classroom Examples: pp. 362–363, Exercises 30 and 46

Simplify. Assume that $y \neq 0$.

a. $\left(\dfrac{1}{6}\right)^{-2}$ **b.** $\left(-\dfrac{3}{5}\right)^{-3}$ **c.** $\dfrac{1}{y^{-5}}$

Solution:

a. $\left(\dfrac{1}{6}\right)^{-2} = 6^2$ Take the reciprocal of the base, and change the sign of the exponent.

       $= 36$ Simplify.

b. $\left(-\dfrac{3}{5}\right)^{-3} = \left(-\dfrac{5}{3}\right)^3$ Take the reciprocal of the base, and change the sign of the exponent.

       $= -\dfrac{125}{27}$ Simplify.

c. $\dfrac{1}{y^{-5}} = \left(\dfrac{1}{y}\right)^{-5}$ Apply the power of a quotient rule.

       $= (y)^5$ Take the reciprocal of the base, and change the sign of the exponent.

       $= y^5$

TIP: Example 3(c) illustrates that $\dfrac{1}{b^{-n}} = b^n$, for $b \neq 0$.

Skill Practice Simplify. Assume that $w \neq 0$.

10. $\left(\dfrac{1}{3}\right)^{-1}$ **11.** $\left(-\dfrac{2}{5}\right)^{-2}$ **12.** $\dfrac{1}{w^{-7}}$

Answers

7. $\dfrac{1}{p^4}$ **8.** $\dfrac{1}{3^3}$ or $\dfrac{1}{27}$

9. $\dfrac{1}{(-5)^2}$ or $\dfrac{1}{25}$ **10.** 3

11. $\dfrac{25}{4}$ **12.** w^7

Classroom Examples: pp. 362–363, Exercises 36, 38, and 42

| Example 4 | **Simplifying Expressions with Negative Exponents** |

Simplify. Assume that $x \neq 0$.

a. $(5x)^{-3}$ **b.** $5x^{-3}$ **c.** $-5x^{-3}$

Solution:

a. $(5x)^{-3} = \left(\dfrac{1}{5x}\right)^3$ Take the reciprocal of the base, and change the sign of the exponent.

$= \dfrac{(1)^3}{(5x)^3}$ Apply the exponent of 3 to each factor within parentheses.

$= \dfrac{1}{125x^3}$ Simplify.

b. $5x^{-3} = 5 \cdot x^{-3}$ Note that the exponent, -3, applies only to x.

$= 5 \cdot \dfrac{1}{x^3}$ Rewrite x^{-3} as $\dfrac{1}{x^3}$.

$= \dfrac{5}{x^3}$ Multiply.

c. $-5x^{-3} = -5 \cdot x^{-3}$ Note that the exponent, -3, applies only to x, and that -5 is a coefficient.

$= -5 \cdot \dfrac{1}{x^3}$ Rewrite x^{-3} as $\dfrac{1}{x^3}$.

$= -\dfrac{5}{x^3}$ Multiply.

Skill Practice Simplify. Assume that $w \neq 0$.

13. $(2w)^{-4}$ **14.** $2w^{-4}$ **15.** $-2w^{-4}$

It is important to note that the definition of b^{-n} is consistent with the other properties of exponents learned thus far. For example, consider the expression

$$\frac{x^4}{x^7} = \frac{\cancel{x} \cdot \cancel{x} \cdot \cancel{x} \cdot \cancel{x}}{\cancel{x} \cdot \cancel{x} \cdot \cancel{x} \cdot \cancel{x} \cdot x \cdot x \cdot x} = \frac{1}{x^3}$$

Subtract exponents.   Hence, $x^{-3} = \dfrac{1}{x^3}$.

By subtracting exponents, we have $\dfrac{x^4}{x^7} = x^{4-7} = x^{-3}$.

3. Properties of Integer Exponents: A Summary

The definitions of b^0 and b^{-n} enable us to extend the properties of exponents. These are summarized in Table 5-1.

Answers

13. $\dfrac{1}{16w^4}$ **14.** $\dfrac{2}{w^4}$

15. $-\dfrac{2}{w^4}$

Table 5-1

<table>
<tr><td colspan="3" align="center">**Properties of Integer Exponents**
Assume that a and b are real numbers ($b \neq 0$) and that m and n represent integers.</td></tr>
<tr><td align="center">**Property**</td><td align="center">**Example**</td><td align="center">**Details/Notes**</td></tr>
<tr><td>Multiplication of Expressions with Like Bases

$b^m b^n = b^{m+n}$</td><td>$b^2 b^4 = b^{2+4} = b^6$</td><td>$b^2 b^4 = (b \cdot b)(b \cdot b \cdot b \cdot b) = b^6$</td></tr>
<tr><td>Division of Expressions with Like Bases

$\dfrac{b^m}{b^n} = b^{m-n}$</td><td>$\dfrac{b^5}{b^2} = b^{5-2} = b^3$</td><td>$\dfrac{b^5}{b^2} = \dfrac{\not{b} \cdot \not{b} \cdot b \cdot b \cdot b}{\not{b} \cdot \not{b}} = b^3$</td></tr>
<tr><td>The Power Rule

$(b^m)^n = b^{m \cdot n}$</td><td>$\left(b^4\right)^2 = b^{4 \cdot 2} = b^8$</td><td>$\left(b^4\right)^2 = (b \cdot b \cdot b \cdot b)(b \cdot b \cdot b \cdot b) = b^8$</td></tr>
<tr><td>Power of a Product

$(ab)^m = a^m b^m$</td><td>$(ab)^3 = a^3 b^3$</td><td>$(ab)^3 = (ab)(ab)(ab)$
$\quad = (a \cdot a \cdot a)(b \cdot b \cdot b) = a^3 b^3$</td></tr>
<tr><td>Power of a Quotient

$\left(\dfrac{a}{b}\right)^m = \dfrac{a^m}{b^m}$</td><td>$\left(\dfrac{a}{b}\right)^3 = \dfrac{a^3}{b^3}$</td><td>$\left(\dfrac{a}{b}\right)^3 = \left(\dfrac{a}{b}\right)\left(\dfrac{a}{b}\right)\left(\dfrac{a}{b}\right) = \dfrac{a \cdot a \cdot a}{b \cdot b \cdot b} = \dfrac{a^3}{b^3}$</td></tr>
<tr><td colspan="3" align="center">**Definitions**
Assume that b is a real number ($b \neq 0$) and that n represents an integer.</td></tr>
<tr><td align="center">**Definition**</td><td align="center">**Example**</td><td align="center">**Details/Notes**</td></tr>
<tr><td>$b^0 = 1$</td><td>$(4)^0 = 1$</td><td>Any nonzero quantity raised to the zero power equals 1.</td></tr>
<tr><td>$b^{-n} = \left(\dfrac{1}{b}\right)^n = \dfrac{1}{b^n}$</td><td>$b^{-5} = \left(\dfrac{1}{b}\right)^5 = \dfrac{1}{b^5}$</td><td>To simplify a negative exponent, take the reciprocal of the base and make the exponent positive.</td></tr>
</table>

Example 5 ## Simplifying Expressions with Exponents

Classroom Example: p. 363, Exercise 74

Simplify the expressions. Write the answers with positive exponents only. Assume all variables are nonzero.

a. $\dfrac{a^3 b^{-2}}{c^{-5}}$ **b.** $\dfrac{x^2 x^{-7}}{x^3}$ **c.** $\dfrac{z^2}{w^{-4} w^4 z^{-8}}$

Solution:

a. $\dfrac{a^3 b^{-2}}{c^{-5}}$

$= \dfrac{a^3}{1} \cdot \dfrac{b^{-2}}{1} \cdot \dfrac{1}{c^{-5}}$

$= \dfrac{a^3}{1} \cdot \dfrac{1}{b^2} \cdot \dfrac{c^5}{1}$ Simplify negative exponents.

$= \dfrac{a^3 c^5}{b^2}$ Multiply.

b. $\dfrac{x^2 x^{-7}}{x^3}$

$$= \dfrac{x^{2+(-7)}}{x^3}$$ Add the exponents in the numerator.

$$= \dfrac{x^{-5}}{x^3}$$ Simplify.

$$= x^{-5-3}$$ Subtract the exponents.

$$= x^{-8}$$

$$= \dfrac{1}{x^8}$$ Simplify the negative exponent.

c. $\dfrac{z^2}{w^{-4} w^4 z^{-8}}$

$$= \dfrac{z^2}{w^{-4+4} z^{-8}}$$ Add the exponents in the denominator.

$$= \dfrac{z^2}{w^0 z^{-8}}$$

$$= \dfrac{z^2}{(1) z^{-8}}$$ Recall that $w^0 = 1$.

$$= z^{2-(-8)}$$ Subtract the exponents.

$$= z^{10}$$ Simplify.

Skill Practice Simplify the expressions. Assume all variables are nonzero.

16. $\dfrac{x^{-6}}{y^4 z^{-8}}$ **17.** $\dfrac{x^3 x^{-8}}{x^4}$ **18.** $\dfrac{p^3}{w^7 w^{-7} z^{-2}}$

Classroom Examples: p. 363, Exercises 82 and 92

| Example 6 | **Simplifying Expressions with Exponents** |

Simplify the expressions. Write the answers with positive exponents only. Assume that all variables are nonzero.

a. $\left(-4ab^{-2}\right)^{-3}$ **b.** $\left(\dfrac{2p^{-4} q^3}{5p^2 q}\right)^{-2}$

Solution:

a. $\left(-4ab^{-2}\right)^{-3}$

$$= (-4)^{-3} a^{-3} (b^{-2})^{-3}$$ Apply the power rule of exponents.

$$= (-4)^{-3} a^{-3} b^6$$

$$= \dfrac{1}{(-4)^3} \cdot \dfrac{1}{a^3} \cdot b^6$$ Simplify the negative exponents.

$$= \dfrac{1}{-64} \cdot \dfrac{1}{a^3} \cdot b^6$$ Simplify.

$$= -\dfrac{b^6}{64a^3}$$ Multiply fractions.

Answers

16. $\dfrac{z^8}{y^4 x^6}$ **17.** $\dfrac{1}{x^9}$ **18.** $p^3 z^2$

b. $\left(\dfrac{2p^{-4}q^3}{5p^2q}\right)^{-2}$ First simplify within the parentheses.

$= \left(\dfrac{2p^{-4-2}q^{3-1}}{5}\right)^{-2}$ Divide expressions with like bases by subtracting exponents.

$= \left(\dfrac{2p^{-6}q^2}{5}\right)^{-2}$ Simplify.

$= \dfrac{(2p^{-6}q^2)^{-2}}{(5)^{-2}}$ Apply the power rule of a quotient.

$= \dfrac{2^{-2}(p^{-6})^{-2}(q^2)^{-2}}{5^{-2}}$ Apply the power rule of a product.

$= \dfrac{2^{-2}p^{12}q^{-4}}{5^{-2}}$ Simplify.

$= \dfrac{5^2p^{12}}{2^2q^4}$ Simplify the negative exponents.

$= \dfrac{25p^{12}}{4q^4}$ Simplify.

> **TIP:** For Example 6(b), the power rule of exponents can be performed first. In that case, the second step would be
>
> $$\dfrac{2^{-2}p^8q^{-6}}{5^{-2}p^{-4}q^{-2}}$$

Skill Practice   Simplify the expressions. Assume all variables are nonzero.

19. $(-5x^{-2}y^3)^{-2}$ **20.** $\left(\dfrac{3x^{-3}y^{-2}}{4xy^{-3}}\right)^{-2}$

| Example 7 | **Simplifying an Expression with Exponents** |

Classroom Example: p. 363, Exercise 96

Simplify the expression $2^{-1} + 3^{-1} + 5^0$. Write the answer with positive exponents only.

Solution:

$2^{-1} + 3^{-1} + 5^0$

$= \dfrac{1}{2} + \dfrac{1}{3} + 1$ Simplify negative exponents. Simplify $5^0 = 1$.

$= \dfrac{3}{6} + \dfrac{2}{6} + \dfrac{6}{6}$ The least common denominator is 6.

$= \dfrac{11}{6}$ Simplify.

Skill Practice   Simplify the expressions.

21. $2^{-1} + 4^{-2} + 3^0$

Answers

19. $\dfrac{x^4}{25y^6}$ **20.** $\dfrac{16x^8}{9y^2}$ **21.** $\dfrac{25}{16}$

Section 5.3 Practice Exercises

For this exercise set, assume all variables represent nonzero real numbers.

For additional exercises, see Classroom Activities 5.3A–5.3B in the *Student's Resource Manual* at www.mhhe.com/moh.

Study Skills Exercise

To help you remember the properties of exponents, write them on 3×5 cards. On each card, write a property on one side and an example using that property on the other side. Keep these cards with you, and when you have a spare moment (such as waiting at the doctor's office), pull out these cards and go over the properties.

Vocabulary and Key Concepts

1. a. The expression b^0 is defined to be _____ 1 _____ provided that $b \neq 0$.

   b. The expression b^{-n} is defined as _____ provided that $b \neq 0$. $\left(\frac{1}{b}\right)^n$ or $\frac{1}{b^n}$

Review Exercises

For Exercises 2–9, simplify.

2. $b^3 b^8$ b^{11}

3. $c^7 c^2$ c^9

4. $\dfrac{x^6}{x^2}$ x^4

5. $\dfrac{y^9}{y^8}$ y

6. $\dfrac{9^4 \cdot 9^8}{9}$ 9^{11}

7. $\dfrac{3^{14}}{3^3 \cdot 3^5}$ 3^6 or 729

8. $(6ab^3c^2)^5$ $6^5 a^5 b^{15} c^{10}$ or $7776 a^5 b^{15} c^{10}$

9. $(7w^7 z^2)^4$ $7^4 w^{28} z^8$ or $2401 w^{28} z^8$

Concept 1: Definition of b^0

10. Simplify.

   a. 8^0 1   **b.** $\dfrac{8^4}{8^4}$ 1

11. Simplify.

   a. d^0 1   **b.** $\dfrac{d^3}{d^3}$ 1

12. Simplify.

   a. m^0 1   **b.** $\dfrac{m^5}{m^5}$ 1

For Exercises 13–24, simplify. **(See Example 1.)**

13. p^0 1

14. k^0 1

15. 5^0 1

16. 2^0 1

17. -4^0 -1

18. -1^0 -1

19. $(-6)^0$ 1

20. $(-2)^0$ 1

21. $(8x)^0$ 1

22. $(-3y^3)^0$ 1

23. $-7x^0$ -7

24. $6y^0$ 6

Concept 2: Definition of b^{-n}

25. Simplify and write the answers with positive exponents.

   a. t^{-5} $\dfrac{1}{t^5}$ **b.** $\dfrac{t^3}{t^8}$ $\dfrac{1}{t^5}$

26. Simplify and write the answers with positive exponents.

   a. 4^{-3} $\dfrac{1}{4^3}$ **b.** $\dfrac{4^2}{4^5}$ $\dfrac{1}{4^3}$

For Exercises 27–46, simplify. **(See Examples 2–4.)**

27. $\left(\dfrac{2}{7}\right)^{-3}$ $\dfrac{343}{8}$

28. $\left(\dfrac{5}{4}\right)^{-1}$ $\dfrac{4}{5}$

29. $\left(-\dfrac{1}{5}\right)^{-2}$ 25

30. $\left(-\dfrac{1}{3}\right)^{-3}$ -27

31. a^{-3} $\dfrac{1}{a^3}$

32. c^{-5} $\dfrac{1}{c^5}$

33. 12^{-1} $\dfrac{1}{12}$

34. 4^{-2} $\dfrac{1}{16}$

35. $(4b)^{-2}$ $\dfrac{1}{16b^2}$

36. $(3z)^{-1}$ $\dfrac{1}{3z}$

37. $6x^{-2}$ $\dfrac{6}{x^2}$

38. $7y^{-1}$ $\dfrac{7}{y}$

✏ Writing   ↔ Translating Expression   △ Geometry   🖩 Scientific Calculator   Video

39. $(-8)^{-2}$ $\dfrac{1}{64}$

40. -8^{-2} $-\dfrac{1}{64}$

41. $-3y^{-4}$ $-\dfrac{3}{y^4}$

42. $-6a^{-2}$ $-\dfrac{6}{a^2}$

43. $(-t)^{-3}$ $-\dfrac{1}{t^3}$

44. $(-r)^{-5}$ $-\dfrac{1}{r^5}$

45. $\dfrac{1}{a^{-5}}$ a^5

46. $\dfrac{1}{b^{-6}}$ b^6

Concept 3: Properties of Integer Exponents: A Summary

For Exercises 47–50, correct the statement.

47. $\dfrac{x^4}{x^{-6}} = x^{4-6} = x^{-2}$ $\dfrac{x^4}{x^{-6}} = x^{4-(-6)} = x^{10}$

48. $\dfrac{y^5}{y^{-3}} = y^{5-3} = y^2$ $\dfrac{y^5}{y^{-3}} = y^{5-(-3)} = y^8$

49. $2a^{-3} = \dfrac{1}{2a^3}$ $2a^{-3} = 2 \cdot \dfrac{1}{a^3} = \dfrac{2}{a^3}$

50. $5b^{-2} = \dfrac{1}{5b^2}$ $5b^{-2} = 5 \cdot \dfrac{1}{b^2} = \dfrac{5}{b^2}$

Mixed Exercises

For Exercises 51–94, simplify each expression. Write the answer with positive exponents only. **(See Examples 5–6.)**

51. $x^{-8}x^4$ $\dfrac{1}{x^4}$

52. $s^5 s^{-6}$ $\dfrac{1}{s}$

53. $a^{-8}a^8$ 1

54. $q^3 q^{-3}$ 1

55. $y^{17}y^{-13}$ y^4

56. $b^{20}b^{-14}$ b^6

57. $(m^{-6}n^9)^3$ $\dfrac{n^{27}}{m^{18}}$

58. $(c^4 d^{-5})^{-2}$ $\dfrac{d^{10}}{c^8}$

59. $(-3j^{-5}k^6)^4$ $\dfrac{81k^{24}}{j^{20}}$

60. $(6xy^{-11})^{-3}$ $\dfrac{y^{33}}{6^3 x^3}$ or $\dfrac{y^{33}}{216x^3}$

61. $\dfrac{p^3}{p^9}$ $\dfrac{1}{p^6}$

62. $\dfrac{q^2}{q^{10}}$ $\dfrac{1}{q^8}$

63. $\dfrac{r^{-5}}{r^{-2}}$ $\dfrac{1}{r^3}$

64. $\dfrac{u^{-2}}{u^{-6}}$ u^4

65. $\dfrac{a^2}{a^{-6}}$ a^8

66. $\dfrac{p^3}{p^{-5}}$ p^8

67. $\dfrac{y^{-2}}{y^6}$ $\dfrac{1}{y^8}$

68. $\dfrac{s^{-4}}{s^3}$ $\dfrac{1}{s^7}$

69. $\dfrac{7^3}{7^2 \cdot 7^8}$ $\dfrac{1}{7^7}$

70. $\dfrac{3^4 \cdot 3}{3^7}$ $\dfrac{1}{9}$

71. $\dfrac{a^2 a}{a^3}$ 1

72. $\dfrac{t^5}{t^2 t^3}$ 1

73. $\dfrac{a^{-1}b^2}{a^3 b^8}$ $\dfrac{1}{a^4 b^6}$

74. $\dfrac{k^{-4}h^{-1}}{k^6 h}$ $\dfrac{1}{k^{10}h^2}$

75. $\dfrac{w^{-8}(w^2)^{-5}}{w^3}$ $\dfrac{1}{w^{21}}$

76. $\dfrac{p^2 p^{-7}}{(p^2)^3}$ $\dfrac{1}{p^{11}}$

77. $\dfrac{3^{-2}}{3}$ $\dfrac{1}{27}$

78. $\dfrac{5^{-1}}{5}$ $\dfrac{1}{25}$

79. $\left(\dfrac{p^{-1}q^5}{p^{-6}}\right)^0$ 1

80. $\left(\dfrac{ab^{-4}}{a^{-5}}\right)^0$ 1

81. $(8x^3 y^0)^{-2}$ $\dfrac{1}{64x^6}$

82. $(3u^2 v^0)^{-3}$ $\dfrac{1}{27u^6}$

83. $(-8y^{-12})(2y^{16}z^{-2})$ $\dfrac{16y^4}{z^2}$

84. $(5p^{-2}q^5)(-2p^{-4}q^{-1})$ $-\dfrac{10q^4}{p^6}$

85. $\dfrac{-18a^{10}b^6}{108a^{-2}b^6}$ $-\dfrac{a^{12}}{6}$

86. $\dfrac{-35x^{-4}y^{-3}}{-21x^2 y^{-3}}$ $\dfrac{5}{3x^6}$

87. $\dfrac{(-4c^{12}d^7)^2}{(5c^{-3}d^{10})^{-1}}$ $80c^{21}d^{24}$

88. $\dfrac{(s^3 t^{-2})^4}{(3s^{-4}t^6)^{-2}}$ $9s^4 t^4$

89. $\dfrac{(2x^3 y^2)^{-3}}{(3x^2 y^4)^{-2}}$ $\dfrac{9y^2}{8x^5}$

90. $\dfrac{(5p^4 q)^{-3}}{(p^3 q^5)^{-4}}$ $\dfrac{q^{17}}{125}$

91. $\left(\dfrac{5cd^{-3}}{10d^5}\right)^{-2}$ $\dfrac{4d^{16}}{c^2}$

92. $\left(\dfrac{4m^{10}n^4}{2m^{12}n^{-2}}\right)^{-1}$ $\dfrac{m^2}{2n^6}$

93. $(2xy^3)\left(\dfrac{9xy}{4x^3 y^2}\right)$ $\dfrac{9y^2}{2x}$

94. $(-3a^3)\left(\dfrac{ab}{27a^4 b^2}\right)$ $-\dfrac{1}{9b}$

For Exercises 95–102, simplify. **(See Example 7.)**

95. $5^{-1} + 2^{-2}$ $\dfrac{9}{20}$

96. $4^{-2} + 8^{-1}$ $\dfrac{3}{16}$

97. $10^0 - 10^{-1}$ $\dfrac{9}{10}$

98. $3^0 - 3^{-2}$ $\dfrac{8}{9}$

99. $2^{-2} + 1^{-2}$ $\dfrac{5}{4}$

100. $4^{-1} + 8^{-1}$ $\dfrac{3}{8}$

101. $4 \cdot 5^0 - 2 \cdot 3^{-1}$ $\dfrac{10}{3}$

102. $2 \cdot 4^0 - 3 \cdot 4^{-1}$ $\dfrac{5}{4}$

Writing   Translating Expression   Geometry   Scientific Calculator   Video

Expanding Your Skills

For Exercises 103–106, determine the missing exponent.

103. $\dfrac{y^4 y^{\square}}{y^{-2}} = y^8$ 2

104. $\dfrac{x^4 x^{\square}}{x^{-1}} = x^9$ 4

105. $\dfrac{w^{-9}}{w^{\square}} = w^2$ −11

106. $\dfrac{a^{-2}}{a^{\square}} = a^6$ −8

Problem Recognition Exercises

Properties of Exponents

For Exercises 1–40, simplify completely. Assume that all variables represent nonzero real numbers.

1. $t^3 t^5$ t^8

2. $2^3 2^5$ 2^8 or 256

3. $\dfrac{y^7}{y^2}$ y^5

4. $\dfrac{p^9}{p^3}$ p^6

5. $(r^2 s^4)^2$ $r^4 s^8$

6. $(ab^3 c^2)^3$ $a^3 b^9 c^6$

7. $\dfrac{w^4}{w^{-2}}$ w^6

8. $\dfrac{m^{-14}}{m^2}$ $\dfrac{1}{m^{16}}$

9. $\dfrac{y^{-7} x^4}{z^{-3}}$ $\dfrac{x^4 z^3}{y^7}$

10. $\dfrac{a^3 b^{-6}}{c^{-8}}$ $\dfrac{a^3 c^8}{b^6}$

11. $\dfrac{x^4 x^{-3}}{x^{-5}}$ x^6

12. $\dfrac{y^{-4}}{y^7 y^{-1}}$ $\dfrac{1}{y^{10}}$

13. $\dfrac{t^{-2} t^4}{t^8 t^{-1}}$ $\dfrac{1}{t^5}$

14. $\dfrac{w^8 w^{-5}}{w^{-2} w^{-2}}$ w^7

15. $\dfrac{1}{p^{-6} p^{-8} p^{-1}}$ p^{15}

16. $p^6 p^8 p$ p^{15}

17. $\dfrac{v^9}{v^{11}}$ $\dfrac{1}{v^2}$

18. $(c^5 d^4)^{10}$ $c^{50} d^{40}$

19. $\left(\dfrac{1}{2}\right)^{-1} + \left(\dfrac{1}{3}\right)^0$ 3

20. $\left(\dfrac{1}{4}\right)^0 - \left(\dfrac{1}{5}\right)^{-1}$ −4

21. $(2^5 b^{-3})^{-3}$ $\dfrac{b^9}{2^{15}}$

22. $(3^{-2} y^3)^{-2}$ $\dfrac{81}{y^6}$

23. $\left(\dfrac{3x}{2y}\right)^{-4}$ $\dfrac{16 y^4}{81 x^4}$

24. $\left(\dfrac{6c}{5d^3}\right)^{-2}$ $\dfrac{25 d^6}{36 c^2}$

25. $(3ab^2)(a^2 b)^3$ $3a^7 b^5$

26. $(4x^2 y^3)^3 (xy^2)$ $64 x^7 y^{11}$

27. $\left(\dfrac{xy^2}{x^3 y}\right)^4$ $\dfrac{y^4}{x^8}$

28. $\left(\dfrac{a^3 b}{a^5 b^3}\right)^5$ $\dfrac{1}{a^{10} b^{10}}$

29. $\dfrac{(t^{-2})^3}{t^{-4}}$ $\dfrac{1}{t^2}$

30. $\dfrac{(p^3)^{-4}}{p^{-5}}$ $\dfrac{1}{p^7}$

31. $\left(\dfrac{2w^2 x^3}{3y^0}\right)^3$ $\dfrac{8 w^6 x^9}{27}$

32. $\left(\dfrac{5 a^0 b^4}{4 c^3}\right)^2$ $\dfrac{25 b^8}{16 c^6}$

33. $\dfrac{q^3 r^{-2}}{s^{-1} t^5}$ $\dfrac{q^3 s}{r^2 t^5}$

34. $\dfrac{n^{-3} m^2}{p^{-3} q^{-1}}$ $\dfrac{m^2 p^3 q}{n^3}$

35. $\dfrac{(y^{-3})^2 (y^5)}{(y^{-3})^{-4}}$ $\dfrac{1}{y^{13}}$

36. $\dfrac{(w^2)^{-4} (w^{-2})}{(w^5)^{-4}}$ w^{10}

37. $\left(\dfrac{-2 a^2 b^{-3}}{a^{-4} b^{-5}}\right)^{-3}$ $-\dfrac{1}{8 a^{18} b^6}$

38. $\left(\dfrac{-3 x^{-4} y^3}{2 x^5 y^{-2}}\right)^{-2}$ $\dfrac{4 x^{18}}{9 y^{10}}$

39. $(5h^{-2} k^0)^3 (5k^{-2})^{-4}$ $\dfrac{k^8}{5 h^6}$

40. $(6 m^3 n^{-5})^{-4} (6 m^0 n^{-2})^5$ $\dfrac{6 n^{10}}{m^{12}}$

Writing   ◀▶ Translating Expression   Geometry   Scientific Calculator   Video

Scientific Notation

1. Writing Numbers in Scientific Notation

In many applications in mathematics, it is necessary to work with very large or very small numbers. For example, the number of movie tickets sold in the United States recently is estimated to be 1,500,000,000. The weight of a flea is approximately 0.00066 lb. To avoid writing numerous zeros in very large or small numbers, scientific notation was devised as a shortcut.

The principle behind scientific notation is to use a power of 10 to express the magnitude of the number. For example, the numbers 4000 and 0.07 can be written as:

$$4000 = 4 \times 1000 = 4 \times 10^3$$

$$0.07 = 7.0 \times 0.01 = 7.0 \times 10^{-2} \qquad \text{Note that } 10^{-2} = \frac{1}{100} = 0.01$$

Definition of Scientific Notation

A positive number expressed in the form: $a \times 10^n$, where $1 \le a < 10$ and n is an integer, is said to be written in **scientific notation**.

To write a positive number in scientific notation, we apply the following guidelines:

1. Move the decimal point so that its new location is to the right of the first nonzero digit. The number should now be greater than or equal to 1 but less than 10. Count the number of places that the decimal point is moved.

2. If the original number is *large* (greater than or equal to 10), use the number of places the decimal point was moved as a *positive* power of 10.

$$450,000 = 4.5 \times 100,000 = 4.5 \times 10^5$$

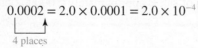
5 places

3. If the original number is *small* (between 0 and 1), use the number of places the decimal point was moved as a *negative* power of 10.

$$0.0002 = 2.0 \times 0.0001 = 2.0 \times 10^{-4}$$

4 places

4. If the original number is greater than or equal to 1 but less than 10, use 0 as the power of 10.

$7.592 = 7.592 \times 10^0$ *Note*: A number between 1 and 10 is seldom written in scientific notation.

5. If the original number is negative, then $-10 < a \le -1$.

$$-450,000 = -4.5 \times 100,000 = -4.5 \times 10^5$$

5 places

Classroom Examples: p. 369,
Exercises 22 and 26

Example 1 **Writing Numbers in Scientific Notation**

Write the numbers in scientific notation.

   a. 53,000   **b.** 0.00053

Solution:

   a. $53,000. = 5.3 \times 10^4$ To write 53,000 in scientific notation, the decimal point must be moved four places to the left. Because 53,000 is larger than 10, a *positive* power of 10 is used.

   b. $0.00053 = 5.3 \times 10^{-4}$ To write 0.00053 in scientific notation, the decimal point must be moved four places to the right. Because 0.00053 is between 0 and 1, a *negative* power of 10 is used.

Skill Practice Write the numbers in scientific notation.

   1. 175,000,000   **2.** 0.000005

Classroom Examples: p. 369,
Exercises 30 and 32

Example 2 **Writing Numbers in Scientific Notation**

Write the numbers in scientific notation.

   a. The number of movie tickets sold in the United States for a recent year is estimated to be 1,500,000,000.

   b. The weight of a flea is approximately 0.00066 lb.

   c. The temperature on a January day in Fargo dropped to −43°F.

   d. A bench is 8.2 ft long.

Solution:

   a. $1,500,000,000 = 1.5 \times 10^9$ **b.** $0.00066 \text{ lb} = 6.6 \times 10^{-4} \text{ lb}$

   c. $-43°\text{F} = -4.3 \times 10^1 \text{ °F}$ **d.** $8.2 \text{ ft} = 8.2 \times 10^0 \text{ ft}$

Skill Practice Write the numbers in scientific notation.

   3. In the year 2011, the population of the Earth was approximately 7,000,000,000.

   4. The weight of a grain of salt is approximately 0.000002 ounce.

2. Writing Numbers in Standard Form

Classroom Examples: p. 370,
Exercises 40 and 50

Example 3 **Writing Numbers in Standard Form**

Write the numbers in standard form.

   a. The mass of a proton is approximately 1.67×10^{-24} g.

   b. The "nearby" star Vega is approximately 1.552×10^{14} miles from Earth.

Solution:

   a. $1.67 \times 10^{-24} \text{ g} = 0.000\ 000\ 000\ 000\ 000\ 000\ 000\ 001\ 67 \text{ g}$

   Because the power of 10 is negative, the value of 1.67×10^{-24} is a decimal number between 0 and 1. Move the decimal point 24 places to the *left*.

   b. $1.552 \times 10^{14} \text{ miles} = 155,200,000,000,000 \text{ miles}$

   Because the power of 10 is a positive integer, the value of 1.552×10^{14} is a large number greater than 10. Move the decimal point 14 places to the *right*.

Answers
1. 1.75×10^8 **2.** 5×10^{-6}
3. 7×10^9 **4.** 2×10^{-6} oz

Skill Practice Write the numbers in standard form.

5. The probability of winning the California Super Lotto Jackpot is 5.5×10^{-8}.
6. The Sun's mass is 2×10^{30} kilograms.

3. Multiplying and Dividing Numbers in Scientific Notation

To multiply or divide two numbers in scientific notation, use the commutative and associative properties of multiplication to group the powers of 10. For example:

$$400 \times 2000 = (4 \times 10^2)(2 \times 10^3) = (4 \cdot 2) \times (10^2 \cdot 10^3) = 8 \times 10^5$$

$$\frac{0.00054}{150} = \frac{5.4 \times 10^{-4}}{1.5 \times 10^2} = \left(\frac{5.4}{1.5}\right) \times \left(\frac{10^{-4}}{10^2}\right) = 3.6 \times 10^{-6}$$

Example 4	**Multiplying and Dividing Numbers in Scientific Notation**

Classroom Examples: p. 370, Exercises 54 and 68

Multiply or divide as indicated.

a. $(8.7 \times 10^4)(2.5 \times 10^{-12})$ **b.** $\dfrac{4.25 \times 10^{13}}{8.5 \times 10^{-2}}$

Solution:

a. $(8.7 \times 10^4)(2.5 \times 10^{-12})$

$= (8.7 \cdot 2.5) \times (10^4 \cdot 10^{-12})$ Regroup factors using the commutative and associative properties of multiplication.

$= 21.75 \times 10^{-8}$ The number 21.75 is not in proper scientific notation because 21.75 is not between 1 and 10.

$= (2.175 \times 10^1) \times 10^{-8}$ Rewrite 21.75 as 2.175×10^1.

$= 2.175 \times (10^1 \times 10^{-8})$ Associative property of multiplication

$= 2.175 \times 10^{-7}$ Simplify.

b. $\dfrac{4.25 \times 10^{13}}{8.5 \times 10^{-2}}$

$= \left(\dfrac{4.25}{8.5}\right) \times \left(\dfrac{10^{13}}{10^{-2}}\right)$ Regroup factors using the commutative and associative properties.

$= 0.5 \times 10^{15}$ The number 0.5×10^{15} is not in proper scientific notation because 0.5 is not between 1 and 10.

$= (5.0 \times 10^{-1}) \times 10^{15}$ Rewrite 0.5 as 5.0×10^{-1}.

$= 5.0 \times (10^{-1} \times 10^{15})$ Associative property of multiplication

$= 5.0 \times 10^{14}$ Simplify.

Skill Practice Multiply or divide as indicated.

7. $(7 \times 10^5)(5 \times 10^3)$ 8. $\dfrac{1 \times 10^{-2}}{4 \times 10^{-7}}$

Answers
5. 0.000 000 055
6. 2,000,000,000,000,000,000,000,000,000,000
7. 3.5×10^9
8. 2.5×10^4

Calculator Connections

Topic: Using Scientific Notation

Both scientific and graphing calculators can perform calculations involving numbers written in scientific notation. Most calculators use an $\boxed{\text{EE}}$ key or an $\boxed{\text{EXP}}$ key to enter the power of 10.

Scientific Calculator

Enter: 2.7 $\boxed{\text{EE}}$ 5 $\boxed{=}$ or   2.7 $\boxed{\text{EXP}}$ 5 $\boxed{=}$ **Result:** | 270000 |

Enter: 7.1 $\boxed{\text{EE}}$ 3 $\boxed{+\circ-}$ $\boxed{=}$ or   7.1 $\boxed{\text{EXP}}$ 3 $\boxed{+\circ-}$ $\boxed{=}$ **Result:** | 0.0071 |

Graphing Calculator

```
2.7E5
            270000
7.1E-3
             .0071
```

We recommend that you use parentheses to enclose each number written in scientific notation when performing calculations. Try using your calculator to perform the calculations from Example 4.

a. $(8.7 \times 10^4)(2.5 \times 10^{-12})$ b. $\dfrac{4.25 \times 10^{13}}{8.5 \times 10^{-2}}$

Scientific Calculator

Enter: $\boxed{(}$ 8.7 $\boxed{\text{EE}}$ 4 $\boxed{)}$ $\boxed{\times}$ $\boxed{(}$ 2.5 $\boxed{\text{EE}}$ 12 $\boxed{+\circ-}$ $\boxed{)}$ $\boxed{=}$ **Result:** | 0.000000218 |

Enter: $\boxed{(}$ 4.25 $\boxed{\text{EE}}$ 13 $\boxed{)}$ $\boxed{\div}$ $\boxed{(}$ 8.5 $\boxed{\text{EE}}$ 2 $\boxed{+\circ-}$ $\boxed{)}$ $\boxed{=}$ **Result:** | 5E14 |

Notice that the answer to part (b) is shown on the calculator in scientific notation. The calculator does not have enough room to display 14 zeros. Also notice that the calculator rounds the answer to part (a). The exact answer is 2.175×10^{-7} or 0.0000002175.

Graphing Calculator

```
(8.7E4)*(2.5E-12
)
          2.175E-7
(4.25E13)/(8.5E-
2)
             5E14
```

Avoiding Mistakes

A display of 5E14 on a calculator does not mean 5^{14}. It is scientific notation and means 5×10^{14}.

Calculator Exercises

Use a calculator to perform the indicated operations:

1. $(5.2 \times 10^6)(4.6 \times 10^{-3})$ 23,920

2. $(2.19 \times 10^{-8})(7.84 \times 10^{-4})$ 1.71696 × 10^{-11}

3. $\dfrac{4.76 \times 10^{-5}}{2.38 \times 10^9}$ 2 × 10^{-14}

4. $\dfrac{8.5 \times 10^4}{4.0 \times 10^{-1}}$ 212,500

5. $\dfrac{(9.6 \times 10^7)(4.0 \times 10^{-3})}{2.0 \times 10^{-2}}$ 19,200,000

6. $\dfrac{(5.0 \times 10^{-12})(6.4 \times 10^{-5})}{(1.6 \times 10^{-8})(4.0 \times 10^2)}$ 5 × 10^{-11}

Section 5.4   Practice Exercises

Vocabulary and Key Concepts

For additional exercises, see Classroom Activities 5.4A–5.4B in the
Student's Resource Manual at www.mhhe.com/moh.

1. A positive number expressed in the form $a \times 10^n$, where $1 \leq a < 10$ and n is an integer is said to be written
   in ___scientific___ ___notation___.

Review Exercises

For Exercises 2–13, simplify each expression. Assume all variables represent nonzero real numbers.

2. $a^3 a^{-4}$ $\frac{1}{a}$

3. $b^5 b^8$ b^{13}

4. $10^3 \cdot 10^{-4}$ $\frac{1}{10}$

5. $10^5 \cdot 10^8$ 10^{13}

6. $\dfrac{x^3}{x^6}$ $\frac{1}{x^3}$

7. $\dfrac{y^2}{y^7}$ $\frac{1}{y^5}$

8. $(c^4 d^2)^3$ $c^{12} d^6$

9. $(x^5 y^{-3})^4$ $\frac{x^{20}}{y^{12}}$

10. $\dfrac{z^9 z^4}{z^3}$ z^{10}

11. $\dfrac{w^{-2} w^5}{w^{-1}}$ w^4

12. $\dfrac{10^9 \cdot 10^4}{10^3}$ 10^{10}

13. $\dfrac{10^{-2} \cdot 10^5}{10^{-1}}$ 10^4

Concept 1: Writing Numbers in Scientific Notation

14. Explain how scientific notation might be valuable in studying astronomy.   Answers may vary.
    For example: Scientific notation would be helpful in writing the distance between Earth and the stars or planets.

15. Explain how you would write the number 0.000 000 000 23 in scientific notation.   Move the decimal point
    between the 2 and 3 and multiply by 10^{-10}; 2.3×10^{-10}.

16. Explain how you would write the number 23,000,000,000,000 in scientific notation.   Move the decimal point
    between the 2 and 3 and multiply by 10^{13}; 2.3×10^{13}.

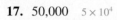

For Exercises 17–28, write the number in scientific notation. **(See Example 1.)**

17. 50,000 5×10^4

18. 900,000 9×10^5

19. 208,000 2.08×10^5

20. 420,000,000 4.2×10^8

21. 6,010,000 6.01×10^6

22. 75,000 7.5×10^4

23. 0.000008 8×10^{-6}

24. 0.003 3×10^{-3}

25. 0.000125 1.25×10^{-4}

26. 0.00000025 2.5×10^{-7}

27. 0.006708 6.708×10^{-3}

28. 0.02004 2.004×10^{-2}

For Exercises 29–34, write each number in scientific notation. **(See Example 2.)**

29. The mass of a proton is approximately
    0.000 000 000 000 000 000 000 000 0017 g. 1.7×10^{-24} g

30. The total combined salaries of the president, vice
    president, senators, and representatives of the
    United States federal government is approximately
    \$85,000,000. $\$8.5 \times 10^7$

31. A renowned foundation has over
    \$27,000,000,000 from which it makes
    contributions to global charities. $\$2.7 \times 10^{10}$

32. One gram is equivalent to 0.0035 oz.
    3.5×10^{-3} oz

33. One of the world's largest tanker disasters spilled
    68,000,000 gal of oil off Portsall, France, causing
    widespread environmental damage over 100 miles
    of Brittany coast.
    6.8×10^7 gal; 1×10^2 miles

34. The human heart pumps about 2100 gal of blood
    per day. That means that it pumps approximately
    767,000 gal per year. 2.1×10^3 gal; 7.67×10^5 gal

Writing   Translating Expression   Geometry   Scientific Calculator   Video

Concept 2: Writing Numbers in Standard Form

35. Explain how you would write the number 3.1×10^{-9} in standard form.   Move the decimal point nine places to the left; 0.000 000 0031.

36. Explain how you would write the number 3.1×10^{9} in standard form.   Move the decimal point nine places to the right; 3,100,000,000.

For Exercises 37–52, write each number in standard form. **(See Example 3.)**

37. 5×10^{-5} 0.00005

38. 2×10^{-7} 0.0000002

39. 2.8×10^{3} 2800

40. 9.1×10^{6} 9,100,000

41. 6.03×10^{-4} 0.000603

42. 7.01×10^{-3} 0.00701

43. 2.4×10^{6} 2,400,000

44. 3.1×10^{4} 31,000

45. 1.9×10^{-2} 0.019

46. 2.8×10^{-6} 0.0000028

47. 7.032×10^{3} 7032

48. 8.205×10^{2} 820.5

49. One picogram (pg) is equal to 1×10^{-12} g.   0.000 000 000 001 g

50. A nanometer (nm) is approximately 3.94×10^{-8} in.   0.000 000 0394 in.

51. A normal diet contains between 1.6×10^{3} Cal and 2.8×10^{3} Cal per day.   1600 Cal and 2800 Cal

52. The total land area of Texas is approximately 2.62×10^{5} square miles.   262,000 square miles

© Ryan McVay/Getty Images RF

Concept 3: Multiplying and Dividing Numbers in Scientific Notation

For Exercises 53–72, multiply or divide as indicated. Write the answers in scientific notation. **(See Example 4.)**

53. $(2.5 \times 10^{6})(2 \times 10^{-2})$ 5×10^{4}

54. $(2 \times 10^{-7})(3 \times 10^{13})$ 6×10^{6}

55. $(1.2 \times 10^{4})(3 \times 10^{7})$ 3.6×10^{11}

56. $(3.2 \times 10^{-3})(2.5 \times 10^{8})$ 8×10^{5}

57. $\dfrac{7.7 \times 10^{6}}{3.5 \times 10^{2}}$ 2.2×10^{4}

58. $\dfrac{9.5 \times 10^{11}}{1.9 \times 10^{3}}$ 5×10^{8}

59. $\dfrac{9 \times 10^{-6}}{4 \times 10^{7}}$ 2.25×10^{-13}

60. $\dfrac{7 \times 10^{-2}}{5 \times 10^{9}}$ 1.4×10^{-11}

61. $80,000,000,000 \times 4000$ 3.2×10^{14}

62. 0.0006×0.03 1.8×10^{-5}

63. $(3.2 \times 10^{-4})(7.6 \times 10^{-7})$ 2.432×10^{-10}

64. $(5.9 \times 10^{12})(3.6 \times 10^{9})$ 2.124×10^{22}

65. $\dfrac{210,000,000,000}{0.007}$ 3×10^{13}

66. $\dfrac{160,000,000,000,000}{0.00008}$ 2×10^{18}

67. $\dfrac{5.7 \times 10^{-2}}{9.5 \times 10^{-8}}$ 6×10^{5}

68. $\dfrac{2.72 \times 10^{-6}}{6.8 \times 10^{-4}}$ 4×10^{-3}

69. $6,000,000,000 \times 0.0000000023$ 1.38×10^{1}

70. $0.000055 \times 40,000$ 2.2×10^{0}

71. $\dfrac{0.0000000003}{6000}$ 5×10^{-14}

72. $\dfrac{420,000}{0.0000021}$ 2×10^{11}

Mixed Exercises

73. If a piece of paper is 3×10^{-3} in. thick, how thick is a stack of 1.25×10^{3} pieces of paper?   3.75 in.

74. A box of staples contains 5×10^{3} staples and weighs 15 oz. How much does one staple weigh? Write your answer in scientific notation. 3×10^{-3} oz

 Writing   Translating Expression   Geometry   Scientific Calculator   Video

75. At one time, Bill Gates owned approximately 1,100,000,000 shares of Microsoft stock. If the stock price was $27 per share, how much was Bill Gates' stock worth? 2.97×10^{10}

76. A state lottery had a jackpot of 5.2×10^7. This week the winner was a group of office employees that included 13 people. How much would each person receive? 4×10^6 or $4,000,000$

77. Dinosaurs became extinct about 65 million years ago.

   a. Write the number 65 million in scientific notation. 6.5×10^7

   b. How many days is 65 million years? 2.3725×10^{10} days

   c. How many hours is 65 million years? 5.694×10^{11} hr

   d. How many seconds is 65 million years? 2.04984×10^{15} sec

78. The Earth is 150,000,000 km from the Sun.

   a. Write the number 150,000,000 in scientific notation. 1.5×10^8

   b. There are 1000 m in a kilometer. How many meters is the Earth from the Sun? 1.5×10^{11} m

   c. There are 100 cm in a meter. How many centimeters is the Earth from the Sun? 1.5×10^{13} cm

Addition and Subtraction of Polynomials

Section 5.5

1. Introduction to Polynomials

Concepts

1. **Introduction to Polynomials**
2. **Addition of Polynomials**
3. **Subtraction of Polynomials**
4. **Polynomials and Applications to Geometry**

One commonly used algebraic expression is called a polynomial. A **polynomial** in one variable, x, is defined as a single term or a sum of terms of the form ax^n, where a is a real number and the exponent, n, is a nonnegative integer. For each term, a is called the **coefficient**, and n is called the **degree of the term**. For example:

Term (Expressed in the Form ax^n)	Coefficient	Degree
$-12z^7$	-12	7
$x^3 \rightarrow$ rewrite as $1x^3$	1	3
$10w \rightarrow$ rewrite as $10w^1$	10	1
$7 \rightarrow$ rewrite as $7x^0$	7	0

If a polynomial has exactly one term, it is categorized as a **monomial**. A two-term polynomial is called a **binomial**, and a three-term polynomial is called a **trinomial**. Usually the terms of a polynomial are written in descending order according to degree. The term with highest degree is called the **leading term**, and its coefficient is called the **leading coefficient**. The **degree of a polynomial** is the greatest degree of all of its terms. A polynomial in one variable is written in **descending order** if the term with highest degree is written first, followed by the term of next highest degree and so on. Thus, for a polynomial written in descending order, the leading term determines the degree of the polynomial.

Instructor Note: Ask students why a constant has degree 0.

	Expression	Descending Order	Leading Coefficient	Degree of Polynomial
Monomials	$-3x^4$	$-3x^4$	-3	4
	17	17	17	0
Binomials	$4y^3 - 6y^5$	$-6y^5 + 4y^3$	-6	5
	$\dfrac{1}{2} - \dfrac{1}{4}c$	$-\dfrac{1}{4}c + \dfrac{1}{2}$	$-\dfrac{1}{4}$	1
Trinomials	$4p - 3p^3 + 8p^6$	$8p^6 - 3p^3 + 4p$	8	6
	$7a^4 - 1.2a^8 + 3a^3$	$-1.2a^8 + 7a^4 + 3a^3$	-1.2	8

 Writing   Translating Expression   Geometry    Scientific Calculator   Video

Classroom Example: p. 378, Exercise 14

> **Example 1** **Identifying the Parts of a Polynomial**

Given the polynomial: $3a - 2a^4 + 6 - a^3$

a. List the terms of the polynomial, and state the coefficient and degree of each term.

b. Write the polynomial in descending order.

c. State the degree of the polynomial and the leading coefficient.

d. Evaluate the polynomial for $a = -2$.

Solution:

a. term: $3a$ coefficient: 3   degree: 1

     term: $-2a^4$ coefficient: -2 degree: 4

     term: 6   coefficient: 6   degree: 0

     term: $-a^3$ coefficient: -1 degree: 3

b. $-2a^4 - a^3 + 3a + 6$

c. The degree of the polynomial is 4 and the leading coefficient is -2.

d. $-2a^4 - a^3 + 3a + 6$

$= -2(-2)^4 - (-2)^3 + 3(-2) + 6$ Substitute -2 for the variable, a. Remember to use parentheses when replacing a variable.

$= -2(16) - (-8) + 3(-2) + 6$ Simplify exponents first.

$= -32 - (-8) + (-6) + 6$ Perform multiplication before addition and subtraction.

$= -24$ Simplify.

Skill Practice

1. Given the polynomial: $5x^3 - x + 8x^4 + 3x^2$

a. Write the polynomial in descending order.

b. State the degree of the polynomial.

c. State the coefficient of the leading term.

d. Evaluate the polynomial for $x = -1$.

Polynomials may have more than one variable. In such a case, the degree of a term is the sum of the exponents of the variables contained in the term. For example, the term, $32x^2y^5z$, has degree 8 because the exponents applied to x, y, and z are 2, 5, and 1, respectively. The following polynomial has a degree of 11 because the highest degree of its terms is 11.

$$32x^2y^5z \quad - \quad 2x^3y \quad + \quad 2x^2yz^8 \quad + \quad 7$$

$\uparrow$	$\uparrow$	$\uparrow$	$\uparrow$
degree	degree	degree	degree
8	4	11	0

Answers

1. a. $8x^4 + 5x^3 + 3x^2 - x$
   b. 4   **c.** 8   **d.** 7

2. Addition of Polynomials

Recall that two terms are *like* terms if they each have the same variables, and the corresponding variables are raised to the same powers.

Like Terms: $3x^2, -7x^2$ $-5yz^3, yz^3$

Unlike Terms: $9z^2, 12z^6$ $\dfrac{1}{3}w^6, \dfrac{2}{5}p^6$ $4y, 7$

Recall that the distributive property is used to add or subtract *like* terms. For example,

$3x^2 + 9x^2 - 2x^2$

$= (3 + 9 - 2)x^2$ Apply the distributive property.

$= (10)x^2$ Simplify.

$= 10x^2$

Avoiding Mistakes

Note that when adding terms, the exponents do *not* change.
$$2x^3 + 3x^3 \neq 5x^6$$

Example 2 **Adding Polynomials**

Add the polynomials. $3x^2y + 5x^2y$

Solution:

$3x^2y + 5x^2y$ The terms are *like* terms.

$= (3 + 5)x^2y$ Apply the distributive property.

$= (8)x^2y$

$= 8x^2y$ Simplify.

Classroom Example: p. 378, Exercise 26

Skill Practice Add the polynomials.

2. $13a^2b^3 + 2a^2b^3$

It is the distributive property that enables us to add *like* terms. We shorten the process by adding the coefficients of *like* terms.

Example 3 **Adding Polynomials**

Add the polynomials. $(-3c^3 + 5c^2 - 7c) + (11c^3 + 6c^2 + 3)$

Solution:

$(-3c^3 + 5c^2 - 7c) + (11c^3 + 6c^2 + 3)$

$= -3c^3 + 11c^3 + 5c^2 + 6c^2 - 7c + 3$ Clear parentheses, and group *like* terms.

$= 8c^3 + 11c^2 - 7c + 3$ Combine *like* terms.

Classroom Example: p. 378, Exercise 28

Avoiding Mistakes

When *adding* like terms, the exponents do not change.

Answer

2. $15a^2b^3$

> **TIP:** Polynomials can also be added by combining *like* terms in columns. The sum of the polynomials from Example 3 is shown here.
>
> $$-3c^3 + 5c^2 - 7c + 0$$
> $$+\ 11c^3 + 6c^2 + 0c + 3$$
> $$\overline{8c^3 + 11c^2 - 7c + 3}$$
>
> Place holders such as 0 and 0c may be used to help line up *like* terms.

Skill Practice Add the polynomials.

3. $(7q^2 - 2q + 4) + (5q^2 + 6q - 9)$

Classroom Example: p. 378, Exercise 30

Example 4   **Adding Polynomials**

Add the polynomials. $(4w^2 - 2x) + (3w^2 - 4x^2 + 6x)$

Solution:

$$(4w^2 - 2x) + (3w^2 - 4x^2 + 6x)$$
$$= 4w^2 + 3w^2 - 4x^2 - 2x + 6x \qquad \text{Clear parentheses and group } like \text{ terms.}$$
$$= 7w^2 - 4x^2 + 4x$$

Skill Practice Add the polynomials.

4. $(5x^2 - 4xy + y^2) + (-3x^2 - 5y^2)$

3. Subtraction of Polynomials

Subtraction of two polynomials requires us to find the opposite of the polynomial being subtracted. To find the opposite of a polynomial, take the opposite of each term. This is equivalent to multiplying the polynomial by -1.

Classroom Example: p. 379, Exercise 46

Example 5   **Finding the Opposite of a Polynomial**

Find the opposite of the polynomials.

  a. $5x$ **b.** $3a - 4b - c$ **c.** $5.5y^4 - 2.4y^3 + 1.1y$

Solution:

	Expression	Opposite	Simplified Form
a.	$5x$	$-(5x)$	$-5x$
b.	$3a - 4b - c$	$-(3a - 4b - c)$	$-3a + 4b + c$
c.	$5.5y^4 - 2.4y^3 + 1.1y$	$-(5.5y^4 - 2.4y^3 + 1.1y)$	$-5.5y^4 + 2.4y^3 - 1.1y$

> **TIP:** Notice that the sign of each term is changed when finding the opposite of a polynomial.

Skill Practice Find the opposite of the polynomials.

5. $x - 3$ **6.** $3y^2 - 2xy + 6x + 2$ **7.** $-2.1w^3 + 4.9w^2 - 1.9w$

Answers

3. $12q^2 + 4q - 5$
4. $2x^2 - 4xy - 4y^2$
5. $-x + 3$
6. $-3y^2 + 2xy - 6x - 2$
7. $2.1w^3 - 4.9w^2 + 1.9w$

Subtraction of two polynomials is similar to subtracting real numbers. Add the opposite of the second polynomial to the first polynomial.

Subtraction of Polynomials

If A and B are polynomials, then $A - B = A + (-B)$.

Example 6   **Subtracting Polynomials**

Classroom Example: p. 379, Exercise 58

Subtract the polynomials. $(-4p^4 + 5p^2 - 3) - (11p^2 + 4p - 6)$

Solution:

$(-4p^4 + 5p^2 - 3) - (11p^2 + 4p - 6)$

$\quad = (-4p^4 + 5p^2 - 3) + (-11p^2 - 4p + 6)$ Add the opposite of the second polynomial.

$\quad = -4p^4 + 5p^2 - 11p^2 - 4p - 3 + 6$ Group *like* terms.

$\quad = -4p^4 - 6p^2 - 4p + 3$ Combine *like* terms.

TIP: Two polynomials can also be subtracted in columns by adding the opposite of the second polynomial to the first polynomial. Place holders (shown in red) may be used to help line up *like* terms.

$$\begin{array}{l} -4p^4 + 0p^3 + 5p^2 + 0p - 3 \\ \underline{-(0p^4 + 0p^3 + 11p^2 + 4p - 6)} \end{array} \xrightarrow{\text{Add the opposite}} \begin{array}{l} -4p^4 + 0p^3 + 5p^2 + 0p - 3 \\ \underline{+-0p^4 - 0p^3 - 11p^2 - 4p + 6} \\ -4p^4 - 6p^2 - 4p + 3 \end{array}$$

The difference of the polynomials is $-4p^4 - 6p^2 - 4p + 3$.

Skill Practice   Subtract the polynomials.

8. $(x^2 + 3x - 2) - (4x^2 + 6x + 1)$

Example 7   **Subtracting Polynomials**

Classroom Example: p. 379, Exercise 60

Subtract the polynomials. $(a^2 - 2ab + 7b^2) - (-8a^2 - 6ab + 2b^2)$

Solution:

$(a^2 - 2ab + 7b^2) - (-8a^2 - 6ab + 2b^2)$

$\quad = (a^2 - 2ab + 7b^2) + (8a^2 + 6ab - 2b^2)$ Add the opposite of the second polynomial.

$\quad = a^2 + 8a^2 - 2ab + 6ab + 7b^2 - 2b^2$ Group *like* terms.

$\quad = 9a^2 + 4ab + 5b^2$ Combine *like* terms.

Skill Practice   Subtract the polynomials.

9. $(-3y^2 + xy + 2x^2) - (-2y^2 - 3xy - 8x^2)$

In Example 8, we illustrate the subtraction of polynomials by first clearing parentheses and then combining like terms.

Answers

8. $-3x^2 - 3x - 3$
9. $-y^2 + 4xy + 10x^2$

Classroom Example: p. 379,
Exercise 72

| **Example 8** | **Subtracting Polynomials** |

Subtract $\frac{1}{3}t^4 + \frac{1}{2}t^2$ from $t^2 - 4$, and simplify the result.

Solution:

To subtract a from b, we write $b - a$. Thus, to subtract $\overset{a}{\overbrace{\frac{1}{3}t^4 + \frac{1}{2}t^2}}$ from $\overset{b}{\overbrace{t^2 - 4}}$, we have

$$\overset{b}{\overbrace{(t^2 - 4)}} - \overset{a}{\overbrace{\left(\frac{1}{3}t^4 + \frac{1}{2}t^2\right)}}$$

Avoiding Mistakes

Example 8 involves subtracting two *expressions*. This is not an equation. Therefore, we cannot clear fractions.

$$= t^2 - 4 - \frac{1}{3}t^4 - \frac{1}{2}t^2 \qquad \text{Apply the distributive property to clear parentheses.}$$

$$= -\frac{1}{3}t^4 + t^2 - \frac{1}{2}t^2 - 4 \qquad \text{Group } like \text{ terms in descending order.}$$

$$= -\frac{1}{3}t^4 + \frac{2}{2}t^2 - \frac{1}{2}t^2 - 4 \qquad \begin{array}{l} \text{The } t^2\text{-terms are the only } like \text{ terms.} \\ \text{Get a common denominator for the } t^2\text{-terms.} \end{array}$$

$$= -\frac{1}{3}t^4 + \frac{1}{2}t^2 - 4 \qquad \text{Add } like \text{ terms.}$$

Skill Practice

10. Subtract $\frac{3}{4}x^2 + \frac{2}{5}$ from $x^2 + 3x$.

4. Polynomials and Applications to Geometry

Classroom Example: p. 380,
Exercise 76

| **Example 9** | **Subtracting Polynomials in Geometry** |

If the perimeter of the triangle in Figure 5-1 can be represented by the polynomial $2x^2 + 5x + 6$, find a polynomial that represents the length of the missing side.

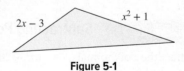

Figure 5-1

Solution:

The missing side of the triangle can be found by subtracting the sum of the two known sides from the perimeter.

$$\begin{pmatrix} \text{Length} \\ \text{of missing} \\ \text{side} \end{pmatrix} = (\text{perimeter}) - \begin{pmatrix} \text{sum of the} \\ \text{two known sides} \end{pmatrix}$$

$$\begin{pmatrix} \text{Length} \\ \text{of missing} \\ \text{side} \end{pmatrix} = (2x^2 + 5x + 6) - [(2x - 3) + (x^2 + 1)]$$

Answer
10. $\frac{1}{4}x^2 + 3x - \frac{2}{5}$

$$= 2x^2 + 5x + 6 - [2x - 3 + x^2 + 1] \quad \text{Clear inner parentheses.}$$

$$= 2x^2 + 5x + 6 - (x^2 + 2x - 2) \quad \text{Combine } \textit{like} \text{ terms within } [\ ].$$

$$= 2x^2 + 5x + 6 - x^2 - 2x + 2 \quad \text{Apply the distributive property.}$$

$$= 2x^2 - x^2 + 5x - 2x + 6 + 2 \quad \text{Group } \textit{like} \text{ terms.}$$

$$= x^2 + 3x + 8 \quad \text{Combine } \textit{like} \text{ terms.}$$

The polynomial $x^2 + 3x + 8$ represents the length of the missing side.

Skill Practice

11. If the perimeter of the triangle is represented by the polynomial $6x - 9$, find the polynomial that represents the missing side.

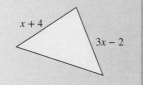

Answer

11. $2x - 11$

Section 5.5   Practice Exercises

For additional exercises, see Classroom Activities 5.5A–5.5C in the *Student's Resource Manual* at www.mhhe.com/moh.

Vocabulary and Key Concepts

1. **a.** A __polynomial__ is a single term or a sum of terms.

   b. For the term ax^n, a is called the __coefficient__ and n is called the __degree__ of the term.

   c. Given the term x, the coefficient of the term is __1__.

   d. A monomial is a polynomial with exactly __one__ term(s).

   e. A __binomial__ is a polynomial with exactly two terms.

   f. A __trinomial__ is a polynomial with exactly three terms.

   g. The term with the highest degree is called the __leading__ term and its coefficient is called the __leading coefficient__.

   h. The degree of a polynomial is the __greatest__ degree of all of its terms.

   i. The degree of a nonzero constant such as 5 is __zero__.

Review Exercises

For Exercises 2–7, simplify each expression.

2. $\dfrac{p^3 \cdot 4p}{p^2}$ $4p^2$

3. $(3x)^2(5x^{-4})$ $\dfrac{45}{x^2}$

4. $(6y^{-3})(2y^9)$ $12y^6$

5. $\dfrac{8t^{-6}}{4t^{-2}}$ $\dfrac{2}{t^4}$

6. $\dfrac{8^3 \cdot 8^{-4}}{8^{-2} \cdot 8^6}$ $\dfrac{1}{8^5}$

7. $\dfrac{3^4 \cdot 3^{-8}}{3^{12} \cdot 3^{-4}}$ $\dfrac{1}{3^{12}}$

8. Explain the difference between 3×10^7 and 3^7. 3×10^7 is in scientific notation in which 10 is raised to the seventh power. 3^7 is not in scientific notation and 3 is being raised to the seventh power.

9. Explain the difference between 4×10^{-2} and 4^{-2}. 4×10^{-2} is in scientific notation in which 10 is raised to the -2 power. 4^{-2} is not in scientific notation and 4 is being raised to the -2 power.

Writing   Translating Expression   Geometry   Scientific Calculator   Video

Concept 1: Introduction to Polynomials

For Exercises 10–12,

    a. write the polynomial in descending order.

    b. identify the leading coefficient.

    c. identify the degree of the polynomial. **(See Example 1.)**

10. $10 - 8a - a^3 + 2a^2 + a^5$
  a. $a^5 - a^3 + 2a^2 - 8a + 10$ b. 1 c. 5

11. $6 + 7x^2 - 7x^5 + 9x$
  a. $-7x^5 + 7x^2 + 9x + 6$ b. -7 c. 5

12. $\frac{1}{2}y + y^2 - 12y^4 + y^3 - 6$
  a. $-12y^4 + y^3 + y^2 + \frac{1}{2}y - 6$ b. -12 c. 4

For Exercises 13–22, categorize each expression as a monomial, a binomial, or a trinomial. Then evaluate the polynomial given $x = -3$, $y = 2$, and $z = -1$. **(See Example 1.)**

13. $10x^2 + 5x$
  Binomial; 75

14. $7z + 13z^2 - 15$
  Trinomial; −9

15. $6x^2$
  Monomial; 54

16. 9
  Monomial; 9

17. $2y - y^4$
  Binomial; −12

18. $7x + 2$
  Binomial; −19

19. $2y^4 - 3y + 1$
  Trinomial; 27

20. 23
  Monomial; 23

21. $-32xyz$
  Monomial; −192

22. $y^4 - x^2$
  Binomial; 7

Concept 2: Addition of Polynomials

23. Explain why the terms $3x$ and $3x^2$ are not *like* terms. The exponents on the x-factors are different.

24. Explain why the terms $4w^3$ and $4z^3$ are not *like* terms. The variables are different.

For Exercises 25–42, add the polynomials. **(See Examples 2–4.)**

25. $23x^2y + 12x^2y$ $35x^2y$

26. $-5ab^3 + 17ab^3$ $12ab^3$

27. $3b^5d^2 + (5b^5d^2 - 9d)$ $8b^5d^2 - 9d$

28. $4c^2d^3 + (3cd - 10c^2d^3)$ $-6c^2d^3 + 3cd$

29. $(7y^2 + 2y - 9) + (-3y^2 - y)$ $4y^2 + y - 9$

30. $(-3w^2 + 4w - 6) + (5w^2 + 2)$ $2w^2 + 4w - 4$

31. $(5x + 3x^2 - x^3) + (2x^2 + 4x - 10)$
  $-x^3 + 5x^2 + 9x - 10$

32. $(t^2 - 4t + t^4) + (3t^4 + 2t + 6)$ $4t^4 + t^2 - 2t + 6$

33. $(6.1y + 3.2x) + (4.8y - 3.2x)$ $10.9y$

34. $(2.7m - 0.5h) + (-3.2m + 0.2h)$ $-0.5m - 0.3h$

35. $6a + 2b - 5c$ $4a - 8c$
  $+\ \underline{-2a - 2b - 3c}$

36. $-13x + 5y + 10z$ $-16x + 2y + 12z$
  $+\ \underline{-3x - 3y +\ 2z}$

37. $\left(\frac{2}{5}a + \frac{1}{4}b - \frac{5}{6}\right) + \left(\frac{3}{5}a - \frac{3}{4}b - \frac{7}{6}\right)$ $a - \frac{1}{2}b - 2$

38. $\left(\frac{5}{9}x + \frac{1}{10}y\right) + \left(-\frac{4}{9}x + \frac{3}{10}y\right)$ $\frac{1}{9}x + \frac{2}{5}y$

39. $\left(z - \frac{8}{3}\right) + \left(\frac{4}{3}z^2 - z + 1\right)$ $\frac{4}{3}z^2 - \frac{5}{3}$

40. $\left(-\frac{7}{5}r + 1\right) + \left(-\frac{3}{5}r^2 + \frac{7}{5}r + 1\right)$ $-\frac{3}{5}r^2 + 2$

41. $7.9t^3$ $+ 2.6t - 1.1$
  $+\ \underline{\quad -3.4t^2 + 3.4t - 3.1}$ $7.9t^3 - 3.4t^2 + 6t - 4.2$

42. $0.34y^2$ $+ 1.23$
  $+\ \underline{\quad\quad 3.42y - 7.56}$ $0.34y^2 + 3.42y - 6.33$

Concept 3: Subtraction of Polynomials

For Exercises 43–48, find the opposite of each polynomial. **(See Example 5.)**

43. $4h - 5$
$-4h + 5$

44. $5k - 12$
$-5k + 12$

45. $-2.3m^2 + 3.1m - 1.5$
$2.3m^2 - 3.1m + 1.5$

46. $-11.8n^2 - 6.7n + 9.3$
$11.8n^2 + 6.7n - 9.3$

47. $3v^3 + 5v^2 + 10v + 22$
$-3v^3 - 5v^2 - 10v - 22$

48. $7u^4 + 3v^2 + 17$
$-7u^4 - 3v^2 - 17$

For Exercises 49–68, subtract the polynomials. **(See Examples 6–7.)**

49. $4a^3b^2 - 12a^3b^2$ $\quad -8a^3b^2$

50. $5yz^4 - 14yz^4$ $\quad -9yz^4$

51. $-32x^3 - 21x^3$ $\quad -53x^3$

52. $-23c^5 - 12c^5$ $\quad -35c^5$

53. $(7a - 7) - (12a - 4)$ $\quad -5a - 3$

54. $(4x + 3v) - (-3x + v)$ $\quad 7x + 2v$

55.
$$4k + 3$$
$$-(-12k - 6)$$
$$\overline{16k + 9}$$

56.
$$3h - 15$$
$$-(8h + 13)$$
$$\overline{-5h - 28}$$

57. $25m^4 - (23m^4 + 14m)$ $\quad 2m^4 - 14m$

58. $3x^2 - (-x^2 - 12)$ $\quad 4x^2 + 12$

59. $(5s^2 - 3st - 2t^2) - (2s^2 + st + t^2)$
$3s^2 - 4st - 3t^2$

60. $(6k^2 + 2kp + p^2) - (3k^2 - 6kp + 2p^2)$
$3k^2 + 8kp - p^2$

61.
$$10r - 6s + 2t$$
$$-(12r - 3s - \ t)$$
$$\overline{-2r - 3s + 3t}$$

62.
$$a - 14b + 7c$$
$$-(-3a - \ 8b + 2c)$$
$$\overline{4a - 6b + 5c}$$

63. $\left(\frac{7}{8}x + \frac{2}{3}y - \frac{3}{10}\right) - \left(\frac{1}{8}x + \frac{1}{3}y\right)$ $\quad \frac{3}{4}x + \frac{1}{3}y - \frac{3}{10}$

64. $\left(r - \frac{1}{12}s\right) - \left(\frac{1}{2}r - \frac{5}{12}s - \frac{4}{11}\right)$ $\quad \frac{1}{2}r + \frac{1}{3}s + \frac{4}{11}$

65. $\left(\frac{2}{3}h^2 - \frac{1}{5}h - \frac{3}{4}\right) - \left(\frac{4}{3}h^2 - \frac{4}{5}h + \frac{7}{4}\right)$ $\quad -\frac{2}{3}h^2 + \frac{3}{5}h - \frac{5}{2}$

66. $\left(\frac{3}{8}p^3 - \frac{5}{7}p^2 - \frac{2}{5}\right) - \left(\frac{5}{8}p^3 - \frac{2}{7}p^2 + \frac{7}{5}\right)$ $\quad -\frac{1}{4}p^3 - \frac{3}{7}p^2 - \frac{9}{5}$

67.
$$4.5x^4 - 3.1x^2 \qquad - 6.7$$
$$-(2.1x^4 \qquad + 4.4x + 1.2)$$
$$\overline{2.4x^4 - 3.1x^2 - 4.4x - 7.9}$$

68.
$$1.3c^2 \qquad + 4.8$$
$$- (4.3c^2 - 2c - 2.2)$$
$$\overline{-3c^2 + 2c + 7}$$

69. Find the difference of $(4b^3 + 6b - 7)$ and $(-12b^2 + 11b + 5)$. $\quad 4b^3 + 12b^2 - 5b - 12$

70. Find the difference of $(-5y^2 + 3y - 21)$ and $(-4y^2 - 5y + 23)$. $\quad -y^2 + 8y - 44$

71. Subtract $\left(\frac{3}{2}x^2 - 5x\right)$ from $\left(-2x^2 + 11\right)$.
(See Example 8.) $\quad -\frac{7}{2}x^2 + 5x - 11$

72. Subtract $\left(a^5 - \frac{1}{3}a^3 + 5a\right)$ from $\left(\frac{3}{4}a^5 + \frac{1}{2}a^4 + 6a\right)$. $\quad -\frac{1}{4}a^5 + \frac{1}{2}a^4 + \frac{1}{3}a^3 + a$

Concept 4: Polynomials and Applications to Geometry

73. Find a polynomial that represents the perimeter of the figure. $\quad 4y^3 + 2y^2 + 2$

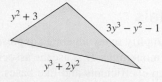

$y^2 + 3$
$3y^3 - y^2 - 1$
$y^3 + 2y^2$

74. Find a polynomial that represents the perimeter of the figure. $\quad 9t^3 + t^2$

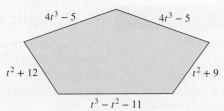

$4t^3 - 5$
$4t^3 - 5$
$t^2 + 12$
$t^2 + 9$
$t^3 - t^2 - 11$

75. If the perimeter of the figure can be represented by the polynomial $5a^2 - 2a + 1$, find a polynomial that represents the length of the missing side.
(See Example 9.) $3a^2 - 3a + 5$

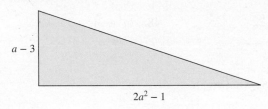

$a - 3$

$2a^2 - 1$

76. If the perimeter of the figure can be represented by the polynomial $6w^3 - 2w - 3$, find a polynomial that represents the length of the missing side. $4w^3 - 6w^2 - 2w$

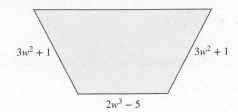

$3w^2 + 1$ $3w^2 + 1$

$2w^3 - 5$

Mixed Exercises

For Exercises 77–92, perform the indicated operation.

77. $(2ab^2 + 9a^2b) + (7ab^2 - 3ab + 7a^2b)$ $9ab^2 - 3ab + 16a^2b$

78. $(8x^2y - 3xy - 6xy^2) + (3x^2y - 12xy)$
$11x^2y - 15xy - 6xy^2$

79.
$$4z^5 \quad + \quad z^3 - 3z + 13$$
$$- (\quad - z^4 - 8z^3 \quad + 15)$$
$4z^5 + z^4 + 9z^3 - 3z - 2$

80.
$$-15t^4 \quad - 23t^2 + 16t$$
$$-(\quad 21t^3 + 18t^2 + \quad t)$$
$-15t^4 - 21t^3 - 41t^2 + 15t$

81. $(9x^4 + 2x^3 - x + 5) + (9x^3 - 3x^2 + 8x + 3) - (7x^4 - x + 12)$
$2x^4 + 11x^3 - 3x^2 + 8x - 4$

82. $(-6y^3 - 9y^2 + 23) - (7y^2 + 2y - 11) + (3y^3 - 25)$
$-3y^3 - 16y^2 - 2y + 9$

83. $(0.2w^2 + 3w + 1.3) - (w^3 - 0.7w + 2)$
$-w^3 + 0.2w^2 + 3.7w - 0.7$

84. $(8.1u^3 - 5.2u^2 + 4) + (2.8u^3 + 6.3u - 7)$
$10.9u^3 - 5.2u^2 + 6.3u - 3$

85. $(7p^2q - 3pq^2) - (8p^2q + pq) + (4pq - pq^2)$
$-p^2q - 4pq^2 + 3pq$

86. $(12c^2d - 2cd + 8cd^2) - (-c^2d + 4cd) - (5cd - 2cd^2)$
$13c^2d - 11cd + 10cd^2$

87. $(5x - 2x^3) + (2x^3 - 5x)$ 0

88. $(p^2 - 4p + 2) - (2 + p^2 - 4p)$ 0

89.
$$2a^2b - 4ab + ab^2$$
$$-(2a^2b + ab - 5ab^2)$$
$-5ab + 6ab^2$

90.
$$-3xy + 7xy^2 + 5x^2y$$
$$+ (-8xy - 11xy^2 + 3x^2y)$$
$-11xy - 4xy^2 + 8x^2y$

91. $[(3y^2 - 5y) - (2y^2 + y - 1)] + (10y^2 - 4y - 5)$
$11y^2 - 10y - 4$

92. $(12c^3 - 5c^2 - 2c) + [(7c^3 - 2c^2 + c) - (4c^3 + 4c)]$
$15c^3 - 7c^2 - 5c$

Expanding Your Skills

93. Write a binomial of degree 3. (Answers may vary.)
For example: $x^3 + 6$

94. Write a trinomial of degree 6. (Answers may vary.)
For example: $5x^6 + x - 4$

95. Write a monomial of degree 5. (Answers may vary.)
For example: $8x^5$

96. Write a monomial of degree 1. (Answers may vary.)
For example: $3x$

97. Write a trinomial with the leading coefficient -6.
(Answers may vary.) For example: $-6x^2 + 2x + 5$

98. Write a binomial with the leading coefficient 13.
(Answers may vary.) For example: $13x + 7$

Section 5.6 Multiplication of Polynomials and Special Products

Concepts

1. Multiplication of Polynomials
2. Special Case Products: Difference of Squares and Perfect Square Trinomials
3. Applications to Geometry

1. Multiplication of Polynomials

The properties of exponents can be used to simplify many algebraic expressions including the multiplication of monomials. To multiply monomials, first use the associative and commutative properties of multiplication to group coefficients and like bases. Then simplify the result by using the properties of exponents.

Example 1 **Multiplying Monomials**

Classroom Example: p. 387, Exercise 18

Multiply the monomials.

a. $(3x^4)(4x^2)$ **b.** $(-4c^5d)(2c^2d^3e)$ **c.** $\left(\frac{1}{3}a^4b^3\right)\left(\frac{3}{4}b^7\right)$

Solution:

a. $(3x^4)(4x^2)$

$= (3 \cdot 4)(x^4 x^2)$ Group coefficients and like bases.

$= 12x^6$ Multiply the coefficients and add the exponents on x.

b. $(-4c^5d)(2c^2d^3e)$

$= (-4 \cdot 2)(c^5 c^2)(dd^3)(e)$ Group coefficients and like bases.

$= -8c^7d^4e$ Simplify.

c. $\left(\frac{1}{3}a^4b^3\right)\left(\frac{3}{4}b^7\right)$

$= \left(\frac{1}{3} \cdot \frac{3}{4}\right)(a^4)(b^3 b^7)$ Group coefficients and like bases.

$= \frac{1}{4}a^4b^{10}$ Simplify.

Skill Practice Multiply the monomials.

1. $(-5y)(6y^3)$ **2.** $(7x^2y)(-2x^3y^4)$ **3.** $\left(\frac{2}{5}w^5z^3\right)\left(\frac{15}{4}w^4\right)$

The distributive property is used to multiply polynomials: $a(b+c) = ab + ac$.

Example 2 **Multiplying a Polynomial by a Monomial**

Classroom Example: p. 387, Exercise 24

Multiply the polynomials.

a. $2t(4t - 3)$ **b.** $-3a^2\left(-4a^2 + 2a - \frac{1}{3}\right)$

Solution:

a. $2t(4t - 3)$

$= (2t)(4t) + (2t)(-3)$ Apply the distributive property by multiplying each term by $2t$.

$= 8t^2 - 6t$ Simplify each term.

b. $-3a^2\left(-4a^2 + 2a - \frac{1}{3}\right)$

$= (-3a^2)(-4a^2) + (-3a^2)(2a) + (-3a^2)\left(-\frac{1}{3}\right)$ Apply the distributive property by multiplying each term by $-3a^2$.

$= 12a^4 - 6a^3 + a^2$ Simplify each term.

Answers

1. $-30y^4$
2. $-14x^5y^5$
3. $\frac{3}{2}w^9z^3$

Skill Practice Multiply the polynomials.

4. $-4a(5a - 3)$ **5.** $-4p\left(2p^2 - 6p + \dfrac{1}{4}\right)$

Thus far, we have illustrated polynomial multiplication involving monomials. Next, the distributive property will be used to multiply polynomials with more than one term.

$$(x + 3)(x + 5) = x(x + 5) + 3(x + 5) \qquad \text{Apply the distributive property.}$$

$$= x(x + 5) + 3(x + 5) \qquad \text{Apply the distributive property again.}$$

$$= (x)(x) + (x)(5) + (3)(x) + (3)(5)$$

$$= x^2 + 5x + 3x + 15$$

$$= x^2 + 8x + 15 \qquad \text{Combine } like \text{ terms.}$$

Note: Using the distributive property results in multiplying each term of the first polynomial by each term of the second polynomial.

$$(x + 3)(x + 5) = (x)(x) + (x)(5) + (3)(x) + (3)(5)$$

$$= x^2 + 5x + 3x + 15$$

$$= x^2 + 8x + 15$$

Classroom Example: p. 387, Exercise 30

| Example 3 | Multiplying a Polynomial by a Polynomial |

Multiply the polynomials. $(c - 7)(c + 2)$

Solution:

Multiply each term in the first polynomial by each term in the second. That is, apply the distributive property.

$$= (c)(c) + (c)(2) + (-7)(c) + (-7)(2)$$

$$= c^2 + 2c - 7c - 14 \qquad \text{Simplify.}$$

$$= c^2 - 5c - 14 \qquad \text{Combine } like \text{ terms.}$$

Instructor Note: Mention to students that the product $(c - 7)(c + 2)$ requires multiple applications of the distributive property. First distribute c, then distribute -7.

TIP: Notice that the product of two *binomials* equals the sum of the products of the **F**irst terms, the **O**uter terms, the **I**nner terms, and the **L**ast terms. The acronym **FOIL** (First Outer Inner Last) can be used as a memory device to multiply two binomials.

| | First | Outer | Inner | Last |

$$(c - 7)(c + 2) = (c)(c) + (c)(2) + (-7)(c) + (-7)(2)$$

$$= c^2 + 2c - 7c - 14$$

$$= c^2 - 5c - 14$$

Outer terms / First terms / Inner terms / Last terms

Skill Practice Multiply the polynomials.

6. $(x + 2)(x + 8)$

Example 4   **Multiplying a Polynomial by a Polynomial**

Multiply the polynomials. $(y - 2)(3y^2 + y - 5)$

Solution:

$(y - 2)(3y^2 + y - 5)$ Multiply each term in the first polynomial by each
                                 term in the second.

$= (y)(3y^2) + (y)(y) + (y)(-5) + (-2)(3y^2) + (-2)(y) + (-2)(-5)$

$= 3y^3 + y^2 - 5y - 6y^2 - 2y + 10$ Simplify each term.

$= 3y^3 - 5y^2 - 7y + 10$ Combine *like* terms.

> **TIP:** Multiplication of polynomials can be performed vertically by a process similar to
> column multiplication of real numbers. For example,
>
> $$\begin{array}{r} 235 \\ \times\ \ 21 \\ \hline 235 \\ 4700 \\ \hline 4935 \end{array}$$
>
> $$\begin{array}{r} 3y^2\ +\ y\ -\ 5 \\ \times\ \qquad\quad y\ -\ 2 \\ \hline -6y^2\ -\ 2y\ +\ 10 \\ 3y^3\ +\ y^2\ -\ 5y\ +\ 0 \\ \hline 3y^3\ -\ 5y^2\ -\ 7y\ +\ 10 \end{array}$$
>
> *Note:*   When multiplying by the column method, it is important to *align like* terms
> vertically before adding terms.

Skill Practice Multiply the polynomials.

7. $(2y + 4)(3y^2 - 5y + 2)$

Classroom Example: p. 387,
Exercise 44

Avoiding Mistakes

It is important to note that the acronym FOIL does not apply to Example 4 because the product does not involve two binomials.

Example 5   **Multiplying Polynomials**

Multiply the polynomials. $2(10x + 3y)(2x - 4y)$

Solution:

$2(10x + 3y)(2x - 4y)$ In this case we are multiplying three polynomials—a
                                 monomial times two binomials. The associative
                                 property of multiplication enables us to choose which
                                 two polynomials to multiply first.

$= 2[(10x + 3y)(2x - 4y)]$ First we will multiply the binomials. Multiply
                                     each term in the first binomial by each term in
                                     the second binomial. That is, apply the
                                     distributive property.

$= 2[(10x)(2x) + (10x)(-4y) + (3y)(2x) + (3y)(-4y)]$

$= 2[20x^2 - 40xy + 6xy - 12y^2]$ Simplify each term.

$= 2(20x^2 - 34xy - 12y^2)$ Combine *like* terms.

$= 40x^2 - 68xy - 24y^2$ Multiply by 2 using the distributive property.

Skill Practice Multiply.

8. $3(4a - 3c)(5a - 2c)$

Classroom Example: p. 387,
Exercise 42

TIP: When multiplying three polynomials, first multiply two of the polynomials. Then multiply the result by the third polynomial.

Answers

7. $6y^3 + 2y^2 - 16y + 8$
8. $60a^2 - 69ac + 18c^2$

2. Special Case Products: Difference of Squares and Perfect Square Trinomials

In some cases the product of two binomials takes on a special pattern.

I. The first special case occurs when multiplying the sum and difference of the same two terms. For example:

$(2x + 3)(2x - 3)$

$= 4x^2 - 6x + 6x - 9$

$= 4x^2 - 9$

Notice that the middle terms are opposites. This leaves only the difference between the square of the first term and the square of the second term. For this reason, the product is called a *difference of squares*.

Note: The binomials $2x + 3$ and $2x - 3$ are called **conjugates**. In one expression, $2x$ and 3 are added, and in the other, $2x$ and 3 are subtracted.

II. The second special case involves the square of a binomial. For example:

$(3x + 7)^2$

$= (3x + 7)(3x + 7)$

$= 9x^2 + 21x + 21x + 49$

$= 9x^2 + 42x + 49$

$= (3x)^2 + 2(3x)(7) + (7)^2$

When squaring a binomial, the product will be a trinomial called a *perfect square trinomial*. The first and third terms are formed by squaring each term of the binomial. The middle term equals twice the product of the terms in the binomial.

Note: The expression $(3x - 7)^2$ also expands to a perfect square trinomial, but the middle term will be negative:

$$(3x - 7)(3x - 7) = 9x^2 - 21x - 21x + 49 = 9x^2 - 42x + 49$$

Special Case Product Formulas

1. $(a + b)(a - b) = a^2 - b^2$ — The product is called a **difference of squares**.

2. $(a + b)^2 = a^2 + 2ab + b^2$
   $(a - b)^2 = a^2 - 2ab + b^2$ — The product is called a **perfect square trinomial**.

Instructor Note: Have students substitute some numbers to confirm that $(a + b)^2 \neq a^2 + b^2$.

You should become familiar with these special case products because they will be used again in the next chapter to factor polynomials.

Classroom Examples: p. 388, Exercises 56 and 62

TIP: The product of two conjugates can be checked by applying the distributive property:

$(x - 9)(x + 9)$

$= x^2 + 9x - 9x - 81$

$= x^2 - 81$

Example 6   Multiplying Conjugates

Multiply the conjugates.

a. $(x - 9)(x + 9)$ **b.** $\left(\dfrac{1}{2}p + 6\right)\left(\dfrac{1}{2}p - 6\right)$

Solution:

a. $(x - 9)(x + 9)$ Apply the formula: $(a + b)(a - b) = a^2 - b^2$.

$ a^2 - b^2$

$= (x)^2 - (9)^2$ Substitute $a = x$ and $b = 9$.

$= x^2 - 81$

b. $\left(\frac{1}{2}p + 6\right)\left(\frac{1}{2}p - 6\right)$ Apply the formula: $(a + b)(a - b) = a^2 - b^2$.

$a^2 - b^2$

$= \left(\frac{1}{2}p\right)^2 - (6)^2$ Substitute $a = \frac{1}{2}p$ and $b = 6$.

$= \frac{1}{4}p^2 - 36$ Simplify each term.

Skill Practice Multiply the conjugates.

9. $(a + 7)(a - 7)$ **10.** $\left(\frac{4}{5}x - 10\right)\left(\frac{4}{5}x + 10\right)$

Example 7 **Squaring Binomials**

Classroom Examples: p. 388, Exercises 72 and 74

Square the binomials.

a. $(3w - 4)^2$ **b.** $(5x^2 + 2)^2$

Solution:

a. $(3w - 4)^2$ Apply the formula: $(a - b)^2 = a^2 - 2ab + b^2$

$a^2 - 2ab + b^2$

$= (3w)^2 - 2(3w)(4) + (4)^2$ Substitute $a = 3w$, $b = 4$.

$= 9w^2 - 24w + 16$ Simplify each term.

TIP: The square of a binomial can be checked by explicitly writing the product of the two binomials and applying the distributive property:

$(3w - 4)^2 = (3w - 4)(3w - 4) = 9w^2 - 12w - 12w + 16$

$= 9w^2 - 24w + 16$

b. $(5x^2 + 2)^2$ Apply the formula: $(a + b)^2 = a^2 + 2ab + b^2$.

$a^2 + 2ab + b^2$

$= (5x^2)^2 + 2(5x^2)(2) + (2)^2$ Substitute $a = 5x^2$, $b = 2$.

$= 25x^4 + 20x^2 + 4$ Simplify each term.

Avoiding Mistakes

The property for squaring two factors is different than the property for squaring two terms: $(ab)^2 = a^2b^2$ but $(a + b)^2 = a^2 + 2ab + b^2$

Skill Practice Square the binomials.

11. $(2x + 3)^2$ **12.** $(5c^2 - 6)^2$

Answers
9. $a^2 - 49$
10. $\frac{16}{25}x^2 - 100$
11. $4x^2 + 12x + 9$
12. $25c^4 - 60c^2 + 36$

3. Applications to Geometry

Classroom Example: p. 389, Exercise 86

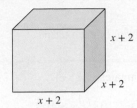 **Example 8** ### Using Special Case Products in an Application of Geometry

Find a polynomial that represents the volume of the cube (Figure 5-2).

$x + 2$

$x + 2$

$x + 2$

Figure 5-2

Solution:

$$\text{Volume} = (\text{length})(\text{width})(\text{height})$$

$$V = (x + 2)(x + 2)(x + 2) \qquad \text{or} \qquad V = (x + 2)^3$$

To expand $(x + 2)(x + 2)(x + 2)$, multiply the first two factors. Then multiply the result by the last factor.

$$V = \underline{(x + 2)(x + 2)}(x + 2)$$

$$= (x^2 + 4x + 4)(x + 2) \longleftarrow$$

> **TIP:** $(x + 2)(x + 2) = (x + 2)^2$ and results in a perfect square trinomial.
> $$(x + 2)^2 = (x)^2 + 2(x)(2) + (2)^2$$
> $$= x^2 + 4x + 4$$

$$= (x^2)(x) + (x^2)(2) + (4x)(x) + (4x)(2) + (4)(x) + (4)(2) \qquad \text{Apply the distributive property.}$$

$$= x^3 + 2x^2 + 4x^2 + 8x + 4x + 8 \qquad \text{Group } like \text{ terms.}$$

$$= x^3 + 6x^2 + 12x + 8 \qquad \text{Combine } like \text{ terms.}$$

The volume of the cube can be represented by

$$V = (x + 2)^3 = x^3 + 6x^2 + 12x + 8$$

Skill Practice

13. Find the polynomial that represents the volume of the cube.

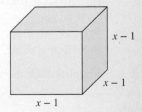

$x - 1$

$x - 1$

$x - 1$

Answer

13. The volume of the cube can be represented by $x^3 - 3x^2 + 3x - 1$.

Section 5.6 Practice Exercises

For additional exercises, see Classroom Activities 5.6A–5.6B in the *Student's Resource Manual* at www.mhhe.com/moh.

Vocabulary and Key Concepts

1. **a.** The conjugate of $5 + 2x$ is ___$5 - 2x$___.

   b. When two conjugates are multiplied the resulting binomial is a difference of ___squares___. This is given by the formula $(a + b)(a - b) =$ ___$a^2 - b^2$___.

   c. When a binomial is squared, the resulting trinomial is a ___perfect___ square trinomial. This is given by the formula $(a + b)^2 =$ ___$a^2 + 2ab + b^2$___.

Review Exercises

For Exercises 2–9, simplify each expression (if possible).

2. $4x + 5x$ $9x$ 3. $2y^2 - 4y^2$ $-2y^2$ 4. $(4x)(5x)$ $20x^2$ 5. $(2y^2)(-4y^2)$ $-8y^4$

6. $-5a^3b - 2a^3b$ $-7a^3b$ 7. $7uvw^2 + uvw^2$ $8uvw^2$ 8. $(-5a^3b)(-2a^3b)$ $10a^6b^2$ 9. $(7uvw^2)(uvw^2)$ $7u^2v^2w^4$

Concept 1: Multiplication of Polynomials

For Exercises 10–18, multiply the expressions. **(See Example 1.)**

10. $8(4x)$ $32x$ 11. $-2(6y)$ $-12y$ 12. $-10(5z)$ $-50z$

13. $7(3p)$ $21p$ 14. $(x^{10})(4x^3)$ $4x^{13}$ 15. $(a^{13}b^4)(12ab^4)$ $12a^{14}b^8$

16. $(4m^3n^7)(-3m^6n)$ $-12m^9n^8$ 17. $(2c^7d)(-c^3d^{11})$ $-2c^{10}d^{12}$ 18. $(-5u^2v)(-8u^3v^2)$ $40u^5v^3$

For Exercises 19–54, multiply the polynomials. **(See Examples 2–5.)**

19. $8pq(2pq - 3p + 5q)$
    $16p^2q^2 - 24p^2q + 40pq^2$

20. $5ab(2ab + 6a - 3b)$
    $10a^2b^2 + 30a^2b - 15ab^2$

21. $(k^2 - 13k - 6)(-4k)$
    $-4k^3 + 52k^2 + 24k$

22. $(h^2 + 5h - 12)(-2h)$
    $-2h^3 - 10h^2 + 24h$

23. $-15pq(3p^2 + p^3q^2 - 2q)$
    $-45p^3q - 15p^4q^3 + 30pq^2$

24. $-4u^2v(2u - 5uv^3 + v)$
    $-8u^3v + 20u^3v^4 - 4u^2v^2$

25. $(y + 10)(y + 9)$
    $y^2 + 19y + 90$

26. $(x + 5)(x + 6)$
    $x^2 + 11x + 30$

27. $(m - 12)(m - 2)$
    $m^2 - 14m + 24$

28. $(n - 7)(n - 2)$
    $n^2 - 9n + 14$

29. $(3p - 2)(4p + 1)$
    $12p^2 - 5p - 2$

30. $(7q + 11)(q - 5)$
    $7q^2 - 24q - 55$

31. $(8 - 4w)(-3w + 2)$
    $12w^2 - 32w + 16$

32. $(-6z + 10)(4 - 2z)$
    $12z^2 - 44z + 40$

33. $(p - 3w)(p - 11w)$
    $p^2 - 14pw + 33w^2$

34. $(y - 7x)(y - 10x)$
    $y^2 - 17xy + 70x^2$

35. $(6x - 1)(2x + 5)$
    $12x^2 + 28x - 5$

36. $(3x + 7)(x - 8)$
    $3x^2 - 17x - 56$

37. $(4a - 9)(1.5a - 2)$
    $6a^2 - 21.5a + 18$

38. $(2.1y - 0.5)(y + 3)$
    $2.1y^2 + 5.8y - 1.5$

39. $(3t - 7)(1 + 3t^2)$
    $9t^3 - 21t^2 + 3t - 7$

40. $(2 - 5w)(2w^2 - 5)$
    $-10w^3 + 4w^2 + 25w - 10$

41. $3(3m + 4n)(m + 2n)$
    $9m^2 + 30mn + 24n^2$

42. $2(7y + z)(3y + 5z)$
    $42y^2 + 76yz + 10z^2$

43. $(5s + 3)(s^2 + s - 2)$
    $5s^3 + 8s^2 - 7s - 6$

44. $(t - 4)(2t^2 - t + 6)$
    $2t^3 - 9t^2 + 10t - 24$

45. $(3w - 2)(9w^2 + 6w + 4)$
    $27w^3 - 8$

46. $(z + 5)(z^2 - 5z + 25)$
    $z^3 + 125$

47. $(p^2 + p - 5)(p^2 + 4p - 1)$
    $p^4 + 5p^3 - 2p^2 - 21p + 5$

48. $(-x^2 - 2x + 4)(x^2 + 2x - 6)$
    $-x^4 - 4x^3 + 6x^2 + 20x - 24$

49. $\begin{array}{r} 3a^2 - 4a + 9 \\ \times\ \underline{2a - 5} \\ 6a^3 - 23a^2 + 38a - 45 \end{array}$

50. $\begin{array}{r} 7x^2 - 3x - 4 \\ \times\ \underline{5x + 1} \\ 35x^3 - 8x^2 - 23x - 4 \end{array}$

51. $\begin{array}{r} 4x^2 - 12xy + 9y^2 \\ \times\ \underline{2x - 3y} \\ 8x^3 - 36x^2y + 54xy^2 - 27y^3 \end{array}$

52. $\begin{array}{r} 25a^2 + 10ab + b^2 \\ \times\ \underline{5a + b} \\ 125a^3 + 75a^2b + 15ab^2 + b^3 \end{array}$

53. $\begin{array}{r} 6x + 2y \\ \times\ \underline{0.2x + 1.2y} \\ 1.2x^2 + 7.6xy + 2.4y^2 \end{array}$

54. $\begin{array}{r} 4.5a + 2b \\ \times\ \underline{2a - 1.8b} \\ 9a^2 - 4.1ab - 3.6b^2 \end{array}$

Concept 2: Special Case Products: Difference of Squares and Perfect Square Trinomials

For Exercises 55–66, multiply the conjugates. **(See Example 6.)**

55. $(y - 6)(y + 6)$
$y^2 - 36$

56. $(x + 3)(x - 3)$
$x^2 - 9$

57. $(3a - 4b)(3a + 4b)$
$9a^2 - 16b^2$

58. $(5y + 7x)(5y - 7x)$
$25y^2 - 49x^2$

59. $(9k + 6)(9k - 6)$
$81k^2 - 36$

60. $(2h - 5)(2h + 5)$
$4h^2 - 25$

61. $\left(\frac{2}{3}t - 3\right)\left(\frac{2}{3}t + 3\right)$
$\frac{4}{9}t^2 - 9$

62. $\left(\frac{1}{4}r - 1\right)\left(\frac{1}{4}r + 1\right)$
$\frac{1}{16}r^2 - 1$

63. $(u^3 + 5v)(u^3 - 5v)$
$u^6 - 25v^2$

64. $(8w^2 - x)(8w^2 + x)$
$64w^4 - x^2$

65. $\left(\frac{2}{3} - p\right)\left(\frac{2}{3} + p\right)$
$\frac{4}{9} - p^2$

66. $\left(\frac{1}{8} - q\right)\left(\frac{1}{8} + q\right)$
$\frac{1}{64} - q^2$

For Exercises 67–78, square the binomials. **(See Example 7.)**

67. $(a + 5)^2$
$a^2 + 10a + 25$

68. $(a - 3)^2$
$a^2 - 6a + 9$

69. $(x - y)^2$
$x^2 - 2xy + y^2$

70. $(x + y)^2$
$x^2 + 2xy + y^2$

71. $(2c + 5)^2$
$4c^2 + 20c + 25$

72. $(5d - 9)^2$
$25d^2 - 90d + 81$

73. $(3t^2 - 4s)^2$
$9t^4 - 24st^2 + 16s^2$

74. $(u^2 + 4v)^2$
$u^4 + 8u^2v + 16v^2$

75. $(7 - t)^2$
$t^2 - 14t + 49$

76. $(4 + w)^2$
$w^2 + 8w + 16$

77. $(3 + 4q)^2$
$16q^2 + 24q + 9$

78. $(2 - 3b)^2$
$9b^2 - 12b + 4$

79. a. Evaluate $(2 + 4)^2$ by working within the parentheses first. 36

  b. Evaluate $2^2 + 4^2$. 20

  c. Compare the answers to parts (a) and (b) and make a conjecture about $(a + b)^2$ and $a^2 + b^2$.
  $(a + b)^2 \neq a^2 + b^2$ in general.

80. a. Evaluate $(6 - 5)^2$ by working within the parentheses first. 1

  b. Evaluate $6^2 - 5^2$. 11

  c. Compare the answers to parts (a) and (b) and make a conjecture about $(a - b)^2$ and $a^2 - b^2$.
  $(a - b)^2 \neq a^2 - b^2$ in general.

Concept 3: Applications to Geometry

81. Find a polynomial expression that represents the area of the rectangle shown in the figure.
$4x^2 - 25$

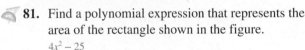
2x + 5
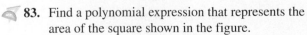
2x − 5

82. Find a polynomial expression that represents the area of the rectangle shown in the figure.
$36 - y^2$

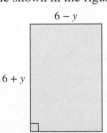
6 − y

6 + y

83. Find a polynomial expression that represents the area of the square shown in the figure.
$16p^2 + 40p + 25$

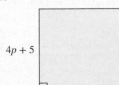
4p + 5

84. Find a polynomial expression that represents the area of the square shown in the figure.
$49q^2 - 42q + 9$

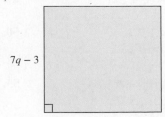
7q − 3

 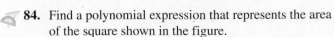

85. Find a polynomial that represents the volume of the cube shown in the figure. **(See Example 8.)**

(Recall: $V = s^3$) $27p^3 - 135p^2 + 225p - 125$

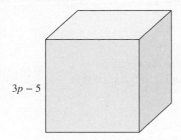

$3p - 5$

 86. Find a polynomial that represents the volume of the rectangular solid shown in the figure.

(Recall: $V = lwh$) $r^3 - 15r^2 + 63r - 49$

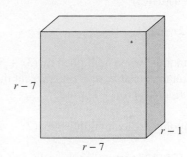

$r - 7$

$r - 1$

$r - 7$

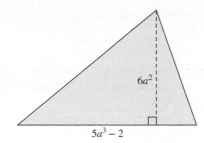 **87.** Find a polynomial that represents the area of the triangle shown in the figure.

(Recall: $A = \frac{1}{2}bh$) $15a^5 - 6a^2$

$6a^2$

$5a^3 - 2$

88. Find a polynomial that represents the area of the triangle shown in the figure. $3t^3 - 12t$

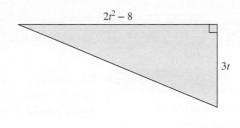

$2t^2 - 8$

$3t$

Mixed Exercises

For Exercises 89–118, multiply the expressions.

89. $(7x + y)(7x - y)$ $49x^2 - y^2$

90. $(9w - 4z)(9w + 4z)$ $81w^2 - 16z^2$

91. $(5s + 3t)^2$ $25s^2 + 30st + 9t^2$

92. $(5s - 3t)^2$ $25s^2 - 30st + 9t^2$

93. $(7x - 3y)(3x - 8y)$
$21x^2 - 65xy + 24y^2$

94. $(5a - 4b)(2a - b)$
$10a^2 - 13ab + 4b^2$

95. $\left(\frac{2}{3}t + 2\right)(3t + 4)$ $2t^2 + \frac{26}{3}t + 8$

96. $\left(\frac{1}{5}s + 6\right)(5s - 3)$ $s^2 + \frac{147}{5}s - 18$

97. $-5(3x + 5)(2x - 1)$
$-30x^2 - 35x + 25$

98. $-4(2k - 5)(4k - 3)$
$-32k^2 + 104k - 60$

99. $(3a - 2)(5a + 1 + 2a^2)$
$6a^3 + 11a^2 - 7a - 2$

100. $(u + 4)(2 - 3u + u^2)$
$u^3 + u^2 - 10u + 8$

101. $(y^2 + 2y + 4)(y - 5)$
$y^3 - 3y^2 - 6y - 20$

102. $(w^2 - w + 6)(w + 2)$
$w^3 + w^2 + 4w + 12$

103. $\left(\frac{1}{3}m - n\right)^2$ $\frac{1}{9}m^2 - \frac{2}{3}mn + n^2$

104. $\left(\frac{2}{5}p - q\right)^2$ $\frac{4}{25}p^2 - \frac{4}{5}pq + q^2$

105. $6w^2(7w - 14)$ $42w^3 - 84w^2$

106. $4v^3(v + 12)$ $4v^4 + 48v^3$

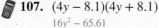 **107.** $(4y - 8.1)(4y + 8.1)$
$16y^2 - 65.61$

108. $(2h + 2.7)(2h - 2.7)$
$4h^2 - 7.29$

109. $(3c^2 + 4)(7c^2 - 8)$
$21c^4 + 4c^2 - 32$

110. $(5k^3 - 9)(k^3 - 2)$
$5k^6 - 19k^3 + 18$

111. $(3.1x + 4.5)^2$
$9.61x^2 + 27.9x + 20.25$

112. $(2.5y + 1.1)^2$
$6.25y^2 + 5.5y + 1.21$

113. $(k - 4)^3$
$k^3 - 12k^2 + 48k - 64$

114. $(h + 3)^3$
$h^3 + 9h^2 + 27h + 27$

115. $(5x + 3)^3$
$125x^3 + 225x^2 + 135x + 27$

116. $(2a - 4)^3$
$8a^3 - 48a^2 + 96a - 64$

117. $(y^2 + 2y + 1)(2y^2 - y + 3)$
$2y^4 + 3y^3 + 3y^2 + 5y + 3$

118. $(2w^2 - w - 5)(3w^2 + 2w + 1)$
$6w^4 + w^3 - 15w^2 - 11w - 5$

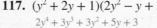

Expanding Your Skills

For Exercises 119–122, multiply the expressions containing more than two factors.

119. $2a(3a - 4)(a + 5)$ $6a^3 + 22a^2 - 40a$ **120.** $5x(x + 2)(6x - 1)$ $30x^3 + 55x^2 - 10x$

121. $(x - 3)(2x + 1)(x - 4)$ $2x^3 - 13x^2 + 17x + 12$ **122.** $(y - 2)(2y - 3)(y + 3)$ $2y^3 - y^2 - 15y + 18$

123. What binomial when multiplied by $(3x + 5)$ will produce a product of $6x^2 - 11x - 35$?
[*Hint:* Let the quantity $(a + b)$ represent the unknown binomial.] Then find a and b such that
$(3x + 5)(a + b) = 6x^2 - 11x - 35$. $2x - 7$

124. What binomial when multiplied by $(2x - 4)$ will produce a product of $2x^2 + 8x - 24$? $x + 6$

For Exercises 125–127, determine the values of k that would create a perfect square trinomial.

125. $x^2 + kx + 25$ $k = 10$ or -10 **126.** $w^2 + kw + 9$ $k = 6$ or -6 **127.** $a^2 + ka + 16$ $k = 8$ or -8

Section 5.7   Division of Polynomials

Concepts

1. Division by a Monomial
2. Long Division

Division of polynomials will be presented in this section as two separate cases: The first case illustrates division by a monomial divisor. The second case illustrates long division by a polynomial with two or more terms.

1. Division by a Monomial

To divide a polynomial by a monomial, divide each individual term in the polynomial by the divisor and simplify the result.

> **Dividing a Polynomial by a Monomial**
> If a, b, and c are polynomials such that $c \neq 0$, then
> $$\frac{a + b}{c} = \frac{a}{c} + \frac{b}{c} \qquad \text{Similarly,} \qquad \frac{a - b}{c} = \frac{a}{c} - \frac{b}{c}$$

Classroom Examples: p. 396,
Exercises 26 and 28

Example 1   Dividing a Polynomial by a Monomial

Divide the polynomials.

a. $\dfrac{5a^3 - 10a^2 + 20}{5a}$ **b.** $(12y^2z^3 - 15yz^2 + 6y^2z) \div (-6y^2z)$

Solution:

a. $\dfrac{5a^3 - 10a^2 + 20}{5a}$

$= \dfrac{5a^3}{5a} - \dfrac{10a^2}{5a} + \dfrac{20}{5a}$ Divide each term in the numerator by $5a$.

$= a^2 - 2a + \dfrac{4}{a}$ Simplify each term using the properties of exponents.

Writing   Translating Expression   Geometry   Scientific Calculator   Video

b. $(12y^2z^3 - 15yz^2 + 6y^2z) \div (-6y^2z)$

$$= \frac{12y^2z^3 - 15yz^2 + 6y^2z}{-6y^2z}$$

$$= \frac{12y^2z^3}{-6y^2z} - \frac{15yz^2}{-6y^2z} + \frac{6y^2z}{-6y^2z} \qquad \text{Divide each term by } -6y^2z.$$

$$= -2z^2 + \frac{5z}{2y} - 1 \qquad \text{Simplify each term.}$$

Instructor Note: In Example 1(b), mention to students that the third term in the quotient is −1 not 0.

Skill Practice Divide the polynomials.

1. $(36a^4 - 48a^3 + 12a^2) \div (6a^3)$

2. $\dfrac{-15x^3y^4 + 25x^2y^3 - 5xy^2}{-5xy^2}$

2. Long Division

If the divisor has two or more terms, a *long division* process similar to the division of real numbers is used. Take a minute to review the long division process for real numbers by dividing 2273 by 5.

$$
\begin{array}{r}
454 \longleftarrow \text{Quotient} \\
5{\overline{\smash{\big)}\,2273}} \\
\underline{-20}\downarrow \\
27 \\
\underline{-25}\downarrow \\
23 \\
\underline{-20} \\
3 \longleftarrow \text{Remainder}
\end{array}
$$

Therefore, $2273 \div 5 = 454\frac{3}{5}$.

A similar procedure is used for long division of polynomials as shown in Example 2.

Example 2 Using Long Division to Divide Polynomials

Divide the polynomials using long division. $(2x^2 - x + 3) \div (x - 3)$

Classroom Example: p. 396, Exercise 34

Solution:

$x - 3{\overline{\smash{\big)}\,2x^2 - x + 3}}$ — Divide the leading term in the dividend by the leading term in the divisor.

$$\frac{2x^2}{x} = 2x$$

This is the first term in the quotient.

$$
\begin{array}{r}
2x \\
x - 3{\overline{\smash{\big)}\,2x^2 - x + 3}} \\
-(2x^2 - 6x)
\end{array}
$$

Multiply $2x$ by the divisor: $2x(x - 3) = 2x^2 - 6x$ and subtract the result.

TIP: Recall that taking the opposite of a polynomial changes the sign of each term of the polynomial.

Answers

1. $6a - 8 + \dfrac{2}{a}$

2. $3x^2y^2 - 5xy + 1$

$$\begin{array}{r} 2x \\ x-3\overline{)2x^2-x+3} \\ \underline{-2x^2+6x} \\ 5x \end{array}$$

Subtract the quantity $2x^2-6x$. To do this, add the opposite.

$$\begin{array}{r} 2x+5 \\ x-3\overline{)2x^2-x+3} \\ \underline{-2x^2+6x}\downarrow \\ 5x+3 \end{array}$$

Bring down the next column, and repeat the process.

Divide the leading term by x: $(5x)/x = 5$. Place 5 in the quotient.

$$\begin{array}{r} 2x+5 \\ x-3\overline{)2x^2-x+3} \\ -2x^2+6x \\ 5x+3 \\ -(5x-15) \end{array}$$

Multiply the divisor by 5: $5(x-3) = 5x - 15$ and subtract the result.

$$\begin{array}{r} 2x+5 \\ x-3\overline{)2x^2-x+3} \\ -2x^2+6x \\ 5x+3 \\ \underline{-5x+15} \\ 18 \end{array}$$

Subtract the quantity $5x - 15$ by adding the opposite.

The remainder is 18.

Summary:

The quotient is $2x+5$
The remainder is   18
The divisor is $x-3$
The dividend is $2x^2-x+3$

The solution to a long division problem is usually written in the form:

$$\text{quotient} + \frac{\text{remainder}}{\text{divisor}}$$

Hence,

$$(2x^2-x+3) \div (x-3) = 2x+5 + \frac{18}{x-3}$$

Skill Practice Divide the polynomials using long division.

3. $(3x^2+2x-5) \div (x+2)$

The division of polynomials can be checked in the same fashion as the division of real numbers. To check Example 2, we use the **division algorithm**:

$$\text{Dividend} = (\text{divisor})(\text{quotient}) + \text{remainder}$$

$$2x^2-x+3 \stackrel{?}{=} (x-3)(2x+5) + (18)$$

$$\stackrel{?}{=} 2x^2+5x-6x-15 + (18)$$

$$= 2x^2-x+3 \ \checkmark$$

Answer

3. $3x-4 + \dfrac{3}{x+2}$

Example 3 **Using Long Division to Divide Polynomials**

Classroom Example: p. 397, Exercise 50

Divide the polynomials using long division: $(3w^3 + 26w^2 - 3) \div (3w - 1)$

Solution:

First note that the dividend has a missing power of w and can be written as $3w^3 + 26w^2 + 0w - 3$. The term $0w$ is a place holder for the missing term. It is helpful to use the place holder to keep the powers of w lined up.

$$
\begin{array}{r}
w^2 \\
3w - 1\overline{)3w^3 + 26w^2 + 0w - 3} \\
-(3w^3 - w^2)
\end{array}
$$

Divide $3w^3 \div 3w = w^2$. This is the first term of the quotient.
Then multiply $w^2(3w - 1) = 3w^3 - w^2$.

$$
\begin{array}{r}
w^2 \\
3w - 1\overline{)3w^3 + 26w^2 + 0w - 3} \\
\underline{-3w^3 + w^2} \\
27w^2 + 0w
\end{array}
$$

Subtract by adding the opposite.

Bring down the next column, and repeat the process

$$
\begin{array}{r}
w^2 + 9w \\
3w - 1\overline{)3w^3 + 26w^2 + 0w - 3} \\
\underline{-3w^3 + w^2} \\
27w^2 + 0w \\
-(27w^2 - 9w)
\end{array}
$$

Divide $27w^2$ by the leading term in the divisor. $27w^2 \div 3w = 9w$
Place $9w$ in the quotient.
Multiply $9w(3w - 1) = 27w^2 - 9w$.

$$
\begin{array}{r}
w^2 + 9w \\
3w - 1\overline{)3w^3 + 26w^2 + 0w - 3} \\
\underline{-3w^3 + w^2} \\
27w^2 + 0w \\
\underline{-27w^2 + 9w} \\
9w - 3
\end{array}
$$

Subtract by adding the opposite.

Bring down the next column, and repeat the process.

$$
\begin{array}{r}
w^2 + 9w + 3 \\
3w - 1\overline{)3w^3 + 26w^2 + 0w - 3} \\
\underline{-3w^3 + w^2} \\
27w^2 + 0w \\
\underline{-27w^2 + 9w} \\
9w - 3 \\
-(9w - 3)
\end{array}
$$

Divide $9w$ by the leading term in the divisor. $9w \div 3w = 3$
Place 3 in the quotient.
Multiply $3(3w - 1) = 9w - 3$.

$$
\begin{array}{r}
w^2 + 9w + 3 \\
3w - 1\overline{)3w^3 + 26w^2 + 0w - 3} \\
\underline{-3w^3 + w^2} \\
27w^2 + 0w \\
\underline{-27w^2 + 9w} \\
9w - 3 \\
\underline{-9w + 3} \\
0
\end{array}
$$

Subtract by adding the opposite.

The remainder is 0.

The quotient is $w^2 + 9w + 3$, and the remainder is 0.

Skill Practice Divide the polynomials using long division.

4. $\dfrac{9x^3 + 11x + 10}{3x + 2}$

Answer

4. $3x^2 - 2x + 5$

In Example 3, the remainder is zero. Therefore, we say that $3w - 1$ divides evenly into $3w^3 + 26w^2 - 3$. For this reason, the divisor and quotient are factors of $3w^3 + 26w^2 - 3$. To check, we have

$$\text{Dividend} = (\text{divisor})(\text{quotient}) + \text{remainder}$$

$$3w^3 + 26w^2 - 3 \overset{?}{=} (3w - 1)(w^2 + 9w + 3) + 0$$

$$\overset{?}{=} 3w^3 + 27w^2 + 9w - w^2 - 9w - 3$$

$$= 3w^3 + 26w^2 - 3 \checkmark$$

Classroom Example: p. 397, Exercise 56

Example 4   **Using Long Division to Divide Polynomials**

Divide the polynomials using long division.

$$\frac{2y + y^4 - 5}{1 + y^2}$$

Solution:

First note that both the dividend and divisor should be written in descending order:

$$\frac{y^4 + 2y - 5}{y^2 + 1}$$

Also note that the dividend and the divisor have missing powers of y. Leave place holders.

$$y^2 + 0y + 1 \overline{)y^4 + 0y^3 + 0y^2 + 2y - 5}$$

$$\begin{array}{r} y^2 \\ y^2 + 0y + 1 \overline{)y^4 + 0y^3 + 0y^2 + 2y - 5} \\ -(y^4 + 0y^3 + y^2) \end{array}$$

Divide $y^4 \div y^2 = y^2$. This is the first term of the quotient.

Multiply $y^2(y^2 + 0y + 1) = y^4 + 0y^3 + y^2$.

$$\begin{array}{r} y^2 \\ y^2 + 0y + 1 \overline{)y^4 + 0y^3 + 0y^2 + 2y - 5} \\ -y^4 - 0y^3 - y^2 \\ \hline -y^2 + 2y - 5 \end{array}$$

Subtract by adding the opposite.

Bring down the next columns.

$$\begin{array}{r} y^2 -1 \\ y^2 + 0y + 1 \overline{)y^4 + 0y^3 + 0y^2 + 2y - 5} \\ -y^4 - 0y^3 - y^2 \\ \hline -y^2 + 2y - 5 \\ -(-y^2 - 0y - 1) \end{array}$$

Divide $-y^2 \div y^2 = -1$.

Multiply $-1(y^2 + 0y + 1) = -y^2 - 0y - 1$.

$$\begin{array}{r} y^2 -1 \\ y^2 + 0y + 1 \overline{)y^4 + 0y^3 + 0y^2 + 2y - 5} \\ -y^4 - 0y^3 - y^2 \\ \hline -y^2 + 2y - 5 \\ y^2 + 0y + 1 \\ \hline 2y - 4 \end{array}$$

Subtract by adding the opposite.

Remainder

Therefore, $\dfrac{y^4 + 2y - 5}{y^2 + 1} = y^2 - 1 + \dfrac{2y - 4}{y^2 + 1}$.

Skill Practice Divide the polynomials using long division.

5. $(4 - x^2 + x^3) \div (2 + x^2)$

Answer

5. $x - 1 + \dfrac{-2x + 6}{x^2 + 2}$

| **Example 5** | **Determining Whether Long Division Is Necessary** |

Classroom Examples: p. 397, Exercises 66 and 70

Determine whether long division is necessary for each division of polynomials.

a. $\dfrac{2p^5 - 8p^4 + 4p - 16}{p^2 - 2p + 1}$

b. $\dfrac{2p^5 - 8p^4 + 4p - 16}{2p^2}$

c. $(3z^3 - 5z^2 + 10) \div (15z^3)$

d. $(3z^3 - 5z^2 + 10) \div (3z + 1)$

Solution:

- Long division is used when the divisor has *two or more terms*.
- If the divisor has *one term*, then divide each term in the dividend by the monomial divisor.

a. $\dfrac{2p^5 - 8p^4 + 4p - 16}{p^2 - 2p + 1}$ The divisor has three terms. Use long division.

b. $\dfrac{2p^5 - 8p^4 + 4p - 16}{2p^2}$ The divisor has one term. Long division is not necessary.

c. $(3z^3 - 5z^2 + 10) \div (15z^3)$ The divisor has one term. Long division is not necessary.

d. $(3z^3 - 5z^2 + 10) \div (3z + 1)$ The divisor has two terms. Use long division.

Skill Practice Divide the polynomials using the appropriate method of division.

6. $\dfrac{6x^3 - x^2 + 3x - 5}{2x + 3}$

7. $\dfrac{9w^3 - 18w^2 + 6w + 12}{3w}$

Answers

6. $3x^2 - 5x + 9 + \dfrac{-32}{2x+3}$

7. $3w^2 - 6w + 2 + \dfrac{4}{w}$

| **Section 5.7** | **Practice Exercises** |

For additional exercises, see Classroom Activities 5.7A–5.7B in the *Student's Resource Manual* at www.mhhe.com/moh.

Vocabulary and Key Concepts

1. The ___division___ algorithm states that: Dividend = (divisor)(___quotient___) + (___remainder___).

Review Exercises

For Exercises 2–11, perform the indicated operations.

2. $(6z^5 - 2z^3 + z - 6) - (10z^4 + 2z^3 + z^2 + z)$
$6z^5 - 10z^4 - 4z^3 - z^2 - 6$

3. $(7a^2 + a - 6) + (2a^2 + 5a + 11)$ $9a^2 + 6a + 5$

4. $(10x + y)(x - 3y)$ $10x^2 - 29xy - 3y^2$

5. $8b^2(2b^2 - 5b + 12)$ $16b^4 - 40b^3 + 96b^2$

6. $(10x + y) + (x - 3y)$ $11x - 2y$

7. $(2w^3 + 5)^2$ $4w^6 + 20w^3 + 25$

8. $\left(\dfrac{4}{3}y^2 - \dfrac{1}{2}y + \dfrac{3}{8}\right) - \left(\dfrac{1}{3}y^2 + \dfrac{1}{4}y - \dfrac{1}{8}\right)$ $y^2 - \dfrac{3}{4}y + \dfrac{1}{2}$

9. $\left(\dfrac{7}{8}w - 1\right)\left(\dfrac{7}{8}w + 1\right)$ $\dfrac{49}{64}w^2 - 1$

10. $(a + 3)(a^2 - 3a + 9)$ $a^3 + 27$

11. $(2x + 1)(5x - 3)$ $10x^2 - x - 3$

Concept 1: Division by a Monomial

12. There are two methods for dividing polynomials. Explain when long division is used.

   Use long division when the divisor is a polynomial with two or more terms.

13. **a.** Divide $\dfrac{15t^3 + 18t^2}{3t}$ $5t^2 + 6t$

   b. Check by multiplying the quotient by the divisor.

14. **a.** Divide $(-9y^4 + 6y^2 - y) \div (3y)$ $-3y^3 + 2y - \dfrac{1}{3}$

   b. Check by multiplying the quotient by the divisor.

For Exercises 15–30, divide the polynomials. **(See Example 1.)**

15. $(6a^2 + 4a - 14) \div (2)$ $3a^2 + 2a - 7$

16. $\dfrac{4b^2 + 16b - 12}{4}$ $b^2 + 4b - 3$

17. $\dfrac{-5x^2 - 20x + 5}{-5}$ $x^2 + 4x - 1$

18. $\dfrac{-3y^3 + 12y - 6}{-3}$ $y^3 - 4y + 2$

19. $\dfrac{3p^3 - p^2}{p}$ $3p^2 - p$

20. $(7q^4 + 5q^2) \div q$ $7q^3 + 5q$

21. $(4m^2 + 8m) \div 4m^2$ $1 + \dfrac{2}{m}$

22. $\dfrac{n^2 - 8}{n}$ $n - \dfrac{8}{n}$

23. $\dfrac{14y^4 - 7y^3 + 21y^2}{-7y^2}$ $-2y^2 + y - 3$

24. $(25a^5 - 5a^4 + 15a^3 - 5a) \div (-5a)$ $-5a^4 + a^3 - 3a^2 + 1$

25. $(4x^3 - 24x^2 - x + 8) \div (4x)$ $x^2 - 6x - \dfrac{1}{4} + \dfrac{2}{x}$

26. $\dfrac{20w^3 + 15w^2 - w + 5}{10w}$ $2w^2 + \dfrac{3}{2}w - \dfrac{1}{10} + \dfrac{1}{2w}$

27. $\dfrac{-a^3b^2 + a^2b^2 - ab^3}{-a^2b^2}$ $a - 1 + \dfrac{b}{a}$

28. $(3x^4y^3 - x^2y^2 - xy^3) \div (-x^2y^2)$ $-3x^2y + 1 + \dfrac{y}{x}$

29. $(6t^4 - 2t^3 + 3t^2 - t + 4) \div (2t^3)$

   $3t - 1 + \dfrac{3}{2t} - \dfrac{1}{2t^2} + \dfrac{2}{t^3}$

30. $\dfrac{2y^3 - 2y^2 + 3y - 9}{2y^2}$ $y - 1 + \dfrac{3}{2y} - \dfrac{9}{2y^2}$

Concept 2: Long Division

31. **a.** Divide $(z^2 + 7z + 11) \div (z + 5)$. $z + 2 + \dfrac{1}{z + 5}$

   b. Check by multiplying the quotient by the divisor and adding the remainder.

32. **a.** Divide $\dfrac{2w^2 - 7w + 3}{w - 4}$. $2w + 1 + \dfrac{7}{w - 4}$

   b. Check by multiplying the quotient by the divisor and adding the remainder.

For Exercises 33–58, divide the polynomials. **(See Examples 2–4.)**

33. $\dfrac{t^2 + 4t + 5}{t + 1}$ $t + 3 + \dfrac{2}{t + 1}$

34. $(3x^2 + 8x + 5) \div (x + 2)$ $3x + 2 + \dfrac{1}{x + 2}$

35. $(7b^2 - 3b - 4) \div (b - 1)$ $7b + 4$

36. $\dfrac{w^2 - w - 2}{w - 2}$ $w + 1$

37. $\dfrac{5k^2 - 29k - 6}{5k + 1}$ $k - 6$

38. $(4y^2 + 25y - 21) \div (4y - 3)$ $y + 7$

39. $(4p^3 + 12p^2 + p - 12) \div (2p + 3)$ $2p^2 + 3p - 4$

40. $\dfrac{12a^3 - 2a^2 - 17a - 5}{3a + 1}$ $4a^2 - 2a - 5$

41. $\dfrac{-k - 6 + k^2}{1 + k}$ $k - 2 + \dfrac{-4}{k + 1}$

42. $(1 + h^2 + 3h) \div (2 + h)$ $h + 1 + \dfrac{-1}{h + 2}$

43. $(4x^3 - 8x^2 + 15x - 16) \div (2x - 3)$
$2x^2 - x + 6 + \dfrac{2}{2x - 3}$

44. $\dfrac{3b^3 + b^2 + 17b - 49}{3b - 5}$ $b^2 + 2b + 9 + \dfrac{-4}{3b - 5}$

45. $\dfrac{3y^3 + 5y^2 + y + 1}{3y - 1}$ $y^2 + 2y + 1 + \dfrac{2}{3y - 1}$

46. $\dfrac{4t^3 + 4t^2 - 9t + 3}{2t + 3}$ $2t^2 - t - 3 + \dfrac{12}{2t + 3}$

47. $\dfrac{9 + a^2}{a + 3}$ $a - 3 + \dfrac{18}{a + 3}$

48. $(3 + m^2) \div (m + 3)$ $m - 3 + \dfrac{12}{m + 3}$

49. $(4x^3 - 3x - 26) \div (x - 2)$ $4x^2 + 8x + 13$

50. $(4y^3 + y + 1) \div (2y + 1)$ $2y^2 - y + 1$

51. $(w^4 + 5w^3 - 5w^2 - 15w + 7) \div (w^2 - 3)$
$w^2 + 5w - 2 + \dfrac{1}{w^2 - 3}$

52. $\dfrac{p^4 - p^3 - 4p^2 - 2p - 15}{p^2 + 2}$ $p^2 - p - 6 + \dfrac{-3}{p^2 + 2}$

53. $\dfrac{2n^4 + 5n^3 - 11n^2 - 20n + 12}{2n^2 + 3n - 2}$ $n^2 + n - 6$

54. $(6y^4 - 5y^3 - 8y^2 + 16y - 8) \div (2y^2 - 3y + 2)$
$3y^2 + 2y - 4$

55. $\dfrac{3y^4 + 2y + 3}{1 + y^2}$ $3y^2 - 3 + \dfrac{2y + 6}{y^2 + 1}$

56. $\dfrac{2x^4 + 6x + 4}{2 + x^2}$ $2x^2 - 4 + \dfrac{6x + 12}{x^2 + 2}$

57. $(5x^3 - 4x - 9) \div (5x^2 + 5x + 1)$
$x - 1 + \dfrac{-8}{5x^2 + 5x + 1}$

58. $\dfrac{3a^3 - 5a + 16}{3a^2 - 6a + 7}$ $a + 2 + \dfrac{2}{3a^2 - 6a + 7}$

59. Show that $(x^3 - 8) \div (x - 2)$ is *not* $(x^2 + 4)$.
Multiply $(x - 2)(x^2 + 4) = x^3 - 2x^2 + 4x - 8$, which does not equal $x^3 - 8$.

60. Explain why $(y^3 + 27) \div (y + 3)$ is *not* $(y^2 + 9)$.
To check, multiply $(y + 3)(y^2 + 9) = y^3 + 3y^2 + 9y + 27$, which does not equal $y^3 + 27$.

Mixed Exercises

For Exercises 61–72, determine which method to use to divide the polynomials: monomial division or long division. Then use that method to divide the polynomials. **(See Example 5.)**

61. $\dfrac{9a^3 + 12a^2}{3a}$ Monomial division; $3a^2 + 4a$

62. $\dfrac{3y^2 + 17y - 12}{y + 6}$ Long division; $3y - 1 + \dfrac{-6}{y + 6}$

63. $(p^3 + p^2 - 4p - 4) \div (p^2 - p - 2)$
Long division; $p + 2$

64. $(q^3 + 1) \div (q + 1)$
Long division; $q^2 - q + 1$

65. $\dfrac{t^4 + t^2 - 16}{t + 2}$
Long division; $t^3 - 2t^2 + 5t - 10 + \dfrac{4}{t + 2}$

66. $\dfrac{-8m^5 - 4m^3 + 4m^2}{-2m^2}$
Monomial division; $4m^3 + 2m - 2$

67. $(w^4 + w^2 - 5) \div (w^2 - 2)$
Long division; $w^2 + 3 + \dfrac{1}{w^2 - 2}$

68. $(2k^2 + 9k + 7) \div (k + 1)$
Long division; $2k + 7$

69. $\dfrac{n^3 - 64}{n - 4}$
Long division; $n^2 + 4n + 16$

70. $\dfrac{15s^2 + 34s + 28}{5s + 3}$
Long division; $3s + 5 + \dfrac{13}{5s + 3}$

71. $(9r^3 - 12r^2 + 9) \div (-3r^2)$
Monomial division; $-3r + 4 - \dfrac{3}{r^2}$

72. $(6x^4 - 16x^3 + 15x^2 - 5x + 10) \div (3x + 1)$
Long division; $2x^3 - 6x^2 + 7x - 4 + \dfrac{14}{3x + 1}$

 Writing   ← → Translating Expression   Geometry   Scientific Calculator   Video

Expanding Your Skills

For Exercises 73–80, divide the polynomials and note any patterns.

73. $(x^2 - 1) \div (x - 1)$
$x + 1$

74. $(x^3 - 1) \div (x - 1)$
$x^2 + x + 1$

75. $(x^4 - 1) \div (x - 1)$
$x^3 + x^2 + x + 1$

76. $(x^5 - 1) \div (x - 1)$
$x^4 + x^3 + x^2 + x + 1$

77. $x^2 \div (x - 1)$ $x + 1 + \dfrac{1}{x - 1}$

78. $x^3 \div (x - 1)$ $x^2 + x + 1 + \dfrac{1}{x - 1}$

79. $x^4 \div (x - 1)$
$x^3 + x^2 + x + 1 + \dfrac{1}{x - 1}$

80. $x^5 \div (x - 1)$
$x^4 + x^3 + x^2 + x + 1 + \dfrac{1}{x - 1}$

Problem Recognition Exercises

Operations on Polynomials

For Exercises 1–40, perform the indicated operations and simplify.

1. a. $6x^2 + 2x^2$ a. $8x^2$
   b. $(6x^2)(2x^2)$ b. $12x^4$

2. a. $8y^3 + y^3$ a. $9y^3$
   b. $(8y^3)(y^3)$ b. $8y^6$

3. a. $(4x + y)^2$
   b. $(4xy)^2$ a. $16x^2 + 8xy + y^2$
      b. $16x^2y^2$

4. a. $(2a + b)^2$
   b. $(2ab)^2$ a. $4a^2 + 4ab + b^2$
      b. $4a^2b^2$

5. a. $(2x + 3) + (4x - 2)$
   b. $(2x + 3)(4x - 2)$
a. $6x + 1$ b. $8x^2 + 8x - 6$

6. a. $(5m^2 + 1) + (m^2 + m)$
   b. $(5m^2 + 1)(m^2 + m)$
a. $6m^2 + m + 1$ b. $5m^4 + 5m^3 + m^2 + m$

7. a. $(3z + 2)^2$
   b. $(3z + 2)(3z - 2)$
a. $9z^2 + 12z + 4$ b. $9z^2 - 4$

8. a. $(6y - 7)^2$
   b. $(6y - 7)(6y + 7)$
a. $36y^2 - 84y + 49$ b. $36y^2 - 49$

9. a. $(2x - 4)(x^2 - 2x + 3)$
   b. $(2x - 4) + (x^2 - 2x + 3)$
a. $2x^3 - 8x^2 + 14x - 12$ b. $x^2 - 1$

10. a. $(3y^2 + 8)(-y^2 - 4)$
    b. $(3y^2 + 8) - (-y^2 - 4)$
a. $-3y^4 - 20y^2 - 32$ b. $4y^2 + 12$

11. a. $x + x$ a. $2x$
    b. $x \cdot x$ b. x^2

12. a. $2c + 2c$ a. $4c$
    b. $2c \cdot 2c$ b. $4c^2$

13. $(7mn)^2$ $49m^2n^2$

14. $(8pq)^2$ $64p^2q^2$

15. $(-2x^4 - 6x^3 + 8x^2) \div (2x^2)$ $-x^2 - 3x + 4$

16. $(-15m^3 + 12m^2 - 3m) \div (-3m)$ $5m^2 - 4m + 1$

17. $(m^3 - 4m^2 - 6) - (3m^2 + 7m) + (-m^3 - 9m + 6)$
$-7m^2 - 16m$

18. $(n^4 + 2n^2 - 3n) + (4n^2 + 2n - 1) - (4n^5 + 6n - 3)$
$-4n^5 + n^4 + 6n^2 - 7n + 2$

19. $(8x^3 + 2x + 6) \div (x - 2)$ $8x^2 + 16x + 34 + \dfrac{74}{x - 2}$

20. $(-4x^3 + 2x^2 - 5) \div (x - 3)$ $-4x^2 - 10x - 30 + \dfrac{-95}{x - 3}$

21. $(2x - y)(3x^2 + 4xy - y^2)$ $6x^3 + 5x^2y - 6xy^2 + y^3$

22. $(3a + b)(2a^2 - ab + 2b^2)$ $6a^3 - a^2b + 5ab^2 + 2b^3$

23. $(x + y^2)(x^2 - xy^2 + y^4)$ $x^3 + y^6$

24. $(m^2 + 1)(m^4 - m^2 + 1)$ $m^6 + 1$

25. $(a^2 + 2b) - (a^2 - 2b)$
$4b$

26. $(y^3 - 6z) - (y^3 + 6z)$
$-12z$

27. $(a^3 + 2b)(a^3 - 2b)$
$a^6 - 4b^2$

28. $(y^3 - 6z)(y^3 + 6z)$
$y^6 - 36z^2$

29. $\dfrac{8p^2 + 4p - 6}{2p - 1}$
$4p + 4 + \dfrac{-2}{2p - 1}$

30. $\dfrac{4v^2 - 8v + 8}{2v + 3}$
$2v - 7 + \dfrac{29}{2v + 3}$

31. $\dfrac{12x^3y^7}{3xy^5}$ $4x^2y^2$

32. $\dfrac{-18p^2q^4}{2pq^3}$ $-9pq$

33. $\left(\dfrac{3}{7}x - \dfrac{1}{2}\right)\left(\dfrac{3}{7}x + \dfrac{1}{2}\right)$
$\dfrac{9}{49}x^2 - \dfrac{1}{4}$

34. $\left(\dfrac{2}{5}y + \dfrac{4}{3}\right)\left(\dfrac{2}{5}y - \dfrac{4}{3}\right)$
$\dfrac{4}{25}y^2 - \dfrac{16}{9}$

35. $\left(\dfrac{1}{9}x^3 + \dfrac{2}{3}x^2 + \dfrac{1}{6}x - 3\right) - \left(\dfrac{4}{3}x^3 + \dfrac{1}{9}x^2 + \dfrac{2}{3}x + 1\right)$
$-\dfrac{11}{9}x^3 + \dfrac{5}{9}x^2 - \dfrac{1}{2}x - 4$

36. $\left(\dfrac{1}{10}y^2 - \dfrac{3}{5}y - \dfrac{1}{15}\right) - \left(\dfrac{7}{5}y^2 + \dfrac{3}{10}y - \dfrac{1}{3}\right)$
$-\dfrac{13}{10}y^2 - \dfrac{9}{10}y + \dfrac{4}{15}$

37. $(0.05x^2 - 0.16x - 0.75) + (1.25x^2 - 0.14x + 0.25)$
$1.3x^2 - 0.3x - 0.5$

38. $(1.6w^3 + 2.8w + 6.1) + (3.4w^3 - 4.1w^2 - 7.3)$ $5w^3 - 4.1w^2 + 2.8w - 1.2$

39. $(3x^2y)(-2xy^5)$ $-6x^3y^6$

40. $(10ab^4)(5a^3b^2)$ $50a^4b^6$

 Writing   Translating Expression   Geometry   Scientific Calculator   Video

Group Activity

The Pythagorean Theorem and a Geometric "Proof"

Estimated Time: 25–30 minutes

Group Size: 2

Right triangles occur in many applications of mathematics. By definition, a right triangle is a triangle that contains a 90° angle. The two shorter sides in a right triangle are referred to as the "legs," and the longest side is called the "hypotenuse." In the triangle shown, the legs are labeled as a and b, and the hypotenuse is labeled as c.

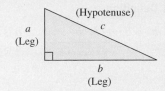

   Right triangles have an important property that the sum of the squares of the two legs of a right triangle equals the square of the hypotenuse. This fact is referred to as the Pythagorean theorem. In symbols, the Pythagorean theorem is stated as:

$$a^2 + b^2 = c^2$$

1. The following triangles are right triangles. Verify that $a^2 + b^2 = c^2$. (The units may be left off when performing these calculations.)

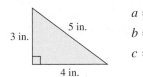

$a = 3$
$b = 4$
$c = 5$

$$a^2 + b^2 = c^2$$
$$(3)^2 + (4)^2 \overset{?}{=} (5)^2$$
$$9 + 16 = 25 \ \checkmark$$

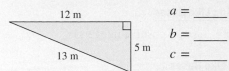

$a = \underline{\quad}$
$b = \underline{\quad}$
$c = \underline{\quad}$

$$a^2 + b^2 = c^2$$
$$(\underline{\quad})^2 + (\underline{\quad})^2 \overset{?}{=} (\underline{\quad})^2$$
$$\underline{\quad} + \underline{\quad} = \underline{\quad} \ \checkmark$$

2. The following geometric "proof" of the Pythagorean theorem uses addition, subtraction, and multiplication of polynomials. Consider the square figure. The length of each side of the large outer square is $(a + b)$. Therefore, the area of the large outer square is $(a + b)^2$.

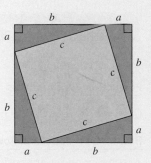

   The area of the large outer square can also be found by adding the area of the inner square (pictured in light gray) plus the area of the four right triangles (pictured in dark gray).

Area of inner square: c^2 Area of the four right triangles: $4 \cdot \left(\frac{1}{2} a b\right)$

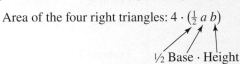

½ Base · Height

3. Now equate the two expressions representing the area of the large outer square:

$$\left(\begin{array}{c}\text{Area of outer}\\ \text{square}\end{array}\right) = \left(\begin{array}{c}\text{area of inner}\\ \text{square}\end{array}\right) + \left(\begin{array}{c}\text{4 times the area}\\ \text{of the right triangles}\end{array}\right)$$

$(a+b)^2 = c^2 + 4 \cdot (\frac{1}{2}ab);$
$a^2 + 2ab + b^2 = c^2 + 2ab;$
$a^2 + b^2 = c^2$

$$\underline{\qquad} = \underline{\qquad} + \underline{\qquad}$$

◀——— Clear parentheses on both sides of the equation.

$$\underline{\qquad} = \underline{\qquad} + \underline{\qquad}$$

◀——— Subtract $2ab$ from both sides.

$$\underline{\qquad} = \underline{\qquad}$$

Chapter 5   Summary

Section 5.1   Multiplying and Dividing Expressions with Common Bases

Key Concepts

Definition

$b^n = \underbrace{b \cdot b \cdot b \cdot b \cdot \ldots \ b}_{n \text{ factors of } b}$ b is the base,
n is the exponent

Multiplying Like Bases

$b^m b^n = b^{m+n}$ $(m, n \text{ positive integers})$

Dividing Like Bases

$\dfrac{b^m}{b^n} = b^{m-n}$ $(b \neq 0, m, n, \text{ positive integers})$

Examples

Example 1

$3^4 = 3 \cdot 3 \cdot 3 \cdot 3 = 81$ 3 is the base
4 is the exponent

Example 2

Compare: $(-5)^2$ versus -5^2

versus $\left[\begin{array}{l} \rule{0pt}{1em} \\ \\ \end{array} \right.$

$(-5)^2 = (-5)(-5) = 25$

$-5^2 = -1(5^2) = -1(5)(5) = -25$

Example 3

Simplify: $x^3 \cdot x^4 \cdot x^2 \cdot x = x^{3+4+2+1} = x^{10}$

Example 4

Simplify: $\dfrac{c^4 d^{10}}{c d^5} = c^{4-1} d^{10-5} = c^3 d^5$

Section 5.2   More Properties of Exponents

Key Concepts

Power Rule for Exponents

$(b^m)^n = b^{mn}$ $(b \neq 0, m, n \text{ positive integers})$

Power of a Product and Power of a Quotient

Assume m and n are positive integers and a and b are real numbers where $b \neq 0$.
Then,

$(ab)^m = a^m b^m$ and $\left(\dfrac{a}{b} \right)^m = \dfrac{a^m}{b^m}$

Examples

Example 1

Simplify: $(x^4)^5 = x^{20}$

Example 2

Simplify: $(4uv^2)^3 = 4^3 u^3 (v^2)^3 = 64 u^3 v^6$

Example 3

Simplify: $\left(\dfrac{p^5 q^3}{5pq^2} \right)^2 = \left(\dfrac{p^{5-1} q^{3-2}}{5} \right)^2 = \left(\dfrac{p^4 q}{5} \right)^2$

$= \dfrac{(p^4)^2 (q)^2}{(5)^2} = \dfrac{p^8 q^2}{25}$

Section 5.3 — Definitions of b^0 and b^{-n}

Key Concepts

Definitions

If b is a nonzero real number and n is an integer, then:

1. $b^0 = 1$

2. $b^{-n} = \left(\dfrac{1}{b}\right)^n = \dfrac{1}{b^n}$

Examples

Example 1

Simplify: $4^0 = 1$

Example 2

Simplify: $y^{-7} = \dfrac{1}{y^7}$

Example 3

Simplify: $\dfrac{8a^0 b^{-2}}{c^{-5} d}$

$$= \frac{8(1)c^5}{b^2 d} = \frac{8c^5}{b^2 d}$$

Section 5.4 — Scientific Notation

Key Concepts

A positive number written in **scientific notation** is expressed in the form:

$a \times 10^n$ where $1 \le a < 10$ and n is an integer.

$35{,}000 = 3.5 \times 10^4$

$0.000\,000\,548 = 5.48 \times 10^{-7}$

Examples

Example 1

Multiply: $(3.5 \times 10^4)(2 \times 10^{-6})$

$$= 7 \times 10^{-2}$$

Example 2

Divide: $\dfrac{2.1 \times 10^{-9}}{8.4 \times 10^3} = 0.25 \times 10^{-9-3}$

$$= 0.25 \times 10^{-12}$$

$$= (2.5 \times 10^{-1}) \times 10^{-12}$$

$$= 2.5 \times 10^{-13}$$

| **Section 5.5** | **Addition and Subtraction of Polynomials** |

Key Concepts

A **polynomial** in one variable is a sum of terms of the form ax^n, where a is a real number and the exponent, n, is a nonnegative integer. For each term, a is called the **coefficient** of the term and n is the **degree of the term**. The term with highest degree is the **leading term**, and its coefficient is called the **leading coefficient**. The **degree of the polynomial** is the largest degree of all its terms.

   To add or subtract polynomials, add or subtract *like* terms.

Examples

Example 1

Given: $4x^5 - 8x^3 + 9x - 5$

Coefficients of each term: 4, −8, 9, −5

Degree of each term: 5, 3, 1, 0

Leading term: $4x^5$

Leading coefficient: 4

Degree of polynomial: 5

Example 2

Perform the indicated operations:

$$(2x^4 - 5x^3 + 1) - (x^4 + 3) + (x^3 - 4x - 7)$$
$$= 2x^4 - 5x^3 + 1 - x^4 - 3 + x^3 - 4x - 7$$
$$= 2x^4 - x^4 - 5x^3 + x^3 - 4x + 1 - 3 - 7$$
$$= x^4 - 4x^3 - 4x - 9$$

| **Section 5.6** | **Multiplication of Polynomials and Special Products** |

Key Concepts

Multiplying Monomials

Use the commutative and associative properties of multiplication to group coefficients and like bases.

Multiplying Polynomials

Multiply each term in the first polynomial by each term in the second polynomial.

Product of Conjugates

The product of conjugates results in a **difference of squares**

$$(a + b)(a - b) = a^2 - b^2$$

Square of a Binomial

The square of a binomial results in a **perfect square trinomial**

$$(a + b)^2 = a^2 + 2ab + b^2$$
$$(a - b)^2 = a^2 - 2ab + b^2$$

Examples

Example 1

Multiply: $(5a^2b)(-2ab^3)$
$$= [5 \cdot (-2)](a^2a)(bb^3)$$
$$= -10a^3b^4$$

Example 2

Multiply: $(x - 2)(3x^2 - 4x + 11)$
$$= 3x^3 - 4x^2 + 11x - 6x^2 + 8x - 22$$
$$= 3x^3 - 10x^2 + 19x - 22$$

Example 3

Multiply: $(3w - 4v)(3w + 4v)$
$$= (3w)^2 - (4v)^2$$
$$= 9w^2 - 16v^2$$

Example 4

Multiply: $(5c - 8d)^2$
$$= (5c)^2 - 2(5c)(8d) + (8d)^2$$
$$= 25c^2 - 80cd + 64d^2$$

Section 5.7 Division of Polynomials

Key Concepts

Division of Polynomials

1. Division by a monomial, use the properties:

$$\frac{a+b}{c} = \frac{a}{c} + \frac{b}{c} \qquad \text{and} \qquad \frac{a-b}{c} = \frac{a}{c} - \frac{b}{c}$$

2. If the divisor has more than one term, use long division.

Examples

Example 1

Divide: $\dfrac{-3x^2 - 6x + 9}{-3x}$

$$= \frac{-3x^2}{-3x} - \frac{6x}{-3x} + \frac{9}{-3x}$$

$$= x + 2 - \frac{3}{x}$$

Example 2

Divide: $(3x^2 - 5x + 1) \div (x + 2)$

$$\begin{array}{r} 3x - 11 \\ x+2\overline{)3x^2 - 5x + 1} \\ -(3x^2 + 6x) \\ \hline -11x + 1 \\ -(-11x - 22) \\ \hline 23 \end{array}$$

$$3x - 11 + \frac{23}{x+2}$$

Chapter 5 Review Exercises

Section 5.1

For Exercises 1–4, identify the base and the exponent.

1. 5^3
Base: 5; exponent: 3

2. x^4
Base: x; exponent: 4

3. $(-2)^0$
Base: −2; exponent: 0

4. y
Base: y; exponent: 1

5. Evaluate the expressions.

   a. 6^2 36   **b.** $(-6)^2$ 36   **c.** -6^2 −36

6. Evaluate the expressions.

   a. 4^3 64   **b.** $(-4)^3$ −64   **c.** -4^3 −64

For Exercises 7–18, simplify and write the answers in exponent form. Assume that all variables represent nonzero real numbers.

7. $5^3 \cdot 5^{10}$ 5^{13}

8. $a^7 a^4$ a^{11}

9. $x \cdot x^6 \cdot x^2$ x^9

10. $6^3 \cdot 6 \cdot 6^5$ 6^9

11. $\dfrac{10^7}{10^4}$ 10^3

12. $\dfrac{y^{14}}{y^8}$ y^6

13. $\dfrac{b^9}{b}$ b^8

14. $\dfrac{7^8}{7}$ 7^7

15. $\dfrac{k^2 k^3}{k^4}$ k

16. $\dfrac{8^4 \cdot 8^7}{8^{11}}$ 1

17. $\dfrac{2^8 \cdot 2^{10}}{2^3 \cdot 2^7}$ 2^8

18. $\dfrac{q^3 q^{12}}{qq^8}$ q^6

19. Explain why $2^2 \cdot 4^4$ does *not* equal 8^6.
Exponents are added only when multiplying factors with the same base. In such a case, the base does not change.

20. Explain why $\frac{10^5}{5^2}$ does *not* equal 2^3.
Exponents are subtracted only when dividing factors with the same base. In such a case, the base does not change.

Writing   Translating Expression   Geometry   Scientific Calculator   Video

For Exercises 21–22, use the formula

$$A = P(1 + r)^t$$

21. Find the amount in an account after 3 years if the initial investment is \$6000, invested at 6% interest compounded annually.   \$7146.10

22. Find the amount in an account after 2 years if the initial investment is \$20,000, invested at 5% interest compounded annually.   \$22,050

Section 5.2

For Exercises 23–40, simplify each expression. Write the answer in exponent form. Assume all variables represent nonzero real numbers.

23. $(7^3)^4$ 7^{12}

24. $(c^2)^6$ c^{12}

25. $(p^4p^2)^3$ p^{18}

26. $(9^5 \cdot 9^2)^4$ 9^{28}

27. $\left(\dfrac{a}{b}\right)^2$ $\dfrac{a^2}{b^2}$

28. $\left(\dfrac{1}{3}\right)^4$ $\dfrac{1}{3^4}$

29. $\left(\dfrac{5}{c^2d^5}\right)^2$ $\dfrac{5^2}{c^4d^{10}}$

30. $\left(-\dfrac{m^2}{4n^6}\right)^5$ $-\dfrac{m^{10}}{4^5n^{30}}$

31. $(2ab^2)^4$ $2^4a^4b^8$

32. $(-x^7y)^2$ $x^{14}y^2$

33. $\left(\dfrac{-3x^3}{5y^2z}\right)^3$ $-\dfrac{3^3x^9}{5^3y^6z^3}$

34. $\left(\dfrac{r^3}{s^2t^6}\right)^5$ $\dfrac{r^{15}}{s^{10}t^{30}}$

35. $\dfrac{a^4(a^2)^8}{(a^3)^3}$ a^{11}

36. $\dfrac{(8^3)^4 \cdot 8^{10}}{(8^4)^5}$ 8^2

37. $\dfrac{(4h^2k)^2(h^3k)^4}{(2hk^3)^2}$ $4h^{14}$

38. $\dfrac{(p^3q)^3(2p^2q^4)^4}{(8p)(pq^3)^2}$ $2p^{14}q^{13}$

39. $\left(\dfrac{2x^4y^3}{4xy^2}\right)^2$ $\dfrac{x^6y^2}{4}$

40. $\left(\dfrac{a^4b^6}{ab^4}\right)^3$ a^9b^6

Section 5.3

For Exercises 41–62, simplify each expression. Write the answers with positive exponents. Assume all variables represent nonzero real numbers.

41. 8^0 1

42. $(-b)^0$ 1

43. $-x^0$ -1

44. 1^0 1

45. $2y^0$ 2

46. $(2y)^0$ 1

47. z^{-5} $\dfrac{1}{z^5}$

48. 10^{-4} $\dfrac{1}{10^4}$

49. $(6a)^{-2}$ $\dfrac{1}{36a^2}$

50. $6a^{-2}$ $\dfrac{6}{a^2}$

51. $4^0 + 4^{-2}$ $\dfrac{17}{16}$

52. $9^{-1} + 9^0$ $\dfrac{10}{9}$

53. $t^{-6}t^{-2}$ $\dfrac{1}{t^8}$

54. r^8r^{-9} $\dfrac{1}{r}$

55. $\dfrac{12x^{-2}y^3}{6x^4y^{-4}}$ $\dfrac{2y^7}{x^6}$

56. $\dfrac{8ab^{-3}c^0}{10a^{-5}b^{-4}c^{-1}}$ $\dfrac{4a^6bc}{5}$

57. $(-2m^2n^{-4})^{-4}$ $\dfrac{n^{16}}{16m^8}$

58. $(3u^{-5}v^2)^{-3}$ $\dfrac{u^{15}}{27v^6}$

59. $\dfrac{(k^{-6})^{-2}(k^3)}{5k^{-6}k^0}$ $\dfrac{k^{21}}{5}$

60. $\dfrac{(3h)^{-2}(h^{-5})^{-3}}{h^{-4}h^8}$ $\dfrac{h^9}{9}$

61. $2 \cdot 3^{-1} - 6^{-1}$ $\dfrac{1}{2}$

62. $2^{-1} - 2^{-2} + 2^0$ $\dfrac{5}{4}$

Section 5.4

63. Write the numbers in scientific notation.

   a. In a recent year there were 97,400,000 packages of M&Ms sold in the United States.
   9.74×10^7

   b. The thickness of a piece of paper is 0.0042 in.
   4.2×10^{-3} in.

64. Write the numbers in standard form.

   a. A pH of 10 means the hydrogen ion concentration is 1×10^{-10} units.   0.000 000 0001

   b. A fundraising event for neurospinal research raised 2.56×10^5.   \$256,000

For Exercises 65–68, perform the indicated operations. Write the answers in scientific notation.

65. $(41 \times 10^{-6})(2.3 \times 10^{11})$
   9.43×10^5

66. $\dfrac{9.3 \times 10^3}{6 \times 10^{-7}}$ 1.55×10^{10}

67. $\dfrac{2000}{0.000008}$ 2.5×10^8

68. $(0.000078)(21,000,000)$ 1.638×10^3

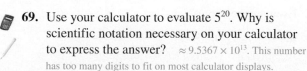 69. Use your calculator to evaluate 5^{20}. Why is scientific notation necessary on your calculator to express the answer? $\approx 9.5367 \times 10^{13}$. This number has too many digits to fit on most calculator displays.

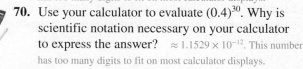 70. Use your calculator to evaluate $(0.4)^{30}$. Why is scientific notation necessary on your calculator to express the answer? $\approx 1.1529 \times 10^{-12}$. This number has too many digits to fit on most calculator displays.

71. The average distance between the Earth and Sun is 9.3×10^7 mi.

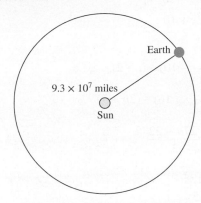

Earth

9.3×10^7 miles

Sun

a. If the Earth's orbit is approximated by a circle, find the total distance the Earth travels around the Sun in one orbit. (*Hint:* The circumference of a circle is given by $C = 2\pi r$.) Express the answer in scientific notation.
$\approx 5.84 \times 10^8$ mi

b. If the Earth makes one complete trip around the Sun in 1 year (365 days = 8.76×10^3 hr), find the average speed that the Earth travels around the Sun in miles per hour. Express the answer in scientific notation.
$\approx 6.67 \times 10^4$ mph

 72. The average distance between the planet Mercury and the Sun is 3.6×10^7 mi.

a. If Mercury's orbit is approximated by a circle, find the total distance Mercury travels around the Sun in one orbit. (*Hint:* The circumference of a circle is given by $C = 2\pi r$.) Express the answer in scientific notation. $\approx 2.26 \times 10^8$ mi

b. If Mercury makes one complete trip around the Sun in 88 days (2.112×10^3 hr), find the average speed that Mercury travels around the Sun in miles per hour. Express the answer in scientific notation.
$\approx 1.07 \times 10^5$ mph

Section 5.5

73. For the polynomial $7x^4 - x + 6$

a. Classify as a monomial, a binomial, or a trinomial.   Trinomial

b. Identify the degree of the polynomial.   4

c. Identify the leading coefficient.   7

74. For the polynomial $2y^3 - 5y^7$

a. Classify as a monomial, a binomial, or a trinomial.   Binomial

b. Identify the degree of the polynomial.   7

c. Identify the leading coefficient.   −5

For Exercises 75–80, add or subtract as indicated.

75. $(4x + 2) + (3x - 5)$ $7x - 3$

76. $(7y^2 - 11y - 6) - (8y^2 + 3y - 4)$ $-y^2 - 14y - 2$

77. $(9a^2 - 6) - (-5a^2 + 2a)$ $14a^2 - 2a - 6$

78. $\left(5x^3 - \dfrac{1}{4}x^2 + \dfrac{5}{8}x + 2\right) + \left(\dfrac{5}{2}x^3 + \dfrac{1}{2}x^2 - \dfrac{1}{8}x\right)$
$\dfrac{15}{2}x^3 + \dfrac{1}{4}x^2 + \dfrac{1}{2}x + 2$

79. $8w^4 \qquad\quad - 6w + 3$ $10w^4 + 2w^3 - 7w + 4$
$+ 2w^4 + 2w^3 - \ w + 1$

80. $-0.02b^5 + b^4 \qquad\quad - 0.7b + 0.3$
$+ \ \underline{0.03b^5 \qquad - 0.1b^3 + \quad b + 0.03}$
$0.01b^5 + b^4 - 0.1b^3 + 0.3b + 0.33$

81. Subtract $(9x^2 + 4x + 6)$ from $(7x^2 - 5x)$.
$-2x^2 - 9x - 6$

82. Find the difference of $(x^2 - 5x - 3)$ and $(6x^2 + 4x + 9)$. $-5x^2 - 9x - 12$

83. Write a trinomial of degree 2 with a leading coefficient of −5. (Answers may vary.)
For example, $-5x^2 + 2x - 4$

84. Write a binomial of degree 5 with leading coefficient 6. (Answers may vary.)
For example, $6x^5 + 8$

 85. Find a polynomial that represents the perimeter of the given rectangle. $6w + 6$

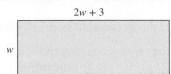

$2w + 3$

w

Section 5.6

For Exercises 86–103, multiply the expressions.

86. $(25x^4y^3)(-3x^2y)$
$-75x^6y^4$

87. $(9a^6)(2a^2b^4)$
$18a^8b^4$

88. $5c(3c^3 - 7c + 5)$
$15c^4 - 35c^2 + 25c$

89. $(x^2 + 5x - 3)(-2x)$
$-2x^3 - 10x^2 + 6x$

90. $(5k - 4)(k + 1)$
$5k^2 + k - 4$

91. $(4t - 1)(5t + 2)$
$20t^2 + 3t - 2$

92. $(q + 8)(6q - 1)$
$6q^2 + 47q - 8$

93. $(2a - 6)(a + 5)$
$2a^2 + 4a - 30$

94. $\left(7a + \dfrac{1}{2}\right)^2$
$49a^2 + 7a + \dfrac{1}{4}$

95. $(b - 4)^2$
$b^2 - 8b + 16$

96. $(4p^2 + 6p + 9)(2p - 3)$ $8p^3 - 27$

97. $(2w - 1)(-w^2 - 3w - 4)$ $-2w^3 - 5w^2 - 5w + 4$

98. $\begin{array}{r} 2x^2 - 3x + 4 \\ \times \underline{\quad 2x - 1} \end{array}$
$4x^3 - 8x^2 + 11x - 4$

99. $\begin{array}{r} 4a^2 + \ a - 5 \\ \times \underline{\quad 3a + 2} \end{array}$
$12a^3 + 11a^2 - 13a - 10$

100. $(b-4)(b+4)$
$b^2 - 16$

101. $\left(\frac{1}{3}r^4 - s^2\right)\left(\frac{1}{3}r^4 + s^2\right)$
$\frac{1}{9}r^8 - s^4$

102. $(-7z^2 + 6)^2$ $49z^4 - 84z^2 + 36$

103. $(2h+3)(h^4 - h^3 + h^2 - h + 1)$
$2h^5 + h^4 - h^3 + h^2 - h + 3$

104. Find a polynomial that represents the area of the given rectangle. $2x^2 + 3x - 20$

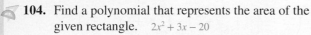

$2x - 5$

$x + 4$

Section 5.7

For Exercises 105–117, divide the polynomials.

105. $\dfrac{20y^3 - 10y^2}{5y}$ $4y^2 - 2y$

106. $(18a^3b^2 - 9a^2b - 27ab^2) \div 9ab$ $2a^2b - a - 3b$

107. $(12x^4 - 8x^3 + 4x^2) \div (-4x^2)$ $-3x^2 + 2x - 1$

108. $\dfrac{10z^7w^4 - 15z^3w^2 - 20zw}{-20z^2w}$ $-\dfrac{z^5w^3}{2} + \dfrac{3zw}{4} + \dfrac{1}{z}$

109. $\dfrac{x^2 + 7x + 10}{x + 5}$ $x + 2$

110. $(2t^2 + t - 10) \div (t - 2)$ $2t + 5$

111. $(2p^2 + p - 16) \div (2p + 7)$ $p - 3 + \dfrac{5}{2p + 7}$

112. $\dfrac{5a^2 + 27a - 22}{5a - 3}$ $a + 6 + \dfrac{-4}{5a - 3}$

113. $\dfrac{b^3 - 125}{b - 5}$ $b^2 + 5b + 25$

114. $(z^3 + 4z^2 + 5z + 20) \div (5 + z^2)$ $z + 4$

115. $(y^4 - 4y^3 + 5y^2 - 3y + 2) \div (y^2 + 3)$
$y^2 - 4y + 2 + \dfrac{9y - 4}{y^2 + 3}$

116. $(3t^4 - 8t^3 + t^2 - 4t - 5) \div (3t^2 + t + 1)$
$t^2 - 3t + 1 + \dfrac{-2t - 6}{3t^2 + t + 1}$

117. $\dfrac{2w^4 + w^3 + 4w - 3}{2w^2 - w + 3}$ $w^2 + w - 1$

Chapter 5 Test

Assume all variables represent nonzero real numbers.

1. Expand the expression using the definition of exponents, then simplify: $\dfrac{3^4 \cdot 3^3}{3^6}$
$\dfrac{(3 \cdot 3 \cdot 3 \cdot 3) \cdot (3 \cdot 3 \cdot 3)}{3 \cdot 3 \cdot 3 \cdot 3 \cdot 3 \cdot 3} = 3$

For Exercises 2–13, simplify each expression. Write the answer with positive exponents only.

2. $9^5 \cdot 9$ 9^6

3. $\dfrac{q^{10}}{q^2}$ q^8

4. $(-7)^0$ 1

5. c^{-3} $\dfrac{1}{c^3}$

6. $3^0 + 2^{-1} - 4^{-1}$ $\dfrac{5}{4}$

7. $4 \cdot 8^{-1} + 16^0$ $\dfrac{3}{2}$

8. $(3a^2b)^3$ $27a^6b^3$

9. $\left(\dfrac{2x}{y^3}\right)^4$ $\dfrac{16x^4}{y^{12}}$

10. $\dfrac{14^3 \cdot 14^9}{14^{10} \cdot 14}$ 14

11. $\dfrac{(s^2t)^3(7s^4t)^4}{(7s^2t^3)^2}$ $49s^{18}t$

12. $(2a^0b^{-6})^2$ $\dfrac{4}{b^{12}}$

13. $\left(\dfrac{6a^{-5}b}{8ab^{-2}}\right)^{-2}$ $\dfrac{16a^{12}}{9b^6}$

14. a. Write the number in scientific notation: $43{,}000{,}000{,}000$ 4.3×10^{10}

b. Write the number in standard form: 5.6×10^{-6} $0.000\,0056$

15. Multiply: $(1.2 \times 10^6)(7 \times 10^{-15})$ 8.4×10^{-9}

16. Divide: $\dfrac{60{,}000}{0.008}$ 7.5×10^6

17. The average amount of water flowing over Niagara Falls is 1.68×10^5 m^3/min.

a. How many cubic meters of water flow over the falls in one day? 2.4192×10^8 m^3

b. How many cubic meters of water flow over the falls in one year? 8.83008×10^{10} m^3

© Alberto Fresco/Alamy RF

 Writing Translating Expression Geometry Scientific Calculator  Video

18. Write the polynomial in descending order:
$4x + 5x^3 - 7x^2 + 11$ $5x^3 - 7x^2 + 4x + 11$

   a. Identify the degree of the polynomial.   3

   b. Identify the leading coefficient of the polynomial.   5

For Exercises 19–28, perform the indicated operations.

19. $(5t^4 - 2t^2 - 17) + (12t^3 + 2t^2 + 7t - 2)$
$5t^4 + 12t^3 + 7t - 19$

20. $(7w^2 - 11w - 6) + (8w^2 + 3w + 4) -$
$(-9w^2 - 5w + 2)$ $24w^2 - 3w - 4$

21. $-2x^3(5x^2 + x - 15)$
$-10x^5 - 2x^4 + 30x^3$

22. $(4a - 3)(2a - 1)$
$8a^2 - 10a + 3$

23. $(4y - 5)(y^2 - 5y + 3)$
$4y^3 - 25y^2 + 37y - 15$

24. $(2 + 3b)(2 - 3b)$
$4 - 9b^2$

25. $(5z - 6)^2$
$25z^2 - 60z + 36$

26. $(5x + 3)(3x - 2)$
$15x^2 - x - 6$

27. $(2x^2 + 5x) - (6x^2 - 7)$ $-4x^2 + 5x + 7$

28. $(y^2 - 5y + 2)(y - 6)$ $y^3 - 11y^2 + 32y - 12$

29. Subtract $(3x^2 - 5x^3 + 2x)$ from $(10x^3 - 4x^2 + 1)$.
$15x^3 - 7x^2 - 2x + 1$

30. Find the perimeter and the area of the rectangle shown in the figure.   Perimeter: $12x - 2$; area: $5x^2 - 13x - 6$

$5x + 2$

$x - 3$

For Exercises 31–35, divide the polynomials.

31. $(-12x^8 + x^6 - 8x^3) \div (4x^2)$ $-3x^6 + \dfrac{x^4}{4} - 2x$

32. $\dfrac{16a^3b - 2a^2b^2 + 8ab}{-4ab}$ $-4a^2 + \dfrac{ab}{2} - 2$

33. $\dfrac{2y^2 - 13y + 21}{y - 3}$ $2y - 7$

34. $(-5w^2 + 2w^3 - 2w + 5) \div (2w + 3)$ $w^2 - 4w + 5 + \dfrac{-10}{2w + 3}$

35. $\dfrac{3x^4 + x^3 + 4x - 33}{x^2 + 4}$ $3x^2 + x - 12 + \dfrac{15}{x^2 + 4}$

Chapters 1–5   Cumulative Review Exercises

For Exercises 1–2, simplify completely.

1. $-5 - \dfrac{1}{2}[4 - 3(-7)]$ $-\dfrac{35}{2}$

2. $|-3^2 + 5|$ 4

3. Translate the phrase into a mathematical expression and simplify:

The difference of the square of five and the square root of four. $5^2 - \sqrt{4}$; 23

4. Solve for x: $\dfrac{1}{2}(x - 6) + \dfrac{2}{3} = \dfrac{1}{4}x$ $\left\{\dfrac{28}{3}\right\}$

5. Solve for y: $-2y - 3 = -5(y - 1) + 3y$ $\{ \ \}$

6. For a point in a rectangular coordinate system, in which quadrant are both the x- and y-coordinates negative?   Quadrant III

7. For a point in a rectangular coordinate system, on which axis is the x-coordinate zero and the y-coordinate nonzero? y-axis

8. In a triangle, one angle measures 23° more than the smallest angle. The third angle measures 10° more than the sum of the other two angles. Find the measure of each angle.   The measures are 31°, 54°, 95°.

9. A snow storm lasts for 9 hr and dumps snow at a rate of $1\frac{1}{2}$ in./hr. If there was already 6 in. of snow on the ground before the storm, the snow depth is given by the equation:

$$y = \dfrac{3}{2}x + 6$$ where y is the snow depth in inches and $x \geq 0$ is the time in hours after the storm began.

   a. Find the snow depth after 4 hr.   12 in.

   b. Find the snow depth at the end of the storm.   19.5 in.

   c. How long had it snowed when the total depth of snow was $14\frac{1}{4}$ in.?   5.5 hr

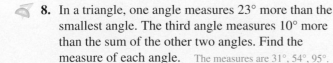

© Brand X Photography/Veer RF

10. Solve the system of equations. $\{(-3, 4)\}$

$$5x + 3y = -3$$
$$3x + 2y = -1$$

11. Solve the inequality. Graph the solution set on the real number line and express the solution in interval notation. $2 - 3(2x + 4) \le -2x - (x - 5)$

$[-5, \infty)$

$$\xleftarrow{\qquad\underset{-5}{[\quad}\qquad\qquad\qquad}\rightarrow$$

For Exercises 12–15, perform the indicated operations.

12. $(2x^2 + 3x - 7) - (-3x^2 + 12x + 8)$ $5x^2 - 9x - 15$

13. $(2y + 3z)(-y - 5z)$ $-2y^2 - 13yz - 15z^2$

14. $(4t - 3)^2$ $16t^2 - 24t + 9$ **15.** $\left(\dfrac{2}{5}a + \dfrac{1}{3}\right)\left(\dfrac{2}{5}a - \dfrac{1}{3}\right)$

$\dfrac{4}{25}a^2 - \dfrac{1}{9}$

For Exercises 16–17, divide the polynomials.

16. $(12a^4b^3 - 6a^2b^2 + 3ab) \div (-3ab)$

$-4a^3b^2 + 2ab - 1$

17. $\dfrac{4m^3 - 5m + 2}{m - 2}$ $4m^2 + 8m + 11 + \dfrac{24}{m - 2}$

For Exercises 18–19, use the properties of exponents to simplify the expressions. Write the answers with positive exponents only. Assume all variables represent nonzero real numbers.

18. $\left(\dfrac{2c^2d^4}{8cd^6}\right)^2$ $\dfrac{c^2}{16d^4}$ **19.** $\dfrac{10a^{-2}b^{-3}}{5a^0b^{-6}}$ $\dfrac{2b^3}{a^2}$

20. Perform the indicated operations, and write the final answer in scientific notation. 4.1×10^3

$$\dfrac{8.2 \times 10^{-2}}{2 \times 10^{-5}}$$

 Writing   Translating Expression   Geometry   Scientific Calculator   Video

Factoring Polynomials

6

Mathematics in the Workplace

Suppose that you are the manager of a laboratory that produces prototype electronics. This week you are tasked with producing 6 laptops over a 6-day period. With these constraints, you could produce 1 laptop per day for 6 days, 2 laptops per day for 3 days, 3 laptops per day for 2 days, or 6 laptops all in 1 day.

Any of these combinations will accomplish your task.

Laptops per day	Number of days	Total number of laptops
1	6	6
2	3	6
3	2	6
6	1	6

© Henrik Jonsson/Getty Images RF

The word **factor** is Latin for "maker." In this example, the numbers 1, 2, 3, and 6 are **factors** of 6 because they *make* a 6 by using multiplication. In other words, we can multiply each of the numbers 1, 2, 3, and 6 by another factor to produce 6. In essence, *factoring* is the process of breaking numbers or expressions into the elements that made them via multiplication.

Factoring (finding the factors) is used extensively in algebra to make expressions easier to use. In this chapter you will learn how to factor both numerical values and polynomials.

Section 6.1 Greatest Common Factor and Factoring by Grouping

1. Identifying the Greatest Common Factor

We have already learned how to multiply two or more polynomials. We now devote our study to a related operation called **factoring**. To factor an integer means to write the integer as a product of two or more integers. To factor a polynomial means to express the polynomial as a product of two or more polynomials.

In the product $2 \cdot 5 = 10$, for example, 2 and 5 are factors of 10.

In the product $(3x + 4)(2x - 1) = 6x^2 + 5x - 4$, the quantities $(3x + 4)$ and $(2x - 1)$ are factors of $6x^2 + 5x - 4$.

We begin our study of factoring by factoring integers. The number 20, for example, can be factored as $1 \cdot 20$ or $2 \cdot 10$ or $4 \cdot 5$ or $2 \cdot 2 \cdot 5$. The product $2 \cdot 2 \cdot 5$ (or equivalently $2^2 \cdot 5$) consists only of prime numbers and is called the **prime factorization**.

The **greatest common factor** (denoted **GCF**) of two or more integers is the largest factor common to each integer. To find the greatest common factor of two or more integers, it is often helpful to express the numbers as a product of prime factors as shown in the next example.

Classroom Example: p. 418, Exercise 6

Example 1 Identifying the Greatest Common Factor

Find the greatest common factor.

a. 24 and 36   **b.** 105, 40, and 60

Solution:

First find the prime factorization of each number. Then find the product of common factors.

a. 2|24   2|36   Factors of 24 = 2 · 2 · 2 · 3 ◄— Common
        2|12   2|18   factors are
        2|6   3|9   Factors of 36 = 2 · 2 · 3 · 3 ◄— circled.
        3   3

The numbers 24 and 36 share two factors of 2 and one factor of 3. Therefore, the greatest common factor is $2 \cdot 2 \cdot 3 = 12$.

b. 5|105   5|40   5|60   Factors of 105 = 3 · 7 · 5
        3|21   2|8   3|12
        7   2|4   2|4   Factors of 40 = 2 · 2 · 2 · 5
                  2   2
                                         Factors of 60 = 2 · 2 · 3 · 5

The greatest common factor is 5.

Skill Practice Find the GCF.

1. 12 and 20   **2.** 45, 75, and 30

Answers
1. 4   **2.** 15

In Example 2, we find the greatest common factor of two or more variable terms.

Classroom Example: p. 418, Exercise 10

Example 2 Identifying the Greatest Common Factor

Find the GCF among each group of terms.

a. $7x^3$, $14x^2$, $21x^4$ **b.** $15a^4b$, $25a^3b^2$

Solution:

List the factors of each term.

a. $7x^3 = \boxed{7 \cdot x \cdot x} \cdot x$
$14x^2 = 2 \cdot \boxed{7 \cdot x \cdot x}$
$21x^4 = 3 \cdot \boxed{7 \cdot x \cdot x} \cdot x \cdot x$
The GCF is $7x^2$.

b. $15a^4b = 3 \cdot \boxed{5 \cdot a \cdot a \cdot a} \cdot a \cdot \boxed{b}$
$25a^3b^2 = 5 \cdot \boxed{5 \cdot a \cdot a \cdot a} \cdot b \cdot \boxed{b}$
The GCF is $5a^3b$.

TIP: Notice in Example 2(b) the expressions $15a^4b$ and $25a^3b^2$ share factors of 5, a, and b. The GCF is the product of the common factors, where each factor is raised to the lowest power to which it occurs in all the original expressions.

$\left.\begin{array}{l} 15a^4b = 3^1 \cdot 5^1 a^4 b^1 \\ 25a^3b^2 = 5^2 a^3 b^2 \end{array}\right\}$ $\left.\begin{array}{l} \text{Lowest power of 5 is 1:} \quad 5^1 \\ \text{Lowest power of } a \text{ is 3:} \quad a^3 \\ \text{Lowest power of } b \text{ is 1:} \quad b^1 \end{array}\right\}$ The GCF is $5a^3b$.

Skill Practice Find the GCF.

3. $10z^3$, $15z^5$, $40z$ **4.** $6w^3y^5$, $21w^4y^2$

Classroom Example: p. 418, Exercise 12

Example 3 Identifying the Greatest Common Factor

Find the GCF of the terms $8c^2d^7e$ and $6c^3d^4$.

Solution:

$\left.\begin{array}{l} 8c^2d^7e = 2^3c^2d^7e \\ 6c^3d^4 = 2 \cdot 3c^3d^4 \end{array}\right\}$ The common factors are 2, c, and d.

$\left.\begin{array}{l} \text{The lowest power of 2 is 1:} \quad 2^1 \\ \text{The lowest power of } c \text{ is 2:} \quad c^2 \\ \text{The lowest power of } d \text{ is 4:} \quad d^4 \end{array}\right\}$ The GCF is $2c^2d^4$.

Skill Practice Find the GCF.

5. $9m^2np^8$, $15n^4p^5$

Sometimes polynomials share a common binomial factor, as shown in Example 4.

Answers

3. $5z$ **4.** $3w^3y^2$
5. $3np^5$

Classroom Example: p. 418, Exercise 14

Example 4 | Identifying the Greatest Common Binomial Factor

Find the greatest common factor of the terms $3x(a + b)$ and $2y(a + b)$.

Solution:

$\left. \begin{array}{l} 3x(a + b) \\ 2y(a + b) \end{array} \right\}$ The only common factor is the binomial $(a + b)$.

The GCF is $(a + b)$.

Skill Practice Find the GCF.

6. $a(x + 2)$ and $b(x + 2)$

2. Factoring out the Greatest Common Factor

The process of factoring a polynomial is the reverse process of multiplying polynomials. Both operations use the distributive property: $ab + ac = a(b + c)$.

Multiply

$$5y(y^2 + 3y + 1) = 5y(y^2) + 5y(3y) + 5y(1)$$
$$= 5y^3 + 15y^2 + 5y$$

Factor

$$5y^3 + 15y^2 + 5y = 5y(y^2) + 5y(3y) + 5y(1)$$
$$= 5y(y^2 + 3y + 1)$$

Factoring out the Greatest Common Factor

Step 1 Identify the GCF of all terms of the polynomial.
Step 2 Write each term as the product of the GCF and another factor.
Step 3 Use the distributive property to remove the GCF.
Note: To check the factorization, multiply the polynomials to remove parentheses.

Classroom Examples: p. 418, Exercises 18 and 22

Example 5 | Factoring out the Greatest Common Factor

Factor out the GCF.

a. $4x - 20$ **b.** $6w^2 + 3w$

Instructor Note: Tell students that an expression is factored only if the GCF has been removed. The expression $2(2x - 10)$ is not factored completely.

Solution:

a. $4x - 20$ The GCF is 4.

$\quad = 4(x) - 4(5)$ Write each term as the product of the GCF and another factor.

$\quad = 4(x - 5)$ Use the distributive property to factor out the GCF.

TIP: Any factoring problem can be checked by multiplying the factors:

Check: $4(x - 5) = 4x - 20$ ✓

Answer

6. $(x + 2)$

b. $6w^2 + 3w$ The GCF is $3w$.

$= 3w(2w) + 3w(1)$ Write each term as the product of $3w$ and another factor.

$= 3w(2w + 1)$ Use the distributive property to factor out the GCF.

Check: $3w(2w + 1) = 6w^2 + 3w$ ✓

Avoiding Mistakes

In Example 5(b), the GCF, $3w$, is equal to one of the terms of the polynomial. In such a case, you must leave a 1 in place of that term after the GCF is factored out.

Skill Practice Factor out the GCF.

7. $6w + 18$ **8.** $21m^3 - 7m$

Example 6   **Factoring out the Greatest Common Factor**

Factor out the GCF.

a. $15y^3 + 12y^4$ **b.** $9a^4b - 18a^5b + 27a^6b$

Classroom Examples: p. 418, Exercises 24 and 34

Solution:

a. $15y^3 + 12y^4$ The GCF is $3y^3$.

$= 3y^3(5) + 3y^3(4y)$ Write each term as the product of $3y^3$ and another factor.

$= 3y^3(5 + 4y)$ Use the distributive property to factor out the GCF.

Check: $3y^3(5 + 4y) = 15y^3 + 12y^4$ ✓

TIP: When factoring out the GCF from a polynomial, the terms within parentheses are found by dividing the original terms by the GCF. For example:

$15y^3 + 12y^4$ The GCF is $3y^3$.

$$\frac{15y^3}{3y^3} = 5 \quad \text{and} \quad \frac{12y^4}{3y^3} = 4y$$

Thus, $15y^3 + 12y^4 = 3y^3(5 + 4y)$

b. $9a^4b - 18a^5b + 27a^6b$ The GCF is $9a^4b$.

$= 9a^4b(1) - 9a^4b(2a) + 9a^4b(3a^2)$ Write each term as the product of $9a^4b$ and another factor.

$= 9a^4b(1 - 2a + 3a^2)$ Use the distributive property to factor out the GCF.

Check: $9a^4b(1 - 2a + 3a^2)$

$= 9a^4b - 18a^5b + 27a^6b$ ✓

Avoiding Mistakes

The GCF is $9a^4b$, not $3a^4b$. The expression $3a^4b(3 - 6a + 9a^2)$ is not factored completely.

Skill Practice Factor out the GCF.

9. $9y^2 - 6y^5$ **10.** $50s^3t - 40st^2 + 10st$

The greatest common factor of the polynomial $2x + 5y$ is 1. If we factor out the GCF, we have $1(2x + 5y)$. A polynomial whose only factors are itself and 1 is called a **prime polynomial**.

Answers

7. $6(w + 3)$
8. $7m(3m^2 - 1)$
9. $3y^2(3 - 2y^3)$
10. $10st(5s^2 - 4t + 1)$

Instructor Note: Remind students that first we put the polynomial in descending order, then determine if the leading coefficient is negative.

3. Factoring out a Negative Factor

Usually it is advantageous to factor out the *opposite* of the GCF when the leading coefficient of the polynomial is negative. This is demonstrated in Example 7. Notice that this *changes the signs* of the remaining terms inside the parentheses.

Classroom Example: p. 418, Exercise 42

Example 7	Factoring out a Negative Factor

Factor out -3 from the polynomial $-3x^2 + 6x - 33$.

Solution:

$-3x^2 + 6x - 33$ The GCF is 3. However, in this case, we will factor out the *opposite* of the GCF, -3.

$= -3(x^2) + (-3)(-2x) + (-3)(11)$ Write each term as the product of -3 and another factor.

$= -3[x^2 + (-2x) + 11]$ Factor out -3.

$= -3(x^2 - 2x + 11)$ Simplify. Notice that each sign within the trinomial has changed.

Check: $-3(x^2 - 2x + 11) = -3x^2 + 6x - 33$ ✓

Skill Practice Factor out -2 from the polynomial.

11. $-2x^2 - 10x + 16$

Classroom Example: p. 418, Exercise 44

Example 8	Factoring out a Negative Factor

Factor out the quantity $-4pq$ from the polynomial $-12p^3q - 8p^2q^2 + 4pq^3$.

Solution:

$-12p^3q - 8p^2q^2 + 4pq^3$ The GCF is $4pq$. However, in this case, we will factor out the *opposite* of the GCF, $-4pq$.

$= -4pq(3p^2) + (-4pq)(2pq) + (-4pq)(-q^2)$ Write each term as the product of $-4pq$ and another factor.

$= -4pq[3p^2 + 2pq + (-q^2)]$ Factor out $-4pq$. Notice that each sign within the trinomial has changed.

$= -4pq(3p^2 + 2pq - q^2)$ To verify that this is the correct factorization and that the signs are correct, multiply the factors.

Check: $-4pq(3p^2 + 2pq - q^2) = -12p^3q - 8p^2q^2 + 4pq^3$ ✓

Skill Practice Factor out $-5xy$ from the polynomial.

12. $-10x^2y + 5xy - 15xy^2$

4. Factoring out a Binomial Factor

The distributive property can also be used to factor out a common factor that consists of more than one term, as shown in Example 9.

Answers
11. $-2(x^2 + 5x - 8)$
12. $-5xy(2x - 1 + 3y)$

Example 9   **Factoring out a Binomial Factor**

Classroom Example: p. 418, Exercise 48

Factor out the GCF. $2w(x + 3) - 5(x + 3)$

Solution:

$2w(x + 3) - 5(x + 3)$ The greatest common factor is the quantity $(x + 3)$.

$= (x + 3)(2w - 5)$ Use the distributive property to factor out the GCF.

Skill Practice Factor out the GCF.

13. $8y(a + b) + 9(a + b)$

5. Factoring by Grouping

When two binomials are multiplied, the product before simplifying contains four terms. For example:

$$(x + 4)(3a + 2b) = (x + 4)(3a) + (x + 4)(2b)$$

$$= (x + 4)(3a) + (x + 4)(2b)$$

$$= 3ax + 12a + 2bx + 8b$$

In Example 10, we learn how to reverse this process. That is, given a four-term polynomial, we will factor it as a product of two binomials. The process is called *factoring by grouping*.

Factoring by Grouping

To factor a four-term polynomial by grouping:

Step 1   Identify and factor out the GCF from all four terms.

Step 2   Factor out the GCF from the first pair of terms. Factor out the GCF from the second pair of terms. (Sometimes it is necessary to factor out the opposite of the GCF.)

Step 3   If the two terms share a common binomial factor, factor out the binomial factor.

Example 10   **Factoring by Grouping**

Classroom Example: p. 419, Exercise 58

Factor by grouping. $3ax + 12a + 2bx + 8b$

Solution:

$3ax + 12a + 2bx + 8b$ **Step 1:** Identify and factor out the GCF from all four terms. In this case, the GCF is 1.

$= 3ax + 12a \mid + 2bx + 8b$ Group the first pair of terms and the second pair of terms.

Answer

13. $(a + b)(8y + 9)$

$$= 3a(x + 4) + 2b(x + 4)$$

Step 2: Factor out the GCF from each pair of terms. *Note:* The two terms now share a common binomial factor of $(x + 4)$.

$$= (x + 4)(3a + 2b)$$

Step 3: Factor out the common binomial factor.

Check: $(x + 4)(3a + 2b) = 3ax + 2bx + 12a + 8b$ ✓

Note: Step 2 results in two terms with a common binomial factor. If the two binomials are different, step 3 cannot be performed. In such a case, the original polynomial may not be factorable by grouping, or different pairs of terms may need to be grouped and inspected.

Skill Practice Factor by grouping.

14. $5x + 10y + ax + 2ay$

TIP: One frequently asked question when factoring is whether the order can be switched between the factors. The answer is yes. Because multiplication is commutative, the order in which the factors are written does not matter.

$$(x + 4)(3a + 2b) = (3a + 2b)(x + 4)$$

Classroom Example: p. 419, Exercise 60

Example 11 Factoring by Grouping

Factor by grouping. $ax + ay - x - y$

Solution:

$$ax + ay - x - y$$

Step 1: Identify and factor out the GCF from all four terms. In this case, the GCF is 1.

$$= ax + ay \mid - x - y$$

Group the first pair of terms and the second pair of terms.

$$= a(x + y) - 1(x + y)$$

Step 2: Factor out a from the first pair of terms.

Factor out -1 from the second pair of terms. (This causes sign changes within the second parentheses.) The terms in parentheses now match.

$$= (x + y)(a - 1)$$

Step 3: Factor out the common binomial factor.

Check: $(x + y)(a - 1) = x(a) + x(-1) + y(a) + y(-1)$

$$= ax - x + ay - y ✓$$

Avoiding Mistakes

In step 2, the expression $a(x + y) - (x + y)$ is not yet factored completely because it is a *difference*, not a product. To factor the expression, you must carry it one step further.

$$a(x + y) - 1(x + y)$$
$$= (x + y)(a - 1)$$

The factored form must be represented as a product.

Skill Practice Factor by grouping.

15. $tu - tv - u + v$

Answers

14. $(x + 2y)(5 + a)$
15. $(u - v)(t - 1)$

Example 12   **Factoring by Grouping**

Factor by grouping. $16w^4 - 40w^3 - 12w^2 + 30w$

Classroom Example: p. 419, Exercise 74

Solution:

$16w^4 - 40w^3 - 12w^2 + 30w$

Step 1: Identify and factor out the GCF from all four terms. In this case, the GCF is $2w$.

$= 2w[8w^3 - 20w^2 - 6w + 15]$

$= 2w[8w^3 - 20w^2 \mid - 6w + 15]$

Group the first pair of terms and the second pair of terms.

$= 2w[4w^2(2w - 5) - 3(2w - 5)]$

Step 2: Factor out $4w^2$ from the first pair of terms.

Factor out -3 from the second pair of terms. (This causes sign changes within the second parentheses.) The terms in parentheses now match.

$= 2w[(2w - 5)(4w^2 - 3)]$

Step 3: Factor out the common binomial factor.

$= 2w(2w - 5)(4w^2 - 3)$

Skill Practice Factor by grouping.

16. $3ab^2 + 6b^2 - 12ab - 24b$

Answer

16. $3b(a + 2)(b - 4)$

Section 6.1   Practice Exercises

Study Skills Exercise

For additional exercises, see Classroom Activities 6.1A–6.1B in the *Student's Resource Manual* at www.mhhe.com/moh.

The final exam is just around the corner. Your old tests and quizzes provide good material to study for the final exam. Use your old tests to make a list of the chapters on which you need to concentrate. Ask your professor for help if there are still concepts that you do not understand.

Vocabulary and Key Concepts

1. **a.** Factoring a polynomial means to write it as a __product__ of two or more polynomials.

   b. The prime factorization of a number consists of only __prime__ factors.

   c. The __greatest__ __common__ __factor__ (GCF) of two or more integers is the largest whole number that is a factor of each integer.

   d. A polynomial whose only factors are 1 and itself is called a __prime__ polynomial.

   e. The first step toward factoring a polynomial is to factor out the __greatest__ __common__ __factor (GCF)__.

   f. To factor a four-term polynomial, we try the process of factoring by __grouping__.

Concept 1: Identifying the Greatest Common Factor

2. List all the factors of 24.   1, 2, 3, 4, 6, 8, 12, 24

For Exercises 3–14, identify the greatest common factor. **(See Examples 1–4.)**

3. 28, 63   7

4. 24, 40   8

5. 42, 30, 60   6

6. 20, 52, 32   4

7. $3xy, 7y$ y

8. $10mn, 11n$ n

9. $12w^3z, 16w^2z$ $4w^2z$

10. $20cd, 15c^3d$ $5cd$

11. $8x^3y^4z^2, 12xy^5z^4, 6x^2y^8z^3$ $2xy^4z^2$

12. $15r^2s^2t^5, 5r^3s^4t^3, 30r^4s^3t^2$ $5r^2s^2t^2$

13. $7(x-y), 9(x-y)$ $(x-y)$

14. $(2a-b), 3(2a-b)$ $(2a-b)$

Concept 2: Factoring out the Greatest Common Factor

15. a. Use the distributive property to multiply $3(x-2y)$. $3x-6y$

   b. Use the distributive property to factor $3x-6y$. $3(x-2y)$

16. a. Use the distributive property to multiply $a^2(5a+b)$. $5a^3+a^2b$

   b. Use the distributive property to factor $5a^3+a^2b$. $a^2(5a+b)$

For Exercises 17–36, factor out the GCF. **(See Examples 5–6.)**

17. $4p+12$ $4(p+3)$

18. $3q-15$ $3(q-5)$

19. $5c^2-10c+15$
$5(c^2-2c+3)$

20. $16d^3+24d^2+32d$
$8d(2d^2+3d+4)$

21. x^5+x^3 $x^3(x^2+1)$

22. y^2-y^3 $y^2(1-y)$

23. $t^4-4t+8t^2$
$t(t^3-4+8t)$

24. $7r^3-r^5+r^4$
$r^3(7-r^2+r)$

25. $2ab+4a^3b$
$2ab(1+2a^2)$

26. $5u^3v^2-5uv$
$5uv(u^2v-1)$

27. $38x^2y-19x^2y^4$
$19x^2y(2-y^3)$

28. $100a^5b^3+16a^2b$
$4a^2b(25a^3b^2+4)$

29. $6x^3y^5-18xy^9z$
$6xy^5(x^2-3y^4z)$

30. $15mp^7q^4+12m^4q^3$
$3mq^3(5p^7q+4m^3)$

31. $5+7y^3$ The expression is prime because it is not factorable.

32. $w^3-5u^3v^2$ The expression is prime because it is not factorable.

33. $42p^3q^2+14pq^2-7p^4q^4$ $7pq^2(6p^2+2-p^3q^2)$

34. $8m^2n^3-24m^2n^2+4m^3n$ $4m^2n(2n^2-6n+m)$

35. $t^5+2rt^3-3t^4+4r^2t^2$ $t^2(t^3+2rt-3t^2+4r^2)$

36. $u^2v+5u^3v^2-2u^2+8uv$ $u(uv+5u^2v^2-2u+8v)$

Concept 3: Factoring out a Negative Factor

37. For the polynomial $-2x^3-4x^2+8x$

   a. Factor out $-2x$.
   $-2x(x^2+2x-4)$

   b. Factor out $2x$.
   $2x(-x^2-2x+4)$

38. For the polynomial $-9y^5+3y^3-12y$

   a. Factor out $-3y$.
   $-3y(3y^4-y^2+4)$

   b. Factor out $3y$.
   $3y(-3y^4+y^2-4)$

39. Factor out -1 from the polynomial.
$-8t^2-9t-2$ $-1(8t^2+9t+2)$

40. Factor out -1 from the polynomial.
$-6x^3-2x-5$ $-1(6x^3+2x+5)$

For Exercises 41–46, factor out the opposite of the greatest common factor. **(See Examples 7–8.)**

41. $-15p^3-30p^2$ $-15p^2(p+2)$

42. $-24m^3-12m^4$ $-12m^3(2+m)$

43. $-3m^4n^2+6m^2n-9mn^2$
$-3mn(m^3n-2m+3n)$

44. $-12p^3t+2p^2t^3+6pt^2$
$-2pt(6p^2-pt^2-3t)$

45. $-7x-6y-2z$ $-1(7x+6y+2z)$

46. $-4a+5b-c$ $-1(4a-5b+c)$

Concept 4: Factoring out a Binomial Factor

For Exercises 47–52, factor out the GCF. **(See Example 9.)**

47. $13(a+6)-4b(a+6)$
$(a+6)(13-4b)$

48. $7(x^2+1)-y(x^2+1)$
$(x^2+1)(7-y)$

49. $8v(w^2-2)+(w^2-2)$
$(w^2-2)(8v+1)$

50. $t(r+2)+(r+2)$
$(r+2)(t+1)$

51. $21x(x+3)+7x^2(x+3)$
$7x(x+3)^2$

52. $5y^3(y-2)-15y(y-2)$
$5y(y-2)(y^2-3)$

Concept 5: Factoring by Grouping

For Exercises 53–72, factor by grouping. **(See Examples 10–11.)**

53. $8a^2 - 4ab + 6ac - 3bc$
$(2a - b)(4a + 3c)$

54. $4x^3 + 3x^2y + 4xy^2 + 3y^3$
$(4x + 3y)(x^2 + y^2)$

55. $3q + 3p + qr + pr$
$(q + p)(3 + r)$

56. $xy - xz + 7y - 7z$
$(y - z)(x + 7)$

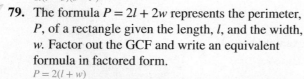 **57.** $6x^2 + 3x + 4x + 2$
$(2x + 1)(3x + 2)$

58. $4y^2 + 8y + 7y + 14$
$(y + 2)(4y + 7)$

59. $2t^2 + 6t - t - 3$
$(t + 3)(2t - 1)$

60. $2p^2 - p - 2p + 1$
$(2p - 1)(p - 1)$

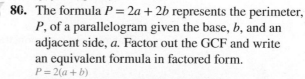 **61.** $6y^2 - 2y - 9y + 3$
$(3y - 1)(2y - 3)$

62. $5a^2 + 30a - 2a - 12$
$(a + 6)(5a - 2)$

63. $b^4 + b^3 - 4b - 4$
$(b + 1)(b^3 - 4)$

64. $8w^5 + 12w^2 - 10w^3 - 15$
$(2w^3 + 3)(4w^2 - 5)$

65. $3j^2k + 15k + j^2 + 5$
$(j^2 + 5)(3k + 1)$

66. $2ab^2 - 6ac + b^2 - 3c$
$(b^2 - 3c)(2a + 1)$

67. $14w^6x^6 + 7w^6 - 2x^6 - 1$
$(2x^6 + 1)(7w^6 - 1)$

68. $18p^4x - 4x - 9p^5 + 2p$
$(9p^4 - 2)(2x - p)$

69. $ay + bx + by + ax$
(*Hint:* Rearrange the terms.)
$(y + x)(a + b)$

70. $2c + 3ay + ac + 6y$
$(c + 3y)(2 + a)$

71. $vw^2 - 3 + w - 3wv$
$(vw + 1)(w - 3)$

72. $2x^2 + 6m + 12 + x^2m$
$(m + 2)(6 + x^2)$

Mixed Exercises

For Exercises 73–78, factor out the GCF first. Then factor by grouping. **(See Example 12.)**

73. $15x^4 + 15x^2y^2 + 10x^3y + 10xy^3$
$5x(x^2 + y^2)(3x + 2y)$

74. $2a^3b - 4a^2b + 32ab - 64b$
$2b(a - 2)(a^2 + 16)$

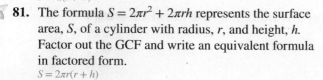 **75.** $4abx - 4b^2x - 4ab + 4b^2$
$4b(a - b)(x - 1)$

76. $p^2q - pq^2 - rp^2q + rpq^2$
$pq(p - q)(1 - r)$

77. $6st^2 - 18st - 6t^4 + 18t^3$
$6t(t - 3)(s - t^2)$

78. $15j^3 - 10j^2k - 15j^2k^2 + 10jk^3$
$5j(3j - 2k)(j - k^2)$

79. The formula $P = 2l + 2w$ represents the perimeter, P, of a rectangle given the length, l, and the width, w. Factor out the GCF and write an equivalent formula in factored form.
$P = 2(l + w)$

80. The formula $P = 2a + 2b$ represents the perimeter, P, of a parallelogram given the base, b, and an adjacent side, a. Factor out the GCF and write an equivalent formula in factored form.
$P = 2(a + b)$

81. The formula $S = 2\pi r^2 + 2\pi rh$ represents the surface area, S, of a cylinder with radius, r, and height, h. Factor out the GCF and write an equivalent formula in factored form.
$S = 2\pi r(r + h)$

82. The formula $A = P + Prt$ represents the total amount of money, A, in an account that earns simple interest at a rate, r, for t years. Factor out the GCF and write an equivalent formula in factored form.
$A = P(1 + rt)$

Expanding Your Skills

83. Factor out $\frac{1}{7}$ from $\frac{1}{7}x^2 + \frac{3}{7}x - \frac{5}{7}$. $\frac{1}{7}(x^2 + 3x - 5)$

84. Factor out $\frac{1}{5}$ from $\frac{6}{5}y^2 - \frac{4}{5}y + \frac{1}{5}$. $\frac{1}{5}(6y^2 - 4y + 1)$

85. Factor out $\frac{1}{4}$ from $\frac{5}{4}w^2 + \frac{3}{4}w + \frac{9}{4}$. $\frac{1}{4}(5w^2 + 3w + 9)$

86. Factor out $\frac{1}{6}$ from $\frac{1}{6}p^2 - \frac{3}{6}p + \frac{5}{6}$. $\frac{1}{6}(p^2 - 3p + 5)$

87. Write a polynomial that has a GCF of $3x$. (Answers may vary.)   For example: $6x^2 + 9x$

88. Write a polynomial that has a GCF of $7y$. (Answers may vary.)   For example: $14y - 21y^3 + 7y^2$

89. Write a polynomial that has a GCF of $4p^2q$. (Answers may vary.)
For example: $16p^4q^2 + 8p^3q - 4p^2q$

90. Write a polynomial that has a GCF of $2ab^2$. (Answers may vary.)   For example: $18a^2b^3 - 2ab^2$

✎ Writing   ⬅➡ Translating Expression   ◢ Geometry   🖩 Scientific Calculator   ▶ Video

Concept

1. Factoring Trinomials with a Leading Coefficient of 1

1. Factoring Trinomials with a Leading Coefficient of 1

We have already learned how to multiply two binomials. We also saw that such a product often results in a trinomial. For example:

Product of first terms

Product of last terms

$$(x + 3)(x + 7) = x^2 + 7x + 3x + 21 = x^2 + 10x + 21$$

Sum of products of inner terms and outer terms

In this section, we want to reverse the process. That is, given a trinomial, we want to *factor* it as a product of two binomials. In particular, we begin our study with the case in which a trinomial has a leading coefficient of 1.

Consider the quadratic trinomial $x^2 + bx + c$. To produce a leading term of x^2, we can construct binomials of the form $(x + \ )(x + \ )$. The remaining terms can be obtained from two integers, p and q, whose product is c and whose sum is b.

Factors of c

$$x^2 + bx + c = (x + p)(x + q) = x^2 + qx + px + pq$$
$$= x^2 + (q + p)x + pq$$

Sum = b Product = c

This process is demonstrated in Example 1.

Classroom Example: p. 424, Exercise 8

| **Example 1** | **Factoring a Trinomial of the Form $x^2 + bx + c$** |

Factor. $x^2 + 4x - 45$

Solution:

$x^2 + 4x - 45 = (x + \square)(x + \square)$ The product of the first terms in the binomials must equal the leading term of the trinomial $x \cdot x = x^2$.

We must fill in the blanks with two integers whose product is -45 and whose sum is 4. The factors must have opposite signs to produce a negative product. The possible factorizations of -45 are:

Product = −45	**Sum**
$-1 \cdot 45$	44
$-3 \cdot 15$	12
$-5 \cdot 9$	4
$-9 \cdot 5$	−4
$-15 \cdot 3$	−12
$-45 \cdot 1$	−44

$$x^2 + 4x - 45 = (x + \square)(x + \square)$$
$$= [x + (-5)](x + 9) \qquad \text{Fill in the blanks with } -5 \text{ and } 9.$$
$$= (x - 5)(x + 9) \qquad \text{Factored form}$$

Check:
$$(x - 5)(x + 9) = x^2 + 9x - 5x - 45$$
$$= x^2 + 4x - 45 \checkmark$$

Skill Practice Factor.

1. $x^2 - 5x - 14$

Multiplication of polynomials is a commutative operation. Therefore, in Example 1, we can express the factorization as $(x - 5)(x + 9)$ or as $(x + 9)(x - 5)$.

| **Example 2** | **Factoring a Trinomial of the Form $x^2 + bx + c$** |

Classroom Example: p. 424, Exercise 14

Factor. $w^2 - 15w + 50$

Solution:

$$w^2 - 15w + 50 = (w + \square)(w + \square) \qquad \text{The product } w \cdot w = w^2.$$

Find two integers whose product is 50 and whose sum is -15. To form a positive product, the factors must be either both positive or both negative. The sum must be negative, so we will choose negative factors of 50.

Product = 50	**Sum**
$(-1)(-50)$	-51
$(-2)(-25)$	-27
$(-5)(-10)$	-15

$$w^2 - 15w + 50 = (w + \square)(w + \square)$$
$$= [w + (-5)][w + (-10)]$$
$$= (w - 5)(w - 10) \qquad \text{Factored form}$$

Check:
$$(w - 5)(w - 10) = w^2 - 10w - 5w + 50$$
$$= w^2 - 15w + 50 \checkmark$$

Skill Practice Factor.

2. $z^2 - 16z + 48$

TIP: Practice will help you become proficient in factoring polynomials. As you do your homework, keep these important guidelines in mind:

- To factor a trinomial, write the trinomial in descending order such as $x^2 + bx + c$.
- For all factoring problems, always factor out the GCF from all terms first.

Answers

1. $(x - 7)(x + 2)$
2. $(z - 4)(z - 12)$

Furthermore, we offer the following rules for determining the signs within the binomial factors.

Sign Rules for Factoring Trinomials

Given the trinomial $x^2 + bx + c$, the signs within the binomial factors are determined as follows:

Case 1 If c is *positive*, then the signs in the binomials must be the same (either both positive or both negative). The correct choice is determined by the middle term. If the middle term is positive, then both signs must be positive. If the middle term is negative, then both signs must be negative.

c is positive.
$$x^2 + 6x + 8$$
$$(x + 2)(x + 4)$$
Same signs

c is positive.
$$x^2 - 6x + 8$$
$$(x - 2)(x - 4)$$
Same signs

Case 2 If c is *negative*, then the signs in the binomials must be different.

c is negative.
$$x^2 + 2x - 35$$
$$(x + 7)(x - 5)$$
Different signs

c is negative.
$$x^2 - 2x - 35$$
$$(x - 7)(x + 5)$$
Different signs

Classroom Examples: p. 425, Exercises 32 and 40

Example 3   Factoring Trinomials

Factor.   **a.** $-8p - 48 + p^2$ **b.** $-40t - 30t^2 + 10t^3$

Solution:

a. $-8p - 48 + p^2$

$= p^2 - 8p - 48$ Write in descending order.

$= (p \ \square)(p \ \square)$ Find two integers whose product is -48 and whose sum is -8. The numbers are -12 and 4.

$= (p - 12)(p + 4)$ Factored form

b. $-40t - 30t^2 + 10t^3$

$= 10t^3 - 30t^2 - 40t$ Write in descending order.

$= 10t(t^2 - 3t - 4)$ Factor out the GCF.

$= 10t(t \ \square)(t \ \square)$ Find two integers whose product is -4 and whose sum is -3. The numbers are -4 and 1.

$= 10t(t - 4)(t + 1)$ Factored form

Skill Practice Factor.

3. $-5w + w^2 - 6$ **4.** $30y^3 + 2y^4 + 112y^2$

Answers

3. $(w - 6)(w + 1)$
4. $2y^2(y + 8)(y + 7)$

Example 4   **Factoring Trinomials**

Factor.   **a.** $-a^2 + 6a - 8$ **b.** $-2c^2 - 22cd - 60d^2$

Solution:

a. $-a^2 + 6a - 8$

$= -1(a^2 - 6a + 8)$

$= -1(a \ \Box)(a \ \Box)$

$= -1(a - 4)(a - 2)$

$= -(a - 4)(a - 2)$

It is generally easier to factor a trinomial with a *positive* leading coefficient. Therefore, we will factor out -1 from all terms.

Find two integers whose product is 8 and whose sum is -6. The numbers are -4 and -2.

b. $-2c^2 - 22cd - 60d^2$

$= -2(c^2 + 11cd + 30d^2)$ Factor out -2.

$= -2(c \ \Box d)(c \ \Box d)$ Notice that the second pair of terms has a factor of d. This will produce a product of d^2.

$= -2(c + 5d)(c + 6d)$ Find two integers whose product is 30 and whose sum is 11. The numbers are 5 and 6.

Skill Practice Factor.

5. $-x^2 + x + 12$ **6.** $-3a^2 + 15ab - 12b^2$

Classroom Examples: p. 425, Exercises 44 and 46

Avoiding Mistakes

Recall that factoring out -1 from a polynomial changes the signs of all terms within parentheses.

Instructor Note: Show students how a sign difference between two expressions can affect the factors used.
1. $x^2 - 10x + 24, x^2 - 10x - 24$
2. $x^2 - 5x + 6, x^2 - 5x - 6$
3. $x^2 - 13x + 30, x^2 - 13x - 30$

To factor a trinomial of the form $x^2 + bx + c$, we must find two integers whose product is c and whose sum is b. If no such integers exist, then the trinomial is prime.

Example 5   **Factoring Trinomials**

Factor. $x^2 - 13x + 14$

Classroom Example: p. 424, Exercise 18

Solution:

$x^2 - 13x + 14$ The trinomial is in descending order. The GCF is 1.

$= (x \ \Box)(x \ \Box)$ Find two integers whose product is 14 and whose sum is -13. No such integers exist.

The trinomial $x^2 - 13x + 14$ is prime.

Skill Practice Factor.

7. $x^2 - 7x + 28$

Answers

5. $-(x - 4)(x + 3)$
6. $-3(a - b)(a - 4b)$
7. Prime

Section 6.2 Practice Exercises

For additional exercises, see Classroom Activity 6.2A in the *Student's Resource Manual* at www.mhhe.com/moh.

Vocabulary and Key Concepts

1. **a.** Given a trinomial $x^2 + bx + c$, if c is positive, then the signs in the binomial factors are either both ___positive___ or both negative.

    b. Given a trinomial $x^2 + bx + c$, if c is negative, then the signs in the binomial factors are (choose one: both positive, both negative, different). different

    c. Which is the correct factored form of $x^2 - 7x - 44$? The product $(x + 4)(x - 11)$ or $(x - 11)(x + 4)$?
    Both are correct.

    d. Which is the complete factorization of $3x^2 + 24x + 36$? The product $(3x + 6)(x + 6)$ or $3(x + 6)(x + 2)$?
    $3(x + 6)(x + 2)$

Review Exercises

For Exercises 2–6, factor completely.

2. $9a^6b^3 - 27a^3b^6 - 3a^2b^2$ $3a^2b^2(3a^4b - 9ab^4 - 1)$

3. $3t(t - 5) - 6(t - 5)$ $3(t - 5)(t - 2)$

4. $4(3x - 2) + 8x(3x - 2)$ $4(3x - 2)(1 + 2x)$

5. $ax + 2bx - 5a - 10b$ $(a + 2b)(x - 5)$

6. $m^2 - mx - 3pm + 3px$ $(m - x)(m - 3p)$

Concept 1: Factoring Trinomials with a Leading Coefficient of 1

For Exercises 7–20, factor completely. **(See Examples 1, 2, and 5.)**

7. $x^2 + 10x + 16$ $(x + 8)(x + 2)$

8. $y^2 + 18y + 80$ $(y + 10)(y + 8)$

9. $z^2 - 11z + 18$ $(z - 9)(z - 2)$

10. $w^2 - 7w + 12$ $(w - 3)(w - 4)$

11. $z^2 - 3z - 18$ $(z - 6)(z + 3)$

12. $w^2 + 4w - 12$ $(w + 6)(w - 2)$

13. $p^2 - 3p - 40$ $(p - 8)(p + 5)$

14. $a^2 - 10a + 9$ $(a - 9)(a - 1)$

15. $t^2 + 6t - 40$ $(t + 10)(t - 4)$

16. $m^2 - 12m + 11$ $(m - 11)(m - 1)$

17. $x^2 - 3x + 20$ Prime

18. $y^2 + 6y + 18$ Prime

19. $n^2 + 8n + 16$ $(n + 4)^2$

20. $v^2 + 10v + 25$ $(v + 5)^2$

For Exercises 21–24, assume that b and c represent positive integers.

21. When factoring a polynomial of the form $x^2 + bx + c$, pick an appropriate combination of signs.   a
    a. (+)(+)   **b.** (−)(−)   **c.** (+)(−)

22. When factoring a polynomial of the form $x^2 + bx - c$, pick an appropriate combination of signs.   c
    a. (+)(+)   **b.** (−)(−)   **c.** (+)(−)

23. When factoring a polynomial of the form $x^2 - bx - c$, pick an appropriate combination of signs.   c
    a. (+)(+)   **b.** (−)(−)   **c.** (+)(−)

24. When factoring a polynomial of the form $x^2 - bx + c$, pick an appropriate combination of signs.   b
    a. (+)(+)   **b.** (−)(−)   **c.** (+)(−)

25. Which is the correct factorization of $y^2 - y - 12$, the product $(y - 4)(y + 3)$ or $(y + 3)(y - 4)$? Explain.
    They are both correct because multiplication of polynomials is a commutative operation.

26. Which is the correct factorization of $x^2 + 14x + 13$, the product $(x + 13)(x + 1)$ or $(x + 1)(x + 13)$? Explain.
They are both correct because multiplication of polynomials is a commutative operation.

27. Which is the correct factorization of $w^2 + 2w + 1$, the product $(w + 1)(w + 1)$ or $(w + 1)^2$? Explain.
The expressions are equal and both are correct.

28. Which is the correct factorization of $z^2 - 4z + 4$, the product $(z - 2)(z - 2)$ or $(z - 2)^2$? Explain.
The expressions are equal and both are correct.

29. In what order should a trinomial be written before attempting to factor it?
It should be written in descending order.

30. Once a polynomial is written in descending order, what is the next step to factor the polynomial?
Factor out the greatest common factor (GCF) from all terms.

For Exercises 31–66, factor completely. Be sure to factor out the GCF when necessary. **(See Examples 3–4.)**

31. $-13x + x^2 - 30$
$(x - 15)(x + 2)$

32. $12y - 160 + y^2$
$(y + 20)(y - 8)$

33. $-18w + 65 + w^2$
$(w - 13)(w - 5)$

34. $17t + t^2 + 72$
$(t + 8)(t + 9)$

35. $22t + t^2 + 72$
$(t + 18)(t + 4)$

36. $10q - 1200 + q^2$
$(q - 30)(q + 40)$

37. $3x^2 - 30x - 72$
$3(x - 12)(x + 2)$

38. $2z^2 + 4z - 198$
$2(z + 11)(z - 9)$

39. $8p^3 - 40p^2 + 32p$
$8p(p - 1)(p - 4)$

40. $5w^4 - 35w^3 + 50w^2$
$5w^2(w - 2)(w - 5)$

41. $y^4z^2 - 12y^3z^2 + 36y^2z^2$
$y^2z^2(y - 6)(y - 6)$ or $y^2z^2(y - 6)^2$

42. $t^4u^2 + 6t^3u^2 + 9t^2u^2$
$t^2u^2(t + 3)(t + 3)$ or $t^2u^2(t + 3)^2$

43. $-x^2 + 10x + 24$
$-(x - 4)(x - 6)$

44. $-y^2 - 12y - 35$
$-(y + 5)(y + 7)$

45. $-5a^2 + 5ax + 30x^2$
$-5(a - 3x)(a + 2x)$

46. $-2m^2 + 10mn + 12n^2$
$-2(m + n)(m - 6n)$

47. $-4 - 2c^2 - 6c$
$-2(c + 2)(c + 1)$

48. $-40d - 30 - 10d^2$
$-10(d + 3)(d + 1)$

49. $x^3y^3 - 19x^2y^3 + 60xy^3$
$xy^3(x - 4)(x - 15)$

50. $y^2z^5 + 17yz^5 + 60z^5$
$z^5(y + 5)(y + 12)$

51. $12p^2 - 96p + 84$
$12(p - 7)(p - 1)$

52. $5w^2 - 40w - 45$
$5(w - 9)(w + 1)$

53. $-2m^2 + 22m - 20$
$-2(m - 10)(m - 1)$

54. $-3x^2 - 36x - 81$
$-3(x + 9)(x + 3)$

55. $c^2 + 6cd + 5d^2$ $(c + 5d)(c + d)$

56. $x^2 + 8xy + 12y^2$ $(x + 6y)(x + 2y)$

57. $a^2 - 9ab + 14b^2$ $(a - 2b)(a - 7b)$

58. $m^2 - 15mn + 44n^2$
$(m - 4n)(m - 11n)$

59. $a^2 + 4a + 18$ Prime

60. $b^2 - 6a + 15$ Prime

61. $2q + q^2 - 63$ $(q - 7)(q + 9)$

62. $-32 - 4t + t^2$ $(t - 8)(t + 4)$

63. $x^2 + 20x + 100$ $(x + 10)^2$

64. $z^2 - 24z + 144$ $(z - 12)^2$

65. $t^2 + 18t - 40$ $(t + 20)(t - 2)$

66. $d^2 + 2d - 99$ $(d + 11)(d - 9)$

67. A student factored a trinomial as $(2x - 4)(x - 3)$. The instructor did not give full credit. Why?   The student forgot to factor out the GCF before factoring the trinomial further. The polynomial is not factored completely, because $(2x - 4)$ has a common factor of 2.

68. A student factored a trinomial as $(y + 2)(5y - 15)$. The instructor did not give full credit. Why?   The student forgot to factor out the GCF before factoring the trinomial further. The polynomial is not factored completely, because $(5y - 15)$ has a common factor of 5.

69. What polynomial factors as $(x - 4)(x + 13)$? $x^2 + 9x - 52$

70. What polynomial factors as $(q - 7)(q + 10)$? $q^2 + 3q - 70$

71. Raul purchased a parcel of land in the country. The given expressions represent the lengths of the boundary lines of his property.

a. Write the perimeter of the land as a polynomial in simplified form. $3x^2 + 9x - 12$

b. Write the polynomial from part (a) in factored form. $3(x + 4)(x - 1)$

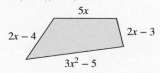

72. Jamison painted a mural in the shape of a triangle on the wall of a building. The given expressions represent the lengths of the sides of the triangle.

a. Write the perimeter of the triangle as a polynomial in simplified form. $2y^2 + 12y + 16$

b. Write the polynomial in factored form.
$2(y + 4)(y + 2)$

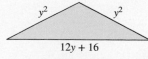

  Writing   ↔ Translating Expression    Geometry   Scientific Calculator   Video

Expanding Your Skills

For Exercises 73–76, factor completely.

73. $x^4 + 10x^2 + 9$
$(x^2 + 1)(x^2 + 9)$

74. $y^4 + 4y^2 - 21$
$(y^2 - 3)(y^2 + 7)$

75. $w^4 + 2w^2 - 15$
$(w^2 + 5)(w^2 - 3)$

76. $p^4 - 13p^2 + 40$
$(p^2 - 8)(p^2 - 5)$

77. Find all integers, b, that make the trinomial $x^2 + bx + 6$ factorable. $7, 5, -7, -5$

78. Find all integers, b, that make the trinomial $x^2 + bx + 10$ factorable. $11, 7, -11, -7$

79. Find a value of c that makes the trinomial $x^2 + 6x + c$ factorable.   For example: $c = -16$

80. Find a value of c that makes the trinomial $x^2 + 8x + c$ factorable.   For example: $c = 7$

Section 6.3 Factoring Trinomials: Trial-and-Error Method

Concept

1. Factoring Trinomials by the Trial-and-Error Method

In this section we will learn how to factor a trinomial of the form $ax^2 + bx + c = 0$ (where $a \neq 0$). The method presented here is called the trial-and-error method.

1. Factoring Trinomials by the Trial-and-Error Method

To understand the basis of factoring trinomials of the form $ax^2 + bx + c$, first consider the multiplication of two binomials:

$$\overset{\text{Product of } 2 \cdot 1}{\qquad} \quad \overset{\text{Product of } 3 \cdot 2}{\qquad}$$

$$(2x + 3)(1x + 2) = 2x^2 + \underline{\mathbf{4x + 3x}} + 6 = 2x^2 + 7x + 6$$

Sum of products of inner terms and outer terms

To factor the trinomial, $2x^2 + 7x + 6$, this operation is reversed.

Factors of 2

$$2x^2 + 7x + 6 = (\square x \quad \square)(\square x \quad \square)$$

Factors of 6

We need to fill in the blanks so that the product of the first terms in the binomials is $2x^2$ and the product of the last terms in the binomials is 6. Furthermore, the factors of $2x^2$ and 6 must be chosen so that the sum of the products of the inner terms and outer terms equals $7x$. To produce the product $2x^2$, we might try the factors $2x$ and x within the binomials:

$$(2x \quad \square)(x \quad \square)$$

To produce a product of 6, the remaining terms in the binomials must either both be positive or both be negative. To produce a positive middle term, we will try positive factors of 6 in the remaining blanks until the correct product is found. The possibilities are $1 \cdot 6, 2 \cdot 3, 6 \cdot 1$, and $3 \cdot 2$.

$(2x + 1)(x + 6) = 2x^2 + 12x + 1x + 6 = 2x^2 + 13x + 6$ Wrong middle term

$(2x + 2)(x + 3) = 2x^2 + 6x + 2x + 6 = 2x^2 + 8x + 6$ Wrong middle term

$(2x + 6)(x + 1) = 2x^2 + 2x + 6x + 6 = 2x^2 + 8x + 6$ Wrong middle term

$(2x + 3)(x + 2) = 2x^2 + 4x + 3x + 6 = 2x^2 + 7x + 6$ Correct!

The correct factorization of $2x^2 + 7x + 6$ is $(2x + 3)(x + 2)$. ✔

As this example shows, we factor a trinomial of the form $ax^2 + bx + c$ by shuffling the factors of a and c within the binomials until the correct product is obtained. However, sometimes it is not necessary to test all the possible combinations of factors. In the previous example, the GCF of the original trinomial is 1. Therefore, any binomial factor whose terms share a common factor *greater than 1* does not need to be considered. In this case, the possibilities $(2x + 2)(x + 3)$ and $(2x + 6)(x + 1)$ cannot work.

$$\underbrace{(2x + 2)(x + 3)}_{\substack{\text{Common} \\ \text{factor of 2}}} \qquad \underbrace{(2x + 6)(x + 1)}_{\substack{\text{Common} \\ \text{factor of 2}}}$$

Trial-and-Error Method to Factor $ax^2 + bx + c$

Step 1 Factor out the GCF.

Step 2 List all pairs of positive factors of a and pairs of positive factors of c. Consider the reverse order for one of the lists of factors.

Step 3 Construct two binomials of the form:

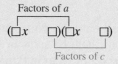

Step 4 Test each combination of factors and signs until the correct product is found.

Step 5 If no combination of factors produces the correct product, the trinomial cannot be factored further and is a *prime polynomial*.

Before we begin Example 1, keep these two important guidelines in mind:

- For any factoring problem you encounter, always factor out the GCF from all terms first.
- To factor a trinomial, write the trinomial in the form $ax^2 + bx + c$.

| **Example 1** | **Factoring a Trinomial by the Trial-and-Error Method** |

Classroom Example: p. 433, Exercise 14

Factor the trinomial by the trial-and-error method. $10x^2 + 11x + 1$

Solution:

$10x^2 + 11x + 1$ **Step 1:** Factor out the GCF from all terms. In this case, the GCF is 1.

The trinomial is written in the form $ax^2 + bx + c$.

To factor $10x^2 + 11x + 1$, two binomials must be constructed in the form:

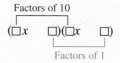

Step 2: To produce the product $10x^2$, we might try $5x$ and $2x$, or $10x$ and $1x$. To produce a product of 1, we will try the factors $(1)(1)$ and $(-1)(-1)$.

Step 3: Construct all possible binomial factors using different combinations of the factors of $10x^2$ and 1.

$(5x + 1)(2x + 1) = 10x^2 + 5x + 2x + 1 = 10x^2 + 7x + 1$ Wrong middle term

$(5x - 1)(2x - 1) = 10x^2 - 5x - 2x + 1 = 10x^2 - 7x + 1$ Wrong middle term

Because the numbers 1 and -1 did not produce the correct trinomial when coupled with $5x$ and $2x$, try using $10x$ and $1x$.

$(10x - 1)(1x - 1) = 10x^2 - 10x - 1x + 1 = 10x^2 - 11x + 1$ Wrong sign on the middle term

$(10x + 1)(1x + 1) = 10x^2 + 10x + 1x + 1 = 10x^2 + 11x + 1$ Correct!

Therefore, $10x^2 + 11x + 1 = (10x + 1)(x + 1)$.

Skill Practice   Factor using the trial-and-error method.

1. $3b^2 + 8b + 4$

In Example 1, the factors of 1 must have the same signs to produce a positive product. Therefore, the binomial factors must both be sums or both be differences. Determining the correct signs is an important aspect of factoring trinomials. We suggest the following guidelines:

> ## Sign Rules for the Trial-and-Error Method
>
> Given the trinomial $ax^2 + bx + c$, $(a > 0)$, the signs can be determined as follows:
>
> - If c *is positive*, then the signs in the binomials must be the same (either both positive or both negative). The correct choice is determined by the middle term. If the middle term is positive, then both signs must be positive. If the middle term is negative, then both signs must be negative.
>
>
>
> c is positive c is positive
> ↓   ↓
> $20x^2 + 43x + 21$ $20x^2 - 43x + 21$
> $(4x + 3)(5x + 7)$ $(4x - 3)(5x - 7)$
> Same signs   Same signs
>
>
>
> - If c *is negative*, then the signs in the binomial must be different. The middle term in the trinomial determines which factor gets the positive sign and which gets the negative sign.
>
>
>
> c is negative c is negative
> ↓   ↓
> $x^2 + 3x - 28$ $x^2 - 3x - 28$
> $(x + 7)(x - 4)$ $(x - 7)(x + 4)$
> Different signs   Different signs
>
>

TIP: Look at the sign on the third term. If it is a sum, the signs will be the same in the two binomials. If it is a difference, the signs in the two binomials will be different: sum–same sign; difference–different signs.

Answer

1. $(3b + 2)(b + 2)$

| Example 2 | **Factoring a Trinomial by the Trial-and-Error Method** |

Classroom Example: p. 433, Exercise 20

Factor the trinomial. $13y - 6 + 8y^2$

Solution:

$13y - 6 + 8y^2$

$= 8y^2 + 13y - 6$ Write the polynomial in descending order.

$(\Box y \quad \Box)(\Box y \quad \Box)$ **Step 1:** The GCF is 1.

Factors of 8	**Factors of 6**	**Step 2:**

$1 \cdot 8$ $1 \cdot 6$

$2 \cdot 4$ $2 \cdot 3$

              $3 \cdot 2$
              $6 \cdot 1$ } (reverse order)

Step 2: List the positive factors of 8 and positive factors of 6. Consider the reverse order in one list of factors.

Step 3: Construct all possible binomial factors using different combinations of the factors of 8 and 6.

$(1y \quad 1)(8y \quad 6)$

$(1y \quad 3)(8y \quad 2)$

$(2y \quad 1)(4y \quad 6)$ Without regard to signs, these factorizations cannot work because the red terms in the binomials share a common factor greater than 1.

$(2y \quad 2)(4y \quad 3)$

$(2y \quad 3)(4y \quad 2)$

$(2y \quad 6)(4y \quad 1)$

Test the remaining factorizations. Keep in mind that to produce a product of -6, the signs within the parentheses must be opposite (one positive and one negative). Also, the sum of the products of the inner terms and outer terms must be combined to form $13y$.

$(1y \quad 6)(8y \quad 1)$ *Incorrect.*   Wrong middle term. Regardless of the signs, the product of inner terms, $48y$, and the product of outer terms, $1y$, cannot be combined to form the middle term $13y$.

$(1y \quad 2)(8y \quad 3)$ *Correct.*   The terms $16y$ and $3y$ can be combined to form the middle term $13y$, provided the signs are applied correctly. We require $+16y$ and $-3y$.

The correct factorization of $8y^2 + 13y - 6$ is $(y + 2)(8y - 3)$.

Skill Practice Factor.

2. $-25w + 6w^2 + 4$

Remember that the first step in any factoring problem is to remove the GCF. By removing the GCF, the remaining terms of the trinomial will have smaller coefficients.

Answer

2. $(6w - 1)(w - 4)$

Classroom Example: p. 434, Exercise 32

Example 3 Factoring a Trinomial by the Trial-and-Error Method

Factor the trinomial by the trial-and-error method. $40x^3 - 104x^2 + 10x$

Solution:

$40x^3 - 104x^2 + 10x$

$= 2x(20x^2 - 52x + 5)$ **Step 1:** The GCF is $2x$.

$= 2x(\square x \quad \square)(\square x \quad \square)$ **Step 2:** List the factors of 20 and factors of 5. Consider the reverse order in one list of factors.

> **TIP:** Notice that when the GCF, $2x$, is removed from the original trinomial, the new trinomial has smaller coefficients. This makes the factoring process simpler. It is easier to list the factors of 20 and 5 than the factors of 40 and 10.

Factors of 20	**Factors of 5**
$1 \cdot 20$	$1 \cdot 5$
$2 \cdot 10$	$5 \cdot 1$
$4 \cdot 5$	

Step 3: Construct all possible binomial factors using different combinations of the factors of 20 and factors of 5. The signs in the parentheses must both be negative.

$= 2x(1x - 1)(20x - 5)$
$= 2x(2x - 1)(10x - 5)$ $\left.\right\}$ *Incorrect.*
$= 2x(4x - 1)(5x - 5)$

Once the GCF has been removed from the original polynomial, the binomial factors cannot contain a GCF greater than 1.

$= 2x(1x - 5)(20x - 1)$ *Incorrect.*   Wrong middle term.

$2x(x - 5)(20x - 1)$
$= 2x(20x^2 - 1x - 100x + 5)$
$= 2x(20x^2 - 101x + 5)$

$= 2x(4x - 5)(5x - 1)$ *Incorrect.*   Wrong middle term.

$2x(4x - 5)(5x - 1)$
$= 2x(20x^2 - 4x - 25x + 5)$
$= 2x(20x^2 - 29x + 5)$

$= 2x(2x - 5)(10x - 1)$ *Correct.*

$2x(2x - 5)(10x - 1)$
$= 2x(20x^2 - 2x - 50x + 5)$
$= 2x(20x^2 - 52x + 5)$
$= 40x^3 - 104x^2 + 10x$

The correct factorization is $2x(2x - 5)(10x - 1)$.

Skill Practice Factor.

3. $8t^3 + 38t^2 + 24t$

Often it is easier to factor a trinomial when the leading coefficient is positive. If the leading coefficient is negative, consider factoring out the opposite of the GCF.

Answer

3. $2t(4t + 3)(t + 4)$

| **Example 4** | **Factoring a Trinomial by the Trial-and-Error Method** |

Classroom Example: p. 434, Exercise 36

Factor. $-45x^2 - 3xy + 18y^2$

Solution:

$-45x^2 - 3xy + 18y^2$

$= -3(15x^2 + xy - 6y^2)$ **Step 1:** Factor out -3 to make the leading coefficient positive.

$= -3(\Box x \quad \Box y)(\Box x \quad \Box y)$ **Step 2:** List the factors of 15 and 6.

Factors of 15	**Factors of 6**	
$1 \cdot 15$	$1 \cdot 6$	
$3 \cdot 5$	$2 \cdot 3$	
	$3 \cdot 2$	
	$6 \cdot 1$	**Step 3:** We will construct all binomial combinations, without regard to signs first.

$\left.\begin{array}{l} -3(x \quad y)(15x \quad 6y) \\ -3(x \quad 2y)(15x \quad 3y) \\ -3(3x \quad 3y)(5x \quad 2y) \\ -3(3x \quad 6y)(5x \quad y) \end{array}\right\}$ *Incorrect.* These combinations cannot work because the binomials in red each contain a common factor.

Test the remaining factorizations. The signs within parentheses must be opposite to produce a product of $-6y^2$. Also, the sum of the products of the inner terms and outer terms must be combined to form $1xy$.

$-3(x \quad 3y)(15x \quad 2y)$ *Incorrect.* Regardless of signs, $45xy$ and $2xy$ cannot be combined to equal xy.

$-3(x \quad 6y)(15x \quad y)$ *Incorrect.* Regardless of signs, $90xy$ and xy cannot be combined to equal xy.

$-3(3x \quad y)(5x \quad 6y)$ *Incorrect.* Regardless of signs, $5xy$ and $18xy$ cannot be combined to equal xy.

$-3(3x \quad 2y)(5x \quad 3y)$ *Correct.* The terms $10xy$ and $9xy$ can be combined to form xy provided that the signs are applied correctly. We require $10xy$ and $-9xy$.

$-3(3x + 2y)(5x - 3y)$ Factored form

Avoiding Mistakes

Do not forget to write the GCF in the final answer.

Skill Practice Factor.

4. $-4x^2 + 26xy - 40y^2$

Recall that a prime polynomial is a polynomial whose only factors are itself and 1. Not every trinomial is factorable by the methods presented in this text.

Answer

4. $-2(2x - 5y)(x - 4y)$

Classroom Example: p. 434, Exercise 52

| Example 5 | **Factoring a Trinomial by the Trial-and-Error Method** |

Factor the trinomial by the trial-and-error method. $2p^2 - 8p + 3$

Solution:

$2p^2 - 8p + 3$ **Step 1:** The GCF is 1.

$= (1p \quad \square)(2p \quad \square)$ **Step 2:** List the factors of 2 and the factors of 3.

Factors of 2	**Factors of 3**	**Step 3:**	Construct all possible binomial factors using different combinations of the factors of 2 and 3. Because the third term in the trinomial is positive, both signs in the binomial must be the same. Because the middle term coefficient is negative, both signs will be negative.
$1 \cdot 2$	$1 \cdot 3$		
	$3 \cdot 1$		

$(p - 1)(2p - 3) = 2p^2 - 3p - 2p + 3$

$\qquad\qquad\qquad\quad = 2p^2 - 5p + 3$ *Incorrect.*   Wrong middle term.

$(p - 3)(2p - 1) = 2p^2 - p - 6p + 3$

$\qquad\qquad\qquad\quad = 2p^2 - 7p + 3$ *Incorrect.*   Wrong middle term.

None of the combinations of factors results in the correct product. Therefore, the polynomial $2p^2 - 8p + 3$ is prime and cannot be factored further.

Skill Practice Factor.

   5. $3a^2 + a + 4$

In Example 6, we use the trial-and-error method to factor a higher degree trinomial into two binomial factors.

Classroom Example: p. 434, Exercise 38

| Example 6 | **Factoring a Higher Degree Trinomial** |

Factor the trinomial. $\qquad 3x^4 + 8x^2 + 5$

Solution:

$3x^4 + 8x^2 + 5$ **Step 1:** The GCF is 1.

$= (\square x^2 + \square)(\square x^2 + \square)$ **Step 2:** To produce the product $3x^4$, we must use $3x^2$ and $1x^2$. To produce a product of 5, we will try the factors $(1)(5)$ and $(5)(1)$.

     Step 3: Construct all possible binomial factors using the combinations of factors of $3x^4$ and 5.

$(3x^2 + 1)(x^2 + 5) = 3x^4 + 15x^2 + 1x^2 + 5 = 3x^4 + 16x^2 + 5$ Wrong middle term.

$(3x^2 + 5)(x^2 + 1) = 3x^4 + 3x^2 + 5x^2 + 5 = 3x^4 + 8x^2 + 5$ Correct!

Therefore, $3x^4 + 8x^2 + 5 = (3x^2 + 5)(x^2 + 1)$

Skill Practice Factor.

   6. $2y^4 - y^2 - 15$

Answers

5. Prime   **6.** $(y^2 - 3)(2y^2 + 5)$

For additional exercises, see Classroom Activity 6.3A in the *Student's Resource Manual* at www.mhhe.com/moh.

Vocabulary and Key Concepts

1. **a.** Which is the correct factored form of $2x^2 - 5x - 12$, the product $(2x + 3)(x - 4)$ or $(x - 4)(2x + 3)$?
    Both are correct.

    b. Which is the complete factorization of $6x^2 - 4x - 10$, the product $(3x - 5)(2x + 2)$ or $2(3x - 5)(x + 1)$?
    $2(3x - 5)(x + 1)$

Review Exercises

For Exercises 2–6, factor completely.

2. $5uv^2 - 10u^2v + 25u^2v^2$
    $5uv(v - 2u + 5uv)$

3. $mn - m - 2n + 2$
    $(n - 1)(m - 2)$

4. $5x - 10 - xy + 2y$
    $(x - 2)(5 - y)$

5. $6a^2 - 30a - 84$
    $6(a - 7)(a + 2)$

6. $10b^2 + 20b - 240$
    $10(b + 6)(b - 4)$

Concept 1: Factoring Trinomials by the Trial-and-Error Method

For Exercises 7–10, assume a, b, and c represent positive integers.

7. When factoring a polynomial of the form $ax^2 + bx + c$, pick an appropriate combination of signs. a

    a. $(\ +\ )(\ +\ )$

    b. $(\ -\ )(\ -\ )$

    c. $(\ +\ )(\ -\ )$

8. When factoring a polynomial of the form $ax^2 - bx - c$, pick an appropriate combination of signs. c

    a. $(\ +\ )(\ +\ )$

    b. $(\ -\ )(\ -\ )$

    c. $(\ +\ )(\ -\ )$

9. When factoring a polynomial of the form $ax^2 - bx + c$, pick an appropriate combination of signs. b

    a. $(\ +\ )(\ +\ )$

    b. $(\ -\ )(\ -\ )$

    c. $(\ +\ )(\ -\ )$

10. When factoring a polynomial of the form $ax^2 + bx - c$, pick an appropriate combination of signs. c

    a. $(\ +\ )(\ +\ )$

    b. $(\ -\ )(\ -\ )$

    c. $(\ +\ )(\ -\ )$

For Exercises 11–28, factor completely by using the trial-and-error method. **(See Examples 1, 2, and 5.)**

11. $3n^2 + 13n + 4$
    $(3n + 1)(n + 4)$

12. $2w^2 + 5w - 3$
    $(2w - 1)(w + 3)$

13. $2y^2 - 3y - 2$
    $(2y + 1)(y - 2)$

14. $2a^2 + 7a + 6$
    $(2a + 3)(a + 2)$

15. $5x^2 - 14x - 3$
    $(5x + 1)(x - 3)$

16. $7y^2 + 9y - 10$
    $(7y - 5)(y + 2)$

17. $12c^2 - 5c - 2$
    $(4c + 1)(3c - 2)$

18. $6z^2 + z - 12$
    $(3z - 4)(2z + 3)$

19. $-12 + 10w^2 + 37w$
    $(10w - 3)(w + 4)$

20. $-10 + 10p^2 + 21p$
    $(2p + 5)(5p - 2)$

21. $-5q - 6 + 6q^2$
    $(3q + 2)(2q - 3)$

22. $17a - 2 + 3a^2$
    Prime

23. $6b - 23 + 4b^2$
    Prime

24. $8 + 7x^2 - 18x$
    $(7x - 4)(x - 2)$

25. $-8 + 25m^2 - 10m$
    $(5m + 2)(5m - 4)$

26. $8q^2 + 31q - 4$
    $(8q - 1)(q + 4)$

27. $6y^2 + 19xy - 20x^2$
    $(6y - 5x)(y + 4x)$

28. $12y^2 - 73yz + 6z^2$
    $(12y - z)(y - 6z)$

For Exercises 29–36, factor completely. Be sure to factor out the GCF first. **(See Examples 3–4.)**

29. $2m^2 - 12m - 80$
$2(m + 4)(m - 10)$

30. $3c^2 - 33c + 72$
$3(c - 8)(c - 3)$

31. $2y^5 + 13y^4 + 6y^3$
$y^3(2y + 1)(y + 6)$

32. $3u^8 - 13u^7 + 4u^6$
$u^6(3u - 1)(u - 4)$

33. $-a^2 - 15a + 34$
$-(a + 17)(a - 2)$

34. $-x^2 - 7x - 10$
$-(x + 2)(x + 5)$

35. $-80m^2 + 100mp + 30p^2$
$-10(4m + p)(2m - 3p)$

36. $-60w^2 - 550wz + 500z^2$
$-10(w + 10z)(6w - 5z)$

For Exercises 37–42, factor the higher degree polynomial. **(See Example 6.)**

37. $x^4 + 10x^2 + 9$
$(x^2 + 1)(x^2 + 9)$

38. $y^4 + 4y^2 - 21$
$(y^2 - 3)(y^2 + 7)$

39. $w^4 + 2w^2 - 15$
$(w^2 + 5)(w^2 - 3)$

40. $p^4 - 13p^2 + 40$
$(p^2 - 8)(p^2 - 5)$

41. $2x^4 - 7x^2 - 15$
$(2x^2 + 3)(x^2 - 5)$

42. $5y^4 + 11y^2 + 2$
$(5y^2 + 1)(y^2 + 2)$

Mixed Exercises

For Exercises 43–82, factor each trinomial completely.

43. $20z - 18 - 2z^2$
$-2(z - 9)(z - 1)$

44. $25t - 5t^2 - 30$
$-5(t - 2)(t - 3)$

45. $42 - 13q + q^2$
$(q - 7)(q - 6)$

46. $-5w - 24 + w^2$
$(w - 8)(w + 3)$

47. $6t^2 + 7t - 3$
$(2t + 3)(3t - 1)$

48. $4p^2 - 9p + 2$
$(4p - 1)(p - 2)$

49. $4m^2 - 20m + 25$
$(2m - 5)^2$

50. $16r^2 + 24r + 9$
$(4r + 3)^2$

51. $5c^2 - c + 2$
Prime

52. $7s^2 + 2s + 9$
Prime

53. $6x^2 - 19xy + 10y^2$
$(2x - 5y)(3x - 2y)$

54. $15p^2 + pq - 2q^2$
$(3p - q)(5p + 2q)$

55. $12m^2 + 11mn - 5n^2$
$(4m + 5n)(3m - n)$

56. $4a^2 + 5ab - 6b^2$
$(4a - 3b)(a + 2b)$

57. $30r^2 + 5r - 10$
$5(3r + 2)(2r - 1)$

58. $36x^2 - 18x - 4$
$2(6x + 1)(3x - 2)$

59. $4s^2 - 8st + t^2$
Prime

60. $6u^2 - 10uv + 5v^2$
Prime

61. $10t^2 - 23t - 5$
$(2t - 5)(5t + 1)$

62. $16n^2 + 14n + 3$
$(8n + 3)(2n + 1)$

63. $14w^2 + 13w - 12$
$(7w - 4)(2w + 3)$

64. $12x^2 - 16x + 5$
$(6x - 5)(2x - 1)$

65. $a^2 - 10a - 24$
$(a - 12)(a + 2)$

66. $b^2 + 6b - 7$
$(b + 7)(b - 1)$

67. $x^2 + 9xy + 20y^2$
$(x + 5y)(x + 4y)$

68. $p^2 - 13pq + 36q^2$
$(p - 9q)(p - 4q)$

69. $a^2 + 21ab + 20b^2$
$(a + 20b)(a + b)$

70. $x^2 - 17xy - 18y^2$
$(x - 18y)(x + y)$

71. $t^2 - 10t + 21$
$(t - 7)(t - 3)$

72. $z^2 - 15z + 36$
$(z - 12)(z - 3)$

73. $5d^3 + 3d^2 - 10d$
$d(5d^2 + 3d - 10)$

74. $3y^3 - y^2 + 12y$
$y(3y^2 - y + 12)$

75. $4b^3 - 4b^2 - 80b$
$4b(b - 5)(b + 4)$

76. $2w^2 + 20w + 42$
$2(w + 7)(w + 3)$

77. $x^2y^2 - 13xy^2 + 30y^2$
$y^2(x - 3)(x - 10)$

78. $p^2q^2 - 14pq^2 + 33q^2$
$q^2(p - 3)(p - 11)$

79. $-12u^3 - 22u^2 + 20u$
$-2u(2u + 5)(3u - 2)$

80. $-18z^4 + 15z^3 + 12z^2$
$-3z^2(3z - 4)(2z + 1)$

81. $8x^4 + 14x^2 + 3$
$(2x^2 + 3)(4x^2 + 1)$

82. $6y^4 - 5y^2 - 4$
$(3y^2 - 4)(2y^2 + 1)$

83. A rock is thrown straight upward from the top of a 40-ft building. Its height in feet after t seconds is given by the polynomial $-16t^2 + 12t + 40$.

  a. Calculate the height of the rock after 1 sec. $(t = 1)$ 36 ft

  b. Write $-16t^2 + 12t + 40$ in factored form. Then evaluate the factored form of the polynomial for $t = 1$. Is the result the same as from part (a)?
  $-4(4t + 5)(t - 2)$; Yes

84. A baseball is thrown straight downward from the top of a 120-ft building. Its height in feet after t seconds is given by $-16t^2 - 8t + 120$.

  a. Calculate the height of the ball after 2 sec. $(t = 2)$ 40 ft

  b. Write $-16t^2 - 8t + 120$ in factored form. Then evaluate the factored form of the polynomial for $t = 2$. Is the result the same as from part (a)?
  $-8(2t - 5)(t + 3)$; Yes

Expanding Your Skills

For Exercises 85–88, the two trinomials look similar but differ by one sign. Factor each trinomial and see how their factored forms differ.

85. a. $x^2 - 10x - 24$ $(x - 12)(x + 2)$

  b. $x^2 - 10x + 24$ $(x - 6)(x - 4)$

86. a. $x^2 - 13x - 30$ $(x - 15)(x + 2)$

  b. $x^2 - 13x + 30$ $(x - 10)(x - 3)$

87. a. $x^2 - 5x - 6$ $(x - 6)(x + 1)$

  b. $x^2 - 5x + 6$ $(x - 2)(x - 3)$

88. a. $x^2 - 10x + 9$ $(x - 9)(x - 1)$

  b. $x^2 + 10x + 9$ $(x + 9)(x + 1)$

Factoring Trinomials: AC-Method

We have already learned how to factor a trinomial of the form $ax^2 + bx + c = 0$ with a leading coefficient of 1. Then we learned the trial-and-error method to factor the more general case in which the leading coefficient is any integer. In this section, we provide an alternative method to factor trinomials, called the ac-method.

Concept

1. Factoring Trinomials by the AC-Method

1. Factoring Trinomials by the AC-Method

The product of two binomials results in a four-term expression that can sometimes be simplified to a trinomial. To factor the trinomial, we want to reverse the process.

Multiply:

Multiply the binomials.   Add the middle terms.

$$(2x + 3)(x + 2) = \longrightarrow 2x^2 + 4x + 3x + 6 = \longrightarrow 2x^2 + 7x + 6$$

Factor:

$$2x^2 + 7x + 6 = \longrightarrow 2x^2 + 4x + 3x + 6 = \longrightarrow (2x + 3)(x + 2)$$

Rewrite the middle term as   Factor by grouping.
a sum or difference of terms.

To factor a quadratic trinomial, $ax^2 + bx + c$, by the ac-method, we rewrite the middle term, bx, as a sum or difference of terms. The goal is to produce a four-term polynomial that can be factored by grouping. The process is outlined as follows.

AC-Method: Factoring $ax^2 + bx + c$ ($a \neq 0$)

Step 1   Factor out the GCF from all terms.

Step 2   Multiply the coefficients of the first and last terms (ac).

Step 3   Find two integers whose product is ac and whose sum is b. (If no pair of integers can be found, then the trinomial cannot be factored further and is a *prime polynomial*.)

Step 4   Rewrite the middle term, bx, as the sum of two terms whose coefficients are the integers found in step 3.

Step 5   Factor the polynomial by grouping.

Instructor Note: Tell students that we use the phrase *prime polynomial* to mean that a polynomial is not factorable over the *integers*.

The ac-method for factoring trinomials is illustrated in Example 1. However, before we begin, keep these two important guidelines in mind:

- For any factoring problem you encounter, always factor out the GCF from all terms first.
- To factor a trinomial, write the trinomial in the form $ax^2 + bx + c$.

Classroom Example: p. 440,
Exercise 14

Example 1 **Factoring a Trinomial by the AC-Method**

Factor the trinomial by the ac-method. $2x^2 + 7x + 6$

Solution:

$2x^2 + 7x + 6$ **Step 1:** Factor out the GCF from all terms. In this case, the GCF is 1. The trinomial is written in the form $ax^2 + bx + c$.

$a = 2, b = 7, c = 6$ **Step 2:** Find the product $ac = (2)(6) = 12$.

$\underline{\quad 12 \quad}$	$\underline{\quad 12 \quad}$
$1 \cdot 12$	$(-1)(-12)$
$2 \cdot 6$	$(-2)(-6)$
$3 \cdot 4$	$(-3)(-4)$

Step 3: List all factors of ac and search for the pair whose sum equals the value of b. That is, list the factors of 12 and find the pair whose sum equals 7.

The numbers 3 and 4 satisfy both conditions: $3 \cdot 4 = 12$ and $3 + 4 = 7$.

$2x^2 + 7x + 6$

$= 2x^2 + 3x + 4x + 6$ **Step 4:** Write the middle term of the trinomial as the sum of two terms whose coefficients are the selected pair of numbers: 3 and 4.

$= 2x^2 + 3x \mid + 4x + 6$ **Step 5:** Factor by grouping.

$= x(2x + 3) + 2(2x + 3)$

$= (2x + 3)(x + 2)$

$\underline{\text{Check}}$: $(2x + 3)(x + 2) = 2x^2 + 4x + 3x + 6$

$= 2x^2 + 7x + 6$ ✓

Skill Practice Factor by the ac-method.

  1. $2x^2 + 5x + 3$

TIP: One frequently asked question is whether the order matters when we rewrite the middle term of the trinomial as two terms (step 4). The answer is no. From the previous example, the two middle terms in step 4 could have been reversed to obtain the same result:

$$2x^2 + 7x + 6$$
$$= 2x^2 + 4x + 3x + 6$$
$$= 2x(x + 2) + 3(x + 2)$$
$$= (x + 2)(2x + 3)$$

This example also points out that the order in which two factors are written does not matter. The expression $(x + 2)(2x + 3)$ is equivalent to $(2x + 3)(x + 2)$ because multiplication is a commutative operation.

Answer

1. $(x + 1)(2x + 3)$

Example 2 **Factoring a Trinomial by the AC-Method**

Factor the trinomial by the ac-method. $-2x + 8x^2 - 3$

Classroom Example: p. 440, Exercise 28

Solution:

$-2x + 8x^2 - 3$ First rewrite the polynomial in the form $ax^2 + bx + c$.

$= 8x^2 - 2x - 3$ **Step 1:** The GCF is 1.

$a = 8, b = -2, c = -3$ **Step 2:** Find the product $ac = (8)(-3) = -24$.

Step 3: List all the factors of -24 and find the pair of factors whose sum equals -2.

$\underline{-24}$	$\underline{-24}$
$-1 \cdot 24$	$-24 \cdot 1$
$-2 \cdot 12$	$-12 \cdot 2$
$-3 \cdot 8$	$-8 \cdot 3$
$-4 \cdot 6$	$-6 \cdot 4$

The numbers -6 and 4 satisfy both conditions: $(-6)(4) = -24$ and $-6 + 4 = -2$.

$= 8x^2 - 2x - 3$ **Step 4:** Write the middle term of the trinomial as two terms whose coefficients are the selected pair of numbers, -6 and 4.

$= 8x^2 - 6x + 4x - 3$

$= 8x^2 - 6x \vdots + 4x - 3$ **Step 5:** Factor by grouping.

$= 2x(4x - 3) + 1(4x - 3)$

$= (4x - 3)(2x + 1)$

$\underline{\text{Check:}}\ (4x - 3)(2x + 1) = 8x^2 + 4x - 6x - 3$

$= 8x^2 - 2x - 3 \checkmark$

Skill Practice Factor by the ac-method.

2. $13w + 6w^2 + 6$

Example 3 **Factoring a Trinomial by the AC-Method**

Factor the trinomial by the ac-method. $10x^3 - 85x^2 + 105x$

Classroom Example: p. 440, Exercise 38

Solution:

$10x^3 - 85x^2 + 105x$ **Step 1:** Factor out the GCF of $5x$.

$= 5x(2x^2 - 17x + 21)$ The trinomial is in the form $ax^2 + bx + c$.

$a = 2, b = -17, c = 21$ **Step 2:** Find the product $ac = (2)(21) = 42$.

Step 3: List all the factors of 42 and find the pair whose sum equals -17.

$\underline{42}$	$\underline{42}$
$1 \cdot 42$	$(-1)(-42)$
$2 \cdot 21$	$(-2)(-21)$
$3 \cdot 14$	$(-3)(-14)$
$6 \cdot 7$	$(-6)(-7)$

The numbers -3 and -14 satisfy both conditions: $(-3)(-14) = 42$ and $-3 + (-14) = -17$.

Answer

2. $(2w + 3)(3w + 2)$

$$= 5x(2x^2 - 17x + 21)$$

Step 4: Write the middle term of the trinomial as two terms whose coefficients are the selected pair of numbers, -3 and -14.

$$= 5x(2x^2 - 3x - 14x + 21)$$

$$= 5x(2x^2 - 3x \mid - 14x + 21)$$

Step 5: Factor by grouping.

$$= 5x[x(2x - 3) - 7(2x - 3)]$$

$$= 5x(2x - 3)(x - 7)$$

Avoiding Mistakes

Be sure to bring down the GCF in each successive step as you factor.

TIP: Notice when the GCF is removed from the original trinomial, the new trinomial has smaller coefficients. This makes the factoring process simpler because the product ac is smaller. It is much easier to list the factors of 42 than the factors of 1050.

Original trinomial	**With the GCF factored out**
$10x^3 - 85x^2 + 105x$	$5x(2x^2 - 17x + 21)$
$ac = (10)(105) = 1050$	$ac = (2)(21) = 42$

Skill Practice Factor by the ac-method.

3. $9y^3 - 30y^2 + 24y$

In most cases, it is easier to factor a trinomial with a positive leading coefficient.

Classroom Example: p. 440, Exercise 36

Example 4 **Factoring a Trinomial by the AC-Method**

Factor the trinomial by the ac-method. $\qquad -18x^2 + 21xy + 15y^2$

Solution:

$-18x^2 + 21xy + 15y^2$

$= -3(6x^2 - 7xy - 5y^2)$

Step 1: Factor out the GCF.

Factor out -3 to make the leading term positive.

Step 2: The product $ac = (6)(-5) = -30$.

Step 3: The numbers -10 and 3 have a product of -30 and a sum of -7.

$= -3[6x^2 - 10xy + 3xy - 5y^2]$

Step 4: Rewrite the middle term, $-7xy$ as $-10xy + 3xy$.

$= -3[6x^2 - 10xy \mid + 3xy - 5y^2]$

Step 5: Factor by grouping.

$= -3[2x(3x - 5y) + y(3x - 5y)]$

$= -3(3x - 5y)(2x + y)$

Factored form

Skill Practice Factor.

4. $-8x^2 - 8xy + 30y^2$

Answers

3. $3y(3y - 4)(y - 2)$
4. $-2(2x - 3y)(2x + 5y)$

Recall that a prime polynomial is a polynomial whose only factors are itself and 1. It also should be noted that not every trinomial is factorable by the methods presented in this text.

Example 5   **Factoring a Trinomial by the AC-Method**

Classroom Example: p. 440, Exercise 26

Factor the trinomial by the ac-method. $2p^2 - 8p + 3$

Solution:

$2p^2 - 8p + 3$

Step 1: The GCF is 1.

Step 2: The product $ac = 6$.

6	6
$1 \cdot 6$	$(-1)(-6)$
$2 \cdot 3$	$(-2)(-3)$

Step 3: List the factors of 6. Notice that no pair of factors has a sum of -8. Therefore, the trinomial cannot be factored.

The trinomial $2p^2 - 8p + 3$ is a prime polynomial.

Skill Practice Factor.

5. $4x^2 + 5x + 2$

In Example 6, we use the ac-method to factor a higher degree trinomial.

Example 6   **Factoring a Higher Degree Trinomial**

Classroom Example: p. 440, Exercise 42

Factor the trinomial by the ac-method. $2x^4 + 5x^2 + 2$

Solution:

$2x^4 + 5x^2 + 2$

Step 1: The GCF is 1.

$a = 2, b = 5, c = 2$

Step 2: Find the product $ac = (2)(2) = 4$.

Step 3: The numbers 1 and 4 have a product of 4 and a sum of 5.

$2x^4 + x^2 + 4x^2 + 2$

Step 4: Rewrite the middle term, $5x^2$, as $x^2 + 4x^2$.

$2x^4 + x^2 \mid + 4x^2 + 2$

Step 5: Factor by grouping.

$x^2(2x^2 + 1) + 2(2x^2 + 1)$

$(2x^2 + 1)(x^2 + 2)$ Factored form

Skill Practice Factor.

6. $3y^4 + 2y^2 - 8$

Section 6.4 Practice Exercises

Vocabulary and Key Concepts

For additional exercises, see Classroom Activity 6.4A in the Student's Resource Manual at www.mhhe.com/moh.

1. a. Which is the correct factored form of $10x^2 - 13x - 3$? The product $(5x + 1)(2x - 3)$ or $(2x - 3)(5x + 1)$?
Both are correct.

   b. Which is the complete factorization of $12x^2 - 15x - 18$? The product $(4x + 3)(3x - 6)$ or $3(4x + 3)(x - 2)$?
$3(4x + 3)(x - 2)$

Review Exercises

For Exercises 2–4, factor completely.

2. $5x(x - 2) - 2(x - 2)$ $(x - 2)(5x - 2)$

3. $8(y + 5) + 9y(y + 5)$ $(y + 5)(8 + 9y)$

4. $6ab + 24b - 12a - 48$ $6(a + 4)(b - 2)$

Concept 1: Factoring Trinomials by the AC-Method

For Exercises 5–12, find the pair of integers whose product and sum are given.

5. Product: 12   Sum: 13   12, 1

6. Product: 12   Sum: 7   3, 4

7. Product: 8   Sum: −9   −8, −1

8. Product: −4   Sum: −3   −4, 1

9. Product: −20   Sum: 1   5, −4

10. Product: −6   Sum: −1   −3, 2

11. Product: −18   Sum: 7   9, −2

12. Product: −72   Sum: −6   −12, 6

For Exercises 13–30, factor the trinomials using the ac-method. **(See Examples 1, 2, and 5.)**

13. $3x^2 + 13x + 4$ $(x + 4)(3x + 1)$

14. $2y^2 + 7y + 6$ $(2y + 3)(y + 2)$

15. $4w^2 - 9w + 2$ $(w - 2)(4w - 1)$

16. $2p^2 - 3p - 2$ $(p - 2)(2p + 1)$

17. $x^2 + 7x - 18$ $(x + 9)(x - 2)$

18. $y^2 - 6y - 40$ $(y - 10)(y + 4)$

19. $2m^2 + 5m - 3$ $(m + 3)(2m - 1)$

20. $6n^2 + 7n - 3$ $(2n + 3)(3n - 1)$

21. $8k^2 - 6k - 9$ $(4k + 3)(2k - 3)$

22. $9h^2 - 3h - 2$ $(3h - 2)(3h + 1)$

23. $4k^2 - 20k + 25$ $(2k - 5)^2$

24. $16h^2 + 24h + 9$ $(4h + 3)^2$

25. $5x^2 + x + 7$ Prime

26. $4y^2 - y + 2$ Prime

27. $10 + 9z^2 - 21z$ $(3z - 5)(3z - 2)$

28. $13x + 4x^2 - 12$
$(4x - 3)(x + 4)$

29. $12y^2 + 8yz - 15z^2$
$(6y - 5z)(2y + 3z)$

30. $20a^2 + 3ab - 9b^2$
$(5a - 3b)(4a + 3b)$

For Exercises 31–38, factor completely. Be sure to factor out the GCF first. **(See Examples 3–4.)**

31. $50y + 24 + 14y^2$
$2(7y + 4)(y + 3)$

32. $-24 + 10w + 4w^2$
$2(2w - 3)(w + 4)$

33. $-15w^2 + 22w + 5$
$-(3w - 5)(5w + 1)$

34. $-16z^2 + 34z + 15$
$-(8z + 3)(2z - 5)$

35. $-12x^2 + 20xy - 8y^2$
$-4(x - y)(3x - 2y)$

36. $-6p^2 - 21pq - 9q^2$
$-3(2p + q)(p + 3q)$

37. $18y^3 + 60y^2 + 42y$
$6y(y + 1)(3y + 7)$

38. $8t^3 - 4t^2 - 40t$
$4t(2t - 5)(t + 2)$

For Exercises 39–44, factor the higher degree polynomial. **(See Example 6.)**

39. $a^4 + 5a^2 + 6$
$(a^2 + 2)(a^2 + 3)$

40. $y^4 - 2y^2 - 35$
$(y^2 + 5)(y^2 - 7)$

41. $6x^4 - x^2 - 15$
$(3x^2 - 5)(2x^2 + 3)$

42. $8t^4 + 2t^2 - 3$
$(4t^2 + 3)(2t^2 - 1)$

43. $8p^4 + 37p^2 - 15$
$(8p^2 - 3)(p^2 + 5)$

44. $2a^4 + 11a^2 + 14$
$(2a^2 + 7)(a^2 + 2)$

 Writing ←→ Translating Expression Geometry Scientific Calculator Video

Mixed Exercises

For Exercises 45–80, factor completely.

45. $20p^2 - 19p + 3$
$(5p - 1)(4p - 3)$

46. $4p^2 + 5pq - 6q^2$
$(p + 2q)(4p - 3q)$

47. $6u^2 - 19uv + 10v^2$
$(3u - 2v)(2u - 5v)$

48. $15m^2 + mn - 2n^2$
$(5m + 2n)(3m - n)$

49. $12a^2 + 11ab - 5b^2$
$(4a + 5b)(3a - b)$

50. $3r^2 - rs - 14s^2$
$(r + 2s)(3r - 7s)$

51. $3h^2 + 19hk - 14k^2$
$(h + 7k)(3h - 2k)$

52. $2u^2 + uv - 15v^2$
$(2u - 5v)(u + 3v)$

53. $2x^2 - 13xy + y^2$ Prime

54. $3p^2 + 20pq - q^2$ Prime

55. $3 - 14z + 16z^2$
$(2z - 1)(8z - 3)$

56. $10w + 1 + 16w^2$
$(8w + 1)(2w + 1)$

57. $b^2 + 16 - 8b$
$(b - 4)^2$

58. $1 + q^2 - 2q$
$(q - 1)^2$

59. $25x - 5x^2 - 30$
$-5(x - 2)(x - 3)$

60. $20a - 18 - 2a^2$
$-2(a - 1)(a - 9)$

61. $-6 - t + t^2$
$(t - 3)(t + 2)$

62. $-6 + m + m^2$
$(m + 3)(m - 2)$

63. $v^2 + 2v + 15$ Prime

64. $x^2 - x - 1$ Prime

65. $72x^2 + 18x - 2$
$2(12x - 1)(3x + 1)$

66. $20y^2 - 78y - 8$
$2(10y + 1)(y - 4)$

67. $p^3 - 6p^2 - 27p$
$p(p + 3)(p - 9)$

68. $w^5 - 11w^4 + 28w^3$
$w^3(w - 4)(w - 7)$

69. $3x^3 + 10x^2 + 7x$
$x(3x + 7)(x + 1)$

70. $4r^3 + 3r^2 - 10r$
$r(4r - 5)(r + 2)$

71. $2p^3 - 38p^2 + 120p$
$2p(p - 15)(p - 4)$

72. $4q^3 - 4q^2 - 80q$
$4q(q - 5)(q + 4)$

73. $x^2y^2 + 14x^2y + 33x^2$
$x^2(y + 3)(y + 11)$

74. $a^2b^2 + 13ab^2 + 30b^2$
$b^2(a + 10)(a + 3)$

75. $-k^2 - 7k - 10$
$-(k + 2)(k + 5)$

76. $-m^2 - 15m + 34$
$-(m - 2)(m + 17)$

77. $-3n^2 - 3n + 90$
$-3(n + 6)(n - 5)$

78. $-2h^2 + 28h - 90$
$-2(h - 9)(h - 5)$

79. $x^4 - 7x^2 + 10$
$(x^2 - 2)(x^2 - 5)$

80. $m^4 + 10m^2 + 21$
$(m^2 + 3)(m^2 + 7)$

81. Is the expression $(2x + 4)(x - 7)$ factored completely? Explain why or why not.
No. $(2x + 4)$ contains a common factor of 2.

82. Is the expression $(3x + 1)(5x - 10)$ factored completely? Explain why or why not.
No. $(5x - 10)$ contains a common factor of 5.

83. Colleen noticed that the number of tables placed in her restaurant affects the number of customers who eat at the restaurant. The number of customers each night is given by $-2x^2 + 40x - 72$, where x is the number of tables set up in the room, and $2 \leq x \leq 18$.

  a. Calculate the number of customers when there are 10 tables set up. ($x = 10$)
  128 customers

  b. Write $-2x^2 + 40x - 72$ in factored form. Then evaluate the factored form of the polynomial for $x = 10$. Is the result the same as from part (a)? $-2(x - 18)(x - 2)$; 128; Yes

84. Roland sells cases for smartphones online. He noticed that for every dollar he discounts the price, he sells two more cases per week. His income in dollars each week is given by $-2d^2 + 30d + 200$, where d is the dollar amount of the discount in price.

  a. Calculate his income if the discount is $2. ($d = 2$) $252

  b. Write $-2d^2 + 30d + 200$ in factored form. Then evaluate the factored form of the polynomial for $d = 2$. Is the result the same as from part (a)?
  $-2(d - 20)(d + 5)$; 252; Yes

85. A formula for finding the sum of the first n even integers is given by $n^2 + n$.

  a. Find the sum of the first 6 even integers $(2 + 4 + 6 + 8 + 10 + 12)$ by evaluating the expression for $n = 6$. 42

  b. Write the polynomial in factored form. Then evaluate the factored form of the expression for $n = 6$. Is the result the same as part (a)?
  $n(n + 1)$; 42; Yes

86. A formula for finding the sum of the squares of the first n integers is given by $\dfrac{2n^3 + 3n^2 + n}{6}$.

  a. Find the sum of the squares of the first 4 integers $(1^2 + 2^2 + 3^2 + 4^2)$ by evaluating the expression for $n = 4$. 30

  b. Write the polynomial in the numerator of the expression in factored form. Then evaluate the factored form of the expression for $n = 4$. Is the result the same as part (a)?
  $\dfrac{n(n + 1)(2n + 1)}{6}$; 30; Yes

Writing ← → Translating Expression   Geometry   Scientific Calculator   Video

Concepts

1. Factoring a Difference of Squares
2. Factoring Perfect Square Trinomials

1. Factoring a Difference of Squares

Up to this point, we have learned several methods of factoring, including:

- Factoring out the greatest common factor from a polynomial
- Factoring a four-term polynomial by grouping
- Factoring trinomials by the ac-method or by the trial-and-error method

In this section, we will learn to factor polynomials that fit two special case patterns: a difference of squares, and a perfect square trinomial. First recall that the product of two conjugates results in a **difference of squares**:

$$(a + b)(a - b) = a^2 - b^2$$

Therefore, to factor a difference of squares, the process is reversed. Identify a and b and construct the conjugate factors.

Factored Form of a Difference of Squares

$$a^2 - b^2 = (a + b)(a - b)$$

To help recognize a difference of squares, we recommend that you become familiar with the first several perfect squares.

Perfect Squares	Perfect Squares	Perfect Squares
$1 = (1)^2$	$36 = (6)^2$	$121 = (11)^2$
$4 = (2)^2$	$49 = (7)^2$	$144 = (12)^2$
$9 = (3)^2$	$64 = (8)^2$	$169 = (13)^2$
$16 = (4)^2$	$81 = (9)^2$	$196 = (14)^2$
$25 = (5)^2$	$100 = (10)^2$	$225 = (15)^2$

It is also important to recognize that a variable expression is a perfect square if its exponent is a multiple of 2. For example:

Perfect Squares

$$x^2 = (x)^2$$
$$x^4 = (x^2)^2$$
$$x^6 = (x^3)^2$$
$$x^8 = (x^4)^2$$
$$x^{10} = (x^5)^2$$

| **Example 1** | **Factoring Differences of Squares** |

Classroom Examples: p. 447, Exercises 16 and 18

Factor the binomials.

a. $y^2 - 25$ **b.** $49s^2 - 4t^4$ **c.** $18w^2z - 2z$

Solution:

a. $y^2 - 25$ The binomial is a difference of squares.

    $= (y)^2 - (5)^2$ Write in the form: $a^2 - b^2$, where $a = y$, $b = 5$.

    $= (y + 5)(y - 5)$ Factor as $(a + b)(a - b)$.

b. $49s^2 - 4t^4$ The binomial is a difference of squares.

    $= (7s)^2 - (2t^2)^2$ Write in the form $a^2 - b^2$, where $a = 7s$ and $b = 2t^2$.

    $= (7s + 2t^2)(7s - 2t^2)$ Factor as $(a + b)(a - b)$.

c. $18w^2z - 2z$ The GCF is $2z$.

    $= 2z(9w^2 - 1)$ $(9w^2 - 1)$ is a difference of squares.

    $= 2z[(3w)^2 - (1)^2]$ Write in the form: $a^2 - b^2$, where $a = 3w$, $b = 1$.

    $= 2z(3w - 1)(3w + 1)$ Factor as $(a - b)(a + b)$.

TIP: Recall that multiplication is commutative. Therefore,
$$a^2 - b^2 = (a + b)(a - b)$$
or $(a - b)(a + b)$.

Skill Practice Factor the binomials.

1. $a^2 - 64$ **2.** $25q^2 - 49w^2$ **3.** $98m^3n - 50mn$

The difference of squares $a^2 - b^2$ factors as $(a + b)(a - b)$. However, the *sum* of squares is not factorable.

Sum of Squares

Suppose a and b have no common factors. Then the **sum of squares** $a^2 + b^2$ is *not* factorable over the real numbers.

That is, $a^2 + b^2$ is prime over the real numbers.

To see why $a^2 + b^2$ is not factorable, consider the product of binomials:

    $(a + b)(a - b) = a^2 - b^2$ Wrong sign

    $(a + b)(a + b) = a^2 + 2ab + b^2$ Wrong middle term

    $(a - b)(a - b) = a^2 - 2ab + b^2$ Wrong middle term

After exhausting all possibilities, we see that if a and b share no common factors, then the sum of squares $a^2 + b^2$ is a prime polynomial.

| **Example 2** | **Factoring Binomials** |

Classroom Example: p. 447, Exercise 24

Factor the binomials, if possible.   **a.** $p^2 - 9$ **b.** $p^2 + 9$

Solution:

a. $p^2 - 9$ Difference of squares

    $= (p - 3)(p + 3)$ Factor as $a^2 - b^2 = (a - b)(a + b)$.

b. $p^2 + 9$ Sum of squares

    Prime (cannot be factored)

Answers

1. $(a + 8)(a - 8)$
2. $(5q + 7w)(5q - 7w)$
3. $2mn(7m + 5)(7m - 5)$

Classroom Example: p. 447,
Exercise 36

Skill Practice Factor the binomials, if possible.

4. $t^2 - 144$ **5.** $t^2 + 144$

Some factoring problems require several steps. Always be sure to factor completely.

Example 3   **Factoring a Difference of Squares**

Factor completely. $w^4 - 81$

Solution:

$w^4 - 81$	The GCF is 1. $w^4 - 81$ is a difference of squares.
$= (w^2)^2 - (9)^2$	Write in the form: $a^2 - b^2$, where $a = w^2$, $b = 9$.
$= (w^2 + 9)(w^2 - 9)$	Factor as $(a + b)(a - b)$.
$= (w^2 + 9)\overbrace{(w + 3)(w - 3)}$	Note that $w^2 - 9$ can be factored further as a difference of squares. (The binomial $w^2 + 9$ is a sum of squares and cannot be factored further.)

Skill Practice Factor completely.

6. $y^4 - 1$

Classroom Example: p. 447,
Exercise 40

Example 4   **Factoring a Polynomial**

Factor completely. $y^3 - 5y^2 - 4y + 20$

Solution:

$y^3 - 5y^2 - 4y + 20$	The GCF is 1. The polynomial has four terms. Factor by grouping.
$= y^3 - 5y^2 \mid - 4y + 20$	
$= y^2(y - 5) - 4(y - 5)$	
$= (y - 5)(y^2 - 4)$	The expression $y^2 - 4$ is a difference of squares and can be factored further as $(y - 2)(y + 2)$.
$= (y - 5)(y - 2)(y + 2)$	

$$\underline{\text{Check:}} \ (y - 5)(y - 2)(y + 2) = (y - 5)(y^2 - 2y + 2y - 4)$$
$$= (y - 5)(y^2 - 4)$$
$$= (y^3 - 4y - 5y^2 + 20)$$
$$= y^3 - 5y^2 - 4y + 20 \ \checkmark$$

Skill Practice Factor completely.

7. $p^3 + 7p^2 - 9p - 63$

2. Factoring Perfect Square Trinomials

Answers

4. $(t - 12)(t + 12)$
5. Prime
6. $(y + 1)(y - 1)(y^2 + 1)$
7. $(p - 3)(p + 3)(p + 7)$

Recall that the square of a binomial always results in a **perfect square trinomial**.

$$(a + b)^2 = (a + b)(a + b) \xrightarrow{\text{Multiply.}} = a^2 + 2ab + b^2$$

$$(a - b)^2 = (a - b)(a - b) \xrightarrow{\text{Multiply.}} = a^2 - 2ab + b^2$$

For example, $(3x + 5)^2 = (3x)^2 + 2(3x)(5) + (5)^2$

$$= 9x^2 + 30x + 25 \text{ (perfect square trinomial)}$$

We now want to reverse this process by factoring a perfect square trinomial. The trial-and-error method or the ac-method can always be used; however, if we recognize the pattern for a perfect square trinomial, we can use one of the following formulas to reach a quick solution.

Factored Form of a Perfect Square Trinomial

$$a^2 + 2ab + b^2 = (a + b)^2$$
$$a^2 - 2ab + b^2 = (a - b)^2$$

For example, $4x^2 + 36x + 81$ is a perfect square trinomial with $a = 2x$ and $b = 9$. Therefore, it factors as

$$4x^2 + 36x + 81 = (2x)^2 + 2(2x)(9) + (9)^2 = (2x + 9)^2$$

$$a^2 \quad + \quad 2(a)(b) + (b)^2 = (a + b)^2$$

To apply the formula to factor a perfect square trinomial, we must first be sure that the trinomial is indeed a perfect square trinomial.

Checking for a Perfect Square Trinomial

Step 1 Determine whether the first and third terms are both perfect squares and have positive coefficients.

Step 2 If this is the case, identify a and b and determine if the middle term equals $2ab$ or $-2ab$.

Example 5 **Factoring Perfect Square Trinomials**

Classroom Examples: p. 448, Exercises 52 and 54

Factor the trinomials completely.

a. $x^2 + 14x + 49$ **b.** $25y^2 - 20y + 4$

Solution:

a. $x^2 + 14x + 49$ The GCF is 1.

- The first and third terms are positive.

- The first term is a perfect square: $x^2 = (x)^2$.

Perfect squares

- The third term is a perfect square: $49 = (7)^2$.

$x^2 + 14x + 49$

- The middle term is twice the product of x and 7: $14x = 2(x)(7)$.

$= (x)^2 + 2(x)(7) + (7)^2$ The trinomial is in the form $a^2 + 2ab + b^2$, where $a = x$ and $b = 7$.

$= (x + 7)^2$ Factor as $(a + b)^2$.

TIP: The sign of the middle term in a perfect square trinomial determines the sign within the binomial of the factored form.

$$a^2 + 2ab + b^2 = (a + b)^2$$
$$a^2 - 2ab + b^2 = (a - b)^2$$

b. $25y^2 - 20y + 4$

The GCF is 1.

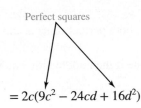

Perfect squares

- The first and third terms are positive.
- The first term is a perfect square: $25y^2 = (5y)^2$.

$25y^2 - 20y + 4$

$= (5y)^2 - 2(5y)(2) + (2)^2$

$= (5y - 2)^2$

- The third term is a perfect square: $4 = (2)^2$.
- In the middle: $-20y = -2(5y)(2)$

Factor as $(a - b)^2$.

Skill Practice Factor completely.

8. $x^2 - 6x + 9$ **9.** $81w^2 + 72w + 16$

Classroom Examples: p. 448,
Exercises 60 and 62

Example 6   **Factoring Perfect Square Trinomials**

Factor the trinomials completely.

   a. $18c^3 - 48c^2d + 32cd^2$ **b.** $5w^2 + 50w + 45$

Solution:

   a. $18c^3 - 48c^2d + 32cd^2$

     $= 2c(9c^2 - 24cd + 16d^2)$

The GCF is $2c$.

Perfect squares

- The first and third terms are positive.
- The first term is a perfect square: $9c^2 = (3c)^2$.

$= 2c(9c^2 - 24cd + 16d^2)$

- The third term is a perfect square: $16d^2 = (4d)^2$.

$= 2c[(3c)^2 - 2(3c)(4d) + (4d)^2]$

$= 2c(3c - 4d)^2$

- In the middle: $-24cd = -2(3c)(4d)$

Factor as $(a - b)^2$.

   b. $5w^2 + 50w + 45$

     $= 5(w^2 + 10w + 9)$

The GCF is 5.

The first and third terms are perfect squares.

$$w^2 = (w)^2 \qquad \text{and} \qquad 9 = (3)^2$$

Perfect squares

$= 5(w^2 + 10w + 9)$

However, the middle term is not 2 times the product of w and 3.

$$10w \neq 2(w)(3)$$

Therefore, this is not a perfect square trinomial.

$= 5(w + 9)(w + 1)$

To factor, use the trial-and-error method.

TIP: If you do not recognize that a trinomial is a perfect square trinomial, you can still use the trial-and-error method or ac-method to factor it.

Skill Practice Factor completely.

10. $5z^3 + 20z^2w + 20zw^2$ **11.** $40x^2 + 130x + 90$

Answers

8. $(x - 3)^2$
9. $(9w + 4)^2$
10. $5z(z + 2w)^2$
11. $10(4x + 9)(x + 1)$

Vocabulary and Key Concepts

For additional exercises, see Classroom Activities 6.5A–6.5B in the *Student's Resource Manual* at www.mhhe.com/moh.

1. **a.** The binomial $x^2 - 16$ is an example of a _____difference_____ of squares. After factoring out the GCF, we factor a difference of squares $a^2 - b^2$ as _____$(a + b)(a - b)$_____.

   b. The binomial $y^2 + 121$ is an example of a _____sum_____ of squares.

   c. A sum of squares with greatest common factor 1 (is/is not) factorable over the real numbers. _____is not_____

   d. The square of a binomial always results in a perfect _____square_____ trinomial.

   e. A perfect square trinomial $a^2 + 2ab + b^2$ factors as _____$(a + b)^2$_____. Likewise, $a^2 - 2ab + b^2$ factors as _____$(a - b)^2$_____.

Review Exercises

For Exercises 2–10, factor completely.

2. $3x^2 + x - 10$
   $(3x - 5)(x + 2)$

3. $6x^2 - 17x + 5$
   $(3x - 1)(2x - 5)$

4. $6a^2b + 3a^3b$
   $3a^2b(2 + a)$

5. $15x^2y^5 - 10xy^6$
   $5xy^5(3x - 2y)$

6. $5p^2q + 20p^2 - 3pq - 12p$
   $p(5p - 3)(q + 4)$

7. $ax + ab - 6x - 6b$
   $(x + b)(a - 6)$

8. $-6x + 5 + x^2$
   $(x - 1)(x - 5)$

9. $6y - 40 + y^2$
   $(y + 10)(y - 4)$

10. $a^2 + 7a + 1$
    Prime

Concept 1: Factoring a Difference of Squares

11. What binomial factors as $(x - 5)(x + 5)$? $x^2 - 25$

12. What binomial factors as $(n - 3)(n + 3)$? $n^2 - 9$

13. What binomial factors as $(2p - 3q)(2p + 3q)$?
    $4p^2 - 9q^2$

14. What binomial factors as $(7x - 4y)(7x + 4y)$?
    $49x^2 - 16y^2$

For Exercises 15–38, factor each binomial completely. **(See Examples 1–3.)**

15. $x^2 - 36$ $(x - 6)(x + 6)$

16. $r^2 - 81$ $(r - 9)(r + 9)$

17. $3w^2 - 300$
    $3(w + 10)(w - 10)$

18. $t^3 - 49t$
    $t(t + 7)(t - 7)$

19. $4a^2 - 121b^2$
    $(2a - 11b)(2a + 11b)$

20. $9x^2 - y^2$
    $(3x - y)(3x + y)$

21. $49m^2 - 16n^2$
    $(7m - 4n)(7m + 4n)$

22. $100a^2 - 49b^2$
    $(10a - 7b)(10a + 7b)$

23. $9q^2 + 16$ Prime

24. $36 + s^2$ Prime

25. $y^2 - 4z^2$
    $(y + 2z)(y - 2z)$

26. $b^2 - 144c^2$
    $(b + 12c)(b - 12c)$

27. $a^2 - b^4$
    $(a - b^2)(a + b^2)$

28. $y^4 - x^2$
    $(y^2 - x)(y^2 + x)$

29. $25p^2q^2 - 1$
    $(5pq - 1)(5pq + 1)$

30. $81s^2t^2 - 1$
    $(9st - 1)(9st + 1)$

31. $c^2 - \dfrac{1}{25}$ $\left(c - \dfrac{1}{5}\right)\left(c + \dfrac{1}{5}\right)$

32. $z^2 - \dfrac{1}{4}$ $\left(z - \dfrac{1}{2}\right)\left(z + \dfrac{1}{2}\right)$

33. $50 - 32t^2$
    $2(5 - 4t)(5 + 4t)$

34. $63 - 7h^2$
    $7(3 - h)(3 + h)$

35. $x^4 - 256$
    $(x + 4)(x - 4)(x^2 + 16)$

36. $y^4 - 625$
    $(y + 5)(y - 5)(y^2 + 25)$

37. $16 - z^4$
    $(2 - z)(2 + z)(4 + z^2)$

38. $81 - a^4$
    $(3 - a)(3 + a)(9 + a^2)$

For Exercises 39–46, factor each polynomial completely. **(See Example 4.)**

39. $x^3 + 5x^2 - 9x - 45$
    $(x + 5)(x + 3)(x - 3)$

40. $y^3 + 6y^2 - 4y - 24$
    $(y + 6)(y + 2)(y - 2)$

41. $c^3 - c^2 - 25c + 25$
    $(c - 1)(c + 5)(c - 5)$

42. $t^3 + 2t^2 - 16t - 32$
    $(t + 2)(t + 4)(t - 4)$

43. $2x^2 - 18 + x^2y - 9y$
    $(x + 3)(x - 3)(2 + y)$

44. $5a^2 - 5 + a^2b - b$
    $(a + 1)(a - 1)(5 + b)$

45. $x^2y^2 - 9x^2 - 4y^2 + 36$
    $(y + 3)(y - 3)(x + 2)(x - 2)$

46. $w^2z^2 - w^2 - 25z^2 + 25$
    $(z + 1)(z - 1)(w + 5)(w - 5)$

Concept 2: Factoring Perfect Square Trinomials

47. Multiply. $(3x + 5)^2$ $9x^2 + 30x + 25$

48. Multiply. $(2y - 7)^2$ $4y^2 - 28y + 49$

Writing   Translating Expression   Geometry   Scientific Calculator   Video

49. a. Which trinomial is a perfect square trinomial?
$x^2 + 4x + 4$ or $x^2 + 5x + 4$
a. $x^2 + 4x + 4$ is a perfect square trinomial.

b. Factor the trinomials from part (a).
b. $x^2 + 4x + 4 = (x + 2)^2$;
$x^2 + 5x + 4 = (x + 1)(x + 4)$

50. a. Which trinomial is a perfect square trinomial?
$x^2 + 13x + 36$ or $x^2 + 12x + 36$
a. $x^2 + 12x + 36$ is a perfect square trinomial.

b. Factor the trinomials from part (a).
b. $x^2 + 13x + 36 = (x + 9)(x + 4)$;
$x^2 + 12x + 36 = (x + 6)^2$

For Exercises 51–68, factor completely. (*Hint:* Look for the pattern of a perfect square trinomial.) **(See Examples 5–6.)**

51. $x^2 + 18x + 81$
$(x + 9)^2$

52. $y^2 - 8y + 16$
$(y - 4)^2$

53. $25z^2 - 20z + 4$
$(5z - 2)^2$

54. $36p^2 + 60p + 25$
$(6p + 5)^2$

55. $49a^2 + 42ab + 9b^2$
$(7a + 3b)^2$

56. $25m^2 - 30mn + 9n^2$
$(5m - 3n)^2$

57. $-2y + y^2 + 1$
$(y - 1)^2$

58. $4 + w^2 - 4w$
$(w - 2)^2$

59. $80z^2 + 120zw + 45w^2$
$5(4z + 3w)^2$

60. $36p^2 - 24pq + 4q^2$
$4(3p - q)^2$

61. $9y^2 + 78y + 25$
$(3y + 25)(3y + 1)$

62. $4y^2 + 20y + 9$
$(2y + 9)(2y + 1)$

63. $2a^2 - 20a + 50$
$2(a - 5)^2$

64. $3t^2 + 18t + 27$
$3(t + 3)^2$

65. $4x^2 + x + 9$
Prime

66. $c^2 - 4c + 16$
Prime

67. $4x^2 + 4xy + y^2$
$(2x + y)^2$

68. $100y^2 + 20yz + z^2$
$(10y + z)^2$

69. The volume of the box shown is given as $3x^3 - 6x^2 + 3x$. Write the polynomial in factored form.
$3x(x - 1)^2$

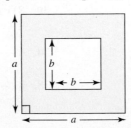

70. The volume of the box shown is given as $20y^3 + 20y^2 + 5y$. Write the polynomial in factored form.
$5y(2y + 1)^2$

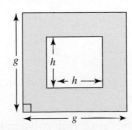

Expanding Your Skills

For Exercises 71–78, factor the difference of squares.

71. $(y - 3)^2 - 9$
$y(y - 6)$

72. $(x - 2)^2 - 4$
$x(x - 4)$

73. $(2p + 1)^2 - 36$
$(2p - 5)(2p + 7)$

74. $(4q + 3)^2 - 25$
$8(2q - 1)(q + 2)$

75. $16 - (t + 2)^2$
$(-t + 2)(t + 6)$ or
$-(t - 2)(t + 6)$

76. $81 - (a + 5)^2$
$(-a + 4)(a + 14)$ or
$-(a - 4)(a + 14)$

77. $(2a - 5)^2 - 100b^2$
$(2a - 5 + 10b)(2a - 5 - 10b)$

78. $(3k + 7)^2 - 49m^2$
$(3k + 7 + 7m)(3k + 7 - 7m)$

79. a. Write a polynomial that represents the area of the shaded region in the figure.
$a^2 - b^2$

b. Factor the expression from part (a).
$(a - b)(a + b)$

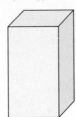

80. a. Write a polynomial that represents the area of the shaded region in the figure. $g^2 - h^2$

b. Factor the expression from part (a).
$(g - h)(g + h)$

Section 6.6 Sum and Difference of Cubes

Concepts

1. **Factoring a Sum or Difference of Cubes**
2. **Factoring Binomials: A Summary**

1. Factoring a Sum or Difference of Cubes

A binomial $a^2 - b^2$ is a difference of squares and can be factored as $(a - b)(a + b)$. Further-more, if a and b share no common factors, then a sum of squares $a^2 + b^2$ is not factorable over the real numbers. In this section, we will learn that both a difference of cubes, $a^3 - b^3$, and a sum of cubes, $a^3 + b^3$, are factorable.

Factored Form of a Sum or Difference of Cubes

Sum of Cubes: $\qquad a^3 + b^3 = (a + b)(a^2 - ab + b^2)$

Difference of Cubes: $\quad a^3 - b^3 = (a - b)(a^2 + ab + b^2)$

Multiplication can be used to confirm the formulas for factoring a sum or difference of cubes:

$$(a + b)(a^2 - ab + b^2) = a^3 - a^2b + ab^2 + a^2b - ab^2 + b^3 = a^3 + b^3 \checkmark$$

$$(a - b)(a^2 + ab + b^2) = a^3 + a^2b + ab^2 - a^2b - ab^2 - b^3 = a^3 - b^3 \checkmark$$

To help you remember the formulas for factoring a sum or difference of cubes, keep the following guidelines in mind:

- The factored form is the product of a binomial and a trinomial.
- The first and third terms in the trinomial are the squares of the terms within the binomial factor.
- Without regard to signs, the middle term in the trinomial is the product of terms in the binomial factor.

Square the first term of the binomial. $\qquad$ Product of terms in the binomial

$$x^3 + 8 = (x)^3 + (2)^3 = (x + 2)[(x)^2 - (x)(2) + (2)^2]$$

Square the last term of the binomial.

- The sign within the binomial factor is the same as the sign of the original binomial.
- The first and third terms in the trinomial are always positive.
- The sign of the middle term in the trinomial is opposite the sign within the binomial.

Same sign $\qquad$ Positive

$$x^3 + 8 = (x)^3 + (2)^3 = (x + 2)[(x)^2 - (x)(2) + (2)^2]$$

Opposite signs

TIP: To help remember the placement of the signs in factoring the sum or difference of cubes, remember SOAP: **S**ame sign, **O**pposite signs, **A**lways **P**ositive.

To help you recognize a sum or difference of cubes, we recommend that you familiarize yourself with the first several perfect cubes:

Perfect Cubes	**Perfect Cubes**
$1 = (1)^3$	$216 = (6)^3$
$8 = (2)^3$	$343 = (7)^3$
$27 = (3)^3$	$512 = (8)^3$
$64 = (4)^3$	$729 = (9)^3$
$125 = (5)^3$	$1000 = (10)^3$

It is also helpful to recognize that a variable expression is a perfect cube if its exponent is a multiple of 3. For example:

Perfect Cubes

$$x^3 = (x)^3$$
$$x^6 = (x^2)^3$$
$$x^9 = (x^3)^3$$
$$x^{12} = (x^4)^3$$

Classroom Example: p. 453,
Exercise 16

| Example 1 | **Factoring a Sum of Cubes** |

Factor. $\quad w^3 + 64$

Solution:

$$w^3 + 64 \qquad\qquad w^3 \text{ and } 64 \text{ are perfect cubes.}$$

$$= (w)^3 + (4)^3 \qquad \text{Write as } a^3 + b^3, \text{ where } a = w, b = 4.$$

$$a^3 + b^3 = (a+b)(a^2 - ab + b^2) \qquad\qquad \text{Apply the formula for a sum of cubes.}$$

$$(w)^3 + (4)^3 = (w+4)[(w)^2 - (w)(4) + (4)^2]$$

$$= (w+4)(w^2 - 4w + 16) \qquad \text{Simplify.}$$

Skill Practice Factor.

1. $p^3 + 125$

Classroom Example: p. 453,
Exercise 26

| Example 2 | **Factoring a Difference of Cubes** |

Factor. $\quad 27p^3 - 1000q^3$

Solution:

$$27p^3 - 1000q^3 \qquad 27p^3 \text{ and } 1000q^3 \text{ are perfect cubes.}$$

$$= (3p)^3 - (10q)^3 \qquad \text{Write as } a^3 - b^3, \text{ where } a = 3p, b = 10q.$$

$$a^3 - b^3 = (a-b)(a^2 + ab + b^2) \qquad\qquad \text{Apply the formula for a difference of cubes.}$$

$$(3p)^3 - (10q)^3 = (3p - 10q)[(3p)^2 + (3p)(10q) + (10q)^2]$$

$$= (3p - 10q)(9p^2 + 30pq + 100q^2) \qquad \text{Simplify.}$$

Skill Practice Factor.

2. $8y^3 - 27z^3$

2. Factoring Binomials: A Summary

After removing the GCF, the next step in any factoring problem is to recognize what type of pattern it follows. Exponents that are divisible by 2 are perfect squares and those divisible by 3 are perfect cubes. The formulas for factoring binomials are summarized in the following box:

Factored Forms of Binomials

Difference of Squares: $a^2 - b^2 = (a+b)(a-b)$

Difference of Cubes: $\quad a^3 - b^3 = (a-b)(a^2 + ab + b^2)$

Sum of Cubes: $\qquad\quad a^3 + b^3 = (a+b)(a^2 - ab + b^2)$

Classroom Examples: p. 454,
Exercises 48 and 52

| Example 3 | **Factoring Binomials** |

Factor completely.

a. $27y^3 + 1$ **b.** $\dfrac{1}{25}m^2 - \dfrac{1}{4}$ **c.** $z^6 - 8w^3$

Answers

1. $(p+5)(p^2 - 5p + 25)$
2. $(2y - 3z)(4y^2 + 6yz + 9z^2)$

Solution:

a. $27y^3 + 1$ Sum of cubes: $27y^3 = (3y)^3$ and $1 = (1)^3$.

   $= (3y)^3 + (1)^3$ Write as $a^3 + b^3$, where $a = 3y$ and $b = 1$.

   $= (3y + 1)[(3y)^2 - (3y)(1) + (1)^2]$ Apply the formula
   $$a^3 + b^3 = (a + b)(a^2 - ab + b^2).$$

   $= (3y + 1)(9y^2 - 3y + 1)$ Simplify.

b. $\dfrac{1}{25}m^2 - \dfrac{1}{4}$ Difference of squares

   $= \left(\dfrac{1}{5}m\right)^2 - \left(\dfrac{1}{2}\right)^2$ Write as $a^2 - b^2$, where $a = \frac{1}{5}m$ and $b = \frac{1}{2}$.

   $= \left(\dfrac{1}{5}m + \dfrac{1}{2}\right)\left(\dfrac{1}{5}m - \dfrac{1}{2}\right)$ Apply the formula $a^2 - b^2 = (a + b)(a - b)$.

c. $z^6 - 8w^3$ Difference of cubes: $z^6 = (z^2)^3$ and $8w^3 = (2w)^3$

   $= (z^2)^3 - (2w)^3$ Write as $a^3 - b^3$, where $a = z^2$ and $b = 2w$.

   $= (z^2 - 2w)[(z^2)^2 + (z^2)(2w) + (2w)^2]$ Apply the formula
   $$a^3 - b^3 = (a - b)(a^2 + ab + b^2).$$

   $= (z^2 - 2w)(z^4 + 2z^2w + 4w^2)$ Simplify.

Each factorization in this example can be checked by multiplying.

Skill Practice Factor completely.

3. $1000x^3 + 1$ **4.** $25p^2 - \dfrac{1}{9}$ **5.** $27a^6 - b^3$

Some factoring problems require more than one method of factoring. In general, when factoring a polynomial, be sure to factor completely.

Example 4   Factoring a Polynomial

Factor completely. $3y^4 - 48$

Classroom Example: p. 454, Exercise 58

Solution:

   $3y^4 - 48$

   $= 3(y^4 - 16)$ Factor out the GCF. The binomial is a difference of squares.

   $= 3[(y^2)^2 - (4)^2]$ Write as $a^2 - b^2$, where $a = y^2$ and $b = 4$.

   $= 3(y^2 + 4)(y^2 - 4)$ Apply the formula
   $$a^2 - b^2 = (a + b)(a - b).$$

   $y^2 + 4$ is a sum of squares and cannot be factored.

   $= 3(y^2 + 4)(y + 2)(y - 2)$ $y^2 - 4$ is a difference of squares and can be factored further.

Skill Practice Factor completely.

6. $2x^4 - 2$

Answers

3. $(10x + 1)(100x^2 - 10x + 1)$

4. $\left(5p - \dfrac{1}{3}\right)\left(5p + \dfrac{1}{3}\right)$

5. $(3a^2 - b)(9a^4 + 3a^2b + b^2)$

6. $2(x^2 + 1)(x - 1)(x + 1)$

Classroom Example: p. 454,
Exercise 64

| **Example 5** | **Factoring a Polynomial** |

Factor completely. $4x^3 + 4x^2 - 25x - 25$

Solution:

$4x^3 + 4x^2 - 25x - 25$ The GCF is 1.

$= 4x^3 + 4x^2 \mid - 25x - 25$ The polynomial has four terms. Factor by grouping.

$= 4x^2(x + 1) - 25(x + 1)$

$= (x + 1)(4x^2 - 25)$ $4x^2 - 25$ is a difference of squares.

$= (x + 1)\overbrace{(2x + 5)(2x - 5)}$

Skill Practice   Factor completely.

7. $x^3 + 6x^2 - 4x - 24$

Classroom Example: p. 454,
Exercise 50

| **Example 6** | **Factoring a Binomial** |

Factor the binomial $x^6 - y^6$ as

a. A difference of cubes   **b.** A difference of squares

Notice that the expressions x^6 and y^6 are both perfect squares and perfect cubes because both exponents are multiples of 2 and of 3. Consequently, $x^6 - y^6$ can be factored initially as either the difference of squares or as the difference of cubes.

Solution:

a. $x^6 - y^6$

      Difference of cubes

$= (x^2)^{\overbrace{3}} - (y^2)^{\overbrace{3}}$ Write as $a^3 - b^3$, where $a = x^2$ and $b = y^2$.

$= (x^2 - y^2)[(x^2)^2 + (x^2)(y^2) + (y^2)^2]$ Apply the formula $a^3 - b^3 = (a - b)(a^2 + ab + b^2)$.

$= (x^2 - y^2)(x^4 + x^2y^2 + y^4)$ Factor $x^2 - y^2$ as a difference of squares.

$= \overbrace{(x + y)(x - y)}(x^4 + x^2y^2 + y^4)$

Avoiding Mistakes

The trinomial $x^4 + x^2y^2 + y^4$ cannot be factored further with the techniques presented in this chapter.

b. $x^6 - y^6$

      Difference of squares

$= (x^3)^{\overbrace{2}} - (y^3)^{\overbrace{2}}$ Write as $a^2 - b^2$, where $a = x^3$ and $b = y^3$.

$= (x^3 + y^3)(x^3 - y^3)$ Apply the formula $a^2 - b^2 = (a + b)(a - b)$.

    Sum of   Difference
    cubes   of cubes

Factor $x^3 + y^3$ as a sum of cubes.

Factor $x^3 - y^3$ as a difference of cubes.

$= \overbrace{(x + y)(x^2 - xy + y^2)} \; \overbrace{(x - y)(x^2 + xy + y^2)}$

Answer

7. $(x + 6)(x + 2)(x - 2)$

In a case such as this, it is recommended that you factor the expression as a difference of squares first because it factors more completely into polynomials of lower degree.

$$x^6 - y^6 = (x + y)(x^2 - xy + y^2)(x - y)(x^2 + xy + y^2)$$

Skill Practice Factor completely.

8. $z^6 - 64$

Answer

8. $(z + 2)(z - 2)(z^2 + 2z + 4)$
$(z^2 - 2z + 4)$

Section 6.6   Practice Exercises

Vocabulary and Key Concepts

For additional exercises, see Classroom Activities 6.6A–6.6B in the *Student's Resource Manual* at www.mhhe.com/moh.

1. a. The binomial $x^3 + 27$ is an example of a _____sum_____ of ____cubes____.

   b. The binomial $c^3 - 8$ is an example of a __difference__ of ____cubes____.

   c. A difference of cubes $a^3 - b^3$ factors as $(a - b)(a^2 + ab + b^2)$.

   d. A sum of cubes $a^3 + b^3$ factors as $(a + b)(a^2 - ab + b^2)$.

Review Exercises

For Exercises 2–10, factor completely.

2. $600 - 6x^2$
$6(10 - x)(10 + x)$

3. $20 - 5t^2$
$5(2 - t)(2 + t)$

4. $ax + bx + 5a + 5b$
$(a + b)(x + 5)$

5. $2t + 2u + st + su$
$(t + u)(2 + s)$

6. $5y^2 + 13y - 6$
$(5y - 2)(y + 3)$

7. $3v^2 + 5v - 12$
$(3v - 4)(v + 3)$

8. $40a^3b^3 - 16a^2b^2 + 24a^3b$
$8a^2b(5ab^2 - 2b + 3a)$

9. $-c^2 - 10c - 25$
$-(c + 5)^2$

10. $-z^2 + 6z - 9$
$-(z - 3)^2$

Concept 1: Factoring a Sum or Difference of Cubes

11. Identify the expressions that are perfect cubes:
$$x^3, 8, 9, y^6, a^4, b^2, 3p^3, 27q^3, w^{12}, r^3s^6$$
$x^3, 8, y^6, 27q^3, w^{12}, r^3s^6$

12. Identify the expressions that are perfect cubes:
$$z^9, -81, 30, 8, 6x^3, y^{15}, 27a^3, b^2, p^3q^2, -1$$
$z^9, 8, y^{15}, 27a^3, -1$

13. Identify the expressions that are perfect cubes:
$$36, t^3, -1, 27, a^3b^6, -9, 125, -8x^2, y^6, 25$$
$t^3, -1, 27, a^3b^6, 125, y^6$

14. Identify the expressions that are perfect cubes:
$$343, 15b^3, z^3, w^{12}, -p^9, -1000, a^2b^3, 3x^3, -8, 60$$
$343, z^3, w^{12}, -p^9, -1000, -8$

For Exercises 15–30, factor the sums or differences of cubes. **(See Examples 1–2.)**

15. $y^3 - 8$
$(y - 2)(y^2 + 2y + 4)$

16. $x^3 + 27$
$(x + 3)(x^2 - 3x + 9)$

17. $1 - p^3$
$(1 - p)(1 + p + p^2)$

18. $q^3 + 1$
$(q + 1)(q^2 - q + 1)$

19. $w^3 + 64$
$(w + 4)(w^2 - 4w + 16)$

20. $8 - t^3$
$(2 - t)(4 + 2t + t^2)$

21. $x^3 - 1000y^3$
$(x - 10y)(x^2 + 10xy + 100y^2)$

22. $8r^3 - 27t^3$
$(2r - 3t)(4r^2 + 6rt + 9t^2)$

 23. $64t^3 + 1$
$(4t + 1)(16t^2 - 4t + 1)$

24. $125r^3 + 1$
$(5r + 1)(25r^2 - 5r + 1)$

25. $1000a^3 + 27$
$(10a + 3)(100a^2 - 30a + 9)$

26. $216b^3 - 125$
$(6b - 5)(36b^2 + 30b + 25)$

27. $n^3 - \dfrac{1}{8}$
$\left(n - \dfrac{1}{2}\right)\left(n^2 + \dfrac{1}{2}n + \dfrac{1}{4}\right)$

28. $\dfrac{8}{27} + m^3$
$\left(\dfrac{2}{3} + m\right)\left(\dfrac{4}{9} - \dfrac{2}{3}m + m^2\right)$

29. $125x^3 + 8y^3$
$(5x + 2y)(25x^2 - 10xy + 4y^2)$

30. $27t^3 + 64u^3$
$(3t + 4u)(9t^2 - 12tu + 16u^2)$

Concept 2: Factoring Binomials: A Summary

For Exercises 31–66, factor completely. **(See Examples 3–6.)**

31. $x^4 - 4$
$(x^2 - 2)(x^2 + 2)$

32. $b^4 - 25$
$(b^2 - 5)(b^2 + 5)$

33. $a^2 + 9$
Prime

34. $w^2 + 36$
Prime

35. $t^3 + 64$
$(t + 4)(t^2 - 4t + 16)$

36. $u^3 + 27$
$(u + 3)(u^2 - 3u + 9)$

37. $g^3 - 4$
Prime

38. $h^3 - 25$
Prime

39. $4b^3 + 108$
$4(b + 3)(b^2 - 3b + 9)$

40. $3c^3 - 24$
$3(c - 2)(c^2 + 2c + 4)$

41. $5p^2 - 125$
$5(p - 5)(p + 5)$

42. $2q^4 - 8$
$2(q^2 - 2)(q^2 + 2)$

43. $\dfrac{1}{64} - 8h^3$
$(\frac{1}{4} - 2h)(\frac{1}{16} + \frac{1}{2}h + 4h^2)$

44. $\dfrac{1}{125} + k^6$
$(\frac{1}{5} + k^2)(\frac{1}{25} - \frac{1}{5}k^2 + k^4)$

45. $x^4 - 16$
$(x - 2)(x + 2)(x^2 + 4)$.

46. $p^4 - 81$
$(p - 3)(p + 3)(p^2 + 9)$

47. $\dfrac{4}{9}x^2 - w^2$
$\left(\frac{2}{3}x - w\right)\left(\frac{2}{3}x + w\right)$

48. $\dfrac{16}{25}y^2 - x^2$
$\left(\frac{4}{5}y - x\right)\left(\frac{4}{5}y + x\right)$

49. $q^6 - 64$
$(q - 2)(q^2 + 2q + 4)(q + 2)(q^2 - 2q + 4)$

50. $a^6 - 1$ $(a - 1)(a^2 + a + 1)$
$(a + 1)(a^2 - a + 1)$
(*Hint:* Factor using the difference of squares first.)

51. $x^9 + 64y^3$
$(x^3 + 4y)(x^6 - 4x^3y + 16y^2)$

52. $125w^3 - z^9$
$(5w - z^3)(25w^2 + 5wz^3 + z^6)$

53. $2x^3 + 3x^2 - 2x - 3$
$(2x + 3)(x - 1)(x + 1)$

54. $3x^3 + x^2 - 12x - 4$
$(3x + 1)(x - 2)(x + 2)$

55. $16x^4 - y^4$
$(2x - y)(2x + y)(4x^2 + y^2)$

56. $1 - t^4$
$(1 - t)(1 + t)(1 + t^2)$

57. $81y^4 - 16$
$(3y - 2)(3y + 2)(9y^2 + 4)$

58. $u^5 - 256u$
$u(u - 4)(u + 4)(u^2 + 16)$

59. $a^3 + b^6$
$(a + b^2)(a^2 - ab^2 + b^4)$

60. $u^6 - v^3$
$(u^2 - v)(u^4 + u^2v + v^2)$

61. $x^4 - y^4$
$(x^2 + y^2)(x - y)(x + y)$

62. $a^4 - b^4$
$(a^2 + b^2)(a - b)(a + b)$

63. $k^3 + 4k^2 - 9k - 36$
$(k + 4)(k - 3)(k + 3)$

64. $w^3 - 2w^2 - 4w + 8$
$(w - 2)^2(w + 2)$

65. $2t^3 - 10t^2 - 2t + 10$
$2(t - 5)(t - 1)(t + 1)$

66. $9a^3 + 27a^2 - 4a - 12$
$(a + 3)(3a - 2)(3a + 2)$

Expanding Your Skills

For Exercises 67–70, factor completely.

67. $\dfrac{64}{125}p^3 - \dfrac{1}{8}q^3$
$\left(\frac{4}{5}p - \frac{1}{2}q\right)\left(\frac{16}{25}p^2 + \frac{2}{5}pq + \frac{1}{4}q^2\right)$

68. $\dfrac{1}{1000}r^3 + \dfrac{8}{27}s^3$
$\left(\frac{1}{10}r + \frac{2}{3}s\right)\left(\frac{1}{100}r^2 - \frac{1}{15}rs + \frac{4}{9}s^2\right)$

69. $a^{12} + b^{12}$
$(a^4 + b^4)(a^8 - a^4b^4 + b^8)$

70. $a^9 - b^9$
$(a - b)(a^2 + ab + b^2)(a^6 + a^3b^3 + b^6)$

Use Exercises 71–72 to investigate the relationship between division and factoring.

71. a. Use long division to divide $x^3 - 8$ by $(x - 2)$.   The quotient is $x^2 + 2x + 4$.

   b. Factor $x^3 - 8$. $(x - 2)(x^2 + 2x + 4)$

72. a. Use long division to divide $y^3 + 27$ by $(y + 3)$.   The quotient is $y^2 - 3y + 9$.

   b. Factor $y^3 + 27$. $(y + 3)(y^2 - 3y + 9)$

73. What trinomial multiplied by $(x - 4)$ gives a difference of cubes? $x^2 + 4x + 16$

74. What trinomial multiplied by $(p + 5)$ gives a sum of cubes? $p^2 - 5p + 25$

75. Write a binomial that when multiplied by $(4x^2 - 2x + 1)$ produces a sum of cubes. $2x + 1$

76. Write a binomial that when multiplied by $(9y^2 + 15y + 25)$ produces a difference of cubes. $3y - 5$

Writing   Translating Expression   Geometry   Scientific Calculator   Video

Problem Recognition Exercises

Factoring Strategy

> **Factoring Strategy**
>
> **Step 1** Factor out the GCF.
>
> **Step 2** Identify whether the polynomial has two terms, three terms, or more than three terms.
>
> **Step 3** If the polynomial has more than three terms, try factoring by grouping.
>
> **Step 4** If the polynomial has three terms, check first for a perfect square trinomial. Otherwise, factor the trinomial with the trial-and-error method or the ac-method.
>
> **Step 5** If the polynomial has two terms, determine if it fits the pattern for
> - A difference of squares: $a^2 - b^2 = (a - b)(a + b)$
> - A sum of squares: $a^2 + b^2$ is prime.
> - A difference of cubes: $a^3 - b^3 = (a - b)(a^2 + ab + b^2)$
> - A sum of cubes: $a^3 + b^3 = (a + b)(a^2 - ab + b^2)$
>
> **Step 6** Be sure to factor the polynomial completely.
>
> **Step 7** Check by multiplying.

1. What is meant by a prime polynomial?   A prime polynomial cannot be factored further.

2. What is the first step in factoring any polynomial?   Factor out the GCF.

3. When factoring a binomial, what patterns can you look for?   Look for a difference of squares: $a^2 - b^2$, a difference of cubes: $a^3 - b^3$, or a sum of cubes: $a^3 + b^3$.

4. What technique should be considered when factoring a four-term polynomial?   Grouping

For Exercises 5–73,

a. Factor out the GCF from each polynomial. Then identify the category in which the remaining polynomial best fits. Choose from
- difference of squares
- sum of squares
- difference of cubes
- sum of cubes
- trinomial (perfect square trinomial)
- trinomial (nonperfect square trinomial)
- four terms-grouping
- none of these

Instructor Note: The following exercises involve a sum or difference of cubes: 10, 12, 13, 18, 23, 28, 29.

b. Factor the polynomial completely.

5. $2a^2 - 162$ a. Difference of squares
    b. $2(a - 9)(a + 9)$

6. $y^2 + 4y + 3$
    a. Nonperfect square trinomial
    b. $(y + 3)(y + 1)$

7. $6w^2 - 6w$ a. None of these
    b. $6w(w - 1)$

8. $16z^4 - 81$
a. Difference of squares
b. $(2z + 3)(2z - 3)(4z^2 + 9)$

9. $3t^2 + 13t + 4$
    a. Nonperfect square trinomial
    b. $(3t + 1)(t + 4)$

10. $5r^3 + 5$ a. Sum of cubes
    b. $5(r + 1)(r^2 - r + 1)$

11. $3ac + ad - 3bc - bd$
a. Four terms-grouping
b. $(3c + d)(a - b)$

12. $x^3 - 125$
    a. Difference of cubes
    b. $(x - 5)(x^2 + 5x + 25)$

13. $y^3 + 8$ a. Sum of cubes
    b. $(y + 2)(y^2 - 2y + 4)$

14. $7p^2 - 29p + 4$
a. Nonperfect square trinomial
b. $(7p - 1)(p - 4)$

15. $3q^2 - 9q - 12$
    a. Nonperfect square trinomial
    b. $3(q - 4)(q + 1)$

16. $-2x^2 + 8x - 8$
    a. Perfect square trinomial
    b. $-2(x - 2)^2$

17. $18a^2 + 12a$
  a. None of these
  b. $6a(3a + 2)$

18. $54 - 2y^3$
  a. Difference of cubes
  b. $2(3 - y)(9 + 3y + y^2)$

19. $4t^2 - 100$
  a. Difference of squares
  b. $4(t - 5)(t + 5)$

20. $4t^2 - 31t - 8$
  a. Nonperfect square trinomial
  b. $(4t + 1)(t - 8)$

21. $10c^2 + 10c + 10$
  a. Nonperfect square trinomial
  b. $10(c^2 + c + 1)$

22. $2xw - 10x + 3yw - 15y$
  a. Four terms-grouping
  b. $(w - 5)(2x + 3y)$

23. $x^3 + 0.001$
  a. Sum of cubes
  b. $(x + 0.1)(x^2 - 0.1x + 0.01)$

24. $4q^2 - 9$
  a. Difference of squares
  b. $(2q - 3)(2q + 3)$

25. $64 + 16k + k^2$
  a. Perfect square trinomial
  b. $(8 + k)^2$

26. $s^2t + 5t + 6s^2 + 30$
  a. Four terms-grouping
  b. $(t + 6)(s^2 + 5)$

27. $2x^2 + 2x - xy - y$
  a. Four terms-grouping
  b. $(x + 1)(2x - y)$

28. $w^3 + y^3$
  a. Sum of cubes
  b. $(w + y)(w^2 - wy + y^2)$

29. $a^3 - c^3$
  a. Difference of cubes
  b. $(a - c)(a^2 + ac + c^2)$

30. $3y^2 + y + 1$
  a. Nonperfect square trinomial
  b. Prime

31. $c^2 + 8c + 9$
  a. Nonperfect square trinomial
  b. Prime

32. $a^2 + 2a + 1$
  a. Perfect square trinomial
  b. $(a + 1)^2$

33. $b^2 + 10b + 25$
  a. Perfect square trinomial
  b. $(b + 5)^2$

34. $-t^2 - 4t + 32$
  a. Nonperfect square trinomial
  b. $-(t + 8)(t - 4)$

35. $-p^3 - 5p^2 - 4p$
  a. Nonperfect square trinomial
  b. $-p(p + 4)(p + 1)$

36. $x^2y^2 - 49$
  a. Difference of squares
  b. $(xy - 7)(xy + 7)$

37. $6x^2 - 21x - 45$
  a. Nonperfect square trinomial
  b. $3(2x + 3)(x - 5)$

38. $20y^2 - 14y + 2$
  a. Nonperfect square trinomial
  b. $2(5y - 1)(2y - 1)$

39. $5a^2bc^3 - 7abc^2$
  a. None of these
  b. $abc^2(5ac - 7)$

40. $8a^2 - 50$
  a. Differenc of squares
  b. $2(2a - 5)(2a + 5)$

41. $t^2 + 2t - 63$
  a. Nonperfect square trinomial
  b. $(t + 9)(t - 7)$

42. $b^2 + 2b - 80$
  a. Nonperfect square trinomial
  b. $(b + 10)(b - 8)$

43. $ab + ay - b^2 - by$
  a. Four terms-grouping
  b. $(b + y)(a - b)$

44. $6x^3y^4 + 3x^2y^5$
  a. None of these
  b. $3x^2y^4(2x + y)$

45. $14u^2 - 11uv + 2v^2$
  a. Nonperfect square trinomial
  b. $(7u - 2v)(2u - v)$

46. $9p^2 - 36pq + 4q^2$
  a. Nonperfect square trinomial
  b. Prime

47. $4q^2 - 8q - 6$
  a. Nonperfect square trinomial
  b. $2(2q^2 - 4q - 3)$

48. $9w^2 + 3w - 15$
  a. Nonperfect square trinomial
  b. $3(3w^2 + w - 5)$

49. $9m^2 + 16n^2$
  a. Sum of squares
  b. Prime

50. $5b^2 - 30b + 45$
  a. Perfect square trinomial
  b. $5(b - 3)^2$

51. $6r^2 + 11r + 3$
  a. Nonperfect square trinomial
  b. $(3r + 1)(2r + 3)$

52. $4s^2 + 4s - 15$
  a. Nonperfect square trinomial
  b. $(2s - 3)(2s + 5)$

53. $16a^4 - 1$
  a. Difference of squares
  b. $(2a - 1)(2a + 1)(4a^2 + 1)$

54. $p^3 + p^2c - 9p - 9c$
  a. Four terms-grouping
  b. $(p + c)(p - 3)(p + 3)$

55. $81u^2 - 90uv + 25v^2$
  a. Perfect square trinomial
  b. $(9u - 5v)^2$

56. $4x^2 + 16$
  a. Sum of squares
  b. $4(x^2 + 4)$

57. $x^2 - 5x - 6$
  a. Nonperfect square trinomial
  b. $(x - 6)(x + 1)$

58. $q^2 + q - 7$
  a. Nonperfect square trinomial
  b. Prime

59. $2ax - 6ay + 4bx - 12by$
  a. Four terms-grouping
  b. $2(x - 3y)(a + 2b)$

60. $8m^3 - 10m^2 - 3m$
  a. Nonperfect square trinomial
  b. $m(4m + 1)(2m - 3)$

61. $21x^4y + 41x^3y + 10x^2y$
  a. Nonperfect square trinomial
  b. $x^2y(3x + 5)(7x + 2)$

62. $2m^4 - 128$
  a. Difference of squares
  b. $2(m^2 - 8)(m^2 + 8)$

63. $8uv - 6u + 12v - 9$
  a. Four terms-grouping
  b. $(4v - 3)(2u + 3)$

64. $4t^2 - 20t + st - 5s$
  a. Four terms-grouping
  b. $(t - 5)(4t + s)$

65. $12x^2 - 12x + 3$
  a. Perfect square trinomial
  b. $3(2x - 1)^2$

66. $p^2 + 2pq + q^2$
  a. Perfect square trinomial
  b. $(p + q)^2$

67. $6n^3 + 5n^2 - 4n$
  a. Nonperfect square trinomial
  b. $n(2n - 1)(3n + 4)$

68. $4k^3 + 4k^2 - 3k$
  a. Nonperfect square trinomial
  b. $k(2k - 1)(2k + 3)$

69. $64 - y^2$
  a. Difference of squares
  b. $(8 - y)(8 + y)$

70. $36b - b^3$
  a. Difference of squares
  b. $b(6 - b)(6 + b)$

71. $b^2 - 4b + 10$
  a. Nonperfect square trinomial
  b. Prime

72. $y^2 + 6y + 8$
  a. Nonperfect square trinomial
  b. $(y + 4)(y + 2)$

73. $c^4 - 12c^2 + 20$
  a. Nonperfect square trinomial
  b. $(c^2 - 10)(c^2 - 2)$

Section 6.7 Solving Equations Using the Zero Product Rule

Concepts

1. Definition of a Quadratic Equation
2. Zero Product Rule
3. Solving Equations by Factoring

1. Definition of a Quadratic Equation

We have already learned to solve linear equations in one variable. These are equations of the form $ax + b = c$ $(a \neq 0)$. A linear equation in one variable is sometimes called a first-degree polynomial equation because the highest degree of all its terms is 1. A second-degree polynomial equation in one variable is called a quadratic equation.

 Writing  Translating Expression   Geometry Scientific Calculator   Video

A Quadratic Equation in One Variable

If a, b, and c are real numbers such that $a \neq 0$, then a **quadratic equation** is an equation that can be written in the form

$$ax^2 + bx + c = 0$$

The following equations are quadratic because they can each be written in the form $ax^2 + bx + c = 0$, $(a \neq 0)$.

$$-4x^2 + 4x = 1 \qquad x(x - 2) = 3 \qquad (x - 4)(x + 4) = 9$$
$$-4x^2 + 4x - 1 = 0 \qquad x^2 - 2x = 3 \qquad x^2 - 16 = 9$$
$$x^2 - 2x - 3 = 0 \qquad x^2 - 25 = 0$$
$$x^2 + 0x - 25 = 0$$

2. Zero Product Rule

One method for solving a quadratic equation is to factor and apply the zero product rule. The **zero product rule** states that if the product of two factors is zero, then one or both of its factors is zero.

Zero Product Rule

If $ab = 0$, then $a = 0$ or $b = 0$.

Example 1   **Applying the Zero Product Rule**

Solve the equation by using the zero product rule. $(x - 4)(x + 3) = 0$

Solution:

$(x - 4)(x + 3) = 0$	Apply the zero product rule.
$x - 4 = 0 \quad$ or $\quad x + 3 = 0$	Set each factor equal to zero.
$x = 4 \quad$ or $\quad x = -3$	Solve each equation for x.

Check: $x = 4$ Check: $x = -3$

$(4 - 4)(4 + 3) \overset{?}{=} 0$ $(-3 - 4)(-3 + 3) \overset{?}{=} 0$

$(0)(7) \overset{?}{=} 0 ✓$ $(-7)(0) \overset{?}{=} 0 ✓$

The solution set is $\{4, -3\}$.

Skill Practice Solve.

1. $(x + 1)(x - 8) = 0$

Classroom Example: p. 462, Exercise 14

Answer

1. $\{-1, 8\}$

Classroom Example: p. 462, Exercise 20

Example 2   **Applying the Zero Product Rule**

Solve the equation by using the zero product rule. $(x + 8)(4x + 1) = 0$

Solution:

$(x + 8)(4x + 1) = 0$	Apply the zero product rule.
$x + 8 = 0$ or $4x + 1 = 0$	Set each factor equal to zero.
$x = -8$ or $4x = -1$	Solve each equation for x.
$x = -8$ or $x = -\dfrac{1}{4}$	The solutions check in the original equation.

The solution set is $\left\{ -8, -\dfrac{1}{4} \right\}$.

Skill Practice Solve.

   2. $(4x - 5)(x + 6) = 0$

Classroom Example: p. 462, Exercise 22

Example 3   **Applying the Zero Product Rule**

Solve the equation using the zero product rule. $x(3x - 7) = 0$

Solution:

$x(3x - 7) = 0$	Apply the zero product rule.
$x = 0$ or $3x - 7 = 0$	Set each factor equal to zero.
$x = 0$ or $3x = 7$	Solve each equation for x.
$x = 0$ or $x = \dfrac{7}{3}$	The solutions check in the original equation.

The solution set is $\left\{ 0, \dfrac{7}{3} \right\}$.

Skill Practice Solve.

   3. $x(4x + 9) = 0$

3. Solving Equations by Factoring

Quadratic equations, like linear equations, arise in many applications in mathematics, science, and business. The following steps summarize the factoring method for solving a quadratic equation.

> **Solving a Quadratic Equation by Factoring**
>
> **Step 1**   Write the equation in the form: $ax^2 + bx + c = 0$.
>
> **Step 2**   Factor the quadratic expression completely.
>
> **Step 3**   Apply the zero product rule. That is, set each factor equal to zero and solve the resulting equations.
>
> *Note:* The solution(s) found in step 3 may be checked by substitution in the original equation.

Answers

2. $\left\{ \dfrac{5}{4}, -6 \right\}$ **3.** $\left\{ 0, -\dfrac{9}{4} \right\}$

| **Example 4** | **Solving a Quadratic Equation** |

Solve the quadratic equation. $\qquad 2x^2 - 9x = 5$

Solution:

$$2x^2 - 9x = 5$$

$$2x^2 - 9x - 5 = 0 \qquad\qquad \text{Write the equation in the form}$$
$$ax^2 + bx + c = 0.$$

$$(2x + 1)(x - 5) = 0 \qquad\qquad \text{Factor the polynomial completely.}$$

$$2x + 1 = 0 \quad \text{ or } \quad x - 5 = 0 \qquad \text{Set each factor equal to zero.}$$

$$2x = -1 \quad \text{ or } \qquad x = 5 \qquad \text{Solve each equation.}$$

$$x = -\frac{1}{2} \quad \text{ or } \qquad x = 5$$

$\underline{\text{Check:}}\ x = -\dfrac{1}{2}$ $\qquad\qquad$ $\underline{\text{Check:}}\ x = 5$

$$2x^2 - 9x = 5 \qquad\qquad 2x^2 - 9x = 5$$

$$2\left(-\frac{1}{2}\right)^2 - 9\left(-\frac{1}{2}\right) \overset{?}{=} 5 \qquad 2(5)^2 - 9(5) \overset{?}{=} 5$$

$$2\left(\frac{1}{4}\right) + \frac{9}{2} \overset{?}{=} 5 \qquad\qquad 2(25) - 45 \overset{?}{=} 5$$

$$\frac{1}{2} + \frac{9}{2} \overset{?}{=} 5 \qquad\qquad\qquad 50 - 45 \overset{?}{=} 5 \checkmark$$

$$\frac{10}{2} \overset{?}{=} 5 \checkmark$$

The solution set is $\left\{ -\dfrac{1}{2}, 5 \right\}$.

Classroom Example: p. 462, Exercise 30

Instructor Note: Show students why $2x^2 - 9x = 5$ cannot be solved as:

$$2x^2 - 9x = 5$$
$$x(2x - 9) = 5$$
$$x = 5 \quad \text{ or } \quad 2x - 9 = 5$$

Skill Practice Solve the quadratic equation.

4. $2y^2 + 19y = -24$

| **Example 5** | **Solving a Quadratic Equation** |

Solve the quadratic equation. $\qquad 4x^2 + 24x = 0$

Solution:

$$4x^2 + 24x = 0 \qquad\qquad \text{The equation is already in the form}$$
$$ax^2 + bx + c = 0. \text{ (Note that } c = 0.)$$

$$4x(x + 6) = 0 \qquad\qquad \text{Factor completely.}$$

$$4x = 0 \quad \text{ or } \quad x + 6 = 0 \qquad \text{Set each factor equal to zero.}$$

$$x = 0 \quad \text{ or } \qquad x = -6 \qquad \text{The solutions check in the original equation.}$$

The solution set is $\{0, -6\}$.

Classroom Example: p. 462, Exercise 46

Skill Practice Solve the quadratic equation.

5. $5s^2 = 45$

Answers

4. $\left\{ -8, -\dfrac{3}{2} \right\}$ $\qquad$ **5.** $\{3, -3\}$

Classroom Example: p. 462,
Exercise 54

Example 6 **Solving a Quadratic Equation**

Solve the quadratic equation. $5x(5x + 2) = 10x + 9$

Solution:

$5x(5x + 2) = 10x + 9$

$25x^2 + 10x = 10x + 9$ Clear parentheses.

$25x^2 + 10x - 10x - 9 = 0$ Set the equation equal to zero.

$25x^2 - 9 = 0$ The equation is in the form
$ax^2 + bx + c = 0$. (Note that $b = 0$.)

$(5x - 3)(5x + 3) = 0$ Factor completely.

$5x - 3 = 0$ or $5x + 3 = 0$ Set each factor equal to zero.

$5x = 3$ or $5x = -3$ Solve each equation.

$\dfrac{5x}{5} = \dfrac{3}{5}$ or $\dfrac{5x}{5} = \dfrac{-3}{5}$

$x = \dfrac{3}{5}$ or $x = -\dfrac{3}{5}$ The solutions check in the original equation.

The solution set is $\left\{ \dfrac{3}{5}, -\dfrac{3}{5} \right\}$.

Skill Practice Solve the quadratic equation.

6. $4z(z + 3) = 4z + 5$

The zero product rule can be used to solve higher degree polynomial equations provided the equations can be set to zero and written in factored form.

Classroom Example: p. 462,
Exercise 38

Example 7 **Solving a Higher Degree Polynomial Equation**

Solve the equation. $-6(y + 3)(y - 5)(2y + 7) = 0$

Solution:

$-6(y + 3)(y - 5)(2y + 7) = 0$ The equation is already in factored
form and equal to zero.

Set each factor equal to zero.

Solve each equation for y.

$-6 \neq 0$ or $y + 3 = 0$ or $y - 5 = 0$ or $2y + 7 = 0$

No solution, $y = -3$ or $y = 5$ or $y = -\dfrac{7}{2}$

Notice that when the constant factor is set equal to zero, the result is a contradiction, $-6 = 0$. The constant factor does not produce a solution to the equation. Therefore, the solution set is $\{-3, 5, -\frac{7}{2}\}$. Each solution can be checked in the original equation.

Answer

6. $\left\{ -\dfrac{5}{2}, \dfrac{1}{2} \right\}$

Skill Practice Solve the equation.

7. $5(p - 4)(p + 7)(2p - 9) = 0$

| **Example 8** | Solving a Higher Degree Polynomial Equation |

Classroom Example: p. 463, Exercise 68

Solve the equation. $w^3 + 5w^2 - 9w - 45 = 0$

Solution:

$w^3 + 5w^2 - 9w - 45 = 0$ This is a higher degree polynomial equation.

$w^3 + 5w^2 \mid - 9w - 45 = 0$ The equation is already set equal to zero. Now factor.

$w^2(w + 5) - 9(w + 5) = 0$ Because there are four terms, try factoring
$(w + 5)(w^2 - 9) = 0$ by grouping.

$(w + 5)(w - 3)(w + 3) = 0$ $w^2 - 9$ is a difference of squares and can be factored further.

$w + 5 = 0$ or $w - 3 = 0$ or $w + 3 = 0$ Set each factor equal to zero.

$w = -5$ or $w = 3$ or $w = -3$ Solve each equation.

The solution set is $\{-5, 3, -3\}$. Each solution checks in the original equation.

Skill Practice Solve the equation.

8. $x^3 + 3x^2 - 4x - 12 = 0$

Answers

7. $\left\{4, -7, \dfrac{9}{2}\right\}$ **8.** $\{-2, -3, 2\}$

Section 6.7   Practice Exercises

Vocabulary and Key Concepts

For additional exercises, see Classroom Activities 6.7A–6.7B in the *Student's Resource Manual* at www.mhhe.com/moh.

1. a. An equation that can be written in the form $ax^2 + bx + c = 0$, $(a \neq 0)$, is called a ___quadratic___ equation.

    b. The zero product rule states that if $ab = 0$, then $a = $ ___0___ or $b = $ ___0___.

Review Exercises

For Exercises 2–7, factor completely.

2. $6a - 8 - 3ab + 4b$
    $(3a - 4)(2 - b)$

3. $4b^2 - 44b + 120$
    $4(b - 5)(b - 6)$

4. $8u^2v^2 - 4uv$
    $4uv(2uv - 1)$

5. $3x^2 + 10x - 8$
    $(3x - 2)(x + 4)$

6. $3h^2 - 75$
    $3(h - 5)(h + 5)$

7. $4x^2 + 16y^2$
    $4(x^2 + 4y^2)$

 Writing   Translating Expression   Geometry   Scientific Calculator   Video

Concept 1: Definition of a Quadratic Equation

For Exercises 8–13, identify the equations as linear, quadratic, or neither.

8. $4 - 5x = 0$
Linear

9. $5x^3 + 2 = 0$
Neither

10. $3x - 6x^2 = 0$
Quadratic

11. $1 - x + 2x^2 = 0$
Quadratic

12. $7x^4 + 8 = 0$
Neither

13. $3x + 2 = 0$
Linear

Concept 2: Zero Product Rule

For Exercises 14–22, solve each equation using the zero product rule. **(See Examples 1–3.)**

14. $(x - 5)(x + 1) = 0$ $\{5, -1\}$

15. $(x + 3)(x - 1) = 0$ $\{-3, 1\}$

16. $(3x - 2)(3x + 2) = 0$ $\left\{\frac{2}{3}, -\frac{2}{3}\right\}$

17. $(2x - 7)(2x + 7) = 0$ $\left\{\frac{7}{2}, -\frac{7}{2}\right\}$

18. $2(x - 7)(x - 7) = 0$ $\{7\}$

19. $3(x + 5)(x + 5) = 0$ $\{-5\}$

20. $(3x - 2)(2x - 3) = 0$ $\left\{\frac{2}{3}, \frac{3}{2}\right\}$

21. $x(5x - 1) = 0$ $\left\{0, \frac{1}{5}\right\}$

22. $x(3x + 8) = 0$ $\left\{0, -\frac{8}{3}\right\}$

23. For a quadratic equation of the form $ax^2 + bx + c = 0$, what must be done before applying the zero product rule?   The polynomial must be factored completely.

24. What are the requirements needed to use the zero product rule to solve a quadratic equation or higher degree polynomial equation?   The equation must have one side equal to zero and the other side factored completely.

Concept 3: Solving Equations by Factoring

For Exercises 25–72, solve each equation. **(See Examples 4–8.)**

25. $p^2 - 2p - 15 = 0$ $\{5, -3\}$

26. $y^2 - 7y - 8 = 0$ $\{8, -1\}$

27. $z^2 + 10z - 24 = 0$ $\{-12, 2\}$

28. $w^2 - 10w + 16 = 0$ $\{8, 2\}$

29. $2q^2 - 7q = 4$ $\left\{4, -\frac{1}{2}\right\}$

30. $4x^2 - 11x = 3$ $\left\{-\frac{1}{4}, 3\right\}$

31. $0 = 9x^2 - 4$ $\left\{\frac{2}{3}, -\frac{2}{3}\right\}$

32. $4a^2 - 49 = 0$ $\left\{\frac{7}{2}, -\frac{7}{2}\right\}$

33. $2k^2 - 28k + 96 = 0$ $\{6, 8\}$

34. $0 = 2t^2 + 20t + 50$ $\{-5\}$

35. $0 = 2m^3 - 5m^2 - 12m$ $\left\{0, -\frac{3}{2}, 4\right\}$

36. $3n^3 + 4n^2 + n = 0$ $\left\{0, -\frac{1}{3}, -1\right\}$

37. $5(3p + 1)(p - 3)(p + 6) = 0$
$\left\{-\frac{1}{3}, 3, -6\right\}$

38. $4(2x - 1)(x - 10)(x + 7) = 0$
$\left\{\frac{1}{2}, 10, -7\right\}$

39. $x(x - 4)(2x + 3) = 0$ $\left\{0, 4, -\frac{3}{2}\right\}$

40. $x(3x + 1)(x + 1) = 0$ $\left\{0, -\frac{1}{3}, -1\right\}$

41. $-5x(2x + 9)(x - 11) = 0$ $\left\{0, -\frac{9}{2}, 11\right\}$

42. $-3x(x + 7)(3x - 5) = 0$
$\left\{0, -7, \frac{5}{3}\right\}$

43. $x^3 - 16x = 0$ $\{0, 4, -4\}$

44. $t^3 - 36t = 0$ $\{0, 6, -6\}$

45. $3x^2 + 18x = 0$ $\{-6, 0\}$

46. $2y^2 - 20y = 0$ $\{0, 10\}$

47. $16m^2 = 9$ $\left\{\frac{3}{4}, -\frac{3}{4}\right\}$

48. $9n^2 = 1$ $\left\{\frac{1}{3}, -\frac{1}{3}\right\}$

49. $2y^3 + 14y^2 = -20y$ $\{0, -5, -2\}$

50. $3d^3 - 6d^2 = 24d$ $\{0, -2, 4\}$

51. $5t - 2(t - 7) = 0$ $\left\{-\frac{14}{3}\right\}$

52. $8h = 5(h - 9) + 6$ $\{-13\}$

53. $2c(c - 8) = -30$ $\{5, 3\}$

54. $3q(q - 3) = 12$ $\{-1, 4\}$

55. $b^3 = -4b^2 - 4b$ $\{0, -2\}$

56. $x^3 + 36x = 12x^2$ $\{0, 6\}$

57. $3(a^2 + 2a) = 2a^2 - 9$ $\{-3\}$

58. $9(k - 1) = -4k^2$ $\left\{\frac{3}{4}, -3\right\}$

59. $2n(n + 2) = 6$ $\{-3, 1\}$

60. $3p(p - 1) = 18$ $\{3, -2\}$

61. $x(2x + 5) - 1 = 2x^2 + 3x + 2$ $\left\{\frac{3}{2}\right\}$

62. $3z(z - 2) - z = 3z^2 + 4$ $\left\{-\frac{4}{7}\right\}$

63. $27q^2 = 9q$ $\left\{0, \frac{1}{3}\right\}$

64. $21w^2 = 14w$ $\left\{0, \frac{2}{3}\right\}$

65. $3(c^2 - 2c) = 0$ $\{0, 2\}$

66. $2(4d^2 + d) = 0$ $\left\{0, -\frac{1}{4}\right\}$

67. $y^3 - 3y^2 - 4y + 12 = 0$ {3, −2, 2}   **68.** $t^3 + 2t^2 - 16t - 32 = 0$ {−2, 4, −4}   **69.** $(x - 1)(x + 2) = 18$ {−5, 4}

70. $(w + 5)(w - 3) = 20$ {−7, 5}   **71.** $(p + 2)(p + 3) = 1 - p$ {−5, −1}   **72.** $(k - 6)(k - 1) = -k - 2$ {4, 2}

Problem Recognition Exercises

Polynomial Expressions Versus Polynomial Equations

For Exercises 1–36, factor the expressions and solve the equations.

1. a. $x^2 + 6x - 7$ $(x + 7)(x - 1)$ **2. a.** $c^2 + 8c + 12$ $(c + 6)(c + 2)$ **3. a.** $2y^2 + 7y + 3$ $(2y + 1)(y + 3)$
  b. $x^2 + 6x - 7 = 0$ {−7, 1}   **b.** $c^2 + 8c + 12 = 0$ {−6, −2}   **b.** $2y^2 + 7y + 3 = 0$ $\left\{-\frac{1}{2}, -3\right\}$

4. a. $3x^2 - 8x + 5$ $(3x - 5)(x - 1)$ **5. a.** $5q^2 + q - 4 = 0$ $\left\{\frac{4}{5}, -1\right\}$ **6. a.** $6a^2 - 7a - 3 = 0$ $\left\{-\frac{1}{3}, \frac{3}{2}\right\}$
  b. $3x^2 - 8x + 5 = 0$ $\left\{\frac{5}{3}, 1\right\}$ **b.** $5q^2 + q - 4$ $(5q - 4)(q + 1)$ **b.** $6a^2 - 7a - 3$ $(3a + 1)(2a - 3)$

7. a. $a^2 - 64 = 0$ {−8, 8}   **8. a.** $v^2 - 100 = 0$ {−10, 10}   **9. a.** $4b^2 - 81$ $(2b + 9)(2b - 9)$
  b. $a^2 - 64$ $(a + 8)(a - 8)$ **b.** $v^2 - 100$ $(v + 10)(v - 10)$ **b.** $4b^2 - 81 = 0$ $\left\{-\frac{9}{2}, \frac{9}{2}\right\}$

10. a. $36t^2 - 49$ $(6t + 7)(6t - 7)$ **11. a.** $8x^2 + 16x + 6 = 0$ $\left\{-\frac{3}{2}, -\frac{1}{2}\right\}$ **12. a.** $12y^2 + 40y + 32 = 0$ $\left\{-\frac{4}{3}, -2\right\}$
  b. $36t^2 - 49 = 0$ $\left\{-\frac{7}{6}, \frac{7}{6}\right\}$ **b.** $8x^2 + 16x + 6$ $2(2x + 3)(2x + 1)$ **b.** $12y^2 + 40y + 32$ $4(3y + 4)(y + 2)$

13. a. $x^3 - 8x^2 - 20x$ $x(x - 10)(x + 2)$ **14. a.** $k^3 + 5k^2 - 14k$ $k(k + 7)(k - 2)$ **15. a.** $b^3 + b^2 - 9b - 9 = 0$ {−1, 3, −3}
  b. $x^3 - 8x^2 - 20x = 0$ {0, 10, −2}   **b.** $k^3 + 5k^2 - 14k = 0$ {0, −7, 2}   **b.** $b^3 + b^2 - 9b - 9$
        $(b + 1)(b - 3)(b + 3)$

16. a. $x^3 - 8x^2 - 4x + 32 = 0$ {8, −2, 2}   **17.** $2s^2 - 6s + rs - 3r$ $(s - 3)(2s + r)$ **18.** $6t^2 + 3t + 10tu + 5u$
  b. $x^3 - 8x^2 - 4x + 32$ $(2t + 1)(3t + 5u)$
        $(x - 8)(x + 2)(x - 2)$

19. $8x^3 - 2x = 0$ $\left\{-\frac{1}{2}, 0, \frac{1}{2}\right\}$ **20.** $2b^3 - 50b = 0$ {−5, 0, 5}   **21.** $2x^3 - 4x^2 + 2x = 0$ {0, 1}

22. $3t^3 + 18t^2 + 27t = 0$ {−3, 0}   **23.** $7c^2 - 2c + 3 = 7(c^2 + c)$ $\left\{\frac{1}{3}\right\}$ **24.** $3z(2z + 4) = -7 + 6z^2$ $\left\{-\frac{7}{12}\right\}$

25. $8w^3 + 27$ $(2w + 3)(4w^2 - 6w + 9)$ **26.** $1000q^3 - 1$ **27.** $5z^2 + 2z = 7$ $\left\{-\frac{7}{5}, 1\right\}$
                                                $(10q - 1)(100q^2 + 10q + 1)$

28. $4h^2 + 25h = -6$ $\left\{-6, -\frac{1}{4}\right\}$ **29.** $3b(b + 6) = b - 10$ $\left\{-\frac{2}{3}, -5\right\}$ **30.** $3y^2 + 1 = y(y - 3)$ $\left\{-1, -\frac{1}{2}\right\}$

31. $5(2x - 3) - 2(3x + 1) = 4 - 3x$ {3}   **32.** $11 - 6a = -4(2a - 3) - 1$ {0}   **33.** $4s^2 = 64$ {−4, 4}

34. $81v^2 = 36$ $\left\{-\frac{2}{3}, \frac{2}{3}\right\}$ **35.** $(x - 3)(x - 4) = 6$ {1, 6}   **36.** $(x + 5)(x + 9) = 21$ {−2, −12}

Section 6.8   Applications of Quadratic Equations

Concepts

1. Applications of Quadratic Equations
2. Pythagorean Theorem

Classroom Example: p. 468, Exercise 16

1. Applications of Quadratic Equations

In this section we solve applications using the Problem-Solving Strategies for Word Problems.

Example 1   Translating to a Quadratic Equation

The product of two consecutive integers is 14 more than 6 times the smaller integer.

Solution:

Let x represent the first (smaller) integer.

Then $x + 1$ represents the second (larger) integer.   Label the variables.

(Smaller integer)(larger integer) $= 6 \cdot$ (smaller integer) $+ 14$ Verbal model

$$x(x + 1) = 6(x) + 14 \qquad \text{Algebraic equation}$$

$$x^2 + x = 6x + 14 \qquad \text{Simplify.}$$

$$x^2 + x - 6x - 14 = 0 \qquad \text{Set one side of the equation equal to zero.}$$

$$x^2 - 5x - 14 = 0$$

$$(x - 7)(x + 2) = 0 \qquad \text{Factor.}$$

$$x - 7 = 0 \quad \text{or} \quad x + 2 = 0 \qquad \text{Set each factor equal to zero.}$$

$$x = 7 \quad \text{or} \quad x = -2 \qquad \text{Solve for } x.$$

Recall that x represents the smaller integer. Therefore, there are two possibilities for the pairs of consecutive integers.

If $x = 7$, then the larger integer is $x + 1$ or $7 + 1 = 8$.

If $x = -2$, then the larger integer is $x + 1$ or $-2 + 1 = -1$.

The integers are 7 and 8, or −2 and −1.

Skill Practice

1. The product of two consecutive odd integers is 9 more than 10 times the smaller integer. Find the pair of integers.

Classroom Example: p. 469, Exercise 18

Example 2   Using a Quadratic Equation in a Geometry Application

A rectangular sign has an area of 40 ft². If the width is 3 feet shorter than the length, what are the dimensions of the sign?

Label the variables.

Solution:

Let x represent the length of the sign. Then $x - 3$ represents the width (Figure 6-1).

The problem gives information about the length of the sides and about the area. Therefore, we can form a relationship by using the formula for the area of a rectangle.

$x - 3$

x

Figure 6-1

© McGraw-Hill Education/Jill Braaten, photographer

Answer

1. The integers are 9 and 11 or −1 and 1.

$A = l \cdot w$	Area equals length times width.
$40 = x(x - 3)$	Set up an algebraic equation.
$40 = x^2 - 3x$	Clear parentheses.
$0 = x^2 - 3x - 40$	Write the equation in the form, $ax^2 + bx + c = 0$.
$0 = (x - 8)(x + 5)$	Factor.
$0 = x - 8 \quad$ or $\quad 0 = x + 5$	Set each factor equal to zero.
$8 = x \quad\quad$ or $\quad -5 \cancel{=} x$	Because x represents the length of a rectangle, reject the negative solution.

The variable x represents the length of the sign. The length is 8 ft.

The expression $x - 3$ represents the width. The width is 8 ft − 3 ft, or 5 ft.

Skill Practice

2. The length of a rectangle is 5 ft more than the width. The area is 36 ft². Find the length and width.

Example 3	**Using a Quadratic Equation in an Application**

Classroom Example: p. 469, Exercise 28

A stone is dropped off a 64-ft cliff and falls into the ocean below. The height of the stone above sea level is given by the equation

$$h = -16t^2 + 64 \qquad \text{where } h \text{ is the stone's height in feet, and } t \text{ is the time in seconds.}$$

Find the time required for the stone to hit the water.

Solution:

When the stone hits the water, its height is zero. Therefore, substitute $h = 0$ into the equation.

$h = -16t^2 + 64$	The equation is quadratic.
$0 = -16t^2 + 64$	Substitute $h = 0$.
$0 = -16(t^2 - 4)$	Factor out the GCF.
$0 = -16(t - 2)(t + 2)$	Factor as a difference of squares.
$-16 \cancel{=} 0 \quad$ or $\quad t - 2 = 0 \quad$ or $\quad t + 2 = 0$	Set each factor to zero.
No solution, $\quad\quad t = 2 \quad$ or $\quad t \cancel{=} -2$	Solve for t.

The negative value of t is rejected because the stone cannot fall for a negative time. Therefore, the stone hits the water after 2 sec.

Skill Practice

3. An object is launched into the air from the ground and its height is given by $h = -16t^2 + 144t$, where h is the height in feet after t seconds. Find the time required for the object to hit the ground.

Answers

2. The width is 4 ft, and the length is 9 ft.
3. The object hits the ground in 9 sec.

2. Pythagorean Theorem

Recall that a right triangle is a triangle that contains a 90° angle. Furthermore, the sum of the squares of the two legs (the shorter sides) of a right triangle equals the square of the hypotenuse (the longest side). This important fact is known as the Pythagorean theorem. The Pythagorean theorem is an enduring landmark of mathematical history from which many mathematical ideas have been built. Although the theorem is named after Pythagoras (sixth century B.C.E.), a Greek mathematician and philosopher, it is thought that the ancient Babylonians were familiar with the principle more than a thousand years earlier.

For the right triangle shown in Figure 6-2, the **Pythagorean theorem** is stated as:

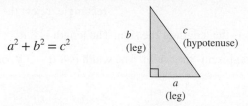

$$a^2 + b^2 = c^2$$

Figure 6-2

In this formula, a and b are the legs of the right triangle and c is the hypotenuse. Notice that the hypotenuse is the longest side of the right triangle and is opposite the 90° angle.

The triangle shown below is a right triangle. Notice that the lengths of the sides satisfy the Pythagorean theorem.

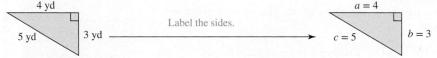

$$a^2 + b^2 = c^2 \qquad \text{Apply the Pythagorean theorem.}$$
$$(4)^2 + (3)^2 = (5)^2 \qquad \text{Substitute } a = 4, b = 3, \text{ and } c = 5.$$
$$16 + 9 = 25$$
$$25 = 25 \ \checkmark$$

Classroom Example: p. 470, Exercise 32

Example 4   **Applying the Pythagorean Theorem**

Find the length of the missing side of the right triangle.

Solution:

Label the triangle.

$$a^2 + b^2 = c^2 \qquad \text{Apply the Pythagorean theorem.}$$
$$a^2 + 6^2 = 10^2 \qquad \text{Substitute } b = 6 \text{ and } c = 10.$$
$$a^2 + 36 = 100 \qquad \text{Simplify. The equation is quadratic.}$$

$$a^2 + 36 - 100 = 100 - 100$$ Subtract 100 from both sides.

$$a^2 - 64 = 0$$ One side is now equal to zero.

$$(a + 8)(a - 8) = 0$$ Factor.

$$a + 8 = 0 \quad \text{or} \quad a - 8 = 0$$ Set each factor equal to zero.

$$a \neq -8 \quad \text{or} \quad a = 8$$ Because x represents the length of a side of a triangle, reject the negative solution.

The third side is 8 ft.

Skill Practice

4. Find the length of the missing side.

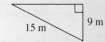

15 m   9 m

Example 5   **Using a Quadratic Equation in an Application**

Classroom Example: p. 470, Exercise 40

A 13-ft board is used as a ramp to unload furniture off a loading platform. If the distance between the top of the board and the ground is 7 ft less than the distance between the bottom of the board and the base of the platform, find both distances.

Solution:

Let x represent the distance between the bottom of the board and the base of the platform. Then $x - 7$ represents the distance between the top of the board and the ground (Figure 6-3).

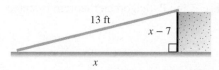

13 ft

$x - 7$

x

Figure 6-3

$$a^2 + b^2 = c^2$$ Pythagorean theorem

$$x^2 + (x - 7)^2 = (13)^2$$

$$x^2 + (x)^2 - 2(x)(7) + (7)^2 = 169$$

$$x^2 + x^2 - 14x + 49 = 169$$

$$2x^2 - 14x + 49 = 169$$ Combine *like* terms.

$$2x^2 - 14x + 49 - 169 = 169 - 169$$ Set the equation equal to zero.

$$2x^2 - 14x - 120 = 0$$ Write the equation in the form $ax^2 + bx + c = 0$.

$$2(x^2 - 7x - 60) = 0$$ Factor.

$$2(x - 12)(x + 5) = 0$$

$$2 \neq 0 \quad \text{or} \quad x - 12 = 0 \quad \text{or} \quad x + 5 = 0$$ Set each factor equal to zero.

$$x = 12 \quad \text{or} \quad x \neq -5$$ Solve both equations for x.

Avoiding Mistakes

Recall that the square of a binomial results in a perfect square trinomial.

$$(a - b)^2 = a^2 - 2ab + b^2$$
$$(x - 7)^2 = x^2 - 2(x)(7) + 7^2$$
$$= x^2 - 14x + 49$$

Don't forget the middle term.

Answer

4. The length of the third side is 12 m.

Recall that x represents the distance between the bottom of the board and the base of the platform. We reject the negative value of x because a distance cannot be negative. Therefore, the distance between the bottom of the board and the base of the platform is 12 ft. The distance between the top of the board and the ground is $x - 7 = 5$ ft.

Skill Practice

5. A 5-yd ladder leans against a wall. The distance from the bottom of the wall to the top of the ladder is 1 yd more than the distance from the bottom of the wall to the bottom of the ladder. Find both distances.

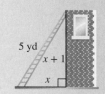

Answer

5. The distance along the wall to the top of the ladder is 4 yd. The distance on the ground from the ladder to the wall is 3 yd.

Section 6.8   Practice Exercises

For additional exercises, see Classroom Activities 6.8A–6.8B in the *Student's Resource Manual* at www.mhhe.com/moh.

Vocabulary and Key Concepts

1. **a.** If x is the smaller of two consecutive integers, then _____ $x + 1$ _____ represents the next greater integer.

   b. If x is the smaller of two consecutive odd integers, then _____ $x + 2$ _____ represents the next greater odd integer.

   c. If x is the smaller of two consecutive even integers, then _____ $x + 2$ _____ represents the next greater even integer.

   d. The area of a rectangle of length L and width W is given by $A =$ _____ LW _____.

   e. The area of a triangle with base b and height h is given by the formula $A =$ _____ $\frac{1}{2}bh$ _____.

   f. Given a right triangle with legs a and b and hypotenuse c, the Pythagorean theorem is stated as _____ $a^2 + b^2 = c^2$ _____.

Review Exercises

For Exercises 2–7, solve the quadratic equations.

2. $(6x + 1)(x + 4) = 0$ $\left\{-\frac{1}{6}, -4\right\}$

3. $9x(3x + 2) = 0$ $\left\{0, -\frac{2}{3}\right\}$

4. $4x^2 - 1 = 0$ $\left\{-\frac{1}{2}, \frac{1}{2}\right\}$

5. $x^2 - 5x = 6$ $\{6, -1\}$

6. $x(x - 20) = -100$ $\{10\}$

7. $6x^2 - 7x - 10 = 0$ $\left\{-\frac{5}{6}, 2\right\}$

8. Explain what is wrong with the following logic. $(x - 3)(x + 2) = 5$
   $$x - 3 = 5 \quad \text{or} \quad x + 2 = 5$$
   A factored expression must be set equal to zero to use the zero product rule.

Concept 1: Applications of Quadratic Equations

9. If eleven is added to the square of a number, the result is sixty. Find all such numbers.
   The numbers are 7 and −7.

10. If a number is added to two times its square, the result is thirty-six. Find all such numbers.
    The numbers are $-\frac{9}{2}$ and 4.

11. If twelve is added to six times a number, the result is twenty-eight less than the square of the number. Find all such numbers.
    The numbers are 10 and −4.

12. The square of a number is equal to twenty more than the number. Find all such numbers.
    The numbers are 5 and −4.

13. The product of two consecutive odd integers is sixty-three. Find all such integers. **(See Example 1.)**
    The numbers are −9 and −7, or 7 and 9.

14. The product of two consecutive even integers is forty-eight. Find all such integers.
    The numbers are 6 and 8, or −8 and −6.

15. The sum of the squares of two consecutive integers is sixty-one. Find all such integers.
    The numbers are 5 and 6, or −6 and −5.

16. The sum of the squares of two consecutive even integers is fifty-two. Find all such integers.
    The numbers are 4 and 6 or −6 and −4.

 Writing   ⟷ Translating Expression   Geometry   Scientific Calculator   Video

17. *Las Meninas* (Spanish for *The Maids of Honor*) is a famous painting by Spanish painter Diego Velázquez. This work is regarded as one of the most important paintings in Western art history. The height of the painting is approximately 2 ft more than its width. If the total area is 99 ft², determine the dimensions of the painting.
(See Example 2.)   The height of the painting is 11 ft and the width is 9 ft.

 18. The width of a rectangular painting is 2 in. less than the length. The area is 120 in.² Find the length and width.   The painting has length 12 in. and width 10 in.

x

x − 2

© IT Stock Free/Alamy RF

 19. The width of a rectangular slab of concrete is 3 m less than the length. The area is 28 m².

   a. What are the dimensions of the rectangle?
     The slab is 7 m by 4 m.

   b. What is the perimeter of the rectangle?
     The perimeter is 22 m.

 20. The width of a rectangular picture is 7 in. less than the length. The area of the picture is 78 in.²

   a. What are the dimensions of the picture?
     The picture is 13 in. by 6 in.

   b. What is the perimeter of the picture?
     The perimeter is 38 in.

 21. The base of a triangle is 3 ft more than the height. If the area is 14 ft², find the base and the height.   The base is 7 ft and the height is 4 ft.

22. The height of a triangle is 15 cm more than the base. If the area is 125 cm², find the base and the height.   The base is 10 cm and the height is 25 cm.

23. The height of a triangle is 7 cm less than 3 times the base. If the area is 20 cm², find the base and the height.   The base is 5 cm and the height is 8 cm.

24. The base of a triangle is 2 ft less than 4 times the height. If the area is 6 ft², find the base and the height.   The base is 6 ft and the height is 2 ft.

25. In a physics experiment, a ball is dropped off a 144-ft platform. The height of the ball above the ground is given by the equation

$h = -16t^2 + 144$ where h is the ball's height in feet, and t is the time in seconds after the ball is dropped ($t \geq 0$).

Find the time required for the ball to hit the ground. (*Hint:* Let $h = 0$.) **(See Example 3.)**
The ball hits the ground in 3 sec.

26. A stone is dropped off a 256-ft cliff. The height of the stone above the ground is given by the equation

 $h = -16t^2 + 256$ where h is the stone's height in feet, and t is the time in seconds after the stone is dropped ($t \geq 0$).

Find the time required for the stone to hit the ground.
The stone hits the ground in 4 sec.

27. An object is shot straight up into the air from ground level with an initial speed of 24 ft/sec. The height of the object (in feet) is given by the equation

$h = -16t^2 + 24t$ where t is the time in seconds after launch ($t \geq 0$).

Find the time(s) when the object is at ground level.
The object is at ground level at 0 sec and 1.5 sec.

 28. A rocket is launched straight up into the air from the ground with initial speed of 64 ft/sec. The height of the rocket (in feet) is given by the equation

$h = -16t^2 + 64t$ where t is the time in seconds after launch ($t \geq 0$).

Find the time(s) when the rocket is at ground level.
The rocket is at ground level at 0 sec and 4 sec.

Concept 2: Pythagorean Theorem

 29. Sketch a right triangle and label the sides with the words *leg* and *hypotenuse*.

 30. State the Pythagorean theorem.
Given a right triangle with legs a and b and hypotenuse c, then $a^2 + b^2 = c^2$.

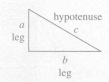

hypotenuse

a

c

leg

b

leg

For Exercises 31–34, find the length of the missing side of the right triangle. **(See Example 4.)**

31.

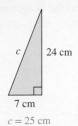

c 24 cm

7 cm

$c = 25$ cm

32.

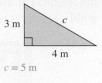

3 m c

4 m

$c = 5$ m

33.

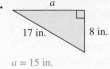

a

17 in.   8 in.

$a = 15$ in.

34.

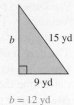

b 15 yd

9 yd

$b = 12$ yd

35. Find the length of the supporting brace.
The brace is 20 in. long.

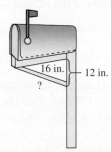

16 in.   12 in.

?

36. Find the height of the airplane above the ground.
The height is 9 km.

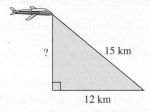

?   15 km

12 km

37. Darcy holds the end of a kite string 3 ft (1 yd) off the ground and wants to estimate the height of the kite. Her friend Jenna is 24 yd away from her, standing directly under the kite as shown in the figure. If Darcy has 30 yd of string out, find the height of the kite (ignore the sag in the string).
The kite is 19 yd high.

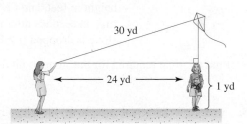

30 yd

← 24 yd →   1 yd

38. Two cars leave the same point at the same time, one traveling north and the other traveling east. After an hour, one car has traveled 48 mi and the other has traveled 64 mi. How many miles apart were they at that time?   They were 80 mi apart.

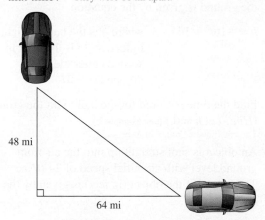

48 mi

64 mi

39. A 17-ft ladder rests against the side of a house. The distance between the top of the ladder and the ground is 7 ft more than the distance between the base of the ladder and the bottom of the house. Find both distances. **(See Example 5.)**   The bottom of the ladder is 8 ft from the house. The distance from the top of the ladder to the ground is 15 ft.

17 ft

$x + 7$

x

40. Two boats leave a marina. One travels east, and the other travels south. After 30 min, the second boat has traveled 1 mi farther than the first boat and the distance between the boats is 5 mi. Find the distance each boat traveled.   The first boat traveled 3 mi; the second boat traveled 4 mi.

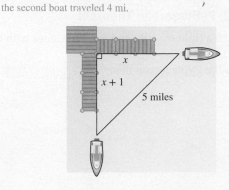

x

$x + 1$ 5 miles

Writing   ←→ Translating Expression   Geometry   Scientific Calculator   Video

41. One leg of a right triangle is 4 m less than the hypotenuse. The other leg is 2 m less than the hypotenuse. Find the length of the hypotenuse.
The hypotenuse is 10 m.

 42. The longer leg of a right triangle is 1 cm less than twice the shorter leg. The hypotenuse is 1 cm greater than twice the shorter leg. Find the length of the shorter leg.   The shorter leg is 8 cm.

Group Activity

Building a Factoring Test

Estimated Time: 30–45 minutes

Group Size: 3

Answers will vary throughout this exercise.

In this activity, each group will make a test for this chapter. Then the groups will trade papers and take the test.

For Exercises 1–8, write a polynomial that has the given conditions. Do not use reference materials such as the textbook or your notes.

1. A trinomial with a GCF not equal to 1. The GCF should include a constant and at least one variable.

1. _____

2. A four-term polynomial that is factorable by grouping.

2. _____

3. A factorable trinomial with a leading coefficient of 1. (The trinomial should factor as a product of two binomials.)

3. _____

4. A factorable trinomial with a leading coefficient not equal to 1. (The trinomial should factor as a product of two binomials.)

4. _____

5. A trinomial that requires the GCF to be removed. The resulting trinomial should factor as a product of two binomials.

5. _____

6. A difference of squares.

6. _____

7. A difference of cubes.

7. _____

8. A sum of cubes.

8. _____

9. Write a quadratic *equation* that has solution set $\{4, -7\}$.

9. _____

10. Write a quadratic *equation* that has solution set $\left\{0, -\dfrac{2}{3}\right\}$.

10. _____

 Writing   Translating Expression   Geometry    Scientific Calculator   Video

Chapter 6   Summary

Section 6.1   Greatest Common Factor and Factoring by Grouping

Key Concepts

The **greatest common factor** (GCF) is the greatest factor common to all terms of a polynomial. To factor out the GCF from a polynomial, use the distributive property.

A four-term polynomial may be factorable by grouping.

Steps to Factoring by Grouping

1. Identify and factor out the GCF from all four terms.
2. Factor out the GCF from the first pair of terms. Factor out the GCF or its opposite from the second pair of terms.
3. If the two terms share a common binomial factor, factor out the binomial factor.

Examples

Example 1

$3x(a + b) - 5(a + b)$ Greatest common factor is $(a + b)$.

$= (a + b)(3x - 5)$

Example 2

$60xa - 30xb - 80ya + 40yb$

$= 10[6xa - 3xb - 8ya + 4yb]$ Factor out the GCF.

$= 10[3x(2a - b) - 4y(2a - b)]$ Factor by grouping.

$= 10(2a - b)(3x - 4y)$

Section 6.2   Factoring Trinomials of the Form $x^2 + bx + c$

Key Concepts

Factoring a Trinomial with a Leading Coefficient of 1

A trinomial of the form $x^2 + bx + c$ factors as

$$x^2 + bx + c = (x \ \Box)(x \ \Box)$$

where the remaining terms are given by two integers whose product is c and whose sum is b.

Example

Example 1

$x^2 - 14x + 45$ The integers -5 and -9 have a product of 45 and a sum of -14.

$= (x \ \Box)(x \ \Box)$

$= (x - 5)(x - 9)$

Section 6.3   Factoring Trinomials: Trial-and-Error Method

Key Concepts

Trial-and-Error Method for Factoring Trinomials in the Form $ax^2 + bx + c$ (where $a \neq 0$)

1. Factor out the GCF from all terms.
2. List the pairs of factors of a and the pairs of factors of c. Consider the reverse order in one of the lists.
3. Construct two binomials of the form

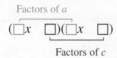

Factors of a

$(\square x \quad \square)(\square x \quad \square)$

Factors of c

4. Test each combination of factors and signs until the product forms the correct trinomial.
5. If no combination of factors produces the correct product, then the trinomial is prime.

Example

Example 1

$10y^2 + 35y - 20$

$= 5(2y^2 + 7y - 4)$

The pairs of factors of 2 are: $2 \cdot 1$
The pairs of factors of -4 are:

$$-1(4) \qquad 1(-4)$$
$$-2(2) \qquad 2(-2)$$
$$-4(1) \qquad 4(-1)$$

$(2y - 2)(y + 2) = 2y^2 + 2y - 4$ No

$(2y - 4)(y + 1) = 2y^2 - 2y - 4$ No

$(2y + 1)(y - 4) = 2y^2 - 7y - 4$ No

$(2y + 2)(y - 2) = 2y^2 - 2y - 4$ No

$(2y + 4)(y - 1) = 2y^2 + 2y - 4$ No

$(2y - 1)(y + 4) = 2y^2 + 7y - 4$ Yes

$10y^2 + 35y - 20 = 5(2y - 1)(y + 4)$

Section 6.4   Factoring Trinomials: AC-Method

Key Concepts

AC-Method for Factoring Trinomials of the Form $ax^2 + bx + c$ (where $a \neq 0$)

1. Factor out the GCF from all terms.
2. Find the product ac.
3. Find two integers whose product is ac and whose sum is b. (If no pair of integers can be found, then the trinomial is prime.)
4. Rewrite the middle term (bx) as the sum of two terms whose coefficients are the integers found in step 3.
5. Factor the polynomial by grouping.

Example

Example 1

$10y^2 + 35y - 20$

$= 5(2y^2 + 7y - 4)$ First factor out the GCF.

Identify the product $ac = (2)(-4) = -8$.

Find two integers whose product is -8 and whose sum is 7. The numbers are 8 and -1.

$5[2y^2 + 8y - 1y - 4]$

$= 5[2y(y + 4) - 1(y + 4)]$

$= 5(y + 4)(2y - 1)$

Section 6.5	Difference of Squares and Perfect Square Trinomials

Key Concepts

Factoring a Difference of Squares

$a^2 - b^2 = (a + b)(a - b)$

Factoring a Perfect Square Trinomial

The factored form of a **perfect square trinomial** is the square of a binomial:

$a^2 + 2ab + b^2 = (a + b)^2$

$a^2 - 2ab + b^2 = (a - b)^2$

Examples

Example 1

$25z^2 - 4y^2$

$= (5z + 2y)(5z - 2y)$

Example 2

Factor: $25y^2 + 10y + 1$

$= (5y)^2 + 2(5y)(1) + (1)^2$

$= (5y + 1)^2$

Section 6.6	Sum and Difference of Cubes

Key Concepts

Factoring a Sum or Difference of Cubes

$a^3 + b^3 = (a + b)(a^2 - ab + b^2)$

$a^3 - b^3 = (a - b)(a^2 + ab + b^2)$

Examples

Example 1

$x^6 + 8y^3$

$= (x^2)^3 + (2y)^3$

$= (x^2 + 2y)(x^4 - 2x^2y + 4y^2)$

Example 2

$m^3 - 64$

$= (m)^3 - (4)^3$

$= (m - 4)(m^2 + 4m + 16)$

Section 6.7   Solving Equations Using the Zero Product Rule

Key Concepts

An equation of the form $ax^2 + bx + c = 0$, where $a \neq 0$, is a **quadratic equation**.

The zero product rule states that if $ab = 0$, then $a = 0$ or $b = 0$. The zero product rule can be used to solve a quadratic equation or a higher degree polynomial equation that is factored and set to zero.

Examples

Example 1

The equation $2x^2 - 17x + 30 = 0$ is a quadratic equation.

Example 2

$3w(w - 4)(2w + 1) = 0$

$3w = 0$ or $w - 4 = 0$ or $2w + 1 = 0$

$w = 0$ or $w = 4$ or $w = -\dfrac{1}{2}$

The solution set is $\left\{ 0, 4, -\dfrac{1}{2} \right\}$.

Example 3

$4x^2 = 34x - 60$

$4x^2 - 34x + 60 = 0$

$2(2x^2 - 17x + 30) = 0$

$2(2x - 5)(x - 6) = 0$

$2 \neq 0$ or $2x - 5 = 0$ or $x - 6 = 0$

$x = \dfrac{5}{2}$ or $x = 6$

The solution set is $\left\{ \dfrac{5}{2}, 6 \right\}$.

Section 6.8   Applications of Quadratic Equations

Key Concepts

Use the zero product rule to solve applications.

Examples

Example 1

Find two consecutive integers such that the sum of their squares is 61.

Let x represent one integer.
Let $x + 1$ represent the next consecutive integer.

$$x^2 + (x + 1)^2 = 61$$

$$x^2 + x^2 + 2x + 1 = 61$$

$$2x^2 + 2x - 60 = 0$$

$$2(x^2 + x - 30) = 0$$

$$2(x - 5)(x + 6) = 0$$

$$x = 5 \quad \text{or} \quad x = -6$$

If $x = 5$, then the next consecutive integer is 6.
If $x = -6$, then the next consecutive integer is -5.

The integers are 5 and 6, or -6 and -5.

Some applications involve the Pythagorean theorem.

$$a^2 + b^2 = c^2$$

Example 2

Find the length of the missing side.

$$x^2 + (7)^2 = (25)^2$$

$$x^2 + 49 = 625$$

$$x^2 - 576 = 0$$

$$(x - 24)(x + 24) = 0$$

$$x = 24 \quad \text{or} \quad x = -24$$

The length of the side is 24 ft.

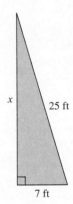

Chapter 6 Review Exercises

Section 6.1

For Exercises 1–4, identify the greatest common factor for each group of terms.

1. $15a^2b^4, 30a^3b, 9a^5b^3$
   $3a^2b$

2. $3(x + 5), x(x + 5)$
   $x + 5$

3. $2c^3(3c - 5), 4c(3c - 5)$
   $2c(3c - 5)$

4. $-2wyz, -4xyz$
   $-2yz$ or $2yz$

For Exercises 5–10, factor out the greatest common factor.

5. $6x^2 + 2x^4 - 8x$
   $2x(3x + x^3 - 4)$

6. $11w^3y^3 - 44w^2y^5$
   $11w^2y^3(w - 4y^2)$

7. $-t^2 + 5t$
   $t(-t + 5)$ or $-t(t - 5)$

8. $-6u^2 - u$
   $u(-6u - 1)$ or $-u(6u + 1)$

9. $3b(b + 2) - 7(b + 2)$ $(b + 2)(3b - 7)$

10. $2(5x + 9) + 8x(5x + 9)$ $2(5x + 9)(1 + 4x)$

For Exercises 11–14, factor by grouping.

11. $7w^2 + 14w + wb + 2b$ $(w + 2)(7w + b)$

12. $b^2 - 2b + yb - 2y$ $(b - 2)(b + y)$

13. $60y^2 - 45y - 12y + 9$ $3(4y - 3)(5y - 1)$

14. $6a - 3a^2 - 2ab + a^2b$ $a(2 - a)(3 - b)$

Section 6.2

For Exercises 15–24, factor completely.

15. $x^2 - 10x + 21$
    $(x - 3)(x - 7)$

16. $y^2 - 19y + 88$
    $(y - 8)(y - 11)$

17. $-6z + z^2 - 72$
    $(z - 12)(z + 6)$

18. $-39 + q^2 - 10q$
    $(q - 13)(q + 3)$

19. $3p^2w + 36pw + 60w$
    $3w(p + 10)(p + 2)$

20. $2m^4 + 26m^3 + 80m^2$
    $2m^2(m + 8)(m + 5)$

21. $-t^2 + 10t - 16$
    $-(t - 8)(t - 2)$

22. $-w^2 - w + 20$
    $-(w - 4)(w + 5)$

23. $a^2 + 12ab + 11b^2$
    $(a + b)(a + 11b)$

24. $c^2 - 3cd - 18d^2$
    $(c - 6d)(c + 3d)$

Section 6.3

For Exercises 25–28, assume that a, b, and c represent positive integers.

25. When factoring a polynomial of the form $ax^2 - bx - c$, should the signs of the binomials be both positive, both negative, or different?
    Different

26. When factoring a polynomial of the form $ax^2 - bx + c$, should the signs of the binomials be both positive, both negative, or different?
    Both negative

27. When factoring a polynomial of the form $ax^2 + bx + c$, should the signs of the binomials be both positive, both negative, or different?
    Both positive

28. When factoring a polynomial of the form $ax^2 + bx - c$, should the signs of the binomials be both positive, both negative, or different?
    Different

For Exercises 29–42, factor each trinomial using the trial-and-error method.

29. $2y^2 - 5y - 12$
    $(2y + 3)(y - 4)$

30. $4w^2 - 5w - 6$
    $(4w + 3)(w - 2)$

31. $10z^2 + 29z + 10$
    $(2z + 5)(5z + 2)$

32. $8z^2 + 6z - 9$
    $(4z - 3)(2z + 3)$

33. $2p^2 - 5p + 1$
    Prime

34. $5r^2 - 3r + 7$
    Prime

35. $10w^2 - 60w - 270$
    $10(w - 9)(w + 3)$

36. $-3y^2 + 18y + 48$
    $-3(y - 8)(y + 2)$

37. $9c^2 - 30cd + 25d^2$
    $(3c - 5d)^2$

38. $x^2 + 12x + 36$
    $(x + 6)^2$

39. $6g^2 + 7gh + 2h^2$
    $(3g + 2h)(2g + h)$

40. $12m^2 - 32mn + 5n^2$
    $(6m - n)(2m - 5n)$

41. $v^4 - 2v^2 - 3$
    $(v^2 + 1)(v^2 - 3)$

42. $x^4 + 7x^2 + 10$
    $(x^2 + 5)(x^2 + 2)$

Section 6.4

For Exercises 43–44, find a pair of integers whose product and sum are given.

43. Product: -5 sum: 4 $5, -1$

44. Product: 15 sum: -8 $-3, -5$

For Exercises 45–58, factor each trinomial using the ac-method.

45. $3c^2 - 5c - 2$
    $(c - 2)(3c + 1)$

46. $4y^2 + 13y + 3$
    $(y + 3)(4y + 1)$

47. $t^2 + 13tw + 12w^2$
    $(t + 12w)(t + w)$

48. $4x^4 + 17x^2 - 15$
    $(4x^2 - 3)(x^2 + 5)$

49. $w^4 + 7w^2 + 10$
    $(w^2 + 5)(w^2 + 2)$

50. $p^2 - 8pq + 15q^2$
    $(p - 3q)(p - 5q)$

51. $-40v^2 - 22v + 6$
    $-2(4v + 3)(5v - 1)$

52. $40s^2 + 30s - 100$
    $10(4s - 5)(s + 2)$

53. $a^3b - 10a^2b^2 + 24ab^3$
    $ab(a - 6b)(a - 4b)$

54. $2z^6 + 8z^5 - 42z^4$
    $2z^4(z + 7)(z - 3)$

55. $m + 9m^2 - 2$
Prime

56. $2 + 6p^2 + 19p$
Prime

57. $49x^2 + 140x + 100$
$(7x + 10)^2$

58. $9w^2 - 6wz + z^2$
$(3w - z)^2$

Section 6.5

For Exercises 59–60, write the formula to factor each binomial, if possible.

59. $a^2 - b^2$
$(a - b)(a + b)$

60. $a^2 + b^2$
Prime

For Exercises 61–76, factor completely.

61. $a^2 - 49$
$(a - 7)(a + 7)$

62. $d^2 - 64$
$(d - 8)(d + 8)$

63. $100 - 81t^2$
$(10 - 9t)(10 + 9t)$

64. $4 - 25k^2$
$(2 - 5k)(2 + 5k)$

65. $x^2 + 16$
Prime

66. $y^2 + 121$
Prime

67. $y^2 + 12y + 36$
$(y + 6)^2$

68. $t^2 + 16t + 64$
$(t + 8)^2$

69. $9a^2 - 12a + 4$
$(3a - 2)^2$

70. $25x^2 - 40x + 16$
$(5x - 4)^2$

71. $-3v^2 - 12v - 12$
$-3(v + 2)^2$

72. $-2x^2 + 20x - 50$
$-2(x - 5)^2$

73. $2c^4 - 18$
$2(c^2 - 3)(c^2 + 3)$

74. $72x^2 - 2y^2$
$2(6x - y)(6x + y)$

75. $p^3 + 3p^2 - 16p - 48$
$(p + 3)(p - 4)(p + 4)$

76. $4k - 8 - k^3 + 2k^2$
$(k - 2)(2 - k)(2 + k)$
or $-(k - 2)^2(2 + k)$

Section 6.6

For Exercises 77–78, write the formula to factor each binomial, if possible.

77. $a^3 + b^3$
$(a + b)(a^2 - ab + b^2)$

78. $a^3 - b^3$
$(a - b)(a^2 + ab + b^2)$

For Exercises 79–92, factor completely.

79. $64 + a^3$
$(4 + a)(16 - 4a + a^2)$

80. $125 - b^3$
$(5 - b)(25 + 5b + b^2)$

81. $p^6 + 8$
$(p^2 + 2)(p^4 - 2p^2 + 4)$

82. $q^6 - \dfrac{1}{27}$
$\left(q^2 - \dfrac{1}{3}\right)\left(q^4 + \dfrac{1}{3}q^2 + \dfrac{1}{9}\right)$

83. $6x^3 - 48$
$6(x - 2)(x^2 + 2x + 4)$

84. $7y^3 + 7$
$7(y + 1)(y^2 - y + 1)$

85. $x^3 - 36x$
$x(x - 6)(x + 6)$

86. $q^4 - 64q$
$q(q - 4)(q^2 + 4q + 16)$

87. $8h^2 + 20$
$4(2h^2 + 5)$

88. $m^2 - 8m$
$m(m - 8)$

89. $x^3 + 4x^2 - x - 4$
$(x + 4)(x + 1)(x - 1)$

90. $5p^4q - 20q^3$
$5q(p^2 - 2q)(p^2 + 2q)$

91. $8n + n^4$
$n(2 + n)(4 - 2n + n^2)$

92. $14m^3 - 14$
$14(m - 1)(m^2 + m + 1)$

Section 6.7

93. For which of the following equations can the zero product rule be applied directly? Explain.
$(x - 3)(2x + 1) = 0$ or $(x - 3)(2x + 1) = 6$
$(x - 3)(2x + 1) = 0$ can be solved directly by the zero product rule because it is a product of factors set equal to zero.

For Exercises 94–109, solve each equation using the zero product rule.

94. $(4x - 1)(3x + 2) = 0$ $\left\{\dfrac{1}{4}, -\dfrac{2}{3}\right\}$

95. $(a - 9)(2a - 1) = 0$ $\left\{9, \dfrac{1}{2}\right\}$

96. $3w(w + 3)(5w + 2) = 0$ $\left\{0, -3, -\dfrac{2}{5}\right\}$

97. $6u(u - 7)(4u - 9) = 0$ $\left\{0, 7, \dfrac{9}{4}\right\}$

98. $7k^2 - 9k - 10 = 0$ $\left\{-\dfrac{5}{7}, 2\right\}$

99. $4h^2 - 23h - 6 = 0$ $\left\{-\dfrac{1}{4}, 6\right\}$

100. $q^2 - 144 = 0$
$\{12, -12\}$

101. $r^2 = 25$
$\{5, -5\}$

102. $5v^2 - v = 0$ $\left\{0, \dfrac{1}{5}\right\}$

103. $x(x - 6) = -8$
$\{4, 2\}$

104. $36t^2 + 60t = -25$ $\left\{-\dfrac{5}{6}\right\}$

105. $9s^2 + 12s = -4$
$\left\{-\dfrac{2}{3}\right\}$

106. $3(y^2 + 4) = 20y$
$\left\{\dfrac{2}{3}, 6\right\}$

107. $2(p^2 - 66) = -13p$
$\left\{\dfrac{11}{2}, -12\right\}$

108. $2y^3 - 18y^2 = -28y$
$\{0, 7, 2\}$

109. $x^3 - 4x = 0$
$\{0, 2, -2\}$

Section 6.8

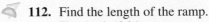

110. The base of a parallelogram is 1 ft longer than twice the height. If the area is 78 ft^2, find the base and height of the parallelogram.
The height is 6 ft, and the base is 13 ft.

111. A ball is tossed into the air from ground level with initial speed of 16 ft/sec. The height of the ball is given by the equation

$$h = -16t^2 + 16t \quad (t \geq 0)$$ where h is the ball's height in feet, and t is the time in seconds.

Find the time(s) when the ball is at ground level.
The ball is at ground level at 0 and 1 sec.

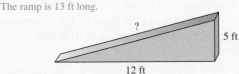

112. Find the length of the ramp.
The ramp is 13 ft long.

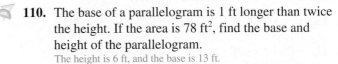

?

5 ft

12 ft

113. A right triangle has one leg that is 2 ft longer than the other leg. The hypotenuse is 2 ft less than twice the shorter leg. Find the lengths of all sides of the triangle.
The legs are 6 ft and 8 ft; the hypotenuse is 10 ft.

114. If the square of a number is subtracted from 60, the result is −4. Find all such numbers.
The numbers are −8 and 8.

115. The product of two consecutive integers is 44 more than 14 times their sum. Find all such integers.
The numbers are 29 and 30, or −2 and −1.

116. The base of a triangle is 1 m longer than twice the height. If the area of the triangle is 18 m^2, find the base and height.
The height is 4 m, and the base is 9 m.

Chapter 6   Test

1. Factor out the GCF. $15x^4 - 3x + 6x^3$
$3x(5x^3 - 1 + 2x^2)$

2. Factor by grouping. $7a - 35 - a^2 + 5a$
$(a - 5)(7 - a)$

3. Factor the trinomial. $6w^2 - 43w + 7$
$(6w - 1)(w - 7)$

4. Factor the difference of squares. $169 - p^2$
$(13 - p)(13 + p)$

5. Factor the perfect square trinomial.
$q^2 - 16q + 64$ $(q - 8)^2$

6. Factor the sum of cubes. $8 + t^3$
$(2 + t)(4 - 2t + t^2)$

For Exercises 7–26, factor completely.

7. $a^2 + 12a + 32$
$(a + 4)(a + 8)$

8. $x^2 + x - 42$
$(x + 7)(x - 6)$

9. $2y^2 - 17y + 8$
$(2y - 1)(y - 8)$

10. $6z^2 + 19z + 8$
$(2z + 1)(3z + 8)$

11. $9t^2 - 100$
$(3t - 10)(3t + 10)$

12. $v^2 - 81$
$(v + 9)(v - 9)$

13. $3a^2 + 27ab + 54b^2$
$3(a + 6b)(a + 3b)$

14. $c^4 - 1$
$(c - 1)(c + 1)(c^2 + 1)$

15. $xy - 7x + 3y - 21$
$(y - 7)(x + 3)$

16. $49 + p^2$ Prime

17. $-10u^2 + 30u - 20$
$-10(u - 2)(u - 1)$

18. $12t^2 - 75$
$3(2t - 5)(2t + 5)$

19. $5y^2 - 50y + 125$
$5(y - 5)^2$

20. $21q^2 + 14q$
$7q(3q + 2)$

21. $2x^3 + x^2 - 8x - 4$
$(2x + 1)(x - 2)(x + 2)$

22. $y^3 - 125$
$(y - 5)(y^2 + 5y + 25)$

23. $m^2n^2 - 81$
$(mn - 9)(mn + 9)$

24. $16a^2 - 64b^2$
$16(a - 2b)(a + 2b)$

25. $64x^3 - 27y^6$
$(4x - 3y^2)(16x^2 + 12xy^2 + 9y^4)$

26. $3x^2y - 6xy - 24y$
$3y(x - 4)(x + 2)$

For Exercises 27–31, solve the equation.

27. $(2x - 3)(x + 5) = 0$
$\left\{\dfrac{3}{2}, -5\right\}$

28. $x^2 - 7x = 0$ $\{0, 7\}$

29. $x^2 - 6x = 16$ $\{8, -2\}$

30. $x(5x + 4) = 1$
$\left\{\dfrac{1}{5}, -1\right\}$

31. $y^3 + 10y^2 - 9y - 90 = 0$
$\{3, -3, -10\}$

32. A tennis court has an area of 312 yd^2. If the length is 2 yd more than twice the width, find the dimensions of the court.
The tennis court is 12 yd by 26 yd.

33. The product of two consecutive odd integers is 35. Find the integers.
The two integers are 5 and 7, or −5 and −7.

34. The height of a triangle is 5 in. less than the length of the base. The area is 42 in^2. Find the base and the height of the triangle.
The base is 12 in., and the height is 7 in.

35. The hypotenuse of a right triangle is 2 ft less than three times the shorter leg. The longer leg is 3 ft less than three times the shorter leg. Find the length of the shorter leg.   The shorter leg is 5 ft.

36. A stone is dropped off a 64-ft cliff. The height of the stone above the ground is given in the equation

$h = -16t^2 + 64$ where h is the stone's height in feet, and t is the time in seconds after the stone is dropped ($t \geq 0$).

Find the time required for the stone to hit the ground.   The stone hits the ground in 2 sec.

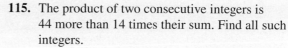

Chapters 1–6   Cumulative Review Exercises

1. Simplify. $\dfrac{|4 - 25 \div (-5) \cdot 2|}{\sqrt{8^2 + 6^2}}$ $\dfrac{7}{5}$

2. Solve. $5 - 2(t + 4) = 3t + 12$ $\{-3\}$

3. Solve for y. $3x - 2y = 8$ $y = \dfrac{3}{2}x - 4$

4. A child's piggy bank has 17 coins in quarters and dimes. The number of dimes is three less than the number of quarters. How many of each type of coin is in the piggy bank?
   There are 10 quarters and 7 dimes.

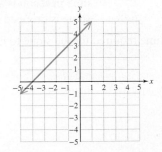

© Ryan McVay/Getty Images RF

5. Solve the inequality. Graph the solution on a number line and write the solution set in interval notation. $[-4, \infty)$

   $-\dfrac{5}{12}x \le \dfrac{5}{3}$

6. Given the equation $y = x + 4$,

   a. Is the equation linear?   Yes

   b. Identify the slope.   1

   c. Identify the y-intercept. $(0, 4)$

   d. Identify the x-intercept. $(-4, 0)$

   e. Graph the line

7. Consider the equation $x = 5$,

   a. Does the equation represent a horizontal or vertical line?   Vertical line

   b. Determine the slope of the line, if it exists.
      Undefined

   c. Identify the x-intercept, if it exists. $(5, 0)$

   d. Identify the y-intercept, if it exists.
      Does not exist

8. Find an equation of the line passing through the point $(-3, 5)$ and having a slope of 3. Write the final answer in slope-intercept form. $y = 3x + 14$

9. Solve the system. $2x - 3y = 4$ $\{(5, 2)\}$
                        $5x - 6y = 13$

For Exercises 10–13, perform the indicated operations.

10. $2\left(\dfrac{1}{3}y^3 - \dfrac{3}{2}y^2 - 7\right) - \left(\dfrac{2}{3}y^3 + \dfrac{1}{2}y^2 + 5y\right)$
    $-\dfrac{7}{2}y^2 - 5y - 14$

11. $(4p^2 - 5p - 1)(2p - 3)$ $8p^3 - 22p^2 + 13p + 3$

12. $(2w - 7)^2$ $4w^2 - 28w + 49$

13. $(r^4 + 2r^3 - 5r + 1) \div (r - 3)$ $r^3 + 5r^2 + 15r + 40 + \dfrac{121}{r - 3}$

14. Simplify. $\dfrac{c^{12}c^{-5}}{c^3}$ c^4

15. Simplify. $\left(\dfrac{6a^2b^{-4}}{2a^4b^{-5}}\right)^{-2}$ $\dfrac{a^4}{9b^2}$

16. Divide. Write the final answer in scientific notation. $\dfrac{8 \times 10^{-3}}{5 \times 10^{-6}}$ 1.6×10^3

For Exercises 17–19, factor completely.

17. $w^4 - 16$ $(w - 2)(w + 2)(w^2 + 4)$

18. $2ax + 10bx - 3ya - 15yb$ $(a + 5b)(2x - 3y)$

19. $4x^2 - 8x - 5$ $(2x - 5)(2x + 1)$

20. Solve. $4x(2x - 1)(x + 5) = 0$ $\left\{0, \dfrac{1}{2}, -5\right\}$

Rational Expressions and Equations

7

CHAPTER OUTLINE

Mathematics in Purchasing Land

Suppose you want to purchase a parcel of land in a remote area. The current owner has a large amount of acreage that she plans to divide into n smaller lots for sale. Her total acreage is worth \$120,000 and the smaller lots are of equal size. Thus, the cost (in dollars) of an individual lot is given by:

$$\frac{120{,}000}{n}$$

This fraction is called a **rational expression**, because it consists of a polynomial in the numerator and a nonzero polynomial in the denominator.

The cost to purchase an individual lot also includes closing costs. If closing costs are \$5600, then the total cost (in dollars) to purchase one lot is given by:

© AlinaMD/Shutterstock RF

$$\frac{120{,}000}{n} + 5600$$

This expression can be simplified by combining the two terms into a single rational expression. In this chapter we will learn how to perform operations on rational expressions and see numerous applications of their use.

Section 7.1 Introduction to Rational Expressions

1. Definition of a Rational Expression

We define a rational number as the ratio of two integers, $\frac{p}{q}$, where $q \neq 0$.

Examples of rational numbers: $\frac{2}{3}, -\frac{1}{5}, 9$

In a similar way, we define a **rational expression** as the ratio of two polynomials, $\frac{p}{q}$, where $q \neq 0$.

Examples of rational expressions: $\frac{3x-6}{x^2-4}, \frac{3}{4}, \frac{6r^5+2r}{7r^3}$

2. Evaluating Rational Expressions

Classroom Examples: p. 489, Exercises 6 and 10

Example 1 **Evaluating Rational Expressions**

Evaluate the rational expression (if possible) for the given values of x. $\frac{x}{x-3}$

a. $x = 6$ b. $x = -3$ c. $x = 0$ d. $x = 3$

Solution:

Substitute the given value for the variable. Use the order of operations to simplify.

a. $\dfrac{x}{x-3}$

$\dfrac{(6)}{(6)-3}$ Substitute $x = 6$.

$= \dfrac{6}{3}$

$= 2$

b. $\dfrac{x}{x-3}$

$\dfrac{(-3)}{(-3)-3}$ Substitute $x = -3$.

$= \dfrac{-3}{-6}$

$= \dfrac{1}{2}$

Instructor Note: Memory Device: Zero **un**der the line is **un**defined.

c. $\dfrac{x}{x-3}$

$\dfrac{(0)}{(0)-3}$ Substitute $x = 0$.

$= \dfrac{0}{-3}$

$= 0$

d. $\dfrac{x}{x-3}$

$\dfrac{(3)}{(3)-3}$ Substitute $x = 3$.

$= \dfrac{3}{0}$ Undefined.

Recall that division by zero is undefined.

Avoiding Mistakes

The numerator is 0 and the denominator is nonzero. Therefore, the fraction is equal to 0.

Avoiding Mistakes

The denominator is 0, therefore the fraction is undefined.

Skill Practice Evaluate the expression for the given values of x. $\dfrac{x-3}{x+5}$

1. $x = 2$ **2.** $x = 0$ **3.** $x = 3$ **4.** $x = -5$

Answers

1. $-\dfrac{1}{7}$ **2.** $-\dfrac{3}{5}$

3. 0   **4.** Undefined

3. Restricted Values of a Rational Expression

From Example 1 we see that not all values of x can be substituted into a rational expression. The values that make the denominator zero must be restricted.

The expression $\dfrac{x}{x-3}$ is undefined for $x = 3$, so we call $x = 3$ a restricted value.

Restricted values of a rational expression are all values that make the expression undefined, that is, make the denominator equal to zero.

Finding the Restricted Values of a Rational Expression

- Set the denominator equal to zero and solve the resulting equation.
- The restricted values are the solutions to the equation.

Example 2 **Finding the Restricted Values of Rational Expressions**

Classroom Examples: p. 490, Exercises 14 and 24

Identify the restricted values for each expression.

a. $\dfrac{y-3}{2y+7}$ **b.** $\dfrac{-5}{x}$

Solution:

a. $\dfrac{y-3}{2y+7}$

$2y + 7 = 0$ Set the denominator equal to zero.

$2y = -7$ Solve the equation.

$\dfrac{2y}{2} = \dfrac{-7}{2}$

$y = -\dfrac{7}{2}$ The restricted value is $y = -\dfrac{7}{2}$.

b. $\dfrac{-5}{x}$

$x = 0$ Set the denominator equal to zero.

The restricted value is $x = 0$.

Instructor Note: Ask the students to find the restricted values of a rational expression that has no variable in the denominator.

For example: $\dfrac{3x+7}{8}$

Skill Practice Identify the restricted values.

5. $\dfrac{a+2}{2a-8}$ **6.** $\dfrac{2}{t}$

Answers

5. $a = 4$ **6.** $t = 0$

Classroom Examples: p. 490, Exercises 18 and 20

Example 3 Finding the Restricted Values of Rational Expressions

Identify the restricted values for each expression.

a. $\dfrac{a+10}{a^2-25}$ b. $\dfrac{2x^3+5}{x^2+9}$

Solution:

a. $\dfrac{a+10}{a^2-25}$

$a^2 - 25 = 0$	Set the denominator equal to zero. The equation is quadratic.
$(a-5)(a+5) = 0$	Factor.
$a-5 = 0$ or $a+5 = 0$	Set each factor equal to zero.
$a = 5$ or $a = -5$	

The restricted values are $a = 5$ and $a = -5$.

b. $\dfrac{2x^3+5}{x^2+9}$

$x^2 + 9 = 0$

$x^2 = -9$

The quantity x^2 cannot be negative for any real number, x. The denominator x^2+9 is the sum of two nonnegative values, and thus cannot equal zero. Therefore, there are no restricted values.

Skill Practice Identify the restricted values.

7. $\dfrac{w-4}{w^2-9}$ 8. $\dfrac{8}{z^4+1}$

4. Simplifying Rational Expressions

Instructor Note: Remind students that we can reduce common factors, but not terms. We know that $\frac{33}{3} = 11$. If we rewrite as $\frac{30+3}{3}$ it is incorrect to write $\frac{30+\cancel{3}}{\cancel{3}}$.

In many cases, it is advantageous to simplify or reduce a fraction to lowest terms. The same is true for rational expressions.

The method for simplifying rational expressions mirrors the process for simplifying fractions. In each case, factor the numerator and denominator. Common factors in the numerator and denominator form a ratio of 1 and can be reduced.

Simplifying a fraction: $\dfrac{21}{35} \xrightarrow{\text{Factor}} \dfrac{3 \cdot 7}{5 \cdot 7} = \dfrac{3}{5} \cdot \dfrac{\overset{1}{\cancel{7}}}{\cancel{7}} = \dfrac{3}{5} \cdot (1) = \dfrac{3}{5}$

Simplifying a rational expression: $\dfrac{2x-6}{x^2-9} \xrightarrow{\text{Factor}} \dfrac{2(x-3)}{(x+3)(x-3)} = \dfrac{2}{(x+3)} \cdot \dfrac{\overset{1}{\cancel{(x-3)}}}{\cancel{(x-3)}} = \dfrac{2}{(x+3)} (1)$

$= \dfrac{2}{x+3}$

Informally, to simplify a rational expression, we simplify the ratio of common factors to 1. Formally, this is accomplished by applying the fundamental principle of rational expressions.

Answers

7. $w = 3, w = -3$
8. There are no restricted values.

Fundamental Principle of Rational Expressions

Let p, q, and r represent polynomials where $q \neq 0$ and $r \neq 0$. Then

$$\frac{pr}{qr} = \frac{p}{q} \cdot \frac{r}{r} = \frac{p}{q} \cdot 1 = \frac{p}{q}$$

TIP: In practice we often shorten the process to reduce a rational expression by dividing out common factors.

$$\frac{p \cdot \overset{1}{\cancel{r}}}{q \cdot \cancel{r}} = \frac{p}{q}$$

Example 4 Simplifying a Rational Expression

Given the expression $\dfrac{2p - 14}{p^2 - 49}$

a. Factor the numerator and denominator.

b. Identify the restricted values.

c. Simplify the rational expression.

Classroom Example: p. 491, Exercise 40

Solution:

a. $\dfrac{2p - 14}{p^2 - 49}$ Factor out the GCF in the numerator.

$= \dfrac{2(p - 7)}{(p + 7)(p - 7)}$ Factor the denominator as a difference of squares.

b. $(p + 7)(p - 7) = 0$ To find the restricted values, set the denominator equal to zero. The equation is quadratic.

$p + 7 = 0$ or $p - 7 = 0$ Set each factor equal to 0.

$p = -7$ or $p = 7$ The restricted values are -7 and 7.

c. $\dfrac{2(p - 7)}{(p + 7)(p - 7)}$ Simplify the ratio of common factors to 1.

$= \dfrac{2}{p + 7}$ (provided $p \neq 7$ and $p \neq -7$)

Avoiding Mistakes

The restricted values of a rational expression are always determined *before* simplifying the expression.

Skill Practice Given $\dfrac{5z + 25}{z^2 + 3z - 10}$

9. Factor the numerator and the denominator.

10. Identify the restricted values.   **11.** Simplify the rational expression.

In Example 4, it is important to note that the expressions

$$\frac{2p - 14}{p^2 - 49} \quad \text{and} \quad \frac{2}{p + 7}$$

are equal for all values of p that make each expression a real number. Therefore,

$$\frac{2p - 14}{p^2 - 49} = \frac{2}{p + 7}$$

for all values of p except $p = 7$ and $p = -7$. (At $p = 7$ and $p = -7$, the original expression is undefined.) This is why the restricted values are determined before the expression is simplified.

Answers

9. $\dfrac{5(z + 5)}{(z + 5)(z - 2)}$

10. $z = -5, z = 2$

11. $\dfrac{5}{z - 2}$ $(z \neq 2, z \neq -5)$

From this point forward, we will write statements of equality between two rational expressions with the assumption that they are equal for all values of the variable for which each expression is defined.

Classroom Example: p. 491, Exercise 44

Example 5 **Simplifying a Rational Expression**

Simplify the rational expression. $\dfrac{18a^4}{9a^5}$

TIP: The expression $\dfrac{18a^4}{9a^5}$ can also be simplified using the properties of exponents.

$\dfrac{18a^4}{9a^5} = 2a^{4-5} = 2a^{-1} = \dfrac{2}{a}$

Solution:

$\dfrac{18a^4}{9a^5}$

$= \dfrac{2 \cdot 3 \cdot 3 \cdot a \cdot a \cdot a \cdot a}{3 \cdot 3 \cdot a \cdot a \cdot a \cdot a \cdot a}$ Factor the numerator and denominator.

$= \dfrac{2 \cdot (3 \cdot 3 \cdot a \cdot a \cdot a \cdot a)}{(3 \cdot 3 \cdot a \cdot a \cdot a \cdot a) \cdot a}$ Simplify common factors.

$= \dfrac{2}{a}$

Skill Practice Simplify the rational expression.

12. $\dfrac{15q^3}{9q^2}$

Classroom Example: p. 491, Exercise 58

Example 6 **Simplifying a Rational Expression**

Simplify the rational expression. $\dfrac{2c - 8}{10c^2 - 80c + 160}$

Avoiding Mistakes

Given the expression

$\dfrac{2c - 8}{10c^2 - 80c + 160}$

do not be tempted to reduce before factoring. The terms $2c$ and $10c^2$ cannot be "canceled" because they are *terms* not factors.

The numerator and denominator must be in factored form before simplifying.

Solution:

$\dfrac{2c - 8}{10c^2 - 80c + 160}$

$= \dfrac{2(c - 4)}{10(c^2 - 8c + 16)}$ Factor out the GCF.

$= \dfrac{2(c - 4)}{10(c - 4)^2}$ Factor the denominator.

$= \dfrac{2(c - 4)}{2 \cdot 5(c - 4)(c - 4)}$ Simplify the ratio of common factors to 1.

$= \dfrac{1}{5(c - 4)}$

Skill Practice Simplify the rational expression.

13. $\dfrac{x^2 - 1}{2x^2 - x - 3}$

Answers

12. $\dfrac{5q}{3}$ **13.** $\dfrac{x - 1}{2x - 3}$

The process to simplify a rational expression is based on the identity property of multiplication. Therefore, this process applies only to factors (remember that factors are multiplied). For example:

$$\frac{3x}{3y} = \frac{3 \cdot x}{3 \cdot y} = \overset{1}{\cancel{\frac{3}{3}}} \cdot \frac{x}{y} = 1 \cdot \frac{x}{y} = \frac{x}{y}$$

$$\underset{\text{Simplify}}{\uparrow}$$

Terms that are added or subtracted cannot be reduced to lowest terms. For example:

$$\frac{x+3}{y+3}$$

$$\underset{\text{Cannot be simplified}}{\uparrow}$$

The objective of simplifying a rational expression is to create an equivalent expression that is simpler to use. Consider the rational expression from Example 6 in its original form and in its reduced form. If we substitute a value c into each expression, we see that the reduced form is easier to evaluate. For example, substitute $c = 3$:

	Original Expression	**Simplified Expression**
	$\dfrac{2c-8}{10c^2 - 80c + 160}$	$\dfrac{1}{5(c-4)}$
Substitute $c = 3$	$= \dfrac{2(3)-8}{10(3)^2 - 80(3) + 160}$	$= \dfrac{1}{5(3-4)}$
	$= \dfrac{6-8}{10(9) - 240 + 160}$	$= \dfrac{1}{5(-1)}$
	$= \dfrac{-2}{90 - 240 + 160}$	$= -\dfrac{1}{5}$
	$= \dfrac{-2}{10} \quad \text{or} \quad -\dfrac{1}{5}$	

5. Simplifying a Ratio of −1

When two factors are identical in the numerator and denominator, they form a ratio of 1 and can be reduced. Sometimes we encounter two factors that are *opposites* and form a ratio of −1. For example:

Simplified Form	**Details/Notes**
$\dfrac{-5}{5} = -1$	The ratio of a number and its opposite is −1.
$\dfrac{100}{-100} = -1$	The ratio of a number and its opposite is −1.
$\dfrac{x+7}{-x-7} = -1$	$\dfrac{x+7}{-x-7} = \dfrac{x+7}{-1(x+7)} = \dfrac{\overset{1}{\cancel{x+7}}}{-1\cancel{(x+7)}} = \dfrac{1}{-1} = -1$
	$\underset{\text{factor out } -1}{\curvearrowright}$
$\dfrac{2-x}{x-2} = -1$	$\dfrac{2-x}{x-2} = \dfrac{-1(-2+x)}{x-2} = \dfrac{-1\overset{1}{\cancel{(x-2)}}}{\cancel{x-2}} = \dfrac{-1}{1} = -1$

Recognizing factors that are opposites is useful when simplifying rational expressions.

> **Avoiding Mistakes**
>
> While the expression $2 - x$ and $x - 2$ are opposites, the expressions $2 - x$ and $2 + x$ are *not*.
> Therefore, $\dfrac{2-x}{2+x}$ does not simplify to −1.

Classroom Example: p. 491,
Exercise 80

Example 7   **Simplifying a Rational Expression**

Simplify the rational expression. $\dfrac{3c - 3d}{d - c}$

Solution:

$$\dfrac{3c - 3d}{d - c}$$

$$= \dfrac{3(c - d)}{d - c} \qquad \text{Factor the numerator and denominator.}$$

Notice that $(c - d)$ and $(d - c)$ are opposites and form a ratio of -1.

$$= \dfrac{3(c - d)^{-1}}{d - c} \qquad \underline{\text{Details:}} \; \dfrac{3(c - d)}{d - c} = \dfrac{3(-1)(-c + d)}{d - c} = \dfrac{-3(d - c)}{d - c)} = -3$$

$$= 3(-1)$$

$$= -3$$

Skill Practice   Simplify the rational expression.

14. $\dfrac{2t - 12}{6 - t}$

TIP: It is important to recognize that a rational expression can be written in several equivalent forms. In particular, two numbers with opposite signs form a negative quotient. Therefore, a number such as $-\frac{3}{4}$ can be written as:

$$-\dfrac{3}{4} \quad \text{or} \quad \dfrac{-3}{4} \quad \text{or} \quad \dfrac{3}{-4}$$

The negative sign can be written in the numerator, in the denominator, or out in front of the fraction. We demonstrate this concept in Example 8.

Classroom Example: p. 491,
Exercises 86

Example 8   **Simplifying a Rational Expression**

Simplify the rational expression. $\dfrac{5 - y}{y^2 - 25}$

Solution:

$$\dfrac{5 - y}{y^2 - 25}$$

$$= \dfrac{5 - y}{(y - 5)(y + 5)} \qquad \text{Factor the numerator and denominator.}$$

Notice that $5 - y$ and $y - 5$ are opposites and form a ratio of -1.

Answer

14. -2

$$= \frac{\overset{-1}{5-y}}{(y-5)(y+5)} \qquad \underline{\text{Details:}} \quad \frac{5-y}{(y-5)(y+5)} = \frac{-1(-5+y)}{(y-5)(y+5)}$$

$$= \frac{-1(y-5)}{(y-5)(y+5)} = \frac{-1}{y+5}$$

$$= \frac{-1}{y+5} \quad \text{or} \quad \frac{1}{-(y+5)} \quad \text{or} \quad -\frac{1}{y+5}$$

Skill Practice Simplify the rational expression.

15. $\dfrac{b-a}{a^2-b^2}$

Answer

15. $-\dfrac{1}{a+b}$

Section 7.1 Practice Exercises

Study Skills Exercise

For additional exercises, see Classroom Activities 7.1A–7.1B in the *Student's Resource Manual* at www.mhhe.com/moh.

> Write an example of how to simplify (reduce) a fraction, multiply two fractions, divide two fractions, add two fractions, and subtract two fractions. Then as you learn about rational expressions, compare the operations on rational expressions with those on fractions. This is a great place to use 3×5 cards again. Write an example of an operation with fractions on one side and the same operation with rational expressions on the other side.

Vocabulary and Key Concepts

1. **a.** A ___rational___ expression is the ratio of two polynomials, $\dfrac{p}{q}$, where $q \neq 0$.

   b. Restricted values of a rational expression are all values that make the ___denominator___ equal to ___zero___.

   c. For polynomials p, q, and r, where $(q \neq 0 \text{ and } r \neq 0)$, $\dfrac{pr}{qr} = $ ___$\dfrac{p}{q}$___.

   d. The ratio $\dfrac{a-b}{a-b} = $ ___1___ whereas the ratio $\dfrac{a-b}{b-a} = $ ___-1___ provided that $a \neq b$.

Concept 2: Evaluating Rational Expressions

For Exercises 2–10, substitute the given number into the expression and simplify (if possible). **(See Example 1.)**

2. $\dfrac{5}{y-4}$; $y=6$ $\dfrac{5}{2}$

3. $\dfrac{t-2}{t^2-4t+8}$; $t=2$ 0

4. $\dfrac{4x}{x-7}$; $x=8$ 32

5. $\dfrac{1}{x-6}$; $x=-2$ $-\dfrac{1}{8}$

6. $\dfrac{w-10}{w+6}$; $w=0$ $-\dfrac{5}{3}$

7. $\dfrac{w-4}{2w+8}$; $w=0$ $-\dfrac{1}{2}$

8. $\dfrac{y-8}{2y^2+y-1}$; $y=8$ 0

9. $\dfrac{(a-7)(a+1)}{(a-2)(a+5)}$; $a=2$ Undefined

10. $\dfrac{(a+4)(a+1)}{(a-4)(a-1)}$; $a=1$ Undefined

© Royalty Free/Corbis RF

11. A bicyclist rides 24 mi against a wind and returns 24 mi with the same wind. His average speed for the return trip traveling with the wind is 8 mph faster than his speed going out against the wind. If x represents the bicyclist's speed going out against the wind, then the total time, t, required for the round trip is given by

$$t = \frac{24}{x} + \frac{24}{x+8}$$ where t is measured in hours.

a. Find the time required for the round trip if the cyclist rides 12 mph against the wind. $3\frac{1}{5}$ hr or 3.2 hr

b. Find the time required for the round trip if the cyclist rides 24 mph against the wind. $1\frac{3}{4}$ hr or 1.75 hr

12. The manufacturer of mountain bikes has a fixed cost of $56,000, plus a variable cost of $140 per bike. The average cost per bike, y (in dollars), is given by the equation:

$$y = \frac{56,000 + 140x}{x}$$ where x represents the number of bikes produced.

a. Find the average cost per bike if the manufacturer produces 1000 bikes.
$196

b. Find the average cost per bike if the manufacturer produces 2000 bikes.
$168

c. Find the average cost per bike if the manufacturer produces 10,000 bikes.
$145.80

© Jeff Maloney/Getty Images RF

Concept 3: Restricted Values of a Rational Expression

For Exercises 13–24, identify the restricted values. **(See Examples 2–3.)**

13. $\dfrac{5}{k+2}$ $k = -2$

14. $\dfrac{-3}{h-4}$ $h = 4$

15. $\dfrac{x+5}{(2x-5)(x+8)}$ $x = \frac{5}{2}, x = -8$

16. $\dfrac{4y+1}{(3y+7)(y+3)}$ $y = -\frac{7}{3}, y = -3$

17. $\dfrac{m+12}{m^2+5m+6}$ $m = -2, m = -3$

18. $\dfrac{c-11}{c^2-5c-6}$ $c = 6, c = -1$

19. $\dfrac{x-4}{x^2+9}$
There are no restricted values.

20. $\dfrac{x+1}{x^2+4}$
There are no restricted values.

21. $\dfrac{y^2-y-12}{12}$
There are no restricted values.

22. $\dfrac{z^2+10z+9}{9}$
There are no restricted values.

23. $\dfrac{t-5}{t}$ $t = 0$

24. $\dfrac{2w+7}{w}$ $w = 0$

25. Construct a rational expression that is undefined for $x = 2$. (Answers will vary.)   For example: $\dfrac{1}{x-2}$

26. Construct a rational expression that is undefined for $x = 5$. (Answers will vary.)   For example: $\dfrac{1}{x-5}$

27. Construct a rational expression that is undefined for $x = -3$ and $x = 7$. (Answers will vary.)   For example: $\dfrac{1}{(x+3)(x-7)}$

28. Construct a rational expression that is undefined for $x = -1$ and $x = 4$. (Answers will vary.)   For example: $\dfrac{1}{(x+1)(x-4)}$

29. Evaluate the expressions for $x = -1$.

a. $\dfrac{3x^2-2x-1}{6x^2-7x-3}$ $\frac{2}{5}$

b. $\dfrac{x-1}{2x-3}$ $\frac{2}{5}$

30. Evaluate the expressions for $x = 4$.

a. $\dfrac{(x+5)^2}{x^2+6x+5}$ $\frac{9}{5}$

b. $\dfrac{x+5}{x+1}$ $\frac{9}{5}$

31. Evaluate the expressions for $x = 1$.

a. $\dfrac{5x+5}{x^2-1}$
Undefined

b. $\dfrac{5}{x-1}$
Undefined

32. Evaluate the expressions for $x = -3$.

a. $\dfrac{2x^2-4x-6}{2x^2-18}$
Undefined

b. $\dfrac{x+1}{x+3}$
Undefined

Concept 4: Simplifying Rational Expressions

For Exercises 33–42,

    a. Identify the restricted values.   **b.** Simplify the rational expression. **(See Example 4.)**

33. $\dfrac{3y+6}{6y+12}$ a. $y=-2$ b. $\frac{1}{2}$

34. $\dfrac{8x-8}{4x-4}$ a. $x=1$ b. 2

35. $\dfrac{t^2-1}{t+1}$ a. $t=-1$ b. $t-1$

36. $\dfrac{r^2-4}{r-2}$ a. $r=2$ b. $r+2$

37. $\dfrac{7w}{21w^2-35w}$ a. $w=0,\ w=\frac{5}{3}$ b. $\frac{1}{3w-5}$

38. $\dfrac{12a^2}{24a^2-18a}$ a. $a=0,\ a=\frac{3}{4}$ b. $\frac{2a}{4a-3}$

39. $\dfrac{9x^2-4}{6x+4}$ a. $x=-\frac{2}{3}$ b. $\frac{3x-2}{2}$

40. $\dfrac{8n-20}{4n^2-25}$ a. $n=\frac{5}{2},\ n=-\frac{5}{2}$ b. $\frac{4}{2n+5}$

41. $\dfrac{a^2+3a-10}{a^2+a-6}$ a. $a=-3,\ a=2$ b. $\frac{a+5}{a+3}$

42. $\dfrac{t^2+3t-10}{t^2+t-20}$ a. $t=-5,\ t=4$ b. $\frac{t-2}{t-4}$

For Exercises 43–72, simplify the rational expression. **(See Examples 5–6.)**

43. $\dfrac{7b^2}{21b}$ $\frac{b}{3}$

44. $\dfrac{15c^3}{3c^5}$ $\frac{5}{c^2}$

45. $\dfrac{-24x^2y^5z}{8xy^4z^3}$ $-\frac{3xy}{z^2}$

46. $\dfrac{60rs^4t^2}{-12r^4s^2t^3}$ $-\frac{5s^2}{r^3t}$

47. $\dfrac{(p-3)(p+5)}{(p+5)(p+4)}$ $\frac{p-3}{p+4}$

48. $\dfrac{(c+4)(c-1)}{(c+4)(c+2)}$ $\frac{c-1}{c+2}$

49. $\dfrac{m+11}{4(m+11)(m-11)}$ $\frac{1}{4(m-11)}$

50. $\dfrac{n-7}{9(n+2)(n-7)}$ $\frac{1}{9(n+2)}$

51. $\dfrac{x(2x+1)^2}{4x^3(2x+1)}$ $\frac{2x+1}{4x^2}$

52. $\dfrac{(p+2)(p-3)^4}{(p+2)^2(p-3)^2}$ $\frac{(p-3)^2}{p+2}$

53. $\dfrac{5}{20a-25}$ $\frac{1}{4a-5}$

54. $\dfrac{7}{14c-21}$ $\frac{1}{2c-3}$

 55. $\dfrac{3x^2-6x}{9xy+18x}$ $\frac{x-2}{3(y+2)}$

56. $\dfrac{6p^2+12p}{2pq-4p}$ $\frac{3(p+2)}{q-2}$

57. $\dfrac{2x+4}{x^2-3x-10}$ $\frac{2}{x-5}$

58. $\dfrac{5z+15}{z^2-4z-21}$ $\frac{5}{z-7}$

59. $\dfrac{a^2-49}{a-7}$ $a+7$

60. $\dfrac{b^2-64}{b-8}$ $b+8$

61. $\dfrac{q^2+25}{q+5}$ Cannot simplify

62. $\dfrac{r^2+36}{r+6}$ Cannot simplify

63. $\dfrac{y^2+6y+9}{2y^2+y-15}$ $\frac{y+3}{2y-5}$

64. $\dfrac{h^2+h-6}{h^2+2h-8}$ $\frac{h+3}{h+4}$

65. $\dfrac{5q^2+5}{q^4-1}$ $\frac{5}{(q+1)(q-1)}$

66. $\dfrac{4t^2+16}{t^4-16}$ $\frac{4}{(t-2)(t+2)}$

67. $\dfrac{ac-ad+2bc-2bd}{2ac+ad+4bc+2bd}$ (*Hint*: Factor by grouping.) $\frac{c-d}{2c+d}$

68. $\dfrac{3pr-ps-3qr+qs}{3pr-ps+3qr-qs}$ (*Hint*: Factor by grouping.) $\frac{p-q}{p+q}$

69. $\dfrac{5x^3+4x^2-45x-36}{x^2-9}$ $5x+4$

 70. $\dfrac{x^2-1}{ax^3-bx^2-ax+b}$ $\frac{1}{ax-b}$

71. $\dfrac{2x^2-xy-3y^2}{2x^2-11xy+12y^2}$ $\frac{x+y}{x-4y}$

72. $\dfrac{2c^2+cd-d^2}{5c^2+3cd-2d^2}$ $\frac{2c-d}{5c-2d}$

Concept 5: Simplifying a Ratio of −1

73. What is the relationship between $x-2$ and $2-x$? They are opposites.

74. What is the relationship between $w+p$ and $-w-p$? They are opposites.

For Exercises 75–86, simplify the rational expressions. **(See Examples 7–8.)**

75. $\dfrac{x-5}{5-x}$ -1

76. $\dfrac{8-p}{p-8}$ -1

77. $\dfrac{-4-y}{4+y}$ -1

78. $\dfrac{z+10}{-z-10}$ -1

79. $\dfrac{3y-6}{12-6y}$ $-\frac{1}{2}$

80. $\dfrac{4q-4}{12-12q}$ $-\frac{1}{3}$

81. $\dfrac{k+5}{5-k}$ Cannot simplify

82. $\dfrac{2+n}{2-n}$ Cannot simplify

83. $\dfrac{10x-12}{10x+12}$ $\frac{5x-6}{5x+6}$

84. $\dfrac{4t-16}{16+4t}$ $\frac{t-4}{4+t}$

 85. $\dfrac{x^2-x-12}{16-x^2}$ $-\frac{x+3}{4+x}$

86. $\dfrac{49-b^2}{b^2-10b+21}$ $-\frac{7+b}{b-3}$

Mixed Exercises

For Exercises 87–100, simplify the rational expressions.

87. $\dfrac{3x^2 + 7x - 6}{x^2 + 7x + 12}$ $\dfrac{3x-2}{x+4}$

88. $\dfrac{y^2 - 5y - 14}{2y^2 - y - 10}$ $\dfrac{y-7}{2y-5}$

89. $\dfrac{3(m-2)}{6(2-m)}$ $-\dfrac{1}{2}$

90. $\dfrac{8(1-x)}{4(x-1)}$ -2

91. $\dfrac{w^2 - 4}{8 - 4w}$ $-\dfrac{w+2}{4}$

92. $\dfrac{15 - 3x}{x^2 - 25}$ $-\dfrac{3}{x+5}$

93. $\dfrac{18st^5}{12st^3}$ $\dfrac{3t^2}{2}$

94. $\dfrac{20a^4b^2}{25ab^2}$ $\dfrac{4a^3}{5}$

95. $\dfrac{4r^2 - 4rs + s^2}{s^2 - 4r^2}$ $-\dfrac{2r-s}{s+2r}$

96. $\dfrac{y^2 - 9z^2}{3z^2 + 2yz - y^2}$ $-\dfrac{y+3z}{z+y}$

97. $\dfrac{3y - 3x}{2x^2 - 4xy + 2y^2}$ $-\dfrac{3}{2(x-y)}$

98. $\dfrac{49p^2 - 28pq + 4q^2}{4q - 14p}$ $-\dfrac{7p-2q}{2}$

99. $\dfrac{2t^2 - 3t}{2t^4 - 13t^3 + 15t^2}$ $\dfrac{1}{t(t-5)}$

100. $\dfrac{4m^3 + 3m^2}{4m^3 + 7m^2 + 3m}$ $\dfrac{m}{m+1}$

Expanding Your Skills

For Exercises 101–104, factor and simplify.

101. $\dfrac{w^3 - 8}{w^2 + 2w + 4}$ $w - 2$

102. $\dfrac{y^3 + 27}{y^2 - 3y + 9}$ $y + 3$

103. $\dfrac{z^2 - 16}{z^3 - 64}$ $\dfrac{z+4}{z^2 + 4z + 16}$

104. $\dfrac{x^2 - 25}{x^3 + 125}$ $\dfrac{x-5}{x^2 - 5x + 25}$

Section 7.2	Multiplication and Division of Rational Expressions

Concepts

1. Multiplication of Rational Expressions
2. Division of Rational Expressions

1. Multiplication of Rational Expressions

Recall that to multiply fractions, we multiply the numerators and multiply the denominators. The same is true for multiplying rational expressions.

> **Multiplication of Rational Expressions**
>
> Let p, q, r, and s represent polynomials, such that $q \neq 0$, $s \neq 0$. Then,
>
> $$\frac{p}{q} \cdot \frac{r}{s} = \frac{pr}{qs}$$

For example:

Multiply the Fractions	**Multiply the Rational Expressions**
$\dfrac{2}{3} \cdot \dfrac{5}{7} = \dfrac{10}{21}$	$\dfrac{2x}{3y} \cdot \dfrac{5z}{7} = \dfrac{10xz}{21y}$

Sometimes it is possible to simplify a ratio of common factors to 1 *before* multiplying. To do so, we must first factor the numerators and denominators of each fraction.

$$\frac{15}{14} \cdot \frac{21}{10} = \frac{3 \cdot \overset{1}{\cancel{5}}}{2 \cdot \cancel{7}} \cdot \frac{3 \cdot \overset{1}{\cancel{7}}}{2 \cdot \cancel{5}} = \frac{9}{4}$$

The same process is also used to multiply rational expressions.

Multiplying Rational Expressions

Step 1   Factor the numerators and denominators of all rational expressions.

Step 2   Simplify the ratios of common factors to 1 and opposite factors to -1.

Step 3   Multiply the remaining factors in the numerator, and multiply the remaining factors in the denominator.

Example 1   **Multiplying Rational Expressions**

Classroom Example: p. 497, Exercise 10

Multiply. $\dfrac{5a^2b}{2} \cdot \dfrac{6a}{10b}$

Solution:

$\dfrac{5a^2b}{2} \cdot \dfrac{6a}{10b}$

$= \dfrac{5 \cdot a \cdot a \cdot b}{2} \cdot \dfrac{2 \cdot 3 \cdot a}{2 \cdot 5 \cdot b}$ Factor into prime factors.

$= \dfrac{\overset{1}{\cancel{5}} \cdot a \cdot a \cdot \overset{1}{\cancel{b}}}{2} \cdot \dfrac{\overset{1}{\cancel{2}} \cdot 3 \cdot a}{\cancel{2} \cdot \cancel{5} \cdot \cancel{b}}$ Simplify.

$= \dfrac{3a^3}{2}$ Multiply remaining factors.

Skill Practice   Multiply.

1. $\dfrac{7a}{3b} \cdot \dfrac{15b}{14a^2}$

Example 2   **Multiplying Rational Expressions**

Classroom Example: p. 497, Exercise 16

Multiply. $\dfrac{3c - 3d}{6c} \cdot \dfrac{2}{c^2 - d^2}$

Solution:

$\dfrac{3c - 3d}{6c} \cdot \dfrac{2}{c^2 - d^2}$

$= \dfrac{3(c - d)}{2 \cdot 3 \cdot c} \cdot \dfrac{2}{(c - d)(c + d)}$ Factor.

$= \dfrac{\overset{1}{\cancel{3}}(\overset{1}{\cancel{c - d}})}{\cancel{2} \cdot \cancel{3} \cdot c} \cdot \dfrac{\overset{1}{\cancel{2}}}{(\cancel{c - d})(c + d)}$ Simplify.

$= \dfrac{1}{c(c + d)}$ Multiply remaining factors.

Avoiding Mistakes

If all the factors in the numerator reduce to a ratio of 1, a factor of 1 is left in the numerator.

Skill Practice   Multiply.

2. $\dfrac{4x - 8}{x + 6} \cdot \dfrac{x^2 + 6x}{2x}$

Answers

1. $\dfrac{5}{2a}$ **2.** $2(x - 2)$

Classroom Example: p. 497, Exercise 20

Example 3   **Multiplying Rational Expressions**

Multiply. $\dfrac{35 - 5x}{5x + 5} \cdot \dfrac{x^2 + 5x + 4}{x^2 - 49}$

Solution:

$$\dfrac{35 - 5x}{5x + 5} \cdot \dfrac{x^2 + 5x + 4}{x^2 - 49}$$

$$= \dfrac{5(7 - x)}{5(x + 1)} \cdot \dfrac{(x + 4)(x + 1)}{(x - 7)(x + 7)}$$ Factor the numerators and denominators completely.

$$= \dfrac{\overset{1}{\cancel{5}}(\overset{-1}{\cancel{7 - x}})}{\cancel{5}(x + 1)} \cdot \dfrac{(x + 4)(\overset{1}{\cancel{x + 1}})}{(\cancel{x - 7})(x + 7)}$$ Simplify the ratios of common factors to 1 or −1.

> **TIP:** The ratio $\dfrac{7 - x}{x - 7} = -1$ because $7 - x$ and $x - 7$ are opposites.

$$= \dfrac{-1(x + 4)}{x + 7}$$ Multiply remaining factors.

$$= \dfrac{-(x + 4)}{x + 7} \quad \text{or} \quad \dfrac{x + 4}{-(x + 7)} \quad \text{or} \quad -\dfrac{x + 4}{x + 7}$$

Skill Practice   Multiply.

3. $\dfrac{p^2 + 4p + 3}{5p + 10} \cdot \dfrac{p^2 - p - 6}{9 - p^2}$

2. Division of Rational Expressions

Recall that to divide two fractions, multiply the first fraction by the reciprocal of the second.

$$\dfrac{21}{10} \div \dfrac{49}{15} \xrightarrow[\text{of the second fraction}]{\text{multiply by the reciprocal}} \dfrac{21}{10} \cdot \dfrac{15}{49} \xrightarrow{\text{factor}} \dfrac{3 \cdot \overset{1}{\cancel{7}}}{2 \cdot \cancel{5}} \cdot \dfrac{3 \cdot \overset{1}{\cancel{5}}}{\cancel{7} \cdot 7} = \dfrac{9}{14}$$

The same process is used to divide rational expressions.

> **Division of Rational Expressions**
>
> Let p, q, r, and s represent polynomials, such that $q \neq 0$, $r \neq 0$, $s \neq 0$. Then,
>
> $$\dfrac{p}{q} \div \dfrac{r}{s} = \dfrac{p}{q} \cdot \dfrac{s}{r} = \dfrac{ps}{qr}$$

Classroom Example: p. 498, Exercise 32

Example 4   **Dividing Rational Expressions**

Divide. $\dfrac{5t - 15}{2} \div \dfrac{t^2 - 9}{10}$

> **Avoiding Mistakes**
>
> When dividing rational expressions, take the reciprocal of the second fraction and change to multiplication *before* reducing like factors.

Solution:

$$\dfrac{5t - 15}{2} \div \dfrac{t^2 - 9}{10}$$

$$= \dfrac{5t - 15}{2} \cdot \dfrac{10}{t^2 - 9}$$ Multiply the first fraction by the reciprocal of the second.

Answer

3. $\dfrac{-(p + 1)}{5}$ or $\dfrac{p + 1}{-5}$ or $-\dfrac{p + 1}{5}$

$$= \frac{5(t-3)}{2} \cdot \frac{2 \cdot 5}{(t-3)(t+3)} \qquad \text{Factor each polynomial.}$$

$$= \frac{5(\overset{1}{\cancel{t-3}})}{\cancel{2}} \cdot \frac{\overset{1}{\cancel{2}} \cdot 5}{(\cancel{t-3})(t+3)} \qquad \text{Simplify the ratio of common factors to 1.}$$

$$= \frac{25}{t+3}$$

Skill Practice Divide.

4. $\dfrac{7y-14}{y+1} \div \dfrac{y^2+2y-8}{2y+2}$

| **Example 5** | **Dividing Rational Expressions** |

Classroom Example: p. 498, Exercise 38

Divide. $\quad \dfrac{p^2-11p+30}{10p^2-250} \div \dfrac{30p-5p^2}{2p+4}$

Solution:

$$\frac{p^2-11p+30}{10p^2-250} \div \frac{30p-5p^2}{2p+4}$$

$$= \frac{p^2-11p+30}{10p^2-250} \cdot \frac{2p+4}{30p-5p^2} \qquad \begin{array}{l}\text{Multiply the first fraction by the}\\\text{reciprocal of the second.}\end{array}$$

Factor the trinomial.
$p^2-11p+30 = (p-5)(p-6)$

$$= \frac{(p-5)(p-6)}{2 \cdot 5(p-5)(p+5)} \cdot \frac{2(p+2)}{5p(6-p)} \qquad \begin{array}{l}\text{Factor out the GCF.}\\ 2p+4 = 2(p+2)\end{array}$$

Factor out the GCF. Then factor the difference of squares.
$10p^2-250 = 10(p^2-25)$
$\qquad\qquad\quad = 2 \cdot 5(p-5)(p+5)$

Factor out the GCF.
$30p-5p^2 = 5p(6-p)$

$$= \frac{\overset{1}{(\cancel{p-5})}\overset{-1}{(\cancel{p-6})}}{\cancel{2} \cdot 5(\cancel{p-5})(p+5)} \cdot \frac{\overset{1}{\cancel{2}}(p+2)}{5p(\cancel{6-p})} \qquad \begin{array}{l}\text{Simplify the ratio of common}\\\text{factors to 1 or }-1.\end{array}$$

$$= -\frac{(p+2)}{25p(p+5)}$$

Skill Practice Divide.

5. $\dfrac{4x^2-9}{2x^2-x-3} \div \dfrac{20x+30}{x^2+7x+6}$

Answers

4. $\dfrac{14}{y+4}$ 5. $\dfrac{x+6}{10}$

Classroom Example: p. 498,
Exercise 28

Example 6   **Dividing Rational Expressions**

Divide. $\dfrac{\dfrac{3x}{4y}}{\dfrac{5x}{6y}}$

Solution:

$$\dfrac{\dfrac{3x}{4y}}{\dfrac{5x}{6y}} \longleftarrow \text{This fraction bar denotes division } (\div).$$

$$= \dfrac{3x}{4y} \div \dfrac{5x}{6y}$$

$$= \dfrac{3x}{4y} \cdot \dfrac{6y}{5x} \qquad \text{Multiply by the reciprocal of the second fraction.}$$

$$= \dfrac{3 \cdot \overset{1}{\cancel{x}}}{\cancel{2} \cdot 2 \cdot \cancel{y}} \cdot \dfrac{\overset{1}{\cancel{2}} \cdot 3 \cdot \overset{1}{\cancel{y}}}{5 \cdot \cancel{x}} \qquad \text{Simplify the ratio of common factors to 1.}$$

$$= \dfrac{9}{10}$$

Skill Practice Divide.

6. $\dfrac{\dfrac{a^3b}{9c}}{\dfrac{4ab}{3c^3}}$

Sometimes multiplication and division of rational expressions appear in the same problem. In such a case, apply the order of operations by multiplying or dividing in order from left to right.

Classroom Example: p. 499,
Exercise 66

Example 7   **Multiplying and Dividing Rational Expressions**

Perform the indicated operations. $\dfrac{4}{c^2 - 9} \div \dfrac{6}{c - 3} \cdot \dfrac{3c}{8}$

Solution:

In this example, division occurs first, before multiplication. Parentheses may be inserted to reinforce the proper order.

$$\left(\dfrac{4}{c^2 - 9} \div \dfrac{6}{c - 3} \right) \cdot \dfrac{3c}{8}$$

$$= \left(\dfrac{4}{c^2 - 9} \cdot \dfrac{c - 3}{6} \right) \cdot \dfrac{3c}{8} \qquad \text{Multiply the first fraction by the reciprocal of the second.}$$

Answer

6. $\dfrac{a^2c^2}{12}$

$$= \left(\frac{2 \cdot 2}{(c-3)(c+3)} \cdot \frac{c-3}{2 \cdot 3} \right) \cdot \frac{3 \cdot c}{2 \cdot 2 \cdot 2}$$

Now that each operation is written as multiplication, factor the polynomials and reduce the common factors.

$$= \frac{\overset{1}{\cancel{2}} \cdot \overset{1}{\cancel{2}}}{(\cancel{c-3})(c+3)} \cdot \frac{(\cancel{c-3})}{2 \cdot \cancel{3}} \cdot \frac{\cancel{3} \cdot c}{\cancel{2} \cdot 2 \cdot 2}$$

$$= \frac{c}{4(c+3)}$$

Simplify.

Skill Practice Perform the indicated operations.

7. $\dfrac{v}{v+2} \div \dfrac{5v^2}{v^2-4} \cdot \dfrac{v}{10}$

Answer

7. $\dfrac{v-2}{50}$

Section 7.2   Practice Exercises

For additional exercises, see Classroom Activities 7.2A–7.2B in the *Student's Resource Manual* at www.mhhe.com/moh.

Review Exercises

For Exercises 1–8, multiply or divide the fractions.

1. $\dfrac{3}{5} \cdot \dfrac{1}{2}$ $\dfrac{3}{10}$

2. $\dfrac{6}{7} \cdot \dfrac{5}{12}$ $\dfrac{5}{14}$

3. $\dfrac{3}{4} \div \dfrac{3}{8}$ 2

4. $\dfrac{18}{5} \div \dfrac{2}{5}$ 9

5. $6 \cdot \dfrac{5}{12}$ $\dfrac{5}{2}$

6. $\dfrac{7}{25} \cdot 5$ $\dfrac{7}{5}$

7. $\dfrac{\frac{21}{4}}{\frac{7}{5}}$ $\dfrac{15}{4}$

8. $\dfrac{\frac{9}{2}}{\frac{3}{4}}$ 6

Concept 1: Multiplication of Rational Expressions

For Exercises 9–24, multiply. **(See Examples 1–3.)**

9. $\dfrac{2xy}{5x^2} \cdot \dfrac{15}{4y}$ $\dfrac{3}{2x}$

10. $\dfrac{7s}{t^2} \cdot \dfrac{t^2}{14s^2}$ $\dfrac{1}{2s}$

11. $\dfrac{6x^3}{9x^6y^2} \cdot \dfrac{18x^4y^7}{4y}$ $3xy^4$

12. $\dfrac{10a^2b}{15b^2} \cdot \dfrac{30b}{2a^3}$ $\dfrac{10}{a}$

13. $\dfrac{4x-24}{20x} \cdot \dfrac{5x}{8}$ $\dfrac{x-6}{8}$

14. $\dfrac{5a+20}{a} \cdot \dfrac{3a}{10}$ $\dfrac{3(a+4)}{2}$ **15.** $\dfrac{3y+18}{y^2} \cdot \dfrac{4y}{6y+36}$ $\dfrac{2}{y}$

16. $\dfrac{2p-4}{6p} \cdot \dfrac{4p^2}{8p-16}$ $\dfrac{p}{6}$

17. $\dfrac{10}{2-a} \cdot \dfrac{a-2}{16}$ $-\dfrac{5}{8}$

18. $\dfrac{w-3}{6} \cdot \dfrac{20}{3-w}$ $-\dfrac{10}{3}$

19. $\dfrac{b^2-a^2}{a-b} \cdot \dfrac{a}{a^2-ab}$ $\dfrac{b+a}{a-b}$

20. $\dfrac{(x-y)^2}{x^2+xy} \cdot \dfrac{x}{y-x}$ $\dfrac{x-y}{x+y}$

21. $\dfrac{y^2+2y+1}{5y-10} \cdot \dfrac{y^2-3y+2}{y^2-1}$ $\dfrac{y+1}{5}$

22. $\dfrac{6a^2-6}{a^2+6a+5} \cdot \dfrac{a^2+5a}{12a}$ $\dfrac{a-1}{2}$

23. $\dfrac{10x}{2x^2+3x+1} \cdot \dfrac{x^2+7x+6}{5x}$ $\dfrac{2(x+6)}{2x+1}$

24. $\dfrac{p-3}{p^2+p-12} \cdot \dfrac{4p+16}{p+1}$ $\dfrac{4}{p+1}$

Concept 2: Division of Rational Expressions

For Exercises 25–38, divide. **(See Examples 4–6.)**

25. $\dfrac{4x}{7y} \div \dfrac{2x^2}{21xy}$ 6

26. $\dfrac{6cd}{5d^2} \div \dfrac{8c^3}{10d}$ $\dfrac{3}{2c^2}$

27. $\dfrac{\dfrac{8m^4n^5}{5n^6}}{\dfrac{24mn}{15m^3}}$ $\dfrac{m^6}{n^2}$

28. $\dfrac{\dfrac{10a^3b}{3a}}{\dfrac{5b}{9ab}}$ $6a^3b$

29. $\dfrac{4a+12}{6a-18} \div \dfrac{3a+9}{5a-15}$ $\dfrac{10}{9}$

30. $\dfrac{8m-16}{3m+3} \div \dfrac{5m-10}{2m+2}$ $\dfrac{16}{15}$

31. $\dfrac{3x-21}{6x^2-42x} \div \dfrac{7}{12x}$ $\dfrac{6}{7}$

32. $\dfrac{4a^2-4a}{9a-9} \div \dfrac{5}{12a}$ $\dfrac{16a^2}{15}$

 33. $\dfrac{m^2-n^2}{9} \div \dfrac{3n-3m}{27m}$ $-m(m+n)$

34. $\dfrac{9-t^2}{15t+15} \div \dfrac{t-3}{5t}$ $-\dfrac{t(3+t)}{3(t+1)}$

35. $\dfrac{3p+4q}{p^2+4pq+4q^2} \div \dfrac{4}{p+2q}$ $\dfrac{3p+4q}{4(p+2q)}$

36. $\dfrac{x^2+2xy+y^2}{2x-y} \div \dfrac{x+y}{5}$ $\dfrac{5(x+y)}{2x-y}$

37. $\dfrac{p^2-2p-3}{p^2-p-6} \div \dfrac{p^2-1}{p^2+2p}$ $\dfrac{p}{p-1}$

38. $\dfrac{4t^2-1}{t^2-5t} \div \dfrac{2t^2+5t+2}{t^2-3t-10}$ $\dfrac{2t-1}{t}$

Mixed Exercises

For Exercises 39–64, multiply or divide as indicated.

39. $(w+3) \cdot \dfrac{w}{2w^2+5w-3}$ $\dfrac{w}{2w-1}$

40. $\dfrac{5t+1}{5t^2-31t+6} \cdot (t-6)$ $\dfrac{5t+1}{5t-1}$

41. $(r-5) \cdot \dfrac{4r}{2r^2-7r-15}$ $\dfrac{4r}{2r+3}$

42. $\dfrac{q+1}{5q^2-28q-12} \cdot (5q+2)$ $\dfrac{q+1}{q-6}$

43. $\dfrac{\dfrac{5t-10}{12}}{\dfrac{4t-8}{8}}$ $\dfrac{5}{6}$

44. $\dfrac{\dfrac{6m+6}{5}}{\dfrac{3m+3}{10}}$ 4

 45. $\dfrac{2a^2+13a-24}{8a-12} \div (a+8)$ $\dfrac{1}{4}$

46. $\dfrac{3y^2+20y-7}{5y+35} \div (3y-1)$ $\dfrac{1}{5}$

47. $\dfrac{y^2+5y-36}{y^2-2y-8} \cdot \dfrac{y+2}{y-6}$ $\dfrac{y+9}{y-6}$

48. $\dfrac{z^2-11z+28}{z-1} \cdot \dfrac{z+1}{z^2-6z-7}$ $\dfrac{z-4}{z-1}$

49. $\dfrac{2t^2+t-1}{t^2+3t+2} \cdot \dfrac{t+2}{2t-1}$ 1

50. $\dfrac{3p^2-2p-8}{3p^2-5p-12} \cdot \dfrac{p-3}{p-2}$ 1

51. $(5t-1) \div \dfrac{5t^2+9t-2}{3t+8}$ $\dfrac{3t+8}{t+2}$

52. $(2q-3) \div \dfrac{2q^2+5q-12}{q-7}$ $\dfrac{q-7}{q+4}$

53. $\dfrac{x^2+2x-3}{x^2-3x+2} \cdot \dfrac{x^2+2x-8}{x^2+4x+3}$ $\dfrac{x+4}{x+1}$

54. $\dfrac{y^2+y-12}{y^2-y-20} \cdot \dfrac{y^2+y-30}{y^2-2y-3}$ $\dfrac{y+6}{y+1}$

 55. $\dfrac{\dfrac{w^2-6w+9}{8}}{\dfrac{9-w^2}{4w+12}}$ $-\dfrac{w-3}{2}$

56. $\dfrac{\dfrac{p^2-6p+8}{24}}{\dfrac{16-p^2}{6p+6}}$ $-\dfrac{(p-2)(p+1)}{4(4+p)}$

57. $\dfrac{5k^2+7k+2}{k^2+5k+4} \div \dfrac{5k^2+17k+6}{k^2+10k+24}$ $\dfrac{k+6}{k+3}$

58. $\dfrac{4h^2-5h+1}{h^2+h-2} \div \dfrac{6h^2-7h+2}{2h^2+3h-2}$ $\dfrac{4h-1}{3h-2}$

 59. $\dfrac{ax+a+bx+b}{2x^2+4x+2} \cdot \dfrac{4x+4}{a^2+ab}$ $\dfrac{2}{a}$

60. $\dfrac{3my+9m+ny+3n}{9m^2+6mn+n^2} \cdot \dfrac{30m+10n}{5y^2+15y}$ $\dfrac{2}{y}$

61. $\dfrac{y^4 - 1}{2y^2 - 3y + 1} \div \dfrac{2y^2 + 2}{8y^2 - 4y}$ $\quad 2y(y+1)$

62. $\dfrac{x^4 - 16}{6x^2 + 24} \div \dfrac{x^2 - 2x}{3x}$ $\quad \dfrac{x+2}{2}$

63. $\dfrac{x^2 - xy - 2y^2}{x + 2y} \div \dfrac{x^2 - 4xy + 4y^2}{x^2 - 4y^2}$ $\quad x + y$

64. $\dfrac{4m^2 - 4mn - 3n^2}{8m^2 - 18n^2} \div \dfrac{3m + 3n}{6m^2 + 15mn + 9n^2}$ $\quad \dfrac{2m+n}{2}$

For Exercises 65–70, multiply or divide as indicated. **(See Example 7.)**

65. $\dfrac{y^3 - 3y^2 + 4y - 12}{y^4 - 16} \cdot \dfrac{3y^2 + 5y - 2}{3y^2 - 10y + 3} \div \dfrac{3}{6y - 12}$ $\quad 2$

66. $\dfrac{x^2 - 25}{3x^2 + 3xy} \cdot \dfrac{x^2 + 4x + xy + 4y}{x^2 + 9x + 20} \div \dfrac{x - 5}{x}$ $\quad \dfrac{1}{3}$

67. $\dfrac{a^2 - 5a}{a^2 + 7a + 12} \div \dfrac{a^3 - 7a^2 + 10a}{a^2 + 9a + 18} \div \dfrac{a + 6}{a + 4}$ $\quad \dfrac{1}{a-2}$

68. $\dfrac{t^2 + t - 2}{t^2 + 5t + 6} \div \dfrac{t - 1}{t} \div \dfrac{5t - 5}{t + 3}$ $\quad \dfrac{t}{5(t-1)}$

69. $\dfrac{p^3 - q^3}{p - q} \cdot \dfrac{p + q}{2p^2 + 2pq + 2q^2}$ $\quad \dfrac{p+q}{2}$

70. $\dfrac{r^3 + s^3}{r - s} \div \dfrac{r^2 + 2rs + s^2}{r^2 - s^2}$ $\quad r^2 - rs + s^2$

Least Common Denominator

1. Least Common Denominator

We have already learned how to simplify, multiply, and divide rational expressions. Our next goal is to add and subtract rational expressions. As with fractions, rational expressions may be added or subtracted only if they have the same denominator.

The **least common denominator (LCD)** of two or more rational expressions is defined as the least common multiple of the denominators. For example, consider the fractions $\frac{1}{20}$ and $\frac{1}{8}$. By inspection, you can probably see that the least common denominator is 40. To understand why, find the prime factorization of both denominators:

$$20 = 2^2 \cdot 5 \quad \text{and} \quad 8 = 2^3$$

A common multiple of 20 and 8 must be a multiple of 5, a multiple of 2^2, and a multiple of 2^3. However, any number that is a multiple of $2^3 = 8$ is automatically a multiple of $2^2 = 4$. Therefore, it is sufficient to construct the least common denominator as the product of unique prime factors, in which each factor is raised to its highest power.

$$\text{The LCD of } \frac{1}{20} \text{ and } \frac{1}{8} \text{ is } 2^3 \cdot 5 = 40.$$

> **Finding the Least Common Denominator of Two or More Rational Expressions**
>
> **Step 1** Factor all denominators completely.
>
> **Step 2** The LCD is the product of unique prime factors from the denominators, in which each factor is raised to the highest power to which it appears in any denominator.

Concepts

1. **Least Common Denominator**
2. **Writing Rational Expressions with the Least Common Denominator**

Classroom Example: p. 504, Exercise 20

| Example 1 | **Finding the Least Common Denominator** |

Find the LCD of the rational expressions.

a. $\dfrac{5}{14}; \dfrac{3}{49}; \dfrac{1}{8}$ **b.** $\dfrac{5}{3x^2z}; \dfrac{7}{x^5y^3}$

Solution:

a. $\dfrac{5}{14}; \dfrac{3}{49}; \dfrac{1}{8}$

$= \dfrac{5}{2 \cdot 7}; \dfrac{3}{7^2}; \dfrac{1}{2^3}$ Factor the denominators.

The LCD is $7^2 \cdot 2^3 = 392$.   The LCD is the product of unique factors, each raised to its highest power.

b. $\dfrac{5}{3x^2z}; \dfrac{7}{x^5y^3}$

$= \dfrac{5}{3 \cdot x^2 \cdot z^1}; \dfrac{7}{x^5 \cdot y^3}$ The denominators are in factored form.

The LCD is the product of $3 \cdot x^5 \cdot y^3 \cdot z^1$ or simply $3x^5y^3z$.

Skill Practice Find the LCD for each set of expressions.

1. $\dfrac{3}{8}; \dfrac{7}{10}; \dfrac{1}{15}$ **2.** $\dfrac{1}{5a^3b^2}; \dfrac{1}{10a^4b}$

Classroom Examples: p. 504, Exercises 26 and 28

| Example 2 | **Finding the Least Common Denominator** |

Find the LCD for each pair of rational expressions.

a. $\dfrac{a+b}{a^2-25}; \dfrac{1}{2a-10}$ **b.** $\dfrac{x-5}{x^2-2x}; \dfrac{1}{x^2-4x+4}$

Solution:

a. $\dfrac{a+b}{a^2-25}; \dfrac{1}{2a-10}$

$= \dfrac{a+b}{(a-5)(a+5)}; \dfrac{1}{2(a-5)}$ Factor the denominators.

The LCD is $2(a-5)(a+5)$.   The LCD is the product of unique factors, each raised to its highest power.

b. $\dfrac{x-5}{x^2-2x}; \dfrac{1}{x^2-4x+4}$

$= \dfrac{x-5}{x(x-2)}; \dfrac{1}{(x-2)^2}$ Factor the denominators.

The LCD is $x(x-2)^2$.   The LCD is the product of unique factors, each raised to its highest power.

Answers

1. 120
2. $10a^4b^2$

Skill Practice Find the LCD.

3. $\dfrac{x}{x^2 - 16}; \dfrac{2}{3x + 12}$ **4.** $\dfrac{6}{t^2 + 5t - 14}; \dfrac{8}{t^2 - 3t + 2}$

2. Writing Rational Expressions with the Least Common Denominator

To add or subtract two rational expressions, the expressions must have the same denominator. Therefore, we must first practice the skill of converting each rational expression into an equivalent expression with the LCD as its denominator.

Writing Equivalent Fractions with Common Denominators

Step 1 Identify the LCD for the expressions.

Step 2 Multiply the numerator and denominator of each fraction by the factors from the LCD that are missing from the original denominators.

Example 3 **Converting to the Least Common Denominator**

Find the LCD of each pair of rational expressions. Then convert each expression to an equivalent fraction with the denominator equal to the LCD.

a. $\dfrac{3}{2ab}; \dfrac{6}{5a^2}$ **b.** $\dfrac{4}{x + 1}; \dfrac{7}{x - 4}$

Solution:

a. $\dfrac{3}{2ab}; \dfrac{6}{5a^2}$ The LCD is $10a^2b$.

$\dfrac{3}{2ab} = \dfrac{3 \cdot 5a}{2ab \cdot 5a} = \dfrac{15a}{10a^2b}$ The first expression is missing the factor $5a$ from the denominator.

$\dfrac{6}{5a^2} = \dfrac{6 \cdot 2b}{5a^2 \cdot 2b} = \dfrac{12b}{10a^2b}$ The second expression is missing the factor $2b$ from the denominator.

b. $\dfrac{4}{x + 1}; \dfrac{7}{x - 4}$ The LCD is $(x + 1)(x - 4)$.

$\dfrac{4}{x + 1} = \dfrac{4(x - 4)}{(x + 1)(x - 4)} = \dfrac{4x - 16}{(x + 1)(x - 4)}$ The first expression is missing the factor $(x - 4)$ from the denominator.

$\dfrac{7}{x - 4} = \dfrac{7(x + 1)}{(x - 4)(x + 1)} = \dfrac{7x + 7}{(x - 4)(x + 1)}$ The second expression is missing the factor $(x + 1)$ from the denominator.

Skill Practice For each pair of expressions, find the LCD, and then convert each expression to an equivalent fraction with the denominator equal to the LCD.

5. $\dfrac{2}{rs^2}; \dfrac{-1}{r^3s}$ **6.** $\dfrac{5}{x - 3}; \dfrac{x}{x + 1}$

Classroom Examples: p. 504, Exercises 34 and 40

Answers

3. $3(x - 4)(x + 4)$
4. $(t + 7)(t - 2)(t - 1)$
5. $\dfrac{2}{rs^2} = \dfrac{2r^2}{r^3s^2}; \dfrac{-1}{r^3s} = \dfrac{-s}{r^3s^2}$
6. $\dfrac{5}{x - 3} = \dfrac{5x + 5}{(x - 3)(x + 1)}$
  $\dfrac{x}{x + 1} = \dfrac{x^2 - 3x}{(x + 1)(x - 3)}$

Classroom Example: p. 505, Exercise 46

Example 4 **Converting to the Least Common Denominator**

Find the LCD of the pair of rational expressions. Then convert each expression to an equivalent fraction with the denominator equal to the LCD.

$$\frac{w+2}{w^2 - w - 12}; \frac{1}{w^2 - 9}$$

Solution:

$\dfrac{w+2}{w^2 - w - 12}; \dfrac{1}{w^2 - 9}$	To find the LCD, factor each denominator.
$\dfrac{w+2}{(w-4)(w+3)}; \dfrac{1}{(w-3)(w+3)}$	The LCD is $(w-4)(w+3)(w-3)$.
$\dfrac{w+2}{(w-4)(w+3)} = \dfrac{(w+2)(w-3)}{(w-4)(w+3)(w-3)}$	The first expression is missing the factor $(w-3)$ from the denominator.
$= \dfrac{w^2 - w - 6}{(w-4)(w+3)(w-3)}$	
$\dfrac{1}{(w-3)(w+3)} = \dfrac{1(w-4)}{(w-3)(w+3)(w-4)}$	The second expression is missing the factor $(w-4)$ from the denominator.
$= \dfrac{w-4}{(w-3)(w+3)(w-4)}$	

Skill Practice Find the LCD. Then convert each expression to an equivalent expression with the denominator equal to the LCD.

7. $\dfrac{z}{z^2 - 4}; \dfrac{-3}{z^2 - z - 2}$

Classroom Example: p. 505, Exercise 48

Example 5 **Converting to the Least Common Denominator**

Convert each expression to an equivalent expression with the denominator equal to the LCD.

$$\frac{3}{x-7} \quad \text{and} \quad \frac{1}{7-x}$$

TIP: In Example 5, the expressions

$$\frac{3}{x-7} \quad \text{and} \quad \frac{1}{7-x}$$

have opposite factors in the denominators. In such a case, you do not need to include *both* factors in the LCD.

Solution:

Notice that the expressions $x - 7$ and $7 - x$ are opposites and differ by a factor of -1. Therefore, we may use either $x - 7$ or $7 - x$ as a common denominator. Each case is shown below.

Converting to the Denominator $x - 7$

$\dfrac{3}{x-7}; \dfrac{1}{7-x}$	Leave the first fraction unchanged because it has the desired LCD.
$\dfrac{1}{7-x} = \dfrac{(-1)1}{(-1)(7-x)}$	Multiply the *second* rational expression by the ratio $\frac{-1}{-1}$ to change its denominator to $x - 7$.
$= \dfrac{-1}{-7+x}$	Apply the distributive property.
$= \dfrac{-1}{x-7}$	

Answer

7. $\dfrac{z^2 + z}{(z-2)(z+2)(z+1)};$

$\dfrac{-3z - 6}{(z-2)(z+2)(z+1)}$

Converting to the Denominator $7 - x$

$\dfrac{3}{x-7}; \dfrac{1}{7-x}$ Leave the second fraction unchanged because it has the desired LCD.

$\dfrac{3}{x-7} = \dfrac{(-1)3}{(-1)(x-7)};$ Multiply the *first* rational expression by the ratio $\frac{-1}{-1}$ to change its denominator to $7 - x$.

$= \dfrac{-3}{-x+7}$ Apply the distributive property.

$= \dfrac{-3}{7-x}$

Skill Practice

8. a. Find the LCD of the expressions. $\dfrac{9}{w-2}; \dfrac{11}{2-w}$

     b. Convert each expression to an equivalent fraction with denominator equal to the LCD.

Answers

8. a. The LCD is $(w - 2)$ or $(2 - w)$.

    b. $\dfrac{9}{w-2} = \dfrac{9}{w-2};$

       $\dfrac{11}{2-w} = \dfrac{-11}{w-2}$

       or

       $\dfrac{9}{w-2} = \dfrac{-9}{2-w};$

       $\dfrac{11}{2-w} = \dfrac{11}{2-w}$

Section 7.3   Practice Exercises

For additional exercises, see Classroom Activities 7.3A–7.3C in the *Student's Resource Manual* at www.mhhe.com/moh.

Vocabulary and Key Concepts

1. The least common denominator (LCD) of two rational expressions is defined as the least common ___multiple___ of the ___denominators___.

Review Exercises

2. Evaluate the expression for the given values of x. $\dfrac{2x}{x+5}$

    a. $x = 1$ $\frac{1}{3}$ **b.** $x = 5$ 1 **c.** $x = -5$ Undefined

For Exercises 3–4, identify the restricted values. Then simplify the expression.

3. $\dfrac{3x+3}{5x^2-5}$ $x = 1, x = -1; \dfrac{3}{5(x-1)}$ **4.** $\dfrac{x+2}{x^2-3x-10}$ $x = -2, x = 5; \dfrac{1}{x-5}$

For Exercises 5–8, multiply or divide as indicated.

5. $\dfrac{a+3}{a+7} \cdot \dfrac{a^2+3a-10}{a^2+a-6}$ $\dfrac{a+5}{a+7}$ **6.** $\dfrac{6(a+2b)}{2(a-3b)} \cdot \dfrac{4(a+3b)(a-3b)}{9(a+2b)(a-2b)}$ $\dfrac{4(a+3b)}{3(a-2b)}$

7. $\dfrac{16y^2}{9y+36} \div \dfrac{8y^3}{3y+12}$ $\dfrac{2}{3y}$ **8.** $\dfrac{5w^2+6w+1}{w^2+5w+6} \div (5w+1)$ $\dfrac{w+1}{(w+2)(w+3)}$

9. Which of the expressions are equivalent to $-\dfrac{5}{x-3}$? Circle all that apply.

    a. $\dfrac{-5}{x-3}$ **b.** $\dfrac{5}{-x+3}$ **c.** $\dfrac{5}{3-x}$ **d.** $\dfrac{5}{-(x-3)}$ a, b, c, d

Writing   Translating Expression   Geometry   Scientific Calculator   Video

10. Which of the expressions are equivalent to $\dfrac{4-a}{6}$? Circle all that apply.

a. $\dfrac{a-4}{-6}$ **b.** $\dfrac{a-4}{6}$ **c.** $\dfrac{-(4-a)}{-6}$ **d.** $-\dfrac{a-4}{6}$ a, c, d

Concept 1: Least Common Denominator

11. Explain why the least common denominator of $\frac{1}{x}$, $\frac{1}{x^3}$, and $\frac{1}{x^4}$ is x^5.

x^5 is the greatest power of x that appears in any denominator.

12. Explain why the least common denominator of $\frac{2}{y^3}$, $\frac{9}{y^6}$, and $\frac{4}{y^5}$ is y^6.

y^6 is the greatest power of y that appears in any denominator.

For Exercises 13–30, identify the LCD. **(See Examples 1–2.)**

13. $\dfrac{4}{15}; \dfrac{5}{9}$ 45

14. $\dfrac{7}{12}; \dfrac{1}{18}$ 36

15. $\dfrac{1}{16}; \dfrac{1}{4}; \dfrac{1}{6}$ 48

16. $\dfrac{1}{2}; \dfrac{11}{12}; \dfrac{3}{8}$ 24

17. $\dfrac{1}{7}; \dfrac{2}{9}$ 63

18. $\dfrac{2}{3}; \dfrac{5}{8}$ 24

19. $\dfrac{1}{3x^2y}; \dfrac{8}{9xy^3}$ $9x^2y^3$

20. $\dfrac{5}{2a^4b^2}; \dfrac{1}{8ab^3}$ $8a^4b^3$

21. $\dfrac{6}{w^2}; \dfrac{7}{y}$ w^2y

22. $\dfrac{2}{r}; \dfrac{3}{s^2}$ rs^2

23. $\dfrac{p}{(p+3)(p-1)}; \dfrac{2}{(p+3)(p+2)}$

$(p+3)(p-1)(p+2)$

24. $\dfrac{6}{(q+4)(q-4)}; \dfrac{q^2}{(q+1)(q+4)}$

$(q+4)(q-4)(q+1)$

25. $\dfrac{7}{3t(t+1)}; \dfrac{10t}{9(t+1)^2}$

$9t(t+1)^2$

26. $\dfrac{13x}{15(x-1)^2}; \dfrac{5}{3x(x-1)}$

$15x(x-1)^2$

27. $\dfrac{y}{y^2-4}; \dfrac{3y}{y^2+5y+6}$

$(y-2)(y+2)(y+3)$

28. $\dfrac{4}{w^2-3w+2}; \dfrac{w}{w^2-4}$

$(w-1)(w-2)(w+2)$

29. $\dfrac{5}{3-x}; \dfrac{7}{x-3}$ $3-x$ or $x-3$

30. $\dfrac{4}{x-6}; \dfrac{9}{6-x}$ $x-6$ or $6-x$

31. Explain why a common denominator of

$$\dfrac{b+1}{b-1} \quad \text{and} \quad \dfrac{b}{1-b}$$

could be either $(b-1)$ or $(1-b)$.

Because $(b-1)$ and $(1-b)$ are opposites, they differ by a factor of -1.

32. Explain why a common denominator of

$$\dfrac{1}{6-t} \quad \text{and} \quad \dfrac{t}{t-6}$$

could be either $(6-t)$ or $(t-6)$.

Because $(6-t)$ and $(t-6)$ are opposites, they differ by a factor of -1.

Concept 2: Writing Rational Expressions with the Least Common Denominator

For Exercises 33–56, find the LCD. Then convert each expression to an equivalent expression with the denominator equal to the LCD. **(See Examples 3–5.)**

33. $\dfrac{6}{5x^2}; \dfrac{1}{x}$ $\dfrac{6}{5x^2}; \dfrac{5x}{5x^2}$

34. $\dfrac{3}{y}; \dfrac{7}{9y^2}$ $\dfrac{27y}{9y^2}; \dfrac{7}{9y^2}$

35. $\dfrac{4}{5x^2}; \dfrac{y}{6x^3}$ $\dfrac{24x}{30x^3}; \dfrac{5y}{30x^3}$

36. $\dfrac{3}{15b^2}; \dfrac{c}{3b^2}$ $\dfrac{3}{15b^2}; \dfrac{5c}{15b^2}$

37. $\dfrac{5}{6a^2b}; \dfrac{a}{12b}$ $\dfrac{10}{12a^2b}; \dfrac{a^3}{12a^2b}$

38. $\dfrac{x}{15y^2}; \dfrac{y}{5xy}$ $\dfrac{x^2}{15xy^2}; \dfrac{3y^2}{15xy^2}$

39. $\dfrac{6}{m+4}; \dfrac{3}{m-1}$

$\dfrac{6m-6}{(m+4)(m-1)}; \dfrac{3m+12}{(m+4)(m-1)}$

40. $\dfrac{3}{n-5}; \dfrac{7}{n+2}$

$\dfrac{3n+6}{(n-5)(n+2)}; \dfrac{7n-35}{(n-5)(n+2)}$

41. $\dfrac{6}{2x-5}; \dfrac{1}{x+3}$

$\dfrac{6x+18}{(2x-5)(x+3)}; \dfrac{2x-5}{(2x-5)(x+3)}$

42. $\dfrac{4}{m+3}; \dfrac{-3}{5m+1}$

$\dfrac{20m+4}{(m+3)(5m+1)}; \dfrac{-3m-9}{(m+3)(5m+1)}$

43. $\dfrac{6}{(w+3)(w-8)}; \dfrac{w}{(w-8)(w+1)}$

$\dfrac{6w+6}{(w+3)(w-8)(w+1)}; \dfrac{w^2+3w}{(w+3)(w-8)(w+1)}$

44. $\dfrac{t}{(t+2)(t+12)}; \dfrac{18}{(t-2)(t+2)}$

$\dfrac{t^2-2t}{(t+2)(t-2)(t+12)}; \dfrac{18t+216}{(t+2)(t-2)(t+12)}$

45. $\dfrac{6p}{p^2-4}$; $\dfrac{3}{p^2+4p+4}$

$\dfrac{6p^2+12p}{(p-2)(p+2)^2}$; $\dfrac{3p-6}{(p-2)(p+2)^2}$

46. $\dfrac{5}{t^2-6t+9}$; $\dfrac{t}{t^2-9}$

$\dfrac{5t+15}{(t-3)^2(t+3)}$; $\dfrac{t^2-3t}{(t-3)^2(t+3)}$

47. $\dfrac{1}{a-4}$; $\dfrac{a}{4-a}$

$\dfrac{1}{a-4}$; $\dfrac{-a}{a-4}$ or $\dfrac{-1}{4-a}$; $\dfrac{a}{4-a}$

48. $\dfrac{3b}{2b-5}$; $\dfrac{2b}{5-2b}$

$\dfrac{3b}{2b-5}$; $\dfrac{-2b}{2b-5}$ or $\dfrac{-3b}{5-2b}$; $\dfrac{2b}{5-2b}$

49. $\dfrac{4}{x-7}$; $\dfrac{y}{14-2x}$

$\dfrac{8}{2(x-7)}$; $\dfrac{-y}{2(x-7)}$ or $\dfrac{-8}{2(7-x)}$; $\dfrac{y}{2(7-x)}$

50. $\dfrac{4}{3x-15}$; $\dfrac{z}{5-x}$

$\dfrac{4}{3(x-5)}$; $\dfrac{-3z}{3(x-5)}$ or $\dfrac{-4}{3(5-x)}$; $\dfrac{3z}{3(5-x)}$

51. $\dfrac{1}{a+b}$; $\dfrac{6}{-a-b}$

$\dfrac{1}{a+b}$; $\dfrac{-6}{a+b}$ or $\dfrac{-1}{-a-b}$; $\dfrac{6}{-a-b}$

52. $\dfrac{p}{-q-8}$; $\dfrac{1}{q+8}$

$\dfrac{p}{-q-8}$; $\dfrac{-1}{-q-8}$ or $\dfrac{-p}{q+8}$; $\dfrac{1}{q+8}$

53. $\dfrac{-3}{24y+8}$; $\dfrac{5}{18y+6}$

$\dfrac{-9}{24(3y+1)}$; $\dfrac{20}{24(3y+1)}$

54. $\dfrac{r}{10r+5}$; $\dfrac{2}{16r+8}$

$\dfrac{8r}{40(2r+1)}$; $\dfrac{10}{40(2r+1)}$

55. $\dfrac{3}{5z}$; $\dfrac{1}{z+4}$

$\dfrac{3z+12}{5z(z+4)}$; $\dfrac{5z}{5z(z+4)}$

56. $\dfrac{-1}{4a-8}$; $\dfrac{5}{4a}$

$\dfrac{-a}{4a(a-2)}$; $\dfrac{5a-10}{4a(a-2)}$

Expanding Your Skills

For Exercises 57–60, find the LCD. Then convert each expression to an equivalent expression with the denominator equal to the LCD.

57. $\dfrac{z}{z^2+9z+14}$; $\dfrac{-3z}{z^2+10z+21}$; $\dfrac{5}{z^2+5z+6}$

$\dfrac{z^2+3z}{(z+2)(z+7)(z+3)}$; $\dfrac{-3z^2-6z}{(z+2)(z+7)(z+3)}$; $\dfrac{5z+35}{(z+2)(z+7)(z+3)}$

58. $\dfrac{6}{w^2-3w-4}$; $\dfrac{1}{w^2+6w+5}$; $\dfrac{-9w}{w^2+w-20}$

$\dfrac{6w+30}{(w-4)(w+1)(w+5)}$; $\dfrac{w-4}{(w-4)(w+1)(w+5)}$; $\dfrac{-9w^2-9w}{(w-4)(w+1)(w+5)}$

59. $\dfrac{3}{p^3-8}$; $\dfrac{p}{p^2-4}$; $\dfrac{5p}{p^2+2p+4}$

$\dfrac{3p+6}{(p-2)(p^2+2p+4)(p+2)}$; $\dfrac{p^3+2p^2+4p}{(p-2)(p^2+2p+4)(p+2)}$; $\dfrac{5p^3-20p}{(p-2)(p^2+2p+4)(p+2)}$

60. $\dfrac{7}{n^3+125}$; $\dfrac{n}{n^2-25}$; $\dfrac{12}{n^2-5n+25}$

$\dfrac{7n-35}{(n+5)(n^2-5n+25)(n-5)}$; $\dfrac{n^3-5n^2+25n}{(n+5)(n^2-5n+25)(n-5)}$; $\dfrac{12n^2-300}{(n+5)(n^2-5n+25)(n-5)}$

Addition and Subtraction of Rational Expressions

Section 7.4

1. Addition and Subtraction of Rational Expressions with the Same Denominator

To add or subtract rational expressions, the expressions must have the same denominator. As with fractions, we add or subtract rational expressions with the same denominator by combining the terms in the numerator and then writing the result over the common denominator. Then, if possible, simplify the expression.

Concepts

1. Addition and Subtraction of Rational Expressions with the Same Denominator
2. Addition and Subtraction of Rational Expressions with Different Denominators
3. Using Rational Expressions in Translations

Addition and Subtraction of Rational Expressions

Let p, q, and r represent polynomials where $q \neq 0$. Then,

1. $\dfrac{p}{q}+\dfrac{r}{q}=\dfrac{p+r}{q}$

2. $\dfrac{p}{q}-\dfrac{r}{q}=\dfrac{p-r}{q}$

Classroom Example: p. 512,
Exercises 8 and 12

> **Example 1**   **Adding and Subtracting Rational Expressions with the Same Denominator**

Add or subtract as indicated.   **a.** $\dfrac{1}{12} + \dfrac{7}{12}$ **b.** $\dfrac{2}{5p} - \dfrac{7}{5p}$

Solution:

a. $\dfrac{1}{12} + \dfrac{7}{12}$ The fractions have the same denominator.

$= \dfrac{1+7}{12}$ Add the terms in the numerators, and write the result over the common denominator.

$= \dfrac{\overset{2}{\cancel{8}}}{\underset{3}{\cancel{12}}}$

$= \dfrac{2}{3}$ Simplify.

b. $\dfrac{2}{5p} - \dfrac{7}{5p}$ The rational expressions have the same denominator.

$= \dfrac{2-7}{5p}$ Subtract the terms in the numerators, and write the result over the common denominator.

$= \dfrac{-5}{5p}$

$= \dfrac{\overset{-1}{\cancel{-5}}}{\cancel{5}p}$ Simplify.

$= -\dfrac{1}{p}$

Skill Practice Add or subtract as indicated.

1. $\dfrac{3}{14} + \dfrac{4}{14}$ **2.** $\dfrac{2}{7d} - \dfrac{9}{7d}$

Classroom Example: p. 512,
Exercise 18

> **Example 2**   **Adding and Subtracting Rational Expressions with the Same Denominator**

Add or subtract as indicated.

a. $\dfrac{2}{3d+5} + \dfrac{7d}{3d+5}$ **b.** $\dfrac{x^2}{x-3} - \dfrac{-5x+24}{x-3}$

Solution:

a. $\dfrac{2}{3d+5} + \dfrac{7d}{3d+5}$ The rational expressions have the same denominator.

$= \dfrac{2+7d}{3d+5}$ Add the terms in the numerators, and write the result over the common denominator.

$= \dfrac{7d+2}{3d+5}$ Because the numerator and denominator share no common factors, the expression is in lowest terms.

Answers

1. $\dfrac{1}{2}$ **2.** $-\dfrac{1}{d}$

b. $\dfrac{x^2}{x-3} - \dfrac{-5x+24}{x-3}$ The rational expressions have the same denominator.

$= \dfrac{x^2 - (-5x+24)}{x-3}$ Subtract the terms in the numerators, and write the result over the common denominator.

$= \dfrac{x^2 + 5x - 24}{x-3}$ Simplify the numerator.

$= \dfrac{(x+8)(x-3)}{(x-3)}$ Factor the numerator and denominator to determine if the rational expression can be simplified.

$= \dfrac{(x+8)\overset{1}{\cancel{(x-3)}}}{\cancel{(x-3)}}$ Simplify.

$= x + 8$

Skill Practice Add or subtract as indicated.

3. $\dfrac{x^2+2}{x+3} + \dfrac{4x+1}{x+3}$ **4.** $\dfrac{4t-9}{2t+1} - \dfrac{t-5}{2t+1}$

2. Addition and Subtraction of Rational Expressions with Different Denominators

To add or subtract two rational expressions with unlike denominators, we must convert the expressions to equivalent expressions with the same denominator. For example, consider adding

$$\frac{1}{10} + \frac{12}{5y}$$

The LCD is $10y$. For each expression, identify the factors from the LCD that are missing from the denominator. Then multiply the numerator and denominator of the expression by the missing factor(s).

$$\underset{\substack{\text{Missing} \\ y}}{\frac{1}{10}} + \underset{\substack{\text{Missing} \\ 2}}{\frac{12}{5y}}$$

$= \dfrac{1 \cdot y}{10 \cdot y} + \dfrac{12 \cdot 2}{5y \cdot 2}$

$= \dfrac{y}{10y} + \dfrac{24}{10y}$ The rational expressions now have the same denominators.

$= \dfrac{y+24}{10y}$ Add the numerators.

After successfully adding or subtracting two rational expressions, always check to see if the final answer is simplified. If necessary, factor the numerator and denominator, and reduce common factors. The expression

$$\frac{y+24}{10y}$$

is in lowest terms because the numerator and denominator do not share any common factors.

Answers

3. $x+1$ **4.** $\dfrac{3t-4}{2t+1}$

> **Adding or Subtracting Rational Expressions**
> **Step 1** Factor the denominators of each rational expression.
> **Step 2** Identify the LCD.
> **Step 3** Rewrite each rational expression as an equivalent expression with the LCD as its denominator.
> **Step 4** Add or subtract the numerators, and write the result over the common denominator.
> **Step 5** Simplify.

Classroom Example: p. 513, Exercise 32

Example 3 **Subtracting Rational Expressions with Different Denominators**

Subtract. $\dfrac{4}{7k} - \dfrac{3}{k^2}$

Solution:

$\dfrac{4}{7k} - \dfrac{3}{k^2}$

Step 1: The denominators are already factored.

Step 2: The LCD is $7k^2$.

$= \dfrac{4 \cdot k}{7k \cdot k} - \dfrac{3 \cdot 7}{k^2 \cdot 7}$ **Step 3:** Write each expression with the LCD.

$= \dfrac{4k}{7k^2} - \dfrac{21}{7k^2}$

$= \dfrac{4k - 21}{7k^2}$ **Step 4:** Subtract the numerators, and write the result over the LCD.

Step 5: The expression is in lowest terms because the numerator and denominator share no common factors.

Avoiding Mistakes

Do not reduce after rewriting the individual fractions with the LCD. You will revert back to the original expression.

Skill Practice Subtract.

5. $\dfrac{4}{3x} - \dfrac{1}{2x^2}$

Classroom Example: p. 513, Exercise 42

Example 4 **Subtracting Rational Expressions with Different Denominators**

Subtract. $\dfrac{2q - 4}{3} - \dfrac{q + 1}{2}$

Solution:

$\dfrac{2q - 4}{3} - \dfrac{q + 1}{2}$

Step 1: The denominators are already factored.

Step 2: The LCD is 6.

$= \dfrac{2(2q - 4)}{2 \cdot 3} - \dfrac{3(q + 1)}{3 \cdot 2}$ **Step 3:** Write each expression with the LCD.

Answer

5. $\dfrac{8x - 3}{6x^2}$

$$= \frac{2(2q-4) - 3(q+1)}{6}$$

Step 4: Subtract the numerators, and write the result over the LCD.

$$= \frac{4q - 8 - 3q - 3}{6}$$

$$= \frac{q - 11}{6}$$

Step 5: The expression is in lowest terms because the numerator and denominator share no common factors.

Skill Practice Subtract.

6. $\dfrac{t}{12} - \dfrac{t-2}{4}$

Example 5	**Adding Rational Expressions with Different Denominators**

Classroom Example: p. 513, Exercise 40

Add. $\dfrac{1}{x-5} + \dfrac{-10}{x^2 - 25}$

Solution:

$$\frac{1}{x-5} + \frac{-10}{x^2 - 25}$$

$$= \frac{1}{x-5} + \frac{-10}{(x-5)(x+5)}$$

Step 1: Factor the denominators.

Step 2: The LCD is $(x-5)(x+5)$.

$$= \frac{1(x+5)}{(x-5)(x+5)} + \frac{-10}{(x-5)(x+5)}$$

Step 3: Write each expression with the LCD.

$$= \frac{1(x+5) + (-10)}{(x-5)(x+5)}$$

Step 4: Add the numerators, and write the result over the LCD.

$$= \frac{x + 5 - 10}{(x-5)(x+5)}$$

$$= \frac{\overset{1}{\cancel{x-5}}}{\cancel{(x-5)}(x+5)}$$

Step 5: Simplify.

$$= \frac{1}{x+5}$$

Skill Practice Add.

7. $\dfrac{1}{x-4} + \dfrac{-8}{x^2 - 16}$

Classroom Example: p. 514, Exercise 68

Example 6   **Subtracting Rational Expressions with Different Denominators**

Subtract. $\dfrac{p+2}{p-1} - \dfrac{2}{p+6} - \dfrac{14}{p^2+5p-6}$

Solution:

$\dfrac{p+2}{p-1} - \dfrac{2}{p+6} - \dfrac{14}{p^2+5p-6}$

$= \dfrac{p+2}{p-1} - \dfrac{2}{p+6} - \dfrac{14}{(p-1)(p+6)}$

Step 1: Factor the denominators.

Step 2: The LCD is $(p-1)(p+6)$.

Step 3: Write each expression with the LCD.

$= \dfrac{(p+2)(p+6)}{(p-1)(p+6)} - \dfrac{2(p-1)}{(p+6)(p-1)} - \dfrac{14}{(p-1)(p+6)}$

$= \dfrac{(p+2)(p+6) - 2(p-1) - 14}{(p-1)(p+6)}$

Step 4: Combine the numerators, and write the result over the LCD.

$= \dfrac{p^2 + 6p + 2p + 12 - 2p + 2 - 14}{(p-1)(p+6)}$

Step 5: Clear parentheses in the numerator.

$= \dfrac{p^2 + 6p}{(p-1)(p+6)}$

Combine *like* terms.

$= \dfrac{p(p+6)}{(p-1)(p+6)}$

Factor the numerator to determine if the expression is in lowest terms.

$= \dfrac{p(\overset{1}{\cancel{p+6}})}{(p-1)(\cancel{p+6})}$

Simplify.

$= \dfrac{p}{p-1}$

Skill Practice   Subtract.

8. $\dfrac{2y}{y-1} - \dfrac{1}{y} - \dfrac{2y+1}{y^2-y}$

When the denominators of two rational expressions are opposites, we can produce identical denominators by multiplying one of the expressions by the ratio $\frac{-1}{-1}$. This is demonstrated in Example 7.

Classroom Example: p. 513, Exercise 52

Example 7   **Adding Rational Expressions with Different Denominators**

Add the rational expressions. $\dfrac{1}{d-7} + \dfrac{5}{7-d}$

Answer

8. $\dfrac{2y-3}{y-1}$

Solution:

$$\frac{1}{d-7} + \frac{5}{7-d}$$

The expressions $d-7$ and $7-d$ are opposites and differ by a factor of -1. Therefore, multiply the numerator and denominator of *either* expression by -1 to obtain a common denominator.

$$= \frac{1}{d-7} + \frac{(-1)5}{(-1)(7-d)}$$

Note that $-1(7-d) = -7+d$ or $d-7$.

$$= \frac{1}{d-7} + \frac{-5}{d-7}$$

Simplify.

$$= \frac{1+(-5)}{d-7}$$

Add the terms in the numerators, and write the result over the common denominator.

$$= \frac{-4}{d-7}$$

Skill Practice Add.

9. $\dfrac{3}{p-8} + \dfrac{1}{8-p}$

3. Using Rational Expressions in Translations

Example 8   Using Rational Expressions in Translations

Classroom Example: p. 514, Exercise 80

Write the English phrase as a mathematical expression. Then simplify by combining the rational expressions.

The difference of the reciprocal of n and the quotient of n and 3

Solution:

The difference of the reciprocal of n and the quotient of n and 3

The difference of

$$\left(\frac{1}{n}\right) - \left(\frac{n}{3}\right)$$

The reciprocal of n The quotient of n and 3

$$\frac{1}{n} - \frac{n}{3}$$

The LCD is $3n$.

$$= \frac{3 \cdot 1}{3 \cdot n} - \frac{n \cdot n}{3 \cdot n}$$

Write each expression with the LCD.

$$= \frac{3 - n^2}{3n}$$

Subtract the numerators.

Skill Practice Write the English phrase as a mathematical expression. Then simplify by combining the rational expressions.

10. The sum of 1 and the quotient of 2 and a

Answers

9. $\dfrac{2}{p-8}$ or $\dfrac{-2}{8-p}$

10. $1 + \dfrac{2}{a}; \dfrac{a+2}{a}$

Section 7.4   Practice Exercises

Review Exercises

For additional exercises, see Classroom Activities 7.4A–7.4B in the *Student's Resource Manual* at www.mhhe.com/moh.

1. For the rational expression $\dfrac{x^2 - 4x - 5}{x^2 - 7x + 10}$

   a. Find the value of the expression (if possible) when $x = 0, 1, -1, 2,$ and 5. $-\frac{1}{2}, -2, 0$, undefined, undefined

   b. Factor the denominator and identify the restricted values. $(x-5)(x-2); x = 5, x = 2$

   c. Simplify the expression. $\frac{x+1}{x-2}$

2. For the rational expression $\dfrac{a^2 + a - 2}{a^2 - 4a - 12}$

   a. Find the value of the expression (if possible) when $a = 0, 1, -2, 2,$ and 6. $\frac{1}{6}, 0$, undefined, $-\frac{1}{4}$, undefined

   b. Factor the denominator, and identify the restricted values. $(a-6)(a+2); a = 6, a = -2$

   c. Simplify the expression. $\frac{a-1}{a-6}$

For Exercises 3–4, multiply or divide as indicated.

3. $\dfrac{2x^2 - x - 3}{2x^2 - 3x - 9} \div \dfrac{x^2 - 1}{4x + 6}$ $\frac{2(2x-3)}{(x-3)(x-1)}$

4. $\dfrac{6t - 1}{5t - 30} \cdot \dfrac{10t - 25}{2t^2 - 3t - 5}$ $\frac{6t-1}{(t+1)(t-6)}$

Concept 1: Addition and Subtraction of Rational Expressions with the Same Denominator

For Exercises 5–26, add or subtract the expressions with the same denominators as indicated. **(See Examples 1–2.)**

5. $\dfrac{7}{8} + \dfrac{3}{8}$ $\frac{5}{4}$

6. $\dfrac{1}{3} + \dfrac{7}{3}$ $\frac{8}{3}$

7. $\dfrac{9}{16} - \dfrac{3}{16}$ $\frac{3}{8}$

8. $\dfrac{14}{15} - \dfrac{4}{15}$ $\frac{2}{3}$

9. $\dfrac{5a}{a+2} - \dfrac{3a - 4}{a+2}$ 2

10. $\dfrac{2b}{b-3} - \dfrac{b-9}{b-3}$ $\frac{b+9}{b-3}$

11. $\dfrac{5c}{c+6} + \dfrac{30}{c+6}$ 5

12. $\dfrac{12}{2+d} + \dfrac{6d}{2+d}$ 6

13. $\dfrac{5}{t-8} - \dfrac{2t+1}{t-8}$ $\frac{-2(t-2)}{t-8}$

14. $\dfrac{7p+1}{2p+1} - \dfrac{p-4}{2p+1}$ $\frac{6p+5}{2p+1}$

15. $\dfrac{9x^2}{3x-7} - \dfrac{49}{3x-7}$ $3x+7$

16. $\dfrac{4w^2}{2w-1} - \dfrac{1}{2w-1}$ $2w+1$

17. $\dfrac{m^2}{m+5} + \dfrac{10m+25}{m+5}$ $m+5$

18. $\dfrac{k^2}{k-3} - \dfrac{6k-9}{k-3}$ $k-3$

19. $\dfrac{2a}{a+2} + \dfrac{4}{a+2}$ 2

20. $\dfrac{5b}{b+4} + \dfrac{20}{b+4}$ 5

21. $\dfrac{x^2}{x+5} - \dfrac{25}{x+5}$ $x-5$

22. $\dfrac{y^2}{y-7} - \dfrac{49}{y-7}$ $y+7$

23. $\dfrac{r}{r^2 + 3r + 2} + \dfrac{2}{r^2 + 3r + 2}$ $\frac{1}{r+1}$

24. $\dfrac{x}{x^2 - x - 12} - \dfrac{4}{x^2 - x - 12}$ $\frac{1}{x+3}$

25. $\dfrac{1}{3y^2 + 22y + 7} - \dfrac{-3y}{3y^2 + 22y + 7}$ $\frac{1}{y+7}$

26. $\dfrac{5}{2x^2 + 13x + 20} + \dfrac{2x}{2x^2 + 13x + 20}$ $\frac{1}{x+4}$

Writing   Translating Expression   Geometry   Scientific Calculator   Video

For Exercises 27–28, find an expression that represents the perimeter of the figure (assume that $x > 0$, $y > 0$, and $t > 0$).

27.

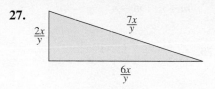

28.

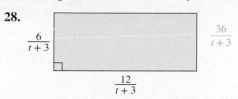

Concept 2: Addition and Subtraction of Rational Expressions with Different Denominators

For Exercises 29–70, add or subtract the expressions with different denominators as indicated. **(See Examples 3–7.)**

29. $\dfrac{5}{4} + \dfrac{3}{2a}$ $\quad \dfrac{5a+6}{4a}$

30. $\dfrac{11}{6p} + \dfrac{-7}{4p}$ $\quad \dfrac{1}{12p}$

31. $\dfrac{4}{5xy^3} + \dfrac{2x}{15y^2}$ $\quad \dfrac{2(6+x^2y)}{15xy^3}$

32. $\dfrac{5}{3a^2b} - \dfrac{7}{6b^2}$ $\quad \dfrac{10b-7a^2}{6a^2b^2}$

33. $\dfrac{2}{s^3t^3} - \dfrac{3}{s^4t}$ $\quad \dfrac{2s-3t^2}{s^4t^3}$

34. $\dfrac{1}{p^2q} - \dfrac{2}{pq^3}$ $\quad \dfrac{q^2-2p}{p^2q^3}$

35. $\dfrac{z}{3z-9} - \dfrac{z-2}{z-3}$ $\quad -\dfrac{2}{3}$

36. $\dfrac{3w-8}{2w-4} - \dfrac{w-3}{w-2}$ $\quad \dfrac{1}{2}$

37. $\dfrac{5}{a+1} + \dfrac{4}{3a+3}$ $\quad \dfrac{19}{3(a+1)}$

38. $\dfrac{2}{c-4} + \dfrac{1}{5c-20}$ $\quad \dfrac{11}{5(c-4)}$

39. $\dfrac{k}{k^2-9} - \dfrac{4}{k-3}$ $\quad \dfrac{-3(k+4)}{(k-3)(k+3)}$

40. $\dfrac{7}{h+2} + \dfrac{2h-3}{h^2-4}$ $\quad \dfrac{9h-17}{(h-2)(h+2)}$

41. $\dfrac{3a-7}{6a+10} - \dfrac{10}{3a^2+5a}$ $\quad \dfrac{a-4}{2a}$

42. $\dfrac{k+2}{8k} - \dfrac{3-k}{12k}$ $\quad \dfrac{5}{24}$

43. $\dfrac{x}{x-4} + \dfrac{3}{x+1}$ $\quad \dfrac{(x+6)(x-2)}{(x-4)(x+1)}$

44. $\dfrac{4}{y-3} + \dfrac{y}{y-5}$ $\quad \dfrac{(y+5)(y-4)}{(y-5)(y-3)}$

45. $\dfrac{3x}{x^2+6x+9} + \dfrac{x}{x^2+5x+6}$ $\quad \dfrac{x(4x+9)}{(x+3)^2(x+2)}$

46. $\dfrac{7x}{x^2+2xy+y^2} + \dfrac{3x}{x^2+xy}$ $\quad \dfrac{10x+3y}{(x+y)^2}$

47. $\dfrac{p}{3} - \dfrac{4p-1}{-3}$ $\quad \dfrac{5p-1}{3}$ or $\dfrac{-5p+1}{-3}$

48. $\dfrac{r}{7} - \dfrac{r-5}{-7}$ $\quad \dfrac{2r-5}{7}$ or $\dfrac{-2r+5}{-7}$

49. $\dfrac{8}{x-3} - \dfrac{1}{3-x}$ $\quad \dfrac{9}{x-3}$ or $\dfrac{-9}{3-x}$

50. $\dfrac{5y}{y-1} - \dfrac{3y}{1-y}$ $\quad \dfrac{8y}{y-1}$ or $\dfrac{-8y}{1-y}$

51. $\dfrac{4n}{n-8} - \dfrac{2n-1}{8-n}$ $\quad \dfrac{6n-1}{n-8}$ or $\dfrac{-6n+1}{8-n}$

52. $\dfrac{m}{m-2} - \dfrac{3m+1}{2-m}$ $\quad \dfrac{4m+1}{m-2}$ or $\dfrac{-4m-1}{2-m}$

53. $\dfrac{5}{x} + \dfrac{3}{x+2}$ $\quad \dfrac{2(4x+5)}{x(x+2)}$

54. $\dfrac{6}{y-1} + \dfrac{9}{y}$ $\quad \dfrac{3(5y-3)}{y(y-1)}$

55. $\dfrac{y}{4y+2} + \dfrac{3y}{6y+3}$ $\quad \dfrac{3y}{2(2y+1)}$

56. $\dfrac{4}{q^2-2q} - \dfrac{5}{3q-6}$ $\quad \dfrac{12-5q}{3q(q-2)}$

57. $\dfrac{4w}{w^2+2w-3} + \dfrac{2}{1-w}$ $\quad \dfrac{2(w-3)}{(w+3)(w-1)}$

58. $\dfrac{z-23}{z^2-z-20} - \dfrac{2}{5-z}$ $\quad \dfrac{3}{z+4}$

59. $\dfrac{3a-8}{a^2-5a+6} + \dfrac{a+2}{a^2-6a+8}$ $\quad \dfrac{4a-13}{(a-3)(a-4)}$

60. $\dfrac{3b+5}{b^2+4b+3} + \dfrac{-b+5}{b^2+2b-3}$ $\quad \dfrac{2b}{(b+1)(b-1)}$

61. $\dfrac{4x}{x^2+4x-5} - \dfrac{x}{x^2+10x+25}$ $\quad \dfrac{3x(x+7)}{(x+5)^2(x-1)}$

62. $\dfrac{x}{x^2+5x+4} - \dfrac{2x}{x^2+8x+16}$ $\quad \dfrac{-x(x-2)}{(x+1)(x+4)^2}$

63. $\dfrac{3y}{2y^2-y-1} - \dfrac{4y}{2y^2-7y-4}$ $\quad \dfrac{-y(y+8)}{(2y+1)(y-1)(y-4)}$

64. $\dfrac{5}{6y^2-7y-3} + \dfrac{4y}{3y^2+4y+1}$ $\quad \dfrac{8y^2-7y+5}{(3y+1)(2y-3)(y+1)}$

65. $\dfrac{3}{2p-1} - \dfrac{4p+4}{4p^2-1}$ $\quad \dfrac{1}{2p+1}$

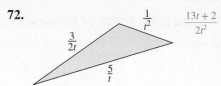 **66.** $\dfrac{1}{3q-2} - \dfrac{6q+4}{9q^2-4}$ $\quad \dfrac{-1}{3q-2}$

67. $\dfrac{m}{m+n} - \dfrac{m}{m-n} + \dfrac{1}{m^2-n^2}$ $\quad \dfrac{-2mn+1}{(m+n)(m-n)}$

68. $\dfrac{x}{x+y} - \dfrac{2xy}{x^2-y^2} + \dfrac{y}{x-y}$ $\quad \dfrac{x-y}{x+y}$

69. $\dfrac{2}{a+b} + \dfrac{2}{a-b} - \dfrac{4a}{a^2-b^2}$ $\quad 0$

70. $\dfrac{-2x}{x^2-y^2} + \dfrac{1}{x+y} - \dfrac{1}{x-y}$ $\quad \dfrac{-2}{x-y}$

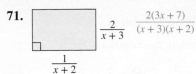 For Exercises 71–72, find an expression that represents the perimeter of the figure (assume that $x > 0$ and $t > 0$).

71. $\dfrac{2}{x+3}$ $\dfrac{2(3x+7)}{(x+3)(x+2)}$

$\dfrac{1}{x+2}$

72. $\dfrac{1}{t^2}$ $\dfrac{13t+2}{2t^2}$

$\dfrac{3}{2t}$ $\dfrac{5}{t}$

Concept 3: Using Rational Expressions in Translations

↔ **73.** Let a number be represented by n. Write the reciprocal of n. $\dfrac{1}{n}$

↔ **74.** Write the reciprocal of the sum of a number and 6. $\dfrac{1}{n+6}$

↔ **75.** Write the quotient of 5 and the sum of a number and 2. $\dfrac{5}{n+2}$

↔ **76.** Write the quotient of 12 and p. $\dfrac{12}{p}$

↔ For Exercises 77–80, translate the English phrases into algebraic expressions. Then simplify by combining the rational expressions. **(See Example 8.)**

 77. The sum of a number and the quantity seven times the reciprocal of the number. $n + \left(7 \cdot \dfrac{1}{n}\right); \dfrac{n^2+7}{n}$

78. The sum of a number and the quantity five times the reciprocal of the number. $n + \left(5 \cdot \dfrac{1}{n}\right); \dfrac{n^2+5}{n}$

79. The difference of the reciprocal of n and the quotient of 2 and n. $\dfrac{1}{n} - \dfrac{2}{n}; \dfrac{-1}{n}$

80. The difference of the reciprocal of m and the quotient of $3m$ and 7. $\dfrac{1}{m} - \dfrac{3m}{7}; \dfrac{7-3m^2}{7m}$

Expanding Your Skills

For Exercises 81–86, perform the indicated operations.

81. $\dfrac{-3}{w^3+27} - \dfrac{1}{w^2-9}$ $\dfrac{-w^2}{(w+3)(w-3)(w^2-3w+9)}$

82. $\dfrac{m}{m^3-1} + \dfrac{1}{(m-1)^2}$ $\dfrac{2m^2+1}{(m-1)^2(m^2+m+1)}$

83. $\dfrac{2p}{p^2+5p+6} - \dfrac{p+1}{p^2+2p-3} + \dfrac{3}{p^2+p-2}$ $\dfrac{p^2-2p+7}{(p+2)(p+3)(p-1)}$

84. $\dfrac{3t}{8t^2+2t-1} - \dfrac{5t}{2t^2-9t-5} + \dfrac{2}{4t^2-21t+5}$ $\dfrac{-17t^2-6t+2}{(2t+1)(4t-1)(t-5)}$

85. $\dfrac{3m}{m^2+3m-10} + \dfrac{5}{4-2m} - \dfrac{1}{m+5}$ $\dfrac{-m-21}{2(m+5)(m-2)}$ or $\dfrac{m+21}{2(m+5)(2-m)}$

86. $\dfrac{2n}{3n^2-8n-3} + \dfrac{1}{6-2n} - \dfrac{3}{3n+1}$ $\dfrac{-5n+17}{2(3n+1)(n-3)}$ or $\dfrac{5n-17}{2(3n+1)(3-n)}$

For Exercises 87–90, simplify by applying the order of operations.

87. $\left(\dfrac{2}{k+1} + 3\right)\left(\dfrac{k+1}{4k+7}\right)$ $\dfrac{3k+5}{4k+7}$

88. $\left(\dfrac{p+1}{3p+4}\right)\left(\dfrac{1}{p+1} + 2\right)$ $\dfrac{2p+3}{3p+4}$

89. $\left(\dfrac{1}{10a} - \dfrac{b}{10a^2}\right) \div \left(\dfrac{1}{10} - \dfrac{b}{10a}\right)$ $\dfrac{1}{a}$

90. $\left(\dfrac{1}{2m} + \dfrac{n}{2m^2}\right) \div \left(\dfrac{1}{4} + \dfrac{n}{4m}\right)$ $\dfrac{2}{m}$

Problem Recognition Exercises

Operations on Rational Expressions

We have learned how to simplify, add, subtract, multiply, and divide rational expressions. The procedure for each operation is different, and it takes considerable practice to determine the correct method to apply for a given problem. The following review exercises give you the opportunity to practice the specific techniques for simplifying rational expressions.

For Exercises 1–20, perform any indicated operations, and simplify the expression.

1. $\dfrac{5}{3x+1} - \dfrac{2x-4}{3x+1}$ $\dfrac{-2x+9}{3x+1}$

2. $\dfrac{\dfrac{w+1}{w^2-16}}{\dfrac{w+1}{w+4}}$ $\dfrac{1}{w-4}$

3. $\dfrac{3}{y} \cdot \dfrac{y^2-5y}{6y-9}$ $\dfrac{y-5}{2y-3}$

4. $\dfrac{-1}{x+3} + \dfrac{2}{2x-1}$ $\dfrac{7}{(x+3)(2x-1)}$

5. $\dfrac{x-9}{9x-x^2}$ $-\dfrac{1}{x}$

6. $\dfrac{1}{p} - \dfrac{3}{p^2+3p} + \dfrac{p}{3p+9}$ $\dfrac{1}{3}$

7. $\dfrac{c^2+5c+6}{c^2+c-2} \div \dfrac{c}{c-1}$ $\dfrac{c+3}{c}$

8. $\dfrac{2x^2-5x-3}{x^2-9} \cdot \dfrac{x^2+6x+9}{10x+5}$ $\dfrac{x+3}{5}$

9. $\dfrac{6a^2b^3}{72ab^7c}$ $\dfrac{a}{12b^4c}$

10. $\dfrac{2a}{a+b} - \dfrac{b}{a-b} - \dfrac{-4ab}{a^2-b^2}$ $\dfrac{2a-b}{a-b}$

11. $\dfrac{p^2+10pq+25q^2}{p^2+6pq+5q^2} \div \dfrac{10p+50q}{2p^2-2q^2}$ $\dfrac{p-q}{5}$

12. $\dfrac{3k-8}{k-5} + \dfrac{k-12}{k-5}$ 4

13. $\dfrac{20x^2+10x}{4x^3+4x^2+x}$ $\dfrac{10}{2x+1}$

14. $\dfrac{w^2-81}{w^2+10w+9} \cdot \dfrac{w^2+w+2zw+2z}{w^2-9w+zw-9z}$ $\dfrac{w+2z}{w+z}$

15. $\dfrac{8x^2-18x-5}{4x^2-25} \div \dfrac{4x^2-11x-3}{3x-9}$ $\dfrac{3}{2x+5}$

16. $\dfrac{xy+7x+5y+35}{x^2+ax+5x+5a}$ $\dfrac{y+7}{x+a}$

17. $\dfrac{a}{a^2-9} - \dfrac{3}{6a-18}$ $\dfrac{1}{2(a+3)}$

18. $\dfrac{4}{y^2-36} + \dfrac{2}{y^2-4y-12}$ $\dfrac{2(3y+10)}{(y-6)(y+6)(y+2)}$

19. $(t^2+5t-24)\left(\dfrac{t+8}{t-3}\right)$ $(t+8)^2$

20. $\dfrac{6b^2-7b-10}{b-2} \cdot 6b+5$

Concepts

1. Simplifying Complex Fractions (Method I)
2. Simplifying Complex Fractions (Method II)

1. Simplifying Complex Fractions (Method I)

A **complex fraction** is an expression containing one or more fractions in the numerator, denominator, or both. For example,

$$\dfrac{\dfrac{1}{ab}}{\dfrac{2}{b}} \quad \text{and} \quad \dfrac{1+\dfrac{3}{4}-\dfrac{1}{6}}{\dfrac{1}{2}+\dfrac{1}{3}}$$

are complex fractions.

Two methods will be presented to simplify complex fractions. The first method (Method I) follows the order of operations to simplify the numerator and denominator separately before dividing. The process is summarized as follows.

> **Simplifying a Complex Fraction (Method I)**
>
> **Step 1** Add or subtract expressions in the numerator to form a single fraction. Add or subtract expressions in the denominator to form a single fraction.
>
> **Step 2** Divide the rational expressions from step 1 by multiplying the numerator of the complex fraction by the reciprocal of the denominator of the complex fraction.
>
> **Step 3** Simplify to lowest terms if possible.

Classroom Example: p. 522, Exercise 12

Example 1 **Simplifying a Complex Fraction (Method I)**

Simplify the expression. $\dfrac{\dfrac{1}{ab}}{\dfrac{2}{b}}$

Solution:

Step 1: The numerator and denominator of the complex fraction are already single fractions.

$\dfrac{\dfrac{1}{ab}}{\dfrac{2}{b}}$ ⟵ This fraction bar denotes division (÷).

$= \dfrac{1}{ab} \div \dfrac{2}{b}$

$= \dfrac{1}{ab} \cdot \dfrac{b}{2}$ **Step 2:** Multiply the numerator of the complex fraction by the reciprocal of $\frac{2}{b}$, which is $\frac{b}{2}$.

$= \dfrac{1}{a\cancel{b}} \cdot \dfrac{\cancel{b}^{1}}{2}$ **Step 3:** Reduce common factors and simplify.

$= \dfrac{1}{2a}$

Skill Practice Simplify the expression.

1. $\dfrac{\dfrac{6x}{y}}{\dfrac{9}{2y}}$

Sometimes it is necessary to simplify the numerator and denominator of a complex fraction before the division can be performed. This is illustrated in Example 2.

Example 2 Simplifying a Complex Fraction (Method I)

Classroom Example: p. 522, Exercise 16

Simplify the expression. $\dfrac{1 + \dfrac{3}{4} - \dfrac{1}{6}}{\dfrac{1}{2} + \dfrac{1}{3}}$

Solution:

$\dfrac{1 + \dfrac{3}{4} - \dfrac{1}{6}}{\dfrac{1}{2} + \dfrac{1}{3}}$

Step 1: Combine fractions in the numerator and denominator separately.

$= \dfrac{1 \cdot \dfrac{12}{12} + \dfrac{3}{4} \cdot \dfrac{3}{3} - \dfrac{1}{6} \cdot \dfrac{2}{2}}{\dfrac{1}{2} \cdot \dfrac{3}{3} + \dfrac{1}{3} \cdot \dfrac{2}{2}}$

The LCD in the numerator is 12. The LCD in the denominator is 6.

$= \dfrac{\dfrac{12}{12} + \dfrac{9}{12} - \dfrac{2}{12}}{\dfrac{3}{6} + \dfrac{2}{6}}$

$= \dfrac{\dfrac{19}{12}}{\dfrac{5}{6}}$

Form a single fraction in the numerator and in the denominator.

$= \dfrac{19}{\overset{2}{12}} \cdot \dfrac{\overset{1}{6}}{5}$

Step 2: Multiply the numerator by the reciprocal of the denominator.

$= \dfrac{19}{10}$

Step 3: Simplify.

Skill Practice Simplify the expression.

2. $\dfrac{\dfrac{3}{4} - \dfrac{1}{6} + 2}{\dfrac{1}{3} + \dfrac{1}{2}}$

Answers

1. $\dfrac{4x}{3}$ 2. $\dfrac{31}{10}$

Classroom Example: p. 522, Exercise 24

| **Example 3** | **Simplifying a Complex Fraction (Method I)** |

Simplify the expression.

$$\dfrac{\dfrac{1}{x} + \dfrac{1}{y}}{x - \dfrac{y^2}{x}}$$

Solution:

$$\dfrac{\dfrac{1}{x} + \dfrac{1}{y}}{x - \dfrac{y^2}{x}}$$

The LCD in the numerator is xy.
The LCD in the denominator is x.

$$= \dfrac{\dfrac{1 \cdot y}{x \cdot y} + \dfrac{1 \cdot x}{y \cdot x}}{\dfrac{x \cdot x}{1 \cdot x} - \dfrac{y^2}{x}}$$

Rewrite the expressions using common denominators.

$$= \dfrac{\dfrac{y}{xy} + \dfrac{x}{xy}}{\dfrac{x^2}{x} - \dfrac{y^2}{x}}$$

$$= \dfrac{\dfrac{y + x}{xy}}{\dfrac{x^2 - y^2}{x}}$$

Form single fractions in the numerator and denominator.

$$= \dfrac{y + x}{xy} \cdot \dfrac{x}{x^2 - y^2}$$

Multiply the numerator by the reciprocal of the denominator.

$$= \dfrac{y + x}{xy} \cdot \dfrac{x}{(x + y)(x - y)}$$

Factor and reduce. Note that $(y + x) = (x + y)$.

$$= \dfrac{1}{y(x - y)}$$

Simplify.

Skill Practice Simplify the expression.

3. $\dfrac{1 - \dfrac{1}{p}}{\dfrac{p}{w} + \dfrac{w}{p}}$

2. Simplifying Complex Fractions (Method II)

We will now simplify the expressions from Examples 2 and 3 again using a second method to simplify complex fractions (Method II). Recall that multiplying the numerator and denominator of a rational expression by the same quantity does not change the value of the expression because we are multiplying by a number equivalent to 1. This is the basis for Method II.

Answer

3. $\dfrac{w(p - 1)}{p^2 + w^2}$

Simplifying a Complex Fraction (Method II)

Step 1   Multiply the numerator and denominator of the complex fraction by the LCD of *all* individual fractions within the expression.

Step 2   Apply the distributive property, and simplify the numerator and denominator.

Step 3   Simplify to lowest terms if possible.

Example 4   **Simplifying a Complex Fraction (Method II)**

Classroom Example: p. 522, Exercise 16

Simplify the expression. $\dfrac{1+\dfrac{3}{4}-\dfrac{1}{6}}{\dfrac{1}{2}+\dfrac{1}{3}}$

Solution:

$$\dfrac{1+\dfrac{3}{4}-\dfrac{1}{6}}{\dfrac{1}{2}+\dfrac{1}{3}}$$

The LCD of the expressions $1, \frac{3}{4}, \frac{1}{6}, \frac{1}{2},$ and $\frac{1}{3}$ is 12.

$$=\dfrac{12\left(1+\dfrac{3}{4}-\dfrac{1}{6}\right)}{12\left(\dfrac{1}{2}+\dfrac{1}{3}\right)}$$

Step 1:   Multiply the numerator and denominator of the complex fraction by 12.

TIP: In step 1, we multiply the original expression by $\frac{12}{12}$, which equals 1.

$$=\dfrac{12\cdot 1+12\cdot\dfrac{3}{4}-12\cdot\dfrac{1}{6}}{12\cdot\dfrac{1}{2}+12\cdot\dfrac{1}{3}}$$

Step 2:   Apply the distributive property.

$$=\dfrac{12\cdot 1+\overset{3}{\cancel{12}}\cdot\dfrac{3}{\cancel{4}}-\overset{2}{\cancel{12}}\cdot\dfrac{1}{\cancel{6}}}{\overset{6}{\cancel{12}}\cdot\dfrac{1}{\cancel{2}}+\overset{4}{\cancel{12}}\cdot\dfrac{1}{\cancel{3}}}$$

Simplify each term.

$$=\dfrac{12+9-2}{6+4}$$

$$=\dfrac{19}{10}$$

Step 3:   Simplify. This is the same result as in Example 2.

Skill Practice   Simplify the expression.

4. $\dfrac{1-\dfrac{3}{5}}{\dfrac{1}{4}-\dfrac{7}{10}+1}$

Answer

4. $\dfrac{8}{11}$

Classroom Example: p. 522, Exercise 26

Example 5 Simplifying a Complex Fraction (Method II)

Simplify the expression.

$$\dfrac{\dfrac{1}{x}+\dfrac{1}{y}}{x-\dfrac{y^2}{x}}$$

Solution:

$$\dfrac{\dfrac{1}{x}+\dfrac{1}{y}}{x-\dfrac{y^2}{x}}$$

The LCD of the expressions $\frac{1}{x}$, $\frac{1}{y}$, x, and $\frac{y^2}{x}$ is xy.

$$=\dfrac{xy\left(\dfrac{1}{x}+\dfrac{1}{y}\right)}{xy\left(x-\dfrac{y^2}{x}\right)}$$

Step 1: Multiply numerator and denominator of the complex fraction by xy.

$$=\dfrac{xy\cdot\dfrac{1}{x}+xy\cdot\dfrac{1}{y}}{xy\cdot x-xy\cdot\dfrac{y^2}{x}}$$

Step 2: Apply the distributive property, and simplify each term.

$$=\dfrac{y+x}{x^2y-y^3}$$

$$=\dfrac{y+x}{y(x^2-y^2)}$$

Step 3: Factor completely, and reduce common factors.

$$=\dfrac{\overset{1}{\cancel{y+x}}}{y\,\cancel{(x+y)}(x-y)}$$

Note that $(y+x)=(x+y)$.

$$=\dfrac{1}{y(x-y)}$$

This is the same result as in Example 3.

Skill Practice Simplify the expression.

5. $\dfrac{\dfrac{z}{3}-\dfrac{3}{z}}{1+\dfrac{3}{z}}$

Answer

5. $\dfrac{z-3}{3}$

Example 6 Simplifying a Complex Fraction (Method II)

Classroom Example: p. 522, Exercise 28

Simplify the expression.
$$\dfrac{\dfrac{1}{k+1} - 1}{\dfrac{1}{k+1} + 1}$$

Solution:

$$\dfrac{\dfrac{1}{k+1} - 1}{\dfrac{1}{k+1} + 1}$$

The LCD of $\dfrac{1}{k+1}$ and 1 is $(k+1)$.

$$= \dfrac{(k+1)\left(\dfrac{1}{k+1} - 1\right)}{(k+1)\left(\dfrac{1}{k+1} + 1\right)}$$

Step 1: Multiply numerator and denominator of the complex fraction by $(k+1)$.

$$= \dfrac{(\overset{1}{\cancel{k+1}}) \cdot \dfrac{1}{\cancel{(k+1)}} - (k+1) \cdot 1}{(\overset{1}{\cancel{k+1}}) \cdot \dfrac{1}{\cancel{(k+1)}} + (k+1) \cdot 1}$$

Step 2: Apply the distributive property.

$$= \dfrac{1 - (k+1)}{1 + (k+1)}$$

Simplify.

$$= \dfrac{1 - k - 1}{1 + k + 1}$$

$$= \dfrac{-k}{k+2}$$

Step 3: The expression is already in lowest terms.

Skill Practice Simplify the expression.

6. $\dfrac{\dfrac{4}{p-3} + 1}{1 + \dfrac{2}{p-3}}$

Answer

6. $\dfrac{p+1}{p-1}$

Section 7.5 Practice Exercises

Vocabulary and Key Concepts

For additional exercises, see Classroom Activities 7.5A–7.5B in the *Student's Resource Manual* at www.mhhe.com/moh.

1. A ___complex___ fraction is an expression containing one or more fractions in the numerator, denominator, or both.

Review Exercises

For Exercises 2–3, simplify the expression.

2. $\dfrac{y(2y+9)}{y^2(2y+9)}$ $\dfrac{1}{y}$

3. $\dfrac{a+5}{2a^2+7a-15}$ $\dfrac{1}{2a-3}$

 Writing Translating Expression  Geometry Scientific Calculator Video

For Exercises 4–6, perform the indicated operations.

4. $\dfrac{2}{w-2}+\dfrac{3}{w}$ $\dfrac{5w-6}{w(w-2)}$

5. $\dfrac{6}{5}-\dfrac{3}{5k-10}$ $\dfrac{3(2k-5)}{5(k-2)}$

6. $\dfrac{x^2-2xy+y^2}{x^4-y^4}\div\dfrac{3x^2y-3xy^2}{x^2+y^2}$ $\dfrac{1}{3xy(x+y)}$

Concepts 1–2: Simplifying Complex Fractions (Methods I and II)

For Exercises 7–34, simplify the complex fractions using either method. **(See Examples 1–6.)**

7. $\dfrac{\frac{7}{18y}}{\frac{2}{9}}$ $\dfrac{7}{4y}$

8. $\dfrac{\frac{a^2}{2a-3}}{\frac{5a}{8a-12}}$ $\dfrac{4a}{5}$

9. $\dfrac{\frac{3x+2y}{2y}}{\frac{6x+4y}{2}}$ $\dfrac{1}{2y}$

10. $\dfrac{\frac{2x-10}{4}}{\frac{x^2-5x}{3x}}$ $\dfrac{3}{2}$

11. $\dfrac{\frac{8a^4b^3}{3c}}{\frac{a^7b^2}{9c}}$ $\dfrac{24b}{a^3}$

12. $\dfrac{\frac{12x^2}{5y}}{\frac{8x^6}{9y^2}}$ $\dfrac{27y}{10x^4}$

13. $\dfrac{\frac{4r^3s}{t^5}}{\frac{2s^7}{r^2t^9}}$ $\dfrac{2r^5t^4}{s^6}$

14. $\dfrac{\frac{5p^4q}{w^4}}{\frac{10p^2}{qw^2}}$ $\dfrac{p^2q^2}{2w^2}$

 15. $\dfrac{\frac{1}{8}+\frac{4}{3}}{\frac{1}{2}-\frac{5}{12}}$ $\dfrac{35}{2}$

16. $\dfrac{\frac{8}{9}-\frac{1}{3}}{\frac{7}{6}+\frac{1}{9}}$ $\dfrac{10}{23}$

 17. $\dfrac{\frac{1}{h}+\frac{1}{k}}{\frac{1}{hk}}$ $k+h$

18. $\dfrac{\frac{1}{b}+1}{\frac{1}{b}}$ $1+b$

19. $\dfrac{\frac{n+1}{n^2-9}}{\frac{2}{n+3}}$ $\dfrac{n+1}{2(n-3)}$

20. $\dfrac{\frac{5}{k-5}}{\frac{k+1}{k^2-25}}$ $\dfrac{5(k+5)}{k+1}$

 21. $\dfrac{2+\frac{1}{x}}{4+\frac{1}{x}}$ $\dfrac{2x+1}{4x+1}$

22. $\dfrac{6+\frac{6}{k}}{1+\frac{1}{k}}$ 6

23. $\dfrac{\frac{m}{7}-\frac{7}{m}}{\frac{1}{7}+\frac{1}{m}}$ $m-7$

24. $\dfrac{\frac{2}{p}+\frac{p}{2}}{\frac{p}{3}-\frac{3}{p}}$ $\dfrac{3(4+p^2)}{2(p-3)(p+3)}$

25. $\dfrac{\frac{1}{5}-\frac{1}{y}}{\frac{7}{10}+\frac{1}{y^2}}$ $\dfrac{2y(y-5)}{7y^2+10}$

26. $\dfrac{\frac{1}{m^2}+\frac{2}{3}}{\frac{1}{m}-\frac{5}{6}}$ $\dfrac{2(3+2m^2)}{m(6-5m)}$

 27. $\dfrac{\frac{8}{a+4}+2}{\frac{12}{a+4}-2}$ $-\dfrac{a+8}{a-2}$ or $\dfrac{a+8}{2-a}$

28. $\dfrac{\frac{2}{w+1}+3}{\frac{3}{w+1}+4}$ $\dfrac{3w+5}{4w+7}$

29. $\dfrac{1-\frac{4}{t^2}}{1-\frac{2}{t}-\frac{8}{t^2}}$ $\dfrac{t-2}{t-4}$

30. $\dfrac{1-\frac{9}{p^2}}{1-\frac{1}{p}-\frac{6}{p^2}}$ $\dfrac{p+3}{p+2}$

31. $\dfrac{t+4+\frac{3}{t}}{t-4-\frac{5}{t}}$ $\dfrac{t+3}{t-5}$

32. $\dfrac{\frac{9}{4m}+\frac{9}{2m^2}}{\frac{3}{2}+\frac{3}{m}}$ $\dfrac{3}{2m}$

33. $\dfrac{\frac{1}{k-6}-1}{\frac{2}{k-6}-2}$ $\dfrac{1}{2}$

34. $\dfrac{\frac{3}{y-3}+4}{8+\frac{6}{y-3}}$ $\dfrac{1}{2}$

 For Exercises 35–38, write the English phrases as algebraic expressions. Then simplify the expressions.

35. The sum of one-half and two-thirds, divided by five $\dfrac{\frac{1}{2}+\frac{2}{3}}{5};\ \dfrac{7}{30}$

36. The quotient of ten and the difference of two-fifths and one-fourth $\dfrac{10}{\frac{2}{5}-\frac{1}{4}};\ \dfrac{200}{3}$

37. The quotient of three and the sum of two-thirds and three-fourths $\dfrac{3}{\frac{2}{3}+\frac{3}{4}};\ \dfrac{36}{17}$

38. The difference of three-fifths and one-half, divided by four $\dfrac{\frac{3}{5}-\frac{1}{2}}{4};\ \dfrac{1}{40}$

 Writing   Translating Expression   Geometry   Scientific Calculator   Video

39. In electronics, resistors oppose the flow of current. For two resistors in parallel, the total resistance is given by

$$R = \dfrac{1}{\dfrac{1}{R_1} + \dfrac{1}{R_2}}$$

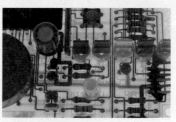

© S. Meltzer/PhotoLink/
Getty Images RF

a. Find the total resistance if $R_1 = 2\ \Omega$ (ohms) and $R_2 = 3\ \Omega$. $\frac{6}{5}\ \Omega$

b. Find the total resistance if $R_1 = 10\ \Omega$ and $R_2 = 15\ \Omega$. $6\ \Omega$

40. Suppose that Joëlle makes a round trip to a location that is d miles away. If the average rate going to the location is r_1 and the average rate on the return trip is given by r_2, the average rate of the entire trip, R, is given by

$$R = \dfrac{2d}{\dfrac{d}{r_1} + \dfrac{d}{r_2}}$$

a. Find the average rate of a trip to a destination 30 mi away when the average rate going there is 60 mph and the average rate returning home is 45 mph. (Round to the nearest tenth of a mile per hour.)
51.4 mph average

b. Find the average rate of a trip to a destination that is 50 mi away if the driver travels at the same rates as in part (a). (Round to the nearest tenth of a mile per hour.)
51.4 mph

c. Compare your answers from parts (a) and (b) and explain the results in the context of the problem.
Because the rates going to and leaving from the destination are the same, the average rate is unchanged. The average rate is not affected by the distance traveled.

Expanding Your Skills

For Exercises 41–50, simplify the complex fractions using either method.

41. $\dfrac{x^{-1} - y^{-1}}{x^{-2} - y^{-2}}$ $\dfrac{xy}{y+x}$

42. $\dfrac{a^{-2} - 1}{a^{-1} - 1}$ $\dfrac{1+a}{a}$

43. $\dfrac{2x^{-1} + 8y^{-1}}{4x^{-1}}$ $\dfrac{y+4x}{2y}$ $\left(Hint:\ 2x^{-1} = \dfrac{2}{x}\right)$

44. $\dfrac{6a^{-1} + 4b^{-1}}{8b^{-1}}$ $\dfrac{3b+2a}{4a}$

45. $\dfrac{(mn)^{-2}}{m^{-2} + n^{-2}}$ $\dfrac{1}{n^2 + m^2}$

46. $\dfrac{(xy)^{-1}}{2x^{-1} + 3y^{-1}}$ $\dfrac{1}{2y + 3x}$

47. $\dfrac{\dfrac{1}{z^2 - 9} + \dfrac{2}{z+3}}{\dfrac{3}{z-3}}$ $\dfrac{2z-5}{3(z+3)}$

48. $\dfrac{\dfrac{5}{w^2 - 25} - \dfrac{3}{w+5}}{\dfrac{4}{w-5}}$ $\dfrac{-3w+20}{4(w+5)}$

49. $\dfrac{\dfrac{2}{x-1} + 2}{\dfrac{2}{x+1} - 2}$ $-\dfrac{x+1}{x-1}$ or $\dfrac{x+1}{1-x}$

50. $\dfrac{\dfrac{1}{y-3} + 1}{\dfrac{2}{y+3} - 1}$ $\dfrac{(y+3)(y-2)}{(y+1)(y-3)}$

For Exercises 51–52, simplify the complex fractions. (*Hint:* Use the order of operations and begin with the fraction on the lower right.)

51. $1 + \dfrac{1}{1+1}$ $\dfrac{3}{2}$

52. $1 + \dfrac{1}{1 + \dfrac{1}{1+1}}$ $\dfrac{5}{3}$

Concepts

1. **Introduction to Rational Equations**
2. **Solving Rational Equations**
3. **Solving Formulas Involving Rational Expressions**

1. Introduction to Rational Equations

Thus far we have studied two specific types of equations in one variable: linear equations and quadratic equations. Recall,

$$ax + b = c, \text{ where } a \neq 0, \text{ is a } \textbf{linear equation}.$$

$$ax^2 + bx + c = 0, \text{ where } a \neq 0, \text{ is a } \textbf{quadratic equation}.$$

We will now study another type of equation called a rational equation.

> ### Definition of a Rational Equation
>
> An equation with one or more rational expressions is called a **rational equation**.

The following equations are rational equations:

$$\frac{1}{x} + \frac{1}{3} = \frac{5}{6} \qquad \frac{6}{t^2 - 7t + 12} + \frac{2t}{t - 3} = \frac{3t}{t - 4}$$

To understand the process of solving a rational equation, first review the process of clearing fractions. We can clear the fractions in an equation by multiplying both sides of the equation by the LCD of all terms.

Classroom Example: p. 532, Exercise 12

Example 1   Solving an Equation Containing Fractions

Solve. $\dfrac{y}{2} + \dfrac{y}{4} = 6$

Solution:

$$\frac{y}{2} + \frac{y}{4} = 6 \qquad \text{The LCD of all terms in the equation is } 4.$$

$$4\left(\frac{y}{2} + \frac{y}{4}\right) = 4(6) \qquad \text{Multiply both sides of the equation by 4 to clear fractions.}$$

$$\overset{2}{\cancel{4}} \cdot \frac{y}{2} + \overset{1}{\cancel{4}} \cdot \frac{y}{4} = 4(6) \qquad \text{Apply the distributive property.}$$

$$2y + y = 24 \qquad \text{Clear fractions.}$$

$$3y = 24 \qquad \text{Solve the resulting equation (linear).}$$

$$y = 8$$

$$\underline{\text{Check:}} \qquad \frac{y}{2} + \frac{y}{4} = 6$$

$$\frac{(8)}{2} + \frac{(8)}{4} \overset{?}{=} 6$$

$$4 + 2 \overset{?}{=} 6$$

The solution set is {8}. $\qquad 6 \overset{?}{=} 6 \checkmark \text{ (True)}$

Skill Practice Solve the equation.

1. $\dfrac{t}{5} - \dfrac{t}{4} = 2$

2. Solving Rational Equations

The same process of clearing fractions is used to solve rational equations when variables are present in the denominator. However, variables in the denominator make it necessary to take note of the restricted values.

Example 2 Solving a Rational Equation

Solve the equation. $\quad \dfrac{x+1}{x} + \dfrac{1}{3} = \dfrac{5}{6}$

Classroom Example: p. 532, Exercise 26

Solution:

$$\dfrac{x+1}{x} + \dfrac{1}{3} = \dfrac{5}{6}$$

The LCD of all the expressions is $6x$. The restricted value is $x = 0$.

$$6x \cdot \left(\dfrac{x+1}{x} + \dfrac{1}{3}\right) = 6x \cdot \left(\dfrac{5}{6}\right)$$

Multiply by the LCD.

$$6\overset{1}{x} \cdot \left(\dfrac{x+1}{x}\right) + \overset{2}{6x} \cdot \left(\dfrac{1}{3}\right) = \overset{1}{6x} \cdot \left(\dfrac{5}{6}\right)$$

Apply the distributive property.

$$6(x+1) + 2x = 5x$$

Clear fractions.

$$6x + 6 + 2x = 5x$$

Solve the resulting equation.

$$8x + 6 = 5x$$

$$3x = -6$$

$$x = -2$$

-2 is not a restricted value.

TIP: The restricted value tells us that $x = 0$ is *not* a possible solution to the equation.

Check: $\quad \dfrac{x+1}{x} + \dfrac{1}{3} = \dfrac{5}{6}$

$$\dfrac{(-2)+1}{(-2)} + \dfrac{1}{3} \overset{?}{=} \dfrac{5}{6}$$

$$\dfrac{-1}{-2} + \dfrac{1}{3} \overset{?}{=} \dfrac{5}{6}$$

$$\dfrac{1}{2} + \dfrac{1}{3} \overset{?}{=} \dfrac{5}{6}$$

$$\dfrac{3}{6} + \dfrac{2}{6} \overset{?}{=} \dfrac{5}{6}$$

The solution set is $\{-2\}$.

$$\dfrac{5}{6} \overset{?}{=} \dfrac{5}{6} \checkmark \text{ (True)}$$

Skill Practice Solve the equation.

2. $\dfrac{3}{4} + \dfrac{5+a}{a} = \dfrac{1}{2}$

Answers

1. $\{-40\}$ **2.** $\{-4\}$

Classroom Example: p. 533, Exercise 32

Example 3   Solving a Rational Equation

Solve the equation. $1 + \dfrac{3a}{a-2} = \dfrac{6}{a-2}$

Solution:

$$1 + \dfrac{3a}{a-2} = \dfrac{6}{a-2}$$

The LCD of all the expressions is $a - 2$. The restricted value is $a = 2$.

$$(a-2)\left(1 + \dfrac{3a}{a-2}\right) = (a-2)\left(\dfrac{6}{a-2}\right)$$

Multiply by the LCD.

$$(a-2)1 + (a-2)\overset{1}{\left(\dfrac{3a}{a-2}\right)} = (a-2)\overset{1}{\left(\dfrac{6}{a-2}\right)}$$

Apply the distributive property.

$$a - 2 + 3a = 6$$

Solve the resulting equation (linear).

$$4a - 2 = 6$$

$$4a = 8$$

$$a = 2$$

2 is a restricted value.

Check: $1 + \dfrac{3a}{a-2} = \dfrac{6}{a-2}$

$$1 + \dfrac{3(2)}{(2)-2} \overset{?}{=} \dfrac{6}{(2)-2}$$

$$1 + \dfrac{6}{0} \overset{?}{=} \dfrac{6}{0}$$

The denominator is 0 when $a = 2$.

Instructor Note: Have students explain the difference between

$$\dfrac{2}{x+3} + \dfrac{5}{x+2}$$

and

$$\dfrac{2}{x+3} = \dfrac{5}{x+2}$$

Because the value $a = 2$ makes the denominator zero in one (or more) of the rational expressions within the equation, the equation is undefined for $a = 2$. No other potential solutions exist for the equation, therefore, the solution set is { }.

Skill Practice Solve the equation.

3. $\dfrac{x}{x+1} - 2 = \dfrac{-1}{x+1}$

Examples 1–3 show that the steps to solve a rational equation mirror the process of clearing fractions. However, there is one significant difference. The solutions of a rational equation must not make the denominator equal to zero for any expression within the equation.

Answer

3. { } (The value −1 does not check.)

The steps to solve a rational equation are summarized as follows.

Solving a Rational Equation

Step 1   Factor the denominators of all rational expressions. Identify the restricted values.

Step 2   Identify the LCD of all expressions in the equation.

Step 3   Multiply both sides of the equation by the LCD.

Step 4   Solve the resulting equation.

Step 5   Check potential solutions in the original equation.

After multiplying by the LCD and then simplifying, the rational equation will be either a linear equation or higher degree equation.

Example 4   Solving a Rational Equation

Classroom Example: p. 532, Exercise 24

Solve the equation. $1 - \dfrac{4}{p} = -\dfrac{3}{p^2}$

Solution:

$$1 - \frac{4}{p} = -\frac{3}{p^2}$$

Step 1:   The denominators are already factored. The restricted value is $p = 0$.

Step 2:   The LCD of all expressions is p^2.

$$p^2\left(1 - \frac{4}{p}\right) = p^2\left(-\frac{3}{p^2}\right)$$

Step 3:   Multiply by the LCD.

$$p^2(1) - \overset{p}{\cancel{p^2}}\left(\frac{4}{\cancel{p}}\right) = \overset{1}{\cancel{p^2}}\left(-\frac{3}{\cancel{p^2}}\right)$$

Apply the distributive property.

$$p^2 - 4p = -3$$

Step 4:   Solve the resulting quadratic equation.

$$p^2 - 4p + 3 = 0$$

Set the equation equal to zero and factor.

$$(p - 3)(p - 1) = 0$$

$$p - 3 = 0 \quad \text{or} \quad p - 1 = 0$$

Set each factor equal to zero.

$$p = 3 \quad \text{or} \quad p = 1$$

Step 5:   Check: $p = 3$ Check: $p = 1$

3 and 1 are not restricted values.

$$1 - \frac{4}{p} = -\frac{3}{p^2} \qquad 1 - \frac{4}{p} = -\frac{3}{p^2}$$

$$1 - \frac{4}{(3)} \overset{?}{=} -\frac{3}{(3)^2} \qquad 1 - \frac{4}{(1)} \overset{?}{=} -\frac{3}{(1)^2}$$

$$\frac{3}{3} - \frac{4}{3} \overset{?}{=} -\frac{3}{9} \qquad 1 - 4 \overset{?}{=} -3$$

The solution set is $\{3, 1\}$.

$$-\frac{1}{3} \overset{?}{=} -\frac{1}{3} \checkmark \qquad -3 \overset{?}{=} -3 \checkmark$$

Skill Practice   Solve the equation.

4. $\dfrac{z}{2} - \dfrac{1}{2z} = \dfrac{12}{z}$

Answer

4. $\{5, -5\}$

Classroom Example: p. 533,
Exercise 42

Example 5   **Solving a Rational Equation**

Solve the equation. $\dfrac{6}{t^2 - 7t + 12} + \dfrac{2t}{t-3} = \dfrac{3t}{t-4}$

Solution:

$$\frac{6}{t^2 - 7t + 12} + \frac{2t}{t-3} = \frac{3t}{t-4}$$

$$\frac{6}{(t-3)(t-4)} + \frac{2t}{t-3} = \frac{3t}{t-4}$$

Step 1:   Factor the denominators. The restricted values are $t = 3$ and $t = 4$.

Step 2:   The LCD is $(t-3)(t-4)$.

Step 3:   Multiply by the LCD on both sides.

$$(t-3)(t-4)\left[\frac{6}{(t-3)(t-4)} + \frac{2t}{t-3}\right] = (t-3)(t-4)\left(\frac{3t}{t-4}\right)$$

$$(t-3)(t-4)\left[\frac{6}{(t-3)(t-4)}\right] + (t-3)(t-4)\left(\frac{2t}{t-3}\right) = (t-3)(t-4)\left(\frac{3t}{t-4}\right)$$

$$6 + 2t(t-4) = 3t(t-3)$$

$$6 + 2t^2 - 8t = 3t^2 - 9t$$

$$0 = 3t^2 - 2t^2 - 9t + 8t - 6$$

$$0 = t^2 - t - 6$$

$$0 = (t-3)(t+2)$$

$$t - 3 = 0 \quad \text{or} \quad t + 2 = 0$$

$$t = 3 \quad \text{or} \quad t = -2$$

3 is a restricted value, but -2 is not restricted.

Check: $t = 3$

3 cannot be a solution to the equation because it will make the denominator zero in the original equation.

$$\frac{6}{t^2 - 7t + 12} + \frac{2t}{t-3} = \frac{3t}{t-4}$$

$$\frac{6}{(3)^2 - 7(3) + 12} + \frac{2(3)}{(3)-3} \overset{?}{=} \frac{3(3)}{(3)-4}$$

$$\frac{6}{0} + \frac{6}{0} \overset{?}{=} \frac{9}{-1}$$

Zero in the denominator

The solution set is $\{-2\}$.

Step 4:   Solve the resulting equation.

Because the resulting equation is quadratic, set the equation equal to zero and factor.

Set each factor equal to zero.

Step 5:   Check the potential solutions in the original equation.

Check: $t = -2$

$$\frac{6}{t^2 - 7t + 12} + \frac{2t}{t-3} = \frac{3t}{t-4}$$

$$\frac{6}{(-2)^2 - 7(-2) + 12} + \frac{2(-2)}{(-2)-3} \overset{?}{=} \frac{3(-2)}{(-2)-4}$$

$$\frac{6}{4 + 14 + 12} + \frac{-4}{-5} \overset{?}{=} \frac{-6}{-6}$$

$$\frac{6}{30} + \frac{4}{5} \overset{?}{=} 1$$

$$\frac{1}{5} + \frac{4}{5} \overset{?}{=} 1 \checkmark \text{ (True)}$$

$t = -2$ is a solution.

Skill Practice   Solve the equation.

5. $\dfrac{-8}{x^2 + 6x + 8} + \dfrac{x}{x+4} = \dfrac{2}{x+2}$

Answer

5. $\{4\}$ (The value -4 does not check.)

Example 6	**Translating to a Rational Equation**

Classroom Example: p. 533, Exercise 48

Ten times the reciprocal of a number is added to four. The result is equal to the quotient of twenty-two and the number. Find the number.

Solution:

Let x represent the number.

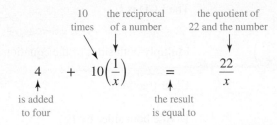

$$4 + \frac{10}{x} = \frac{22}{x}$$

Step 1: The denominators are already factored. The restricted value is $x = 0$.

Step 2: The LCD is x.

$$x\left(4 + \frac{10}{x}\right) = x\left(\frac{22}{x}\right)$$

Step 3: Multiply both sides by the LCD.

$$4x + 10 = 22$$

Apply the distributive property.

$$4x = 12$$

Step 4: Solve the resulting linear equation.

$x = 3$ is a potential solution.

Step 5: 3 is not a restricted value. Substituting $x = 3$ into the original equation verifies that it is a solution.

The number is 3.

Skill Practice

6. The quotient of ten and a number is two less than four times the reciprocal of the number. Find the number.

3. Solving Formulas Involving Rational Expressions

A rational equation may have more than one variable. To solve for a specific variable within a rational equation, we can still apply the principle of clearing fractions.

Classroom Example: p. 533,
Exercise 54

Example 7   Solving Formulas Involving Rational Equations

Solve for k. $F = \dfrac{ma}{k}$

Solution:

To solve for k, we must clear fractions so that k no longer appears in the denominator.

$$F = \frac{ma}{k} \qquad\qquad \text{The LCD is } k.$$

$$k \cdot (F) = k \cdot \left(\frac{ma}{k}\right) \qquad\qquad \text{Multiply both sides of the equation by the LCD.}$$

$$kF = ma \qquad\qquad \text{Clear fractions.}$$

$$\frac{kF}{F} = \frac{ma}{F} \qquad\qquad \text{Divide both sides by } F.$$

$$k = \frac{ma}{F}$$

Skill Practice

7. Solve for t. $C = \dfrac{rt}{d}$

Classroom Example: p. 533,
Exercise 56

Example 8   Solving Formulas Involving Rational Equations

A formula to find the height of a trapezoid given its area and the lengths of the two parallel sides is $h = \dfrac{2A}{B+b}$. Solve for b, the length of one of the parallel sides.

Solution:

To solve for b, we must clear fractions so that b no longer appears in the denominator.

$$h = \frac{2A}{B+b} \qquad\qquad \text{The LCD is } (B+b).$$

$$h(B+b) = \left(\frac{2A}{B+b}\right) \cdot (B+b) \qquad\qquad \text{Multiply both sides of the equation by the LCD.}$$

$$hB + hb = 2A \qquad\qquad \text{Apply the distributive property.}$$

$$hb = 2A - hB \qquad\qquad \text{Subtract } hB \text{ from both sides to isolate the } b \text{ term.}$$

$$\frac{hb}{h} = \frac{2A - hB}{h} \qquad\qquad \text{Divide by } h.$$

$$b = \frac{2A - hB}{h}$$

Avoiding Mistakes

Algebra is case-sensitive. The variables B and b represent different values.

Answers

7. $t = \dfrac{Cd}{r}$

8. $x = \dfrac{3 + 2y}{y}$ or $x = \dfrac{3}{y} + 2$

Skill Practice

8. Solve the formula for x. $y = \dfrac{3}{x-2}$

TIP: The solution to Example 8 can be written in several forms. The quantity

$$\frac{2A - hB}{h}$$

can be left as a single rational expression or can be split into two fractions and simplified.

$$b = \frac{2A - hB}{h} = \frac{2A}{h} - \frac{hB}{h} = \frac{2A}{h} - B$$

Example 9   Solving Formulas Involving Rational Equations

Classroom Example: p. 533, Exercise 64

Solve for z. $y = \dfrac{x - z}{x + z}$

Solution:

To solve for z, we must clear fractions so that z no longer appears in the denominator.

$y = \dfrac{x - z}{x + z}$ LCD is $(x + z)$.

$y(x + z) = \left(\dfrac{x - z}{x + z}\right)(x + z)$ Multiply both sides of the equation by the LCD.

$yx + yz = x - z$ Apply the distributive property.

$yz + z = x - yx$ Collect z terms on one side of the equation and collect terms not containing z on the other side.

$z(y + 1) = x - yx$ Factor out z.

$z = \dfrac{x - yx}{y + 1}$ Divide by $y + 1$ to solve for z.

Skill Practice

9. Solve for h. $\dfrac{b}{x} = \dfrac{a}{h} + 1$

Answer

9. $h = \dfrac{ax}{b - x}$ or $\dfrac{-ax}{x - b}$

Section 7.6   Practice Exercises

For additional exercises, see Classroom Activities 7.6A–7.6B in the *Student's Resource Manual* at www.mhhe.com/moh.

Vocabulary and Key Concepts

1. a. The equation $4x + 7 = -18$ is an example of a ___linear___ equation, whereas $3y^2 - 4y - 7 = 0$ is an example of a ___quadratic___ equation.

b. The equation $\dfrac{6}{x + 2} + \dfrac{1}{4} = \dfrac{2}{3}$ is an example of a ___rational___ equation.

c. After solving a rational equation, check each potential solution to determine if it makes the ___denominator___ equal to zero in one or more of the rational expressions. If so, that potential solution is not part of the solution set.

 Writing   Translating Expression   Geometry   Scientific Calculator   Video

Review Exercises

For Exercises 2–7, perform the indicated operations.

2. $\dfrac{2}{x-3} - \dfrac{3}{x^2-x-6}$ $\dfrac{2x+1}{(x-3)(x+2)}$

3. $\dfrac{2x-6}{4x^2+7x-2} \div \dfrac{x^2-5x+6}{x^2-4}$ $\dfrac{2}{4x-1}$

4. $\dfrac{2y}{y-3} + \dfrac{4}{y^2-9}$ $\dfrac{2(y+2)(y+1)}{(y-3)(y+3)}$

5. $\dfrac{h-\dfrac{1}{h}}{\dfrac{1}{5}-\dfrac{1}{5h}}$ $5(h+1)$

6. $\dfrac{w-4}{w^2-9} \cdot \dfrac{w-3}{w^2-8w+16}$ $\dfrac{1}{(w+3)(w-4)}$

7. $1+\dfrac{1}{x}-\dfrac{12}{x^2}$ $\dfrac{(x+4)(x-3)}{x^2}$

Concept 1: Introduction to Rational Equations

For Exercises 8–13, solve the equations by first clearing the fractions. **(See Example 1.)**

8. $\dfrac{1}{3}z + \dfrac{2}{3} = -2z + 10$ $\{4\}$

9. $\dfrac{5}{2} + \dfrac{1}{2}b = 5 - \dfrac{1}{3}b$ $\{3\}$

10. $\dfrac{3}{2}p + \dfrac{1}{3} = \dfrac{2p-3}{4}$ $\left\{-\dfrac{13}{12}\right\}$

11. $\dfrac{5}{3} - \dfrac{1}{6}k = \dfrac{3k+5}{4}$ $\left\{\dfrac{5}{11}\right\}$

12. $\dfrac{2x-3}{4} + \dfrac{9}{10} = \dfrac{x}{5}$ $\left\{-\dfrac{1}{2}\right\}$

13. $\dfrac{4y+2}{3} - \dfrac{7}{6} = -\dfrac{y}{6}$ $\left\{\dfrac{1}{3}\right\}$

Concept 2: Solving Rational Equations

14. For the equation

$$\dfrac{1}{w} - \dfrac{1}{2} = -\dfrac{1}{4}$$

  a. Identify the restricted values. $w = 0$

  b. Identify the LCD of the fractions in the equation. $4w$

  c. Solve the equation. $\{4\}$

15. For the equation

$$\dfrac{3}{z} - \dfrac{4}{5} = -\dfrac{1}{5}$$

  a. Identify the restricted values. $z = 0$

  b. Identify the LCD of the fractions in the equation. $5z$

  c. Solve the equation. $\{5\}$

16. Identify the LCD of all the denominators in the equation.

$$\dfrac{x+1}{x^2+2x-3} = \dfrac{1}{x+3} - \dfrac{1}{x-1}$$ $(x+3)(x-1)$

For Exercises 17–46, solve the equations. **(See Examples 2–5.)**

17. $\dfrac{1}{8} = \dfrac{3}{5} + \dfrac{5}{y}$ $\left\{-\dfrac{200}{19}\right\}$

18. $\dfrac{2}{7} - \dfrac{1}{x} = \dfrac{2}{3}$ $\left\{-\dfrac{21}{8}\right\}$

19. $\dfrac{7}{4a} = \dfrac{3}{a-5}$ $\{-7\}$

20. $\dfrac{2}{x+4} = \dfrac{5}{3x}$ $\{20\}$

21. $\dfrac{5}{6x} + \dfrac{7}{x} = 1$ $\left\{\dfrac{47}{6}\right\}$

22. $\dfrac{14}{3x} - \dfrac{5}{x} = 2$ $\left\{-\dfrac{1}{6}\right\}$

23. $1 - \dfrac{2}{y} = \dfrac{3}{y^2}$ $\{3, -1\}$

24. $1 - \dfrac{2}{m} = \dfrac{8}{m^2}$ $\{4, -2\}$

25. $\dfrac{a+1}{a} = 1 + \dfrac{a-2}{2a}$ $\{4\}$

26. $\dfrac{7b-4}{5b} = \dfrac{9}{5} - \dfrac{4}{b}$ $\{8\}$

27. $\dfrac{w}{5} - \dfrac{w+3}{w} = -\dfrac{3}{w}$
  $\{5\}$ (The value 0 does not check.)

28. $\dfrac{t}{12} + \dfrac{t+3}{3t} = \dfrac{1}{t}$
  $\{-4\}$ (The value 0 does not check.)

29. $\dfrac{2}{m+3} = \dfrac{5}{4m+12} - \dfrac{3}{8}$
$\{-5\}$

30. $\dfrac{2}{4n-4} - \dfrac{7}{4} = \dfrac{-3}{n-1}$ $\{3\}$

31. $\dfrac{p}{p-4} - 5 = \dfrac{4}{p-4}$
$\{\ \}$ (The value 4 does not check.)

32. $\dfrac{-5}{q+5} = \dfrac{q}{q+5} + 2$
$\{\ \}$ (The value -5 does not check.)

33. $\dfrac{2t}{t+2} - 2 = \dfrac{t-8}{t+2}$ $\{4\}$

34. $\dfrac{4w}{w-3} - 3 = \dfrac{3w-1}{w-3}$ $\{5\}$

35. $\dfrac{x^2-x}{x-2} = \dfrac{12}{x-2}$ $\{4, -3\}$

36. $\dfrac{x^2+9}{x+4} = \dfrac{-10x}{x+4}$ $\{-9, -1\}$

37. $\dfrac{x^2+3x}{x-1} = \dfrac{4}{x-1}$
$\{-4\}$ (The value 1 does not check.)

38. $\dfrac{2x^2-21}{2x-3} = \dfrac{-11x}{2x-3}$
$\{-7\}$ (The value $\frac{3}{2}$ does not check.)

 39. $\dfrac{2x}{x+4} - \dfrac{8}{x-4} = \dfrac{2x^2+32}{x^2-16}$
$\{\ \}$ (The value -4 does not check.)

40. $\dfrac{4x}{x+3} - \dfrac{12}{x-3} = \dfrac{4x^2+36}{x^2-9}$
$\{\ \}$ (The value -3 does not check.)

41. $\dfrac{x}{x+6} = \dfrac{72}{x^2-36} + 4$ $\{4\}$ (The value -6 does not check.)

42. $\dfrac{y}{y+4} = \dfrac{32}{y^2-16} + 3$ $\{2\}$ (The value -4 does not check.)

43. $\dfrac{5}{3x-3} - \dfrac{2}{x-2} = \dfrac{7}{x^2-3x+2}$ $\{-25\}$

44. $\dfrac{6}{5a+10} - \dfrac{1}{a-5} = \dfrac{4}{a^2-3a-10}$ $\{60\}$

45. $\dfrac{w}{w-3} = \dfrac{17}{w^2-7w+12} + \dfrac{1}{w-4}$ $\{-2, 7\}$

46. $\dfrac{y}{y+6} = \dfrac{-6}{y^2+7y+6} + \dfrac{2}{y+1}$ $\{-2, 3\}$

For Exercises 47–50, translate to a rational equation and solve. **(See Example 6.)**

 **47.** The reciprocal of a number is added to three. The result is the quotient of 25 and the number. Find the number.   The number is 8.

 48. The difference of three and the reciprocal of a number is equal to the quotient of 20 and the number. Find the number.   The number is 7.

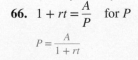 **49.** If a number added to five is divided by the difference of the number and two, the result is three-fourths. Find the number.   The number is -26.

 50. If twice a number added to three is divided by the number plus one, the result is three-halves. Find the number.   The number is -3.

Concept 3: Solving Formulas Involving Rational Expressions

For Exercises 51–68, solve for the indicated variable. **(See Examples 7–9.)**

51. $K = \dfrac{ma}{F}$ for m $m = \dfrac{FK}{a}$

52. $K = \dfrac{ma}{F}$ for a $a = \dfrac{FK}{m}$

53. $K = \dfrac{IR}{E}$ for E $E = \dfrac{IR}{K}$

54. $K = \dfrac{IR}{E}$ for R $R = \dfrac{KE}{I}$

55. $I = \dfrac{E}{R+r}$ for R
$R = \dfrac{E-Ir}{I}$ or $R = \dfrac{E}{I} - r$

56. $I = \dfrac{E}{R+r}$ for r
$r = \dfrac{E-IR}{I}$ or $r = \dfrac{E}{I} - R$

57. $h = \dfrac{2A}{B+b}$ for B
$B = \dfrac{2A-hb}{h}$ or $B = \dfrac{2A}{h} - b$

58. $\dfrac{C}{\pi r} = 2$ for r
$r = \dfrac{C}{2\pi}$

59. $\dfrac{V}{\pi h} = r^2$ for h
$h = \dfrac{V}{r^2 \pi}$

60. $\dfrac{V}{lw} = h$ for w
$w = \dfrac{V}{lh}$

61. $x = \dfrac{at+b}{t}$ for t
$t = \dfrac{b}{x-a}$ or $t = \dfrac{-b}{a-x}$

 62. $\dfrac{T+mf}{m} = g$ for m
$m = \dfrac{T}{g-f}$ or $m = \dfrac{-T}{f-g}$

63. $\dfrac{x-y}{xy} = z$ for x
$x = \dfrac{y}{1-yz}$ or $x = \dfrac{-y}{yz-1}$

64. $\dfrac{w-n}{wn} = P$ for w
$w = \dfrac{n}{1-Pn}$ or $w = \dfrac{-n}{Pn-1}$

65. $a + b = \dfrac{2A}{h}$ for h $h = \dfrac{2A}{a+b}$

66. $1 + rt = \dfrac{A}{P}$ for P
$P = \dfrac{A}{1+rt}$

67. $\dfrac{1}{R} = \dfrac{1}{R_1} + \dfrac{1}{R_2}$ for R
$R = \dfrac{R_1 R_2}{R_2 + R_1}$

68. $\dfrac{b+a}{ab} = \dfrac{1}{f}$ for b
$b = \dfrac{fa}{a-f}$ or $b = \dfrac{-fa}{f-a}$

/ Writing   ↔ Translating Expression   △ Geometry   ▤ Scientific Calculator   ▶ Video

Problem Recognition Exercises

Comparing Rational Equations and Rational Expressions

Often adding or subtracting rational expressions is confused with solving rational equations. When adding rational expressions, we combine the terms to simplify the expression. When solving an equation, we clear the fractions and find numerical solutions, if possible. Both processes begin with finding the LCD, but the LCD is used differently in each process. Compare these two examples.

Example 1:

Add. $\dfrac{4}{x} + \dfrac{x}{3}$ (The LCD is $3x$.)

$$= \frac{3}{3} \cdot \left(\frac{4}{x}\right) + \left(\frac{x}{3}\right) \cdot \frac{x}{x}$$

$$= \frac{12}{3x} + \frac{x^2}{3x}$$

$$= \frac{12 + x^2}{3x} \quad \text{The answer is a} \atop \text{rational expression.}$$

Example 2:

Solve. $\dfrac{4}{x} + \dfrac{x}{3} = -\dfrac{8}{3}$ (The LCD is $3x$.)

$$\frac{3x}{1}\left(\frac{4}{x} + \frac{x}{3}\right) = \frac{3x}{1}\left(-\frac{8}{3}\right)$$

$$12 + x^2 = -8x$$

$$x^2 + 8x + 12 = 0$$

$$(x + 2)(x + 6) = 0$$

$$x + 2 = 0 \text{ or } x + 6 = 0$$

$$x = -2 \text{ or } x = -6 \quad \text{The answer is} \atop \text{the set } \{-2, -6\}.$$

For Exercises 1–20, solve the equations and simplify the expressions.

1. $\dfrac{y}{2y + 4} - \dfrac{2}{y^2 + 2y}$ $\dfrac{y-2}{2y}$

2. $\dfrac{1}{x + 2} + 2 = \dfrac{x + 11}{x + 2}$ $\{6\}$

3. $\dfrac{5t}{2} - \dfrac{t - 2}{3} = 5$ $\{2\}$

4. $3 - \dfrac{2}{a - 5}$ $\dfrac{3a - 17}{a - 5}$

5. $\dfrac{7}{6p^2} + \dfrac{2}{9p} + \dfrac{1}{3p^2}$ $\dfrac{4p + 27}{18p^2}$

6. $\dfrac{3b}{b + 1} - \dfrac{2b}{b - 1}$ $\dfrac{b(b - 5)}{(b - 1)(b + 1)}$

7. $4 + \dfrac{2}{h - 3} = 5$ $\{5\}$

8. $\dfrac{2}{w + 1} + \dfrac{3}{(w + 1)^2}$ $\dfrac{2w + 5}{(w + 1)^2}$

9. $\dfrac{1}{x - 6} - \dfrac{3}{x^2 - 6x} = \dfrac{4}{x}$ $\{7\}$

10. $\dfrac{3}{m} - \dfrac{6}{5} = -\dfrac{3}{m}$ $\{5\}$

11. $\dfrac{7}{2x + 2} + \dfrac{3x}{4x + 4}$ $\dfrac{3x + 14}{4(x + 1)}$

12. $\dfrac{10}{2t - 1} - 1 = \dfrac{t}{2t - 1}$ $\left\{\dfrac{11}{3}\right\}$

13. $\dfrac{3}{5x} + \dfrac{7}{2x} = 1$ $\left\{\dfrac{41}{10}\right\}$

14. $\dfrac{7}{t^2 - 5t} - \dfrac{3}{t - 5}$ $\dfrac{7 - 3t}{t(t - 5)}$

15. $\dfrac{5}{2a - 1} + 4$ $\dfrac{8a + 1}{2a - 1}$

16. $p - \dfrac{5p}{p - 2} = -\dfrac{10}{p - 2}$
$\{5\}$ (The value of 2 does not check.)

17. $\dfrac{3}{u} + \dfrac{12}{u^2 - 3u} = \dfrac{u + 1}{u - 3}$
$\{-1\}$ (The value 3 does not check.)

18. $\dfrac{5}{4k} - \dfrac{2}{6k}$ $\dfrac{11}{12k}$

19. $\dfrac{-2h}{h^2 - 9} + \dfrac{3}{h - 3}$ $\dfrac{h + 9}{(h - 3)(h + 3)}$

20. $\dfrac{3y}{y^2 - 5y + 4} = \dfrac{2}{y - 4} + \dfrac{3}{y - 1}$ $\{7\}$

Writing   Translating Expression   Geometry   Scientific Calculator   Video

Applications of Rational Equations and Proportions

1. Solving Proportions

In this section, we look at how rational equations can be used to solve a variety of applications. The first type of rational equation that will be applied is called a proportion.

Concepts

1. **Solving Proportions**
2. **Applications of Proportions and Similar Triangles**
3. **Distance, Rate, and Time Applications**
4. **Work Applications**

Definition of Ratio and Proportion

1. The **ratio** of a to b is $\dfrac{a}{b}$ ($b \neq 0$) and can also be expressed as $a{:}b$ or $a \div b$.

2. An equation that equates two ratios or rates is called a **proportion**. Therefore, if $b \neq 0$ and $d \neq 0$, then $\dfrac{a}{b} = \dfrac{c}{d}$ is a proportion.

A proportion can be solved by multiplying both sides of the equation by the LCD and clearing fractions.

Example 1 Solving a Proportion

Solve the proportion. $\dfrac{3}{11} = \dfrac{123}{w}$

Classroom Example: p. 543, Exercise 10

Solution:

$$\frac{3}{11} = \frac{123}{w} \qquad \text{The LCD is } 11w.$$

$$11w\left(\frac{3}{11}\right) = 11w\left(\frac{123}{w}\right) \qquad \text{Multiply by the LCD and clear fractions.}$$

$$3w = 11 \cdot 123 \qquad \text{Solve the resulting equation (linear).}$$

$$3w = 1353$$

$$\frac{3w}{3} = \frac{1353}{3}$$

$$w = 451$$

$$\underline{\text{Check}}: w = 451$$

$$\frac{3}{11} = \frac{123}{w}$$

$$\frac{3}{11} \stackrel{?}{=} \frac{123}{(451)}$$

The solution set is $\{451\}$.

$$\frac{3}{11} \stackrel{?}{=} \frac{3}{11} \checkmark \text{ (True)} \qquad \text{Simplify to lowest terms.}$$

Skill Practice Solve the proportion.

1. $\dfrac{10}{b} = \dfrac{2}{33}$

Answer

1. $\{165\}$

2. Applications of Proportions and Similar Triangles

Classroom Example: p. 543, Exercise 28

Example 2   Using a Proportion in an Application

For a recent year, the population of Alabama was approximately 4.2 million. At that time, Alabama had seven representatives in the U.S. House of Representatives. In the same year, North Carolina had a population of approximately 7.2 million. If representation in the House is based on population in equal proportions for each state, how many representatives did North Carolina have?

© Brand X Pictures/
PunchStock RF

Solution:

Let x represent the number of representatives for North Carolina.

Set up a proportion by writing two equivalent ratios.

$$\frac{\text{Population of Alabama}}{\text{number of representatives}} \rightarrow \frac{4.2}{7} = \frac{7.2}{x} \leftarrow \frac{\text{Population of North Carolina}}{\text{number of representatives}}$$

$$\frac{4.2}{7} = \frac{7.2}{x}$$

$$7x \cdot \frac{4.2}{7} = 7x \cdot \frac{7.2}{x} \qquad \text{Multiply by the LCD, } 7x.$$

$$4.2x = (7.2)(7) \qquad \text{Solve the resulting linear equation.}$$

$$4.2x = 50.4$$

$$\frac{4.2x}{4.2} = \frac{50.4}{4.2}$$

$$x = 12 \qquad \text{North Carolina had 12 representatives.}$$

> **TIP:** The equation from Example 2 could have been solved by first equating the cross products:
>
> $$\frac{4.2}{7} \diagdown\!\!\!\!\times \frac{7.2}{x}$$
>
> $$4.2x = (7.2)(7)$$
> $$4.2x = 50.4$$
> $$x = 12$$

Skill Practice

2. A university has a ratio of students to faculty of 105 to 2. If the student population at the university is 15,750, how many faculty members are needed?

Proportions are used in geometry with **similar triangles**. Two triangles are similar if their angles have equal measure. In such a case, the lengths of the corresponding sides are proportional. In Figure 7-1, triangle ABC is similar to triangle XYZ. Therefore, the following ratios are equivalent.

$$\frac{a}{x} = \frac{b}{y} = \frac{c}{z}$$

Figure 7-1

Answer

2. 300 faculty members are needed.

Example 3	Using Similar Triangles to Find an Unknown Side in a Triangle

Classroom Example: p. 544, Exercise 34

In Figure 7-2, triangle *XYZ* is similar to triangle *ABC*.

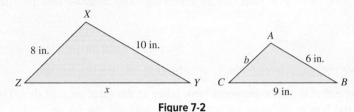

Figure 7-2

a. Solve for *x*.   **b.** Solve for *b*.

Solution:

a. The lengths of the upper right sides of the triangles are given. These form a known ratio of $\frac{10}{6}$. Because the triangles are similar, the ratio of the other corresponding sides must also be equal to $\frac{10}{6}$. To solve for *x*, we have

Bottom side from large triangle	$\rightarrow$	$\dfrac{x}{9 \text{ in.}} = \dfrac{10 \text{ in.}}{6 \text{ in.}}$	$\leftarrow$	Right side from large triangle
bottom side from small triangle	$\rightarrow$		$\leftarrow$	right side from small triangle

$\dfrac{x}{9} = \dfrac{10}{6}$ The LCD is 18.

$\overset{2}{\cancel{18}} \cdot \left(\dfrac{x}{\cancel{9}} \right) = \overset{3}{\cancel{18}} \cdot \left(\dfrac{10}{\cancel{6}} \right)$ Multiply by the LCD.

$2x = 30$ Clear fractions.

$x = 15$ Divide by 2.

The length of side *x* is 15 in.

b. To solve for *b*, the ratio of the upper left sides of the triangles must equal $\frac{10}{6}$.

Left side from large triangle	$\rightarrow$	$\dfrac{8 \text{ in.}}{b} = \dfrac{10 \text{ in.}}{6 \text{ in.}}$	$\leftarrow$	Right side from large triangle
left side from small triangle	$\rightarrow$		$\leftarrow$	right side from small triangle

$\dfrac{8}{b} = \dfrac{10}{6}$ The LCD is 6*b*.

$6\cancel{b} \cdot \left(\dfrac{8}{\cancel{b}} \right) = \cancel{6}b \cdot \left(\dfrac{10}{\cancel{6}} \right)$ Multiply by the LCD.

$48 = 10b$ Clear fractions.

$\dfrac{48}{10} = \dfrac{10b}{10}$

$4.8 = b$

The length of side *b* is 4.8 in.

Skill Practice

3. Triangle *ABC* is similar to triangle *XYZ*. Solve for the lengths of the missing sides.

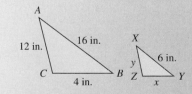

Answer

3. $x = 1.5$ in., and $y = 4.5$ in.

Classroom Example: p. 544,
Exercise 38

| Example 4 | **Using Similar Triangles in an Application** |

A tree that is 20 ft from a house is to be cut down. Use the following information and similar triangles to find the height of the tree to determine if it will hit the house.

    The shadow cast by a yardstick is 2 ft long. The shadow cast by the tree is 11 ft long.

Solution:

Let x represent the height of the tree.

Step 1:	Read the problem.
Step 2:	Label the variables.

We will assume that the measurements were taken at the same time of day. Therefore, the angle of the Sun is the same on both objects, and we can set up similar triangles (Figure 7-3).

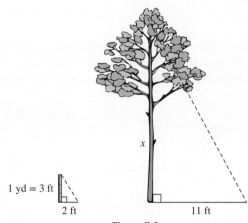

1 yd = 3 ft   2 ft x 11 ft

Figure 7-3

Step 3:   Create a verbal model.

$$\boxed{\frac{\text{Height of yardstick}}{\text{height of tree}}} \longrightarrow \frac{3 \text{ ft}}{x} = \frac{2 \text{ ft}}{11 \text{ ft}} \longleftarrow \boxed{\frac{\text{Length of yardstick's shadow}}{\text{length of tree's shadow}}}$$

$$\frac{3}{x} = \frac{2}{11}$$

Step 4:   Write a mathematical equation.

$$11x\left(\frac{3}{x}\right) = \left(\frac{2}{11}\right)11x$$

Step 5:   Multiply by the LCD.

$$33 = 2x$$

Solve the equation.

$$\frac{33}{2} = \frac{2x}{2}$$

$$16.5 = x$$

Step 6:   Interpret the results, and write the answer in words.

The tree is 16.5 ft high. The tree is less than 20 ft high so it will not hit the house.

Skill Practice

   4. The Sun casts a 3.2-ft shadow of a 6-ft man. At the same time, the Sun casts an 80-ft shadow of a building. How tall is the building?

Answer

4. The building is 150 ft tall.

3. Distance, Rate, and Time Applications

In Examples 5 and 6, we use the familiar relationship among the variables distance, rate, and time. Recall that $d = rt$.

Example 5	Using a Rational Equation in a Distance, Rate, and Time Application

Classroom Example: p. 545, Exercise 42

A small plane flies 440 mi with the wind from Memphis, TN, to Oklahoma City, OK. In the same amount of time, the plane flies 340 miles against the wind from Oklahoma City to Little Rock, AR (see Figure 7-4). If the wind speed is 30 mph, find the speed of the plane in still air.

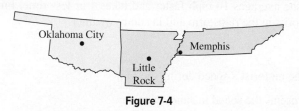

Figure 7-4

Solution:

Let x represent the speed of the plane in still air.

Then $x + 30$ is the speed of the plane with the wind.

$x - 30$ is the speed of the plane against the wind.

Organize the given information in a chart.

	Distance	Rate	Time
With the wind	440	$x + 30$	$\dfrac{440}{x + 30}$
Against the wind	340	$x - 30$	$\dfrac{340}{x - 30}$

Because $d = rt$, then $t = \dfrac{d}{r}$

The plane travels with the wind for the same amount of time as it travels against the wind, so we can equate the two expressions for time.

$$\left(\begin{array}{c}\text{Time with} \\ \text{the wind}\end{array}\right) = \left(\begin{array}{c}\text{time against} \\ \text{the wind}\end{array}\right)$$

$$\frac{440}{x + 30} = \frac{340}{x - 30} \qquad \text{The LCD is}$$
$$(x + 30)(x - 30).$$

$$(x + 30)(x - 30) \cdot \frac{440}{x + 30} = (x + 30)(x - 30) \cdot \frac{340}{x - 30}$$

$$440(x - 30) = 340(x + 30)$$

$$440x - 13{,}200 = 340x + 10{,}200 \qquad \text{Solve the resulting}$$
$$\text{linear equation.}$$

$$100x = 23{,}400$$

$$x = 234$$

The plane's speed in still air is 234 mph.

TIP: The equation

$$\frac{440}{x + 30} = \frac{340}{x - 30}$$

is a proportion. The fractions can also be cleared by equating the cross products.

$$\frac{440}{x + 30} \diagup\!\!\!\!\diagdown \frac{340}{x - 30}$$

$$440(x - 30) = 340(x + 30)$$

Skill Practice

5. Alison paddles her kayak in a river where the current is 2 mph. She can paddle 20 mi with the current in the same amount of time that she can paddle 10 mi against the current. Find the speed of the kayak in still water.

Example 6   **Using a Rational Equation in a Distance, Rate, and Time Application**

A motorist drives 100 mi between two cities in a bad rainstorm. For the return trip in sunny weather, she averages 10 mph faster and takes $\frac{1}{2}$ hr less time. Find the average speed of the motorist in the rainstorm and in sunny weather.

Solution:

Let x represent the motorist's speed during the rain.

Then $x + 10$ represents the speed in sunny weather.

	Distance	Rate	Time
Trip during rainstorm	100	x	$\dfrac{100}{x}$
Trip during sunny weather	100	$x + 10$	$\dfrac{100}{x + 10}$

Because $d = rt$, then $t = \dfrac{d}{r}$

Because the same distance is traveled in $\frac{1}{2}$ hr less time, the difference between the time of the trip during the rainstorm and the time during sunny weather is $\frac{1}{2}$ hr.

$$\left(\begin{array}{c}\text{Time during}\\ \text{the rainstorm}\end{array}\right) - \left(\begin{array}{c}\text{time during}\\ \text{sunny weather}\end{array}\right) = \left(\frac{1}{2}\,\text{hr}\right) \qquad \text{Verbal model}$$

$$\frac{100}{x} - \frac{100}{x + 10} = \frac{1}{2} \qquad \begin{array}{l}\text{Mathematical}\\ \text{equation}\end{array}$$

$$2x(x + 10)\left(\frac{100}{x} - \frac{100}{x + 10}\right) = 2x(x + 10)\left(\frac{1}{2}\right) \qquad \begin{array}{l}\text{Multiply by}\\ \text{the LCD.}\end{array}$$

$$2x(x + 10)\left(\frac{100}{x}\right) - 2x(x + 10)\left(\frac{100}{x + 10}\right) = 2x(x + 10)\left(\frac{1}{2}\right) \qquad \begin{array}{l}\text{Apply the}\\ \text{distributive}\\ \text{property.}\end{array}$$

$$200(x + 10) - 200x = x(x + 10) \qquad \text{Clear fractions.}$$

$$200x + 2000 - 200x = x^2 + 10x \qquad \begin{array}{l}\text{Solve the}\\ \text{resulting}\\ \text{equation}\\ \text{(quadratic).}\end{array}$$

$$2000 = x^2 + 10x$$

$$0 = x^2 + 10x - 2000 \qquad \begin{array}{l}\text{Set the}\\ \text{equation equal}\\ \text{to zero.}\end{array}$$

$$0 = (x - 40)(x + 50) \qquad \text{Factor.}$$

$$x = 40 \quad \text{or} \quad x \cancel{=} -50$$

Avoiding Mistakes

The equation

$$\frac{100}{x} - \frac{100}{x + 10} = \frac{1}{2}$$

is not a proportion because the left-hand side has more than one fraction. Do not try to multiply the cross products. Instead, multiply by the LCD to clear fractions.

Answer

5. The speed of the kayak is 6 mph.

Because a rate of speed cannot be negative, reject $x = -50$. Therefore, the speed of the motorist in the rainstorm is 40 mph. Because $x + 10 = 40 + 10 = 50$, the average speed for the return trip in sunny weather is 50 mph.

Skill Practice

6. Harley rode his mountain bike 12 mi to the top of a mountain and the same distance back down. His speed going up was 8 mph slower than coming down. The ride up took 2 hr longer than the ride coming down. Find his speed in each direction.

4. Work Applications

Example 7 demonstrates how work rates are related to a portion of a job that can be completed in one unit of time.

| **Example 7** | **Using a Rational Equation in a Work Problem** |

Classroom Example: p. 545, Exercise 52

A new printing press can print the morning edition in 2 hr, whereas the old printer requires 4 hr. How long would it take to print the morning edition if both printers work together?

Solution:

One method to solve this problem is to add rates.

Let x represent the time required for both printers working together to complete the job.

$$\left(\begin{array}{c}\text{Rate} \\ \text{of old printer}\end{array}\right) + \left(\begin{array}{c}\text{rate} \\ \text{of new printer}\end{array}\right) = \left(\begin{array}{c}\text{rate of} \\ \text{both working together}\end{array}\right)$$

$$\frac{1 \text{ job}}{4 \text{ hr}} \quad + \quad \frac{1 \text{ job}}{2 \text{ hr}} \quad = \quad \frac{1 \text{ job}}{x \text{ hr}}$$

$$\frac{1}{4} + \frac{1}{2} = \frac{1}{x}$$

$$4x\left(\frac{1}{4} + \frac{1}{2}\right) = 4x\left(\frac{1}{x}\right) \qquad \text{The LCD is } 4x.$$

$$\overset{1}{\cancel{4}}x \cdot \frac{1}{\cancel{4}} + \overset{2}{\cancel{4}}x \cdot \frac{1}{\cancel{2}} = 4\cancel{x} \cdot \frac{1}{\cancel{x}} \qquad \text{Apply the distributive property.}$$

$$x + 2x = 4 \qquad \text{Solve the resulting linear equation.}$$

$$3x = 4$$

$$x = \frac{4}{3} \qquad \begin{array}{l}\text{The time required to print the morning} \\ \text{edition using both printers is } 1\frac{1}{3} \text{ hr.}\end{array}$$

© Getty Images RF

Animation

Skill Practice

7. The computer at a bank can process and prepare the bank statements in 30 hr. A new faster computer can do the job in 20 hr. If the bank uses both computers together, how long will it take to process the statements?

Answers

6. Uphill speed was 4 mph; downhill speed was 12 mph.
7. 12 hr

An alternative approach to Example 7 is to determine the portion of the job that each printer can complete in 1 hr and extend that rate to the portion of the job completed in x hours.

- The old printer can perform the job in 4 hr. Therefore, it completes $\frac{1}{4}$ of the job in 1 hr and $\frac{1}{4}x$ jobs in x hours.

- The new printer can perform the job in 2 hr. Therefore, it completes $\frac{1}{2}$ of the job in 1 hr and $\frac{1}{2}x$ jobs in x hours.

The sum of the portions of the job completed by each printer must equal one whole job.

$$\left(\begin{array}{c}\text{Portion of job} \\ \text{completed by} \\ \text{old printer}\end{array}\right) + \left(\begin{array}{c}\text{portion of job} \\ \text{completed by} \\ \text{new printer}\end{array}\right) = \left(\begin{array}{c}1 \\ \text{whole} \\ \text{job}\end{array}\right)$$

$\frac{1}{4}x + \frac{1}{2}x = 1$ The LCD is 4.

$4\left(\frac{1}{4}x + \frac{1}{2}x\right) = 4(1)$ Multiply by the LCD.

$x + 2x = 4$ Solve the resulting linear equation.

$3x = 4$

$x = \frac{4}{3}$ The time required using both printers is $1\frac{1}{3}$ hr.

Section 7.7   Practice Exercises

For additional exercises, see Classroom Activities 7.7A–7.7C in the *Student's Resource Manual* at www.mhhe.com/moh.

Vocabulary and Key Concepts

1. **a.** An equation that equates two ratios is called a ___proportion___.

    b. Given similar triangles, the lengths of corresponding sides are ___proportional___.

Review Exercises

For Exercises 2–7, determine whether each of the following is an equation or an expression. If it is an equation, solve it. If it is an expression, perform the indicated operation.

2. $\frac{b}{5} + 3 = 9$ Equation; {30}

3. $\frac{m}{m-1} - \frac{2}{m+3}$ Expression; $\frac{m^2+m+2}{(m-1)(m+3)}$

4. $\frac{2}{a+5} + \frac{5}{a^2-25}$ Expression; $\frac{2a-5}{(a+5)(a-5)}$

5. $\frac{3y+6}{20} \div \frac{4y+8}{8}$ Expression; $\frac{3}{10}$

6. $\frac{z^2+z}{24} \cdot \frac{8}{z+1}$ Expression; $\frac{z}{3}$

7. $\frac{3}{p+3} = \frac{12p+19}{p^2+7p+12} - \frac{5}{p+4}$ Equation; {2}

8. Determine whether 1 is a solution to the equation. $\frac{1}{x-1} + \frac{1}{2} = \frac{2}{x^2-1}$ No

Writing   Translating Expression   Geometry   Scientific Calculator   Video

Concept 1: Solving Proportions

For Exercises 9–22, solve the proportions. **(See Example 1.)**

9. $\dfrac{8}{5} = \dfrac{152}{p}$ {95}

10. $\dfrac{6}{7} = \dfrac{96}{y}$ {112}

11. $\dfrac{19}{76} = \dfrac{z}{4}$ {1}

12. $\dfrac{15}{135} = \dfrac{w}{9}$ {1}

13. $\dfrac{5}{3} = \dfrac{a}{8}$ $\left\{\dfrac{40}{3}\right\}$

14. $\dfrac{b}{14} = \dfrac{3}{8}$ $\left\{\dfrac{21}{4}\right\}$

15. $\dfrac{2}{1.9} = \dfrac{x}{38}$ {40}

16. $\dfrac{16}{1.3} = \dfrac{30}{p}$ {2.4375}

17. $\dfrac{y+1}{2y} = \dfrac{2}{3}$ {3}

18. $\dfrac{w-2}{4w} = \dfrac{1}{6}$ {6}

 19. $\dfrac{9}{2z-1} = \dfrac{3}{z}$ {−1}

20. $\dfrac{1}{t} = \dfrac{1}{4-t}$ {2}

21. $\dfrac{8}{9a-1} = \dfrac{5}{3a+2}$ {1}

22. $\dfrac{4p+1}{3} = \dfrac{2p-5}{6}$ $\left\{-\dfrac{7}{6}\right\}$

23. Charles' law describes the relationship between the initial and final temperature and volume of a gas held at a constant pressure.

$$\dfrac{V_i}{V_f} = \dfrac{T_i}{T_f}$$

   a. Solve the equation for V_f. $V_f = \dfrac{V_i T_f}{T_i}$

   b. Solve the equation for T_f. $T_f = \dfrac{T_i V_f}{V_i}$

24. The relationship between the area, height, and base of a triangle is given by the proportion

$$\dfrac{A}{b} = \dfrac{h}{2}$$

   a. Solve the equation for A. $A = \dfrac{hb}{2}$

   b. Solve the equation for b. $b = \dfrac{2A}{h}$

Concept 2: Applications of Proportions and Similar Triangles

For Exercises 25–32, solve using proportions.

25. Toni drives her car 132 mi on the highway on 4 gal of gas. At this rate how many miles can she drive on 9 gal of gas?
   (See Example 2.)   Toni can drive 297 mi on 9 gal of gas.

26. Tim takes his pulse for 10 sec and counts 12 beats. How many beats per minute is this?   This is 72 beats/min.

27. It is recommended that 7.8 mL of Grow-It-Right plant food be mixed with 2 L of water for feeding house plants. How much plant food should be mixed with 1 gal of water to maintain the same concentration? (1 gal ≈ 3.8 L.) Express the answer in milliliters.   Mix 14.82 mL of plant food.

28. According to the website for the state of Virginia, 0.8 million tons of clothing is reused or recycled out of 7 million tons of clothing discarded. If 17.5 million tons of clothing is discarded, how many tons will be reused or recycled?
   2 million tons of clothing will be reused or recycled.

© Duncan Smith/Getty Images RF

 29. Andrew is on a low-carbohydrate diet. If his diet book tells him that an 8-oz serving of pineapple contains 19.2 g of carbohydrate, how many grams of carbohydrate does a 5-oz serving contain?
   5 oz contains 12 g of carbohydrate.

30. Cooking oatmeal requires 1 cup of water for every $\frac{1}{2}$ cup of oats. How many cups of water will be required for $\frac{3}{4}$ cup of oats?   1.5 cups of water will be necessary.

31. According to a building code, a wheelchair ramp must be at least 12 ft long for each foot of height. If the height of a newly constructed ramp is to be $1\frac{2}{3}$ ft, find the minimum acceptable length.   The minimum length is 20 ft.

32. A map has a scale of 50 mi/in. If two cities measure 6.5 in. apart, how many miles does this represent?
   This represents 325 mi.

For Exercises 33–36, triangle *ABC* is similar to triangle *XYZ*. Solve for *x* and *y*. **(See Example 3.)**

33.

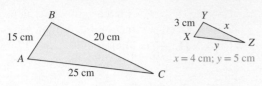

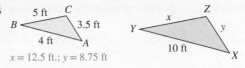

3 cm

x

y

$x = 4$ cm; $y = 5$ cm

34.

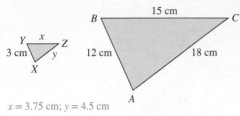

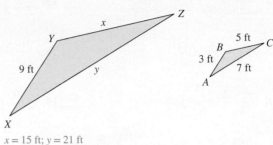

$x = 12.5$ ft.; $y = 8.75$ ft

35.

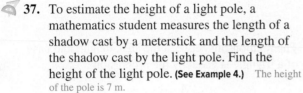

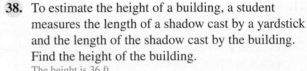

$x = 3.75$ cm; $y = 4.5$ cm

36.

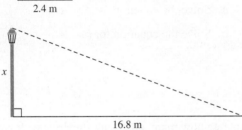

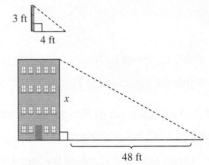

$x = 15$ ft; $y = 21$ ft

37. To estimate the height of a light pole, a mathematics student measures the length of a shadow cast by a meterstick and the length of the shadow cast by the light pole. Find the height of the light pole. **(See Example 4.)** The height of the pole is 7 m.

1 m

2.4 m

x

16.8 m

38. To estimate the height of a building, a student measures the length of a shadow cast by a yardstick and the length of the shadow cast by the building. Find the height of the building.
The height is 36 ft.

3 ft

4 ft

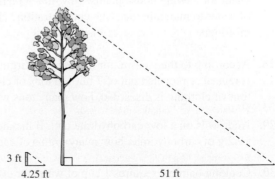

x

48 ft

39. A 6-ft-tall man standing 54 ft from a light post casts an 18-ft shadow. What is the height of the light post? The light post is 24 ft high.

6 ft

54 ft

18 ft

40. For a science project at school, a student must measure the height of a tree. The student measures the length of the shadow of the tree and then measures the length of the shadow cast by a yardstick. Find the height of the tree.
The tree is 36 ft high.

3 ft

4.25 ft

51 ft

Concept 3: Distance, Rate, and Time Applications

41. A boat travels 54 mi upstream against the current in the same amount of time it takes to travel 66 mi downstream with the current. If the current is 2 mph, what is the speed of the boat in still water? (Use $t = \frac{d}{r}$ to complete the table.)
(See Example 5.) The speed of the boat is 20 mph.

	Distance	Rate	Time
With the current (downstream)			
Against the current (upstream)			

 42. A plane flies 630 mi with the wind in the same amount of time that it takes to fly 455 mi against the wind. If this plane flies at the rate of 217 mph in still air, what is the speed of the wind? (Use $t = \frac{d}{r}$ to complete the table.) The wind speed is 35 mph.

	Distance	Rate	Time
With the wind			
Against the wind			

43. The jet stream is a fast-flowing air current found in the atmosphere at around 36,000 ft above the surface of the Earth. During one summer day, the speed of the jet stream is 35 mph. A plane flying with the jet stream can fly 700 mi in the same amount of time that it would take to fly 500 mi against the jet stream. What is the speed of the plane in still air? The plane flies 210 mph in still air.

44. A fishing boat travels 9 mi downstream with the current in the same amount of time that it travels 3 mi upstream against the current. If the speed of the current is 6 mph, what is the speed at which the boat travels in still water? The speed of the boat in still water is 12 mph.

45. An athlete in training rides his bike 20 mi and then immediately follows with a 10-mi run. The total workout takes him 2.5 hr. He also knows that he bikes about twice as fast as he runs. Determine his biking speed and his running speed. He runs 8 mph and bikes 16 mph.

46. Devon can cross-country ski 5 km/hr faster than his sister Shanelle. Devon skis 45 km in the same amount of time Shanelle skis 30 km. Find their speeds. Shanelle skis 10 km/hr and Devon skis 15 km/hr.

© PhotoLink/Getty Images RF

47. Floyd can walk 2 mph faster than his wife, Rachel. It takes Rachel 3 hr longer than Floyd to hike a 12-mi trail through a park. Find their speeds.
(See Example 6.) Floyd walks 4 mph and Rachel walks 2 mph.

48. Janine bikes 3 mph faster than her sister, Jessica. Janine can ride 36 mi in 1 hr less time than Jessica can ride the same distance. Find each of their speeds. Janine's speed is 12 mph and Jessica's is 9 mph.

49. Sergio rode his bike 4 mi. Then he got a flat tire and had to walk back 4 mi. It took him 1 hr longer to walk than it did to ride. If his rate walking was 9 mph less than his rate riding, find the two rates. Sergio rode 12 mph and walked 3 mph.

50. Amber jogs 10 km in $\frac{3}{4}$ hr less time than she can walk the same distance. If her walking rate is 3 km/hr less than her jogging rate, find her rates jogging and walking (in km/hr). Amber jogs 8 km/hr and walks 5 km/hr.

Concept 4: Work Applications

51. If the cold-water faucet is left on, the sink will fill in 10 min. If the hot-water faucet is left on, the sink will fill in 12 min. How long would it take to fill the sink if both faucets are left on?
(See Example 7.) $5\frac{5}{11}$ $(5.\overline{45})$ min

52. The CUT-IT-OUT lawn mowing company consists of two people: Tina and Bill. If Tina cuts a lawn by herself, she can do it in 4 hr. If Bill cuts the same lawn himself, it takes him an hour longer than Tina. How long would it take them if they worked together? $2\frac{2}{9}$ $(2.\overline{2})$ hr

53. A manuscript needs to be printed. One printer can do the job in 50 min, and another printer can do the job in 40 min. How long would it take if both printers were used? $22\frac{2}{9}$ $(22.\overline{2})$ min

54. A pump can empty a small pond in 4 hr. Another more efficient pump can do the job in 3 hr. How long would it take to empty the pond if both pumps were used? $1\frac{5}{7}$ (approximately 1.7) hr

55. A pipe can fill a reservoir in 16 hr. A drainage pipe can drain the reservoir in 24 hr. How long would it take to fill the reservoir if the drainage pipe were left open by mistake? (*Hint:* The rate at which water drains should be negative.)   48 hr

56. A hole in the bottom of a child's plastic swimming pool can drain the pool in 60 min. If the pool had no hole, a hose could fill the pool in 40 min. How long would it take the hose to fill the pool with the hole?   120 min

57. Tim and Al are bricklayers. Tim can construct an outdoor grill in 5 days. If Al helps Tim, they can build it in only 2 days. How long would it take Al to build the grill alone? $3\frac{1}{3}$ (3.$\overline{3}$) days

58. Norma is a new and inexperienced secretary. It takes her 3 hr to prepare a mailing. If her boss helps her, the mailing can be completed in 1 hr. How long would it take the boss to do the job by herself? $1\frac{1}{2}$ (1.5) hr

Expanding Your Skills

For Exercises 59–62, solve using proportions.

59. The ratio of smokers to nonsmokers in a restaurant is 2 to 7. There are 100 more nonsmokers than smokers. How many smokers and nonsmokers are in the restaurant?
There are 40 smokers and 140 nonsmokers.

60. The ratio of fiction to nonfiction books sold in a bookstore is 5 to 3. One week there were 180 more fiction books sold than nonfiction. Find the number of fiction and nonfiction books sold during that week.   There are 450 fiction and 270 nonfiction books sold.

61. There are 440 students attending a biology lecture. The ratio of male to female students at the lecture is 6 to 5. How many men and women are attending the lecture?   There are 240 men and 200 women.

62. The ratio of dogs to cats at the humane society is 5 to 8. The total number of dogs and cats is 650. How many dogs and how many cats are at the humane society?   There are 250 dogs and 400 cats.

Section 7.8   Variation

Concepts

1. Definition of Direct and Inverse Variation
2. Translations Involving Variation
3. Applications of Variation

1. Definition of Direct and Inverse Variation

In this section, we introduce the concept of variation. Direct and inverse variation models can show how one quantity varies in proportion to another.

> **Definition of Direct and Inverse Variation**
>
> Let k be a nonzero constant real number. Then the following statements are equivalent:
>
> **1.** y varies **directly** as x.
> y is directly proportional to x. $\Big\}$ $y = kx$
>
> **2.** y varies **inversely** as x.
> y is inversely proportional to x. $\Big\}$ $y = \dfrac{k}{x}$
>
> *Note:* The value of k is called the constant of variation.

For a car traveling 30 mph, the equation $d = 30t$ indicates that the distance traveled is *directly proportional* to the time of travel. For positive values of k, when two variables are directly related, as one variable increases, the other variable will also increase. Likewise, if one variable decreases, the other will decrease. In the equation $d = 30t$, the longer the time of the trip, the greater the distance traveled. The shorter the time of the trip, the shorter the distance traveled.

For positive values of k, when two positive variables are *inversely related*, as one variable increases, the other will decrease, and vice versa. Consider a car traveling between Toronto and Montreal, a distance of 500 km. The time required to make the trip is inversely proportional to the speed of travel: $t = 500/r$. As the rate of speed, r, increases, the quotient $500/r$ will decrease. Thus, the time will decrease. Similarly, as the rate of speed decreases, the trip will take longer.

2. Translations Involving Variation

The first step in using a variation model is to write an English phrase as an equivalent mathematical equation.

| **Example 1** | **Translating to a Variation Model** |

Classroom Examples: p. 552, Exercises 12 and 14

Write each expression as an equivalent mathematical model.

a. The circumference of a circle varies directly as the radius.

b. At a constant temperature, the volume of a gas varies inversely as the pressure.

c. The length of time of a meeting is directly proportional to the *square* of the number of people present.

Solution:

a. Let C represent circumference and r represent radius. The variables are directly related, so use the model $C = kr$.

b. Let V represent volume and P represent pressure. Because the variables are inversely related, use the model $V = \frac{k}{P}$.

c. Let t represent time, and let N be the number of people present at a meeting. Because t is directly related to N^2, use the model $t = kN^2$.

Skill Practice Write each expression as an equivalent mathematical model.

1. The distance, d, driven in a particular time varies directly with the speed of the car, s.
2. The weight of an individual kitten, w, varies inversely with the number of kittens in the litter, n.
3. The value of v varies inversely as the square root of b.

Sometimes a variable varies directly as the product of two or more other variables. In this case, we have joint variation.

Definition of Joint Variation

Let k be a nonzero constant real number. Then the following statements are equivalent:

$\left.\begin{array}{l} y \text{ varies } \textbf{jointly} \text{ as } w \text{ and } z. \\ y \text{ is jointly proportional to } w \text{ and } z. \end{array}\right\} \quad y = kwz$

Answers

1. $d = ks$ 2. $w = \dfrac{k}{n}$

3. $v = \dfrac{k}{\sqrt{b}}$

Classroom Examples: p. 552,
Exercises 18 and 22

| **Example 2** | **Translating to a Variation Model** |

Write each expression as an equivalent mathematical model.

a. y varies jointly as u and the square root of v.

b. The gravitational force of attraction between two planets varies jointly as the product of their masses and inversely as the square of the distance between them.

Solution:

a. $y = ku\sqrt{v}$

b. Let m_1 and m_2 represent the masses of the two planets. Let F represent the gravitational force of attraction and d represent the distance between the planets.

The variation model is: $F = \dfrac{km_1m_2}{d^2}$

Skill Practice Write each expression as an equivalent mathematical model.

4. The value of q varies jointly as u and v.

5. The value of x varies directly as the square of y and inversely as z.

3. Applications of Variation

Consider the variation models $y = kx$ and $y = \frac{k}{x}$. In either case, if values for x and y are known, we can solve for k. Once k is known, we can use the variation equation to find y if x is known, or to find x if y is known. This concept is the basis for solving many applications involving variation.

Finding a Variation Model

Step 1 Write a general variation model that relates the variables given in the problem. Let k represent the constant of variation.

Step 2 Solve for k by substituting known values of the variables into the model from step 1.

Step 3 Substitute the value of k into the original variation model from step 1.

Classroom Examples: pp. 552–553,
Exercises 24 and 32

| **Example 3** | **Solving an Application Involving Direct Variation** |

The variable z varies directly as w. When w is 16, z is 56.

a. Write a variation model for this situation. Use k as the constant of variation.

b. Solve for the constant of variation.

c. Find the value of z when w is 84.

Answers

4. $q = kuv$ **5.** $x = \dfrac{ky^2}{z}$

Solution:

a. $z = kw$

b. $z = kw$

$56 = k(16)$ Substitute known values for z and w. Then solve for the unknown value of k.

$\dfrac{56}{16} = \dfrac{k(16)}{16}$ To isolate k, divide both sides by 16.

$\dfrac{7}{2} = k$ Simplify $\dfrac{56}{16}$ to $\dfrac{7}{2}$.

c. With the value of k known, the variation model can now be written as $z = \dfrac{7}{2}w$.

$z = \dfrac{7}{2}(84)$ To find z when $w = 84$, substitute $w = 84$ into the equation.

$z = 294$

Skill Practice   The variable t varies directly as the square of v. When v is 8, t is 32.

6. Write a variation model for this relationship.
7. Solve for the constant of variation.
8. Find t when $v = 10$.

> **Example 4**   Solving an Application Involving Direct Variation

Classroom Example: p. 553, Exercise 42

The speed of a racing canoe in still water varies directly as the square root of the length of the canoe.

a. If a 16-ft canoe can travel 6.2 mph in still water, find a variation model that relates the speed of a canoe to its length.

b. Find the speed of a 25-ft canoe.

Solution:

a. Let s represent the speed of the canoe and L represent the length. The general variation model is $s = k\sqrt{L}$. To solve for k, substitute the known values for s and L.

$s = k\sqrt{L}$

$6.2 = k\sqrt{16}$ Substitute $s = 6.2$ mph and $L = 16$ ft.

$6.2 = k \cdot 4$

$\dfrac{6.2}{4} = \dfrac{4k}{4}$ Solve for k.

$k = 1.55$

$s = 1.55\sqrt{L}$ Substitute $k = 1.55$ into the model $s = k\sqrt{L}$.

Animation

Answers

6. $t = kv^2$ **7.** $\dfrac{1}{2}$ **8.** 50

b. $s = 1.55\sqrt{L}$

$\qquad = 1.55\sqrt{25}$ Find the speed when $L = 25$ ft.

$\qquad = 7.75$ mph The speed is 7.75 mph.

Skill Practice

9. The amount of water needed by a mountain hiker varies directly as the time spent hiking. The hiker needs 2.4 L for a 4-hr hike. How much water will be needed for a 5-hr hike?

Classroom Example: p. 554, Exercise 50

| Example 5 | **Solving an Application Involving Inverse Variation** |

The loudness of sound measured in decibels (dB) varies inversely as the square of the distance between the listener and the source of the sound. If the loudness of sound is 17.92 dB at a distance of 10 ft from a home theater speaker, what is the decibel level 20 ft from the speaker?

Solution:

Let L represent the loudness of sound in decibels and d represent the distance in feet. The inverse relationship between decibel level and the square of the distance is modeled by

$$L = \frac{k}{d^2}$$

$17.92 = \dfrac{k}{(10)^2}$ Substitute $L = 17.92$ dB and $d = 10$ ft.

$17.92 = \dfrac{k}{100}$

$(17.92)100 = \dfrac{k}{100} \cdot 100$ Solve for k (clear fractions).

$k = 1792$

$L = \dfrac{1792}{d^2}$ Substitute $k = 1792$ into the original

$\qquad\qquad$ model $L = \dfrac{k}{d^2}$.

With the value of k known, we can find L for any value of d.

$L = \dfrac{1792}{(20)^2}$ Find the loudness when $d = 20$ ft.

$\qquad = 4.48$ dB The loudness is 4.48 dB.

Notice that the loudness of sound is 17.92 dB at a distance 10 ft from the speaker. When the distance from the speaker is increased to 20 ft, the decibel level decreases to 4.48 dB. This is consistent with an inverse relationship. For $k > 0$, as one variable is increased, the other is decreased. It also seems reasonable that the farther one moves away from the source of a sound, the softer the sound becomes.

Skill Practice

10. The yield on a bond varies inversely as the price. The yield on a particular bond is 5% when the price is $100. Find the yield when the price is $125.

Answers

9. 3 L $\qquad$ **10.** 4%

| **Example 6** | **Solving an Application Involving Joint Variation** |

Classroom Example: p. 555, Exercise 58

The kinetic energy of an object varies jointly as the weight of the object at sea level and as the square of its velocity. During a hurricane, a 0.5-lb stone traveling at 60 mph has 81 J (joules) of kinetic energy. Suppose the wind speed doubles to 120 mph. Find the kinetic energy.

Solution:

Let E represent the kinetic energy, let w represent the weight, and let v represent the velocity of the stone. The variation model is

$E = kwv^2$

$81 = k(0.5)(60)^2$ Substitute $E = 81$ J, $w = 0.5$ lb, and $v = 60$ mph.

$81 = k(0.5)(3600)$ Simplify exponents.

$81 = k(1800)$

$\dfrac{81}{1800} = \dfrac{k(1800)}{1800}$ Divide by 1800.

$0.045 = k$ Solve for k.

With the value of k known, the model $E = kwv^2$ can now be written as $E = 0.045wv^2$. We now find the kinetic energy of a 0.5-lb stone traveling at 120 mph.

$E = 0.045(0.5)(120)^2$

$\quad = 324$

The kinetic energy of a 0.5-lb stone traveling at 120 mph is 324 J.

Skill Practice

11. The amount of simple interest earned in an account varies jointly as the interest rate and time of the investment. An account earns $72 in 4 years at 2% interest. How much interest would be earned in 3 years at a rate of 5%?

In Example 6, when the velocity increased by 2 times, the kinetic energy increased by 4 times (note that 324 J $= 4 \cdot 81$ J). This factor of 4 occurs because the kinetic energy is proportional to the *square* of the velocity. When the velocity increased by 2 times, the kinetic energy increased by 2^2 times.

Answer

11. $135

| **Section 7.8** | **Practice Exercises** |

Vocabulary and Key Concepts

For additional exercises, see Classroom Activities 7.8A–7.8C in the *Student's Resource Manual* at www.mhhe.com/moh.

1. a. Let k be a nonzero constant. If y varies directly as x, then $y =$ _____ kx _____ where k is the constant of variation.

b. Let k be a nonzero constant. If y varies inversely as x, then $y =$ _____ $\dfrac{k}{x}$ _____ where k is the constant of variation.

c. Let k be a nonzero constant. If y varies jointly as x and w, then $y =$ _____ kxw _____ where k is the constant of variation.

/ Writing   ↔ Translating Expression   Geometry   Scientific Calculator   Video

Review Exercises

For Exercises 2–7, perform the indicated operation, or solve the equation.

2. $\dfrac{5p}{p+2} + \dfrac{10}{p+2}$ 5

3. $\dfrac{2y}{3} - \dfrac{3y-1}{5} = 1$ $\{12\}$

4. $\dfrac{3}{q-1} \cdot \dfrac{2q^2+3q-5}{6q+24}$ $\dfrac{2q+5}{2(q+4)}$

5. $\dfrac{a}{4} + \dfrac{3}{a} = 2$ $\{6, 2\}$

6. $\dfrac{3}{b^2+5b-14} - \dfrac{2}{b^2-49}$ $\dfrac{b-17}{(b+7)(b-7)(b-2)}$

7. $\dfrac{a+\dfrac{a}{b}}{\dfrac{a}{b}-a}$ $\dfrac{b+1}{1-b}$

Concept 1: Definition of Direct and Inverse Variation

8. In the equation $r = kt$, does r vary directly or inversely with t?   Directly

9. In the equation $w = \dfrac{k}{v}$, does w vary directly or inversely with v?   Inversely

10. In the equation $P = \dfrac{k \cdot c}{v}$, does P vary directly or inversely as v?   Inversely

Concept 2: Translations Involving Variation

For Exercises 11–22, write a variation model. Use k as the constant of variation. **(See Examples 1–2.)**

11. T varies directly as q. $T = kq$

12. W varies directly as z. $W = kz$

13. b varies inversely as c. $b = \dfrac{k}{c}$

14. m varies inversely as t. $m = \dfrac{k}{t}$

15. Q is directly proportional to x and inversely proportional to y. $Q = \dfrac{kx}{y}$

16. d is directly proportional to p and inversely proportional to n. $d = \dfrac{kp}{n}$

17. c varies jointly as s and t. $c = kst$

18. w varies jointly as p and f. $w = kpf$

19. L varies jointly as w and the square root of v. $L = kw\sqrt{v}$

20. q varies jointly as v and the square root of w. $q = kv\sqrt{w}$

21. x varies directly as the square of y and inversely as z. $x = \dfrac{ky^2}{z}$

22. a varies directly as n and inversely as the square of d. $a = \dfrac{kn}{d^2}$

Concept 3: Applications of Variation

For Exercises 23–28, find the constant of variation, k. **(See Example 3.)**

23. y varies directly as x and when x is 4, y is 18. $k = \dfrac{9}{2}$

24. m varies directly as x and when x is 8, m is 22. $k = \dfrac{11}{4}$

25. p varies inversely as q and when q is 16, p is 32. $k = 512$

26. T varies inversely as x and when x is 40, T is 200. $k = 8000$

27. y varies jointly as w and v. When w is 50 and v is 0.1, y is 8.75. $k = 1.75$

28. N varies jointly as t and p. When t is 1 and p is 7.5, N is 330. $k = 44$

For Exercises 29–40, solve for the indicated variable. **(See Example 3.)**

29. x varies directly as p. If $x = 50$ when $p = 10$, find x when p is 14. $x = 70$

30. y is directly proportional to z. If $y = 12$ when $z = 36$, find y when z is 21. $y = 7$

31. b is inversely proportional to c. If b is 4 when c is 3, find b when $c = 2$. $b = 6$

32. q varies inversely as w. If q is 8 when w is 50, find q when w is 125. $q = 3.2$

33. Z varies directly as the square of w. If $Z = 14$ when $w = 4$, find Z when $w = 8$. $Z = 56$

34. m varies directly as the square of x. If $m = 200$ when $x = 20$, find m when x is 32. $m = 512$

35. Q varies inversely as the square of p. If $Q = 4$ when $p = 3$, find Q when $p = 2$. $Q = 9$

36. z is inversely proportional to the square of t. If $z = 15$ when $t = 4$, find z when $t = 10$. $z = 2.4$

37. L varies jointly as a and the square root of b. If $L = 72$ when $a = 8$ and $b = 9$, find L when $a = \frac{1}{2}$ and $b = 36$. $L = 9$

38. Y varies jointly as the cube of x and the square root of w. $Y = 128$ when $x = 2$ and $w = 16$. Find Y when $x = \frac{1}{2}$ and $w = 64$. $Y = 4$

39. B varies directly as m and inversely as n. $B = 20$ when $m = 10$ and $n = 3$. Find B when $m = 15$ and $n = 12$. $B = \dfrac{15}{2}$

40. R varies directly as s and inversely as t. $R = 14$ when $s = 2$ and $t = 9$. Find R when $s = 4$ and $t = 3$. $R = 84$

For Exercises 41–58, use a variation model to solve for the unknown value. **(See Examples 4–6.)**

41. The weight of a person's heart varies directly as the person's actual weight. For a 150-lb man, his heart would weigh 0.75 lb.

   a. Approximate the weight of a 184-lb man's heart.   The heart weighs 0.92 lb.

   b. How much does your heart weigh?
     Answers will vary.

42. The number of calories, C, in beer varies directly with the number of ounces, n. If 12 oz of beer contains 153 calories, how many calories are in 40 oz of beer?   There are 510 calories in 40 oz of beer.

43. The amount of medicine that a physician prescribes for a patient varies directly as the weight of the patient. A physician prescribes 3 g of a medicine for a 150-lb person.

   a. How many grams should be prescribed for a 180-lb person?   3.6 g

   b. How many grams should be prescribed for a 225-lb person?   4.5 g

   c. How many grams should be prescribed for a 120-lb person?   2.4 g

44. The number of turkeys needed for a banquet is directly proportional to the number of guests that must be fed. Master Chef Rico knows that he needs to cook 3 turkeys to feed 42 guests.

   a. How many turkeys should he cook to feed 70 guests?   5 turkeys

   b. How many turkeys should he cook to feed 140 guests?   10 turkeys

   c. How many turkeys should be cooked to feed 700 guests at an inaugural ball?   50 turkeys

   d. How many turkeys should be cooked for a wedding with 100 guests?   8 turkeys; Note that the answer was rounded up, because it would be better for the chef to have too much food than too little.

© Terry Vine/Blend Images LLC RF

© BananaStock/PunchStock RF

 Writing   Translating Expression    Geometry   Scientific Calculator   Video

45. The unit cost of producing CDs is inversely proportional to the number of CDs produced. If 5000 CDs are produced, the cost per CD is $0.48.

   a. What would be the unit cost if 6000 CDs were produced? $0.40

   b. What would be the unit cost if 8000 CDs were produced? $0.30

   c. What would be the unit cost if 2400 CDs were produced? $1.00

46. An author self-publishes a book and finds that the number of books she can sell per month varies inversely as the price of the book. The author can sell 1500 books per month when the price is set at $8 per book.

   a. How many books would she expect to sell if the price were $12? 1000 books

   b. How many books would she expect to sell if the price were $15? 800 books

   c. How many books would she expect to sell if the price were $6? 2000 books

47. The amount of pollution entering the atmosphere over a given time varies directly as the number of people living in an area. If 80,000 people cause 56,800 tons of pollutants, how many tons enter the atmosphere in a city with a population of 500,000? 355,000 tons

© Patrick Clark/Getty Images RF

48. The area of a picture projected on a wall varies directly as the square of the distance from the projector to the wall. If a 10-ft distance produces a 16-ft^2 picture, what is the area of a picture produced when the projection unit is moved to a distance 20 ft from the wall? 64 ft^2

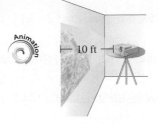

49. The stopping distance of a car varies directly as the square of the speed of the car. If a car traveling 40 mph has a stopping distance of 109 ft, find the stopping distance of a car that travels 25 mph. (Round the answer to one decimal place.) 42.6 ft

50. The intensity of a light source varies inversely as the square of the distance from the source. If the intensity of a light bulb is 400 lumens/m^2 (lux) at a distance of 5 m, determine the intensity at 8 m. 156.25 lumens/m^2

51. The power in an electric circuit varies jointly as the current and the square of the resistance. If the power is 144 W (watts) when the current is 4 A (amperes) and the resistance is 6 Ω (ohms), find the power when the current is 3 A and the resistance is 10 Ω. 300 W

52. Some bodybuilders claim that, within safe limits, the number of repetitions that a person can complete on a given weight-lifting exercise is inversely proportional to the amount of weight lifted. Roxanne can bench press 45 lb for 15 repetitions.

   a. How many repetitions can Roxanne bench with 60 lb of weight? 11 repetitions

   b. How many repetitions can Roxanne bench with 75 lb of weight? 9 repetitions

   c. How many repetitions can Roxanne bench with 100 lb of weight? 6 repetitions

53. The current in a wire varies directly as the voltage and inversely as the resistance. If the current is 9 A when the voltage is 90 V (volts) and the resistance is 10 Ω (ohms), find the current when the voltage is 185 V and the resistance is 10 Ω. 18.5 A

54. The resistance of a wire varies directly as its length and inversely as the square of its diameter. A 40-ft wire 0.1 in. in diameter has a resistance of 4 Ω. What is the resistance of a 50-ft wire with a diameter of 0.2 in.? 1.25 Ω

55. The weight of a medicine ball varies directly as the cube of its radius. A ball with a radius of 3 in. weighs 4.32 lb. How much would a medicine ball weigh if its radius is 5 in.?   20 lb

56. The surface area of a cube varies directly as the square of the length of an edge. The surface area is 24 ft² when the length of an edge is 2 ft. Find the surface area of a cube with an edge that is 5 ft.
150 ft²

57. The amount of simple interest earned in an account varies jointly as the amount of principal invested and the amount of time the money is invested. If $2500 in principal earns $500 in interest after 4 years, then how much interest will be earned on $7000 invested for 10 years? $3500

58. The amount of simple interest earned in an account varies jointly as the amount of principal invested and the amount of time the money is invested. If $6000 in principal earns $840 in interest after 2 years, then how much interest will be earned on $4500 invested for 8 years? $2520

Group Activity

Computing Monthly Mortgage Payments

Materials: A calculator

Estimated Time: 15–20 minutes

Group Size: 3

When a person borrows money to buy a house, the bank usually requires a down payment of between 0% and 20% of the cost of the house. The bank then issues a loan for the remaining balance on the house. The loan to buy a house is called a *mortgage.* Monthly payments are made to pay off the mortgage over a period of years.

A formula to calculate the monthly payment, P, for a loan is given by:

$$P = \frac{\frac{Ar}{12}}{1 - \frac{1}{\left(1 + \frac{r}{12}\right)^{12t}}}$$ where

P is the monthly payment
A is the original amount of the mortgage
r is the annual interest rate written as a decimal
t is the term of the loan in years

Suppose a person wants to buy a $200,000 house. The bank requires a down payment of 20%, and the loan is issued for 30 years at 7.5% interest for 30 years.

1. Find the amount of the down payment. ____$40,000____

2. Find the amount of the mortgage. ____$160,000____

3. Find the monthly payment (to the nearest cent). ____$1118.74____

4. Multiply the monthly payment found in question 3 by the total number of months in a 30-year period. Interpret what this value means in the context of the problem.
$402,746.40; This is the total amount owed to the bank.

5. How much total interest was paid on the loan for the house? ____$242,746.40____

6. What was the total amount paid to the bank (include the down payment). ____$442,746.40____

 Writing    Translating Expression   Geometry   Scientific Calculator   Video

Chapter 7 Summary

Section 7.1 Introduction to Rational Expressions

Key Concepts

A **rational expression** is a ratio of the form $\frac{p}{q}$ where p and q are polynomials and $q \neq 0$.

Restricted values of a rational expression are those values that, when substituted for the variable, make the expression undefined. To find restricted values, set the denominator equal to 0 and solve the equation.

Simplifying a Rational Expression

Factor the numerator and denominator completely, and reduce factors whose ratio is equal to 1 or to -1. A rational expression written in lowest terms will still have the same restricted values as the original expression.

Examples

Example 1

$$\frac{x+2}{x^2 - 5x - 14} \text{ is a rational expression.}$$

Example 2

To find the restricted values of $\dfrac{x+2}{x^2 - 5x - 14}$ factor the

denominator: $\dfrac{x+2}{(x+2)(x-7)}$

The restricted values are $x = -2$ and $x = 7$.

Example 3

Simplify the rational expression. $\dfrac{x+2}{x^2 - 5x - 14}$

$\dfrac{\overset{1}{\cancel{x+2}}}{\cancel{(x+2)}(x-7)}$ Simplify.

$= \dfrac{1}{x-7}$ (provided $x \neq 7$, $x \neq -2$).

Section 7.2 Multiplication and Division of Rational Expressions

Key Concepts

Multiplying Rational Expressions

Multiply the numerators and multiply the denominators. That is, if $q \neq 0$ and $s \neq 0$, then

$$\frac{p}{q} \cdot \frac{r}{s} = \frac{pr}{qs}$$

Factor the numerator and denominator completely. Then reduce factors whose ratio is 1 or -1.

Examples

Example 1

Multiply. $\dfrac{b^2 - a^2}{a^2 - 2ab + b^2} \cdot \dfrac{a^2 - 3ab + 2b^2}{2a + 2b}$

$= \dfrac{\overset{-1}{\cancel{(b-a)}}(b+a)}{\cancel{(a-b)}(a-b)} \cdot \dfrac{(a-2b)\overset{1}{\cancel{(a-b)}}}{2(a+b)}$

$= -\dfrac{a - 2b}{2}$ or $\dfrac{2b - a}{2}$

Dividing Rational Expressions

Multiply the first expression by the reciprocal of the second expression. That is, for $q \neq 0$, $r \neq 0$, and $s \neq 0$,

$$\frac{p}{q} \div \frac{r}{s} = \frac{p}{q} \cdot \frac{s}{r} = \frac{ps}{qr}$$

Example 2

Divide. $\dfrac{x-2}{15} \div \dfrac{x^2 + 2x - 8}{20x}$

$$= \frac{x-2}{15} \cdot \frac{20x}{x^2 + 2x - 8}$$

$$= \frac{\overset{1}{(x-2)}}{\underset{3}{15}} \cdot \frac{\overset{4}{20x}}{\underset{1}{(x-2)(x+4)}}$$

$$= \frac{4x}{3(x+4)}$$

Section 7.3   Least Common Denominator

Key Concepts

Finding the Least Common Denominator (LCD) of Two or More Rational Expressions

1. Factor all denominators completely.
2. The LCD is the product of unique factors from the denominators, where each factor is raised to its highest power.

Examples

Example 1

Identify the LCD. $\dfrac{1}{8x^3y^2z}; \dfrac{5}{6xy^4}$

1. Write the denominators as a product of prime factors:

$$\frac{1}{2^3 x^3 y^2 z}; \frac{5}{2 \cdot 3 x y^4}$$

2. The LCD is $2^3 \cdot 3x^3y^4z$ or $24x^3y^4z$

Converting a Rational Expression to an Equivalent Expression with a Different Denominator

Multiply numerator and denominator of the rational expression by the missing factors necessary to create the desired denominator.

Example 2

Convert $\dfrac{-3}{x-2}$ to an equivalent expression with the indicated denominator:

$$\frac{-3}{x-2} = \frac{}{5(x-2)(x+2)}$$

Multiply numerator and denominator by the missing factors from the denominator.

$$\frac{-3 \cdot 5(x+2)}{(x-2) \cdot 5(x+2)} = \frac{-15x - 30}{5(x-2)(x+2)}$$

| Section 7.4 | Addition and Subtraction of Rational Expressions |

Key Concepts

To add or subtract rational expressions, the expressions must have the same denominator.

Steps to Add or Subtract Rational Expressions

1. Factor the denominators of each rational expression.
2. Identify the LCD.
3. Rewrite each rational expression as an equivalent expression with the LCD as its denominator.
4. Add or subtract the numerators, and write the result over the common denominator.
5. Simplify.

Example

Example 1

Add. $\dfrac{c-2}{c+1} + \dfrac{12c-3}{2c^2-c-3}$

$= \dfrac{c-2}{c+1} + \dfrac{12c-3}{(2c-3)(c+1)}$

The LCD is $(2c-3)(c+1)$.

$= \dfrac{(2c-3)(c-2)}{(2c-3)(c+1)} + \dfrac{12c-3}{(2c-3)(c+1)}$

$= \dfrac{2c^2-4c-3c+6+12c-3}{(2c-3)(c+1)}$

$= \dfrac{2c^2+5c+3}{(2c-3)(c+1)}$

$= \dfrac{(2c+3)\cancel{(c+1)}}{(2c-3)\cancel{(c+1)}} = \dfrac{2c+3}{2c-3}$

| Section 7.5 | Complex Fractions |

Key Concepts

Complex fractions can be simplified by using Method I or Method II.

Method I

1. Add or subtract expressions in the numerator to form a single fraction. Add or subtract expressions in the denominator to form a single fraction.
2. Divide the rational expressions from step 1 by multiplying the numerator of the complex fraction by the reciprocal of the denominator of the complex fraction.
3. Simplify to lowest terms, if possible.

Examples

Example 1

Simplify. $\dfrac{1 - \dfrac{4}{w^2}}{1 - \dfrac{1}{w} - \dfrac{6}{w^2}} = \dfrac{1 \cdot \dfrac{w^2}{w^2} - \dfrac{4}{w^2}}{1 \cdot \dfrac{w^2}{w^2} - \dfrac{1}{w} \cdot \dfrac{w}{w} - \dfrac{6}{w^2}}$

$= \dfrac{\dfrac{w^2}{w^2} - \dfrac{4}{w^2}}{\dfrac{w^2}{w^2} - \dfrac{w}{w^2} - \dfrac{6}{w^2}} = \dfrac{\dfrac{w^2-4}{w^2}}{\dfrac{w^2-w-6}{w^2}}$

$= \dfrac{w^2-4}{w^2} \cdot \dfrac{w^2}{w^2-w-6}$

$= \dfrac{(w-2)\cancel{(w+2)}}{\cancel{w^2}} \cdot \dfrac{\cancel{w^2}}{(w-3)\cancel{(w+2)}}$

$= \dfrac{w-2}{w-3}$

Method II

1. Multiply the numerator and denominator of the complex fraction by the LCD of all individual fractions within the expression.
2. Apply the distributive property, and simplify the result.
3. Simplify to lowest terms, if possible.

Example 2

Simplify. $\dfrac{1 - \dfrac{4}{w^2}}{1 - \dfrac{1}{w} - \dfrac{6}{w^2}} = \dfrac{w^2\left(1 - \dfrac{4}{w^2}\right)}{w^2\left(1 - \dfrac{1}{w} - \dfrac{6}{w^2}\right)}$

$= \dfrac{w^2 - 4}{w^2 - w - 6} = \dfrac{(w-2)(w+2)}{(w-3)(w+2)}$

$= \dfrac{w-2}{w-3}$

Section 7.6   Rational Equations

Key Concepts

An equation with one or more rational expressions is called a **rational equation**.

Steps to Solve a Rational Equation

1. Factor the denominators of all rational expressions. Identify the restricted values.
2. Identify the LCD of all expressions in the equation.
3. Multiply both sides of the equation by the LCD.
4. Solve the resulting equation.
5. Check each potential solution in the original equation.

Examples

Example 1

Solve. $\dfrac{1}{w} - \dfrac{1}{2w-1} = \dfrac{-2w}{2w-1}$ The restricted values are $w = 0$ and $w = \frac{1}{2}$.

The LCD is $w(2w-1)$.

$w(2w-1)\dfrac{1}{w} - w(2w-1)\dfrac{1}{2w-1}$

$\qquad = w(2w-1)\dfrac{-2w}{2w-1}$

$(2w-1)(1) - w(1) = w(-2w)$

$2w - 1 - w = -2w^2$ Quadratic equation

$2w^2 + w - 1 = 0$

$(2w-1)(w+1) = 0$

$\qquad w = \frac{1}{2} \qquad$ or $\qquad w = -1$
$\qquad$ Does not check. $\qquad\qquad$ Checks.

The solution set is $\{-1\}$.

Example 2

Solve for I. $q = \dfrac{VQ}{I}$

$I \cdot q = \dfrac{VQ}{I} \cdot I$

$Iq = VQ$

$I = \dfrac{VQ}{q}$

| Section 7.7 | Applications of Rational Equations and Proportions |

Key Concepts and Examples

Solving Proportions

An equation that equates two rates or ratios is called a **proportion**:

$$\frac{a}{b} = \frac{c}{d} \quad (b \neq 0, d \neq 0)$$

To solve a proportion, multiply both sides of the equation by the LCD.

Examples

Example 1

A 90-g serving of a particular ice cream contains 10 g of fat. How much fat does 400 g of the same ice cream contain?

$$\frac{10 \text{ g fat}}{90 \text{ g ice cream}} = \frac{x \text{ grams fat}}{400 \text{ g ice cream}}$$

$$\frac{10}{90} = \frac{x}{400} \qquad \text{The LCD is 3600.}$$

$$\overset{40}{3600} \cdot \left(\frac{10}{90}\right) = \left(\frac{x}{400}\right) \cdot \overset{9}{3600}$$

$$400 = 9x$$

$$x = \frac{400}{9} \approx 44.4 \text{ g}$$

Examples 2 and 3 give applications of rational equations.

Example 2

Two cars travel from Los Angeles to Las Vegas. One car travels an average of 8 mph faster than the other car. If the faster car travels 189 mi in the same amount of time that the slower car travels 165 mi, what is the average speed of each car?

Let r represent the speed of the slower car.
Let $r + 8$ represent the speed of the faster car.

	Distance	Rate	Time
Slower car	165	r	$\dfrac{165}{r}$
Faster car	189	$r + 8$	$\dfrac{189}{r + 8}$

$$\frac{165}{r} = \frac{189}{r + 8}$$

$$165(r + 8) = 189r$$

$$165r + 1320 = 189r$$

$$1320 = 24r$$

$$55 = r$$

The slower car travels 55 mph, and the faster car travels $55 + 8 = 63$ mph.

Example 3

Beth and Cecelia have a house cleaning business. Beth can clean a particular house in 5 hr by herself. Cecelia can clean the same house in 4 hr. How long would it take if they cleaned the house together?

Let x be the number of hours it takes for both Beth and Cecelia to clean the house.

Beth's rate is $\dfrac{1 \text{ job}}{5 \text{ hr}}$. Cecelia's rate is $\dfrac{1 \text{ job}}{4 \text{ hr}}$.

The rate together is $\dfrac{1 \text{ job}}{x \text{ hr}}$.

$$\frac{1}{5} + \frac{1}{4} = \frac{1}{x} \qquad \text{Add the rates.}$$

$$20x\left(\frac{1}{5} + \frac{1}{4}\right) = 20x\left(\frac{1}{x}\right)$$

$$4x + 5x = 20$$

$$9x = 20$$

$$x = \frac{20}{9}$$

It takes $\dfrac{20}{9}$ hr or $2\dfrac{2}{9}$ hr working together.

Section 7.8   Variation

Key Concepts

Direct Variation

y varies directly as x.
y is directly proportional to x. $\left.\right\}$ $y = kx$

Inverse Variation

y varies inversely as x.
y is inversely proportional to x. $\left.\right\}$ $y = \dfrac{k}{x}$

Joint Variation

y varies jointly as w and z.
y is jointly proportional to w and z. $\left.\right\}$ $y = kwz$

Steps to Find a Variation Model

1. Write a general variation model that relates the variables given in the problem. Let k represent the constant of variation.
2. Solve for k by substituting known values of the variables into the model from step 1.
3. Substitute the value of k into the original variation model from step 1.

Examples

Example 1

t varies directly as the square root of x.

$t = k\sqrt{x}$

Example 2

W is inversely proportional to the cube of x.

$W = \dfrac{k}{x^3}$

Example 3

y is jointly proportional to x and the square of z.

$y = kxz^2$

Example 4

C varies directly as the square root of d and inversely as t. If $C = 12$ when d is 9 and t is 6, find C if d is 16 and t is 12.

Step 1. $C = \dfrac{k\sqrt{d}}{t}$

Step 2. $12 = \dfrac{k\sqrt{9}}{6} \Rightarrow 12 = \dfrac{k \cdot 3}{6} \Rightarrow k = 24$

Step 3. $C = \dfrac{24\sqrt{d}}{t} \Rightarrow C = \dfrac{24\sqrt{16}}{12} \Rightarrow C = 8$

Chapter 7   Review Exercises

Section 7.1

1. For the rational expression $\dfrac{t-2}{t+9}$

   a. Evaluate the expression (if possible) for
   $t = 0, 1, 2, -3, -9$ $-\dfrac{2}{9}, -\dfrac{1}{10}, 0, -\dfrac{5}{6}$, undefined

   b. Identify the restricted values. $t = -9$

2. For the rational expression $\dfrac{k+1}{k-5}$

   a. Evaluate the expression for $k = 0, 1, 5, -1, -2$
   $-\dfrac{1}{5}, -\dfrac{1}{2}$, undefined, $0, \dfrac{1}{7}$

   b. Identify the restricted values. $k = 5$

3. Which of the rational expressions are equal to -1?   a, c, d

   a. $\dfrac{2-x}{x-2}$ b. $\dfrac{x-5}{x+5}$

   c. $\dfrac{-x-7}{x+7}$ d. $\dfrac{x^2-4}{4-x^2}$

For Exercises 4–13, identify the restricted values. Then simplify the expressions.

4. $\dfrac{x-3}{(2x-5)(x-3)}$
   $x = \dfrac{5}{2}, x = 3; \dfrac{1}{2x-5}$

5. $\dfrac{h+7}{(3h+1)(h+7)}$
   $h = -\dfrac{1}{3}, h = -7; \dfrac{1}{3h+1}$

 Writing   Translating Expression   Geometry   Scientific Calculator   Video

6. $\dfrac{4a^2 + 7a - 2}{a^2 - 4}$
$a = 2, a = -2; \dfrac{4a - 1}{a - 2}$

7. $\dfrac{2w^2 + 11w + 12}{w^2 - 16}$
$w = 4, w = -4; \dfrac{2w + 3}{w - 4}$

8. $\dfrac{z^2 - 4z}{8 - 2z}$ $z = 4; -\dfrac{z}{2}$

9. $\dfrac{15 - 3k}{2k^2 - 10k}$
$k = 0, k = 5; -\dfrac{3}{2k}$

10. $\dfrac{2b^2 + 4b - 6}{4b + 12}$
$b = -3; \dfrac{b - 1}{2}$

11. $\dfrac{3m^2 - 12m - 15}{9m + 9}$
$m = -1; \dfrac{m - 5}{3}$

12. $\dfrac{n + 3}{n^2 + 6n + 9}$
$n = -3; \dfrac{1}{n + 3}$

13. $\dfrac{p + 7}{p^2 + 14p + 49}$
$p = -7; \dfrac{1}{p + 7}$

Section 7.2

For Exercises 14–27, multiply or divide as indicated.

14. $\dfrac{3y^3}{3y - 6} \cdot \dfrac{y - 2}{y}$ y^2

15. $\dfrac{2u + 10}{u} \cdot \dfrac{u^3}{4u + 20}$ $\dfrac{u^2}{2}$

16. $\dfrac{11}{v - 2} \cdot \dfrac{2v^2 - 8}{22}$ $v + 2$

17. $\dfrac{8}{x^2 - 25} \cdot \dfrac{3x + 15}{16}$ $\dfrac{3}{2(x - 5)}$

18. $\dfrac{4c^2 + 4c}{c^2 - 25} \div \dfrac{8c}{c^2 - 5c}$
$\dfrac{c(c + 1)}{2(c + 5)}$

19. $\dfrac{q^2 - 5q + 6}{2q + 4} \div \dfrac{2q - 6}{q + 2}$
$\dfrac{q - 2}{4}$

20. $\left(\dfrac{-2t}{t + 1}\right)(t^2 - 4t - 5)$
$-2t(t - 5)$

21. $(s^2 - 6s + 8)\left(\dfrac{4s}{s - 2}\right)$
$4s(s - 4)$

22. $\dfrac{\dfrac{a^2 + 5a + 1}{7a - 7}}{\dfrac{a^2 + 5a + 1}{a - 1}}$ $\dfrac{1}{7}$

23. $\dfrac{\dfrac{n^2 + n + 1}{n^2 - 4}}{\dfrac{n^2 + n + 1}{n + 2}}$ $\dfrac{1}{n - 2}$

24. $\dfrac{5h^2 - 6h + 1}{h^2 - 1} \div \dfrac{16h^2 - 9}{4h^2 + 7h + 3} \cdot \dfrac{3 - 4h}{30h - 6}$ $-\dfrac{1}{6}$

25. $\dfrac{3m - 3}{6m^2 + 18m + 12} \cdot \dfrac{2m^2 - 8}{m^2 - 3m + 2} \div \dfrac{m + 3}{m + 1}$ $\dfrac{1}{m + 3}$

26. $\dfrac{x - 2}{x^2 - 3x - 18} \cdot \dfrac{6 - x}{x^2 - 4}$ $\dfrac{-1}{(x + 3)(x + 2)}$

27. $\dfrac{4y^2 - 1}{1 + 2y} \div \dfrac{y^2 - 4y - 5}{5 - y}$ $\dfrac{2y - 1}{y + 1}$

Section 7.3

28. Determine the LCD.

$$\dfrac{6}{n^2 - 9}; \dfrac{5}{n^2 - n - 6}$$
LCD $= (n + 3)(n - 3)(n + 2)$

29. Determine the LCD.

$$\dfrac{8}{m^2 - 16}; \dfrac{7}{m^2 - m - 12}$$
LCD $= (m + 4)(m - 4)(m + 3)$

30. State two possible LCDs that could be used to add the fractions. $c - 2$ or $2 - c$

$$\dfrac{7}{c - 2} + \dfrac{4}{2 - c}$$

31. State two possible LCDs that could be used to subtract the fractions. $3 - x$ or $x - 3$

$$\dfrac{10}{3 - x} - \dfrac{5}{x - 3}$$

For Exercises 32–37, write each fraction as an equivalent fraction with the LCD as its denominator.

32. $\dfrac{2}{5a}; \dfrac{3}{10b}$ $\dfrac{4b}{10ab}; \dfrac{3a}{10ab}$

33. $\dfrac{7}{4x}; \dfrac{11}{6y}$ $\dfrac{21y}{12xy}; \dfrac{22x}{12xy}$

34. $\dfrac{1}{x^2 y^4}; \dfrac{3}{xy^5}$ $\dfrac{y}{x^2 y^5}; \dfrac{3x}{x^2 y^5}$

35. $\dfrac{5}{ab^3}; \dfrac{3}{ac^2}$ $\dfrac{5c^2}{ab^3 c^2}; \dfrac{3b^3}{ab^3 c^2}$

36. $\dfrac{5}{p + 2}; \dfrac{p}{p - 4}$ $\dfrac{5p - 20}{(p + 2)(p - 4)}; \dfrac{p^2 + 2p}{(p + 2)(p - 4)}$

37. $\dfrac{6}{q}; \dfrac{1}{q + 8}$ $\dfrac{6q + 48}{q(q + 8)}; \dfrac{q}{q(q + 8)}$

Section 7.4

For Exercises 38–49, add or subtract as indicated.

38. $\dfrac{h + 3}{h + 1} + \dfrac{h - 1}{h + 1}$ 2

39. $\dfrac{b - 6}{b - 2} + \dfrac{b + 2}{b - 2}$ 2

40. $\dfrac{a^2}{a - 5} - \dfrac{25}{a - 5}$ $a + 5$

41. $\dfrac{x^2}{x + 7} - \dfrac{49}{x + 7}$ $x - 7$

42. $\dfrac{y}{y^2 - 81} + \dfrac{2}{9 - y}$
$\dfrac{-y - 18}{(y - 9)(y + 9)}$ or $\dfrac{y + 18}{(9 - y)(y + 9)}$

43. $\dfrac{3}{4 - t^2} + \dfrac{t}{2 - t}$
$\dfrac{t^2 + 2t + 3}{(2 - t)(2 + t)}$

44. $\dfrac{4}{3m} - \dfrac{1}{m + 2}$
$\dfrac{m + 8}{3m(m + 2)}$

45. $\dfrac{5}{2r + 12} - \dfrac{1}{r}$ $\dfrac{3(r - 4)}{2r(r + 6)}$

46. $\dfrac{4p}{p^2 + 6p + 5} - \dfrac{3p}{p^2 + 5p + 4}$ $\dfrac{p}{(p + 4)(p + 5)}$

47. $\dfrac{3q}{q^2 + 7q + 10} - \dfrac{2q}{q^2 + 6q + 8}$ $\dfrac{q}{(q + 5)(q + 4)}$

48. $\dfrac{1}{h} + \dfrac{h}{2h+4} - \dfrac{2}{h^2+2h}$ $\frac{1}{2}$

49. $\dfrac{x}{3x+9} - \dfrac{3}{x^2+3x} + \dfrac{1}{x}$ $\frac{1}{3}$

Section 7.5

For Exercises 50–57, simplify the complex fractions.

50. $\dfrac{\dfrac{a-4}{3}}{\dfrac{a-2}{3}}$ $\frac{a-4}{a-2}$

51. $\dfrac{\dfrac{z+5}{z}}{\dfrac{z-5}{3}}$ $\frac{3(z+5)}{z(z-5)}$

52. $\dfrac{\dfrac{2-3w}{2}}{\dfrac{2}{w}-3}$ $\frac{w}{2}$

53. $\dfrac{\dfrac{2}{y}+6}{\dfrac{3y+1}{4}}$ $\frac{8}{y}$

54. $\dfrac{\dfrac{y}{x}-\dfrac{x}{y}}{\dfrac{1}{x}+\dfrac{1}{y}}$ $y-x$

55. $\dfrac{\dfrac{b}{a}-\dfrac{a}{b}}{\dfrac{1}{b}-\dfrac{1}{a}}$ $-(b+a)$

56. $\dfrac{\dfrac{6}{p+2}+4}{\dfrac{8}{p+2}-4}$ $-\frac{2p+7}{2p}$

57. $\dfrac{\dfrac{25}{k+5}+5}{\dfrac{5}{k+5}-5}$ $-\frac{k+10}{k+4}$

Section 7.6

For Exercises 58–65, solve the equations.

58. $\dfrac{2}{x} + \dfrac{1}{2} = \dfrac{1}{4}$ $\{-8\}$

59. $\dfrac{1}{y} + \dfrac{3}{4} = \dfrac{1}{4}$ $\{-2\}$

60. $\dfrac{2}{h-2} + 1 = \dfrac{h}{h+2}$ $\{0\}$

61. $\dfrac{w}{w-1} = \dfrac{3}{w+1} + 1$ $\{2\}$

62. $\dfrac{t+1}{3} - \dfrac{t-1}{6} = \dfrac{1}{6}$ $\{-2\}$

63. $\dfrac{w+1}{w-3} - \dfrac{3}{w} = \dfrac{12}{w^2-3w}$ $\{-1\}$ (The value 3 does not check.)

64. $\dfrac{1}{z+2} = \dfrac{4}{z^2-4} - \dfrac{1}{z-2}$ $\{\ \}$ (The value 2 does not check.)

65. $\dfrac{y+1}{y+3} = \dfrac{y^2-11y}{y^2+y-6} - \dfrac{y-3}{y-2}$ $\{-11, 1\}$

66. Four times a number is added to 5. The sum is then divided by 6. The result is $\frac{7}{2}$. Find the number.
The number is 4.

67. Solve the formula $\dfrac{V}{h} = \dfrac{\pi r^2}{3}$ for h. $h = \frac{3V}{\pi r^2}$

68. Solve the formula $\dfrac{A}{b} = \dfrac{h}{2}$ for b. $b = \frac{2A}{h}$

Section 7.7

For Exercises 69–70, solve the proportions.

69. $\dfrac{m+2}{8} = \dfrac{m}{3}$ $\left\{\frac{6}{5}\right\}$

70. $\dfrac{12}{a} = \dfrac{5}{8}$ $\left\{\frac{96}{5}\right\}$

71. A bag of popcorn states that it contains 4 g of fat per serving. If a serving is 2 oz, how many grams of fat are in a 5-oz bag?   It contains 10 g of fat.

72. Bud goes 10 mph faster on his motorcycle than Ed goes on his motorcycle. If Bud travels 105 mi in the same amount of time that Ed travels 90 mi, what are the rates of the two bikers?
Ed travels 60 mph, and Bud travels 70 mph.

© Glow Images RF

73. There are two pumps set up to fill a small swimming pool. One pump takes 24 min by itself to fill the pool, but the other takes 56 min by itself. How long would it take if both pumps work together?
Together the pumps would fill the pool in 16.8 min.

74. Triangle XYZ is similar to triangle ABC. Find the values of x and b. $x = 11$; $b = 26$

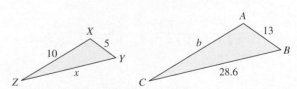

Section 7.8

75. The force F applied to a spring varies directly with the distance d that the spring is stretched.

  a. Write a variation model using k as the constant of variation. $F = kd$

  b. When 6 lb of force is applied, the spring stretches 2 ft. Find k. $k = 3$

  c. How much force is required to stretch the spring 4.2 ft? 12.6 lb

76. Suppose y varies inversely with the cube of x, and $y = 9$ when $x = 2$. Find y when $x = 3$. $y = \dfrac{8}{3}$

77. Suppose y varies jointly with x and the square root of z, and $y = 3$ when $x = 3$ and $z = 4$. Find y when $x = 8$ and $z = 9$. $y = 12$

 78. The distance, d, that one can see to the horizon varies directly as the square root of the height above sea level. If a person 25 m above sea level can see 30 km, how far can a person see if she is 64 m above sea level? 48 km

© Royalty Free/Corbis RF

Chapter 7 Test

For Exercises 1–2,

  a. Identify the restricted values.

  b. Simplify the rational expression.

1. $\dfrac{5(x-2)(x+1)}{30(2-x)}$
  a. $x = 2$ b. $-\dfrac{x+1}{6}$

2. $\dfrac{7a^2 - 42a}{a^3 - 4a^2 - 12a}$
  a. $a = 6, a = -2, a = 0$ b. $\dfrac{7}{a+2}$

3. Identify the rational expressions that are equal to -1. b, c, d

  a. $\dfrac{x+4}{x-4}$ **b.** $\dfrac{7-2x}{2x-7}$

  c. $\dfrac{9x^2+16}{-9x^2-16}$ **d.** $-\dfrac{x+5}{x+5}$

4. Find the LCD of the following pairs of rational expressions.

  a. $\dfrac{x}{3(x+3)}; \dfrac{7}{5(x+3)}$
    $15(x+3)$

  b. $\dfrac{-2}{3x^2y}; \dfrac{4}{xy^2}$
    $3x^2y^2$

For Exercises 5–11, perform the indicated operation.

5. $\dfrac{2}{y^2+4y+3} + \dfrac{1}{3y+9}$ $\dfrac{y+7}{3(y+3)(y+1)}$

6. $\dfrac{9-b^2}{5b+15} \div \dfrac{b-3}{b+3}$ $\dfrac{b+3}{5}$

7. $\dfrac{w^2-4w}{w^2-8w+16} \cdot \dfrac{w-4}{w^2+w}$ $\dfrac{1}{w+1}$

8. $\dfrac{t}{t-2} - \dfrac{8}{t^2-4}$ $\dfrac{t+4}{t+2}$

9. $\dfrac{1}{x+4} + \dfrac{2}{x^2+2x-8} + \dfrac{x}{x-2}$ $\dfrac{x(x+5)}{(x+4)(x-2)}$

10. $\dfrac{2y}{y-6} - \dfrac{7}{6-y}$ $\dfrac{2y+7}{y-6}$ or $\dfrac{-2y-7}{6-y}$

11. $\dfrac{1 - \dfrac{4}{m}}{m - \dfrac{16}{m}}$ $\dfrac{1}{m+4}$

For Exercises 12–16, solve the equation.

12. $\dfrac{3}{a} + \dfrac{5}{2} = \dfrac{7}{a}$ $\left\{\dfrac{8}{5}\right\}$

13. $\dfrac{p}{p-1} + \dfrac{1}{p} = \dfrac{p^2+1}{p^2-p}$ $\{2\}$

14. $\dfrac{3}{c-2} - \dfrac{1}{c+1} = \dfrac{7}{c^2-c-2}$ $\{1\}$

15. $\dfrac{4x}{x-4} = 3 + \dfrac{16}{x-4}$ $\{\ \}$ (The value 4 does not check.)

16. $\dfrac{y^2+7y}{y-2} - \dfrac{36}{2y-4} = 4$ $\{-5\}$ (The value 2 does not check.)

17. Solve the formula $\dfrac{C}{2} = \dfrac{A}{r}$ for r. $r = \dfrac{2A}{C}$

18. Solve the proportion.

$$\frac{y+7}{-4} = \frac{1}{4} \quad \{-8\}$$

19. A recipe for vegetable soup calls for $\frac{1}{2}$ cup of carrots for six servings. How many cups of carrots are needed to prepare 15 servings?
$1\frac{1}{4}$ (1.25) cups of carrots

20. A motorboat can travel 28 mi downstream in the same amount of time as it can travel 18 mi upstream. Find the speed of the current if the boat can travel 23 mph in still water.
The speed of the current is 5 mph.

21. Two printers working together can complete a job in 2 hr. If one printer requires 6 hr to do the job alone, how many hours would the second printer need to complete the job alone?   It would take the second printer 3 hr to do the job working alone.

22. Triangle *XYZ* is similar to triangle *ABC*. Find the values of *a* and *b*. $a = 5.6$; $b = 12$

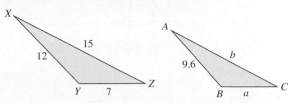

23. The amount of medication prescribed for a patient varies directly as the patient's weight. If a 160-lb person is prescribed 6 mL of a medicine, how much medicine would be prescribed for a 220-lb person?   8.25 mL

24. The number of drinks sold at a concession stand varies inversely as price. If the price is set at \$1.25 per drink, then 400 drinks are sold. If the price is set at \$2.50 per drink, then how many drinks are sold?   200 drinks are sold.

Chapters 1–7   Cumulative Review Exercises

For Exercises 1–2, simplify completely.

1. $\left(\frac{1}{2}\right)^{-4} + 2^4$ 32

2. $|3 - 5| + |-2 + 7|$
7

3. Solve. $\frac{1}{2} - \frac{3}{4}(y - 1) = \frac{5}{12}$ $\left\{\frac{10}{9}\right\}$

4. Complete the table.

Set-Builder Notation	Graph	Interval Notation
$\{x \mid x \geq -1\}$	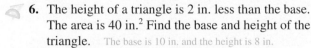	$[-1, \infty)$
$\{x \mid x < 5\}$		$(-\infty, 5)$

 5. The perimeter of a rectangular swimming pool is 104 m. The length is 1 m more than twice the width. Find the length and width.
The width is 17 m and the length is 35 m.

6. The height of a triangle is 2 in. less than the base. The area is 40 in.^{2} Find the base and height of the triangle.   The base is 10 in. and the height is 8 in.

7. Simplify. $\left(\frac{4x^{-1}y^{-2}}{z^4}\right)^{-2}(2y^{-1}z^3)^3$ $\frac{x^2yz^{17}}{2}$

8. The length and width of a rectangle are given in terms of *x*.

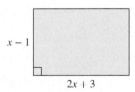

   a. Write a polynomial that represents the perimeter of the rectangle. $6x + 4$

   b. Write a polynomial that represents the area of the rectangle. $2x^2 + x - 3$

9. Factor completely. $25x^2 - 30x + 9$ $(5x - 3)^2$

10. Factor completely. $10cd + 5d - 6c - 3$
$(2c + 1)(5d - 3)$

11. Identify the restricted values of the expression.

$$\frac{x+3}{(x-5)(2x+1)} \qquad x = 5, x = -\frac{1}{2}$$

12. Solve the system.

$\begin{aligned} x + 2y &= -7 \\ 6x + 3y &= -6 \end{aligned}$ $\{(1, -4)\}$

13. Divide. $\frac{2x - 6}{x^2 - 16} \div \frac{10x^2 - 90}{x^2 - x - 12}$ $\frac{1}{5(x+4)}$

Writing   Translating Expression   Geometry   Scientific Calculator   Video

14. Simplify.

$$\frac{\dfrac{3}{4}-\dfrac{1}{x}}{\dfrac{1}{3x}-\dfrac{1}{4}} \quad -3$$

15. Solve. $\dfrac{7}{y^2-4}=\dfrac{3}{y-2}+\dfrac{2}{y+2}$ {1}

16. Solve the proportion.

$$\frac{2b-5}{6}=\frac{4b}{7} \quad \left\{-\frac{7}{2}\right\}$$

17. Determine the x- and y-intercepts.

   a. $-2x+4y=8$ x-intercept: (−4, 0);
                                y-intercept: (0, 2)

   b. $y=5x$
      x-intercept: (0,0);
      y-intercept: (0,0)

18. Determine the slope

   a. of the line containing the points (0, −6) and (−5, 1). $m=-\dfrac{7}{5}$

   b. of the line $y=-\dfrac{2}{3}x-6$. $m=-\dfrac{2}{3}$

   c. of a line parallel to a line having a slope of 4. $m=4$

   d. of a line perpendicular to a line having a slope of 4. $m=-\dfrac{1}{4}$

19. Find an equation of a line passing through the point (1, 2) and having a slope of 5. Write the answer in slope-intercept form. $y=5x-3$

20. A group of teenagers buys 2 large popcorns and 6 drinks at the movie theater for $27. A couple buys 1 large popcorn and 2 drinks for $10.50. Find the price for 1 large popcorn and the price for 1 drink.   One large popcorn costs $4.50, and one drink costs $3.00.

© BananaStock/PunchStock RF

Radicals

8

Mathematics in History

We are familiar with the idea of a power or exponent. For example, when we raise 3 to the power of 2, we understand that we will multiply 3 by itself.

$$3^2 = 3 \cdot 3 = 9$$

By virtue of this, when asked, "What number multiplied by itself equals 9?" we would quickly answer, "3."

What would we answer if asked, "What number multiplied by itself equals 7?" For many thousands of years since the beginning of civilization, this *undoing* operation was difficult to answer!

© Digital Vision/Getty Images RF

It was mathematicians in the 16th century who formally developed the tools to do such a thing and adopted the **radical sign** $\sqrt{}$. This symbol represents the nonnegative number that when multiplied by itself results in the number inside the symbol. For example,

$$\sqrt{9} = 3 \quad \text{and} \quad \sqrt{7} \approx 2.645$$

This operation represented by a radical sign is called **extracting a square root**. In this chapter, we will perform operations on radical expressions and use radicals in applications.

Section 8.1	Introduction to Roots and Radicals

Concepts

1. **Definition of a Square Root**
2. **Definition of an *n*th-Root**
3. **Translations Involving *n*th-Roots**
4. **Pythagorean Theorem**

1. Definition of a Square Root

Recall that to square a number means to multiply the number by itself: $b^2 = b \cdot b$. To find a square root of a number, we reverse the process of squaring a number. For example, finding a square root of 49 is equivalent to asking: "What number when squared equals 49?"

One obvious answer to this question is 7 because $(7)^2 = 49$. But -7 will also work because $(-7)^2 = 49$.

> ### Definition of a Square Root
>
> b is a **square root** of a if $b^2 = a$.

Classroom Examples: p. 576, Exercises 2 and 4

Example 1	Identifying the Square Roots of a Number

Identify the square roots of each number.

    a. 9   **b.** 121   **c.** 0   **d.** -4

Solution:

    a. 3 is a square root of 9 because $(3)^2 = 9$.
       -3 is a square root of 9 because $(-3)^2 = 9$.

    b. 11 is a square root of 121 because $(11)^2 = 121$.
       -11 is a square root of 121 because $(-11)^2 = 121$.

    c. 0 is a square root of 0 because $(0)^2 = 0$.

    d. There are no real numbers that when squared will equal a negative number. Therefore, there are no real-valued square roots of -4.

TIP: All positive real numbers have two real-valued square roots: one positive and one negative. Zero has only one square root, which is 0 itself. Finally, for any negative real number, there are no real-valued square roots.

Skill Practice Identify the square roots of each number.

    1. 64   **2.** -36 **3.** 36   **4.** $\dfrac{25}{16}$

Recall that the positive square root of a real number can be denoted with a radical sign, $\sqrt{}$.

> ### Notation for Positive and Negative Square Roots
>
> Let a represent a positive real number. Then,
>
>   **1.** $\sqrt{a}$ is the **positive square root** of a. The positive square root is also called the **principal square root**.
>   **2.** $-\sqrt{a}$ is the **negative square root** of a.
>   **3.** $\sqrt{0} = 0$

Answers

1. $8; -8$
2. There are no real-valued square roots.
3. $6; -6$ **4.** $\dfrac{5}{4}; -\dfrac{5}{4}$

Example 2 **Simplifying Square Roots**

Classroom Examples: p. 576,
Exercises 18, 20, and 24

Simplify the square roots.

a. $\sqrt{36}$ **b.** $\sqrt{225}$ **c.** $\sqrt{1}$ **d.** $\sqrt{\dfrac{9}{4}}$ **e.** $\sqrt{0.49}$

Solution:

a. $\sqrt{36}$ denotes the positive square root of 36. $\sqrt{36} = 6$

b. $\sqrt{225}$ denotes the positive square root of 225. $\sqrt{225} = 15$

c. $\sqrt{1}$ denotes the positive square root of 1. $\sqrt{1} = 1$

d. $\sqrt{\dfrac{9}{4}}$ denotes the positive square root of $\dfrac{9}{4}$. $\sqrt{\dfrac{9}{4}} = \dfrac{3}{2}$

e. $\sqrt{0.49}$ denotes the positive square root. $\sqrt{0.49} = 0.7$

Skill Practice Simplify the square roots.

5. $\sqrt{81}$ **6.** $\sqrt{144}$ **7.** $\sqrt{0}$ **8.** $\sqrt{\dfrac{1}{4}}$ **9.** $\sqrt{0.09}$

The numbers 36, 225, 1, $\frac{9}{4}$, and 0.49 are **perfect squares** because their square roots are rational numbers. Radicals that cannot be simplified to rational numbers are irrational numbers. Recall that an irrational number cannot be written as a terminating or repeating decimal. For example, the symbol $\sqrt{13}$ is used to represent the exact value of the square root of 13. The symbol $\sqrt{42}$ is used to represent the exact value of the square root of 42. These values are irrational numbers but can be approximated by rational numbers by using a calculator.

$$\sqrt{13} \approx 3.605551275 \qquad \sqrt{42} \approx 6.480740698$$

Note: The only way to denote the *exact* values of the square root of 13 and the square root of 42 is $\sqrt{13}$ and $\sqrt{42}$.

A negative number cannot have a real number as a square root because no real number when squared is negative. For example, $\sqrt{-25}$ is *not a real number* because there is no real number, b, for which $(b)^2 = -25$.

> **TIP:** Before using a calculator to evaluate a square root, try estimating the value first.
>
> $\sqrt{13}$ must be a number between 3 and 4 because $\sqrt{9} < \sqrt{13} < \sqrt{16}$.
>
> $\sqrt{42}$ must be a number between 6 and 7 because $\sqrt{36} < \sqrt{42} < \sqrt{49}$.

Example 3 **Simplifying Square Roots if Possible**

Classroom Examples: p. 576,
Exercises 36 and 38

Simplify the square roots, if possible.

a. $\sqrt{-100}$ **b.** $-\sqrt{100}$ **c.** $\sqrt{-64}$

Solution:

a. $\sqrt{-100}$ Not a real number

b. $-\sqrt{100}$

    $-1 \cdot \sqrt{100}$ The expression $-\sqrt{100}$ is equivalent to $-1 \cdot \sqrt{100}$.

    $-1 \cdot 10 = -10$

c. $\sqrt{-64}$ Not a real number

Skill Practice Simplify the square roots, if possible.

10. $\sqrt{-25}$ **11.** $-\sqrt{25}$ **12.** $\sqrt{-4}$

Answers

5. 9   **6.** 12

7. 0   **8.** $\dfrac{1}{2}$

9. 0.3   **10.** Not a real number

11. −5   **12.** Not a real number

2. Definition of an *n*th-Root

To find a square root of a number, we reverse the process of squaring a number. This concept can be extended to finding a third root (called a cube root), a fourth root, and in general, an *n*th-root.

> ### Definition of an *n*th-Root
>
> b is an **nth-root** of a if $b^n = a$.

The radical sign, $\sqrt{}$, is used to denote the principal square root of a number. The symbol, $\sqrt[n]{}$, is used to denote the principal *n*th-root of a number.

In the expression $\sqrt[n]{a}$, n is called the **index** of the radical, and a is called the **radicand**. For a square root, the index is 2, but it is usually not written ($\sqrt[2]{a}$ is denoted simply as $\sqrt{a}$). A radical with an index of 3 is called a **cube root**, $\sqrt[3]{a}$.

> ### Definition of $\sqrt[n]{a}$
>
> 1. If n is a positive *even* integer and $a > 0$, then $\sqrt[n]{a}$ is the principal (positive) *n*th-root of a.   Example: $\sqrt[4]{81} = 3$
> 2. If $n > 1$ is a positive *odd* integer, then $\sqrt[n]{a}$ is the *n*th-root of a.   Example: $\sqrt[3]{-125} = -5$
> 3. If $n > 1$ is a positive integer, then $\sqrt[n]{0} = 0$.   Example: $\sqrt[6]{0} = 0$

For the purpose of simplifying radicals, it is helpful to know the following patterns:

Perfect cubes	Perfect fourth powers	Perfect fifth powers
$1^3 = 1$	$1^4 = 1$	$1^5 = 1$
$2^3 = 8$	$2^4 = 16$	$2^5 = 32$
$3^3 = 27$	$3^4 = 81$	$3^5 = 243$
$4^3 = 64$	$4^4 = 256$	$4^5 = 1024$
$5^3 = 125$	$5^4 = 625$	$5^5 = 3125$

Classroom Examples: pp. 576–577, Exercises 52 and 62

Example 4   Simplifying *n*th-Roots

Simplify the expressions, if possible.

a. $\sqrt[3]{8}$ **b.** $\sqrt[4]{16}$ **c.** $\sqrt[5]{32}$ **d.** $\sqrt[3]{-64}$

e. $\sqrt[3]{\dfrac{125}{27}}$ **f.** $\sqrt{0.01}$ **g.** $\sqrt[4]{-81}$

Solution:

a. $\sqrt[3]{8} = 2$ Because $(2)^3 = 8$

b. $\sqrt[4]{16} = 2$ Because $(2)^4 = 16$

c. $\sqrt[5]{32} = 2$ Because $(2)^5 = 32$

d. $\sqrt[3]{-64} = -4$ Because $(-4)^3 = -64$

e. $\sqrt[3]{\dfrac{125}{27}} = \dfrac{5}{3}$ Because $\left(\dfrac{5}{3}\right)^3 = \dfrac{125}{27}$

> **TIP:** Even-indexed roots of negative numbers are not real numbers. Odd-indexed roots of negative numbers are negative.

f. $\sqrt{0.01} = 0.1$ Because $(0.1)^2 = 0.01$

Note: $\sqrt{0.01}$ is equivalent to $\sqrt{\dfrac{1}{100}} = \dfrac{1}{10}$, or 0.1.

g. $\sqrt[4]{-81}$ is not a real number because no real number raised to the fourth power equals -81.

Avoiding Mistakes

When evaluating $\sqrt[n]{a}$, where n is *even*, always choose the principal (positive) root.

$$\sqrt[4]{16} = 2 \quad (\text{not} -2)$$
$$\sqrt{0.01} = 0.1 \quad (\text{not} -0.1)$$

Skill Practice Simplify the expressions, if possible.

13. $\sqrt[3]{27}$ **14.** $\sqrt[4]{1}$ **15.** $\sqrt[3]{216}$ **16.** $\sqrt[5]{-32}$

17. $\sqrt[4]{\dfrac{16}{625}}$ **18.** $\sqrt{0.25}$ **19.** $\sqrt[4]{-1}$

Example 4(g) illustrates that an *n*th-root of a negative number is not a real number if the index is even because no real number raised to an even power is negative.

Finding an *n*th-root of a variable expression is similar to finding an *n*th-root of a numerical expression. However, for roots with an even index, particular care must be taken to obtain a nonnegative value.

Definition of $\sqrt[n]{a^n}$

1. If n is a positive odd integer, then $\sqrt[n]{a^n} = a$
     Example: cube root $\sqrt[3]{a^3} = a$

2. If n is a positive even integer, then $\sqrt[n]{a^n} = |a|$
     Example: square root $\sqrt{a^2} = |a|$

If n is an even integer, then $\sqrt[n]{a^n} = |a|$. However, if the variable a is assumed to be nonnegative, then the absolute value bars may be omitted, that is, $\sqrt[n]{a^n} = a$ provided $a \geq 0$. In the following examples and exercises, we will make the assumption that the variables within a radical expression are positive real numbers. In such a case, the absolute value bars are not needed to evaluate $\sqrt[n]{a^n}$.

It is helpful to become familiar with the patterns associated with perfect squares and perfect cubes involving variable expressions.

The following powers of x are perfect squares:

Perfect squares

$(x^1)^2 = x^2$
$(x^2)^2 = x^4$
$(x^3)^2 = x^6$
$(x^4)^2 = x^8$
. . .

TIP: Any expression raised to an even power (multiple of 2) is a perfect square.

The following powers of x are perfect cubes:

Perfect cubes

$(x^1)^3 = x^3$
$(x^2)^3 = x^6$
$(x^3)^3 = x^9$
$(x^4)^3 = x^{12}$
. . .

TIP: Any expression raised to a power that is a multiple of 3 is a perfect cube.

Answers

13. 3   **14.** 1   **15.** 6
16. -2 **17.** $\dfrac{2}{5}$ **18.** 0.5
19. Not a real number

Classroom Examples: p. 577, Exercises 74 and 80

Example 5	**Simplifying *n*th-Roots**

Simplify the expressions. Assume that the variables are positive real numbers.

    a. $\sqrt{c^6}$ **b.** $\sqrt[3]{d^{15}}$ **c.** $\sqrt{a^2b^2}$ **d.** $\sqrt[3]{64z^6}$

Solution:

    a. $\sqrt{c^6}$ The expression c^6 is a perfect square.

       $\sqrt{c^6} = c^3$ This is because $\sqrt{(c^3)^2} = c^3$.

    b. $\sqrt[3]{d^{15}}$ The expression d^{15} is a perfect cube.

       $\sqrt[3]{d^{15}} = d^5$ This is because $\sqrt[3]{(d^5)^3} = d^5$.

    c. $\sqrt{a^2b^2} = ab$ This is because $\sqrt{a^2b^2} = \sqrt{(ab)^2} = ab$.

    d. $\sqrt[3]{64z^6} = 4z^2$ This is because $\sqrt[3]{(4z^2)^3} = 4z^2$.

Skill Practice Simplify the expressions. Assume the variables represent positive real numbers.

    20. $\sqrt{y^{10}}$ **21.** $\sqrt[3]{x^{12}}$ **22.** $\sqrt{x^4y^2}$ **23.** $\sqrt{25c^4}$

3. Translations Involving *n*th-Roots

It is important to understand the vocabulary and language associated with *n*th-roots. For instance, you must be able to distinguish between the square of a number and the square *root* of a number. The following example offers practice translating between English form and algebraic form.

Classroom Example: p. 577, Exercise 90

Example 6	**Writing an English Phrase in Algebraic Form**

Write each English phrase as an algebraic expression.

    a. The difference of the square of x and the principal square root of 7

    b. The quotient of 1 and the cube root of z

Solution:

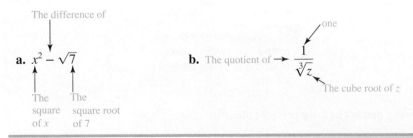

Skill Practice Write the English phrases as algebraic expressions.

    24. The product of the square of y and the principal square root of x

    25. The sum of 2 and the cube root of y

Answers

20. y^5 **21.** x^4

22. x^2y **23.** $5c^2$

24. $y^2\sqrt{x}$ **25.** $2 + \sqrt[3]{y}$

4. Pythagorean Theorem

Recall that the **Pythagorean theorem** relates the lengths of the three sides of a right triangle (Figure 8-1).

$$a^2 + b^2 = c^2$$

The principal square root can be used to solve for an unknown side of a right triangle if the lengths of the other two sides are known.

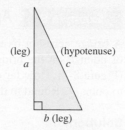

Figure 8-1

Example 7	**Applying the Pythagorean Theorem**

Classroom Example: p. 577, Exercise 96

Use the Pythagorean theorem and the definition of the principal square root of a number to find the length of the unknown side.

Solution:

Label the sides of the triangle.

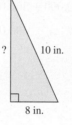

$a^2 + b^2 = c^2$

$a^2 + (8)^2 = (10)^2$ Apply the Pythagorean theorem.

$a^2 + 64 = 100$ Simplify.

$a^2 = 36$ This equation is quadratic. One method for solving the equation is to set the equation equal to zero, factor, and apply the zero product rule. However, we can also use the definition of a square root to solve for a.

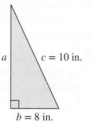

$a = \sqrt{36}$ or $a = -\sqrt{36}$ By definition, a must be one of the square roots
        (Reject   of 36 (either 6 or −6). However, because a
$a = 6$ negative   represents a distance, choose the *positive*
        value)   (principal) square root of 36.

The third side is 6 in. long.

Skill Practice Use the Pythagorean theorem to find the length of the unknown side.

26.

Answer

26. 5 cm

Classroom Example: p. 578,
Exercise 102

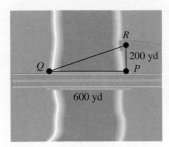

Figure 8-2

| Example 8 | **Applying the Pythagorean Theorem** |

A bridge across a river is 600 yd long. A boat ramp at point R is 200 yd due north of point P on the bridge, such that the line segments $\overline{PQ}$ and $\overline{PR}$ form a right angle (Figure 8-2). How far does a kayak travel if it leaves from the boat ramp and paddles to point Q? Round to the nearest yard.

Solution:

Label the triangle:

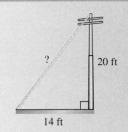

$$a^2 + b^2 = c^2$$

$$(200)^2 + (600)^2 = c^2 \qquad \text{Apply the Pythagorean theorem.}$$

$$40{,}000 + 360{,}000 = c^2 \qquad \text{Simplify.}$$

$$400{,}000 = c^2 \qquad \begin{array}{l}\text{By definition, } c \text{ must be one of the square} \\ \text{roots of 400,000. Because the value of } c \text{ is a}\end{array}$$

$$c = \sqrt{400{,}000} \qquad \begin{array}{l}\text{distance, choose the positive square root of} \\ 400{,}000.\end{array}$$

$$c \approx 632 \qquad \begin{array}{l}\text{A calculator can be used to approximate the} \\ \text{positive square root of 400,000.}\end{array}$$

The kayak must travel approximately 632 yd.

Skill Practice

27. A wire is attached to the top of a 20-ft pole. How long is the wire if it reaches a point on the ground 14 ft from the base of the pole? Round to the nearest tenth of a foot.

Answers

27. The wire is 24.4 ft long.

Calculator Connections

Topic: Evaluating Square Roots and Higher Order Roots on a Calculator

A calculator can be used to approximate the value of a radical expression. To evaluate a square root, use the $\sqrt{}$ key. For example, evaluate: $\sqrt{25}$, $\sqrt{60}$, $\sqrt{\frac{13}{3}}$

Scientific Calculator

Enter: 25 $\boxed{\sqrt{x}}$ **Result:** $\boxed{5}$

Enter: 60 $\boxed{\sqrt{x}}$ **Result:** $\boxed{7.745966692}$

Enter: 13 $\boxed{\div}$ 3 $\boxed{=}$ $\boxed{\sqrt{x}}$ **Result:** $\boxed{2.081665999}$

Graphing Calculator

On the graphing calculator, the radicand is enclosed in parentheses.

```
√(25)
                5
√(60)
      7.745966692
√(13/3)
      2.081665999
```

TIP: The values $\sqrt{60}$ and $\sqrt{\frac{13}{3}}$ are approximated on the calculator to 10 digits. However, $\sqrt{60}$ and $\sqrt{\frac{13}{3}}$ are actually irrational numbers. Their decimal forms are nonterminating and nonrepeating. The only way to represent the exact answers is by writing the radical forms, $\sqrt{60}$ and $\sqrt{\frac{13}{2}}$.

To evaluate cube roots, your calculator may have a $\boxed{\sqrt[3]{}}$ key. Otherwise, for cube roots and roots of higher index (fourth roots, fifth roots, and so on), try using the $\boxed{\sqrt[x]{y}}$ key or $\boxed{\sqrt[x]{}}$ key. For example, evaluate $\sqrt[3]{64}$, $\sqrt[4]{81}$, and $\sqrt[3]{162}$:

Scientific Calculator

Enter:	64 $\boxed{2^{nd}}$ $\boxed{\sqrt[x]{y}}$ 3 $\boxed{=}$		Result:	4
Enter:	81 $\boxed{2^{nd}}$ $\boxed{\sqrt[x]{y}}$ 4 $\boxed{=}$		Result:	3
Enter:	162 $\boxed{2^{nd}}$ $\boxed{\sqrt[x]{y}}$ 3 $\boxed{=}$		Result:	5.451361778

Graphing Calculator

On a graphing calculator, the index is usually entered first.

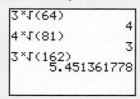

```
3*√(64)
              4
4*√(81)
              3
3*√(162)
      5.451361778
```

Calculator Exercises

Estimate the value of each radical. Then use a calculator to approximate the radical to three decimal places.

1. $\sqrt{5}$ 2.236

2. $\sqrt{17}$ 4.123

3. $\sqrt{50}$ 7.071

4. $\sqrt{96}$ 9.798

5. $\sqrt{33}$ 5.745

6. $\sqrt{145}$ 12.042

7. $\sqrt{80}$ 8.944

8. $\sqrt{170}$ 13.038

9. $\sqrt[3]{7}$ 1.913

10. $\sqrt[3]{28}$ 3.037

11. $\sqrt[3]{65}$ 4.021

12. $\sqrt[3]{124}$ 4.987

Section 8.1 Practice Exercises

Vocabulary and Key Concepts

For additional exercises, see Classroom Activities 8.1A–8.1D in the *Student's Resource Manual* at www.mhhe.com/moh.

1. a. If $b^2 = a$, then _____b_____ is a square root of _____a_____.

   b. The symbol $\sqrt{a}$ denotes the _____principal_____ or positive square root of a.

   c. A number is a perfect square if its square root is a _____rational_____ number.

   d. b is an nth root of a if _____b^n_____ = _____a_____.

   e. Given the symbol $\sqrt[n]{a}$, n is called the _____index_____ and a is called the _____radicand_____.

   f. The symbol $\sqrt[3]{a}$ denotes the _____cube_____ root of a.

   g. The expression $\sqrt{-4}$ (is/is not) a real number. The expression $-\sqrt{4}$ (is/is not) a real number.   is not; is

   h. The expression $\sqrt[n]{a^n} = |a|$ if n is (even/odd). The expression $\sqrt[n]{a^n} = a$ if n is (even/odd).   even; odd

   i. Given a right triangle with legs a and b and hypotenuse c, the Pythagorean theorem is stated as _____$a^2 + b^2 = c^2$_____.

 Writing   Translating Expression   Geometry   Scientific Calculator   Video

Concept 1: Definition of a Square Root

For Exercises 2–9, determine the square roots. **(See Example 1.)**

2. 4   2, −2

3. 144   12, −12

4. −64   There are no real-valued square roots of −64.

5. −49   There are no real-valued square roots of −49.

6. 81   9, −9

7. 0   0

8. $\dfrac{16}{9}$ $\dfrac{4}{3}$, $-\dfrac{4}{3}$

9. $\dfrac{1}{25}$ $\dfrac{1}{5}$, $-\dfrac{1}{5}$

10. a. What is the principal square root of 64?   8

    b. What is the negative square root of 64?   −8

11. a. What is the principal square root of 169?   13

    b. What is the negative square root of 169?   −13

12. Does every number have two square roots? Explain.

No, only positive numbers have two square roots. Zero has only one square root, and negative numbers have no real-valued square roots.

13. Which number has only one square root?   0

14. Which of the following are perfect squares?

0, 1, 4, 15, 30, 49, 72, 81, 144, 300, 625, 900

0, 1, 4, 49, 81, 144, 625, 900

15. Which of the following are perfect squares?

8, 9, 12, 16, 25, 36, 42, 64, 95, 121, 140, 169

9, 16, 25, 36, 64, 121, 169

For Exercises 16–31, simplify the square roots. **(See Example 2.)**

16. $\sqrt{16}$ 4

17. $\sqrt{4}$ 2

18. $\sqrt{81}$ 9

19. $\sqrt{49}$ 7

20. $\sqrt{0.25}$ 0.5

21. $\sqrt{0.16}$ 0.4

22. $\sqrt{0.64}$ 0.8

23. $\sqrt{0.09}$ 0.3

24. $\sqrt{\dfrac{1}{9}}$ $\dfrac{1}{3}$

25. $\sqrt{\dfrac{25}{16}}$ $\dfrac{5}{4}$

26. $\sqrt{\dfrac{49}{121}}$ $\dfrac{7}{11}$

27. $\sqrt{\dfrac{1}{144}}$ $\dfrac{1}{12}$

28. $\sqrt{64+36}$ 10

29. $\sqrt{16+9}$ 5

30. $\sqrt{169-144}$ 5

31. $\sqrt{225-144}$ 9

32. Explain the difference between $\sqrt{-16}$ and $-\sqrt{16}$.

$\sqrt{-16}$ is not a real number, and $-\sqrt{16}$ simplifies to −4, which is a real number.

33. Using the definition of a square root, explain why $\sqrt{-16}$ does not have a real-valued square root.

There is no real value of b for which $b^2 = -16$.

34. Evaluate. $-\sqrt{|-25|}$ −5

For Exercises 35–46, simplify the square roots, if possible. **(See Example 3.)**

35. $-\sqrt{4}$ −2

36. $-\sqrt{1}$ −1

37. $\sqrt{-4}$
Not a real number

38. $\sqrt{-1}$
Not a real number

39. $\sqrt{-\dfrac{4}{49}}$
Not a real number

40. $-\sqrt{-\dfrac{9}{25}}$
Not a real number

41. $-\sqrt{-\dfrac{1}{36}}$
Not a real number

42. $-\sqrt{\dfrac{1}{36}}$ $-\dfrac{1}{6}$

43. $-\sqrt{400}$ −20

44. $-\sqrt{121}$ −11

45. $\sqrt{-900}$
Not a real number

46. $\sqrt{-169}$
Not a real number

Concept 2: Definition of an nth-Root

47. Which of the following are perfect cubes?

0, 1, 3, 9, 27, 36, 42, 90, 125   0, 1, 27, 125

48. Which of the following are perfect cubes?

6, 8, 16, 20, 30, 64, 111, 150, 216   8, 64, 216

49. Does −27 have a real-valued cube root?
Yes, −3

50. Does −8 have a real-valued cube root?
Yes, −2

For Exercises 51–66, simplify the nth roots, if possible. **(See Example 4.)**

51. $\sqrt[3]{27}$ 3

52. $\sqrt[3]{-27}$ −3

53. $\sqrt[3]{64}$ 4

54. $\sqrt[3]{-64}$ −4

55. $-\sqrt[4]{16}$ −2

56. $-\sqrt[4]{81}$ −3

57. $\sqrt[4]{-1}$ Not a real number

58. $\sqrt[4]{0}$ 0

59. $\sqrt[4]{-256}$ Not a real number

60. $\sqrt[4]{-625}$ Not a real number

61. $\sqrt[5]{-\dfrac{1}{32}}$ $-\dfrac{1}{2}$

62. $-\sqrt[5]{\dfrac{1}{32}}$ $-\dfrac{1}{2}$

63. $-\sqrt[6]{1}$ -1

64. $\sqrt[6]{64}$ 2

65. $\sqrt[6]{0}$ 0

66. $\sqrt[6]{-1}$ Not a real number

67. Determine which of the expressions are perfect squares. Then state a rule for determining perfect squares based on the exponent of the expression.

$x^2, a^3, y^4, z^5, (ab)^6, (pq)^7, w^8x^8, c^9d^9, m^{10}, n^{11}$
$x^2, y^4, (ab)^6, w^8x^8, m^{10}$ The expression is a perfect square if the exponent is even.

68. Determine which of the expressions are perfect cubes. Then state a rule for determining perfect cubes based on the exponent of the expression.

$a^2, b^3, c^4, d^5, e^6, (xy)^7, (wz)^8, (pq)^9, t^{10}s^{10}, m^{11}n^{11}, u^{12}v^{12}$
$b^3, e^6, (pq)^9, u^{12}v^{12}$ The expression is a perfect cube if the exponent is a multiple of 3.

For Exercises 69–88, simplify the expressions. Assume the variables represent positive real numbers. **(See Example 5.)**

69. $\sqrt{(4)^2}$ 4

70. $\sqrt{(8)^2}$ 8

71. $\sqrt[3]{(5)^3}$ 5

72. $\sqrt[3]{(7)^3}$ 7

73. $\sqrt{y^{12}}$ y^6

74. $\sqrt{z^{20}}$ z^{10}

75. $\sqrt{a^8b^{30}}$ a^4b^{15}

76. $\sqrt{t^{50}s^{60}}$ $t^{25}s^{30}$

77. $\sqrt[3]{q^{24}}$ q^8

78. $\sqrt[3]{x^{33}}$ x^{11}

79. $\sqrt[3]{8w^6}$ $2w^2$

80. $\sqrt[3]{-27x^{27}}$ $-3x^9$

81. $\sqrt{(5x)^2}$ $5x$

82. $\sqrt{(6w)^2}$ $6w$

83. $-\sqrt{25x^2}$ $-5x$

84. $-\sqrt{36w^2}$ $-6w$

85. $\sqrt[3]{(5p^2)^3}$ $5p^2$

86. $\sqrt[3]{(2k^4)^3}$ $2k^4$

87. $\sqrt[3]{125p^6}$ $5p^2$

88. $\sqrt[3]{8k^{12}}$ $2k^4$

Concept 3: Translations Involving *n*th-Roots

For Exercises 89–92, write each English phrase as an algebraic expression. **(See Example 6.)**

89. The sum of the principal square root of q and the square of p $\sqrt{q}+p^2$

90. The product of the principal square root of 11 and the cube of x
$\sqrt{11} \cdot x^3$

91. The quotient of 6 and the cube root of x $\dfrac{6}{\sqrt[3]{x}}$

92. The difference of the square of y and 1 y^2-1

Concept 4: Pythagorean Theorem

For Exercises 93–98, find the length of the third side of each triangle using the Pythagorean theorem. Round the answer to the nearest tenth if necessary. **(See Example 7.)**

93.

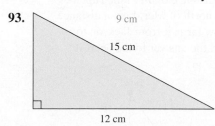

9 cm
15 cm
12 cm

94.

10 in.
8 in.
6 in.

95.

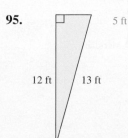

5 ft
12 ft   13 ft

96.

3 m
5 m
4 m

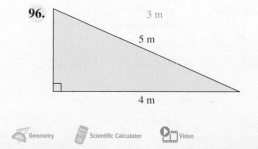

/ Writing   ←→ Translating Expression   Geometry   Scientific Calculator   Video

97.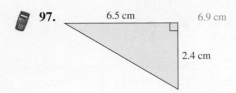
6.5 cm   6.9 cm
2.4 cm

98.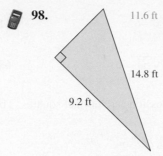
11.6 ft
14.8 ft
9.2 ft

99. Find the length of the diagonal of the square tile shown in the figure. Round the answer to the nearest tenth of an inch.   17.0 in.

12 in.
12 in.

100. A baseball diamond is 90 ft on a side. Find the distance between home plate and second base. Round the answer to the nearest tenth of a foot.   127.3 ft

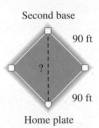
Second base
90 ft
?
90 ft
Home plate

101. A new television is listed as being 42 in. This distance is the diagonal distance across the screen. If the screen measures 28 in. in height, what is the actual width of the screen? Round to the nearest tenth of an inch. **(See Example 8.)**   31.3 in.

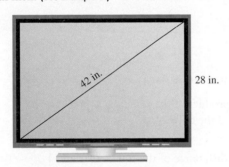
42 in.
28 in.

102. A marine biologist wants to track the migration of a pod of whales. He receives a radio signal from a tagged humpback whale and determines that the whale is 21 mi east and 37 mi north of his laboratory. Find the direct distance between the whale and the laboratory. Round to the nearest tenth of a mile.   42.5 mi

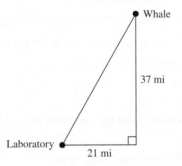
Whale
37 mi
Laboratory
21 mi

103. On a map, the cities Asheville, North Carolina, Roanoke, Virginia, and Greensboro, North Carolina, form a right triangle (see the figure). The distance between Asheville and Roanoke is 300 km. The distance between Roanoke and Greensboro is 134 km. How far is it from Greensboro to Asheville? Round the answer to the nearest kilometer.   268 km

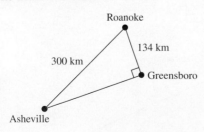
Roanoke
134 km
300 km
Greensboro
Asheville

104. Jackson, Mississippi, is west of Meridian, Mississippi, a distance of 141 km. Tupelo, Mississippi, is north of Meridian, a distance of 209 km. How far is it from Jackson to Tupelo? Round the answer to the nearest kilometer.   252 km

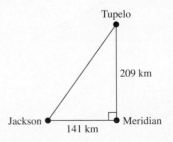
Tupelo
209 km
Jackson
141 km
Meridian

 Writing   Translating Expression   Geometry   Scientific Calculator   Video

Expanding Your Skills

105. For what values of x will $\sqrt{x}$ be a real number?
$x \geq 0$

106. For what values of x will $\sqrt{-x}$ be a real number?
$x \leq 0$

107. Under what conditions will $\sqrt{a - b}$ be a real number? $a \geq b$

108. Under what conditions will $\sqrt{m - n}$ be a real number? $m \geq n$

Simplifying Radicals

1. Multiplication Property of Radicals

You may have already recognized certain properties of radicals involving a product.

Concepts

1. **Multiplication Property of Radicals**
2. **Simplifying Radicals Using the Order of Operations**
3. **Simplifying Cube Roots**

Multiplication Property of Radicals

Let a and b represent real numbers such that $\sqrt[n]{a}$ and $\sqrt[n]{b}$ are both real. Then,

$$\sqrt[n]{ab} = \sqrt[n]{a} \cdot \sqrt[n]{b} \qquad \textbf{Multiplication property of radicals}$$

The multiplication property of radicals indicates that a product within a radicand can be written as a product of radicals provided the roots are real numbers.

$$\sqrt{100} = \sqrt{25} \cdot \sqrt{4}$$

The reverse process is also true. A product of radicals can be written as a single radical provided the roots are real numbers and they have the same indices.

$$\overset{\text{Same index}}{\sqrt{2} \cdot \sqrt{18}} = \sqrt{36}$$

In algebra, it is customary to simplify radical expressions as much as possible.

Simplified Form of a Radical

Consider any radical expression where the radicand is written as a product of prime factors. The expression is in **simplified form** if all of the following conditions are met:

1. The radicand has no factor raised to a power greater than or equal to the index.
2. There are no radicals in the denominator of a fraction.
3. The radicand does not contain a fraction.

The expression $\sqrt{x^2}$ is not simplified because it fails condition 1. Because x^2 is a perfect square, $\sqrt{x^2}$ is easily simplified.

$$\sqrt{x^2} = x \quad \text{(for } x \geq 0)$$

However, how is an expression such as $\sqrt{x^7}$ simplified? This and many other radical expressions are simplified using the multiplication property of radicals. Examples 1–3 illustrate how nth powers can be removed from the radicands of square roots.

Classroom Example: p. 586, Exercise 24

Example 1	**Using the Multiplication Property to Simplify a Radical Expression**

Use the multiplication property of radicals to simplify the expression $\sqrt{x^7}$. Assume $x \geq 0$.

Solution:

The expression $\sqrt{x^7}$ is equivalent to $\sqrt{x^6 \cdot x}$. By applying the multiplication property of radicals, we have

$$\sqrt{x^6 \cdot x} = \sqrt{x^6} \cdot \sqrt{x} \qquad x^6 \text{ is a perfect square because } (x^3)^2 = x^6$$

$$= x^3 \cdot \sqrt{x} \qquad \text{Simplify.}$$

$$= x^3\sqrt{x}$$

Skill Practice   Use the multiplication property of radicals to simplify the expression. Assume $x \geq 0$.

   1. $\sqrt{x^5}$

In Example 1, the expression x^7 is not a perfect square. Therefore, to simplify $\sqrt{x^7}$, it was necessary to write the expression as the product of the largest perfect square and a remaining, or "leftover," factor: $\sqrt{x^7} = \sqrt{x^6 \cdot x}$.

Classroom Example: p. 586, Exercise 30

Example 2	**Using the Multiplication Property to Simplify Radicals**

Use the multiplication property of radicals to simplify the expressions. Assume the variables represent positive real numbers.

   a. $\sqrt{a^{15}}$ **b.** $\sqrt{x^2 y^5}$ **c.** $\sqrt{s^9 t^{11}}$

Solution:

The goal is to rewrite each radicand as the product of the largest perfect square and a leftover factor.

   a. $\sqrt{a^{15}}$

$$= \sqrt{a^{14} \cdot a} \qquad a^{14} \text{ is the largest perfect square in the radicand.}$$

$$= \sqrt{a^{14}} \cdot \sqrt{a} \qquad \text{Apply the multiplication property of radicals.}$$

$$= a^7\sqrt{a} \qquad \text{Simplify.}$$

   b. $\sqrt{x^2 y^5}$

$$= \sqrt{x^2 y^4 \cdot y} \qquad x^2 y^4 \text{ is the largest perfect square in the radicand.}$$

$$= \sqrt{x^2 y^4} \cdot \sqrt{y} \qquad \text{Apply the multiplication property of radicals.}$$

$$= xy^2\sqrt{y} \qquad \text{Simplify.}$$

   c. $\sqrt{s^9 t^{11}}$

$$= \sqrt{s^8 t^{10} \cdot st} \qquad s^8 t^{10} \text{ is the largest perfect square in the radical.}$$

$$= \sqrt{s^8 t^{10}} \cdot \sqrt{st} \qquad \text{Apply the multiplication property of radicals.}$$

$$= s^4 t^5 \sqrt{st} \qquad \text{Simplify.}$$

Answer

1. $x^2\sqrt{x}$

Skill Practice Simplify the expressions. Assume the variables represent positive real numbers.

2. $\sqrt{y^{11}}$ **3.** $\sqrt{x^8 y^{13}}$ **4.** $\sqrt{u^3 w^9}$

Each expression in Example 2 involves a radicand that is a product of variable factors. If a numerical factor is present, sometimes it is necessary to factor the coefficient before simplifying the radical.

Example 3 **Using the Multiplication Property to Simplify Radicals**

Classroom Examples: p. 586, Exercises 22 and 50

Use the multiplication property of radicals to simplify the expressions. Assume the variables represent positive real numbers.

a. $\sqrt{50}$ **b.** $5\sqrt{24a^6}$ **c.** $-\sqrt{81x^4 y^3}$

Solution:

The goal is to rewrite each radicand as the product of the largest perfect square and a leftover factor.

a. Write the radicand as a product of prime factors. From the prime factorization, the largest perfect square is easily identified.

$\sqrt{50} = \sqrt{5^2 \cdot 2}$ Factor the radicand. 5^2 is the largest perfect square.

$\begin{array}{r} 2\overline{)50} \\ 5\overline{)25} \\ 5 \end{array}$

$= \sqrt{5^2} \cdot \sqrt{2}$ Apply the multiplication property of radicals.

$= 5\sqrt{2}$ Simplify.

> **TIP:** The expression $\sqrt{50}$ can also be written as:
> $\sqrt{25 \cdot 2}$
> $= \sqrt{25} \cdot \sqrt{2}$
> $= 5\sqrt{2}$

b. $5\sqrt{24a^6} = 5\sqrt{2^3 \cdot 3 \cdot a^6}$ Write the radicand as a product of prime factors: $24 = 2^3 \cdot 3$.

$= 5\sqrt{2^2 a^6 \cdot 2 \cdot 3}$ $2^2 a^6$ is the largest perfect square in the radicand.

$= 5\sqrt{2^2 a^6} \cdot \sqrt{2 \cdot 3}$ Apply the multiplication property of radicals.

$= 5 \cdot 2a^3 \sqrt{6}$ Simplify the radical.

$= 10a^3 \sqrt{6}$ Simplify the coefficient of the radical.

c. $-\sqrt{81x^4 y^3} = -\sqrt{3^4 x^4 y^3}$ Write the radical as a product of prime factors. *Note:* $81 = 3^4$.

$= -\sqrt{3^4 x^4 y^2 \cdot y}$ $3^4 x^4 y^2$ is the largest square in the radicand.

$= -\sqrt{3^4 x^4 y^2} \cdot \sqrt{y}$ Apply the multiplication property of radicals.

$= -3^2 x^2 y \cdot \sqrt{y}$ Simplify the radical.

$= -9x^2 y \sqrt{y}$ Simplify the coefficient of the radical.

Skill Practice Simplify the expressions. Assume the variables represent positive real numbers.

5. $\sqrt{12}$ **6.** $\sqrt{60x^2}$ **7.** $7\sqrt{18t^{10}}$

Answers

2. $y^5 \sqrt{y}$ **3.** $x^4 y^6 \sqrt{y}$
4. $uw^4 \sqrt{uw}$ **5.** $2\sqrt{3}$
6. $2x\sqrt{15}$ **7.** $21t^5 \sqrt{2}$

> **Avoiding Mistakes**
>
> The multiplication property of radicals enables us to simplify a product within a radical. That is,
> $$\sqrt{x^2y^2} = \sqrt{x^2} \cdot \sqrt{y^2} = xy \qquad \text{(for } x \geq 0 \text{ and } y \geq 0\text{)}$$
> However, this rule does not apply to *terms* that are added or subtracted *within* the radical. That is,
> $$\sqrt{x^2 + y^2} \neq \sqrt{x^2} + \sqrt{y^2} \qquad \text{and} \qquad \sqrt{x^2 - y^2} \neq \sqrt{x^2} - \sqrt{y^2}$$
> For example: $\qquad \sqrt{(16) \cdot (9)} = 4 \cdot 3,$ however, $\sqrt{16 + 9} \neq 4 + 3.$

2. Simplifying Radicals Using the Order of Operations

Often a radical can be simplified by applying the order of operations. In Example 4, the first step will be to simplify the expression within the radicand.

Classroom Example: p. 586, Exercise 62

Example 4   Simplifying Radicals Using the Order of Operations

Simplify the expressions. Assume the variables represent positive real numbers.

a. $\sqrt{\dfrac{a^5}{a^3}}$ b. $\sqrt{\dfrac{6}{96}}$ c. $\sqrt{\dfrac{27x^5}{3x}}$

Solution:

a. $\sqrt{\dfrac{a^5}{a^3}}$ The radical contains a fraction. However, the fraction can be simplified.

$= \sqrt{a^2}$ Reduce the fraction to lowest terms.

$= a$ Simplify the radical.

b. $\sqrt{\dfrac{6}{96}}$ The radical contains a fraction that can be simplified.

$= \sqrt{\dfrac{1}{16}}$ Reduce the fraction to lowest terms.

$= \dfrac{1}{4}$ Simplify.

c. $\sqrt{\dfrac{27x^5}{3x}}$ The fraction within the radicand can be simplified.

$= \sqrt{9x^4}$ Reduce to lowest terms.

$= 3x^2$ Simplify.

Skill Practice Simplify the expressions. Assume the variables represent positive real numbers.

8. $\sqrt{\dfrac{y^{11}}{y^3}}$ 9. $\sqrt{\dfrac{8}{50}}$ 10. $\sqrt{\dfrac{32z^3}{2z}}$

Answers

8. y^4 9. $\dfrac{2}{5}$ 10. $4z$

Example 5   ### Simplifying Radical Expressions

Simplify the expressions.

a. $\dfrac{5\sqrt{20}}{2}$ **b.** $\dfrac{2-\sqrt{36}}{12}$

Classroom Example: p. 586, Exercise 70

Solution:

a. $\dfrac{5\sqrt{20}}{2} = \dfrac{5\sqrt{4\cdot5}}{2}$ Following the order of operations, first simplify the radical. 4 is the largest perfect square factor in the radicand.

$= \dfrac{5\sqrt{4}\cdot\sqrt{5}}{2}$ Apply the multiplication property of radicals.

$= \dfrac{5\cdot2\sqrt{5}}{2}$ Simplify the radical.

$= \dfrac{\overset{5}{\cancel{10}}\sqrt{5}}{\cancel{2}}$ Simplify to lowest terms.

$= 5\sqrt{5}$

b. $\dfrac{2-\sqrt{36}}{12}$

$= \dfrac{2-6}{12}$ Following the order of operations, first simplify the radical.

$= \dfrac{-4}{12}$ Next, simplify the numerator.

$= -\dfrac{1}{3}$ Simplify to lowest terms.

Avoiding Mistakes

$\dfrac{5\sqrt{20}}{2}$ cannot be simplified as written because 20 is under the radical and 2 is not under the radical. To reduce to lowest terms, the radical must be simplified first, $\dfrac{10\sqrt{5}}{2}$. Then factors outside the radical can be simplified.

Skill Practice Simplify the expressions.

11. $\dfrac{7\sqrt{18}}{3}$ **12.** $\dfrac{5+\sqrt{49}}{6}$

3. Simplifying Cube Roots

To simplify a cube root, we write the radicand as a product of the largest perfect cube times another factor. Then apply the multiplication property of radicals.

Example 6   ### Simplifying Cube Roots

Use the multiplication property of radicals to simplify the expressions.

a. $\sqrt[3]{z^5}$ **b.** $\sqrt[3]{-80}$

Classroom Example: p. 587, Exercise 80

Solution:

a. $\sqrt[3]{z^5}$

$= \sqrt[3]{z^3\cdot z^2}$ z^3 is the largest perfect cube in the radicand.

$= \sqrt[3]{z^3}\cdot\sqrt[3]{z^2}$ Apply the multiplication property of radicals.

$= z\sqrt[3]{z^2}$ Simplify.

Answers
11. $7\sqrt{2}$ **12.** 2

TIP: In Example 6(b), rather than factoring −80 as a product of prime factors, we factored as

$$-80 = -1 \cdot 8 \cdot 10$$

because −1 and 8 are easily recognized as perfect cubes.

b. $\sqrt[3]{-80}$

$= \sqrt[3]{-1 \cdot 8 \cdot 10}$ Factor the radicand. −1 and 8 are perfect cubes.

$= \sqrt[3]{-1} \cdot \sqrt[3]{8} \cdot \sqrt[3]{10}$ Apply the multiplication property of radicals.

$= -1 \cdot 2 \cdot \sqrt[3]{10}$ Simplify.

$= -2\sqrt[3]{10}$

Skill Practice Simplify.

13. $\sqrt[3]{y^4}$ **14.** $\sqrt[3]{-24}$

Classroom Example: p. 587, Exercise 82

Example 7   Simplifying Cube Roots

Simplify the expressions.

a. $\sqrt[3]{\dfrac{a^{16}}{a}}$ **b.** $\sqrt[3]{\dfrac{2}{16}}$

Solution:

a. $\sqrt[3]{\dfrac{a^{16}}{a}}$ The radical contains a fraction that can be simplified.

$= \sqrt[3]{a^{15}}$ Reduce to lowest terms.

$= a^5$ Simplify.

b. $\sqrt[3]{\dfrac{2}{16}}$ The radical contains a fraction that can be simplified.

$= \sqrt[3]{\dfrac{1}{8}}$ Reduce to lowest terms.

$= \dfrac{1}{2}$ Simplify.

Skill Practice Simplify.

15. $\sqrt[3]{\dfrac{x^{12}}{x^6}}$ **16.** $\sqrt[3]{\dfrac{81}{3}}$

Answers

13. $y\sqrt[3]{y}$ **14.** $-2\sqrt[3]{3}$

15. x^2 **16.** 3

Calculator Connections

Topic: Verifying Simplified Radicals

A calculator can support the multiplication property of radicals. For example, use a calculator to evaluate $\sqrt{50}$ and its simplified form $5\sqrt{2}$.

Scientific Calculator

| Enter: | 50 $\boxed{\sqrt{x}}$ | Result: | $\boxed{7.071067812}$ |
| Enter: | 2 $\boxed{\sqrt{x}}$ $\boxed{\times}$ 5 $\boxed{=}$ | Result: | $\boxed{7.071067812}$ |

Instructor Note: This is a good opportunity to work on estimation skills. $\sqrt{50}$ is between what two integers? Closer to which one?

Graphing Calculator

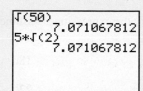

TIP: The decimal approximation for $\sqrt{50}$ and $5\sqrt{2}$ agree for the first 10 digits. This in itself does not make $\sqrt{50} = 5\sqrt{2}$. It is the multiplication property of radicals that guarantees that the expressions are equal.

Calculator Exercises

Simplify the radical expressions algebraically. Then use a calculator to approximate the original expression and its simplified form.

1. $\sqrt{125}$
$5\sqrt{5}$; 11.18033989

2. $\sqrt{18}$
$3\sqrt{2}$; 4.242640687

3. $\sqrt[3]{54}$
$3\sqrt[3]{2}$; 3.77976315

4. $\sqrt[3]{108}$
$3\sqrt[3]{4}$; 4.762203156

Section 8.2 Practice Exercises

Vocabulary and Key Concepts

For additional exercises, see Classroom Activities 8.2A–8.2B in the *Student's Resource Manual* at www.mhhe.com/moh.

1. a. The multiplication property of radicals indicates that if both $\sqrt[n]{a}$ and $\sqrt[n]{b}$ are real numbers, then $\sqrt[n]{ab} = \underline{\quad \sqrt[n]{a} \quad} \cdot \sqrt[n]{b}$.

b. Explain why the radical is not in simplified form. $\sqrt{x^3}$ The exponent within the radicand is not less than the index.

c. On a calculator, $\sqrt{2}$ is given as 1.414213562. Is this decimal number the exact value of $\sqrt{2}$?
No. $\sqrt{2}$ is an irrational number; therefore its decimal form is a nonterminating, nonrepeating decimal.

Review Exercises

2. Which of the following are perfect squares? 2, 4, 6, 16, 20, 25, x^2, x^3, x^{15}, x^{20}, x^{25} 4, 16, 25, x^2, x^{20}

3. Which of the following are perfect cubes? 3, 6, 8, 9, 12, 27, y^3, y^8, y^9, y^{12}, y^{27} 8, 27, y^3, y^9, y^{12}, y^{27}

4. Which of the following are perfect fourth powers? 4, 16, 20, 25, 81, w^4, w^{16}, w^{20}, w^{25}, w^{81} 16, 18, w^4, w^{16}, w^{20}

For Exercises 5–12, simplify the expressions, if possible. Assume the variables represent positive real numbers.

5. $-\sqrt{25}$ -5

6. $\sqrt{-25}$ Not a real number

7. $-\sqrt[3]{27}$ -3

8. $\sqrt[3]{-27}$ -3

9. $\sqrt{a^8}$ a^4

10. $\sqrt[3]{b^{15}}$ b^5

11. $\sqrt{4x^2y^4}$ $2xy^2$

12. $\sqrt{9p^{10}}$ $3p^5$

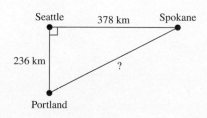

13. On a map, Seattle, Washington, is 378 km west of Spokane, Washington. Portland, Oregon, is 236 km south of Seattle. Approximate the distance between Portland and Spokane to the nearest kilometer.   446 km

14. A new roof is needed on a shed. How many square feet of tar paper would be needed to cover the top of the roof?   1040 ft²

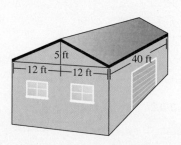

Writing   Translating Expression   Geometry   Scientific Calculator   Video

Concept 1: Multiplication Property of Radicals

For Exercises 15–50, use the multiplication property of radicals to simplify the expressions. Assume the variables represent positive real numbers. **(See Examples 1–3.)**

15. $\sqrt{18}$ $3\sqrt{2}$

16. $\sqrt{75}$ $5\sqrt{3}$

17. $\sqrt{28}$ $2\sqrt{7}$

18. $\sqrt{40}$ $2\sqrt{10}$

19. $6\sqrt{20}$ $12\sqrt{5}$

20. $10\sqrt{27}$ $30\sqrt{3}$

21. $-2\sqrt{50}$ $-10\sqrt{2}$

22. $-11\sqrt{8}$ $-22\sqrt{2}$

23. $\sqrt{a^5}$ $a^2\sqrt{a}$

24. $\sqrt{b^9}$ $b^4\sqrt{b}$

25. $\sqrt{w^{22}}$ w^{11}

26. $\sqrt{p^{18}}$ p^9

27. $\sqrt{m^4n^5}$ $m^2n^2\sqrt{n}$

28. $\sqrt{c^2d^9}$ $cd^4\sqrt{d}$

29. $x\sqrt{x^{13}y^{10}}$ $x^7y^5\sqrt{x}$

30. $v\sqrt{u^{10}v^7}$ $u^5v^4\sqrt{v}$

31. $3\sqrt{t^{10}}$ $3t^5$

32. $-4\sqrt{m^8n^4}$ $-4m^4n^2$

33. $\sqrt{8x^3}$ $2x\sqrt{2x}$

34. $\sqrt{27y^5}$ $3y^2\sqrt{3y}$

35. $\sqrt{16z^3}$ $4z\sqrt{z}$

36. $\sqrt{9y^5}$ $3y^2\sqrt{y}$

37. $-\sqrt{45w^6}$ $-3w^3\sqrt{5}$

38. $-\sqrt{56v^8}$ $-2v^4\sqrt{14}$

39. $\sqrt{z^{25}}$ $z^{12}\sqrt{z}$

40. $\sqrt{25p^{49}}$ $5p^{24}\sqrt{p}$

41. $-\sqrt{15z^{11}}$ $-z^5\sqrt{15z}$

42. $-\sqrt{6k^{15}}$ $-k^7\sqrt{6k}$

43. $5\sqrt{104a^2b^7}$ $10ab^3\sqrt{26b}$

44. $3\sqrt{88m^4n^{11}}$ $6m^2n^5\sqrt{22n}$

45. $\sqrt{26pq}$ $\sqrt{26pq}$

46. $\sqrt{15a}$ $\sqrt{15a}$

47. $m\sqrt{m^{10}n^{16}}$ m^6n^8

48. $c^2\sqrt{c^4d^{12}}$ c^4d^6

49. $-\sqrt{48a^3b^5c^4}$ $-4ab^2c^2\sqrt{3ab}$

50. $-\sqrt{18xy^4z^3}$ $-3y^2z\sqrt{2xz}$

Concept 2: Simplifying Radicals Using the Order of Operations

For Exercises 51–70, use the order of operations, if necessary, to simplify the expressions. Assume the variables represent positive real numbers. **(See Examples 4–5.)**

51. $\sqrt{\dfrac{a^9}{a}}$ a^4

52. $\sqrt{\dfrac{x^5}{x}}$ x^2

53. $\sqrt{\dfrac{y^{15}}{y^5}}$ y^5

54. $\sqrt{\dfrac{c^{31}}{c^{11}}}$ c^{10}

55. $\sqrt{\dfrac{5}{20}}$ $\dfrac{1}{2}$

56. $\sqrt{\dfrac{3}{75}}$ $\dfrac{1}{5}$

57. $\sqrt{\dfrac{40}{10}}$ 2

58. $\sqrt{\dfrac{80}{5}}$ 4

59. $\sqrt{\dfrac{32x^3}{8x}}$ $2x$

60. $\sqrt{\dfrac{200b^{11}}{2b^5}}$ $10b^3$

61. $\sqrt{\dfrac{50p^7}{2p}}$ $5p^3$

62. $\sqrt{\dfrac{45t^9}{5t^5}}$ $3t^2$

63. $\dfrac{3\sqrt{20}}{2}$ $3\sqrt{5}$

64. $\dfrac{5\sqrt{18}}{3}$ $5\sqrt{2}$

65. $\dfrac{5\sqrt{24}}{10}$ $\sqrt{6}$

66. $\dfrac{2\sqrt{27}}{6}$ $\sqrt{3}$

67. $\dfrac{10+\sqrt{4}}{3}$ 4

68. $\dfrac{-1+\sqrt{25}}{4}$ 1

69. $\dfrac{20-\sqrt{36}}{2}$ 7

70. $\dfrac{3-\sqrt{81}}{3}$ -2

For Exercises 71–74, find the exact length of the third side of each triangle using the Pythagorean theorem. Write the answer as a simplified radical.

71.

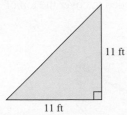

11 ft

11 ft

$11\sqrt{2}$ ft

72.

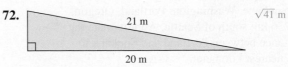

21 m

20 m

$\sqrt{41}$ m

73.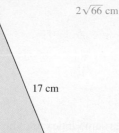

2√66 cm

17 cm

5 cm

74.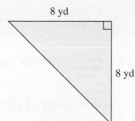

8 yd

8√2 yd

8 yd

Concept 3: Simplifying Cube Roots

For Exercises 75–86, simplify the cube roots. **(See Examples 6–7.)**

75. $\sqrt[3]{a^8}$ $a^2\sqrt[3]{a^2}$

76. $\sqrt[3]{8v^3}$ $2v$

77. $7\sqrt[3]{16z^3}$ $14z\sqrt[3]{2}$

78. $5\sqrt[3]{54t^6}$ $15t^2\sqrt[3]{2}$

79. $\sqrt[3]{16a^5b^6}$ $2ab^2\sqrt[3]{2a^2}$

80. $\sqrt[3]{81p^9q^{11}}$ $3p^3q^3\sqrt[3]{3q^2}$

81. $\sqrt[3]{\dfrac{z^4}{z}}$ z

82. $\sqrt[3]{\dfrac{w^8}{w^2}}$ w^2

83. $\sqrt[3]{-\dfrac{32}{4}}$ -2

84. $\sqrt[3]{-\dfrac{128}{2}}$ -4

85. $-\sqrt[3]{40}$ $-2\sqrt[3]{5}$

86. $-\sqrt[3]{54}$ $-3\sqrt[3]{2}$

Mixed Exercises

For Exercises 87–110, simplify the expressions. Assume the variables represent positive real numbers.

87. $\sqrt{\dfrac{3}{27}}$ $\dfrac{1}{3}$

88. $\sqrt{\dfrac{5}{125}}$ $\dfrac{1}{5}$

89. $\sqrt{16a^3}$ $4a\sqrt{a}$

90. $\sqrt{125x^6}$ $5x^3\sqrt{5}$

91. $\sqrt{\dfrac{4x^3}{x}}$ $2x$

92. $\sqrt{\dfrac{9z^5}{z}}$ $3z^2$

93. $\sqrt{8p^2q}$ $2p\sqrt{2q}$

94. $\sqrt{6cd^3}$ $d\sqrt{6cd}$

95. $-\sqrt{32}$ $-4\sqrt{2}$

96. $-\sqrt{64}$ -8

97. $\sqrt{52u^4v^7}$ $2u^2v^3\sqrt{13v}$

98. $\sqrt{44p^8q^{10}}$ $2p^4q^5\sqrt{11}$

99. $\sqrt{216}$ $6\sqrt{6}$

100. $\sqrt{250}$ $5\sqrt{10}$

101. $\sqrt[3]{216}$ 6

102. $\sqrt[3]{250}$ $5\sqrt[3]{2}$

103. $\sqrt[3]{16a^3}$ $2a\sqrt[3]{2}$

104. $\sqrt[3]{125x^6}$ $5x^2$

105. $\sqrt[3]{\dfrac{x^5}{x^2}}$ x

106. $\sqrt[3]{\dfrac{y^{11}}{y^2}}$ y^3

107. $\dfrac{-6\sqrt{20}}{12}$ $-\sqrt{5}$

108. $\dfrac{-5\sqrt{32}}{10}$ $-2\sqrt{2}$

 **109.** $\dfrac{-4-\sqrt{25}}{18}$ $-\dfrac{1}{2}$

110. $\dfrac{8-\sqrt{100}}{2}$ -1

Expanding Your Skills

For Exercises 111–114, simplify the expressions. Assume the variables represent positive real numbers.

111. $\sqrt{(-2-5)^2+(-4+3)^2}$ $5\sqrt{2}$

112. $\sqrt{(-1-7)^2+[1-(-1)]^2}$ $2\sqrt{17}$

113. $\sqrt{x^2+10x+25}$ $x+5$

114. $\sqrt{x^2+6x+9}$ $x+3$

Section 8.3 Addition and Subtraction of Radicals

Concepts

1. Definition of *Like* Radicals
2. Addition and Subtraction of Radicals

1. Definition of *Like* Radicals

> ### Definition of *Like* Radicals
>
> Two radical terms are called *like* **radicals** if they have the same index and the same radicand.

Like radicals can be added or subtracted by using the distributive property.

$$9\sqrt{2y} + 4\sqrt{2y} = (9+4)\sqrt{2y} = 13\sqrt{2y}$$

Same index

Distributive property

Same radicand

2. Addition and Subtraction of Radicals

Classroom Examples: p. 591, Exercises 14 and 18

Example 1   Adding and Subtracting Radicals

Add or subtract the radicals as indicated. Assume all variables represent positive real numbers.

a. $\sqrt{5} + \sqrt{5}$ **b.** $6\sqrt[3]{15} + 3\sqrt[3]{15} + \sqrt[3]{15}$ **c.** $\sqrt{xy} - 6\sqrt{xy} + 4\sqrt{xy}$

Solution:

a. $\sqrt{5} + \sqrt{5}$

$\quad = 1\sqrt{5} + 1\sqrt{5}$ *Note:* $\sqrt{5} = 1\sqrt{5}$

$\quad = (1+1)\sqrt{5}$ Apply the distributive property.

$\quad = 2\sqrt{5}$ Simplify.

b. $6\sqrt[3]{15} + 3\sqrt[3]{15} + \sqrt[3]{15}$ The radicals have the same radicand and same index.

$\quad = 6\sqrt[3]{15} + 3\sqrt[3]{15} + 1\sqrt[3]{15}$ *Note:* $\sqrt[3]{15} = 1\sqrt[3]{15}$

$\quad = (6+3+1)\sqrt[3]{15}$ Apply the distributive property.

$\quad = 10\sqrt[3]{15}$

c. $\sqrt{xy} - 6\sqrt{xy} + 4\sqrt{xy}$ The radicals have the same radicand and same index.

$\quad = 1\sqrt{xy} - 6\sqrt{xy} + 4\sqrt{xy}$ *Note:* $\sqrt{xy} = 1\sqrt{xy}$

$\quad = (1 - 6 + 4)\sqrt{xy}$ Apply the distributive property.

$\quad = -1\sqrt{xy}$ Simplify.

$\quad = -\sqrt{xy}$

Skill Practice   Add or subtract the radicals as indicated. Assume the variables represent positive real numbers.

1. $3\sqrt{2} + 7\sqrt{2}$ **2.** $8\sqrt[3]{x} - \sqrt[3]{x}$ **3.** $4\sqrt{ab} - 2\sqrt{ab} - 9\sqrt{ab}$

Avoiding Mistakes

The process of adding *like* radicals with the distributive property is similar to adding *like* terms. The numerical coefficients are added and the radical factor is unchanged.

$\sqrt{5} + \sqrt{5}$

$= 1\sqrt{5} + 1\sqrt{5}$

$= 2\sqrt{5}$ Correct

Be careful: $\sqrt{5} + \sqrt{5} \neq \sqrt{10}$

In general,

$$\sqrt{x} + \sqrt{y} \neq \sqrt{x+y}$$

Answers

1. $10\sqrt{2}$ **2.** $7\sqrt[3]{x}$ **3.** $-7\sqrt{ab}$

Sometimes it is necessary to simplify radicals before adding or subtracting.

Example 2 **Simplifying Radicals before Adding or Subtracting**

Classroom Example: p. 591, Exercise 30

Add or subtract the radicals as indicated.

a. $\sqrt{20} + 7\sqrt{5}$ **b.** $\sqrt{50} - \sqrt{8}$

Solution:

a. $\sqrt{20} + 7\sqrt{5}$ Because the radicands are different, try simplifying the radicals first.

$= \sqrt{4 \cdot 5} + 7\sqrt{5}$ Factor the radicand.

$= 2\sqrt{5} + 7\sqrt{5}$ The terms are *like* radicals.

$= (2 + 7)\sqrt{5}$ Apply the distributive property.

$= 9\sqrt{5}$ Simplify.

b. $\sqrt{50} - \sqrt{8}$ Because the radicands are different, try simplifying the radicals first.

$= \sqrt{25 \cdot 2} - \sqrt{4 \cdot 2}$ Factor the radicands.

$= 5\sqrt{2} - 2\sqrt{2}$ The terms are *like* radicals.

$= (5 - 2)\sqrt{2}$ Apply the distributive property.

$= 3\sqrt{2}$ Simplify.

Instructor Note: We can only decide if radicals are *like* if they are simplified. For example, $\sqrt{8}$ and $\sqrt{18}$ simplify as $\sqrt{8} = 2\sqrt{2}$ and $\sqrt{18} = 3\sqrt{2}$. Thus $2\sqrt{2}$ and $3\sqrt{2}$ are *like* radicals.

Skill Practice Add or subtract the radicals as indicated.

4. $4\sqrt{18} + \sqrt{8}$ **5.** $\sqrt{50} - \sqrt{98}$

Example 3 **Simplifying Radicals before Adding or Subtracting**

Classroom Examples: p. 591, Exercises 36 and 54

Add or subtract the radicals as indicated. Assume the variables represent positive real numbers.

a. $-4\sqrt{3x^2} - x\sqrt{27} + 5x\sqrt{3}$ **b.** $a\sqrt{8a^5} + 6\sqrt{2a^7} + \sqrt{9a}$

Solution:

a. $-4\sqrt{3x^2} - x\sqrt{27} + 5x\sqrt{3}$ Simplify each radical.

$= -4\sqrt{3x^2} - x\sqrt{9 \cdot 3} + 5x\sqrt{3}$ Factor the radicands.

$= -4x\sqrt{3} - 3x\sqrt{3} + 5x\sqrt{3}$ The terms are *like* radicals.

$= (-4x - 3x + 5x)\sqrt{3}$ Apply the distributive property.

$= -2x\sqrt{3}$ Simplify.

Answers

4. $14\sqrt{2}$ **5.** $-2\sqrt{2}$

Avoiding Mistakes

In Example 3(b), notice that the radical expression

$$8a^3\sqrt{2a} + 3\sqrt{a}$$

cannot be simplified further because the two terms have different radicands.

b. $a\sqrt{8a^5} + 6\sqrt{2a^7} + \sqrt{9a}$ Simplify each radical.

$= a\sqrt{4a^4 \cdot 2a} + 6\sqrt{a^6 \cdot 2a} + \sqrt{9 \cdot a}$ Factor the radicands.

$= a \cdot 2a^2\sqrt{2a} + 6 \cdot a^3\sqrt{2a} + 3\sqrt{a}$ Simplify the radicals.

$= 2a^3\sqrt{2a} + 6a^3\sqrt{2a} + 3\sqrt{a}$ The first two terms are *like* radicals.

$= (2a^3 + 6a^3)\sqrt{2a} + 3\sqrt{a}$ Apply the distributive property.

$= 8a^3\sqrt{2a} + 3\sqrt{a}$

Skill Practice Add or subtract the radicals as indicated. Assume the variables represent positive real numbers.

6. $4x\sqrt{12} - \sqrt{27x^2}$ **7.** $\sqrt{28y^3} - y\sqrt{63y} + \sqrt{700}$

It is important to realize that only *like* radicals can be added or subtracted. The next example provides extra practice for recognizing *unlike* radicals.

Classroom Examples: p. 592, Exercises 64, 66, and 68

Example 4   **Recognizing *Unlike* Radicals**

Explain why the radicals cannot be simplified further by adding or subtracting.

a. $2\sqrt{x} - 5\sqrt{y}$ **b.** $7 + 4\sqrt{5}$

Solution:

a. $2\sqrt{x} - 5\sqrt{y}$ The radicands are not the same.

b. $7 + 4\sqrt{5}$ One term has a radical, and one does not.

Answers

6. $5x\sqrt{3}$
7. $-y\sqrt{7y} + 10\sqrt{7}$
8. One term has a radical and one does not.
9. The radicands are not the same.

Skill Practice Explain why the radicals cannot be simplified further.

8. $12 - 7\sqrt{5}$ **9.** $2\sqrt{3} - 3\sqrt{2}$

Section 8.3   Practice Exercises

Vocabulary and Key Concepts

For additional exercises, see Classroom Activity 8.3A in the *Student's Resource Manual* at www.mhhe.com/moh.

1. Two radical terms are called *like* radicals if they have the same ___index___ and the same ___radicand___.

Review Exercises

For Exercises 2–9, simplify each expression. Assume the variables represent positive real numbers.

2. $\sqrt{25w^2}$ $5w$ **3.** $\sqrt[3]{8y^3}$ $2y$ **4.** $\sqrt[3]{4z^4}$ $z\sqrt[3]{4z}$ **5.** $\sqrt{36x^3}$ $6x\sqrt{x}$

6. $\sqrt{\dfrac{9a^6}{a^2}}$ $3a^2$ **7.** $\sqrt{\dfrac{12x^3}{3x}}$ $2x$ **8.** $\dfrac{\sqrt{25c^6}}{16}$ $\dfrac{5c^3}{16}$ **9.** $\sqrt{-25}$ Not a real number

Concept 1: Definition of *Like* Radicals

10. How do you determine whether two radicals are *like* or *unlike*?

Two radicals are *like* if they have the same radicand and same index.

11. Write two radicals that are considered *unlike*. For example, $2\sqrt{3}, 6\sqrt[3]{3}$

12. Which pairs of radicals are *like* radicals? b

   a. $2\sqrt{x}$ and $8\sqrt[3]{x}$

   b. $\sqrt{5}$ and $-3\sqrt{5}$

   c. $3a\sqrt{3}$ and $3a\sqrt{2}$

13. Which pairs of radicals are *like* radicals? c

   a. $13\sqrt{5b}$ and $13b\sqrt{5}$

   b. $\sqrt[4]{x^2y}$ and $\sqrt[3]{x^2y}$

   c. $-2\sqrt[3]{y^2}$ and $6\sqrt[3]{y^2}$

Concept 2: Addition and Subtraction of Radicals

For Exercises 14–28, add or subtract the expressions, if possible. Assume the variables represent positive real numbers.
(See Example 1.)

14. $8\sqrt{6} + 2\sqrt{6}$ $10\sqrt{6}$

15. $3\sqrt{2} + 5\sqrt{2}$ $8\sqrt{2}$

16. $4\sqrt{3} - 2\sqrt{3} + 5\sqrt{3}$ $7\sqrt{3}$

17. $5\sqrt{7} - 3\sqrt{7} + 2\sqrt{7}$ $4\sqrt{7}$

18. $\sqrt[3]{11} + \sqrt[3]{11}$ $2\sqrt[3]{11}$

19. $\sqrt[3]{10} + \sqrt[3]{10}$ $2\sqrt[3]{10}$

20. $12\sqrt{x} - 3\sqrt{x}$ $9\sqrt{x}$

21. $15\sqrt{y} - 4\sqrt{y}$ $11\sqrt{y}$

22. $-3\sqrt{a} + 2\sqrt{a} + \sqrt{a}$ 0

23. $5\sqrt{c} - 6\sqrt{c} + \sqrt{c}$ 0

24. $7x\sqrt{11} - 9x\sqrt{11}$ $-2x\sqrt{11}$

25. $8y\sqrt{15} - 3y\sqrt{15}$ $5y\sqrt{15}$

26. $9\sqrt{2} - 9\sqrt{5}$ $9\sqrt{2} - 9\sqrt{5}$

27. $x\sqrt{y} - y\sqrt{x}$ $x\sqrt{y} - y\sqrt{x}$

28. $a\sqrt{b} + b\sqrt{a}$ $a\sqrt{b} + b\sqrt{a}$

Mixed Exercises

For Exercises 29–58, simplify. Then add or subtract the expressions, if possible. Assume the variables represent positive real numbers. **(See Examples 2 and 3.)**

29. $2\sqrt{12} + \sqrt{48}$ $8\sqrt{3}$

30. $5\sqrt{32} + 2\sqrt{50}$ $30\sqrt{2}$

31. $4\sqrt{45} - 6\sqrt{20}$ 0

32. $8\sqrt{54} - 4\sqrt{24}$ $16\sqrt{6}$

33. $\frac{1}{2}\sqrt{8} + \frac{1}{3}\sqrt{18}$ $2\sqrt{2}$

34. $\frac{1}{4}\sqrt{32} - \frac{1}{5}\sqrt{50}$ 0

35. $6p\sqrt{20p^2} + p^2\sqrt{80}$ $16p^2\sqrt{5}$

36. $2q\sqrt{48} + \sqrt{27q^2}$ $11q\sqrt{3}$

37. $-2\sqrt{2k} + 6\sqrt{8k}$ $10\sqrt{2k}$

38. $5\sqrt{27x} - 4\sqrt{12x}$ $7\sqrt{3x}$

39. $11\sqrt{a^4b} - a^2\sqrt{b} - 9a\sqrt{a^2b}$ $a^2\sqrt{b}$

40. $-7\sqrt{x^4y} + 5x^2\sqrt{y} - 6x\sqrt{x^2y}$ $-8x^2\sqrt{y}$

41. $4\sqrt{5} - \sqrt{5}$ $3\sqrt{5}$

42. $-3\sqrt{10} - \sqrt{10}$ $-4\sqrt{10}$

43. $\frac{5}{6}z\sqrt{6} + \frac{7}{9}z\sqrt{6}$ $\frac{29}{18}z\sqrt{6}$

44. $\frac{3}{4}a\sqrt{b} + \frac{1}{6}a\sqrt{b}$ $\frac{11}{12}a\sqrt{b}$

45. $1.1\sqrt{10} - 5.6\sqrt{10} + 2.8\sqrt{10}$ $-1.7\sqrt{10}$

46. $0.25\sqrt{x} + 1.50\sqrt{x} - 0.75\sqrt{x}$ $\sqrt{x}$

47. $4\sqrt{x^3} - 2x\sqrt{x}$ $2x\sqrt{x}$

48. $8\sqrt{y^9} - 2y^2\sqrt{y^5}$ $6y^4\sqrt{y}$

49. $4\sqrt{7} + \sqrt{63} - 2\sqrt{28}$ $3\sqrt{7}$

50. $8\sqrt{3} - 2\sqrt{27} + \sqrt{75}$ $7\sqrt{3}$

51. $\sqrt{16w} + \sqrt{24w} + \sqrt{40w}$ $4\sqrt{w} + 2\sqrt{6w} + 2\sqrt{10w}$

52. $\sqrt{54y} + \sqrt{81y} - \sqrt{12y}$ $3\sqrt{6y} + 9\sqrt{y} - 2\sqrt{3y}$

53. $\sqrt{x^6y} + 5x^2\sqrt{x^2y}$ $6x^3\sqrt{y}$

54. $7\sqrt{a^5b^2} - a^2\sqrt{ab^2}$ $6a^2b\sqrt{a}$

55. $4\sqrt{6} + 2\sqrt{3} - 8\sqrt{6}$ $2\sqrt{3} - 4\sqrt{6}$

56. $-7\sqrt{y} - \sqrt{z} + 2\sqrt{z}$ $-7\sqrt{y} + \sqrt{z}$

57. $x\sqrt{8} - 2\sqrt{18x^2} + \sqrt{2x}$ $-4x\sqrt{2} + \sqrt{2x}$

58. $5\sqrt{p^5} - 2p\sqrt{p} + p\sqrt{16p^3}$ $9p^2\sqrt{p} - 2p\sqrt{p}$

 Writing   Translating Expression   Geometry   Scientific Calculator   Video

For Exercises 59–60, find the exact perimeter of each figure.

59.

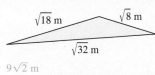

$\sqrt{18}$ m $\sqrt{8}$ m

$\sqrt{32}$ m

$9\sqrt{2}$ m

60.

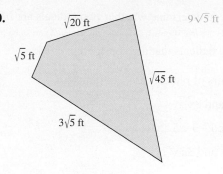

$\sqrt{20}$ ft $9\sqrt{5}$ ft

$\sqrt{5}$ ft

$\sqrt{45}$ ft

$3\sqrt{5}$ ft

61. Find the exact perimeter of a rectangle whose width is $2\sqrt{3}$ in. and whose length is $3\sqrt{12}$ in.
  $16\sqrt{3}$ in.

62. Find the exact perimeter of a square whose side length is $5\sqrt{8}$ cm. $40\sqrt{2}$ cm

For Exercises 63–68, determine the reason why the following radical expressions cannot be combined by addition or subtraction. **(See Example 4.)**

63. $\sqrt{5} + 5\sqrt{2}$
  Radicands are not the same.

64. $3\sqrt{10} + 10\sqrt{3}$
  Radicands are not the same.

65. $3 + 5\sqrt{7}$
  One term has a radical. One does not.

66. $-2 + 5\sqrt{11}$
  One term has a radical. One does not.

67. $5\sqrt{2} + \sqrt[3]{2}$
  The indices are different.

68. $\sqrt[4]{6} - 3\sqrt{6}$
  The indices are different.

Expanding Your Skills

69. Find the slope of the line through the points $(4, 2\sqrt{3})$ and $(1, \sqrt{3})$. $\dfrac{\sqrt{3}}{3}$

70. Find the slope of the line through the points $(7, 4\sqrt{5})$ and $(2, 3\sqrt{5})$. $\dfrac{\sqrt{5}}{5}$

71. A golfer hits a golf ball at an angle of 30° with an initial velocity of 46.0 meters/second (m/sec). The horizontal position of the ball, x (measured in meters), depends on the number of seconds, t, after the ball is struck according to the equation:

$$x = 23t\sqrt{3}$$

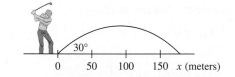

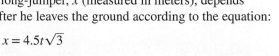

a. What is the horizontal position of the ball after 2 sec? Round the answer to the nearest meter.   80 m

b. What is the horizontal position of the ball after 4 sec? Round the answer to the nearest meter.   159 m

72. A long-jumper leaves the ground at an angle of 30° at a speed of 9 m/sec. The horizontal position of the long-jumper, x (measured in meters), depends on the number of seconds, t, after he leaves the ground according to the equation:

$$x = 4.5t\sqrt{3}$$

a. What is the horizontal position of the long-jumper after 0.5 sec? Round the answer to the nearest hundredth of a meter.   3.90 m

b. What is the horizontal position of the long-jumper after 0.75 sec? Round the answer to the nearest hundredth of a meter.   5.85 m

© Digital Vision/Getty Images RF

Multiplication of Radicals

1. Multiplication Property of Radicals

In this section, we will learn how to multiply radicals that have the same index. Recall the multiplication property of radicals.

Concepts

1. **Multiplication Property of Radicals**
2. **Expressions of the Form $(\sqrt[n]{a})^n$**
3. **Special Case Products**

> ### Multiplication Property of Radicals
> Let a and b represent real numbers such that $\sqrt[n]{a}$ and $\sqrt[n]{b}$ are both real. Then,
> $$\sqrt[n]{a} \cdot \sqrt[n]{b} = \sqrt[n]{ab}$$

To multiply two radical expressions, use the multiplication property of radicals along with the commutative and associative properties of multiplication.

Example 1   **Multiplying Radical Expressions**

Multiply the expressions and simplify the result. Assume the variables represent positive real numbers.

a. $\sqrt{3} \cdot \sqrt{2}$ **b.** $(5\sqrt{3})(2\sqrt{15})$ **c.** $(6a\sqrt{ab})\left(\frac{1}{3}a\sqrt{a}\right)$

Classroom Examples: p. 598, Exercises 8, 18, and 22

Solution:

a. $\sqrt{3} \cdot \sqrt{2} = \sqrt{6}$ Multiplication property of radicals

b. $(5\sqrt{3})(2\sqrt{15}) = (5 \cdot 2)(\sqrt{3} \cdot \sqrt{15})$ Regroup factors.

$\qquad\qquad\quad = 10\sqrt{45}$ Multiplication property of radicals

$\qquad\qquad\quad = 10\sqrt{9 \cdot 5}$ Simplify the radical.

$\qquad\qquad\quad = 10 \cdot 3\sqrt{5}$

$\qquad\qquad\quad = 30\sqrt{5}$

c. $(6a\sqrt{ab})\left(\frac{1}{3}a\sqrt{a}\right) = \left(6a \cdot \frac{1}{3}a\right)(\sqrt{ab} \cdot \sqrt{a})$ Regroup factors.

$\qquad\qquad\qquad = 2a^2\sqrt{a^2b}$ Multiplication property of radicals

$\qquad\qquad\qquad = 2a^2 \cdot a\sqrt{b}$ Simplify the radical.

$\qquad\qquad\qquad = 2a^3\sqrt{b}$

Skill Practice Multiply the expressions and simplify the result. Assume the variables represent positive real numbers.

1. $\sqrt{2} \cdot \sqrt{5}$ **2.** $(-5z\sqrt{6})(4z\sqrt{2})$ **3.** $(9y\sqrt{x})\left(\frac{1}{3}y\sqrt{xy}\right)$

Answers

1. $\sqrt{10}$ **2.** $-40z^2\sqrt{3}$

3. $3xy^2\sqrt{y}$

When multiplying radical expressions with more than one term, we use the distributive property.

Classroom Examples: p. 598, Exercises 38 and 42

Example 2   **Multiplying Radical Expressions with Multiple Terms**

Multiply the expressions. Assume the variables represent positive real numbers.

   a. $\sqrt{5}(4 + 3\sqrt{5})$ **b.** $(\sqrt{x} - 10)(\sqrt{y} + 4)$ **c.** $(2\sqrt{3} - \sqrt{5})(\sqrt{3} + 6\sqrt{5})$

Solution:

a. $\sqrt{5}(4 + 3\sqrt{5})$

$$= \sqrt{5}(4) + \sqrt{5}(3\sqrt{5})$$ Apply the distributive property.

$$= 4\sqrt{5} + 3\sqrt{5}^2$$ Multiplication property of radicals

$$= 4\sqrt{5} + 3 \cdot 5$$ Simplify the radical.

$$= 4\sqrt{5} + 15$$

b. $(\sqrt{x} - 10)(\sqrt{y} + 4)$

$$= \sqrt{x}(\sqrt{y}) + \sqrt{x}(4) - 10(\sqrt{y}) - 10(4)$$ Apply the distributive property.

$$= \sqrt{xy} + 4\sqrt{x} - 10\sqrt{y} - 40$$ Simplify.

c. $(2\sqrt{3} - \sqrt{5})(\sqrt{3} + 6\sqrt{5})$

$$= 2\sqrt{3}(\sqrt{3}) + 2\sqrt{3}(6\sqrt{5}) - \sqrt{5}(\sqrt{3}) - \sqrt{5}(6\sqrt{5})$$ Apply the distributive property.

$$= 2\sqrt{3}^2 + 12\sqrt{15} - \sqrt{15} - 6\sqrt{5}^2$$ Multiplication property of radicals

$$= 2 \cdot 3 + 11\sqrt{15} - 6 \cdot 5$$ Simplify radicals. Combine *like* radicals.

$$= 6 + 11\sqrt{15} - 30$$

$$= -24 + 11\sqrt{15}$$ Combine *like* terms.

Skill Practice Multiply the expressions and simplify the result. Assume the variables represent positive real numbers.

   4. $\sqrt{7}(2\sqrt{7} - 4)$ **5.** $(\sqrt{x} + 2)(\sqrt{x} - 3)$ **6.** $(2\sqrt{a} + 4\sqrt{6})(\sqrt{a} - 3\sqrt{6})$

2. Expressions of the Form $(\sqrt[n]{a})^n$

The multiplication property of radicals can be used to simplify an expression of the form $(\sqrt{a})^2$, where $a \geq 0$.

$$(\sqrt{a})^2 = \sqrt{a} \cdot \sqrt{a} = \sqrt{a^2} = a$$

This logic can be applied to nth-roots. If $\sqrt[n]{a}$ is a real number, then $(\sqrt[n]{a})^n = a$.

Answers
4. $14 - 4\sqrt{7}$
5. $x - \sqrt{x} - 6$
6. $2a - 2\sqrt{6a} - 72$

Example 3 **Simplifying Radical Expressions**

Classroom Examples: p. 599, Exercises 46 and 52

Simplify the expressions.

a. $(\sqrt{7})^2$ **b.** $(\sqrt[3]{x})^3$ **c.** $(3\sqrt{2})^2$

Solution:

a. $(\sqrt{7})^2 = 7$ **b.** $(\sqrt[3]{x})^3 = x$ **c.** $(3\sqrt{2})^2 = 3^2 \cdot (\sqrt{2})^2 = 9 \cdot 2 = 18$

Skill Practice Simplify the expressions.

7. $(\sqrt{13})^2$ **8.** $(\sqrt[3]{y})^3$ **9.** $(2\sqrt{11})^2$

3. Special Case Products

From Example 2, you may have noticed a similarity between multiplying radical expressions and multiplying polynomials.

Recall that the square of a binomial results in a perfect square trinomial.

$$(a + b)^2 = a^2 + 2ab + b^2$$

$$(a - b)^2 = a^2 - 2ab + b^2$$

The same patterns occur when squaring a radical expression with two terms.

Example 4 **Squaring a Two-Term Radical Expression**

Classroom Example: p. 599, Exercise 54

Multiply the radical expression. Assume the variables represent positive real numbers.

$$(\sqrt{x} + \sqrt{y})^2$$

Solution:

$(\sqrt{x} + \sqrt{y})^2$ This expression is in the form $(a + b)^2$, where $a = \sqrt{x}$ and $b = \sqrt{y}$.

$\overbrace{a^2 + 2ab + b^2}$

$= (\sqrt{x})^2 + 2(\sqrt{x})(\sqrt{y}) + (\sqrt{y})^2$ Apply the formula $(a + b)^2 = a^2 + 2ab + b^2$.

$= x + 2\sqrt{xy} + y$ Simplify.

Skill Practice Multiply the radical expression. Assume $p \geq 0$.

10. $(\sqrt{p} + 3)^2$

TIP: The product $(\sqrt{x} + \sqrt{y})^2$ can also be found using the distributive property.

$(\sqrt{x} + \sqrt{y})^2 = (\sqrt{x} + \sqrt{y})(\sqrt{x} + \sqrt{y}) = \sqrt{x} \cdot \sqrt{x} + \sqrt{x} \cdot \sqrt{y} + \sqrt{y} \cdot \sqrt{x} + \sqrt{y} \cdot \sqrt{y}$

$= \sqrt{x^2} + \sqrt{xy} + \sqrt{xy} + \sqrt{y^2}$

$= x + 2\sqrt{xy} + y$

Answers

7. 13   **8.** y **9.** 44

10. $p + 6\sqrt{p} + 9$

Classroom Example: p. 599,
Exercise 58

| **Example 5** | **Squaring a Two-Term Radical Expression** |

Multiply the radical expression. $(\sqrt{2} - 4\sqrt{3})^2$

Solution:

$(\sqrt{2} - 4\sqrt{3})^2$ This expression is in the form $(a - b)^2$, where $a = \sqrt{2}$ and $b = 4\sqrt{3}$.

$\overbrace{a^2 - 2ab + b^2}$

$(\sqrt{2})^2 - 2(\sqrt{2})(4\sqrt{3}) + (4\sqrt{3})^2$ Apply the formula $(a - b)^2 = a^2 - 2ab + b^2$.

$= 2 - 8\sqrt{6} + 16 \cdot 3$ Simplify.

$= 2 - 8\sqrt{6} + 48$

$= 50 - 8\sqrt{6}$

Skill Practice   Multiply the radical expression.

11. $(\sqrt{5} - 3\sqrt{2})^2$

Recall that the product of two conjugate binomials results in a difference of squares.

$$(a + b)(a - b) = a^2 - b^2$$

The same pattern occurs when multiplying two conjugate radical expressions.

Classroom Example: p. 599,
Exercise 62

| **Example 6** | **Multiplying Conjugate Radical Expressions** |

Multiply the radical expressions. $(\sqrt{5} + 4)(\sqrt{5} - 4)$

Solution:

$(\sqrt{5} + 4)(\sqrt{5} - 4)$ This expression is in the form $(a + b)(a - b)$, where $a = \sqrt{5}$ and $b = 4$.

$\overbrace{a^2 - b^2}$

$= (\sqrt{5})^2 - (4)^2$ Apply the formula $(a + b)(a - b) = a^2 - b^2$.

$= 5 - 16$ Simplify.

$= -11$

Skill Practice   Multiply the radical expressions.

12. $(\sqrt{6} - 3)(\sqrt{6} + 3)$

Answers

11. $23 - 6\sqrt{10}$ **12.** -3

TIP: The product $(\sqrt{5} + 4)(\sqrt{5} - 4)$ can also be found using the distributive property.

$$(\sqrt{5} + 4)(\sqrt{5} - 4) = \sqrt{5} \cdot \sqrt{5} + \sqrt{5} \cdot (-4) + 4 \cdot \sqrt{5} + 4 \cdot (-4)$$
$$= 5 - 4\sqrt{5} + 4\sqrt{5} - 16$$
$$= 5 - 16$$
$$= -11$$

Example 7   **Multiplying Conjugate Radical Expressions**

Classroom Example: p. 599, Exercise 70

Multiply the radical expressions. Assume the variables represent positive real numbers.

$$(2\sqrt{c} - 3\sqrt{d})(2\sqrt{c} + 3\sqrt{d})$$

Solution:

$(2\sqrt{c} - 3\sqrt{d})(2\sqrt{c} + 3\sqrt{d})$ This expression is in the form $(a - b)(a + b)$, where $a = 2\sqrt{c}$ and $b = 3\sqrt{d}$.

$= (2\sqrt{c})^2 - (3\sqrt{d})^2$ Apply the formula $(a + b)(a - b) = a^2 - b^2$.

$= 4c - 9d$

Skill Practice Multiply the radical expressions. Assume the variables represent positive real numbers.

13. $(5\sqrt{a} + \sqrt{b})(5\sqrt{a} - \sqrt{b})$

Answer

13. $25a - b$

Section 8.4   Practice Exercises

For additional exercises, see Classroom Activities 8.4A–8.4B in the *Student's Resource Manual* at www.mhhe.com/moh.

Vocabulary and Key Concepts

1. a. The multiplication property of radicals indicates that $\sqrt[n]{a} \cdot \sqrt[n]{b} = \underline{\sqrt[n]{ab}}$ provided that both $\sqrt[n]{a}$ and $\sqrt[n]{b}$ are real numbers.

   b. If $\sqrt[n]{a}$ is a real number, then $(\sqrt[n]{a})^n = \underline{a}$.

   c. Two binomials $(x + \sqrt{2})$ and $(x - \sqrt{2})$ are called $\underline{conjugates}$ of each other, and their product is $(x)^2 - (\sqrt{2})^2$.

Review Exercises

For Exercises 2–6, perform the indicated operations and simplify. Assume the variables represent positive real numbers.

2. $\sqrt{25} + \sqrt{16} - \sqrt{36}$ **3.** $\sqrt{100} - \sqrt{4} + \sqrt{9}$ **4.** $6x\sqrt{18} + 2\sqrt{2x^2}$ **5.** $10\sqrt{zw^4} - w^2\sqrt{49z}$

    3   11 $20x\sqrt{2}$ $3w^2\sqrt{z}$

6. $2\sqrt{16x^2 y} + x\sqrt{25y} - \sqrt{64x^2}$ $13x\sqrt{y} - 8x$

Concept 1: Multiplication Property of Radicals

For Exercises 7–26, multiply the expressions. **(See Example 1.)**

7. $\sqrt{5} \cdot \sqrt{3}$ $\quad \sqrt{15}$

8. $\sqrt{7} \cdot \sqrt{6}$ $\quad \sqrt{42}$

9. $\sqrt{47} \cdot \sqrt{47}$ $\quad 47$

10. $\sqrt{59} \cdot \sqrt{59}$ $\quad 59$

11. $\sqrt{b} \cdot \sqrt{b}$ $\quad b$

12. $\sqrt{t} \cdot \sqrt{t}$ $\quad t$

13. $(2\sqrt{15})(3\sqrt{p})$ $\quad 6\sqrt{15p}$

14. $(4\sqrt{2})(5\sqrt{q})$ $\quad 20\sqrt{2q}$

15. $\sqrt{10} \cdot \sqrt{5}$ $\quad 5\sqrt{2}$

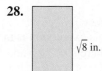 **16.** $\sqrt{2} \cdot \sqrt{10}$ $\quad 2\sqrt{5}$

17. $(-\sqrt{7})(-2\sqrt{14})$ $\quad 14\sqrt{2}$

18. $(-6\sqrt{2})(-\sqrt{22})$ $\quad 12\sqrt{11}$

19. $(3x\sqrt{2})(\sqrt{14})$ $\quad 6x\sqrt{7}$

20. $(4y\sqrt{3})(\sqrt{6})$ $\quad 12y\sqrt{2}$

21. $\left(\frac{1}{6}x\sqrt{xy}\right)(24x\sqrt{x})$ $\quad 4x^3\sqrt{y}$

22. $\left(\frac{1}{4}u\sqrt{uv}\right)(8u\sqrt{v})$ $\quad 2u^2v\sqrt{u}$

23. $(6w\sqrt{5})(w\sqrt{8})$ $\quad 12w^2\sqrt{10}$

24. $(t\sqrt{2})(5\sqrt{6t})$ $\quad 10t\sqrt{3t}$

25. $(-2\sqrt{3})(4\sqrt{5})$ $\quad -8\sqrt{15}$

26. $(-\sqrt{7})(2\sqrt{3})$ $\quad -2\sqrt{21}$

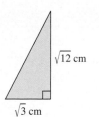 For Exercises 27–28, find the exact perimeter and exact area of the rectangles.

27.

Perimeter: $6\sqrt{5}$ ft; area: 10 ft^2

$\sqrt{5}$ ft

$\sqrt{20}$ ft

28.

Perimeter: $6\sqrt{2}$ in.; area: 4 in.^{2}

$\sqrt{8}$ in.

$\sqrt{2}$ in.

 For Exercises 29–30, find the exact area of the triangles.

29.

3 cm^2

$\sqrt{12}$ cm

$\sqrt{3}$ cm

30.

7 m^2

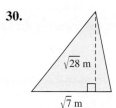

$\sqrt{28}$ m

$\sqrt{7}$ m

For Exercises 31–44, multiply the expressions. Assume the variables represent positive real numbers. **(See Example 2.)**

31. $\sqrt{3w} \cdot \sqrt{3w}$ $\quad 3w$

32. $\sqrt{6p} \cdot \sqrt{6p}$ $\quad 6p$

33. $(8\sqrt{5y})(-2\sqrt{2})$ $\quad -16\sqrt{10y}$

34. $(4\sqrt{5x})(7\sqrt{3})$ $\quad 28\sqrt{15x}$

35. $\sqrt{2}(\sqrt{6} - \sqrt{3})$ $\quad 2\sqrt{3} - \sqrt{6}$

36. $\sqrt{5}(\sqrt{10} + \sqrt{7})$ $\quad 5\sqrt{2} + \sqrt{35}$

37. $4\sqrt{x}(\sqrt{x} + 5)$ $\quad 4x + 20\sqrt{x}$

38. $2\sqrt{y}(3 - \sqrt{y})$ $\quad 6\sqrt{y} - 2y$

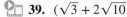 **39.** $(\sqrt{3} + 2\sqrt{10})(4\sqrt{3} - \sqrt{10})$ $\quad -8 + 7\sqrt{30}$

40. $(8\sqrt{7} - \sqrt{5})(\sqrt{7} + 3\sqrt{5})$ $\quad 41 + 23\sqrt{35}$

41. $(\sqrt{a} - 3b)(9\sqrt{a} - b)$ $\quad 9a - 28b\sqrt{a} + 3b^2$

42. $(11\sqrt{m} + 4n)(\sqrt{m} + n)$ $\quad 11m + 15n\sqrt{m} + 4n^2$

43. $(p + 2\sqrt{p})(8p + 3\sqrt{p} - 4)$
$\quad 8p^2 + 19p\sqrt{p} + 2p - 8\sqrt{p}$

44. $(5x - \sqrt{x})(x + 5\sqrt{x} + 6)$
$\quad 5x^2 + 24x\sqrt{x} + 25x - 6\sqrt{x}$

Concept 2: Expressions of the Form $(\sqrt[n]{a})^n$

For Exercises 45–52, simplify the expressions. Assume the variables represent positive real numbers. **(See Example 3.)**

45. $(\sqrt{10})^2$ 10

46. $(\sqrt{23})^2$ 23

47. $(\sqrt[3]{4})^3$ 4

48. $(\sqrt[3]{29})^3$ 29

49. $(\sqrt[3]{t})^3$ t

50. $(\sqrt[3]{xy})^3$ xy

51. $(4\sqrt{c})^2$ $16c$

52. $(10\sqrt{2pq})^2$ $200pq$

Concept 3: Special Case Products

For Exercises 53–60, multiply the radical expressions. Assume the variables represent positive real numbers.
(See Examples 4–5.)

53. $(\sqrt{13} + 4)^2$
$29 + 8\sqrt{13}$

54. $(6 - \sqrt{11})^2$
$47 - 12\sqrt{11}$

55. $(\sqrt{a} - 2)^2$
$a - 4\sqrt{a} + 4$

56. $(\sqrt{p} + 3)^2$
$p + 6\sqrt{p} + 9$

57. $(2\sqrt{a} - 3)^2$
$4a - 12\sqrt{a} + 9$

58. $(3\sqrt{w} + 4)^2$
$9w + 24\sqrt{w} + 16$

59. $(\sqrt{10} - \sqrt{11})^2$
$21 - 2\sqrt{110}$

60. $(\sqrt{3} - \sqrt{2})^2$
$5 - 2\sqrt{6}$

For Exercises 61–72, multiply the radical expressions. Assume the variables represent positive real numbers.
(See Examples 6–7.)

61. $(\sqrt{5} + 2)(\sqrt{5} - 2)$ 1

62. $(\sqrt{3} - 4)(\sqrt{3} + 4)$ -13

63. $(\sqrt{x} + \sqrt{y})(\sqrt{x} - \sqrt{y})$ $x - y$

64. $(\sqrt{a} + \sqrt{b})(\sqrt{a} - \sqrt{b})$ $a - b$

65. $(\sqrt{10} - \sqrt{11})(\sqrt{10} + \sqrt{11})$ -1

66. $(\sqrt{3} - \sqrt{2})(\sqrt{3} + \sqrt{2})$ 1

67. $(6\sqrt{m} + 5\sqrt{n})(6\sqrt{m} - 5\sqrt{n})$
$36m - 25n$

68. $(3\sqrt{p} - 4\sqrt{w})(3\sqrt{p} + 4\sqrt{w})$
$9p - 16w$

69. $(8\sqrt{x} - 2\sqrt{y})(8\sqrt{x} + 2\sqrt{y})$
$64x - 4y$

70. $(4\sqrt{s} + 11\sqrt{t})(4\sqrt{s} - 11\sqrt{t})$
$16s - 121t$

71. $(5\sqrt{3} - \sqrt{2})(5\sqrt{3} + \sqrt{2})$ 73

72. $(2\sqrt{7} - 4\sqrt{3})(2\sqrt{7} + 4\sqrt{3})$
-20

Mixed Exercises

For Exercises 73–84, multiply the expressions in parts (a) and (b) and compare the process used. Assume the variables represent positive real numbers.

73. a. $3(x + 2)$ $3x + 6$

 b. $\sqrt{3}(\sqrt{x} + \sqrt{2})$ $\sqrt{3x} + \sqrt{6}$

74. a. $-5(6 + y)$ $-30 - 5y$

 b. $-\sqrt{5}(\sqrt{6} + \sqrt{y})$ $-\sqrt{30} - \sqrt{5y}$

75. a. $(2a + 3)^2$ $4a^2 + 12a + 9$

 b. $(2\sqrt{a} + 3)^2$ $4a + 12\sqrt{a} + 9$

76. a. $(6 - z)^2$ $36 - 12z + z^2$

 b. $(\sqrt{6} - z)^2$ $6 - 2z\sqrt{6} + z^2$

77. a. $(b - 5)(b + 5)$ $b^2 - 25$

 b. $(\sqrt{b} - 5)(\sqrt{b} + 5)$ $b - 25$

78. a. $(3w - 1)(3w + 1)$ $9w^2 - 1$

 b. $(3\sqrt{w} - 1)(3\sqrt{w} + 1)$
 $9w - 1$

79. a. $(x - 2y)^2$ $x^2 - 4xy + 4y^2$

 b. $(\sqrt{x} - 2\sqrt{y})^2$ $x - 4\sqrt{xy} + 4y$

80. a. $(5c + 2d)^2$ $25c^2 + 20cd + 4d^2$

 b. $(5\sqrt{c} + 2\sqrt{d})^2$ $25c + 20\sqrt{cd} + 4d$

81. a. $(p - q)(p + q)$ $p^2 - q^2$

 b. $(\sqrt{p} - \sqrt{q})(\sqrt{p} + \sqrt{q})$ $p - q$

82. a. $(t - 3)(t + 3)$ $t^2 - 9$

 b. $(\sqrt{t} - \sqrt{3})(\sqrt{t} + \sqrt{3})$ $t - 3$

83. a. $(y - 3)^2$ $y^2 - 6y + 9$

 b. $(\sqrt{x - 2} - 3)^2$ $x - 6\sqrt{x - 2} + 7$

84. a. $(p + 4)^2$ $p^2 + 8p + 16$

 b. $(\sqrt{x + 1} + 4)^2$ $x + 8\sqrt{x + 1} + 17$

Writing   Translating Expression   Geometry   Scientific Calculator   Video

Section 8.5   Division of Radicals and Rationalization

1. Simplified Form of a Radical

Recall the conditions for a radical to be simplified.

Simplified Form of a Radical

Consider any radical expression where the radicand is written as a product of prime factors. The expression is in simplified form if all of the following conditions are met:

1. The radicand has no factor raised to a power greater than or equal to the index.
2. There are no radicals in the denominator of a fraction.
3. The radicand does not contain a fraction.

The basis of the second and third conditions, which restrict radicals from the denominator of an expression, are largely historical. In some cases, removing a radical from the denominator of a fraction will create an expression that is computationally simpler.

The process to remove a radical from the denominator is called **rationalizing the denominator**. In this section, we will show three approaches that can be used to achieve the second and third conditions of a simplified radical.

1. Rationalizing by applying the division property of radicals.
2. Rationalizing when the denominator contains a single radical term.
3. Rationalizing when the denominator contains two terms involving square roots.

2. Division Property of Radicals

The multiplication property of radicals enables a product within a radical to be separated and written as a product of radicals. We now state a similar property for radicals involving quotients.

Division Property of Radicals

Let a and b represent real numbers such that $\sqrt[n]{a}$ and $\sqrt[n]{b}$ are both real. Then,

$$\sqrt[n]{\frac{a}{b}} = \frac{\sqrt[n]{a}}{\sqrt[n]{b}} \qquad b \neq 0$$

The division property of radicals indicates that a quotient within a radicand can be written as a quotient of radicals provided the roots are real numbers. For example:

$$\sqrt{\frac{4}{9}} = \frac{\sqrt{4}}{\sqrt{9}}$$

The reverse process is also true. A quotient of radicals can be written as a single radical provided that the roots are real numbers and they have the same indices.

$$\text{Same index} \quad \boxed{\begin{array}{c} \overset{\downarrow 3}{\sqrt[3]{125}} \\ \overset{\downarrow 3}{\sqrt[3]{8}} \end{array}} = \sqrt[3]{\frac{125}{8}}$$

In Examples 1 and 2, we will apply the division property of radicals to simplify radical expressions.

Example 1 **Using the Division Property to Simplify Radicals**

Classroom Example: p. 607, Exercise 16

Use the division property of radicals to simplify the expressions. Assume the variables represent positive real numbers.

a. $\sqrt{\dfrac{a^{10}}{b^4}}$ b. $\sqrt{\dfrac{20x^3}{9}}$

Solution:

a. $\sqrt{\dfrac{a^{10}}{b^4}}$ The radicand contains an irreducible fraction.

$= \dfrac{\sqrt{a^{10}}}{\sqrt{b^4}}$ Apply the division property to rewrite as a quotient of radicals.

$= \dfrac{a^5}{b^2}$ Simplify the radicals.

b. $\sqrt{\dfrac{20x^3}{9}}$ The radicand contains an irreducible fraction.

$= \dfrac{\sqrt{20x^3}}{\sqrt{9}}$ Apply the division property to rewrite as a quotient of radicals.

$= \dfrac{\sqrt{(4x^2)(5x)}}{\sqrt{9}}$ Factor the radicand in the numerator as a perfect square and another factor.

$= \dfrac{2x\sqrt{5x}}{3}$ Simplify the radicals in the numerator and denominator. The expression is simplified since it now satisfies all conditions.

Skill Practice Simplify the expressions.

1. $\sqrt{\dfrac{c^4}{49}}$ 2. $\sqrt{\dfrac{12b^5}{25}}$

Example 2 **Using the Division Property to Simplify Radicals**

Classroom Examples: p. 607, Exercises 24 and 26

Use the division property of radicals to simplify the expressions. Assume the variables represent positive real numbers.

a. $\dfrac{\sqrt[3]{9}}{\sqrt[3]{72}}$ b. $\dfrac{\sqrt{7y^3}}{\sqrt{y}}$

Solution:

a. $\dfrac{\sqrt[3]{9}}{\sqrt[3]{72}}$ There is a radical in the denominator of the fraction.

$= \sqrt[3]{\dfrac{9}{72}}$ Apply the division property to write the quotient under a single radical.

$= \sqrt[3]{\dfrac{1}{8}}$ Simplify to lowest terms.

$= \dfrac{1}{2}$ Simplify the radical.

Answers

1. $\dfrac{c^2}{7}$ 2. $\dfrac{2b^2\sqrt{3b}}{5}$

b. $\dfrac{\sqrt{7y^3}}{\sqrt{y}}$ There is a radical in the denominator of the fraction.

$= \sqrt{\dfrac{7y^3}{y}}$ Apply the division property to write the quotient under a single radical.

$= \sqrt{7y^2}$ Simplify the fraction.

$= y\sqrt{7}$ Simplify the radical.

Skill Practice Simplify the expressions.

3. $\dfrac{\sqrt[3]{250}}{\sqrt[3]{2}}$ **4.** $\dfrac{\sqrt{10z^9}}{\sqrt{z}}$

3. Rationalizing the Denominator: One Term

Examples 1 and 2 show that radical expressions can sometimes be simplified by using the division property of radicals. However, there are cases where other methods are needed. For example:

$$\dfrac{2}{\sqrt{2}} \quad \text{and} \quad \dfrac{2}{\sqrt{5} + \sqrt{3}} \quad \text{are two such cases.}$$

To begin, recall that the nth-root of a perfect nth power is easily simplified. For example:

$$\sqrt{x^2} = x \quad x \geq 0$$

Classroom Example: p. 607, Exercise 32

Example 3   **Rationalizing the Denominator: One Term**

Simplify the expression. $\dfrac{2}{\sqrt{2}}$

Solution:

A square root of a perfect square is needed in the denominator to remove the radical.

$\dfrac{2}{\sqrt{2}} = \dfrac{2}{\sqrt{2}} \cdot \dfrac{\sqrt{2}}{\sqrt{2}}$ Multiply the numerator and denominator by $\sqrt{2}$ because $\sqrt{2} \cdot \sqrt{2} = \sqrt{2^2}$.

$= \dfrac{2\sqrt{2}}{\sqrt{2^2}}$ Multiply the radicals.

$= \dfrac{2\sqrt{2}}{2}$ Simplify.

$= \dfrac{\overset{1}{2}\sqrt{2}}{2}$ Simplify the fraction to lowest terms.

$= \sqrt{2}$

Skill Practice Simplify the expression.

5. $\dfrac{3}{\sqrt{5}}$

Answers

3. 5   **4.** $z^4\sqrt{10}$ **5.** $\dfrac{3\sqrt{5}}{5}$

Classroom Example: p. 607, Exercise 40

| Example 4 | **Rationalizing the Denominator: One Term** |

Simplify the expression. Assume x represents a positive real number.

$$\sqrt{\dfrac{x}{5}}$$

Solution:

$\sqrt{\dfrac{x}{5}}$ The radicand contains an irreducible fraction.

$=\dfrac{\sqrt{x}}{\sqrt{5}}$ Apply the division property of radicals.

$=\dfrac{\sqrt{x}}{\sqrt{5}}\cdot\dfrac{\sqrt{5}}{\sqrt{5}}$ Multiply the numerator and denominator by $\sqrt{5}$ because $\sqrt{5}\cdot\sqrt{5}=\sqrt{5^2}$.

$=\dfrac{\sqrt{5x}}{\sqrt{5^2}}$ Multiply the radicals.

$=\dfrac{\sqrt{5x}}{5}$ Simplify the radicals.

Skill Practice Simplify the expression.

6. $\sqrt{\dfrac{7}{10}}$

Avoiding Mistakes

In the expression $\dfrac{\sqrt{5x}}{5}$, do not try to "cancel" the factor of $\sqrt{5}$ from the numerator with the factor of 5 in the denominator. $\sqrt{5}$ and 5 are not equal.

Classroom Example: p. 607, Exercise 44

| Example 5 | **Rationalizing the Denominator: One Term** |

Simplify the expression. Assume w represents a positive real number.

$$\dfrac{14\sqrt{w}}{\sqrt{7}}$$

Solution:

$\dfrac{14\sqrt{w}}{\sqrt{7}}$ The fraction contains a radical in the denominator.

$=\dfrac{14\sqrt{w}}{\sqrt{7}}\cdot\dfrac{\sqrt{7}}{\sqrt{7}}$ Multiply the numerator and denominator by $\sqrt{7}$ because $\sqrt{7}\cdot\sqrt{7}=\sqrt{7^2}$.

$=\dfrac{14\sqrt{7w}}{\sqrt{7^2}}$ Multiply the radicals.

$=\dfrac{14\sqrt{7w}}{7}$ Simplify.

$=\dfrac{\overset{2}{14}\sqrt{7w}}{\underset{1}{7}}$ Simplify to lowest terms.

$=2\sqrt{7w}$

Skill Practice Simplify the expression.

7. $\dfrac{6y}{\sqrt{3}}$

TIP: In the expression

$$\dfrac{14\sqrt{7w}}{7}$$

the factor of 14 and the factor of 7 may be reduced because both factors are outside the radical.

$$\dfrac{14\sqrt{7w}}{7}=\dfrac{14}{7}\cdot\sqrt{7w}$$
$$=2\sqrt{7w}$$

Answers

6. $\dfrac{\sqrt{70}}{10}$ 7. $2y\sqrt{3}$

Classroom Example: p. 607, Exercise 46

| Example 6 | **Rationalizing the Denominator: One Term** |

Simplify the expression. Assume w represents a positive real number.

$$\sqrt{\frac{w}{12}}$$

Solution:

$\sqrt{\dfrac{w}{12}}$ The radical contains an irreducible fraction.

$= \dfrac{\sqrt{w}}{\sqrt{12}}$ Apply the division property of radicals.

$= \dfrac{\sqrt{w}}{\sqrt{4 \cdot 3}}$ Factor 12 to simplify the radical.

$= \dfrac{\sqrt{w}}{2\sqrt{3}}$ The $\sqrt{3}$ in the denominator needs to be rationalized.

$= \dfrac{\sqrt{w}}{2\sqrt{3}} \cdot \dfrac{\sqrt{3}}{\sqrt{3}}$ Multiply the numerator and denominator by $\sqrt{3}$ because $\sqrt{3} \cdot \sqrt{3} = \sqrt{3^2}$.

$= \dfrac{\sqrt{3w}}{2\sqrt{3^2}}$ Multiply the radicals.

$= \dfrac{\sqrt{3w}}{2 \cdot 3}$ Simplify.

$= \dfrac{\sqrt{3w}}{6}$ This cannot be simplified further because 3 is inside the radical and 6 is not.

Skill Practice Simplify the expression.

8. $\sqrt{\dfrac{z}{18}}$

4. Rationalizing the Denominator: Two Terms

Recall from the multiplication of polynomials that the product of two conjugates results in a difference of squares.

$$(a + b)(a - b) = a^2 - b^2$$

If either a or b has a square root factor, the expression will simplify without a radical; that is, the expression is rationalized. For example,

$$(\sqrt{5} - \sqrt{3})(\sqrt{5} + \sqrt{3}) = (\sqrt{5})^2 - (\sqrt{3})^2$$
$$= 5 - 3$$
$$= 2$$

Multiplying a binomial by its conjugate is the basis for rationalizing a denominator with two terms involving square roots.

Answer

8. $\dfrac{\sqrt{2z}}{6}$

| Example 7 | **Rationalizing the Denominator: Two Terms** |

Classroom Example: p. 608, Exercise 58

Simplify the expression by rationalizing the denominator. $\dfrac{2}{\sqrt{6}+2}$

Solution:

$\dfrac{2}{\sqrt{6}+2}$ To rationalize a denominator with two terms, multiply the numerator and denominator by the conjugate of the denominator.

$= \dfrac{2}{(\sqrt{6}+2)} \cdot \dfrac{(\sqrt{6}-2)}{(\sqrt{6}-2)}$

Conjugates

The denominator is in the form $(a+b)(a-b)$, where $a=\sqrt{6}$ and $b=2$.

$= \dfrac{2(\sqrt{6}-2)}{(\sqrt{6})^2-(2)^2}$ In the denominator, apply the formula $(a+b)(a-b)=a^2-b^2$.

$= \dfrac{2(\sqrt{6}-2)}{6-4}$ Simplify.

$= \dfrac{2(\sqrt{6}-2)}{2}$

$= \dfrac{\overset{1}{2}(\sqrt{6}-2)}{\underset{1}{2}}$ Simplify to lowest terms.

$= \sqrt{6}-2$

Skill Practice Simplify the expression by rationalizing the denominator.

9. $\dfrac{6}{\sqrt{3}-1}$

| Example 8 | **Rationalizing the Denominator: Two Terms** |

Classroom Example: p. 608, Exercise 66

Simplify the expression by rationalizing the denominator. $\dfrac{\sqrt{x}+\sqrt{2}}{\sqrt{x}-\sqrt{2}}$

Solution:

$\dfrac{\sqrt{x}+\sqrt{2}}{\sqrt{x}-\sqrt{2}} = \dfrac{(\sqrt{x}+\sqrt{2})}{(\sqrt{x}-\sqrt{2})} \cdot \dfrac{(\sqrt{x}+\sqrt{2})}{(\sqrt{x}+\sqrt{2})}$ Multiply the numerator and denominator by the conjugate of the denominator.

Conjugates

$= \dfrac{(\sqrt{x}+\sqrt{2})^2}{(\sqrt{x}-\sqrt{2})(\sqrt{x}+\sqrt{2})}$

$= \dfrac{(\sqrt{x})^2+2(\sqrt{x})(\sqrt{2})+(\sqrt{2})^2}{(\sqrt{x})^2-(\sqrt{2})^2}$ Simplify using special case products.

$= \dfrac{x+2\sqrt{2x}+2}{x-2}$ Simplify the radicals.

Skill Practice Simplify the expression by rationalizing the denominator.

10. $\dfrac{\sqrt{y}-\sqrt{5}}{\sqrt{y}+\sqrt{5}}$

Answers

9. $3\sqrt{3}+3$

10. $\dfrac{y-2\sqrt{5y}+5}{y-5}$

5. Simplifying Quotients That Contain Radicals

Sometimes a radical expression within a quotient must be reduced to lowest terms. This is demonstrated in Example 9.

Classroom Example: p. 608, Exercise 72

Example 9 **Simplifying a Radical Quotient to Lowest Terms**

Simplify the expression. $\dfrac{4 - \sqrt{20}}{10}$

Solution:

$$\dfrac{4 - \sqrt{20}}{10}$$ First simplify $\sqrt{20}$ by writing the radicand as a product of prime factors.

$$= \dfrac{4 - \sqrt{4 \cdot 5}}{10}$$

$$= \dfrac{4 - 2\sqrt{5}}{10}$$ Simplify the radical.

$$= \dfrac{2(2 - \sqrt{5})}{2 \cdot 5}$$ Factor out the GCF.

$$= \dfrac{\overset{1}{\cancel{2}}(2 - \sqrt{5})}{\cancel{2} \cdot 5}$$ Simplify to lowest terms.

$$= \dfrac{2 - \sqrt{5}}{5}$$

Avoiding Mistakes

Remember that it is not correct to reduce *terms* within a rational expression. In the expression

$$\dfrac{4 - 2\sqrt{5}}{10}$$

do not try to reduce the 4 and the 10. Only common *factors* can be canceled.

Skill Practice Simplify the expression.

11. $\dfrac{6 - \sqrt{24}}{12}$

Answer

11. $\dfrac{3 - \sqrt{6}}{6}$

Section 8.5   Practice Exercises

Vocabulary and Key Concepts

For additional exercises, see Classroom Activities 8.5A–8.5C in the *Student's Resource Manual* at www.mhhe.com/moh.

1. a. In the simplified form of a radical, the radicand has no factor raised to a power greater than or equal to the ___index___ .

    b. In the simplified form of a radical, there are no radicals in the ___denominator___ of a fraction.

    c. The process of removing a radical from the denominator of a fraction is called ___rationalizing___ the denominator.

    d. The division property of radicals indicates that $\sqrt[n]{\dfrac{a}{b}} = \underset{\rule{2cm}{0.4pt}}{\dfrac{\sqrt[n]{a}}{\sqrt[n]{b}}}$ provided that both $\sqrt[n]{a}$ and $\sqrt[n]{b}$ are real numbers and that $b \neq 0$.

    e. To rationalize the denominator for the expression $\dfrac{\sqrt{x} + 3}{\sqrt{x} - 2}$, multiply the numerator and denominator by the conjugate of the ___denominator___ .

Review Exercises

For Exercises 2–10, perform the indicated operations. Assume the variables represent positive real numbers.

2. $x\sqrt{45} + 4\sqrt{20x^2}$ $11x\sqrt{5}$

3. $(2\sqrt{y} + 3)(3\sqrt{y} + 7)$
    $6y + 23\sqrt{y} + 21$

4. $(4\sqrt{w} - 2)(2\sqrt{w} - 4)$
    $8w - 20\sqrt{w} + 8$

5. $4\sqrt{3} + \sqrt{5} \cdot \sqrt{15}$ $9\sqrt{3}$

6. $\sqrt{7} \cdot \sqrt{21} + 2\sqrt{27}$ $13\sqrt{3}$

7. $(5 - \sqrt{a})^2$ $25 - 10\sqrt{a} + a$

8. $(\sqrt{z} + 3)^2$ $z + 6\sqrt{z} + 9$

9. $(\sqrt{2} + \sqrt{7})(\sqrt{2} - \sqrt{7})$ -5

10. $(\sqrt{3} + 5)(\sqrt{3} - 5)$ -22

Concept 2: Division Property of Radicals

For Exercises 11–30, use the division property of radicals, if necessary, to simplify the expressions. Assume the variables represent positive real numbers. **(See Examples 1–2.)**

11. $\sqrt{\dfrac{3}{16}}$ $\dfrac{\sqrt{3}}{4}$

12. $\sqrt{\dfrac{7}{25}}$ $\dfrac{\sqrt{7}}{5}$

13. $\sqrt{\dfrac{a^4}{b^4}}$ $\dfrac{a^2}{b^2}$

14. $\sqrt{\dfrac{y^6}{z^2}}$ $\dfrac{y^3}{z}$

15. $\sqrt{\dfrac{c^3}{4}}$ $\dfrac{c\sqrt{c}}{2}$

16. $\sqrt{\dfrac{d^5}{9}}$ $\dfrac{d^2\sqrt{d}}{3}$

17. $\sqrt[3]{\dfrac{x^2}{27}}$ $\dfrac{\sqrt[3]{x^2}}{3}$

18. $\sqrt[3]{\dfrac{c^2}{8}}$ $\dfrac{\sqrt[3]{c^2}}{2}$

19. $\sqrt[3]{\dfrac{y^5}{27y^3}}$ $\dfrac{\sqrt[3]{y^2}}{3}$

20. $\sqrt[3]{\dfrac{7ac}{64c^4}}$ $\dfrac{\sqrt[3]{7a}}{4c}$

21. $\sqrt{\dfrac{200}{81}}$ $\dfrac{10\sqrt{2}}{9}$

22. $\sqrt{\dfrac{80}{49}}$ $\dfrac{4\sqrt{5}}{7}$

23. $\dfrac{\sqrt{8}}{\sqrt{50}}$ $\dfrac{2}{5}$

24. $\dfrac{\sqrt{21}}{\sqrt{12}}$ $\dfrac{\sqrt{7}}{2}$

25. $\dfrac{\sqrt{p}}{\sqrt{4p^3}}$ $\dfrac{1}{2p}$

26. $\dfrac{\sqrt{9t}}{\sqrt{t^5}}$ $\dfrac{3}{t^2}$

27. $\dfrac{\sqrt[3]{z^5}}{\sqrt[3]{z^2}}$ z

28. $\dfrac{\sqrt[3]{a^7}}{\sqrt[3]{a}}$ a^2

29. $\dfrac{\sqrt[3]{24x^5}}{\sqrt[3]{3x^4}}$ $2\sqrt[3]{x}$

30. $\dfrac{\sqrt[3]{2y^8}}{\sqrt[3]{54y^7}}$ $\dfrac{\sqrt[3]{y}}{3}$

Concept 3: Rationalizing the Denominator: One Term

For Exercises 31–50, rationalize the denominators. Assume the variable expressions represent positive real numbers.
(See Examples 3–6.)

31. $\dfrac{1}{\sqrt{6}}$ $\dfrac{\sqrt{6}}{6}$

32. $\dfrac{5}{\sqrt{2}}$ $\dfrac{5\sqrt{2}}{2}$

33. $\dfrac{15}{\sqrt{5}}$ $3\sqrt{5}$

34. $\dfrac{14}{\sqrt{7}}$ $2\sqrt{7}$

35. $\dfrac{6}{\sqrt{x+1}}$ $\dfrac{6\sqrt{x+1}}{x+1}$

36. $\dfrac{8}{\sqrt{y-3}}$ $\dfrac{8\sqrt{y-3}}{y-3}$

37. $\sqrt{\dfrac{6}{x}}$ $\dfrac{\sqrt{6x}}{x}$

38. $\sqrt{\dfrac{8}{y}}$ $\dfrac{2\sqrt{2y}}{y}$

39. $\sqrt{\dfrac{3}{7}}$ $\dfrac{\sqrt{21}}{7}$

40. $\sqrt{\dfrac{5}{11}}$ $\dfrac{\sqrt{55}}{11}$

41. $\dfrac{10}{\sqrt{6y}}$ $\dfrac{5\sqrt{6y}}{3y}$

42. $\dfrac{15}{\sqrt{3w}}$ $\dfrac{5\sqrt{3w}}{w}$

43. $\dfrac{9}{2\sqrt{6}}$ $\dfrac{3\sqrt{6}}{4}$

44. $\dfrac{15}{4\sqrt{10}}$ $\dfrac{3\sqrt{10}}{8}$

45. $\sqrt{\dfrac{p}{27}}$ $\dfrac{\sqrt{3p}}{9}$

46. $\sqrt{\dfrac{x}{32}}$ $\dfrac{\sqrt{2x}}{8}$

47. $\dfrac{5}{\sqrt{20}}$ $\dfrac{\sqrt{5}}{2}$

48. $\dfrac{8}{\sqrt{24}}$ $\dfrac{2\sqrt{6}}{3}$

49. $\sqrt{\dfrac{x^2}{y^3}}$ $\dfrac{x\sqrt{y}}{y^2}$

50. $\sqrt{\dfrac{a}{b^5}}$ $\dfrac{\sqrt{ab}}{b^3}$

Concept 4: Rationalizing the Denominator: Two Terms

For Exercises 51–52, multiply the conjugates.

51. $(\sqrt{2} + 3)(\sqrt{2} - 3)$ -7

52. $(\sqrt{3} + \sqrt{7})(\sqrt{3} - \sqrt{7})$ -4

53. What is the conjugate of $\sqrt{5} - \sqrt{3}$? Multiply $\sqrt{5} - \sqrt{3}$ by its conjugate. $\sqrt{5} + \sqrt{3}; 2$

54. What is the conjugate of $\sqrt{7} + \sqrt{2}$? Multiply $\sqrt{7} + \sqrt{2}$ by its conjugate. $\sqrt{7} - \sqrt{2}; 5$

55. What is the conjugate of $\sqrt{x} + 10$? Multiply $\sqrt{x} + 10$ by its conjugate. $\sqrt{x} - 10; x - 100$

56. What is the conjugate of $12 - \sqrt{y}$? Multiply $12 - \sqrt{y}$ by its conjugate. $12 + \sqrt{y}; 144 - y$

For Exercises 57–68, rationalize the denominators. Assume the variable expressions represent positive real numbers. **(See Examples 7–8.)**

57. $\dfrac{4}{\sqrt{2} + 3}$ $\dfrac{4\sqrt{2} - 12}{-7}$ or $\dfrac{12 - 4\sqrt{2}}{7}$

58. $\dfrac{6}{4 - \sqrt{3}}$ $\dfrac{24 + 6\sqrt{3}}{13}$

59. $\dfrac{1}{\sqrt{5} - \sqrt{2}}$ $\dfrac{\sqrt{5} + \sqrt{2}}{3}$

60. $\dfrac{2}{\sqrt{3} + \sqrt{7}}$ $\dfrac{\sqrt{3} - \sqrt{7}}{-2}$ or $\dfrac{\sqrt{7} - \sqrt{3}}{2}$

61. $\dfrac{\sqrt{8}}{\sqrt{3} + 1}$ $\sqrt{6} - \sqrt{2}$

62. $\dfrac{\sqrt{18}}{1 - \sqrt{2}}$ $-3\sqrt{2} - 6$

63. $\dfrac{1}{\sqrt{x} - \sqrt{3}}$ $\dfrac{\sqrt{x} + \sqrt{3}}{x - 3}$

64. $\dfrac{1}{\sqrt{y} + \sqrt{5}}$ $\dfrac{\sqrt{y} - \sqrt{5}}{y - 5}$

65. $\dfrac{2 - \sqrt{3}}{2 + \sqrt{3}}$ $7 - 4\sqrt{3}$

66. $\dfrac{\sqrt{3} - \sqrt{2}}{\sqrt{3} + \sqrt{2}}$ $5 - 2\sqrt{6}$

67. $\dfrac{\sqrt{5} + 4}{2 - \sqrt{5}}$ $-13 - 6\sqrt{5}$

68. $\dfrac{3 + \sqrt{2}}{\sqrt{2} - 5}$ $\dfrac{17 + 8\sqrt{2}}{-23}$ or $-\dfrac{17 + 8\sqrt{2}}{23}$

Concept 5: Simplifying Quotients That Contain Radicals

For Exercises 69–76, simplify the expression. **(See Example 9.)**

69. $\dfrac{10 - \sqrt{50}}{5}$ $2 - \sqrt{2}$

70. $\dfrac{4 + \sqrt{12}}{2}$ $2 + \sqrt{3}$

71. $\dfrac{21 + \sqrt{98}}{14}$ $\dfrac{3 + \sqrt{2}}{2}$

72. $\dfrac{3 - \sqrt{18}}{6}$ $\dfrac{1 - \sqrt{2}}{2}$

73. $\dfrac{2 - \sqrt{28}}{2}$ $1 - \sqrt{7}$

74. $\dfrac{5 + \sqrt{75}}{5}$ $1 + \sqrt{3}$

75. $\dfrac{14 + \sqrt{72}}{6}$ $\dfrac{7 + 3\sqrt{2}}{3}$

76. $\dfrac{15 - \sqrt{125}}{10}$ $\dfrac{3 - \sqrt{5}}{2}$

Recall that a radical is simplified if

1. The radicand has no factor raised to a power greater than or equal to the index.

2. There are no radicals in the denominator of a fraction.

3. The radicand does not contain a fraction.

For Exercises 77–80, state which condition(s) fails. Then simplify the radical.

77. a. $\sqrt{8x^9}$

    a. Condition 1 fails; $2x^4\sqrt{2x}$

b. $\dfrac{5}{\sqrt{5x}}$

    b. Condition 2 fails; $\dfrac{\sqrt{5x}}{x}$

c. $\sqrt{\dfrac{1}{3}}$

    c. Condition 3 fails; $\dfrac{\sqrt{3}}{3}$

78. a. $\sqrt{\dfrac{7}{2}}$

    a. Condition 3 fails; $\dfrac{\sqrt{14}}{2}$

b. $\sqrt{18y^6}$

    b. Condition 1 fails; $3y^3\sqrt{2}$

c. $\dfrac{2}{\sqrt{4x}}$

    c. Conditions 1 and 2 fail; $\dfrac{\sqrt{x}}{x}$

79. a. $\dfrac{3}{\sqrt{x} + 1}$

    a. Condition 2 fails; $\dfrac{3\sqrt{x} - 3}{x - 1}$

b. $\sqrt{\dfrac{9w^2}{t}}$

    b. Conditions 1 and 3 fail; $\dfrac{3w\sqrt{t}}{t}$

c. $\sqrt{24a^5b^9}$

    c. Condition 1 fails; $2a^2b^4\sqrt{6ab}$

80. a. $\sqrt{\dfrac{12}{z^3}}$

    a. Conditions 1 and 3 fail; $\dfrac{2\sqrt{3z}}{z^2}$

b. $\dfrac{4}{\sqrt{a} - \sqrt{b}}$

    b. Condition 2 fails; $\dfrac{4\sqrt{a} + 4\sqrt{b}}{a - b}$

c. $\sqrt[3]{27m^3n^7}$

    c. Condition 1 fails; $3mn^2\sqrt[3]{n}$

Mixed Exercises

For Exercises 81–96, simplify the radical expressions, if possible. Assume the variables represent positive real numbers.

81. $\sqrt{45}$ $3\sqrt{5}$

82. $-\sqrt{108y^4}$ $-6y^2\sqrt{3}$

83. $-\sqrt{\dfrac{18w^2}{25}}$ $-\dfrac{3w\sqrt{2}}{5}$

84. $\sqrt{\dfrac{8a^2}{7}}$ $\dfrac{2a\sqrt{14}}{7}$

85. $\sqrt{-36}$ Not a real number

86. $\sqrt{54b^5}$ $3b^2\sqrt{6b}$

87. $\sqrt{\dfrac{s^2}{t}}$ $\dfrac{s\sqrt{t}}{t}$

88. $\dfrac{x+\sqrt{y}}{x-\sqrt{y}}$ $\dfrac{x^2+2x\sqrt{y}+y}{x^2-y}$

89. $\dfrac{\sqrt{2m^5}}{\sqrt{8m}}$ $\dfrac{m^2}{2}$

90. $\dfrac{\sqrt{10w}}{\sqrt{5w^3}}$ $\dfrac{\sqrt{2}}{w}$

91. $\sqrt{\dfrac{81}{t^3}}$ $\dfrac{9\sqrt{t}}{t^2}$

92. $-\sqrt{a^3bc^6}$ $-ac^3\sqrt{ab}$

93. $\dfrac{3}{\sqrt{11}+\sqrt{5}}$ $\dfrac{\sqrt{11}-\sqrt{5}}{2}$

94. $\dfrac{4}{\sqrt{10}+\sqrt{2}}$ $\dfrac{\sqrt{10}-\sqrt{2}}{2}$

95. $\dfrac{\sqrt{a}+\sqrt{b}}{\sqrt{a}-\sqrt{b}}$ $\dfrac{a+2\sqrt{ab}+b}{a-b}$

96. $\dfrac{\sqrt{x}+1}{\sqrt{x}-1}$ $\dfrac{x+2\sqrt{x}+1}{x-1}$

Expanding Your Skills

97. Find the slope of the line through the points $(5\sqrt{2}, 3)$ and $(\sqrt{2}, 6)$. $-\dfrac{3\sqrt{2}}{8}$

98. Find the slope of the line through the points $(4\sqrt{5}, -1)$ and $(6\sqrt{5}, -5)$. $-\dfrac{2\sqrt{5}}{5}$

99. Find the slope of the line through the points $(\sqrt{3}, -1)$ and $(4\sqrt{3}, 0)$. $\dfrac{\sqrt{3}}{9}$

100. Find the slope of the line through the points $(-2\sqrt{7}, -5)$ and $(\sqrt{7}, 2)$. $\dfrac{\sqrt{7}}{3}$

Problem Recognition Exercises

Operations on Radicals

For Exercises 1–10, simplify each expression. Assume that all variable expressions represent positive real numbers.

1. a. $(\sqrt{3})(\sqrt{6})$ $3\sqrt{2}$ **b.** $\sqrt{3}+\sqrt{6}$ Cannot be simplified further.   **c.** $\dfrac{\sqrt{6}}{\sqrt{3}}$ $\sqrt{2}$

2. a. $\dfrac{\sqrt{14}}{\sqrt{2}}$ $\sqrt{7}$ **b.** $(\sqrt{2})(\sqrt{14})$ $2\sqrt{7}$ **c.** $\sqrt{2}+\sqrt{14}$ Cannot be simplified further.

3. a. $(3\sqrt{z})^2$ $9z$ **b.** $(3+\sqrt{z})^2$ $9+6\sqrt{z}+z$ **c.** $(3+\sqrt{z})(3-\sqrt{z})$ $9-z$

4. a. $(4-\sqrt{x})^2$ $16-8\sqrt{x}+x$ **b.** $(4-\sqrt{x})(4+\sqrt{x})$ $16-x$ **c.** $(4\sqrt{x})^2$ $16x$

5. a. $\dfrac{12}{\sqrt{2x}}$ $\dfrac{6\sqrt{2x}}{x}$ **b.** $\sqrt{\dfrac{12}{2x}}$ $\dfrac{\sqrt{6x}}{x}$ **c.** $\dfrac{12}{\sqrt{2}+x}$ $\dfrac{12\sqrt{2}-12x}{2-x^2}$

6. a. $\dfrac{15}{3-\sqrt{y}}$ $\dfrac{45+15\sqrt{y}}{9-y}$ **b.** $\dfrac{15}{\sqrt{3y}}$ $\dfrac{5\sqrt{3y}}{y}$ **c.** $\sqrt{\dfrac{15}{3y}}$ $\dfrac{\sqrt{5y}}{y}$

7. a. $(2\sqrt{5}+1)+(\sqrt{5}-2)$ $3\sqrt{5}-1$ **b.** $(2\sqrt{5}+1)(\sqrt{5}-2)$ $8-3\sqrt{5}$ **c.** $2\sqrt{5}(\sqrt{5}-2)$ $10-4\sqrt{5}$

8. a. $(4\sqrt{3}-5)(\sqrt{3}+4)$ $-8+11\sqrt{3}$ **b.** $4\sqrt{3}(\sqrt{3}+4)$ $12+16\sqrt{3}$ **c.** $(4\sqrt{3}-5)-(\sqrt{3}+4)$ $3\sqrt{3}-9$

9. a. $\sqrt{16a^{15}}$ $4a^7\sqrt{a}$ **b.** $\sqrt[3]{16a^{15}}$ $2a^5\sqrt[3]{2}$

10. a. $\sqrt[3]{27y^9}$ $3y^3$ **b.** $\sqrt{27y^9}$ $3y^4\sqrt{3y}$

Writing   Translating Expression   Geometry   Scientific Calculator   Video

Section 8.6 Radical Equations

Concepts

1. Solving Radical Equations
2. Translations Involving Radical Equations
3. Applications of Radical Equations

1. Solving Radical Equations

Radical Equation

An equation with one or more radicals containing a variable is called a **radical equation**.

For example, $\sqrt{x} = 5$ is a radical equation. Recall that $(\sqrt[n]{a})^n = a$ provided $\sqrt[n]{a}$ is a real number. The basis to solve a radical equation is to eliminate the radical by raising both sides of the equation to a power equal to the index of the radical.

To solve the equation $\sqrt{x} = 5$, square both sides of the equation.

$$\sqrt{x} = 5$$
$$(\sqrt{x})^2 = (5)^2$$
$$x = 25$$

By raising each side of a radical equation to a power equal to the index of the radical, a new equation is produced. However, it is important to note that the new equation may have **extraneous solutions**; that is, some or all of the solutions to the new equation may *not* be solutions to the original radical equation. For this reason, it is necessary to check *all* potential solutions in the original equation. For example, consider the equation $\sqrt{x} = -10$. This equation has no solution because by definition, the principal square root of x must be a nonnegative number. However, if we square both sides of the equation, it appears as though a solution exists.

$$\sqrt{x} = -10$$
$$(\sqrt{x})^2 = (-10)^2$$
$$x = 100$$

The value 100 does not check in the original equation $\sqrt{x} = -10$. Therefore, 100 is an extraneous solution.

Check: $\sqrt{x} = -10$

$\sqrt{100} \stackrel{?}{=} -10$ false

Solving a Radical Equation

Step 1 Isolate the radical. If an equation has more than one radical, choose one of the radicals to isolate.

Step 2 Raise each side of the equation to a power equal to the index of the radical.

Step 3 Solve the resulting equation.

Step 4 Check the potential solutions in the original equation.*

*In solving radical equations, extraneous solutions *potentially occur* only when each side of the equation is raised to an even power.

Example 1   Solving a Radical Equation

Solve the equation. $\sqrt{2x+1}+5=8$

Classroom Example: p. 616, Exercise 28

Solution:

$$\sqrt{2x+1}+5=8$$

$$\sqrt{2x+1}=3$$ Isolate the radical by subtracting 5 from both sides.

$$(\sqrt{2x+1})^2=(3)^2$$ Raise both sides to a power equal to the index of the radical.

$$2x+1=9$$ Simplify both sides.

$$2x=8$$ Solve the resulting equation (the equation is linear).

$$x=4$$

Check: Check 4 as a potential solution.

$$\sqrt{2x+1}+5=8$$

$$\sqrt{2(4)+1}+5\overset{?}{=}8$$

$$\sqrt{8+1}+5\overset{?}{=}8$$

$$\sqrt{9}+5\overset{?}{=}8$$

$$3+5\overset{?}{=}8\ \checkmark$$ The answer checks.

The solution set is $\{4\}$.

Skill Practice Solve the equation.

1. $\sqrt{p-4}-2=4$

Example 2   Solving a Radical Equation

Solve the equation. $8+\sqrt{x+2}=7$

Classroom Example: p. 616, Exercise 30

Solution:

$$8+\sqrt{x+2}=7$$

$$\sqrt{x+2}=-1$$ Isolate the radical by subtracting 8 from both sides.

$$(\sqrt{x+2})^2=(-1)^2$$ Raise both sides to a power equal to the index of the radical.

$$x+2=1$$ Simplify.

$$x=-1$$ Solve the resulting equation.

TIP: After isolating the radical in Example 2, the equation shows a square root equated to a negative number.

$$\sqrt{x+2}=-1$$

By definition, a principal square root of any real number must be nonnegative. Therefore, there can be no solution to this equation.

Answer

1. $\{40\}$

Check: Check −1 as a potential solution.

$$8 + \sqrt{x+2} = 7$$

$$8 + \sqrt{(-1)+2} \overset{?}{=} 7$$

$$8 + \sqrt{1} \overset{?}{=} 7$$

$$8 + 1 \neq 7 \qquad \text{The value } -1 \text{ does not check. It is an extraneous solution.}$$

The solution set is { }.

Skill Practice Solve the equation.

2. $\sqrt{2y+5} + 7 = 4$

Classroom Example: p. 616, Exercise 42

| Example 3 | **Solving a Radical Equation** |

Solve the equation. $p + 4 = \sqrt{p+6}$

Solution:

$$p + 4 = \sqrt{p+6} \qquad \text{The radical is already isolated.}$$

$$(p+4)^2 = (\sqrt{p+6})^2 \qquad \text{Raise both sides to a power equal to the index.}$$

$$p^2 + 8p + 16 = p + 6$$

$$p^2 + 7p + 10 = 0 \qquad \text{Solve the resulting equation (the equation is quadratic).}$$

$$(p+5)(p+2) = 0 \qquad \text{Set the equation equal to zero and factor.}$$

$$p + 5 = 0 \quad \text{or} \quad p + 2 = 0 \qquad \text{Set each factor equal to zero.}$$

$$p = -5 \quad \text{or} \qquad p = -2 \qquad \text{Solve for } p.$$

> **Avoiding Mistakes**
>
> Recall that
> $(a+b)^2 = a^2 + 2ab + b^2$
> Hence,
> $(p+4)^2$
> $= (p)^2 + 2(p)(4) + (4)^2$
> $= p^2 + 8p + 16$

Check: $p = -5$ Check: $p = -2$

$$p + 4 = \sqrt{p+6} \qquad\qquad p + 4 = \sqrt{p+6}$$

$$(-5) + 4 \overset{?}{=} \sqrt{(-5)+6} \qquad\qquad (-2) + 4 \overset{?}{=} \sqrt{(-2)+6}$$

$$-1 \overset{?}{=} \sqrt{1} \qquad\qquad\qquad 2 \overset{?}{=} \sqrt{4}$$

$$-1 \neq 1 \quad \text{Does not check.} \qquad\qquad 2 \overset{?}{=} 2 \checkmark \text{ The solution checks.}$$

The solution set is $\{-2\}$. The value -5 does not check.

Skill Practice Solve the equation.

3. $\sqrt{x+34} = x + 4$

Answers

2. { } (The value 2 does not check.)
3. {2} (The value −9 does not check.)

| Example 4 | **Solving a Radical Equation** |

Solve the equation. $2\sqrt[3]{2x-3} - \sqrt[3]{x+6} = 0$

Classroom Example: p. 616, Exercise 46

Solution:

$$2\sqrt[3]{2x-3} - \sqrt[3]{x+6} = 0$$

$$2\sqrt[3]{2x-3} = \sqrt[3]{x+6} \qquad \text{Isolate one of the radicals.}$$

$$(2\sqrt[3]{2x-3})^3 = (\sqrt[3]{x+6})^3 \qquad \text{Raise both sides to a power equal to the index.}$$

$$(2)^3(\sqrt[3]{2x-3})^3 = (\sqrt[3]{x+6})^3 \qquad \text{On the left-hand side, be sure to cube both factors, } (2)^3 \text{ and } (\sqrt[3]{2x-3})^3.$$

$$8(2x-3) = x+6 \qquad \text{Solve the resulting equation.}$$

$$16x - 24 = x + 6$$

$$15x = 30$$

$$x = 2$$

Check:

$$2\sqrt[3]{2x-3} - \sqrt[3]{x+6} = 0 \qquad \text{Check the potential solution, 2.}$$

$$2\sqrt[3]{2(2)-3} - \sqrt[3]{2+6} \stackrel{?}{=} 0$$

$$2\sqrt[3]{4-3} - \sqrt[3]{8} \stackrel{?}{=} 0$$

$$2\sqrt[3]{1} - 2 \stackrel{?}{=} 0$$

$$2 - 2 \stackrel{?}{=} 0 \checkmark \qquad \text{The solution checks.}$$

The solution set is $\{2\}$.

Skill Practice Solve the equation.

4. $\sqrt[3]{4p+1} - \sqrt[3]{p+16} = 0$

2. Translations Involving Radical Equations

| Example 5 | **Translating to a Radical Equation** |

The principal square root of the sum of a number and three is equal to seven. Find the number.

Classroom Example: p. 616, Exercise 48

Solution:

Let x represent the number. Label the variable.

$$\sqrt{x+3} = 7 \qquad \text{Write the verbal model as an algebraic equation.}$$

$$(\sqrt{x+3})^2 = (7)^2 \qquad \text{The radical is already isolated. Square both sides.}$$

$$x + 3 = 49 \qquad \text{The resulting equation is linear.}$$

$$x = 46 \qquad \text{Solve for } x.$$

Answer

4. $\{5\}$

Check: Check 46 as a potential solution.

$$\sqrt{x+3} = 7$$

$$\sqrt{46+3} \stackrel{?}{=} 7$$

$$\sqrt{49} \stackrel{?}{=} 7$$

$$7 \stackrel{?}{=} 7 \checkmark \qquad \text{The solution checks.}$$

The number is 46.

Skill Practice

5. The principal square root of the sum of a number and 5 is 2. Find the number.

3. Applications of Radical Equations

Classroom Example: p. 616, Exercise 54

Example 6   Using a Radical Equation in an Application

For a small company, the weekly sales, y, of its product are related to the money spent on advertising, x, according to the equation:

$$y = 100\sqrt{x}$$

a. Find the amount in sales if the company spends $100 on advertising.

b. Find the amount in sales if the company spends $625 on advertising.

c. Find the amount the company spent on advertising if its sales for 1 week totaled $2000.

Solution:

a. $y = 100\sqrt{x}$

$\quad = 100\sqrt{100} \qquad$ Substitute $x = 100$.

$\quad = 100(10)$

$\quad = 1000$

The amount in sales is $1000.

b. $y = 100\sqrt{x}$

$\quad = 100\sqrt{625} \qquad$ Substitute $x = 625$.

$\quad = 100(25)$

$\quad = 2500$

The amount in sales is $2500.

c. $\qquad y = 100\sqrt{x}$

$\quad 2000 = 100\sqrt{x} \qquad$ Substitute $y = 2000$.

$\quad \dfrac{2000}{100} = \dfrac{100\sqrt{x}}{100} \qquad$ Isolate the radical. Divide both sides by 100.

$\quad 20 = \sqrt{x} \qquad$ Simplify.

$\quad (20)^2 = (\sqrt{x})^2 \qquad$ Raise both sides to a power equal to the index.

$\quad 400 = x \qquad$ Simplify both sides.

Answer

5. The number is −1.

Check: Check 400 as a potential solution.

$$y = 100\sqrt{x}$$

$$2000 \overset{?}{=} 100\sqrt{400}$$

$$2000 \overset{?}{=} 100(20)$$

$$2000 \overset{?}{=} 2000 \checkmark$$ The solution checks.

The amount spent on advertising was $400.

Skill Practice

6. If the small company mentioned in Example 6 changes its advertising media, the equation relating money spent on advertising, x, to weekly sales, y, is $y = 100\sqrt{2x}$.

   a. Use the given equation to find the amount in sales if the company spends $200 on advertising.

   b. Find the amount spent on advertising if the sales for 1 week totaled $3000.

Answer

6. a. $2000   b. $450

Section 8.6   Practice Exercises

For additional exercises, see Classroom Activities 8.6A–8.6B in the *Student's Resource Manual* at www.mhhe.com/moh.

Vocabulary and Key Concepts

1. **a.** An equation with one or more radicals containing a variable is called a ____radical____ equation.

   b. A potential solution that does not check in the original equation is called an ____extraneous____ solution.

   c. What is the first step to solve the equation $\sqrt{x+2} - 3 = 7$?   Isolate the radical by adding 3 to both sides of the equation.

   d. To solve the equation $\sqrt[3]{x-4} = 2$, raise both sides of the equation to the ____third____ power.

Review Exercises

For Exercises 2–5, rationalize the denominators.

2. $\dfrac{1}{\sqrt{3} - \sqrt{7}}$ $\dfrac{\sqrt{3}+\sqrt{7}}{-4}$ or $-\dfrac{\sqrt{3}+\sqrt{7}}{4}$

3. $\dfrac{1}{\sqrt{2} + \sqrt{10}}$ $\dfrac{\sqrt{2}-\sqrt{10}}{-8}$ or $\dfrac{\sqrt{10}-\sqrt{2}}{8}$

4. $\dfrac{6}{\sqrt{6}}$ $\sqrt{6}$

5. $\dfrac{2\sqrt{2}}{\sqrt{3}}$ $\dfrac{2\sqrt{6}}{3}$

6. Simplify the expression. $\dfrac{10 - \sqrt{75}}{5}$ $2 - \sqrt{3}$

For Exercises 7–10, multiply the expressions.

7. $(x+4)^2$ $x^2 + 8x + 16$

8. $(3-y)^2$ $9 - 6y + y^2$

9. $(\sqrt{x}+4)^2$ $x + 8\sqrt{x} + 16$

10. $(\sqrt{3} - \sqrt{y})^2$ $3 - 2\sqrt{3y} + y$

For Exercises 11–14, multiply the expressions. Assume the variable expressions represent positive real numbers.

11. $(\sqrt{2x-3})^2$ $2x - 3$

12. $(\sqrt{m+6})^2$ $m + 6$

13. $(t+1)^2$ $t^2 + 2t + 1$

14. $(y-4)^2$ $y^2 - 8y + 16$

/ Writing   ←→ Translating Expression   ◁ Geometry   ▤ Scientific Calculator   Video

Concept 1: Solving Radical Equations

For Exercises 15–47, solve the equations. Be sure to check all of the potential answers. **(See Examples 1–4.)**

15. $\sqrt{t} = 6$ {36}

16. $\sqrt{p} = 5$ {25}

17. $\sqrt{x+1} = 4$ {15}

18. $\sqrt{x-3} = 7$ {52}

19. $\sqrt{y-4} = -5$ { } (The value 29 does not check.)

20. $\sqrt{p+6} = -1$ { } (The value −5 does not check.)

21. $\sqrt{5-t} = 0$ {5}

22. $\sqrt{13+m} = 0$ {−13}

23. $\sqrt{2n+10} = 3$ $\left\{-\frac{1}{2}\right\}$

24. $\sqrt{1-q} = 15$ {−224}

25. $\sqrt{6w} - 8 = -2$ {6}

26. $\sqrt{2z} - 11 = -3$ {32}

27. $\sqrt{5a-4} - 2 = 4$ {8}

28. $\sqrt{3b+4} - 3 = 2$ {7}

29. $\sqrt{2x-3} + 7 = 3$ { } (The value $\frac{19}{2}$ does not check.)

30. $\sqrt{8y+1} + 5 = 1$ { } (The value $\frac{15}{8}$ does not check.)

31. $5\sqrt{c} = \sqrt{10c+15}$ {1}

32. $4\sqrt{x} = \sqrt{10x+6}$ {1}

33. $\sqrt{x^2-x} = \sqrt{12}$ {4, −3}

34. $\sqrt{x^2+5x} = \sqrt{150}$ {−15, 10}

35. $\sqrt{9y^2-8y+1} = 3y+1$ {0}

36. $\sqrt{4x^2+2x+20} = 2x$ { } (The value −10 does not check.)

37. $\sqrt{x^2+4x+16} = x$ { } (The value −4 does not check.)

38. $\sqrt{x^2+3x-2} = 4$ {−6, 3}

39. $\sqrt{2k^2-3k-4} = k$ {4} (The value −1 does not check.)

40. $\sqrt{6t+7} = t+2$ {3, −1}

41. $\sqrt{y+1} = y+1$ {0, −1}

42. $\sqrt{3p+3} + 5 = p$ {11} (The value 2 does not check.)

43. $\sqrt{2m+1} + 7 = m$ {12} (The value 4 does not check.)

44. $\sqrt[3]{3y+7} = \sqrt[3]{2y-1}$ {−8}

45. $\sqrt[3]{p-5} - \sqrt[3]{2p+1} = 0$ {−6}

46. $\sqrt[3]{2x-8} - \sqrt[3]{-x+1} = 0$ {3}

47. $\sqrt[3]{a-3} = \sqrt[3]{5a+1}$ {−1}

Concept 2: Translations Involving Radical Equations

For Exercises 48–53, write the English sentence as a radical equation and solve the equation. **(See Example 5.)**

48. The square root of the sum of a number and 8 equals 12. Find the number.
$\sqrt{x+8} = 12$;   136

49. The square root of the sum of a number and 10 equals 1. Find the number.
$\sqrt{x+10} = 1$;   −9

50. The square root of a number is 2 less than the number. Find the number.
$\sqrt{x} = x-2$;   4

51. The square root of twice a number is 4 less than the number. Find the number.
$\sqrt{2x} = x-4$;   8

52. The cube root of the sum of a number and 4 is −5. Find the number.
$\sqrt[3]{x+4} = -5$;   −129

53. The cube root of the sum of a number and 1 is 2. Find the number.
$\sqrt[3]{x+1} = 2$;   7

Concept 3: Applications of Radical Equations

54. Ignoring air resistance, the time, t (in seconds), required for an object to fall x feet is given by the equation:

$$t = \frac{\sqrt{x}}{4}$$

a. Find the time required for an object to fall 64 ft.   2 sec

b. Find the distance an object will fall in 4 sec.
256 ft

55. Ignoring air resistance, the velocity, v (in feet per second: ft/sec), of an object in free fall depends on the distance it has fallen, x (in feet), according to the equation:

$$v = 8\sqrt{x}$$

a. Find the velocity of an object that has fallen 100 ft.   80 ft/sec

b. Find the distance that an object has fallen if its velocity is 136 ft/sec. **(See Example 6.)**   289 ft

56. The speed of a car, s (in miles per hour), before the brakes were applied can be approximated by the length of its skid marks, x (in feet), according to the equation:

$$s = 4\sqrt{x}$$

a. Find the speed of a car before the brakes were applied if its skid marks are 324 ft long.

72 mph

b. How long would you expect the skid marks to be if the car had been traveling the speed limit of 60 mph? 225 ft

© Image Source/Getty Images RF

57. The height of a sunflower plant, y (in inches), can be determined by the time, t (in weeks), after the seed has germinated according to the equation:

$$y = 8\sqrt{t} \quad 0 \le t \le 40$$

a. Find the height of the plant after 4 weeks.

16 in.

b. In how many weeks will the plant be 40 in. tall? 25 weeks

© Royalty Free/Corbis RF

Expanding Your Skills

For Exercises 58–61, solve the equations. First isolate one of the radical terms. Then square both sides. The resulting equation will still have a radical. Repeat the process by isolating the radical and squaring both sides again.

58. $\sqrt{t+8} = \sqrt{t} + 2$ {1}

59. $\sqrt{5x-9} = \sqrt{5x} - 3$ $\left\{\dfrac{9}{5}\right\}$

60. $\sqrt{z+1} + \sqrt{2z+3} = 1$

{−1} (The value 3 does not check.)

61. $\sqrt{2m+6} = 1 + \sqrt{7-2m}$

$\left\{\dfrac{3}{2}\right\}$ (The value −1 does not check.)

Rational Exponents

Section 8.7

1. Definition of $a^{1/n}$

We have already presented several properties for simplifying expressions with integer exponents. In this section, the properties are expanded to include expressions with rational exponents. We begin by defining expressions of the form $a^{1/n}$.

> ### Definition of $a^{1/n}$
>
> Let a be a real number, and let n be an integer such that $n > 1$. If $\sqrt[n]{a}$ is a real number, then
>
> $$a^{1/n} = \sqrt[n]{a}$$

Concepts

1. Definition of $a^{1/n}$
2. Definition of $a^{m/n}$
3. Converting between Rational Exponents and Radical Notation
4. Properties of Rational Exponents
5. Applications of Rational Exponents

Classroom Examples: p. 622,
Exercises 8, 10, 14, and 18

> **Example 1** Evaluating Expressions of the Form $a^{1/n}$
>
> Convert the expression to radical notation and simplify, if possible.
>
> **a.** $9^{1/2}$ **b.** $125^{1/3}$
>
> **c.** $16^{1/4}$ **d.** $-25^{1/2}$
>
> **e.** $(-25)^{1/2}$ **f.** $25^{-1/2}$
>
> **Solution:**
>
> **a.** $9^{1/2} = \sqrt{9} = 3$
>
> **b.** $125^{1/3} = \sqrt[3]{125} = 5$
>
> **c.** $16^{1/4} = \sqrt[4]{16} = 2$
>
> **d.** $-25^{1/2}$ is equivalent to $-1 \cdot (25^{1/2})$
>
> $\quad = -1 \cdot \sqrt{25}$
>
> $\quad = -5$
>
> **e.** $(-25)^{1/2}$ is not a real number because $\sqrt{-25}$ is not a real number.
>
> **f.** $25^{-1/2} = \dfrac{1}{25^{1/2}} = \dfrac{1}{\sqrt{25}} = \dfrac{1}{5}$

Avoiding Mistakes

In Example 1(f), the expression is first rewritten with a positive exponent before it can be simplified.

Skill Practice Convert the expression to radical notation and simplify.

1. $36^{1/2}$ **2.** $(-27)^{1/3}$ **3.** $81^{1/4}$ **4.** $(-16)^{1/4}$ **5.** $(16)^{-1/4}$ **6.** $-16^{1/4}$

2. Definition of $a^{m/n}$

If $\sqrt[n]{a}$ is a real number, then we can define an expression of the form $a^{m/n}$ in such a way that the multiplication property of exponents holds true. For example:

$$16^{3/4} = \begin{cases} (16^{1/4})^3 = (\sqrt[4]{16})^3 = (2)^3 = 8 \\ (16^3)^{1/4} = \sqrt[4]{16^3} = \sqrt[4]{4096} = 8 \end{cases}$$

TIP: In simplifying the expression $a^{m/n}$ it is usually easier to take the root first. That is, simplify as $(\sqrt[n]{a})^m$.

> **Definition of $a^{m/n}$**
>
> Let a be a real number, and let m and n be positive integers such that m and n share no common factors and $n > 1$. If $\sqrt[n]{a}$ is a real number, then
>
> $$a^{m/n} = (a^{1/n})^m = (\sqrt[n]{a})^m \quad \text{and} \quad a^{m/n} = (a^m)^{1/n} = \sqrt[n]{a^m}$$

The rational exponent in the expression $a^{m/n}$ is essentially performing two operations. The numerator of the exponent raises the base to the mth-power. The denominator takes the nth-root.

Answers

1. $\sqrt{36}$; 6 **2.** $\sqrt[3]{-27}$; -3
3. $\sqrt[4]{81}$; 3
4. $\sqrt[4]{-16}$; Not a real number
5. $\dfrac{1}{\sqrt[4]{16}}$; $\dfrac{1}{2}$ **6.** $-\sqrt[4]{16}$; -2

| Example 2 | Evaluating Expressions of the Form $a^{m/n}$ |

Convert each expression to radical notation and simplify.

a. $125^{2/3}$ **b.** $100^{-3/2}$ **c.** $(81)^{3/4}$

Solution:

a. $125^{2/3} = (\sqrt[3]{125})^2$ Take the cube root of 125, and square the result.

$\quad\quad = (5)^2$ Simplify.

$\quad\quad = 25$

b. $100^{-3/2} = \dfrac{1}{100^{3/2}}$ Take the reciprocal of the base.

$\quad\quad = \dfrac{1}{(\sqrt{100})^3}$ Take the square root of 100, and cube the result.

$\quad\quad = \dfrac{1}{(10)^3}$ Simplify.

$\quad\quad = \dfrac{1}{1000}$

c. $(81)^{3/4} = (\sqrt[4]{81})^3$ Take the fourth root of 81, and cube the result.

$\quad\quad = (3)^3$ Simplify.

$\quad\quad = 27$

Skill Practice Convert each expression to radical notation and simplify.

7. $16^{3/4}$ **8.** $8^{-2/3}$ **9.** $9^{3/2}$

Classroom Examples: p. 622, Exercises 34 and 36

3. Converting between Rational Exponents and Radical Notation

| Example 3 | Converting Rational Exponents to Radical Notation |

Convert the expressions to radical notation. Assume the variables represent positive real numbers. Write the answers with positive exponents only.

a. $x^{3/5}$ **b.** $(2a^2)^{1/3}$ **c.** $5y^{1/4}$ **d.** $p^{-1/2}$

Solution:

a. $x^{3/5} = \sqrt[5]{x^3}$ or $(\sqrt[5]{x})^3$

b. $(2a^2)^{1/3} = \sqrt[3]{2a^2}$

c. $5y^{1/4} = 5\sqrt[4]{y}$ The exponent $\frac{1}{4}$ applies only to y.

d. $p^{-1/2} = \dfrac{1}{\sqrt{p}}$

Classroom Examples: p. 623, Exercises 42, 46, and 48

Skill Practice Convert each expression to radical notation. Write the answers with positive exponents only. Assume the variables represent positive real numbers.

10. $y^{4/3}$ **11.** $(5x)^{1/2}$ **12.** $10a^{3/5}$ **13.** $z^{-2/3}$

Answers

7. $(\sqrt[4]{16})^3$; 8

8. $\dfrac{1}{(\sqrt[3]{8})^2}$; $\dfrac{1}{4}$ **9.** $(\sqrt{9})^3$; 27

10. $(\sqrt[3]{y})^4$ **11.** $\sqrt{5x}$

12. $10(\sqrt[5]{a})^3$ **13.** $\dfrac{1}{(\sqrt[3]{z})^2}$

Classroom Example: p. 623, Exercise 52

Example 4 Converting Radical Notation to Rational Exponents

Convert each expression to an equivalent expression using rational exponents. Assume that the variables represent positive real numbers.

a. $\sqrt[4]{c^3}$ b. $\sqrt{11p}$ c. $11\sqrt{p}$

Solution:

a. $\sqrt[4]{c^3} = c^{3/4}$ b. $\sqrt{11p} = (11p)^{1/2}$ c. $11\sqrt{p} = 11p^{1/2}$

Skill Practice Convert each expression to an equivalent expression using rational exponents.

14. $\sqrt[5]{y^2}$ 15. $\sqrt{2x}$ 16. $2\sqrt{x}$

4. Properties of Rational Exponents

The properties of integer exponents also apply to rational exponents.

Operations with Exponents

Let a and b be real numbers. Let m and n be rational numbers such that a^m, a^n, and b^n are defined. Then,

Description	Property	Example
1. Multiplying like bases	$a^m a^n = a^{m+n}$	$x^{1/3}x^{4/3} = x^{5/3}$
2. Dividing like bases	$\dfrac{a^m}{a^n} = a^{m-n}$	$\dfrac{x^{3/5}}{x^{1/5}} = x^{2/5}$
3. The power rule	$(a^m)^n = a^{mn}$	$(2^{1/3})^{1/2} = 2^{1/6}$
4. Power of a product	$(ab)^m = a^m b^m$	$(xy)^{1/2} = x^{1/2}y^{1/2}$
5. Power of a quotient	$\left(\dfrac{a}{b}\right)^m = \dfrac{a^m}{b^m}$ $(b \neq 0)$	$\left(\dfrac{4}{25}\right)^{1/2} = \dfrac{4^{1/2}}{25^{1/2}} = \dfrac{2}{5}$

Negative and Zero Exponents

Description	Definition	Example
1. Negative exponents	$a^{-m} = \left(\dfrac{1}{a}\right)^m = \dfrac{1}{a^m}$ $(a \neq 0)$	$(8)^{-1/3} = \left(\dfrac{1}{8}\right)^{1/3} = \dfrac{1}{2}$
2. Zero exponent	$a^0 = 1$ $(a \neq 0)$	$5^0 = 1$

Classroom Examples: p. 623, Exercises 58, 66, and 74

Example 5 Simplifying Expressions with Rational Exponents

Use the properties of exponents to simplify the expressions. Write the final answers with positive exponents only. Assume the variables represent positive real numbers.

a. $x^{2/3}x^{1/3}$ b. $\dfrac{y^{1/10}}{y^{4/5}}$ c. $(z^4)^{1/2}$ d. $(s^4 t^8)^{1/4}$

Solution:

a. $x^{2/3}x^{1/3} = x^{(2/3)+(1/3)}$ Add exponents.

$\qquad\qquad = x^{3/3}$ Simplify.

$\qquad\qquad = x$

Answers

14. $y^{2/5}$ 15. $(2x)^{1/2}$ 16. $2x^{1/2}$

b. $\dfrac{y^{1/10}}{y^{4/5}} = y^{(1/10) - (4/5)}$ Subtract exponents.

$ = y^{(1/10) - (8/10)}$ The common denominator is 10.

$ = y^{-7/10}$ Simplify.

$ = \dfrac{1}{y^{7/10}}$ Write with a positive exponent.

c. $(z^4)^{1/2} = z^{(4)\cdot(1/2)}$ Multiply exponents.

$ = z^2$ Simplify.

d. $(s^4 t^8)^{1/4} = s^{4/4} t^{8/4}$ Multiply exponents.

$ = st^2$

Skill Practice Use the properties of exponents to simplify the expressions. Write the answers with positive exponents only. Assume the variables represent positive real numbers.

17. $a^{3/4} \cdot a^{5/4}$ **18.** $\dfrac{t^{2/3}}{t^2}$ **19.** $(w^{1/3})^{-12}$ **20.** $(y^9 z^{15})^{1/3}$

5. Applications of Rational Exponents

Example 6   **Using Rational Exponents in an Application**

Classroom Example: p. 624, Exercise 84

Suppose P dollars in principal is invested in an account that earns interest annually. If after t years the investment grows to A dollars, then the annual rate of return, r, on the investment is given by

$$r = \left(\frac{A}{P}\right)^{1/t} - 1$$

Find the annual rate of return on \$8000 that grew to \$11,220.41 after 5 years (round to the nearest tenth of a percent).

Solution:

$r = \left(\dfrac{A}{P}\right)^{1/t} - 1$ where $A = \$11,220.41$, $P = \$8000$, and $t = 5$. Hence,

$r = \left(\dfrac{11,220.41}{8000}\right)^{1/5} - 1$

$ = (1.40255125)^{1/5} - 1$

$ \approx 1.069999927 - 1$

$ \approx 0.070$ or 7.0%

There is a 7.0% annual rate of return.

Skill Practice

21. The formula for finding the radius of a circle given the area is

$$r = \left(\frac{A}{\pi}\right)^{1/2}$$

Find the radius of a circle given that the area is 12.56 in.^{2} Use 3.14 for π.

Answers

17. a^2 **18.** $\dfrac{1}{t^{4/3}}$

19. $\dfrac{1}{w^4}$ **20.** $y^3 z^5$

21. The radius is 2 in.

Section 8.7   Practice Exercises

For additional exercises, see Classroom Activities 8.7A–8.7C in the *Student's Resource Manual* at www.mhhe.com/moh.

For the exercises in this set, assume that the variables represent positive real numbers unless otherwise stated.

Vocabulary and Key Concepts

1. **a.** If n is an integer greater than 1, then $a^{1/n} = \underline{\quad \sqrt[n]{a} \quad}$.

   b. If m and n are positive integers that share no common factors and $n > 1$, then in radical notation $a^{m/n} = \underline{\quad \sqrt[n]{a^m} \text{ or } (\sqrt[n]{a})^m \quad}$.

Review Exercises

2. Given $\sqrt[3]{125}$

   a. Identify the index. 3   **b.** Identify the radicand. 125

For Exercises 3–6, simplify the radicals.

3. $(\sqrt[4]{81})^3$ 27

4. $(\sqrt[4]{16})^3$ 8

5. $\sqrt[3]{(a+1)^3}$ $a+1$

6. $\sqrt[5]{(x+y)^5}$ $x+y$

Concept 1: Definition of $a^{1/n}$

For Exercises 7–18, simplify the expression. **(See Example 1.)**

7. $81^{1/2}$ 9

8. $25^{1/2}$ 5

9. $125^{1/3}$ 5

10. $8^{1/3}$ 2

11. $81^{1/4}$ 3

12. $16^{1/4}$ 2

13. $(-8)^{1/3}$ -2

14. $(-9)^{1/2}$ Not a real number

15. $-8^{1/3}$ -2

16. $-9^{1/2}$ -3

17. $36^{-1/2}$ $\dfrac{1}{6}$

18. $16^{-1/2}$ $\dfrac{1}{4}$

For Exercises 19–30, write the expressions in radical notation.

19. $x^{1/3}$ $\sqrt[3]{x}$

20. $y^{1/4}$ $\sqrt[4]{y}$

21. $(4a)^{1/2}$ $\sqrt{4a}$ or $2\sqrt{a}$

22. $(36x)^{1/2}$ $\sqrt{36x}$ or $6\sqrt{x}$

23. $(yz)^{1/5}$ $\sqrt[5]{yz}$

24. $(cd)^{1/4}$ $\sqrt[4]{cd}$

25. $(u^2)^{1/3}$ $\sqrt[3]{u^2}$

26. $(v^3)^{1/4}$ $\sqrt[4]{v^3}$

27. $5q^{1/2}$ $5\sqrt{q}$

28. $6p^{1/2}$ $6\sqrt{p}$

29. $\left(\dfrac{x}{9}\right)^{1/2}$ $\sqrt{\dfrac{x}{9}}$ or $\dfrac{\sqrt{x}}{3}$

30. $\left(\dfrac{y}{8}\right)^{1/3}$ $\sqrt[3]{\dfrac{y}{8}}$ or $\dfrac{\sqrt[3]{y}}{2}$

Concept 2: Definition of $a^{m/n}$

31. Explain how to interpret the expression $a^{m/n}$ as a radical. $a^{m/n} = \sqrt[n]{a^m}$ or $(\sqrt[n]{a})^m$, provided the roots exist.

32. Explain why $(\sqrt[3]{8})^4$ is easier to evaluate than $\sqrt[3]{8^4}$. It is easier to evaluate a cube root of a smaller number.

For Exercises 33–40, simplify. **(See Example 2.)**

33. $16^{3/4}$ 8

34. $32^{2/5}$ 4

35. $27^{-2/3}$ $\dfrac{1}{9}$

36. $4^{-5/2}$ $\dfrac{1}{32}$

37. $(-8)^{5/3}$ -32

38. $(-27)^{2/3}$ 9

39. $\left(\dfrac{1}{4}\right)^{-1/2}$ 2

40. $\left(\dfrac{1}{9}\right)^{3/2}$ $\dfrac{1}{27}$

/ Writing   ⬌ Translating Expression   ◿ Geometry   🖩 Scientific Calculator   ▶ Video

Concept 3: Converting between Rational Exponents and Radical Notation

For Exercises 41–48, convert each expression to radical notation. **(See Example 3.)**

41. $y^{9/2}$ $(\sqrt{y})^9$

42. $b^{4/9}$ $\sqrt[9]{b^4}$

43. $(c^5d)^{1/3}$ $\sqrt[3]{c^5d}$

44. $(a^2b)^{1/8}$ $\sqrt[8]{a^2b}$

45. $(qr)^{-1/5}$ $\dfrac{1}{\sqrt[5]{qr}}$

46. $(3x)^{-1/4}$ $\dfrac{1}{\sqrt[4]{3x}}$

47. $6y^{2/3}$ $6\sqrt[3]{y^2}$

48. $2q^{5/6}$ $2\sqrt[6]{q^5}$

For Exercises 49–56, write the expressions using rational exponents rather than radical notation. **(See Example 4.)**

49. $\sqrt[3]{y^2}$ $y^{2/3}$

50. $\sqrt[5]{b^2}$ $b^{2/5}$

51. $5\sqrt{x}$ $5x^{1/2}$

52. $7\sqrt[3]{z}$ $7z^{1/3}$

53. $\sqrt[3]{xy}$ $(xy)^{1/3}$

54. $\sqrt[5]{ab}$ $(ab)^{1/5}$

55. $\sqrt[4]{m^3n}$ $(m^3n)^{1/4}$

56. $\sqrt[5]{u^3v^4}$ $(u^3v^4)^{1/5}$

Concept 4: Properties of Rational Exponents

For Exercises 57–80, simplify the expressions using the properties of rational exponents. Write the final answers with positive exponents only. **(See Example 5.)**

57. $x^{1/4}x^{3/4}$ x

58. $2^{3/5}2^{2/5}$ 2

59. $(y^{1/5})^{10}$ y^2

60. $(x^{1/2})^8$ x^4

61. $6^{-1/5}6^{6/5}$ 6

62. $a^{-1/3}a^{2/3}$ $a^{1/3}$

63. $(a^{1/3}a^{1/4})^{12}$ a^7

64. $(x^{2/3}x^{1/2})^6$ x^7

65. $\dfrac{y^{5/3}}{y^{1/3}}$ $y^{4/3}$

66. $\dfrac{z^2}{z^{1/2}}$ $z^{3/2}$

67. $\dfrac{2^{4/3}}{2^{1/3}}$ 2

68. $\dfrac{5^{6/5}}{5^{1/5}}$ 5

69. $(x^{-2}y^{1/3})^{1/2}$ $\dfrac{y^{1/6}}{x}$

70. $(a^3b^{-4})^{1/3}$ $\dfrac{a}{b^{4/3}}$

71. $\left(\dfrac{w^{-2}}{z^{-4}}\right)^{-3/2}$ $\dfrac{w^3}{z^6}$

72. $\left(\dfrac{x^{-8}}{y^{-4}}\right)^{-1/4}$ $\dfrac{x^2}{y}$

73. $(5a^2c^{-1/2}d^{1/2})^2$ $\dfrac{25a^4d}{c}$

74. $(2x^{-1/3}y^2z^{5/3})^3$ $\dfrac{8y^6z^5}{x}$

75. $\left(\dfrac{x^{-2/3}}{y^{-3/4}}\right)^{12}$ $\dfrac{y^9}{x^8}$

76. $\left(\dfrac{m^{-1/4}}{n^{-1/2}}\right)^{-4}$ $\dfrac{m}{n^2}$

77. $\left(\dfrac{16w^{-2}z}{2wz^{-8}}\right)^{1/3}$ $\dfrac{2z^3}{w}$

78. $\left(\dfrac{50p^{-1}q}{2pq^{-3}}\right)^{1/2}$ $\dfrac{5q^2}{p}$

79. $(25x^2y^4z^3)^{1/2}$ $5xy^2z^{3/2}$

80. $(8a^6b^3c^2)^{2/3}$ $4a^4b^2c^{4/3}$

Concept 5: Applications of Rational Exponents

81. a. If the area, A, of a square is known, then the length of its sides, s, can be computed by the formula: $s = A^{1/2}$. Compute the length of the sides of a square having an area of 100 in.^{2} 10 in.

b. Compute the length of the sides of a square having an area of 72 in.^{2} Round your answer to the nearest 0.01 in. 8.49 in.

82. The radius, r, of a sphere of volume, V, is given by

$$r = \left(\frac{3V}{4\pi}\right)^{1/3}$$

Find the radius of a spherical ball having a volume of 55 in.^{3} Round your answer to the nearest 0.01 in. 2.36 in.

Writing ↔ Translating Expression Geometry Scientific Calculator Video

For Exercises 83–84, use the following information.

If P dollars in principal grows to A dollars after t years with annual interest, then the rate of return r is given by

$$r = \left(\frac{A}{P}\right)^{1/t} - 1$$

 83. a. In one account, $10,000 grows to $16,802 after 5 years. Compute the interest rate to the nearest tenth of a percent. **(See Example 6.)**
   10.9%

   b. In another account $10,000 grows to $18,000 after 7 years. Compute the interest rate to the nearest tenth of a percent.   8.8%

   c. Which account produced a higher average yearly return?   The account in part (a)

84. a. In one account, $5000 grows to $23,304.79 in 20 years. Compute the interest rate to the nearest whole percent.   8%

   b. In another account, $6000 grows to $34,460.95 in 30 years. Compute the interest rate to the nearest whole percent.   6%

   c. Which account produced a higher average yearly return?   The account in part (a)

Expanding Your Skills

85. Is $(a + b)^{1/2}$ the same as $a^{1/2} + b^{1/2}$? Explain why or why not by giving an example.
   No, for example, $(36 + 64)^{1/2} \neq 36^{1/2} + 64^{1/2}$

For Exercises 86–91, simplify the expressions. Write the final answers with positive exponents only.

86. $\left(\frac{1}{8}\right)^{2/3} + \left(\frac{1}{4}\right)^{1/2}$ $\frac{3}{4}$

87. $\left(\frac{1}{8}\right)^{-2/3} + \left(\frac{1}{4}\right)^{-1/2}$ 6

88. $\left(\frac{1}{16}\right)^{-1/4} - \left(\frac{1}{49}\right)^{-1/2}$ -5

89. $\left(\frac{1}{16}\right)^{1/4} - \left(\frac{1}{49}\right)^{1/2}$ $\frac{5}{14}$

90. $\left(\frac{x^2 y^{-1/3} z^{2/3}}{x^{2/3} y^{1/4} z}\right)^{12}$ $\frac{x^{16}}{y^7 z^4}$

91. $\left(\frac{a^2 b^{1/2} c^{-2}}{a^{-3/4} b^0 c^{1/8}}\right)^8$ $\frac{a^{22} b^4}{c^{17}}$

Group Activity

Calculating Standard Deviation

Materials: Pencil, paper, calculator

Estimated Time: 15 minutes

Group Size: 5 or 6

In statistics, the standard deviation of a set of data measures how much the data values differ from the mean of the values. A large standard deviation means that the values are more spread out. A smaller standard deviation means the values are more "clustered" around the mean.

Consider a set of data consisting of the following eight values: 6, 10, 13, 22, 30, 4, 12, 23

1. Calculate the mean (average) of the values.   The data points have a mean (or average) value of 15.

2. To calculate the standard deviation, first compute the difference of each data point from the mean, and square the result. The process is shown here:

$(6 - 15)^2 = 81$ $(10 - 15)^2 = 25$ $(13 - 15)^2 = 4$ $(22 - 15)^2 = 49$

$(30 - 15)^2 = 225$ $(4 - 15)^2 = 121$ $(12 - 15)^2 = 9$ $(23 - 15)^2 = 64$

3. Next divide the sum of these values by one less than the number of values. Then take the square root. The result is the standard deviation. Round to the nearest whole number.

$$\sqrt{\frac{81 + 25 + 4 + 49 + 225 + 121 + 9 + 64}{8 - 1}}$$

9; The standard deviation is approximately 9.

4. Next, find the standard deviation of the ages of the members of your group. First list the ages. Then follow steps 1–3. Round the standard deviation to the tenths place.   Answers will vary.

5. Compare the standard deviation of your group of ages to that of several other groups in the class. Which group has ages that are closest to its mean?   Answers will vary.

Chapter 8   Summary

Section 8.1   Introduction to Roots and Radicals

Key Concepts

b is a **square root** of a if $b^2 = a$.

The expression $\sqrt{a}$ represents the **principal square root** of a.

b is an nth-root of a if $b^n = a$.

1. If n is a positive *even* integer and $a > 0$, then $\sqrt[n]{a}$ is the principal (positive) nth-root of a.
2. If $n > 1$ is a positive *odd* integer, then $\sqrt[n]{a}$ is the nth-root of a.
3. If $n > 1$ is any positive integer, then $\sqrt[n]{0} = 0$.

$\sqrt[n]{a^n} = |a|$ if n is even.

$\sqrt[n]{a^n} = a$ if n is odd.

$\sqrt[n]{a}$ is not a real number if a is *negative* and n is even.

Pythagorean Theorem

The Pythagorean theorem states that the sum of the squares of the two legs of a right triangle equals the square of the hypotenuse.

$a^2 + b^2 = c^2$

Examples

Example 1

The square roots of 16 are 4 and -4 because $(4)^2 = 16$ and $(-4)^2 = 16$.

$\sqrt{16} = 4$ Because $4^2 = 16$

$\sqrt[3]{125} = 5$ Because $5^3 = 125$

$\sqrt[3]{-8} = -2$ Because $(-2)^3 = -8$

Example 2

$\sqrt{y^2} = |y|$ $\sqrt[3]{y^3} = y$

Example 3

$\sqrt{-16}$ is not a real number.

Example 4

Find the length of the unknown side.

$a^2 + b^2 = c^2$

$(8)^2 + b^2 = (17)^2$

$64 + b^2 = 289$

$b^2 = 225$

$b = \sqrt{225}$ Because b denotes a length, b must be the positive square root of 225.

$b = 15$

The third side is 15 cm.

b | $c = 17$ cm
$a = 8$ cm

Section 8.2 Simplifying Radicals

Key Concepts

Multiplication Property of Radicals

If $\sqrt[n]{a}$ and $\sqrt[n]{b}$ are both real, then

$$\sqrt[n]{ab} = \sqrt[n]{a} \cdot \sqrt[n]{b}$$

Simplifying Radicals

Consider a radical expression whose radicand is written as a product of prime factors. Then the radical is in simplified form if each of the following criteria are met:

1. The radicand has no factor raised to a power greater than or equal to the index.
2. There are no radicals in the denominator of a fraction.
3. The radicand does not contain a fraction.

Examples

Example 1

$$\sqrt{3} \cdot \sqrt{5} = \sqrt{3 \cdot 5} = \sqrt{15}$$

Example 2

$$\sqrt{\frac{b^7}{b^3}} = \sqrt{b^4} = b^2$$

Example 3

$$\sqrt[3]{16x^5y^7} = \sqrt[3]{8x^3y^6 \cdot 2x^2y}$$
$$= \sqrt[3]{8x^3y^6} \cdot \sqrt[3]{2x^2y}$$
$$= 2xy^2\sqrt[3]{2x^2y}$$

Section 8.3 Addition and Subtraction of Radicals

Key Concepts

Two radical terms are *like* radicals if they have the same index and the same radicand.

Use the distributive property to add or subtract *like* radicals.

Examples

Example 1

Like radicals. $\sqrt[3]{5z}, \quad 6\sqrt[3]{5z}$

Example 2

$$3\sqrt{7} - 10\sqrt{7} + \sqrt{7}$$
$$= (3 - 10 + 1)\sqrt{7}$$
$$= -6\sqrt{7}$$

Section 8.4	Multiplication of Radicals

Key Concepts

Multiplication Property of Radicals

$\sqrt[n]{a} \cdot \sqrt[n]{b} = \sqrt[n]{ab}$ provided $\sqrt[n]{a}$ and $\sqrt[n]{b}$ are both real.

Examples

Example 1

$$(6\sqrt{5})(4\sqrt{3}) = (6 \cdot 4)(\sqrt{5} \cdot \sqrt{3})$$
$$= 24\sqrt{15}$$

Example 2

$$3\sqrt{2}(\sqrt{2} + 5\sqrt{7} - \sqrt{6}) = 3\sqrt{4} + 15\sqrt{14} - 3\sqrt{12}$$
$$= 3\sqrt{4} + 15\sqrt{14} - 3\sqrt{4 \cdot 3}$$
$$= 3 \cdot 2 + 15\sqrt{14} - 3 \cdot 2\sqrt{3}$$
$$= 6 + 15\sqrt{14} - 6\sqrt{3}$$

Special Case Products

$(a + b)(a - b) = a^2 - b^2$

$(a + b)^2 = a^2 + 2ab + b^2$

$(a - b)^2 = a^2 - 2ab + b^2$

Example 3

$$(4\sqrt{x} + \sqrt{2})(4\sqrt{x} - \sqrt{2}) = (4\sqrt{x})^2 - (\sqrt{2})^2$$
$$= 16x - 2$$

Example 4

$$(\sqrt{x} - \sqrt{5y})^2 = (\sqrt{x})^2 - 2(\sqrt{x})(\sqrt{5y}) + (\sqrt{5y})^2$$
$$= x - 2\sqrt{5xy} + 5y$$

Section 8.5	Division of Radicals and Rationalization

Key Concepts

Division Property of Radicals

If $\sqrt[n]{a}$ and $\sqrt[n]{b}$ are both real, then

$$\sqrt[n]{\frac{a}{b}} = \frac{\sqrt[n]{a}}{\sqrt[n]{b}} \quad b \neq 0$$

Rationalizing the Denominator with One Term

Multiply the numerator and denominator by an appropriate expression to create an nth-root of an nth-power in the denominator.

Rationalizing a Two-Term Denominator Involving Square Roots

Multiply the numerator and denominator by the conjugate of the denominator.

Examples

Example 1

$$\sqrt{\frac{w}{16}} = \frac{\sqrt{w}}{\sqrt{16}} = \frac{\sqrt{w}}{4}$$

Example 2

$$\frac{10}{\sqrt{5}} = \frac{10}{\sqrt{5}} \cdot \frac{\sqrt{5}}{\sqrt{5}} = \frac{10\sqrt{5}}{\sqrt{5^2}} = \frac{10\sqrt{5}}{5} = 2\sqrt{5}$$

Example 3

$$\frac{\sqrt{2}}{\sqrt{x} - \sqrt{3}} = \frac{\sqrt{2}}{(\sqrt{x} - \sqrt{3})} \cdot \frac{(\sqrt{x} + \sqrt{3})}{(\sqrt{x} + \sqrt{3})}$$
$$= \frac{\sqrt{2x} + \sqrt{6}}{x - 3}$$

Section 8.6 Radical Equations

Key Concepts

An equation with one or more radicals containing a variable is a **radical equation**.

Steps for Solving a Radical Equation

1. Isolate the radical. If an equation has more than one radical, choose one of the radicals to isolate.
2. Raise each side of the equation to a power equal to the index of the radical.
3. Solve the resulting equation.
4. Check the potential solutions in the original equation.

Note: Raising both sides of an equation to an even power may result in extraneous solutions.

Examples

Example 1

Solve. $\sqrt{2x - 4} + 3 = 7$

Step 1:	$\sqrt{2x - 4} = 4$	Isolate the radical.
Step 2:	$(\sqrt{2x - 4})^2 = (4)^2$	Square both sides.
Step 3:	$2x - 4 = 16$	Solve the resulting equation.
	$2x = 20$	
	$x = 10$	

Step 4:

Check:

$$\sqrt{2x - 4} + 3 = 7$$

$$\sqrt{2(10) - 4} + 3 \stackrel{?}{=} 7$$

$$\sqrt{20 - 4} + 3 \stackrel{?}{=} 7$$

$$\sqrt{16} + 3 \stackrel{?}{=} 7$$

$$4 + 3 \stackrel{?}{=} 7 \checkmark \qquad \text{The solution checks.}$$

The solution set is $\{10\}$.

Section 8.7 Rational Exponents

Key Concepts

If $\sqrt[n]{a}$ is a real number, then

- $a^{1/n} = \sqrt[n]{a}$
- $a^{m/n} = (\sqrt[n]{a})^m = \sqrt[n]{a^m}$

Examples

Example 1

$$121^{1/2} = \sqrt{121} = 11$$

Example 2

$$27^{2/3} = (\sqrt[3]{27})^2 = (3)^2 = 9$$

Example 3

$$8^{-1/3} = \frac{1}{8^{1/3}} = \frac{1}{\sqrt[3]{8}} = \frac{1}{2}$$

Chapter 8   Review Exercises

Section 8.1

For Exercises 1–4, state the principal square root and the negative square root.

1. 196
Principal square root: 14;
negative square root: −14

2. 1.44
Principal square root: 1.2;
negative square root: −1.2

3. 0.64
Principal square root: 0.8;
negative square root: −0.8

4. 225
Principal square root: 15;
negative square root: −15

 5. Explain why $\sqrt{-64}$ is *not* a real number.
There is no real number b such that $b^2 = -64$.

6. Explain why $\sqrt[3]{-64}$ *is* a real number.
$\sqrt[3]{-64} = -4$ because $(-4)^3 = -64$.

 For Exercises 7–18, simplify the expressions, if possible. Assume all variables represent positive real numbers.

7. $-\sqrt{144}$ −12
8. $-\sqrt{25}$ −5
9. $\sqrt{-144}$
    Not a real number

10. $\sqrt{-25}$
Not a real number
11. $\sqrt{y^2}$ y
12. $\sqrt[3]{a^3}$ a

13. $\sqrt[3]{8p^3}$ $2p$
14. $-\sqrt[3]{125}$ −5
15. $-\sqrt[4]{625}$ −5

16. $\sqrt[3]{p^{12}}$ p^4
17. $\sqrt[3]{\dfrac{64}{t^6}}$ $\dfrac{4}{t^2}$
18. $\sqrt[3]{\dfrac{-27}{w^3}}$ $-\dfrac{3}{w}$

19. The radius, r, of a circle can be found from the area of the circle according to the formula:

$$r = \sqrt{\dfrac{A}{\pi}}$$

a. What is the radius of a circular garden whose area is 160 m²? Round to the nearest tenth of a meter.   7.1 m

b. What is the radius of a circular fountain whose area is 1600 ft²? Round to the nearest tenth of a foot.   22.6 ft

20. Suppose a ball is thrown with an initial velocity of 76 ft/sec at an angle of 30° (see figure). Then the horizontal position of the ball, x (measured in feet), depends on the number of seconds, t, after the ball is thrown according to the equation:

$$x = 38t\sqrt{3}$$

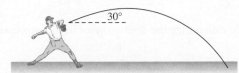

a. What is the horizontal position of the ball after 1 sec? Round your answer to the nearest tenth of a foot.   65.8 ft

b. What is the horizontal position of the ball after 2 sec? Round your answer to the nearest tenth of a foot.   131.6 ft

For Exercises 21–22, write the English phrases as algebraic expressions.

21. The square of b plus the principal square root of 5 $b^2 + \sqrt{5}$

22. The difference of the cube root of y and the fourth root of x $\sqrt[3]{y} - \sqrt[4]{x}$

For Exercises 23–24, write the algebraic expressions as English phrases. (Answers may vary.)

23. $\dfrac{2}{\sqrt{p}}$ The quotient of 2 and the principal square root of p

24. $8\sqrt{q}$ The product of 8 and the principal square root of q

25. A hedge extends 5 ft from the wall of a house. A 13-ft ladder is placed at the edge of the hedge. How far up the house is the tip of the ladder?
12 ft

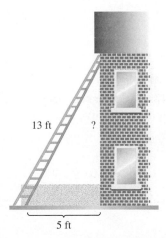

26. Nashville, Tennessee, is north of Birmingham, Alabama, a distance of 182 miles. Augusta, Georgia, is east of Birmingham, a distance of 277 miles. How far is it from Augusta to Nashville? Round the answer to the nearest mile.
331 mi

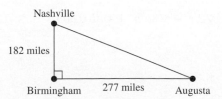

Section 8.2

For Exercises 27–32, use the multiplication property of radicals to simplify. Assume the variables represent positive real numbers.

27. $\sqrt{x^{17}}$ $x^8\sqrt{x}$ **28.** $\sqrt[3]{40}$ $2\sqrt[3]{5}$ **29.** $\sqrt{28}$ $2\sqrt{7}$

30. $5\sqrt{18x^3}$ **31.** $\sqrt[3]{27y^{10}}$ **32.** $2\sqrt{27y^{10}}$
$15x\sqrt{2x}$ $3y^3\sqrt[3]{y}$ $6y^5\sqrt{3}$

For Exercises 33–42, use order of operations to simplify. Assume the variables represent positive real numbers.

33. $\sqrt{\dfrac{c^5}{c^3}}$ c **34.** $\sqrt{\dfrac{t^9}{t^3}}$ t^3

35. $\sqrt{\dfrac{200y^5}{2y}}$ $10y^2$ **36.** $\sqrt{\dfrac{18x^3}{2x}}$ $3x$

37. $\sqrt[3]{\dfrac{48x^4}{6x}}$ $2x$ **38.** $\sqrt[3]{\dfrac{128a^{17}}{2a^2}}$ $4a^5$

39. $\dfrac{5\sqrt{12}}{2}$ $5\sqrt{3}$ **40.** $\dfrac{2\sqrt{45}}{6}$ $\sqrt{5}$

41. $\dfrac{12-\sqrt{49}}{5}$ 1 **42.** $\dfrac{20+\sqrt{100}}{5}$ 6

Section 8.3

For Exercises 43–50, add or subtract as indicated. Assume the variables represent positive real numbers.

43. $8\sqrt{6}-\sqrt{6}$ $7\sqrt{6}$

44. $1.6\sqrt{y}-1.4\sqrt{y}+0.6\sqrt{y}$ $0.8\sqrt{y}$

45. $x\sqrt{20}-2\sqrt{45x^2}$ $-4x\sqrt{5}$

46. $y\sqrt{64y}+3\sqrt{y^3}$ $11y\sqrt{y}$

47. $3\sqrt{75}-4\sqrt{28}+\sqrt{7}$ $15\sqrt{3}-7\sqrt{7}$

48. $2\sqrt{50}-4\sqrt{20}-6\sqrt{2}$ $4\sqrt{2}-8\sqrt{5}$

49. $7\sqrt{3x^9}-3x^4\sqrt{75x}$ $-8x^4\sqrt{3x}$

50. $3a^2\sqrt{2b^3}-\sqrt{8a^4b^3}+4a^2b\sqrt{50b}$ $21a^2b\sqrt{2b}$

51. Find the exact perimeter of the triangle. $12\sqrt{2}$ ft

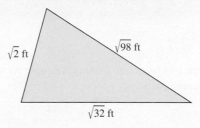

52. Find the exact perimeter of a square whose sides are $3\sqrt{48}$ m. $48\sqrt{3}$ m

Section 8.4

For Exercises 53–62, multiply the expressions. Assume the variables represent positive real numbers.

53. $\sqrt{5}\cdot\sqrt{125}$ 25 **54.** $\sqrt{10p}\cdot\sqrt{6}$
$2\sqrt{15p}$

55. $(5\sqrt{6})(7\sqrt{2x})$ $70\sqrt{3x}$ **56.** $(3\sqrt{y})(-2z\sqrt{11y})$
$-6yz\sqrt{11}$

57. $8\sqrt{m}(\sqrt{m}+3)$ **58.** $\sqrt{2}(\sqrt{7}+8)$
$8m+24\sqrt{m}$ $\sqrt{14}+8\sqrt{2}$

59. $(5\sqrt{2}+\sqrt{13})(-\sqrt{2}-3\sqrt{13})$
$-49-16\sqrt{26}$

60. $(\sqrt{p}+2\sqrt{q})(4\sqrt{p}-\sqrt{q})$
$4p+7\sqrt{pq}-2q$

61. $(8\sqrt{w}-\sqrt{z})(8\sqrt{w}+\sqrt{z})$ **62.** $(2x-\sqrt{y})^2$
$64w-z$ $4x^2-4x\sqrt{y}+y$

63. Find the exact volume of the box. $10\sqrt{3}$ m³

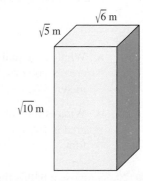

Section 8.5

For Exercises 64–67, use the division property of radicals to write the radicals in simplified form. Assume all variables are positive real numbers.

64. $\dfrac{\sqrt[3]{x^7}}{\sqrt[3]{x^4}}$ **65.** $\dfrac{\sqrt{a^{11}}}{\sqrt{a}}$ **66.** $\dfrac{\sqrt{250c}}{\sqrt{10}}$ **67.** $\dfrac{\sqrt{96y^3}}{\sqrt{6y^2}}$
x a^5 $5\sqrt{c}$ $4\sqrt{y}$

68. To rationalize the denominator in the expression

$$\frac{6}{\sqrt{a}+5}$$

which quantity would you multiply by in the numerator and denominator?   b

a. $\sqrt{a}+5$ **b.** $\sqrt{a}-5$ **c.** $\sqrt{a}$ **d.** -5

69. To rationalize the denominator in the expression

$$\frac{w}{\sqrt{w} - 4}$$

which quantity would you multiply by in the numerator and denominator?   b

a. $\sqrt{w} - 4$ **b.** $\sqrt{w} + 4$

c. $\sqrt{w}$ **d.** 4

For Exercises 70–75, rationalize the denominators. Assume the variables represent positive real numbers.

70. $\dfrac{11}{\sqrt{7}}$ $\dfrac{11\sqrt{7}}{7}$ **71.** $\sqrt{\dfrac{18}{y}}$ $\dfrac{3\sqrt{2y}}{y}$ **72.** $\dfrac{\sqrt{24}}{\sqrt{6x^7}}$ $\dfrac{2\sqrt{x}}{x^4}$

73. $\dfrac{10}{\sqrt{7} - \sqrt{2}}$ **74.** $\dfrac{6}{\sqrt{w} + 2}$ $\dfrac{6\sqrt{w} - 12}{w - 4}$ **75.** $\dfrac{\sqrt{7} + 3}{\sqrt{7} - 3}$

$2\sqrt{7} + 2\sqrt{2}$ $\qquad\qquad\qquad -8 - 3\sqrt{7}$

76. The velocity of an object, v (in meters per second: m/sec) depends on the kinetic energy, E (in joules: J), and mass, m (in kilograms: kg), of the object according to the formula:

$$v = \sqrt{\frac{2E}{m}}$$

a. What is the exact velocity of a 3-kg object whose kinetic energy is 100 J? $\dfrac{10\sqrt{6}}{3}$ m/sec

b. What is the exact velocity of a 5-kg object whose kinetic energy is 162 J? $\dfrac{18\sqrt{5}}{5}$ m/sec

Section 8.6

For Exercises 77–85, solve the equations. Be sure to check the potential solutions.

77. $\sqrt{p + 6} = 12$
{138}

78. $\sqrt{k + 1} = -7$
{ } (The value 48 does not check.)

79. $\sqrt{3x - 17} - 10 = 0$ {39}

80. $\sqrt{14n + 10} = 4\sqrt{n}$ {5}

81. $\sqrt{2z + 2} = \sqrt{3z - 5}$ {7}

82. $\sqrt{5y - 5} - \sqrt{4y + 1} = 0$ {6}

83. $\sqrt{2m + 5} = m + 1$ {2} (The value −2 does not check.)

84. $\sqrt{3n - 8} - n + 2 = 0$ {3, 4}

85. $\sqrt[3]{2y + 13} = -5$ {−69}

86. The length of the sides of a cube is related to the volume of the cube according to the formula: $x = \sqrt[3]{V}$.

a. What is the volume of the cube if the side length is 21 in.?   9261 in.^{3}

b. What is the volume of the cube if the side length is 15 cm?   3375 cm^3

Section 8.7

For Exercises 87–92, simplify the expressions.

87. $(-27)^{1/3}$ −3   **88.** $121^{1/2}$ 11   **89.** $-16^{1/4}$ −2

90. $(-16)^{1/4}$ **91.** $4^{-3/2}$ $\dfrac{1}{8}$ **92.** $\left(\dfrac{1}{9}\right)^{-3/2}$ 27
Not a real number

For Exercises 93–96, write the expression in radical notation. Assume the variables represent positive real numbers.

93. $z^{1/5}$ $\sqrt[5]{z}$ **94.** $q^{2/3}$ $\sqrt[3]{q^2}$

95. $(w^3)^{1/4}$ $\sqrt[4]{w^3}$ **96.** $\left(\dfrac{b}{121}\right)^{1/2}$

$\sqrt{\dfrac{b}{121}} = \dfrac{\sqrt{b}}{11}$

For Exercises 97–100, write the expression using rational exponents rather than radical notation. Assume the variables represent positive real numbers.

97. $\sqrt[5]{a^2}$ $a^{2/5}$ **98.** $5\sqrt[3]{m^2}$ $5m^{2/3}$

99. $\sqrt[5]{a^2 b^4}$ $(a^2 b^4)^{1/5}$ **100.** $\sqrt{6}$ $6^{1/2}$

For Exercises 101–106, simplify using the properties of rational exponents. Write the answer with positive exponents only. Assume the variables represent positive real numbers.

101. $y^{2/3} y^{4/3}$ y^2 **102.** $a^{1/3} a^{1/2}$ $a^{5/6}$

103. $\dfrac{6^{4/5}}{6^{1/5}}$ $6^{3/5}$ **104.** $\left(\dfrac{b^4 b^0}{b^{1/4}}\right)^4$ b^{15}

105. $(64a^3 b^6)^{1/3}$ $4ab^2$ **106.** $(5^{1/2})^{3/2}$ $5^{3/4}$

107. The radius, r, of a right circular cylinder can be found if the volume, V, and height, h, are known. The radius is given by

$$r = \left(\frac{V}{\pi h}\right)^{1/2}$$

Find the radius of a right circular cylinder whose volume is 150.8 cm^3 and whose height is 12 cm. Round the answer to the nearest tenth of a centimeter.   2.0 cm

Chapter 8 Test

1. For a right triangle with legs of lengths x and y, and a hypotenuse of length z, state the Pythagorean theorem. $x^2 + y^2 = z^2$

For Exercises 2–7, simplify the radicals, if possible. Assume the variables represent positive real numbers.

2. $\sqrt{242x^2}$ $11x\sqrt{2}$

3. $\sqrt[3]{48y^4}$ $2y\sqrt[3]{6y}$

4. $\sqrt{-64}$ Not a real number

5. $\sqrt{\dfrac{5a^6}{81}}$ $\dfrac{a^3\sqrt{5}}{9}$

6. $\dfrac{9}{\sqrt{6}}$ $\dfrac{3\sqrt{6}}{2}$

7. $\dfrac{2}{\sqrt{5}+6}$ $\dfrac{2\sqrt{5}-12}{-31}$ or $\dfrac{12-2\sqrt{5}}{31}$

8. Write the English phrases as algebraic expressions and simplify.

   a. The sum of the principal square root of twenty-five and the cube of five $\sqrt{25}+5^3$; 130

   b. The difference of the square of four and the principal square root of 16 $4^2-\sqrt{16}$; 12

9. A baseball player hits the ball at an angle of 30° with an initial velocity of 112 ft/sec. The horizontal position of the ball, x (measured in feet), depends on the number of seconds, t, after the ball is struck according to the equation:

$$x = 56t\sqrt{3}$$

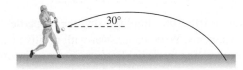

   What is the horizontal position of the ball after 1 sec? Round the answer to the nearest foot.   97 ft

For Exercises 10–19, perform the indicated operations. Assume the variables represent positive real numbers.

10. $6\sqrt{z} - 3\sqrt{z} + 5\sqrt{z}$ $8\sqrt{z}$

11. $\sqrt{3}(4\sqrt{2} - 5\sqrt{3})$ $4\sqrt{6} - 15$

12. $\sqrt{50t^2} - t\sqrt{288}$ $-7t\sqrt{2}$

13. $\sqrt{360} + \sqrt{250} - \sqrt{40}$ $9\sqrt{10}$

14. $(6\sqrt{2} - \sqrt{5})(\sqrt{2} + 4\sqrt{5})$ $-8 + 23\sqrt{10}$

15. $(3\sqrt{5} - 1)^2$ $46 - 6\sqrt{5}$

16. $\dfrac{\sqrt{2m^3n}}{\sqrt{72m^5}}$ $\dfrac{\sqrt{n}}{6m}$

17. $(4 - 3\sqrt{x})(4 + 3\sqrt{x})$ $16 - 9x$

18. $\sqrt{\dfrac{2}{11}}$ $\dfrac{\sqrt{22}}{11}$

19. $\dfrac{6}{\sqrt{7} - \sqrt{3}}$ $\dfrac{3\sqrt{7} + 3\sqrt{3}}{2}$

20. A triathlon consists of a swim, followed by a bike ride, followed by a run. The swim begins on a beach at point A. The swimmers must swim 50 yd to a buoy at point B, then 200 yd to a buoy at point C, and then return to point A on the beach. How far is the distance from point C to point A? (Round to the nearest yard.)   206 yd

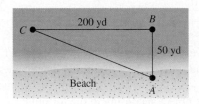

For Exercises 21–23, solve the equations.

21. $\sqrt{2x+7} + 6 = 2$ $\{\ \}$ (The value $\frac{9}{2}$ does not check.)

22. $\sqrt{1 - 7x} = 1 - x$ $\{0, -5\}$

23. $\sqrt[3]{x+6} = \sqrt[3]{2x-8}$ $\{14\}$

24. The height, y (in inches), of a tomato plant can be approximated by the time, t (in weeks), after the seed has germinated according to the equation:

$$y = 6\sqrt{t}$$

   a. Use the equation to find the height of the plant after 4 weeks.   12 in.

   b. Use the equation to find the time required for the plant to reach a height of 30 in. Verify your answer from the graph.   25 weeks

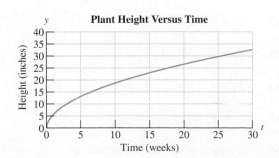

For Exercises 25–26, simplify the expression.

25. $10,000^{3/4}$ 1000

26. $\left(\dfrac{1}{8}\right)^{-1/3}$ 2

For Exercises 27–28, write the expressions in radical notation. Assume the variables represent positive real numbers.

27. $x^{3/5}$ $\sqrt[5]{x^3}$ or $(\sqrt[5]{x})^3$

28. $5y^{1/2}$ $5\sqrt{y}$

29. Write the expression using rational exponents: $\sqrt[4]{ab^3}$. (Assume $a \geq 0$ and $b \geq 0$.) $(ab^3)^{1/4}$

For Exercises 30–32, simplify using the properties of rational exponents. Write the final answer with positive exponents only. Assume the variables represent positive real numbers.

30. $p^{1/4} \cdot p^{2/3}$ $p^{11/12}$

31. $\dfrac{5^{4/5}}{5^{1/5}}$ $5^{3/5}$

32. $(9m^2n^4)^{1/2}$ $3mn^2$

Chapters 1–8   Cumulative Review Exercises

1. Simplify. $\dfrac{|-3 - 12 \div 6 + 2|}{\sqrt{5^2 - 4^2}}$ 1

2. Solve.
$2 - 5(2y + 4) - (-3y - 1) = -(y + 5)$ $\{-2\}$

3. Simplify. Write the final answer with positive exponents only.

$$\left(\frac{1}{3}\right)^0 - \left(\frac{1}{4}\right)^{-2}$$ -15

4. Perform the indicated operations.
$2(x - 3) - (3x + 4)(3x - 4)$ $-9x^2 + 2x + 10$

5. Divide. $\dfrac{14x^3y - 7x^2y^2 + 28xy^2}{7x^2y^2}$ $\dfrac{2x}{y} - 1 + \dfrac{4}{x}$

6. Factor completely. $50c^2 + 40c + 8$ $2(5c + 2)^2$

7. Solve. $10x^2 = x + 2$ $\left\{-\dfrac{2}{5}, \dfrac{1}{2}\right\}$

8. Perform the indicated operation.

$$\frac{5a^2 + 2ab - 3b^2}{10a + 10b} \div \frac{25a^2 - 9b^2}{50a + 30b}$$ 1

9. Solve for z. $\dfrac{1}{5} + \dfrac{z}{z - 5} = \dfrac{5}{z - 5}$
$\{\ \}$ (The value 5 does not check.)

10. Simplify.

$$\frac{\dfrac{5}{4} + \dfrac{2}{x}}{\dfrac{4}{x} - \dfrac{4}{x^2}}$$ $\dfrac{x(5x + 8)}{16(x - 1)}$

11. Graph. $3y = 6$

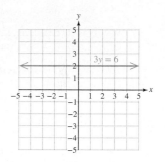

12. The equation $y = 210x + 250$ represents the cost, y (in dollars), of renting office space for x months.

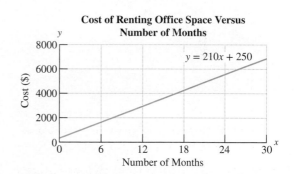

Cost of Renting Office Space Versus Number of Months

a. Find y when x is 3. Interpret the result in the context of the problem. $y = 880$; the cost of renting the office space for 3 months is $880.

b. Find x when y is $2770. Interpret the result in the context of the problem. $x = 12$; the cost of renting office space for 12 months is $2770.

c. What is the slope of the line? Interpret the meaning of the slope in the context of the problem. $m = 210$; the cost increases at a rate of $210 per month.

d. What is the y-intercept? Interpret the meaning of the y-intercept in the context of the problem. $(0, 250)$; the initial deposit to rent the office space is $250.

13. Write an equation of the line passing through the points $(2, -1)$ and $(-3, 4)$. Write the answer in slope-intercept form. $y = -x + 1$

14. Solve the system of equations using the addition method. If the system has no solution or infinitely many solutions, so state.

$$3x - 5y = 23$$
$$2x + 4y = -14$$ $\{(1, -4)\}$

15. Graph the solution to the inequality.
$-2x - y > 3$

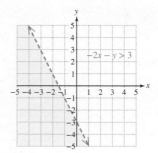

16. How many liters (L) of 20% acid solution must be mixed with a 50% acid solution to obtain 12 L of a 30% acid solution?
   8 L of 20% solution should be mixed with 4 L of 50% solution.

17. Simplify. $\sqrt{99}$ $3\sqrt{11}$

18. Perform the indicated operation.

$$5x\sqrt{3} + \sqrt{12x^2}$$ $7x\sqrt{3}$

19. Rationalize the denominator. $\dfrac{\sqrt{x}}{\sqrt{x} - \sqrt{y}}$ $\dfrac{x + \sqrt{xy}}{x - y}$

20. Solve. $\sqrt{2y - 1} - 4 = -1$ $\{5\}$

Quadratic Equations, Complex Numbers, and Functions

9

Mathematics in Communication

Imagine that every time you call your friend's cell phone your call is sent to the wrong device, and a different person answers the call. Suppose that one time you reach a single mother in Alabama, another time perhaps a small-business owner in Georgia, and the next time a person in Kansas.

This sort of haphazard **relationship** between a phone number and a target would create chaos and seriously compromise the future of your cell phone provider. Instead, when we dial 10 digits in our phone we know that the call will be routed to *only one* person's device—the person to whom the phone number is registered.

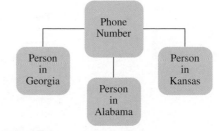

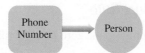

This example illustrates the importance of a **function**. That is, every item in a first set of items (in this case, a phone number being dialed) is associated with one and only one element in a second set of items (in this case, the phone of the proper recipient). Functions are relationships that take an input value and perform an operation on that value to produce a unique and predictable output value. The study of functions concludes this chapter and is the springboard for mathematics at the next level.

© koya979/Shutterstock RF

Section 9.1   The Square Root Property

Concepts

1. **Review of the Zero Product Rule**
2. **Solving Quadratic Equations Using the Square Root Property**

1. Review of the Zero Product Rule

We have learned that an equation that can be written in the form $ax^2 + bx + c = 0$, where $a \neq 0$, is a quadratic equation. One method to solve a quadratic equation is to factor the equation and apply the zero product rule. Recall that the zero product rule states that if $a \cdot b = 0$, then $a = 0$ or $b = 0$. This is reviewed in Examples 1–3.

Classroom Example: p. 640, Exercise 6

Example 1   Solving a Quadratic Equation Using the Zero Product Rule

Solve the equation by factoring and applying the zero product rule.

$$2x^2 - 7x - 30 = 0$$

Solution:

$$2x^2 - 7x - 30 = 0 \qquad \text{The equation is in the form } ax^2 + bx + c = 0.$$

$$(2x + 5)(x - 6) = 0 \qquad \text{Factor.}$$

$$2x + 5 = 0 \quad \text{or} \quad x - 6 = 0 \qquad \text{Set each factor equal to zero.}$$

$$2x = -5 \quad \text{or} \quad x = 6 \qquad \text{Solve the resulting equations.}$$

$$x = -\frac{5}{2}$$

The solution set is $\left\{ -\dfrac{5}{2}, 6 \right\}$.

Skill Practice Solve the quadratic equation by using the zero product rule.

1. $2x^2 + 3x - 20 = 0$

Classroom Example: p. 640, Exercise 10

Example 2   Solving a Quadratic Equation Using the Zero Product Rule

Solve the equation by factoring and applying the zero product rule.

$$2x(x + 4) = x^2 - 15$$

Solution:

$$2x(x + 4) = x^2 - 15$$

$$2x^2 + 8x = x^2 - 15 \qquad \text{Clear parentheses and combine like terms.}$$

$$x^2 + 8x + 15 = 0 \qquad \text{Set one side of the equation equal to zero. The equation is now in the form } ax^2 + bx + c = 0.$$

$$(x + 5)(x + 3) = 0 \qquad \text{Factor.}$$

$$x + 5 = 0 \quad \text{or} \quad x + 3 = 0 \qquad \text{Set each factor equal to zero.}$$

$$x = -5 \quad \text{or} \quad x = -3 \qquad \text{Solve each equation.}$$

The solution set is $\{-5, -3\}$.

Answer

1. $\left\{ -4, \dfrac{5}{2} \right\}$

TIP: The solutions to an equation can be checked in the original equation.

Check: $x = -5$

$$2x(x + 4) = x^2 - 15$$

$$2(-5)(-5 + 4) \overset{?}{=} (-5)^2 - 15$$

$$-10(-1) \overset{?}{=} 25 - 15$$

$$10 \overset{?}{=} 10 \checkmark$$

Check: $x = -3$

$$2x(x + 4) = x^2 - 15$$

$$2(-3)(-3 + 4) \overset{?}{=} (-3)^2 - 15$$

$$-6(1) \overset{?}{=} 9 - 15$$

$$-6 \overset{?}{=} -6 \checkmark$$

Skill Practice Solve the quadratic equation by using the zero product rule.

2. $y(y - 1) = 2y + 10$

Example 3   **Solving a Quadratic Equation Using the Zero Product Rule**

Classroom Example: p. 640, Exercise 12

Solve the equation by factoring and applying the zero product rule.

$$x^2 = 25$$

Solution:

$$x^2 = 25$$

$$x^2 - 25 = 0 \qquad\qquad \text{Set one side of the equation equal to zero.}$$

$$(x - 5)(x + 5) = 0 \qquad\qquad \text{Factor.}$$

$$x - 5 = 0 \quad \text{or} \quad x + 5 = 0 \qquad \text{Set each factor equal to zero.}$$

$$x = 5 \quad \text{or} \qquad x = -5$$

The solution set is $\{5, -5\}$.

Skill Practice Solve the quadratic equation by using the zero product rule.

3. $t^2 = 49$

---

2. Solving Quadratic Equations Using the Square Root Property

In Examples 1–3, the quadratic equations were all factorable. In this chapter, we learn techniques to solve *all* quadratic equations, factorable and nonfactorable. The first technique uses the **square root property**.

Square Root Property

For any real number, k, if $x^2 = k$, then $x = \pm\sqrt{k}$.
The solution set is $\{\sqrt{k}, -\sqrt{k}\}$.

Note: The expression $\pm\sqrt{k}$ is read as "plus or minus the square root of k."

Answers

2. $\{5, -2\}$

3. $\{7, -7\}$

Classroom Example: p. 640, Exercise 22

Example 4   Solving a Quadratic Equation Using the Square Root Property

Use the square root property to solve the equation.

$$x^2 = 25$$

Solution:

$x^2 = 25$	The equation is in the form $x^2 = k$.
$x = \pm\sqrt{25}$	Apply the square root property.
$x = \pm 5$	

The solution set is $\{5, -5\}$. Note that this result is the same as in Example 3.

Skill Practice   Use the square root property to solve the equation.

4. $c^2 = 64$

Classroom Example: p. 640, Exercise 28

Example 5   Solving a Quadratic Equation Using the Square Root Property

Use the square root property to solve the equation.

$$2x^2 - 10 = 0$$

Solution:

$2x^2 - 10 = 0$	To apply the square root property, the equation must be in the form $x^2 = k$, that is, we must isolate x^2.
$2x^2 = 10$	Add 10 to both sides.
$x^2 = 5$	Divide both sides by 2. The equation is in the form $x^2 = k$.
$x = \pm\sqrt{5}$	Apply the square root property.

Avoiding Mistakes

Remember to use the $\pm$ symbol when applying the square root property.

Check: $x = \sqrt{5}$

$$2x^2 - 10 = 0$$
$$2(\sqrt{5})^2 - 10 \overset{?}{=} 0$$
$$2(5) - 10 \overset{?}{=} 0$$
$$10 - 10 \overset{?}{=} 0 \checkmark$$

Check: $x = -\sqrt{5}$

$$2x^2 - 10 = 0$$
$$2(-\sqrt{5})^2 - 10 \overset{?}{=} 0$$
$$2(5) - 10 \overset{?}{=} 0$$
$$10 - 10 \overset{?}{=} 0 \checkmark$$

The solution set is $\{\sqrt{5}, -\sqrt{5}\}$.

Skill Practice   Use the square root property to solve the equation.

5. $3x^2 - 36 = 0$

Answers

4. $\{8, -8\}$ **5.** $\{2\sqrt{3}, -2\sqrt{3}\}$

Example 6	Solving a Quadratic Equation Using the Square Root Property

Classroom Example: p. 640, Exercis 36

Use the square root property to solve the equation.

$$(t - 4)^2 = 12$$

Solution:

$(t - 4)^2 = 12$ The equation is in the form $x^2 = k$, where $x = (t - 4)$.

$t - 4 = \pm\sqrt{12}$ Apply the square root property.

$t - 4 = \pm\sqrt{4 \cdot 3}$ Simplify the radical.

$t - 4 = \pm 2\sqrt{3}$

$t = 4 \pm 2\sqrt{3}$ Solve for t.

Check: $t = 4 + 2\sqrt{3}$ Check: $t = 4 - 2\sqrt{3}$

$(t - 4)^2 = 12$ $(t - 4)^2 = 12$

$(4 + 2\sqrt{3} - 4)^2 \overset{?}{=} 12$ $(4 - 2\sqrt{3} - 4)^2 \overset{?}{=} 12$

$(2\sqrt{3})^2 \overset{?}{=} 12$ $(-2\sqrt{3})^2 \overset{?}{=} 12$

$4 \cdot 3 \overset{?}{=} 12$ $4 \cdot 3 \overset{?}{=} 12$

$12 \overset{?}{=} 12 \checkmark$ $12 \overset{?}{=} 12 \checkmark$

The solution set is $\{4 \pm 2\sqrt{3}\}$.

Skill Practice Use the square root property to solve the equation.

6. $(p + 3)^2 = 8$

Example 7	Solving a Quadratic Equation Using the Square Root Property

Classroom Example: p. 640, Exercise 42

Use the square root property to solve the equation.

$$y^2 = -4$$

Solution:

$y^2 = -4$ The equation is in the form $y^2 = k$.

$y = \pm\sqrt{-4}$

The expression $\sqrt{-4}$ is not a real number. Thus, the equation, $y^2 = -4$, has no real-valued solutions.

Skill Practice Use the square root property to solve the equation.

7. $z^2 = -9$

Answers

6. $\{-3 \pm 2\sqrt{2}\}$

7. The equation has no real-valued solutions.

Section 9.1 Practice Exercises

For additional exercises, see Classroom Activity 9.1A in the *Student's Resource Manual* at www.mhhe.com/moh.

Vocabulary and Key Concepts

1. **a.** The zero product rule states that if $ab = 0$, then $a = $ _____0_____ or $b = $ _____0_____.

   b. To apply the zero product rule, one side of the equation must be equal to _____0_____ and the other side must be written in factored form.

   c. The square root property states that for any real number k, if $x^2 = k$, then $x = $ _____$\sqrt{k}$_____ or $x = $ _____$-\sqrt{k}$_____.

   d. To apply the square root property to the equation $x^2 + 4 = 13$, first subtract _____4_____ from both sides. The solution set is _____$\{3, -3\}$_____.

Concept 1: Review of the Zero Product Rule

2. Identify the equations as linear or quadratic.

   a. $2x - 5 = 3(x + 2) - 1$ Linear

   b. $2x(x - 5) = 3(x + 2) - 1$ Quadratic

   c. $ax^2 + bx + c = 0$
   (a, b, and c are real numbers, and $a \neq 0$)
   Quadratic

3. Identify the equations as linear or quadratic.

   a. $ax + b = c$
   (a, b, and c are real numbers, and $a \neq 0$)
   Linear

   b. $\frac{1}{2}p - \frac{3}{4}p^2 = 0$ Quadratic

   c. $\frac{1}{2}(p - 3) = 5$ Linear

For Exercises 4–19, solve using the zero product rule. **(See Examples 1–3.)**

4. $(3z - 2)(4z + 5) = 0$ $\left\{\frac{2}{3}, -\frac{5}{4}\right\}$

5. $(t + 5)(2t - 1) = 0$ $\left\{-5, \frac{1}{2}\right\}$

6. $r^2 + 7r + 12 = 0$ $\{-4, -3\}$

7. $y^2 - 2y - 35 = 0$ $\{7, -5\}$

8. $10x^2 = 13x - 4$ $\left\{\frac{1}{2}, \frac{4}{5}\right\}$

9. $6p^2 = -13p - 2$ $\left\{-2, -\frac{1}{6}\right\}$

10. $2m(m - 1) = 3m - 3$ $\left\{\frac{3}{2}, 1\right\}$

11. $2x^2 + 10x = -7(x + 3)$ $\left\{-7, -\frac{3}{2}\right\}$

12. $x^2 = 4$ $\{2, -2\}$

13. $c^2 = 144$ $\{12, -12\}$

14. $(x - 1)^2 = 16$ $\{-3, 5\}$

15. $(x - 3)^2 = 25$ $\{8, -2\}$

16. $3p^2 + 4p = 15$ $\left\{\frac{5}{3}, -3\right\}$

17. $4a^2 + 7a = 2$ $\left\{\frac{1}{4}, -2\right\}$

18. $(x + 2)(x + 3) = 2$ $\{-4, -1\}$

19. $(x + 2)(x + 6) = 5$ $\{-1, -7\}$

Concept 2: Solving Quadratic Equations Using the Square Root Property

20. The symbol "$\pm$" is read as . . .   Plus or minus

For Exercises 21–44, solve the equations using the square root property. **(See Examples 4–7.)**

21. $x^2 = 49$ $\{7, -7\}$

22. $x^2 = 16$ $\{4, -4\}$

23. $k^2 - 100 = 0$ $\{10, -10\}$

24. $m^2 - 64 = 0$ $\{8, -8\}$

25. $p^2 = -24$ There are no real-valued solutions.

26. $q^2 = -50$ There are no real-valued solutions.

27. $3w^2 - 9 = 0$ $\{\sqrt{3}, -\sqrt{3}\}$

28. $4v^2 - 24 = 0$ $\{\sqrt{6}, -\sqrt{6}\}$

29. $(a - 5)^2 = 16$ $\{9, 1\}$

30. $(b + 3)^2 = 1$ $\{-2, -4\}$

31. $(y - 5)^2 = 36$ $\{11, -1\}$

32. $(y + 4)^2 = 4$ $\{-2, -6\}$

33. $(x - 11)^2 = 5$ $\{11 \pm \sqrt{5}\}$

34. $(z - 2)^2 = 7$ $\{2 \pm \sqrt{7}\}$

35. $(a + 1)^2 = 18$ $\{-1 \pm 3\sqrt{2}\}$

36. $(b - 1)^2 = 12$ $\{1 \pm 2\sqrt{3}\}$

37. $\left(t - \frac{1}{4}\right)^2 = \frac{7}{16}$ $\left\{\frac{1}{4} \pm \frac{\sqrt{7}}{4}\right\}$

38. $\left(t - \frac{1}{3}\right)^2 = \frac{1}{9}$ $\left\{\frac{2}{3}, 0\right\}$

39. $\left(x - \frac{1}{2}\right)^2 + 5 = 20$ $\left\{\frac{1}{2} \pm \sqrt{15}\right\}$

40. $\left(x + \frac{5}{2}\right)^2 - 3 = 18$ $\left\{-\frac{5}{2} \pm \sqrt{21}\right\}$

41. $(p - 3)^2 = -16$ There are no real-valued solutions.

42. $(t + 4)^2 = -9$ There are no real-valued solutions.

43. $12t^2 = 75$ $\left\{\frac{5}{2}, -\frac{5}{2}\right\}$

44. $8p^2 = 18$ $\left\{\frac{3}{2}, -\frac{3}{2}\right\}$

45. Check the solution $-3 + \sqrt{5}$ in the equation $(x + 3)^2 = 5$.   The solution checks.

46. Check the solution $-5 - \sqrt{7}$ in the equation $(p + 5)^2 = 7$.   The solution checks.

For Exercises 47–48, answer true or false. If a statement is false, explain why.

47. The only solution to the equation $x^2 = 64$ is 8.
False. −8 is also a solution.

48. There are two real solutions to every quadratic equation of the form $x^2 = k$, where $k \geq 0$ is a real number.   False. If $k = 0$, there is only one solution.

 49. Ignoring air resistance, the distance, d (in feet), that an object drops in t seconds is given by the equation

$$d = 16t^2$$

  a. Find the distance traveled in 2 sec.   64 ft

  b. Find the time required for the object to fall 200 ft. Round to the nearest tenth of a second.
    3.5 sec

  c. Find the time required for an object to fall from the top of the Empire State Building in New York City if the building is 1250 ft high. Round to the nearest tenth of a second.   8.8 sec

 50. Ignoring air resistance, the distance, d (in meters), that an object drops in t seconds is given by the equation

$$d = 4.9t^2$$

  a. Find the distance traveled in 5 sec.   122.5 m

  b. Find the time required for the object to fall 50 m. Round to the nearest tenth of a second.
    3.2 sec

  c. Find the time required for an object to fall from the top of the TD Canada Trust Tower in Toronto, Canada, if the building is 261 m high. Round to the nearest tenth of a second.   7.3 sec

 51. A right triangle has legs of equal length. If the hypotenuse is 10 m long, find the length (in meters) of each leg. Round the answer to the nearest tenth of a meter.   7.1 m

 52. The diagonal of a square computer monitor screen is 24 in. long. Find the length of the sides to the nearest tenth of an inch.   17.0 in.

 53. The area of a circular wading pool is approximately 200 ft². Find the radius to the nearest tenth of a foot.   8.0 ft

© Doug Menuez/Getty Images RF

 54. According to the International Swimming Federation, the volume of an eight-lane Olympic size pool should be 2500 m³. The length of the pool is twice the width, and the depth is 2 m. Use a calculator to find the length and width of the pool.   The length is 50 m and the width is 25 m.

© Ryan McVay/Getty Images RF

Concepts

1. Completing the Square
2. Solving Quadratic Equations by Completing the Square

1. Completing the Square

In an earlier example, we used the square root property to solve an equation in which the square of a binomial was equal to a constant.

$$\underbrace{(t-4)^2}_{\substack{\text{Square of a} \\ \text{binomial}}} = \overset{\displaystyle \uparrow}{\underset{\text{Constant}}{12}}$$

Furthermore, any equation $ax^2 + bx + c = 0$ $(a \neq 0)$ can be rewritten as the square of a binomial equal to a constant by using a process called **completing the square**.

We begin our discussion of completing the square with some vocabulary. For a trinomial $ax^2 + bx + c$ $(a \neq 0)$, the term ax^2 is called the **quadratic term**. The term bx is called the **linear term**, and the term c is called the **constant term**.

Next, notice that the square of a binomial is the factored form of a perfect square trinomial.

Perfect Square Trinomial   Factored Form

$$x^2 + 10x + 25 \longrightarrow (x+5)^2$$
$$t^2 - 6t + 9 \longrightarrow (t-3)^2$$
$$p^2 - 14p + 49 \longrightarrow (p-7)^2$$

Furthermore, for a perfect square trinomial with a leading coefficient of 1, the constant term is the square of half the coefficient of the linear term. For example:

$$x^2 + 10x + 25 \qquad t^2 - 6t + 9 \qquad p^2 - 14p + 49$$
$$\left[\tfrac{1}{2}(10)\right]^2 = [5]^2 = 25 \qquad \left[\tfrac{1}{2}(-6)\right]^2 = [-3]^2 = 9 \qquad \left[\tfrac{1}{2}(-14)\right]^2 = [-7]^2 = 49$$

In general, an expression of the form $x^2 + bx$ will result in a perfect square trinomial if the square of half the linear term coefficient, $\left(\tfrac{1}{2}b\right)^2$, is added to the expression.

Classroom Examples: p. 646, Exercises 6, 14, and 16

Example 1   Completing the Square

Determine the value of n that makes the polynomial a perfect square trinomial. Then factor the expression as the square of a binomial.

a. $x^2 + 12x + n$ **b.** $x^2 - 22x + n$ **c.** $x^2 + 5x + n$ **d.** $x^2 - \dfrac{3}{5}x + n$

Solution:

The expressions are in the form $x^2 + bx$. Add the square of half the linear term coefficient, $\left(\tfrac{1}{2}b\right)^2$.

a. $x^2 + 12x + n$

$x^2 + 12x + 36 \qquad n = \left[\tfrac{1}{2}(12)\right]^2 = (6)^2 = 36$

$(x+6)^2 \qquad$ Factored form

b. $x^2 - 22x + n$

$x^2 - 22x + 121 \qquad n = \left[\frac{1}{2}(-22)\right]^2 = (-11)^2 = 121$

$(x - 11)^2 \qquad$ Factored form

c. $x^2 + 5x + n$

$x^2 + 5x + \frac{25}{4} \qquad n = \left[\frac{1}{2}(5)\right]^2 = \left(\frac{5}{2}\right)^2 = \frac{25}{4}$

$\left(x + \frac{5}{2}\right)^2 \qquad$ Factored form

d. $x^2 - \frac{3}{5}x + n$

$x^2 - \frac{3}{5}x + \frac{9}{100} \qquad n = \left[\frac{1}{2}\left(-\frac{3}{5}\right)\right]^2 = \left(-\frac{3}{10}\right)^2 = \frac{9}{100}$

$\left(x - \frac{3}{10}\right)^2 \qquad$ Factored form

Skill Practice Determine the value of n that makes the polynomial a perfect square trinomial. Then factor the expression as the square of a binomial.

1. $q^2 + 8q + n$ **2.** $t^2 - 10t + n$

3. $v^2 + 3v + n$ **4.** $y^2 + \frac{1}{4}y + n$

2. Solving Quadratic Equations by Completing the Square

A quadratic equation can be solved by completing the square and applying the square root property. The following steps outline the procedure.

Solving a Quadratic Equation in the Form $ax^2 + bx + c = 0$ $(a \neq 0)$ by Completing the Square and Applying the Square Root Property

Step 1 Divide both sides by a to make the leading coefficient 1.

Step 2 Isolate the variable terms on one side of the equation.

Step 3 Complete the square by adding the square of one-half the linear term coefficient to both sides of the equation. Then factor the resulting perfect square trinomial.

Step 4 Apply the square root property, and solve for x.

Answers

1. $n = 16; (q + 4)^2$

2. $n = 25; (t - 5)^2$

3. $n = \frac{9}{4}; \left(v + \frac{3}{2}\right)^2$

4. $n = \frac{1}{64}; \left(y + \frac{1}{8}\right)^2$

Classroom Example: p. 646, Exercise 22

Example 2 Solving a Quadratic Equation by Completing the Square and Applying the Square Root Property

Solve the quadratic equation by completing the square and applying the square root property.

$$x^2 + 6x - 8 = 0$$

Solution:

$x^2 + 6x - 8 = 0$	The equation is in the form $ax^2 + bx + c = 0$.
	Step 1: The leading coefficient is already 1.
$x^2 + 6x = 8$	**Step 2:** Isolate the variable terms on one side.
$x^2 + 6x + 9 = 8 + 9$	**Step 3:** To complete the square, add $\left[\frac{1}{2}(6)\right]^2 = (3)^2 = 9$ to both sides.
$(x + 3)^2 = 17$	Factor the perfect square trinomial.
$x + 3 = \pm\sqrt{17}$	**Step 4:** Apply the square root property.
$x = -3 \pm \sqrt{17}$	Solve for x.

The solution set is $\{-3 \pm \sqrt{17}\}$.

Skill Practice Solve the equation by completing the square and applying the square root property.

5. $t^2 + 4t + 2 = 0$

Classroom Example: p. 647, Exercise 26

Example 3 Solving a Quadratic Equation by Completing the Square and Applying the Square Root Property

Solve the quadratic equation by completing the square and applying the square root property.

$$2x^2 - 16x - 24 = 0$$

Solution:

$2x^2 - 16x - 24 = 0$	The equation is in the form $ax^2 + bx + c = 0$.
$\dfrac{2x^2}{2} - \dfrac{16x}{2} - \dfrac{24}{2} = \dfrac{0}{2}$	**Step 1:** Divide both sides by the leading coefficient, 2.
$x^2 - 8x - 12 = 0$	
$x^2 - 8x = 12$	**Step 2:** Isolate the variable terms on one side.
$x^2 - 8x + 16 = 12 + 16$	**Step 3:** To complete the square, add $\left[\frac{1}{2}(-8)\right]^2 = 16$ to both sides of the equation.
$(x - 4)^2 = 28$	Factor the perfect square trinomial.

Answers

5. $\{-2 \pm \sqrt{2}\}$

$$x - 4 = \pm\sqrt{28}$$ **Step 4:** Apply the square root property.

$$x - 4 = \pm 2\sqrt{7}$$ Simplify the radical.

$$x = 4 \pm 2\sqrt{7}$$ Solve for x.

The solution set is $\{4 \pm 2\sqrt{7}\}$.

Skill Practice Solve the equation by completing the square and applying the square root property.

6. $3y^2 - 6y - 51 = 0$

| **Example 4** | **Solving a Quadratic Equation by Completing the Square and Applying the Square Root Property** |

Classroom Example: p. 647, Exercise 36

Solve the quadratic equation by completing the square and applying the square root property.

$$x(2x - 5) - 3 = 0$$

Solution:

$$x(2x - 5) - 3 = 0$$ Clear parentheses.

$$2x^2 - 5x - 3 = 0$$ The equation is in the form $ax^2 + bx + c = 0$.

$$\frac{2x^2}{2} - \frac{5x}{2} - \frac{3}{2} = \frac{0}{2}$$ **Step 1:** Divide both sides by the leading coefficient, 2.

$$x^2 - \frac{5}{2}x - \frac{3}{2} = 0$$

$$x^2 - \frac{5}{2}x = \frac{3}{2}$$ **Step 2:** Isolate the variable terms on one side.

$$x^2 - \frac{5}{2}x + \frac{25}{16} = \frac{3}{2} + \frac{25}{16}$$ **Step 3:** Add $\left[\frac{1}{2}\left(-\frac{5}{2}\right)\right]^2 = \left(-\frac{5}{4}\right)^2 = \frac{25}{16}$ to both sides.

$$\left(x - \frac{5}{4}\right)^2 = \frac{24}{16} + \frac{25}{16}$$ Factor the perfect square trinomial. Rewrite the right-hand side with a common denominator and simplify.

$$\left(x - \frac{5}{4}\right)^2 = \frac{49}{16}$$

$$x - \frac{5}{4} = \pm\sqrt{\frac{49}{16}}$$ **Step 4:** Apply the square root property.

$$x - \frac{5}{4} = \pm\frac{7}{4}$$ Simplify the radical.

$$x = \frac{5}{4} \pm \frac{7}{4}$$ Solve for x.

Answers

6. $\{1 \pm 3\sqrt{2}\}$

We have

$$x = \begin{cases} \dfrac{5}{4} + \dfrac{7}{4} = \dfrac{12}{4} = 3 \\[2mm] \dfrac{5}{4} - \dfrac{7}{4} = -\dfrac{2}{4} = -\dfrac{1}{2} \end{cases}$$

The solution set is $\left\{ 3, -\dfrac{1}{2} \right\}$.

Skill Practice Solve the equation by completing the square and applying the square root property.

7. $5x(x + 2) = 6 + 3x$

Answers

7. $\left\{ \dfrac{3}{5}, -2 \right\}$

Section 9.2   Practice Exercises

Vocabulary and Key Concepts

For additional exercises, see Classroom Activity 9.2A in the *Student's Resource Manual* at www.mhhe.com/moh.

1. a. The process to create a perfect square trinomial is called _____completing_____ the square.

b. Fill in the blank to complete the square for the trinomial $x^2 + 20x + $ _____100_____.

c. To use completing the square to solve the equation $5x^2 + 3x + 1 = 0$, the first step is to divide both sides of the equation by _____5_____ so that the coefficient on the x^2 term is _____1_____.

d. Given the trinomial $y^2 + 8y + 16$, the coefficient of the linear term is _____8_____.

Review Exercises

For Exercises 2–4, solve each quadratic equation using the square root property.

2. $x^2 = 21$ $\{\sqrt{21}, -\sqrt{21}\}$

3. $(x - 5)^2 = 21$ $\{5 \pm \sqrt{21}\}$

4. $(x - 5)^2 = -21$
There are no real-valued solutions.

Concept 1: Completing the Square

For Exercises 5–16, find the value of n so that the expression is a perfect square trinomial. Then factor the trinomial.
(See Example 1.)

5. $y^2 + 4y + n$
$n = 4; (y + 2)^2$

6. $w^2 - 6w + n$
$n = 9; (w - 3)^2$

7. $p^2 - 12p + n$
$n = 36; (p - 6)^2$

8. $q^2 + 16q + n$
$n = 64; (q + 8)^2$

9. $x^2 - 9x + n$
$n = \dfrac{81}{4}; \left(x - \dfrac{9}{2}\right)^2$

 10. $a^2 - 5a + n$
$n = \dfrac{25}{4}; \left(a - \dfrac{5}{2}\right)^2$

11. $d^2 + \dfrac{5}{3}d + n$
$n = \dfrac{25}{36}; \left(d + \dfrac{5}{6}\right)^2$

12. $t^2 + \dfrac{1}{4}t + n$
$n = \dfrac{1}{64}; \left(t + \dfrac{1}{8}\right)^2$

13. $m^2 - \dfrac{1}{5}m + n$
$n = \dfrac{1}{100}; \left(m - \dfrac{1}{10}\right)^2$

14. $x^2 - \dfrac{5}{7}x + n$
$n = \dfrac{25}{196}; \left(x - \dfrac{5}{14}\right)^2$

15. $u^2 + u + n$
$n = \dfrac{1}{4}; \left(u + \dfrac{1}{2}\right)^2$

16. $v^2 - v + n$
$n = \dfrac{1}{4}; \left(v - \dfrac{1}{2}\right)^2$

Concept 2: Solving Quadratic Equations by Completing the Square

For Exercises 17–36, solve each equation by completing the square and applying the square root property.
(See Examples 2–4.)

17. $x^2 + 4x = 12$
$\{2, -6\}$

18. $x^2 - 2x = 8$
$\{4, -2\}$

19. $y^2 + 6y = -5$
$\{-1, -5\}$

20. $t^2 + 10t = 11$
$\{1, -11\}$

21. $x^2 = 2x + 1$
$\{1 \pm \sqrt{2}\}$

22. $x^2 = 6x - 2$
$\{3 \pm \sqrt{7}\}$

23. $3x^2 - 6x - 15 = 0$
$\{1 \pm \sqrt{6}\}$

 24. $5x^2 + 10x - 30 = 0$
$\{-1 \pm \sqrt{7}\}$

25. $4p^2 + 16p = -4$
$\{-2 \pm \sqrt{3}\}$

26. $2t^2 - 12t = 12$
$\{3 \pm \sqrt{15}\}$

27. $w^2 + w - 3 = 0$
$\left\{-\dfrac{1}{2} \pm \dfrac{\sqrt{13}}{2}\right\}$

28. $z^2 - 3z - 5 = 0$
$\left\{\dfrac{3}{2} \pm \dfrac{\sqrt{29}}{2}\right\}$

29. $x(x + 2) = 40$
$\{-1 \pm \sqrt{41}\}$

30. $y(y - 4) = 10$
$\{2 \pm \sqrt{14}\}$

31. $a^2 - 4a - 1 = 0$
$\{2 \pm \sqrt{5}\}$

32. $c^2 - 2c - 9 = 0$
$\{1 \pm \sqrt{10}\}$

33. $2r^2 + 12r + 16 = 0$
$\{-2, -4\}$

34. $3p^2 + 12p + 9 = 0$
$\{-3, -1\}$

35. $h(h - 11) = -24$
$\{3, 8\}$

36. $k(k - 8) = -7$
$\{1, 7\}$

Mixed Exercises

For Exercises 37–64, solve each quadratic equation by using the zero product rule or the square root property. (*Hint:* For some exercises, you may have to factor or complete the square first.)

37. $y^2 = 121$
$\{11, -11\}$

38. $x^2 = 81$
$\{9, -9\}$

39. $(p + 2)^2 = 2$
$\{-2 \pm \sqrt{2}\}$

40. $(q - 6)^2 = 3$
$\{6 \pm \sqrt{3}\}$

41. $(k + 13)(k - 5) = 0$
$\{-13, 5\}$

42. $(r - 10)(r + 12) = 0$
$\{10, -12\}$

43. $(x - 13)^2 = 0$
$\{13\}$

44. $(p + 14)^2 = 0$
$\{-14\}$

45. $z^2 - 8z - 20 = 0$
$\{10, -2\}$

46. $b^2 - 14b + 48 = 0$
$\{8, 6\}$

47. $(x - 3)^2 = 16$
$\{7, -1\}$

48. $(x + 2)^2 = 49$
$\{5, -9\}$

49. $a^2 - 8a + 1 = 0$
$\{4 \pm \sqrt{15}\}$

50. $x^2 + 12x - 4 = 0$
$\{-6 \pm 2\sqrt{10}\}$

51. $2y^2 + 4y = 10$
$\{-1 \pm \sqrt{6}\}$

52. $3z^2 - 48z = 6$
$\{8 \pm \sqrt{66}\}$

53. $x^2 - 9x - 22 = 0$
$\{11, -2\}$

54. $y^2 + 11y + 18 = 0$
$\{-9, -2\}$

55. $5h(h - 7) = 0$
$\{0, 7\}$

56. $-2w(w + 9) = 0$
$\{0, -9\}$

57. $8t^2 + 2t - 3 = 0$
$\left\{\dfrac{1}{2}, -\dfrac{3}{4}\right\}$

58. $18a^2 - 21a + 5 = 0$
$\left\{\dfrac{5}{6}, \dfrac{1}{3}\right\}$

59. $t^2 = 14$
$\{\sqrt{14}, -\sqrt{14}\}$

60. $s^2 = 17$
$\{\sqrt{17}, -\sqrt{17}\}$

61. $c^2 + 9 = 0$
There are no real-valued solutions.

62. $k^2 + 25 = 0$
There are no real-valued solutions.

63. $4x^2 - 8x = -4$
$\{1\}$

64. $3x^2 + 12x = -12$
$\{-2\}$

Expanding Your Skills

For Exercises 65–66, solve by completing the square.

65. To comply with airline regulations, a piece of luggage must be checked to the luggage compartment of the plane if its combined linear measurement of length, width, and height is over 45 in. Katie's suitcase has a total volume of 4200 in.³ Its length is 30 in., and its width is 4 in. greater than the height. Find the dimensions of the suitcase. Will this suitcase need to be checked? Explain.   The suitcase is 10 in. by 14 in. by 30 in. The bag must be checked because 10 in. + 14 in. + 30 in. = 54 in., which is greater than 45 in.

66. Luggage that is checked to the baggage compartment of an airplane must not exceed the dimensional requirements set by the carrier. Most carriers do not allow bags that exceed 30 in. in any dimension. They also require that the combined length, width, and height of the bag not exceed 62 in. Suppose a suitcase has a total volume of 5040 in.³ If the length is 28 in. and the width is 8 in. greater than the height, find the dimensions of the bag. Can this bag be checked to the luggage compartment of the plane? Explain.   The dimensions are 10 in., 18 in., and 28 in. The bag can be checked because 10 in. + 18 in. + 28 in. = 56 in., which is under the 62-in. limit.

© Corbis Premium RF/Alamy RF

 Writing   ⟵⟶ Translating Expression   Geometry   Scientific Calculator   Video

Section 9.3   Quadratic Formula

1. Derivation of the Quadratic Formula

If we solve a general quadratic equation $ax^2 + bx + c = 0$ $(a \neq 0)$ by completing the square and using the square root property, the result is a formula that gives the solutions for x in terms of a, b, and c.

$$ax^2 + bx + c = 0$$

Begin with a quadratic equation in standard form.

$$\frac{ax^2}{a} + \frac{b}{a}x + \frac{c}{a} = \frac{0}{a}$$

Divide by the leading coefficient.

$$x^2 + \frac{b}{a}x = -\frac{c}{a}$$

Isolate the terms containing x.

$$x^2 + \frac{b}{a}x + \left(\frac{1}{2} \cdot \frac{b}{a}\right)^2 = \left(\frac{1}{2} \cdot \frac{b}{a}\right)^2 - \frac{c}{a}$$

Add the square of $\frac{1}{2}$ the linear term coefficient to both sides of the equation.

$$\left(x + \frac{b}{2a}\right)^2 = \frac{b^2}{4a^2} - \frac{c}{a}$$

Factor the left side as a perfect square.

$$\left(x + \frac{b}{2a}\right)^2 = \frac{b^2}{4a^2} - \frac{c}{a} \cdot \frac{(4a)}{(4a)}$$

On the right side, write the fractions with the common denominator, $4a^2$.

$$\left(x + \frac{b}{2a}\right)^2 = \frac{b^2 - 4ac}{4a^2}$$

Combine the fractions.

$$x + \frac{b}{2a} = \pm\sqrt{\frac{b^2 - 4ac}{4a^2}}$$

Apply the square root property.

$$x + \frac{b}{2a} = \frac{\pm\sqrt{b^2 - 4ac}}{2a}$$

Simplify the denominator.

$$x = -\frac{b}{2a} \pm \frac{\sqrt{b^2 - 4ac}}{2a}$$

Subtract $\frac{b}{2a}$ from both sides.

$$= \frac{-b \pm \sqrt{b^2 - 4ac}}{2a}$$

Combine fractions.

Quadratic Formula

For any quadratic equation of the form $ax^2 + bx + c = 0$ $(a \neq 0)$, the solutions are

$$x = \frac{-b \pm \sqrt{b^2 - 4ac}}{2a}$$

2. Solving Quadratic Equations Using the Quadratic Formula

| Example 1 | **Solving a Quadratic Equation Using the Quadratic Formula** |

Classroom Example: p. 655, Exercise 18

Solve the quadratic equation using the quadratic formula. $3x^2 - 7x = -2$

Solution:

$3x^2 - 7x = -2$

$3x^2 - 7x + 2 = 0$ Write the equation in the form $ax^2 + bx + c = 0$.

$a = 3, b = -7, c = 2$ Identify a, b, and c.

$x = \dfrac{-b \pm \sqrt{b^2 - 4ac}}{2a}$

$x = \dfrac{-(-7) \pm \sqrt{(-7)^2 - 4(3)(2)}}{2(3)}$ Apply the quadratic formula.

$x = \dfrac{7 \pm \sqrt{49 - 24}}{6}$ Simplify.

$= \dfrac{7 \pm \sqrt{25}}{6}$

$= \dfrac{7 \pm 5}{6}$

There are two rational solutions.

$x = \begin{cases} \dfrac{7 + 5}{6} = \dfrac{12}{6} = 2 \\ \dfrac{7 - 5}{6} = \dfrac{2}{6} = \dfrac{1}{3} \end{cases}$

The solution set is $\left\{ 2, \dfrac{1}{3} \right\}$.

Instructor Note: Ask students what type of solutions we get when $b^2 - 4ac$ is a perfect square, zero, a nonperfect square, or a negative number

TIP: If the solutions to a quadratic equation are rational numbers, then the original equation could have been solved by factoring and using the zero product rule.

Skill Practice Solve by using the quadratic formula.

1. $5x^2 - 9x + 4 = 0$

Answer

1. $\left\{ 1, \dfrac{4}{5} \right\}$

Classroom Example: p. 655, Exercise 16

Example 2 **Solving a Quadratic Equation Using the Quadratic Formula**

Solve the quadratic equation using the quadratic formula.

$$4x(x - 5) + 25 = 0$$

Solution:

$$4x(x - 5) + 25 = 0$$

$$4x^2 - 20x + 25 = 0 \qquad \text{Write the equation in the form } ax^2 + bx + c = 0.$$

$$a = 4, b = -20, c = 25 \qquad \text{Identify } a, b, \text{ and } c.$$

$$x = \frac{-b \pm \sqrt{b^2 - 4ac}}{2a}$$

$$x = \frac{-(-20) \pm \sqrt{(-20)^2 - 4(4)(25)}}{2(4)} \qquad \text{Apply the quadratic formula.}$$

$$= \frac{20 \pm \sqrt{400 - 400}}{8} \qquad \text{Simplify.}$$

$$= \frac{20 \pm \sqrt{0}}{8} \qquad \text{Simplify the radical.}$$

$$= \frac{\overset{5}{\cancel{20}}}{\underset{2}{\cancel{8}}} \qquad \text{Simplify the fraction.}$$

$$= \frac{5}{2}$$

The solution set is $\left\{ \dfrac{5}{2} \right\}$.

TIP: When using the quadratic formula, if the radical term results in the square root of zero, there will be only one rational solution.

Skill Practice Solve by using the quadratic formula.

2. $x(x + 6) = -9$

Classroom Example: p. 655, Exercise 32

Example 3 **Solving a Quadratic Equation Using the Quadratic Formula**

Solve the quadratic equation using the quadratic formula.

$$\frac{1}{4}w^2 - \frac{1}{2}w - \frac{5}{4} = 0$$

Solution:

It is easier to work with the quadratic formula with integer values of a, b, and c. Therefore, for the first step, we will choose to clear fractions.

$$\frac{1}{4}w^2 - \frac{1}{2}w - \frac{5}{4} = 0 \qquad \text{Multiply each side of the equation by the LCD, which is 4.}$$

$$4\left(\frac{1}{4}w^2 - \frac{1}{2}w - \frac{5}{4} \right) = 4(0) \qquad \text{Use the distributive property to clear fractions.}$$

$$w^2 - 2w - 5 = 0 \qquad \text{The equation is in the form } ax^2 + bx + c = 0.$$

Answer

2. $\{-3\}$

$a = 1, b = -2, c = -5$ Identify a, b, and c.

$$w = \frac{-b \pm \sqrt{b^2 - 4ac}}{2a}$$

$$w = \frac{-(-2) \pm \sqrt{(-2)^2 - 4(1)(-5)}}{2(1)}$$ Apply the quadratic formula.

$$= \frac{2 \pm \sqrt{4 + 20}}{2}$$ Simplify.

$$= \frac{2 \pm \sqrt{24}}{2}$$

$$= \frac{2 \pm 2\sqrt{6}}{2}$$ The solutions are irrational numbers.

$$= \frac{\cancel{2}(1 \pm \sqrt{6})}{\cancel{2}}$$ Factor and simplify.

$$= 1 \pm \sqrt{6}$$

The solution set is $\{1 \pm \sqrt{6}\}$.

> **Avoiding Mistakes**
>
> The fraction bar must extend under the term $-b$ as well as the radical.

Skill Practice Solve by using the quadratic formula.

3. $\dfrac{1}{6}t^2 + \dfrac{2}{3}t - \dfrac{1}{3} = 0$

3. Review of the Methods for Solving a Quadratic Equation

Three methods have been presented for solving quadratic equations.

Methods for Solving a Quadratic Equation

- Factor and use the zero product rule.
- Use the square root property. Complete the square if necessary.
- Use the quadratic formula.

    The zero product rule can be used only if one side of the equation is zero, and the expression on the other side is factorable. The square root property and the quadratic formula can be used to solve any quadratic equation. Before solving a quadratic equation, take a minute to analyze it. Each problem must be examined individually before choosing the most efficient method to find its solutions.

Answer

3. $\{-2 \pm \sqrt{6}\}$

Classroom Examples: p. 655, Exercises 34, 38, and 44

| **Example 4** | **Solving Quadratic Equations Using Any Method** |

Solve the quadratic equations using any method.

a. $(x+1)^2 = 5$ **b.** $t^2 - t - 30 = 0$ **c.** $2x^2 + 5x + 1 = 0$

Solution:

a. $(x+1)^2 = 5$ ... Because the equation is the square of a binomial equal to a constant, the square root property can be applied easily.

$x + 1 = \pm\sqrt{5}$ ... Apply the square root property.

$x = -1 \pm \sqrt{5}$ ... Isolate x.

The solution set is $\{-1 \pm \sqrt{5}\}$.

b. $t^2 - t - 30 = 0$ ... The expression factors.

$(t-6)(t+5) = 0$ ... Factor and apply the zero product rule.

$t = 6$ or $t = -5$

The solution set is $\{6, -5\}$.

c. $2x^2 + 5x + 1 = 0$ ... The expression does not factor. Because the equation is already in the form $ax^2 + bx + c = 0$, use the quadratic formula.

$a = 2, b = 5, c = 1$ ... Identify a, b, and c.

$x = \dfrac{-(5) \pm \sqrt{(5)^2 - 4(2)(1)}}{2(2)}$ ... Apply the quadratic formula.

$x = \dfrac{-5 \pm \sqrt{25 - 8}}{4}$ ... Simplify.

$x = \dfrac{-5 \pm \sqrt{17}}{4}$

The solution set is $\left\{\dfrac{-5 \pm \sqrt{17}}{4}\right\}$.

Skill Practice Solve the equations using any method.

4. $p^2 + 7p + 12 = 0$ **5.** $5y^2 + 7y + 1 = 0$ **6.** $(w-8)^2 = 3$

4. Applications of Quadratic Equations

Classroom Example: p. 655, Exercise 58

| **Example 5** | **Solving a Quadratic Equation in an Application** |

The length of a box is 2 in. longer than the width. The height of the box is 4 in. and the volume of the box is 200 in.^{3} Find the exact dimensions of the box. Then use a calculator to approximate the dimensions to the nearest tenth of an inch.

Answers

4. $\{-3, -4\}$ **5.** $\left\{\dfrac{-7 \pm \sqrt{29}}{10}\right\}$

6. $\{8 \pm \sqrt{3}\}$

Solution:

Label the box as follows (Figure 9-1):

Width = x

Length = $x + 2$

Height = 4

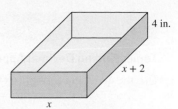

4 in.

$x + 2$

x

Figure 9-1

The volume of a box is given by the formula: $V = lwh$

$$V = l \cdot w \cdot h$$

$$200 = (x + 2)(x)(4)$$ Substitute $V = 200$, $l = x + 2$, $w = x$, and $h = 4$.

$$200 = (x + 2)4x$$

$$200 = 4x^2 + 8x$$

$$0 = 4x^2 + 8x - 200$$

$$4x^2 + 8x - 200 = 0$$ The equation is in the form $ax^2 + bx + c = 0$.

$$\frac{4x^2}{4} + \frac{8x}{4} - \frac{200}{4} = \frac{0}{4}$$ The coefficients are all divisible by 4. Dividing by 4 will create smaller values of a, b, and c to be used in the quadratic formula.

$$x^2 + 2x - 50 = 0$$ $a = 1$, $b = 2$, $c = -50$

$$x = \frac{-2 \pm \sqrt{(2)^2 - 4(1)(-50)}}{2(1)}$$ Apply the quadratic formula.

$$= \frac{-2 \pm \sqrt{4 + 200}}{2}$$ Simplify.

$$= \frac{-2 \pm \sqrt{204}}{2}$$

$$= \frac{-2 \pm 2\sqrt{51}}{2}$$ Simplify the radical. $\sqrt{204} = \sqrt{4 \cdot 51} = 2\sqrt{51}$

$$= \frac{\overset{1}{\cancel{2}}(-1 \pm \sqrt{51})}{\underset{1}{\cancel{2}}}$$ Factor and simplify.

$$= -1 \pm \sqrt{51}$$

Because the width of the box must be positive, use $x = -1 + \sqrt{51}$.

The width is $(-1 + \sqrt{51})$ in. ≈ 6.1 in.

The length is $x + 2$: $(-1 + \sqrt{51} + 2)$ in. or $(1 + \sqrt{51})$ in. ≈ 8.1 in.

The height is 4 in.

Avoiding Mistakes

We do not use the solution $x = -1 - \sqrt{51}$ because it is a negative number, that is,

$$-1 - \sqrt{51} \approx -8.1$$

The width of an object cannot be negative.

Skill Practice

7. The length of a rectangle is 2 in. longer than the width. The area is 10 in.² Find the exact values of the length and width. Then use a calculator to approximate the dimensions to the nearest tenth of an inch.

Answer

7. The width is $(-1 + \sqrt{11})$ in. or approximately 2.3 in. The length is $(1 + \sqrt{11})$ in. or approximately 4.3 in.

Calculator Connections

Topic: Finding Decimal Approximations to the Solutions of a Quadratic Equation

Use the quadratic formula to verify that the solutions to the equation $x^2 + 7x + 4 = 0$ are:

$$x = \frac{-7 + \sqrt{33}}{2} \quad \text{and} \quad x = \frac{-7 - \sqrt{33}}{2}$$

A calculator can be used to obtain decimal approximations for the irrational solutions of a quadratic equation.

Scientific Calculator

Enter: 7 $\boxed{+/-}$ $\boxed{+}$ 33 $\boxed{\sqrt{}}$ $\boxed{=}$ $\boxed{\div}$ 2 $\boxed{=}$ **Result:** $\boxed{-0.627718677}$

Enter: 7 $\boxed{+/-}$ $\boxed{-}$ 33 $\boxed{\sqrt{}}$ $\boxed{=}$ $\boxed{\div}$ 2 $\boxed{=}$ **Result:** $\boxed{-6.372281323}$

Graphing Calculator

```
(-7+√(33))/2
          -.6277186767
(-7-√(33))/2
          -6.372281323
```

Calculator Exercises

Use a calculator to obtain a decimal approximation of each expression.

1. $\dfrac{-5 + \sqrt{17}}{4}$ and $\dfrac{-5 - \sqrt{17}}{4}$ −0.2192235936; −2.280776406

2. $\dfrac{-40 + \sqrt{1920}}{-32}$ and $\dfrac{-40 - \sqrt{1920}}{-32}$ −0.1193063938; 2.619306394

Section 9.3   **Practice Exercises**

For additional exercises, see Classroom Activities 9.3A–9.3B in the *Student's Resource Manual* at www.mhhe.com/moh.

Vocabulary and Key Concepts

1. **a.** For the equation $ax^2 + bx + c = 0$ $(a \neq 0)$, the quadratic formula gives the solutions as $x = \dfrac{-b \pm \sqrt{b^2 - 4ac}}{2a}$.

   b. To apply the quadratic formula, a quadratic equation must be written in the form, $\underline{ax^2 + bx + c = 0}$, where $a \neq 0$.

   c. To apply the quadratic formula to solve the equation $5x^2 - 24x - 36 = 0$, the value of a is $\underline{\quad 5 \quad}$, the value of b is $\underline{\quad -24 \quad}$, and the value of c is $\underline{\quad -36 \quad}$.

   d. To apply the quadratic formula to solve the equation $2x^2 - 5x - 6 = 0$, the value of $-b$ is $\underline{\quad 5 \quad}$ and the value of the radicand is $\underline{\quad 73 \quad}$.

Review Exercises

For Exercises 2–5, apply the square root property to solve the equation.

2. $z^2 = 169$ $\{13, -13\}$ 3. $p^2 = 1$ $\{1, -1\}$ 4. $(x - 4)^2 = 28$ $\{4 \pm 2\sqrt{7}\}$ 5. $(y + 3)^2 = 7$ $\{-3 \pm \sqrt{7}\}$

For Exercises 6–8, solve the equations by completing the square.

6. $p^2 + 10p + 2 = 0$ $\{-5 \pm \sqrt{23}\}$ 7. $3a^2 - 12a - 12 = 0$ $\{2 \pm 2\sqrt{2}\}$ 8. $x^2 - 5x + 1 = 0$ $\left\{ \dfrac{5}{2} \pm \dfrac{\sqrt{21}}{2} \right\}$

Concept 1: Derivation of the Quadratic Formula

For Exercises 9–14, write each equation in the form $ax^2 + bx + c = 0$. Then identify the values of a, b, and c.

9. $2x^2 - x = 5$
   $2x^2 - x - 5 = 0; a = 2, b = -1, c = -5$

10. $5(x^2 + 2) = -3x$
    $5x^2 + 3x + 10 = 0; a = 5, b = 3, c = 10$

11. $-3x(x - 4) = -2x$
    $-3x^2 + 14x + 0 = 0; a = -3, b = 14, c = 0$

12. $x(x - 2) = 3(x + 1)$
    $x^2 - 5x - 3 = 0; a = 1, b = -5, c = -3$

13. $x^2 - 9 = 0$
    $x^2 + 0x - 9 = 0; a = 1, b = 0, c = -9$

14. $x^2 + 25 = 0$
    $x^2 + 0x + 25 = 0; a = 1, b = 0, c = 25$

✎ Writing   ↔ Translating Expression   ◁ Geometry   🖩 Scientific Calculator   ▶ Video

Concept 2: Solving Quadratic Equations Using the Quadratic Formula

For Exercises 15–32, solve each equation using the quadratic formula. **(See Examples 1–3.)**

15. $t^2 + 16t + 64 = 0$ $\{-8\}$

16. $y^2 - 10y + 25 = 0$ $\{5\}$

17. $6k^2 - k - 2 = 0$ $\left\{\frac{2}{3}, -\frac{1}{2}\right\}$

18. $3n^2 + 5n - 2 = 0$ $\left\{-2, \frac{1}{3}\right\}$

19. $5t^2 - t = 3$ $\left\{\frac{1 \pm \sqrt{61}}{10}\right\}$

20. $2a^2 + 5a = 1$ $\left\{\frac{-5 \pm \sqrt{33}}{4}\right\}$

21. $x(x - 2) = 1$ $\{1 \pm \sqrt{2}\}$

22. $2y(y - 3) = -1$ $\left\{\frac{3 \pm \sqrt{7}}{2}\right\}$

23. $2p^2 = -10p - 11$ $\left\{\frac{-5 \pm \sqrt{3}}{2}\right\}$

24. $z^2 = 4z + 1$ $\{2 \pm \sqrt{5}\}$

25. $-4y^2 - y + 1 = 0$ $\left\{\frac{1 \pm \sqrt{17}}{-8}\right\}$ or $\left\{\frac{-1 \pm \sqrt{17}}{8}\right\}$

26. $-5z^2 - 3z + 4 = 0$ $\left\{\frac{3 \pm \sqrt{89}}{-10}\right\}$ or $\left\{\frac{-3 \pm \sqrt{89}}{10}\right\}$

27. $2x(x + 1) = 3 - x$ $\left\{\frac{-3 \pm \sqrt{33}}{4}\right\}$

28. $3m(m - 2) = -m + 1$ $\left\{\frac{5 \pm \sqrt{37}}{6}\right\}$

29. $0.2y^2 = -1.5y - 1$ $\left\{\frac{-15 \pm \sqrt{145}}{4}\right\}$

30. $0.2t^2 = t + 0.5$ $\left\{\frac{5 \pm \sqrt{35}}{2}\right\}$

31. $\frac{2}{3}x^2 + \frac{4}{9}x = \frac{1}{3}$ $\left\{\frac{-2 \pm \sqrt{22}}{6}\right\}$

32. $\frac{1}{2}x^2 + \frac{1}{6}x = 1$ $\left\{\frac{-1 \pm \sqrt{73}}{6}\right\}$

Concept 3: Review of the Methods for Solving a Quadratic Equation

For Exercises 33–56, choose any method to solve the quadratic equations. **(See Example 4.)**

33. $16x^2 - 9 = 0$ $\left\{\frac{3}{4}, -\frac{3}{4}\right\}$

34. $\frac{1}{4}x^2 + 5x + 13 = 0$ $\{-10 \pm 4\sqrt{3}\}$

35. $(x - 5)^2 = -21$
There are no real-valued solutions.

36. $2x^2 + x + 5 = 0$
There are no real-valued solutions.

37. $\frac{1}{9}x^2 + \frac{8}{3}x + 11 = 0$ $\{-12 \pm 3\sqrt{5}\}$

38. $7x^2 = 12x$ $\left\{0, \frac{12}{7}\right\}$

39. $2x^2 - 6x - 3 = 0$ $\left\{\frac{3 \pm \sqrt{15}}{2}\right\}$

40. $4(x + 1)^2 = -15$
There are no real-valued solutions.

41. $9x^2 = 11x$ $\left\{0, \frac{11}{9}\right\}$

42. $25x^2 - 4 = 0$ $\left\{\frac{2}{5}, -\frac{2}{5}\right\}$

43. $(2y - 3)^2 = 5$ $\left\{\frac{3 \pm \sqrt{5}}{2}\right\}$

44. $(6z + 1)^2 = 7$ $\left\{\frac{-1 \pm \sqrt{7}}{6}\right\}$

45. $0.4x^2 = 0.2x + 1$ $\left\{\frac{1 \pm \sqrt{41}}{4}\right\}$

46. $0.6x^2 = 0.1x + 0.8$ $\left\{\frac{1 \pm \sqrt{193}}{12}\right\}$

47. $9z^2 - z = 0$ $\left\{0, \frac{1}{9}\right\}$

48. $16p^2 - p = 0$ $\left\{0, \frac{1}{16}\right\}$

49. $r^2 - 52 = 0$ $\{2\sqrt{13}, -2\sqrt{13}\}$

50. $y^2 - 32 = 0$ $\{4\sqrt{2}, -4\sqrt{2}\}$

51. $-2.5t(t - 4) = 1.5$ $\left\{\frac{-10 \pm \sqrt{85}}{-5}\right\}$ or $\left\{\frac{10 \pm \sqrt{85}}{5}\right\}$

52. $1.6p(p - 2) = 0.8$ $\left\{\frac{2 \pm \sqrt{6}}{2}\right\}$

53. $(m - 3)(m + 2) = 9$ $\left\{\frac{1 \pm \sqrt{61}}{2}\right\}$

54. $(h - 6)(h - 1) = 12$ $\left\{\frac{7 \pm \sqrt{73}}{2}\right\}$

55. $x^2 + x + 3 = 0$
There are no real-valued solutions.

56. $3x^2 - 20x + 12 = 0$ $\left\{6, \frac{2}{3}\right\}$

Concept 4: Applications of Quadratic Equations

57. In a rectangle, the length is 1 m less than twice the width and the area is 100 m². Approximate the dimensions to the nearest tenth of a meter. **(See Example 5.)**
The width is 7.3 m. The length is 13.6 m.

58. In a triangle, the height is 2 cm more than the base. The area is 72 cm². Approximate the base and height to the nearest tenth of a centimeter.
The base is 11.0 cm. The height is 13.0 cm.

59. The volume of a rectangular storage area is 240 ft³. The length is 2 ft more than the width. The height is 6 ft. Approximate the dimensions to the nearest tenth of a foot.
The length is 7.4 ft. The width is 5.4 ft. The height is 6 ft.

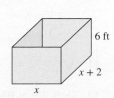

60. In a right triangle, one leg is 2 ft shorter than the other leg. The hypotenuse is 12 ft. Approximate the lengths of the legs to the nearest tenth of a foot.
The legs are 9.4 ft and 7.4 ft.

61. In a rectangle, the length is 4 ft longer than the width. The area is 72 ft². Approximate the dimensions to the nearest tenth of a foot.
The width is 6.7 ft. The length is 10.7 ft.

Writing   Translating Expression   Geometry   Scientific Calculator   Video

62. In a triangle, the base is 4 cm less than twice the height. The area is 60 cm². Approximate the base and height to the nearest tenth of a centimeter.
The height is 8.8 cm. The base is 13.6 cm.

63. In a right triangle, one leg is 3 m longer than the other leg. The hypotenuse is 13 m. Approximate the lengths of the legs to the nearest tenth of a meter.
The legs are 10.6 m and 7.6 m.

64. A bad punter on a football team kicks a football approximately straight upward with an initial velocity of 90 ft/sec. The height h (in feet) of the ball t seconds after being kicked is given by $h = -16t^2 + 90t + 4$. Find the time(s) at which the ball is at a height of 90 ft. Round to one decimal place.
The ball will be at a 90-ft height 1.2 sec and 4.4 sec after being kicked.

65. A professional basketball player has a 48-in. vertical leap. Suppose that he jumps from ground level with an initial velocity of 16 ft/sec. His height h (in feet) at a time t seconds after he leaves the ground is given by $h = -16t^2 + 16t$. Use the equation to determine the times at which the basketball player would be at a height of 1 ft. Round to two decimal places.
He will be 1 ft off the ground 0.07 sec after leaving the ground (on the way up) and after 0.93 sec (on the way back down).

Problem Recognition Exercises

Solving Different Types of Equations

For Exercises 1–2, solve the equations using each of the three methods.

   a. Factoring and applying the zero product rule

   b. Completing the square and applying the square root property

   c. Applying the quadratic formula

1. $6x^2 + 7x - 3 = 0$ $\left\{\frac{1}{3}, -\frac{3}{2}\right\}$ **2.** $y^2 + 14y + 49 = 0$ $\{-7\}$

For Exercises 3–16,

   a. Identify the type of equation as
      • linear
      • quadratic
      • rational
      • radical
   b. Solve the equation.

3. $x(x - 8) = 6$ a. Quadratic   b. $\{4 \pm \sqrt{22}\}$ **4.** $2 - 6y = -y^2$ a. Quadratic   b. $\{3 \pm \sqrt{7}\}$

5. $3(k - 6) = 2k - 5$ a. Linear   b. $\{13\}$ **6.** $13x + 4 = 5(x - 4)$ a. Linear   b. $\{-3\}$

7. $8x^2 - 22x + 5 = 0$ a. Quadratic   b. $\left\{\frac{5}{2}, \frac{1}{4}\right\}$ **8.** $9w^2 - 15w + 4 = 0$ a. Quadratic   b. $\left\{\frac{4}{3}, \frac{1}{3}\right\}$

9. $\frac{2}{x - 1} - \frac{5}{4} = -\frac{1}{x + 1}$ a. Rational   b. $\left\{-\frac{3}{5}, 3\right\}$ **10.** $\frac{5}{p - 2} = 7 - \frac{10}{p + 2}$ a. Rational   b. $\left\{-\frac{6}{7}, 3\right\}$

11. $\sqrt{2y - 2} = y - 1$ a. Radical   b. $\{1, 3\}$ **12.** $\sqrt{5p - 1} = p + 1$ a. Radical   b. $\{1, 2\}$

13. $(w + 1)^2 = 100$ a. Quadratic   b. $\{9, -11\}$ **14.** $(u - 5)^2 = 64$ a. Quadratic   b. $\{13, -3\}$

15. $\frac{2}{x + 1} = \frac{5}{4}$ a. Rational   b. $\left\{\frac{3}{5}\right\}$ **16.** $\frac{7}{t - 1} = \frac{21}{2}$ a. Rational   b. $\left\{\frac{5}{3}\right\}$

 Writing   Translating Expression   Geometry   Scientific Calculator   Video

Complex Numbers

1. Definition of *i*

We have learned that there are no real-valued square roots of a negative number. For example, $\sqrt{-9}$ is not a real number because no real number when squared equals -9. However, the square roots of a negative number are defined over another set of numbers called the imaginary numbers. The foundation of the set of *imaginary numbers* is the definition of the imaginary number, *i*.

> #### Definition of *i*
>
> $$i = \sqrt{-1}$$
>
> *Note:* From the definition of *i*, it follows that $i^2 = -1$.

2. Simplifying Expressions in Terms of *i*

Using the imaginary number *i*, we can define the square root of any negative real number.

> #### Definition of $\sqrt{-b}$, $b > 0$
> Let *b* be a real number such that $b > 0$, then $\sqrt{-b} = i\sqrt{b}$.

> **Example 1**   **Simplifying Expressions in Terms of *i***
>
> Simplify the expressions in terms of *i*.
>
>   **a.** $\sqrt{-25}$ **b.** $-3\sqrt{-81}$ **c.** $\sqrt{-13}$
>
> **Solution:**
>
>   **a.** $\sqrt{-25} = i\sqrt{25} = 5i$
>
>   **b.** $-3\sqrt{-81} = -3 \cdot i\sqrt{81} = -3(9i) = -27i$
>
>   **c.** $\sqrt{-13} = i\sqrt{13}$

Skill Practice Simplify in terms of *i*.

   1. $\sqrt{-144}$ **2.** $-2\sqrt{-100}$ **3.** $\sqrt{-7}$

Concepts

1. Definition of *i*
2. Simplifying Expressions in Terms of *i*
3. Definition of a Complex Number
4. Addition, Subtraction, and Multiplication of Complex Numbers
5. Division of Complex Numbers
6. Quadratic Equations with Imaginary Solutions

Classroom Examples: p. 664, Exercises 2, 4, and 6

Avoiding Mistakes

In an expression such as $i\sqrt{13}$ the *i* is usually written in front of the square root. The expression $\sqrt{13}\,i$ is also correct but may be misinterpreted as $\sqrt{13i}$ (with *i* incorrectly placed under the radical).

In our initial study of radical expressions, we presented the multiplication and division properties of radicals as follows:

   If *a* and *b* represent real numbers such that $\sqrt[n]{a}$ and $\sqrt[n]{b}$ are both real, then

$$\sqrt[n]{ab} = \sqrt[n]{a} \cdot \sqrt[n]{b} \quad \text{and} \quad \sqrt[n]{\frac{a}{b}} = \frac{\sqrt[n]{a}}{\sqrt[n]{b}} \quad b \neq 0$$

The conditions that $\sqrt[n]{a}$ and $\sqrt[n]{b}$ must both be real numbers prevent us from applying the multiplication and division properties of radicals for square roots with negative radicands. Therefore, to multiply or divide radicals with negative radicands, write the radicals in terms of the imaginary number *i* first. This is demonstrated in Example 2.

Answers

1. $12i$ **2.** $-20i$ **3.** $i\sqrt{7}$

Classroom Examples: p. 664, Exercises 12 and 14

Example 2 **Simplifying Expressions in Terms of *i***

Simplify the expressions.

a. $\dfrac{\sqrt{-100}}{\sqrt{-25}}$ **b.** $\sqrt{-16} \cdot \sqrt{-4}$

Solution:

a. $\dfrac{\sqrt{-100}}{\sqrt{-25}}$

$= \dfrac{10i}{5i}$ Simplify each radical in terms of *i* *before* dividing.

$= 2$ Simplify.

b. $\sqrt{-16} \cdot \sqrt{-4}$

$= (4i)(2i)$ Simplify each radical in terms of *i* *first* before multiplying.

$= 8i^2$

$= 8(-1)$ Substitute i^2 with -1.

$= -8$

Avoiding Mistakes

In Example 2(b), the radical expressions were written in terms of *i* first before multiplying. If we had mistakenly applied the multiplication property first we would obtain an incorrect answer.

Be careful:

$\sqrt{-16} \cdot \sqrt{-4} \neq \sqrt{64}$

Skill Practice Simplify.

4. $\dfrac{\sqrt{-36}}{\sqrt{-4}}$ **5.** $\sqrt{-1} \cdot \sqrt{-9}$

For Example 3, recall that $i^2 = -1$.

Classroom Examples: p. 664, Exercises 22 and 26

Example 3 **Simplifying Expressions Involving i^2**

Simplify the expressions.

a. $7i^2$ **b.** $2i^2 - 3$ **c.** $-i^2 + 11$

Solution:

Substitute i^2 with -1.

a. $7i^2$

$= 7(-1)$

$= -7$

b. $2i^2 - 3$

$= 2(-1) - 3$

$= -2 - 3$

$= -5$

c. $-i^2 + 11$

$= -(-1) + 11$

$= 1 + 11$

$= 12$

Skill Practice Simplify.

6. $2i^2$ **7.** $3 + 4i^2$ **8.** $9 - i^2$

3. Definition of a Complex Number

We have already learned the definitions of the integers, rational numbers, irrational numbers, and real numbers. In this section, we define the complex numbers.

Answers

4. 3   **5.** −3   **6.** −2
7. −1   **8.** 10

Complex Numbers

A **complex number** is a number of the form $a + bi$, where a and b are real numbers and $i = \sqrt{-1}$.

Notes:

- If $b = 0$, then the complex number, $a + bi$ is a real number.
- If $b \neq 0$, then we say that $a + bi$ is an **imaginary number**.
- The complex number $a + bi$ is said to be written in **standard form**. The quantities a and b are called the **real** and **imaginary parts**, respectively.
- The complex numbers $(a - bi)$ and $(a + bi)$ are called **complex conjugates**.

From the definition of a complex number, it follows that all real numbers are complex numbers and all imaginary numbers are complex numbers. Figure 9-2 illustrates the relationship among the sets of numbers we have learned so far.

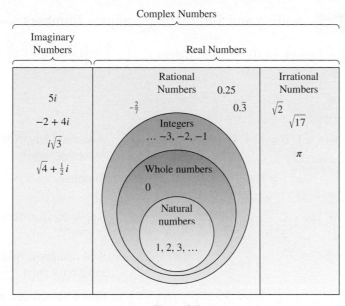

Figure 9-2

Example 4	Identifying the Real and Imaginary Parts of a Complex Number

Classroom Examples: p. 664, Exercises 30, 32, and 34

Identify the real and imaginary parts of the complex numbers.

a. $7 + 4i$ **b.** -6 **c.** $-\dfrac{1}{2}i$

Solution:

a. $7 + 4i$ The real part is 7, and the imaginary part is 4.

b. -6

$= -6 + 0i$ Rewrite -6 in the form $a + bi$.
The real part is -6, and the imaginary part is 0.

TIP: Example 4(b) illustrates that a real number is also a complex number.

TIP: Example 4(c) illustrates that an imaginary number is also a complex number.

c. $-\dfrac{1}{2}i$

$= 0 + -\dfrac{1}{2}i$ Rewrite $-\frac{1}{2}i$ in the form $a + bi$.
                The real part is 0, and the imaginary part is $-\frac{1}{2}$.

Skill Practice Identify the real and the imaginary part.

9. $-3 + 2i$ **10.** $6i$ **11.** $-\dfrac{3}{4}$

4. Addition, Subtraction, and Multiplication of Complex Numbers

The operations for addition, subtraction, and multiplication of real numbers also apply to imaginary numbers. To add or subtract complex numbers, combine the real parts and combine the imaginary parts. The commutative, associative, and distributive properties that apply to real numbers also apply to complex numbers.

Classroom Examples: pp. 664–665, Exercises 38 and 52

Example 5   **Adding and Subtracting Complex Numbers**

a. Add. $(2 - 3i) + (4 + 17i)$ **b.** Subtract. $\left(-\dfrac{3}{2} + \dfrac{1}{3}i\right) - \left(2 - \dfrac{2}{3}i\right)$

Solution:

Real parts

a. $(2 - 3i) + (4 + 17i) = (2 + 4) + (-3 + 17)i$ Add real parts. Add imaginary parts.

Imaginary parts

$= 6 + 14i$ Simplify.

b. $\left(-\dfrac{3}{2} + \dfrac{1}{3}i\right) - \left(2 - \dfrac{2}{3}i\right) = -\dfrac{3}{2} + \dfrac{1}{3}i - 2 + \dfrac{2}{3}i$ Apply the distributive property.

$= \left(-\dfrac{3}{2} - 2\right) + \left(\dfrac{1}{3} + \dfrac{2}{3}\right)i$ Add real parts. Add imaginary parts.

$= \left(-\dfrac{3}{2} - \dfrac{4}{2}\right) + \left(\dfrac{3}{3}\right)i$ Find common denominators and simplify.

$= -\dfrac{7}{2} + i$

Skill Practice

12. Add. $(-3 + 4i) + (-5 - 6i)$ **13.** Subtract. $\left(\dfrac{1}{2} - \dfrac{4}{5}i\right) - \left(\dfrac{1}{3} + \dfrac{7}{10}i\right)$

Answers

9. real part: -3; imaginary part: 2
10. real part: 0; imaginary part: 6
11. real part: $-\dfrac{3}{4}$; imaginary part: 0
12. $-8 - 2i$ **13.** $\dfrac{1}{6} - \dfrac{3}{2}i$

| Example 6 | **Multiplying Complex Numbers** |

Classroom Examples: p. 665, Exercises 50 and 56

Multiply.

a. $(5 - 2i)(3 + 4i)$ **b.** $(2 + 7i)(2 - 7i)$

Solution:

a. $(5 - 2i)(3 + 4i)$

$= (5)(3) + (5)(4i) + (-2i)(3) + (-2i)(4i)$ Apply the distributive property.

$= 15 + 20i - 6i - 8i^2$ Simplify.

$= 15 + 14i - 8(-1)$ Recall $i^2 = -1$.

$= 15 + 14i + 8$

$= 23 + 14i$ Write the answer in the form $a + bi$.

b. $(2 + 7i)(2 - 7i)$ The expressions $(2 + 7i)$ and $(2 - 7i)$ are complex conjugates.

The product is a difference of squares.

$(a + b)(a - b) = a^2 - b^2$

$(2 + 7i)(2 - 7i) = (2)^2 - (7i)^2$ Apply the formula, where $a = 2$ and $b = 7i$.

$= 4 - 49i^2$ Simplify.

$= 4 - 49(-1)$ Recall $i^2 = -1$.

$= 4 + 49$

$= 53$

TIP: The complex numbers $2 + 7i$ and $2 - 7i$ can also be multiplied by using the distributive property.

$(2 + 7i)(2 - 7i) = 4 - 14i + 14i - 49i^2$

$= 4 - 49(-1)$

$= 4 + 49$

$= 53$

Skill Practice Multiply.

14. $(2 - 7i)(3 + 5i)$ **15.** $(5 - i)(5 + i)$

Answers

14. $41 - 11i$ **15.** 26

5. Division of Complex Numbers

Example 6(b) illustrates that the product of a complex number and its complex conjugate produces a real number. Consider the complex numbers $a + bi$ and $a - bi$, where a and b are real numbers. Then,

$$(a + bi)(a - bi) = (a)^2 - (bi)^2$$
$$= a^2 - b^2 i^2$$
$$= a^2 - b^2(-1)$$
$$= a^2 + b^2 \quad \text{(real number)}$$

To divide by a complex number, multiply the numerator and denominator by the complex conjugate of the denominator. This produces a real number in the denominator so that the resulting expression can be written in the form $a + bi$.

Classroom Example: p. 665, Exercise 80

Example 7	**Dividing by a Complex Number**

Divide the complex numbers. Write the answer in the form $a + bi$.

$$\frac{2 + 3i}{4 - 5i}$$

Solution:

$$\frac{2 + 3i}{4 - 5i}$$

> **TIP:** Dividing by a complex number mimics the same process as rationalizing a denominator with two terms.

$$\frac{(2 + 3i)}{(4 - 5i)} \cdot \frac{(4 + 5i)}{(4 + 5i)} = \frac{(2)(4) + (2)(5i) + (3i)(4) + (3i)(5i)}{(4)^2 - (5i)^2}$$

Multiply the numerator and denominator by the complex conjugate of the denominator.

$$= \frac{8 + 10i + 12i + 15i^2}{16 - 25i^2}$$

Simplify the numerator and denominator.

$$= \frac{8 + 22i + 15(-1)}{16 - 25(-1)}$$

Recall $i^2 = -1$.

$$= \frac{8 + 22i - 15}{16 + 25}$$

$$= \frac{-7 + 22i}{41}$$

Simplify.

$$= -\frac{7}{41} + \frac{22}{41}i$$

Write in the form $a + bi$.

Skill Practice Divide. Write the answer in $a + bi$ form.

16. $\dfrac{3 - i}{7 + 5i}$

Answer

16. $\dfrac{8}{37} - \dfrac{11}{37}i$

6. Quadratic Equations with Imaginary Solutions

We have solved quadratic equations using the square root property and the quadratic formula. For some equations, we saw that the solutions were not real numbers. We now have the tools to solve these types of equations.

Example 8 — Solving a Quadratic Equation Using the Square Root Property

Classroom Example: p. 665, Exercise 86

Solve the equation using the square root property. $(x+2)^2 = -9$

Solution:

$$(x+2)^2 = -9$$

$$x + 2 = \pm\sqrt{-9} \qquad \text{Apply the square root property.}$$

$$x + 2 = \pm 3i \qquad \text{Write } \sqrt{-9} \text{ as } i\sqrt{9} \text{ and simplify to } 3i.$$

$$x = -2 \pm 3i \qquad \text{Solve for } x.$$

The solution set is $\{-2 \pm 3i\}$.

Skill Practice Solve the quadratic equation.

17. $(y-5)^2 = -16$

Example 9 — Solving a Quadratic Equation Using the Quadratic Formula

Classroom Example: p. 665, Exercise 90

Solve the equation using the quadratic formula. $2x^2 + 4x + 5 = 0$

Solution:

$$2x^2 + 4x + 5 = 0 \qquad\qquad a = 2, b = 4, c = 5$$

$$x = \frac{-4 \pm \sqrt{(4)^2 - 4(2)(5)}}{2(2)} \qquad \text{Apply the quadratic formula.}$$

$$= \frac{-4 \pm \sqrt{-24}}{4} \qquad\qquad \text{Simplify.}$$

$$= \frac{-4 \pm i\sqrt{24}}{4} \qquad\qquad \text{Write } \sqrt{-24} \text{ as } i\sqrt{24}.$$

$$= \frac{-4 \pm 2i\sqrt{6}}{4} \qquad\qquad \sqrt{24} = \sqrt{4 \cdot 6} = 2\sqrt{6}$$

$$= -\frac{\overset{1}{\cancel{4}}}{\underset{1}{\cancel{4}}} \pm \frac{\overset{1}{\cancel{2}}i\sqrt{6}}{\underset{2}{\cancel{4}}} \qquad\qquad \text{Simplify the terms.}$$

$$= -1 \pm \frac{\sqrt{6}}{2}i \qquad\qquad \text{Write the answer in standard form.}$$

The solution set is $\left\{ -1 \pm \dfrac{\sqrt{6}}{2}i \right\}$.

Skill Practice Solve the quadratic equation.

18. $3x^2 + 2x + 3 = 0$

Answers

17. $\{5 \pm 4i\}$

18. $\left\{ -\dfrac{1}{3} \pm \dfrac{2\sqrt{2}}{3}i \right\}$

Section 9.4 Practice Exercises

Vocabulary and Key Concepts

For additional exercises, see Classroom Activities 9.4A–9.4D in the *Student's Resource Manual* at www.mhhe.com/moh.

1. **a.** A square root of a negative number is not a real number, but rather is an ___imaginary___ number.

   b. $i =$ ___$\sqrt{-1}$___, and $i^2 =$ ___-1___.

   c. A complex number is a number of the form ___$a + bi$___, where a and b are real numbers and $i =$ ___$\sqrt{-1}$___.

   d. Given a complex number $a + bi$, a is called the ___real___ part, and ___b___ is called the imaginary part.

   e. The complex conjugate of $a - bi$ is ___$a + bi$___.

   f. Answer true or false. All real numbers are complex numbers.   True

   g. Answer true or false. All imaginary numbers are complex numbers.   True

Concept 2: Simplifying Expressions in Terms of i

For Exercises 2–8, simplify each expression in terms of i. **(See Example 1.)**

2. $\sqrt{-49}$ $7i$

3. $\sqrt{-36}$ $6i$

4. $\sqrt{-15}$ $i\sqrt{15}$

5. $\sqrt{-21}$ $i\sqrt{21}$

6. $-5\sqrt{-12}$ $-10i\sqrt{3}$

7. $-4\sqrt{-48}$ $-16i\sqrt{3}$

8. $\sqrt{-1}$ i

For Exercises 9–20, perform the indicated operations. Remember to write the radicals in terms of i first. **(See Example 2.)**

9. $\sqrt{-100} \cdot \sqrt{-4}$ -20

10. $\sqrt{-9} \cdot \sqrt{-25}$ -15

11. $\sqrt{-3} \cdot \sqrt{-12}$ -6

12. $\sqrt{-8} \cdot \sqrt{-2}$ -4

13. $\dfrac{\sqrt{-81}}{\sqrt{-9}}$ 3

14. $\dfrac{\sqrt{-64}}{\sqrt{-16}}$ 2

15. $\dfrac{\sqrt{-50}}{\sqrt{-2}}$ 5

16. $\dfrac{\sqrt{-45}}{\sqrt{-5}}$ 3

17. $\sqrt{-9} + \sqrt{-121}$ $14i$

18. $\sqrt{-36} - \sqrt{-49}$ $-i$

19. $\sqrt{-1} - \sqrt{-144} - \sqrt{-169}$ $-24i$

20. $\sqrt{-4} + \sqrt{-64} + \sqrt{-81}$ $19i$

For Exercises 21–28, simplify the expressions involving i^2. **(See Example 3.)**

21. $10i^2$ -10

22. $12i^2$ -12

23. $6 + i^2$ 5

24. $-3 + i^2$ -4

25. $-i^2 - 4$ -3

26. $-i^2 + 1$ 2

27. $-5i^2$ 5

28. $-9i^2$ 9

Concept 3: Definition of a Complex Number

For Exercises 29–34, identify the real part and the imaginary part of the complex number. **(See Example 4.)**

29. $-3 - 2i$ Real part: -3; Imaginary part: -2

30. $5 + i$ Real part: 5; Imaginary part: 1

31. 4 Real part: 4; Imaginary part: 0

32. -6 Real part: -6; Imaginary part: 0

33. $\frac{2}{7}i$ Real part: 0; Imaginary part: $\frac{2}{7}$

34. $0.52i$ Real part: 0; Imaginary part: 0.52

Concept 4: Addition, Subtraction, and Multiplication of Complex Numbers

35. Explain how to add or subtract complex numbers.   Add or subtract the real parts. Add or subtract the imaginary parts.

36. Explain how to multiply complex numbers.   Multiply using the distributive property, remembering to replace i^2 with -1.

For Exercises 37–66, perform the indicated operations. Write the answers in standard form, $a + bi$. **(See Examples 5–6.)**

37. $(2 + 7i) + (-8 + i)$ $-6 + 8i$

38. $(6 - i) + (4 + 2i)$ $10 + i$

39. $(3 - 4i) + (7 - 6i)$ $10 - 10i$

Writing   Translating Expression   Geometry   Scientific Calculator   Video

40. $(-4 - 15i) - (-3 - 17i)$
$-1 + 2i$

41. $4i - (9 + i) + 15$ $6 + 3i$

42. $10i - (1 - 5i) - 8$ $-9 + 15i$

43. $(5 - 6i) - (9 - 8i) - (3 - i)$
$-7 + 3i$

44. $(1 - i) - (5 - 19i) - (24 + 19i)$
$-28 - i$

45. $(2 - i)(7 - 7i)$ $7 - 21i$

46. $(1 + i)(8 - i)$ $9 + 7i$

47. $(13 - 5i) - (2 + 4i)$ $11 - 9i$

48. $(1 + 8i) + (-6 + 3i)$ $-5 + 11i$

49. $(5 + 3i)(3 + 2i)$ $9 + 19i$

50. $(9 + i)(8 + 2i)$ $70 + 26i$

51. $\left(\dfrac{1}{2} + \dfrac{1}{5}i\right) - \left(\dfrac{3}{4} + \dfrac{2}{5}i\right)$ $-\dfrac{1}{4} - \dfrac{1}{5}i$

52. $\left(\dfrac{5}{6} + \dfrac{1}{8}i\right) + \left(\dfrac{1}{3} - \dfrac{3}{8}i\right)$ $\dfrac{7}{6} - \dfrac{1}{4}i$

53. $8.4i - (3.5 - 9.7i)$ $-3.5 + 18.1i$

54. $(4.2 - 3i) - (10 - 18.2i)$
$-5.8 + 15.2i$

55. $(3 - 2i)(3 + 2i)$ 13

56. $(18 + i)(18 - i)$ 325

57. $(10 - 2i)(10 + 2i)$ 104

58. $(3 - 5i)(3 + 5i)$ 34

59. $\left(\dfrac{1}{2} - i\right)\left(\dfrac{1}{2} + i\right)$ $\dfrac{5}{4}$

60. $\left(\dfrac{1}{3} - i\right)\left(\dfrac{1}{3} + i\right)$ $\dfrac{10}{9}$

61. $(6 - i)^2$ $35 - 12i$

62. $(4 + 3i)^2$ $7 + 24i$

63. $(5 + 2i)^2$ $21 + 20i$

64. $(7 - 6i)^2$ $13 - 84i$

65. $(4 - 7i)^2$ $-33 - 56i$

66. $(3 - i)^2$ $8 - 6i$

67. What is the conjugate of $7 - 4i$? Multiply $7 - 4i$ by its conjugate. $7 + 4i;\ 65$

68. What is the conjugate of $-3 - i$? Multiply $-3 - i$ by its conjugate. $-3 + i;\ 10$

69. What is the conjugate of $\frac{3}{2} + \frac{2}{5}i$? Multiply $\frac{3}{2} + \frac{2}{5}i$ by its conjugate. $\dfrac{3}{2} - \dfrac{2}{5}i;\ \dfrac{241}{100}$

70. What is the conjugate of $-1.3 + 5.7i$? Multiply $-1.3 + 5.7i$ by its conjugate. $-1.3 - 5.7i;\ 34.18$

71. What is the conjugate of $4i$? Multiply $4i$ by its conjugate. $-4i;\ 16$

72. What is the conjugate of $-8i$? Multiply $-8i$ by its conjugate. $8i;\ 64$

Concept 5: Division of Complex Numbers

For Exercises 73–84, divide the complex numbers. Write the answers in standard form, $a + bi$. **(See Example 7.)**

73. $\dfrac{-3i}{2 + i}$ $-\dfrac{3}{5} - \dfrac{6}{5}i$

74. $\dfrac{6i}{3 - 2i}$ $-\dfrac{12}{13} + \dfrac{18}{13}i$

75. $\dfrac{4i}{5 - i}$ $-\dfrac{2}{13} + \dfrac{10}{13}i$

76. $\dfrac{6i}{3 + i}$ $\dfrac{3}{5} + \dfrac{9}{5}i$

77. $\dfrac{4 + i}{4 - i}$ $\dfrac{15}{17} + \dfrac{8}{17}i$

78. $\dfrac{1 - 5i}{1 + 5i}$ $-\dfrac{12}{13} - \dfrac{5}{13}i$

79. $\dfrac{4 + 3i}{2 + 5i}$ $\dfrac{23}{29} - \dfrac{14}{29}i$

80. $\dfrac{1 + 7i}{3 + 2i}$ $\dfrac{17}{13} + \dfrac{19}{13}i$

81. $\dfrac{2}{7 - 4i}$ $\dfrac{14}{65} + \dfrac{8}{65}i$

82. $\dfrac{-3}{-3 - i}$ $\dfrac{9}{10} - \dfrac{3}{10}i$

83. $\dfrac{5}{1 + i}$ $\dfrac{5}{2} - \dfrac{5}{2}i$

84. $\dfrac{6}{1 - i}$ $3 + 3i$

Concept 6: Quadratic Equations with Imaginary Solutions

For Exercises 85–92, solve the quadratic equations. **(See Examples 8–9.)**

85. $(x + 4)^2 = -25$ $\{-4 \pm 5i\}$

86. $(x + 2)^2 = -49$ $\{-2 \pm 7i\}$

87. $(p - 3)^2 = -8$ $\{3 \pm 2i\sqrt{2}\}$

88. $(m - 6)^2 = -40$ $\{6 \pm 2i\sqrt{10}\}$

89. $x^2 - 2x + 4 = 0$ $\{1 \pm i\sqrt{3}\}$

90. $x^2 - 4x + 6 = 0$ $\{2 \pm i\sqrt{2}\}$

91. $6y^2 + 3y + 2 = 0$ $\left\{-\dfrac{1}{4} \pm \dfrac{\sqrt{39}}{12}i\right\}$

92. $2x^2 + 5x + 12 = 0$ $\left\{-\dfrac{5}{4} \pm \dfrac{\sqrt{71}}{4}i\right\}$

Writing   ◄ ➡ Translating Expression   Geometry   Scientific Calculator   Video

Expanding Your Skills

For Exercises 93–105, answer true or false. If an answer is false, explain why.

93. Every complex number is a real number.
False. For example: $2 + 3i$ is not a real number.

94. Every real number is a complex number.   True

95. Every imaginary number is a complex number.
True

96. $\sqrt{-64}$ is an imaginary number.   True

97. $\sqrt[3]{-64}$ is an imaginary number.
False. $\sqrt[3]{-64} = -4$.

98. The product $(2 + 3i)(2 - 3i)$ is a real number.
True

99. The product $(1 + 4i)(1 - 4i)$ is an imaginary number.   False. $(1 + 4i)(1 - 4i) = 17$.

100. The imaginary part of the complex number $2 - 3i$ is 3.   False. The imaginary part is -3.

101. The imaginary part of the complex number $4 - 5i$ is -5.   True

102. i^2 is a real number.   True

103. i^4 is an imaginary number.   False. $i^4 = 1$.

104. i^3 is a real number.   False. $i^3 = -i$.

105. i^4 is a real number.   True

Section 9.5 Graphing Quadratic Equations

Concepts

1. Definition of a Quadratic Equation in Two Variables
2. Vertex of a Parabola
3. Graphing a Parabola
4. Applications of Quadratic Equations

1. Definition of a Quadratic Equation in Two Variables

We have already learned how to graph the solutions to linear equations in two variables. Now suppose we want to graph the *nonlinear* equation, $y = x^2$. To begin, we create a table of points representing several solutions to the equation (Table 9-1). These points form the curve shown in Figure 9-3.

Table 9-1

x	y
-3	9
-2	4
-1	1
0	0
1	1
2	4
3	9

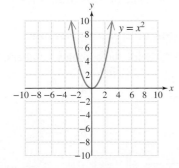

Figure 9-3

The equation $y = x^2$ is a special type of equation called a quadratic equation, and its graph is in the shape of a **parabola**.

> ### Quadratic Equation in Two Variables
>
> Let a, b, and c represent real numbers such that $a \neq 0$. Then an equation of the form $y = ax^2 + bx + c$ is called a **quadratic equation in two variables**.

The graph of a quadratic equation is a parabola that opens upward or downward. The leading coefficient, a, determines the direction of the parabola. For the quadratic equation $y = ax^2 + bx + c$,

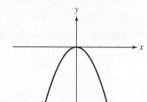

If $a > 0$, the parabola opens *upward*.
For example: $y = x^2$.

$$y = 1x^2 \quad (a = 1)$$

If $a < 0$, the parabola opens *downward*.
For example: $y = -x^2$.

$$y = -1x^2 \quad (a = -1)$$

If a parabola opens upward, the **vertex** is the lowest point on the graph. If a parabola opens downward, the **vertex** is the highest point on the graph. For a parabola defined by $y = ax^2 + bx + c$, the **axis of symmetry** is the vertical line that passes through the vertex. Notice that the graph of the parabola is its own mirror image to the left and right of the axis of symmetry.

Here are four quadratic equations and their graphs.

$y = 0.5x^2 + 2x + 3$
$a > 0$
Vertex $(-2, 1)$
Axis of symmetry: $x = -2$

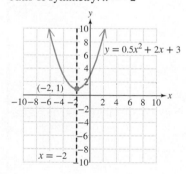

$y = x^2 - 6x + 9$
$a > 0$
Vertex $(3, 0)$
Axis of symmetry: $x = 3$

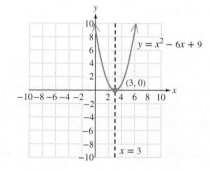

$y = -x^2 + 4x$
$a < 0$
Vertex $(2, 4)$
Axis of symmetry: $x = 2$

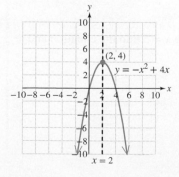

$y = -2x^2 - 4$
$a < 0$
Vertex $(0, -4)$
Axis of symmetry: $x = 0$

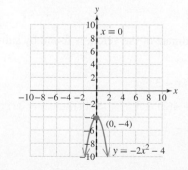

2. Vertex of a Parabola

Quadratic equations arise in many applications of mathematics and applied sciences. For example, an object thrown through the air follows a parabolic path. The mirror inside a reflecting telescope is parabolic in shape. In applications, it is often advantageous to analyze the graph of a parabola. In particular, we want to find the location of the x- and y-intercepts and the vertex.

To find the vertex of a parabola defined by $y = ax^2 + bx + c$ $(a \neq 0)$, we use the following steps:

Finding the Vertex of a Parabola

Step 1 The x-coordinate of the vertex of the parabola defined by
$y = ax^2 + bx + c$ $(a \neq 0)$ is given by

$$x = \frac{-b}{2a}$$

Step 2 To find the corresponding y-coordinate of the vertex, substitute the value of the x-coordinate found in step 1 and solve for y.

Classroom Example: p. 675, Exercise 46

> **Example 1** **Analyzing a Quadratic Equation**

Given the equation $y = -x^2 + 4x - 3$,

   a. Determine whether the parabola opens upward or downward.

   b. Find the vertex of the parabola.

   c. Find the x-intercept(s).

   d. Find the y-intercept.

   e. Sketch the parabola.

Solution:

   a. The equation $y = -x^2 + 4x - 3$ is written in the form $y = ax^2 + bx + c$, where $a = -1$, $b = 4$, and $c = -3$. Because the value of a is negative, the parabola opens *downward*.

   b. The x-coordinate of the vertex is given by $x = \dfrac{-b}{2a}$.

$$x = \frac{-b}{2a} = \frac{-(4)}{2(-1)} \qquad \text{Substitute } b = 4 \text{ and } a = -1.$$

$$= \frac{-4}{-2} \qquad \text{Simplify.}$$

$$= 2$$

The y-coordinate of the vertex is found by substituting $x = 2$ into the equation and solving for y.

$$y = -x^2 + 4x - 3$$

$$= -(2)^2 + 4(2) - 3 \qquad \text{Substitute } x = 2.$$

$$= -4 + 8 - 3$$

$$= 1$$

The vertex is (2, 1).　Because the parabola opens downward, the vertex is the maximum point on the graph of the parabola.

c. To find the x-intercept(s), substitute $y = 0$ and solve for x.

$y = -x^2 + 4x - 3$

$0 = -x^2 + 4x - 3$　　Substitute $y = 0$. The resulting equation is quadratic.

$0 = -1(x^2 - 4x + 3)$　　Factor out -1.

$0 = -1(x - 3)(x - 1)$　　Factor the trinomial.

$x - 3 = 0$　or　$x - 1 = 0$　　Apply the zero product rule.

$x = 3$　　or　$x = 1$

The x-intercepts are (3, 0) and (1, 0).

d. To find the y-intercept, substitute $x = 0$ and solve for y.

$y = -x^2 + 4x - 3$

$\quad = -(0)^2 + 4(0) - 3$　　　　Substitute $x = 0$.

$\quad = -3$

The y-intercept is (0, −3).

e. Using the results of parts (a)–(d), we have a parabola that opens downward with vertex (2, 1), x-intercepts at (3, 0) and (1, 0), and y-intercept at (0, −3) (Figure 9-4).

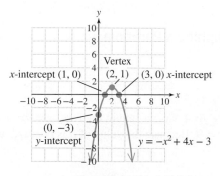

Figure 9-4

> **TIP:** Because of the symmetry of a parabola, the x-coordinate of the vertex will be halfway between the x-intercepts.

Skill Practice

1. Given $y = -x^2 - 4x$, perform parts (a)–(e), as in Example 1.

---

3. Graphing a Parabola

To sketch a quadratic equation in two variables, determine the vertex and x- and y-intercepts. Furthermore, notice that the parabola defining the graph of a quadratic equation is symmetric with respect to the axis of symmetry.

Answers

1. a. downward
b. (−2, 4)
c. (0, 0) and (−4, 0)
d. (0, 0)
e.

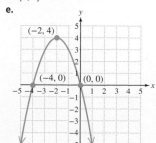

To analyze a parabola, we recommend the following guidelines.

Graphing a Parabola

Given a quadratic equation defined by $y = ax^2 + bx + c$ $(a \neq 0)$, consider the following guidelines to graph the parabola.

Step 1 Determine whether the parabola opens upward or downward.
- If $a > 0$, the parabola opens upward.
- If $a < 0$, the parabola opens downward.

Step 2 Find the vertex.
- The x-coordinate is given by $x = \dfrac{-b}{2a}$
- To find the y-coordinate, substitute the x-coordinate of the vertex into the equation and solve for y.

Step 3 Find the x-intercept(s) by substituting $y = 0$ and solving the quadratic equation for x.
- *Note:* If the solutions to the equation in step 3 are not real numbers, then there are no x-intercepts.

Step 4 Find the y-intercept by substituting $x = 0$ and solving the equation for y.

Step 5 Plot the vertex and x- and y-intercepts. If necessary, find and plot additional points near the vertex. Then use the symmetry of the parabola to sketch the curve through the points. (*Note:* The axis of symmetry is the vertical line that passes through the vertex.)

Classroom Example: p. 675, Exercise 42

Example 2   Graphing a Parabola

Graph $y = x^2 - 6x + 9$.

Solution:

1. The equation $y = x^2 - 6x + 9$ is written in the form $y = ax^2 + bx + c$, where $a = 1$, $b = -6$, and $c = 9$. Because the value of a is positive, the parabola opens upward.

2. The x-coordinate of the vertex is given by

$$x = \frac{-b}{2a} = \frac{-(-6)}{2(1)} = 3$$

Substituting $x = 3$ into the equation, we have

$$y = (3)^2 - 6(3) + 9$$
$$= 9 - 18 + 9$$
$$= 0$$

The vertex is $(3, 0)$.

3. To find the x-intercept(s), substitute $y = 0$ and solve for x.

$$y = x^2 - 6x + 9 \longrightarrow 0 = x^2 - 6x + 9$$

$$0 = (x - 3)^2 \qquad \text{Factor.}$$

$$x = 3 \qquad \text{Apply the zero product rule.}$$

The x-intercept is $(3, 0)$.

4. To find the y-intercept, substitute $x = 0$ and solve for y.

$$y = x^2 - 6x + 9 \longrightarrow y = (0)^2 - 6(0) + 9$$
$$= 9$$

The y-intercept is $(0, 9)$.

5. Sketch the parabola through the x- and y-intercepts and vertex (Figure 9-5).

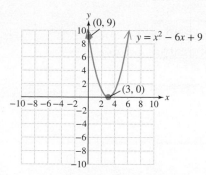

Figure 9-5

TIP: Using the symmetry of the parabola, we know that the points to the right of the vertex must mirror the points to the left of the vertex. For example, because the point $(0, 9)$ is on the parabola, then the point $(6, 9)$ is also on the parabola.

Skill Practice

2. Graph $y = x^2 - 2x + 1$.

Example 3   **Graphing a Parabola**

Graph $y = -x^2 - 4$.

Classroom Example: p. 675, Exercise 50

Solution:

1. The equation $y = -x^2 - 4$ is written in the form $y = ax^2 + bx + c$, where $a = -1$, $b = 0$, and $c = -4$. Because the value of a is negative, the parabola opens downward.

2. The x-coordinate of the vertex is given by

$$x = \frac{-b}{2a} = \frac{-(0)}{2(-1)} = 0$$

Substituting $x = 0$ into the equation, we have

$$y = -(0)^2 - 4$$
$$= -4$$

The vertex is $(0, -4)$.

3. Substituting $y = 0$ into the equation $y = -x^2 - 4$ results in an equation with no real solutions. Therefore, the graph of $y = -x^2 - 4$ has no x-intercepts.

$$y = -x^2 - 4$$
$$0 = -x^2 - 4$$
$$x^2 = -4$$
$$x = \pm\sqrt{-4} \quad \text{Not a real}$$
$$\text{number}$$

4. The vertex is $(0, -4)$. This is also the y-intercept.

TIP: The vertex is below the x-axis and the parabola opens downward. Therefore, there can be no x-intercepts. A quick sketch shows this.

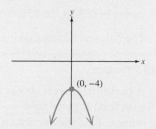

Answers

2.

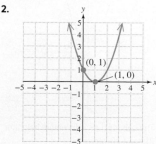

5. Sketch the parabola through the y-intercept and vertex (Figure 9-6).

To verify the proper shape of the graph, find additional points to the right or left of the vertex and use the symmetry of the parabola to sketch the curve.

x	y
1	-5
2	-8
3	-13

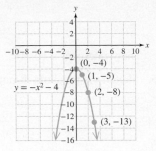

Figure 9-6

Skill Practice

3. Graph $y = x^2 + 1$.

---

4. Applications of Quadratic Equations

Classroom Example: p. 676, Exercise 56

| Example 4 | **Using a Quadratic Equation in an Application** |

A golfer hits a ball at an angle of 30°. The height of the ball y (in feet) can be represented by

$y = -16x^2 + 60x$ where x is the time in seconds after the ball was hit (Figure 9–7).

Find the maximum height of the ball. In how many seconds will the ball reach its maximum height?

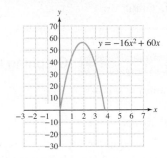

Figure 9-7

Solution:

The equation is written in the form $y = ax^2 + bx + c$, where $a = -16$, $b = 60$, and $c = 0$. Because a is negative, the parabola opens downward. Therefore, the maximum height of the ball occurs at the vertex of the parabola.

The x-coordinate of the vertex is given by

$$x = \frac{-b}{2a} = \frac{-(60)}{2(-16)} = \frac{-60}{-32} = \frac{15}{8} = 1.875$$

Substituting $x = 1.875$ into the equation, we have

$$y = -16(1.875)^2 + 60(1.875)$$
$$= -56.25 + 112.5$$
$$= 56.25$$

Answers

3.

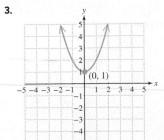

4. The ball reaches a maximum height of 31 ft in 1.25 sec.

The vertex is $(1.875, 56.25)$.

The ball reaches its maximum height of 56.25 ft after 1.875 sec.

Skill Practice

4. A basketball player shoots a basketball at an angle of 45°. The height of the ball y (in feet) is given by $y = -16x^2 + 40x + 6$ where x is the time (in seconds) after release. Find the maximum height of the ball and the time required to reach that height.

Calculator Connections

Topic: Finding the Maximum or Minimum Point of a Parabola

Some graphing calculators have *Minimum* and *Maximum* features that enable the user to approximate the minimum and maximum values of an equation. Otherwise, *Zoom* and *Trace* can be used.

For example, the maximum value of the equation from Example 4, $y = -16x^2 + 60x$, can be found using the *Maximum* feature.

The minimum value of the equation from Example 2, $y = x^2 - 6x + 9$, can be found using the *Minimum* feature.

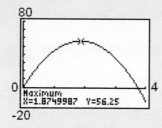 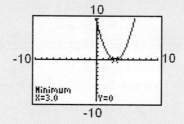

Calculator Exercises

Find the maximum or minimum point for each parabola. Identify the point as a maximum or a minimum.

1. $y = x^2 + 4x + 7$
   $(-2, 3)$; minimum

2. $y = x^2 - 20x + 105$
   $(10, 5)$; minimum

3. $y = -x^2 - 3x - 4.85$
   $(-1.5, -2.6)$; maximum

4. $y = -x^2 + 3.5x - 0.5625$
   $(1.75, 2.5)$; maximum

5. $y = 2x^2 - 10x + \dfrac{25}{2}$
   $\left(\dfrac{5}{2}, 0\right)$; minimum

6. $y = 3x^2 + 16x + \dfrac{64}{3}$
   $\left(-\dfrac{8}{3}, 0\right)$; minimum

Section 9.5   Practice Exercises

Vocabulary and Key Concepts

For additional exercises, see Classroom Activities 9.5A–9.5D, in the *Student's Resource Manual* at www.mhhe.com/moh.

1. a. The graph of a quadratic equation, $y = ax^2 + bx + c$, is a ___parabola___.

   b. The parabola defined by $y = ax^2 + bx + c$ $(a \neq 0)$ will open upward if a ___>___ 0 and will open downward if a ___<___ 0.

   c. If a parabola opens upward, the vertex is the (highest/lowest) point on the graph. If a parabola opens downward, the vertex is the (highest/lowest) point on the graph.   lowest; highest

   d. The axis of ___symmetry___ is a vertical line that passes through the vertex.

   e. The formula, $x = \dfrac{-b}{2a}$, gives the x-coordinate of the ___vertex___.

Review Exercises

For Exercises 2–8, solve each quadratic equation using any one of the following methods: factoring, the square root property, or the quadratic formula.

2. $3(y^2 + 1) = 10y$ $\left\{\dfrac{1}{3}, 3\right\}$

3. $3 + a(a + 2) = 18$ $\{-5, 3\}$

4. $4t^2 - 7 = 0$ $\left\{\dfrac{\sqrt{7}}{2}, -\dfrac{\sqrt{7}}{2}\right\}$

5. $2z^2 + 4z - 10 = 0$
   $\{-1 \pm \sqrt{6}\}$

6. $(b + 1)^2 = 6$
   $\{-1 \pm \sqrt{6}\}$

7. $(x - 5)^2 = 12$
   $\{5 \pm 2\sqrt{3}\}$

8. $3p^2 - 12p - 12 = 0$
   $\{2 \pm 2\sqrt{2}\}$

Concept 1: Definition of a Quadratic Equation in Two Variables

For Exercises 9–20, identify each equation as linear, quadratic, or neither.

9. $y = -8x + 3$ Linear

10. $y = 5x - 12$ Linear

11. $y = 4x^2 - 8x + 22$
   Quadratic

12. $y = x^2 + 10x - 3$
   Quadratic

Writing   Translating Expression   Geometry   Scientific Calculator   Video

13. $y = -5x^3 - 8x + 14$
Neither

14. $y = -3x^4 + 7x - 11$
Neither

15. $y = 15x$ Linear

16. $y = -9x$ Linear

17. $y = -21x^2$ Quadratic

18. $y = 3x^2$ Quadratic

19. $y = -x^3 + 1$ Neither

20. $y = 7x^4 - 4$ Neither

Concept 2: Vertex of a Parabola

21. How do you determine whether the graph of $y = ax^2 + bx + c$ $(a \neq 0)$ opens upward or downward?
If $a > 0$ the graph opens upward; if $a < 0$ the graph opens downward.

For Exercises 22–25, identify a and determine if the parabola opens upward or downward. **(See Example 1.)**

22. $y = x^2 - 15$
$a = 1$; upward

23. $y = 2x^2 + 23$
$a = 2$; upward

24. $y = -3x^2 + x - 18$
$a = -3$; downward

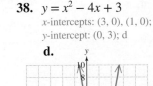 **25.** $y = -10x^2 - 6x - 20$
$a = -10$; downward

26. How do you find the vertex of a parabola?   Find the x-coordinate by $\dfrac{-b}{2a}$. Then substitute the value of x into the equation and solve for y.

For Exercises 27–34, find the vertex of the parabola. **(See Example 1.)**

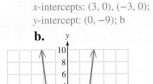 **27.** $y = 2x^2 + 4x - 6$
$(-1, -8)$

28. $y = x^2 - 4x - 4$
$(2, -8)$

29. $y = -x^2 + 2x - 5$
$(1, -4)$

30. $y = 2x^2 - 4x - 6$
$(1, -8)$

31. $y = x^2 - 2x + 3$
$(1, 2)$

32. $y = -x^2 + 4x - 2$
$(2, 2)$

33. $y = 3x^2 - 4$
$(0, -4)$

34. $y = 4x^2 - 1$
$(0, -1)$

Concept 3: Graphing a Parabola

For Exercises 35–38, find the x- and y-intercepts. Then match each equation with a graph. **(See Example 1.)**

35. $y = x^2 - 7$
x-intercepts: $(\sqrt{7}, 0), (-\sqrt{7}, 0)$;
y-intercept: $(0, -7)$; c

36. $y = x^2 - 9$
x-intercepts: $(3, 0), (-3, 0)$;
y-intercept: $(0, -9)$; b

37. $y = x^2 + 6x + 5$
x-intercepts: $(-1, 0), (-5, 0)$;
y-intercept: $(0, 5)$; a

38. $y = x^2 - 4x + 3$
x-intercepts: $(3, 0), (1, 0)$;
y-intercept: $(0, 3)$; d

a.
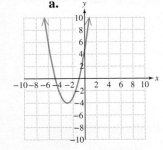

b.
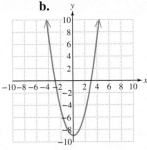

c.
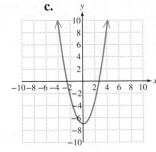

d.
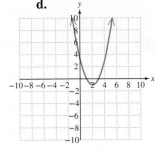

For Exercises 39–50, **(See Examples 1–3.)**

  a. Determine whether the parabola opens upward or downward.

  b. Find the vertex.

  c. Find the x-intercept(s), if possible.

  d. Find the y-intercept.

  e. Sketch the graph.

39. $y = x^2 - 9$
a. Upward
b. $(0, -9)$
c. $(3, 0), (-3, 0)$
d. $(0, -9)$

e.
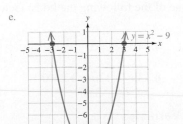

40. $y = x^2 - 4$
a. Upward
b. $(0, -4)$
c. $(2, 0), (-2, 0)$
d. $(0, -4)$

e.
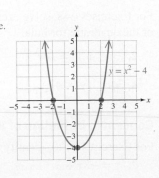

41. $y = x^2 - 2x - 8$
 a. Upward
 b. $(1, -9)$
 c. $(4, 0), (-2, 0)$
 d. $(0, -8)$
 e.
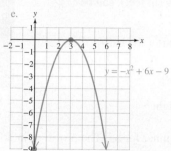

42. $y = x^2 + 2x - 24$
 a. Upward
 b. $(-1, -25)$
 c. $(4, 0), (-6, 0)$
 d. $(0, -24)$
 e.
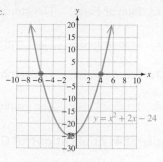

43. $y = -x^2 + 6x - 9$
 a. Downward
 b. $(3, 0)$
 c. $(3, 0)$
 d. $(0, -9)$
 e.
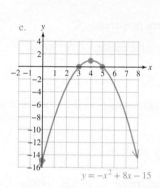

44. $y = -x^2 + 10x - 25$
 a. Downward
 b. $(5, 0)$
 c. $(5, 0)$
 d. $(0, -25)$
 e.
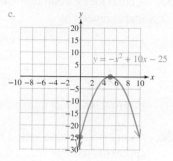

45. $y = -x^2 + 8x - 15$
 a. Downward
 b. $(4, 1)$
 c. $(3, 0), (5, 0)$
 d. $(0, -15)$
 e.
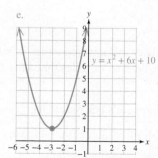

46. $y = -x^2 - 4x + 5$
 a. Downward
 b. $(-2, 9)$
 c. $(-5, 0), (1, 0)$
 d. $(0, 5)$
 e.
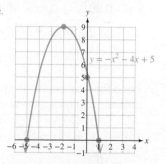

47. $y = x^2 + 6x + 10$
 a. Upward
 b. $(-3, 1)$
 c. none
 d. $(0, 10)$
 e.
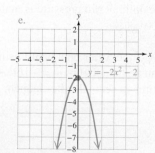

48. $y = x^2 + 4x + 5$
 a. Upward
 b. $(-2, 1)$
 c. none
 d. $(0, 5)$
 e.
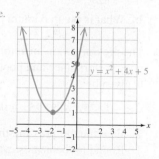

49. $y = -2x^2 - 2$
 a. Downward
 b. $(0, -2)$
 c. none
 d. $(0, -2)$
 e.

50. $y = -x^2 - 5$
 a. Downward
 b. $(0, -5)$
 c. none
 d. $(0, -5)$
 e.
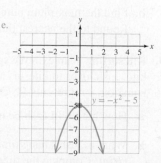

Writing   Translating Expression   Geometry   Scientific Calculator   Video

51. True or False: The graph of $y = -5x^2$ has a maximum value but no minimum value.   True

52. True or False: The graph of $y = -4x^2 + 9x - 6$ opens upward.   False

53. True or False: The graph of $y = 1.5x^2 - 6x - 3$ opens downward.   False

54. True or False: The graph of $y = 2x^2 - 5x + 4$ has a maximum value but no minimum value.   False

Concept 4: Applications of Quadratic Equations

55. A child kicks a ball into the air, and the height of the ball, y (in feet), can be approximated by

$$y = -16t^2 + 40t + 3 \qquad \text{where } t \text{ is the number of seconds after the ball was kicked.}$$

   a. Find the maximum height of the ball. **(See Example 4.)**   28 ft

   b. How long will it take the ball to reach its maximum height?   1.25 sec

© McGraw-Hill Education/
Ken Cavanagh, photographer

56. A concession stand sells a hamburger/drink combination dinner for $5. The profit, y (in dollars), can be approximated by

$$y = -0.001x^2 + 3.6x - 400 \qquad \text{where } x \text{ is the number of dinners prepared.}$$

   a. Find the number of dinners that should be prepared to maximize profit.   1800 dinners

   b. What is the maximum profit? $2840

57. For a fund-raising activity, a charitable organization produces calendars to sell in the community. The profit, y (in dollars), can be approximated by

$$y = -\frac{1}{40}x^2 + 10x - 500 \qquad \text{where } x \text{ is the number of calendars produced.}$$

   a. Find the number of calendars that should be produced to maximize the profit.   200 calendars

   b. What is the maximum profit? $500

58. The pressure, x, in an automobile tire can affect its wear. Both over-inflated and under-inflated tires can lead to poor performance and poor mileage. For one particular tire, the number of miles that a tire lasts, y (in thousands), is given by

$$y = -0.875x^2 + 57.25x - 900 \qquad \text{where } x \text{ is the tire pressure in pounds per square inch (psi).}$$

   a. Find the tire pressure that will yield the maximum number of miles that a tire will last. Round to the nearest whole unit.   33 psi

   b. Find the maximum number of miles that a tire will last if the optimal tire pressure is maintained. Round to the nearest thousand miles.   36 thousand miles or 36,000 miles

59. Kitesurfing is an extreme sport where athletes are propelled across the water on a board using the power of a kite. Josh loves to kitesurf and the height of one of his jumps can be modeled by $y = -16t^2 + 32t$. In this equation, y represents Josh's height in feet and t represents the time in seconds after launch.

   a. How high will Josh be in 0.5 sec?
     Josh will be 12 ft high in 0.5 sec.

   b. What is Josh's hang time?
     (*Hint:* Compute the time required for him to land.)
     Josh will land in 2 sec.

   c. What is Josh's maximum height?
     The maximum height is 16 ft.

Courtesy Rick Iossi

Introduction to Functions

1. Definition of a Relation

Table 9-2 gives the number of points scored by LeBron James corresponding to the number of minutes that he played per game for six games.

Table 9-2

Minutes Played, x	Number of Points, y	
38	33	→ (38, 33)
44	52	→ (44, 52)
40	16	→ (40, 16)
41	47	→ (41, 47)
33	26	→ (33, 26)
38	30	→ (38, 30)

Each ordered pair from Table 9-2 shows a correspondence, or relationship, between the number of minutes played and the number of points scored by LeBron James. The set of ordered pairs: {(38, 33), (44, 52), (40, 16), (41, 47), (33, 26), (38, 30)} defines a relationship between the number of minutes played and the number of points scored.

Definition of a Relation in x and y

Any set of ordered pairs, (x, y), is called a **relation** in x and y. Furthermore:

- The set of first components in the ordered pairs is called the **domain** of the relation.
- The set of second components in the ordered pairs is called the **range** of the relation.

Example 1 **Finding the Domain and Range of a Relation**

Find the domain and range of the relation linking the number of minutes played to the number of points scored by James in six games of the season (Table 9-2).

$$\{(38, 33), (44, 52), (40, 16), (41, 47), (33, 26), (38, 30)\}$$

Solution:

Domain: {38, 44, 40, 41, 33}   The domain is the set of first coordinates. Notice that repeated values are not listed more than once the set.

Range: {33, 52, 16, 47, 26, 30}   The range is the set of second coordinates.

The domain consists of the number of minutes played. The range represents the corresponding number of points.

Classroom Example: p. 686, Exercise 6

Skill Practice

1. Find the domain and range of the relation. {(0, 1), (4, 5), (−6, 8), (4, 13), (−8, 8)}

Answer

1. Domain: {0, 4, −6, −8}; Range: {1, 5, 8, 13}

Classroom Example: p. 687, Exercise 12

Example 2 Finding the Domain and Range of a Relation

The three women represented in Figure 9-8 each have children. Molly has one child, Peggy has two children, and Joanne has three children.

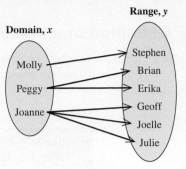

Figure 9-8

a. If the set of mothers is given as the domain and the set of children is the range, write a set of ordered pairs defining the relation given in Figure 9-8.

b. Write the domain and range of the relation.

Solution:

a. {(Molly, Stephen), (Peggy, Brian), (Peggy, Erika), (Joanne, Geoff), (Joanne, Joelle), (Joanne, Julie)}

b. Domain: {Molly, Peggy, Joanne}
   Range: {Stephen, Brian, Erika, Geoff, Joelle, Julie}

Skill Practice Given the relation represented by the figure:

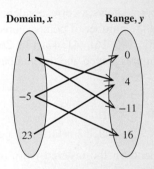

2. Write the relation as a set of ordered pairs.
3. Write the domain and range of the relation.

2. Definition of a Function

In mathematics, a special type of relation, called a function, is used extensively.

Definition of a Function

Given a relation in x and y, we say "y is a **function** of x" if for each element x in the domain, there is exactly one value of y in the range.

Note: This means that no two ordered pairs may have the same first coordinate and different second coordinates.

In Example 2, the relation linking the set of mothers with their respective children is *not* a function. The domain elements, "Peggy" and "Joanne," each have more than one child. Because these x values in the domain have more than one corresponding y value in the range, the relation is not a function.

Answers

2. {(1, 4), (1, −11), (−5, 0), (−5, 16), (23, 4)}
3. Domain: {1, −5, 23};
   Range: {0, 4, −11, 16}

To understand the difference between a relation that is a function and one that is not a function, consider Example 3.

Classroom Example: p. 687, Exercise 16

| **Example 3** | **Determining Whether a Relation Is a Function** |

Determine whether the following relations are functions.

a. $\{(2, -3), (4, 1), (3, -1), (2, 4)\}$ **b.** $\{(-3, 1), (0, 2), (4, -3), (1, 5), (-2, 1)\}$

Solution:

a. This relation is defined by the set of ordered pairs.

same x-values

$$\{(2, -3), (4, 1), (3, -1), (2, 4)\}$$

different y-values

When $x = 2$, there are two possibilities for y: $y = -3$ and $y = 4$

This relation is *not* a function because for $x = 2$, there is more than one corresponding element in the range.

b. This relation is defined by the set of ordered pairs: $\{(-3, 1), (0, 2), (4, -3), (1, 5), (-2, 1)\}$. Notice that no two ordered pairs have the same value of x but different values of y. Therefore, this relation *is* a function.

Skill Practice Determine whether the following relations are functions. If the relation is not a function, state why.

4. $\{(0, -7), (4, 9), (-2, -7), \left(\dfrac{1}{3}, \dfrac{1}{2}\right), (4, 10)\}$

5. $\{(-8, -3), (4, -3), (-12, 7), (-1, -1)\}$

3. Vertical Line Test

A relation that is not a function has at least one domain element, x, paired with more than one range element, y. For example, the ordered pairs (2, 1) and (2, 4) do not make a function. On a graph, these two points are aligned vertically in the xy-plane, and a vertical line drawn through one point also intersects the other point (Figure 9-9). Thus, if a vertical line drawn through a graph of a relation intersects the graph in more than one point, the relation cannot be a function. This idea is stated formally as the **vertical line test**.

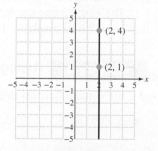

Figure 9-9

Instructor Note: Point out that a relation is not a function if any x value has more than one y value associated with it. However, it is fine for a single y value to have more than one x value associated with it.

Answers

4. Not a function because the domain element, 4, has two different y-values: (4, 9) and (4, 10).

5. Function

Using the Vertical Line Test

Consider a relation defined by a set of points (x, y) on a rectangular coordinate system. Then the graph defines y as a function of x if no vertical line intersects the graph in more than one point.

If any vertical line drawn through the graph of a relation intersects the relation in more than one point, then the relation does *not* define y as a function of x.

The vertical line test can be demonstrated by graphing the ordered pairs from the relations in Example 3 (Figure 9-10 and Figure 9-11).

$$\{(2, -3), (4, 1), (3, -1), (2, 4)\} \qquad \{(-3, 1), (0, 2), (4, -3), (1, 5), (-2, 1)\}$$

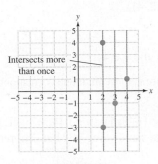

Figure 9-10

Not a Function

A vertical line intersects in more than one point.

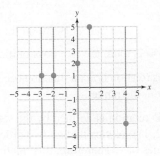

Figure 9-11

Function

No vertical line intersects more than once.

The relations in Examples 1, 2, and 3 consist of a finite number of ordered pairs. A relation may, however, consist of an *infinite* number of points defined by an equation or by a graph. For example, the equation $y = x + 1$ defines infinitely many ordered pairs whose y-coordinate is one more than its x-coordinate. These ordered pairs cannot all be listed but can be depicted in a graph.

The vertical line test is especially helpful in determining whether a relation is a function based on its graph.

Classroom Examples: p. 687, Exercises 20 and 22

Example 4 **Using the Vertical Line Test**

Use the vertical line test to determine whether the following relations are functions.

a.

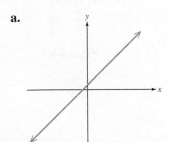

b.

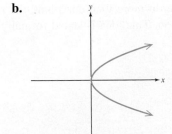

Solution:

a.

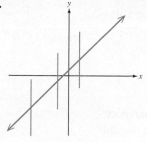

Function

No vertical line intersects
more than once.

b.

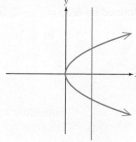

Not a Function

A vertical line intersects
in more than one point.

Skill Practice Use the vertical line test to determine if the following relations are functions.

6.

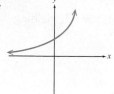

7.

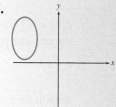

4. Function Notation

A function is defined as a relation with the added restriction that each value of the domain corresponds to only one value in the range. In mathematics, functions are often given by rules or equations to define the relationship between two or more variables. For example, the equation, $y = x + 1$ defines the set of ordered pairs such that the y-value is one more than the x-value.

When a function is defined by an equation, we often use **function notation**. For example, the equation $y = x + 1$ may be written in function notation as

$$f(x) = x + 1$$

where f is the name of the function, x is an input value from the domain of the function, and $f(x)$ is the function value (or y-value) corresponding to x.

The notation $f(x)$ is read as "f of x" or "the value of the function, f, at x."

A function may be evaluated at different values of x by substituting values of x from the domain into the function. For example, for the function defined by $f(x) = x + 1$ we can evaluate f at $x = 3$ by using substitution.

$f(x) = x + 1$

$f(3) = (3) + 1$

$f(3) = 4$ This is read as "f of 3 equals 4."

Thus, when $x = 3$, the corresponding function value is 4. This can also be interpreted as an ordered pair, $(3, 4)$.

The names of functions are often given by either lowercase letters or uppercase letters such as f, g, h, p, k, M, and so on.

> **Avoiding Mistakes**
>
> The notation $f(x)$ is read as "f of x" and does *not* imply multiplication.

Answers

6. Function   **7.** Not a function

Classroom Example: p. 688, Exercise 30

Example 5 **Evaluating a Function**

Given the function defined by $h(x) = x^2 - 2$, find the function values.

a. $h(0)$ **b.** $h(1)$ **c.** $h(2)$ **d.** $h(-1)$ **e.** $h(-2)$

Solution:

a. $h(x) = x^2 - 2$

$h(0) = (0)^2 - 2$ Substitute $x = 0$ into the function.

$= 0 - 2$

$= -2$ $h(0) = -2$ means that when $x = 0$, $y = -2$, yielding the ordered pair $(0, -2)$.

b. $h(x) = x^2 - 2$

$h(1) = (1)^2 - 2$ Substitute $x = 1$ into the function.

$= 1 - 2$

$= -1$ $h(1) = -1$ means that when $x = 1$, $y = -1$, yielding the ordered pair $(1, -1)$.

c. $h(x) = x^2 - 2$

$h(2) = (2)^2 - 2$ Substitute $x = 2$ into the function.

$= 4 - 2$

$= 2$ $h(2) = 2$ means that when $x = 2$, $y = 2$, yielding the ordered pair $(2, 2)$.

Avoiding Mistakes

Remember to use parentheses when substituting values into a function. In Example 5(d) for instance, this will ensure that $(-1)^2$ is evaluated rather than -1^2.

d. $h(x) = x^2 - 2$

$h(-1) = (-1)^2 - 2$ Substitute $x = -1$ into the function.

$= 1 - 2$

$= -1$ $h(-1) = -1$ means that when $x = -1$, $y = -1$, yielding the ordered pair $(-1, -1)$.

e. $h(x) = x^2 - 2$

$h(-2) = (-2)^2 - 2$ Substitute $x = -2$ into the function.

$= 4 - 2$

$= 2$ $h(-2) = 2$ means that when $x = -2$, $y = 2$, yielding the ordered pair $(-2, 2)$.

The rule $h(x) = x^2 - 2$ is equivalent to the equation $y = x^2 - 2$. The function values $h(0)$, $h(1)$, $h(2)$, $h(-1)$, and $h(-2)$ correspond to the y-values in the ordered pairs $(0, -2)$, $(1, -1)$, $(2, 2)$, $(-1, -1)$, and $(-2, 2)$, respectively. These points can be used to sketch a graph of the function (Figure 9-12).

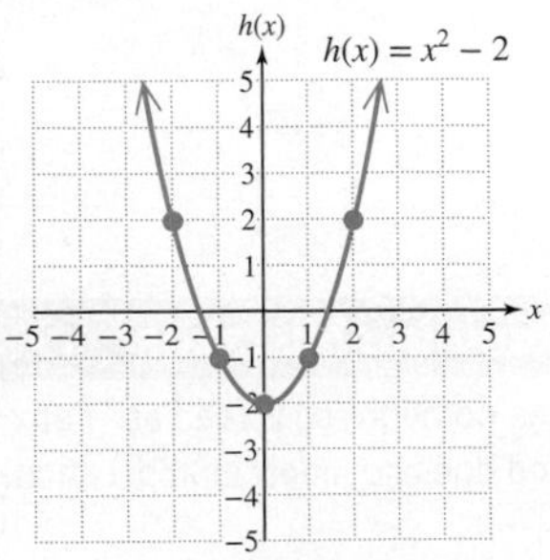

Figure 9-12

Skill Practice Given the function defined by $f(x) = x^2 - 5x$, find the function values.

8. $f(1)$ **9.** $f(0)$ **10.** $f(-3)$ **11.** $f(2)$ **12.** $f(-1)$

Answers

8. -4 **9.** 0 **10.** 24

11. -6 **12.** 6

5. Domain and Range of a Function

A function is a relation, and it is often necessary to determine its domain and range. Consider a function defined by the equation $y = f(x)$. The **domain** of f is the set of all x-values that when substituted into the function produce a real number. The **range** of f is the set of all y-values corresponding to the values of x in the domain.

    For Examples 6 and 7, we find the domain and range based on the graph of the function.

Classroom Example: p. 688, Exercise 38

Example 6	Finding the Domain and Range of a Function

Find the domain and range from the graph of the function.

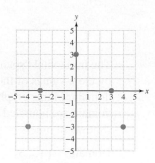

Solution:

From the figure, the function defines the set of ordered pairs:

$$\{(-4, -3), (-3, 0), (0, 3), (3, 0), (4, -3)\}$$

    The domain is the set of all x-coordinates: $\{-4, -3, 0, 3, 4\}$

    The range is the set of all y-coordinates: $\{-3, 0, 3\}$

Skill Practice Find the domain and range from the graph of the function.

13.

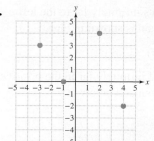

Answer

13. Domain: $\{-3, -1, 2, 4\}$
     Range: $\{-2, 0, 3, 4\}$

Classroom Example: p. 689,
Exercise 46

Animation

Example 7   **Finding the Domain and Range of a Function**

Find the domain and range of the functions based on the graph of the function. Express the answers in interval notation.

a.

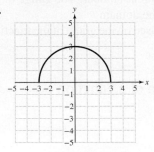

b.

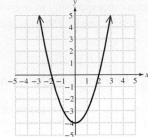

Solution:

a.

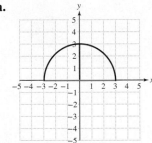

The horizontal "span" of the graph is determined by the x-values of the points. This is the domain. In this graph, the x-values in the domain are bounded between -3 and 3. (Shown in blue.)

Domain: $[-3, 3]$

The vertical "span" of the graph is determined by the y-values of the points. This is the range.

The y-values in the range are bounded between 0 and 3. (Shown in red.)

Range: $[0, 3]$

b.

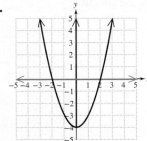

The function extends infinitely far to the left and right. The domain is shown in blue.

Domain: $(-\infty, \infty)$

The y-values extend infinitely far in the positive direction, but are bounded below at $y = -4$. (Shown in red.)

Range: $[-4, \infty)$

Skill Practice   Find the domain and range of the functions based on the graph of the function.

14.

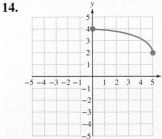

15.

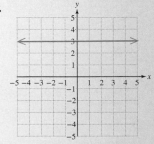

Answers

14. Domain: $[0, 5]$
Range: $[2, 4]$
15. Domain: $(-\infty, \infty)$
Range: $\{3\}$

6. Applications of Functions

| Example 8 | Using a Function in an Application |

Classroom Example: p. 690, Exercise 56

The score a student receives on an exam is a function of the number of hours the student spends studying. The function defined by

$$P(x) = \frac{100x^2}{40 + x^2} \quad (x \geq 0)$$

indicates that a student's percentage score after studying for x hours will be $P(x)$.

a. Evaluate $P(0)$, $P(10)$, and $P(20)$.

b. Interpret the function values from part (a) in the context of this problem.

Solution:

a. $P(x) = \dfrac{100x^2}{40 + x^2}$

$P(0) = \dfrac{100(0)^2}{40 + (0)^2}$ $P(10) = \dfrac{100(10)^2}{40 + (10)^2}$ $P(20) = \dfrac{100(20)^2}{40 + (20)^2}$

$P(0) = \dfrac{0}{40}$ $P(10) = \dfrac{10{,}000}{140}$ $P(20) = \dfrac{40{,}000}{440}$

$P(0) = 0$ $P(10) = \dfrac{500}{7} \approx 71.4$ $P(20) = \dfrac{1000}{11} \approx 90.9$

b. $P(0) = 0$ means that for 0 hr spent studying, the student will receive 0% on the exam.

$P(10) \approx 71.4$ means that for 10 hr spent studying, the student will receive approximately 71.4% on the exam.

$P(20) \approx 90.9$ means that for 20 hr spent studying, the student will receive approximately 90.9% on the exam.

The graph of $P(x) = \dfrac{100x^2}{40 + x^2}$ is shown in Figure 9-13.

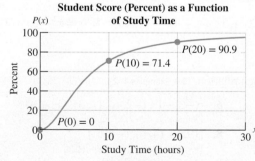

Student Score (Percent) as a Function of Study Time

Figure 9-13

Skill Practice The function defined by $S(x) = 6x^2$ $(x \geq 0)$ indicates the surface area of the cube whose side is length x (in inches).

16. Evaluate $S(5)$.

17. Interpret the function value, $S(5)$.

Answers

16. 150

17. For a cube 5 in. on a side, the surface area is 150 in.^{2}

Calculator Connections

Topic: Graphing Functions

A graphing calculator can be used to graph a function. We replace $f(x)$ by y and enter the defining expression into the calculator. For example:

$$f(x) = \frac{1}{4}x^3 - x^2 - x + 4 \quad \text{becomes} \quad y = \frac{1}{4}x^3 - x^2 - x + 4$$

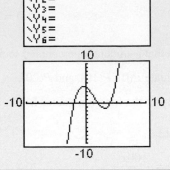

Calculator Exercises

Use a graphing calculator to graph the following functions.

1. $f(x) = x^2 - 5x + 2$

2. $g(x) = -x^2 + 4x + 5$

3. $m(x) = \frac{1}{3}x^3 + x^2 - 3x - 1$

4. $n(x) = x^3 - 9x$

Section 9.6 Practice Exercises

For additional exercises, see Classroom Activities 9.6A–9.6D in the *Student's Resource Manual* at www.mhhe.com/moh.

Vocabulary and Key Concepts

1. a. A set of ordered pairs (x, y) is called a ____relation____ in x and y.

    b. The ____domain____ of a relation is the set of first components in the ordered pairs.

    c. The ____range____ of a relation is the set of second components in the ordered pairs.

    d. Given a relation in x and y, we say that y is a ____function____ of x if for each element x in the domain, there is exactly one value of y in the range.

    e. If a ____vertical____ line intersects the graph of a relation in more than one point, the relation is not a function.

    f. Function notation for the relation $y = 7x - 4$ is $f(x) = $ ____$7x - 4$____.

Review Exercises

For Exercises 2–4, find the vertex of each parabola.

2. $y = -3x^2 + 2x + 2$ $\left(\frac{1}{3}, \frac{7}{3}\right)$

3. $y = 4x^2 - 2x + 3$ $\left(\frac{1}{4}, \frac{11}{4}\right)$

4. $y = x^2 - 5x + 2$ $\left(\frac{5}{2}, -\frac{17}{4}\right)$

Concept 1: Definition of a Relation

For Exercises 5–14, determine the domain and range of each relation. **(See Examples 1–2.)**

5. $\{(4, 2), (3, 7), (4, 1), (0, 6)\}$
   Domain: $\{4, 3, 0\}$; range: $\{2, 7, 1, 6\}$

6. $\{(-3, -1), (-2, 6), (1, 3), (1, -2)\}$
   Domain: $\{-3, -2, 1\}$; range: $\{-1, 6, 3, -2\}$

7. $\{(\frac{1}{2}, 3), (0, 3), (1, 3)\}$
   Domain: $\{\frac{1}{2}, 0, 1\}$; range: $\{3\}$

8. $\{(9, 6), (4, 6), (-\frac{1}{3}, 6)\}$
   Domain: $\{9, 4, -\frac{1}{3}\}$; range: $\{6\}$

9. $\{(0, 0), (5, 0), (-8, 2), (8, 5)\}$
   Domain: $\{0, 5, -8, 8\}$; range: $\{0, 2, 5\}$

10. $\{(\frac{1}{2}, -\frac{1}{2}), (-4, 0), (0, -\frac{1}{2}), (\frac{1}{2}, 0)\}$
    Domain: $\{\frac{1}{2}, -4, 0\}$; range: $\{-\frac{1}{2}, 0\}$

11.

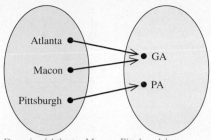

Domain: {Atlanta, Macon, Pittsburgh};
range: {GA, PA}

12.

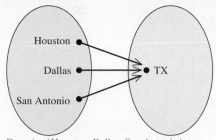

Domain: {Houston, Dallas, San Antonio};
range: {TX}

13.

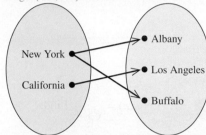

Domain: {New York, California};
range: {Albany, Los Angeles, Buffalo}

14.

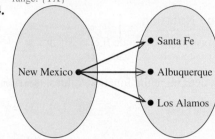

Domain: {New Mexico};
range: {Santa Fe, Albuquerque, Los Alamos}

Concept 2: Definition of a Function

15. How can you determine if a set of ordered pairs represents a function?

The relation is a function if each element in the domain has exactly one corresponding element in the range.

16. Refer back to Exercises 6, 8, 10, 12, and 14. Identify which relations are functions.

The relations in Exercises 8 and 12 are functions.

17. Refer back to Exercises 5, 7, 9, 11, and 13. Identify which relations are functions. **(See Example 3.)**

The relations in Exercises 7, 9, and 11 are functions.

Concept 3: Vertical Line Test

18. How can you tell from the graph of a relation if the relation is a function?

The graph represents a function if no vertical line intersects the graph more than once.

For Exercises 19–27, use the vertical line test to determine if the relation defines y as a function of x. **(See Example 4.)**

19. Yes

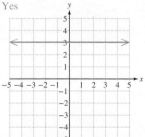

20. Yes

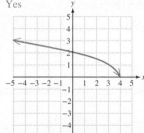

21. No

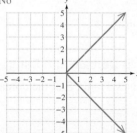

22. No

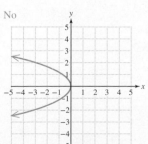

23. No

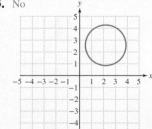

24. No

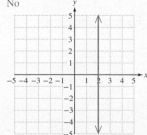

25.   Yes
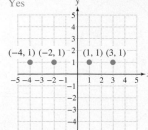
(−4, 1) (−2, 1)   (1, 1) (3, 1)

26.   No
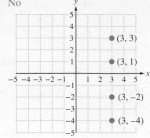
(3, 3)
(3, 1)
(3, −2)
(3, −4)

27.   Yes
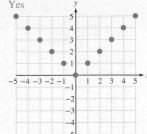

Concept 4: Function Notation

28. Explain how you would evaluate $f(x) = 3x^2$ at $x = -1$.
Substitute −1 for x and simplify the result.

For Exercises 29–36, determine the function values. **(See Example 5.)**

29. Let $f(x) = 2x - 5$. Find:

   a. $f(0)$ −5

   b. $f(2)$ −1

   c. $f(-3)$ −11

30. Let $g(x) = x^2 + 1$. Find:

   a. $g(0)$ 1

   b. $g(-1)$ 2

   c. $g(3)$ 10

31. Let $h(x) = \dfrac{1}{x + 4}$. Find:

   a. $h(1)$ $\frac{1}{5}$

   b. $h(0)$ $\frac{1}{4}$

   c. $h(-2)$ $\frac{1}{2}$

32. Let $p(x) = \sqrt{x + 4}$. Find:

   a. $p(0)$ 2

   b. $p(-4)$ 0

   c. $p(5)$ 3

33. Let $m(x) = |5x - 7|$. Find:

   a. $m(0)$ 7

   b. $m(1)$ 2

   c. $m(2)$ 3

34. Let $w(x) = |2x - 3|$. Find:

   a. $w(0)$ 3

   b. $w(1)$ 1

   c. $w(2)$ 1

35. Let $n(x) = \sqrt{x - 2}$. Find:

   a. $n(2)$ 0

   b. $n(3)$ 1

   c. $n(6)$ 2

36. Let $t(x) = \dfrac{1}{x - 3}$. Find:

   a. $t(1)$ $-\frac{1}{2}$

   b. $t(-1)$ $-\frac{1}{4}$

   c. $t(2)$ −1

Concept 5: Domain and Range of a Function

For Exercises 37–40, find the domain and range from the graphs of the functions. **(See Example 6.)**

37.
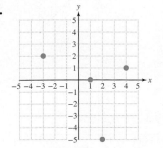
Domain: {−3, 1, 2, 4};
range: {−5, 0, 1, 2}

38.
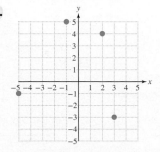
Domain: {−5, −1, 2, 3};
range: {−3, −1, 4, 5}

39.

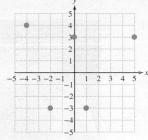

Domain: {−4, −2, 0, 1, 5};
range: {−3, 3, 4}

40.

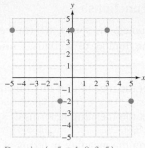

Domain: {−5, −1, 0, 3, 5};
range: {−2, 4,}

For Exercises 41–44, match the domain and range given with a possible graph.

41. Domain: $(-\infty, \infty)$

Range: $[1, \infty)$ b

42. Domain: $[-4, 4]$

Range: $[-2, 2]$ a

43. Domain: $[-2, \infty)$

Range: $(-\infty, \infty)$ c

44. Domain: $(-\infty, \infty)$

Range: $(-\infty, \infty)$ d

a.

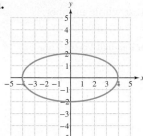

b.

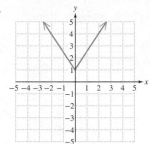

c.

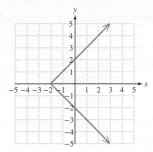

d.

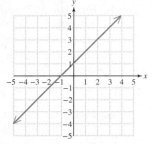

For Exercises 45–48, find the domain and range from the graph of each relation. Express the answers in interval notation.
(See Example 7.)

45.

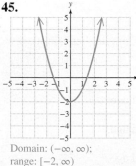

Domain: $(-\infty, \infty)$;
range: $[-2, \infty)$

46.

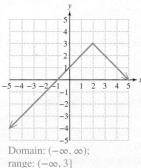

Domain: $(-\infty, \infty)$;
range: $(-\infty, 3]$

47.

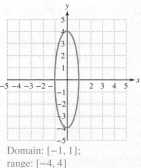

Domain: $[-1, 1]$;
range: $[-4, 4]$

48.

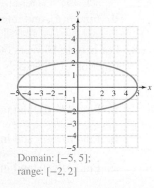

Domain: $[-5, 5]$;
range: $[-2, 2]$

For Exercises 49–52, write each expression as an English phrase.

49. $f(6) = 2$
The function value
at $x = 6$ is 2.

50. $f(-2) = -14$
The function value
at $x = -2$ is −14.

51. $g\left(\dfrac{1}{2}\right) = \dfrac{1}{4}$ The function
value at $x = \frac{1}{2}$ is $\frac{1}{4}$.

52. $h(k) = k^2$
The function value at
$x = k$ is k^2.

53. Consider a function defined by $y = f(x)$. The function value $f(2) = 7$ corresponds to what ordered pair?
(2, 7)

54. Consider a function defined by $y = f(x)$. The function value $f(-3) = -4$ corresponds to what ordered pair?
(−3, −4)

Writing   Translating Expression   Geometry   Scientific Calculator   Video

Concept 6: Applications of Functions

55. In the absence of air resistance, the speed, $s(t)$ (in feet per second: ft/sec), of an object in free fall is a function of the number of seconds, t, after it was dropped. **(See Example 8.)**

$$s(t) = 32t$$

a. Find $s(1)$, and interpret the meaning of this function value in terms of speed and time.
$s(1) = 32$. The speed of an object 1 sec after being dropped is 32 ft/sec.

b. Find $s(2)$, and interpret the meaning in terms of speed and time.
$s(2) = 64$. The speed of an object 2 sec after being dropped is 64 ft/sec.

c. Find $s(10)$, and interpret the meaning in terms of speed and time.
$s(10) = 320$. The speed of an object 10 sec after being dropped is 320 ft/sec.

d. A ball dropped from the top of the Willis Tower in Chicago falls for approximately 9.2 sec. How fast was the ball going the instant before it hit the ground? 294.4 ft/sec

© IMS Communications Ltd./
Capstone Design/FlatEarth
Images RF

56. The number of people diagnosed with skin cancer, $N(x)$, can be approximated by $N(x) = 45,625(1 + 0.029x)$. For this function, x represents the number of years since 2003. (*Source:* Centers for Disease Control)

a. Evaluate $N(0)$ and interpret its meaning in the context of this problem. $N(0) = 45,625$; This means that in the year 2003 (when $x = 0$), the number of people diagnosed with skin cancer was approximately 45,625.

b. Evaluate $N(7)$ and interpret its meaning in the context of this problem. Round to the nearest whole number.
$N(7) = 54,887$; This means that in the year 2010 (when $x = 7$), the number of people diagnosed with skin cancer was approximately 54,887.

57. A punter kicks a football straight up with an initial velocity of 64 ft/sec. The height of the ball, $h(t)$ (in feet), is a function of the number of seconds, t, after the ball is kicked.

$$h(t) = -16t^2 + 64t + 3$$

a. Find $h(0)$, and interpret the meaning of the function value in terms of time and height.
$h(0) = 3$. The initial height of the ball is 3 ft.

b. Find $h(1)$, and interpret the meaning in terms of time and height.
$h(1) = 51$. The height of the ball 1 sec after being kicked is 51 ft.

c. Find $h(2)$, and interpret the meaning in terms of time and height.
$h(2) = 67$. The height of the ball 2 sec after being kicked is 67 ft.

d. Find $h(4)$, and interpret the meaning in terms of time and height.
$h(4) = 3$. The height of the ball 4 sec after being kicked is 3 ft.

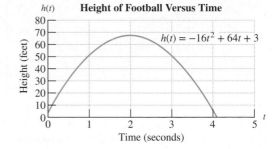

Height of Football Versus Time

$h(t) = -16t^2 + 64t + 3$

58. For people 16 years old and older, the maximum recommended heart rate, $M(x)$ (in beats per minute: beats/min), is a function of a person's age, x (in years).

$$M(x) = 220 - x \quad \text{for } x \geq 16$$

a. Find $M(16)$, and interpret the meaning in terms of maximum recommended heart rate and age.
$M(16) = 204$. A 16-year-old adult's maximum recommended heart rate is 204 beats/min.

b. Find $M(30)$, and interpret the meaning in terms of maximum recommended heart rate and age.
$M(30) = 190$. A 30-year-old adult's maximum recommended heart rate is 190 beats/min.

c. Find $M(60)$, and interpret the meaning in terms of maximum recommended heart rate and age.
$M(60) = 160$. A 60-year-old adult's maximum recommended heart rate is 160 beats/min.

d. Find your own maximum recommended heart rate. Answers will vary.

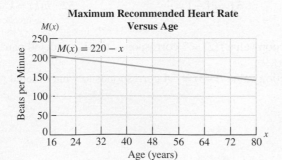

Maximum Recommended Heart Rate Versus Age

$M(x) = 220 - x$

© Imagesource/Jupiterimages RF

 Writing   Translating Expression   Geometry   Scientific Calculator Video

59. An electrician charges $75 to visit, diagnose, and give an estimate for repairing a refrigerator. If Helena decides to have her refrigerator fixed, she will then be charged an additional $50 per hour for labor costs. The equation for the total cost, $C(x)$, of fixing the refrigerator can be modeled by the linear function $C(x) = 75 + 50x$, where x is the number of hours it takes the electrician to fix the refrigerator.

   a. Find the total cost for an estimate and 3 hr of labor.   The cost is $225.

   b. If Helena spent $200 on fixing her refrigerator, how many hours of labor was she charged for?
   She was charged for 2.5 hr.

   c. What is the domain of $C(x)$?   Domain: $[0, \infty)$

   d. What does the y-intercept represent?   The y-intercept represents the cost of the estimate.

Group Activity

Maximizing Volume

Materials: A calculator, a ruler, and a sheet of $8\frac{1}{2}$ by 11 in. paper.

Estimated Time: 25–30 minutes

Group Size: 3

Antonio is going to build a custom gutter system for his house. He plans to use rectangular strips of aluminum that are $8\frac{1}{2}$ in. wide and 72 in. long. Each piece of aluminum will be turned up at a distance of x in. from the sides to form a gutter. See Figure 9-14.

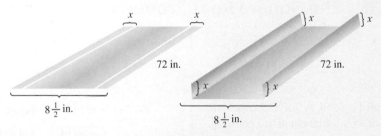

Figure 9-14

Antonio wants to maximize the volume of water that the gutters can hold. To do this, he must determine the distance, x, that should be turned up to form the height of the gutter.

1. To familiarize yourself with this problem, we will simulate the gutter problem using an $8\frac{1}{2}$ by 11 in. piece of paper. For each value of x in the table, turn up the sides of the paper x in. from the edge. Then measure the base, the height, and the length of the paper "gutter," and calculate the volume.

Height, x	Base	Length	Volume
0.5 in.	7.5 in.	11 in.	41.25 in.^{3}
1.0 in.	6.5 in.	11 in.	71.5 in.^{3}
1.5 in.	5.5 in.	11 in.	90.75 in.^{3}
2.0 in.	4.5 in.	11 in.	99 in.^{3}
2.5 in.	3.5 in.	11 in.	96.25 in.^{3}
3.0 in.	2.5 in.	11 in.	82.5 in.^{3}
3.5 in.	1.5 in.	11 in.	57.75 in.^{3}

 Writing   Translating Expression    Geometry   Scientific Calculator   Video

2. From the table, estimate the dimensions for the maximum volume.   Height 2.0 in., base 4.5 in., length 11 in.

3. Now follow these steps to find the optimal distance, x, that you should fold the paper to make the greatest volume within the paper gutter.

   a. If the height of the paper gutter is x in., write an expression for the base of the paper gutter. $8.5 - 2x$

   b. Write a function for the volume of the paper gutter. $V = (8.5 - 2x)(x)(11)$ or $V = -22x^2 + 93.5x$

   c. Find the vertex of the parabola defined by the function in part (b).
   (2.125, 99.34375)

   d. Interpret the meaning of the vertex from part (c).
   Fold the paper 2.125 in. from the edge. This will produce a maximum volume of 99.34375 in.[3]

4. Use the concept from the paper gutter to write a function for the volume of Antonio's aluminum gutter that is 72 in. long.
$V = (8.5 - 2x)(x)(72)$ or $V = -144x^2 + 612x$

5. Now find the optimal distance, x, that he should fold the aluminum sheet to make the greatest volume within the 72-in.-long aluminum gutter. What is the maximum volume?
Fold the aluminum 2.125 in. from the edge. This will produce a maximum volume of 650.25 in.[3]

Chapter 9   Summary

Section 9.1   The Square Root Property

Key Concepts

Square Root Property

If $x^2 = k$, then $x = \pm\sqrt{k}$.

The square root property can be used to solve a quadratic equation written as a square of a binomial equal to a constant.

Example

Example 1

$$(x - 5)^2 = 13$$

$$x - 5 = \pm\sqrt{13} \qquad \text{Square root property}$$

$$x = 5 \pm \sqrt{13} \qquad \text{Solve for } x.$$

The solution set is $\{5 \pm \sqrt{13}\}$.

| **Section 9.2** | **Completing the Square** |

Key Concepts

Solving a Quadratic Equation of the Form

$ax^2 + bx + c = 0 \ (a \neq 0)$ **by Completing the**

Square and Applying the Square Root Property

1. Divide both sides by a to make the leading coefficient 1.
2. Isolate the variable terms on one side of the equation.
3. Complete the square by adding the square of $\frac{1}{2}$ the linear term coefficient to both sides of the equation. Then factor the resulting perfect square trinomial.
4. Apply the square root property and solve for x.

Example

Example 1

$2x^2 - 8x - 6 = 0$

Step 1: $\quad \dfrac{2x^2}{2} - \dfrac{8x}{2} - \dfrac{6}{2} = \dfrac{0}{2}$

$\qquad\qquad\quad x^2 - 4x - 3 = 0$

Step 2: $\quad x^2 - 4x = 3$

Step 3: $\quad x^2 - 4x + 4 = 3 + 4 \quad$ Note that

$\qquad\qquad\qquad\qquad\qquad\qquad [\frac{1}{2}(-4)]^2 = (-2)^2 = 4$

$\qquad\qquad\quad (x - 2)^2 = 7$

Step 4: $\qquad\quad x - 2 = \pm\sqrt{7}$

$\qquad\qquad\qquad\quad x = 2 \pm \sqrt{7}$

The solution set is $\{2 \pm \sqrt{7}\}$.

| **Section 9.3** | **Quadratic Formula** |

Key Concepts

The solutions to a quadratic equation of the form $ax^2 + bx + c = 0 \ (a \neq 0)$ are given by the **quadratic formula**:

$$x = \frac{-b \pm \sqrt{b^2 - 4ac}}{2a}$$

Three Methods for Solving a Quadratic Equation

1. Factoring
2. Completing the square and applying the square root property
3. Using the quadratic formula

Example

Example 1

$$3x^2 = 2x + 4$$

$3x^2 - 2x - 4 = 0 \qquad a = 3, b = -2, c = -4$

$x = \dfrac{-(-2) \pm \sqrt{(-2)^2 - 4(3)(-4)}}{2(3)}$

$\ = \dfrac{2 \pm \sqrt{4 + 48}}{6}$

$\ = \dfrac{2 \pm \sqrt{52}}{6}$

$\ = \dfrac{2 \pm 2\sqrt{13}}{6} \qquad$ Simplify the radical.

$\ = \dfrac{2(1 \pm \sqrt{13})}{6} \qquad$ Factor.

$\ = \dfrac{1 \pm \sqrt{13}}{3} \qquad$ Simplify.

The solution set is $\left\{ \dfrac{1 \pm \sqrt{13}}{3} \right\}$.

Section 9.4 — Complex Numbers

Key Concepts

$i = \sqrt{-1}$ and $i^2 = -1$
For a real number $b > 0$, $\sqrt{-b} = i\sqrt{b}$

A complex number is in the form $a + bi$, where a and b are real numbers. The value a is called the real part, and the value b is called the imaginary part.

To add or subtract complex numbers, combine the real parts and combine the imaginary parts.

Multiply complex numbers by using the distributive property.

Divide complex numbers by multiplying the numerator and denominator by the conjugate of the denominator.

Examples

Example 1

$$\sqrt{-4} \cdot \sqrt{-9}$$
$$= (2i)(3i)$$
$$= 6i^2$$
$$= -6$$

Example 2

$$(3 - 5i) - (2 + i) + (3 - 2i)$$
$$= 3 - 5i - 2 - i + 3 - 2i$$
$$= 4 - 8i$$

Example 3

$$(1 + 6i)(2 + 4i)$$
$$= 2 + 4i + 12i + 24i^2$$
$$= 2 + 16i + 24(-1)$$
$$= -22 + 16i$$

Example 4

$$\frac{3}{2 - 5i}$$

$$= \frac{3}{2 - 5i} \cdot \frac{(2 + 5i)}{(2 + 5i)} = \frac{6 + 15i}{4 - 25i^2} = \frac{6 + 15i}{4 - 25(-1)}$$

$$= \frac{6 + 15i}{29} \quad \text{or} \quad \frac{6}{29} + \frac{15}{29}i$$

Section 9.5 — Graphing Quadratic Equations

Key Concepts

Let a, b, and c represent real numbers such that $a \neq 0$. Then an equation of the form $y = ax^2 + bx + c$ is called a **quadratic equation in two variables**.

The graph of a quadratic equation is called a **parabola**.

The leading coefficient, a, of a quadratic equation, $y = ax^2 + bx + c$, determines if the parabola will open upward or downward. If $a > 0$, then the parabola opens upward. If $a < 0$, then the parabola opens downward.

Examples

Example 1

$y = x^2 - 4x - 3$ is a quadratic equation. Its graph is in the shape of a parabola.

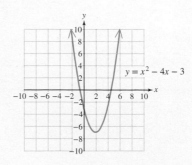

Finding the Vertex of a Parabola

1. For the equation $y = ax^2 + bx + c$ $(a \neq 0)$, the x-coordinate of the vertex is

$$x = \frac{-b}{2a}$$

2. To find the corresponding y-coordinate of the vertex, substitute the value of the x-coordinate found in step 1 and solve for y.

   If a parabola opens upward, the vertex is the lowest point on the graph. If a parabola opens downward, the vertex is the highest point on the graph.

Example 2

Find the vertex of the parabola defined by $y = x^2 - 4x - 3$.

$$x = \frac{-b}{2a} = \frac{-(-4)}{2(1)} = \frac{4}{2} = 2$$

$$y = (2)^2 - 4(2) - 3 = -7 \qquad \text{The vertex is } (2, -7).$$

For the equation $y = x^2 - 4x - 3$, $a > 0$. Therefore, the parabola opens upward. The vertex $(2, -7)$ represents the minimum point on the graph.

Section 9.6   Introduction to Functions

Key Concepts

Any set of ordered pairs, (x, y), is called a **relation** in x and y.

    The **domain** of a relation is the set of first components in the ordered pairs in the relation. The **range** of a relation is the set of second components in the ordered pairs.

    Given a relation in x and y, we say "y is a **function** of x" if for each element x in the domain, there is exactly one value y in the range.

Examples

Example 1

Find the domain and range of the relation.

$\{(0, 0), (1, 1), (2, 4), (3, 9), (-1, 1), (-2, 4), (-3, 9)\}$

Domain: $\{0, 1, 2, 3, -1, -2, -3\}$

Range: $\{0, 1, 4, 9\}$

Example 2

Function: $\{(1, 3), (2, 5), (6, 3)\}$

Not a function: $\{(1, 3), (2, 5), (1, -2)\}$

different y-values for the same x-value

Vertical Line Test for Functions

Consider any relation defined by a set of points (x, y) on a rectangular coordinate system. Then the graph defines y as a function of x if no vertical line intersects the graph in more than one point.

Example 3

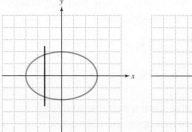

Not a Function
Vertical line intersects more than once.

Function
No vertical line intersects more than once.

Function Notation

$f(x)$ is the value of the function, f, at x.

Example 4

Given $f(x) = -3x^2 + 5x$, find $f(-2)$.

$$f(-2) = -3(-2)^2 + 5(-2)$$
$$= -12 - 10$$
$$= -22$$

Chapter 9   Review Exercises

Section 9.1

For Exercises 1–4, identify each equation as linear or quadratic.

1. $5x - 10 = 3x - 6$
  Linear
2. $(x + 6)^2 = 6$
  Quadratic
3. $x(x - 4) = 5x - 2$
  Quadratic
4. $3(x + 6) = 18(x - 1)$
  Linear

For Exercises 5–12, solve each equation using the square root property.

5. $x^2 = 25$ $\{5, -5\}$
6. $x^2 - 19 = 0$
  $\{\sqrt{19}, -\sqrt{19}\}$
7. $x^2 + 49 = 0$ The equation has no real-valued solutions.
8. $x^2 = -48$ The equation has no real-valued solutions.
9. $(x + 1)^2 = 14$
  $\{-1 \pm \sqrt{14}\}$
10. $(x - 2)^2 = 60$
  $\{2 \pm 2\sqrt{15}\}$
11. $\left(x - \dfrac{1}{8}\right)^2 = \dfrac{3}{64}$
  $\left\{\dfrac{1}{8} \pm \dfrac{\sqrt{3}}{8}\right\}$
12. $(2x - 3)^2 = 20$
  $\left\{\dfrac{3 \pm 2\sqrt{5}}{2}\right\}$

Section 9.2

For Exercises 13–16, determine the value of n that makes the polynomial a perfect square trinomial.

13. $x^2 + 12x + n$ $n = 36$
14. $x^2 - 18x + n$ $n = 81$
15. $x^2 - 5x + n$ $n = \dfrac{25}{4}$
16. $x^2 + 7x + n$ $n = \dfrac{49}{4}$

For Exercises 17–20, solve each quadratic equation by completing the square and applying the square root property.

17. $x^2 + 8x + 3 = 0$
  $\{-4 \pm \sqrt{13}\}$
18. $x^2 - 2x - 4 = 0$
  $\{1 \pm \sqrt{5}\}$
19. $2x^2 - 6x - 6 = 0$
  $\left\{\dfrac{3}{2} \pm \dfrac{\sqrt{21}}{2}\right\}$
20. $3x^2 - 7x - 3 = 0$
  $\left\{\dfrac{7}{6} \pm \dfrac{\sqrt{85}}{6}\right\}$

 21. A right triangle has legs of equal length. If the hypotenuse is 15 ft long, find the length of each leg. Round the answer to the nearest tenth of a foot.   10.6 ft

22. A can in the shape of a right circular cylinder holds approximately 362 cm³ of liquid. If the height of the can is 12.1 cm, find the radius of the can. Round to the nearest tenth of a centimeter. (*Hint:* The volume of a right circular cylinder is given by: $V = \pi r^2 h$)
  3.1 cm

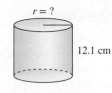
r = ?
12.1 cm

Section 9.3

23. Write the quadratic formula from memory.
  For $ax^2 + bx + c = 0$, $x = \dfrac{-b \pm \sqrt{b^2 - 4ac}}{2a}$

For Exercises 24–33, find the real solutions for each quadratic equation using the quadratic formula.

24. $5x^2 + x - 7 = 0$
  $\left\{\dfrac{-1 \pm \sqrt{141}}{10}\right\}$
25. $x^2 + 4x + 4 = 0$
  $\{-2\}$
26. $3x^2 - 2x + 2 = 0$
  The equation has no real-valued solutions.
27. $2x^2 - x - 3 = 0$
  $\left\{\dfrac{3}{2}, -1\right\}$
28. $\dfrac{1}{8}x^2 + x = \dfrac{5}{2}$
  $\{-10, 2\}$
29. $\dfrac{1}{6}x^2 + x + \dfrac{1}{3} = 0$
  $\{-3 \pm \sqrt{7}\}$
30. $1.2x^2 + 6x = 7.2$ $\{1, -6\}$
31. $0.01x^2 - 0.02x - 0.04 = 0$ $\{1 \pm \sqrt{5}\}$
32. $(x + 6)(x + 2) = 10$ $\{-4 \pm \sqrt{14}\}$
33. $(x - 1)(x - 7) = -18$
  The equation has no real-valued solutions.

34. One number is two more than another number. Their product is 11.25. Find the numbers.
The numbers are −2.5 and −4.5, or 2.5 and 4.5.

35. The base of a parallelogram is 1 cm longer than the height, and the area is 24 cm². Find the values of the base and height of the parallelogram. Use a calculator to approximate the values to the nearest tenth of a centimeter.
The height is approximately 4.4 cm. The base is approximately 5.4 cm.

36. An astronaut on the moon tosses a rock upward with an initial velocity of 25 ft/sec. The height of the rock, $h(t)$ (in feet), is determined by the number of seconds, t, after the rock is released according to the equation.

$$h(t) = -2.7t^2 + 25t + 5$$

Find the time required for the rock to hit the ground. [*Hint:* At ground level, $h(t) = 0$.] Round to the nearest tenth of a second.   9.5 sec

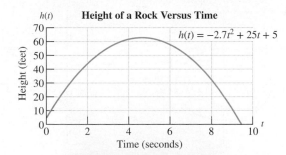

Source: NASA Headquarters – Greatest Images of NASA (NASA-HQ-GRIN)

Section 9.4

37. Define a complex number.
$a + bi$, where a and b are real numbers and $i = \sqrt{-1}$

38. Define an imaginary number. $a + bi$, where $b \neq 0$

For Exercises 39–42, rewrite each expression in terms of i.

39. $\sqrt{-16}$ $4i$

40. $-\sqrt{-5}$ $-i\sqrt{5}$

41. $\sqrt{-75} \cdot \sqrt{-3}$ -15

42. $\dfrac{-\sqrt{-24}}{\sqrt{6}}$ $-2i$

For Exercises 43–46, simplify completely.

43. $-6i^2$ 6

44. $-8i^2$ 8

45. $12 - i^2$ 13

46. $9 - 2i^2$ 11

For Exercises 47–50, perform the indicated operations. Write the final answer in the form $a + bi$.

47. $(-3 + i) - (2 - 4i)$ $-5 + 5i$

48. $(1 + 6i)(3 - i)$ $9 + 17i$

49. $(4 - 3i)(4 + 3i)$ $25 + 0i$

50. $(5 - i)^2$ $24 - 10i$

For Exercises 51–52, write each expression in the form $a + bi$, and determine the real and imaginary parts.

51. $\dfrac{17 - 4i}{-4}$ $-\dfrac{17}{4} + i$
Real part: $-\dfrac{17}{4}$; Imaginary part: 1

52. $\dfrac{-16 - 8i}{8}$ $-2 - i$
Real part: -2; Imaginary part: -1

For Exercises 53–54, divide and simplify. Write the final answer in the form $a + bi$.

53. $\dfrac{2 - i}{3 + 2i}$ $\dfrac{4}{13} - \dfrac{7}{13}i$

54. $\dfrac{10 + 5i}{2 - i}$ $3 + 4i$

For Exercises 55–58, solve each quadratic equation.

55. $(x + 12)^2 = -20$
$\{-12 \pm 2i\sqrt{5}\}$

56. $(x - 7)^2 = -18$
$\{7 \pm 3i\sqrt{2}\}$

57. $4x^2 - x + 2 = 0$
$\left\{\dfrac{1}{8} \pm \dfrac{\sqrt{31}}{8}i\right\}$

58. $2x^2 + 3x + 2 = 0$
$\left\{-\dfrac{3}{4} \pm \dfrac{\sqrt{7}}{4}i\right\}$

Section 9.5

For Exercises 59–62, given $y = ax^2 + bx + c$, identify a and determine if the parabola opens upward or downward.

59. $y = x^2 - 3x + 1$
$a = 1$; upward

60. $y = -x^2 + 8x + 2$
$a = -1$; downward

61. $y = -2x^2 + x - 12$
$a = -2$; downward

62. $y = 5x^2 - 2x - 6$
$a = 5$; upward

For Exercises 63–66, find the vertex for each parabola.

63. $y = 3x^2 + 6x + 4$
Vertex: $(-1, 1)$

64. $y = -x^2 + 8x + 3$
Vertex: $(4, 19)$

65. $y = -2x^2 + 12x - 5$
Vertex: $(3, 13)$

66. $y = 2x^2 + 2x - 1$
Vertex: $\left(-\frac{1}{2}, -\frac{3}{2}\right)$

For Exercises 67–70,

  a. Determine whether the graph of the parabola opens upward or downward.

  b. Find the vertex.

  c. Find the x-intercept(s) if possible.

  d. Find the y-intercept.

  e. Sketch the graph.

67. $y = x^2 + 2x - 3$
  a. Upward
  b. $(-1, -4)$
  c. $(-3, 0), (1, 0)$
  d. $(0, -3)$

e.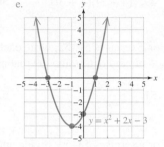

68. $y = x^2 - 2x$
  a. Upward
  b. $(1, -1)$
  c. $(0, 0), (2, 0)$
  d. $(0, 0)$

e.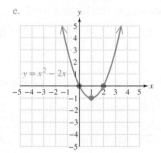

69. $y = -3x^2 + 12x - 9$
  a. Downward
  b. $(2, 3)$
  c. $(1, 0), (3, 0)$
  d. $(0, -9)$

e.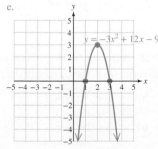

70. $y = -8x^2 - 16x - 12$
  a. Downward
  b. $(-1, -4)$
  c. No x-intercepts
  d. $(0, -12)$

e.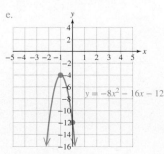

71. An object is launched into the air from ground level with an initial velocity of 256 ft/sec. The height of the object, y (in feet), can be approximated by the function

$$y = -16t^2 + 256t \qquad \text{where } t \text{ is the number of seconds after launch.}$$

  a. Find the maximum height of the object.
    1024 ft

  b. Find the time required for the object to reach its maximum height.   8 sec

Section 9.6

For Exercises 72–77, state the domain and range of each relation. Then determine whether the relation is a function.

72. $\{(6, 3), (10, 3), (-1, 3), (0, 3)\}$
Domain: $\{6, 10, -1, 0\}$; range: $\{3\}$; function

73. $\{(2, 0), (2, 1), (2, -5), (2, 2)\}$
Domain: $\{2\}$; range: $\{0, 1, -5, 2\}$; not a function

74.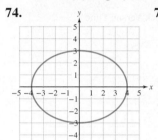

75.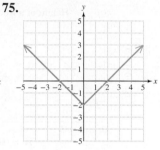

Domain: $[-4, 4]$; range: $[-3, 3]$; not a function

Domain: $(-\infty, \infty)$; range: $[-2, \infty)$; function

76. $\{(4, 23), (3, -2), (-6, 5), (4, 6)\}$
Domain: $\{4, 3, -6\}$; range: $\{23, -2, 5, 6\}$; not a function

77. $\{(3, 0), (-4, \frac{1}{2}), (0, 3), (2, -12)\}$
Domain: $\{3, -4, 0, 2\}$; range: $\{0, \frac{1}{2}, 3, -12\}$; function

78. Given the function defined by $f(x) = x^3$, find:

  a. $f(0)$ 0   **b.** $f(2)$ 8   **c.** $f(-3)$ -27

  d. $f(-1)$ -1 **e.** $f(4)$ 64

79. Given the function defined by $g(x) = \dfrac{x}{5 - x}$, find:

  a. $g(0)$ 0   **b.** $g(4)$ 4   **c.** $g(-1)$ $-\dfrac{1}{6}$

  d. $g(3)$ $\dfrac{3}{2}$ **e.** $g(-5)$ $-\dfrac{1}{2}$

80. The landing distance that a certain plane will travel on a runway is determined by the initial landing speed at the instant the plane touches down. The following function relates landing distance, $D(x)$, to initial landing speed, x, where $x \geq 15$.

$$D(x) = \frac{1}{10}x^2 - 3x + 22 \qquad \text{where } D(x) \text{ is in feet and } x \text{ is in feet per second.}$$

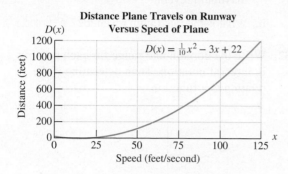

Distance Plane Travels on Runway Versus Speed of Plane

$$D(x) = \frac{1}{10}x^2 - 3x + 22$$

a. Find $D(90)$, and interpret the meaning of the function value in terms of landing speed and length of the runway. $D(90) = 562$. A plane traveling 90 ft/sec when it touches down will require 562 ft of runway.

b. Find $D(110)$, and interpret the meaning in terms of landing speed and length of the runway. $D(110) = 902$. A plane traveling 110 ft/sec when it touches down will require 902 ft of runway.

Chapter 9 Test

1. Solve the equation by applying the square root property. $\{-1 \pm \sqrt{14}\}$

$$(x + 1)^2 = 14$$

2. Solve the equation by completing the square and applying the square root property. $\{4 \pm \sqrt{21}\}$

$$x^2 - 8x - 5 = 0$$

3. Solve the equation by using the quadratic formula.

$$3x^2 - 5x = -1 \qquad \left\{\frac{5 \pm \sqrt{13}}{6}\right\}$$

For Exercises 4–10, solve the equations using any method.

4. $5x^2 + x - 2 = 0$
$\left\{\frac{-1 \pm \sqrt{41}}{10}\right\}$

5. $(c - 12)^2 = 12$
$\{12 \pm 2\sqrt{3}\}$

6. $y^2 + 14y - 1 = 0$
$\{-7 \pm 5\sqrt{2}\}$

7. $3t^2 = 30$
$\{\sqrt{10}, -\sqrt{10}\}$

8. $4x(3x + 2) = 15$ $\left\{\frac{5}{6}, -\frac{3}{2}\right\}$

9. $6p^2 - 11p = 0$
$\left\{0, \frac{11}{6}\right\}$

10. $\frac{1}{4}x^2 - \frac{3}{2}x = \frac{11}{4}$ $\{3 \pm 2\sqrt{5}\}$

11. The surface area, S, of a sphere is given by the formula $S = 4\pi r^2$, where r is the radius of the sphere. Find the radius of a sphere whose surface area is 201 in.^{2} Round to the nearest tenth of an inch. 4.0 in.

$S = 201$ in.^{2}

12. The height of a triangle is 2 m longer than twice the base, and the area is 24 m^2. Find the values of the base and height. Use a calculator to approximate the base and height to the nearest tenth of a meter.
The base is 4.4 m. The height is 10.8 m.

For Exercises 13–15, simplify the expressions in terms of i.

13. $\sqrt{-100}$
$10i$

14. $\sqrt{-23}$
$i\sqrt{23}$

15. $\sqrt{-9} \cdot \sqrt{-49}$
-21

For Exercises 16–17, simplify.

16. $2i^2$ $\;-2$

17. $5 - 3i^2$ $\;8$

For Exercises 18–21, perform the indicated operation. Write the answer in standard form, $a + bi$.

18. $(2 - 7i) - (-3 - 4i)$
$5 - 3i$

19. $(8 + i)(-2 - 3i)$
$-13 - 26i$

20. $(10 - 11i)(10 + 11i)$
221

21. $\dfrac{1}{10 - 11i}$ $\;\dfrac{10}{221} + \dfrac{11}{221}i$

For Exercises 22–23, solve the quadratic equations with complex solutions.

22. $(x + 14)^2 = -81$
$\{-14 \pm 9i\}$

23. $x^2 + x + 7 = 0$
$\left\{-\dfrac{1}{2} \pm \dfrac{3\sqrt{3}}{2}i\right\}$

24. Explain how to determine if a parabola opens upward or downward.   For $y = ax^2 + bx + c$, if $a > 0$ the parabola opens upward, if $a < 0$ the parabola opens downward.

For Exercises 25–27, find the vertex of the parabola.

25. $y = x^2 - 10x + 25$
(5, 0)

26. $y = 3x^2 - 6x + 8$
(1, 5)

27. $y = -x^2 - 16$ (0, −16)

28. Suppose a parabola opens upward and the vertex is located at (−4, 3). How many x-intercepts does the parabola have?
The parabola has no x-intercepts.

29. Given the parabola, $y = x^2 + 6x + 8$

   a. Determine whether the parabola opens upward or downward.   Opens upward

   b. Find the vertex of the parabola.   Vertex: (−3, −1)

   c. Find the x-intercepts.
    x-intercepts: (−2, 0) and (−4, 0)

   d. Find the y-intercept. y-intercept: (0, 8)

   e. Graph the parabola.

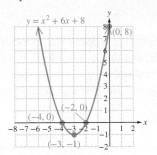

30. Graph the parabola and label the vertex, x-intercepts, and y-intercept.
$$y = -x^2 + 25$$
Vertex: (0, 25); x-intercepts: (−5, 0), (5, 0); y-intercept: (0, 25)

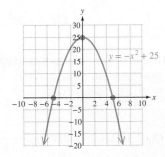

31. A sports franchise in Atlanta knows that if the home basketball team charges x dollars per ticket for a game, then the total revenue, y (in dollars), can be approximated by

$$y = -400x^2 + 20{,}000x \qquad \text{where } x \text{ is the price per ticket.}$$

   a. Find the ticket price that will produce the maximum revenue. $25 per ticket

   b. What is the maximum revenue? $250,000

32. Write the domain and range for each relation in interval notation. Then determine if the relation is a function.

   a.

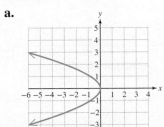

Domain: $(-\infty, 0]$; range: $(-\infty, \infty)$; not a function

   b.

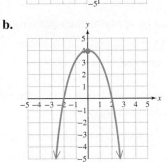

Domain: $(-\infty, \infty)$; range: $(-\infty, 4]$; function

33. For the function defined by $f(x) = \dfrac{1}{x+2}$, find the function values: $f(0)$, $f(-2)$, $f(6)$

$f(0) = \dfrac{1}{2}$, $f(-2)$ is undefined, $f(6) = \dfrac{1}{8}$

34. The number of diagonals, $D(x)$, of a polygon is a function of the number of sides, x, of the polygon according to the equation

$$D(x) = \frac{1}{2}x(x-3)$$

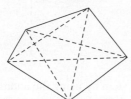

a. Find $D(5)$ and interpret the meaning of the function value. Verify your answer by counting the number of diagonals in the pentagon in the figure.
$D(5) = 5$; a five-sided polygon has five diagonals.

b. Find $D(10)$ and interpret its meaning.
$D(10) = 35$; a 10-sided polygon has 35 diagonals.

c. If a polygon has 20 diagonals, how many sides does it have? (*Hint:* Substitute $D(x) = 20$ and solve for x. Try clearing fractions first.)
8 sides

Chapters 1–9   Cumulative Review Exercises

1. Solve. $3x - 5 = 2(x - 2)$ {1}

2. Solve for h. $A = \frac{1}{2}bh$ $h = \dfrac{2A}{b}$

3. Solve. $\frac{1}{2}y - \frac{5}{6} = \frac{1}{4}y + 2$ $\left\{ \frac{34}{3} \right\}$

4. a. Determine whether 2 is a solution to the inequality. $-3x + 4 < x + 8$
Yes, 2 is a solution.

b. Graph the solution to the inequality $-3x + 4 < x + 8$. Then write the solution in set-builder notation and in interval notation.
$\{x \mid x > -1\}; (-1, \infty)$ ←———→
-1

5. The graph depicts the death rate from 60 U.S. cities versus the median education level of the people living in that city. The death rate, y, is measured in number of deaths per 100,000 people. The median education level, x, is a type of "average" and is measured by grade level. (*Source:* U.S. Bureau of the Census)

The death rate can be predicted from the median education level according to the equation.

$$y = -37.6x + 1353 \quad \text{where } 9 \le x \le 13$$

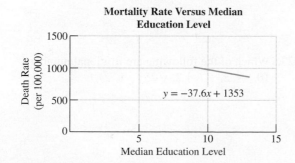

Mortality Rate Versus Median Education Level

$y = -37.6x + 1353$

a. From the graph, does it appear that the death rate increases or decreases as the median education level increases?   Decreases

b. What is the slope of the line? Interpret the slope in the context of the death rate and education level. $m = -37.6$. For each additional increase in education level, the death rate decreases by approximately 38 deaths per 100,000 people.

c. For a city in the United States with a median education level of 12, what would be the expected death rate?   901.8 per 100,000

d. If the death rate of a certain city is 977 per 100,000 people, what would be the approximate median education level?
10th grade

6. Simplify completely. Write the final answer with positive exponents only.

$$\left(\frac{2a^2b^{-3}}{c} \right)^{-1} \cdot \left(\frac{4a^{-1}}{b^2} \right)^2 \quad \frac{8c}{a^4b}$$

7. Approximately 5.2×10^7 disposable diapers are thrown into the trash each day in the United States and Canada. How many diapers are thrown away each year? 1.898×10^{10} diapers

8. In 1989, the Hipparcos satellite found the distance between Earth and the star, Polaris, to be approximately 2.53×10^{15} mi. If 1 light-year is approximately 5.88×10^{12} miles, how many light-years is Polaris from Earth?
Approximately 430 light-years

9. Perform the indicated operations.
$(2x - 3)^2 - 4(x - 1)$ $4x^2 - 16x + 13$

10. Divide using long division.
$(2y^4 - 4y^3 + y - 5) \div (y - 2)$ $2y^3 + 1 - \dfrac{3}{y-2}$

 Writing ←→ Translating Expression   Geometry   Scientific Calculator   Video

11. Factor. $2x^2 - 9x - 35$ $\quad (2x+5)(x-7)$

12. Factor completely. $2xy + 8xa - 3by - 12ab$
$(y+4a)(2x-3b)$

13. The base of a triangle is 1 m more than the height. If the area is 36 m², find the base and height.
The base is 9 m, and the height is 8m.

14. Simplify to lowest terms. $\dfrac{5x+10}{x^2-4}$ $\quad \dfrac{5}{x-2}$

15. Multiply. $\dfrac{x^2+10x+9}{x^2-81} \cdot \dfrac{18-2x}{x^2+2x+1}$ $\quad \dfrac{2}{x+1}$

16. Perform the indicated operation.

$$\dfrac{x^2}{x-5} - \dfrac{10x-25}{x-5} \quad x-5$$

17. Simplify completely.

$$\dfrac{\dfrac{1}{x+1} - \dfrac{1}{x-1}}{\dfrac{x}{x^2-1}} \quad \dfrac{2}{x}$$

18. Solve. $1 - \dfrac{1}{y} = \dfrac{12}{y^2}$ $\quad \{4, -3\}$

19. Write an equation of the line passing through the point $(-2, 3)$ and having a slope of $\frac{1}{2}$. Write the final answer in slope-intercept form.
$y = \dfrac{1}{2}x + 4$

For Exercises 20–21,

  a. Find the x-intercept (if it exists).

  b. Find the y-intercept (if it exists).

  c. Find the slope (if it exists).

  d. Graph the line.

20. $2x - 4y = 12$ d.
  a. $(6, 0)$
  b. $(0, -3)$
  c. $\frac{1}{2}$

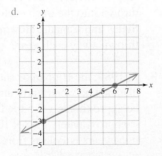

21. $4x + 12 = 0$ d.
  a. $(-3, 0)$
  b. No y-intercept
  c. Slope is undefined.

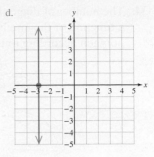

22. Solve the system by using the addition method. If the system has no solution or infinitely many solutions, so state.

$$\begin{aligned} \dfrac{1}{2}x - \dfrac{1}{4}y &= \dfrac{1}{6} \\ 12x - 3y &= 8 \end{aligned} \quad \left\{ \left(1, \dfrac{4}{3}\right) \right\}$$

23. Solve the system by using the substitution method. If the system has no solution or infinitely many solutions, so state.

$$\begin{aligned} 2x - y &= 8 \quad \{(5, 2)\} \\ 4x - 4y &= 3x - 3 \end{aligned}$$

24. In a right triangle, one acute angle is 2° more than three times the other acute angle. Find the measure of each angle.   The angles are 22° and 68°.

25. A bank of 27 coins contains only dimes and quarters. The total value of the coins is \$4.80. Find the number of dimes and the number of quarters.   There are 13 dimes and 14 quarters.

26. Sketch the inequality. $x - y \leq 4$

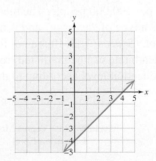

27. Which of the following are irrational numbers?
$\{0, -\frac{2}{3}, \pi, \sqrt{7}, 1.2, \sqrt{25}\}$ $\pi, \sqrt{7}$

For Exercises 28–29, simplify the radicals.

28. $\sqrt{\dfrac{1}{7}}$ $\dfrac{\sqrt{7}}{7}$

29. $\dfrac{\sqrt{16x^4}}{\sqrt{2x}}$ $2x\sqrt{2x}$

30. Perform the indicated operation. $(4\sqrt{3} + \sqrt{x})^2$
$48 + 8\sqrt{3x} + x$

31. Add the radicals. $-3\sqrt{2x} + \sqrt{50x}$ $2\sqrt{2x}$

32. Rationalize the denominator.

$$\dfrac{4}{2 - \sqrt{a}}$$ $\dfrac{8 + 4\sqrt{a}}{4 - a}$

33. Solve. $\sqrt{x + 11} = x + 5$
{−2} (The value −7 does not check.)

34. Factor completely. $8c^3 - y^3$
$(2c - y)(4c^2 + 2cy + y^2)$

35. Which graph defines y as a function of x?   b

a. **b.**

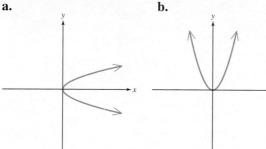

36. Given the functions defined by $f(x) = -\frac{1}{2}x + 4$ and $g(x) = x^2$, find

a. $f(6)$ **b.** $g(-2)$ **c.** $f(0) + g(3)$
   1   4   13

37. Find the domain and range of the function.
$\{(2, 4), (-1, 3), (9, 2), (-6, 8)\}$
Domain: {2, −1, 9, −6}; range: {4, 3, 2, 8}

38. Find the slope of the line passing through the points $(3, -1)$ and $(-4, -6)$. $m = \dfrac{5}{7}$

39. Find the slope of the line defined by
$-4x - 5y = 10$. $m = -\dfrac{4}{5}$

40. What value of n would make the expression a perfect square trinomial? $n = 25$

$$x^2 + 10x + n$$

41. Solve the quadratic equation by completing the square and applying the square root property.
$2x^2 + 12x + 6 = 0$ $\{-3 \pm \sqrt{6}\}$

42. Solve the quadratic equation by using the quadratic formula. $2x^2 + 12x + 6 = 0$ $\{-3 \pm \sqrt{6}\}$

43. Graph the parabola defined by the equation. Label the vertex, x-intercepts, and y-intercept.

$$y = x^2 + 4x + 4$$

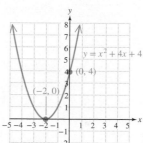

For Exercises 44–45, simplify completely.

44. $-10i^2 + 6$ 16   **45.** $3i(4i - 1)$ $-12 - 3i$

Additional Topics Appendix

Decimals and Percents

1. Introduction to a Place Value System

In a *place value* number system, each digit in a numeral has a particular value determined by its location in the numeral (Figure A-1).

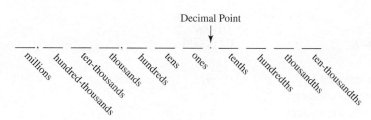

Figure A-1

For example, the number 197.215 represents

$$(1 \times 100) + (9 \times 10) + (7 \times 1) + \left(2 \times \frac{1}{10}\right) + \left(1 \times \frac{1}{100}\right) + \left(5 \times \frac{1}{1000}\right)$$

Each of the digits 1, 9, 7, 2, 1, and 5 is multiplied by 100, 10, 1, $\frac{1}{10}$, $\frac{1}{100}$, and $\frac{1}{1000}$, respectively, depending on its location in the numeral 197.215.

By obtaining a common denominator and adding fractions, we have

$$197.215 = 100 + 90 + 7 + \frac{200}{1000} + \frac{10}{1000} + \frac{5}{1000}$$

$$= 197 + \frac{215}{1000} \quad \text{or} \quad 197\frac{215}{1000}$$

Because 197.215 is equal to the mixed number $197\frac{215}{1000}$, we read 197.215 as one hundred ninety-seven *and* two hundred fifteen thousandths. The decimal point is read as the word *and*.

If there are no digits to the right of the decimal point, we usually omit the decimal point. For example, the number 7125. is written simply as 7125 without a decimal point.

2. Converting Fractions to Decimals

We have learned that a fraction represents part of a whole unit. Likewise, the digits to the right of the decimal point represent a fraction of a whole unit. In this section, we will learn how to convert a fraction to a decimal number and vice versa.

> **Converting a Fraction to a Decimal**
>
> To convert a fraction to a decimal, divide the numerator of the fraction by the denominator of the fraction.

Note: Zeros may be inserted to the right of the decimal point in the dividend until the digits in the quotient either terminate or repeat.

Classroom Examples: p. A-8, Exercises 14 and 16

| Example 1 | **Converting Fractions to Decimals** |

Convert each fraction to a decimal.   **a.** $\dfrac{7}{40}$ **b.** $\dfrac{2}{3}$

Solution:

a. $\dfrac{7}{40} = 0.175$

$$\begin{array}{r} 0.175 \\ 40\overline{)7.000} \\ \underline{40} \\ 300 \\ \underline{280} \\ 200 \\ \underline{200} \\ 0 \end{array}$$

The number 0.175 is said to be a *terminating decimal* because there are no nonzero digits to the right of the last digit, 5.

b. $\dfrac{2}{3} = 0.666\ldots$

$$\begin{array}{r} 0.666\ldots \\ 3\overline{)2.00000} \\ \underline{18} \\ 20 \\ \underline{18} \\ 20 \\ \underline{18} \\ 2\ldots \end{array}$$

The pattern $0.666\ldots$ continues indefinitely. Therefore, we say that this is a *repeating decimal.*

TIP: If a fraction in lowest terms has a denominator whose prime factorization includes *only* 2's and/or 5's, it will terminate. If it contains any other factors, it will repeat.

For a repeating decimal, a horizontal bar is often used to denote the repeating pattern after the decimal point. Hence, $\frac{2}{3} = 0.\overline{6}$.

Skill Practice   Convert to a decimal.

1. $\dfrac{5}{8}$ **2.** $\dfrac{1}{6}$

Sometimes it is useful to round a decimal number to a desired place value.

Rounding Decimals to a Place Value to the Right of the Decimal Point

Step 1   Identify the digit one position to the right of the given place value.

Step 2   If the digit in step 1 is 5 or greater, add 1 to the digit in the given place value. Then discard the digits to its right.

Step 3   If the digit in step 1 is less than 5, discard it and any digits to its right.

Classroom Examples: p. A-8, Exercises 20 and 26

| Example 2 | **Rounding Decimal Numbers** |

Round the numbers to the indicated place value.

a. 0.175 to the hundredths place   **b.** $0.5\overline{4}$ to the thousandths place

Solution:

hundredths place
↓

a. $0.17\boxed{5}$ The digit to the right of the hundredths place is 5 or greater.

≈ 0.18 Round up. Add 1 to the hundredths place digit, and discard digits to its right.

Answers

1. 0.625   **2.** $0.1\overline{6}$

b. $0.\overline{54} = 0.545454\ldots$ This is a repeating decimal. We write out several digits.

thousandths place
↓

$0.545\boxed{4}54\ldots$ The digit to the right of the thousandths place is less than 5. Therefore, discard digits to the right of the thousandths place.

≈ 0.545

Skill Practice Round to the indicated place value.

3. 0.624 hundredths place   **4.** $1.\overline{62}$ ten-thousandths place

3. Converting Decimals to Fractions

For a terminating decimal, use the word name to write the number as a fraction or mixed number.

| **Example 3** | **Converting Terminating Decimals to Fractions** |

Classroom Example: p. A-9, Exercise 30

Convert each decimal to a fraction.

a. 0.0023   **b.** 50.06

Solution:

a. 0.0023 is read as twenty-three ten-thousandths. Thus,

$$0.0023 = \frac{23}{10,000}$$

b. 50.06 is read as fifty and six hundredths. Thus,

$$50.06 = 50\frac{6}{100}$$

$$= 50\frac{3}{50}$$ Simplify the fraction to lowest terms.

$$= \frac{2503}{50}$$ Write the mixed number as a fraction.

Skill Practice Convert to a fraction.

5. 0.107   **6.** 11.25

Repeating decimals also can be written as fractions. However, the procedure to convert a repeating decimal to a fraction requires some knowledge of algebra. Table A-1 shows some common repeating decimals and an equivalent fraction for each.

Table A-1

$0.\overline{1} = \dfrac{1}{9}$	$0.\overline{4} = \dfrac{4}{9}$	$0.\overline{7} = \dfrac{7}{9}$
$0.\overline{2} = \dfrac{2}{9}$	$0.\overline{5} = \dfrac{5}{9}$	$0.\overline{8} = \dfrac{8}{9}$
$0.\overline{3} = \dfrac{3}{9} = \dfrac{1}{3}$	$0.\overline{6} = \dfrac{6}{9} = \dfrac{2}{3}$	$0.\overline{9} = \dfrac{9}{9} = 1$

Answers

3. 0.62   **4.** 1.6263

5. $\dfrac{107}{1000}$ **6.** $\dfrac{45}{4}$

4. Converting Percents to Decimals and Fractions

The concept of percent (%) is widely used in a variety of mathematical applications. The word *percent* means "per 100." Therefore, we can write percents as fractions.

$$6\% = \frac{6}{100}$$ A sales tax of 6% means that 6 cents in tax is charged for every 100 cents spent.

$$91\% = \frac{91}{100}$$ The fact that 91% of the population is right-handed means that 91 people out of 100 are right-handed.

The quantity $91\% = \dfrac{91}{100}$ can be written as $91 \times \dfrac{1}{100}$ or as 91×0.01.

Notice that the % symbol implies "division by 100" or, equivalently, "multiplication by $\frac{1}{100}$." Thus, we have the following rule to convert a percent to a fraction (or to a decimal).

Converting a Percent to a Decimal or Fraction

Replace the % symbol by $\div\ 100$ (or equivalently $\times\ \dfrac{1}{100}$ or $\times\ 0.01$).

Classroom Example: p. A-9, Exercise 40

Example 4   Converting Percents to Decimals

Convert the percents to decimals.

   a. 78%   **b.** 412%   **c.** 0.045%

TIP: Multiplying by 0.01 is equivalent to dividing by 100. This has the effect of moving the decimal point two places to the left.

Solution:

   a. $78\% = 78 \times 0.01 = 0.78$

   b. $412\% = 412 \times 0.01 = 4.12$

   c. $0.045\% = 0.045 \times 0.01 = 0.00045$

Skill Practice Convert the percent to a decimal.

   7. 29%   **8.** 3.5%   **9.** 100%

Classroom Example: p. A-9, Exercise 44

Example 5   Converting Percents to Fractions

Convert the percents to fractions.

   a. 52%   **b.** $33\frac{1}{3}\%$ **c.** 6.5%

Solution:

   a. $52\% = 52 \times \dfrac{1}{100}$ Replace the % symbol by $\frac{1}{100}$.

           $= \dfrac{52}{100}$ Multiply.

           $= \dfrac{13}{25}$ Simplify to lowest terms.

Answers

7. 0.29   **8.** 0.035   **9.** 1.00

b. $33\frac{1}{3}\% = 33\frac{1}{3} \times \frac{1}{100}$ Replace the % symbol by $\frac{1}{100}$.

$\quad = \frac{100}{3} \times \frac{1}{100}$ Write the mixed number as a fraction $33\frac{1}{3} = \frac{100}{3}$.

$\quad = \frac{100}{300}$ Multiply the fractions.

$\quad = \frac{1}{3}$ Simplify to lowest terms.

c. $6.5\% = 6.5 \times \frac{1}{100}$ Replace the % symbol by $\frac{1}{100}$.

$\quad = \frac{65}{10} \times \frac{1}{100}$ Write 6.5 as an improper fraction.

$\quad = \frac{65}{1000}$ Multiply the fractions.

$\quad = \frac{13}{200}$ Simplify to lowest terms.

Skill Practice   Convert the percent to a fraction.

10. 30%   **11.** $120\frac{1}{2}\%$ **12.** 2.5%

5. Converting Decimals and Fractions to Percents

To convert a percent to a decimal or fraction, we replace the % symbol by ÷ 100. To convert a decimal or fraction to a percent, we reverse this process.

> **Converting Decimals and Fractions to Percents**
> Multiply the fraction or decimal by 100%. (Note that 100% = 1)

Example 6   **Converting Decimals to Percents**

Convert the decimals to percents.

a. 0.92   **b.** 10.80   **c.** 0.005

Solution:

a. $0.92 = 0.92 \times 100\% = 92\%$ Multiply by 100%.

b. $10.80 = 10.80 \times 100\% = 1080\%$ Multiply by 100%.

c. $0.005 = 0.005 \times 100\% = 0.5\%$ Multiply by 100%.

Skill Practice   Convert the decimals to percents.

13. 0.56   **14.** 4.36   **15.** 0.002

Classroom Examples: p. A-9,
Exercises 52 and 56

Answers

10. $\frac{3}{10}$ **11.** $\frac{241}{200}$ **12.** $\frac{1}{40}$

13. 56%   **14.** 436%   **15.** 0.2%

Classroom Examples: p. A-9,
Exercises 64 and 72

Example 7	Converting Fractions to Percents

Convert the fractions to percents.

a. $\dfrac{2}{5}$ **b.** $\dfrac{5}{3}$

Solution:

TIP: Notice that 100% = 1. So by multiplying a number by 100%, we are not changing the value of the number.

a. $\dfrac{2}{5} = \dfrac{2}{5} \times 100\%$ Multiply by 100%.

$= \dfrac{2}{5} \times \dfrac{100}{1}\%$ Write the whole number as a fraction.

$= \dfrac{2}{\cancel{5}} \times \dfrac{\overset{20}{\cancel{100}}}{1}\%$ Multiply fractions.

$= \dfrac{40}{1}\%$ or 40% Simplify.

b. $\dfrac{5}{3} = \dfrac{5}{3} \times 100\%$ Multiply by 100%.

$= \dfrac{5}{3} \times \dfrac{100}{1}\%$ Write the whole number as a fraction.

$= \dfrac{500}{3}\%$ Multiply fractions.

$= \dfrac{500}{3}\%$ or $166.\overline{6}\%$ The value $\frac{500}{3}$ can be written in decimal form by dividing 500 by 3.

Skill Practice   Convert the fractions to percents.

16. $\dfrac{7}{8}$ **17.** $\dfrac{5}{6}$

6. Applications of Percents

Many applications involving percents involve finding a percent of some base number. For example, suppose a textbook is discounted 25%. If the book originally costs $60, find the amount of the discount.

In this example, we must find 25% of $60. In this context, the word *of* means multiply.

25% of $60
$\downarrow \quad \downarrow \quad \downarrow$
$0.25 \times 60 = 15$ The amount of the discount is $15.

Note that the *decimal form* of a percent is always used in calculations. Therefore, 25% was converted to 0.25 *before* multiplying by $60.

Answers
16. 87.5%   **17.** $83.\overline{3}\%$ or $83\frac{1}{3}\%$

| **Example 8** | **Applying Percentages** |

Classroom Example: p. A-10, Exercise 76

Shauna received a raise, so now her new salary is 105% of her old salary. Find Shauna's new salary if her old salary was $36,000 per year.

Solution:

The new salary is 105% of $36,000.

$$1.05 \times 36,000 = 37,800 \qquad \text{The new salary is } \$37,800 \text{ per year.}$$

Skill Practice

18. The sales tax rate for a certain county is 6%. Find the amount of sales tax on a $52.00 fishing pole.

In some applications, it is necessary to convert a fractional part of a whole to a percent of the whole.

| **Example 9** | **Finding a Percentage** |

Classroom Example: p. A-10, Exercise 80

A small college accepts approximately 520 students each year from 3500 applicants. What percent does 520 represent? Round to the nearest tenth of a percent.

Solution:

$$\frac{520}{3500} \approx 0.149 \qquad$$ Convert the fractional part of the total number of applicants to decimal form. (*Note:* Rounding the decimal form of the quotient to the thousandths place gives us the nearest tenth of a percent.)

$$= 0.149 \times 100\% \qquad$$ Convert the decimal to a percent.

$$= 14.9\% \qquad$$ Simplify.

Approximately 14.9% of the applicants are accepted.

Skill Practice

19. Eduardo answered 66 questions correctly on a test with 75 questions. What percent of the questions does 66 represent?

Answers

18. $3.12   19. 88%

Calculator Connections

Topic: Approximating Repeating Decimals on a Calculator

Calculators can display only a limited number of digits on the calculator screen. Therefore, repeating decimals and terminating decimals with a large number of digits will be truncated or rounded to fit the calculator display. For example, the fraction $\frac{2}{3} = 0.\overline{6}$ may be entered into the calculator as $\boxed{2} \; \boxed{\div} \; \boxed{3}$. The result may appear as 0.6666666667 or as 0.6666666666. The fraction $\frac{2}{11}$ equals the repeating decimal $0.\overline{18}$. However, the calculator converts $\frac{2}{11}$ to the terminating decimal 0.1818181818.

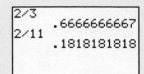

```
2/3   .6666666667
2/11
      .1818181818
```

Calculator Exercises

Without using a calculator, find a repeating decimal to represent each of the following fractions. Then use a calculator to confirm your answer.

1. $\dfrac{4}{9}$ $0.\overline{4}$
2. $\dfrac{7}{11}$ $0.\overline{63}$
3. $\dfrac{3}{22}$ $0.1\overline{36}$
4. $\dfrac{5}{13}$ $0.\overline{384615}$

Section A.1   Practice Exercises

Concept 1: Introduction to a Place Value System

For additional exercises, see Classroom Activities A.1A–A.1D in the *Student's Resource Manual* at www.mhhe.com/moh.

For Exercises 1–8, write the name of the place value for the underlined digit.

1. 481.24
   Tens
2. 1345.42
   Tens
3. 2912.032
   Hundreds
4. 4208.03
   Hundreds
5. 2.381
   Tenths
6. 8.249
   Tenths
7. 21.413
   Hundredths
8. 82.794
   Hundredths

9. The first 10 Roman numerals are: I, II, III, IV, V, VI, VII, VIII, IX, X. Is this numbering system a place value system? Explain your answer.   No, the symbols I, V, X, and so on each represent certain values but the values are not dependent on the position of the symbol within the number.

10. Write the number in decimal form. $3(100) + 7(10) + 6 + \dfrac{1}{100} + \dfrac{5}{1000}$ 376.015

Concept 2: Converting Fractions to Decimals

For Exercises 11–18, convert each fraction to a terminating decimal or a repeating decimal. **(See Example 1.)**

11. $\dfrac{7}{10}$
    0.7
12. $\dfrac{9}{10}$
    0.9
13. $\dfrac{9}{25}$
    0.36
14. $\dfrac{3}{25}$
    0.12

15. $\dfrac{11}{9}$
    $1.\overline{2}$
16. $\dfrac{16}{9}$
    $1.\overline{7}$
17. $\dfrac{7}{33}$
    $0.\overline{21}$
18. $\dfrac{2}{11}$
    $0.\overline{18}$

For Exercises 19–26, round each decimal to the given place value. **(See Example 2.)**

19. 214.059; tenths   214.1

20. 1004.165; hundredths   1004.17

21. 39.26849; thousandths   39.268

22. 0.059499; thousandths   0.059

23. 39,918.2; thousands   40,000

24. 599,621.5; thousands   600,000

25. $0.7\overline{2}$; hundredths   0.73

26. $0.3\overline{4}$; thousandths   0.343

 Writing   Translating Expression   Geometry   Scientific Calculator   Video

Concept 3: Converting Decimals to Fractions

For Exercises 27–38, convert each decimal to a fraction or a mixed number. **(See Example 3.)**

27. 0.45 $\frac{9}{20}$

28. 0.65 $\frac{13}{20}$

29. 0.181 $\frac{181}{1000}$

30. 0.273 $\frac{273}{1000}$

31. 2.04 $\frac{51}{25}$ or $2\frac{1}{25}$

32. 6.02 $\frac{301}{50}$ or $6\frac{1}{50}$

33. 13.007 $\frac{13{,}007}{1000}$ or $13\frac{7}{1000}$

34. 12.003 $\frac{12{,}003}{1000}$ or $12\frac{3}{1000}$

35. $0.\overline{5}$ (*Hint:* Refer to Table A-1) $\frac{5}{9}$

36. $0.\overline{8}$ $\frac{8}{9}$

37. $1.\overline{1}$ $\frac{10}{9}$ or $1\frac{1}{9}$

38. $2.\overline{3}$ $\frac{7}{3}$ or $2\frac{1}{3}$

Concept 4: Converting Percents to Decimals and Fractions

For Exercises 39–48, convert each percent to a decimal and to a fraction. **(See Examples 4–5.)**

39. The sale price is 30% off of the original price. $0.3, \frac{3}{10}$

40. An HMO (health maintenance organization) pays 80% of all doctors' bills. $0.8, \frac{4}{5}$

41. The building will be 75% complete by spring. $0.75, \frac{3}{4}$

42. Chen plants roses in 25% of his garden. $0.25, \frac{1}{4}$

© Image Plan/Corbis RF

43. The bank pays $3\frac{3}{4}$% interest on a checking account. $0.0375, \frac{3}{80}$

44. A credit union pays $4\frac{1}{2}$% interest on a savings account. $0.045, \frac{9}{200}$

45. Kansas received 15.7% of its annual rainfall in 1 week. $0.157, \frac{157}{1000}$

46. Social Security withholds 6.2% of an employee's gross pay. $0.062, \frac{31}{500}$

47. The world population in 2008 was 270% of the world population in 1950. $2.7, \frac{27}{10}$

48. The cost of a home is 140% of its cost 10 years ago. $1.40, \frac{7}{5}$

Concept 5: Converting Decimals and Fractions to Percents

49. Explain how to convert a decimal to a percent.
Multiply by 100%.

50. Explain how to convert a percent to a decimal.
Replace the % symbol by ÷ 100, by × $\frac{1}{100}$, or by × 0.01.

For Exercises 51–62, convert each decimal to a percent. **(See Example 6.)**

51. 0.05 5%

52. 0.06 6%

53. 0.90 90%

54. 0.70 70%

55. 1.2 120%

56. 4.8 480%

57. 7.5 750%

58. 9.3 930%

59. 0.135 13.5%

60. 0.536 53.6%

61. 0.003 0.3%

62. 0.002 0.2%

For Exercises 63–74, convert each fraction to a percent. **(See Example 7.)**

63. $\frac{3}{50}$ 6%

64. $\frac{23}{50}$ 46%

65. $\frac{9}{2}$ 450%

66. $\frac{7}{4}$ 175%

67. $\frac{5}{8}$ 62.5%

68. $\frac{1}{8}$ 12.5%

69. $\frac{5}{16}$ 31.25%

70. $\frac{7}{16}$ 43.75%

71. $\frac{5}{6}$ $83.\overline{3}$%

72. $\frac{4}{15}$ $26.\overline{6}$%

73. $\frac{14}{15}$ $93.\overline{3}$%

74. $\frac{5}{18}$ $27.\overline{7}$%

Concept 6: Applications of Percents

75. A suit that costs $140 is discounted by 30%. How much is the discount? **(See Example 8.)** $42

76. Louise completed 40% of her task that takes a total of 60 hr to finish. How many hours did she complete? 24 hr

 77. Tom's federal taxes amount to 27% of his income. If Tom earns $12,500 per quarter, how much will he pay in taxes for that quarter? $3375

78. A tip of $7 is left for a meal that costs $56. What percent of the cost does the tip represent? 12.5%

79. Jamie paid $5.95 in sales tax on a textbook that costs $85. Find the percent of the sales tax. **(See Example 9.)** 7%

80. Sue saves $37.50 each week out of her paycheck of $625. What percent of her paycheck does her savings represent? 6%

For Exercises 81–84, refer to the graph. The pie graph shows a family budget based on a net income of $2400 per month.

81. Determine the amount spent on rent. $792

82. Determine the amount spent on car payments. $408

83. Determine the amount spent on utilities. $192

84. How much more money is spent than saved? $1488

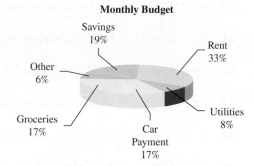

Monthly Budget

Savings 19%
Rent 33%
Other 6%
Utilities 8%
Groceries 17%
Car Payment 17%

85. By the end of the year, Felipe will have 75% of his mortgage paid. If the mortgage was originally for $90,000, how much will have been paid at the end of the year? $67,500

86. An account earns 4% interest in 1 year. If Mr. Patel has $12,000 invested in the account how much interest will he receive at the end of the year? $480

Section A.2 — Mean, Median, and Mode

Concepts

1. Mean
2. Median
3. Mode
4. Weighted Mean

1. Mean

When given a list of numerical data, it is often desirable to obtain a single number that represents the central value of the data. In this section, we discuss three such values called the mean, median, and mode.

Instructor Note: Ask students to compute the mean of their quiz scores or test scores for this class, as a warm-up exercise.

Definition of a Mean

The **mean** (or average) of a set of numbers is the sum of the values divided by the number of values. We can write this as a formula.

$$\text{Mean} = \frac{\text{sum of the values}}{\text{number of values}}$$

Example 1	**Finding the Mean of a Data Set**

Classroom Example: p. A-16, Exercise 10

A small business employs five workers. Their yearly salaries are

$42,000 $36,000 $45,000 $35,000 $38,000

a. Find the mean yearly salary for the five employees.

b. Suppose the owner of the business makes $218,000 per year. Find the mean salary for all six individuals (that is, include the owner's salary).

Solution:

a. Mean salary of five employees

$$= \frac{42,000 + 36,000 + 45,000 + 35,000 + 38,000}{5}$$

$$= \frac{196,000}{5} \qquad \text{Add the data values.}$$

$$= 39,200 \qquad \text{Divide.}$$

The mean salary for employees is $39,200.

b. Mean of all six individuals

$$= \frac{42,000 + 36,000 + 45,000 + 35,000 + 38,000 + 218,000}{6}$$

$$= \frac{414,000}{6}$$

$$= 69,000$$

The mean salary with the owner's salary included is $69,000.

> **Avoiding Mistakes**
>
> When computing a mean remember to add the data first before dividing.

Skill Practice   Housing prices for five homes in one neighborhood are given.

$108,000 $149,000 $164,000 $118,000 $144,000

1. Find the mean of these five prices.

2. Suppose a new home is built in the neighborhood for $1.3 million ($1,300,000). Find the mean price of all six homes.

2. Median

In Example 1, you may have noticed that the mean salary was greatly affected by the unusually high value of $218,000. For this reason, you may want to use a different measure of "center" called the median. The median is the "middle" number in an ordered list of numbers.

Finding the Median

To compute the **median** of a list of numbers, first arrange the numbers in order from least to greatest.

- If the number of data values in the list is *odd*, then the median is the middle number in the list.
- If the number of data values is *even*, there is no single middle number. Therefore, the median is the mean (average) of the two middle numbers in the list.

Answers

1. $136,600   **2.** $330,500

Classroom Examples: p. A-17, Exercises 16 and 18

Example 2   Finding the Median of a Data Set

Consider the salaries of the five workers from Example 1.

$42,000 $36,000 $45,000 $35,000 $38,000

a. Find the median salary for the five workers.

b. Find the median salary including the owner's salary of $218,000.

Solution:

a. 35,000   36,000   38,000   42,000   45,000   Arrange the data in order.

Because there are five data values (an *odd* number), the median is the middle number.

The median is $38,000.

b. Now consider the scores of all six individuals (including the owner). Arrange the data in order.

35,000   36,000   38,000   42,000   45,000   218,000

$$\frac{38,000 + 42,000}{2}$$ There are six data values (an *even* number). The median is the average of the two middle numbers.

$$= \frac{80,000}{2}$$ Add the two middle numbers.

$$= 40,000$$ Divide.

The median of all six salaries is $40,000.

Skill Practice Housing prices for five homes in one neighborhood are given.

$108,000 $149,000 $164,000 $118,000 $144,000

3. Find the median price of these five houses.

4. Suppose a new home is built in the neighborhood for $1,300,000. Find the median price of all six homes. Compare this price with the mean in Skill Practice Exercise 2.

In Examples 1 and 2, the mean of all six salaries is $69,000, whereas the median is $40,000. These examples show that the median is a better representation for a central value when the data list has an unusually high (or low) value.

Classroom Example: p. A-17, Exercise 24

Example 3   Determining the Median of a Data Set

The average monthly temperatures (in °C) for the South Pole are given in the table. Find the median temperature. (*Source:* NOAA)

Jan.	Feb.	March	April	May	June	July	Aug.	Sept.	Oct.	Nov.	Dec.
−2.9	−9.5	−19.2	−20.7	−21.7	−23.0	−25.7	−26.1	−24.6	−18.9	−9.7	−3.4

Answers
3. $144,000   **4.** $146,500

Solution:

First arrange the numbers in order from least to greatest.

$$-26.1 \quad -25.7 \quad -24.6 \quad -23.0 \quad -21.7 \quad -20.7 \quad -19.2 \quad -18.9 \quad -9.7 \quad -9.5 \quad -3.4 \quad -2.9$$

$$\text{Median} = \frac{-20.7 + (-19.2)}{2} = -19.95$$

There are 12 data values (an *even* number). Therefore, the median is the average of the two middle numbers. The median temperature at the South Pole is $-19.95°$C.

> **TIP:** Note that the median may not be one of the original data values.

Skill Practice The gain or loss for a stock is given for an 8-day period. Find the median gain or loss.

5. $-2.4 \qquad -2.0 \qquad 1.25 \qquad 0.6 \qquad -1.8 \qquad -0.4 \qquad 0.6 \qquad -0.9$

3. Mode

A third representative value for a list of data is called the mode.

Definition of a Mode

The **mode** of a set of data is the value or values that occur most often.

- If two values occur most often we say the data are bimodal.
- If more than two values occur most often, we say there is no mode.

Example 4   **Finding the Mode of a Data Set**

Classroom Example: p. A-17, Exercise 28

The student-to-teacher ratio is given for elementary schools for ten selected states. For example, California has a student-to-teacher ratio of 20.6. This means that there are approximately 20.6 students per teacher in California elementary schools. (*Source: National Center for Education Statistics*)

ME	ND	WI	NH	RI	IL	IN	MS	CA	UT
12.5	13.4	14.1	14.5	14.8	16.1	16.1	16.1	20.6	21.9

Find the mode of the student-to-teacher ratio for these states.

Solution:

The data value 16.1 appears most often. Therefore, the mode is 16.1 students per teacher.

Skill Practice The monthly rainfall amounts (in inches) for Houston, Texas, are given. Find the mode. (*Source:* NOAA)

6. 4.5   3.0   3.2   3.5   5.1   6.8
　　 4.3   4.5   5.6   5.3   4.5   3.8

Answers

5. −0.65   **6.** 4.5 in.

Classroom Example: p. A-17,
Exercise 32

| Example 5 | Finding the Mode of a Data Set |

Find the mode of the list of average monthly temperatures for Albany, New York. Values are in °F.

Jan.	Feb.	March	April	May	June	July	Aug.	Sept.	Oct.	Nov.	Dec.
22	25	35	47	58	66	71	69	61	49	39	26

Solution:

No data value occurs most often. There is no mode for this set of data.

Skill Practice

7. Find the mode of the weights in pounds of babies born one day at Brackenridge Hospital in Austin, Texas.

   7.2   8.1   6.9   9.3   8.3   7.7   7.9   6.4   7.5

Classroom Example: p. A-18,
Exercise 36

| Example 6 | Finding the Mode of a Data Set |

The grades for a quiz in college algebra are as follows. The scores are out of a possible 10 points.

9	4	6	9	9	8	2	1	4	9
5	10	10	5	7	7	9	8	7	3
9	7	10	7	10	1	7	4	5	6

Solution:

Sometimes arranging the data in order makes it easier to find the repeated values.

1	1	2	3	4	4	4	5	5	5
6	6	7	7	7	7	7	7	8	8
9	9	9	9	9	9	10	10	10	10

The score of 7 occurs 6 times. The score of 9 occurs 6 times. There are two modes, 7 and 9, because these scores both occur more than any other score. We say that these data are *bimodal*.

Skill Practice

8. The ages of children participating in an after-school sports program are given. Find the mode(s).

13	15	17	15	14	15	16	16
15	16	12	13	15	14	16	15
15	16	16	13	16	13	14	18

TIP: To remember the difference between median and mode, think of the *median* of a highway that goes down the *middle*. Think of the word *mode* as sounding similar to the word *most*.

4. Weighted Mean

Sometimes data values in a list appear multiple times. In such a case, we can compute a weighted mean. In Example 7, we demonstrate how to use a weighted mean to compute a grade point average (GPA). To compute GPA, each grade is assigned a numerical value. For example, an "A" is worth 4 points, a "B" is worth 3 points, and so on. Then each grade for a course is "weighted" by the number of credit-hours that the course is worth.

Answers

7. There is no mode.
8. There are two modes, 15 and 16.

| Example 7 | **Using a Weighted Mean to Compute GPA** |

Classroom Example: p. A-19, Exercise 44

At a certain college, the grades A–F are assigned numerical values as follows.

$$A = 4.0 \qquad B+ = 3.5 \qquad B = 3.0 \qquad C+ = 2.5$$
$$C = 2.0 \qquad D+ = 1.5 \qquad D = 1.0 \qquad F = 0.0$$

Elmer takes the following classes with the grades as shown. Determine Elmer's GPA.

Course	Grade	Number of Credit-Hours
Prealgebra	A = 4 pts	3
Study Skills	C = 2 pts	1
First Aid	B+ = 3.5 pts	2
English I	D = 1.0 pt	4

Solution:

The data in the table can be visualized as follows.

| 4 pts | 4 pts | 4 pts | 2 pts | 3.5 pts | 3.5 pts | 1 pt | 1 pt | 1 pt | 1 pt |
| A | A | A | C | B+ | B+ | D | D | D | D |

3 of these   1 of these   2 of these   4 of these

The number of grade points earned for each course is the product of the grade for the course and the number of credit-hours for the course. For example:

Grade points for Prealgebra: (4 pts)(3 credit-hours) = 12 points

Course	Grade	Number of Credit-Hours (Weights)	Product Number of Grade Points
Prealgebra	A = 4 pts	3	(4 pts)(3 credit-hours) = 12 pts
Study Skills	C = 2 pts	1	(2 pts)(1 credit-hour) = 2 pts
First Aid	B+ = 3.5 pts	2	(3.5 pts)(2 credit-hours) = 7 pts
English I	D = 1.0 pt	4	(1 pt)(4 credit-hours) = 4 pts
		Total hours: 10	Total grade points: 25 pts

To determine GPA, we will add the number of grade points earned for each course and then divide by the total number of credit-hours taken.

$$\text{Mean} = \frac{25}{10} = 2.5 \qquad \text{Elmer's GPA for this term is 2.5.}$$

Skill Practice

9. Clyde received the following grades for the semester. Use the numerical values assigned to grades from Example 7 to find Clyde's GPA.

Course	Grade	Credit-Hours
Math	B+	4
Science	C	3
Speech	A	3

In Example 7, notice that the value of each grade is "weighted" by the number of credit-hours. The grade of "A" for Prealgebra is weighted three times. The grade of "C" for the study skills course is weighted one time. The grade that hurt Elmer's GPA was the

Answer
9. 3.2

"D" in English. Not only did he receive a low grade, but the course was weighted heavily (4 credit-hours). In Exercise 47, we recompute Elmer's GPA with a "B" in English to see how this grade affects his GPA.

Section A.2   Practice Exercises

For additional exercises, see Classroom Activities A.2A–A.2B in the *Student's Resource Manual* at www.mhhe.com/moh.

Concept 1: Mean

For Exercises 1–7, find the mean of each set of numbers. **(See Example 1.)**

1. 93, 96, 88, 72, 91   88

2. 4, 6, 5, 10, 4, 5, 8   6

3. 3, 8, 5, 7, 4, 2, 7, 4   5

4. 0, 5, 7, 4, 7, 2, 4, 3   4

5. 7, 6, 5, 10, 8, 4, 8, 6, 0   6

6. −10, −13, −18, −20, −15   −15.2

7. −22, −14, −12, −16, −15   −15.8

8. Compute the mean of your test scores for this class up to this point.   Answers will vary.

9. The flight times in hours for six flights between New York and Los Angeles are given. Find the mean flight time. Round to the nearest tenth of an hour.   5.8 hr

    5.5, 6.0, 5.8, 5.8, 6.0, 5.6

10. A nurse takes the temperature of a patient every 10 min and records the temperatures as follows: 98°F, 98.4°F, 98.9°F, 100.1°F, and 99.2°F. Find the patient's mean temperature.   98.92°F

11. The number of Calories for six different chicken sandwiches and chicken salads is given in the table.

    a. What is the mean number of Calories for a chicken sandwich? Round to the nearest whole unit.   397 Cal

    b. What is the mean number of Calories for a salad with chicken? Round to the nearest whole unit.   386 Cal

    c. What is the difference in the means?   There is only an 11-Cal difference in the means.

Chicken Sandwiches	Salads with Chicken
360	310
370	325
380	350
400	390
400	440
470	500

12. The heights of the players from two NBA teams are given in the table. All heights are in inches.

    a. Find the mean height for the players on the Philadelphia 76ers.   79 in.

    b. Find the mean height for the players on the Milwaukee Bucks.   77.9 in.

    c. What is the difference in the mean heights?
    The mean height for the 76ers is slightly higher by 1.1 in.

Philadelphia 76ers' Height (in.)	Milwaukee Bucks' Height (in.)
83	70
83	83
72	82
79	72
77	82
84	85
75	75
76	75
82	78
79	77

Writing   ← → Translating Expression   Geometry   Scientific Calculator   Video

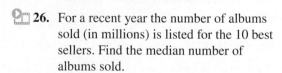

13. Zach received the following scores for his first four tests: 98%, 80%, 78%, 90%.

   a. Find Zach's mean test score.   86.5%

   b. Zach got a 59% on his fifth test. Find the mean of all five tests.   81%

   c. How did the low score of 59% affect the overall mean of five tests?
   The low score of 59% decreased Zach's average by 5.5%.

14. The prices of four steam irons are $50, $30, $25, and $45.

   a. Find the mean of these prices. $37.50

   b. An iron that costs $140 is added to the list. What is the mean of all five irons? $58.00

   c. How does the expensive iron affect the mean?
   By including the iron for $140, the mean increased by $20.50.

Concept 2: Median

For Exercises 15–20, find the median for each set of numbers. **(See Examples 2–3.)**

15. 16, 14, 22, 13, 20, 19, 17
   17

16. 32, 35, 22, 36, 30, 31, 38
   32

17. 109, 118, 111, 110, 123, 100
   110.5

18. 134, 132, 120, 135, 140, 118
   133

19. −58, −55, −50, −40, −40, −55
   −52.5

20. −82, −90, −99, −82, −88, −87
   −87.5

21. The infant mortality rates for five countries are given in the table. Find the median.
   3.93 deaths per 1000

Country	Infant Mortality Rate (Deaths per 1000)
Sweden	3.93
Japan	4.10
Finland	3.82
Andorra	4.09
Singapore	3.87

22. The snowfall amounts for 5 winter months in Burlington, Vermont, are given in the table. Find the median.   16.8 in.

Month	Snowfall (in.)
November	6.6
December	18.1
January	18.8
February	16.8
March	12.4

23. Jonas Slackman played 8 golf tournaments, each with 72-holes of golf. His scores for the tournaments are given. Find the median score.

   −3, −5, 1, 4, −8, 2, 8, −1   0

24. Andrew Strauss recorded the daily low temperature (in °C) at his home in Virginia for 8 days in January. Find the median temperature.

   5, 6, −5, 1, −4, −11, −8, −5   −4.5°C

25. The number of passengers (in millions) on 9 leading airlines for a recent year is listed. Find the median number of passengers. (*Source:* International Airline Transport Association)

   48.3, 42.4, 91.6, 86.8, 46.5, 71.2, 45.4, 56.4, 51.7
   51.7 million passengers

26. For a recent year the number of albums sold (in millions) is listed for the 10 best sellers. Find the median number of albums sold.

   2.7, 3.0, 4.8, 7.4, 3.4, 2.6, 3.0, 3.0, 3.9, 3.2
   3.1 million albums

Concept 3: Mode

For Exercises 27–32, find the mode(s) for each set of numbers. **(See Examples 4–5.)**

27. 4, 5, 3, 8, 4, 9, 4, 2, 1, 4   4

28. 12, 14, 13, 17, 19, 18, 19, 17, 17   17

29. −28, −21, −24, −23, −24, −30, −21
   −21 and −24

30. −45, −42, −40, −41, −49, −49, −42
   −42 and −49

31. 90, 89, 91, 77, 88   No mode

32. 132, 253, 553, 255, 552, 234   No mode

33. The table gives the monthly precipitation for Portland, Oregon for selected months. Find the mode.   *3.66 in.*

Month	Rainfall (in.)
January	4.88
February	3.66
March	3.66
April	2.72
May	2.48
June	1.69

34. The table gives the number of hazardous waste sites for selected states. Find the mode.   *39*

State	Number of Sites
Florida	51
New Jersey	112
Michigan	67
Wisconsin	39
California	96
Pennsylvania	94
Illinois	39
New York	90

35. The unemployment rates for nine countries are given. Find the mode. **(See Example 6.)**

6.3%, 7.0%, 5.8%, 9.1%, 5.2%, 8.8%, 8.4%, 5.8%, 5.2%   *5.2% and 5.8%*

36. The list gives the number of children who were absent from class for an 11-day period. Find the mode.

4, 1, 6, 2, 4, 4, 4, 2, 2, 3, 2   *2 children and 4 children*

Mixed Exercises

37. Six test scores for Jonathan's history class are listed. Find the mean and median. Round to the nearest tenth if necessary. Does the mean or median give a better overall score for Jonathan's performance?

92%, 98%, 43%, 98%, 97%, 85%
Mean: 85.5%; median: 94.5%; The median gives Jonathan a better overall score.

38. Nora's math test results are listed. Find the mean and median. Round to the nearest tenth if necessary. Does the mean or median give a better overall score for Nora's performance?

52%, 85%, 89%, 90%, 83%, 89%
Mean: 81.3%; median: 87%; The median gives Nora a better overall score.

39. Listed below are monthly costs for seven health insurance companies for a self-employed person, 55 years of age, and in good health. Find the mean, median, and mode (if one exists). Round to the nearest dollar. (*Source:* eHealth Insurance Company, 2007)

$312, $225, $221, $256, $308, $280, $147
Mean: $250; median: $256; mode: There is no mode.

40. The salaries for seven Associate Professors at a large university are listed. Find the mean, median, and mode (if one exists). Round to the nearest dollar.

$104,000, $107,000, $67,750, $82,500, $73,500, $88,300, $104,000
Mean: $89,579; median: $88,300; mode: $104,000

41. The prices of 10 single-family, three-bedroom homes for sale in Santa Rosa, California, are listed for a recent year. Find the mean, median, and mode (if one exists).

$850,000, $835,000, $839,000, $829,000, $850,000, $850,000, $850,000, $847,000, $1,850,000, $825,000
Mean: $942,500; median: $848,500; mode: $850,000

42. The prices of 10 single-family, three-bedroom homes for sale in Boston, Massachusetts, are listed for a recent year. Find the mean, median, and mode (if one exists).

$300,000, $2,495,000, $2,120,000, $220,000, $194,000, $391,000, $315,000, $330,000, $435,000, $250,000
Mean: $705,000; median: $322,500; mode: There is no mode.

Concept 4: Weighted Mean

For Exercises 43–46, use the following numerical values assigned to grades to compute GPA. Round each GPA to the hundredths place. **(See Example 7.)**

A = 4.0	B+ = 3.5	B = 3.0	C+ = 2.5
C = 2.0	D+ = 1.5	D = 1.0	F = 0.0

Writing   Translating Expression   Geometry   Scientific Calculator   Video

43. Compute the GPA for the following grades. Round to the nearest hundredth. 2.38

Course	Grade	Number of Credit-Hours (Weights)
Intermediate Algebra	B	4
Theater	C	1
Music Appreciation	A	3
World History	D	5

44. Compute the GPA for the following grades. Round to the nearest hundredth. 3.59

Course	Grade	Number of Credit-Hours (Weights)
General Psychology	B+	3
Beginning Algebra	A	4
Student Success	A	1
Freshman English	B	3

45. Compute the GPA for the following grades. Round to the nearest hundredth. 2.77

Course	Grade	Number of Credit-Hours (Weights)
Business Calculus	B+	3
Biology	C	4
Library Research	F	1
American Literature	A	3

46. Compute the GPA for the following grades. Round to the nearest hundredth. 2.73

Course	Grade	Number of Credit-Hours (Weights)
University Physics	C+	5
Calculus I	A	4
Computer Programming	D	3
Swimming	A	1

47. Refer to the table given in Example 7. Replace the grade of "D" in English I with a grade of "B" and compute the GPA. How did Elmer's GPA differ with a better grade in the 4-hour English class?
3.3; Elmer's GPA improved from 2.5 to 3.3.

Expanding Your Skills

48. There are 20 students enrolled in a 12th-grade math class. The graph displays the number of students by age. First complete the table, and then find the mean.

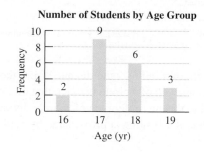

Number of Students by Age Group

Age (yr)	Number of Students	Product
16	2	32
17	9	153
18	6	108
19	3	57
Total:	20	350

The mean age is 17.5 years.

49. A survey was made in a neighborhood of 37 houses. The graph represents the number of residents who live in each house. Complete the table and determine the mean number of residents per house.

Number of Houses for Each Number of Residents

Number of Residents in Each House	Number of Houses	Product
1	3	3
2	9	18
3	10	30
4	9	36
5	6	30
Total:	37	117

The mean number of residents is approximately 3.2.

 Writing   Translating Expression Geometry Scientific Calculator   Video

Concepts

1. Perimeter
2. Area
3. Volume
4. Angles
5. Triangles

1. Perimeter

In this section, we present several facts and formulas that may be used throughout the text in applications of geometry. One of the most important uses of geometry involves the measurement of objects of various shapes. We begin with an introduction to perimeter, area, and volume for several common shapes and objects.

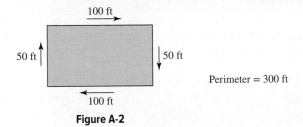

100 ft

50 ft   50 ft

Perimeter = 300 ft

100 ft

Figure A-2

Perimeter is defined as the distance around a figure. If we were to put up a fence around a field, the perimeter would determine the amount of fencing. For example, in Figure A-2 the distance around the field is 300 ft. For a polygon (a closed figure constructed from line segments), the perimeter is the sum of the lengths of the sides. For a circle, the distance around the outside is called the circumference.

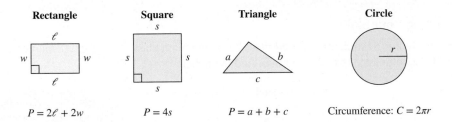

Rectangle	Square	Triangle	Circle
$P = 2\ell + 2w$	$P = 4s$	$P = a + b + c$	Circumference: $C = 2\pi r$

For a circle, r represents the length of a radius—the distance from the center to any point on the circle. The length of a diameter, d, of a circle is twice that of a radius. Thus, $d = 2r$. The number π is a constant equal to the circumference of a circle divided by the length of a diameter. That is, $\pi = \frac{C}{d}$. The value of π is often approximated by 3.14 or $\frac{22}{7}$.

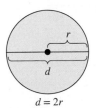

$d = 2r$

Classroom Examples: p. A-30, Exercises 4 and 10

Example 1   Finding Perimeter and Circumference

Find the perimeter or circumference as indicated. Use 3.14 for π.

a. Perimeter of the rectangle   **b.** Circumference of the circle

3.1 ft

5.5 ft

6 cm

Solution:

a. $P = 2\ell + 2w$

$\qquad = 2(5.5 \text{ ft}) + 2(3.1 \text{ ft})$ $\qquad$ Substitute $\ell = 5.5$ ft and $w = 3.1$ ft.

$\qquad = 11 \text{ ft} + 6.2 \text{ ft}$

$\qquad = 17.2 \text{ ft}$ $\qquad$ The perimeter is 17.2 ft.

b. $C = 2\pi r$

$\qquad \approx 2(3.14)(6 \text{ cm})$ $\qquad$ Substitute 3.14 for π and $r = 6$ cm.

$\qquad = 6.28(6 \text{ cm})$

$\qquad = 37.68 \text{ cm}$ $\qquad$ The circumference is 37.68 cm.

> **TIP:** If a calculator is used to find the circumference of a circle, use the π key to get a more accurate answer.

Skill Practice

1. Find the perimeter of the square.

2. Find the circumference. Use 3.14 for π.

7.25 in.

2 in.

2. Area

The area of a geometric figure is the number of square units that can be enclosed within the figure. In applications, we would find the area of a region if we were laying carpet or putting down sod for a lawn. For example, the rectangle shown in Figure A-3 encloses 6 square inches (6 in.^{2}).

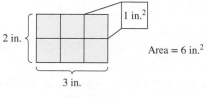

2 in.

1 in.^{2}

Area = 6 in.^{2}

3 in.

Figure A-3

The formulas used to compute the area for several common geometric shapes are given here:

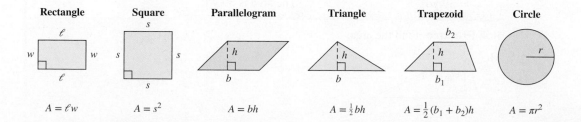

Rectangle	Square	Parallelogram	Triangle	Trapezoid	Circle
$A = \ell w$	$A = s^2$	$A = bh$	$A = \frac{1}{2}bh$	$A = \frac{1}{2}(b_1 + b_2)h$	$A = \pi r^2$

Answers

1. 29 in. $\qquad$ **2.** 12.56 in.

Classroom Example: p. A-30, Exercise 18

TIP: The units of area are given in square units such as square inches (in.²), square feet (ft²), square yards (yd²), square centimeters (cm²), and so on.

| Example 2 | **Finding Area** |

Find the area.

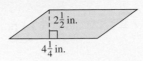

Solution:

$$A = bh$$ The figure is a parallelogram.

$$= (4\tfrac{1}{4} \text{ in.})(2\tfrac{1}{2} \text{ in.})$$ Substitute $b = 4\tfrac{1}{4}$ in. and $h = 2\tfrac{1}{2}$ in.

$$= \left(\frac{17}{4} \text{ in.}\right)\left(\frac{5}{2} \text{ in.}\right)$$

$$= \frac{85}{8} \text{ in.}^2 \text{ or } 10\tfrac{5}{8} \text{ in.}^2$$

Skill Practice Find the area.

3.

Classroom Example: p. A-31, Exercise 24

| Example 3 | **Finding Area** |

Find the area.

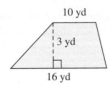

Solution:

$$A = \frac{1}{2}(b_1 + b_2)h$$ The figure is a trapezoid.

$$= \frac{1}{2}(16 \text{ yd} + 10 \text{ yd})(3 \text{ yd})$$ Substitute $b_1 = 16$ yd, $b_2 = 10$ yd, and $h = 3$ yd.

$$= \frac{1}{2}(26 \text{ yd})(3 \text{ yd})$$

$$= (13 \text{ yd})(3 \text{ yd})$$

$$= 39 \text{ yd}^2$$ The area is 39 yd².

Skill Practice Find the area.

4.

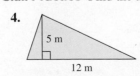

Answers

3. $\dfrac{9}{16}$ cm²   4. 30 m²

Example 4	**Finding the Area of a Circle**

Classroom Example: p. A-31, Exercise 22

Find the area of a circular fountain if the diameter is 50 ft. Use 3.14 for π.

50 ft

Solution:

$A = \pi r^2$ We need the radius, which is $\frac{1}{2}$ the diameter.
$r = \frac{1}{2}(50) = 25$ ft

$\approx (3.14)(25 \text{ ft})^2$ Substitute 3.14 for π and $r = 25$ ft.

$= (3.14)(625 \text{ ft}^2)$

$= 1962.5 \text{ ft}^2$ The area of the fountain is 1962.5 ft^2.

Skill Practice Find the area of the circular region. Use 3.14 for π.

5.

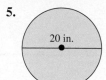

20 in.

3. Volume

The volume of a solid is the number of cubic units that can be enclosed within a solid. The solid shown in Figure A-4 contains 18 cubic inches (18 in.^{3}). In applications, volume might refer to the amount of water in a swimming pool.

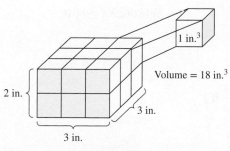

2 in.

3 in.

3 in.

1 in.^{3}

Volume $= 18$ in.^{3}

Figure A-4

The formulas used to compute the volume of several common solids are given here:

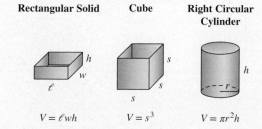

Rectangular Solid	Cube	Right Circular Cylinder
h w ℓ	s s s	h r
$V = \ell wh$	$V = s^3$	$V = \pi r^2 h$

Answer

5. 314 in.^{2}

> **TIP:** Notice that the volume formulas for the three figures just shown are given by the product of the area of the base and the height of the figure:
>
> $$V = \ell w h \qquad\qquad V = s \cdot s \cdot s \qquad\qquad V = \pi r^2 h$$
>
> Area of　　　　　　Area of　　　　　　Area of
> Rectangular Base　Square Base　　　Circular Base

Two additional geometric solids often used in geometry are the right circular cone and the sphere.

Right Circular Cone　　　**Sphere**

$$V = \tfrac{1}{3}\pi r^2 h \qquad\qquad V = \tfrac{4}{3}\pi r^3$$

Classroom Example: p. A-31, Exercise 32

Example 5　**Finding Volume**

Find the volume.

$1\frac{1}{2}$ ft

$1\frac{1}{2}$ ft

$1\frac{1}{2}$ ft

Solution:

$$V = s^3 \qquad\qquad \text{The object is a cube.}$$

$$= \left(1\tfrac{1}{2}\ \text{ft}\right)^3 \qquad\qquad \text{Substitute } s = 1\tfrac{1}{2}\ \text{ft.}$$

$$= \left(\frac{3}{2}\ \text{ft}\right)^3$$

$$= \left(\frac{3}{2}\ \text{ft}\right)\left(\frac{3}{2}\ \text{ft}\right)\left(\frac{3}{2}\ \text{ft}\right)$$

$$= \frac{27}{8}\ \text{ft}^3,\ \text{or}\ 3\frac{3}{8}\ \text{ft}^3$$

> **TIP:** The units of volume are cubic units such as cubic inches (in.³), cubic feet (ft³), cubic yards (yd³), cubic centimeters (cm³), and so on.

Skill Practice Find the volume.

6.

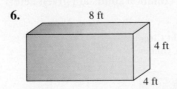

8 ft

4 ft

4 ft

Answer

6. 128 ft³

Example 6 **Finding Volume**

Classroom Example: p. A-31, Exercise 36

Find the volume. Round to the nearest whole unit.

$h = 12$ cm
$r = 4$ cm

Solution:

$V = \dfrac{1}{3}\pi r^2 h$ The object is a right circular cone.

$\approx \dfrac{1}{3}(3.14)(4 \text{ cm})^2(12 \text{ cm})$ Substitute 3.14 for π, $r = 4$ cm, and $h = 12$ cm.

$= \dfrac{1}{3}(3.14)(16 \text{ cm}^2)(12 \text{ cm})$

$= 200.96 \text{ cm}^3$

$\approx 201 \text{ cm}^3$ Round to the nearest whole unit.

Skill Practice Find the volume. Use 3.14 for π. Round to the nearest whole unit.

7. $r = 2$ in.

Example 7 **Finding Volume in an Application**

Classroom Example: p. A-32, Exercise 52

An underground gas tank is in the shape of a right circular cylinder. Find the volume of the tank. Use 3.14 for π.

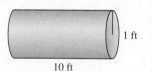

1 ft

10 ft

Solution:

$V = \pi r^2 h$

$\approx (3.14)(1 \text{ ft})^2(10 \text{ ft})$ Substitute 3.14 for π, $r = 1$ ft, and $h = 10$ ft.

$= (3.14)(1 \text{ ft}^2)(10 \text{ ft})$

$= 31.4 \text{ ft}^3$ The tank holds 31.4 ft^3 of gasoline.

Skill Practice

8. Find the volume of soda in the can. Use 3.14 for π. Round to the nearest whole unit.

12 cm

6 cm

Answers

7. 33 in.^{3} **8.** 339 cm^3

4. Angles

Applications involving angles and their measure come up often in the study of algebra, trigonometry, calculus, and applied sciences. The most common unit to measure an angle is the degree (°). Several angles and their corresponding degree measure are shown in Figure A-5.

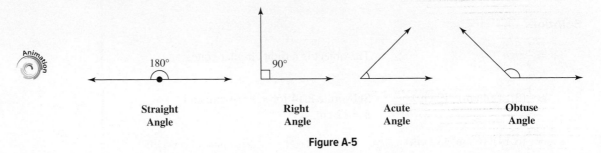

| Straight | Right | Acute | Obtuse |
| Angle | Angle | Angle | Angle |

Figure A-5

- An angle that measures 90° is a **right angle** (right angles are often marked with a square or corner symbol, □).
- An angle that measures 180° is called a **straight angle**.
- An angle that measures between 0° and 90° is called an **acute angle**.
- An angle that measures between 90° and 180° is called an **obtuse angle**.
- Two angles with the same measure are **congruent angles**.

The measure of an angle will be denoted by the symbol m written in front of the angle. Therefore, the measure of $\angle A$ is denoted $m(\angle A)$.

- Two angles are **complementary** if the sum of their measures is 90°.
- Two angles are **supplementary** if the sum of their measures is 180°.

$$m(\angle x) + m(\angle y) = 90°$$

$$m(\angle x) + m(\angle y) = 180°$$

When two lines intersect, four angles are formed (Figure A-6). In Figure A-6, $\angle a$ and $\angle b$ are a pair of vertical angles. Another set of vertical angles is the pair $\angle c$ and $\angle d$. An important property of vertical angles is that the measures of two vertical angles are *equal*. In the figure, $m(\angle a) = m(\angle b)$ and $m(\angle c) = m(\angle d)$.

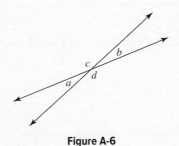

Figure A-6

Parallel lines are lines that lie in the same plane and do not intersect. In Figure A-7, the lines L_1 and L_2 are parallel lines. If a line intersects two parallel lines, the line forms eight angles with the parallel lines.

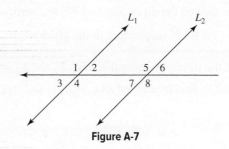

Figure A-7

The measures of angles 1–8 in Figure A-7 have the following special properties.

L_1 and L_2 are Parallel	Name of Angles	Property
	The following pairs of angles are called alternate interior angles:	Alternate interior angles are equal in measure.
	$\angle 2$ and $\angle 7$	$m(\angle 2) = m(\angle 7)$
	$\angle 4$ and $\angle 5$	$m(\angle 4) = m(\angle 5)$
	The following pairs of angles are called alternate exterior angles:	Alternate exterior angles are equal in measure.
	$\angle 1$ and $\angle 8$	$m(\angle 1) = m(\angle 8)$
	$\angle 3$ and $\angle 6$	$m(\angle 3) = m(\angle 6)$
	The following pairs of angles are called corresponding angles:	Corresponding angles are equal in measure.
	$\angle 1$ and $\angle 5$	$m(\angle 1) = m(\angle 5)$
	$\angle 2$ and $\angle 6$	$m(\angle 2) = m(\angle 6)$
	$\angle 3$ and $\angle 7$	$m(\angle 3) = m(\angle 7)$
	$\angle 4$ and $\angle 8$	$m(\angle 4) = m(\angle 8)$

Example 8   **Finding the Measures of Angles in a Diagram**

Classroom Examples: p. A-34, Exercises 76 and 78

Find the measure of each angle and explain how the angle is related to the given angle of 70°.

a. $\angle a$ **b.** $\angle b$

c. $\angle c$ **d.** $\angle d$

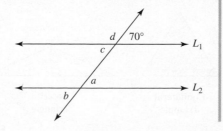

Solution:

 a. $m(\angle a) = 70°$ $\angle a$ is a corresponding angle to the given angle of 70°.

 b. $m(\angle b) = 70°$ $\angle b$ and the given angle of 70° are alternate exterior angles.

 c. $m(\angle c) = 70°$ $\angle c$ and the given angle of 70° are vertical angles.

 d. $m(\angle d) = 110°$ $\angle d$ is the supplement of the given angle of 70°.

Skill Practice Refer to the figure. Assume that lines L_1 and L_2 are parallel.

 9. Given that $m(\angle 3) = 23°$, find $m(\angle 2)$, $m(\angle 4)$, $m(\angle 7)$, and $m(\angle 8)$.

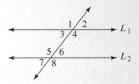

5. Triangles

Triangles are categorized by the measures of the angles (Figure A-8) and by the number of equal sides or angles (Figure A-9).

- An **acute triangle** is a triangle in which all three angles are acute.
- A **right triangle** is a triangle in which one angle is a right angle.
- An **obtuse triangle** is a triangle in which one angle is obtuse.

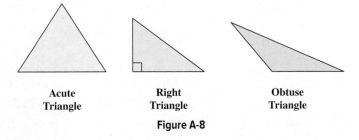

Figure A-8

- An **equilateral triangle** is a triangle in which all three sides (and all three angles) are equal in measure.
- An **isosceles triangle** is a triangle in which two sides are equal in measure (the angles opposite the equal sides are also equal in measure).
- A **scalene triangle** is a triangle in which no sides (or angles) are equal in measure.

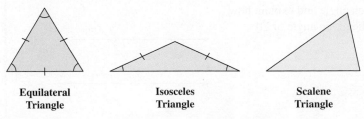

Figure A-9

Answer

9. $m(\angle 2) = 23°$; $m(\angle 4) = 157°$; $m(\angle 7) = 23°$; $m(\angle 8) = 157°$

The following important property is true for all triangles.

> ## Sum of the Angles in a Triangle
> The sum of the measures of the angles of a triangle is 180°.

Animation

| **Example 9** | **Finding the Measures of Angles in a Diagram** |

Classroom Example: p. A-35, Exercise 98

Find the measure of each angle in the figure.

a. $\angle a$

b. $\angle b$

c. $\angle c$

d. $\angle d$

e. $\angle e$

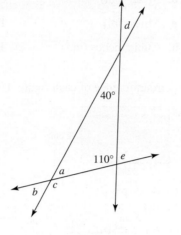

Solution:

a. $m(\angle a) = 30°$ The sum of the angles in a triangle is 180°.

b. $m(\angle b) = 30°$ $\angle a$ and $\angle b$ are vertical angles and have equal measures.

c. $m(\angle c) = 150°$ $\angle c$ and $\angle a$ are supplementary angles ($\angle c$ and $\angle b$ are also supplementary).

d. $m(\angle d) = 40°$ $\angle d$ and the given angle of 40° are vertical angles.

e. $m(\angle e) = 70°$ $\angle e$ and the given angle of 110° are supplementary angles.

Skill Practice For Exercises 10–14, refer to the figure. Find the measure of the indicated angle.

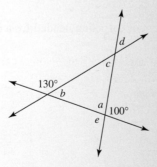

10. $\angle a$ **11.** $\angle b$ **12.** $\angle c$ **13.** $\angle d$ **14.** $\angle e$

Answers

10. 80°   **11.** 50°   **12.** 50°
13. 50°   **14.** 100°

Section A.3   Practice Exercises

Concept 1: Perimeter

For additional exercises, see Classroom Activities A.3A–A.3C in the *Student's Resource Manual* at www.mhhe.com/moh.

1. Would you measure area or perimeter to determine the amount of decorative fence needed to enclose a garden?   Perimeter

2. Identify which of the following units could be measures of perimeter.   b, e, i

   a. Square inches (in.^{2})   **b.** Meters (m)   **c.** Cubic feet (ft^3)

   d. Cubic meters (m^3)   **e.** Miles (mi)   **f.** Square centimeters (cm^2)

   g. Square yards (yd^2)   **h.** Cubic inches (in.^{3})   **i.** Kilometers (km)

For Exercises 3–10, find the perimeter or circumference of each figure. Use 3.14 for π. **(See Example 1.)**

3.
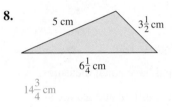
32 m
6 m
10 m

4.
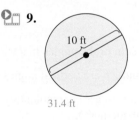
108 cm
22 cm
32 cm

5.
17.2 mi
4.3 mi
4.3 mi

6.
1 ft
0.25 ft
0.25 ft

7.
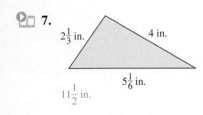
$2\frac{1}{3}$ in.   4 in.
$5\frac{1}{6}$ in.
$11\frac{1}{2}$ in.

8.
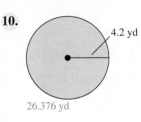
5 cm $3\frac{1}{2}$ cm
$6\frac{1}{4}$ cm
$14\frac{3}{4}$ cm

9.
10 ft
31.4 ft

10.
4.2 yd
26.376 yd

Concept 2: Area

11. Identify which of the following units could be measures of area.   a, f, g

   a. Square inches (in.^{2})   **b.** Meters (m)   **c.** Cubic feet (ft^3)

   d. Cubic meters (m^3)   **e.** Miles (mi)   **f.** Square centimeters (cm^2)

   g. Square yards (yd^2)   **h.** Cubic inches (in.^{3})   **i.** Kilometers (km)

12. Would you measure area or perimeter to determine the amount of carpeting needed for a room?   Area

For Exercises 13–26, find the area. Use 3.14 for π. **(See Examples 2–4.)**

13.
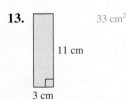
33 cm^2
11 cm
3 cm

14.
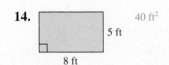
40 ft^2
5 ft
8 ft

15.
16.81 m^2
4,1 m
4.1 m

16.
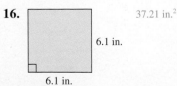
37.21 in.^{2}
6.1 in.
6.1 in.

17.
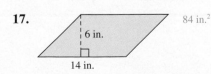
84 in.^{2}
6 in.
14 in.

18.
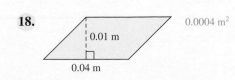
0.0004 m^2
0.01 m
0.04 m

 Writing   Translating Expression   Geometry   Scientific Calculator   Video

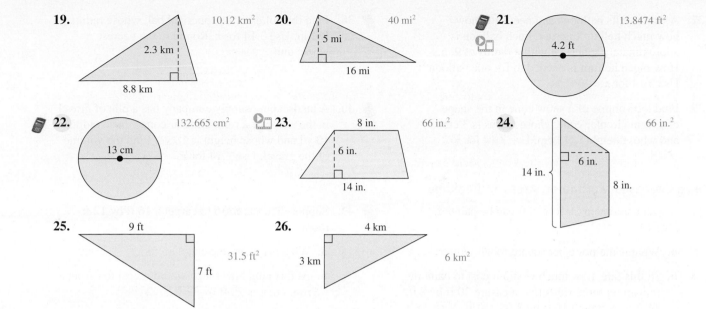

19. 10.12 km²   2.3 km   8.8 km

20. 40 mi²   5 mi   16 mi

21. 13.8474 ft²   4.2 ft

22. 132.665 cm²   13 cm

23. 66 in.²   8 in.   6 in.   14 in.

24. 66 in.²   6 in.   14 in.   8 in.

25. 9 ft   31.5 ft²   7 ft

26. 4 km   3 km   6 km²

Concept 3: Volume

27. Identify which of the following units could be measures of volume.   c, d, h

     a. Square inches (in.²)   **b.** Meters (m)   **c.** Cubic feet (ft³)

     d. Cubic meters (m³)   **e.** Miles (mi)   **f.** Square centimeters (cm²)

     g. Square yards (yd²)   **h.** Cubic inches (in.³)   **i.** Kilometers (km)

28. Would you measure perimeter, area, or volume to determine the amount of water needed to fill a swimming pool?   Volume

For Exercises 29–36, find the volume of each figure. Use 3.14 for π. **(See Examples 5–7.)**

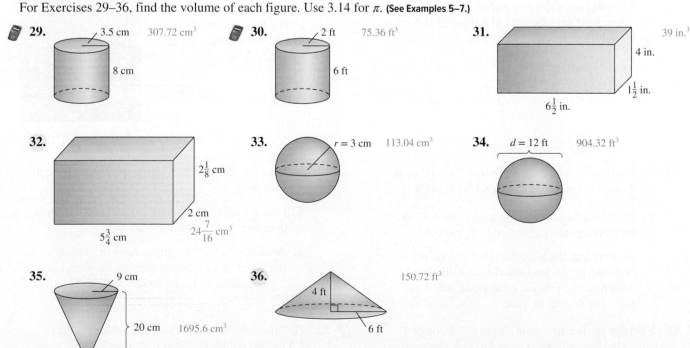

29. 3.5 cm   307.72 cm³   8 cm

30. 2 ft   75.36 ft³   6 ft

31. 39 in.³   4 in. $1\frac{1}{2}$ in. $6\frac{1}{2}$ in.

32. $2\frac{1}{8}$ cm   2 cm $5\frac{3}{4}$ cm $24\frac{7}{16}$ cm³

33. $r = 3$ cm   113.04 cm³

34. $d = 12$ ft   904.32 ft³

35. 9 cm   20 cm   1695.6 cm³

36. 4 ft   6 ft   150.72 ft³

37. A florist sells balloons and needs to know how much helium to order. Each balloon is approximately spherical with a radius of 9 in. How much helium is needed to fill one balloon? Use 3.14 for π.   3052.08 in.³

38. Find the volume of a spherical ball whose radius is 2 in. Use 3.14 for π. Round to the nearest whole unit.   33 in.³

39. Find the volume of a snow cone in the shape of a right circular cone whose radius is 3 cm and whose height is 12 cm. Use 3.14 for π. 113.04 cm³

40. A landscaping supply company has a pile of gravel in the shape of a right circular cone whose radius is 10 yd and whose height is 18 yd. Find the volume of the gravel. Use 3.14 for π.   1884 yd³

Mixed Exercises: Perimeter, Area, and Volume

41. A wall measuring 20 ft by 8 ft can be painted for $40.

   a. What is the price per square foot? $0.25/ft²

   b. At this rate, how much would it cost to paint the remaining three walls that measure 20 ft by 8 ft, 16 ft by 8 ft, and 16 ft by 8 ft? $104

42. Suppose it costs $336 to carpet a 16 ft by 12 ft room.

   a. What is the price per square foot? $1.75/ft²

   b. At this rate, how much would it cost to carpet a room that is 20 ft by 32 ft? $1120

43. If you were to purchase wood to frame a photograph, would you measure the perimeter or area of the photograph?   Perimeter

44. If you were to purchase sod (grass) for your front yard, would you measure the perimeter or area of the yard?   Area

45. How much fencing is needed to enclose a triangularly shaped garden whose sides measure 12 ft, 22 ft, and 20 ft?   54 ft

46. A regulation soccer field is 100 yd long by 60 yd wide. Find the perimeter of the field.   320 yd

47. **a.** An American football field is 360 ft long by 160 ft wide. What is the area of the field? 57,600 ft²

   b. How many pieces of sod, each 1 ft wide and 3 ft long, are needed to sod an entire field? (*Hint:* First find the area of a piece of sod.) 19,200 pieces

48. The Transamerica Pyramid in San Francisco is a tower with triangular sides (excluding the "wings"). Each side measures 145 ft wide with a height of 853 ft. What is the area of each side?   61,842.5 ft²

© R. Morley/PhotoLink/ Getty Images RF

49. **a.** Find the area of a circular pizza that is 8 in. in diameter (the radius is 4 in.). Use 3.14 for π. 50.24 in.²

   b. Find the area of a circular pizza that is 12 in. in diameter (the radius is 6 in.).   113.04 in.²

   c. Assume that the 8-in. diameter and 12-in. diameter pizzas are both the same thickness. Which would provide more pizza, two 8-in. pizzas or one 12-in. pizza?   One 12-in. pizza

50. **a.** Find the area of a rectangular pizza that is 12 in. by 8 in.   96 in.²

   b. Find the area of a circular pizza that has a 16-in. diameter. Use 3.14 for π.   200.96 in.²

   c. Assume that the two pizzas have the same thickness. Which would provide more pizza? Two rectangular pizzas or one circular pizza? One circular pizza

51. Find the volume of a soup can in the shape of a right circular cylinder if its radius is 3.2 cm and its height is 9 cm. Use 3.14 for π.   289.3824 cm³

52. Find the volume of a coffee mug whose radius is 2.5 in. and whose height is 6 in. Use 3.14 for π. 117.75 in.³

Concept 4: Angles

For Exercises 53–58, answer true or false. If an answer is false, explain why.

53. The sum of the measures of two right angles equals the measure of a straight angle.   True

54. Two right angles are complementary.
False; they are supplementary.

55. Two right angles are supplementary.   True

56. Two acute angles cannot be supplementary.
True

57. Two obtuse angles cannot be supplementary.
True

58. An obtuse angle and an acute angle can be supplementary.   True

59. If possible, find two acute angles that are supplementary.   Not possible

60. If possible, find two acute angles that are complementary. Answers may vary.   For example: 30°, 60°

61. If possible, find an obtuse angle and an acute angle that are supplementary. Answers may vary.
For example: 100°, 80°

62. If possible, find two obtuse angles that are supplementary.   Not possible

63. What angle is its own complement?   45°

64. What angle is its own supplement?   90°

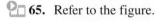 **65.** Refer to the figure.

    a. State all the pairs of vertical angles.   ∠1 and ∠3, ∠2 and ∠4

    b. State all the pairs of supplementary angles.
    ∠1 and ∠2, ∠2 and ∠3, ∠3 and ∠4, ∠1 and ∠4

    c. If the measure of ∠4 is 80°, find the measures of ∠1, ∠2, and ∠3.
    $m(\angle 1) = 100°$, $m(\angle 2) = 80°$, $m(\angle 3) = 100°$

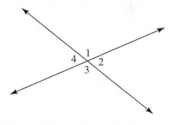

66. Refer to the figure.

    a. State all the pairs of vertical angles.
    ∠b and ∠d, ∠a and ∠c

    b. State all the pairs of supplementary angles.
    ∠a and ∠b, ∠b and ∠c, ∠c and ∠d, ∠a and ∠d

    c. If the measure of ∠a is 25°, find the measures of ∠b, ∠c, and ∠d.
    $m(\angle b) = 155°$, $m(\angle c) = 25°$, $m(\angle d) = 155°$

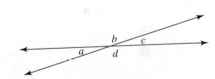

For Exercises 67–70, the measure of an angle is given. Find the measure of its complement.

67. 33°   57°   **68.** 87°   3°   **69.** 12°   78°   **70.** 45°   45°

For Exercises 71–74, the measure of an angle is given. Find the measure of its supplement.

71. 33°   147°   **72.** 87°   93°   **73.** 122°   58°   **74.** 90°   90°

For Exercises 75–82, refer to the figure. Assume that L_1 and L_2 are parallel lines. **(See Example 8.)**

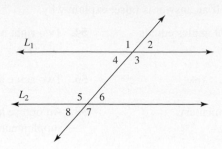

75. $m(\angle 5) = m(\angle \underline{\quad 7 \quad})$ Reason: Vertical angles have equal measures.

76. $m(\angle 5) = m(\angle \underline{\quad 3 \quad})$ Reason: Alternate interior angles have equal measures.

77. $m(\angle 5) = m(\angle \underline{\quad 1 \quad})$ Reason: Corresponding angles have equal measures.

78. $m(\angle 7) = m(\angle \underline{\quad 3 \quad})$ Reason: Corresponding angles have equal measures.

79. $m(\angle 7) = m(\angle \underline{\quad 1 \quad})$ Reason: Alternate exterior angles have equal measures.

80. $m(\angle 7) = m(\angle \underline{\quad 5 \quad})$ Reason: Vertical angles have equal measures.

81. $m(\angle 3) = m(\angle \underline{\quad 5 \quad})$ Reason: Alternate interior angles have equal measures.

82. $m(\angle 3) = m(\angle \underline{\quad 1 \quad})$ Reason: Vertical angles have equal measures.

83. Find the measures of angles a–g in the figure. Assume that L_1 and L_2 are parallel.
$m(\angle a) = 45°, m(\angle b) = 135°, m(\angle c) = 45°, m(\angle d) = 135°,$
$m(\angle e) = 45°, m(\angle f) = 135°, m(\angle g) = 45°$

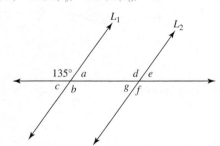

84. Find the measures of angles a–g in the figure. Assume that L_1 and L_2 are parallel.
$m(\angle a) = 65°, m(\angle b) = 115°, m(\angle c) = 115°, m(\angle d) = 65°,$
$m(\angle e) = 115°, m(\angle f) = 115°, m(\angle g) = 65°$

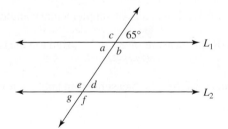

Concept 5: Triangles

For Exercises 85–88, identify the triangle as equilateral, isosceles, or scalene.

85.

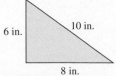

6 in.   10 in.
8 in.
Scalene

86.

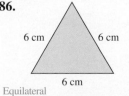

6 cm   6 cm
6 cm
Equilateral

87.

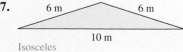

6 m   6 m
10 m
Isosceles

88.

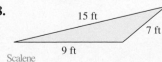

15 ft
7 ft
9 ft
Scalene

89. True or False? If a triangle is equilateral, then it is not scalene.   True

90. True or False? If a triangle is isosceles, then it is also scalene.   False; an isosceles triangle has two sides that are the same length. A scalene triangle has three sides of different length.

91. Can a triangle be both a right triangle and an obtuse triangle? Explain.   No, a 90° angle plus an angle greater than 90° would make the sum of the angles greater than 180°.

92. Can a triangle be both a right triangle and an isosceles triangle? Explain.   Yes, the sides forming the right angle can be equal.

For Exercises 93–96, find the measure of each missing angle.

93.

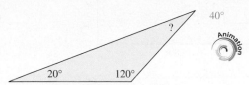

94.

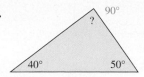

95.

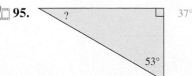

96.

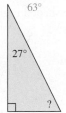

97. Refer to the figure. Find the measures of angles *a–j*. **(See Example 9.)** $m(\angle a) = 80°$, $m(\angle b) = 80°$, $m(\angle c) = 100°$, $m(\angle d) = 100°$, $m(\angle e) = 65°$, $m(\angle f) = 115°$, $m(\angle g) = 115°$, $m(\angle h) = 35°$, $m(\angle i) = 145°$, $m(\angle j) = 145°$

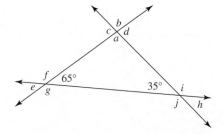

98. Refer to the figure. Find the measures of angles *a–j*.

$m(\angle a) = 120°$, $m(\angle b) = 60°$, $m(\angle c) = 60°$, $m(\angle d) = 40°$, $m(\angle e) = 40°$, $m(\angle f) = 140°$, $m(\angle g) = 140°$, $m(\angle h) = 160°$, $m(\angle i) = 20°$, $m(\angle j) = 160°$

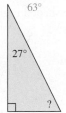

99. Refer to the figure. Find the measures of angles *a–k*. Assume that L_1 and L_2 are parallel.

$m(\angle a) = 70°$, $m(\angle b) = 65°$, $m(\angle c) = 65°$, $m(\angle d) = 110°$, $m(\angle e) = 70°$, $m(\angle f) = 110°$, $m(\angle g) = 115°$, $m(\angle h) = 115°$, $m(\angle i) = 65°$, $m(\angle j) = 70°$, $m(\angle k) = 65°$

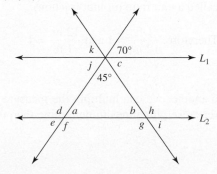

100. Refer to the figure. Find the measures of angles *a–k*. Assume that L_1 and L_2 are parallel.

$m(\angle a) = 75°$, $m(\angle b) = 65°$, $m(\angle c) = 40°$, $m(\angle d) = 65°$, $m(\angle e) = 65°$, $m(\angle f) = 40°$, $m(\angle g) = 40°$, $m(\angle h) = 140°$, $m(\angle i) = 140°$, $m(\angle j) = 115°$, $m(\angle k) = 115°$

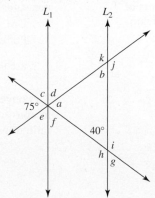

Expanding Your Skills

For Exercises 101–102, find the perimeter.

101.

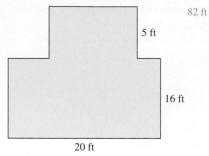

82 ft

5 ft

16 ft

20 ft

102.

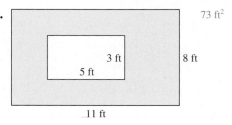

78.4 ft

5.1 ft

4.2 ft

14.1 ft

15.8 ft

For Exercises 103–106, find the area of the shaded region. Use 3.14 for π.

103.

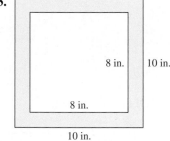

36 in.²

8 in.   10 in.

8 in.

10 in.

104.

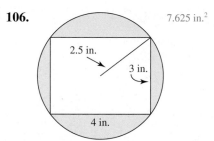

73 ft²

3 ft   8 ft

5 ft

11 ft

105.

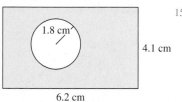

15.2464 cm²

1.8 cm

4.1 cm

6.2 cm

106.

7.625 in.²

2.5 in.

3 in.

4 in.

Section A.4   Converting Units of Measurement

Concepts

1. Converting U.S. and Metric Units of Measure
2. Converting Units of Measure Involving Temperature

1. Converting U.S. and Metric Units of Measure

The procedure for multiplying rational expressions and canceling like factors can be applied in converting units of measurement. A conversion factor is a ratio of equivalent measures. If this ratio is equal to 1, it is called a unit ratio (or unit fraction).

For example, note that 1 ft = 12 in. Therefore, $\dfrac{1\ \text{ft}}{12\ \text{in.}} = 1$ and $\dfrac{12\ \text{in.}}{1\ \text{ft}} = 1$

are unit ratios.

To convert from one unit of measure to another we can multiply the measure we want to convert from, by a unit ratio. We offer these guidelines to determine the proper unit ratio to use.

Choosing a Unit Ratio as a Conversion Factor

To choose the correct unit ratio for unit conversion,

- The unit of measure in the numerator should be the new unit you want to convert *to*.
- The unit of measure in the denominator should be the original unit you want to convert *from*.

Table A-2 summarizes some common metric and U.S. customary equivalents. These equivalents can be used to form unit ratios.

Table A-2

Length	Weight/Mass (on Earth)	Capacity
1 in. = 2.54 cm	1 lb ≈ 0.45 kg	1 qt ≈ 0.95 L
1 ft ≈ 0.305 m	1 oz ≈ 28 g	1 fl oz ≈ 30 mL ≈ 30 cc
1 yd ≈ 0.914 m		
1 mi ≈ 1.61 km		

In Example 1 we use a unit ratio to convert from kilometers to miles.

Example 1 **Converting Metric Units to U.S. Customary Units**

Classroom Example: p. A-39, Exercise 4

A 5K road race is a race that is 5 kilometers in length. From Table A-2, use the fact that 1 mi ≈ 1.61 km to convert this distance to miles. Round to one decimal place.

Solution:

The unit ratios that relate miles to kilometers are $\frac{1 \text{ mi}}{1.61 \text{ km}} = 1$ or $\frac{1.61 \text{ km}}{1 \text{ mi}} = 1$.

We are converting *to* miles *from* kilometers. The unit ratio with miles in the numerator and kilometers in the denominator is $\frac{1 \text{ mi}}{1.61 \text{ km}}$.

$5 \text{ km} \approx \frac{5 \text{ km}}{1} \cdot \frac{1 \text{ mi}}{1.61 \text{ km}}$ Set up a unit ratio to convert kilometers to miles.

$\approx \frac{5}{1.61} \text{ mi}$ Multiply fractions. Like units are canceled.

$\approx 3.1 \text{ mi}$ Divide and round to one decimal place.

The 5K race is approximately 3.1 mi.

Skill Practice

1. Use the fact that 1 mi ≈ 1.61 km to convert 90 mi to kilometers.

Answer

1. 144.9 km

Classroom Examples: pp. A-39–A-40, Exercises 6, 8, and 10

| Example 2 | Converting Between U.S. Customary and Metric Measurements |

Convert the units as indicated. Round to two decimal places, if necessary.

a. 48 oz ≈ _____ g   **b.** 200 m ≈ _____ ft

c. 5.2 L ≈ _____ qt

Solution:

a. $48 \text{ oz} \approx \dfrac{48 \text{ oz}}{1} \cdot \dfrac{28 \text{ g}}{1 \text{ oz}}$ From Table A-2 we know that 1 oz ≈ 28 g.

$\approx 48 \cdot 28 \text{ g}$ Multiply. Cancel like units.

$\approx 1344 \text{ g}$

b. $200 \text{ m} \approx \dfrac{200 \text{ m}}{1} \cdot \dfrac{1 \text{ ft}}{0.305 \text{ m}}$ From Table A-2 we know that 1 ft ≈ 0.305 m.

$\approx \dfrac{200}{0.305} \text{ ft}$ Multiply fractions. Cancel like units.

$\approx 655.74 \text{ ft}$ Divide and round to two decimal places.

c. $5.2 \text{ L} \approx \dfrac{5.2 \text{ L}}{1} \cdot \dfrac{1 \text{ qt}}{0.95 \text{ L}}$ From Table A-2 we know that 1 qt ≈ 0.95 L.

$\approx \dfrac{5.2}{0.95} \text{ qt}$ Multiply fractions. Cancel like units.

$\approx 5.47 \text{ qt}$ Divide and round to two decimal places.

Skill Practice

2. Fill in the blank. Round to two decimal places, if necessary.
   a. 170 lb ≈ _____ kg   **b.** 10 m ≈ _____ ft
   c. 100 cc ≈ _____ fl oz

2. Converting Units of Measure Involving Temperature

In the United States, the Fahrenheit scale is used most often to measure temperature. The symbol °F stands for "degrees Fahrenheit." Another scale used to measure temperature is the Celsius temperature scale. The symbol °C stands for "degrees Celsius." Use the following formulas to convert back and forth between Fahrenheit and Celsius scales.

Conversions for Temperature Scale

To convert from °C to °F: To convert from °F to °C:

$$F = \frac{9}{5}C + 32 \qquad\qquad\qquad C = \frac{5}{9}(F - 32)$$

Answers

2. a. 76.5 kg   **b.** 32.79 ft   **c.** 3.33 fl oz

Example 3	**Converting Units of Temperature**

Classroom Examples: p. A-40, Exercises 14 and 16

a. A warm day in Quebec, Canada, is 25°C. Convert this temperature to °F.

b. A refrigerator should maintain a temperature of not more than 41°F. Convert this temperature to °C.

Solution:

a. To convert from degrees Celsius to degrees Fahrenheit, use the formula $F = \frac{9}{5}C + 32$.

$$F = \frac{9}{5}C + 32$$

$$= \frac{9}{5}(25) + 32 \qquad \text{Substitute } C = 25.$$

$$= 45 + 32 \qquad \text{Perform multiplication first.}$$

$$= 77$$

The temperature in Quebec is equivalent to 77°F.

b. To convert from degrees Fahrenheit to degrees Celsius, use the formula $C = \frac{5}{9}(F - 32)$.

$$C = \frac{5}{9}(F - 32)$$

$$= \frac{5}{9}(41 - 32) \qquad \text{Substitute } F = 41.$$

$$= \frac{5}{9}(9) \qquad \text{Perform the operation within parentheses first.}$$

$$= 5$$

The refrigerator should maintain a temperature of not more than 5°C.

Skill Practice

3. a. Convert 100°F to degrees Celsius. Round to the nearest degree.
   b. Convert −10°C to degrees Fahrenheit.

Answers

3. a. 38°F   **b.** 14°C

Section A.4   Practice Exercises

For additional exercises, see Classroom Activity A.4A in the *Student's Resource Manual* at www.mhhe.com/moh.

Concepts 1 and 2: Converting U.S. and Metric Units of Measurement and Temperature

For Exercises 1–16 convert the units as indicated. Round to one decimal place, if necessary. **(See Examples 1–3.)**

1. 60 m ≈ _____196.7_____ ft

2. 32 cm = _____12.6_____ in.

 3. 400 ft ≈ _____122_____ m

4. 25 mi ≈ _____40.3_____ km

 5. 0.3 lb ≈ _____0.1_____ kg

6. 16 oz ≈ _____448_____ g

7. 8.8 g ≈ _____0.3_____ oz

8. 18 kg ≈ _____40_____ lb

9. 64 fl oz ≈ _____1920_____ mL

10. 1.5 qt ≈ _____1.4_____ L

11. 5.2 L ≈ _____5.5_____ qt

12. 960 cc ≈ _____32_____ fl oz

13. −20°C = _____−4_____ °F

14. 99°C = _____210.2_____ °F

15. 40°F = _____4.4_____ °C

16. −13°F = _____−25_____ °C

Mixed Exercises

For Exercises 17–32, round answers to one decimal place, if necessary.

17. One of the events in the Winter Olympics is speed skating for a distance of 1500 m. Convert this distance to yards.   1641.1 yd

18. In a recent ski jump event in the Winter Olympics, the winner jumped a distance of 108 m. Convert this distance to feet.   354.1 ft

19. The speed of sound is approximately 1126 ft/sec. Convert this speed to meters per second.
343.4 m/sec

20. A tourist from France notices the speed limit on an interstate highway in the United States is 70 mph. What is the speed limit in km per hour?   112.7 km/hr

21. Tyler works with a nutritionist who recommends that males should consume approximately 3 L of liquids per day. How many quarts per day is this amount?   3.2 qt

22. Joanie's nutritionist recommends that females should consume 2.2 L of liquids per day. How many quarts is this amount?   2.3 qt

23. A 2-L bottle of juice sells for $1.99. A half-gallon (2-qt) container of juice sells for $1.79. Compare the price per quart to determine which is the better buy.   The 2-L bottle costs $0.95 per quart. The half-gallon bottle costs $0.90 per quart. The half-gallon is the better buy.

24. The capacity of an automobile gas tank is 16 gal. Determine the capacity of the tank in L. (*Hint:* 1 gal = 4 qt)   60.8 L

25. The recommended daily intake of protein can vary according to gender, weight, and activity level. The recommended amount for a male weighing 90 kg is 72 g protein. Determine his weight in lb.   200 lb

26. A roll of U.S. quarters weighs approximately 227 g. Determine this weight in ounces.   8.1 oz

27. One of the smallest miniature horses on record weighed only 20 lb. Convert this weight to kg.   9 kg

28. On average, a Chihuahua weighs about 5 lb. Convert this weight to kg.   2.3 kg

29. In January, the average temperature at the top of Mt. Everest is about −31°F. What is the temperature in °C?   −35°C

30. Jessica set her oven at 185°F to keep her pizza warm. Convert the temperature to °C.   85°C

31. Technical information for most smartphones states that smartphones should not be operated at temperatures above 35°C. What is this temperature in °F?   95°F

32. The high temperature in London, England, on a typical September day is 18°C, and the low is 13°C. Convert these temperatures to degrees Fahrenheit.   64.4°F, 55.4°F

Student Answer Appendix

Chapter 1

Section 1.1 Practice Exercises, pp. 17–21

1. a. product **b.** factors **c.** numerator; b **d.** lowest
e. 1; 4 **f.** reciprocals **g.** multiple **h.** least
3. Numerator: 2; denominator: 3; proper
5. Numerator: 5; denominator: 2; improper
7. Numerator: 4; denominator: 4; improper
9. Numerator: 5; denominator: 1; improper
11. $\frac{3}{4}$ **13.** $\frac{4}{3}$ **15.** $\frac{1}{6}$ **17.** $\frac{2}{2}$ **19.** $\frac{5}{2}$ or $2\frac{1}{2}$ **21.** $\frac{6}{2}$ or 3
23. The set of whole numbers includes the number 0 and the set of natural numbers does not.
25. For example: $\frac{2}{4}$ **27.** Prime **29.** Composite
31. Composite **33.** Prime **35.** $2 \times 2 \times 3 \times 3$
37. $2 \times 3 \times 7$ **39.** $2 \times 5 \times 11$ **41.** $3 \times 3 \times 3 \times 5$
43. $\frac{1}{5}$ **45.** $\frac{8}{3}$ or $2\frac{2}{3}$ **47.** $\frac{7}{8}$ **49.** $\frac{3}{4}$ **51.** $\frac{5}{8}$ **53.** $\frac{4}{3}$ or $1\frac{1}{3}$
55. False: When adding or subtracting fractions, it is necessary to have a common denominator.
57. $\frac{4}{3}$ or $1\frac{1}{3}$ **59.** $\frac{2}{3}$ **61.** $\frac{9}{2}$ or $4\frac{1}{2}$ **63.** $\frac{3}{5}$ **65.** $\frac{5}{3}$ or $1\frac{2}{3}$
67. $\frac{90}{13}$ or $6\frac{12}{13}$ **69.** \$704 **71.** 35 students **73.** 8 pieces
75. 8 jars **77.** $\frac{3}{7}$ **79.** $\frac{1}{2}$ **81.** 30 **83.** 40 **85.** $\frac{7}{8}$
87. $\frac{43}{40}$ or $1\frac{3}{40}$ **89.** $\frac{3}{26}$ **91.** $\frac{29}{36}$ **93.** $\frac{6}{5}$ or $1\frac{1}{5}$ **95.** $\frac{35}{48}$
97. $\frac{7}{24}$ **99.** $\frac{51}{28}$ or $1\frac{23}{28}$ **101.** $\frac{11}{12}$ cup sugar **103.** $\frac{7}{50}$ in.
105. $\frac{46}{5}$ or $9\frac{1}{5}$ **107.** $\frac{1}{6}$ **109.** $\frac{11}{54}$ **111.** $\frac{7}{2}$ or $3\frac{1}{2}$
113. $\frac{13}{8}$ or $1\frac{5}{8}$ **115.** $\frac{59}{12}$ or $4\frac{11}{12}$ **117.** $\frac{1}{8}$ **119.** $8\frac{19}{24}$ in.
121. $1\frac{7}{12}$ hr **123.** $2\frac{1}{4}$ lb **125.** 25 in.

Section 1.2 Calculator Connections, pp. 30–31

1. ≈ 3.464101615 **2.** ≈ 9.949874371
3. ≈ 12.56637061 **4.** ≈ 1.772453851

Section 1.2 Practice Exercises, pp. 31–34

1. a. variable **b.** constants **c.** set **d.** inequalities **e.** a is less than b **f.** c is greater than or equal to d **g.** 5 is not equal to 6
h. opposites **i.** $|a|$; 0 **3.** $8\frac{1}{4}$ or $\frac{33}{4}$ **5.** 15 **7.** 3
9. 4 **11.** 5.3 **13.** $\frac{1}{80}$ **15. a.** \$3.87 **b.** \$10.32
c. \$12.90 **17. a.** 375 calories **b.** 630 calories **c.** 270 calories
19.

21. a; rational **23.** b; rational **25.** a; rational
27. c; irrational **29.** a; rational **31.** a; rational
33. b; rational **35.** c; irrational **37.** For example: $\pi, -\sqrt{2}, \sqrt{3}$
39. For example: $-4, -1, 0$ **41.** For example: $-\frac{3}{4}, \frac{1}{2}, 0.206$

43. $-\frac{3}{2}, -4, 0.\overline{6}, 0, 1$ **45.** 1 **47.** $-4, 0, 1$ **49. a.** >
b. > **c.** < **d.** > **51.** -18 **53.** 6.1 **55.** $\frac{5}{8}$ **57.** $-\frac{7}{3}$
59. 3 **61.** $-\frac{7}{3}$ **63.** 8 **65.** -72.1 **67.** 2 **69.** 1.5
71. -1.5 **73.** $\frac{3}{2}$ **75.** -10 **77.** $-\frac{1}{2}$
79. False, $|n|$ is never negative. **81.** True **83.** False
85. True **87.** False **89.** False **91.** False
93. True **95.** True **97.** False **99.** True
101. True **103.** True **105.** For all $a < 0$

Section 1.3 Calculator Connections, pp. 41–42

1. 2 **2.** 91 **3.** 84 **4.** 12 **5.** 49 **6.** 18

1–3.

```
(4+6)/(8-3)
              2
110-5*(2+1)-4
             91
100-2*(5-3)^3
             84
```

4–6.

```
3+(4-1)2
              12
(12-6+1)2
              49
3*8-√(32+22)
              18
```

7. 4 **8.** 27 **9.** 0.5

7–9.

```
√(18-2)
              4
(4*3-3*3)^3
             27
(20-32)/(26-22)
             .5
```

Section 1.3 Practice Exercises, pp. 42–45

1. a. quotient; product; sum; difference **b.** base; exponent; power **c.** 8^2 **d.** p^4 **e.** radical; square **f.** order of operations
3. 56 **5.** -19 **7.** $\left(\frac{1}{6}\right)^4$ **9.** a^3b^2 **11.** $(5c)^5$
13. a. x **b.** Yes, 1 **15.** $x \cdot x \cdot x$ **17.** $2b \cdot 2b \cdot 2b$
19. $10 \cdot y \cdot y \cdot y \cdot y \cdot y$ **21.** $2 \cdot w \cdot z \cdot z$ **23.** 36 **25.** $\frac{1}{49}$
27. 0.008 **29.** 64 **31.** 9 **33.** 2 **35.** 12 **37.** 4
39. $\frac{1}{3}$ **41.** $\frac{5}{9}$ **43.** 20 **45.** 60 **47.** 8 **49.** 78 **51.** 0
53. $\frac{7}{6}$ **55.** 45 **57.** 16 **59.** 15 **61.** 19 **63.** 75 **65.** 3
67. 39 **69.** 26 **71.** $\frac{5}{12}$ **73.** $\frac{5}{2}$ **75. a.** 0.12
b. Yes. $0.12 < 0.20$ **77.** $57,600 \text{ ft}^2$ **79.** 21 ft^2 **81.** $3x$
83. $\frac{x}{7}$ or $x \div 7$ **85.** $2 - a$ **87.** $2y + x$ **89.** $4(x + 12)$
91. $3 - Q$ **93.** $2y^3$; 16 **95.** $|z - 8|$; 2 **97.** $5\sqrt{x}$; 10
99. $yz - x$; 16 **101.** 1 **103.** $\frac{1}{4}$
105. a. $36 \div 4 \cdot 3$ Division must be performed before
      $= 9 \cdot 3$ multiplication.
      $= 27$
b. $36 - 4 + 3$ Subtraction must be performed before
      $= 32 + 3$ addition.
      $= 35$

107. This is acceptable, provided division and multiplication are performed in order from left to right, and subtraction and addition are performed in order from left to right.

Section 1.4 Practice Exercises, pp. 51–53

1. a. negative **b.** b **3.** >   **5.** >   **7.** >   **9.** −6
11. 3   **13.** 3   **15.** −3   **17.** −17   **19.** 7   **21.** −19
23. −23   **25.** −5   **27.** −3   **29.** 0   **31.** 0   **33.** −5
35. −3   **37.** 0   **39.** −23   **41.** −6   **43.** −3   **45.** 21.3
47. $-\dfrac{3}{14}$ **49.** $-\dfrac{1}{6}$ **51.** $-\dfrac{15}{16}$ **53.** $\dfrac{1}{20}$ **55.** -2.4 or $-\dfrac{12}{5}$
57. $\dfrac{1}{4}$ or 0.25   **59.** 0   **61.** $-\dfrac{7}{8}$ **63.** −1   **65.** $\dfrac{11}{9}$
67. −23.08   **69.** −0.002117   **71.** To add two numbers with different signs, subtract the smaller absolute value from the larger absolute value and apply the sign of the number with the larger absolute value.   **73.** −1   **75.** 10   **77.** 5   **79.** 1
81. $-6 + (-10); -16$ **83.** $-3 + 8; 5$ **85.** $-21 + 17; -4$
87. $3(-14 + 20); 18$ **89.** $[-7 + (-2)] + 5; -4$
91. $-5 + 13 + (-11); -3°F$
93. $-8 + 1 + 2 + (-5) = -10$; Amara lost 10 lb.
95. a. $52.23 + (-52.95)$ **b.** Yes   **97.** −6; She was 6 below par.

Section 1.5 Calculator Connections, p. 59

1. −13   **2.** −2   **3.** 711
1–3.

```
-8+(-5)
               -13
4+(-5)+(-1)
               -2
627-(-84)
              711
```

4. −0.18   **5.** −17.7   **6.** −990   **7.** −17   **8.** 38
4–6.

```
-0.06-0.12
              -.18
-3.2+(-14.5)
             -17.7
-472+(-518)
             -990
```

7–8.

```
-12-9+4
              -17
209-108+(-63)
               38
```

Section 1.5 Practice Exercises, pp. 59–62

1. a. $-b$ **b.** positive   **3.** x^2 **5.** $-b + 2$ **7.** 9   **9.** −3
11. −12   **13.** 4   **15.** −2   **17.** 8   **19.** −8   **21.** 2
23. 6   **25.** 40   **27.** −40   **29.** 0   **31.** −20   **33.** −24
35. 25   **37.** −5   **39.** $-\dfrac{3}{2}$ **41.** $\dfrac{41}{24}$ **43.** $\dfrac{2}{5}$ **45.** $-\dfrac{2}{3}$
47. 9.2   **49.** −5.72   **51.** −10   **53.** −14   **55.** −20
57. −173.188   **59.** 3.243   **61.** $6 - (-7); 13$
63. $3 - 18; -15$ **65.** $-5 - (-11); 6$ **67.** $-1 - (-13); 12$
69. $-32 - 20; -52$ **71.** $200 + 400 + 600 + 800 - 1000; \$1000$
73. 152°F   **75.** 19,881 m   **77.** 13   **79.** −9   **81.** 5
83. −25   **85.** −2   **87.** $-\dfrac{11}{30}$ **89.** $-\dfrac{29}{9}$ **91.** −2
93. −11   **95.** 2   **97.** −7   **99.** 5   **101.** 5   **103.** 3

Chapter 1 Problem Recognition Exercises, p. 62

1. Add their absolute values and apply a negative sign.
2. Subtract the smaller absolute value from the larger absolute value. Apply the sign of the number with the larger absolute value.
3. a. 6 **b.** −6 **c.** −22 **d.** 22 **e.** −22   **4. a.** −2 **b.** −8
c. −8 **d.** −2 **e.** 8   **5. a.** 0 **b.** 0 **c.** 50 **d.** 0 **e.** −50
6. a. $-\dfrac{1}{6}$ **b.** $\dfrac{1}{6}$ **c.** $-\dfrac{7}{6}$ **d.** $\dfrac{7}{6}$ **e.** $-\dfrac{7}{6}$ **7. a.** −3.6 **b.** 10.6
c. 3.6 **d.** 3.6 **e.** −10.6 **8. a.** 4 **b.** −4 **c.** 6 **d.** 6
9. a. −270 **b.** −110 **c.** −90 **d.** −270   **10. a.** 402
b. −108 **c.** 484 **d.** 24

Section 1.6 Calculator Connections, pp. 69–70

1. −30   **2.** −2   **3.** 625   **4.** 625   **5.** −625   **6.** −5.76
1–3.

```
-6(5)
               -30
-5.2/2.6
                -2
(-5)(-5)(-5)(-5)
              625
```

4–6.

```
(-5)^4
              625
-5^4
             -625
-2.4²
            -5.76
```

7. 5.76   **8.** −1   **9.** 4   **10.** −36
7–8.

```
(-2.4)²
            5.76
(-1)(-1)(-1)
              -1
```

9–10.

```
-8.4/-2.1
                4
90/(-5)(2)
              -36
```

Section 1.6 Practice Exercises, pp. 70–74

1. a. $\dfrac{1}{a}$ **b.** 0 **c.** 0 **d.** undefined **e.** positive **f.** negative
g. $1; -\frac{3}{2}$ **h.** All of these   **3.** True   **5.** False   **7.** −56
9. 143   **11.** −12.76   **13.** $\dfrac{3}{4}$ **15.** 36   **17.** −36
19. $-\dfrac{27}{125}$ **21.** 0.0016   **23.** −6   **25.** $\dfrac{15}{17}$ **27.** 1
29. $-\dfrac{1}{5}$ **31.** $(-2)(-7) = 14$ **33.** $-5 \cdot 0 = 0$
35. No number multiplied by zero equals 6.   **37.** $(-6)(4) = -24$
39. 6   **41.** −6   **43.** −8   **45.** 8   **47.** 0   **49.** Undefined
51. 0   **53.** 0   **55.** $-\dfrac{3}{2}$ **57.** $\dfrac{3}{10}$ **59.** −2   **61.** −7.912
63. 0.092   **65.** −6   **67.** 2.1   **69.** 9   **71.** −9   **73.** $-\dfrac{64}{27}$
75. 340   **77.** $\dfrac{14}{9}$ **79.** −30   **81.** 96   **83.** 2   **85.** −1
87. $-\dfrac{4}{33}$ **89.** $-\dfrac{4}{7}$ **91.** −24   **93.** $-\dfrac{1}{20}$ **95.** −23
97. 12   **99.** $\dfrac{9}{7}$ **101.** Undefined   **103.** −48   **105.** −6
107. −1   **109.** 7   **111.** −4   **113.** −40   **115.** $\dfrac{7}{2}$
117. No. The first expression is equivalent to $10 \div (5x)$. The second is $10 \div 5 \cdot x$.   **119.** $-3.75(0.3); -1.125$
121. $\dfrac{16}{5} \div \left(-\dfrac{8}{9}\right); -\dfrac{18}{5}$ **123.** $-0.4 + 6(-0.42); -2.92$
125. $-\dfrac{1}{4} - 6\left(-\dfrac{1}{3}\right); \dfrac{7}{4}$ **127.** $3(-2) + 3 = -3$; loss of \$3
129. $2(5) + 3(-3) = 1$; Lorne was one sale above quota for the week.
131. $\dfrac{12 + (-15) + 4 + (-9) + 3}{5} = -1$; The average loss was 1 oz.
133. a. −10 **b.** 24 **c.** In part (a), we subtract; in part (b), we multiply.

Chapter 1 Problem Recognition Exercises, p. 74

1. a. −4 **b.** 32 **c.** −12 **d.** 2   **2. a.** 10 **b.** 14 **c.** −24
d. −6   **3. a.** −27 **b.** −324 **c.** −4 **d.** −45   **4. a.** 30
b. 24 **c.** −81 **d.** −9   **5. a.** 50 **b.** −15 **c.** $\dfrac{1}{2}$ **d.** 5
6. a. −5 **b.** −24 **c.** −16 **d.** −80   **7. a.** 64 **b.** 12 **c.** $\dfrac{1}{4}$
d. −20   **8. a.** −7 **b.** −24 **c.** −63 **d.** −18   **9. a.** −400
b. 85 **c.** −16 **d.** 75   **10. a.** 7 **b.** 294 **c.** $\dfrac{2}{3}$ **d.** −35
11. a. 8 **b.** 4 **c.** 4 **d.** 8   **12. a.** −16 **b.** 2 **c.** −2 **d.** 16

Section 1.7 Practice Exercises, pp. 84–87

1. a. constant **b.** coefficient **c.** 1; 1 **d.** like **3.** 7 **5.** 18

7. $-\dfrac{27}{5}$ or -5.4 **9.** 0 **11.** $\dfrac{1}{3}$ **13.** $\dfrac{11}{15}$ **15.** $-8+5$

17. $x+8$ **19.** $4(5)$ **21.** $-12x$ **23.** $x+(-3); -3+x$

25. $4p+(-9); -9+4p$ **27.** $x+(4+9); x+13$

29. $(-5\cdot3)x; -15x$ **31.** $\left(\dfrac{6}{11}\cdot\dfrac{11}{6}\right)x; x$ **33.** $\left(-4\cdot-\dfrac{1}{4}\right)t; t$

35. $(-8+2)+y; -6+y$ **37.** $(-5\cdot2)x; -10x$

39. Reciprocal **41.** 0 **43.** $30x+6$ **45.** $-2a-16$

47. $15c-3d$ **49.** $-7y+14$ **51.** $-\dfrac{2}{3}x+4$ **53.** $\dfrac{1}{3}m-1$

55. $-2p-10$ **57.** $6w+10z-16$ **59.** $4x+8y-4z$

61. $6w-x+3y$ **63.** $6+2x$ **65.** $24z$ **67.** $-14x$

69. $-4-4x$ **71.** b **73.** i **75.** g **77.** d **79.** h

81.

Term	Coefficient
$2x$	2
$-y$	-1
$18xy$	18
5	5

83.

Term	Coefficient
$-x$	-1
$8y$	8
$-9x^2y$	-9
-3	-3

85. The variable factors are different. **87.** The variables are the same and raised to the same power. **89.** For example: $5y, -2x, 6$

91. $-6p$ **93.** $-6y^2$ **95.** $7x^3y-4$ **97.** $3t-\dfrac{7}{5}$

99. $-6x+22$ **101.** $4w$ **103.** $-3x+17$ **105.** $10t-44w$

107. $-18u$ **109.** $-2t+7$ **111.** $51a-27$ **113.** -6

115. $4q-\dfrac{1}{3}$ **117.** $6n$ **119.** $2x+18$ **121.** $2.2c-0.32$

123. $-2x-34$ **125.** $9z-35$ **127.** Equivalent

129. Not equivalent. The terms are not *like* terms and cannot be combined. **131.** Not equivalent; subtraction is not commutative. **133.** Equivalent **135. a.** 55 **b.** 210

Chapter 1 Review Exercises, pp. 94–96

1. Improper **2.** Proper **3.** Improper **4.** Improper

5. $2\times2\times2\times2\times7$ **6.** $\dfrac{6}{5}$ **7.** $\dfrac{35}{36}$ **8.** $\dfrac{13}{16}$ **9.** $\dfrac{2}{7}$

10. $\dfrac{6}{5}$ or $1\dfrac{1}{5}$ **11.** 3 **12.** $\dfrac{17}{10}$ or $1\dfrac{7}{10}$ **13.** 357 million km^2

14. a. 7, 1 **b.** 7, -4, 0, 1 **c.** 7, 0, 1 **d.** 7, $\frac{1}{3}$, -4, 0, $-0.\overline{2}$, 1

e. $-\sqrt{3}$, π **f.** 7, $\frac{1}{3}$, -4, 0, $-\sqrt{3}$, $-0.\overline{2}$, π, 1 **15.** $\dfrac{1}{2}$ **16.** 6

17. $\sqrt{7}$ **18.** 0 **19.** False **20.** False **21.** True

22. True **23.** True **24.** True **25.** False **26.** True

27. True **28.** 0 **29.** 60 **30.** 3 **31.** 4 **32.** $x\cdot\dfrac{2}{3}$ or $\dfrac{2}{3}x$

33. $\dfrac{7}{y}$ or $7\div y$ **34.** $2+3b$ **35.** $a-5$ **36.** $5k+2$

37. $13z-7$ **38.** 216 **39.** 225 **40.** 6 **41.** $\dfrac{1}{10}$

42. $\dfrac{1}{16}$ **43.** $\dfrac{27}{8}$ **44.** 13 **45.** 11 **46.** 7 **47.** 10

48. 2 **49.** 4 **50.** 15 **51.** -17 **52.** $\dfrac{11}{63}$ **53.** $-\dfrac{5}{22}$

54. $-\dfrac{14}{15}$ **55.** $-\dfrac{27}{10}$ **56.** -2.15 **57.** -4.28 **58.** 3

59. 8 **60.** 4 **61.** When a and b are both negative or when a and b have different signs and the number with the larger absolute value is negative. **62.** No. He is still overdrawn by $8. **63.** -12 **64.** 33 **65.** -1 **66.** -17

67. $-\dfrac{29}{18}$ **68.** $-\dfrac{19}{24}$ **69.** -1.2 **70.** -4.25 **71.** -10.2

72. 12.09 **73.** $\dfrac{10}{3}$ **74.** $-\dfrac{17}{20}$ **75.** -1 **76.** If $a<b$

77. $-7-(-18); 11$ **78.** $-6-41; -47$ **79.** $7-13; -6$

80. $[20-(-7)]-5; 22$ **81.** $[6+(-12)]-21; -27$

82. 175°F **83.** -170 **84.** -91 **85.** -2 **86.** 3

87. $-\dfrac{1}{6}$ **88.** $-\dfrac{8}{11}$ **89.** 0 **90.** Undefined **91.** 0

92. 2.25 **93.** $-\dfrac{3}{2}$ **94.** $\dfrac{1}{4}$ **95.** -30 **96.** 450 **97.** $\dfrac{1}{4}$

98. $-\dfrac{1}{7}$ **99.** -2 **100.** $\dfrac{18}{7}$ **101.** 17 **102.** 6

103. $-\dfrac{7}{120}$ **104.** 4.4 **105.** $-\dfrac{1}{3}$ **106.** -1 **107.** -2

108. 11 **109.** 36 **110.** -6 **111.** 70.6 **112.** True

113. False, any nonzero real number raised to an even power is positive. **114.** True **115.** True **116.** False, the product of two negative numbers is positive. **117.** True **118.** True

119. For example: $2+3=3+2$ **120.** For example: $(2+3)+4=2+(3+4)$ **121.** For example: $5+(-5)=0$

122. For example: $7+0=7$ **123.** For example: $5\cdot2=2\cdot5$

124. For example: $(8\cdot2)10=8(2\cdot10)$

125. For example: $3\cdot\dfrac{1}{3}=1$ **126.** For example: $8\cdot1=8$

127. $5x-2y=5x+(-2y)$, then use the commutative property of addition.

128. $3a-9y=3a+(-9y)$, then use the commutative property of addition. **129.** $3y, 10x, -12, xy$ **130.** 3, 10, -12, 1

131. $8a-b-10$ **132.** $-7p-11q+16$ **133.** $-8z-18$

134. $20w-40y+5$ **135.** $p-2w$ **136.** $-h+14m$

137. $-14q-1$ **138.** $-5.7b+2.4$ **139.** $4x+24$

140. $50y+105$

Chapter 1 Test, pp. 96–97

1. $\dfrac{15}{4}$ **2.** $\dfrac{3}{2}$ **3.** $3\dfrac{1}{16}$ **4.** $2\dfrac{3}{8}$

5. Rational, all repeating decimals are rational numbers.

6. a. $(4x)(4x)(4x)$ **b.** $4\cdot x\cdot x\cdot x$

7.

8. a. False **b.** True **c.** True **d.** True **9. a.** Commutative property of multiplication **b.** Identity property of addition **c.** Associative property of addition **d.** Inverse property of multiplication **e.** Associative property of multiplication

10. a. Twice the difference of a and b **b.** b subtracted from twice a

11. $\dfrac{\sqrt{c}}{d^2}$ or $\sqrt{c}\div d^2$ **12.** $12-(-4); 16$ **13.** $6-8; -2$

14. $\dfrac{10}{-12}; -\dfrac{5}{6}$ **15.** $-\dfrac{7}{8}$ **16.** -12 **17.** 28 **18.** -12

19. -32 **20.** 96 **21.** 0 **22.** Undefined **23.** 6 **24.** 4.66

25. -28 **26.** $\dfrac{2}{3}$ **27.** $\dfrac{1}{3}$ **28.** 9 **29.** -8 **30.** The difference is 9.5°C. **31. a.** $5+2+(-10)+4$ **b.** He gained 1 yd.

32. $-6k-8$ **33.** $-12x+2y-4$ **34.** $-4p-23$

35. $-12m-24p+21$ **36.** $4p-\dfrac{4}{3}$ **37.** 5 **38.** 18

39. -6 **40.** -32

Chapter 2

Section 2.1 Practice Exercises, pp. 109–111

1. a. equation **b.** solution **c.** linear **d.** solution set
3. Expression **5.** Equation **7.** Substitute the value into the equation and determine if the right-hand side is equal to the left-hand side. **9.** No **11.** Yes **13.** Yes **15.** $\{-1\}$
17. $\{20\}$ **19.** $\{-17\}$ **21.** $\{-16\}$ **23.** $\{0\}$ **25.** $\{1.3\}$
27. $\left\{\dfrac{11}{2}\right\}$ or $\left\{5\dfrac{1}{2}\right\}$ **29.** $\{-2\}$ **31.** $\{-2.13\}$ **33.** $\{-3.2675\}$
35. $\{9\}$ **37.** $\{-4\}$ **39.** $\{0\}$ **41.** $\{-15\}$ **43.** $\left\{-\dfrac{4}{5}\right\}$
45. $\{-10\}$ **47.** $\{4\}$ **49.** $\{41\}$ **51.** $\{-127\}$
53. $\{-2.6\}$ **55.** $-8 + x = 42$; The number is 50.
57. $x - (-6) = 18$; The number is 12.
59. $x \cdot 7 = -63$ or $7x = -63$; The number is -9.
61. $x - 3.2 = 2.1$; The number is 5.3.
63. $\dfrac{x}{12} = \dfrac{1}{3}$; The number is 4.
65. $x + \dfrac{5}{8} = \dfrac{13}{8}$; The number is 1. **67. a.** $\{10\}$ **b.** $\left\{-\dfrac{1}{9}\right\}$
69. a. $\{-12\}$ **b.** $\left\{\dfrac{22}{3}\right\}$ **71.** $\{-36\}$ **73.** $\{16\}$ **75.** $\{2\}$
77. $\left\{-\dfrac{7}{4}\right\}$ **79.** $\{11\}$ **81.** $\{-36\}$ **83.** $\left\{\dfrac{7}{2}\right\}$ **85.** $\{4\}$
87. $\{3.6\}$ **89.** $\{0.4084\}$ **91.** Yes **93.** No **95.** Yes
97. Yes **99.** For example: $y + 9 = 15$ **101.** For example: $2p = -8$ **103.** For example: $5a + 5 = 5$ **105.** $\{-1\}$ **107.** $\{7\}$

Section 2.2 Practice Exercises, pp. 118–120

1. a. conditional **b.** contradiction **c.** empty or null
d. identity **3.** $4w + 8$ **5.** $6y - 22$ **7.** $\{19\}$ **9.** $\{-3\}$
11. $\{-5\}$ **13.** $\{2\}$ **15.** $\{6\}$ **17.** $\left\{\dfrac{5}{2}\right\}$ **19.** $\{-42\}$
21. $\left\{-\dfrac{3}{4}\right\}$ **23.** $\{5\}$ **25.** $\{-4\}$ **27.** $\{-26\}$ **29.** $\{10\}$
31. $\{-8\}$ **33.** $\left\{-\dfrac{7}{3}\right\}$ **35.** $\{0\}$ **37.** $\{-3\}$ **39.** $\{-2\}$
41. $\left\{\dfrac{9}{2}\right\}$ **43.** $\left\{-\dfrac{1}{3}\right\}$ **45.** $\{10\}$ **47.** $\{-6\}$ **49.** $\{0\}$
51. $\{-2\}$ **53.** $\left\{-\dfrac{25}{4}\right\}$ **55.** $\left\{\dfrac{10}{3}\right\}$ **57.** $\{-0.25\}$
59. $\{\ \}$; contradiction **61.** $\{-15\}$; conditional equation
63. The set of real numbers; identity **65.** One solution
67. Infinitely many solutions **69.** $\{7\}$ **71.** $\left\{\dfrac{1}{2}\right\}$ **73.** $\{0\}$
75. The set of real numbers **77.** $\{-46\}$ **79.** $\{2\}$
81. $\left\{\dfrac{13}{2}\right\}$ **83.** $\{-5\}$ **85.** $\{\ \}$ **87.** $\{2.205\}$ **89.** $\{10\}$
91. $\{-1\}$ **93.** $a = 15$ **95.** $a = 4$
97. For example: $5x + 2 = 2 + 5x$

Section 2.3 Practice Exercises, pp. 126–127

1. a. clearing fractions **b.** clearing decimals **3.** $\{-2\}$
5. $\{-5\}$ **7.** $\{\ \}$ **9.** 18, 36 **11.** 100; 1000; 10,000
13. 30, 60 **15.** $\{4\}$ **17.** $\{-12\}$ **19.** $\left\{-\dfrac{15}{4}\right\}$
21. $\{8\}$ **23.** $\{3\}$ **25.** $\{15\}$ **27.** $\{\ \}$
29. The set of real numbers **31.** $\{5\}$ **33.** $\{2\}$ **35.** $\{-15\}$
37. $\{6\}$ **39.** $\{107\}$ **41.** The set of real numbers **43.** $\{67\}$
45. $\{90\}$ **47.** $\{4\}$ **49.** $\{4\}$ **51.** $\{\ \}$ **53.** $\{-0.25\}$
55. $\{-6\}$ **57.** $\left\{\dfrac{8}{3}\right\}$ or $\left\{2\dfrac{2}{3}\right\}$ **59.** $\{-11\}$ **61.** $\left\{\dfrac{1}{10}\right\}$
63. $\{-2\}$ **65.** $\{-1\}$ **67.** $\{2\}$

Chapter 2 Problem Recognition Exercises, p. 128

1. Expression; $-4b + 18$ **2.** Expression; $20p - 30$
3. Equation; $\{-8\}$ **4.** Equation; $\{-14\}$

5. Equation; $\left\{\dfrac{1}{3}\right\}$ **6.** Equation; $\left\{-\dfrac{4}{3}\right\}$
7. Expression; $6z - 23$ **8.** Expression; $-x - 9$
9. Equation; $\left\{\dfrac{7}{9}\right\}$ **10.** Equation; $\left\{-\dfrac{13}{10}\right\}$
11. Equation; $\{20\}$ **12.** Equation; $\{-3\}$
13. Equation; $\left\{\dfrac{1}{2}\right\}$ **14.** Equation; $\{-6\}$
15. Expression; $\dfrac{5}{8}x + \dfrac{7}{4}$ **16.** Expression; $-26t + 18$
17. Equation; $\{\ \}$ **18.** Equation; $\{\ \}$
19. Equation; $\left\{\dfrac{23}{12}\right\}$ **20.** Equation; $\left\{\dfrac{5}{8}\right\}$
21. Equation; The set of real numbers
22. Equation; The set of real numbers
23. Equation; $\left\{\dfrac{1}{2}\right\}$ **24.** Equation; $\{0\}$
25. Expression; 0 **26.** Expression; -1
27. Expression; $2a + 13$ **28.** Expression; $8q + 3$
29. Equation; $\{10\}$ **30.** Equation; $\left\{-\dfrac{1}{20}\right\}$
31. Expression; $-\dfrac{1}{6}y + \dfrac{1}{3}$ **32.** Expression; $\dfrac{7}{10}x + \dfrac{2}{5}$

Section 2.4 Practice Exercises, pp. 135–138

1. a. consecutive **b.** even **c.** odd **d.** 1 **e.** 2 **f.** 2
3. $x - 5,682,080$ **5.** $10x$ **7.** $3x - 20$ **9.** The number is -4.
11. The number is -3. **13.** The number is 5. **15.** The number is -5. **17.** The number is 3. **19. a.** $x + 1, x + 2$
b. $x - 1, x - 2$ **21.** The integers are -34 and -33.
23. The integers are 13 and 15. **25.** The sides are 14 in., 15 in., 16 in., 17 in., and 18 in. **27.** The integers are -54, -52, and -50.
29. The integers are 13, 15, and 17. **31.** The lengths of the pieces are 33 cm and 53 cm. **33.** Karen's music library has 23 playlists and Claran's library has 35 playlists. **35.** There were 201 Republicans and 232 Democrats. **37.** There were 190 passenger cars and 230 SUVs sold. **39.** The Congo River is 4370 km long, and the Nile River is 6825 km. **41.** The area of Africa is 30,065,000 km². The area of Asia is 44,579,000 km².
43. They walked 12.3 mi on the first day and 8.2 mi on the second.
45. The pieces are 9 in., 17 in., and 22 in. **47.** The integers are 42, 43, and 44. **49.** The winner earned $14 million and the runner-up earned $5 million. **51.** The number is 11. **53.** The page numbers are 470 and 471. **55.** The number is 10. **57.** The deepest point in the Arctic Ocean is 5122 m.
59. The number is $\dfrac{7}{16}$. **61.** The number is 2.5.

Section 2.5 Practice Exercises, pp. 143–146

1. a. simple **b.** 100 **3.** The numbers are 21 and 22.
5. 12.5% **7.** 85% **9.** 0.75 **11.** 1050.8
13. 885 **15.** 2200 **17.** Molly will have to pay $106.99.
19. Approximately 231,000 cases **21.** 2% **23.** Javon's taxable income was $84,000. **25.** Aidan would earn $9 more in the CD.
27. Bob borrowed $1200. **29.** The rate is 6%.
31. Perry needs to invest $3302. **33. a.** $7.44 **b.** $54.56
35. The original price was $470.59. **37.** The discount rate is 12%.
39. The original dosage was 15 cc. **41.** The tax rate is 5%.
43. The original cost was $11.58 per pack.
45. The original price was $210,000. **47.** Alina made $4600 that month. **49.** Diane sold $645 over $200 worth of merchandise.

Section 2.6 Calculator Connections, p. 152

1. 140.056 **2.** 31.831 **3.** 1.273 **4.** 0.455
1–2. 3–4.

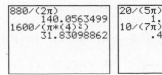

Section 2.6 Practice Exercises, pp. 152–156

1. $\{-5\}$ **3.** $\{0\}$ **5.** $\{-2\}$ **7.** $\{\ \}$ **9.** $a = P - b - c$

11. $y = x + z$ **13.** $q = p - 250$ **15.** $b = \dfrac{A}{h}$

17. $t = \dfrac{PV}{nr}$ **19.** $x = 5 + y$ **21.** $y = -3x - 19$

23. $y = \dfrac{-2x + 6}{3}$ or $y = -\dfrac{2}{3}x + 2$

25. $x = \dfrac{y + 9}{-2}$ or $x = -\dfrac{1}{2}y - \dfrac{9}{2}$

27. $y = \dfrac{-4x + 12}{-3}$ or $y = \dfrac{4}{3}x - 4$

29. $y = \dfrac{-ax + c}{b}$ or $y = -\dfrac{a}{b}x + \dfrac{c}{b}$

31. $t = \dfrac{A - P}{Pr}$ or $t = \dfrac{A}{Pr} - \dfrac{1}{r}$

33. $c = \dfrac{a - 2b}{2}$ or $c = \dfrac{a}{2} - b$ **35.** $y = 2Q - x$

37. $a = MS$ **39.** $R = \dfrac{P}{I^2}$

41. The length is 7 ft and the width is 5 ft.
43. The length is 120 yd and the width is 30 yd.
45. The length is 195 m and the width is 100 m.
47. The sides are 22 m, 22 m, and 27 m.
49. "Adjacent supplementary angles form a straight angle." The words *Supplementary* and *Straight* both begin with the same letter.
51. The angles are 23.5° and 66.5°.
53. The angles are 34.8° and 145.2°.
55. $x = 20$; the vertical angles measure 37°.
57. The measures of the angles are 30°, 60°, and 90°.
59. The measures of the angles are 42°, 54°, and 84°.
61. $x = 17$; the measures of the angles are 34° and 56°.

63. **a.** $A = lw$ **b.** $w = \dfrac{A}{l}$ **c.** The width is 29.5 ft.

65. **a.** $P = 2l + 2w$ **b.** $l = \dfrac{P - 2w}{2}$ **c.** The length is 103 m.

67. **a.** $C = 2\pi r$ **b.** $r = \dfrac{C}{2\pi}$ **c.** The radius is approximately 140 ft.

69. **a.** 415.48 m^2 **b.** 10,386.89 m^3

Section 2.7 Practice Exercises, pp. 161–165

1. **a.** $x = \dfrac{c + by}{a}$ **3.** $y = \dfrac{18 - 7x}{x}$ **5.** $\{4\}$

7. $200 - t$ **9.** $100 - x$ **11.** $3000 - y$
13. 53 tickets were sold at $3 and 28 tickets were sold at $2.
15. There were 17 songs purchased at $1.29 and 8 songs purchased at $1.49.
17. Amber bought 21 books at $6.99 and 11 books at $9.99.   **19.** $x + 7$
21. $d + 2000$ **23.** Mix 500 gal of 5% ethanol fuel mixture.
25. The pharmacist needs to use 21 mL of the 1% saline solution.
27. The contractor needs to mix 6.75 oz of 50% acid solution.
29. **a.** 300 mi   **b.** $5x$ **c.** $5(x + 12)$ or $5x + 60$ **31.** She walks 4 mph to the lake.   **33.** Bryan hiked 6 mi up the canyon.
35. The plane travels 600 mph in still air.   **37.** The slower car travels 48 mph and the faster car travels 52 mph.   **39.** The speeds of the vehicles are 40 mph and 50 mph.   **41.** Sarah walks 3.5 mph and Jeanette runs 7 mph.   **43.** **a.** 2 lb   **b.** $0.10x$
c. $0.10(x + 3) = 0.10x + 0.30$ **45.** Mix 10 lb of coffee sold at $12 per pound and 40 lb of coffee sold at $8 per pound.   **47.** The boats will meet in $\frac{3}{4}$ hr (45 min).   **49.** Sam purchased 16 packages of wax and 5 bottles of sunscreen.   **51.** 2.5 quarts of 85% chlorine solution
53. 20 L of water must be added.   **55.** The Japanese bullet train travels 300 km/hr and the Acela Express travels 240 km/hr.

Section 2.8 Practice Exercises, pp. 176–180

1. **a.** linear inequality   **b.** inequality   **c.** set-builder; interval

3. $\{-3\}$ **5.**   **7.**   **9.**

11.   **13.**   **15.**

Set-Builder Notation	Graph	Interval Notation
17. $\{x \mid x \geq 6\}$		$[6, \infty)$
19. $\{x \mid x \leq 2.1\}$		$(-\infty, 2.1]$
21. $\{x \mid -2 < x \leq 7\}$		$(-2, 7]$

Set-Builder Notation	Graph	Interval Notation
23. $\left\{x \mid x > \dfrac{3}{4}\right\}$		$\left(\dfrac{3}{4}, \infty\right)$
25. $\{x \mid -1 < x < 8\}$		$(-1, 8)$
27. $\{x \mid x \leq -14\}$		$(-\infty, -14]$

Set-Builder Notation	Graph	Interval Notation
29. $\{x \mid x \geq 18\}$		$[18, \infty)$
31. $\{x \mid x < -0.6\}$		$(-\infty, -0.6)$
33. $\{x \mid -3.5 \leq x < 7.1\}$		$[-3.5, 7.1)$

35. **a.** $\{3\}$
b. $\{x \mid x > 3\}$; $(3, \infty)$

37. **a.** $\{13\}$
b. $\{p \mid p \leq 13\}$; $(-\infty, 13]$

39. **a.** $\{-3\}$
b. $\{c \mid c < -3\}$; $(-\infty, -3)$

41. **a.** $\left\{-\dfrac{3}{2}\right\}$
b. $\left\{z \mid z \geq -\dfrac{3}{2}\right\}$; $\left[-\dfrac{3}{2}, \infty\right)$

43. $(-1, 4]$

45. $(-3, 5)$

47. $[2, 6]$

49. **a.** $\{x \mid x \leq 1\}$ **b.** $(-\infty, 1]$

51. **a.** $\{q \mid q > 10\}$
b. $(10, \infty)$

53. **a.** $\{x \mid x > 3\}$
b. $(3, \infty)$

55. **a.** $\{c \mid c > 2\}$
b. $(2, \infty)$

57. **a.** $\{c \mid c < -2\}$
b. $(-\infty, -2)$

59. **a.** $\{h \mid h \geq 14\}$
b. $[14, \infty)$

61. **a.** $\{x \mid x \geq -24\}$
b. $[-24, \infty)$

63. **a.** $\{p \mid -3 \leq p < 3\}$
b. $[-3, 3)$

65. **a.** $\left\{h \mid 0 < h < \dfrac{5}{2}\right\}$
b. $\left(0, \dfrac{5}{2}\right)$

67. **a.** $\{x \mid 10 < x < 12\}$
b. $(10, 12)$

69. **a.** $\{x \mid -8 \leq x < 24\}$
b. $[-8, 24)$

71. a. $\{y|y > -9\}$

b. $(-9, \infty)$

73. a. $\left\{x\middle|x \ge -\dfrac{15}{2}\right\}$

b. $\left[-\dfrac{15}{2}, \infty\right)$

75. a. $\{x|x < -3\}$

b. $(-\infty, -3)$

77. a. $\left\{b\middle|b \ge -\dfrac{1}{3}\right\}$

b. $\left[-\dfrac{1}{3}, \infty\right)$

79. a. $\{n|n > -3\}$
b. $(-3, \infty)$

81. a. $\{x|x < 7\}$
b. $(-\infty, 7)$

83. a. $\{x|x \le -5\}$
b. $(-\infty, -5]$

85. a. $\{z|z \le -2\}$
b. $(-\infty, -2]$

87. a. $\left\{a\middle|a > -\dfrac{2}{3}\right\}$

b. $\left(-\dfrac{2}{3}, \infty\right)$

89. a. $\left\{p\middle|p \le \dfrac{15}{4}\right\}$

b. $\left(-\infty, \dfrac{15}{4}\right]$

91. a. $\{y|y < -9\}$
b. $(-\infty, -9)$

93. a. $\{a|a > -3\}$
b. $(-3, \infty)$

95. a. $\{x|x \le 0\}$ **b.** $(-\infty, 0]$

97. No   **99.** Yes   **101.** $L \ge 10$ **103.** $w > 75$
105. $t \le 72$ **107.** $L \ge 8$ **109.** $2 < h < 5$
111. More than 10.2 in. of rain is needed.   **113.** Trevor needs at least 86 to get a B in the course.   **115. a.** $1539 **b.** 200 birdhouses cost $1440. It is cheaper to purchase 200 birdhouses because the discount is greater.   **117.** Company A is better if more than 400 flyers are printed.   **119.** Madison needs to babysit a minimum of 39.5 hr.

Chapter 2 Review Exercises, pp. 187–190

1. a. Equation   **b.** Expression   **c.** Equation   **d.** Equation
2. A linear equation can be written in the form $ax + b = c$, $a \ne 0$.
3. a. No   **b.** Yes   **c.** No   **d.** Yes   **4. a.** No   **b.** Yes
5. $\{-8\}$ **6.** $\{15\}$ **7.** $\left\{\dfrac{21}{4}\right\}$ **8.** $\{70\}$ **9.** $\left\{-\dfrac{21}{5}\right\}$

10. $\{-60\}$ **11.** $\left\{-\dfrac{10}{7}\right\}$ **12.** $\{27\}$ **13.** The number is 60.

14. The number is $\dfrac{7}{24}$.   **15.** The number is -8.

16. The number is -2.   **17.** $\{1\}$ **18.** $\left\{-\dfrac{3}{5}\right\}$ **19.** $\{2\}$

20. $\{-6\}$ **21.** $\{-3\}$ **22.** $\{18\}$ **23.** $\left\{\dfrac{3}{4}\right\}$ **24.** $\{-3\}$

25. $\{0\}$ **26.** $\left\{\dfrac{1}{8}\right\}$ **27.** $\{2\}$ **28.** $\{6\}$ **29.** A contradiction has no solution and an identity is true for all real numbers.
30. Identity   **31.** Conditional equation   **32.** Contradiction
33. Identity   **34.** Contradiction   **35.** Conditional equation
36. $\{6\}$ **37.** $\{22\}$ **38.** $\{13\}$ **39.** $\{-27\}$ **40.** $\{-10\}$
41. $\{-7\}$ **42.** $\left\{\dfrac{5}{3}\right\}$ **43.** $\left\{-\dfrac{9}{4}\right\}$ **44.** $\{2.5\}$ **45.** $\{-4\}$
46. $\{-4.2\}$ **47.** $\{2.5\}$ **48.** $\{-312\}$ **49.** $\{200\}$ **50.** $\{\ \}$
51. $\{\ \}$ **52.** The set of real numbers   **53.** The set of real numbers   **54.** The number is 30.   **55.** The number is 11.
56. The number is -7.   **57.** The number is -10.
58. The integers are 66, 68, and 70.   **59.** The integers are 27, 28, and 29.   **60.** The sides are 25 in., 26 in., and 27 in.

61. The sides are 36 cm, 37 cm, 38 cm, 39 cm, and 40 cm.
62. In 1975 the minimum salary was $16,000.
63. Indiana has 6.2 million people and Kentucky has 4.1 million.
64. 23.8   **65.** 28.8   **66.** 12.5%   **67.** 95%   **68.** 160
69. 1750   **70.** The dinner was $40 before tax and tip.
71. a. $840 **b.** $3840   **72.** He invested $12,000.
73. The novel originally cost $29.50.   **74.** $K = C + 273$

75. $C = K - 273$ **76.** $s = \dfrac{P}{4}$ **77.** $s = \dfrac{P}{3}$

78. $x = \dfrac{y - b}{m}$ **79.** $x = \dfrac{c - a}{b}$ **80.** $y = \dfrac{-2x - 2}{5}$

81. $b = \dfrac{Q - 4a}{4}$ or $b = \dfrac{Q}{4} - a$ **82.** The height is 7 m.

83. a. $h = \dfrac{3V}{\pi r^2}$ **b.** The height is 5.1 in.   **84.** The angles are $22°$, $78°$, and $80°$.   **85.** The angles are $50°$ and $40°$.
86. The length is 5 ft and the width is 4 ft.   **87.** $x = 20$. The angle measure is $65°$.   **88.** The measure of angle y is $53°$.
89. The truck travels 45 km/hr in bad weather and 60 km/hr in good weather.   **90.** Gus rides 15 mph and Winston rides 18 mph.
91. The cars will be 327.6 mi apart after 2.8 hr (2 hr and 48 min).
92. They meet in 2.25 hr (2 hr and 15 min).   **93.** 2 lb of 24% fat content beef is needed.   **94.** 20 lb of the 40% solder should be used.
95. $(-2, \infty)$

96. $\left(-\infty, \dfrac{1}{2}\right]$

97. $(-1, 4]$

98. a. $637 **b.** 300 plants cost $1410, and 295 plants cost $1416. 295 plants cost more.
99. $\{c|c < 17\}$; $(-\infty, 17)$

100. $\left\{w\middle|w > -\dfrac{1}{3}\right\}$; $\left(-\dfrac{1}{3}, \infty\right)$

101. $\{x|x \le -6\}$; $(-\infty, -6]$

102. $\left\{y\middle|y \le -\dfrac{14}{5}\right\}$; $\left(-\infty, -\dfrac{14}{5}\right]$

103. $\{a|a \ge 49\}$; $[49, \infty)$

104. $\{t|t < 34.5\}$; $(-\infty, 34.5)$

105. $\{k|k > 18\}$; $(18, \infty)$

106. $\left\{h\middle|h \le \dfrac{5}{2}\right\}$; $\left(-\infty, \dfrac{5}{2}\right]$

107. $\{x|x < -1\}$; $(-\infty, -1)$

108. $\{b|-3 < b \le 7\}$; $(-3, 7]$

109. $\{z|-6 \le z \le 5\}$; $[-6, 5]$

110. More than 2.5 in. is required.
111. Collette can have at most 18 wings.

Chapter 2 Test, pp. 190–191

1. b, d   **2. a.** $5x + 7$ **b.** $\{9\}$ **3.** $\left\{\dfrac{7}{3}\right\}$ **4.** $\{\ \}$

5. $\{-16\}$ **6.** $\left\{\dfrac{13}{4}\right\}$ **7.** $\{12\}$ **8.** $\left\{\dfrac{20}{21}\right\}$ **9.** $\left\{-\dfrac{16}{9}\right\}$

10. $\{15\}$ **11.** The set of real numbers   **12.** $\{-3\}$

13. $\{-47\}$ **14.** $y = -3x - 4$ **15.** $r = \dfrac{C}{2\pi}$ **16.** 90

17. a. $(-\infty, 0)$ **b.** $[-2, 5)$

18. $\{x|x > -2\}$; $(-2, \infty)$ **19.** $\{x|x \le -4\}$; $(-\infty, -4]$

20. $\left\{y \mid y > -\frac{3}{2}\right\}; \left(-\frac{3}{2}, \infty\right)$

21. $\{p \mid -5 \le p \le 1\}; [-5, 1]$

22. The cost was $82.00.   **23.** Clarita originally borrowed $5000.
24. The numbers are 18 and 13.   **25.** One family travels 55 mph
and the other travels 50 mph.   **26.** More than 26.5 in. is required.
27. The sides are 61 in., 62 in., 63 in., 64 in., and 65 in.
28. Matthew needs 30 lb of macadamia nuts.
29. Each basketball ticket was $36.32, and each hockey ticket
was $40.64.   **30.** The measures of the angles are 32° and 58°.
31. The field is 110 m long and 75 m wide.   **32.** $y = 30$;
The measures of the angles are 30°, 39°, and 111°.

Chapters 1–2 Cumulative Review Exercises, pp. 191–192

1. $\frac{1}{2}$ **2.** -7 **3.** $-\frac{5}{12}$ **4.** 16 **5.** 4
6. $\sqrt{5^2 - 9}; 4$ **7.** $-14 + 12; -2$ **8.** $-7x^2y, 4xy, -6$
9. $9x + 13$ **10.** $\{4\}$ **11.** $\{-7.2\}$
12. The set of real numbers   **13.** $\{-8\}$ **14.** $\left\{-\frac{4}{7}\right\}$
15. $\{-80\}$ **16.** The numbers are 77 and 79.   **17.** The
cost before tax was $350.00.   **18.** The height is $\frac{41}{6}$ cm or $6\frac{5}{6}$ cm.
19. $\{x \mid x > -2\}; (-2, \infty)$ **20.** $\{x \mid -1 \le x \le 9\}; [-1, 9]$

Chapter 3

Section 3.1 Practice Exercises, pp. 198–203

1. a. x; y-axis   **b.** ordered   **c.** origin; $(0, 0)$ **d.** quadrants
e. negative   **f.** III   **3. a.** Month 10   **b.** 30   **c.** Between months 3
and 5 and between months 10 and 12   **d.** Months 8 and 9
e. Month 3   **f.** 80   **5. a.** On day 1 the price per share was $89.25.
b. $1.75   **c.** $-$2.75

7. 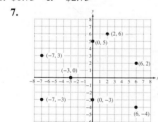   **9.**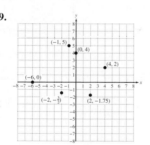

11. IV   **13.** II   **15.** III   **17.** I
19. $(0, -5)$ lies on the y-axis.   **21.** $\left(\frac{7}{8}, 0\right)$ is located on the x-axis.
23. $A(-4, 2)$, $B\left(\frac{1}{2}, 4\right)$, $C(3, -4)$, $D(-3, -4)$, $E(0, -3)$, $F(5, 0)$
25. a. $A(400, 200)$, $B(200, -150)$, $C(-300, -200)$,
$D(-300, 250)$, $E(0, 450)$ **b.** 450 m   **27. a.** $(250, 225)$, $(175, 193)$,
$(315, 330)$, $(220, 209)$, $(450, 570)$, $(400, 480)$, $(190, 185)$; the ordered
pair $(250, 225)$ means that 250 people produce $225 in popcorn sales.
b.

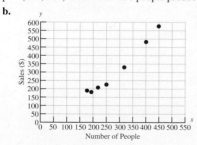

29. a. $(1, -10.2)$, $(2, -9.0)$, $(3, -2.5)$, $(4, 5.7)$, $(5, 13.0)$, $(6, 18.3)$,
$(7, 20.9)$, $(8, 19.6)$, $(9, 14.8)$, $(10, 8.7)$, $(11, 2.0)$, $(12, -6.9)$
b.

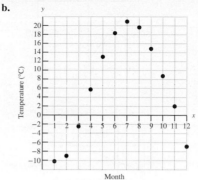

31. a.

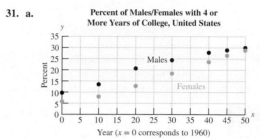

b. Increasing   **c.** Increasing

Section 3.2 Calculator Connections, pp. 211–212

1.   **2.**

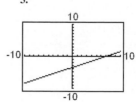   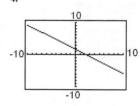

3.   **4.**

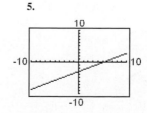   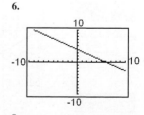

5.   **6.**

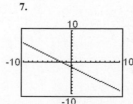   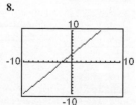

7.   **8.**

9.

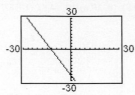

10.

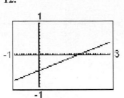

11.

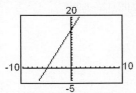

12.

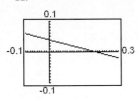

31.

x	y
0	4.6
1	3.4
2	2.2

33.

35.

37.

39.

41.

43.

Section 3.2 Practice Exercises, pp. 212–218

1. a. $Ax + By = C$ **b.** x-intercept **c.** y-intercept
d. vertical **e.** horizontal   **3.** $(-2, -2)$; quadrant III
5. $(-5, 0)$; x-axis   **7.** $(-3, 2)$; quadrant II   **9.** Yes
11. Yes   **13.** No   **15.** No   **17.** Yes

19.

x	y
1	-3
-2	0
-3	1
-4	2

21.

x	y
-2	3
-1	0
-4	9

23.

x	y
0	4
2	0
3	-2

25.

x	y
0	-2
5	-5
10	-8

27.

x	y
0	-2
-3	-2
5	-2

29.

x	y
3/2	-1
3/2	2
3/2	-3

45. y-axis   **47.** x-intercept: $(-1, 0)$; y-intercept: $(0, -3)$
49. x-intercept: $(-4, 0)$; y-intercept: $(0, 1)$
51. x-intercept: $\left(-\dfrac{9}{4}, 0\right)$;   **53.** x-intercept: $\left(\dfrac{8}{3}, 0\right)$;
y-intercept: $(0, 3)$ y-intercept: $(0, 2)$

55. x-intercept: $(-4, 0)$;   **57.** x-intercept: $(0, 0)$;
y-intercept: $(0, 8)$ y-intercept: $(0, 0)$

59. x-intercept: $(10, 0)$;   **61.** x-intercept: $(0, 0)$;
y-intercept: $(0, 5)$ y-intercept: $(0, 0)$

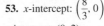

63. True   **65.** True
67. a. Horizontal line   **69. a.** Vertical line
c. no x-intercept;   **c.** x-intercept: $(4, 0)$;
y-intercept: $(0, -1)$ no y-intercept

71. a. Horizontal line **c.** no *x*-intercept; *y*-intercept: (0, 3)

73. a. Vertical line **c.** All points on the *y*-axis are *y*-intercepts; *x*-intercept: (0, 0)

83. For example: (−3, 1) and (−3, 4)

85.

75. A horizontal line may not have an *x*-intercept. A vertical line may not have a *y*-intercept.   **77.** a, b, d   **79. a.** $y = 11,190$
b. $x = 3$ **c.** (1, 11190) One year after purchase the value of the car is $11,190. (3, 9140) Three years after purchase the value of the car is $9140.

Section 3.3 Practice Exercises, pp. 225–231

1. a. slope; $\dfrac{y_2 - y_1}{x_2 - x_1}$ **b.** parallel   **c.** right   **d.** −1
e. undefined; horizontal
3. *x*-intercept: (6, 0); *y*-intercept: (0, −2)

5. *x*-intercept: none; *y*-intercept: $\left(0, \dfrac{3}{2}\right)$

7. $m = \dfrac{1}{3}$ **9.** $m = \dfrac{6}{11}$ **11.** undefined   **13.** positive
15. Negative   **17.** Zero   **19.** Undefined   **21.** Positive
23. Negative   **25.** $m = \dfrac{1}{2}$ **27.** $m = -3$ **29.** $m = 0$
31. The slope is undefined.   **33.** $\dfrac{1}{3}$ **35.** −3   **37.** $\dfrac{3}{5}$
39. Zero   **41.** Undefined   **43.** $\dfrac{28}{5}$ **45.** $-\dfrac{7}{8}$
47. −0.45 or $-\dfrac{9}{20}$ **49.** −0.15 or $-\dfrac{3}{20}$ **51. a.** −2   **b.** $\dfrac{1}{2}$
53. a. 0   **b.** undefined   **55. a.** $\dfrac{4}{5}$ **b.** $-\dfrac{5}{4}$ **57.** Perpendicular
59. Parallel   **61.** Neither   **63.** $l_1: m = 2, l_2: m = 2$; parallel
65. $l_1: m = 5, l_2: m = -\dfrac{1}{5}$; perpendicular
67. $l_1: m = \dfrac{1}{4}, l_2: m = 4$; neither   **69.** The average rate of change is −$160 per year.   **71. a.** $m = 47$ **b.** The number of male prisoners increased at a rate of 47 thousand per year during this time period.   **73. a.** 1 mi   **b.** 2 mi   **c.** 3 mi   **d.** $m = 0.2$; The distance between a lightning strike and an observer increases by 0.2 mi for every additional second between seeing lightning and hearing thunder.
75. $m = \dfrac{3}{4}$ **77.** $m = 0$
79. For example: (4, −4) and (−2, 0)

81. For example: (3, 5) and (1, −1)

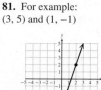

87.

89.

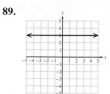

91. $\dfrac{3m - 3n}{2b}$ or $\dfrac{-3m + 3n}{-2b}$ **93.** $\left(\dfrac{c}{a}, 0\right)$
95. For example: (7, 1)

Section 3.4 Calculator Connections, p. 237

1. Perpendicular

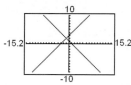

2. Parallel

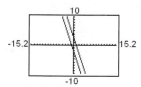

3. Neither

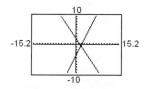

4. The lines may appear parallel; however, they are not parallel because the slopes are different.   **5.** The lines may appear to coincide on a graph; however, they are not the same line because the *y*-intercepts are different.   **6.** The line may appear to be horizontal, but it is not. The slope is 0.001 rather than 0.

Section 3.4 Practice Exercises, pp. 238–242

1. a. $y = mx + b$ **b.** standard   **3.** *x*-intercept: (10, 0); *y*-intercept: (0, −2)   **5.** *x*-intercept: none; *y*-intercept: (0, −3)
7. *x*-intercept: (0, 0); *y*-intercept: (0, 0)   **9.** *x*-intercept: (4, 0); *y*-intercept: none   **11.** $m = -2$; *y*-intercept: (0, 3)   **13.** $m = 1$; *y*-intercept: (0, −2)   **15.** $m = -1$; *y*-intercept: (0, 0)
17. $m = \dfrac{3}{4}$; *y*-intercept: (0, −1)   **19.** $m = \dfrac{2}{5}$; *y*-intercept: $\left(0, -\dfrac{4}{5}\right)$ **21.** $m = 3$; *y*-intercept: (0, −5)
23. $m = -1$; *y*-intercept: (0, 6)   **25.** Undefined slope; no *y*-intercept   **27.** $m = 0$; *y*-intercept: $\left(0, -\dfrac{1}{4}\right)$
29. $m = \dfrac{2}{3}$; *y*-intercept: (0, 0)
31.

33.

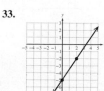

35. b   **37.** e   **39.** c

41. $y = -2x + 9$

43. $y = \frac{1}{2}x - 3$

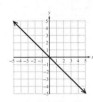

45. $y = -\frac{1}{2}x + \frac{3}{2}$

47. $y = -x$

49. $y = \frac{4}{5}x$

51. $y = -\frac{2}{3}$

53. Perpendicular   **55.** Neither   **57.** Parallel

59. Perpendicular   **61.** Parallel   **63.** Perpendicular

65. Neither   **67.** Parallel   **69.** $y = -\frac{1}{3}x + 2$

71. $y = 10x - 19$ **73.** $y = 6x - 8$ **75.** $y = \frac{1}{2}x - 3$

77. $y = -11$ **79.** $y = 5x$ **81.** $y = -2x + 3$

83. $y = -\frac{1}{3}x + 2$ **85.** $y = 4x - 1$ **87. a.** $m = 1203$;

The slope represents the rate of increase in the number of cases of Lyme disease per year.   **b.** (0, 10006); In the year 1993 there were 10,006 cases reported.   **c.** 30,457 cases   **d.** $x = 27$; the year 2020

89. $y = -\frac{A}{B}x + \frac{C}{B}$; the slope is $-\frac{A}{B}$.

91. $m = -\frac{6}{7}$ **93.** $m = \frac{11}{8}$

Chapter 3 Problem Recognition Exercises, p. 242

1. a, c, d   **2.** b, f, h   **3.** a   **4.** f   **5.** b, f   **6.** c

7. c, d   **8.** f   **9.** e   **10.** g   **11.** b   **12.** h

13. g   **14.** e   **15.** c   **16.** b, h   **17.** e   **18.** e

19. b, f, h   **20.** c, d

Section 3.5 Practice Exercises, pp. 247–251

1. a. $Ax + By = C$ **b.** horizontal   **c.** vertical   **d.** slope; y-intercept   **e.** $y - y_1 = m(x - x_1)$

3.

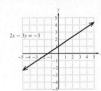

5.

7. 9   **9.** 0   **11.** $y = 3x + 7$ or $3x - y = -7$

13. $y = -4x - 14$ or $4x + y = -14$ **15.** $y = -\frac{1}{2}x - \frac{1}{2}$ or $x + 2y = -1$ **17.** $y = 2x - 2$ or $2x - y = 2$

19. $y = -x - 4$ or $x + y = -4$

21. $y = -0.2x - 2.86$ or $20x + 100y = -286$

23. $y = -2x + 1$ **25.** $y = 2x + 4$ **27.** $y = \frac{1}{2}x - 1$

29. $y = 4x + 13$ or $4x - y = -13$

31. $y = -\frac{3}{2}x + 6$ or $3x + 2y = 12$

33. $y = -2x - 8$ or $2x + y = -8$ **35.** $y = -\frac{1}{5}x - 6$ or $x + 5y = -30$ **37.** iv   **39.** vi   **41.** iii   **43.** $y = 1$

45. $x = 2$ **47.** $y = 2$ **49.** $y = \frac{1}{4}x + 8$ or $x - 4y = -32$

51. $y = 3x - 8$ or $3x - y = 8$ **53.** $y = 4.5x - 25.6$ or $45x - 10y = 256$ **55.** $x = -6$ **57.** $y = -2$

59. $x = -4$ **61.** $y = 2x - 4$ **63.** $y = -\frac{1}{2}x + 1$

Section 3.6 Calculator Connections, p. 255

1. 13.3   **2.** −42.3

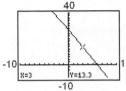   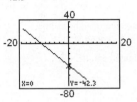

3. 345   **4.** 95

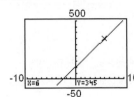   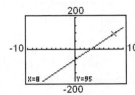

Section 3.6 Practice Exercises, pp. 255–259

1. $m = -\frac{5}{2}$ **3.** x-intercept: (6, 0); y-intercept: (0, 5)

5. x-intercept: (−2, 0); y-intercept: (0, −4)   **7.** x-intercept: none; y-intercept: (0, −9)   **9. a.** $3.00   **b.** $7.90   **c.** The y-intercept is (0, 1.6). This indicates that the minimum wage was $1.60 per hour in the year 1970.   **d.** The slope is 0.14. This indicates that the minimum wage has risen approximately $0.14 per year during this period.

11. a. $m = \frac{2}{7}$ **b.** $m = \frac{4}{7}$ **c.** $m = \frac{2}{7}$ means that Grindel's weight increased at a rate of 2 oz in 7 days. $m = \frac{4}{7}$ means that Frisco's weight increased at a rate of 4 oz in 7 days.   **d.** Frisco gained weight more rapidly.   **13. a.** $106.95   **b.** $201.95   **c.** (0, 11.95). For 0 kilowatt-hours used, the cost consists of only the fixed monthly tax of $11.95.   **d.** $m = 0.095$. The cost increases by $0.095 for each kilowatt-hour used.   **15. a.** $m = -1.0$ **b.** $y = -1.0x + 1051$ **c.** The minimum pressure was approximately 921 mb.

17. a. $m = 21.5$ **b.** The slope means that the consumption of wind energy in the United States increased by 21.5 trillion Btu per year.   **c.** $y = 21.5x + 57$ **d.** 272 trillion Btu   **19. a.** $y = 0.10x + 5000$ **b.** $6130   **21. a.** $y = 90x + 105$ **b.** $1185.00   **23. a.** $y = 0.8x + 100$ **b.** $260.00

Chapter 3 Review Exercises, pp. 265–269

1.

2. $A(4, -3)$; $B(-3, -2)$; $C(\frac{5}{2}, 5)$; $D(-4, 1)$; $E(-\frac{1}{2}, 0)$; $F(0, -5)$

3. III   **4.** II   **5.** IV   **6.** I   **7.** IV   **8.** III   **9.** x-axis

10. y-axis   **11. a.** On day 1, the price was $26.25.   **b.** Day 2

c. $2.25   **12. a.** In 2003 (8 years after 1995), there was only one space shuttle launch. (This was the year that the Columbia and its crew were lost.)

b.

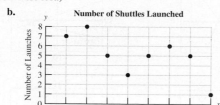

Number of Shuttles Launched

Year ($x = 0$ corresponds to 1995)

13. No   **14.** No   **15.** Yes   **16.** Yes

17.

x	y
2	1
3	4
1	-2

18.

x	y
0	2
-2	7/3
-6	3

19.

x	y
0	-1
3	1
-6	-5

20.

x	y
0	-3
-3	3
1	-5

21.

22.

23.

24.

25. Vertical   **26.** Vertical

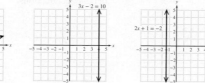

27. Horizontal   **28.** Horizontal

29. x-intercept: $(-3, 0)$; y-intercept: $\left(0, \frac{3}{2}\right)$

30. x-intercept: $(3, 0)$; y-intercept: $(0, 6)$ **31.** x-intercept: $(0, 0)$; y-intercept: $(0, 0)$ **32.** x-intercept: $(0, 0)$; y-intercept: $(0, 0)$

33. x-intercept: none; y-intercept: $(0, -4)$ **34.** x-intercept: none; y-intercept: $(0, 2)$ **35.** x-intercept: $\left(-\frac{5}{2}, 0\right)$; y-intercept: none

36. x-intercept: $\left(\frac{1}{3}, 0\right)$; y-intercept: none   **37.** $m = \frac{12}{5}$

38. -2 **39.** $-\frac{2}{3}$ **40.** 8   **41.** Undefined   **42.** 0

43. a. -5 **b.** $\frac{1}{5}$ **44. a.** 0 **b.** Undefined

45. $m_1 = \frac{2}{3}$; $m_2 = \frac{2}{3}$; parallel   **46.** $m_1 = 8$; $m_2 = 8$; parallel

47. $m_1 = -\frac{5}{12}$; $m_2 = \frac{12}{5}$; perpendicular   **48.** m_1 is undefined; $m_2 = 0$; perpendicular   **49. a.** $m = 35$ **b.** The number of kilowatt-hours increased at a rate of 35 kilowatt-hours per day.

50. a. $m = -994$ **b.** The number of new cars sold in Maryland decreased at a rate of 994 cars per year.

51. $y = \frac{5}{2}x - 5$; $m = \frac{5}{2}$; y-intercept: $(0, -5)$

52. $y = -\frac{3}{4}x + 3$; $m = -\frac{3}{4}$; y-intercept: $(0, 3)$

53. $y = \frac{1}{3}x$; $m = \frac{1}{3}$; y-intercept: $(0, 0)$ **54.** $y = \frac{12}{5}$; $m = 0$; y-intercept: $\left(0, \frac{12}{5}\right)$ **55.** $y = -\frac{5}{2}$; $m = 0$; y-intercept: $\left(0, -\frac{5}{2}\right)$

56. $y = x$; $m = 1$; y-intercept: $(0, 0)$ **57.** Neither   **58.** Perpendicular   **59.** Parallel   **60.** Parallel

61. Perpendicular   **62.** Neither   **63.** $y = -\frac{4}{3}x - 1$ or $4x + 3y = -3$ **64.** $y = 5x$ or $5x - y = 0$ **65.** $y = -\frac{4}{3}x - 6$ or $4x + 3y = -18$ **66.** $y = 5x - 3$ or $5x - y = 3$

67. For example: $y = 3x + 2$ **68.** For example: $5x + 2y = -4$

69. $m = \frac{y_2 - y_1}{x_2 - x_1}$ **70.** $y - y_1 = m(x - x_1)$

71. For example: $x = 6$ **72.** For example: $y = -5$

73. $y = -6x + 2$ or $6x + y = 2$ **74.** $y = \frac{2}{3}x + \frac{5}{3}$ or $2x - 3y = -5$ **75.** $y = \frac{1}{4}x - 4$ or $x - 4y = 16$ **76.** $y = -5$

77. $y = \frac{6}{5}x + 6$ or $6x - 5y = -30$ **78.** $y = 4x + 31$ or $4x - y = -31$ **79. a.** 47.8 in.   **b.** The slope is 2.4 and indicates that the average height for girls increases at a rate of 2.4 in. per year.

80. a. $m = 137$ **b.** The number of prescriptions increased by 137 million per year during this time period.   **c.** $y = 137x + 2825$ **d.** 4880 million   **81. a.** $y = 20x + 55$ **b.** $235

82. a. $y = 8x + 700$ **b.** $1340

Chapter 3 Test, pp. 269–271

1. a. II **b.** IV **c.** III   **2.** 0   **3.** 0

4. a. (4, 14) After 4 days the bamboo plant is 14 in. tall. (8, 28), (12, 42), (16, 56), (20, 70)

b.

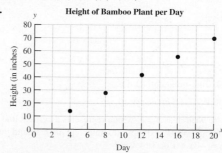

Height of Bamboo Plant per Day

c. Approximately 35 in.

5. a. No **b.** Yes **c.** Yes **d.** Yes

6.

x	y
0	−2
4	−1
6	−½

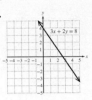

7.

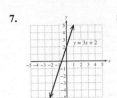

8.

9.

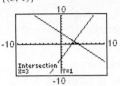

10.

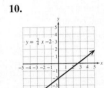

11. Horizontal

12. Vertical

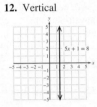

13. x-intercept: $\left(-\dfrac{3}{2}, 0\right)$; y-intercept: (0, 2)

14. x-intercept: (0, 0); y-intercept: (0, 0)   **15.** x-intercept: (4, 0); y-intercept: none   **16.** x-intercept: none; y-intercept: (0, 3)

17. $\dfrac{2}{5}$ **18. a.** $\dfrac{1}{3}$ **b.** $\dfrac{4}{3}$ **19. a.** $-\dfrac{1}{4}$ **b.** 4

20. a. Undefined **b.** 0   **21. a.** $m = 0.6$ **b.** The cost of renting a truck increases at a rate of $0.60 per mile.   **22.** Parallel

23. Perpendicular   **24.** $y = -\dfrac{1}{3}x + 1$ or $x + 3y = 3$

25. $y = -\dfrac{7}{2}x + 15$ or $7x + 2y = 30$ **26.** $y = \dfrac{1}{4}x + \dfrac{1}{2}$ or $x - 4y = -2$ **27.** $y = 3x + 8$ or $3x - y = -8$ **28.** $y = -6$

29. $y = -x - 3$ or $x + y = -3$ **30. a.** $y = 1.5x + 10$ **b.** $25

31. a. $m = 20$; The slope indicates that there is an increase of 20 thousand medical doctors per year. **b.** $y = 20x + 414$ **c.** 1114 thousand or, equivalently, 1,114,000

Chapters 1–3 Cumulative Review Exercises, pp. 271–272

1. a. Rational **b.** Rational **c.** Irrational **d.** Rational

2. a. $\dfrac{2}{3}; \dfrac{2}{3}$ **b.** −5.3; 5.3   **3.** 69   **4.** −13   **5.** 18

6. $\dfrac{3}{4} \div \left(-\dfrac{7}{8}\right); -\dfrac{6}{7}$ **7.** (−2.1)(−6); 12.6   **8.** The associative property of addition   **9.** {4}   **10.** {5}   **11.** $\left\{-\dfrac{9}{2}\right\}$

12. {−2}   **13.** 9241 mi²   **14.** $a = \dfrac{c - b}{3}$

15.

16. x-intercept: (−2, 0); y-intercept: (0, 1)

17. $y = -\dfrac{3}{2}x - 6$; slope: $-\dfrac{3}{2}$; y-intercept: (0, −6)

18. $2x + 3 = 5$ can be written as $x = 1$, which represents a vertical line. A vertical line of the form $x = k \, (k \neq 0)$ has an x-intercept of $(k, 0)$ and no y-intercept.   **19.** $y = -3x + 1$ or $3x + y = 1$

20. $y = \frac{2}{3}x + 6$ or $2x - 3y = -18$

Chapter 4

Section 4.1 Calculator Connections, pp. 278–279

1. {(2, 1)}

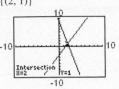

2. {(6, −1)}

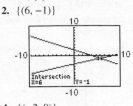

3. {(3, 1)}

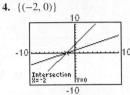

4. {(−2, 0)}

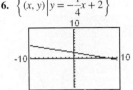

5. { }

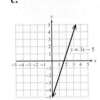

6. $\left\{(x, y)\,\middle|\, y = -\dfrac{1}{4}x + 2\right\}$

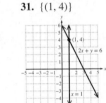

Section 4.1 Practice Exercises, pp. 279–284

1. a. system **b.** solution **c.** intersect **d.** consistent **e.** { } **f.** dependent **g.** independent   **3.** Yes   **5.** No   **7.** Yes

9. No   **11.** b   **13.** d

15. a.

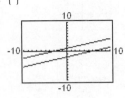

b.

c.

17. c   **19.** a   **21.** a   **23.** b   **25.** c

27. {(3, 1)}   **29.** {(1, −2)}   **31.** {(1, 4)}

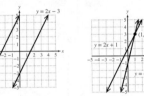

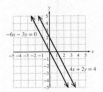

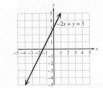

33. No solution; { }; inconsistent system   **35.** Infinitely many solutions; $\{(x, y)\,|\,-2x + y = 3\}$; dependent equations

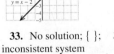

37. $\{(-3, 6)\}$ **39.** $\{(2, -2)\}$ **41.** No solution; { }; inconsistent system

43. $\{(4, 2)\}$ **45.** $\{(1, 2)\}$

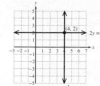

47. Infinitely many solutions; $\{(x, y) \mid y = 0.5x + 2\}$; dependent equations   **49.** $\{(-3, -1)\}$

51. The same cost occurs when $2500 of merchandise is purchased.   **53.** The point of intersection is below the x-axis and cannot have a positive y-coordinate.   **55.** For example: $4x + y = 9; -2x - y = -5$ **57.** For example: $2x + 2y = 1$

Section 4.2 Practice Exercises, pp. 292–294

1. $y = 2x - 4; y = 2x - 4$; coinciding lines
3. $y = -\dfrac{2}{3}x + 2; y = x - 5$; intersecting lines
5. $y = 4x - 4; y = 4x - 13$; parallel lines   **7.** $\{(3, -6)\}$
9. $\{(0, 4)\}$ **11. a.** y in the second equation is easiest to isolate because its coefficient is 1.   **b.** $\{(1, 5)\}$ **13.** $\{(5, 2)\}$ **15.** $\{(10, 5)\}$
17. $\left\{\left(\dfrac{1}{2}, 3\right)\right\}$ **19.** $\{(5, 3)\}$ **21.** $\{(1, 0)\}$ **23.** $\{(1, 4)\}$
25. No solution; { }; inconsistent system   **27.** Infinitely many solutions; $\{(x, y) \mid 2x - 6y = -2\}$; dependent equations   **29.** $\{(5, -7)\}$
31. $\left\{\left(-5, \dfrac{3}{2}\right)\right\}$ **33.** $\{(2, -5)\}$ **35.** $\{(-4, 6)\}$ **37.** $\{(0, 2)\}$
39. Infinitely many solutions; $\{(x, y) \mid y = 0.25x + 1\}$; dependent equations   **41.** $\{(1, 1)\}$ **43.** No solution; { }; inconsistent system
45. $\{(-1, 5)\}$ **47.** $\{(-6, -4)\}$ **49.** The numbers are 48 and 58.   **51.** The numbers are 13 and 39.   **53.** The angles are 165° and 15°.   **55.** The angles are 70° and 20°.   **57.** The angles are 42° and 48°.   **59.** For example: $(0, 3), (1, 5), (-1, 1)$

Section 4.3 Practice Exercises, pp. 301–303

1. No   **3.** No   **5.** Yes   **7. a.** True   **b.** False, multiply the second equation by 5.   **9. a.** x would be easier.   **b.** $\{(0, -3)\}$
11. $\{(4, -1)\}$ **13.** $\{(4, 3)\}$ **15.** $\{(2, 3)\}$ **17.** $\{(1, -4)\}$
19. $\{(1, -1)\}$ **21.** $\{(-4, -6)\}$ **23.** $\left\{\left(\dfrac{7}{9}, \dfrac{5}{9}\right)\right\}$
25. The system will have no solution. The lines are parallel.
27. There are infinitely many solutions. The lines coincide.
29. The system will have one solution. The lines intersect at a point whose x-coordinate is 0.   **31.** No solution; { }; inconsistent system
33. Infinitely many solutions; $\{(x, y) \mid x + 2y = 2\}$; dependent equations   **35.** $\{(1, 4)\}$ **37.** $\{(-1, -2)\}$ **39.** $\{(2, 1)\}$
41. No solution; { }; inconsistent system   **43.** $\{(2, 3)\}$ **45.** $\{(3.5, 2.5)\}$
47. $\left\{\left(\dfrac{1}{3}, 2\right)\right\}$ **49.** $\{(-2, 5)\}$ **51.** $\left\{\left(-\dfrac{1}{2}, 1\right)\right\}$ **53.** $\{(0, 3)\}$

55. { }   **57.** $\{(1, 4)\}$ **59.** $\{(4, 0)\}$ **61.** Infinitely many solutions; $\{(a, b) \mid 9a - 2b = 8\}$ **63.** $\left\{\left(\dfrac{7}{16}, -\dfrac{7}{8}\right)\right\}$
65. The numbers are 17 and 19.   **67.** The numbers are -1 and 3.
69. The angles are 46° and 134°.   **71.** $\{(1, 3)\}$ **73.** One line within the system of equations would have to "bend" for the system to have exactly two points of intersection. This is not possible.
75. $A = -5, B = 2$

Chapter 4 Problem Recognition Exercises, p. 304

1. Infinitely many solutions. The equations represent the same line.
2. No solution. The equations represent parallel lines.
3. One solution. The equations represent intersecting lines.
4. One solution. The equations represent intersecting lines.
5. No solution. The equations represent parallel lines.
6. Infinitely many solutions. The equations represent the same line.
7. The y variable will easily be eliminated by adding the equations. The solution set is $\{(-3, 1)\}$.   **8.** The x variable will easily be eliminated by multiplying the second equation by 2 and then adding the equations. The solution set is $\{(3, 4)\}$.   **9.** Because the variable x is already isolated in the first equation, the expression $-3y + 4$ can be easily substituted for x in the second equation. The solution set is $\{(-2, 2)\}$.   **10.** Because the variable y is already isolated in the second equation, the expression $x - 8$ can be easily substituted for y in the first equation. The solution set is $\{(8, 0)\}$.   **11.** $\{(5, 0)\}$
12. $\{(1, -7)\}$ **13.** $\{(4, -5)\}$ **14.** $\{(2, 3)\}$ **15.** $\{(2, 0)\}$
16. $\{(8, 10)\}$ **17.** $\left\{\left(2, -\dfrac{5}{7}\right)\right\}$ **18.** $\left\{\left(-\dfrac{14}{3}, -4\right)\right\}$
19. No solution; { }; inconsistent system   **20.** No solution; { }; inconsistent system
21. $\{(-1, 0)\}$ **22.** $\{(5, 0)\}$ **23.** Infinitely many solutions; $\{(x, y) \mid y = 2x - 14\}$; dependent equations   **24.** Infinitely many solutions; $\{(x, y) \mid x = 5y - 9\}$; dependent equations
25. $\{(2200, 1000)\}$ **26.** $\{(3300, 1200)\}$ **27.** $\{(5, -7)\}$
28. $\{(2, -1)\}$ **29.** $\left\{\left(\dfrac{2}{3}, \dfrac{1}{2}\right)\right\}$ **30.** $\left\{\left(\dfrac{1}{4}, -\dfrac{3}{2}\right)\right\}$

Section 4.4 Practice Exercises, pp. 310–314

1. $\{(-1, 4)\}$ **3.** $\left\{\left(\dfrac{5}{2}, 1\right)\right\}$ **5.** The numbers are 4 and 16.
7. The angles are 80° and 10°.   **9.** A video game costs $21.50 and a DVD costs $15.   **11.** Technology stock costs $16 per share, and the mutual fund costs $11 per share.   **13.** Mylee bought thirty-five 47-cent stamps and fifteen 34-cent stamps.   **15.** Shanelle invested $3500 in the 10% account and $6500 in the 7% account.
17. $9000 was borrowed at 6%, and $3000 was borrowed at 9%.
19. Invest $12,000 in the bond fund and $18,000 in the stock fund.
21. 15 gal of the 50% mixture should be mixed with 10 gal of the 40% mixture.   **23.** 12 gal of the 45% disinfectant solution should be mixed with 8 gal of the 30% disinfectant solution.
25. She should mix 20 mL of the 13% solution with 30 mL of the 18% solution.   **27.** Chad needs 1 L of pure antifreeze and 5 L of the 40% solution.   **29.** The speed of the boat in still water is 6 mph, and the speed of the current is 2 mph.   **31.** The speed of the plane in still air is 300 mph, and the wind is 20 mph.   **33.** The speed of the plane in still air is 525 mph and the speed of the wind is 75 mph.   **35.** There are 17 dimes and 22 nickels.   **37.** 2 qt of water should be mixed.   **39. a.** 835 free throws and 1597 field goals   **b.** 4029 points   **c.** Approximately 50 points per game
41. The speed of the plane in still air is 160 mph, and the wind is 40 mph.   **43.** $15,000 was invested in the 5.5% account, and $45,000 was invested in the 6.5% account.   **45.** 12 lb of candy should be mixed with 8 lb of nuts.   **47.** Dallas scored 30 points, and Buffalo scored 13 points.   **49.** There were 300 women and 200 men in the survey.

Section 4.5 Practice Exercises, pp. 321–326

1. a. linear **b.** intersect or overlap

3.

5. When the inequality symbol is ≤ or ≥ **7.** All of the points in the shaded region are solutions to the inequality.

9. a **11.** True **13.** False **15.** True

17. For example: **19.** For example: **21.** For example:
$(0, 5), (2, 7), (-1, 8)$ $(1, -1), (3, 0), (-2, -9)$ $(0, 0), (0, 2), (-1, -3)$

23. **25.** **27.**

29. **31.** **33.**

 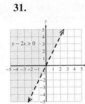

35. **37.** **39.**

41. a. The set of ordered pairs above the line $x + y = 4$, for example: $(6, 3), (-2, 8), (0, 5)$ **b.** The set of ordered pairs on the line $x + y = 4$, for example: $(0, 4), (4, 0), (2, 2)$ **c.** The set of ordered pairs below the line $x + y = 4$, for example: $(0, 0), (-2, 1), (3, 0)$

43. **45.** **47.**

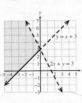

49. **51.** **53.**

55. **57.** **59.**

  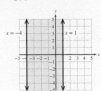

Chapter 4 Review Exercises, pp. 333–336

1. Yes **2.** No **3.** No **4.** Yes **5.** Intersecting lines (The lines have different slopes.) **6.** Intersecting lines (The lines have different slopes.) **7.** Parallel lines (The lines have the same slope but different y-intercepts.) **8.** Intersecting lines (The lines have different slopes.) **9.** Coinciding lines (The lines have the same slope and same y-intercept.) **10.** Intersecting lines (The lines have different slopes.)

11. $\{(0, -2)\}$ **12.** $\{(-2, 3)\}$

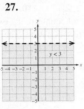

13. Infinitely many solutions; $\{(x, y) \mid 2x + y = 5\}$; dependent equations

14. Infinitely many solutions; $\{(x, y) \mid -x + 5y = -5\}$; dependent equations

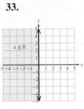

15. $\{(1, -1)\}$ **16.** $\{(-1, 2)\}$

17. No solution; { }; inconsistent system

18. No solution; { }; inconsistent system

19. Using 39 minutes per month, the cost per month for each company is $9.75. **20.** $\left\{\left(\dfrac{2}{3}, -2\right)\right\}$ **21.** $\{(-4, 1)\}$

22. { } **23.** Infinitely many solutions; $\{(x, y) \mid y = -2x + 2\}$

24. a. x in the first equation is easiest to isolate because its coefficient is 1. **b.** $\left\{\left(6, \dfrac{5}{2}\right)\right\}$ **25. a.** y in the second equation is easiest to isolate because its coefficient is 1. **b.** $\left\{\left(\dfrac{9}{2}, 3\right)\right\}$

26. $\{(5, -4)\}$ **27.** $\{(0, 4)\}$ **28.** Infinitely many solutions; $\{(x, y) \mid x - 3y = 9\}$ **29.** { } **30.** The numbers are 50 and 8.

31. The angles are 41° and 49°. **32.** The angles are $115\frac{1}{3}°$ and $64\frac{2}{3}°$.

33. b. $\{(-1, 4)\}$ **34. b.** $\{(-3, -2)\}$ **35. b.** $\{(2, 2)\}$

36. $\{(2, -1)\}$ **37.** $\{(-6, 2)\}$ **38.** $\left\{\left(-\dfrac{1}{2}, \dfrac{1}{3}\right)\right\}$

39. $\left\{\left(\dfrac{1}{4}, -\dfrac{2}{5}\right)\right\}$ **40.** Infinitely many solutions;
$\{(x, y)\,|\,-4x - 6y = -2\}$ **41.** $\{\ \}$ **42.** $\{(-4, -2)\}$

43. $\{(1, 0)\}$ **44. a.** Use the substitution method because y is already isolated in the second equation.   **b.** $\{(5, -3)\}$

45. a. Use the addition method. The substitution method would be cumbersome because isolating either variable from either equation would result in fractional coefficients.   **b.** $\{(-2, -1)\}$

46. There were 8 adult tickets and 52 children's tickets sold.

47. He should invest \$75,000 at 12% and \$525,000 at 4%.

48. 20 gal of whole milk should be mixed with 40 gal of low fat milk.

49. The speed of the boat is 18 mph, and that of the current is 2 mph.

50. The plane's speed in still air is 320 mph. The wind speed is 30 mph.

51. A hot dog costs \$4.50 and a drink costs \$3.50.   **52.** The score was 72 on the first round and 82 on the second round.

53. For example: **54.** For example:
$(1, -1), (0, -4), (2, 0)$ $(5, 5), (4, 0), (0, 7)$

55. For example: **56.** For example:
$(-4, 0), (-2, -2), (1, -4)$ $(0, 0), (0, -5), (-1, 1)$

57.   **58.**   **59.**

        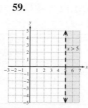

60.   **61.**   **62.**

        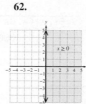

63.   **64.**

65.   **66.**

Chapter 4 Test, pp. 336–337

1. $y = -\dfrac{5}{2}x - 3; y = -\dfrac{5}{2}x + 3;$ Parallel lines   **2. a.** No solution
b. Infinitely many solutions   **c.** One solution

3. $\{(3, 2)\}$ **4.** Infinitely many solutions; $\{(x, y)\,|\,2x + 4y = 12\}$

    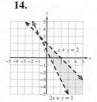

5. $\{(-2, 0)\}$ **6.** $\left\{\left(2, -\dfrac{1}{3}\right)\right\}$ **7.** $\{(-5, 4)\}$ **8.** $\{\ \}$

9. $\{(3, -5)\}$ **10.** $\{(-1, 2)\}$ **11.** Infinitely many solutions; $\{(x, y)\,|\,10x + 2y = -8\}$ **12.** $\{(1, -2)\}$

13.   **14.**

    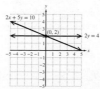

15. Swoopes scored 614 points and Jackson scored 597.   **16.** CDs cost \$8 each and DVDs cost \$11 each.   **17.** 12 mL of the 50% acid solution should be mixed with 24 mL of the 20% solution.

18. a. \$18 was required.   **b.** They used 24 quarters and 12 \$1 bills.

19. \$1200 was borrowed at 10%, and \$3800 was borrowed at 8%.

20. They would be the same cost for a distance of approximately 24 mi.   **21.** The plane travels 500 mph in still air, and the wind speed is 45 mph.   **22.** The cake has 340 calories, and the ice cream has 120 calories.   **23.** 60 mL of 10% solution and 40 mL of 25% solution

Chapters 1–4 Cumulative Review Exercises, pp. 337–338

1. $\dfrac{11}{6}$ **2.** $\left\{-\dfrac{21}{2}\right\}$ **3.** $\{\ \}$ **4.** $y = \dfrac{3}{2}x - 3$

5. $x = 5z + m$ **6.** $\left[\dfrac{3}{11}, \infty\right)$

7. The angles are 37°, 33°, and 110°.   **8.** The rates of the hikers are 2 mph and 4 mph.   **9.** Jesse Ventura received approximately 762,200 votes.   **10.** 36% of the goal has been achieved.   **11.** The angles are 36.5° and 53.5°.

12. Slope: $-\dfrac{5}{3}$; y-intercept: $(0, -2)$

13. a. $-\dfrac{2}{3}$ **b.** $\dfrac{3}{2}$ **14.** $y = -3x + 3$

15. c. $\{(0, 2)\}$ **16.** $\{(0, 2)\}$

17. a.   **b.**

c. Part (a) represents the solutions to an equation. Part (b) represents the solutions to a strict inequality.   **18.** 20 gal of the 15% solution should be mixed with 40 gal of the 60% solution.

19. x is 27°; y is 63°   **20. a.** 1.4   **b.** Between 1920 and 1990, the winning speed in the Indianapolis 500 increased on average by 1.4 mph per year.

Chapter 5

Section 5.1 Calculator Connections, p. 346

1.–3.

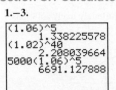

4.–6.

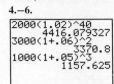

Section 5.1 Practice Exercises, pp. 347–349

1. a. exponent **b.** base; exponent **c.** 1 **d.** $I = Prt$
e. compound interest   **3.** Base: x; exponent: 4   **5.** Base: 3;
exponent: 5   **7.** Base: -1; exponent: 4   **9.** Base: 13; exponent: 1
11. Base: 10; exponent: 3   **13.** Base: t; exponent: 6
15. v **17.** 1   **19.** $(-6b)^2$ **21.** $-6b^2$ **23.** $(y+2)^4$
25. $\dfrac{-2}{t^3}$ **27.** No; $-5^2 = -25$ and $(-5)^2 = 25$ **29.** Yes; -32
31. Yes; $\dfrac{1}{8}$ **33.** Yes; -16 **35.** 16   **37.** -1 **39.** $\dfrac{1}{9}$
41. $-\dfrac{4}{25}$ **43.** 48   **45.** 4   **47.** 9   **49.** 50   **51.** -100
53. 400   **55.** 1   **57.** 1   **59.** 1000   **61.** -800
63. a. $(x \cdot x \cdot x \cdot x)(x \cdot x \cdot x) = x^7$ **b.** $(5 \cdot 5 \cdot 5 \cdot 5)(5 \cdot 5 \cdot 5) = 5^7$
65. z^8 **67.** a^9 **69.** 4^{14} **71.** $\left(\dfrac{2}{3}\right)^4$ **73.** c^{14} **75.** x^{18}
77. a. $\dfrac{p \cdot p \cdot p \cdot p \cdot p \cdot p \cdot p \cdot p}{p \cdot p \cdot p} = p^5$
b. $\dfrac{8 \cdot 8 \cdot 8 \cdot 8 \cdot 8 \cdot 8 \cdot 8 \cdot 8}{8 \cdot 8 \cdot 8} = 8^5$ **79.** x^2 **81.** a^9 **83.** 7^7
85. 5^7 **87.** y **89.** h^4 **91.** 7^7 **93.** 10^9 **95.** $6x^7$
97. $40a^5b^5$ **99.** s^9t^{16} **101.** $-30v^8$ **103.** $16m^{20}n^{10}$
105. $2cd^4$ **107.** z^4 **109.** $\dfrac{25hjk^4}{12}$ **111.** $-8p^7q^9r^6$
113. $-3stu$ **115.** \$5724.50   **117.** \$4764.06   **119.** 201 in.^{2}
121. 268 in.^{3} **123.** x^{2n+1} **125.** p^{2m+3} **127.** z **129.** r^3

Section 5.2 Practice Exercises, pp. 353–354

1. 4^9 **3.** a^{20} **5.** d^9 **7.** 7^6 **9.** When multiplying
expressions with the same base, add the exponents. When raising an
expression with an exponent to a power, multiply the exponents.
11. 5^{12} **13.** 12^6 **15.** y^{14} **17.** w^{25} **19.** a^{36} **21.** y^{14}
23. They are both equal to 2^6.   **25.** $4^{3^2} = 4^9$; $(4^3)^2 = 4^6$; the
expression 4^{3^2} is greater than $(4^3)^2$.   **27.** $25w^2$ **29.** $s^4r^4t^4$
31. $\dfrac{16}{r^4}$ **33.** $\dfrac{x^5}{y^5}$ **35.** $81a^4$ **37.** $-27a^3b^3c^3$ **39.** $-\dfrac{64}{x^3}$
41. $\dfrac{a^2}{b^2}$ **43.** $6^3u^6v^{12}$ or $216u^6v^{12}$ **45.** $5x^8y^4$ **47.** $-h^{28}$
49. m^{12} **51.** $\dfrac{4^5}{r^5s^{20}}$ or $\dfrac{1024}{r^5s^{20}}$ **53.** $\dfrac{3^5p^5}{q^{15}}$ or $\dfrac{243p^5}{q^{15}}$ **55.** y^{14}
57. x^{31} **59.** $200a^{14}b^9$ **61.** $16p^8q^{16}$ **63.** $-m^{35}n^{15}$
65. $25a^{18}b^6$ **67.** $\dfrac{4c^2d^6}{9}$ **69.** $\dfrac{c^{27}d^{31}}{2}$ **71.** $-\dfrac{27a^9b^3}{c^6}$
73. $16b^{26}$ **75.** x^{2m} **77.** $125a^{6n}$ **79.** $\dfrac{m^{2b}}{n^{3b}}$ **81.** $\dfrac{3^na^{3n}}{5^nb^{4n}}$

Section 5.3 Practice Exercises, pp. 362–364

1. a. 1 **b.** $\left(\dfrac{1}{b}\right)^n$ or $\dfrac{1}{b^n}$ **3.** c^9 **5.** y **7.** 3^6 or 729
9. $7^4w^{28}z^8$ or $2401w^{28}z^8$ **11. a.** 1 **b.** 1   **13.** 1   **15.** 1
17. -1 **19.** 1   **21.** 1   **23.** -7 **25. a.** $\dfrac{1}{t^5}$ **b.** $\dfrac{1}{t^5}$

27. $\dfrac{343}{8}$ **29.** 25   **31.** $\dfrac{1}{a^3}$ **33.** $\dfrac{1}{12}$ **35.** $\dfrac{1}{16b^2}$ **37.** $\dfrac{6}{x^2}$
39. $\dfrac{1}{64}$ **41.** $-\dfrac{3}{y^4}$ **43.** $-\dfrac{1}{t^3}$ **45.** a^5
47. $\dfrac{x^4}{x^{-6}} = x^{4-(-6)} = x^{10}$ **49.** $2a^{-3} = 2 \cdot \dfrac{1}{a^3} = \dfrac{2}{a^3}$ **51.** $\dfrac{1}{x^4}$
53. 1   **55.** y^4 **57.** $\dfrac{n^{27}}{m^{18}}$ **59.** $\dfrac{81k^{24}}{j^{20}}$ **61.** $\dfrac{1}{p^6}$ **63.** $\dfrac{1}{r^3}$
65. a^8 **67.** $\dfrac{1}{y^8}$ **69.** $\dfrac{1}{7^7}$ **71.** 1   **73.** $\dfrac{1}{a^4b^6}$ **75.** $\dfrac{1}{w^{21}}$
77. $\dfrac{1}{27}$ **79.** 1   **81.** $\dfrac{1}{64x^6}$ **83.** $\dfrac{16y^4}{z^2}$ **85.** $-\dfrac{a^{12}}{6}$
87. $80c^{21}d^{24}$ **89.** $\dfrac{9y^2}{8x^5}$ **91.** $\dfrac{4d^{16}}{c^2}$ **93.** $\dfrac{9y^2}{2x}$ **95.** $\dfrac{9}{20}$
97. $\dfrac{9}{10}$ **99.** $\dfrac{5}{4}$ **101.** $\dfrac{10}{3}$ **103.** 2   **105.** -11

Chapter 5 Problem Recognition Exercises, p. 364

1. t^8 **2.** 2^8 or 256   **3.** y^5 **4.** p^6 **5.** r^4s^8 **6.** $a^3b^9c^6$
7. w^6 **8.** $\dfrac{1}{m^{16}}$ **9.** $\dfrac{x^4z^3}{y^7}$ **10.** $\dfrac{a^3c^8}{b^6}$ **11.** x^6 **12.** $\dfrac{1}{y^{10}}$
13. $\dfrac{1}{t^5}$ **14.** w^7 **15.** p^{15} **16.** p^{15} **17.** $\dfrac{1}{v^2}$ **18.** $c^{50}d^{40}$
19. 3   **20.** -4 **21.** $\dfrac{b^9}{2^{15}}$ **22.** $\dfrac{81}{y^6}$ **23.** $\dfrac{16y^4}{81x^4}$ **24.** $\dfrac{25d^6}{36c^2}$
25. $3a^7b^5$ **26.** $64x^7y^{11}$ **27.** $\dfrac{y^4}{x^8}$ **28.** $\dfrac{1}{a^{10}b^{10}}$ **29.** $\dfrac{1}{t^2}$ **30.** $\dfrac{1}{p^7}$
31. $\dfrac{8w^6x^9}{27}$ **32.** $\dfrac{25b^8}{16c^6}$ **33.** $\dfrac{q^3s}{r^2t^5}$ **34.** $\dfrac{m^2p^3q}{n^3}$ **35.** $\dfrac{1}{y^{13}}$
36. w^{10} **37.** $-\dfrac{1}{8a^{18}b^6}$ **38.** $\dfrac{4x^{18}}{9y^{10}}$ **39.** $\dfrac{k^8}{5h^6}$ **40.** $\dfrac{6n^{10}}{m^{12}}$

Section 5.4 Calculator Connections, p. 368

1.–2.

```
(5.2E6)*(4.6E-3)
              23920
(2.19E-8)*(7.84E
-4)
        1.71696E-11
```

3.–4.

```
(4.76E-5)/(2.38E
9)
              2E-14
(8.5E4)/(4.0E-1)
             212500
```

5.

```
((9.6E7)*(4.0E-3
))/(2.0E-2)
           19200000
```

6.

```
((5.0E-12)*(6.4E
-5))/((1.6E-8)*(
4.0E2))
               5E-11
```

Section 5.4 Practice Exercises, pp. 369–371

1. scientific notation   **3.** b^{13} **5.** 10^{13} **7.** $\dfrac{1}{y^5}$ **9.** $\dfrac{x^{20}}{y^{12}}$
11. w^4 **13.** 10^4 **15.** Move the decimal point between the
2 and 3 and multiply by 10^{-10}; 2.3×10^{-10}.   **17.** 5×10^4
19. 2.08×10^5 **21.** 6.01×10^6 **23.** 8×10^{-6}
25. 1.25×10^{-4} **27.** 6.708×10^{-3} **29.** 1.7×10^{-24} g
31. $\$2.7 \times 10^{10}$ **33.** 6.8×10^7 gal; 1×10^2 miles
35. Move the decimal point nine places to the left; 0.000 000 0031.
37. 0.00005   **39.** 2800   **41.** 0.000603   **43.** 2,400,000
45. 0.019   **47.** 7032   **49.** 0.000 000 000 001 g
51. 1600 Cal and 2800 Cal   **53.** 5×10^4 **55.** 3.6×10^{11}
57. 2.2×10^4 **59.** 2.25×10^{-13} **61.** 3.2×10^{14}
63. 2.432×10^{-10} **65.** 3×10^{13} **67.** 6×10^5 **69.** 1.38×10^1
71. 5×10^{-14} **73.** 3.75 in.   **75.** $\$2.97 \times 10^{10}$
77. a. 6.5×10^7 **b.** 2.3725×10^{10} days **c.** 5.694×10^{11} hr
d. 2.04984×10^{15} sec

Section 5.5 Practice Exercises, pp. 377–380

1. a. polynomial **b.** coefficient; degree **c.** 1 **d.** one
e. binomial **f.** trinomial **g.** leading; leading coefficient **h.** greatest
i. zero   **3.** $\dfrac{45}{x^2}$ **5.** $\dfrac{2}{t^4}$ **7.** $\dfrac{1}{3^{12}}$ **9.** 4×10^{-2} is in scientific
notation in which 10 is raised to the -2 power. 4^{-2} is not in scientific notation and 4 is being raised to the -2 power.

11. a. $-7x^5 + 7x^2 + 9x + 6$ **b.** -7 **c.** 5   **13.** Binomial; 75
15. Monomial; 54   **17.** Binomial; -12 **19.** Trinomial; 27
21. Monomial; -192 **23.** The exponents on the x-factors are
different.   **25.** $35x^2y$ **27.** $8b^5d^2 - 9d$ **29.** $4y^2 + y - 9$
31. $-x^3 + 5x^2 + 9x - 10$ **33.** $10.9y$ **35.** $4a - 8c$
37. $a - \dfrac{1}{2}b - 2$ **39.** $\dfrac{4}{3}z^2 - \dfrac{5}{3}$ **41.** $7.9t^3 - 3.4t^2 + 6t - 4.2$
43. $-4h + 5$ **45.** $2.3m^2 - 3.1m + 1.5$
47. $-3v^3 - 5v^2 - 10v - 22$ **49.** $-8a^3b^2$ **51.** $-53x^3$
53. $-5a - 3$ **55.** $16k + 9$ **57.** $2m^4 - 14m$
59. $3s^2 - 4st - 3t^2$ **61.** $-2r - 3s + 3t$ **63.** $\dfrac{3}{4}x + \dfrac{1}{3}y - \dfrac{3}{10}$
65. $-\dfrac{2}{3}h^2 + \dfrac{3}{5}h - \dfrac{5}{2}$ **67.** $2.4x^4 - 3.1x^2 - 4.4x - 7.9$
69. $4b^3 + 12b^2 - 5b - 12$ **71.** $-\dfrac{7}{2}x^2 + 5x - 11$
73. $4y^3 + 2y^2 + 2$ **75.** $3a^2 - 3a + 5$
77. $9ab^2 - 3ab + 16a^2b$ **79.** $4z^5 + z^4 + 9z^3 - 3z - 2$
81. $2x^4 + 11x^3 - 3x^2 + 8x - 4$ **83.** $-w^3 + 0.2w^2 + 3.7w - 0.7$
85. $-p^2q - 4pq^2 + 3pq$ **87.** 0   **89.** $-5ab + 6ab^2$
91. $11y^2 - 10y - 4$ **93.** For example: $x^3 + 6$
95. For example: $8x^5$ **97.** For example: $-6x^2 + 2x + 5$

Section 5.6 Practice Exercises, pp. 387–390

1. a. $5 - 2x$ **b.** squares; $a^2 - b^2$ **c.** perfect; $a^2 + 2ab + b^2$
3. $-2y^2$ **5.** $-8y^4$ **7.** $8uvw^2$ **9.** $7u^2v^2w^4$ **11.** $-12y$
13. $21p$ **15.** $12a^{14}b^8$ **17.** $-2c^{10}d^{12}$
19. $16p^2q^2 - 24p^2q + 40pq^2$ **21.** $-4k^3 + 52k^2 + 24k$
23. $-45p^3q - 15p^4q^3 + 30pq^2$ **25.** $y^2 + 19y + 90$
27. $m^2 - 14m + 24$ **29.** $12p^2 - 5p - 2$
31. $12w^2 - 32w + 16$ **33.** $p^2 - 14pw + 33w^2$
35. $12x^2 + 28x - 5$ **37.** $6a^2 - 21.5a + 18$
39. $9t^3 - 21t^2 + 3t - 7$ **41.** $9m^2 + 30mn + 24n^2$
43. $5s^3 + 8s^2 - 7s - 6$ **45.** $27w^3 - 8$
47. $p^4 + 5p^3 - 2p^2 - 21p + 5$ **49.** $6a^3 - 23a^2 + 38a - 45$
51. $8x^3 - 36x^2y + 54xy^2 - 27y^3$ **53.** $1.2x^2 + 7.6xy + 2.4y^2$
55. $y^2 - 36$ **57.** $9a^2 - 16b^2$ **59.** $81k^2 - 36$
61. $\dfrac{4}{9}t^2 - 9$ **63.** $u^6 - 25v^2$ **65.** $\dfrac{4}{9} - p^2$
67. $a^2 + 10a + 25$ **69.** $x^2 - 2xy + y^2$ **71.** $4c^2 + 20c + 25$
73. $9t^4 - 24st^2 + 16s^2$ **75.** $t^2 - 14t + 49$ **77.** $16q^2 + 24q + 9$
79. a. 36 **b.** 20 **c.** $(a + b)^2 \neq a^2 + b^2$ in general.
81. $4x^2 - 25$ **83.** $16p^2 + 40p + 25$
85. $27p^3 - 135p^2 + 225p - 125$ **87.** $15a^5 - 6a^2$
89. $49x^2 - y^2$ **91.** $25s^2 + 30st + 9t^2$ **93.** $21x^2 - 65xy + 24y^2$
95. $2t^2 + \dfrac{26}{3}t + 8$ **97.** $-30x^2 - 35x + 25$
99. $6a^3 + 11a^2 - 7a - 2$ **101.** $y^3 - 3y^2 - 6y - 20$
103. $\dfrac{1}{9}m^2 - \dfrac{2}{3}mn + n^2$ **105.** $42w^3 - 84w^2$ **107.** $16y^2 - 65.61$
109. $21c^4 + 4c^2 - 32$ **111.** $9.61x^2 + 27.9x + 20.25$
113. $k^3 - 12k^2 + 48k - 64$ **115.** $125x^3 + 225x^2 + 135x + 27$
117. $2y^4 + 3y^3 + 3y^2 + 5y + 3$ **119.** $6a^3 + 22a^2 - 40a$
121. $2x^3 - 13x^2 + 17x + 12$ **123.** $2x - 7$
125. $k = 10$ or -10 **127.** $k = 8$ or -8

Section 5.7 Practice Exercises, pp. 395–398

1. division; quotient; remainder   **3.** $9a^2 + 6a + 5$
5. $16b^4 - 40b^3 + 96b^2$ **7.** $4w^6 + 20w^3 + 25$ **9.** $\dfrac{49}{64}w^2 - 1$
11. $10x^2 - x - 3$ **13.** $5t^2 + 6t$ **15.** $3a^2 + 2a - 7$
17. $x^2 + 4x - 1$ **19.** $3p^2 - p$ **21.** $1 + \dfrac{2}{m}$
23. $-2y^2 + y - 3$ **25.** $x^2 - 6x - \dfrac{1}{4} + \dfrac{2}{x}$ **27.** $a - 1 + \dfrac{b}{a}$
29. $3t - 1 + \dfrac{3}{2t} - \dfrac{1}{2t^2} + \dfrac{2}{t^3}$ **31. a.** $z + 2 + \dfrac{1}{z + 5}$
33. $t + 3 + \dfrac{2}{t + 1}$ **35.** $7b + 4$ **37.** $k - 6$ **39.** $2p^2 + 3p - 4$
41. $k - 2 + \dfrac{-4}{k + 1}$ **43.** $2x^2 - x + 6 + \dfrac{2}{2x - 3}$
45. $y^2 + 2y + 1 + \dfrac{2}{3y - 1}$ **47.** $a - 3 + \dfrac{18}{a + 3}$
49. $4x^2 + 8x + 13$ **51.** $w^2 + 5w - 2 + \dfrac{1}{w^2 - 3}$
53. $n^2 + n - 6$ **55.** $3y^2 - 3 + \dfrac{2y + 6}{y^2 + 1}$
57. $x - 1 + \dfrac{-8}{5x^2 + 5x + 1}$
59. Multiply $(x - 2)(x^2 + 4) = x^3 - 2x^2 + 4x - 8$,
which does not equal $x^3 - 8$.   **61.** Monomial division; $3a^2 + 4a$
63. Long division; $p + 2$
65. Long division; $t^3 - 2t^2 + 5t - 10 + \dfrac{4}{t + 2}$
67. Long division; $w^2 + 3 + \dfrac{1}{w^2 - 2}$
69. Long division; $n^2 + 4n + 16$
71. Monomial division; $-3r + 4 - \dfrac{3}{r^2}$ **73.** $x + 1$
75. $x^3 + x^2 + x + 1$ **77.** $x + 1 + \dfrac{1}{x - 1}$
79. $x^3 + x^2 + x + 1 + \dfrac{1}{x - 1}$

Chapter 5 Problem Recognition Exercises, p. 398

1. a. $8x^2$ **b.** $12x^4$ **2. a.** $9y^3$ **b.** $8y^6$ **3. a.** $16x^2 + 8xy + y^2$
b. $16x^2y^2$ **4. a.** $4a^2 + 4ab + b^2$ **b.** $4a^2b^2$ **5. a.** $6x + 1$
b. $8x^2 + 8x - 6$ **6. a.** $6m^2 + m + 1$ **b.** $5m^4 + 5m^3 + m^2 + m$
7. a. $9z^2 + 12z + 4$ **b.** $9z^2 - 4$ **8. a.** $36y^2 - 84y + 49$
b. $36y^2 - 49$ **9. a.** $2x^3 - 8x^2 + 14x - 12$ **b.** $x^2 - 1$
10. a. $-3y^4 - 20y^2 - 32$ **b.** $4y^2 + 12$ **11. a.** $2x$ **b.** x^2
12. a. $4c$ **b.** $4c^2$ **13.** $49m^2n^2$ **14.** $64p^2q^2$ **15.** $-x^2 - 3x + 4$
16. $5m^2 - 4m + 1$ **17.** $-7m^2 - 16m$
18. $-4n^5 + n^4 + 6n^2 - 7n + 2$ **19.** $8x^2 + 16x + 34 + \dfrac{74}{x - 2}$
20. $-4x^2 - 10x - 30 + \dfrac{-95}{x - 3}$ **21.** $6x^3 + 5x^2y - 6xy^2 + y^3$
22. $6a^3 - a^2b + 5ab^2 + 2b^3$ **23.** $x^3 + y^6$ **24.** $m^6 + 1$
25. $4b$ **26.** $-12z$ **27.** $a^6 - 4b^2$ **28.** $y^6 - 36z^2$
29. $4p + 4 + \dfrac{-2}{2p - 1}$ **30.** $2v - 7 + \dfrac{29}{2v + 3}$ **31.** $4x^2y^2$
32. $-9pq$ **33.** $\dfrac{9}{49}x^2 - \dfrac{1}{4}$ **34.** $\dfrac{4}{25}y^2 - \dfrac{16}{9}$
35. $-\dfrac{11}{9}x^3 + \dfrac{5}{9}x^2 - \dfrac{1}{2}x - 4$ **36.** $-\dfrac{13}{10}y^2 - \dfrac{9}{10}y + \dfrac{4}{15}$
37. $1.3x^2 - 0.3x - 0.5$ **38.** $5w^3 - 4.1w^2 + 2.8w - 1.2$
39. $-6x^3y^6$ **40.** $50a^4b^6$

Chapter 5 Review Exercises, pp. 403–406

1. Base: 5; exponent: 3   **2.** Base: x; exponent: 4
3. Base: -2; exponent: 0   **4.** Base: y; exponent: 1
5. a. 36 **b.** 36 **c.** -36 **6. a.** 64 **b.** -64 **c.** -64
7. 5^{13} **8.** a^{11} **9.** x^9 **10.** 6^9 **11.** 10^3 **12.** y^6
13. b^8 **14.** 7^7 **15.** k **16.** 1 **17.** 2^8 **18.** q^6
19. Exponents are added only when multiplying factors with the same base. In such a case, the base does not change.
20. Exponents are subtracted only when dividing factors with the same base. In such a case, the base does not change.   **21.** $7146.10
22. $$22,050 **23.** 7^{12} **24.** c^{12} **25.** p^{18} **26.** 9^{28}
27. $\dfrac{a^2}{b^2}$ **28.** $\dfrac{1}{3^4}$ **29.** $\dfrac{5^2}{c^4 d^{10}}$ **30.** $\dfrac{m^{10}}{4^5 n^{30}}$ **31.** $2^4 a^4 b^8$
32. $x^{14} y^2$ **33.** $-\dfrac{3^3 x^9}{5^3 y^6 z^3}$ **34.** $\dfrac{r^{15}}{s^{10} t^{30}}$ **35.** a^{11} **36.** 8^2
37. $4h^{14}$ **38.** $2p^{14} q^{13}$ **39.** $\dfrac{x^6 y^2}{4}$ **40.** $a^9 b^6$ **41.** 1
42. 1 **43.** -1 **44.** 1 **45.** 2 **46.** 1 **47.** $\dfrac{1}{z^5}$
48. $\dfrac{1}{10^4}$ **49.** $\dfrac{1}{36a^2}$ **50.** $\dfrac{6}{a^2}$ **51.** $\dfrac{17}{16}$ **52.** $\dfrac{10}{9}$
53. $\dfrac{1}{t^8}$ **54.** $\dfrac{1}{r}$ **55.** $\dfrac{2y^7}{x^6}$ **56.** $\dfrac{4a^6 bc}{5}$ **57.** $\dfrac{n^{16}}{16 m^8}$
58. $\dfrac{u^{15}}{27 v^6}$ **59.** $\dfrac{k^{21}}{5}$ **60.** $\dfrac{h^9}{9}$ **61.** $\dfrac{1}{2}$ **62.** $\dfrac{5}{4}$
63. a. 9.74×10^7 **b.** 4.2×10^{-3} in.   **64. a.** 0.000 000 0001
b. $256,000 **65.** 9.43×10^5 **66.** 1.55×10^{10}
67. 2.5×10^8 **68.** 1.638×10^3 **69.** $\approx 9.5367 \times 10^{13}$.
This number has too many digits to fit on most calculator displays.
70. $\approx 1.1529 \times 10^{-12}$. This number has too many digits to fit on most calculator displays.   **71. a.** $\approx 5.84 \times 10^8$ mi
b. $\approx 6.67 \times 10^4$ mph   **72. a.** $\approx 2.26 \times 10^8$ mi
b. $\approx 1.07 \times 10^5$ mph   **73. a.** Trinomial **b.** 4 **c.** 7
74. a. Binomial **b.** 7 **c.** -5 **75.** $7x - 3$
76. $-y^2 - 14y - 2$ **77.** $14a^2 - 2a - 6$
78. $\dfrac{15}{2} x^3 + \dfrac{1}{4} x^2 + \dfrac{1}{2} x + 2$ **79.** $10w^4 + 2w^3 - 7w + 4$
80. $0.01b^5 + b^4 - 0.1b^3 + 0.3b + 0.33$ **81.** $-2x^2 - 9x - 6$
82. $-5x^2 - 9x - 12$ **83.** For example, $-5x^2 + 2x - 4$
84. For example, $6x^5 + 8$ **85.** $6w + 6$ **86.** $-75 x^6 y^4$
87. $18 a^8 b^4$ **88.** $15 c^4 - 35 c^2 + 25 c$ **89.** $-2x^3 - 10 x^2 + 6x$
90. $5k^2 + k - 4$ **91.** $20 t^2 + 3t - 2$ **92.** $6q^2 + 47q - 8$
93. $2a^2 + 4a - 30$ **94.** $49 a^2 + 7a + \dfrac{1}{4}$ **95.** $b^2 - 8b + 16$
96. $8p^3 - 27$ **97.** $-2w^3 - 5w^2 - 5w + 4$
98. $4x^3 - 8x^2 + 11x - 4$ **99.** $12 a^3 + 11 a^2 - 13a - 10$
100. $b^2 - 16$ **101.** $\dfrac{1}{9} r^8 - s^4$ **102.** $49 z^4 - 84 z^2 + 36$
103. $2h^5 + h^4 - h^3 + h^2 - h + 3$ **104.** $2x^2 + 3x - 20$
105. $4y^2 - 2y$ **106.** $2a^2 b - a - 3b$ **107.** $-3x^2 + 2x - 1$
108. $-\dfrac{z^5 w^3}{2} + \dfrac{3zw}{4} + \dfrac{1}{z}$ **109.** $x + 2$ **110.** $2t + 5$
111. $p - 3 + \dfrac{5}{2p + 7}$ **112.** $a + 6 + \dfrac{-4}{5a - 3}$
113. $b^2 + 5b + 25$ **114.** $z + 4$ **115.** $y^2 - 4y + 2 + \dfrac{9y - 4}{y^2 + 3}$
116. $t^2 - 3t + 1 + \dfrac{-2t - 6}{3t^2 + t + 1}$ **117.** $w^2 + w - 1$

Chapter 5 Test, pp. 406–407

1. $\dfrac{(3 \cdot 3 \cdot 3 \cdot 3) \cdot (3 \cdot 3 \cdot 3)}{3 \cdot 3 \cdot 3 \cdot 3 \cdot 3 \cdot 3} = 3$ **2.** 9^6 **3.** q^8 **4.** 1
5. $\dfrac{1}{c^3}$ **6.** $\dfrac{5}{4}$ **7.** $\dfrac{3}{2}$ **8.** $27 a^6 b^3$ **9.** $\dfrac{16 x^4}{y^{12}}$ **10.** 14

11. $49 s^{18} t$ **12.** $\dfrac{4}{b^{12}}$ **13.** $\dfrac{16 a^{12}}{9 b^6}$ **14. a.** 4.3×10^{10}
b. 0.000 0056 **15.** 8.4×10^{-9} **16.** 7.5×10^6
17. a. 2.4192×10^8 m^3 **b.** 8.83008×10^{10} m^3
18. $5x^3 - 7x^2 + 4x + 11$ **a.** 3 **b.** 5
19. $5t^4 + 12 t^3 + 7t - 19$ **20.** $24 w^2 - 3w - 4$
21. $-10 x^5 - 2x^4 + 30 x^3$ **22.** $8a^2 - 10a + 3$
23. $4y^3 - 25 y^2 + 37y - 15$ **24.** $4 - 9b^2$ **25.** $25 z^2 - 60z + 36$
26. $15 x^2 - x - 6$ **27.** $-4x^2 + 5x + 7$
28. $y^3 - 11 y^2 + 32y - 12$ **29.** $15 x^3 - 7x^2 - 2x + 1$
30. Perimeter: $12x - 2$; area: $5x^2 - 13x - 6$
31. $-3x^6 + \dfrac{x^4}{4} - 2x$ **32.** $-4a^2 + \dfrac{ab}{2} - 2$ **33.** $2y - 7$
34. $w^2 - 4w + 5 + \dfrac{-10}{2w + 3}$ **35.** $3x^2 + x - 12 + \dfrac{15}{x^2 + 4}$

Chapters 1–5 Cumulative Review Exercises, pp. 407–408

1. $-\dfrac{35}{2}$ **2.** 4   **3.** $5^2 - \sqrt{4}$; 23   **4.** $\left\{ \dfrac{28}{3} \right\}$
5. { }   **6.** Quadrant III   **7.** y-axis
8. The measures are $31°$, $54°$, $95°$.   **9. a.** 12 in. **b.** 19.5 in.
c. 5.5 hr   **10.** $\{(-3, 4)\}$
11. $[-5, \infty)$ ⟶
        -5
12. $5x^2 - 9x - 15$ **13.** $-2y^2 - 13 yz - 15 z^2$
14. $16 t^2 - 24t + 9$ **15.** $\dfrac{4}{25} a^2 - \dfrac{1}{9}$
16. $-4a^3 b^2 + 2ab - 1$ **17.** $4m^2 + 8m + 11 + \dfrac{24}{m - 2}$
18. $\dfrac{c^2}{16 d^4}$ **19.** $\dfrac{2b^3}{a^2}$ **20.** 4.1×10^3

Chapter 6

Section 6.1 Practice Exercises, pp. 417–419

1. a. product   **b.** prime   **c.** greatest common factor
d. prime   **e.** greatest common factor (GCF)   **f.** grouping
3. 7 **5.** 6 **7.** y **9.** $4w^2 z$ **11.** $2xy^4 z^2$ **13.** $(x - y)$
15. a. $3x - 6y$ **b.** $3(x - 2y)$ **17.** $4(p + 3)$ **19.** $5(c^2 - 2c + 3)$
21. $x^3(x^2 + 1)$ **23.** $t(t^3 - 4 + 8t)$ **25.** $2ab(1 + 2a^2)$
27. $19 x^2 y(2 - y^3)$ **29.** $6xy^5(x^2 - 3y^4 z)$ **31.** The expression is prime because it is not factorable.   **33.** $7pq^2(6p^2 + 2 - p^3 q^2)$
35. $t^2(t^3 + 2rt - 3t^2 + 4r^2)$ **37. a.** $-2x(x^2 + 2x - 4)$
b. $2x(-x^2 - 2x + 4)$ **39.** $-1(8t^2 + 9t + 2)$ **41.** $-15 p^2(p + 2)$
43. $-3mn(m^3 n - 2m + 3n)$ **45.** $-1(7x + 6y + 2z)$
47. $(a + 6)(13 - 4b)$ **49.** $(w^2 - 2)(8v + 1)$ **51.** $7x(x + 3)^2$
53. $(2a - b)(4a + 3c)$ **55.** $(q + p)(3 + r)$ **57.** $(2x + 1)(3x + 2)$
59. $(t + 3)(2t - 1)$ **61.** $(3y - 1)(2y - 3)$ **63.** $(b + 1)(b^3 - 4)$
65. $(j^2 + 5)(3k + 1)$ **67.** $(2x^6 + 1)(7w^6 - 1)$
69. $(y + x)(a + b)$ **71.** $(vw + 1)(w - 3)$ **73.** $5x(x^2 + y^2)(3x + 2y)$
75. $4b(a - b)(x - 1)$ **77.** $6t(t - 3)(s - t^2)$ **79.** $P = 2(l + w)$
81. $S = 2\pi r(r + h)$ **83.** $\dfrac{1}{7}(x^2 + 3x - 5)$ **85.** $\dfrac{1}{4}(5w^2 + 3w + 9)$
87. For example: $6x^2 + 9x$ **89.** For example: $16 p^4 q^2 + 8p^3 q - 4p^2 q$

Section 6.2 Practice Exercises, pp. 424–426

1. a. positive   **b.** different   **c.** Both are correct.
d. $3(x + 6)(x + 2)$ **3.** $3(t - 5)(t - 2)$ **5.** $(a + 2b)(x - 5)$
7. $(x + 8)(x + 2)$ **9.** $(z - 9)(z - 2)$ **11.** $(z - 6)(z + 3)$
13. $(p - 8)(p + 5)$ **15.** $(t + 10)(t - 4)$ **17.** Prime
19. $(n + 4)^2$ **21.** a   **23.** c   **25.** They are both correct because multiplication of polynomials is a commutative operation.
27. The expressions are equal and both are correct.
29. It should be written in descending order.   **31.** $(x - 15)(x + 2)$
33. $(w - 13)(w - 5)$ **35.** $(t + 18)(t + 4)$ **37.** $3(x - 12)(x + 2)$
39. $8p(p - 1)(p - 4)$ **41.** $y^2 z^2(y - 6)(y - 6)$ or $y^2 z^2(y - 6)^2$
43. $-(x - 4)(x - 6)$ **45.** $-5(a - 3x)(a + 2x)$
47. $-2(c + 2)(c + 1)$ **49.** $xy^3(x - 4)(x - 15)$

51. $12(p-7)(p-1)$ **53.** $-2(m-10)(m-1)$
55. $(c+5d)(c+d)$ **57.** $(a-2b)(a-7b)$ **59.** Prime
61. $(q-7)(q+9)$ **63.** $(x+10)^2$ **65.** $(t+20)(t-2)$
67. The student forgot to factor out the GCF before factoring the trinomial further. The polynomial is not factored completely, because $(2x-4)$ has a common factor of 2.   **69.** $x^2+9x-52$
71. a. $3x^2+9x-12$ **b.** $3(x+4)(x-1)$ **73.** $(x^2+1)(x^2+9)$
75. $(w^2+5)(w^2-3)$ **77.** $7, 5, -7, -5$
79. For example: $c=-16$

Section 6.3 Practice Exercises, pp. 433–434

1. a. Both are correct.   **b.** $2(3x-5)(x+1)$ **3.** $(n-1)(m-2)$
5. $6(a-7)(a+2)$ **7.** a   **9.** b   **11.** $(3n+1)(n+4)$
13. $(2y+1)(y-2)$ **15.** $(5x+1)(x-3)$ **17.** $(4c+1)(3c-2)$
19. $(10w-3)(w+4)$ **21.** $(3q+2)(2q-3)$ **23.** Prime
25. $(5m+2)(5m-4)$ **27.** $(6y-5x)(y+4x)$
29. $2(m+4)(m-10)$ **31.** $y^3(2y+1)(y+6)$
33. $-(a+17)(a-2)$ **35.** $-10(4m+p)(2m-3p)$
37. $(x^2+1)(x^2+9)$ **39.** $(w^2+5)(w^2-3)$ **41.** $(2x^2+3)(x^2-5)$
43. $-2(z-9)(z-1)$ **45.** $(q-7)(q-6)$ **47.** $(2t+3)(3t-1)$
49. $(2m-5)^2$ **51.** Prime   **53.** $(2x-5y)(3x-2y)$
55. $(4m+5n)(3m-n)$ **57.** $5(3r+2)(2r-1)$ **59.** Prime
61. $(2t-5)(5t+1)$ **63.** $(7w-4)(2w+3)$
65. $(a-12)(a+2)$ **67.** $(x+5y)(x+4y)$ **69.** $(a+20b)(a+b)$
71. $(t-7)(t-3)$ **73.** $d(5d^2+3d-10)$
75. $4b(b-5)(b+4)$ **77.** $y^2(x-3)(x-10)$
79. $-2u(2u+5)(3u-2)$ **81.** $(2x^2+3)(4x^2+1)$
83. a. 36 ft   **b.** $-4(4t+5)(t-2)$; Yes   **85. a.** $(x-12)(x+2)$
b. $(x-6)(x-4)$ **87. a.** $(x-6)(x+1)$ **b.** $(x-2)(x-3)$

Section 6.4 Practice Exercises, pp. 440–441

1. a. Both are correct.   **b.** $3(4x+3)(x-2)$ **3.** $(y+5)(8+9y)$
5. $12, 1$ **7.** $-8, -1$ **9.** $5, -4$ **11.** $9, -2$
13. $(x+4)(3x+1)$ **15.** $(w-2)(4w-1)$ **17.** $(x+9)(x-2)$
19. $(m+3)(2m-1)$ **21.** $(4k+3)(2k-3)$ **23.** $(2k-5)^2$
25. Prime   **27.** $(3z-5)(3z-2)$ **29.** $(6y-5z)(2y+3z)$
31. $2(7y+4)(y+3)$ **33.** $-(3w-5)(5w+1)$
35. $-4(x-y)(3x-2y)$ **37.** $6y(y+1)(3y+7)$
39. $(a^2+2)(a^2+3)$ **41.** $(3x^2-5)(2x^2+3)$
43. $(8p^2-3)(p^2+5)$ **45.** $(5p-1)(4p-3)$
47. $(3u-2v)(2u-5v)$ **49.** $(4a+5b)(3a-b)$
51. $(h+7k)(3h-2k)$ **53.** Prime   **55.** $(2z-1)(8z-3)$
57. $(b-4)^2$ **59.** $-5(x-2)(x-3)$ **61.** $(t-3)(t+2)$
63. Prime   **65.** $2(12x-1)(3x+1)$ **67.** $p(p+3)(p-9)$
69. $x(3x+7)(x+1)$ **71.** $2p(p-15)(p-4)$
73. $x^2(y+3)(y+11)$ **75.** $-(k+2)(k+5)$
77. $-3(n+6)(n-5)$ **79.** $(x^2-2)(x^2-5)$
81. No. $(2x+4)$ contains a common factor of 2.
83. a. 128 customers   **b.** $-2(x-18)(x-2)$; 128; Yes
85. a. 42   **b.** $n(n+1)$; 42; Yes

Section 6.5 Practice Exercises, pp. 447–448

1. a. difference; $(a+b)(a-b)$ **b.** sum   **c.** is not
d. square   **e.** $(a+b)^2$; $(a-b)^2$ **3.** $(3x-1)(2x-5)$
5. $5xy^5(3x-2y)$ **7.** $(x+b)(a-6)$ **9.** $(y+10)(y-4)$
11. x^2-25 **13.** $4p^2-9q^2$ **15.** $(x-6)(x+6)$
17. $3(w+10)(w-10)$ **19.** $(2a-11b)(2a+11b)$
21. $(7m-4n)(7m+4n)$ **23.** Prime   **25.** $(y+2z)(y-2z)$
27. $(a-b^2)(a+b^2)$ **29.** $(5pq-1)(5pq+1)$ **31.** $\left(c-\dfrac{1}{5}\right)\left(c+\dfrac{1}{5}\right)$
33. $2(5-4t)(5+4t)$ **35.** $(x+4)(x-4)(x^2+16)$
37. $(2-z)(2+z)(4+z^2)$ **39.** $(x+5)(x+3)(x-3)$
41. $(c-1)(c+5)(c-5)$ **43.** $(x+3)(x-3)(2+y)$
45. $(y+3)(y-3)(x+2)(x-2)$ **47.** $9x^2+30x+25$
49. a. x^2+4x+4 is a perfect square trinomial.
b. $x^2+4x+4=(x+2)^2$; $x^2+5x+4=(x+1)(x+4)$
51. $(x+9)^2$ **53.** $(5z-2)^2$ **55.** $(7a+3b)^2$ **57.** $(y-1)^2$
59. $5(4z+3w)^2$ **61.** $(3y+25)(3y+1)$ **63.** $2(a-5)^2$
65. Prime   **67.** $(2x+y)^2$ **69.** $3x(x-1)^2$ **71.** $y(y-6)$

73. $(2p-5)(2p+7)$ **75.** $(-t+2)(t+6)$ or $-(t-2)(t+6)$
77. $(2a-5+10b)(2a-5-10b)$ **79. a.** a^2-b^2
b. $(a-b)(a+b)$

Section 6.6 Practice Exercises, pp. 453–454

1. a. sum; cubes   **b.** difference; cubes   **c.** $(a-b)(a^2+ab+b^2)$
d. $(a+b)(a^2-ab+b^2)$ **3.** $5(2-t)(2+t)$ **5.** $(t+u)(2+s)$
7. $(3v-4)(v+3)$ **9.** $-(c+5)^2$ **11.** $x^3, 8, y^6, 27q^3, w^{12}, r^3s^6$
13. $t^3, -1, 27, a^3b^6, 125, y^6$ **15.** $(y-2)(y^2+2y+4)$
17. $(1-p)(1+p+p^2)$ **19.** $(w+4)(w^2-4w+16)$
21. $(x-10y)(x^2+10xy+100y^2)$ **23.** $(4t+1)(16t^2-4t+1)$
25. $(10a+3)(100a^2-30a+9)$ **27.** $\left(n-\dfrac{1}{2}\right)\left(n^2+\dfrac{1}{2}n+\dfrac{1}{4}\right)$
29. $(5x+2y)(25x^2-10xy+4y^2)$ **31.** $(x^2-2)(x^2+2)$
33. Prime   **35.** $(t+4)(t^2-4t+16)$ **37.** Prime
39. $4(b+3)(b^2-3b+9)$ **41.** $5(p-5)(p+5)$
43. $\left(\dfrac{1}{4}-2h\right)\left(\dfrac{1}{16}+\dfrac{1}{2}h+4h^2\right)$ **45.** $(x-2)(x+2)(x^2+4)$
47. $\left(\dfrac{2}{3}x-w\right)\left(\dfrac{2}{3}x+w\right)$
49. $(q-2)(q^2+2q+4)(q+2)(q^2-2q+4)$
51. $(x^3+4y)(x^6-4x^3y+16y^2)$ **53.** $(2x+3)(x-1)(x+1)$
55. $(2x-y)(2x+y)(4x^2+y^2)$ **57.** $(3y-2)(3y+2)(9y^2+4)$
59. $(a+b^2)(a^2-ab^2+b^4)$ **61.** $(x^2+y^2)(x-y)(x+y)$
63. $(k+4)(k-3)(k+3)$ **65.** $2(t-5)(t-1)(t+1)$
67. $\left(\dfrac{4}{5}p-\dfrac{1}{2}q\right)\left(\dfrac{16}{25}p^2+\dfrac{2}{5}pq+\dfrac{1}{4}q^2\right)$ **69.** $(a^4+b^4)(a^8-a^4b^4+b^8)$
71. a. The quotient is x^2+2x+4.   **b.** $(x-2)(x^2+2x+4)$
73. $x^2+4x+16$ **75.** $2x+1$

Chapter 6 Problem Recognition Exercises, pp. 455–456

1. A prime polynomial cannot be factored further.
2. Factor out the GCF.   **3.** Look for a difference of squares: a^2-b^2, a difference of cubes: a^3-b^3, or a sum of cubes: a^3+b^3.   **4.** Grouping
5. a. Difference of squares   **b.** $2(a-9)(a+9)$
6. a. Nonperfect square trinomial   **b.** $(y+3)(y+1)$
7. a. None of these   **b.** $6w(w-1)$
8. a. Difference of squares   **b.** $(2z+3)(2z-3)(4z^2+9)$
9. a. Nonperfect square trinomial   **b.** $(3t+1)(t+4)$
10. a. Sum of cubes   **b.** $5(r+1)(r^2-r+1)$
11. a. Four terms-grouping   **b.** $(3c+d)(a-b)$
12. a. Difference of cubes   **b.** $(x-5)(x^2+5x+25)$
13. a. Sum of cubes   **b.** $(y+2)(y^2-2y+4)$
14. a. Nonperfect square trinomial   **b.** $(7p-1)(p-4)$
15. a. Nonperfect square trinomial   **b.** $3(q-4)(q+1)$
16. a. Perfect square trinomial   **b.** $-2(x-2)^2$
17. a. None of these   **b.** $6a(3a+2)$
18. a. Difference of cubes   **b.** $2(3-y)(9+3y+y^2)$
19. a. Difference of squares   **b.** $4(t-5)(t+5)$
20. a. Nonperfect square trinomial   **b.** $(4t+1)(t-8)$
21. a. Nonperfect square trinomial   **b.** $10(c^2+c+1)$
22. a. Four terms-grouping   **b.** $(w-5)(2x+3y)$
23. a. Sum of cubes   **b.** $(x+0.1)(x^2-0.1x+0.01)$
24. a. Difference of squares   **b.** $(2q-3)(2q+3)$
25. a. Perfect square trinomial   **b.** $(8+k)^2$
26. a. Four terms-grouping   **b.** $(t+6)(s^2+5)$
27. a. Four terms-grouping   **b.** $(x+1)(2x-y)$
28. a. Sum of cubes   **b.** $(w+y)(w^2-wy+y^2)$
29. a. Difference of cubes   **b.** $(a-c)(a^2+ac+c^2)$
30. a. Nonperfect square trinomial   **b.** Prime
31. a. Nonperfect square trinomial   **b.** Prime
32. a. Perfect square trinomial   **b.** $(a+1)^2$
33. a. Perfect square trinomial   **b.** $(b+5)^2$
34. a. Nonperfect square trinomial   **b.** $-(t+8)(t-4)$
35. a. Nonperfect square trinomial   **b.** $-p(p+4)(p+1)$
36. a. Difference of squares   **b.** $(xy-7)(xy+7)$
37. a. Nonperfect square trinomial   **b.** $3(2x+3)(x-5)$
38. a. Nonperfect square trinomial   **b.** $2(5y-1)(2y-1)$

39. a. None of these   **b.** $abc^2(5ac - 7)$
40. a. Difference of squares   **b.** $2(2a - 5)(2a + 5)$
41. a. Nonperfect square trinomial   **b.** $(t + 9)(t - 7)$
42. a. Nonperfect square trinomial   **b.** $(b + 10)(b - 8)$
43. a. Four terms-grouping   **b.** $(b + y)(a - b)$
44. a. None of these   **b.** $3x^2y^4(2x + y)$
45. a. Nonperfect square trinomial   **b.** $(7u - 2v)(2u - v)$
46. a. Nonperfect square trinomial   **b.** Prime
47. a. Nonperfect square trinomial   **b.** $2(2q^2 - 4q - 3)$
48. a. Nonperfect square trinomial   **b.** $3(3w^2 + w - 5)$
49. a. Sum of squares   **b.** Prime
50. a. Perfect square trinomial   **b.** $5(b - 3)^2$
51. a. Nonperfect square trinomial   **b.** $(3r + 1)(2r + 3)$
52. a. Nonperfect square trinomial   **b.** $(2s - 3)(2s + 5)$
53. a. Difference of squares   **b.** $(2a - 1)(2a + 1)(4a^2 + 1)$
54. a. Four terms-grouping   **b.** $(p + c)(p - 3)(p + 3)$
55. a. Perfect square trinomial   **b.** $(9u - 5v)^2$
56. a. Sum of squares   **b.** $4(x^2 + 4)$
57. a. Nonperfect square trinomial   **b.** $(x - 6)(x + 1)$
58. a. Nonperfect square trinomial   **b.** Prime
59. a. Four terms-grouping   **b.** $2(x - 3y)(a + 2b)$
60. a. Nonperfect square trinomial   **b.** $m(4m + 1)(2m - 3)$
61. a. Nonperfect square trinomial   **b.** $x^2y(3x + 5)(7x + 2)$
62. a. Difference of squares   **b.** $2(m^2 - 8)(m^2 + 8)$
63. a. Four terms-grouping   **b.** $(4v - 3)(2u + 3)$
64. a. Four terms-grouping   **b.** $(t - 5)(4t + s)$
65. a. Perfect square trinomial   **b.** $3(2x - 1)^2$
66. a. Perfect square trinomial   **b.** $(p + q)^2$
67. a. Nonperfect square trinomial   **b.** $n(2n - 1)(3n + 4)$
68. a. Nonperfect square trinomial   **b.** $k(2k - 1)(2k + 3)$
69. a. Difference of squares   **b.** $(8 - y)(8 + y)$
70. a. Difference of squares   **b.** $b(6 - b)(6 + b)$
71. a. Nonperfect square trinomial   **b.** Prime
72. a. Nonperfect square trinomial   **b.** $(y + 4)(y + 2)$
73. a. Nonperfect square trinomial   **b.** $(c^2 - 10)(c^2 - 2)$

Section 6.7 Practice Exercises, pp. 461–463

1. a. quadratic   **b.** $0; 0$ **3.** $4(b - 5)(b - 6)$
5. $(3x - 2)(x + 4)$ **7.** $4(x^2 + 4y^2)$ **9.** Neither   **11.** Quadratic
13. Linear   **15.** $\{-3, 1\}$ **17.** $\left\{\dfrac{7}{2}, -\dfrac{7}{2}\right\}$ **19.** $\{-5\}$
21. $\left\{0, \dfrac{1}{5}\right\}$ **23.** The polynomial must be factored completely.
25. $\{5, -3\}$ **27.** $\{-12, 2\}$ **29.** $\left\{4, -\dfrac{1}{2}\right\}$ **31.** $\left\{\dfrac{2}{3}, -\dfrac{2}{3}\right\}$
33. $\{6, 8\}$ **35.** $\left\{0, -\dfrac{3}{2}, 4\right\}$ **37.** $\left\{-\dfrac{1}{3}, 3, -6\right\}$ **39.** $\left\{0, 4, -\dfrac{3}{2}\right\}$
41. $\left\{0, -\dfrac{9}{2}, 11\right\}$ **43.** $\{0, 4, -4\}$ **45.** $\{-6, 0\}$
47. $\left\{\dfrac{3}{4}, -\dfrac{3}{4}\right\}$ **49.** $\{0, -5, -2\}$ **51.** $\left\{-\dfrac{14}{3}\right\}$ **53.** $\{5, 3\}$
55. $\{0, -2\}$ **57.** $\{-3\}$ **59.** $\{-3, 1\}$ **61.** $\left\{\dfrac{3}{2}\right\}$
63. $\left\{0, \dfrac{1}{3}\right\}$ **65.** $\{0, 2\}$ **67.** $\{3, -2, 2\}$ **69.** $\{-5, 4\}$
71. $\{-5, -1\}$

Chapter 6 Problem Recognition Exercises, p. 463

1. a. $(x + 7)(x - 1)$ **b.** $\{-7, 1\}$ **2. a.** $(c + 6)(c + 2)$
b. $\{-6, -2\}$ **3. a.** $(2y + 1)(y + 3)$ **b.** $\left\{-\dfrac{1}{2}, -3\right\}$
4. a. $(3x - 5)(x - 1)$ **b.** $\left\{\dfrac{5}{3}, 1\right\}$ **5. a.** $\left\{\dfrac{4}{5}, -1\right\}$
b. $(5q - 4)(q + 1)$ **6. a.** $\left\{-\dfrac{1}{3}, \dfrac{3}{2}\right\}$ **b.** $(3a + 1)(2a - 3)$
7. a. $\{-8, 8\}$ **b.** $(a + 8)(a - 8)$ **8. a.** $\{-10, 10\}$
b. $(v + 10)(v - 10)$ **9. a.** $(2b + 9)(2b - 9)$ **b.** $\left\{-\dfrac{9}{2}, \dfrac{9}{2}\right\}$
10. a. $(6t + 7)(6t - 7)$ **b.** $\left\{-\dfrac{7}{6}, \dfrac{7}{6}\right\}$

11. a. $\left\{-\dfrac{3}{2}, -\dfrac{1}{2}\right\}$ **b.** $2(2x + 3)(2x + 1)$
12. a. $\left\{-\dfrac{4}{3}, -2\right\}$ **b.** $4(3y + 4)(y + 2)$
13. a. $x(x - 10)(x + 2)$ **b.** $\{0, 10, -2\}$
14. a. $k(k + 7)(k - 2)$ **b.** $\{0, -7, 2\}$
15. a. $\{-1, 3, -3\}$ **b.** $(b + 1)(b - 3)(b + 3)$
16. a. $\{8, -2, 2\}$ **b.** $(x - 8)(x + 2)(x - 2)$
17. $(s - 3)(2s + r)$ **18.** $(2t + 1)(3t + 5u)$
19. $\left\{-\dfrac{1}{2}, 0, \dfrac{1}{2}\right\}$ **20.** $\{-5, 0, 5\}$ **21.** $\{0, 1\}$
22. $\{-3, 0\}$ **23.** $\left\{\dfrac{1}{3}\right\}$ **24.** $\left\{-\dfrac{7}{12}\right\}$
25. $(2w + 3)(4w^2 - 6w + 9)$ **26.** $(10q - 1)(100q^2 + 10q + 1)$
27. $\left\{-\dfrac{7}{5}, 1\right\}$ **28.** $\left\{-6, -\dfrac{1}{4}\right\}$ **29.** $\left\{-\dfrac{2}{3}, -5\right\}$ **30.** $\left\{-1, -\dfrac{1}{2}\right\}$
31. $\{3\}$ **32.** $\{0\}$ **33.** $\{-4, 4\}$ **34.** $\left\{-\dfrac{2}{3}, \dfrac{2}{3}\right\}$
35. $\{1, 6\}$ **36.** $\{-2, -12\}$

Section 6.8 Practice Exercises, pp. 468–471

1. a. $x + 1$ **b.** $x + 2$ **c.** $x + 2$ **d.** LW **e.** $\dfrac{1}{2}bh$
f. $a^2 + b^2 = c^2$ **3.** $\left\{0, -\dfrac{2}{3}\right\}$ **5.** $\{6, -1\}$ **7.** $\left\{-\dfrac{5}{6}, 2\right\}$
9. The numbers are 7 and -7.   **11.** The numbers are 10 and -4.
13. The numbers are -9 and -7, or 7 and 9.   **15.** The numbers are 5 and 6, or -6 and -5.   **17.** The height of the painting is 11 ft and the width is 9 ft.   **19. a.** The slab is 7 m by 4 m.   **b.** The perimeter is 22 m.   **21.** The base is 7 ft and the height is 4 ft.
23. The base is 5 cm and the height is 8 cm.   **25.** The ball hits the ground in 3 sec.
27. The object is at ground level at 0 sec and 1.5 sec.
29. 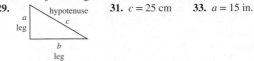   **31.** $c = 25$ cm   **33.** $a = 15$ in.

35. The brace is 20 in. long.   **37.** The kite is 19 yd high.
39. The bottom of the ladder is 8 ft from the house. The distance from the top of the ladder to the ground is 15 ft.
41. The hypotenuse is 10 m.

Chapter 6 Review Exercises, pp. 477–479

1. $3a^2b$ **2.** $x + 5$ **3.** $2c(3c - 5)$ **4.** $-2yz$ or $2yz$
5. $2x(3x + x^3 - 4)$ **6.** $11w^2y^3(w - 4y^2)$
7. $t(-t + 5)$ or $-t(t - 5)$ **8.** $u(-6u - 1)$ or $-u(6u + 1)$
9. $(b + 2)(3b - 7)$ **10.** $2(5x + 9)(1 + 4x)$
11. $(w + 2)(7w + b)$ **12.** $(b - 2)(b + y)$
13. $3(4y - 3)(5y - 1)$ **14.** $a(2 - a)(3 - b)$
15. $(x - 3)(x - 7)$ **16.** $(y - 8)(y - 11)$
17. $(z - 12)(z + 6)$ **18.** $(q - 13)(q + 3)$
19. $3w(p + 10)(p + 2)$ **20.** $2m^2(m + 8)(m + 5)$
21. $-(t - 8)(t - 2)$ **22.** $-(w - 4)(w + 5)$
23. $(a + b)(a + 11b)$ **24.** $(c - 6d)(c + 3d)$
25. Different   **26.** Both negative   **27.** Both positive
28. Different   **29.** $(2y + 3)(y - 4)$ **30.** $(4w + 3)(w - 2)$
31. $(2z + 5)(5z + 2)$ **32.** $(4z - 3)(2z + 3)$ **33.** Prime
34. Prime   **35.** $10(w - 9)(w + 3)$ **36.** $-3(y - 8)(y + 2)$
37. $(3c - 5d)^2$ **38.** $(x + 6)^2$ **39.** $(3g + 2h)(2g + h)$
40. $(6m - n)(2m - 5n)$ **41.** $(v^2 + 1)(v^2 - 3)$
42. $(x^2 + 5)(x^2 + 2)$ **43.** $5, -1$ **44.** $-3, -5$
45. $(c - 2)(3c + 1)$ **46.** $(y + 3)(4y + 1)$ **47.** $(t + 12w)(t + w)$
48. $(4x^2 - 3)(x^2 + 5)$ **49.** $(w^2 + 5)(w^2 + 2)$
50. $(p - 3q)(p - 5q)$ **51.** $-2(4v + 3)(5v - 1)$
52. $10(4s - 5)(s + 2)$ **53.** $ab(a - 6b)(a - 4b)$
54. $2z^4(z + 7)(z - 3)$ **55.** Prime   **56.** Prime   **57.** $(7x + 10)^2$
58. $(3w - z)^2$ **59.** $(a - b)(a + b)$ **60.** Prime
61. $(a - 7)(a + 7)$ **62.** $(d - 8)(d + 8)$ **63.** $(10 - 9t)(10 + 9t)$
64. $(2 - 5k)(2 + 5k)$ **65.** Prime   **66.** Prime   **67.** $(y + 6)^2$
68. $(t + 8)^2$ **69.** $(3a - 2)^2$ **70.** $(5x - 4)^2$ **71.** $-3(v + 2)^2$

72. $-2(x-5)^2$ **73.** $2(c^2-3)(c^2+3)$ **74.** $2(6x-y)(6x+y)$
75. $(p+3)(p-4)(p+4)$ **76.** $(k-2)(2-k)(2+k)$ or
$-(k-2)^2(2+k)$ **77.** $(a+b)(a^2-ab+b^2)$
78. $(a-b)(a^2+ab+b^2)$ **79.** $(4+a)(16-4a+a^2)$
80. $(5-b)(25+5b+b^2)$ **81.** $(p^2+2)(p^4-2p^2+4)$
82. $\left(q^2-\dfrac{1}{3}\right)\left(q^4+\dfrac{1}{3}q^2+\dfrac{1}{9}\right)$ **83.** $6(x-2)(x^2+2x+4)$
84. $7(y+1)(y^2-y+1)$ **85.** $x(x-6)(x+6)$
86. $q(q-4)(q^2+4q+16)$ **87.** $4(2h^2+5)$ **88.** $m(m-8)$
89. $(x+4)(x+1)(x-1)$ **90.** $5q(p^2-2q)(p^2+2q)$
91. $n(2+n)(4-2n+n^2)$ **92.** $14(m-1)(m^2+m+1)$
93. $(x-3)(2x+1)=0$ can be solved directly by the zero product rule
because it is a product of factors set equal to zero.
94. $\left\{\dfrac{1}{4},-\dfrac{2}{3}\right\}$ **95.** $\left\{9,\dfrac{1}{2}\right\}$ **96.** $\left\{0,-3,-\dfrac{2}{5}\right\}$
97. $\left\{0,7,\dfrac{9}{4}\right\}$ **98.** $\left\{-\dfrac{5}{7},2\right\}$ **99.** $\left\{-\dfrac{1}{4},6\right\}$
100. $\{12,-12\}$ **101.** $\{5,-5\}$ **102.** $\left\{0,\dfrac{1}{5}\right\}$ **103.** $\{4,2\}$
104. $\left\{-\dfrac{5}{6}\right\}$ **105.** $\left\{-\dfrac{2}{3}\right\}$ **106.** $\left\{\dfrac{2}{3},6\right\}$ **107.** $\left\{\dfrac{11}{2},-12\right\}$
108. $\{0,7,2\}$ **109.** $\{0,2,-2\}$ **110.** The height is 6 ft, and the
base is 13 ft.   **111.** The ball is at ground level at 0 and 1 sec.
112. The ramp is 13 ft long.   **113.** The legs are 6 ft and 8 ft; the
hypotenuse is 10 ft.   **114.** The numbers are -8 and 8.
115. The numbers are 29 and 30, or -2 and -1.
116. The height is 4 m, and the base is 9 m.

Chapter 6 Test, p. 479

1. $3x(5x^3-1+2x^2)$ **2.** $(a-5)(7-a)$
3. $(6w-1)(w-7)$ **4.** $(13-p)(13+p)$
5. $(q-8)^2$ **6.** $(2+t)(4-2t+t^2)$ **7.** $(a+4)(a+8)$
8. $(x+7)(x-6)$ **9.** $(2y-1)(y-8)$ **10.** $(2z+1)(3z+8)$
11. $(3t-10)(3t+10)$ **12.** $(v+9)(v-9)$ **13.** $3(a+6b)(a+3b)$
14. $(c-1)(c+1)(c^2+1)$ **15.** $(y-7)(x+3)$ **16.** Prime
17. $-10(u-2)(u-1)$ **18.** $3(2t-5)(2t+5)$ **19.** $5(y-5)^2$
20. $7q(3q+2)$ **21.** $(2x+1)(x-2)(x+2)$
22. $(y-5)(y^2+5y+25)$ **23.** $(mn-9)(mn+9)$
24. $16(a-2b)(a+2b)$ **25.** $(4x-3y^2)(16x^2+12xy^2+9y^4)$
26. $3y(x-4)(x+2)$ **27.** $\left\{\dfrac{3}{2},-5\right\}$ **28.** $\{0,7\}$ **29.** $\{8,-2\}$
30. $\left\{\dfrac{1}{5},-1\right\}$ **31.** $\{3,-3,-10\}$
32. The tennis court is 12 yd by 26 yd.   **33.** The two integers are 5 and 7,
or -5 and -7.   **34.** The base is 12 in., and the height is 7 in.
35. The shorter leg is 5 ft.   **36.** The stone hits the ground in 2 sec.

Chapters 1–6 Cumulative Review Exercises, p. 480

1. $\dfrac{7}{5}$ **2.** $\{-3\}$ **3.** $y=\dfrac{3}{2}x-4$
4. There are 10 quarters and 7 dimes.
5. $[-4,\infty)$ **6. a.** Yes   **b.** 1
c. $(0,4)$ **d.** $(-4,0)$ **e.**
7. a. Vertical line   **b.** Undefined   **c.** $(5,0)$ **d.** Does not exist
8. $y=3x+14$ **9.** $\{(5,2)\}$ **10.** $-\dfrac{7}{2}y^2-5y-14$
11. $8p^3-22p^2+13p+3$ **12.** $4w^2-28w+49$
13. $r^3+5r^2+15r+40+\dfrac{121}{r-3}$ **14.** c^4 **15.** $\dfrac{a^4}{9b^2}$
16. 1.6×10^3 **17.** $(w-2)(w+2)(w^2+4)$
18. $(a+5b)(2x-3y)$ **19.** $(2x-5)(2x+1)$
20. $\left\{0,\dfrac{1}{2},-5\right\}$

Chapter 7

Section 7.1 Practice Exercises, pp. 489–492

1. a. rational   **b.** denominator; zero   **c.** $\dfrac{p}{q}$ **d.** $1;-1$ **3.** 0
5. $-\dfrac{1}{8}$ **7.** $-\dfrac{1}{2}$ **9.** Undefined   **11. a.** $3\frac{1}{5}$ hr or 3.2 hr
b. $1\frac{3}{4}$ hr or 1.75 hr   **13.** $k=-2$ **15.** $x=\dfrac{5}{2},x=-8$
17. $m=-2,m=-3$ **19.** There are no restricted values.
21. There are no restricted values.   **23.** $t=0$
25. For example: $\dfrac{1}{x-2}$ **27.** For example: $\dfrac{1}{(x+3)(x-7)}$
29. a. $\dfrac{2}{5}$ **b.** $\dfrac{2}{5}$ **31. a.** Undefined   **b.** Undefined
33. a. $y=-2$ **b.** $\dfrac{1}{2}$ **35. a.** $t=-1$ **b.** $t-1$
37. a. $w=0,w=\dfrac{5}{3}$ **b.** $\dfrac{1}{3w-5}$ **39. a.** $x=-\dfrac{2}{3}$ **b.** $\dfrac{3x-2}{2}$
41. a. $a=-3,a=2$ **b.** $\dfrac{a+5}{a+3}$ **43.** $\dfrac{b}{3}$ **45.** $-\dfrac{3xy}{z^2}$
47. $\dfrac{p-3}{p+4}$ **49.** $\dfrac{1}{4(m-11)}$ **51.** $\dfrac{2x+1}{4x^2}$ **53.** $\dfrac{1}{4a-5}$
55. $\dfrac{x-2}{3(y+2)}$ **57.** $\dfrac{2}{x-5}$ **59.** $a+7$ **61.** Cannot simplify
63. $\dfrac{y+3}{2y-5}$ **65.** $\dfrac{5}{(q+1)(q-1)}$ **67.** $\dfrac{c-d}{2c+d}$ **69.** $5x+4$
71. $\dfrac{x+y}{x-4y}$ **73.** They are opposites.   **75.** -1 **77.** -1
79. $-\dfrac{1}{2}$ **81.** Cannot simplify   **83.** $\dfrac{5x-6}{5x+6}$ **85.** $\dfrac{x+3}{4+x}$
87. $\dfrac{3x-2}{x+4}$ **89.** $-\dfrac{1}{2}$ **91.** $-\dfrac{w+2}{4}$ **93.** $\dfrac{3t^2}{2}$ **95.** $-\dfrac{2r-s}{s+2r}$
97. $-\dfrac{3}{2(x-y)}$ **99.** $\dfrac{1}{t(t-5)}$ **101.** $w-2$ **103.** $\dfrac{z+4}{z^2+4z+16}$

Section 7.2 Practice Exercises, pp. 497–499

1. $\dfrac{3}{10}$ **3.** 2   **5.** $\dfrac{5}{2}$ **7.** $\dfrac{15}{4}$ **9.** $\dfrac{3}{2x}$ **11.** $3xy^4$
13. $\dfrac{x-6}{8}$ **15.** $\dfrac{2}{y}$ **17.** $-\dfrac{5}{8}$ **19.** $\dfrac{b+a}{a-b}$ **21.** $\dfrac{y+1}{5}$
23. $\dfrac{2(x+6)}{2x+1}$ **25.** 6   **27.** $\dfrac{m^6}{n^2}$ **29.** $\dfrac{10}{9}$ **31.** $\dfrac{6}{7}$
33. $-m(m+n)$ **35.** $\dfrac{3p+4q}{4(p+2q)}$ **37.** $\dfrac{p}{p-1}$ **39.** $\dfrac{w}{2w-1}$
41. $\dfrac{4r}{2r+3}$ **43.** $\dfrac{5}{6}$ **45.** $\dfrac{1}{4}$ **47.** $\dfrac{y+9}{y-6}$ **49.** 1
51. $\dfrac{3t+8}{t+2}$ **53.** $\dfrac{x+4}{x+1}$ **55.** $-\dfrac{w-3}{2}$ **57.** $\dfrac{k+6}{k+3}$ **59.** $\dfrac{2}{a}$
61. $2y(y+1)$ **63.** $x+y$ **65.** 2   **67.** $\dfrac{1}{a-2}$ **69.** $\dfrac{p+q}{2}$

Section 7.3 Practice Exercises, pp. 503–505

1. multiple; denominators   **3.** $x=1,x=-1;\dfrac{3}{5(x-1)}$
5. $\dfrac{a+5}{a+7}$ **7.** $\dfrac{2}{3y}$ **9.** a, b, c, d   **11.** x^5 is the greatest
power of x that appears in any denominator.   **13.** 45   **15.** 48
17. 63   **19.** $9x^2y^3$ **21.** w^2y **23.** $(p+3)(p-1)(p+2)$
25. $9t(t+1)^2$ **27.** $(y-2)(y+2)(y+3)$ **29.** $3-x$ or $x-3$
31. Because $(b-1)$ and $(1-b)$ are opposites, they differ
by a factor of -1.   **33.** $\dfrac{6}{5x^2};\dfrac{5x}{5x^2}$ **35.** $\dfrac{24x}{30x^3};\dfrac{5y}{30x^3}$

37. $\dfrac{10}{12a^2b}; \dfrac{a^3}{12a^2b}$ **39.** $\dfrac{6m-6}{(m+4)(m-1)}; \dfrac{3m+12}{(m+4)(m-1)}$

41. $\dfrac{6x+18}{(2x-5)(x+3)}; \dfrac{2x-5}{(2x-5)(x+3)}$

43. $\dfrac{6w+6}{(w+3)(w-8)(w+1)}; \dfrac{w^2+3w}{(w+3)(w-8)(w+1)}$

45. $\dfrac{6p^2+12p}{(p-2)(p+2)^2}; \dfrac{3p-6}{(p-2)(p+2)^2}$

47. $\dfrac{1}{a-4}; \dfrac{-a}{a-4}$ or $\dfrac{-1}{4-a}; \dfrac{a}{4-a}$

49. $\dfrac{8}{2(x-7)}; \dfrac{-y}{2(x-7)}$ or $\dfrac{-8}{2(7-x)}; \dfrac{y}{2(7-x)}$

51. $\dfrac{1}{a+b}; \dfrac{-6}{a+b}$ or $\dfrac{-1}{-a-b}; \dfrac{6}{-a-b}$

53. $\dfrac{-9}{24(3y+1)}; \dfrac{20}{24(3y+1)}$ **55.** $\dfrac{3z+12}{5z(z+4)}; \dfrac{5z}{5z(z+4)}$

57. $\dfrac{z^2+3z}{(z+2)(z+7)(z+3)}; \dfrac{-3z^2-6z}{(z+2)(z+7)(z+3)};$
$\dfrac{5z+35}{(z+2)(z+7)(z+3)}$

59. $\dfrac{3p+6}{(p-2)(p^2+2p+4)(p+2)}; \dfrac{p^3+2p^2+4p}{(p-2)(p^2+2p+4)(p+2)};$
$\dfrac{5p^3-20p}{(p-2)(p^2+2p+4)(p+2)}$

Section 7.4 Practice Exercises, pp. 512–514

1. a. $-\dfrac{1}{2}, -2, 0$, undefined, undefined

b. $(x-5)(x-2); x=5, x=2$ **c.** $\dfrac{x+1}{x-2}$ **3.** $\dfrac{2(2x-3)}{(x-3)(x-1)}$

5. $\dfrac{5}{4}$ **7.** $\dfrac{3}{8}$ **9.** 2   **11.** 5   **13.** $\dfrac{-2(t-2)}{t-8}$ **15.** $3x+7$

17. $m+5$ **19.** 2   **21.** $x-5$ **23.** $\dfrac{1}{r+1}$ **25.** $\dfrac{1}{y+7}$

27. $\dfrac{15x}{y}$ **29.** $\dfrac{5a+6}{4a}$ **31.** $\dfrac{2(6+x^2y)}{15xy^3}$ **33.** $\dfrac{2s-3t^2}{s^4t^3}$

35. $-\dfrac{2}{3}$ **37.** $\dfrac{19}{3(a+1)}$ **39.** $\dfrac{-3(k+4)}{(k-3)(k+3)}$ **41.** $\dfrac{a-4}{2a}$

43. $\dfrac{(x+6)(x-2)}{(x-4)(x+1)}$ **45.** $\dfrac{x(4x+9)}{(x+3)^2(x+2)}$

47. $\dfrac{5p-1}{3}$ or $\dfrac{-5p+1}{-3}$ **49.** $\dfrac{9}{x-3}$ or $\dfrac{-9}{3-x}$

51. $\dfrac{6n-1}{n-8}$ or $\dfrac{-6n+1}{8-n}$ **53.** $\dfrac{2(4x+5)}{x(x+2)}$ **55.** $\dfrac{3y}{2(2y+1)}$

57. $\dfrac{2(w-3)}{(w+3)(w-1)}$ **59.** $\dfrac{4a-13}{(a-3)(a-4)}$ **61.** $\dfrac{3x(x+7)}{(x+5)^2(x-1)}$

63. $\dfrac{-y(y+8)}{(2y+1)(y-1)(y-4)}$ **65.** $\dfrac{1}{2p+1}$ **67.** $\dfrac{-2mn+1}{(m+n)(m-n)}$

69. 0   **71.** $\dfrac{2(3x+7)}{(x+3)(x+2)}$ **73.** $\dfrac{1}{n}$ **75.** $\dfrac{5}{n+2}$

77. $n+\left(7\cdot\dfrac{1}{n}\right); \dfrac{n^2+7}{n}$ **79.** $\dfrac{1}{n}-\dfrac{2}{n}; -\dfrac{1}{n}$

81. $\dfrac{-w^2}{(w+3)(w-3)(w^2-3w+9)}$ **83.** $\dfrac{p^2-2p+7}{(p+2)(p+3)(p-1)}$

85. $\dfrac{-m-21}{2(m+5)(m-2)}$ or $\dfrac{m+21}{2(m+5)(2-m)}$ **87.** $\dfrac{3k+5}{4k+7}$ **89.** $\dfrac{1}{a}$

Chapter 7 Problem Recognition Exercises, p. 515

1. $\dfrac{-2x+9}{3x+1}$ **2.** $\dfrac{1}{w-4}$ **3.** $\dfrac{y-5}{2y-3}$ **4.** $\dfrac{7}{(x+3)(2x-1)}$

5. $-\dfrac{1}{x}$ **6.** $\dfrac{1}{3}$ **7.** $\dfrac{c+3}{c}$ **8.** $\dfrac{x+3}{5}$ **9.** $\dfrac{a}{12b^4c}$

10. $\dfrac{2a-b}{a-b}$ **11.** $\dfrac{p-q}{5}$ **12.** 4   **13.** $\dfrac{10}{2x+1}$ **14.** $\dfrac{w+2z}{w+z}$

15. $\dfrac{3}{2x+5}$ **16.** $\dfrac{y+7}{x+a}$ **17.** $\dfrac{1}{2(a+3)}$

18. $\dfrac{2(3y+10)}{(y-6)(y+6)(y+2)}$ **19.** $(t+8)^2$ **20.** $6b+5$

Section 7.5 Practice Exercises, pp. 521–523

1. complex   **3.** $\dfrac{1}{2a-3}$ **5.** $\dfrac{3(2k-5)}{5(k-2)}$ **7.** $\dfrac{7}{4y}$ **9.** $\dfrac{1}{2y}$

11. $\dfrac{24b}{a^3}$ **13.** $\dfrac{2r^5t^4}{s^6}$ **15.** $\dfrac{35}{2}$ **17.** $k+h$ **19.** $\dfrac{n+1}{2(n-3)}$

21. $\dfrac{2x+1}{4x+1}$ **23.** $m-7$ **25.** $\dfrac{2y(y-5)}{7y^2+10}$ **27.** $-\dfrac{a+8}{a-2}$ or $\dfrac{a+8}{2-a}$

29. $\dfrac{t-2}{t-4}$ **31.** $\dfrac{t+3}{t-5}$ **33.** $\dfrac{1}{2}$ **35.** $\dfrac{\frac{1}{2}+\frac{2}{3}}{5}; \dfrac{7}{30}$

37. $\dfrac{3}{\frac{2}{3}+\frac{3}{4}}; \dfrac{36}{17}$ **39. a.** $\dfrac{6}{5}\,\Omega$ **b.** $6\,\Omega$ **41.** $\dfrac{xy}{y+x}$ **43.** $\dfrac{y+4x}{2y}$

45. $\dfrac{1}{n^2+m^2}$ **47.** $\dfrac{2z-5}{3(z+3)}$ **49.** $-\dfrac{x+1}{x-1}$ or $\dfrac{x+1}{1-x}$ **51.** $\dfrac{3}{2}$

Section 7.6 Practice Exercises, pp. 531–533

1. a. linear; quadratic   **b.** rational   **c.** denominator   **3.** $\dfrac{2}{4x-1}$

5. $5(h+1)$ **7.** $\dfrac{(x+4)(x-3)}{x^2}$ **9.** $\{3\}$ **11.** $\left\{\dfrac{5}{11}\right\}$

13. $\left\{\dfrac{1}{3}\right\}$ **15. a.** $z=0$ **b.** $5z$ **c.** $\{5\}$ **17.** $\left\{-\dfrac{200}{19}\right\}$

19. $\{-7\}$ **21.** $\left\{\dfrac{47}{6}\right\}$ **23.** $\{3,-1\}$ **25.** $\{4\}$

27. $\{5\}$ (The value 0 does not check.)   **29.** $\{-5\}$

31. $\{\ \}$ (The value 4 does not check.)   **33.** $\{4\}$ **35.** $\{4,-3\}$

37. $\{-4\}$; (The value 1 does not check.)   **39.** $\{\ \}$ (The value -4 does not check.)   **41.** $\{4\}$ (The value -6 does not check.)

43. $\{-25\}$ **45.** $\{-2,7\}$ **47.** The number is 8.

49. The number is -26.   **51.** $m=\dfrac{FK}{a}$ **53.** $E=\dfrac{IR}{K}$

55. $R=\dfrac{E-Ir}{I}$ or $R=\dfrac{E}{I}-r$

57. $B=\dfrac{2A-hb}{h}$ or $B=\dfrac{2A}{h}-b$ **59.** $h=\dfrac{V}{r^2\pi}$

61. $t=\dfrac{b}{x-a}$ or $t=\dfrac{-b}{a-x}$ **63.** $x=\dfrac{y}{1-yz}$ or $x=\dfrac{-y}{yz-1}$

65. $h=\dfrac{2A}{a+b}$ **67.** $R=\dfrac{R_1R_2}{R_2+R_1}$

Chapter 7 Problem Recognition Exercises, p. 534

1. $\dfrac{y-2}{2y}$ **2.** $\{6\}$ **3.** $\{2\}$ **4.** $\dfrac{3a-17}{a-5}$ **5.** $\dfrac{4p+27}{18p^2}$

6. $\dfrac{b(b-5)}{(b-1)(b+1)}$ **7.** $\{5\}$ **8.** $\dfrac{2w+5}{(w+1)^2}$ **9.** $\{7\}$ **10.** $\{5\}$

11. $\dfrac{3x+14}{4(x+1)}$ **12.** $\left\{\dfrac{11}{3}\right\}$ **13.** $\left\{\dfrac{41}{10}\right\}$ **14.** $\dfrac{7-3t}{t(t-5)}$

15. $\dfrac{8a+1}{2a-1}$ **16.** $\{5\}$ (The value of 2 does not check.)

17. $\{-1\}$ (The value of 3 does not check.)   **18.** $\dfrac{11}{12k}$

19. $\dfrac{h+9}{(h-3)(h+3)}$ **20.** $\{7\}$

Section 7.7 Practice Exercises, pp. 542–546

1. a. proportion **b.** proportional

3. Expression; $\dfrac{m^2 + m + 2}{(m-1)(m+3)}$ **5.** Expression; $\dfrac{3}{10}$

7. Equation; $\{2\}$ **9.** $\{95\}$ **11.** $\{1\}$ **13.** $\left\{\dfrac{40}{3}\right\}$ **15.** $\{40\}$

17. $\{3\}$ **19.** $\{-1\}$ **21.** $\{1\}$ **23. a.** $V_f = \dfrac{V_i T_f}{T_i}$ **b.** $T_f = \dfrac{T_i V_f}{V_i}$

25. Toni can drive 297 mi on 9 gal of gas. **27.** Mix 14.82 mL of plant food. **29.** 5 oz contains 12 g of carbohydrate.

31. The minimum length is 20 ft. **33.** $x = 4$ cm; $y = 5$ cm

35. $x = 3.75$ cm; $y = 4.5$ cm **37.** The height of the pole is 7 m.

39. The light post is 24 ft high. **41.** The speed of the boat is 20 mph. **43.** The plane flies 210 mph in still air. **45.** He runs 8 mph and bikes 16 mph. **47.** Floyd walks 4 mph and Rachel walks 2 mph. **49.** Sergio rode 12 mph and walked 3 mph.

51. $5\frac{5}{11}$ ($5.\overline{45}$) min **53.** $22\frac{2}{9}$ ($22.\overline{2}$) min **55.** 48 hr

57. $3\frac{1}{3}$ ($3.\overline{3}$) days **59.** There are 40 smokers and 140 nonsmokers.

61. There are 240 men and 200 women.

Section 7.8 Practice Exercises, pp. 551–555

1. a. kx **b.** $\dfrac{k}{x}$ **c.** kxw **3.** $\{12\}$ **5.** $\{6, 2\}$ **7.** $\dfrac{b+1}{1-b}$

9. Inversely **11.** $T = kq$ **13.** $b = \dfrac{k}{c}$ **15.** $Q = \dfrac{kx}{y}$

17. $c = kst$ **19.** $L = kw\sqrt{v}$ **21.** $x = \dfrac{ky^2}{z}$ **23.** $k = \dfrac{9}{2}$

25. $k = 512$ **27.** $k = 1.75$ **29.** $x = 70$ **31.** $b = 6$

33. $Z = 56$ **35.** $Q = 9$ **37.** $L = 9$ **39.** $B = \dfrac{15}{2}$

41. a. The heart weighs 0.92 lb. **b.** Answers will vary.

43. a. 3.6 g **b.** 4.5 g **c.** 2.4 g **45. a.** \$0.40 **b.** \$0.30

c. \$1.00 **47.** 355,000 tons **49.** 42.6 ft **51.** 300 W

53. 18.5 A **55.** 20 lb **57.** \$3500

Chapter 7 Review Exercises pp. 561–564

1. a. $-\dfrac{2}{9}, -\dfrac{1}{10}, 0, -\dfrac{5}{6}$, undefined **b.** $t = -9$

2. a. $-\dfrac{1}{5}, -\dfrac{1}{2}$, undefined, $0, \dfrac{1}{7}$ **b.** $k = 5$ **3.** a, c, d

4. $x = \dfrac{5}{2}, x = 3; \dfrac{1}{2x-5}$ **5.** $h = -\dfrac{1}{3}, h = -7; \dfrac{1}{3h+1}$

6. $a = 2, a = -2; \dfrac{4a-1}{a-2}$ **7.** $w = 4, w = -4; \dfrac{2w+3}{w-4}$

8. $z = 4; -\dfrac{z}{2}$ **9.** $k = 0, k = 5; -\dfrac{3}{2k}$ **10.** $b = -3; \dfrac{b-1}{2}$

11. $m = -1; \dfrac{m-5}{3}$ **12.** $n = -3; \dfrac{1}{n+3}$ **13.** $p = -7; \dfrac{1}{p+7}$

14. y^2 **15.** $\dfrac{u^2}{2}$ **16.** $v + 2$ **17.** $\dfrac{3}{2(x-5)}$ **18.** $\dfrac{c(c+1)}{2(c+5)}$

19. $\dfrac{q-2}{4}$ **20.** $-2t(t-5)$ **21.** $4s(s-4)$ **22.** $\dfrac{1}{7}$

23. $\dfrac{1}{n-2}$ **24.** $-\dfrac{1}{6}$ **25.** $\dfrac{1}{m+3}$ **26.** $\dfrac{-1}{(x+3)(x+2)}$

27. $\dfrac{2y-1}{y+1}$ **28.** LCD $= (n+3)(n-3)(n+2)$

29. LCD $= (m+4)(m-4)(m+3)$ **30.** $c - 2$ or $2 - c$

31. $3 - x$ or $x - 3$ **32.** $\dfrac{4b}{10ab}; \dfrac{3a}{10ab}$ **33.** $\dfrac{21y}{12xy}; \dfrac{22x}{12xy}$

34. $\dfrac{y}{x^2y^5}; \dfrac{3x}{x^2y^5}$ **35.** $\dfrac{5c^2}{ab^3c^2}; \dfrac{3b^3}{ab^3c^2}$

36. $\dfrac{5p-20}{(p+2)(p-4)}; \dfrac{p^2+2p}{(p+2)(p-4)}$ **37.** $\dfrac{6q+48}{q(q+8)}; \dfrac{q}{q(q+8)}$

38. 2 **39.** 2 **40.** $a + 5$ **41.** $x - 7$

42. $\dfrac{-y-18}{(y-9)(y+9)}$ or $\dfrac{y+18}{(9-y)(y+9)}$ **43.** $\dfrac{t^2+2t+3}{(2-t)(2+t)}$

44. $\dfrac{m+8}{3m(m+2)}$ **45.** $\dfrac{3(r-4)}{2r(r+6)}$ **46.** $\dfrac{p}{(p+4)(p+5)}$

47. $\dfrac{q}{(q+5)(q+4)}$ **48.** $\dfrac{1}{2}$ **49.** $\dfrac{1}{3}$ **50.** $\dfrac{a-4}{a-2}$

51. $\dfrac{3(z+5)}{z(z-5)}$ **52.** $\dfrac{w}{2}$ **53.** $\dfrac{8}{y}$ **54.** $y - x$ **55.** $-(b+a)$

56. $-\dfrac{2p+7}{2p}$ **57.** $\dfrac{k+10}{k+4}$ **58.** $\{-8\}$ **59.** $\{-2\}$

60. $\{0\}$ **61.** $\{2\}$ **62.** $\{-2\}$ **63.** $\{-1\}$ (The value 3 does not check.) **64.** $\{\ \}$ (The value 2 does not check.) **65.** $\{-11, 1\}$

66. The number is 4. **67.** $h = \dfrac{3V}{\pi r^2}$ **68.** $b = \dfrac{2A}{h}$

69. $\left\{\dfrac{6}{5}\right\}$ **70.** $\left\{\dfrac{96}{5}\right\}$ **71.** It contains 10 g of fat.

72. Ed travels 60 mph, and Bud travels 70 mph.

73. Together the pumps would fill the pool in 16.8 min.

74. $x = 11; b = 26$ **75. a.** $F = kd$ **b.** $k = 3$ **c.** 12.6 lb

76. $y = \dfrac{8}{3}$ **77.** $y = 12$ **78.** 48 km

Chapter 7 Test, pp. 564–565

1. a. $x = 2$ **b.** $-\dfrac{x+1}{6}$ **2. a.** $a = 6, a = -2, a = 0$

b. $\dfrac{7}{a+2}$ **3.** b, c, d **4. a.** $15(x+3)$ **b.** $3x^2y^2$

5. $\dfrac{y+7}{3(y+3)(y+1)}$ **6.** $-\dfrac{b+3}{5}$ **7.** $\dfrac{1}{w+1}$ **8.** $\dfrac{t+4}{t+2}$

9. $\dfrac{x(x+5)}{(x+4)(x-2)}$ **10.** $\dfrac{2y+7}{y-6}$ or $\dfrac{-2y-7}{6-y}$ **11.** $\dfrac{1}{m+4}$

12. $\left\{\dfrac{8}{5}\right\}$ **13.** $\{2\}$ **14.** $\{1\}$ **15.** $\{\ \}$ (The value 4 does not check.) **16.** $\{-5\}$ (The value 2 does not check.)

17. $r = \dfrac{2A}{C}$ **18.** $\{-8\}$ **19.** $1\frac{1}{4}$ (1.25) cups of carrots

20. The speed of the current is 5 mph. **21.** It would take the second printer 3 hr to do the job working alone.

22. $a = 5.6; b = 12$ **23.** 8.25 mL **24.** 200 drinks are sold.

Chapters 1–7 Cumulative Review Exercises, pp. 565–566

1. 32 **2.** 7 **3.** $\left\{\dfrac{10}{9}\right\}$

4.

Set-Builder Notation	Graph	Interval Notation
$\{x \mid x \geq -1\}$	$\xrightarrow{\quad}$ -1	$[-1, \infty)$
$\{x \mid x < 5\}$	$\xleftarrow{\quad}$ 5	$(-\infty, 5)$

5. The width is 17 m and the length is 35 m.

6. The base is 10 in. and the height is 8 in. **7.** $\dfrac{x^2yz^{17}}{2}$

8. a. $6x + 4$ **b.** $2x^2 + x - 3$ **9.** $(5x-3)^2$

10. $(2c+1)(5d-3)$ **11.** $x = 5, x = -\dfrac{1}{2}$ **12.** $\{(1, -4)\}$

13. $\dfrac{1}{5(x+4)}$ **14.** -3 **15.** $\{1\}$ **16.** $\left\{-\dfrac{7}{2}\right\}$

17. a. x-intercept: $(-4, 0)$; y-intercept: $(0, 2)$
b. x-intercept: $(0, 0)$; y-intercept: $(0, 0)$

18. a. $m = -\dfrac{7}{5}$ **b.** $m = -\dfrac{2}{3}$ **c.** $m = 4$ **d.** $m = -\dfrac{1}{4}$

19. $y = 5x - 3$ **20.** One large popcorn costs \$4.50, and one drink costs \$3.00.

Chapter 8

Section 8.1 Calculator Connections, pp. 574–575

1. 2.236   **2.** 4.123   **3.** 7.071   **4.** 9.798
5. 5.745   **6.** 12.042   **7.** 8.944   **8.** 13.038
9. 1.913   **10.** 3.037   **11.** 4.021   **12.** 4.987

Section 8.1 Practice Exercises, pp. 575–579

1. a. b; a **b.** principal   **c.** rational   **d.** b^n; a **e.** index; radicand
f. cube   **g.** is not; is   **h.** even; odd   **i.** $a^2 + b^2 = c^2$ **3.** 12, −12
5. There are no real-valued square roots of −49.   **7.** 0

9. $\dfrac{1}{5}, -\dfrac{1}{5}$ **11. a.** 13   **b.** −13   **13.** 0

15. 9, 16, 25, 36, 64, 121, 169   **17.** 2   **19.** 7   **21.** 0.4

23. 0.3   **25.** $\dfrac{5}{4}$ **27.** $\dfrac{1}{12}$ **29.** 5   **31.** 9

33. There is no real value of b for which $b^2 = -16$.   **35.** −2
37. Not a real number   **39.** Not a real number
41. Not a real number   **43.** −20   **45.** Not a real number
47. 0, 1, 27, 125   **49.** Yes, −3   **51.** 3   **53.** 4   **55.** −2

57. Not a real number   **59.** Not a real number   **61.** $-\dfrac{1}{2}$

63. −1   **65.** 0   **67.** $x^2, y^4, (ab)^6, w^8x^8, m^{10}$ The expression is a
perfect square if the exponent is even.   **69.** 4   **71.** 5   **73.** y^6
75. a^4b^{15} **77.** q^8 **79.** $2w^2$ **81.** $5x$ **83.** $-5x$ **85.** $5p^2$

87. $5p^2$ **89.** $\sqrt{q} + p^2$ **91.** $\dfrac{6}{\sqrt[3]{x}}$ **93.** 9 cm   **95.** 5 ft

97. 6.9 cm   **99.** 17.0 in.   **101.** 31.3 in.   **103.** 268 km
105. $x \geq 0$ **107.** $a \geq b$

Section 8.2 Calculator Connections, pp. 584–585

1.
```
√(125)
        11.18033989
5*√(5)
        11.18033989
```

2.
```
√(18)
        4.242640687
3*√(2)
        4.242640687
```

3.
```
³√(54)
        3.77976315
3*³√(2)
        3.77976315
```

4.
```
³√(108)
        4.762203156
3*³√(4)
        4.762203156
```

Section 8.2 Practice Exercises, pp. 585–587

1. a. $\sqrt[n]{a}$ **b.** The exponent within the radicand is not less than the
index.   **c.** No. $\sqrt{2}$ is an irrational number; therefore its decimal form is a
nonterminating, nonrepeating decimal.   **3.** 8, 27, y^3, y^9, y^{12}, y^{27}
5. −5   **7.** −3   **9.** a^4 **11.** $2xy^2$ **13.** 446 km
15. $3\sqrt{2}$ **17.** $2\sqrt{7}$ **19.** $12\sqrt{5}$ **21.** $-10\sqrt{2}$
23. $a^2\sqrt{a}$ **25.** w^{11} **27.** $m^2n^2\sqrt{n}$ **29.** $x^7y^5\sqrt{x}$ **31.** $3t^5$
33. $2x\sqrt{2x}$ **35.** $4z\sqrt{z}$ **37.** $-3w^3\sqrt{5}$ **39.** $z^{12}\sqrt{z}$
41. $-z^5\sqrt{15z}$ **43.** $10ab^3\sqrt{26b}$ **45.** $\sqrt{26pq}$ **47.** m^6n^8

49. $-4ab^2c^2\sqrt{3ab}$ **51.** a^4 **53.** y^5 **55.** $\dfrac{1}{2}$ **57.** 2   **59.** $2x$

61. $5p^3$ **63.** $3\sqrt{5}$ **65.** $\sqrt{6}$ **67.** 4   **69.** 7   **71.** $11\sqrt{2}$ ft
73. $2\sqrt{66}$ cm   **75.** $a^2\sqrt[3]{a^2}$ **77.** $14z\sqrt[3]{2}$ **79.** $2ab^2\sqrt[3]{2a^2}$

81. z **83.** −2   **85.** $-2\sqrt[3]{5}$ **87.** $\dfrac{1}{3}$ **89.** $4a\sqrt{a}$ **91.** $2x$

93. $2p\sqrt{2q}$ **95.** $-4\sqrt{2}$ **97.** $2u^2v^3\sqrt{13v}$ **99.** $6\sqrt{6}$

101. 6   **103.** $2a\sqrt[3]{2}$ **105.** x **107.** $-\sqrt{5}$ **109.** $-\dfrac{1}{2}$

111. $5\sqrt{2}$ **113.** $x + 5$

Section 8.3 Practice Exercises, pp. 590–592

1. index, radicand   **3.** $2y$ **5.** $6x\sqrt{x}$ **7.** $2x$ **9.** Not a
real number   **11.** For example, $2\sqrt{3}, 6\sqrt[3]{3}$ **13.** c   **15.** $8\sqrt{2}$

17. $4\sqrt{7}$ **19.** $2\sqrt[3]{10}$ **21.** $11\sqrt{y}$ **23.** 0   **25.** $5y\sqrt{15}$
27. $x\sqrt{y} - y\sqrt{x}$ **29.** $8\sqrt{3}$ **31.** 0   **33.** $2\sqrt{2}$ **35.** $16p^2\sqrt{5}$

37. $10\sqrt{2k}$ **39.** $a^2\sqrt{b}$ **41.** $3\sqrt{5}$ **43.** $\dfrac{29}{18}z\sqrt{6}$

45. $-1.7\sqrt{10}$ **47.** $2x\sqrt{x}$ **49.** $3\sqrt{7}$
51. $4\sqrt{w} + 2\sqrt{6w} + 2\sqrt{10w}$ **53.** $6x^3\sqrt{y}$ **55.** $2\sqrt{3} - 4\sqrt{6}$
57. $-4x\sqrt{2} + \sqrt{2x}$ **59.** $9\sqrt{2}$ m   **61.** $16\sqrt{3}$ in.
63. Radicands are not the same.   **65.** One term has a radical. One
does not.   **67.** The indices are different.

69. $\dfrac{\sqrt{3}}{3}$ **71. a.** 80 m   **b.** 159 m

Section 8.4 Practice Exercises, pp. 597–599

1. a. $\sqrt[n]{ab}$ **b.** a **c.** conjugates   **3.** 11   **5.** $3w^2\sqrt{z}$
7. $\sqrt{15}$ **9.** 47   **11.** b **13.** $6\sqrt{15p}$ **15.** $5\sqrt{2}$
17. $14\sqrt{2}$ **19.** $6x\sqrt{7}$ **21.** $4x^3\sqrt{y}$ **23.** $12w^2\sqrt{10}$
25. $-8\sqrt{15}$ **27.** Perimeter: $6\sqrt{5}$ ft; area: 10 ft^2 **29.** 3 cm^2
31. $3w$ **33.** $-16\sqrt{10y}$ **35.** $2\sqrt{3} - \sqrt{6}$ **37.** $4x + 20\sqrt{x}$
39. $-8 + 7\sqrt{30}$ **41.** $9a - 28b\sqrt{a} + 3b^2$
43. $8p^2 + 19p\sqrt{p} + 2p - 8\sqrt{p}$ **45.** 10   **47.** 4   **49.** t
51. $16c$ **53.** $29 + 8\sqrt{13}$ **55.** $a - 4\sqrt{a} + 4$
57. $4a - 12\sqrt{a} + 9$ **59.** $21 - 2\sqrt{110}$ **61.** 1
63. $x - y$ **65.** −1   **67.** $36m - 25n$ **69.** $64x - 4y$
71. 73   **73. a.** $3x + 6$ **b.** $\sqrt{3x} + \sqrt{6}$ **75. a.** $4a^2 + 12a + 9$
b. $4a + 12\sqrt{a} + 9$ **77. a.** $b^2 - 25$ **b.** $b - 25$
79. a. $x^2 - 4xy + 4y^2$ **b.** $x - 4\sqrt{xy} + 4y$ **81. a.** $p^2 - q^2$
b. $p - q$ **83. a.** $y^2 - 6y + 9$ **b.** $x - 6\sqrt{x - 2} + 7$

Section 8.5 Practice Exercises, pp. 606–609

1. a. index   **b.** denominator   **c.** rationalizing   **d.** $\dfrac{\sqrt[n]{a}}{\sqrt[n]{b}}$
e. denominator   **3.** $6y + 23\sqrt{y} + 21$ **5.** $9\sqrt{3}$

7. $25 - 10\sqrt{a} + a$ **9.** −5   **11.** $\dfrac{\sqrt{3}}{4}$ **13.** $\dfrac{a^2}{b^2}$ **15.** $\dfrac{c\sqrt{c}}{2}$

17. $\dfrac{\sqrt[3]{x^2}}{3}$ **19.** $\dfrac{\sqrt[3]{y^2}}{3}$ **21.** $\dfrac{10\sqrt{2}}{9}$ **23.** $\dfrac{2}{5}$ **25.** $\dfrac{1}{2p}$

27. z **29.** $2\sqrt[3]{x}$ **31.** $\dfrac{\sqrt{6}}{6}$ **33.** $3\sqrt{5}$ **35.** $\dfrac{6\sqrt{x + 1}}{x + 1}$

37. $\dfrac{\sqrt{6x}}{x}$ **39.** $\dfrac{\sqrt{21}}{7}$ **41.** $\dfrac{5\sqrt{6y}}{3y}$ **43.** $\dfrac{3\sqrt{6}}{4}$ **45.** $\dfrac{\sqrt{3p}}{9}$

47. $\dfrac{\sqrt{5}}{2}$ **49.** $\dfrac{x\sqrt{y}}{y^2}$ **51.** −7   **53.** $\sqrt{5} + \sqrt{3}$; 2

55. $\sqrt{x} - 10$; $x - 100$ **57.** $\dfrac{4\sqrt{2} - 12}{-7}$ or $\dfrac{12 - 4\sqrt{2}}{7}$

59. $\dfrac{\sqrt{5} + \sqrt{2}}{3}$ **61.** $\sqrt{6} - \sqrt{2}$ **63.** $\dfrac{\sqrt{x} + \sqrt{3}}{x - 3}$

65. $7 - 4\sqrt{3}$ **67.** $-13 - 6\sqrt{5}$ **69.** $2 - \sqrt{2}$

71. $\dfrac{3 + \sqrt{2}}{2}$ **73.** $1 - \sqrt{7}$ **75.** $\dfrac{7 + 3\sqrt{2}}{3}$

77. a. Condition 1 fails; $2x^4\sqrt{2x}$ **b.** Condition 2 fails; $\dfrac{\sqrt{5x}}{x}$

c. Condition 3 fails; $\dfrac{\sqrt{3}}{3}$ **79. a.** Condition 2 fails; $\dfrac{3\sqrt{x} - 3}{x - 1}$

b. Conditions 1 and 3 fail; $\dfrac{3w\sqrt{t}}{t}$ **c.** Condition 1 fails; $2a^2b^4\sqrt{6ab}$

81. $3\sqrt{5}$ **83.** $-\dfrac{3w\sqrt{2}}{5}$ **85.** Not a real number   **87.** $\dfrac{s\sqrt{t}}{t}$

89. $\dfrac{m^2}{2}$ **91.** $\dfrac{9\sqrt{t}}{t^2}$ **93.** $\dfrac{\sqrt{11} - \sqrt{5}}{2}$ **95.** $\dfrac{a + 2\sqrt{ab} + b}{a - b}$

97. $-\dfrac{3\sqrt{2}}{8}$ **99.** $\dfrac{\sqrt{3}}{9}$

Chapter 8 Problem Recognition Exercises, p. 609

1. a. $3\sqrt{2}$ **b.** Cannot be simplified further. **c.** $\sqrt{2}$ **2. a.** $\sqrt{7}$
b. $2\sqrt{7}$ **c.** Cannot be simplified further. **3. a.** $9z$
b. $9 + 6\sqrt{z} + z$ **c.** $9 - z$ **4. a.** $16 - 8\sqrt{x} + x$ **b.** $16 - x$ **c.** $16x$
5. a. $\dfrac{6\sqrt{2x}}{x}$ **b.** $\dfrac{\sqrt{6x}}{x}$ **c.** $\dfrac{12\sqrt{2} - 12x}{2 - x^2}$ **6. a.** $\dfrac{45 + 15\sqrt{y}}{9 - y}$
b. $\dfrac{5\sqrt{3y}}{y}$ **c.** $\dfrac{\sqrt{5y}}{y}$ **7. a.** $3\sqrt{5} - 1$ **b.** $8 - 3\sqrt{5}$ **c.** $10 - 4\sqrt{5}$
8. a. $-8 + 11\sqrt{3}$ **b.** $12 + 16\sqrt{3}$ **c.** $3\sqrt{3} - 9$ **9. a.** $4a^7\sqrt{a}$
b. $2a^5\sqrt[3]{2}$ **10. a.** $3y^3$ **b.** $3y^4\sqrt[3]{3y}$

Section 8.6 Practice Exercises, pp. 615–617

1. a. radical **b.** extraneous **c.** Isolate the radical by adding 3 to both sides of the equation. **d.** third
3. $\dfrac{\sqrt{2} - \sqrt{10}}{-8}$ or $\dfrac{\sqrt{10} - \sqrt{2}}{8}$ **5.** $\dfrac{2\sqrt{6}}{3}$ **7.** $x^2 + 8x + 16$
9. $x + 8\sqrt{x} + 16$ **11.** $2x - 3$ **13.** $t^2 + 2t + 1$ **15.** $\{36\}$
17. $\{15\}$ **19.** $\{\ \}$ (The value 29 does not check.) **21.** $\{5\}$
23. $\left\{-\dfrac{1}{2}\right\}$ **25.** $\{6\}$ **27.** $\{8\}$
29. $\{\ \}$ (The value $\frac{19}{2}$ does not check.) **31.** $\{1\}$ **33.** $\{4, -3\}$ **35.** $\{0\}$
37. $\{\ \}$ (The value -4 does not check.)
39. $\{4\}$ (The value -1 does not check.) **41.** $\{0, -1\}$
43. $\{12\}$ (The value 4 does not check.) **45.** $\{-6\}$ **47.** $\{-1\}$
49. $\sqrt{x + 10} = 1$; -9 **51.** $\sqrt{2x} = x - 4$; 8 **53.** $\sqrt[3]{x+1} = 2$; 7
55. a. 80 ft/sec **b.** 289 ft **57. a.** 16 in. **b.** 25 weeks
59. $\left\{\dfrac{9}{5}\right\}$ **61.** $\left\{\dfrac{3}{2}\right\}$ (The value -1 does not check.)

Section 8.7 Practice Exercises, pp. 622–624

1. a. $\sqrt[n]{a}$ **b.** $\sqrt[n]{a^m}$ or $(\sqrt[n]{a})^m$ **3.** 27 **5.** $a + 1$ **7.** 9
9. 5 **11.** 3 **13.** -2 **15.** -2 **17.** $\dfrac{1}{6}$ **19.** $\sqrt[3]{x}$
21. $\sqrt{4a}$ or $2\sqrt{a}$ **23.** $\sqrt[5]{yz}$ **25.** $\sqrt[3]{u^2}$ **27.** $5\sqrt{q}$
29. $\sqrt{\dfrac{x}{9}}$ or $\dfrac{\sqrt{x}}{3}$ **31.** $a^{m/n} = \sqrt[n]{a^m}$ or $(\sqrt[n]{a})^m$, provided the
roots exist. **33.** 8 **35.** $\dfrac{1}{9}$ **37.** -32 **39.** 2 **41.** $(\sqrt{y})^9$
43. $\sqrt[3]{c^5 d}$ **45.** $\dfrac{1}{\sqrt[5]{qr}}$ **47.** $6\sqrt[3]{y^2}$ **49.** $y^{2/3}$ **51.** $5x^{1/2}$
53. $(xy)^{1/3}$ **55.** $(m^3 n)^{1/4}$ **57.** x **59.** y^2 **61.** 6 **63.** a^7
65. $y^{4/3}$ **67.** 2 **69.** $\dfrac{y^{1/6}}{x}$ **71.** $\dfrac{w^3}{z^6}$ **73.** $\dfrac{25a^4 d}{c}$ **75.** $\dfrac{y^9}{x^8}$
77. $\dfrac{2z^3}{w}$ **79.** $5xy^2 z^{3/2}$ **81. a.** 10 in. **b.** 8.49 in.
83. a. 10.9% **b.** 8.8% **c.** The account in part (a)
85. No, for example, $(36 + 64)^{1/2} \neq 36^{1/2} + 64^{1/2}$
87. 6 **89.** $\dfrac{5}{14}$ **91.** $\dfrac{a^{22} b^4}{c^{17}}$

Chapter 8 Review Exercises, pp. 629–631

1. Principal square root: 14; negative square root: -14
2. Principal square root: 1.2; negative square root: -1.2
3. Principal square root: 0.8; negative square root: -0.8
4. Principal square root: 15; negative square root: -15
5. There is no real number b such that $b^2 = -64$.
6. $\sqrt[3]{-64} = -4$ because $(-4)^3 = -64$. **7.** -12 **8.** -5
9. Not a real number **10.** Not a real number **11.** y **12.** a
13. $2p$ **14.** -5 **15.** -5 **16.** p^4 **17.** $\dfrac{4}{t^2}$ **18.** $-\dfrac{3}{w}$
19. a. 7.1 m **b.** 22.6 ft **20. a.** 65.8 ft **b.** 131.6 ft
21. $b^2 + \sqrt{5}$ **22.** $\sqrt[3]{y} - \sqrt[4]{x}$ **23.** The quotient of 2 and the
principal square root of p **24.** The product of 8 and the principal
square root of q **25.** 12 ft **26.** 331 mi
27. $x^8\sqrt{x}$ **28.** $2\sqrt[3]{5}$ **29.** $2\sqrt{7}$ **30.** $15x\sqrt{2x}$ **31.** $3y^3\sqrt[3]{y}$
32. $6y^5\sqrt{3}$ **33.** c **34.** t^3 **35.** $10y^2$ **36.** $3x$ **37.** $2x$
38. $4a^5$ **39.** $5\sqrt{3}$ **40.** $\sqrt{5}$ **41.** 1 **42.** 6 **43.** $7\sqrt{6}$

44. $0.8\sqrt{y}$ **45.** $-4x\sqrt{5}$ **46.** $11y\sqrt{y}$ **47.** $15\sqrt{3} - 7\sqrt{7}$
48. $4\sqrt{2} - 8\sqrt{5}$ **49.** $-8x^4\sqrt{3x}$ **50.** $21a^2 b\sqrt{2b}$
51. $12\sqrt{2}$ ft **52.** $48\sqrt{3}$ m **53.** 25 **54.** $2\sqrt{15p}$
55. $70\sqrt{3x}$ **56.** $-6yz\sqrt{11}$ **57.** $8m + 24\sqrt{m}$
58. $\sqrt{14} + 8\sqrt{2}$ **59.** $-49 - 16\sqrt{26}$ **60.** $4p + 7\sqrt{pq} - 2q$
61. $64w - z$ **62.** $4x^2 - 4x\sqrt{y} + y$ **63.** $10\sqrt{3}$ m³ **64.** x
65. a^5 **66.** $5\sqrt{c}$ **67.** $4\sqrt{y}$ **68.** b **69.** b **70.** $\dfrac{11\sqrt{7}}{7}$
71. $\dfrac{3\sqrt{2y}}{y}$ **72.** $\dfrac{2\sqrt{x}}{x^4}$ **73.** $2\sqrt{7} + 2\sqrt{2}$ **74.** $\dfrac{6\sqrt{w} - 12}{w - 4}$
75. $-8 - 3\sqrt{7}$ **76. a.** $\dfrac{10\sqrt{6}}{3}$ m/sec **b.** $\dfrac{18\sqrt{5}}{5}$ m/sec
77. $\{138\}$ **78.** $\{\ \}$ (The value 48 does not check.) **79.** $\{39\}$
80. $\{5\}$ **81.** $\{7\}$ **82.** $\{6\}$ **83.** $\{2\}$ (The value -2
does not check.) **84.** $\{3, 4\}$ **85.** $\{-69\}$ **86. a.** 9261 in.³
b. 3375 cm³ **87.** -3 **88.** 11 **89.** -2
90. Not a real number **91.** $\dfrac{1}{8}$ **92.** 27 **93.** $\sqrt[5]{z}$ **94.** $\sqrt[3]{q^2}$
95. $\sqrt[4]{w^3}$ **96.** $\sqrt{\dfrac{b}{121}} = \dfrac{\sqrt{b}}{11}$ **97.** $a^{2/5}$ **98.** $5m^{2/3}$
99. $(a^2 b^4)^{1/5}$ **100.** $6^{1/2}$ **101.** y^2 **102.** $a^{5/6}$ **103.** $6^{3/5}$
104. b^{15} **105.** $4ab^2$ **106.** $5^{3/4}$ **107.** 2.0 cm

Chapter 8 Test, pp. 632–633

1. $x^2 + y^2 = z^2$ **2.** $11x\sqrt{2}$ **3.** $2y\sqrt[3]{6y}$
4. Not a real number **5.** $\dfrac{a^3\sqrt{5}}{9}$ **6.** $\dfrac{3\sqrt{6}}{2}$
7. $\dfrac{2\sqrt{5} - 12}{-31}$ or $\dfrac{12 - 2\sqrt{5}}{31}$ **8. a.** $\sqrt{25} + 5^3$; 130
b. $4^2 - \sqrt{16}$; 12 **9.** 97 ft **10.** $8\sqrt{z}$ **11.** $4\sqrt{6} - 15$
12. $-7t\sqrt{2}$ **13.** $9\sqrt{10}$ **14.** $-8 + 23\sqrt{10}$ **15.** $46 - 6\sqrt{5}$
16. $\dfrac{\sqrt{n}}{6m}$ **17.** $16 - 9x$ **18.** $\dfrac{\sqrt{22}}{11}$ **19.** $\dfrac{3\sqrt{7} + 3\sqrt{3}}{2}$
20. 206 yd **21.** $\{\ \}$ (The value $\frac{9}{2}$ does not check.) **22.** $\{0, -5\}$
23. $\{14\}$ **24. a.** 12 in. **b.** 25 weeks **25.** 1000 **26.** 2
27. $\sqrt[5]{x^3}$ or $(\sqrt[5]{x})^3$ **28.** $5\sqrt{y}$ **29.** $(ab^3)^{1/4}$ **30.** $p^{11/12}$
31. $5^{3/5}$ **32.** $3mn^2$

Chapters 1–8 Cumulative Review Exercises, pp. 633–634

1. 1 **2.** $\{-2\}$ **3.** -15 **4.** $-9x^2 + 2x + 10$
5. $\dfrac{2x}{y} - 1 + \dfrac{4}{x}$ **6.** $2(5c + 2)^2$ **7.** $\left\{-\dfrac{2}{5}, \dfrac{1}{2}\right\}$ **8.** 1
9. $\{\ \}$ (The value 5 does not check.)
10. $\dfrac{x(5x + 8)}{16(x - 1)}$ **11.**

12. a. $y = 880$; the cost of renting the office space for
3 months is $880. **b.** $x = 12$; the cost of renting office space for
12 months is $2770. **c.** $m = 210$; the cost increases at a rate of
$210 per month. **d.** $(0, 250)$; the initial deposit to rent the office space is
$250. **13.** $y = -x + 1$ **14.** $\{(1, -4)\}$
15.

16. 8 L of 20% solution should be mixed with 4 L of 50% solution.

17. $3\sqrt{11}$ **18.** $7x\sqrt{3}$ **19.** $\dfrac{x + \sqrt{xy}}{x - y}$ **20.** $\{5\}$

Chapter 9

Section 9.1 Practice Exercises, pp. 640–641

1. a. $0; 0$ **b.** 0 **c.** $\sqrt{k}; -\sqrt{k}$ **d.** $4; \{3, -3\}$

3. a. Linear   **b.** Quadratic   **c.** Linear   **5.** $\left\{-5, \frac{1}{2}\right\}$

7. $\{7, -5\}$ **9.** $\left\{-2, -\frac{1}{6}\right\}$ **11.** $\left\{-7, -\frac{3}{2}\right\}$ **13.** $\{12, -12\}$

15. $\{8, -2\}$ **17.** $\left\{\frac{1}{4}, -2\right\}$ **19.** $\{-1, -7\}$ **21.** $\{7, -7\}$

23. $\{10, -10\}$ **25.** There are no real-valued solutions.
27. $\left\{\sqrt{3}, -\sqrt{3}\right\}$ **29.** $\{9, 1\}$ **31.** $\{11, -1\}$ **33.** $\{11 \pm \sqrt{5}\}$

35. $\{-1 \pm 3\sqrt{2}\}$ **37.** $\left\{\frac{1}{4} \pm \frac{\sqrt{7}}{4}\right\}$ **39.** $\left\{\frac{1}{2} \pm \sqrt{15}\right\}$

41. There are no real-valued solutions.   **43.** $\left\{\frac{5}{2}, -\frac{5}{2}\right\}$

45. The solution checks.   **47.** False. -8 is also a solution.
49. a. 64 ft   **b.** 3.5 sec   **c.** 8.8 sec   **51.** 7.1 m   **53.** 8.0 ft

Section 9.2 Practice Exercises, pp. 646–647

1. a. completing   **b.** 100   **c.** 5; 1   **d.** 8

3. $\{5 \pm \sqrt{21}\}$ **5.** $n = 4; (y + 2)^2$ **7.** $n = 36; (p - 6)^2$

9. $n = \frac{81}{4}; \left(x - \frac{9}{2}\right)^2$ **11.** $n = \frac{25}{36}; \left(d + \frac{5}{6}\right)^2$

13. $n = \frac{1}{100}; \left(m - \frac{1}{10}\right)^2$ **15.** $n = \frac{1}{4}; \left(u + \frac{1}{2}\right)^2$ **17.** $\{2, -6\}$
19. $\{-1, -5\}$ **21.** $\{1 \pm \sqrt{2}\}$ **23.** $\{1 \pm \sqrt{6}\}$

25. $\{-2 \pm \sqrt{3}\}$ **27.** $\left\{-\frac{1}{2} \pm \frac{\sqrt{13}}{2}\right\}$ **29.** $\{-1 \pm \sqrt{41}\}$

31. $\{2 \pm \sqrt{5}\}$ **33.** $\{-2, -4\}$ **35.** $\{3, 8\}$ **37.** $\{11, -11\}$
39. $\{-2 \pm \sqrt{2}\}$ **41.** $\{-13, 5\}$ **43.** $\{13\}$ **45.** $\{10, -2\}$
47. $\{7, -1\}$ **49.** $\{4 \pm \sqrt{15}\}$ **51.** $\{-1 \pm \sqrt{6}\}$

53. $\{11, -2\}$ **55.** $\{0, 7\}$ **57.** $\left\{\frac{1}{2}, -\frac{3}{4}\right\}$ **59.** $\{\sqrt{14}, -\sqrt{14}\}$

61. There are no real-valued solutions.   **63.** $\{1\}$
65. The suitcase is 10 in. by 14 in. by 30 in. The bag must be checked because 10 in. + 14 in. + 30 in. = 54 in., which is greater than 45 in.

Section 9.3 Calculator Connections, p. 654

1.
```
(-5+√(17))/4
          -.2192235936
(-5-√(17))/4
          -2.280776406
```
2.
```
(-40+√(1920))/-3
2
          -.1193063938
(-40-√(1920))/-3
2
          2.619306394
```

Section 9.3 Practice Exercises, pp. 654–656

1. a. $\dfrac{-b \pm \sqrt{b^2 - 4ac}}{2a}$ **b.** $ax^2 + bx + c = 0$ **c.** $5; -24; -36$
d. $5; 73$ **3.** $\{1, -1\}$ **5.** $\{-3 \pm \sqrt{7}\}$ **7.** $\{2 \pm 2\sqrt{2}\}$

9. $2x^2 - x - 5 = 0; a = 2, b = -1, c = -5$
11. $-3x^2 + 14x + 0 = 0; a = -3, b = 14, c = 0$
13. $x^2 + 0x - 9 = 0; a = 1, b = 0, c = -9$

15. $\{-8\}$ **17.** $\left\{\frac{2}{3}, -\frac{1}{2}\right\}$ **19.** $\left\{\frac{1 \pm \sqrt{61}}{10}\right\}$

21. $\{1 \pm \sqrt{2}\}$ **23.** $\left\{\frac{-5 \pm \sqrt{3}}{2}\right\}$ **25.** $\left\{\frac{1 \pm \sqrt{17}}{-8}\right\}$ or

$\left\{\frac{-1 \pm \sqrt{17}}{8}\right\}$ **27.** $\left\{\frac{-3 \pm \sqrt{33}}{4}\right\}$ **29.** $\left\{\frac{-15 \pm \sqrt{145}}{4}\right\}$

31. $\left\{\frac{-2 \pm \sqrt{22}}{6}\right\}$ **33.** $\left\{\frac{3}{4}, -\frac{3}{4}\right\}$

35. There are no real-valued solutions.

37. $\{-12 \pm 3\sqrt{5}\}$ **39.** $\left\{\frac{3 \pm \sqrt{15}}{2}\right\}$ **41.** $\left\{0, \frac{11}{9}\right\}$

43. $\left\{\frac{3 \pm \sqrt{5}}{2}\right\}$ **45.** $\left\{\frac{1 \pm \sqrt{41}}{4}\right\}$ **47.** $\left\{0, \frac{1}{9}\right\}$

49. $\{2\sqrt{13}, -2\sqrt{13}\}$ **51.** $\left\{\frac{-10 \pm \sqrt{85}}{-5}\right\}$ or $\left\{\frac{10 \pm \sqrt{85}}{5}\right\}$

53. $\left\{\frac{1 \pm \sqrt{61}}{2}\right\}$ **55.** There are no real-valued solutions.

57. The width is 7.3 m. The length is 13.6 m.   **59.** The length is 7.4 ft. The width is 5.4 ft. The height is 6 ft.   **61.** The width is 6.7 ft. The length is 10.7 ft.   **63.** The legs are 10.6 m and 7.6 m.
65. He will be 1 ft off the ground 0.07 sec after leaving the ground (on the way up) and after 0.93 sec (on the way back down).

Chapter 9 Problem Recognition Exercises, p. 656

1. $\left\{\frac{1}{3}, -\frac{3}{2}\right\}$ **2.** $\{-7\}$ **3. a.** Quadratic   **b.** $\{4 \pm \sqrt{22}\}$

4. a. Quadratic   **b.** $\{3 \pm \sqrt{7}\}$ **5. a.** Linear   **b.** $\{13\}$

6. a. Linear   **b.** $\{-3\}$ **7. a.** Quadratic   **b.** $\left\{\frac{5}{2}, \frac{1}{4}\right\}$

8. a. Quadratic   **b.** $\left\{\frac{4}{3}, \frac{1}{3}\right\}$ **9. a.** Rational   **b.** $\left\{-\frac{3}{5}, 3\right\}$

10. a. Rational   **b.** $\left\{-\frac{6}{7}, 3\right\}$ **11. a.** Radical   **b.** $\{1, 3\}$

12. a. Radical   **b.** $\{1, 2\}$ **13. a.** Quadratic   **b.** $\{9, -11\}$

14. a. Quadratic   **b.** $\{13, -3\}$ **15. a.** Rational   **b.** $\left\{\frac{3}{5}\right\}$

16. a. Rational   **b.** $\left\{\frac{5}{3}\right\}$

Section 9.4 Practice Exercises, pp. 664–666

1. a. imaginary   **b.** $\sqrt{-1}; -1$ **c.** $a + bi; \sqrt{-1}$ **d.** real; b
e. $a + bi$ **f.** True   **g.** True   **3.** $6i$ **5.** $i\sqrt{21}$ **7.** $-16i\sqrt{3}$
9. -20 **11.** -6 **13.** 3   **15.** 5   **17.** $14i$ **19.** $-24i$
21. -10 **23.** 5   **25.** -3 **27.** 5   **29.** Real part: -3;
Imaginary part: -2 **31.** Real part: 4; Imaginary part: 0
33. Real part: 0; Imaginary part: $\frac{2}{7}$ **35.** Add or subtract the real
parts. Add or subtract the imaginary parts.   **37.** $-6 + 8i$
39. $10 - 10i$ **41.** $6 + 3i$ **43.** $-7 + 3i$ **45.** $7 - 21i$

47. $11 - 9i$ **49.** $9 + 19i$ **51.** $-\frac{1}{4} - \frac{1}{5}i$ **53.** $-3.5 + 18.1i$

55. 13   **57.** 104   **59.** $\frac{5}{4}$ **61.** $35 - 12i$ **63.** $21 + 20i$

65. $-33 - 56i$ **67.** $7 + 4i; 65$ **69.** $\frac{3}{2} - \frac{2}{5}i; \frac{241}{100}$ **71.** $-4i; 16$

73. $-\frac{3}{5} - \frac{6}{5}i$ **75.** $-\frac{2}{13} + \frac{10}{13}i$ **77.** $\frac{15}{17} + \frac{8}{17}i$

79. $\frac{23}{29} - \frac{14}{29}i$ **81.** $\frac{14}{65} + \frac{8}{65}i$ **83.** $\frac{5}{2} - \frac{5}{2}i$ **85.** $\{-4 \pm 5i\}$

87. $\{3 \pm 2i\sqrt{2}\}$ **89.** $\{1 \pm i\sqrt{3}\}$ **91.** $\left\{-\frac{1}{4} \pm \frac{\sqrt{39}}{12}i\right\}$

93. False. For example: $2 + 3i$ is not a real number.
95. True   **97.** False. $\sqrt[3]{-64} = -4$.
99. False. $(1 + 4i)(1 - 4i) = 17$ **101.** True
103. False. $i^4 = 1$ **105.** True

Section 9.5 Calculator Connections, p. 673

1. $(-2, 3)$; minimum

2. $(10, 5)$; minimum

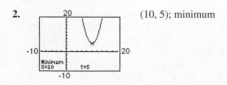

3. 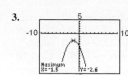 $(-1.5, -2.6)$; maximum

4. 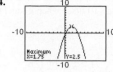 $(1.75, 2.5)$; maximum

5. 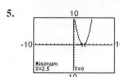 $\left(\frac{5}{2}, 0\right)$; minimum

6. 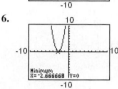 $\left(-\frac{8}{3}, 0\right)$; minimum

Section 9.5 Practice Exercises, pp. 673–676

1. a. parabola **b.** $>$; $<$ **c.** lowest; highest **d.** symmetry **e.** vertex   **3.** $\{-5, 3\}$ **5.** $\{-1 \pm \sqrt{6}\}$ **7.** $\{5 \pm 2\sqrt{3}\}$
9. Linear   **11.** Quadratic   **13.** Neither   **15.** Linear
17. Quadratic   **19.** Neither   **21.** If $a > 0$ the graph opens upward; if $a < 0$ the graph opens downward.   **23.** $a = 2$; upward
25. $a = -10$; downward   **27.** $(-1, -8)$ **29.** $(1, -4)$ **31.** $(1, 2)$
33. $(0, -4)$ **35.** x-intercepts: $(\sqrt{7}, 0), (-\sqrt{7}, 0)$; y-intercept: $(0, -7)$; c   **37.** x-intercepts: $(-1, 0), (-5, 0)$; y-intercept: $(0, 5)$; a
39. a. Upward **b.** $(0, -9)$ **c.** $(3, 0), (-3, 0)$ **d.** $(0, -9)$
e.

41. a. Upward **b.** $(1, -9)$ **c.** $(4, 0), (-2, 0)$ **d.** $(0, -8)$
e.
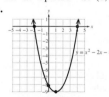

43. a. Downward **b.** $(3, 0)$ **c.** $(3, 0)$ **d.** $(0, -9)$
e.
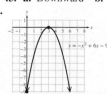

45. a. Downward **b.** $(4, 1)$ **c.** $(3, 0), (5, 0)$ **d.** $(0, -15)$
c.
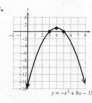

47. a. Upward **b.** $(-3, 1)$ **c.** none **d.** $(0, 10)$
e.
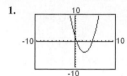

49. a. Downward **b.** $(0, -2)$ **c.** none **d.** $(0, -2)$
e.
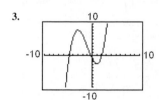

51. True   **53.** False   **55. a.** 28 ft **b.** 1.25 sec
57. a. 200 calendars **b.** \$500   **59. a.** Josh will be 12 ft high in 0.5 sec. **b.** Josh will land in 2 sec. **c.** The maximum height is 16 ft.

Section 9.6 Calculator Connections, p. 686

1.

2.

3.

4.

Section 9.6 Practice Exercises, pp. 686–691

1. a. relation **b.** domain **c.** range **d.** function **e.** vertical
f. $7x - 4$ **3.** $\left(\frac{1}{4}, \frac{11}{4}\right)$ **5.** Domain: $\{4, 3, 0\}$; range: $\{2, 7, 1, 6\}$
7. Domain: $\{\frac{1}{2}, 0, 1\}$; range: $\{3\}$ **9.** Domain: $\{0, 5, -8, 8\}$; range: $\{0, 2, 5\}$ **11.** Domain: $\{$Atlanta, Macon, Pittsburgh$\}$; range: $\{$GA, PA$\}$ **13.** Domain: $\{$New York, California$\}$; range: $\{$Albany, Los Angeles, Buffalo$\}$ **15.** The relation is a function if each element in the domain has exactly one corresponding element in the range.   **17.** The relations in Exercises 7, 9, and 11 are functions.
19. Yes   **21.** No   **23.** No   **25.** Yes   **27.** Yes
29. a. -5 **b.** -1 **c.** -11 **31. a.** $\frac{1}{5}$ **b.** $\frac{1}{4}$ **c.** $\frac{1}{2}$ **33. a.** 7 **b.** 2 **c.** 3   **35. a.** 0 **b.** 1 **c.** 2   **37.** Domain: $\{-3, 1, 2, 4\}$; range: $\{-5, 0, 1, 2\}$ **39.** Domain: $\{-4, -2, 0, 1, 5\}$; range: $\{-3, 3, 4\}$ **41.** b
43. c   **45.** Domain: $(-\infty, \infty)$; range: $[-2, \infty)$ **47.** Domain: $[-1, 1]$; range: $[-4, 4]$ **49.** The function value at $x = 6$ is 2.
51. The function value at $x = \frac{1}{2}$ is $\frac{1}{4}$.   **53.** $(2, 7)$
55. a. $s(1) = 32$. The speed of an object 1 sec after being dropped is 32 ft/sec. **b.** $s(2) = 64$. The speed of an object 2 sec after being dropped is 64 ft/sec. **c.** $s(10) = 320$. The speed of an object 10 sec after being dropped is 320 ft/sec. **d.** 294.4 ft/sec   **57. a.** $h(0) = 3$. The initial height of the ball is 3 ft. **b.** $h(1) = 51$. The height of the ball 1 sec after being kicked is 51 ft. **c.** $h(2) = 67$. The height of the ball 2 sec after being kicked is 67 ft. **d.** $h(4) = 3$. The height of the ball 4 sec after being kicked is 3 ft.   **59. a.** The cost is \$225.
b. She was charged for 2.5 hr. **c.** Domain: $[0, \infty)$ **d.** The y-intercept represents the cost of the estimate.

Chapter 9 Review Exercises, pp. 696–699

1. Linear   **2.** Quadratic   **3.** Quadratic   **4.** Linear
5. $\{5, -5\}$ **6.** $\{\sqrt{19}, -\sqrt{19}\}$ **7.** The equation has no real-valued solutions.   **8.** The equation has no real-valued solutions.

9. $\{-1 \pm \sqrt{14}\}$ **10.** $\{2 \pm 2\sqrt{15}\}$ **11.** $\left\{\dfrac{1}{8} \pm \dfrac{\sqrt{3}}{8}\right\}$

12. $\left\{\dfrac{3 \pm 2\sqrt{5}}{2}\right\}$ **13.** $n = 36$ **14.** $n = 81$ **15.** $n = \dfrac{25}{4}$

16. $n = \dfrac{49}{4}$ **17.** $\{-4 \pm \sqrt{13}\}$ **18.** $\{1 \pm \sqrt{5}\}$

19. $\left\{\dfrac{3}{2} \pm \dfrac{\sqrt{21}}{2}\right\}$ **20.** $\left\{\dfrac{7}{6} \pm \dfrac{\sqrt{85}}{6}\right\}$ **21.** 10.6 ft   **22.** 3.1 cm

23. For $ax^2 + bx + c = 0$, $x = \dfrac{-b \pm \sqrt{b^2 - 4ac}}{2a}$

24. $\left\{\dfrac{-1 \pm \sqrt{141}}{10}\right\}$ **25.** $\{-2\}$ **26.** The equation has no real-valued solutions.   **27.** $\left\{\dfrac{3}{2}, -1\right\}$ **28.** $\{-10, 2\}$

29. $\{-3 \pm \sqrt{7}\}$ **30.** $\{1, -6\}$ **31.** $\{1 \pm \sqrt{5}\}$

32. $\{-4 \pm \sqrt{14}\}$ **33.** The equation has no real-valued solutions.

34. The numbers are -2.5 and -4.5, or 2.5 and 4.5.

35. The height is approximately 4.4 cm The base is approximately 5.4 cm.   **36.** 9.5 sec   **37.** $a + bi$, where a and b are real numbers and $i = \sqrt{-1}$ **38.** $a + bi$, where $b \neq 0$ **39.** $4i$

40. $-i\sqrt{5}$ **41.** -15 **42.** $-2i$ **43.** 6 **44.** 8 **45.** 13

46. 11 **47.** $-5 + 5i$ **48.** $9 + 17i$ **49.** $25 + 0i$

50. $24 - 10i$ **51.** $-\dfrac{17}{4} + i$; Real part: $-\dfrac{17}{4}$; Imaginary part: 1

52. $-2 - i$; Real part: -2; Imaginary part: -1 **53.** $\dfrac{4}{13} - \dfrac{7}{13}i$

54. $3 + 4i$ **55.** $\{-12 \pm 2i\sqrt{5}\}$ **56.** $\{7 \pm 3i\sqrt{2}\}$

57. $\left\{\dfrac{1}{8} \pm \dfrac{\sqrt{31}}{8}i\right\}$ **58.** $\left\{-\dfrac{3}{4} \pm \dfrac{\sqrt{7}}{4}i\right\}$ **59.** $a = 1$; upward

60. $a = -1$; downward   **61.** $a = -2$; downward   **62.** $a = 5$; upward   **63.** Vertex: $(-1, 1)$ **64.** Vertex: $(4, 19)$

65. Vertex: $(3, 13)$ **66.** Vertex: $\left(-\dfrac{1}{2}, -\dfrac{3}{2}\right)$

67. a. Upward
b. $(-1, -4)$
c. $(-3, 0), (1, 0)$
d. $(0, -3)$
e.

68. a. Upward
b. $(1, -1)$
c. $(0, 0), (2, 0)$
d. $(0, 0)$
e.

69. a. Downward   **b.** $(2, 3)$ **c.** $(1, 0), (3, 0)$ **d.** $(0, -9)$
e.

70. a. Downward   **b.** $(-1, -4)$ **c.** No x-intercepts
d. $(0, -12)$
e.

71. a. 1024 ft   **b.** 8 sec   **72.** Domain: $\{6, 10, -1, 0\}$; range: $\{3\}$; function   **73.** Domain: $\{2\}$; range: $\{0, 1, -5, 2\}$; not a function

74. Domain: $[-4, 4]$; range: $[-3, 3]$; not a function   **75.** Domain: $(-\infty, \infty)$; range: $[-2, \infty)$; function   **76.** Domain: $\{4, 3, -6\}$; range: $\{23, -2, 5, 6\}$; not a function   **77.** Domain: $\{3, -4, 0, 2\}$; range: $\{0, \frac{1}{2}, 3, -12\}$; function   **78. a.** 0 **b.** 8 **c.** -27 **d.** -1 **e.** 64

79. a. 0 **b.** 4 **c.** $-\dfrac{1}{6}$ **d.** $\dfrac{3}{2}$ **e.** $-\dfrac{1}{2}$ **80. a.** $D(90) = 562$. A plane traveling 90 ft/sec when it touches down will require 562 ft of runway.   **b.** $D(110) = 902$. A plane traveling 110 ft/sec when it touches down will require 902 ft of runway.

Chapter 9 Test, pp. 699–701

1. $\{-1 \pm \sqrt{14}\}$ **2.** $\{4 \pm \sqrt{21}\}$ **3.** $\left\{\dfrac{5 \pm \sqrt{13}}{6}\right\}$

4. $\left\{\dfrac{-1 \pm \sqrt{41}}{10}\right\}$ **5.** $\{12 \pm 2\sqrt{3}\}$ **6.** $\{-7 \pm 5\sqrt{2}\}$

7. $\{\sqrt{10}, -\sqrt{10}\}$ **8.** $\left\{\dfrac{5}{6}, -\dfrac{3}{2}\right\}$ **9.** $\left\{0, \dfrac{11}{6}\right\}$

10. $\{3 \pm 2\sqrt{5}\}$ **11.** 4.0 in.   **12.** The base is 4.4 m. The height is 10.8 m.   **13.** $10i$ **14.** $i\sqrt{23}$ **15.** -21 **16.** -2

17. 8 **18.** $5 - 3i$ **19.** $-13 - 26i$ **20.** 221

21. $\dfrac{10}{221} + \dfrac{11}{221}i$ **22.** $\{-14 \pm 9i\}$ **23.** $\left\{-\dfrac{1}{2} \pm \dfrac{3\sqrt{3}}{2}i\right\}$

24. For $y = ax^2 + bx + c$, if $a > 0$ the parabola opens upward, if $a < 0$ the parabola opens downward.   **25.** $(5, 0)$ **26.** $(1, 5)$

27. $(0, -16)$ **28.** The parabola has no x-intercepts.

29. a. Opens upward   **b.** Vertex: $(-3, -1)$ **c.** x-intercepts: $(-2, 0)$ and $(-4, 0)$ **d.** y-intercept: $(0, 8)$
e.

30. Vertex: $(0, 25)$; x-intercepts: $(-5, 0)$, $(5, 0)$; y-intercept: $(0, 25)$

31. a. \$25 per ticket   **b.** \$250,000

32. a. Domain: $(-\infty, 0]$; range: $(-\infty, \infty)$; not a function
b. Domain: $(-\infty, \infty)$; range: $(-\infty, 4]$; function

33. $f(0) = \dfrac{1}{2}$, $f(-2)$ is undefined, $f(6) = \dfrac{1}{8}$

34. a. $D(5) = 5$; a five-sided polygon has five diagonals.
b. $D(10) = 35$; a 10-sided polygon has 35 diagonals.   **c.** 8 sides

Chapters 1–9 Cumulative Review Exercises, pp. 701–703

1. $\{1\}$ **2.** $h = \dfrac{2A}{b}$ **3.** $\left\{\dfrac{34}{3}\right\}$ **4. a.** Yes, 2 is a solution.   **b.** $\{x \mid x > -1\}$; $(-1, \infty)$

5. a. Decreases   **b.** $m = -37.6$. For each additional increase in education level, the death rate decreases by approximately 38 deaths per 100,000 people.   **c.** 901.8 per 100,000   **d.** 10th grade

6. $\dfrac{8c}{a^4 b}$ **7.** 1.898×10^{10} diapers   **8.** Approximately 430 light-years

9. $4x^2 - 16x + 13$ **10.** $2y^3 + 1 - \dfrac{3}{y - 2}$

11. $(2x + 5)(x - 7)$ **12.** $(y + 4a)(2x - 3b)$

13. The base is 9 m, and the height is 8 m.   **14.** $\dfrac{5}{x - 2}$

15. $-\dfrac{2}{x+1}$ **16.** $x-5$ **17.** $-\dfrac{2}{x}$ **18.** $\{4,-3\}$

19. $y=\dfrac{1}{2}x+4$ **20. a.** $(6,0)$ **b.** $(0,-3)$ **c.** $\frac{1}{2}$
d.

21. a. $(-3,0)$ **b.** No y-intercept   **c.** Slope is undefined.
d.

22. $\left\{\left(1,\dfrac{4}{3}\right)\right\}$ **23.** $\{(5,2)\}$ **24.** The angles are
$22°$ and $68°$.   **25.** There are 13 dimes and 14 quarters.
26.

27. $\pi,\sqrt{7}$ **28.** $\dfrac{\sqrt{7}}{7}$ **29.** $2x\sqrt{2x}$ **30.** $48+8\sqrt{3x}+x$
31. $2\sqrt{2x}$ **32.** $\dfrac{8+4\sqrt{a}}{4-a}$ **33.** $\{-2\}$ (The value -7 does
not check.)   **34.** $(2c-y)(4c^2+2cy+y^2)$ **35.** b   **36. a.** 1
b. 4   **c.** 13   **37.** Domain: $\{2,-1,9,-6\}$; range: $\{4,3,2,8\}$
38. $m=\dfrac{5}{7}$ **39.** $m=-\dfrac{4}{5}$ **40.** $n=25$ **41.** $\{-3\pm\sqrt{6}\}$
42. $\{-3\pm\sqrt{6}\}$
43.   **44.** 16   **45.** $-12-3i$

Additional Topics Appendix

Section A.1 Calculator Connections, p. A-8

1. $0.\overline{4}$ **2.** $0.\overline{63}$ **3.** $0.1\overline{36}$ **4.** $0.\overline{384615}$
1.–2.   **3.–4.**

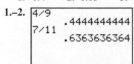

Section A.1 Practice Exercises, pp. A-8–A-10

1. Tens   **3.** Hundreds   **5.** Tenths   **7.** Hundredths
9. No, the symbols I, V, X, and so on each represent certain values
but the values are not dependent on the position of the symbol
within the number.   **11.** 0.7   **13.** 0.36   **15.** $1.\overline{2}$
17. $0.\overline{21}$ **19.** 214.1   **21.** 39.268   **23.** 40,000   **25.** 0.73
27. $\dfrac{9}{20}$ **29.** $\dfrac{181}{1000}$ **31.** $\dfrac{51}{25}$ or $2\dfrac{1}{25}$ **33.** $\dfrac{13,007}{1000}$ or $13\dfrac{7}{1000}$
35. $\dfrac{5}{9}$ **37.** $\dfrac{10}{9}$ or $1\dfrac{1}{9}$ **39.** $0.3,\dfrac{3}{10}$ **41.** $0.75,\dfrac{3}{4}$

43. $0.0375,\dfrac{3}{80}$ **45.** $0.157,\dfrac{157}{1000}$ **47.** $2.7,\dfrac{27}{10}$
49. Multiply by 100%.   **51.** 5%   **53.** 90%   **55.** 120%
57. 750%   **59.** 13.5%   **61.** 0.3%   **63.** 6%   **65.** 450%
67. 62.5%   **69.** 31.25%   **71.** $83.\overline{3}\%$ **73.** $93.\overline{3}\%$
75. $42   **77.** $3375   **79.** 7%   **81.** $792
83. $192   **85.** $67,500

Section A.2 Practice Exercises, pp. A-16–A-19

1. 88   **3.** 5   **5.** 6   **7.** -15.8 **9.** 5.8 hr
11. a. 397 Cal   **b.** 386 Cal   **c.** There is only an 11-Cal difference
in the means.   **13. a.** 86.5%   **b.** 81%   **c.** The low score of 59%
decreased Zach's average by 5.5%.   **15.** 17   **17.** 110.5
19. -52.5 **21.** 3.93 deaths per 1000   **23.** 0
25. 51.7 million passengers   **27.** 4   **29.** -21 and -24
31. No mode   **33.** 3.66 in.   **35.** 5.2% and 5.8%
37. Mean: 85.5%; median: 94.5%; The median gives Jonathan a
better overall score.   **39.** Mean: $250; median: $256;
mode: There is no mode.   **41.** Mean: $942,500; median: $848,500;
mode: $850,000   **43.** 2.38   **45.** 2.77   **47.** 3.3; Elmer's GPA
improved from 2.5 to 3.3.
49.

Number of Residents in Each House	Number of Houses	Product
1	3	3
2	9	18
3	10	30
4	9	36
5	6	30
Total:	37	117

The mean number of residents is approximately 3.2.

Section A.3 Practice Exercises, pp. A-30–A-36

1. Perimeter   **3.** 32 m   **5.** 17.2 mi   **7.** $11\dfrac{1}{2}$ in.
9. 31.4 ft   **11.** a, f, g   **13.** 33 cm^2 **15.** 16.81 m^2
17. 84 in.^{2} **19.** 10.12 km^2 **21.** 13.8474 ft^2 **23.** 66 in.^{2}
25. 31.5 ft^2 **27.** c, d, h   **29.** 307.72 cm^3 **31.** 39 in.^{3}
33. 113.04 cm^3 **35.** 1695.6 cm^3 **37.** 3052.08 in.^{3}
39. 113.04 cm^3 **41. a.** $0.25/ft^2 **b.** $104   **43.** Perimeter
45. 54 ft   **47. a.** 57,600 ft^2 **b.** 19,200 pieces
49. a. 50.24 in.^{2} **b.** 113.04 in.^{2} **c.** One 12-in. pizza
51. 289.3824 cm^3 **53.** True   **55.** True   **57.** True
59. Not possible   **61.** For example: 100°, 80°
63. 45°   **65. a.** $\angle 1$ and $\angle 3$, $\angle 2$ and $\angle 4$
b. $\angle 1$ and $\angle 2$, $\angle 2$ and $\angle 3$, $\angle 3$ and $\angle 4$, $\angle 1$ and $\angle 4$
c. $m(\angle 1)=100°$, $m(\angle 2)=80°$, $m(\angle 3)=100°$
67. 57°   **69.** 78°   **71.** 147°   **73.** 58°   **75.** 7   **77.** 1
79. 1   **81.** 5   **83.** $m(\angle a)=45°$, $m(\angle b)=135°$, $m(\angle c)=45°$,
$m(\angle d)=135°$, $m(\angle e)=45°$, $m(\angle f)=135°$, $m(\angle g)=45°$ **85.** Scalene
87. Isosceles   **89.** True   **91.** No, a 90° angle plus an angle
greater than 90° would make the sum of the angles greater than 180°.
93. 40°   **95.** 37°   **97.** $m(\angle a)=80°$, $m(\angle b)=80°$, $m(\angle c)=100°$,
$m(\angle d)=100°$, $m(\angle e)=65°$, $m(\angle f)=115°$, $m(\angle g)=115°$, $m(\angle h)=35°$,
$m(\angle i)=145°$, $m(\angle j)=145°$
99. $m(\angle a)=70°$, $m(\angle b)=65°$, $m(\angle c)=65°$,
$m(\angle d)=110°$, $m(\angle e)=70°$, $m(\angle f)=110°$, $m(\angle g)=115°$,
$m(\angle h)=115°$, $m(\angle i)=65°$, $m(\angle j)=70°$, $m(\angle k)=65°$
101. 82 ft   **103.** 36 in.^{2} **105.** 15.2464 cm^2

Section A.4 Practice Exercises, pp. A-39–A-40

1. 196.7 ft   **3.** 122 m   **5.** 0.1 kg   **7.** 0.3 oz   **9.** 1920 mL
11. 5.5 qt   **13.** $-4°$F   **15.** 4.4°C   **17.** 1641.1 yd
19. 343.4 m/sec   **21.** 3.2 qt   **23.** The 2-L bottle costs
$0.95 per quart. The half-gallon bottle costs $0.90 per quart.
The half-gallon is the better buy.   **25.** 200 lb   **27.** 9 kg
29. $-35°$C   **31.** 95°F

Application Index

Note: Page numbers preceded by the letter *A* indicate Appendix pages.

Subject Index

Note: Page numbers preceded by the letter *A* indicate Appendix pages.

Perimeter and Circumference

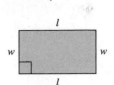

Rectangle

$P = 2l + 2w$

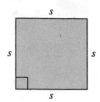

Square

$P = 4s$

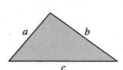

Triangle

$P = a + b + c$

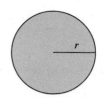

Circle

Circumference: $C = 2\pi r$

Area

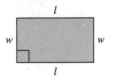

Rectangle

$A = lw$

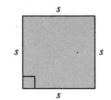

Square

$A = s^2$

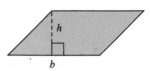

Parallelogram

$A = bh$

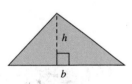

Triangle

$A = \frac{1}{2}bh$

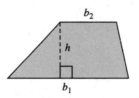

Trapezoid

$A = \frac{1}{2}(b_1 + b_2)h$

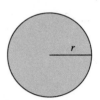

Circle

$A = \pi r^2$

Volume

Rectangular Solid

$V = lwh$

Cube

$V = s^3$

Right Circular Cylinder

$V = \pi r^2 h$

Right Circular Cone

$V = \frac{1}{3}\pi r^2 h$

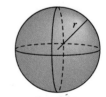

Sphere

$V = \frac{4}{3}\pi r^3$

Angles

- Two angles are **complementary** if the sum of their measures is 90°.

- Two angles are **supplementary** if the sum of their measures is 180°.

- $\angle a$ and $\angle c$ are vertical angles, and $\angle b$ and $\angle d$ are vertical angles. The measures of vertical angles are equal.

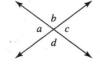

- The sum of the measures of the angles of a triangle is 180°.

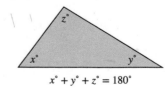

$x° + y° + z° = 180°$